Lecture Notes in Mechanical Engineering

Lecture Notes in Mechanical Engineering (LNME) publishes the latest developments in Mechanical Engineering - quickly, informally and with high quality. Original research reported in proceedings and post-proceedings represents the core of LNME. Volumes published in LNME embrace all aspects, subfields and new challenges of mechanical engineering. Topics in the series include:

- Engineering Design
- Machinery and Machine Elements
- Mechanical Structures and Stress Analysis
- Automotive Engineering
- Engine Technology
- Aerospace Technology and Astronautics
- Nanotechnology and Microengineering
- Control, Robotics, Mechatronics
- MEMS
- Theoretical and Applied Mechanics
- Dynamical Systems, Control
- Fluid Mechanics
- Engineering Thermodynamics, Heat and Mass Transfer
- Manufacturing
- Precision Engineering, Instrumentation, Measurement
- Materials Engineering
- Tribology and Surface Technology

To submit a proposal or request further information, please contact the Springer Editor in your country:

China: Li Shen at li.shen@springer.com
India: Dr. Akash Chakraborty at akash.chakraborty@springernature.com
Rest of Asia, Australia, New Zealand: Swati Meherishi at swati.meherishi@springer.com
All other countries: Dr. Leontina Di Cecco at Leontina.dicecco@springer.com

To submit a proposal for a monograph, please check our Springer Tracts in Mechanical Engineering at http://www.springer.com/series/11693 or contact Leontina.dicecco@springer.com

Indexed by SCOPUS. The books of the series are submitted for indexing to Web of Science.

More information about this series at http://www.springer.com/series/11236

Caterina Rizzi · Angelo Oreste Andrisano ·
Francesco Leali · Francesco Gherardini ·
Fabio Pini · Alberto Vergnano
Editors

Design Tools and Methods in Industrial Engineering

Proceedings of the International Conference on Design Tools and Methods in Industrial Engineering, ADM 2019, September 9–10, 2019, Modena, Italy

Springer

Editors
Caterina Rizzi
Dipartimento di Ingegneria Gestionale, dell'Informazione e della Produzione
Università di Bergamo
Dalmine, Italy

Angelo Oreste Andrisano
Dipartimento Di Ingegneria "Enzo Ferrari"
Università Di Modena E Reggio Emilia
Modena, Italy

Francesco Leali
Dipartimento Di Ingegneria "Enzo Ferrari"
Università Di Modena E Reggio Emilia
Modena, Italy

Francesco Gherardini
Dipartimento Di Ingegneria "Enzo Ferrari"
Università Di Modena E Reggio Emilia
Modena, Italy

Fabio Pini
Dipartimento Di Ingegneria "Enzo Ferrari"
Università Di Modena E Reggio Emilia
Modena, Italy

Alberto Vergnano
Dipartimento Di Ingegneria "Enzo Ferrari"
Università Di Modena E Reggio Emilia
Modena, Italy

ISSN 2195-4356 ISSN 2195-4364 (electronic)
Lecture Notes in Mechanical Engineering
ISBN 978-3-030-31153-7 ISBN 978-3-030-31154-4 (eBook)
https://doi.org/10.1007/978-3-030-31154-4

This Springer imprint is published by the registered company Springer Nature Switzerland AG
The registered company address is: Gewerbestrasse 11, 6330 Cham, Switzerland

Preface

The ADM 2019 International Conference, organized by the Italian Association of Design Methods and Tools for Industrial Engineering (ADM), brings the tradition of dissemination meetings back to life. The event in fact represents the reopening of the biennial international conferences organized by ADM in the years in which the JCM Conference (joint conference of ADM, INGEGRAF and S.mart) is not held.

The opportunity arose from the intention of all members of the Italian Association of Design Methods and Tools for Industrial Engineering to express gratitude to Prof. Angelo O. Andrisano, Rector of the University of Modena and Reggio Emilia.

As a member, adviser and President of the ADM board, Prof. Andrisano has with his skills, excellent professionalism, vision and passion offered a significant contribution to raising the works of the association to the highest levels.

As a very appreciated rector, Prof. Andrisano has moreover brought prestige to the association, stimulating the pride of all its members.

The ADM 2019 International Conference has been made possible with will and support of all the members of the ADM board, as well as the sensitivity of all the members of the Scientific Council, in particular the Coordinator, Prof. Caterina Rizzi. Moreover, a fundamental contribution was given by the Scientific Societies, INGEGRAF and S.mart, whom I sincerely thank.

I also thank all the members of the scientific and the organizing committees and Eng. Francesco Gherardini, who has effectively taken care of significant organizational aspects.

Special thanks go to all the authors and all the Italian and international reviewers for the high scientific level achieved by the papers, which has led to the publication of the research works in this prestigious book.

Finally, I thank all the ADM members. Their contribution has been fundamental for the organization of the ADM 2019 International Conference.

Vincenzo Nigrelli
ADM President

Letter to the Authors

Dear Attendees and Authors,

It is a great honour and a pleasure to welcome you to the ADM 2019 International Conference, held in Modena, Italy, 9–10 September 2019. The conference is hosted by the University of Modena and Reggio Emilia, a globally renowned university located in the heart of the Italian Motor Valley.

The conference aims at sharing knowledge, experience and up-to-date scientific information in the areas of design and manufacturing, with links between industry and academia. It provides a forum for researchers, educators and professional engineers for the dissemination and exchange of their latest research results.

Through the exchange of ideas, ADM 2019 intends to facilitate the creation of multidisciplinary cooperation and developments and discoveries for new product design and manufacture, health care, transportation and environment.

This book collects more than 80 scientific papers across a wide range of session topics that cover a broad spectrum of themes including theoretical issues, methods, tools, processes and case studies. The topics span from technical representation and geometric modelling to virtual/augmented reality, virtual ergonomics, human factors, human–machine interactions, knowledge management, additive manufacturing and many other subjects applied in different contexts, such as automotive, agriculture, cultural heritage and health care. They provide opportunities for interaction and for understanding how the application of emerging technologies impact on critical engineering issues activities such as product design, manufacturing, management and integration of information throughout the life cycle.

We have also the privilege to host three outstanding keynote speakers. The first is Prof. Angelo Oreste Andrisano (Rector of the University of Modena and Reggio Emilia, Italy) at the opening plenary session with a speech dealing with four decades of changes in the university and in our scientific sector. The others are two experts from industry: Dr. Massimo Giannozzi (Materials Engineering Manager, F1 Team Ferrari), with a speech on materials and processes for automotive design with a focus on additive manufacturing technologies, and Dr. Cécile Doan (Head of CATIA Strategy, Market Development and Finance—Dassault Systèmes,

Vice President) with a speech on simulation-driven design in industry: tools and future trends.

To conclude, I would like to thank all the authors for their valuable contribution to the book and each and every one of you for making ADM 2019 successful with your expertise, commitment and active engagement. Special thanks go to the Program Chair, Prof. Francesco Leali, and to the members of the Organizing Committee for their tremendous efforts for making this conference possible.

We sincerely hope that you will enjoy the conference, the Gala Dinner @ MEF—Museo Enzo Ferrari—and the beautiful Modena city.

Caterina Rizzi
Conference Chair

Introduction

This book collects the proceedings of the ADM 2019 International Conference, entitled "Design Tools and Methods in Industrial Engineering", held in Modena, Italy, on 9–10 September 2019.

This is not the first time that an event organized by ADM (formerly known as Italian Association of Machines Design, today renamed the Italian Association of Design Methods for Industrial Engineering) or by our scientific sector (ING-IND/15) takes place in Modena. I am pleased to remember the "ADM day" held at the Military Academy of Modena in 2008, as well as the ING-IND/15 workshop organized in 2009 in the hills of Bologna, not far from Modena.

However, this conference is particularly significant, for the reasons I wish to summarize here.

Firstly, the ADM Conference constitutes a new international, permanent event held in Italy, aimed at disseminating cutting-edge research to a national and international audience. For several years, our association has had a collaborative relationship with the Spanish members of INGEGRAF and the French members of S.mart (formerly AIP-PRIMECA), with whom we organize the International Joint Conference on Mechanics, Design Engineering and Advanced Manufacturing (JCM) every two years. This event is hosted alternately by one of our three countries. Therefore, the ADM International Conference was born from the need for an event "in the middle", to strengthen the international research network and to meet our colleagues in Italy. As a matter of fact, the ADM 2019 scientific committee was joined by Spanish and French colleagues, as well as German, Swedish, Norwegian and American ones, thanks to partnerships created by our "Enzo Ferrari" Department of Engineering (Modena) over the years.

Secondly, I am honoured to play the role of conference chair, a position that has led me to be one of the editors of this book. In more than 45 years of experience in this sector, I have seen many developments and evolutions. From the first use of CAD in the industrial sector, today we have reached a high multidisciplinary level and integration with information technology and electronics. The integration of novel tools and approaches in the industrial world is also due to the activity of our scientific sector, capable of developing and integrating innovative tools and methodologies with the traditional themes of industrial engineering.

Finally, this conference is held in 2019, an important year for me, marking the conclusion of my mandate as Rector of the University of Modena and Reggio Emilia, as well as that of Full Professor for our scientific sector. As the Dean of the ADM association and the scientific sector, I enthusiastically accepted the proposal of Francesco Leali, my collaborator for years and now my colleague, to nominate the University of Modena and Reggio Emilia, and in particular our "Enzo Ferrari" Department of Engineering, as the location for the ADM 2019 International Conference. I would like to thank both him and my departmental colleagues, in particular the researchers of my lab, LaPIS lab, for organizing this event at the "Enzo Ferrari" Department of Engineering, so dear to me, having been one of its first promoters and founders.

Equally, I would like to thank the ADM President Prof. Vincenzo Nigrelli, and Prof. Caterina Rizzi, Coordinator of the ADM Scientific Council, whose members, in turn, I gratefully thank. Further thanks go to the publisher, Springer, who honoured us by publishing the proceedings of this first ADM International Conference in the series "Lecture Notes in Mechanical Engineering".

Therefore, as editor of this first book, I hope that it will be the first of a new series of international conference proceedings underlining the leading role of our scientific sector in the international scene. The colleagues who will be hosting the next ADM conferences will have the burden, but above all the honour, of continuing this project that today we have started in Modena.

Angelo Oreste Andrisano
Honorary Conference Chair

Organization

ADM 2019 is organized by ADM—Italian Association of Design Methods and Tools for Industrial Engineering—in cooperation with the "Enzo Ferrari" Department of Engineering of the University of Modena and Reggio Emilia, Italy.

Organizing Committee

Francesco Leali	Università di Modena e Reggio Emilia, Italy
Francesco Gherardini	Università di Modena e Reggio Emilia, Italy
Fabio Pini	Università di Modena e Reggio Emilia, Italy
Alberto Vergnano	Università di Modena e Reggio Emilia, Italy

Scientific Committee

Conference Chair

Caterina Rizzi (ADM Scientific Committee Coordinator)	Università di Bergamo, Italy

Honorary Chair

Angelo O. Andrisano	Rector of Università di Modena e Reggio Emilia, Italy

Conference Program Chair

Francesco Leali	Università di Modena e Reggio Emilia, Italy

Members

Monica Bordegoni	Politecnico di Milano, Italy (ADM)
Rainer Börret	Aalen University, Germany
Pierre Castagna	Université de Nantes, France (S.mart)
Francisco Cavas-Martínez (INGEGRAF Scientific Council Director)	Universidad Politécnica de Cartagena, Spain
Aitor Cazón	TECNUM, Spain (INGEGRAF)
Vincent Cheutet	INSA Lyon, France (S.mart)
Paolo Di Stefano	Università dell'Aquila, Italy (ADM)
Benoit Eynard	Université de Technologie de Compiègne, France (S.mart)
Salvatore Gerbino	Università della Campania "Luigi Vanvitelli", Italy (ADM)
Francesco Gherardini	Università di Modena e Reggio Emilia, Italy (ADM)
Tim Grunwald	Fraunhofer Institute for Production Technology IPT, Germany
Sebastian Hähnel	Fraunhofer Institute for Production Technology IPT, Germany
Rafael Hidalgo Fernández	Universidad de Córdoba, Spain (INGEGRAF)
Carlos León Robles	Universidad de Granada, Spain (INGEGRAF)
Ruben Lostado Lorza	Universidad de la Rioja, Spain (INGEGRAF)
Cristina Manchado del Val	Universidad de Cantabria, Spain (INGEGRAF)
Ferruccio Mandorli	Università Politecnica delle Marche, Italy (ADM)
Cristina Martín Doñate	Univesidad de Jaén, Spain (INGEGRAF)
Rikardo Minguez Gabiña	Universidad del País Vasco, Spain (INGEGRAF)
Ramón Mirálbes Buil	Universidad de Zaragoza, Spain (INGEGRAF)
Maurizio Muzzupappa	Università della Calabria, Italy (ADM)
Vincenzo Nigrelli (ADM President)	Università di Palermo, Italy
Marcello Pellicciari	Università di Modena e Reggio Emilia, Italy (ADM)
Margherita Peruzzini	Università di Modena e Reggio Emilia, Italy (ADM)
Fabio Pini	Università di Modena e Reggio Emilia, Italy (ADM)
David Ranz	Universidad de Zaragoza, Spain (INGEGRAF)
José Ignacio Rojas Sola	Universidad de Jaén, Spain (INGEGRAF)
Terje Rølvåg	Norwegian University of Science and Technology, Norway
Bengt-Göran Rosén	Halmstad University, Sweden
Lionel Roucoules (S.mart Scientific Council Director)	ENSAM Aix-en-Provence, France

Max Schneckenburger	Aalen University, Germany
Eneko Solaberrieta	Universidad del País Vasco, Spain (INGEGRAF)
Miguel Suffo Pino	Universidad de Cádiz, Spain (INGEGRAF)
Alberto Vergnano	Università di Modena e Reggio Emilia, Italy (ADM)
Enrico Vezzetti	Politecnico di Torino, Italy (ADM)
Matteo Zallio	Stanford University, USA
Marc Zolghadri (S.mart Deputy Director)	SupMéca, France

Reviewers

Ambu, Rita
Bartalucci, Chiara
Bianconi, Francesco
Bici, Michele
Borgianni, Yuri
Buonamici, Francesco
Calì, Michele
Campana, Francesca
Cappetti, Nicola
Carfagni, Monica
Caruso, Giandomenico
Cascini, Gaetano
Ceruti, Alessandro
Cocconcelli, Marco
Concheri, Gianmaria
Cristofolini, Ilaria
Cucinotta, Filippo
De Crescenzio, Francesca
De Giorgi, Marta
De Napoli, Luigi
Di Angelo, Luca
Favi, Claudio
Ferro, Paolo
Fiorineschi, Lorenzo
Furferi, Rocco
Gadola, Marco
Germani, Michele
Governi, Lapo
Grazioso, Stanislao
Grigolato, Luca
Guardiani, Emanuele
Ingrassia, Tommaso
Liverani, Alfredo
Mandolini, Marco
Marconi, Marco
Martorelli, Massimo
Marzullo, Domenico
Meneghello, Roberto
Montorsi, Monia
Morabito, Anna Eva
Motyl, Barbara
Mozzillo, Rocco
Naddeo, Alessandro
Neri, Paolo
Olivetti, Elena Carlotta
Paoli, Alessandro
Papetti, Alessandra
Parras-Burgos, Dolores
Piazzolla, Pietro
Raffaeli, Roberto
Razionale, Armando Viviano
Redaelli, Davide Felice
Regazzoni, Daniele
Rizzuti, Sergio
Rosa, Francesco
Rosso, Stefano
Rossoni, Marco
Rotini, Federico
Savio, Gianpaolo
Schneckenburger, Max
Servi, Michaela
Speranza, Domenico
Strozzi, Matteo
Tamburrino, Francesco
Uberti, Stefano
Uccheddu, Francesca
Vitali, Andrea
Zippo, Antonio

Contents

Experimental Methods in Product Development

Knowledge and Product Data Management

Engineering Methods in Human-Related Applications

Geometric Modelling and Analysis

Shape and Texture Analysis of Radiomic Data for Computer-Assisted Diagnosis and Prognostication: An Overview

Francesco Bianconi[1(✉)], Mario Luca Fravolini[1], Isabella Palumbo[2], and Barbara Palumbo[3]

[1] Department of Engineering, Università degli Studi di Perugia, Via Goffredo Duranti 93, 06135 Perugia, Italy
bianco@ieee.org, mario.fravolini@unipg.it

[2] Section of Radiation Oncology, Department of Surgical and Biomedical Sciences, Università degli Studi di Perugia, Piazza Lucio Severi 1, 06132 Perugia, Italy
isabella.palumbo@unipg.it

[3] Section of Nuclear Medicine and Health Physics, Department of Surgical and Biomedical Sciences, Università degli Studi di Perugia, Piazza Lucio Severi 1, 06132 Perugia, Italy
barbara.palumbo@unipg.it

Abstract. There is increasing evidence that shape and texture descriptors from imaging data could be used as image biomarkers for computer-assisted diagnosis and prognostication in a number of clinical conditions. It is believed that such quantitative features may help uncover patterns that would otherwise go unnoticed to the human eye, this way offering significant advantages against traditional visual interpretation. The objective of this paper is to provide an overview of the steps involved in the process – from image acquisition to feature extraction and classification. A significant part of the work deals with the description of the most common texture and shape features used in the literature; overall issues, perspectives and directions for future research are also discussed.

Keywords: Shape · Texture · Radiomics · Computer-assisted medicine

1 Introduction

Recent technological advances including new imaging modalities as well as storing, sharing and computing resources have facilitated the collection of very large amounts of three-dimensional medical data [1]. In this scenario shape and texture analysis of such data has been receiving increasing attention during the last few years. The overall objective is that of extracting quantitative parameters from the imaging data (*biomarkers*) capable of correlating with clinical features such as disease phenotype and/or survival. The whole process, usually referred to as *radiomics*, can be regarded as an improvement on the traditional

C. Rizzi et al. (Eds.): ADM 2019, LNME, pp. 3–14, 2020.
https://doi.org/10.1007/978-3-030-31154-4_1

practice wherein medical images were mostly used as pictures for qualitative visual interpretation only [2,3]. In the management of oncologic disorders, for instance, a number studies have supported the use of radiomics for a variety of tasks including prediction of outcome [4,5] and response to treatment [6,7]; discrimination between benign, malignant, primary and metastatic lesions [8–10]; and classification of hystologic subtypes [11].

Radiomics, however, is still a new discipline and definitely far from being mature. There are significant obstacles that prevent the application on a large scale – chief among them the lack of large enough datasets for building models and classifiers, and the absence of standards establishing how the biomarkers should be computed [12]. The objective of this paper is to provide an overview of the steps involved, discuss the open issues and indicate directions for future research. A significant part of the paper deals with the description of the most common texture and shape features used in the literature.

2 Methods

The flow-chart of Fig. 1 summarises the overall workflow in radiomics. Image acquisition is always the first step and can optionally be followed by a post-processing phase. Segmentation is then required to separate the region of interest (ROI) from the background. Feature extraction is the core of the procedure and consists of extracting a set of meaningful parameters (*features*) from the ROI. The features can undergo some post-processing step as for instance selection and/or reduction. Finally, the resulting data are fed to some classifier or regression model suitable for the required task.

2.1 Image Acquisition

There are three main classes of medical imaging modalities providing three-dimensional data [13]: Computed Tomography (CT), Positron Emission Tomography (PET) and Magnetic Resonance Imaging (MRI).

Computed Tomography is based on the unlike absorption of X-rays by different tissue types, therefore the signal is proportional to the tissue density in this case [14]. Positron Emission Tomography estimates the metabolic activity of the tissue by measuring the radioactive decay of some specific radio-tracers. Those used in PET contain isotopes (e.g. ^{11}C, ^{15}O and ^{18}F) which emit positrons through $\beta+$ decay. The positrons collide and annihilate with the electrons in the tissue, this way emitting two γ rays 180° apart that are detected by the sensors [15]. Finally, in Magnetic Resonance Imaging the signal comes from positrons (hydrogen nuclei) contained in water and lipids. The signal in this case is proportional relaxation time – i.e. the time to return to the equilibrium magnetization state once the external magnetisation field is switched off [16].

In all the imaging modalities the scanning usually proceeds axially (head to feet), this way producing, as a result, a variable number of axial cross-sections with fixed size (slices). A three-dimensional voxel model is eventually reconstructed by piling up all the slices.

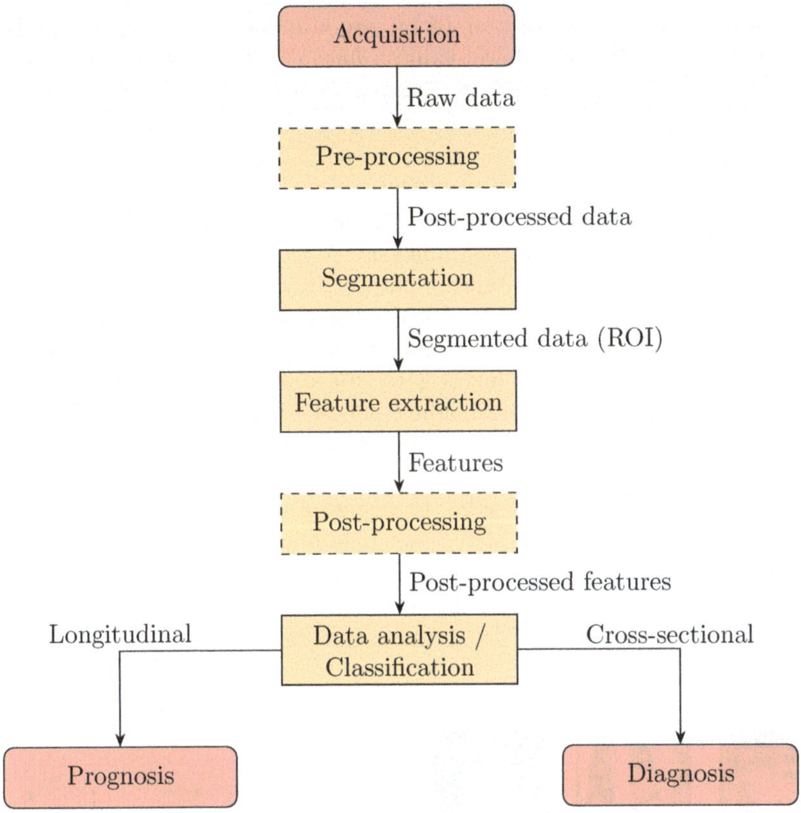

Fig. 1. The overall pipeline. Dashed lines indicate optional steps.

2.2 Pre-processing

Pre-processing usually involves one or more of the following operations: (1) *windowing* (rescaling), (2) *filtering* and (3) *resampling*. Although frequently overlooked, pre-processing is a fundamental step in the pipeline with significant effects on the overall results, as for instance shown in [17,18].

Windowing consists of applying an upper and lower threshold to the raw intensity values returned by the scans, this way excluding from the analysis those values that fall outside the range. In CT, for instance, windowing is routinely used to exclude from the analysis those anatomic parts (e.g. bones) that are reputed not relevant to the disease investigated.

Filtering can be carried out either to reduce noise and/or highlight features at different spatial scales. A variety of methods can be used for this purpose, as for instance Butterworth smoothing [18], Gaussian [19] and Laplacian of Gaussian [7,20] filters.

Resampling involves changing the number of bits (*bit depth*) used for encoding the intensity values. The bit depth of the raw data depends on the scanning device and settings used (12 and 16 bit are standard values in the practice). These are usually reduced to lower values (*downasmpling*) before feature extraction: eight, six and four bit and are common choices [6,18,21,22].

2.3 Segmentation

The objective of segmentation is that of identifying the part of the scan (ROI) that is considered relevant to the analysis. A ROI usually represents a clinically relevant region, as for instance a potentially cancerous lesion (Fig. 2). Segmentation is a crucial step, for it provides the input to the subsequent phases. Unfortunately, this is also a tedious and time-consuming procedure. Although a number of methods have been investigated for automatising the process – these include, among the others thresholding [23,24], region growing [23–25], edge detection [23,24] and convolutional networks [26,27] – segmentation remains by and large a manual procedure in which the experience and sensitivity of the physician play a major role. Besides, the decision whether to include or exclude dubious areas such as necrosis, atelectasis, inflammation and/or oedema is essentially the clinician's responsibility and, as such, hard to automatise.

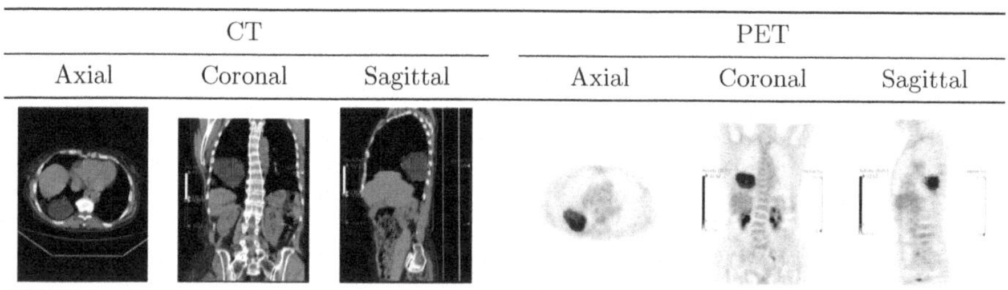

Fig. 2. CT (left) and PET (right) scans of a lung lesion with manually delineated ROI.

2.4 Feature Extraction

Feature extraction can be considered the 'core' of the whole procedure and consists of computing meaningful parameters from the regions of interest. There are two main strategies to feature extraction: the 'hand-designed' or 'hand-crafted' paradigm one the hand, and Deep Learning on the other.

In the hand designed approach the functions for feature extraction (also indicated as *image descriptors*) are mostly designed by hand, the design process being based on some prior knowledge about filtering, perceptual models and/or relatively intuitive visual properties (e.g. coarseness, business, contrast, etc.) This model-driven, 'a priori' paradigm is independent on the data to analyse. By contrast, Deep Learning is a data-driven, 'a posteriori' strategy in which the descriptors are essentially shaped by the data. The feature extractors, in this case, are based on sets of combinable blocks (layers) of which only the overall skeleton is defined a priori, and their behaviour depends on lots of free parameters whose values need to be determined by training over huge sets of data. In this paper we are mostly concerned with the hand-design paradigm; for an overview of Deep Learning and its potential applications in the field we refer the reader to Refs. [28–30].

Regardless the method used, there are some desirable properties that one would always expect from features. First, they should be *discriminative*, i.e.: they should enable good separation among the classes involved in the problem investigated (e.g. classification benign vs. malignant). Second, they should be *interpretable* on the basis of some physical characteristics (e.g. round/elongated, coarse/fine, etc.) Third, they should be *few*: this, again, facilitates interpretation, limits the computational overhead and reduces the chances of overfitting. Here below we briefly review some of the most common shape and texture features used in radiomics.

Shape Features. Shape features have been investigated as potential biomarkers for a range of diseases. In oncologic disorders, for instance, lesions presenting ill-defined ('spiculated') borders are considered suggestive of malignancy, aggressiveness and in general worse prognosis; whereas those with regular, well defined margins are more frequently indicative of benign or less aggressive lesions [31–33]. For a quantitative evaluation of shape different parameters have been proposed – among them *compactness*, *spherical disproportion*, *sphericity* and *surface-to-volume ratio* (Eqs. 1–4). In formulas, indicated with A the surface area of the ROI, V the volume and R the radius of a sphere with volume V we have:

$$\text{compactness} = \frac{V}{\sqrt{\pi} A^{2/3}} \tag{1}$$

$$\text{spherical dispr.} = \frac{A}{4\pi R^2} \tag{2}$$

$$\text{sphericity} = \frac{\pi^{1/3} (6V)^{2/3}}{A} \tag{3}$$

$$\text{surface-to-vol. ratio} = \frac{A}{V} \tag{4}$$

Compactness, spherical disproportion and surface-to-volume ratio from CT, for instance, were found predictive of malignancy in lung lesions [34]; surface-to-volume ratio from MRI showed potential to differentiate between clinically significant and non-significant prostate cancer [31]; and functional sphericity from PET images correlated with clinical outcome in non-small-cell lung cancer [32].

Texture Features

Basic Statistics. These are parameters that can be computed directly from the raw data with no further processing. Resampling is not required. They include: *mean*, *maximum*, *median*, *range*, *standard deviation*, *skewness* and *kurtosis* (for definitions and formulae see also [35]). All these features are by definition invariant to geometric transformations of the input data such as rotation, mirroring, scaling and/or voxel permutation. Most of these features are also rather intuitive and their implementation straightforward.

Histogram-Based Features. This kind of features are derived from the probability distribution (histogram) of the intensity levels within the ROI. Features like *energy* (Eq. 5 – sometimes also referred to as *uniformity*) and *entropy* (Eq. 6) are routinely used for assessing the 'heterogeneity' of tumour lesions. There is indeed evidence that higher heterogeneity may correlate with worse overall prognosis and response to treatment [36–39]. Histogram-based statistics are invariant to geometric transformations of the input data – just as basic statistics are – but they heavily depend on the resampling scheme used. In formulas, given N the number of quantisation levels and p the probability of occurrence of the i-th intensity level, we have:

$$\text{energy}_{\text{H}} = \sum_{i=0}^{N-1} \left[p\left(i\right)\right]^2 \tag{5}$$

$$\text{entropy}_{\text{H}} = \sum_{i=0}^{N-1} p\left(i\right) \log_2 \left[p\left(i\right)\right] \tag{6}$$

where entropy is expressed in bits in this case. Subscript 'H' is used to indicate that the features are computed from histograms and to differentiate them from those computed from co-occurrence matrices (see below).

Grey-Level Co-occurrence Matrices. Co-occurrence matrices (GLCM) represent the two-dimensional joint distribution of the intensity levels between pairs of voxels separated by a given displacement vector. By changing the orientation and the length of the vector GLCM can probe the local signal variation at different scales and orientations. Co-occurrence matrices, a classic tools in texture analysis, were originally designed for planar images [40] but their extension to three-dimensional data is straightforward [41]. In this case there are 26 possible orientations for a given scale and as many GLCM, of which, however, only 13 non-redundant. A GLCM with values mainly clustered around the main diagonal will indicate a texture with low variability; a highly dispersed matrix will be characteristic of a variable texture. To capture this behaviour one usually extracts some global parameters from the GLCM, as for instance *contrast*, *energy*, *entropy* and *homogeneity* (Eqs. 7–10). Again, these have shown potential as clinical biomarkers in a number of studies [5,6,22,42]. Indicated with i and j the indices of the two voxels separated by a given displacement vector, we have:

$$\text{contr.}_{\text{CM}} = \frac{\sum_{i=0}^{N-1} \sum_{j=0}^{N-1} \left(i-j\right)^2 p\left(i,j\right)}{\left(N-1\right)^2} \tag{7}$$

$$\text{energy}_{\text{CM}} = \sum_{i=0}^{N-1} \sum_{i=0}^{N-1} [p(i,j)]^2 \tag{8}$$

$$\text{entr.}_{\text{CM}} = \sum_{i=0}^{N-1} \sum_{j=0}^{N-1} p(i,j) \log_2 [p(i,j)] \tag{9}$$

$$\text{hom.}_{\text{CM}} = \sum_{i=0}^{N-1} \sum_{j=0}^{N-1} \frac{p(i,j)}{1+|i-j|} \tag{10}$$

Other Texture Features. Texture analysis has been an area of intense research for more than forty years, and, as a results, the amount of available methods is huge. Among those that have received attention in the field of radiomics are: neighbourhood grey-tone difference matrices (NGTDM [6,21,22,43]), grey-level run-length matrices (GLRLM [21,22,44]), Local Binary Patterns (LBP [17,45]), Laws' masks [46,47] and wavelets [48,49]. For definitions and further details we refer the reader to the given references.

2.5 Post-processing

The features returned by the extraction phase can undergo further processing to (a) reduce their number and (b) increase their discrimination capability. The main strategies to achieve this goal are *feature selection* and *feature generation* [50]. The first aims at identifying the most discriminative features so as to reduce their overall number while retaining as much information as possible. This is particularly important in radiomics, where some shape and texture features tend to be highly correlated to each other, as recently shown in [51]. The second consists of generating new features from the original ones via some suitable transformation, as for instance Principal Component Analysis (PCA) and Independent Component Analysis (ICA) [50].

2.6 Data Analysis/Classification

The last step consists of feeding the features to a classifier to make predictions about the disease type (*computer-assisted diagnosis*) and/or the clinical outlook (*prognostication*). To this end suitable machine learning models and large enough sets of labelled data (*train set*) are required. As for the model, one can choose among a number of different solutions (e.g. linear classifiers, Support Vector Machines, Classification Trees, neural networks and/or a combination of them [50,52]): the main problem is selecting the right model for the specific task. Getting the right data for training, however, can be rather hard, for it requires finding large enough sets of manually classified/annotated clinical records. For prognostication the data need also to be longitudinal, which implies following up on a cohort of patients for a long period of time.

3 Discussion

A number of recent studies have advocated the use of shape and texture descriptors as potential image biomarkers in the diagnosis and treatment of a range of disorders. Still, clear advantages for patient management are yet to be demonstrated above and beyond traditional imaging techniques and basic biomarkers [37]. From a technical standpoint there are at least three major hurdles that still limit the applicability of radiomics on a large scale. The first is the lack of standardisation in all the steps detailed in Sect. 2. Differences in the imaging acquisition parameters, feature definition and naming conventions, pre- and post-processing procedures are all sources of variations that reduce reproducibility, increase artefacts and eventually lead to biased results [36]. To overcome this problem, an important standardisation initiative is in progress [12]. The second is the lack of large enough datasets of imaging data to train the classifiers. To fill this gap, international, open access repositories are being developed [53]. The third is that some steps of the pipeline (e.g. lesion segmentation) still rely too heavily on human intervention, which may potentially lead to biased results and low reproducibility [54].

4 Conclusions

Shape and texture features from three-dimensional biomedical data have been attracting much research interest in recent years. It is believed that such quantitative imaging features may help uncover patterns that would otherwise go unnoticed to the human eye. In this paper we have provided an overview about the methods, issues and perspectives. On the whole it is clear that radiomics has the potential to improve the patient management in a number of diseases, but there are still significant obstacles along the way – chief among them the absence of standardisation and the lack of large datasets of clinical data.

Acknowledgements. This work was partially supported by the Department of Engineering at the University of Perugia, Italy, under the Fundamental Research programme 2017.

Appendix: Tools and Packages for Radiomics

A number of applications are already available for managing radiomics data. Among them, TexRAD (Feedback Medical Ltd, Cambridge, UK) is a commercial tool enabling lesion segmentation, feature extraction and statistical interpretation over radiomic data [55]. LIFEx (IMIV/CEA, Orsay, France) is a freeware package allowing textural analysis and radiomic feature measurements from PET, CT, ultrasound and MR images [56]. Finally, Pyradiomics (Computational Imaging & Bioinformatics Lab, Harvard Medical School, Cambridge, USA) is an open-source set of python libraries for the extraction of radiomics data from medical images as well as image loading and preprocessing [57].

References

1. Chalkidou, A., O'Doherty, M., Marsden, P.: False discovery rates in PET and CT studies with texture features: a systematic review. PLoS ONE **10**(5), article no. e0124165 (2015)
2. Gillies, R., Kinahan, P., Hricak, H.: Radiomics: images are more than pictures, they are data. Radiology **278**(2), 563–577 (2016)
3. Keek, S., Leijenaar, R., Jochems, A., Woodruff, H.: A review on radiomics and the future of theranostics for patient selection in precision medicine. Br. J. Radiol. **91**(1091), article no. 20170926 (2018)
4. Wang, X., Sun, W., Liang, H., Mao, X., Lu, Z.: Radiomics signatures of computed tomography imaging for predicting risk categorization and clinical stage of thymomas. BioMed Res. Int. **2019**, 10 (2019). article ID 3616852
5. Song, J., Liu, Z., Zhong, W., Huang, Y., Ma, Z., Dong, D., Liang, C., Tian, J.: Non-small cell lung cancer: quantitative phenotypic analysis of CT images as a potential marker of prognosis. Sci. Rep. **6**, article no. 38282 (2016)
6. Pyka, T., Bundschuh, R., Andratschke, N., Mayer, B., Specht, H., Papp, L., Zsótér, N., Essler, M.: Textural features in pre-treatment [F18]-FDG-PET/CT are correlated with risk of local recurrence and disease-specific survival in early stage NSCLC patients receiving primary stereotactic radiation therapy. Radiat. Oncol. **10**(1), article no. 100 (2015)
7. Ravanelli, M., Farina, D., Morassi, M., Roca, E., Cavalleri, G., Tassi, G., Maroldi, R.: Texture analysis of advanced non-small cell lung cancer (NSCLC) on contrast-enhanced computed tomography: prediction of the response to the first-line chemotherapy. Eur. Radiol. **23**(12), 3450–3455 (2013)
8. Andersen, M., Harders, S., Ganeshan, B., Thygesen, J., Madsen, H., Rasmussen, F.: CT texture analysis can help differentiate between malignant and benign lymph nodes in the mediastinum in patients suspected for lung cancer. Acta Radiol. **57**(6), 669–676 (2016)
9. Dennie, C., Thornhill, R., Sethi-Virmani, V., Souza, C., Bayanati, H., Gupta, A., Maziak, D.: Role of quantitative computed tomography texture analysis in the differentiation of primary lung cancer and granulomatous nodules. Quant. Imaging Med. Surg. **6**(1), 6–15 (2016)
10. Kirienko, M., Cozzi, L., Rossi, A., Voulaz, E., Antunovic, L., Fogliata, A., Chiti, A., Sollini, M.: Ability of FDG PET and CT radiomics features to differentiate between primary and metastatic lung lesions. Eur. J. Nucl. Med. Mol. Imaging **45**(10), 1649–1660 (2018)
11. Zhu, X., Dong, D., Chen, Z., Fang, M., Zhang, L., Song, J., Yu, D., Zang, Y., Liu, Z., Shi, J., Tian, J.: Radiomic signature as a diagnostic factor for histologic subtype classification of non-small cell lung cancer. Eur. Radiol. **28**(7), 2772–2778 (2018)
12. Hatt, M., Vallieres, M., Visvikis, D., Zwanenburg, A.: IBSI: an international community radiomics standardization initiative. J. Nucl. Med. **59**(1 supp.), 287 (2018)
13. Barrie Smith, N., Webb, A.: Introduction to Medical Imaging: Physics, Engineering and Clinical Applications. Cambridge University Press, Cambridge (2011)
14. Goldman, L.: Principles of CT and CT technology. J. Nucl. Med. Technol. **35**(3), 115–128 (2007)
15. Basu, S., Kwee, T., Surti, S., Akin, E., Yoo, D., Alavi, A.: Fundamentals of PET and PET/CT imaging. Ann. N. Y. Acad. Sci. **1228**(1), 1–18 (2011)

16. Grover, V., Tognarelli, J., Crossey, M., Cox, I., Taylor-Robinson, S., McPhail, M.: Magnetic resonance imaging: principles and techniques: lessons for clinicians. J. Clin. Exp. Hepatol. **5**(3), 246–255 (2015)
17. Bagher-Ebadian, H., Siddiqui, F., Liu, C., Movsas, B., Chetty, I.: On the impact of smoothing and noise on robustness of CT and CBCT radiomics features for patients with head and neck cancers. Med. Phys. **44**(5), 1755–1770 (2017)
18. Fave, X., Zhang, L., Yang, J., Mackin, D., Balter, P., Gomez, D., Followill, D., Jones, A., Stingo, F., Court, L.: Impact of image preprocessing on the volume dependence and prognostic potential of radiomics features in non-small cell lung cancer. Transl. Cancer Res. **5**(4), 349–363 (2016)
19. Brunese, L., Greco, B., Setola, F., Lassandro, F., Guarracino, M., De Rimini, M., Piccolo, S., De Rosa, N., Muto, R., Bianco, A., Muto, P., Grassi, R., Rotondo, A.: Non-small cell lung cancer evaluated with quantitative contrast-enhanced CT and PET-CT: net enhancement and standardized uptake values are related to tumour size and histology. Med. Sci. Monit. **19**(1), 95–101 (2013)
20. Ganeshan, B., Goh, V., Mandeville, H., Ng, Q., Hoskin, P., Miles, K.: Non-small cell lung cancer: histopathologic correlates for texture parameters at CT. Radiology **266**(1), 326–336 (2013)
21. Bianconi, F., Fravolini, M., Bello-Cerezo, R., Minestrini, M., Scialpi, M., Palumbo, B.: Evaluation of shape and textural features from CT as prognostic biomarkers in non-small cell lung cancer. Anticancer Res. **38**(4), 2155–2160 (2018)
22. Yip, S., McCall, K., Aristophanous, M., Chen, A., Aerts, H., Berbeco, R.: Comparison of texture features derived from static and respiratory-gated PET images in non-small cell lung cancer. PLoS ONE **9**(12), article no. e115510 (2014)
23. Zheng, L., Lei, Y.: A review of image segmentation methods for lung nodule detection based on computed tomography images. In: Proceedings of the 2nd International Conference on Electronic Information Technology and Computer Engineering. MATEC Web of Conferences, Shanghai, China, vol. 232, October 2018. art. no. 02001
24. Jayanthi, M.: Comparative study of different techniques used for medical image segmentation of liver from abdominal CT scan. In: Proceedings of the 2016 IEEE International Conference on Wireless Communications, Signal Processing and Networking (WiSPNET), Chennai, India, pp. 1462–1465, March 2016. art. no. 7566379
25. Sandor, T., Metcalf, D., Kim, Y.J.: Segmentation of brain CT images using the concept of region growing. Int. J. Biomed. Comput. **29**(2), 133–147 (1991)
26. Chen, L., Bentley, P., Mori, K., Misawa, K., Fujiwara, M., Rueckert, D.: DRINet for medical image segmentation. IEEE Trans. Med. Imaging **37**(11), 2453–2462 (2018). art. no. 8357580
27. Chlebus, G., Schenk, A., Moltz, J., van Ginneken, B., Hahn, H., Meine, H.: Automatic liver tumor segmentation in CT with fully convolutional neural networks and object-based postprocessing. Sci. Rep. **8**(1), article no. 15497 (2018)
28. Vial, A., Stirling, D., Field, M., Ros, M., Ritz, C., Carolan, M., Holloway, L., Miller, A.: The role of deep learning and radiomic feature extraction in cancer-specific predictive modelling: a review. Transl. Cancer Res. **7**(3), 803–816 (2018)
29. Lecun, Y., Bengio, Y., Hinton, G.: Deep learning. Nature **521**(7553), 436–444 (2015)
30. Xie, D., Zhang, L., Bai, L.: Deep learning in visual computing and signal processing. Appl. Comput. Intell. Soft Comput., article no. 1320780, February 2017

31. Cuocolo, R., Stanzione, A., Ponsiglione, A., Romeo, V., Verde, F., Creta, M., La Rocca, R., Longo, N., Pace, L., Imbriaco, M.: Clinically significant prostate cancer detection on MRI: a radiomic shape features study. Eur. J. Radiol. **116**, 144–149 (2019)
32. Hatt, M., Laurent, B., Fayad, H., Jaouen, V., Visvikis, D., Le Rest, C.: Tumour functional sphericity from PET images: prognostic value in NSCLC and impact of delineation method. Eur. J. Nucl. Med. Mol. Imaging **45**(4), 630–641 (2018)
33. Limkin, E., Reuzé, S., Carré, A., Sun, R., Schernberg, A., Alexis, A., Deutsch, E., Ferté, C., Robert, C.: The complexity of tumor shape, spiculatedness, correlates with tumor radiomic shape features. Sci. Rep. **9**(1), article no. 4329, (2019)
34. Wang, J., Liu, X., Dong, D., Song, J., Xu, M., Zang, Y., Tian, J.: Prediction of malignant and benign of lung tumor using a quantitative radiomic method. In: Proceedings of the Annual International Conference of the IEEE Engineering in Medicine and Biology Society (EMBS), Orlando, United States, pp. 1272–1275, October 2016. art. no. 7590938
35. Newbold, P., Carlson, W., Thorne, B.: Statistics for Business and Economics, 6th edn. Pearson Education International (2007)
36. Bashir, U., Siddique, M., McLean, E., Goh, V., Cook, G.: Imaging heterogeneity in lung cancer: techniques, applications, and challenges. Am. J. Roentgenol. **207**(3), 534–543 (2016)
37. O'Connor, J., Rose, C., Waterton, J., Carano, R., Parker, G., Jackson, A.: Imaging intratumor heterogeneity: role in therapy response, resistance, and clinical outcome. Clin. Cancer Res. **21**(2), 249–257 (2015)
38. Win, T., Miles, K., Janes, S., Ganeshan, B., Shastry, M., Endozo, R., Meagher, M., Shortman, R., Wan, S., Kayani, I., Ell, P., Groves, A.: Tumor heterogeneity and permeability as measured on the CT component of PET/CT predict survival in patients with non-small cell lung cancer. Clin. Cancer Res. **19**(13), 3591–3599 (2013)
39. Ganeshan, B., Panayiotou, E., Burnand, K., Dizdarevic, S., Miles, K.: Tumour heterogeneity in non-small cell lung carcinoma assessed by CT texture analysis: a potential marker of survival. Eur. Radiol. **22**(4), 796–802 (2012)
40. Haralick, R.M., Shanmugam, K., Dinstein, I.: Textural features for image classification. IEEE Trans. Syst. Man Cybern. **3**(6), 610–621 (1973)
41. Ben Othmen, E., Sayadi, M., Fniaech, F.: 3D gray level co-occurrence matrices for volumetric texture classification. In: Proceedings of the 3rd International Conference on Systems and Control (ICSC), Algiers, Algeria, pp. 833–837, October 2013
42. Aerts, H., Velazquez, E., Leijenaar, R., Parmar, C., Grossmann, P., Cavalho, S., Bussink, J., Monshouwer, R., Haibe-Kains, B., Rietveld, D., Hoebers, F., Rietbergen, M., Leemans, C., Dekker, A., Quackenbush, J., Gillies, R., Lambin, P.: Decoding tumour phenotype by noninvasive imaging using a quantitative radiomics approach. Nat. Commun. **5**, article no. 4006 (2014)
43. Adamasun, M., King, R.: Textural features corresponding to textural properties. IEEE Trans. Syst. Man Cybern. **19**(5), 1264–1274 (1989)
44. Galloway, M.: Texture analysis using gray level run lengths. Comput. Graph. Image Process. **4**(2), 172–179 (1975)
45. Ojala, T., Pietikäinen, M., Mäenpää, T.: Multiresolution gray-scale and rotation invariant texture classification with local binary patterns. IEEE Trans. Pattern Anal. Mach. Intell. **24**(7), 971–987 (2002)

46. Hawkins, S., Korecki, J., Balagurunathan, Y., Gu, Y., Kumar, V., Basu, S., Hall, L., Goldgof, D., Gatenby, R., Gillies, R.: Predicting outcomes of nonsmall cell lung cancer using CT image features. IEEE Access **2**, 1418–1426 (2014). art. no. 6966732
47. Laws, K.: Rapid texture identification. In: Wiener, T. (ed.) Image Processing for Missile Guidance. SPIE Proceedings, vol. 0238 (1980)
48. Balagurunathan, Y., Gu, Y., Wang, H., Kumar, V., Grove, O., Hawkins, S., Kim, J., Goldgof, D., Hall, L., Gatenby, R., Gillies, R.: Reproducibility and prognosis of quantitative features extracted from CT images. Transl. Oncol. **7**(1), 72–87 (2014)
49. Daubechies, I.: CBMS-NSF Regional Conference Series in Applied Mathematics, Ten Lectures on Wavelets, vol. 61. Society for Industrial and Applied Mathematics (1992)
50. Theodoridis, S., Koutroumbas, K.: Pattern Recognition, 3rd edn. Academic Press, Burlington (2006)
51. Bianconi, F., Palumbo, I., Fravolini, M., Chiari, R., Minestrini, M., Brunese, L., Palumbo, B.: Texture analysis on [18F]FDG PET/CT in non-small-cell lung cancer: correlations between PET features, CT features, and histological types. Molecular Imaging and Biology (2019, in press)
52. Duda, R., Hart, P., Stork, D.: Pattern Classification, 2nd edn. Wiley, New York (2001)
53. The Cancer Imaging Archive (TCIA). http://www.cancerimagingarchive.net/. Accessed 04 June 2019
54. Rizzo, S., Botta, F., Raimondi, S., Origgi, D., Fanciullo, C., Morganti, A., Bellomi, M.: Radiomics: the facts and the challenges of image analysis. Eur. Radiol. Exp. **2**(1), 36 (2018)
55. Feedback Medical Ltd: TexRAD (Texture + Radiology). https://fbkmed.com/texrad-landing-2/. Accessed 20 July 2019
56. Nioche, C., Orlhac, F., Boughdad, S., Reuze, S., Goya-Outi, J., Robert, C., Pellot-Barakat, C., Soussan, M., Frouin, F., Buvat, I.: LIFEx: a freeware for radiomic feature calculation in multimodality imaging to accelerate advances in the characterization of tumor heterogeneity. Cancer Res. **78**(16), 4786–4789 (2018)
57. van Griethuysenand, J., Fedorovand, A., Parmarand, C., Hosnyand, A., Vivek Narayanand, A., Beets-Tanand, R., Fillion-Robinand, J.C., Pieperand, S., Aerts, H.: Computational radiomics system to decode the radiographic phenotype. Cancer Res. **77**(21), e104–7 (2017)

Mandible Morphing Through Principal Components Analysis

Giulia Pascoletti[1], Michele Calì[2(✉)], Cristina Bignardi[3], Paolo Conti[1], and Elisabetta M. Zanetti[1]

[1] Department of Engineering, University of Perugia, Via Duranti 67, 06125 Perugia, Italy

[2] Department of Electric, Electronics and Computer Engineering, University of Catania, Viale A. Doria 6, 95125 Catania, Italy
michele.cali@diei.unict.it

[3] PolitoBIOMed Lab, Department of Mechanical and Aerospace Engineering, Politecnico di Torino, Corso Duca degli Abruzzi 24, 10129 Turin, Italy

Abstract. The aim of this research is to develop patient-specific 3D mandible models, based on a limited number of measurements taken on the patient. Twenty Computed Tomography scans were used to build the respective 3D cad models of the mandible. Fifteen of these models were given as an input to a Principal Component Analysis software, and eight 'principal' mandible morphologies were produced. The following step was to identify the most efficient landmarks to 'weight' these morphologies when building a patient-specific model. Two further mandible computed tomography scans (a 'normal' mandible and a 'severely resorbed' one) were used to test the full procedure and to assess its accuracy.

The accuracy of the 3D morphed surface resulted to range between 0.025 and 3.235 mm for the 'normal' mandible and between 0.012 and 1.149 mm for the 'severely resorbed' one having used eight landmarks to morph a 'standard' mandible.

This work demonstrates how patient-specific models can be obtained registering the position of a limited number of points (on panoramic x-ray or on the physical model), reaching a good accuracy. This allows performing patient-specific planning and numerical simulations even for those cases where a computed tomography scan would not be available. In fact, this procedure can be interfaced with mesh morphing algorithms to automatically build finite element models. The accuracy of the procedure can be further improved, widening the mandibles computed tomography scans database and optimizing landmarks position.

Keywords: Morphing · Principal Component Analysis · Patient-specific models

1 Introduction

The use of Computer Aided Surgery [1] or, more in general, the pre-operative planning of interventions and of surgical procedures requires the creation of patient-specific models [2]. The bone geometry can be used to produce physical models through additive manufacturing techniques [3], or it can be coupled to the mechanical properties distribution,

C. Rizzi et al. (Eds.): ADM 2019, LNME, pp. 15–23, 2020.
https://doi.org/10.1007/978-3-030-31154-4_2

as often done for biomechanical considerations both in relation to orthopaedics and dentistry applications [4–7]. This work is focused on the second applications, being finalized to the creation of patient-specific models of the mandibular bone.

The first step for the creation of these models is the definition of the respective morphology. Bone morphology can be obtained with optimal accuracy from CT (Computed Tomography) scans (through reverse engineering), however CT scans are not always available: for example, in the dental field, panoramic x-rays are still the standard imaging approach for most applications. Secondly, statistical analyses, simulating a wide range of possible morphologies, are often required and building a so large CT dataset could not be a viable perspective. This work represents an alternative to CT to build patient- specific morphologies, and it is based on the creation of a parametric mandibular model. This 3D model can be deformed or 'morphed', based on features observable on x-rays and panoramic x-ray views [8], from 3D scanning [9], or features assigned from a given statistical distribution [10].

In the following, the methodology used for mandible parametrization is explained in details and the respective accuracy is tested on two mandibles.

2 Materials and Methods

A database of twenty CT scans of mandibular bones was used. Three CT scans were discarded since their quality was not satisfactory (CT scans of a full head) or they did not represent a complete mandible. Two CT scans were set apart and were used to test the accuracy of the set up methodology. Fifteen CT scans were used to perform Principal Component Analysis (PCA in the following). CT sample details are reported in Table 1. Mandibles of patients with different age and gender were used in order to create a database as wide as possible.

2.1 3D Mandible Models from CT Scans

The first step of the modelling procedure required the digital reconstruction of mandibles from CT scans.

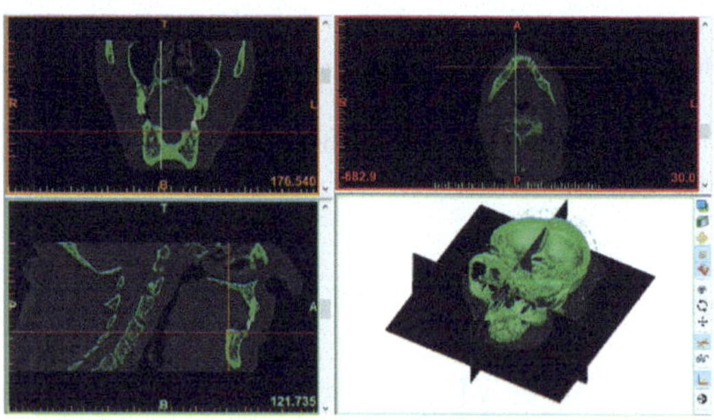

Fig. 1. The green areas refer to bone volumes, obtained through suitable thresholds: frontal (top left), transversal (top right), lateral (bottom left), (d) isometric (bottom right) views.

Table 1. CT Samples details

Sample #	Gender	Age [years]
1	M	80
2	F	55
3	F	23
4	F	21
5	M	31
6	F	89
7	M	17
8	M	47
9	F	20
10	M	75
11	F	67
12	F	24
13	M	52
14	M	54
15	M	26
16	M	20
17	F	29
18	F	38
19	M	40
20	F	44

MIMICS Software® by Materialise was used to build 3D models from CT scans (each one contained about 600 images). Images were segmented using a threshold filter in order to keep only bones and eliminate all other tissues (Fig. 1), based on the respective grey level (Hounsfield Units, HU). For each mandible, the segmentation mask has been created starting from the default 'Bone Threshold' ($HU_{min} = 226$ – $Hu_{max} = 3071$) and adjusting these values for the specific CT data. Indeed, each mandible has little different HU value and therefore a unique segmentation mask could not be defined. In the next step, 'growing region' algorithms were used to isolate the only mandibular bone from other bones. Polylines were calculated as the contours of bone areas in the transverse sections, and both the outer and inner contours were built automatically. Bone volumes were created from polylines and the STL model of the associated external geometry was computed, and finally the surface mesh was exported to ANSYS Mechanical software® by ANSYS. Sometimes some residual teeth volumes remained since teeth cannot be perfectly differentiated from bone, based on the HU only. In these cases teeth volumes were removed based on curvature discontinuity at the apical bone region and patches were created by extrusion of the adjacent bone regions.

After having meshed the first mandible ('template' mesh), the mesh of all other mandibles was obtained through Radial Basis Function (RBF) morph algorithm [11, 12].

In this way all mandibles shared iso-topological meshes: a certain node number identifies the same bone area [13] (for example the top of the coronoid process) on all mandibles.

2.2 Principal Components Analysis

The Principal Component Analysis (PCA) is a *statistical* technique where an orthogonal transformation is used to convert a set of observations of possibly correlated variables into a set of values of linearly uncorrelated variables called principal components. When the 'set of observations' are the coordinates of surface points of a volume (the mandible in this case), the principal components are the principal 'morphing modes' of that volume: a morphing mode is a set of nodal displacement which defines the transition from the 'average' shape to another shape. Morphing modes are called 'principal' when they are uncorrelated.

The principal component analysis was here performed by means of MorphoJ software, which is freely accessible online [14]. A detailed description of this analysis and its theoretical fundament can be found in [15]. As a result, the principal 'modes' of deformation of the 'average' mandible can be identified; these modes are used to describe whichever mandible as a linear combination of principal modes of deformation of the 'average' mandible, with proper 'weights':

$$\{y'\} = \{y\} + a_1 \cdot \{\Delta y_1\} + a_2 \cdot \{\Delta y_2\} + \cdots + a_n \cdot \{\Delta y_n\} \tag{1}$$

where:

- y' is the vector containing all node coordinates for the 'current' mandible
- y is the vector containing all node coordinates for the 'average' mandible
- Δy_1, Δy_2, Δy_n are vectors containing node-by-node coordinate variations, respectively for deformation modes '1', '2', …'n' (where 'n' is the number of CT scans used to perform the principal components analysis).
- a_1, a_2, a_n are the weights to be given respectively to deformation (morphing modes) *'1', '2', ... 'n'*.

The number of principal components can be reduced to a number 'm' lower than 'n', if an approximation is tolerated: in this work, 'm' number has been defined considering that principal components had to cover at least 85% variability of experimental results.

The estimation of 'm' weight factors a_i can be performed on the basis of a subset of coordinates ('landmark' coordinates): strictly speaking 'm' coordinates would be enough to estimate 'm' weights. However a higher number of coordinates allows reaching a more robust regression. In this work eight landmarks have been considered (with three coordinates per landmark, Fig. 2), referring to anatomical points which are commonly used and which can be retrieved from panoramic x-rays.

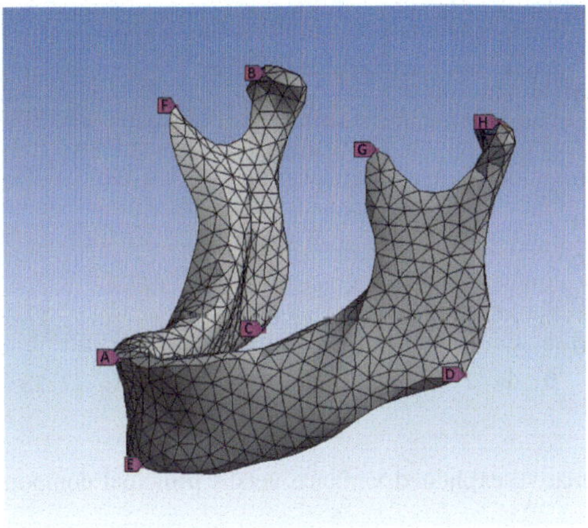

Fig. 2. Landmarks used for mandible morphing.

2.3 Evaluation of Reconstruction Accuracy

The whole procedure was tested with 3-MATIC Software® by Materialise, considering two mandibles which do not belong to PCA database. The first mandible will be called 'Normal' in the following since it is a mandible which does not present any peculiar features. The second one is a 'Severely resorbed' mandible and it was selected in order to test the procedure on a mandible whose shape is significantly different from mandibles used to set up PCA database; as such, this mandible could give an idea of the minimum performance of the set up methodology.

3 Results and Discussion

3.1 Principal Component Analysis

According to the results of Principal Component Analysis, eight Components are sufficient to cover over 85% variability of experimental results (for the 15 mandibles belonging to PCA dataset), as illustrated in Fig. 3.

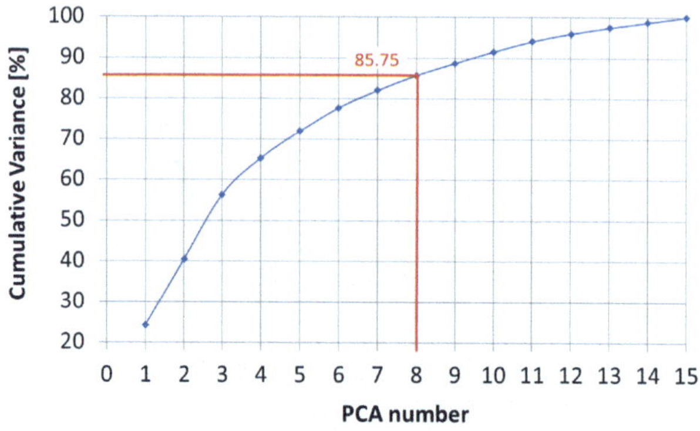

Fig. 3. Cumulative explained variance versus principal components number.

The principal morphing modes are depicted in Fig. 4.

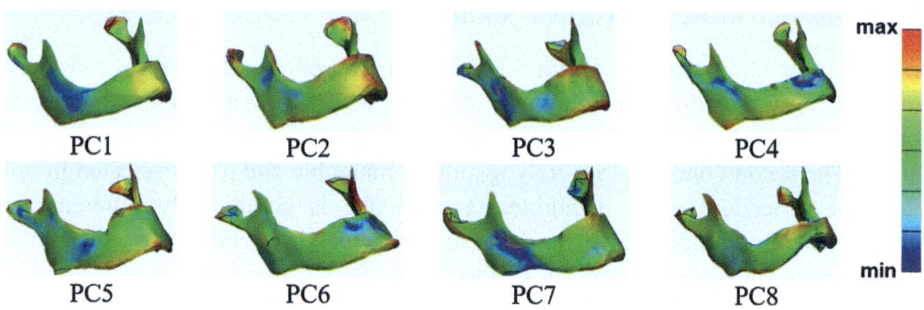

Fig. 4. First eight principals morphing modes; a qualitative colour map is used to visualize differences from the 'standard' geometry.

3.2 Reconstruction Accuracy

The whole procedure was tested on the 'Normal' variable, producing an accuracy ranging from 0.025 to 3.235 mm. Its spatial distribution is uneven: the largest errors can be found in lateral alveolar parts (Fig. 5).

The second test was performed on a 'severely resorbed mandible'. Results are reported in Fig. 6. In this case the accuracy ranges between 0.012 and 2.587 mm, with an average error equal to 0.840 mm. The most critical areas are still the lateral alveolar parts, but also the central alveolar part and the chin area are affected by a larger error in this case.

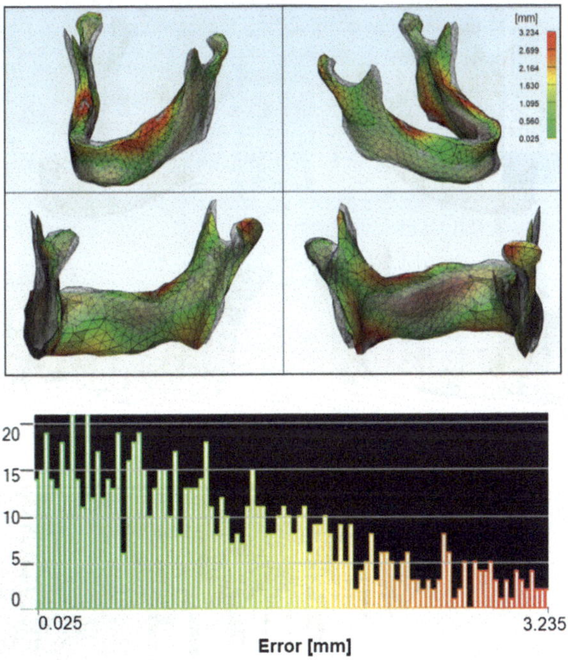

Fig. 5. Results obtained on a ‘normal’ mandible. Top: 3D views of the actual mandible (in colors) and of the fitted mandible (in transparency). Bottom: Error distribution. Red color identifies region with the maximum deviation (namely with the maximum error), green region identifies region with the minimum deviation (namely with the minimum error).

As general remarks, it can be concluded that the full methodology is characterized by an average accuracy which is below 1 mm. The reconstruction error has resulted to be particularly high in the alveolar areas, but this is likely to be due to the difficult differentiation between the bone tissue and the dental tissue: smoothing of this area seems to be advisable. The methodology accuracy is not affected by the mandibular resorption. The methodology can be further optimized, widening the mandibular database and making an effort to optimize landmarks position. The final aim is being able to study the load distribution in the bone and in bone-implant system with numerical models [16], supported by experimental tests [17]: the next step is considering inner volume nodes as well, reading their density from CT and assigning the respective elastic properties, according to functions reported in literature [18].

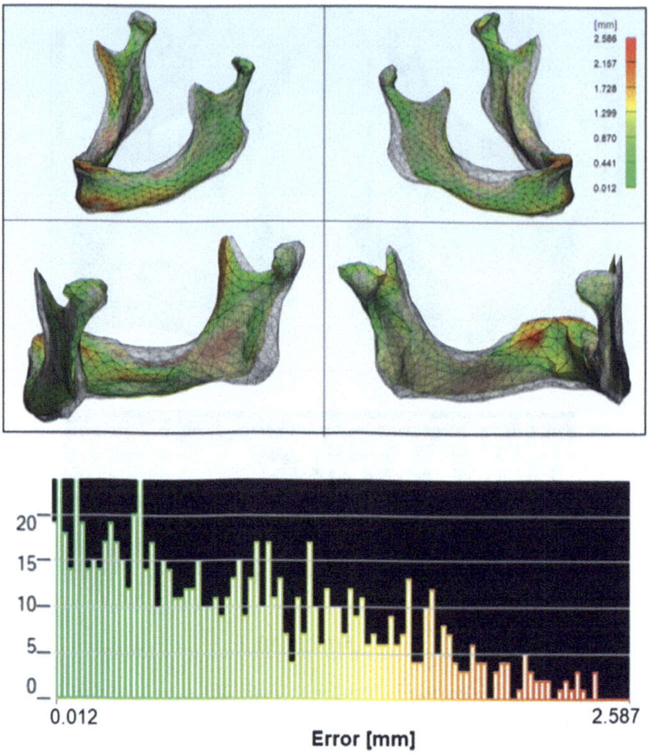

Fig. 6. Results obtained on a 'resorbed' mandible. Top: 3D views of the actual mandible (in colours) and of the fitted mandible (in transparency). Bottom: Error distribution.

4 Conclusions

A methodology was set up to build patient-specific mandibular models. This methodology is able to produce patient-specific 3D models of the mandibular bone, given the coordinates of eight landmarks, measured on the patient-specific geometry. The methodology proved to produce a good average accuracy (less than 1 mm); further efforts will be devoted to optimize the choice of the landmarks set and to calculate not only the outer bone morphology but also the distribution of mechanical properties.

References

1. Gargallo-Albiol, J., Barootchi, S., Salomó-Coll, O., Wang, H.L.: Advantages and disadvantages of implant navigation surgery. A systematic review. Annals of Anatomy-Anatomischer Anzeiger (2019)
2. Alves, P.V., Bolognese, A.M., Zhao, L.: Three-dimensional computerized orthognathic surgical treatment planning. Clin. Plast. Surg. **34**(3), 427–436 (2007)

3. Kamio, T., Hayashi, K., Onda, T., Takaki, T., Shibahara, T., Yakushiji, T., Kato, H.: Utilizing a low-cost desktop 3D printer to develop a "one-stop 3D printing lab" for oral and maxillofacial surgery and dentistry fields. 3D printing in medicine **4**(1), 6 (2018)
4. Zanetti, E.M., Bignardi, C.: Structural analysis of skeletal body elements: numerical and experimental methods. In: Biomechanical Systems Technology: Volume 3: Muscular Skeletal Systems, pp. 185–225 (2009)
5. Calì, M., Zanetti, E.M., Oliveri, S.M., Asero, R., Ciaramella, S., Martorelli, M., Bignardi, C.: Influence of thread shape and inclination on the biomechanical behaviour of plateau implant systems. Dent. Mater. **34**(3), 460–469 (2018)
6. Zanetti, E.M., Ciaramella, S., Calì, M., Pascoletti, G., Martorelli, M., Asero, R., Watts, D.C.: Modal analysis for implant stability assessment: sensitivity of this methodology for different implant designs. Dent. Mater. **34**(8), 1235–1245 (2018)
7. Pascoletti, G., Cianetti, F., Putame, G., Terzini, M., Zanetti, E.M.: numerical simulation of an intramedullary elastic nail: expansion phase and load-bearing behavior. Front. Bioeng. Biotechnol. **21**(6), 174 (2018)
8. Shah, N., Bansal, N., Logani, A.: Recent advances in imaging technologies in dentistry **6** (10), 794–807 (2014)
9. Logozzo, S., Kilpelä, A., Mäkynen, A., Zanetti, E.M., Franceschini, G.: Recent advances in dental optics–Part II: experimental tests for a new intraoral scanner. Opt. Lasers Eng. **54**, 187–196 (2014)
10. Yates, K.M., Untaroiu, C.D.: Finite element modeling of the human kidney for probabilistic occupant models: statistical shape analysis and mesh morphing. J. Biomech. **74**, 50–56 (2018)
11. Valentini, P.P., Biancolini, M.E.: Interactive sculpting using augmented-reality, mesh morphing, and force feedback: force-feedback capabilities in an augmented reality environment. IEEE Consum. Electron. Mag. **7**(2), 83–90 (2018)
12. Biancolini, M.E., Valentini, P.P.: Virtual human bone modelling by interactive sculpting, mesh morphing and force-feedback. Int. J. Interact. Des. Manuf. (IJIDeM) **12**(4), 1223–1234 (2018)
13. Zanetti, E.M., Crupi, V., Bignardi, C., Calderale, P.M.: Radiograph-based femur morphing method. Med. Biol. Eng. Comput. **43**(2), 181–188 (2005)
14. Klingenberg, C.P.: MorphoJ: an integrated software package for geometric morphometrics. Mol. Ecol. Resour. **11**(2), 353–357 (2011)
15. Dunteman, G.H.: Principal Components Analysis (No. 69). Sage (1989)
16. Menicucci, G., Ceruti, P., Barabino, E., Screti, A., Bignardi, C., Preti, G.: A preliminary in vivo trial of load transfer in mandibular implant-retained overdentures anchored in 2 different ways: allowing and counteracting free rotation. Int. J. Prosthodont. **19**(6), 574–576 (2006)
17. Vitale, M.C., Chiesa, M., Coltellaro, F., Bignardi, C., Celozzi, M., Poggio, C.: FEM analysis of different dental root canal-post systems in young permanent teeth. Eur. J. Paediatr. Dent. **9** (3), 111–117 (2008)
18. Xin, P., Nie, P., Jiang, B., Deng, S., Hu, G., Shen, S.G.: Material assignment in finite element modeling: heterogeneous properties of the mandibular bone. J. Craniofac. Surg. **24** (2), 405–410 (2013)

Flying Shape Sails Analysis by Radial Basis Functions Mesh Morphing

Michele Calì[1(✉)], Domenico Speranza[2], Ubaldo Cella[3], and Marco Evangelos Biancolini[3]

[1] Department of Electric, Electronics and Computer Engineering, University of Catania, Catania, Italy
mcali@dii.unict.it
[2] Department of Civil and Mechanical Engineering, University of Cassino and Southern Lazio, Cassino, Italy
[3] Department of Enterprise Engineering "Mario Lucertini", University of Rome "Tor Vergata", Rome, Italy

Abstract. High fidelity calculation tools are well established in the nautical design sector where advanced numerical simulations are adopted for the prediction of the interaction of boat parts with surrounding fluids. The capability to couple such tools with efficient shape parametrization procedures offers the possibility to further improve the performance speeding up the design process. Radial Basis Functions (RBF) Mesh Morphing (MM) allows to quickly modify the shape within numerical domains without the need of updating the underlying CAD representation. The validity of this approach, widely adopted in aeronautical and automotive fields, is demonstrated in this paper by applying the method to the analysis of the flying shape of a symmetric spinnaker also investigating the importance of panel arrangement on sail characteristics. The performance, in terms of drive and side forces, is evaluated for different morphed geometries by RANS (Reynolds Averaged Navier Stokes) analyses. The RBF setup proved to be efficient and robust in generating a good quality of the morphed domain within the full range of amplification from the undeformed to the flying shape geometry.

Keywords: Geometric parametrization · Mesh morphing · Symmetric spinnaker · Radial basis functions · Drive and side force · RANS analysis

1 Introduction

Many researches have been carried out to improve the performance of sailing yachts and sailing boats not only within the context of America's Cup and major international sailing races, but also for a variety of other applications where the design is driven by competition. Higher performance computing environment and very large domains are often adopted to study the interaction of boat parts with surrounding fluids. Methods coupling Computational Fluid Dynamics (CFD) and Finite Element Analysis (FEA) offer the possibility to investigate the fluid-structure interaction allowing to analyze and design the flying shape of sails. In literature, it is possible to find several

C. Rizzi et al. (Eds.): ADM 2019, LNME, pp. 24–36, 2020.
https://doi.org/10.1007/978-3-030-31154-4_3

papers that deal with sail aeroelastic analysis, shapes parametrization as well as materials and manufacturing production technologies [1, 2].

A comparison among different design shapes, arrangement of panels and materials can be effective only through a realistic sail flying shape evaluation [3, 4]. Various tools and methods have been adopted to evaluate sail flying shape performance during navigation or in real and virtual wind tunnel, such as: Coordinate Measuring Machine [5]; the optical profilometry technique that measures free-surface deformations using fringe pattern projection; cameras in wind tunnel tests [6]; laser measuring tools like "LIDAR"; a "Time-Of-Flight" radar to detect flying shapes [7]. Some other studies, instead, deal with the sail performance by recording pressure and stress measurements: total aerodynamic forces and loads by means of sailing dynamometers [8]; combined sail and rig shape recognition system using imaging techniques – VSPARS [9]; determination of the pressure distributions on the sails [10, 11]; real-time velocity prediction program; influence of rigging; aerodynamic force deduction using pressure and shape measurement [12, 13]. Nevertheless, none of the above mentioned methods allow to make a quantitative shared evaluation of the influence of fiber orientation and panel arrangement on sail performance. In general, CFD analyses are relative to isotropic homogeneous membranes and the simulation does not consider any effect of panel deformation and arrangement.

This paper contains the main results of the authors' research works about aerodynamic analysis of the sails performances using Fluid Structure Interaction (FSI) analysis and Radial Basis Functions (RBF) - Mesh Morphing (MM). The results obtained adapting the numerical model according to the experimentally acquired flying shape of nylon SK75 tri-radial symmetric spinnaker are presented. The RBF morphing action is integrated within the CFD solver and allows to efficiently setup an aeroelastic model.

The proposed mesh morphing parametrization represents the main novelty of the method and allows to setup complex shape variation [14–19] by exploiting the feature of a mathematical framework being one of the most efficient in facing the smoothing problem though only recently available out from research environments. RBF Morph is, in fact, the first commercial software with offers a mesh morphing tool based on Radial Basis Functions [20].

The remainder of this article is organized in 4 sections, introduction excluded, described as follows: in Sect. 2, a background knowledge on RBF-MM method is discussed with a special concern of the parametrization strategy used in the FSI problem analysis; in Sect. 3, the application of RBF-MM method for a symmetric spinnaker shape adaption onto the flying shape is presented; in Sect. 4, the principal results of the studies are shown. Final considerations and conclusions are discussed in Sect. 5.

2 Geometric Parametrization Based on RBF-MM

The geometric parametrization based on RBF-MM implements shape modifiers, amplified by intensity parameters directly on the computational domain. New geometric configurations result from the displacement of a set of mesh regions (e.g. walls, boundaries or discrete points within the volume) through the use of algorithms, based on RBFs, which smoothly propagate the imposed displacement to the surrounding volume.

The advantages of this approach are: no need to regenerate the grid; the robustness of the procedure; the capability to support different mesh typologies thanks to its meshless nature and the high parallelizable smoothing process.

The mesh morphing process we adopt to make the computational domain parametric with respect to the shape is comprised of the following steps:

- setup: that is the manual definition of the domain boundaries of the morphing action, the selection of the source points where mesh regions are imposed, and the definition of the required movements of the points driving the shape deformation.
- fitting: the solution of RBF system and the storing of computed coefficients into a file ready to apply the shape modification.
- smoothing by applying the prescribed displacement to the grid surfaces and then smoothly propagating the deformation to the surrounding domain volume combining several RBF solutions as defined by a proper Amplification Factor (AF) constituting the parametric configuration of the computational domain.

The RBF interpolation functions drive mesh morphing of the discretised domain of computational models by applying intended displacements to a set of purposely generated points, called source points. During the mesh morphing the nodes of the moving walls are first moved according to the defined kinematics; then the RBF action is extended in the volume, smoothing the grid up to the prescribed surrounding limits. The obtained RBF solution is sent to the RANS solver. This approach is characterized as follows: in general morphing requires less computational time than unstructured grids re-meshing; preservation of mesh consistency and topology; exact control of node position; low remeshing noise; it can be integrated in the computational fluid dynamic solver; it is capable of handling large grids.

2.1 RBF Background Theory

The effectiveness of RBF-MM in the nautical field has been previously demonstrated in [14, 17, 19]. The RBF background theory [20] is here briefly summarized. The interpolation of a scalar function known at N discrete locations (source points) can be achieved adopting an RBF expansion, which includes weighted distance interactions with all the source points, and an optional polynomial term that is usually introduced to guarantee uniqueness and compatibility with rigid modes:

$$s(x) = \sum\nolimits_{i=1}^{N} \gamma_i \varphi(\|x - x_s\|) + h(x) \qquad (1)$$

where x is the vector indicating the position of a generic node of the surface and/or volume mesh, $\boldsymbol{x}_s$ is the i_{th} source node position vector and $\|\cdot\|$ is the Euclidean norm.

The RBF fitting solution is possible when the RBF coefficients vector γ_i and the weights of the polynomial corrector vector β_i are found so that, at source points, the interpolant function has the specified (known) values of displacement g_i, and orthogonality conditions are satisfied:

$$s(\boldsymbol{x}_{s_i}) = g_{s_i}, \quad 1 \le i \le N \tag{2}$$

$$\sum_{i=1}^{N} \gamma_i p(\boldsymbol{x}_{s_i}) = 0 \tag{3}$$

for all polynomials q with a degree less than or equal to that of polynomial h. A unique RBF interpolant exists if the RBF is conditionally positive definite. Moreover, if the order is less than or equal to 2, a linear polynomial applies:

$$h(\boldsymbol{x}) = \beta_0 + \beta_1 x + \beta_2 y + \beta_3 z \tag{4}$$

allowing for rigid body translations recover. In this case, the interpolant has the form:

$$s(x) = \sum_{i=1}^{N} \gamma_i \varphi(\|\boldsymbol{x} - \boldsymbol{x}_{ki}\|) + \beta_0 + \beta_1 x + \beta_2 y + \beta_3 z \tag{5}$$

and γ_i and β_i values are obtained by this solution of the system

$$\begin{pmatrix} M & P_s \\ P_s^T & 0 \end{pmatrix} \begin{pmatrix} \gamma \\ \beta \end{pmatrix} = \begin{pmatrix} g_s \\ 0 \end{pmatrix} \tag{6}$$

where M is the interpolation matrix whose elements are derived by calculating all the radial interactions between source points as follows:

$$M_{ij} = \varphi\left(\left\|\boldsymbol{x}_{s_i} - \boldsymbol{x}_{s_j}\right\|\right), 1 \le i \le N, 1 \le j \le N \tag{7}$$

and P_s is a constraint matrix that arises balancing the polynomial contribution:

$$P_s = \begin{pmatrix} 1 & x_{s_1} & y_{s_1} & z_{s_1} \\ 1 & x_{s_2} & y_{s_2} & z_{s_2} \\ \vdots & \vdots & \vdots & \vdots \\ 1 & x_{s_N} & y_{s_N} & z_{s_N} \end{pmatrix} \tag{8}$$

assuming that source points are not contained in the same plane (otherwise the interpolation matrix would be singular). For MM application each component of the deformation field is handled as a scalar so satisfying the displacement field prescribed at source points:

$$\begin{cases} s_x(\boldsymbol{x}) = \sum_{i=1}^{N} \gamma_i^x \varphi(\|\boldsymbol{x} - \boldsymbol{x}_{s_i}\|) + \beta_0^x + \beta_1^x x + \beta_2^x y + \beta_3^x z \\ s_y(\boldsymbol{x}) = \sum_{i=1}^{N} \gamma_i^y \varphi(\|\boldsymbol{x} - \boldsymbol{x}_{s_i}\|) + \beta_0^y + \beta_1^y x + \beta_2^y y + \beta_3^y z \\ s_z(\boldsymbol{x}) = \sum_{i=1}^{N} \gamma_i^z \varphi(\|\boldsymbol{x} - \boldsymbol{x}_{s_i}\|) + \beta_0^z + \beta_1^z x + \beta_2^z y + \beta_3^z z \end{cases} \tag{9}$$

A great flexibility using RBF can be achieved acting on the radial functions that can be compactly or globally supported. Common radial functions are summarized in Table 1.

In our work flow we adopt the distance function (global supported and bi-harmonic in 3D) for volume mesh adaption because it allows to get very good quality and it is accelerated so that it can handle RBF problem beyond 1 million of centres (source points). We use Wendland functions (C0, C2 and C4) for surface sculpting because their high level of continuity allows to control a surface using just a few control points.

Table 1. Radial functions with global and compact support.

RBF with global support	$\varphi(r)$
Spline type (R_n)	r^n, n odd
Thin plate spline (TPS_n)	$r^n \log(r)$, n even
Multiquadric (MQ)	$\sqrt{1+r^2}$
Inverse multiquadric (IMQ)	$\frac{1}{\sqrt{1+r^2}}$
Inverse quadratic (IQ)	$\frac{1}{1+r^2}$
Gaussian (GS)	e^{-r^2}
RBF with compact support	$\varphi(r) = f(\xi),\ \xi \leq 1,\ \xi = \frac{r}{R_{sup}}$
Wendland (C^0)	$(1-\xi)^2$
Wendland (C^2)	$(1-\xi)^4(4\xi+1)$
Wendland (C^4)	$(1-\xi)^6\left(\frac{35}{3}\xi^2+6\xi+1\right)$

According to Eq. (9), every node of the volume mesh to be morphed by the RBF field has a new position computed in a meshless way using as input just its original position and the RBF interpolation:

$$\boldsymbol{x}_{node_{new}} = \boldsymbol{x}_{node} + \begin{bmatrix} s_x(\boldsymbol{x}_{node}) \\ s_y(\boldsymbol{x}_{node}) \\ s_z(\boldsymbol{x}_{node}) \end{bmatrix} \tag{10}$$

3 Case Study: Spinnaker Heavy Runner S4

The effectiveness of the proposed RBF-MM approach is demonstrated in this paper adopting the commercial software ANSYS Fluent® and *RBF Morph*^{TM} to study the effect of flying shape of the nylon SK75 tri-radial symmetric spinnaker described in [21–23].

The spinnaker, type heavy runner S4, is an integral part of Elan 31, a 9 m long cruiser-racer, produced by the sail-maker Banks Sails Naples with a surface area of almost 60 m^2. The sail is fixed to the yacht only by the head, the tack and by two sheets (length of about 6 m) connecting the sail's clews to the boat (Fig. 1).

The aeroelastic numerical model, described in [16], was adopted to evaluate the impact of variations of longitudinal and transverse sail curvatures on aerodynamic

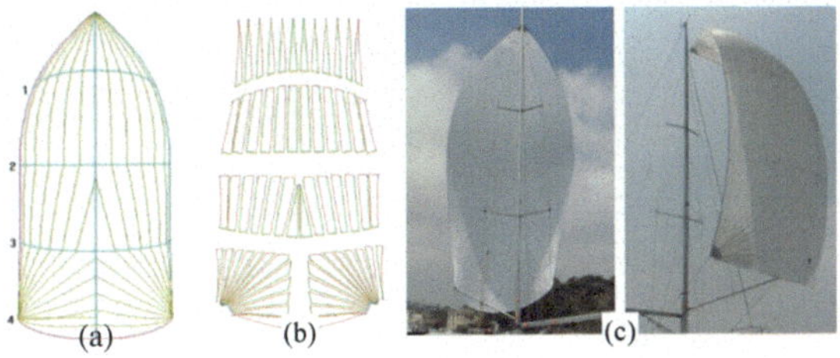

Fig. 1. Symmetric spinnaker: (a) design shape; (b) arrangement of panels; (c) flying shape.

performances. In particular, the values of drive and side forces on the sail were calculated for different longitudinal and transverse sail curvatures.

3.1 Curvatures Sail Parametrization

The surface of the sail studied has noticeably variable main curvatures (Fig. 2(a)) and even the directions of the surface normal are noticeably changeable along the main directions of the sail (Fig. 2(b)).

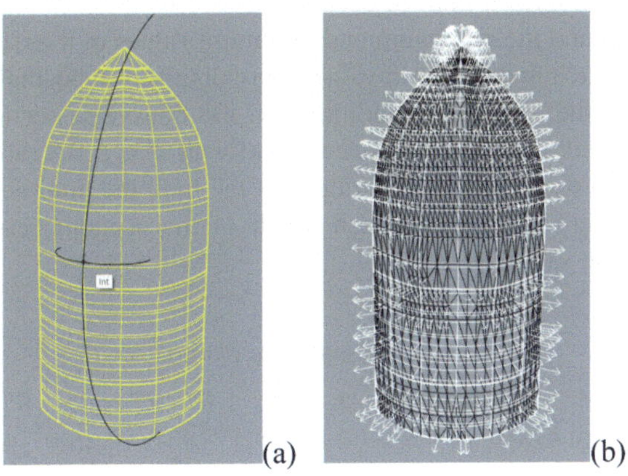

Fig. 2. Symmetric spinnaker: (a) principal curvature; (b) surface normal vectors.

Employing the SolidWorks® (Ver. 2018) commercial software parametrization occurred using the projections along the longitudinal and transverse planes of the sail of four radial sections (Sec. I, Sec. II, Sec. III and Sec. IV) as reported in Fig. 3(a) performed on half of the sail surface. Each of the obtained eight profiles was parametrized through the curvature radii of the two circumference arcs with which the profile was approximated. In Fig. 3 the radii of such circumference arcs approximating each profile and the points of tangency to each pair of arcs are illustrated.

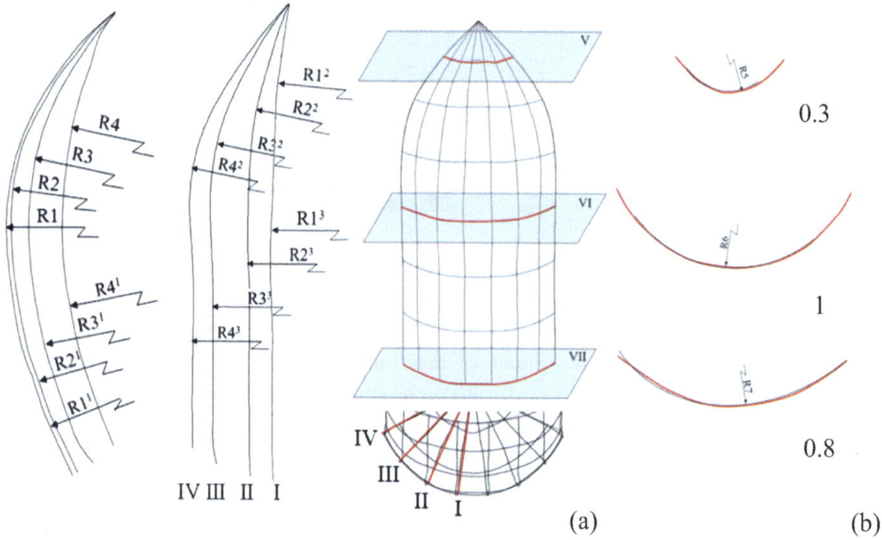

Fig. 3. (a) Curvature profile parametrization; (b) Constraint conditions on cross sections.

The values of the sixteen radii in correspondence with the initial geometry of the sail (Sail on design) and the maximum and minimum values of the radii are reported in meters in Table 2. Such max and min values were obtained imposing in the parametric modelling software the constraint conditions on the transverse sections V, VI e VII. These sections, indeed, must remain concave and with 0.3, 1, 0.8 ratios (Fig. 3(b)). In correspondence with the max and min values of the radii the minimum and maximum NURBS surfaces were obtained through a loft of the sections respectively, named narrow sail and wide sail (Fig. 4).

Table 2. Profile curvature radii (meters).

	R_1	R'_1	R_2	R'_2	R''_2	R_3	R'_3	R''_3	R'''_3	R_4	R'_4	R''_4	R'''_4
On design	4.7	19.5	5.6	19.7	8.9	7.6	13.3	7.7	–	10.1	15.1	4.5	–
Decreased	6.1	16.1	7.0	10.7	18.0	10.8	11.6	7.3	–	17.1	11.1	9.9	–
Increased	6.3	14.7	8.0	10.8	11.8	8.3	15.6	6.8	17.7	5.5	20.9	6.4	11.9

In the co-simulation environment CFD ↔ RBF-MM the amplification factor discussed in Sect. 2 allows to determine with continuity the variation of the mesh between the defined geometric limits (narrow sail and wide sail). The parametrization process starts with determining the optimal number of source points and an initial pre-alignment of narrow and wide sails to the sail on design (Fig. 5).

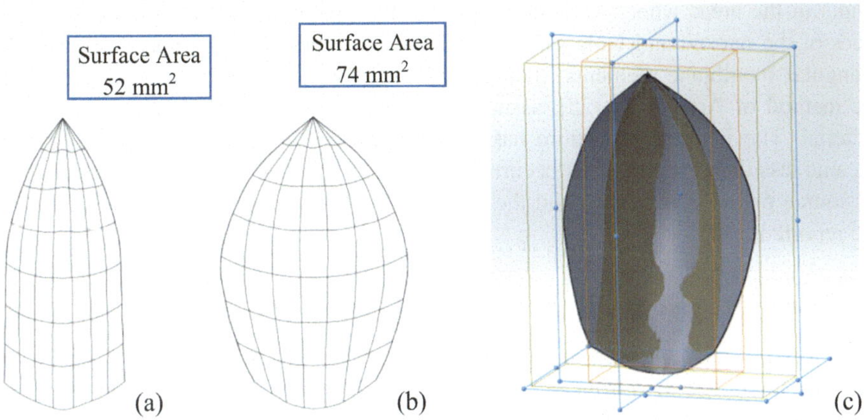

Fig. 4. (a) Narrow sail; (b) Wide sail; (c) Sails initial alignment.

3.2 Determination of Source Points

The narrow sail and the wide sail were initially aligned to the *on design* sail through the centring of the parallelepipeds which enveloped them. In particular, the six planes which enveloped the narrow and large sails were centred to the principal axes of the *on design* sail (Fig. 4(b)). As described in Sect. 2, the first step to deform (morph) the design mesh regions into desired geometric variation was the definition of source points driving the shape deformation. The number and position of such source points were crucial for the quality of morphing that was realized. The number and optimal position of the source points were defined through an iterative process. This started from an initial set of 42 source points, taken in correspondence of the nodes of the NURBS surfaces which had principal curvatures parallel to the longitudinal and transversal directions of the design sail.

Evaluating the accuracy derived by morphing the on design sail towards the narrow and the wide sails, the source points were incremented introducing further driving

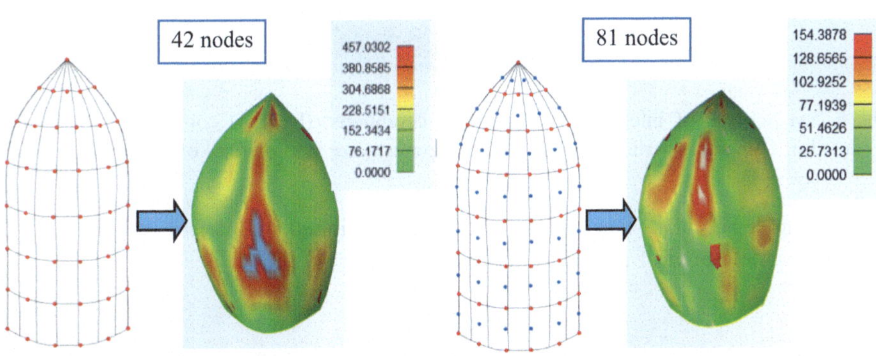

Fig. 5. Results of iterative process to establish number and position of source points.

points in the areas where deviations between the morphed surface and theoretic surfaces of the narrow and wide sails had maximum values. The sail was discretized by triangular membrane elements; their structural deformation calculation, according to the method of Arcaro [24, 25], assumed homogeneous and orthotropic linear elastic material. The iterative procedure has been completed when the maximum deviation became less than 80 mm. This occurred using the 81 nodes. In Fig. 5 the initial set of 42 source points (red colour) and the final set of 81 source points (red + blue colour) are reported.

3.3 Fluid Dynamic Model

A 70 m tall structured multiblock hexahedral mesh was generated. The domain was extended about 100 m in front and on the sides of the isolated sail. A flat inlet boundary was located 70 m behind the sail (Fig. 6(b)), a detail of the mesh in proximity of the sail is reported in (Fig. 6(c)).

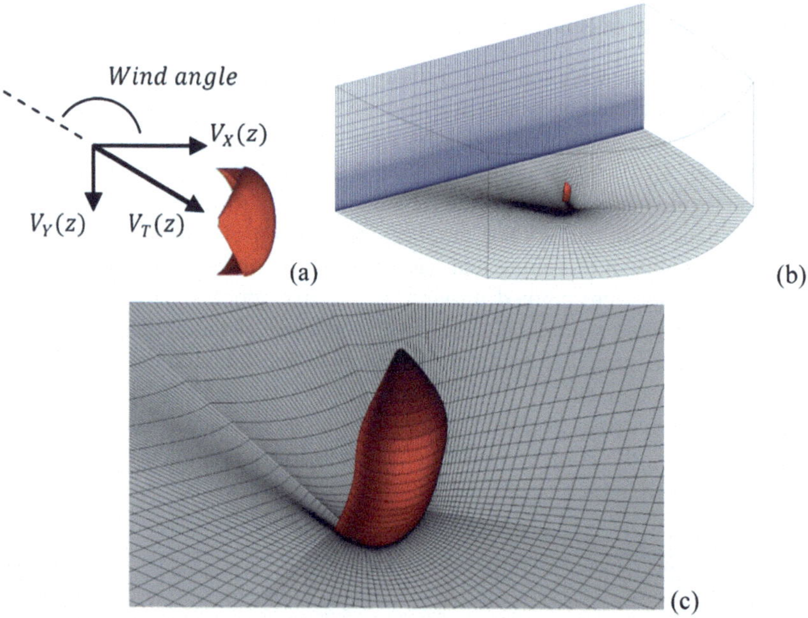

Fig. 6. (a) Scheme of inlet velocity boundary conditions; (b) CFD computational domain; (c) detail of CFD computational domain showing mesh size in proximity of the sail.

The outlet boundary was dimensioned and positioned in order to include the sail wake in a wide range of angles to be investigated.

The vertical distribution of the cell layers from the ground was defined in order to better represent the wind boundary layer profile which was modelled according to the following formula:

$$V_T(z) = V_T(Z_{ref}) \left(\frac{Z}{Z_{ref}} \right)^{\alpha} \quad (11)$$

where $V_T(z)$ is the wind speed and Z_{ref} is the reference eight (usually 10 m). The parameter α ranged between 1/14 and 1/7. The angle between the wind and the sail was varied, in the computations, updating the components of the wind velocity vectors at the inlet boundary according to the scheme reported in Fig. 6(a). The mesh size was 4.5 million of cells. The height of the first layer of cells on the wall was dimensioned in order to adopt wall function formulations maintaining a range of Y^+ between 30 and 100 in the attached regions and for a wind speed ranging from 7 to 20 knots.

4 Results and Discussion

Varying the value of the Amplification Factor (AF) between 0 and 1, the mesh of the sail becomes deformed changing from the geometry of the narrow sail to the geometry of the wide sail. The mesh obtained in correspondence of the value of AF 0.46 reproduces the geometry of the sail on design. In Fig. 7 the values of drive (F_R) and side (F_{LAT}) forces were compared during the mesh morphing ($0 \leq$ AF ≤ 1) for *wind* = 20 knots and Angle variable between 135° and 180°. In Fig. 8 the streamline flow around the narrow and the wide sails flying shapes and the pressure values on the upper surface of the sails were reported for *wind* = 20 knots and incidence angle 180°.

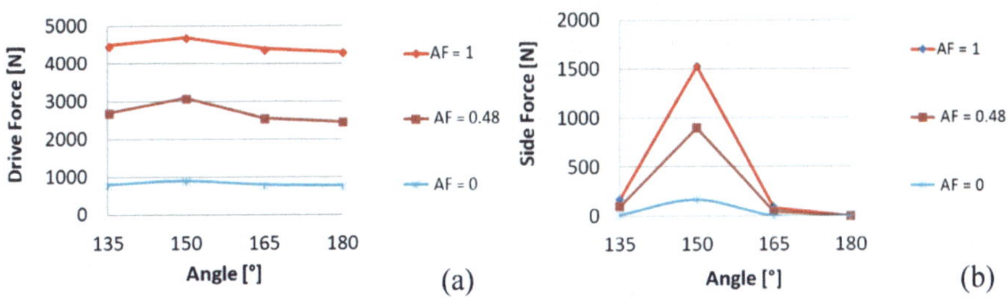

Fig. 7. (a) F_R and (b) F_{LAT} forces versus the amplification factor (*wind* = 20 knots).

Figures 7 and 8 show that the separated flow generates a gradually increasing lift force moving from *angle* = 165° to *angle* = 135° thus significantly contributing to the formation of the complete F_R. The outlined results allow to make a comparison of sail performance providing important information about the optimal surface area of the sail. The results overall show a good compliance with the data recorded in literature [13] and with the experimental tabular data for the forces as well as the flying shapes made available on www.yru-kiel.de or www.fluidengsol.com. In particular, the values of shear stress, drive and side forces show off only irrelevant difference with Renzsch [6] data (less than 2%).

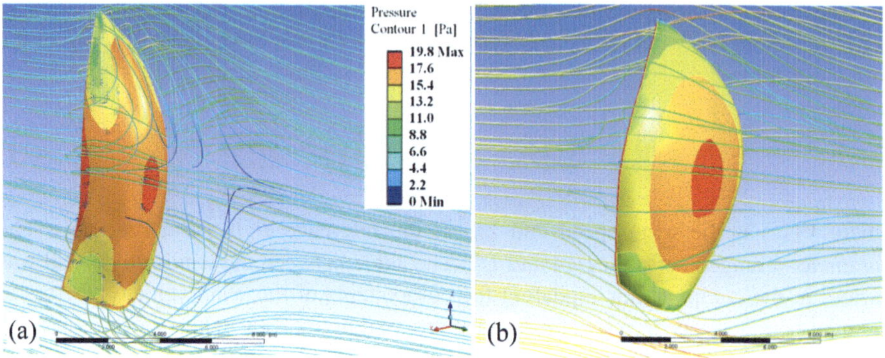

Fig. 8. Streamlines around the spinnaker in 1/2 vertical plane and pressure on the upper surface for: (a) narrow sail; (b) wide sail (*wind* = 20 knots; *angle* = 180°).

5 Conclusions

The potentialities of the Radial Basis Functions (RBF) Mesh Morphing (MM) applied to a sail's flying shape optimization were discussed and demonstrated. In particular, adopting the described method the aerodynamic performance of a symmetric spinnaker in terms of drive force and side force was evaluated for different morphed geometries. The main advantages of the proposed method can be summarized as follows: mesh morphing requires in general (depending on the RBF setup and mesh dimension) less computational time than re-meshing an unstructured grid (example of RBF mesh morphing computational requirements is reported in Sect. 3.1.1 of [26]); preservation of mesh consistency and topology; integration in the computational fluid dynamic solver establishing an iterative optimization process; exact control of node position; capability of handling large grids; low remeshing noise.

Funding Statement. This research was supported by a benefit obtained with the found for the athenaeum research in Catania, research programme 2019/2021.

References

1. Braun, J.B., Imas, L.: High fidelity CFD simulations in racing yacht aerodynamic analysis. In: Proceedings of the 3rd High Performance Yacht Design Conference, Auckland, pp. 168–175 (2008)
2. Manikandan, A., Rajkumar, R.: Evaluation of mechanical properties of synthetic fiber reinforced polymer composites by mixture design analysis. Polym. Polym. Compos. **24**(7), 455–462 (2016)
3. Pearson, W.E.: Textiles to composites: 3D moulding and automated fibre placement for flexible membranes. Mar. Appl. Adv. Fibre Reinf. Compos. **305**, 305–334 (2015)
4. Richter, H.J., Horrigan, K.C., Braun, J.B. Computational fluid dynamics for downwind sails. In: Proceedings of 16th Chesapeake Sailing Yacht Symposium, Annapolis, pp. 19–28 (2003)

5. Ranzenbach, R., Kleene, J.: Utility of flying shapes in the development of offwind sail design database. In: The High Performance Yacht Design Conference, Auckland (2002)
6. Renzsch, H., Graf, K.: An experimental validation case for fluid-structure interaction simulations of downwind sails. In: Proceedings of the 21st Chesapeake Sailing Yacht Symposium, Annapolis, p. 8 (2013)
7. Fossati, F., Bayati, I., Orlandini, F., Muggiasca, S., Vandone, A., Mainetti, G., Sala, R., Bertorello, C., Begovic, E.: A novel full scale laboratory for yacht engineering research. Ocean Eng. **104**, 219–237 (2015)
8. Hochkirch, K.: Design and construction of a full-scale measurement system for the analysis of sailing performance. Technical University of Berlin, Ph.D. thesis (2000)
9. Le Pelley, D.J., Modral, O.: VSPARS: a combined sail and rig shape recognition system using imaging techniques. In: Proceedings of 3rd High Performance Yacht Design Conference, Auckland, pp. 57–66 (2008)
10. Graves, W., Barbera, T., Braun, J.B., Imas, L.: Measurement and simulation of pressure distribution on full size sails. In: 3rd High Performance Yacht Design Conference, Auckland, vol. 2, p. 4 (2008)
11. Augier, B., Bot, P., Hauville, F., Durand, M.: Experimental validation of unsteady models for fluid structure interaction: application to yacht sails and rigs. J. Wind Eng. Ind. Aerodyn. **101**, 53–66 (2012)
12. Le Pelley, D.J., Morris, D., Richards, P.J.: Aerodynamic force deduction on yacht sails using pressure and shape measurement in real time. In: 4th High Performance Yacht Design Conference, Auckland, pp. 28–37 (2012)
13. Motta, D., Flay, R.G.J., Richards, P.J., Le Pelley, D.J., Deparday, J., Bot, P.: Experimental investigation of asymmetric spinnaker aerodynamics using pressure and sail shape measurements. Ocean Eng. **90**, 104–118 (2014)
14. Viola, I.M., Biancolini, M.E., Sacher, M., Cella, U.: A CFD–based wing sail optimization method coupled to a VPP. In: 5th High Performance Yacht Design International Conference, 8–12 March 2015, Auckland (2015)
15. Groth, C., Cella, U., Costa, E., Biancolini, M.E.: Fast high fidelity CFD/CSM fluid structure interaction using RBF mesh morphing and modal superposition method. Aircr. Eng. Aerosp. Technol. J. (2019). https://doi.org/10.1108/aeat-09-2018-0246
16. Biancolini, M.E., Cella, U., Groth, C., Genta, M.: Static aeroelastic analysis of an aircraft wind-tunnel model by means of modal RBF mesh updating. ASCE's J. Aerosp. Eng. **29**(6) (2016). https://doi.org/10.1061/(asce)as.1943-5525.0000627
17. Biancolini, M.E., Cella, U., Clarich, A., Franchini, F.: Multi-objective optimization of A-Class Catamaran foils adopting a geometric parameterization based on RBF mesh morphing. In: Evolutionary and Deterministic Methods for Design Optimization and Control With Applications to Industrial and Societal Problems. Computational Methods in Applied Sciences Series, vol. 49, Springer (2018). https://doi.org/10.1007/978-3-319-89890-2_30
18. Cella, U., Groth, C., Biancolini, M.E.: Geometric parameterization strategies for shape optimization using RBF mesh morphing. In: Advances on Mechanics, Design Engineering and Manufacturing. Lecture Notes series in Mechanical Engineering, pp 537–545 (2016). https://doi.org/10.1007/978-3-319-45781-9_54
19. Biancolini, M.E., Viola, I.M, Riotte, M.: Sails trim optimisation using CFD and RBF mesh morphing. Comput. Fluids 93C, 46–60 (2014)
20. Biancolini, M.E.: Fast Radial Basis Functions for Engineering Applications. Springer International Publishing (2018)
21. Calì, M., Oliveri, S.M., Cella, U., Martorelli, M., Gloria, A., Speranza, D.: Mechanical characterization and modeling of downwind sailcloth in fluid-structure interaction analysis. Ocean Eng. **165**, 488–504 (2018)

22. Calì, M., Oliveri, S.M., Gloria, A., Martorelli, M., Speranza, D.: Comparison of commonly used sail cloths through photogrammetric acquisitions, experimental tests and numerical aerodynamic simulations. Procedia Manuf. **11**, 1651–1658 (2017)
23. Calì, M., Speranza, D., Martorelli, M.: Dynamic spinnaker performance through digital photogrammetry, numerical analysis and experimental tests. In: Advances on Mechanics, Design Engineering and Manufacturing. Lecture Notes in Mechanical Engineering, pp. 585–595 (2017)
24. Arcaro, V.F.: Analysis of orthotropic membrane structures, UNICAMP/FEC (2004). http://www.arcaro.org/tension/main.htm
25. Renzsch, H., Müller, O., Graf, K.: FLEXSAIL–a fluid structure interaction program for the investigation of spinnakers. In: Proceedings of International Conference on Innovations in High Performance Sailing Yachts, Lorient (2008)
26. Cella, U.: Setup and validation of high fidelity aeroelastic analysis methods based on RBF mesh morphing, Ph.D. thesis, University of Rome "Tor Vergata"

Effect of Cell Shape on Nanoindentation Measurements

Antonio Boccaccio(✉), Michele Fiorentino, Vito Modesto Manghisi, Giuseppe Monno, and Antonio E. Uva

Politecnico di Bari, Bari, Italy
antonio.boccaccio@poliba.it

Abstract. The aim of this study is to understand if the shape of a cell can affect the characterization process of the mechanical properties via nanoindentation measurements. The characterization of the cell material by atomic force microscopy, in fact, traditionally implements the Hertz contact theory that is based on hypotheses not satisfied in the contact Atomic Force Microscope tip/cell and that do not take into account the actual cell morphology. In previous experimental studies, the mechanical properties of colorectal cancer cells differently shaped (rounded or elongated cells) and sized were determined via nanoindentation measurements. Implementing the Hertz theory, the authors found that differences in mechanical properties exist between the different cell lines with different shape. At this point, the question that can be raised is the following. Is it possible to state that this difference depends on the differences intrinsically existing between the mechanical properties of the investigated cells? Or, this difference can be justified with the difference in cells shape? In other words, the differences seen with the Hertz theory can depend on the fact that the cell shape was not taken into account. To respond to this question, the nanoindentation process of the different colorectal cancer cells was simulated via the finite element method. The finite element models reproducing the cells morphology were integrated into a numerical optimization algorithm that cyclically perturbs the cell mechanical properties until the difference between the force-indentation curve retrieved numerically and that obtained experimentally becomes smaller than an *a priori* fixed ε value. Once this occurs, the optimization algorithm stops and gives in output the optimal cell material properties. Interestingly, we found that the mechanical properties obtained via the Hertz contact theory are significantly different with respect to those computed with the proposed approach. Furthermore, we found that the material properties of the rounded cells are intrinsically different with respect to those of the elongated ones. The proposed approach provides new insights on the cell mechanobiology and on the effect of cell shape on the specific tasks in cancer growth and invasion.

Keywords: Cell shape · Cell modelling · Cell material characterization

C. Rizzi et al. (Eds.): ADM 2019, LNME, pp. 37–44, 2020.
https://doi.org/10.1007/978-3-030-31154-4_4

1 Introduction

Atomic Force Microscopy (AFM) is a commonly utilized technique that allows characterizing the mechanical properties of cell materials. The AFM device consists in a tip, which can have different geometries fixed to a cantilever beam that, after indenting the material, is subjected to bending moment. Known the displacement experienced by the tip during the indentation and known the force applied, one can determine the mechanical properties of the material under investigation. The technique was successfully utilized, for instance, to characterize entire cells [1–5] as well as cell components such as the cell membrane [6, 7]. The effects related to the shape of the tip on visco-hyperelastic materials [8] as well as to the residual stress included in cell material when subjected to AFM measurements [9] were previously investigated.

The theory utilized to determine, from the force-indentation curve, the material properties is the Hertz contact theory. It is based on strict hypotheses such as, the linear elastic behavior; the small deformation regimen; the homogeneity of the nanoindented material; the indented material has the shape of an infinite plate and behaves as a flat surface [10]. However, it is possible to state that none of these hypotheses is satisfied in the indentation process of cell materials. In fact, the deformation of the cell material can be very significant, the mechanical behavior can be described with hyperelastic constitutive models, the cell material is not homogeneous but includes different subcellular components with different mechanical properties. Finally, the cell material does not have a flat shape but typically presents a "spherical" or a "spherical-like" shape. An important limitation of the Hertz contact theory is that it allows determining the sole mechanical properties of the entire cell and is not capable of separately distinguishing the properties of the different subcellular components. All these aspects lead us to conclude that the results obtained with the Hertz contact theory are much approximated as the hypotheses of the theory are very far from being satisfied in the real conditions. Our hypotheses are that all these aspects such as the heterogeneity of the cell material, the actual cell geometry and the actual deformation regimen occurring in the indentation process do affect the nanoindentation measurements.

In a previous experimental study [11], the mechanical properties of three different lines of colorectal cancer cells were determined and the Hertz contact theory was implemented to measure the cell Young's modulus. In this study, the heterogeneous behavior, and the actual cell geometry were neglected. It was found that the different cell lines have different material properties. The fact that the hypotheses of the Hertz contact theory are not satisfied, leads us to consider as questionable the obtained results. The question that can be raised is: does this difference in mechanical properties depend on the actual and intrinsic differences between the cell material properties? Such differences, in fact, can be justified with the banal argument that the investigated cells present different geometries, which are not taken into account from the Hertz theory. This study attempts to give a response to this question.

In this study we developed the finite element models of three different lines of colorectal cancer cells presenting different shapes and studied in the above mentioned article [11]: rounded, such as the SW480R and SW620, and elongated, such as the SW480E cells. These finite element models were included into a numerical optimization algorithm that iteratively perturbs the material properties until a good fitting between the force-indentation curve obtained experimentally and that computed via the FEM model, is reached.

2 Materials and Methods

The finite element models of the three different cell lines of the colorectal cancer were built and utilized to simulate the experiments carried out by Palmieri et al. [2, 11], where SW480 and SW620 human colon carcinoma-derived cell lines were purchased from the American Type Culture Collection (ATCC; Manassas, VA) and submitted to nanoindentation measurements. The JPK Nanowizard AFM system was utilized to characterize the cells.

The general purpose finite element software Abaqus (version 6.12) was utilized to build the cells model. In particular, exploiting the symmetry of the model, only 1/4 of the entire cell was modelled. Three different subcellular components were modelled: the cytoskeleton, the nucleus and the cell cortex, which presents a thickness of about 200 nm [12]. The nucleus was hypothesized to possess a spherical shape with diameter 5 μm.

The SW620 cell presents a rather smooth surface (Fig. 1). To build its model, simple CAD tools available in Abaqus were utilized.

Fig. 1. Model of the SW620 cancer cell. The model includes the nucleus (red), the cytoskeleton (grey) and the cell cortex (green). Details of the AFM tip are shown on the right side.

The cells SW480R and SW480E, instead, present an important surface roughness that cannot be neglected in nanoindentation measurements (Figs. 2 and 3). The nodes of the external surface of these cells were properly modulated according to a rough surface; further details on the modelling of the rough surface are given in previous studies [13, 14]. The AFM tip was modelled as a rigid surface with the shape of a blunt cone and with a cone aperture angle of $\alpha = 40°$. The radius of the tip was fixed equal to 10 nm. The nodes of the inferior surface were clamped, while a contact interaction was established between the AFM tip and the superior surface of cell. On the lateral planes, symmetry constraints were applied. The finite element mesh was properly refined near the contact region between the AFM tip and the cell.

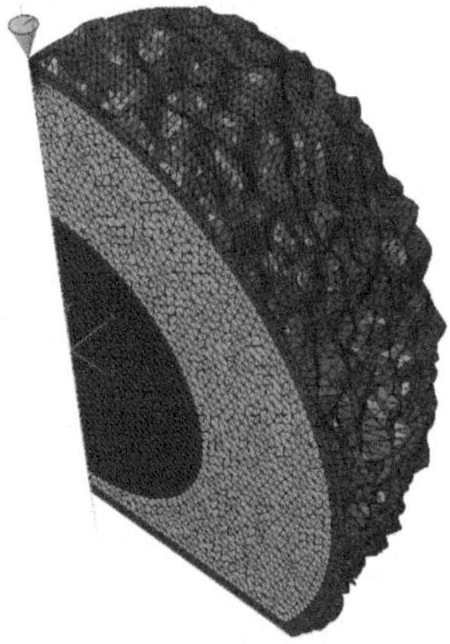

Fig. 2. Model of the SW480R cell. The cell membrane includes nodes that have been modulated according to a random surface. The figure shows also the blunt cone AFM tip.

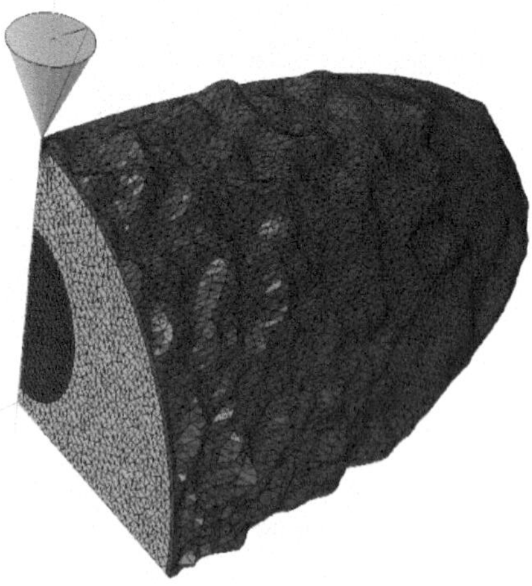

Fig. 3. Model of the SW480E cell.

The cell material was modelled with the Arruda-Boyce hyperelastic constitutive law. The actual values of the cell material properties, i.e. the shear modulus μ and the distensibility λ were determined via a numerical optimization algorithm. The algorithm, in each cycle, determines the numerical force-indentation curve and compares it with that obtained experimentally [11]. The algorithm perturbs the material properties so many times until the "distance" δ between the numerical curve and that experimentally obtained becomes smaller than a *a priori* fixed ε value (Fig. 4).

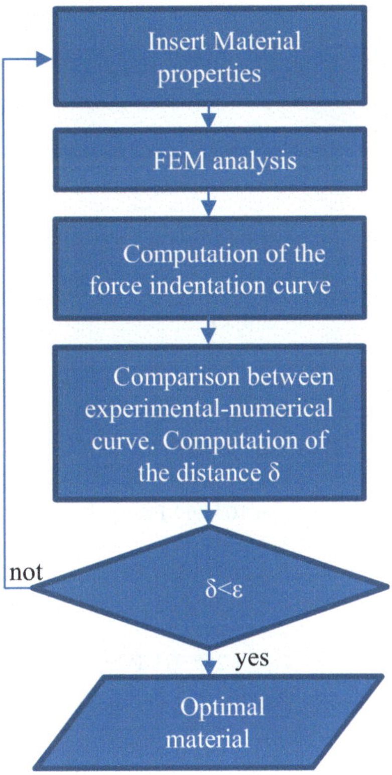

Fig. 4. Schematic of the optimization algorithm implemented to determine the optimal material properties of cells.

3 Results and Discussion

The implementation of the numerical optimization algorithm allowed determining with high accuracy the cell material properties. The Fig. 5 shows, for instance, the typical experimental force-indentation curve fitted with the numerical one. A proof of the goodness of the optimization algorithm is given by the very high correlation coefficients that were found ($R^2 > 0.99$ in all the cell models investigated). In the Table 1 reported below the values obtained with the proposed optimization algorithm are listed. Interestingly, it was found that the cell cortex is always stiffer than the cytoskeleton.

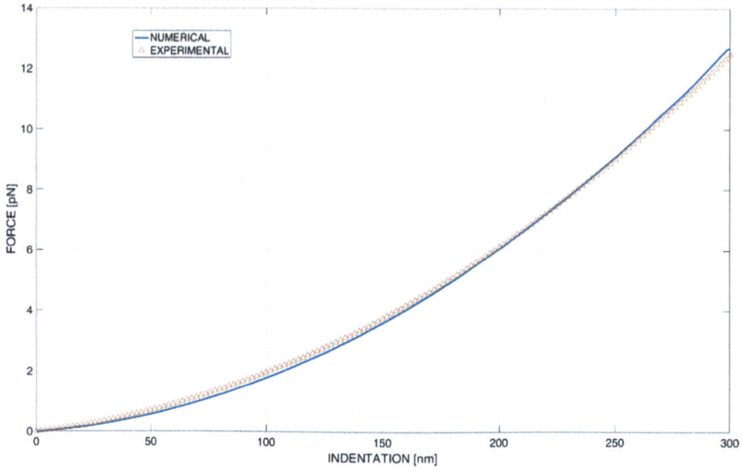

Fig. 5. Typical experimental force-indentation curve fitted with the numerical one.

Table 1. Optimal values of the material properties determined via the proposed optimization algorithm.

Cell line		Arruda-Boyce	Hertz
SW620	Cytoskeleton	μ = 56.586 Pa λ = 2.6499 E_{AB} = 169.760 Pa	E_H = 480.000 Pa [11]
	Cortex	μ = 104.869 Pa λ = 2.9017 E_{AB} = 314.609 Pa	
SW480R	Cytoskeleton	μ = 81.990 Pa λ = 2.6499 E_{AB} = 245.971 Pa	E_H = 540.000 Pa [11]
	Cortex	μ = 184.985 Pa λ = 2.8999 E_{AB} = 554.956 Pa	
SW480E	Cytoskeleton	μ = 429.538 Pa λ = 2.6507 E_{AB} = 1288.615 Pa	E_H = 1060.000 Pa [11]
	Cortex	μ = 731.706 Pa λ = 2.9067 E_{AB} = 2195.118 Pa	

In Table 1, the equivalent Young's modulus E_{AB} is also reported. It can be computed as $E_{AB} = 2\ (1 + \nu)\ \mu$, where ν is the Poisson's ratio that can be set equal to 0.04999 to account for material incompressibility. The E_H represents the Young's modulus reported in Palmieri et al. [11], computed with the Hertz contact theory. It is interesting to see that, in general, the Young's modulus computed with the Hertz theory is different with respect to that computed with the proposed approach. It is worthy to note that with the Hertz theory, only the "average" Young's modulus of the entire cell can be computed. It appears that the rounded cells SW620 and SW480R possess comparable mechanical properties, which are significantly smaller than those possessed by SW480E. This allow us to conclude that the material properties of the rounded cells are intrinsically different with respect to those of the elongated ones. Regarding the effect of the cell shape, we carried out preliminary analyses where we fixed the material properties and changed the cell diameter. We found that the more the radius of curvature increases, the more the indentation force increases. At the limits, the highest value of force can be obtained in the case of planar surface that possess an infinite radius of curvature. This result is consistent with our expectations. In fact, in a planar surface the amounts of material to be moved by the AFM tip (while penetrates) are larger than those to be moved in the case of surface with a finite radius of curvature.

4 Conclusions

In conclusion, we proposed a numerical approach that takes into account the cell morphology, the hyperelastic behavior and the actual deformation regimen. The approach led us to results significantly different with respect to those computed with the Hertz theory. Furthermore, the proposed algorithm allowed us to separately determine the mechanical properties of different subcellular components. The proposed approach provides new insights on the cell mechanobiology and on the effect of cell shape on the specific tasks in cancer growth and invasion.

References

1. Costa, K.D.: Single-cell elastography: probing for disease with the atomic force microscope. Dis. Markers **19**, 139–154 (2004)
2. Palmieri, V., Lucchetti, D., Maiorana, A., Papi, M., Maulucci, G., Calapà, F., Ciasca, G., Giordano, R., Sgambato, A., De Spirito, M.: Mechanical and structural comparison between primary tumor and lymph node metastasis cells in colorectal cancer. Soft Matter **11**, 5719–5726 (2015)
3. Chen, Q., Xiao, P., Chen, J.-N., Cai, J.-Y., Cai, X.-F., Ding, H., Pan, Y.-L.: AFM studies of cellular mechanics during osteogenic differentiation of human amniotic fluid-derived stem cells. Anal. Sci. **26**, 1033–1037 (2010)
4. Liang, X., Shi, X., Ostrovidov, S., Wu, H., Nakajima, K.: Probing stem cell differentiation using atomic force microscopy. Appl. Surf. Sci. **366**, 254–259 (2016)
5. Kuznetsova, T.G., Starodubtseva, M.N., Yegorenkov, N.I., Chizhik, S.A., Zhdanov, R.I.: Atomic force microscopy probing of cell elasticity. Micron **38**, 824–833 (2007)
6. Boccaccio, A., Frassanito, M.C., Lamberti, L., Brunelli, R., Maulucci, G., Monaci, M., Papi, M., Pappalettere, C., Parasassi, T., Sylla, L., et al.: Nanoscale characterization of the biomechanical hardening of bovine zona pellucida. J. R. Soc. Interface **9**, 2871–2882 (2012)
7. Boccaccio, A., Lamberti, L., Papi, M., De Spirito, M., Douet, C., Goudet, G., Pappalettere, C.: A hybrid characterization framework to determine the visco-hyperelastic properties of a porcine zona pellucida. Interface Focus **4**, 20130066 (2014)

8. Boccaccio, A., Lamberti, L., Papi, M., De Spirito, M., Pappalettere, C.: Effect of AFM probe geometry on visco-hyperelastic characterization of soft materials. Nanotechnology **26**, 325701 (2015)
9. Boccaccio, A., Papi, M., De Spirito, M., Lamberti, L., Pappalettere, C.: Effect of the residual stress on soft sample nanoindentation. Appl. Phys. Lett. **102**, 133704 (2013)
10. Liu, K.-K.: Deformation behaviour of soft particles: a review. J. Phys. D Appl. Phys. **39**, R189 (2006)
11. Palmieri, V., Lucchetti, D., Maiorana, A., Papi, M., Maulucci, G., Ciasca, G., Svelto, M., De Spirito, M., Sgambato, A.: Biomechanical investigation of colorectal cancer cells. Appl. Phys. Lett. **105**, 123701 (2014)
12. Vargas-Pinto, R., Gong, H., Vahabikashi, A., Johnson, M.: The effect of the endothelial cell cortex on atomic force microscopy measurements. Biophys. J. **105**, 300–309 (2013)
13. Boccaccio, A., Uva, A.E., Papi, M., Fiorentino, M., De Spirito, M., Monno, G.: Nanoindentation characterisation of human colorectal cancer cells considering cell geometry, surface roughness and hyperelastic constitutive behaviour. Nanotechnology **28**, 045703 (2016)
14. Garcia, N., Stoll, E.: Monte Carlo calculation for electromagnetic-wave scattering from random rough surfaces. Phys. Rev. Lett. **52**, 1798 (1984)

Industrial Design and Ergonomics

Nature Inspired Redesign of the Visual Appearance of an Industrial Product

Dolores Parras-Burgos(✉), Francisco J. F. Cañavate, Francisco Cavas-Martínez, and Daniel G. Fernández-Pacheco

Department of Graphical Expression, Universidad Politécnica de Cartagena, Cartagena, Spain
dolores.parras@upct.es

Abstract. One of the specific characteristics of design is shaping the visual appearance of products, but this feature is not valued in the same way as the functional aspect in some industrial areas. In this sense, this paper describes the real case of an industrial product (a filter housing) that was initially designed considering only functional and technical aspects. Over the years this product became a reference product in the agronomic field and also one of the most copied filters in the world. The company then opted for a new strategy: redesigning the external shape of the filter without losing any functional characteristics. The process of redesigning this new product was based on the use of analogies with nature as source of inspiration. As a result, an helicoidal shape describing the movement of the water inside the filter was the innovative and creative element that permitted to offer a new identity for this product, achieving an industrial design registration in some of the most influential countries in the world.

Keywords: Design process · Product design · Industrial design · Creativity · Bioinspiration

1 Introduction

In the twentieth century, designers became aware of the importance of the shape as an intrinsic value of the object, emerging the binomial shape-function [1]. In this line, the following question may arise: "Should design be based on functionality or a pleasant aesthetic?". According to Víctor Papanek [2], the answer is simple "aesthetic value is an inherent part of the function".

Design provides distinguishing functions, optimizes costs and offers a correct interpretation of the shape and aesthetics of the product in its social environment and its work context. These are aspects that must be taken into account in order to achieve a more efficient product adapted to the consumer´s needs [3].

Some studies evidence that those companies which take the design stage to a higher strategic level, positively contribute to innovation and to their own performance [4, 5]. Those technology companies that emphasise using an aesthetic design while developing new products sell more, are more successful at introducing products into new countries, have a more favourable company image [6–12], and also have the chance to obtain a bigger market share [13–16].

C. Rizzi et al. (Eds.): ADM 2019, LNME, pp. 47–58, 2020.
https://doi.org/10.1007/978-3-030-31154-4_5

Some authors define creativity as the degree to which a new product is distinguished from the products of competitors [17], which they stress is a relevant performance criterion in design [18], as well as essential management to effectively design the innovation of new products [19]. Researchers have acknowledged that design based on analogies is a relevant method to create novel ideas, and is considered a fundamental process to develop creativity [20, 21]. Analogical reasoning is a task performed by reasoning systems based on associative resemblance. It is considered an active method to store diagrams that is based on identifying the connections, parallelisms or resemblances in those elements typically perceived as being different. Some studies have discovered that analogical reasoning often acts as a basic cognitive tool for design problem solving [22].

One way to perform this analogical reasoning is to use the "observation" method. Observation is a key skill for learning process, and observation in nature can generate new insights in product design [23].

During the design process, designers attempt to achieve specific intentions through products' visual appearance. Indeed designers can seek their inspiration in a wide variety of means, such as images, objects and texts [9, 24, 25], although which methods prove more effective is not known for sure [26]. What comes over quite clearly is that a support system is needed to favour the creative process when designing and developing products [27, 28].

Biomimetic designs are widely used to design products in order to emphasise an emotional interaction. Indeed, understanding the psychological effects of biomimetic products is becoming a relevant theme in the product development stage with strong affective qualities. Some studies indicate that consumers show distinct levels of emotional response to products with different biomimetic levels [29]. Using nature as a source of inspiration involves highly potential development [30, 31], is becoming a dominant paradigm for robotics, material science and other technological disciplines, and has a socio-economic impact on this decade and the future [32].

This article describes the redesign process of filter case inspired in the shape of water with an helicoidal effect. This design was a revolution in such products used in the agronomic domain because, until then, only functional aspects of designs were taken into account, and not their aesthetic aspects.

2 Background of the Product

The case described in this article has been developed by the AZUD company (Alcantarilla, Murcia, Spain). This company bases its technological development on the implementation of Industry 4.0 in the management of all its processes, and has divisions for irrigation, industrial filtration and water treatment. It also has successful references in more than 150 countries, with its own subsidiaries in India, Mexico, Brazil and China.

Until 1998, in the area of filtration, metal filters were commercialised, which were formed by a metal case whose configuration was based on: (i) a main base whose interior held the disc filtering element; (ii) an upper lid made of thermoplastic material, which was joined to the case by a stainless-steel clamp. The case had an aligned inlet and outlet so that water could pass through the filter (Fig. 1a).

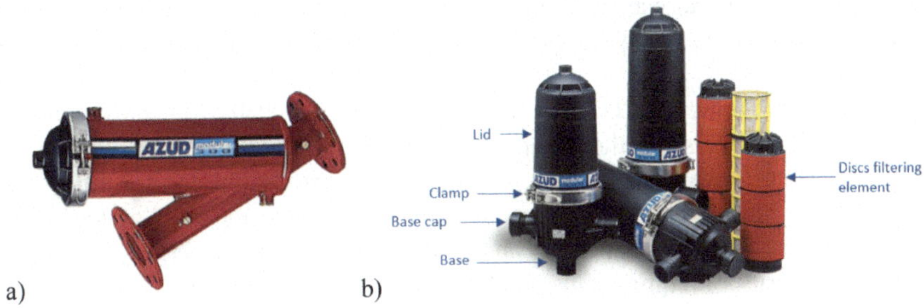

Fig. 1. (a) Metal filter, (b) Classic range filters.

However, in 1998 AZUD launched a new range (called Classic, Fig. 1b) of 2″ and 3″ filters manufactured in thermoplastic to remove 90% of metal filters, and to solve some technical and functional problems it faced. The design requirements defined for this design, and to develop these new filters, were as follows:

- The new plastic case had to support pressures of up to 10 bars without breaking. The main causes of filters breaking are the high pressures, stress and vibrations that filters in filtration systems must withstand.
- To facilitate access to the disc filtering element. In the metal model, access was gained from the top of the case, which made its handling very difficult. The clamp that was then used involved a system with a bolt to open/close it, which made its use complicated as tools were required.
- To design a base with three outlets housing to configure different water inlets and outlets so that the same product could be used in different arrangements (90° and 180°).

The Classic filters range that was launched to the market was based on a modular system that enabled a wide range of flows and configurations using a minimum number of components, and it thus adapted to every customer's requirements. The main filter components were the case (lid, base, clamp and base lid) and the disc filtering element (held inside it). To reach this modular range, two top sizes were defined, with two outlets housing sizes in the base (2″ and 3″) and outlets housing with three connection types (BSP thread, NPT thread, and grooved connection), achieving this way different configurations.

The Classic filters range was available on the market for 10 years, and was sold in more than 35 countries all over the world. AZUD revolutionised the market by launching a range of product novelties thanks to its manufacturing process using technical plastic materials, its modular configuration and the patent of HELIX system inside filters.

The Classic filter case was designed by bearing in mind only functional aspects, which was because simple smooth lines were defined, with ridges and structures to provide reliable resistance. At that time, no importance was attached to its aesthetic design because its registration was not considered possible, and it becoming the most copied filter in the world was never contemplated at that time. The patents that AZUD owned for the filter basically focused on how it worked instead of on its outward appearance.

Today, reproducing and copying products already available on the market is a widespread practice, also in the agronomic domain. Exact copies of the filter design started appearing in several countries, so in 2007 the company opted for a new commercial strategy in relation to this product, investing in the redesign of the filter to differentiate itself from its competitors and all the copies that had emerged. This was no easy task as a series of requirements and limitations came into play that had to be taken into account to start this project:

- The new design had to be representative of the company and easily identifiable by customers, standing out its aesthetic and visual appearance.
- Redesign only the product visible part that, in this case, was only the outer area of the lid and the filter base, by maintaining the same inner geometry.
- The modular structure of all the components had to be maintained to offer the same existing product combinations.
- The product functional aspects should not vary, and its efficiency and operation were essential requirements.
- The weight and sizes of these two components (lid and base) should not vary so as to not increase the cost of the material and its production.

3 Methodology

Various methodologies to guide the designer in the different phases of the design process can be found in the literature [33], although they all have very similar arguments. Therefore, it can be deduced that there does not need to be a single method to develop a design process and obtain the solution that was being sought. As Simon indicates in his work [34], the design method is not fixed but rather flexible and adaptable to the circumstances of the environment and the designer himself. This section describes all aspects related to the methodology, the design process and the manufacturing process developed in this work.

3.1 Methodological Scheme

The design challenge of the present project was important because obtaining an apparently simple new product image was necessary, but one that moved away from its initial aesthetics without losing the brand's essence. To develop this project, several aspects that had to be taken into account are stressed. Firstly, the company's strategy: the company detects the requirement or problem to be solved, establishes the requirements or specifications that the new design must contain, and contributes all the material resources, production processes, human resources, etc. Secondly, design resources, where the designer(s)'s task involves obtaining this new product by conducting studies and resorting to various information sources that offer the data and inspiration required to develop the conceptual design (Fig. 2).

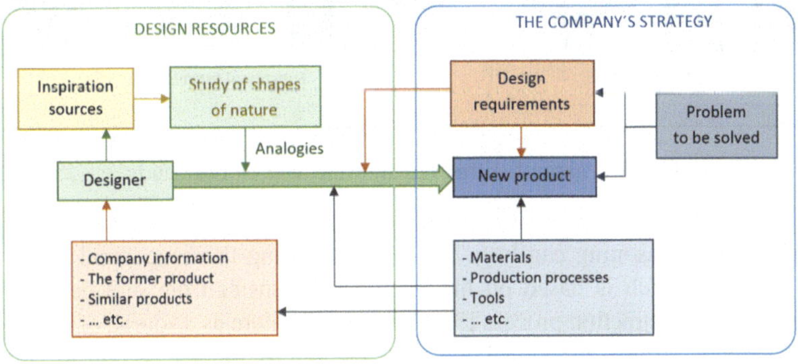

Fig. 2. The methodological scheme.

3.2 Design Process

Taking into account the methodological scheme proposed in Fig. 2, the design process for the development of the new product is described (Fig. 3).

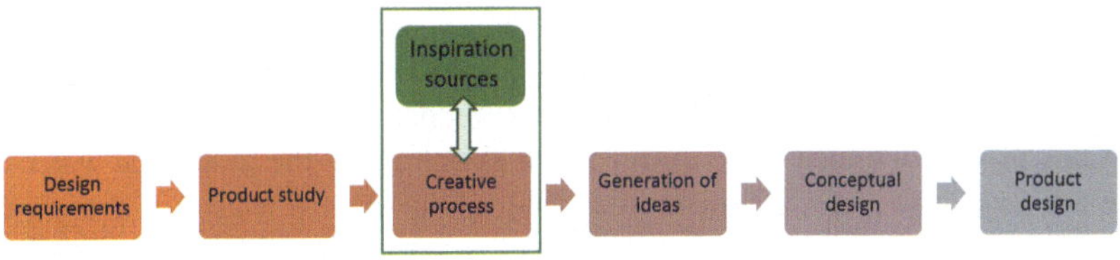

Fig. 3. Design process scheme.

The first two phases that are proposed are (Fig. 4): (i) the definition of the design specifications, and (ii) the study performed to specify the dimensions and functions that must be maintained with the new product. In this sense, it is determined that (Fig. 4, right): (i) the inner part of the body and the lid must have the same geometry (a and b), (ii) the type of assembly of the two pieces must be maintained in the new product (c), (iii) the inlets and outlets connections must be the same to obtain the same modularity (d).

Having defined the design specifications and conducting a product study, the creative phase commences. During the creative process, designers have to bear in mind all the information they have previously acquired but must open their minds to any type of idea without limiting it in an initial phase. Using sources of inspiration is a resource widely used by designers, who can seek analogies from different domains that inspire them.

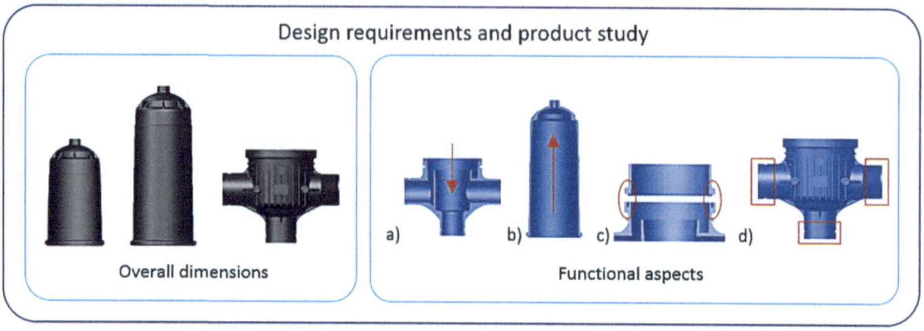

Fig. 4. Phase of design requirements and product study.

Thus, analogical reasoning can be a key element during the conceptual phase of the design process, which is based on the knowledge transfer from one design state to another, from an abstraction process [35]. The use of nature as a source of inspiration is widely researched from a functional point of view, but its application considering shape and aesthetics aspects has not yet been sufficiently studied in the industrial field, although it has been successfully applied in other areas [36–39].

In this case, nature, and more specifically water, were taken into account in this stage. Water is an element of nature that can take very suggestive and varied forms, being able to extract very interesting concepts during this phase of conceptual design (Fig. 5, left).

On the other hand, several researches in cognitive psychology and design thinking have shown that the generation of internal representations in images and external representations through sketches are fundamental in solving design problems [40]. For this reason, illustrations and sketches can be performed to help during this phase and to reflect all the suggestions that inspiration can provide.

In this project, the idea arose from a small sketch about the shape that water takes inside a filter (Fig. 5, left). It represented nothing important when it was drawn, but it became very important in the analysis and review stages (Fig. 5, right). A key part in

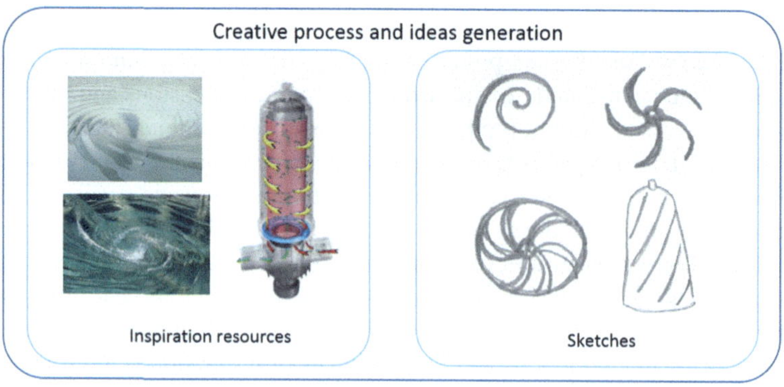

Fig. 5. Phase of creative process and ideas generation.

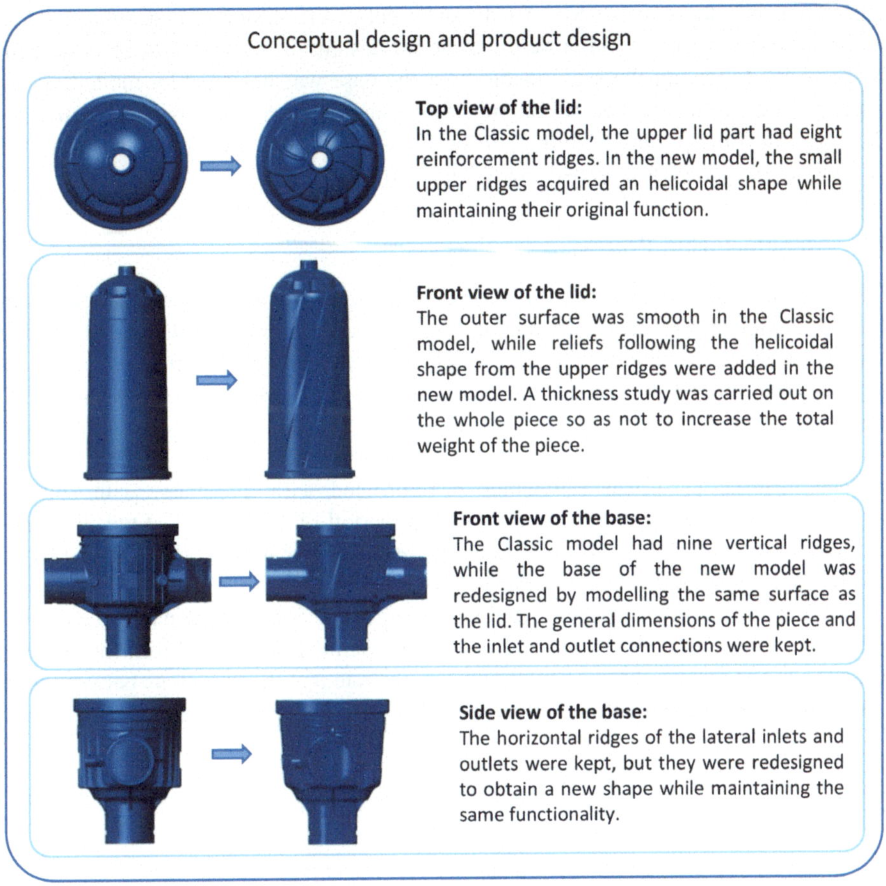

Fig. 6. Phase of conceptual design and product design.

this product is the helix, or clogging-retardant device, which was patented by AZUD. This part has an helicoidal effect on water so that any particles in suspension do not stick so quickly to the filtering discs. This helicoidal effect on water was the key part of this new design. Furthermore, this new concept of visual appearance inspired by an element of nature, such as the helicoidal movement of water, also reflects an internal functional feature of the filter. In this way, it was sought to achieve an original and different aesthetic that was not found in the traditional design schemes of products in the agronomic field.

The Pro/Engineer computer-aided design software (PTC Inc, Boston, Massachusetts, USA) was used to model the new designs. Once the design concept was defined, it had to be adapted to the product itself (Fig. 6). Firstly, the upper lid part was defined by taking into account the helicoidal shape concept. In this way, much importance was attached to the small upper ridges as they took this shape and covered the whole dome. Then the outer lid structure, a fluid shape that continued from the upper area, had to be produced. To do so, reliefs were designed that enhanced the helicoidal effect of ridges. After defining the lid design, the same had to be done with the base. By following the same inspiration, the base was designed by modelling the same surface as the lid. The lateral ridges of holes were completely transformed by changing from the projecting ridges in the Classic model to a bas-relief shape in the new model by completely transforming their structure.

3.3 Manufacturing Process

One of the initial requirements of the project was not to vary the weight of the pieces of the new model with respect to its predecessors so as not to increase the manufacturing cost. To validate the new designs, the weights and volumes of both the designed base and lid were studied. To do so, these two parameters were calculated with the 3D models and the Pro/Engineer software and were compared with the Classic model values.

The material used was Polyamide 6 reinforced with fiberglass, with suitable additives to work outdoors for an approximate period of 10 years. The pieces manufactured with this technical material must be treated eliminating its humidity 100% so that the superficial and functional aspect was the correct one.

Table 1 shows the values obtained by entering in the software the parameters of the reference material for the manufacture of these parts (Polyamide 6 reinforced with fiberglass). The results of the calculations demonstrate that the variations between both models were not significant and, thus, validating the proposed designs.

Table 1. Comparing weights and volumes between the Classic model base and the new one.

	Classic model		New model	
	Weight (kg)	Volume (cm^3)	Weight (kg)	Volume (cm^3)
BASE 2″	1.89	1.47	2.04	1.58
BASE 3″	2.06	1.60	2.24	1.74
LID 2″	1.72	1.34	1.77	1.37
LID 3″	2.33	1.80	2.36	1.83

Concerning the manufacturing process used (injection moulding), all the technical aspects of the Classic model were taken into account: the pieces were designed to house a large sprue that provided a good filling and the mould had the necessary gas outlets so that the piece was completely compacted.

The exterior shape of the new model was designed using a relief with smooth transitions on all surfaces. An exhaustive study of all sections of each piece was carried out to verify that there was a homogeneous thickness throughout the model so as not to produce hot spots in the injection process. These hot spots could cause defects and structural problems in the piece, as well as affect the functional characteristics of the product. Another important aspect that was reviewed during the study of the new surfaces was the existence of undercuts, since they can damage the extraction of the pieces from the mould, causing delays and machine stalls, as well as damaging the quality and the final cost of the pieces. In this sense, the entire design was validated so that their morphology did not affect the injection process.

In short, all the new morphological changes designed on the outside of the pieces did not affect their manufacturing process, the same production levels were obtained as the previous model, and the profitability of the product was not affected by any aesthetic aspect of the new design. For confidentiality reasons, more rigorous data on manufacturing and comparative costs between the two models cannot be provided.

4 Results

The final result was a product with an innovative aesthetic appearance, maintaining all the technical functions of its predecessor, thus fulfilling all the objectives set at the beginning of the project (Fig. 7).

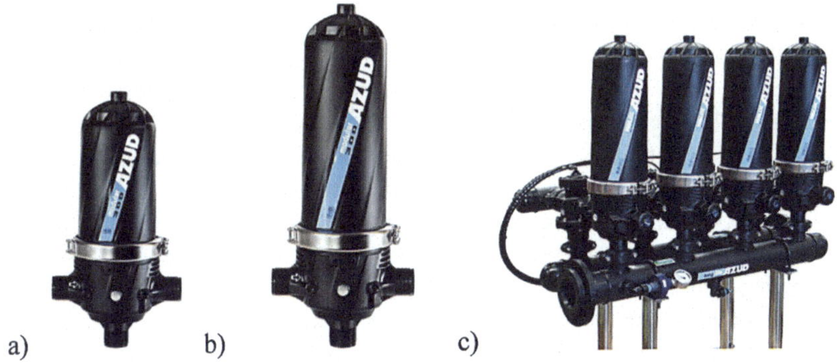

Fig. 7. (a) Front view of the 2″ filter, (b) Front view of the 3″ filter, and (c) Example of filtration equipment with new filter design.

In addition, the new AZUD filter innovated not only in its aesthetic appearance but also in other concepts that were established on this type of products and that were also original, such as: a new sticker location, the engraving of a slogan ("The Original") and the logo on the piece itself, identification plates for the filtration degree, etc. (Fig. 8).

Fig. 8. Details of the new filter.

When a so innovative and different design is developed, the company must study the strategy to determine the countries in which it will be protected, as well as the modality of registration.

With the AZUD filter, a very clear strategy was adopted: to protect it in the most important countries, including those in the EU, China and India, among others. Moreover, not only the filter design was registered in all its versions, but also the various configurations of the equipment and of the variants that had been developed over the years.

The industrial design registration protects it from unloyal competitors that attempt to copy its products, but this is hard to achieve worldwide. Some countries tend to place more bureaucratic obstacles and higher costs, which means that very similar, or almost identical, products can appear. Registering a product worldwide is very difficult and costly. To do so, it is most important to conduct market research, assess markets and decide which countries are more worthwhile to register a product in. A design considered to be obvious cannot be registered, but it is very important to protect it if its design is distinguishable from the rest. This is a guarantee for customers and protects against competitors.

5 Conclusions

The redesign of a product is a challenge for any company. The case that has been shown in this paper describes how a technical and innovative product can be reinvented from a functional perspective by designing new shapes from an aesthetic point of view. The image of this novel product offers: new perspectives in the market, differentiation from competitors and an indisputable brand image. The inspiration in the helicoidal shape of the water provides a dynamic and original design in a type of product that breaks with previously established. This new nature-based design approach has allowed to obtain an industrial design registration in relevant countries worldwide, offering customers a high-quality guarantee against imitations.

Acknowledgment. We thank Sistema AZUD S.A. company for all the provided data and information that enabled to write this manuscript. Special thanks go to Manuel García, Marta García, Pablo Jumilla, María Teresa Munuera, José Miguel López and Yusimí Rodríguez for their contributions and suggestions.

References

1. Gay, A., Samar, L.: El diseño industrial en la historia. Ediciones Tec, Córdoba (Argentina) (2004)
2. Papanek, V.: Diseñar para el mundo real. Ecología humana y cambio social. Pol.len Edicions, Barcelona (2014)
3. Lecuona, M.: Manual sobre gestión de diseño para empresas que abren nuevos mercados. Barcelona Centro de Diseño (BCD), Barcelona (2007)
4. Gemser, G., Leenders, M.A.A.M.: How integrating industrial design in the product development process impacts on company performance. J. Prod. Innov. Manage. **18**, 28–38 (2001)
5. Micheli, P., Perks, H., Beverland, M.B.: Elevating design in the organization. J. Prod. Innov. Manage. **35**, 629–651 (2018)
6. Bloch, P.H.: Seeking the ideal form: product design and consumer response. J. Market. **59**, 16–29 (1995)
7. Candi, M.: Benefits of aesthetic design as an element of new service development. J. Prod. Innov. Manag. **27**, 1047–1064 (2010)
8. Coates, D.: Watches Tell More than Time: Product Design, Information, and the Quest for Elegance, p. 281. McGraw-Hill, Michigan (2003)
9. Crilly, N., Moultrie, J., Clarkson, P.J.: Shaping things: intended consumer response and the other determinants of product form. Des. Stud. **30**, 224–254 (2009)
10. Jagtap, S.: Attributes and emotions in product form design: a survey of professional industrial designers. In: International Conference on Research into Design, pp. 705–714. Springer (2017)
11. Townsend, J.D., Kang, W., Montoya, M.M., Calantone, R.J.: Brand-specific design effects: form and function. J. Prod. Innov. Manag. **30**, 994–1008 (2013)
12. Veryzer, R.W.: Key factors affecting customer evaluation of discontinuous new products. J. Prod. Innov. Manag. **15**, 136–150 (1998)
13. Chakrabarti, A., Morgenstern, S., Knaab, H.: Identification and application of requirements and their impact on the design process: a protocol study. Res. Eng. Des. **15**, 22–39 (2004)
14. Eisentraut, R., Badke-Schaub, P.: Creativity: a personality trait or an illusion. In: International Workshop: Engineering Design and Creativity, Pilsen, Czech Republic. State Scientific Library, Pilsen (1995)
15. Sarkar, P., Chakrabarti, A.: Assessing design creativity. Des. Stud. **32**, 348–383 (2011)
16. Westwood, A.R., Sekine, Y.: Fostering creativity and innovation in an industrial R&D laboratory. Res.-Technol. Manage. **31**, 16–20 (1988)
17. Im, S., Workman Jr., J.P.: Market orientation, creativity, and new product performance in high-technology firms. J. Market. **68**, 114–132 (2004)
18. Christiaans, H.H.: Creativity as a design criterion. Commun. Res. J. **14**, 41–54 (2002)
19. Cropley, D.H., Kaufman, J.C., Cropley, A.J.: Measuring creativity for innovation management. J. Technol. Manage. Innov. **6**, 13–30 (2011)

20. Kalogerakis, K., Lüthje, C., Herstatt, C.: Developing innovations based on analogies: experience from design and engineering consultants. J. Prod. Innov. Manag. **27**, 418–436 (2010)
21. Keshwani, S., Chakrabarti, A.: Towards automatic classification of description of analogies into sapphire constructs. In: Research into Design for Communities, vol. 2, pp. 643–655. Springer (2017)
22. Daugherty, J., Mentzer, N.: Analogical reasoning in the engineering design process and technology education applications. J. Technol. Educ. **19**, 7–21 (2008)
23. Verma, S.K., Punekar, R.M.: Observing nature—what designers can learn from biologists. In: Research into Design for a Connected World, pp. 887–899. Springer (2019)
24. Eckert, C., Stacey, M.: Sources of inspiration: a language of design. Des. Stud. **21**, 523–538 (2000)
25. Gonçalves, M., Cardoso, C., Badke-Schaub, P.: What inspires designers? Preferences on inspirational approaches during idea generation. Des. Stud. **35**, 29–53 (2014)
26. Jagtap, S.: Inspiration in product form design: a survey of professional industrial designers. In: International Conference on Research into Design, pp. 715–725. Springer (2017)
27. Parras-Burgos, D., Fernández-Pacheco, D.G., Cavas-Martínez, F., Martínez, J.N., Cañavate, F.J.: An agent-based approach for the application of nature's forms to product conceptual design. PLoS ONE **13**, e0208930 (2018)
28. Wang, K., Nickerson, J.V.: A literature review on individual creativity support systems. Comput. Hum. Behav. **74**, 139–151 (2017)
29. Wu, T.Y., Chen, H.K.: Products with biomimetic shapes convey emotions more effectively. In: 4th International Conference on Design, User Experience and Usability, pp. 559–566. Springer, Heidelberg (2015)
30. Bar-Cohen, Y.: Biomimetics: mimicking and inspired-by biology. In: SPIE Smart Structures Conference, San Diego, CA, pp. 1–8 (2005)
31. Bar-Cohen, Y.: Biomimetics-using nature to inspire human innovation. Bioinspir. Biomimetics **1**, 1–12 (2006)
32. Lepora, N.F., Verschure, P., Prescott, T.J.: The state of the art in biomimetics. Bioinspir. Biomimetics **8**, 11 (2013)
33. Adams, K.M.: Design methodologies. In: Nonfunctional Requirements in Systems Analysis and Design, vol. 28, pp. 15–43. Springer, Cham (2015)
34. Simón, G.: La trama del diseño: Porqué necesitamos métodos para diseñar. DR Editorial Designio, México (2009)
35. Goel, A.K.: Design, analogy and creativity. IEEE Expert Intell. Syst. Appl. **12**, 62–70 (1997)
36. Avramović, D., Vladić, G., Kašiković, N., Ivan, P.: Aplicability of golden ratio rule in modern product design. J. Graph. Eng. Des. **4**, 29–35 (2013)
37. Ding, W., Yang, L., Zhao, Y.: The conception design of flower-shaped seat in park landscape. In: International Conference on Materials and Products Manufacturing Technology, ICMPMT 2011, Chengdu, pp. 111–114 (2011)
38. Tavsan, C., Tavsan, F., Sonmez, E.: Biomimicry in architectural design education. Procedia-Soc. Behav. Sci. **182**, 489–496 (2015)
39. Tavsan, F.; Sonmez, E.: Biomimicry in furniture design. In: 7th World Conference on Educational Sciences, Athens, Greece, pp. 2285–2292 (2015)
40. Goldschmidt, G., Smolkov, M.: Variances in the impact of visual stimuli on design problem solving performance. Des. Stud. **27**, 549–569 (2006)

Perceived Comfort and Muscular Activity: A Virtual Assessment of Possible Correlations

Nicola Cappetti(✉), Alessandro Naddeo, Vittorio Maria Soldovieri, Ivan Vitillo, and Iolanda Fiorillo

Department of Industrial Engineering, University of Salerno, 840884 Fisciano (SA), Italy
{ncappetti,anaddeo}@unisa.it

Abstract. Generally, comfort may be defined as the "level of well-being" perceived by humans in a working environment. The state-of-the-art about the evaluation of comfort/discomfort shows the need for an objective method to evaluate the "effect in the internal body" and "perceived effects" in main systems of comfort perception. Some medical studies show that each human joint has its own natural Rest Posture in which human muscles are completely relaxed or at the minimum strain level. Basing on this assumption, in other studies postural comfort curves for each Degree of Freedom of human upper and lower-limbs joints have been studied and CaMAN software has been developed to have a direct interface with these curves. In this paper, the upper limbs' postural comfort curves have been compared with the results gained by the biomechanics virtual simulation built on ANYBODY Software. A detailed study has been conducted on the upper body muscular activation during upper limbs movement with and without load. Postures have been analysed both by ANYBODY and CaMAN in order to correlate the postural (dis)comfort perception with the calculated muscular activity.

Keywords: Postural comfort · ANYBODY software · Human Man Interface simulation

1 Introduction

The feeling of postural comfort is strongly related to human body conditions during the execution of a task. A simplified model of comfort perception [1], based on Cappetti-Naddeo model [2], explained the four aspects - User Biomechanics/Posture, Physiologic factor, Environment contribute, Cognitive factor - that strongly affect the global comfort perception. For example, temperature and loads can modify physiological conditions when a subject makes a muscular effort; also, either environment or psychological conditions can raise or decrease a comfort perception.

Many scientific papers deal with the comfort experience during movements with or without loads by using electromyography (EMG) [3, 4] and correlation between EMG and comfort judgements based on assessment questionnaires [5, 6].

Interpreting EMG-based analyses in combination with comfort is complicated as the following elements influence the studies:

C. Rizzi et al. (Eds.): ADM 2019, LNME, pp. 59–70, 2020.
https://doi.org/10.1007/978-3-030-31154-4_6

- it is hard to measure all muscular activities in order to make a complete analysis of the body;
- there is the need to evaluate the cross-talking effects of different activities;
- it could be harsh to clean comfort judgement from non-muscular aspects;
- it is often not possible to generalize the correlations found to other activities, thus, new EMG experiments could be required.

The availability of sophisticated numerical methods for multibody analysis and the deep knowledge of the human body makes it now possible to develop software dedicated to human body simulations. ANYBODY Technology [15] developed a software that is specific for simulations of musculoskeletal systems. There are studies that affirmed the capability of ANYBODY to simulate the functioning of the human body during the execution of specific tasks by comparing the ANYBODY outcome with EMG results [7–10]. However, regarding the correlation of the ANYBODY software to comfort or discomfort not much has been published. In this paper, the authors studied the correlation between muscular loads and postural comfort. Some simple upper limb movements were simulated in ANYBODY and the results were compared with the corresponding comfort rate available in CaMAN software [11].

Upper limbs were chosen because, in scientific literature, they seems to be always involved in tasks performed by humans, both in working environment [12, 13] and in product use [4, 5, 14].

2 Materials and Methods

2.1 Comfort Curves with CaMAN Software

The CaMAN software is based on the comfort curves studied by Naddeo et al. [2, 11, 12]: in those papers it is stated that for each Range of Motion (ROM) of Upper Limbs, there is a unique comfort curve, that is, a bi-univocal correspondence between joint angles and comfort indexes (Fig. 1) exists. Furthermore, looking the Fig. 1, it is possible to notice in the correspondence of Rest Posture Range, the comfort level is higher.

CaMAN software has, as input, the position of the upper limbs of a standing human and, as output, the comfort indexes for each joint angle and the mean postural comfort values.

With this software, it is possible to estimate a Postural Comfort score for a standing human changing the position of the upper limbs within the Range Of Motion. Considered movements were: shoulder flexion/extension; shoulder abduction/adduction; elbow flexion/extension; elbow pronation/supination; wrist flexion/extension; wrist radio/ulnar deviation.

2.2 Human Body Simulation with ANYBODY

Considering a standing human that assumes a determined posture, through the CaMAN and its comfort curve it is possible to obtain the comfort analysis under the postural point of view. Instead, to evaluate muscular stresses of the same posture, ANYBODY simulation software was used. The ANYBODY Modelling System [13] is a multibody software that allows to create detailed musculoskeletal models and run computer simulations that integrate the human body into a 3D environment to evaluate the human/product interaction. As results, individual muscle forces, joint contact-forces and moments, metabolism, elastic energy in tendons and antagonistic muscle action can be calculated.

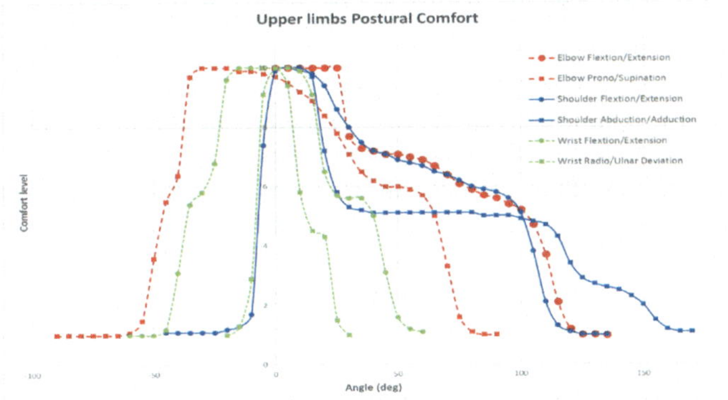

Fig. 1. Upper limbs postural comfort curves over angles. The comfort indexes are between 1° (min value) and 10° (max value). The angle domains represent the Range of Motions.

ANYBODY System provides also a Model Repository that is an open library of components and examples ready to be used in the ANYBODY Modelling System™ with different detailed human parts or whole body models ready to make simulation examples of activities of daily living and validation studies.

In this study, the Standing Human model (Fig. 2) was used; it is a parametric model of a standing human configured for the 50th percentile European male (176 cm high and 75 kg weight). Its feet were bounded to the ground and, during simulations, the body posture was re-evaluated and modified firstly controlling the ankles in order to respect these bounds and to position the Centre of Gravity (CoG) vertically over the CoG of the ground contact.

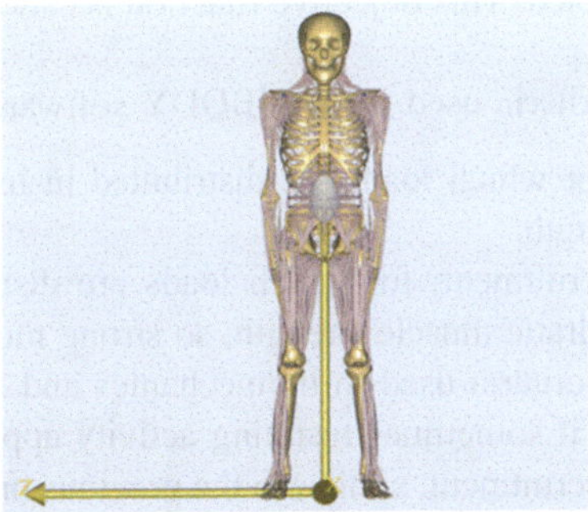

Fig. 2. Standing human in ANYBODY

2.3 Muscle Activity and Muscle Recruitment

In an ANYBODY model, each muscle is characterised by points of connections with bones and by its maximum performable force that corresponds to the maximum muscle traction. The activity of a muscle is defined as the ratio between its instantaneous force and its maximum force.

Moreover, in the human body, the determination of muscular activities equilibrating a load is an undetermined problem because many combinations of muscles' activities can give the same result so a rule is needed. In a real body, the brain assumes the important role to distribute the load on muscles with unknown rules, depending on the contest, such as optimising a specific performance, minimising muscular stresses or consumed energy, protecting articulations and joints, etc.

Therefore, the modelling of these mechanisms is based on assumptions about optimal condition in a given global state. The main problem is that there are more muscles than necessary to drive the degrees of freedom of the system, which implies that there are infinite solutions of muscle recruitment patterns than acceptable from a dynamical point of view. This problem is often referred to as the "redundancy problem of the muscle recruitment".

Musculoskeletal models can be divided into two groups: forward and inverse dynamic models. Forward dynamic computes the motion based on a predicted muscular activation. While this is attractive in the view of the detailed modelling of various physical phenomena, it is a very computationally demanding optimal control problem and requires a costly optimization to make the model performs a specific task. Inverse dynamic computes the muscle activation based on a specified task, i.e., known motion. This puts many restrictions on the model, but it is computationally much more efficient. The ANYBODY Modelling System has sufficient numerical efficiency to allow ergonomic design optimization on inexpensive computers but allows only for inverse dynamic analysis of the models.

The solution of the muscle recruitment problem in the inverse dynamics approach is generally formulated as an optimization problem: the objective function represents the assumed criterion of the recruitment strategy of the real human central nervous system, stated in terms of the muscle forces and it is minimized with respect to all unknown forces in the problem, i.e. muscle forces and joint reactions. With this objective function, the dynamic equilibrium equations, which enter as constraints into the optimization, should be respected. This objective function is called "Muscular Recruitment criteria".

The most important criteria used in ANYBODY software are:

- Linear Recruitment, for which loads are distributed in inversely linear proportion respect to muscle strength.
- Quadratic Muscle Recruitment, for which loads are distributed in inversely proportion respect to quadratic muscle strength, so strong muscles are much involved. This is the most used criteria used in biomechanics and it agrees on experimental tests with EMG, even if sometimes resulting activity appears not physiological.
- Polynomial Muscle Recruitment, similar to the previous one, for which an exponent from 3 (default) to 5 has to be selected for weighing the muscle strength.

- Min/Max Muscle Recruitment, for which all muscles can contribute by becoming active muscles. Even if it corresponds to the minimum fatigue criterion, the results are often not-physiological.

Doing the Muscle Recruitment comparison (Fig. 3), by graphics inspection, the mean activity tendency of simple movements for upper limbs does not change significantly. So, in these simulations, default criteria (Quadratic Muscle Recruitment) has been used.

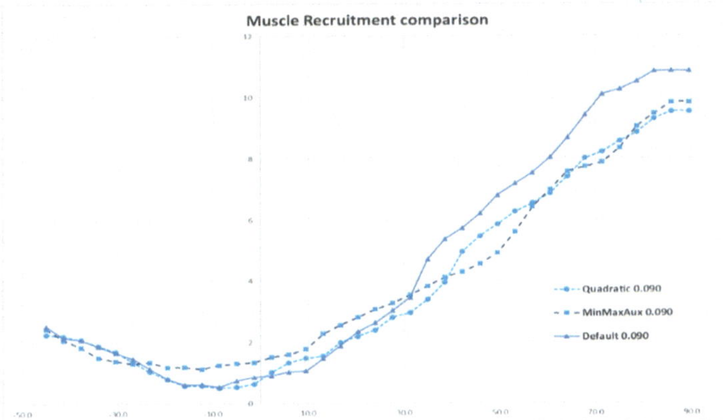

Fig. 3. Muscle recruitment comparison (in ordinate, the % of recruitment) - example of shoulders.

3 Simulations and Results

Based on Standing Human model with default percentile and default Muscle Recruitment Criteria a set of ANYBODY analysis was carried out where the movement of each upper limb considered in CaMAN was simulated and the results obtained were compared with the Comfort curves. The starting posture of upper limbs of the Standing Human model was set to ideal Rest Posture [11] and one of them at a time was varied in time from minimum to maximum articular range listed in Table 1.

Table 1. Upper limbs joints range.

Joint	Movement	[min deg., max deg.]
Shoulder	Extension/Flexion	[−60, 180]
	Adduction/Abduction	[0, 180]
Elbow	Extension/Flexion	[−10, 150]
	Supination/Pronation	[−90, 90]
Wrist	Extension/Flexion	[−85, 80]
	Radio-Ulnar deviation	[−25, 45]

From ANYBODY simulation, some results were pulled out as Muscular Activities, elongation and forces. Figures 4, 5, 6 show Activities of the most active muscles during examined tasks.

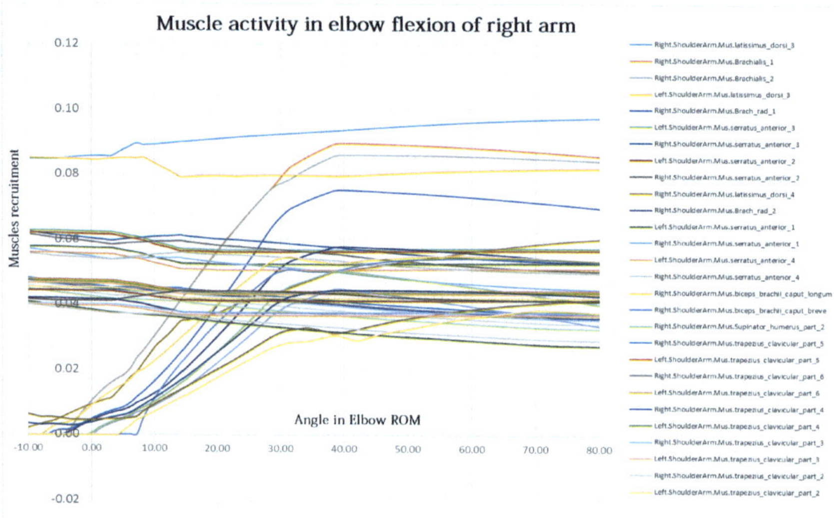

Fig. 4. Most active muscles in Elbow flexion of the right arm

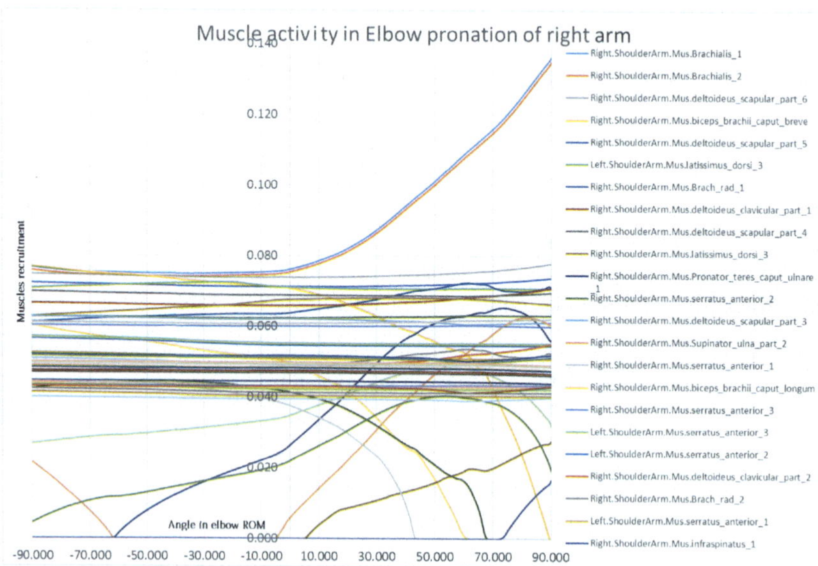

Fig. 5. Most active muscles in Elbow pronation of right arm.

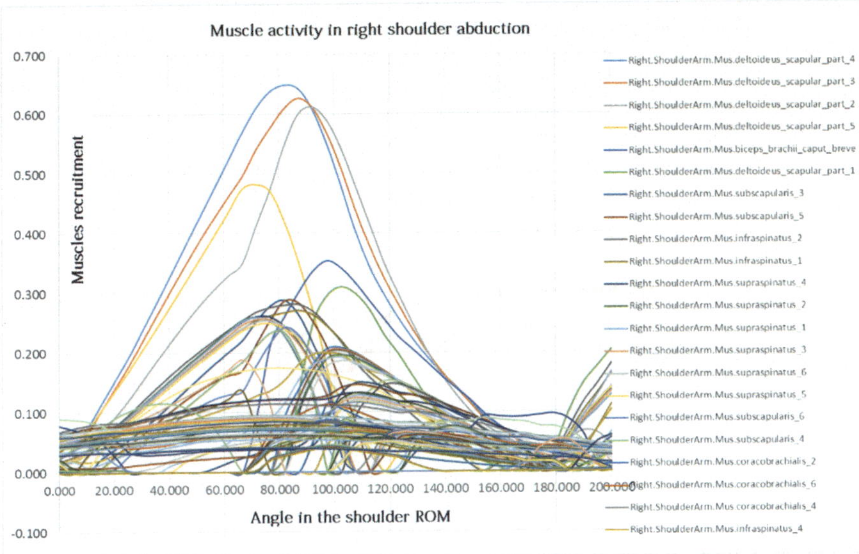

Fig. 6. Most active muscles in right Shoulder Abduction.

3.1 Comfort-Muscle Activity Correlations

It is not useful to compare information about each muscle with Comfort curves because Postural Comfort score states a global subjective evaluation of Muscular condition. As a matter of fact, a simple movement like elbow flexion produces a variation of Centre of Gravity that produces some compensation loads on muscles of abdomen and legs. So the comparison with Comfort curves was performed by the composition of muscular curves: Mean Activity, Envelope and Number of Active Muscles.

- Mean Activity is the curve "Angle-vs-Muscle Activity" where for each Angle the mean Activity value of the body muscles is selected. Since the assumption discomfort is caused by high muscular activity, the Mean Activity was parametrized on a threshold value in order to ignore muscles that are poorly loaded.
- Envelope is the curve "Angle-vs-Muscle Activity" where for each Angle the biggest Activity value of the body muscles is selected.
- Number of Active Muscles counts the muscles with a max activity value greater than the threshold.

In Figs. 7, 8, 9 and 10 resulting curves for shoulder abduction are shown.

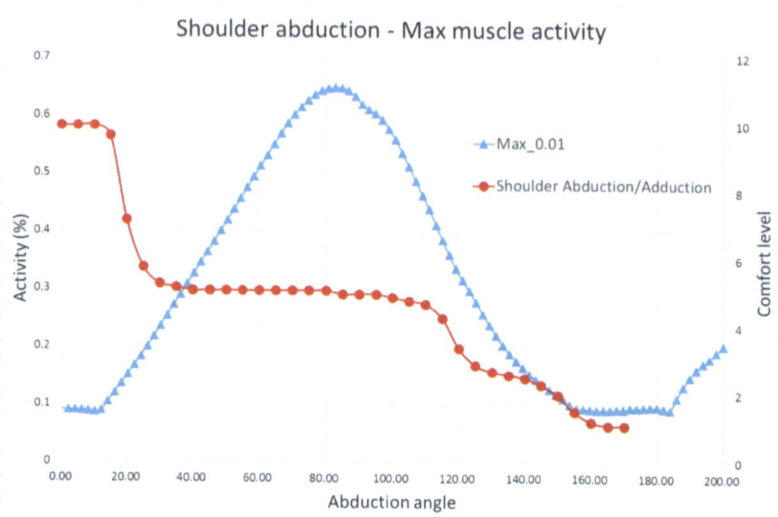

Fig. 7. Activity Envelope in shoulder abduction movement.

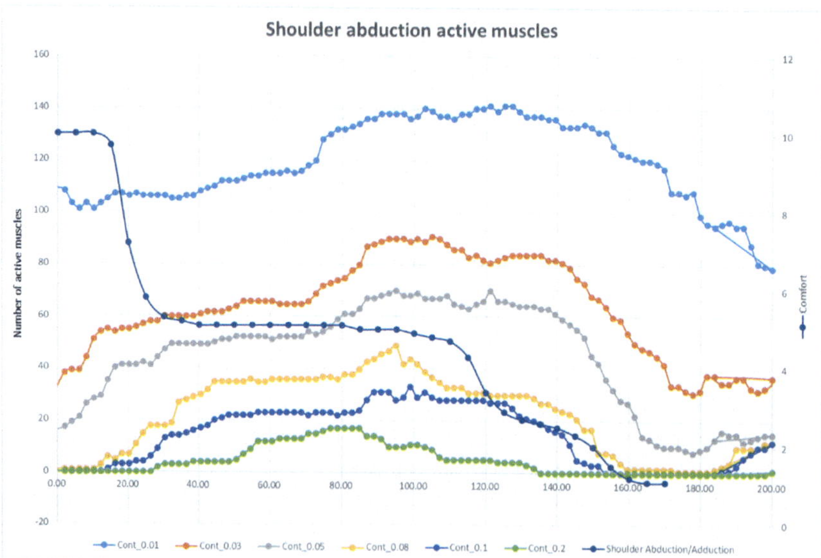

Fig. 8. Active muscles count in shoulder abduction movement.

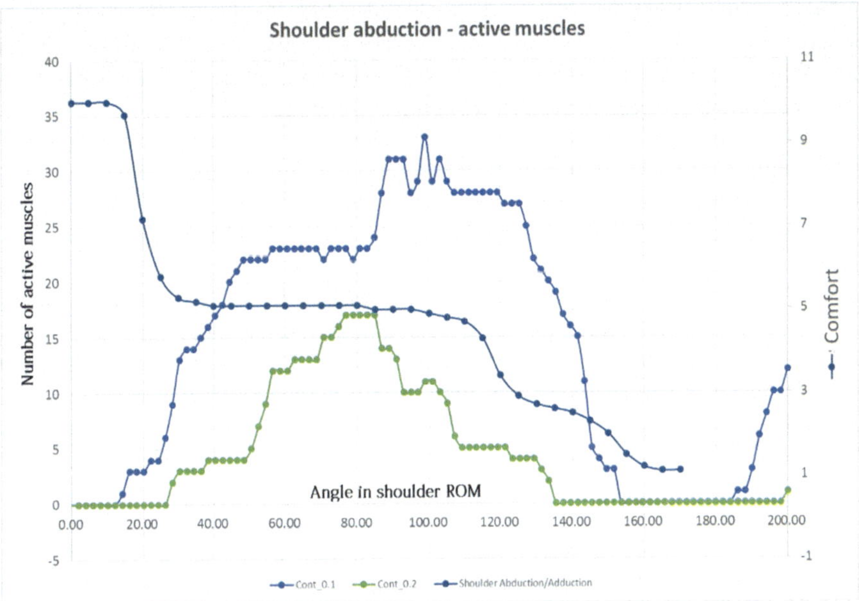

Fig. 9. Number of active muscles in shoulder abduction and threshold 0.01

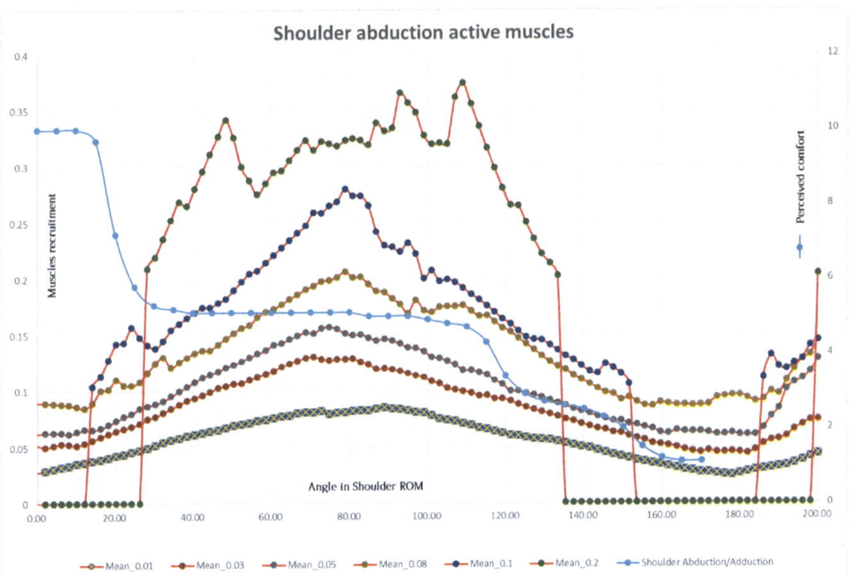

Fig. 10. Mean muscles activity in shoulder abduction movement.

Graphs show that a too low loads level and a low number of involved muscles correspond to a high comfort value, corresponding to the Rest Posture of a human body. When a movement starting from Rest Posture takes place, the number of active muscles increases and the activity of each muscle change.

The Envelope, that is the most used ANYBODY's result, shows a maximum in the angles range where the comfort is constant [30°, 110°] (Fig. 7). Looking in Fig. 8, when the shoulder reached the maximum abduction [110°, 170°] the Envelope decreased to the initial level, the comfort drops down, while the counting of active muscles shows an increasing trend. This last parameter considers both the muscles naturally used to keep the global posture stable unchanging the CoG and the muscles used to perform the movements. Cutting the counting by a threshold, the curves change in trend and in absolute values. A threshold between 0.1 and 0.2 (Fig. 9) seems to shows a change in correspondence of the Comfort curve slope/value change. Also in Fig. 10 there is a change of mean activity from 0.08 threshold, where after the principal peak around 90°, a secondary peak appears at 100° where the Comfort starts to decrease.

Unexpectedly the Comfort level decrease even if both number of active muscles and muscle stress level decrease when the maximum extension is reached. In the particular case of shoulder abduction, the maximum extension corresponds to the arm in the upper vertical position, when its weight loads the articulation of the shoulder and not the muscles maintaining the position, and the ligaments are at their maximum extension.

So the Discomfort sensation at the extremities of movement's ranges is due to the high ligaments elongation and to local pressure on the cartilage of articulations and not to the muscular stress. Figures 11 and 12 show the same graphs for Elbow pronation, where the same considerations and value can be deduced. In addition, other articulations' movements were analysed and results are coherent.

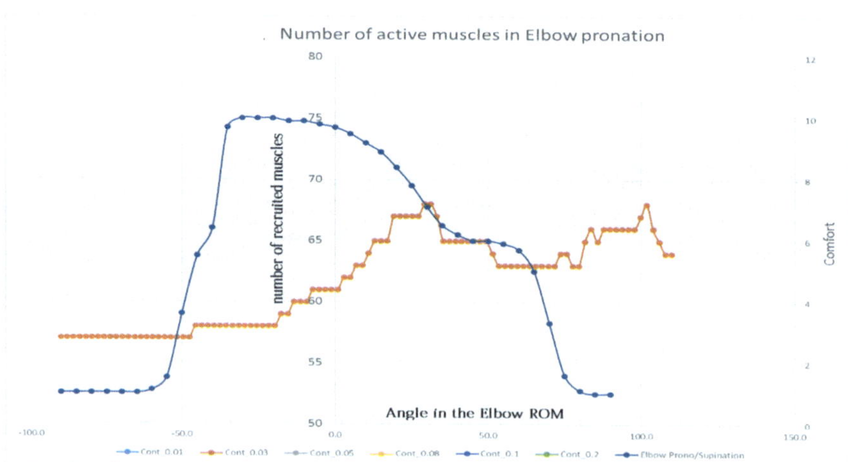

Fig. 11. Number of active muscles in elbow pronation

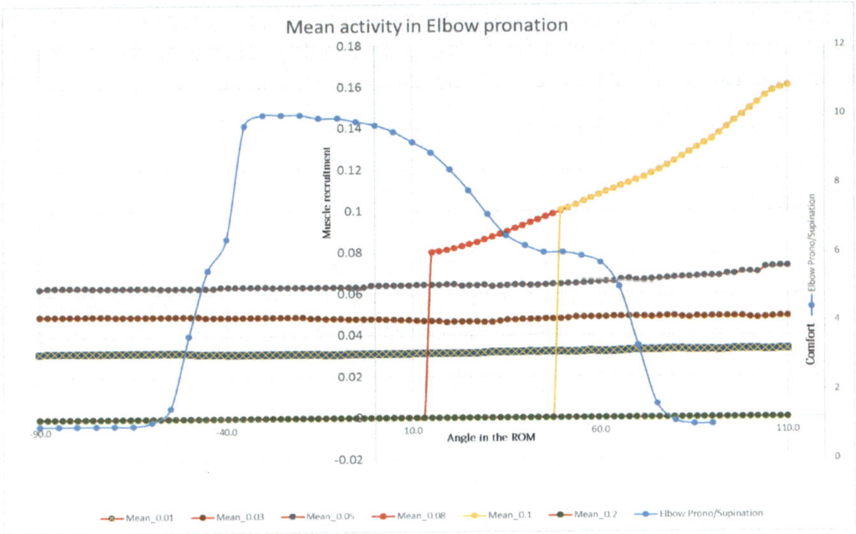

Fig. 12. Mean muscles activity in elbow pronation movement.

4 Conclusions

Physical comfort and discomfort are strongly related to efforts and pain of the muscular-skeletal system, thus, muscular activity could be a relevant indicator of perceived comfort and discomfort. The aim of this paper is the understanding of the influence of muscular activity on postural comfort perception. Therefore, a study based on simulations of a Standing Human Body model by ANYBODY software was performed with the aim to analyse the relation between Postural Comfort and muscles' stresses.

Mean Activity, Envelope and Number of Active Muscles were qualified as the software parameters representing a factor that significantly influences the Comfort experience.

Other parameters, such as ligament extension and joints loads, need a deep investigation because of their influence on Comfort in the extremes of the articular joint range.

However, results showed that:

- when the body is in Rest Posture, the comfort is higher and active muscles and mean activity are at the lower level;
- when the Number of active muscle and also the Mean activity increase, due to movements or external loads, the comfort decrease;
- when the ligaments extension or the articular loads increase, the comfort decrease;
- changes in Mean activity and Number of active muscle could be usefully observed with a 0.1 threshold value.

Additional simulations are suggested in order to verify the hypothesis on ligaments and articular joints. An experimental activity by comfort questionnaires and simulations comparison is suitable to check rules when several joints movements are combined in a single action.

References

1. Naddeo, A., Cappetti, N., Vallone, M., Califano, R.: New trend line of research about comfort evaluation: proposal of a framework for weighing and evaluating contributes coming from cognitive, postural and physiologic comfort perceptions. In: Vink, P. (ed.) Advances in Social and Organizational Factors, Cracovia, pp. 503–515 (2014)
2. Naddeo, A., Cappetti, N., D'Oria, C.: Proposal of a new quantitative method for postural comfort evaluation. Int. J. Ind. Ergon. **48**, 25–35 (2015)
3. Southard, S.A., Mirka, G.A.: An evaluation of backpack harness systems in non-neutral torso postures. Appl. Ergon. **38**, 541–547 (2007)
4. Kong, Y.K., Lowe, B.D.: Evaluation of handle diameters and orientations in a maximum torque task. Int. J. Ind. Ergon. **35**, 1073–1084 (2005)
5. Fernandez, J.E., Agarwal, R., Landwehr, H.R., Poonawala, M.F., Garcia, D.T.: The effects of arm supports during light assembly and computer work tasks. Int. J. Ind. Ergon. **24**, 493–502 (1999)
6. Jianghong, Z., Long, T.: An evaluation of comfort of a bus seat. Appl. Ergon. **25**, 386–392 (1994)
7. Hejazi, S., Rouhi, G., Rasmussen, J.: The effects of gastrocnemius–soleus muscle forces on ankle biomechanics during triple arthrodesis. Comput. Meth. Biomech. Biomed. Eng. **20**, 130–141 (2017)
8. Wibawa, A.D., Verdonschot, N., Halbertsma, J.P.K., Burgerhof, J.G.M., Diercks, R.L., Verkerke, G.J.: A trap motion in validating muscle activity prediction from musculoskeletal model using EMG. Int. J. Bio-Sci. Bio-Technol. **8**, 61–72 (2016)
9. Ji, Z., Wang, H., Jiang, G., Li, L.: Analysis of muscle activity utilizing bench presses in the AnyBody simulation modelling system. Model. Simul. Eng. **2016**, 1–7 (2016)
10. Wu, J.Z., An, K.-N., Cutlip, R.G., Krajnak, K., Welcome, D., Dong, R.G.: Analysis of musculoskeletal loading in an index finger during tapping. J. Biomech. **41**, 668–676 (2008)
11. Apostolico, A., Cappetti, N., D'Oria, C., Naddeo, A., Sestri, M.: Postural comfort evaluation: experimental identification of range of rest posture for human articular joints. Int. J. Interact. Des. Manuf. **8**, 109–120 (2014)
12. Naddeo, A., Cappetti, N., Califano, R., Vallone, M.: Manual assembly workstation redesign based on a new quantitative method for postural comfort evaluation. Appl. Mech. Mater. **459**, 368–379 (2013)
13. Trapanese, S., Naddeo, A., Cappetti, N.: A preventive evaluation of perceived postural comfort in car-cockpit design: differences between the postural approach and the accurate muscular simulation under different load conditions in the case of steering-wheel usage. SAE Technical Paper 2016, April 2016, no. 2016-01-1434 (2016)
14. Vallone, M., Naddeo, A., Cappetti, N., Califano, R.: Comfort driven redesign methods: an application to matresses production systems. Open Mech. Eng. J. **9**(1), 492–507 (2015)
15. Anybody Technology. https://www.anybodytech.com/downloads/documentation/. Accessed 21 Nov 2018

Experimental Comfort Assessment of a T-Shirt for Roadrunner

Enrico Avagnale[1], Rosaria Califano[2(✉)], and Iolanda Fiorillo[2]

[1] MGS Maintenance Global Services SRL, Leghorn, Italy
[2] University of Salerno, 84084 Fisciano, SA, Italy
rocalifano@unisa.it

Abstract. Ergonomic design principles can be applied to improve human interaction. In sport activities, ergonomic planning can be a key-point of development both for sportsman behaviour and for sport facilities/devices' design, in order to reduce injuries and muscular-skeletal disease. In some sport activities, wears and shirts plays a role not only as "dress" for sport but also as "containers" for some sports' devices like instruments, chronometers, bottle-holder and so on. In this study, we have focused our attention on long-running people (half-marathon/marathon) that usually use the running T-shirt also as a bottle holder. In this paper, a new prototype of running T-shirt has been realized and compared with the standard one. A comparative ergonomic and comfort assessment has been done. Interaction between human and T-shirt/bottle has been modelled in DELMIA® DHM environment. The video analysis and the acquisition of human joint's angles were processed using the KINOVEA® software. The data were evaluated using the CA_Man® software. Results are related to the subjective ratings expressed by the subject s that filled a questionnaire.

Keywords: Comfort evaluation · Digital human modelling · Ergonomics

1 Introduction

After the Second World War, the concept of 'fitting something to humans' (not only to workers) became a central component of user-centred design (UCD) and human-centred design (HCD). UCD defines humans as 'users' while HCD takes the term 'human' to include all the elements that can interact with humans, even independently of their will, e.g. the environment or emotions [1]. UCD is a broad term used to describe design processes in which end-users influence the way that a design takes shape [2]. Ergonomic studies are the basis for identifying ergonomic parameters for product improvement. The most used method for ergonomic analysis is based on the following three steps: (1) direct and indirect (video-recorded) observations of users [3, 4]; (2) collection of information; (3) data synthesis and determination of ergonomic indices [5].

In medical field static human posture is evaluated both from tomographic images [6] and from optical 3D scanners (markerless [7] or marker based [8]).

C. Rizzi et al. (Eds.): ADM 2019, LNME, pp. 71–81, 2020.
https://doi.org/10.1007/978-3-030-31154-4_7

In sports' field of research, several researchers were involved in ergonomic/comfort assessment of devices for human. Nevertheless, looking to the literature, most of the papers dealing with sportsman dress, and sportsman shirts are related to general comfort or thermal comfort.

The first work you can find in literature is Ruckman and Green [9] in which a study about skin irritation due to fabric consumption has been investigated; results show how different fabrics can affect the tactile comfort while running. In [10], the physiological and psychological responses during and after high-intensity exercise in a warm and humid environment in subjects wearing shirts of different fabrics has been investigated; results are comparative and allow to assess a methodology for evaluating T-shirts' comfort under the physiological point of view. In [11], authors reports on novel ventilation design in sports T-shirt, which combines spacer and mesh structure, and experimental evidence on the advantages of design in improving thermal comfort. In [12], Sperlich et al. have investigated on thermoregulatory, cardiorespiratory, metabolic, and perceptual responses while running in a hot environment and wearing T-shirts made from different fibre types; also in this paper, results show how different fabrics can affect the tactile comfort while running. In [13], Carneiro et al. made a comparative thermos-physiological study in sport bras for running and demonstrated the effects of raw materials on physiological behaviour of fabrics. Cao et al. [14] in 2016, proposed a customized body mapping of thermal behaviour where the body is in contact with sportswear while Herten et al. [15] in 2016, analysed the effect of functional materials on body-thermoregulatory activity. Finally, in 2017, Ziemele et al. [16] investigated the concept of comfort in sportswear under both physiological and physical point of view, analysing, for the physical aspect, the easiness of body-movements during sport-activities. As demonstrated in this literature review, most of papers deals with thermal-comfort related issue. Only the last one investigate the physical comfort while using sportswear. This article focuses on the correct use of sportswear in running competitions. During the running challenges, professional runners are used to drink water and minerals through a little bottle they attach to their sportswear. During drinking operation several action are done and easiness of them are very important during a run. In this study, several quasi-static positions have been analyses in order to assess the ergonomics of the bottle use both in the standard position and in the new-designed position. The static positions have been recorded by a camera through a video recording and have been analysed; Environmental parameters and some interaction parameters between the runner and the sportswear have not been taken into account as described in throughout the test.

1.1 The Current State of Art and Alternative Solution

The current T-shirt running sport-activity have a minimal design on the back side in order to achieve lightness and usability; usually is possible to find a pocket with a zipper or a spring-loaded pocket (Fig. 1) (T-shirt borrowed from cycling) in order to put inside the required objects. This result to be not so ergonomically compatible with the act of running, especially if you want to bring a bottle to hydrate. Several alternatives have been presented on the market but used the same architype of "belt-pocket". Our idea is to place the pocket in an alternative location, in order to achieve more ergonomics and more comfortable conceived position. The proposed solution presents a high pocket just behind the neck, in the center-back part of the shoulder as shown in Fig. 2.

Fig. 1. Running T-shirt used now

Fig. 2. Alternative solution

2 Materials and Methods

Fifteen runners, all volunteers, participated in the experiment. None had a history of muscular-skeletal diseases. Ten subjects were non-professional runners while five subjects had a previous or a current experience as professional runners. The subjects were informed of the nature of the tests and their written consent, in accordance with ethical standards of the University of Salerno, was obtained. Subjects were asked to use a specific dress code in order to avoid clothing influence on comfort perception [17].

Before each test, subjects have been measured in order to acquire all anthropometric data and were informed about the procedure of testing.

Each participant had to run for a 400 m race. In the middle of the race (after about 200 m) each subject had to take a half liter bottle from the pocket of the T-shirt, drink the water and, then, reposition the bottle in the T-shirt. This operation had to be done both with a standard T-shirt and with the prototyped one, in alternative way, in a randomized sequence.

During the bottle-using operation, two cameras have been placed on a mobile support linked to a bicycle that followed the runner on the right (for righty people) or on the left (for lefty people) side, in order to acquire the video of the bottle-using operation. The video-taping of the operations was carried out using two Go-Pro cameras, one placed lateral side to the participants and the other placed behind them, both at a distance of about one meter (Fig. 3).

After each test, subjects were asked to answer to a comfort questionnaire. Questionnaire was composed of two parts: one for the standard sportswear and one for the non-standard sportswear. For each solution, information about comfort expectation [17], neck comfort, shoulder comfort, elbow comfort, wrist comfort, overall comfort, reachability and repositioning was gathered.

Fig. 3. Behind and lateral video-taping

Questionnaire was composed by nine questions with a continuous band going to the worst perception (extremely in discomfort) to the best perception (extremely in comfort). In order to answer to a question, the subject had to put a cross on the band as shown in [18].

2.1 Human Modelling and Data Acquisition

For the postural analysis, body angles (for upper limb joints) were required. To obtain them, some angles acquired by cameras have been detected through Kinovea® software (see Fig. 4) and other missing angles through DELMIA® Digital Human modeller software.

A virtual mannequin has been realized inside the virtual environment of DELMIA® software to model each subject and the movements thy do for performing the action while running. Human builder and Human Task Simulation features were used to perform these analyses. The following anthropometric measurements of subjects were used to model the mannequin: Height; Arm length; Forearm length; Shoulders height; Shoulder width; Chest height; Hips height; Chest width; Weight.

Angles were used to define the mannequin's movements directly in DELMIA® as shown in [19] and [20].

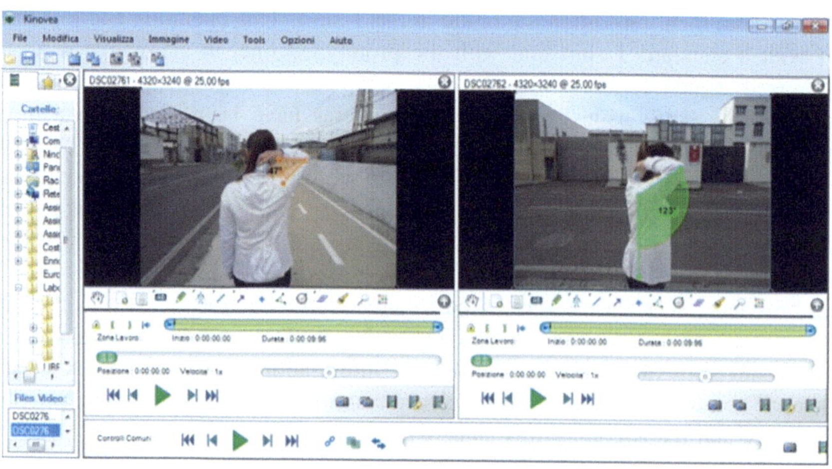

Fig. 4. Movement and postural analyses

Some angles measurements were performed on the subjects in order to check and verify the DELMIA® simulation's results. The error was always under 3% like in [3]. After modelling every task for each mannequin (Fig. 5), angles were saved as a function in time through the feature "Data read" in Task Simulation Environment. These values were used to perform the Comfort analyses.

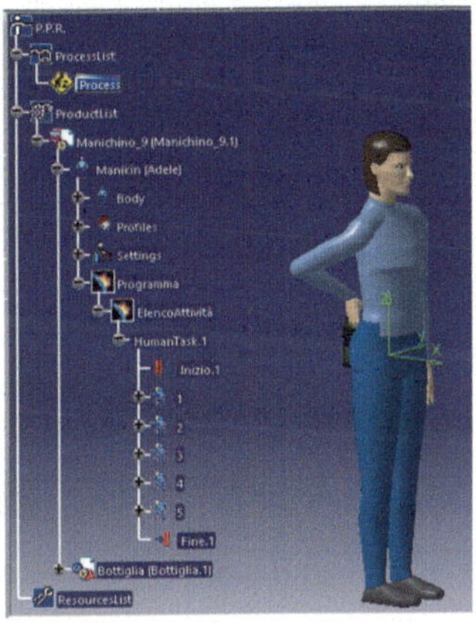

Fig. 5. Modelling tasks of DELMIA® software

To obtain objective comfort indexes from the collected angles, the CAMan® software [21] has been used.

The CAMan® software is based on experimental studies conducted by Naddeo et al. [22] to give a comfort index according to postural angles assumed, especially the angles of the human joint. As far as upper limbs, it considers:

neck: frontal flexion, rotation and lateral flexion
shoulder: frontal flexion, abduction/adduction
elbow: flexion/extension, pronation/supination
wrist: flexion/extension, radio/ulnar deviation

For each joint, curves of postural comfort over angles are used. Comfort indexes are rating on an 11-point scale where 0 = "no comfort" and 10 = "maximum comfort".

Statistical Analyses were performed on the acquired variables and Pearson Index was used to check possible correlations among variables.

2.2 Ergonomic Assessment and Comfort Evaluation

Ergonomic assessment was made by hypothesizing a half-marathon and the use of the bottle for several time during the run. OCRA checklist was used for comfort assessment. All data gave us a very good result in terms of ergonomics both for the standard and for the new T-shirt.

After the ergonomic assessment, a Comfort assessment was done.

The results exported in Excel® were analysed used CA_Man® software, in order to calculate the overall (upper limbs) and the local comfort index in according to the actions listed in the Test paragraph. An example of result is shown in Fig. 6.

Comfort Indexes standard sportswear (low pocket)			
	Tasks		Mean Value
	Extracting	Positioning	
Neck: FF+R	8.4677	8.2382	
Neck: FL	9.9	9.9	9.126
Arm L: FF	9.9	9.9	
Arm L: ad-ab	10	10	
Arm R: FF	1.1	1.1	
Arm R: ad-ab	8.916941133	8.287032967	7.400
Forearm L: ps	1	1	
Forearm L: F-E	10	10	
Forearm R: ps	1	1	
Forearm R: F-E	5.600345992	5.640091575	4.405
Hand L: dev RU	10	10	
Hand L: FE	10	10	
Hand R: dev RU	4.300298351	4.3	
Hand R: FE	1	1	6.325
Global	6.513	6.455	6.484

Fig. 6. Comfort index for each tasks

The answers to the questionnaire were listed as shown in the Fig. 7:

Comfort Questionnaire standard sportswear (low pocket)							
Expected Comfort	Neck Comfort	Shoulder Comfort	Elbow Comfort	Wrist Comfort	Ease of extracting	Ease of repositioning	Overall Comfort
9	9	8	6	7	7	7	8

Fig. 7. Subjective votes

3 Results and Conclusions

In the Figs. 8, 9, 10 and 11, the results of correlations are shown:

Correlations																
		Expected Comfort	Neck Comfort	Shoulder Comfort	Elbow Comfort	Wrist Comfort	Ease of extraction	Ease of repositioning	Overall Comfort	CA_Man® Neck	CA_Man® Shoulder	CA_Man® Elbow	CA_Man® Wrist	CA_Man® Extraction	CA_Man® Repositioning	CA_Man® Global
Expected_Comfort	Correlazione di Pearson	1	0.307	0.394	-0.209	-0.041	-0.173	,529*	-0.059	-0.501	-0.152	-0.259	-0.279	-0.302	-0.352	-0.43
	Sig. (2-code)		0.265	0.146	0.454	0.885	0.538	0.043	0.835	0.057	0.588	0.351	0.313	0.274	0.198	0.109
	N	15	15	15	15	15	15	15	15	15	15	15	15	15	15	15
Neck_Comfort	Correlazione di Pearson	0.307	1	,600*	-,600*	-0.341	0.125	0.24	0.164	-0.324	0.359	-0.355	-,869**	-,540*	-0.369	-,821*
	Sig. (2-code)	0.265		0.018	0.018	0.213	0.657	0.389	0.56	0.239	0.189	0.194	0.006	0.038	0.176	0.014
	N	15	15	15	15	15	15	15	15	15	15	15	15	15	15	15
Shoulder_Comfort	Correlazione di Pearson	0.394	,600*	1	-0.423	-,564*	-0.12	0.154	0.269	-0.131	0.33	-0.104	-,898**	-0.366	-0.353	-0.482
	Sig. (2-code)	0.146	0.018		0.116	0.028	0.67	0.584	0.333	0.642	0.23	0.713	0.004	0.18	0.197	0.069
	N	15	15	15	15	15	15	15	15	15	15	15	15	15	15	15
Elbow_Comfort	Correlazione di Pearson	-0.209	-,600*	-0.423	1	0.018	-0.18	-0.154	0.124	-0.039	-0.274	,526*	0.436	0.345	0.497	,551*
	Sig. (2-code)	0.454	0.018	0.116		0.949	0.521	0.584	0.658	0.89	0.322	0.043	0.104	0.207	0.059	0.033
	N	15	15	15	15	15	15	15	15	15	15	15	15	15	15	15
Wrist_Comfort	Correlazione di Pearson	-0.041	-0.341	-,564*	0.018	1	0.057	0.155	-,608*	0.002	-0.233	-0.07	,750**	0.394	0.262	0.447
	Sig. (2-code)	0.885	0.213	0.028	0.949		0.84	0.582	0.016	0.994	0.404	0.803	0.001	0.146	0.346	0.095
	N	15	15	15	15	15	15	15	15	15	15	15	15	15	15	15
Ease of grip	Correlazione di Pearson	-0.173	0.125	-0.12	-0.18	0.057	1	0.12	-0.02	-0.257	0.078	0.171	-0.067	-0.163	0.297	0.048
	Sig. (2-code)	0.538	0.657	0.67	0.521	0.84		0.67	0.942	0.355	0.781	0.541	0.812	0.561	0.283	0.864
	N	15	15	15	15	15	15	15	15	15	15	15	15	15	15	15
Ease of placing	Correlazione di Pearson	,529*	0.24	0.154	-0.154	0.155	0.12	1	-0.17	-,531*	-0.141	-0.032	-0.113	-0.164	-0.098	-0.181
	Sig. (2-code)	0.043	0.389	0.584	0.584	0.582	0.67		0.544	0.042	0.615	0.91	0.687	0.56	0.729	0.519
	N	15	15	15	15	15	15	15	15	15	15	15	15	15	15	15
Overall_Comfort	Correlazione di Pearson	-0.059	0.164	0.269	0.124	-,608*	-0.02	-0.17	1	-0.073	-0.32	0.169	-0.487	-0.09	-0.449	-0.328
	Sig. (2-code)	0.835	0.56	0.333	0.658	0.016	0.942	0.544		0.795	0.246	0.547	0.065	0.751	0.093	0.232
	N	15	15	15	15	15	15	15	15	15	15	15	15	15	15	15

*Correlation is significant at the 0.05 level (moderate)
**Correlation is significant at the 0.01 level (strong)

Fig. 8. Pearson correlation for standard sportswear, non-expert users

Correlations																
		Expected Comfort	Neck Comfort	Shoulder Comfort	Elbow Comfort	Wrist Comfort	Ease of extraction	Ease of repositioning	Overall Comfort	CA_Man® Neck	CA_Man® Shoulder	CA_Man® Elbow	CA_Man® Wrist	CA_Man® Extraction	CA_Man® Repositioning	CA_Man® Global Comfort
Expected_Comfort	Correlazione di Pearson	1	0.363	-0.16	0.326	0.313	0.326	0.139	0.326	0.194	-0.472	0.106	0.052	-0.218	0.013	-0.13
	Sig. (2-code)		0.184	0.568	0.235	0.257	0.235	0.622	0.235	0.489	0.075	0.708	0.853	0.436	0.963	0.645
	N	15	15	15	15	15	15	15	15	15	15	15	15	15	15	15
Neck_Comfort	Correlazione di Pearson	0.363	1	0.295	,582*	-0.419	0.092	0.039	0.208	0.28	-,623*	,531*	-0.237	-0.04	-0.353	-0.278
	Sig. (2-code)	0.184		0.286	0.023	0.12	0.743	0.89	0.457	0.313	0.013	0.042	0.396	0.887	0.197	0.315
	N	15	15	15	15	15	15	15	15	15	15	15	15	15	15	15
Shoulder_Comfort	Correlazione di Pearson	-0.16	0.295	1	0.35	-0.157	-0.09	-0.432	0.069	-0.123	-0.395	0.149	0.172	0.11	-0.121	-0.015
	Sig. (2-code)	0.568	0.286		0.2	0.577	0.75	0.108	0.807	0.662	0.145	0.596	0.54	0.697	0.667	0.958
	N	15	15	15	15	15	15	15	15	15	15	15	15	15	15	15
Elbow_Comfort	Correlazione di Pearson	0.326	,582*	0.35	1	-0.32	-0.106	0.03	0.423	0.014	-,597*	0.459	-0.272	-0.43	-0.167	-0.4
	Sig. (2-code)	0.235	0.023	0.2		0.244	0.707	0.916	0.116	0.96	0.019	0.085	0.327	0.11	0.551	0.14
	N	15	15	15	15	15	15	15	15	15	15	15	15	15	15	15
Wrist_Comfort	Correlazione di Pearson	0.313	-0.419	-0.157	-0.32	1	0.341	0.23	0.191	-0.168	0.271	-,613*	617*	0.241	0.401	0.441
	Sig. (2-code)	0.257	0.12	0.577	0.244		0.213	0.409	0.496	0.549	0.329	0.015	0.014	0.387	0.139	0.1
	N	15	15	15	15	15	15	15	15	15	15	15	15	15	15	15
Ease of grip	Correlazione di Pearson	0.326	0.092	-0.09	-0.106	0.341	1	-0.019	-0.007	0.171	0.125	-0.212	-0.088	0.318	-0.413	-0.09
	Sig. (2-code)	0.235	0.743	0.75	0.707	0.213		0.946	0.979	0.542	0.657	0.449	0.755	0.249	0.126	0.75
	N	15	15	15	15	15	15	15	15	15	15	15	15	15	15	15
Ease of placing	Correlazione di Pearson	0.139	0.039	-0.432	0.03	0.23	-0.019	1	0.076	0.361	0.213	-0.336	-0.145	0.051	-0.21	-0.115
	Sig. (2-code)	0.622	0.89	0.108	0.916	0.409	0.946		0.788	0.186	0.446	0.22	0.606	0.856	0.452	0.682
	N	15	15	15	15	15	15	15	15	15	15	15	15	15	15	15
Overall_Comfort	Correlazione di Pearson	0.326	0.208	0.069	0.423	0.191	-0.007	0.076	1	0.151	-,836*	0.167	-0.189	-,598*	-0.008	-0.462
	Sig. (2-code)	0.235	0.457	0.807	0.116	0.496	0.979	0.788		0.592	0.011	0.551	0.5	0.019	0.729	0.083
	N	15	15	15	15	15	15	15	15	15	15	15	15	15	15	15

*Correlation is significant at the 0.05 level (moderate)
**Correlation is significant at the 0.01 level (strong)

Fig. 9. Pearson correlation for non-standard sportswear (high pocket), non-expert users

Correlations																
		Expected Comfort	Neck Comfort	Shoulder Comfort	Elbow Comfort	Wrist Comfort	Ease of extraction	Ease of repositioning	Overall Comfort	CA_Man® Neck	CA_Man® Shoulder	CA_Man® Elbow	CA_Man® Wrist	CA_Man® Extraction	CA_Man® Repositioning	CA_Man® Global Comfort
Expected_Comfort	Correlazione di Pearson	1	-0.764	-0.873	-0.327	0.764	-0.327	0.764	-0.607	-0.768	-0.374	-0.596	0.605	-0.153	-0.013	-0.091
	Sig. (2-code)		0.133	0.053	0.591	0.133	0.591	0.133	0.278	0.129	0.535	0.289	0.28	0.806	0.984	0.884
	N	5	5	5	5	5	5	5	5	5	5	5	5	5	5	5
Neck_Comfort	Correlazione di Pearson	-0.764	1	0.667	-0.167	-1,000**	-0.167	-1,000**	0.6	,904*	0.434	0.133	-,964**	-0.513	-0.411	-0.503
	Sig. (2-code)	0.133		0.219	0.789	0	0.789	0	0.285	0.035	0.466	0.831	0.008	0.377	0.492	0.388
	N	5	5	5	5	5	5	5	5	5	5	5	5	5	5	5
Shoulder_Comfort	Correlazione di Pearson	-0.873	0.667	1	0.167	-0.667	0.167	-0.667	0.218	0.735	0.765	0.343	-0.447	0.21	0.05	0.142
	Sig. (2-code)	0.053	0.219		0.789	0.219	0.789	0.219	0.724	0.157	0.132	0.572	0.451	0.735	0.936	0.82
	N	5	5	5	5	5	5	5	5	5	5	5	5	5	5	5
Elbow_Comfort	Correlazione di Pearson	-0.327	-0.167	0.167	1	0.167	0.167	0.167	-0.055	0.153	-0.245	,938*	0.287	0.641	,892*	0.852
	Sig. (2-code)	0.591	0.789	0.789		0.789	0.789	0.789	0.931	0.806	0.691	0.019	0.639	0.243	0.042	0.067
	N	5	5	5	5	5	5	5	5	5	5	5	5	5	5	5
Wrist_Comfort	Correlazione di Pearson	0.764	-1,000**	-0.667	0.167	1	0.167	1,000**	-0.6	-,904*	-0.434	-0.133	,964**	0.513	0.411	0.503
	Sig. (2-code)	0.133	0	0.219	0.789		0.789	0	0.285	0.035	0.466	0.831	0.008	0.377	0.492	0.388
	N	5	5	5	5	5	5	5	5	5	5	5	5	5	5	5
Ease of grip	Correlazione di Pearson	-0.327	-0.167	0.167	0.167	0.167	1	0.167	0.491	-0.349	-0.245	0.274	0.239	0.641	-0.081	0.291
	Sig. (2-code)	0.591	0.789	0.789	0.789	0.789		0.789	0.401	0.565	0.691	0.656	0.699	0.243	0.897	0.634
	N	5	5	5	5	5	5	5	5	5	5	5	5	5	5	5
Ease of placing	Correlazione di Pearson	0.764	-1,000**	-0.667	0.167	1,000**	0.167	1	-0.6	-,904*	-0.434	-0.133	,964**	0.513	0.411	0.503
	Sig. (2-code)	0.133	0	0.219	0.789	0	0.789		0.285	0.035	0.466	0.831	0.008	0.377	0.492	0.388
	N	5	5	5	5	5	5	5	5	5	5	5	5	5	5	5
Overall_Comfort	Correlazione di Pearson	-0.607	0.6	0.218	-0.055	-0.6	0.491	-0.6	1	0.322	-0.299	0.271	-0.654	-0.214	-0.5	-0.397
	Sig. (2-code)	0.278	0.285	0.724	0.931	0.285	0.401	0.285		0.597	0.625	0.66	0.231	0.73	0.391	0.508
	N	5	5	5	5	5	5	5	5	5	5	5	5	5	5	5

*Correlation is significant at the 0.05 level (moderate)
**Correlation is significant at the 0.01 level (strong)

Fig. 10. Pearson correlation for standard sportswear, expert users (professional roadrunners)

Correlations																
		Expected Comfort	Neck Comfort	Shoulder Comfort	Elbow Comfort	Wrist Comfort	Ease of extraction	Ease of repositioning	Overall Comfort	CA_Man® Neck	CA_Man® Shoulder	CA_Man® Elbow	CA_Man® Wrist	CA_Man® Extraction	CA_Man® Repositioning	CA_Man® Global Comfort
Expected_Comfort	Correlazione di Pearson	1	0.363	-0.16	0.326	0.313	0.326	0.139	0.326	0.194	-0.472	0.106	0.052	-0.218	0.013	-0.13
	Sig. (2-code)		0.87	0.272	0.495	0.735	0.239	0.053	0.239	0.942	0.253	0.775	0.5	0.333	0.337	0.34
	N	5	5	5	5	5	5	5	5	5	5	5	5	5	5	5
Comfort_Collo	Correlazione di Pearson	-0.102	1	-0.25	0.687	-,986**	0.395	0.134	-0.395	,996**	-0.173	,912*	-,879*	0.678	-0.686	-0.69
	Sig. (2-code)	0.87		0.685	0.2	0.002	0.51	0.83	0.51	0	0.781	0.031	0.05	0.209	0.201	0.197
	N	5	5	5	5	5	5	5	5	5	5	5	5	5	5	5
Comfort_Spalla	Correlazione di Pearson	-0.612	-0.25	1	-0.875	0.086	0.791	-0.535	-0.791	-0.331	,997**	-0.626	-0.243	0.542	-0.533	-0.529
	Sig. (2-code)	0.272	0.685		0.052	0.891	0.111	0.353	0.111	0.586	0	0.259	0.694	0.345	0.355	0.36
	N	5	5	5	5	5	5	5	5	5	5	5	5	5	5	5
Comfort_Gomito	Correlazione di Pearson	0.408	0.687	-0.875	1	-0.557	-0.395	0.468	0.395	0.747	-0.834	,925*	-0.257	-0.068	0.057	0.052
	Sig. (2-code)	0.495	0.2	0.052		0.329	0.51	0.427	0.51	0.147	0.079	0.024	0.676	0.914	0.928	0.934
	N	5	5	5	5	5	5	5	5	5	5	5	5	5	5	5
Comfort_Polso	Correlazione di Pearson	0.21	-,986**	0.086	-0.557	1	-0.542	-0.046	0.542	-,968**	0.007	-0.831	,946*	-0.791	0.797	0.8
	Sig. (2-code)	0.735	0.002	0.891	0.329		0.345	0.942	0.345	0.007	0.991	0.081	0.015	0.111	0.106	0.104
	N	5	5	5	5	5	5	5	5	5	5	5	5	5	5	5
Facilita_Estrazione	Correlazione di Pearson	-0.645	0.395	0.791	-0.395	-0.542	1	-0.423	-1,000**	0.316	0.836	-0.017	-0.786	,943*	-,940*	-,938*
	Sig. (2-code)	0.239	0.51	0.111	0.51	0.345		0.478	0	0.605	0.077	0.979	0.115	0.016	0.018	0.018
	N	5	5	5	5	5	5	5	5	5	5	5	5	5	5	5
Facilita_Riposiz	Correlazione di Pearson	0.873	0.134	-0.535	0.468	-0.046	-0.423	1	0.423	0.177	-0.533	0.334	0.13	-0.29	0.285	0.283
	Sig. (2-code)	0.053	0.83	0.353	0.427	0.942	0.478		0.478	0.776	0.355	0.582	0.835	0.636	0.642	0.645
	N	5	5	5	5	5	5	5	5	5	5	5	5	5	5	5
Comfort_Generale	Correlazione di Pearson	0.645	-0.395	-0.791	0.395	0.542	-1,000**	0.423	1	-0.316	-0.836	0.017	0.786	-,943*	,940*	,938*
	Sig. (2-code)	0.239	0.51	0.111	0.51	0.345	0	0.478		0.605	0.077	0.979	0.115	0.016	0.018	0.018
	N	5	5	5	5	5	5	5	5	5	5	5	5	5	5	5

*Correlation is significant at the 0.05 level (moderate)
**Correlation is significant at the 0.01 level (strong)

Fig. 11. Pearson correlation for non-standard sportswear (high pocket), expert users (professional road runners)

The tables show a very good correlation among variables acquired through questionnaire and the same calculated by using CA_Man® comfort assessment tool. In particular, strong correlations have been found for average comfort index for neck, elbow and wrist, both for standard and for the new type of T-shirt. As shown in Figs. 8, 9, 10 and 11, correlation among the same variables are higher in the case of only experts' data.

About the comparison among the two kind of T-shirts, several values of perceived comfort are higher in the new (the first) prototype use even if the difference is not so valuable (see Fig. 12). By analysing the mean values in time (see Figs. 13 and 14), during all the duration of the operation of extraction, use and repositioning, the mean values of the overall comfort in the case of new T-shirt are always higher than the case of use of the standard one, thus demonstrating the validity of the idea.

Finally, the effect of expectation on comfort rating can be explained by the Fig. 15.

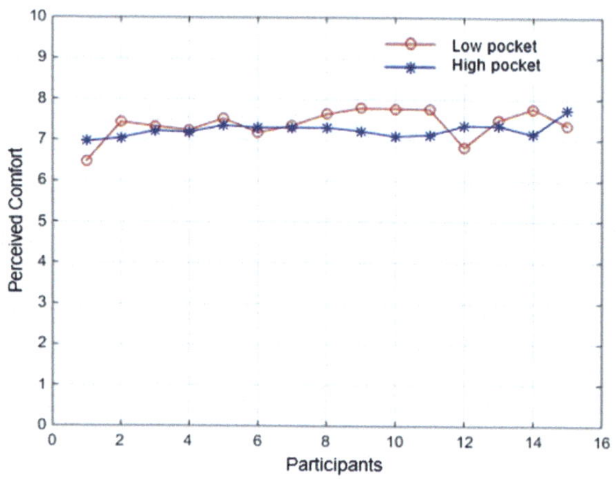

Fig. 12. Comparison of the two different solutions for each participant in terms of perceived comfort.

It is very interesting to highlight that the expert has always a greater value of mean comfort perception for high-pocket use towards the standard-pocket use, while the curve comfort/expectation respects the theory of expectation [14].

In conclusion, the results show clearly the prototype achieved higher value of postural comfort, thus it represent the best solution for roadrunners.

Furthermore, the results show even that CA_Man® tool is able to predict the perceived comfort, with the best results in case of experts' evaluations. Though the comfort index local are similar for both product (Fig. 12), the prototype have the mean value of the comfort index much more high because the movement whom you do is much more ergonomically compatible in its entire complex (Figs. 13 and 14).

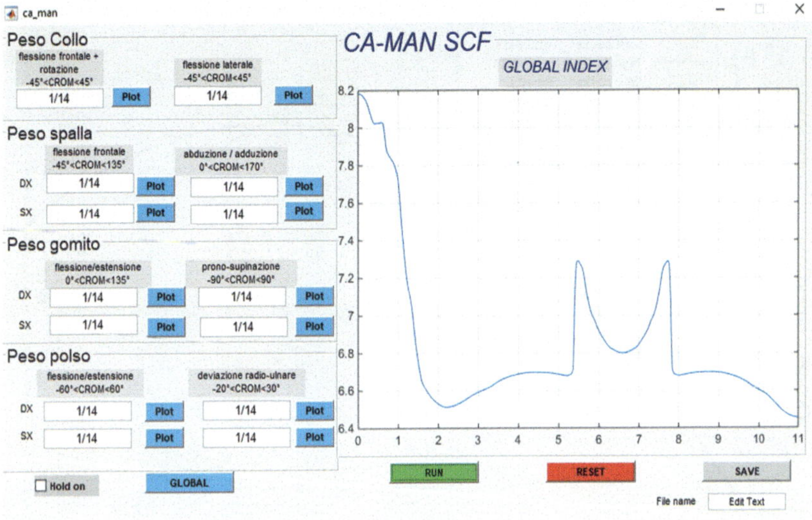

Fig. 13. Global Mean Comfort Index over time during extraction and repositioning in standard pocket use

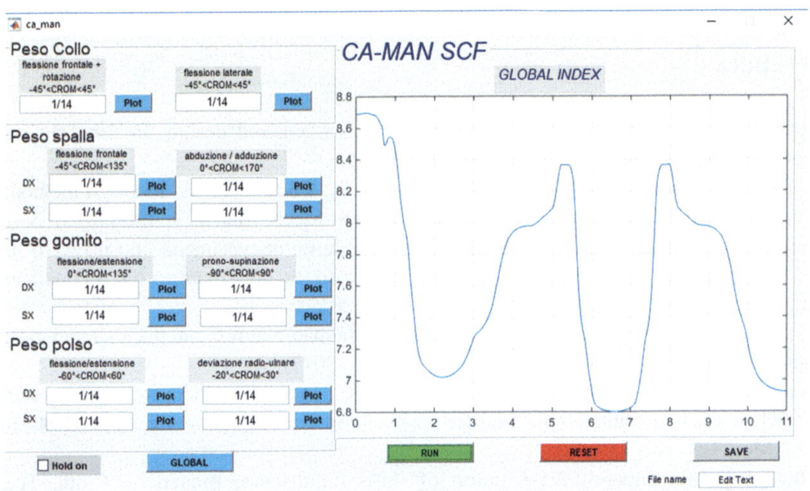

Fig. 14. Global Mean Comfort Index over time during extraction and repositioning in not standard high pocket use

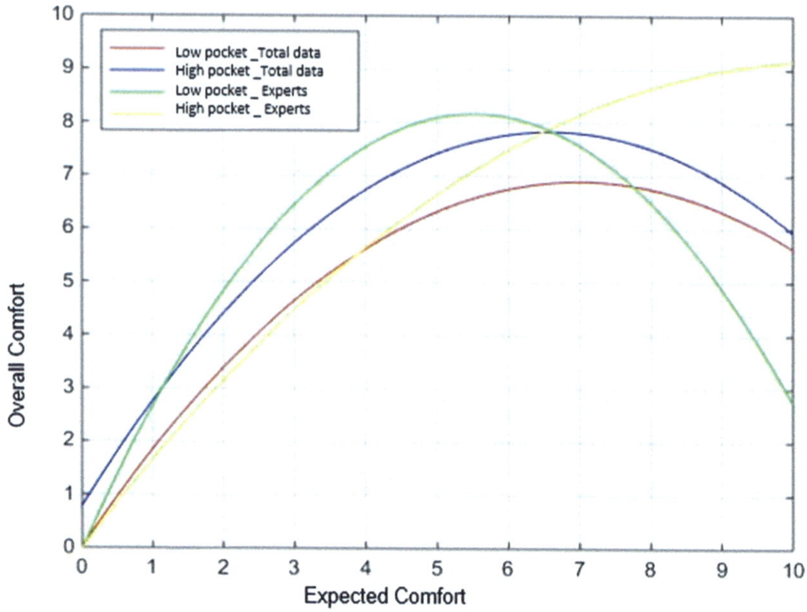

Fig. 15. Comfort curves expected comfort-perceived comfort

References

1. Abras, C., Maloney-Krichmar, D., Preece, J.: User-centered design. In: Encyclopedia of Human-Computer Interaction, pp. 445–456 (2004)
2. Vallone, M., Naddeo, A., Cappetti, N., Califano, R.: Comfort driven redesign methods: An application to mattresses production systems. Open Mech. Eng. J. **9**, 492 (2015)
3. Naddeo, A., Barba, S., Ferrero, F.I.F.: Propuesta de un nuevo método no invasivo para el análisis postural con aplicaciones de fotogrametría 4d. In: XI Congreso Ibero-Americano de Ingegnieria Mecanica (CIBIM2013), pp. 11–14 (2013)
4. Califano, R., Naddeo, A., Vink, P.: The effect of human-mattress interface's temperature on perceived thermal comfort. Appl. Ergon. **58**, 334 (2017)
5. Naddeo, A., Cappetti, N., Califano, R., Vallone, M.: Manual assembly workstation redesign based on a new quantitative method for postural comfort evaluation. In: Applied Mechanics and Materials, vol. 459 (2013)
6. Ruckman, J.E., Green, J.A.: Comfort of shirts for distance runners. J. Cloth. Technol. Manage. **12**, 1–25 (1996)
7. Puggelli, L., Uccheddu, F., Volpe, Y., Furferi, R., Di Feo, D.: Accuracy assessment of CT-based 3D bone surface reconstruction. In: Advances on Mechanics, Design Engineering and Manufacturing II, pp. 487–496 (2019)
8. Furferi, R., Governi, L., Uccheddu, F., Volpe, Y.: A RGB-D based instant body-scanning solution for compact box installation. In: Advances on Mechanics, Design Engineering and Manufacturing, pp. 819–828 (2017)
9. Ferrara, P., Piva, A., Argenti, F., Kusuno, J., Niccolini, M., Ragaglia, M., Uccheddu, F.: Wide-angle and long-range real time pose estimation: a comparison between monocular and stereo vision systems. J. Vis. Commun. Image Represent. **48**, 159–168 (2017)

10. Brazaitis, M., Kamandulis, S., Skurvydas, A., Daniusevičiūtė, L.: The effect of two kinds of T-shirts on physiological and psychological thermal responses during exercise and recovery. Appl. Ergon. **42**, 46–51 (2010)
11. Sun, C., Au, J.S., Fan, J., Zheng, R.: Novel ventilation design of combining spacer and mesh structure in sports T-shirt significantly improves thermal comfort. Appl. Ergon. **48**, 138–147 (2015)
12. Sperlich, B., Born, D., Denise Lefter, M., Holmberg, H.C.: Exercising in a hot environment: which T-shirt to wear? Wilderness Environ. Med. **24**, 211 (2013)
13. Carneiro, L.P., Miranda, T.M.R., Catarino, A.: A comparative thermophysiological study in sport bras for running. In: IOP Conference Series: Materials Science and Engineering (2017)
14. Cao, M., Li, Y., Guo, Y., Yao, L., Pan, Z.: Customized body mapping to facilitate the ergonomic design of sportswear. IEEE Comput. Graph. Appl. **36**, 70–77 (2016)
15. Herten, A., Csapo, R., Kofler, P., Bottoni, G., Hasler, M., Bechtold, T.: Effects of functional shirts with different fiber compositions on thermoregulation in well-trained runners. Proc. Inst. Mech. Eng. Part P J. Sport Eng. Technol. **231**, 75–82 (2016)
16. Ziemele, I., Šroma, I., Kakarāne, A.: Comfort in Sportswear. Key Engineering Materials (2018)
17. Naddeo, A., Cappetti, N., Califano, R., Vallone, M.: The role of expectation in comfort perception: the mattresses' evaluation experience. In: 6th International Conference on Applied Human Factors and Ergonomics (2015)
18. Bouwens, J.M.A., Fasulo, L., Hiemstra-van Mastrigt, S., Schultheis, U.W., Naddeo, A., Vink, P.: Effect of in-seat exercising on comfort perception of airplane passengers. Appl. Ergon. **73**, 7–12 (2018)
19. Califano, R., Cozzitorto, P., Delmastro, M., De Vito, C., Sellitto, G., Vallone, M.: Virtual ergonomic analysis and redesign methods: an application to lunch payment station at university of Salerno. Int. J. Appl. Eng. Res. **11**, 7114 (2016)
20. Bruno, F., Califano, R., Greco, R., Ricco, G., Vallone, M., Valter, N.: Virtual ergonomic analysis and redesign of a dentist's workcycle. ARPN J. Eng. Appl. Sci. (2016)
21. Naddeo, A., D'Ambrosio, D., Antonini, B.: Task analysis and comfort evaluation through simulations: differences between subjective perceptions and simulated data in the case of car-hood lifting. In: Advances in Intelligent Systems and Computing (2018)
22. Naddeo, A., Cappetti, N., D'Oria, C.: Proposal of a new quantitative method for postural comfort evaluation. Int. J. Ind. Ergon. **48**, 25 (2015)

Virtual Reality and Interactive Design

Dynamic Projection for the Design of an Adaptive Museum Guide

Alma Leopardi(✉), Silvia Ceccacci, and Maura Mengoni

Department of Industrial Engineering and Mathematical Sciences, Università Politecnica delle Marche, Via Brecce Bianche, 12, 60131 Ancona, Italy
a.leopardi@pm.univpm.it,
{s.ceccacci,m.mengoni}@univpm.it

Abstract. Very often historical buildings used as museums are characterized by rooms completely covered with decorations (e.g. frescoes, inlay, etc.) rich in details and symbolic contents. Providing adequate information to the visitors to allow them to fully appreciate the artworks is of paramount importance in this context. X-reality technologies have the potential to provide an effective response to the need to combine the educational mission of museums with the ability to involve visitors emotionally, allowing the public to learn new knowledge in a playful way. This study introduces a Museum Guide System, based on Dynamic Projection, to increase the involvement of visitors to the Studiolo by Federico da Montefeltro at Urbino, minimizing technology intrusiveness by ensuring a complete fusion of digital contents with the physical environment. The system is able to track the visitors, to detect their profile and to offer information and multimedia contents tailored to the characteristics of the audience (e.g., children, adults).

Keywords: Spatial Augmented Reality · Dynamic projection · Adaptive system · Museum Guide System · Digital Cultural Heritage

1 Introduction

Nowadays, museums are much more than places dedicated to the conservation and exhibition of collections and artworks. They should be seen as privileged means of communication and play a central role in making culture accessible to the mass audience [1].

One of the keys to attracting a wide audience of visitors is the use of new innovative technologies and new interaction paradigms. When we talk about "new technologies", we conventionally mean mainly visual technologies, which place the image at the center of communication, and interactive technologies. Among these in particular X-reality technologies (e.g., holography, augmented reality, immersive virtual reality, etc.) are today highly attractive because, despite having reached a good level of maturity, they are still far from being widespread enough to be defined as a mass phenomenon. These technologies have the potential to provide an effective response to the need to combine the educational mission of museums with the ability to involve visitors emotionally, allowing the public to learn new knowledge in a playful way.

C. Rizzi et al. (Eds.): ADM 2019, LNME, pp. 85–94, 2020.
https://doi.org/10.1007/978-3-030-31154-4_8

In addition to attracting and fascinating visitors, they also allow curators to modulate the cultural proposal by structuring different information paths for different user profiles (e.g., a training/educational path to the average user, a scientific tool to the professional user) [2].

In recent years, X-reality technologies have reached a good level of international consensus, and the tendency of museums to enrich their collections of objects traditionally exhibited with modern technological systems is growing. However, even today, it is rather rare to find virtual reality installations in museums. This is due to several factors, including: cost of equipment needed for the dedicated space, accessibility and usability of systems, user acceptance, quality of sensory experience, and so on. Moreover, although on the one hand the application of such technologies is contributing to making culture accessible to the mass public, on the other hand it has initiated a process of desacralization of the museum institution, extending its boundaries beyond the modern entertainment industry. This has also contributed to creating a certain diffidence towards technological innovations, especially in the case of historical buildings used as museums (e.g., the Ducal Palace of Urbino, Villa Borghese in Rome, Palace of Te in Mantua, etc.). In fact, in these cases, the structure-container plays an important role in attracting the public and determining the visitor's experience.

However, where the rooms of museum are completely covered of decorations full of details and symbolic contents (e.g., frescoes, marquetry, etc.), providing adequate information to the visitors to allow them to fully appreciate the artworks is of paramount importance. This is the case of the Studiolo of Federico da Montefeltro at Urbino, a tiny inner sanctum (3.6×3.35 m) for study and reflection, whose four walls are entirely clad up to about 2.22 m by elaborately detailed wood inlay, made with over forty different types of wood. Such wooden intarsia depicts an assortment of cupboards, left open in a studiedly casual manner to reveal their contents, in which find place an assortment of objects (e.g., books, musical instruments and scores, armor pieces, weaponry), animals and other figures, all with a strong symbolism that leads back to the deeds and power of the Duke Federico da Montefeltro. However, the allegorical meanings that go beyond the mere figurative representation cannot be known and appreciated by the common visitors, who therefore pass, without being able to understand and appreciate the artwork in its entirety.

Guide systems represents a good solution to overcome this issue, but often they are not satisfactory and suitable for certain categories of users (e.g. children).

In this context, the paradigm of Digital Cultural Heritage (DCH), represents a valid instrument to improve cultural heritage fruition for people visiting museum, in an inclusive way. Applications based on X-reality technologies, developed by using proper storytelling and gamification techniques, can offer new entertaining, informative and enjoyable experience to the visitors and increase their level of engagement and interest [3, 4]. Moreover, they can be very helpful for combining educational purposes with the capability to emotionally involve visitors [5].

In the last two decades, several X-reality applications have been proposed, characterized by different user workspaces, interaction devices, level of immersion, interaction, presence and usability. However, the majority of them pose a variety of practical limitations, such as, development complexity, high installation and maintenance costs, inaccessibility of immersive VR technologies, low usability and high maintenance [6]. These issues strongly reduce the possibility achieve a commercially effective mass entertainment. Moreover, during the development of the contents, demographic and cultural characteristics of the target groups are generally neglected, so that the resulted applications do not provide effective learning environments [6].

In this context, this study aims to define an application able to increase the involvement of visitors at the Studiolo of Federico da Montefeltro, according to their behavior and characteristics (e.g., age, abilities, skills, etc.), to improve learning, and finally push coming back to the museum.

In order to avoid technology intrusiveness and ensure a complete fusion of digital content with the physical environment, so as to create an integrated space, where digital and real coexist, a complete technological framework is defined, which exploits Spatial Augmented Reality (SAR), based on dynamic projection.

2 Research Background

Today, one of the bigger challenges for museum marketing is how to improve the user experience and engage the visitors with artifacts and collections.

To achieve this goal, Virtual Reality (VR) and Augmented Reality (AR) are the most promising tools, as they allow to attract visitors and customize their experience, presenting artwork collections in a more appealing and exciting manner, through the generation of virtual museum exhibitions in various ways.

In particular, while VR allows to simulate real scenarios, AR focuses in enhancing physically-based reality perception through computer-generated sensory outputs [7].

In the last decade, many cultural tourism sites such as art galleries, museum or cultural heritage sites have discovered AR and VR. They have enhanced their visitor experiences with innovations ranging from virtual enhancements to re-live historical sites and events, engage with content in museums, or to visit remote destinations in virtual environments [1, 3]. Moreover, in order to enrich traditional guiding applications with personalized contents, several solutions have been proposed to convey information by using innovative communication channels and new storytelling methods, which exploit various forms of multimedia presentations, VR, AR and games [4, 8].

AR technology can be considered as the optimal solution for guiding systems, in order to engaging visitors and improve their experience while exploring artifacts, because it has the advantage to not isolating the visitors from the environment and allows simple information presentation with storytelling, including references to exhibits in order to enhance comprehension [9].

Today, the majority of proposed AR applications are based on mobile devices [10]. However, smartphone and tablet may distract the visitors from the exhibition and negatively impact on the interaction with artworks, so that users prefer more natural ways of interaction [11]. To this end, several solutions based on smart glasses have been proposed [12, 13], which provide immersive experience without the need to carry a device and to switch focus between the device and the exhibition.

Another emerging technology with the potential to transform learning is Spatial Augmented Reality (SAR). SAR allows real and virtual objects to effectively coexist in the same space by realizing the so called "digital fusion". Such paradigm, powered by video mapping techniques, is adopted in various museum context to improve the visitor perception and learning of geometric details of artefacts, that are difficult to distinguish due to aging effects such as erosion, e.g. Danish rune stone [14], Egyptian sculptures [15].

In the last years, new projection technologies are emerging, such as dynamic projection. This technology is characterized by the use of model high tech mirror (Mirror Head) that is digitally operated and used as an add-on feature from projectors to move either rapidly or slowly the projected pictures, videos and texts onto any imaginable surface. In this way, projections can be moved around from one surface to another or remain statically projected onto one specific surface. This technology has so far been used only in contemporary art exhibitions or commercial fairs and other public contexts, for entertainment purposes. For example, "Gallery Invasion" by Skullmapping [16] was an experimental dynamic projection application made for an art gallery in Leuven (Belgium), which transforms a static exhibition into a lively space, filled with different creatures. However, to the best of our knowledge, it has never been used in museums to improve visitor understanding of the allegorical decorations characterizing the rooms of historical buildings.

Finally, despite the implementation of adaptive systems, able to customize information services in museums have a great potential, many challenges have not yet addressed that still limit their wide adoption. These challenges are mainly due to the lack of tools for localization with minimal infrastructure, non-intrusive devices and methods for dynamic profile extraction, measuring visitor engagement and for group profiling [11].

3 A SAR Application for Enhance Experience with Renaissance Visual Artworks

The basic idea consists in the implementation of museum digital guide system, through the projection of multimedia contents on the wooden inlay of Federico's Studiolo. Given the inestimable value of the artwork, in this preliminary phase, it was decided to proceed with a full-scale reconstruction in the laboratory of a corner, the one between the north and east walls. The choice of these two walls is motivate by richness of symbolism that characterize the decoration of this part. Moreover, one of the two walls, the east wall, has a pillar in the middle, and this can generate critical issues from the technological point of view. The reconstruction was made with panels of hardboard, where some high definition photos of the wooden inlay are applied.

3.1 The Proposed Technological Architecture

The proposed system architecture (Fig. 1) is composed by several modules: tracking of visitor, detection of visitor profile, management of the multimedia contents that best suits the considered visitor profile and visualization of the contents.

The color RGB image provided by the Kinect, can be easily processed through proper machine learning algorithms in order to automatic detect visitor profile (i.e., gender and age) and engagement (e.g., by analyzing eye-gaze behavior and emotions).

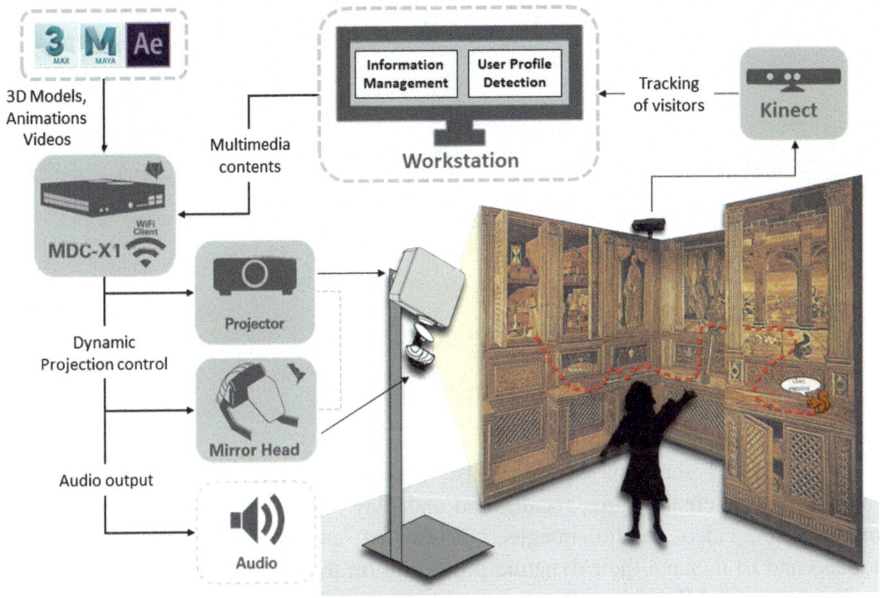

Fig. 1. The system architecture

Several methods are proposed in literature based on image processing and Convolution Neural Networks that allows to determine user gender and age and perform user eye-tracking and emotion detection [17, 18].

To ensure proper management of information services, according to the visitor requirements, it is possible to adopt a decision support system similar of that proposed by [19], which exploits artificial intelligence algorithms (machine learning) based on inductive inference, allowing knowledge related to the user, the context and the information services to be mapped, structured and managed though semantic models.

When visitors enter in the field of interaction, the system detects and recognizes the corresponding user profile and accordingly sends proper commands to the MDC-X1 server in order activate the most suitable multimedia contents. To recognize the user's profile, it is possible to proceed in two ways: by counting the users present in the acquisition field, belonging to each category, and by identifying the profile of the first user entering the room.

The MDC-X1 server is based on the Linux system; it is configured specifically to communicate movements and contents to the "Mirror Head" made by Dynamic Projection.

The Mirror Head (Fig. 2) is a modern high-tech mirror that is digitally operated and used as an add-on feature for projectors. It includes: mirror, motor-drive-arm, tilt motor and mirror head control unit. The device represents the core of the dynamic projection technology and allows to either rapidly or slowly move the projected multimedia contents onto any surface and to move them around from one surface to another.

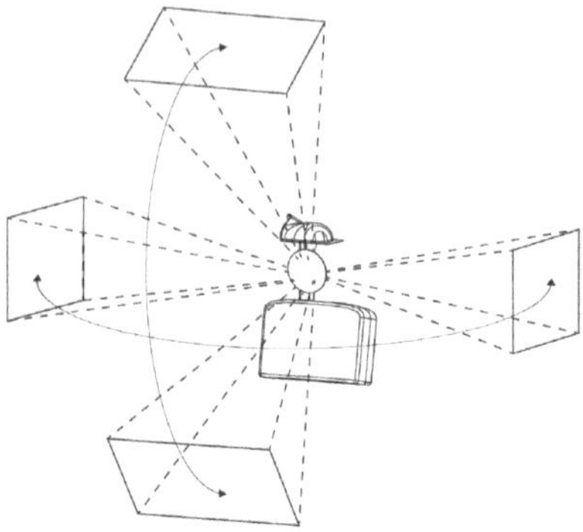

Fig. 2. The Mirror Head

The MDC-X1 server is equipped also with a particular software, called MDX, which allows to create, store, recall, load and play several typologies of multimedia contents (i.e., videos, audio, images, custom text, camera input, capture cards, 3D models) and to manage their dynamic projection on the desired surface.

The software MDX also allows to combine the visual elements in a proper way, through the direct interaction with the mirror head, in order to manage the movements and to make videos and photos straightening, based on the different angulations that the projector can find with the walls.

The multimedia contents are projected through the visualization module (projector, Mirror Head and MDC-X1 server) on the surfaces, overlapping with what is represented on them. Regarding the part for visualization and fruition of contents, besides the "Mirror Head", an Optoma ZH510T projector is implemented. It is a full HD projector, with resolution of 1920 × 1080, 5400 ansilumen. Its high brightness allows to project also on painted or decorated wall. Finally, audio output is provided through several speakers.

3.2 Use Case Scenarios

The Duke's Studiolo, thanks to the richness of hidden meanings, lends itself to become the setting of a story, whose plots come out from the wood inlay decorations. In order to make the allegories hidden in the Studiolo accessible, two typologies of visitors have been identified, based on the age: school-age children and adults. The choice of these two age groups lies in the fact that they cover the majority of the visitors of the Ducal Palace of Urbino during the year.

Through storytelling techniques, two scenarios have been hypothesized, whose structure and protagonists are common. The points of interest in both the scenarios (Fig. 3) were selected for their strong symbolism, importance, allegorical meaning and position in the Studiolo.

The common thread of both stories is represented by the dual nature of Federico's character: warrior and man of letters. However, the narrative interweaving in the two

Fig. 3. The considered points of interest: from the top left

cases develops differently: the story uses a more playful and less formal narrative for the children, while it is enriched with historical details and anecdotes for the adults.

In fact, in the adult scenario, Federico himself describes his personality as a Warrior, through the symbols of armour and ostrich and the honours received (e.g., the ermine, the garter, etc.). The story then focuses on the life of the "Man of Letters", through the animation of the books, the clavichord from which plays a music that accompanies the song of Hope, one of the three theological virtues that most inspired Federico.

In the children's scenario (Fig. 4), the subject narrator is the squirrel, the Federico's alter ego, who accompanies the young visitors throughout the story. Also in this case, the main traits of Federico's personality are retraced, but with less emphasis on his life as a warrior (always represented by the armour and the animals), before moving on his life of a literary man, through the use of more evocative sounds and animations to better involve the children in the representation.

To animate the scene, several multimedia contents are used: 3D model animations are adopted for the animals (e.g., squirrel, ermine and ostrich) and some objects (e.g., books, clavichord, armour pieces, etc.), while glow effects are used to highlight some details (e.g., letters on books). In addition, cinematographic techniques are used to film

Fig. 4. The children scenario

actors who impersonate Federico and the “Hope”. The visualization module overlapping the digital contents in the surfaces reconstruct in laboratory.

4 Conclusion and Further Challenges

The proposed application, based on dynamic projection, aims to increase the involvement of visitors to the Studiolo by Federico da Montefeltro, offering content tailored to the characteristics of the audience and minimize technology intrusiveness by ensuring a complete fusion of digital contents with the physical environment.

This is a very ambitious objective and many studies still need to be carried out to assess the actual potential of the proposed system.

In particular, several technological limits need to be deepen understood and addressed, both in the case of reconstruction in the laboratory and of the real place. For example, the dynamic projection has so far been used to project information only on monochrome surfaces not decorated and opaque. However, the walls of the Duke’s Studiolo are characterized by a semi-gloss wooden coating treated with wax, and they present a wide range of brown colors, varying from very dark to very light shades. Several tests must be carried out to ensure a good performance of the projected image depending on the ambient lighting conditions. Moreover, the projection distance results very limited by the small size of the Studiolo. This could cause problems in focusing distance and consequently limit the maximum size of the projected image. It will also be necessary to evaluate the maximum opening angle of the mirror-head that can be used for projection with a single projector device. Also, the possibility of making the system interactive, allowing you to provide specific information based on the location of the visitor.

Furthermore, several studies have to be carried out to proper evaluate the impact of the proposed system on the visitor experience.

Although all the application based on dynamic projection-based have so far become immediately viral, the degree of acceptance of the system in the considered context needs to be assessed. In fact, for example, the projected images could affect the visualization of the decoration and therefore it is necessary to assess whether and how any disturbing effects may affect the visitor experience.

Finally, another very important aspect that needs to be deepen investigated is the capacity of the proposed system to increase visitor learning. In particular, various studies should be carried out with the aim of comparing performance of the proposed system with that of traditional audio-guides provided by the museum. It will also be interesting to compare the proposed guiding system, based on dynamic projection technology, with a traditional AR application visualized on mobile devices.

Acknowledgements. This research are founded by "Università Politecnica delle Marche" in the framework of Strategic University projects. The project CIVITAS (ChaIn for excellence of reflectiVe Societies to exploit dIgital culTural heritAge and museumS) involves 7 departments and several researchers of the Polytechnic University of Marche.

References

1. Carrozzino, M., Bergamasco, M.: Beyond virtual museums: experiencing immersive virtual reality in real museums. J. Cult. Heritage **11**(4), 452–458 (2010)
2. Gonizzi Barsanti, S., Caruso, G., Micoli, L.L., Covarrubias Rodriguez, M., Guidi, G.: 3D visualization of cultural heritage artefacts with virtual reality devices. In: 25th International CIPA Symposium 2015, vol. 40, no. 5W7, pp. 165–172 (2015)
3. Sylaiou, S., Mania, K., Karoulis, A., White, M.: Exploring the relationship between presence and enjoyment in a virtual museum. Int. J. Hum. Comput. Stud. **68**(5), 243–253 (2010)
4. Mortara, M., Catalano, C.E., Bellotti, F., Fiucci, G., Houry-Panchetti, M., Petridis, P.: Learning cultural heritage by serious games. J. Cult. Heritage **15**(3), 318–325 (2014)
5. Barbieri, L., Bruno, F., Muzzupappa, M.: Virtual museum system evaluation through user studies. J. Cult. Heritage **26**, 101–108 (2017)
6. Tan, B.K., Rahaman, H.: Virtual heritage: reality and criticism. In: CAAD Futures, pp. 143–156. Les Presses de l'Université de Montréal, June 2009
7. Gavish, N., Gutiérrez, T., Webel, S., Rodríguez, J., Peveri, M., Bockholt, U., Tecchia, F.: Evaluating virtual reality and augmented reality training for industrial maintenance and assembly tasks. Interact. Learn. Environ. **23**(6), 778–798 (2015)
8. Wojciechowski, R., Walczak, K., White, M., Cellary, W.: Building virtual and augmented reality museum exhibitions. In: Proceedings of the Ninth International Conference on 3D Web Technology, pp. 135–144, April 2004
9. Dieck, M.C., Jung, T.H., Dieck, D.: Enhancing art gallery visitors' learning experience using wearable augmented reality: generic learning outcomes perspective. Curr. Issues Tourism **21** (17), 2014–2034 (2018)

10. Moro, S., Rita, P., Ramos, P., Esmerado, J.: Analysing recent augmented and virtual reality developments in tourism. J. Hospitality Tourism Technol. (2019). https://doi.org/10.1108/JHTT-07-2018-0059
11. Kosmopoulos, D., Styliaras, G.: A survey on developing personalized content services in museums. Pervasive Mob. Comput. **47**, 54–77 (2018)
12. Dieck, M.C., Jung, T., Han, D.I.: Mapping requirements for the wearable smart glasses augmented reality museum application. J. Hospitality Tourism Technol. **7**(3), 230–253 (2016)
13. Mason, M.: The MIT museum glassware prototype: visitor experience exploration for designing smart glasses. J. Comput. Cult. Heritage (JOCCH) **9**(3), 12 (2016)
14. Basballe, D.A., Halskov, K.: Projections on museum exhibits: engaging visitors in the museum setting. In: Proceedings of the 22nd Conference of the Computer-Human Interaction Special Interest Group of Australia on Computer-Human Interaction, pp. 80–87, November 2010
15. Ridel, B., Reuter, P., Laviole, J., Mellado, N., Couture, N., Granier, X.: The revealing flashlight: interactive spatial augmented reality for detail exploration of cultural heritage artifacts. J. Comput. Cult. Heritage (JOCCH) **7**(2), 6 (2014)
16. Gallery Invasion. https://skullmapping.com/project/gallery-invasion/. Accessed 14 June 2019
17. Generosi, A., Ceccacci, S., Mengoni, M.: A deep learning-based system to track and analyze customer behavior in retail store. In: 2018 IEEE 8th International Conference on Consumer Electronics-Berlin (ICCE-Berlin), pp. 1–6, September 2019
18. Generosi, A., Altieri, A., Ceccacci, S., Foresi, G., Talipu, A., Turri, G., Mengoni, M., Giraldi, L.: MoBeTrack: a toolkit to analyze user experience of mobile apps in the wild. In: 2019 IEEE International Conference on Consumer Electronics (ICCE), pp. 1–2 (2019)
19. Ceccacci, S., Generosi, A., Giraldi, L., Mengoni, M.: Tool to make shopping experience responsive to customer emotions. Int. J. Autom. Technol. **12**(3), 319–326 (2018)

Enhancing Spatial Navigation in Robot-Assisted Surgery: An Application

Marco Gribaudo[1], Sandro Moos[2], Pietro Piazzolla[2](✉), Francesco Porpiglia[3], Enrico Vezzetti[2], and Maria Grazia Violante[2]

[1] Department of Electronics, Information and Bioengineering, Politecnico di Milano, Milan, Italy

[2] Department of Management and Production Engineering, Politecnico di Torino, Turin, Italy
pietro.piazzolla@polito.it

[3] Division of Urology, Department of Oncology, School of Medicine, University of Turin, Turin, Italy

Abstract. Robot-assisted radical prostatectomy (RARP) has become a standardized practice in contemporary prostate cancer surgical procedures. Nowadays, the use of tailored surgical procedures in urologic surgery aims to maximize treatment efficacy while minimizing its impact on patient functions and health-related quality-of-life parameters. Augmented reality (AR) technology appears as a significant development in Image Guided Surgery (IGS) integrating surgical navigation with 3D virtual organ models registered on to the real patient's anatomy. In particular, AR provides surgeons the ability to see through solid objects; as such, it has been exploited in different surgical specialties. In this paper, we present the development of a software system augmenting the spatial navigation of the surgical environment allowed by surgical robots. The application is able to visualize the 3D virtual model of the organ (prostate and kidneys) targeted by the surgical procedure, overlay it over its real counterpart, as captured by the endoscope camera, using of registration and tracking techniques in real time, and stream the augmentation to the surgeon.

Keywords: Augmented reality (AR) · Image Guided Surgery (IGS) · Robotic surgery · Minimally invasive surgery (MIS)

1 Introduction

Image-guided surgical navigation and the remotely controlled robotics are the main technologies supporting minimally invasive approaches in surgery. In minimally invasive surgery, it is very important the use of images registered to the patient. They give a great support for both the planning phase and the guidance during the intervention. Even if the well-known advantages, such as a faster recovery and a reduced risk of wound-related complications, minimally invasive approaches introduce also new difficulties for surgeons. The vision of the organs and the mobility in handling surgical tools are limited by the indirect and restricted access to the operation area in

C. Rizzi et al. (Eds.): ADM 2019, LNME, pp. 95–105, 2020.
https://doi.org/10.1007/978-3-030-31154-4_9

the patient's body. The use of 2D images obtained by means of a camera inserted in the body and the CT slides traditionally consulted by surgeons are not able to estimate the depth of anatomical structures without moving the camera. For this reason, there are difficulties in hand-eye coordination from the surgeons to synchronize the instrument movements shown on a screen with the hand movements.

In this paper, we present the development of an application able to visualize the 3D virtual model of prostate and kidneys targeted by the surgical procedure, overlay it over its real counterpart as captured by the endoscope camera using registration and tracking techniques in real time, and stream the augmentation to the surgeon. The developed application is currently used and tested by the Urology unity of the San Luigi Hospital, in Orbassano (To) – Italy, and exploits the Tile-Pro visualization system of the Da Vinci surgical console. The paper is structured in two parts: in the first one, there is the review of the recent literature in the field of robotic surgery, Minimal invasive surgical (MIS) and Augmented Reality in surgery. In the second, there is the description of the performed application and the conclusions.

2 Brief Overview of Related Works

Since their introduction, surgical robots occupy a powerful role in today's modern surgical operations thanks to their technical advantages for the surgeon such as: better visualization with stereoscopic views, elimination of hand tremor, greater precision and improved maneuvering thanks to up seven degrees of freedom of the robotic wrist. Moreover, the improved kinematics that scale and transform the external movements of the surgical hands into limited internal movements of the "robotic hands" permits to the surgeon to perform complex technical tasks in a limited space work and to work in an ergonomic environment with less stress, achieving higher levels of concentration [1].

The birth of robotic surgery is justified by the increasing demand for greater surgical precision and safer operations and by the increasingly adopting of minimal invasive surgical (MIS) technologies. Minimal invasive surgery when compared to an open approach has been well recognized in many surgical specialties and for many different procedures [1]. The advantages of these minimally invasive approaches are the reduced wound access trauma, shorter hospital stay, less postoperative wound complications (ranging from wound infections to incisional hernias) and less disfigurement [1].

Minimally invasive Robotic-assisted Surgery (RAS) has become an opportunity to integrate current technology advancements in traditional minimally invasive surgery [2]. In urology robotic surgery was introduced into the field of minimally invasive surgery more than 15 years ago and is now used in many departments worldwide [3]. Urological procedures that are being performed robotically now are: radical prostatectomy (RP), radical cystectomy (RC), renal procedures – mainly partial nephrectomy (PN) and pyeloplasty, as well as ureteric re-implantation and adrenalectomy [3].

Even if over the past decades, minimally invasive surgery (MIS) has increasingly become more popular than open surgery thanks to greater clinical benefits, it has a loss of direct vision and tactile feedback upon the scene for the surgeon [1]. The introduction of the augmented reality (AR) to MIS is a viable solution to lighten this drawback.

Augmented reality combines a real world and virtual objects which are artificially generated digital content by a computer [4]. Augmented reality visualization allows the fusion of three-dimensional medical images, such as those from trans rectal ultrasound or computed tomography/MRI, with live camera images in real-time [5]. As another aspect of AR is a registration between the real world and virtual objects, it aims to estimate three dimensional (3D) position of virtual objects related to the real world [6].

Augmented reality (AR) is used in many fields including education, manufacturing, and entertainment but with advances in optics, computer systems, and surgical instruments and with increasing surgery intervention using laparoscopy, endoscopy, or catheterized, AR application to medicine takes an important role in many medical applications [7–13]. In fact since AR can allow the user to see 3D virtual objects superimposed upon the real world, in medicine, a surgeon can see hidden organs inside a body and improve the perception of treatment procedure by interacting with the real world [7].

The paper of Bernhardt, S., et al. describes the status of augmented reality in laparoscopic surgery as of 2016. According the authors, the first intervention assisted by AR was performed in neurosurgery [14–17]. Other interventions in neurosurgery followed such as the works of Peters [18]; Marcus et al. [19]; Cabrilo et al. [20]. The presence of the skull, which serves a rigid common referential between various acquisitions, justifies the rapid development of the AR in the neurosurgery field. While the brain is not rigid, its intraoperative deformation (also known as "brain shift") is limited and thus has been assumed rigid for AR purposes. Neurosurgery is the only discipline where AR has become a standard thanks to two decades of widespread practice. In any case, other surgical fields such as otolaryngology [21–24] in maxillofacial surgery [25] and in orthopedics [26] in dental surgery [27] have used AR. As in the case of the brain surgery, the rapid development of AR in these fields is obtained by the rigidity or semi-rigidity of the structures of interest thanks to their proximity to bones, which limits deformations between the preoperative acquisition/planning and the intraoperative state [28, 29].

In the recent years, augmented reality (AR) has obtained an increasing development in laparoscopic urology too. The first report was by Marescaux et al. in a laparoscopic adrenalectomy in 2004 [30]. In other works, AR assists intraoperative decision making providing the facilitating of the accurate anatomic relationships of the target organ and the facilitating correct tumor resection, ensuring negative surgical margins [31, 32]. The most common operation in which AR is used in urology is the partial nephrectomy. In this operation the understanding of hilar anatomy and 3D location of the tumor is very important [33]. Other publications reporting 3D image-guided surgery indicate that AR with visual cues to the subsurface anatomy could be a replacement in the case of minimally invasive surgery in the field of urology [34]. Early experience centered on laparoscopic prostatectomy involved the use of simultaneous transrectal ultrasound with navigational fiducial markers to overlay images of the prostate and neurovascular bundle [35]. This evolved to utilization of magnetic resonance and computed tomography (CT) images by software platforms to create maps which can be superimposed on real-time intraoperative images [36]. Porpiglia et al. recently demonstrated the accuracy of the superimposed AR image with the endoscopic view during RARP [36, 37].

3 The Software Applications

The architecture of the developed application, implemented in two different versions, each relaying on its own registration and tracking system is shown in Fig. 1. The first application, named pViewer, allows positioning of the virtual anatomic model over the real one by means of a 3D mouse. This requires a dedicated professional operating the device during most of the procedure. The second application, pTracker, introduced a semi-automatic tracking system, active during critical parts of the procedure that minimizes human intervention to those situations where the automation fails.

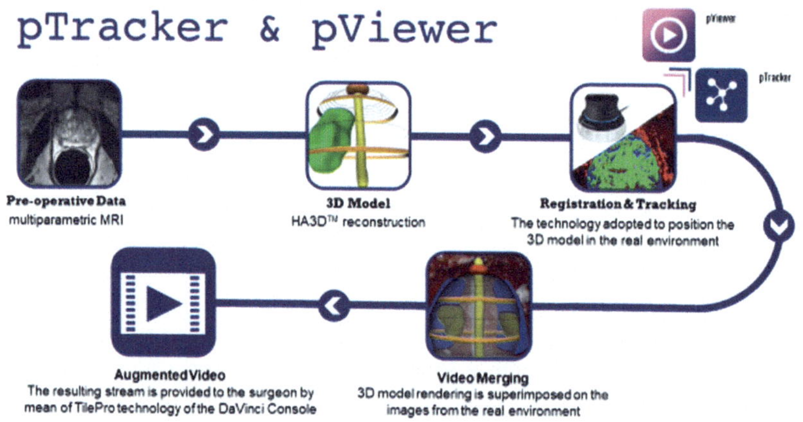

Fig. 1. Architecture of the developed applications, pViewer and pTracker, currently under testing in robot-assisted radical prostatectomy in-vivo procedures.

The virtual prostate or kidney model is generated from high-resolution pre-operatory medical imaging techniques, like e.g.: MRI. This 3D reconstruction accurately reproduce the prostate or the kidney of the patient undergoing surgery, along with surrounding structures, and it is modelled by bio-engineers using the HA3DTM trademarked technique (https://www.medics3d.com/).

Both the software programs are able to load and then display this 3D model of patient's prostate, controlling its translation, rotation and scale transformation values. The base of the urethra or prostate apex is the starting point for all transformations applied. This allows replicating the movements and rotations of the real prostate in this type of surgical procedure, because the chosen point is the last to be severed during the procedure.

To maximize surgeon's awareness about the intra-operatory environment, both applications allow isolating specific parts of the 3D model by modifying their transparency value. The visibility control of the surfaces allows each element of the model to become fully transparent, semi-transparent (maximizing the see-through effect) or fully opaque. The applied shader code leaves no shadows or highlighted zones over the surfaces to ease the subsequent video augmentation procedure.

After the position of the 3D model is determined in real time, the Video Mixer software (https://www.vmix.com/) takes the resulting video stream as input along with the in-vivo stream from the endoscope. The mixer merges the virtual and the real videos and sends the augmented result to the DaVinci remote console monitor, operated by the surgeon who can display it as needed by mean of the TilePro™ multi-input technology.

In the following Sections, we will detail the differences between the two software.

3.1 pViewer

This application is the first developed and has been implemented using the Unity platform (https://unity3d.com/) and C Sharp (https://docs.microsoft.com/en-us/dotnet/csharp/language-reference/language-specification/). The tracking of the target organ is human-assisted. The professional operating the pViewer application overlays the 3D model to the patient's organ in real time, using a 6 Degrees of freedom 3D mouse. Using this device, it is possible to pan, zoom and rotate the 3D model with one hand only, with the perceived feeling of manipulating the virtual prostate in the same way the real organ could have been manipulated.

Even if it does not leverage on any technology for automatic tracking and positioning, it has been tested during several in-vivo surgical procedures. The results of these tests, in term of benefits gained thanks to the AR use are the subject of the works Porpiglia et al. [36, 37]. In the first study, we demonstrate that the HA3D virtual reconstruction of the prostate based on multiparametric magnetic resonance imaging (mpMRI) data and the pViewer real-time superimposed imaging allowed in 80% of the tested subjects to correctly localize the cancer lesion, using biopsies. In the second study, which took into consideration the 3D model deformation, we demonstrate that prostate deformation is correctly simulated during surgery and lesion location is correctly identified, rising the correct localization percentage to 100% of the tested cases.

Moreover, up to date one of the main challenges of the use of AR in surgery is about problems associated with the deformation of the organs. This can be introduced by different sources: soft tissue physical characteristic, breathing of patient, and continuous contact of surgical instruments. The challenge concerns how to simulate deformations in real-time (or at least interactive-time) and the estimation of the in vivo mechanical parameters that can deliver realistic patient-specific tissues behavior. The interaction model applied to the pViewer application allows us to address the problem obtaining a high level of visual efficacy. During the in-vivo surgery, the professional operating the application can also modify the shape of the organ based on the visual information received from the endoscopic camera. Two non-linear parametric deformation of the 3D model can be selected: bend or stretch, the effects of which can be summed together [38]. The use of Barr's formulas gave us fast and intuitive manipulations with little computational effort, suited for an interactive-time application.

3.2 pTracker

As the second step toward a full automatic marker-less tracking system for AR in robot-assisted surgery we developed the pTracker application. The application code is written in C++ (https://isocpp.org/) in order to have a more direct management of the computer's resources because of the computational cost of the techniques used in it. The goal of this application is to determine for each frame of the endoscope video stream the correct pose of the 3D organ model. This is a known problem in the computer vision and augmented reality literature: the Perspective-n-Point (PnP), briefly summarized in Fig. 2. The vertices defining the 3D prostate mesh are the set of world space points to be positioned in the coordinate system of the camera that looks at the scene. The values of these vertices are known, but the matrix that transforms them in camera space not. The goal of solving the PnP problem is to obtain that matrix. The variables required to solve the algorithm are: the position of each of the mesh vertices p_i, their projection u_i onto the 2D space of image framed by the camera as well as the position of the camera c, at the origin of its coordinate system. Other required camera parameters, usually addressed as intrinsic are the focal f and the principal point (u, v), which usually coincide with the center of the image.

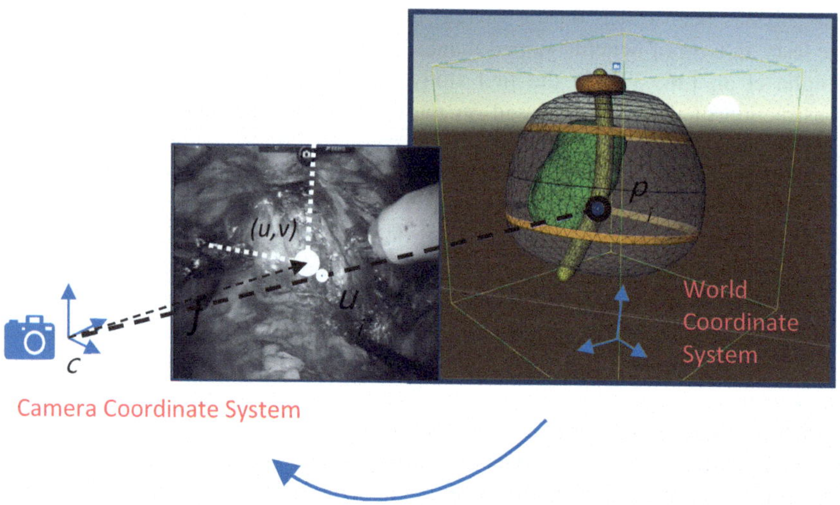

Fig. 2. The Perspective-n-Point problem: given a set on n 3D points in world coordinates and their corresponding 2D projections over an image, determine the pose of the camera able to align them.

Different techniques have been proposed in the literature during the years to solve the PnP problem. We addressed this pose estimation problem testing several of them before opting for the method in [39], refining its results, as suggested by different sources, with the use of the RANSAC model [40].

The 2D projection of the mesh vertices over the camera image is an input parameters required to solve the pose estimation problem. This projection potentially changes every frame of a video stream. To obtain the correspondence between 3D vertices and their 2D representation automatically, a set of features or keypoints are extracted from the image. These points act like markers and can be recognized in each subsequent frame thanks to additional information, or descriptors, automatically extracted with them. In Fig. 3-b, we show in red and green the whole set of keypoints extracted in a given frame using the ORB feature detector algorithm [41]. Before the PnP algorithm can be used, a correspondence need to be determined between the 3D mesh points and these keypoints, which are going to be used as their 2D projection. This is performed during a registration step shown in Fig. 3-a, where a simplified representation of the prostate model is manually positioned over the endoscope image. At the end of this procedure, a subset of points (in green in Fig. 3-b) is associated to the prostate vertexes to act as their 2D projection and then used by the pose estimation algorithm.

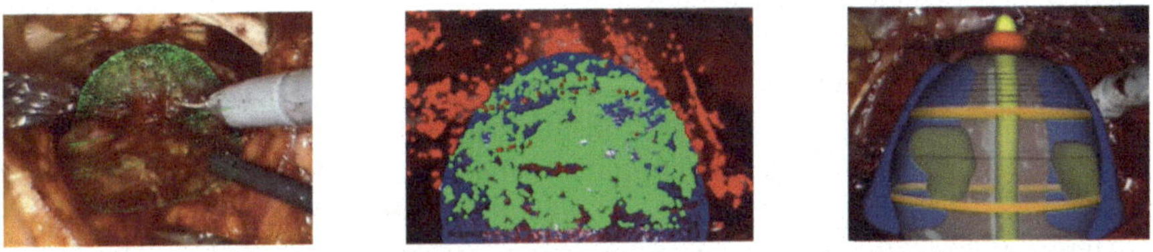

a. b. c.

Fig. 3. (a) The positioning of the virtual organ over the real one; (b) The registration procedure associates the keypoints found with the vertexes of the 3D model; (c) Pose is estimated and the video stream augmented.

At the current state of development of the application, the registration step has to be repeated when the intraoperative condition changes (e.g.: removal of organs) or when there is a sudden change in viewport, and thus in surgery environment illumination. This is because the endoscope has a light source embedded in it and when moved it will determine a change of highlights in the image. Changing the texture will change the results of the ORB algorithm and the keypoints associated with the vertices during the registration are no longer recognizable in the new texture. We are currently investigating different solutions to improve the robustness of the application when it faces texture changes, by testing different filters for the image and with the introduction of an extra light source in the surgery environment able to reduce light changes. In Fig. 3-c, we show the final augmentation stream sent back to the surgeon at the end of the overlay process.

While pViewer is currently in use during in-vivo procedures, pTracker is undergoing an extensive testing campaign. In particular, since we determined in which parts of the surgical procedure the AR will introduce the most of its benefits, we are investigating alternative tracking methods more specifics for the features available for detection in those stages, i.e.: when the prostate has been removed, a plastic catheter is introduced in the empty cave. In that stage, AR could be beneficial to show the position where the tumor mass was before removal in order to proceed with biopsies.

The presence of a distinguishable feature such as the form of the catheter, normally not available in this context, can potentially introduce the use of alternative tracking methods requiring less computational cost.

4 Conclusions

The tremendous progress made by the use of robots in surgery was in a relatively short period of time with improvements for both the patient and surgeon. From the point of view of the patient, a reduction in hospitalization time, fewer complications, and less scarring are obtained while from the surgeon there are benefits in ergonomics thanks to consoles that allow them to sit while operating, joystick controls which compensate for the fulcrum effect from the use of laparoscopic tools, and 3D glasses which display improved views of the operation [2, 42].

Urology has long been recognized as an avid adopter of new technologies and innovations in surgical practice. In concert with the exponential and rapid improvements in laparoscopic techniques and instrumentations over the last two decades, urologists' enthusiasm to implement minimally invasive approaches has led to the near extinction of open surgical approaches in several different urological diseases [2, 43].

Augmented reality (AR) has been an active area of research and development in laparoscopic urology in recent years. It is an advanced form of image-guided surgery and implies the use of enhanced visual information to the normal surgical field to supplement the lack of tactile feedback. This information is typically extrapolated from preoperative or, less commonly, intraoperative 3D imaging from US, computerized tomography, or magnetic resonance imaging [44–47].

AR can be effectively used for preoperative planning as for example tailoring individually preferred incisions and cutting planes and for timely completion of the actual surgery. Moreover, AR increases safety achieved by improving the surgeon's spatial perception of the surgical environment in order to avoid unnecessary manipulations or inadvertent injuries to inner organs.

In the paper a software system that augments the spatial navigation of the surgical environment has been developed. This application permits to visualize the 3D virtual model of the prostate and kidneys targeted by the surgical procedure, overlay it over its real counterpart, as captured by the endoscope camera, using of registration and tracking techniques in real time, and stream the augmentation to the surgeon.

Our work on AR in robot-assisted surgery is an ongoing research aiming to complete automatic tracking of the organ. To this end, in a future implementation of the application, we plan to introduce a machine learning based system able to correctly position the 3D model. The pViewer application will be used to create a large dataset of tagged image to train the algorithm: during the in-vivo procedure we will store in a dedicated database the frame images along with position, rotation and scale parameters of their 3D model overlaid.

References

1. Ashrafian, H., et al.: The evolution of robotic surgery: surgical and anaesthetic aspects. BJ Br. J. Anaesthesia **119**(suppl_1), i72–i84 (2017)
2. Peters, B.S., et al.: Review of emerging surgical robotic technology. Surg. Endosc. **32**(4), 1636–1655 (2018)
3. Schwaibold, H., Wiesend, F., Bach, C.: The age of robotic surgery – is laparoscopy dead? Arab J. Urol. **16**(3), 262–269 (2018)
4. Fischer, J., et al.: Medical augmented reality based on commercial image guided surgery. In: Proceedings of the Tenth Eurographics Conference on Virtual Environments, pp. 83–86. Eurographics Association, Grenoble (2004)
5. Nakamoto, M., et al.: Current progress on augmented reality visualization in endoscopic surgery. Curr. Opin. Urol. **22**(2), 121–126 (2012)
6. Lamata, P., et al.: Augmented reality for minimally invasive surgery: overview and some recent advances. In: Augmented Reality. Intech (2010)
7. Ha, H.-G., Hong, J.: Augmented reality in medicine. Hanyang Med. Rev. **36**(4), 242–247 (2016)
8. Azuma, R., et al.: Recent advances in augmented reality. IEEE Comput. Graph. Appl. **21**(6), 34–47 (2001)
9. Sielhorst, T., Feuerstein, M., Navab, N.: Advanced medical displays: a literature review of augmented reality. J. Display Technol. **4**(4), 451–467 (2008)
10. Violante, M.G., Vezzetti, E.: Design and implementation of 3D web-based interactive medical devices for educational purposes. Int. J. Interact. Des. Manuf. (IJIDeM) **11**(1), 31–44 (2018)
11. Violante, M.G., Vezzetti, E., Piazzolla, P.: Interactive virtual technologies in engineering education: why not 360 videos? Int. J. Interact. Des. Manuf. (IJIDeM) **13**(2), 729–742 (2018)
12. Silva, R., Oliveira, J.C., Giraldi, G.A.: Introduction to augmented reality. National laboratory for scientific computation, Av. Getulio Vargas (2003)
13. Carmigniani, J., et al.: Augmented reality technologies, systems and applications. Multimed. Tools Appl. **51**(1), 341–377 (2011)
14. Bernhardt, S., et al.: The status of augmented reality in laparoscopic surgery as of 2016. Med. Image Anal. **37**, 66–90 (2017)
15. Amparore, D., Checcucci, E., Gribaudo, M., Piazzolla, P., Porpiglia, F., Vezzetti, E.: Nonlinear-optimization using SQP for 3D deformable prostate model pose estimation in minimally invasive surgery. In: Proceedings of the Computer Vision Conference (CVC 2019), pp. 477–496 (2019)
16. Roberts, D.W., et al.: A frameless stereotaxic integration of computerized tomographic imaging and the operating microscope. J. Neurosurg. **65**(4), 545 (1986)
17. Kelly, P.J., et al.: Computer-assisted stereotaxic laser resection of intra-axial brain neoplasms. J. Neurosurg. **64**(3), 427 (1986)
18. Peters, T.M.: Image-guidance for surgical procedures. Phys. Med. Biol. **51**(14), R505–R540 (2006)
19. Marcus, H.J., et al.: Comparative effectiveness and safety of image guidance systems in eurosurgery: a preclinical randomized study. J. Neurosurg. **123**(2), 307–313 (2015)
20. Cabrilo, I., Schaller, K., Bijlenga, P.: Augmented reality-assisted bypass surgery: embracing minimal invasiveness. World Neurosurg. **83**(4), 596–602 (2015)
21. Winne, C., et al.: Overlay visualization in endoscopic ENT surgery. Int. J. Comput. Assist. Radiol. Surg. **6**(3), 401–406 (2011)

22. Mirota, D.J., et al.: Evaluation of a system for high-accuracy 3D image-based registration of endoscopic video to C-arm cone-beam CT for image-guided skull base surgery. IEEE Trans. Med. Imaging **32**(7), 1215–1226 (2013)
23. Dubach, P., et al.: Image-guided otorhinolaryngology. In: Jolesz, F.A. (ed.) Intraoperative Imaging and Image-Guided Therapy, pp. 845–856 (2014)
24. Liu, X., et al.: Application of single-image camera calibration for ultrasound augmented laparoscopic visualization. In: Medical Imaging 2015: Image-Guided Procedures, Robotic Interventions, and Modeling. International Society for Optics and Photonics (2015)
25. Zinser, M.J., et al.: A paradigm shift in orthognathic surgery? A comparison of navigation, computer-aided designed/computer-aided manufactured splints, and "classic" intermaxillary splints to surgical transfer of virtual orthognathic planning. J. Oral Maxillofac. Surg. **71**(12), 2151 (2013)
26. Wengert, C., et al.: Markerless Endoscopic Registration and Referencing. Springer, Heidelberg (2006)
27. Edwards, P.J., et al.: Design and evaluation of a system for microscope-assisted guided interventions (MAGI). IEEE Trans. Med. Imaging **19**(11), 1082–1093 (2000)
28. Shuhaiber, J.H.: Augmented reality in surgery. Arch. Surg. **139**(2), 170–174 (2004)
29. Mezger, U., Jendrewski, C., Bartels, M.: Navigation in surgery. Langenbeck's Arch. Surg. **398**(4), 501–514 (2013)
30. Marescaux, J., et al.: Augmented-reality-assisted laparoscopic adrenalectomy. JAMA **292** (18), 2214–2215 (2004)
31. Ukimura, O., Gill, I.S.: Image-fusion, augmented reality, and predictive surgical navigation. Urol. Clin. North Am. **36**(2), 115–123 (2009)
32. Hughes-Hallett, A., et al.: Augmented reality partial nephrectomy: examining the current status and future perspectives. Urology **83**(2), 266–273 (2014)
33. Altamar, H.O., et al.: Kidney deformation and intraprocedural registration: a study of elements of image-guided kidney surgery. J. Endourol. **25**(3), 511–517 (2011)
34. Hamacher, A., et al.: Application of virtual, augmented, and mixed reality to urology. Int. Neurourol. J. **20**(3), 172–181 (2016)
35. Simpfendorfer, T., et al.: Augmented reality visualization during laparoscopic radical prostatectomy. J. Endourol. **25**(12), 1841–1845 (2011)
36. Porpiglia, F., et al.: Augmented-reality robot-assisted radical prostatectomy using hyper accuracy three-dimensional reconstruction (HA3D™) technology: a radiological and pathological study. BJU Int. **123**, 834 (2018)
37. Porpiglia, F., et al.: 3D elastic augmented-reality robot-assisted radical prostatectomy using hyper-accuracy three-dimensional reconstruction (HA3D™) technology to identify suspicious capsular involvement during dissection of periprostatic tissue: a step further in image guided surgery. J. Eur. Urol. **123**(5), 834–845 (2019)
38. Sederberg, T.W., Parry, S.R.: Free-form deformation of solid geometric models. ACM SIGGRAPH Comput. Graph. **20**(4), 151–160 (1986)
39. Ke, T., Roumeliotis, S.I.: An efficient algebraic solution to the perspective-three-point problem. In: Proceedings of the IEEE Conference on Computer Vision and Pattern Recognition (2017)
40. Fischler, M.A., Bolles, R.C.: Random sample consensus: a paradigm for model fitting with applications to image analysis and automated cartography. Commun. ACM **24**(6), 381–395 (1981)
41. Rublee, E., et al.: ORB: An efficient alternative to SIFT or SURF (2011)
42. Armijo, P.R., et al.: Growth in robotic-assisted procedures is from conversion of laparoscopic procedures and not from open surgeons' conversion: a study of trends and costs. Surg. Endosc. **32**(4), 2106–2113 (2018)

43. Abdul-Muhsin, H., Humphreys M.: Advances in laparoscopic urologic surgery techniques. F1000Research **5**, 716 (2016)
44. Marescaux, J., et al.: Augmented-reality–assisted laparoscopic adrenalectomy. JAMA **292** (18), 2211–2215 (2004)
45. Sugimoto, M., et al.: Image overlay navigation by markerless surface registration in gastrointestinal, hepatobiliary and pancreatic surgery. J. Hepato-Biliary-Pancreat. Sci. **17**(5), 629–636 (2010)
46. Teber, D., et al.: Augmented reality: a new tool to improve surgical accuracy during laparoscopic partial nephrectomy? Preliminary in vitro and in vivo results. Eur. Urol. **56**(2), 332–338 (2009)
47. Su, L.M., et al.: Augmented reality during robot-assisted laparoscopic partial nephrectomy: toward real-time 3D-CT to stereoscopic video registration. Urology **73**(4), 896–900 (2009)

Informing the Use of Visual Assets in Industrial Augmented Reality

Michele Gattullo[1], Giulia Wally Scurati[2], Alessandro Evangelista[1](✉), Francesco Ferrise[2], Michele Fiorentino[1], and Antonio Emmanuele Uva[1]

[1] Department of Mechanics, Mathematics, and Management, Polytechnic Institute of Bari, 70126 Bari, Italy
{michele.gattullo,alessandro.evangelista}@poliba.it
[2] Department of Mechanical Engineering, Polytechnic University of Milan, 20156 Milan, Italy

Abstract. Augmented Reality is considered one of the most promising technologies for Industry 4.0. Augmented Reality allows to display the needed information at the right time and to locate it in the desired space, superimposing it to the real world. In this way, it could simplify the work of industrial operators in a variety of tasks, from planning and commissioning to assembly and maintenance. Despite the growing research interest and efforts for the development of Augmented Reality applications and technologies for the industrial domain, Augmented Reality is still rarely used in real industrial procedures, often remaining at a conceptual level. In fact, developers fail to completely answer the actual needs of industries and workers because of a lack of knowledge of the context and the absence of guidelines to drive the design and development of Augmented Reality applications. One of the open issues is the choice of the proper visualization methods to display technical information in Augmented Reality, which is not trivial. Both the paper-based documentation, based on text and illustrations, and digital documentation, including CAD models and image or video-based tutorials, present different criticalities when adapted to the Augmented Reality technology. Other visual material has been developed specifically to be used in Augmented Reality, but it is difficult to use because of a lack of standards which are instead present for the former paper-based and digital documentation. In this work, we categorize and compare different Augmented Reality visual assets, illustrating their advantages and disadvantages and providing directions for future research.

Keywords: Augmented Reality · Industry 4.0 · Graphical User Interface · Technical documentation

1 Introduction

Pushed by the fourth industrial revolution (Industry 4.0), Augmented Reality (AR) is a candidate to be implemented in industrial facilities for bridging virtual information and real environment [1]. AR is a valid technology for technical communication since it

C. Rizzi et al. (Eds.): ADM 2019, LNME, pp. 106–117, 2020.
https://doi.org/10.1007/978-3-030-31154-4_10

allows the reduction of the cognitive load when performing procedural tasks as for maintenance or assembly procedures [2–5].

The number of AR applications in the industrial domain is continuously increasing in the last decades. Despite this, AR is still rarely used in real industrial procedures, but it often remains at a conceptual level in the laboratories [6]. One of the main reasons is the scarce collaboration between developers, academic researchers and the industrial world [7]. Therefore, the literature is full of various concept AR applications, but poor of guidelines about the use of AR techniques in the industrial domain. Among the open issues, there is the choice of the proper visualization methods to display technical information in AR.

Technical documentation evolved in the last years, starting from paper-based, characterized by text and illustrations, often black and white. Digital documentation, through the use of computer graphics, allows the use of new channels to convey information: interactivity of the contents, color, and animation. Examples of digital contents are CAD models and multimedia as image-based or video-based tutorials. Furthermore, AR allows the registration of graphical visual assets directly on the real world, with technical advantages as objects localization.

However, the authors of AR technical documentation need guidelines on how exploiting these novel technologies, considering both their opportunities and their limitations (e.g., occlusion). In fact, AR offers many options for the GUI designers, but currently, there are no specific standards to follow for industrial technical assets. Existing standards are generalist, like the Augmented Reality Markup Language (ARML) [8] and the Augmented Reality Application Format (ARAF) [9]. However, these standards can be used as a starting point for future implementations, for example for the nomenclature. In this work, we took inspiration from the Separation of Concerns principle, which is at the basis of ARML 2.0 [10], for the analysis of AR interfaces. According to this principle, an augmented scene is composed of three main components: (1) a real-world object (called feature), (2) its projected location on the augmented scene (called Anchor), (3) a virtual model associated to it (Visual Asset). The IEEE is developing a family of standards for virtual and augmented reality [11], that address aspects such as safety, how different technologies should be defined, and how virtual and real objects should work together.

Literature shows that the use of technical visual assets in existing industrial AR is varied and, in some cases, not optimized, because interface design is left to programmer discretion without a dedicated study. In many AR applications, there are no detailed descriptions of the interface, as well as motivations for the choice of the visualization methods. Thus, some visual assets are not used in an optimal way, while others are completely neglected.

According to the authors, there is a lack in the literature of a specific study about technical visual assets that can be used in an AR industrial interface. This study aims to orientate technical writers in the choice of visual assets, supporting them in the designing of future AR manuals.

In this work, we report a list of visual assets used in industrial AR applications (Fig. 1), deriving from a literature review on this topic [12]. We also provide a heuristic evaluation of visual assets, considering the most common issues that we experienced in these years during the development of test cases for industrial companies.

2 Visual Assets

2.1 Text

The traditional way to convey information is using text, since language, both written and oral, is the oldest and most solid way humans use to communicate technically. Text can provide any kind of information: through descriptions, it is possible to procure correct understanding and visualization of the problem, through commands it is possible to give instructions on how to proceed. Text instructions have been used since from the first paper manuals and are still used in AR applications. Authoring of text is very simple, but the need for translation of manuals in every language is a challenge for the use of text in the new AR interfaces.

We included in this category both 2D text and 3D text, as well as both texts, displayed with and without labels. In fact, the amount of information is not affected by the way the text is rendered.

2.2 Signs

We applied the definition of Peirce [13]: "a sign is a thing which serves to convey knowledge of some other thing, which it is said to stand for or represent." Signs are of three classes: icons, indices, and symbols. An icon has a physical resemblance to the thing or concept being represented, like the camera icon in a smartphone, or gloves in an obligation symbol. A symbol may have no resemblance between the signifier and the signified, so the connection between them must be culturally learned: numbers and alphabets are good examples, or the shutdown symbol on a device, or the information symbol. An Index shows evidence of what is being represented, as a skull with two crossed bones is used to warn of toxic material.

Signs are regulated by standards that could be either International Standards, as the ISO 3864 for safety symbols, or internal practices. The information contained in signs is very focused, and this is what distinguishes signs by photographs, where an amount of information could be present.

2.3 Photographs

In this class, we put just photographs of real-world as acquired by a camera. Use of photographs is very common in manuals, especially digital manuals and instructional websites, as iFixit.com [14]. In fact, authoring is very easy since it requires just taking a picture of a real object.

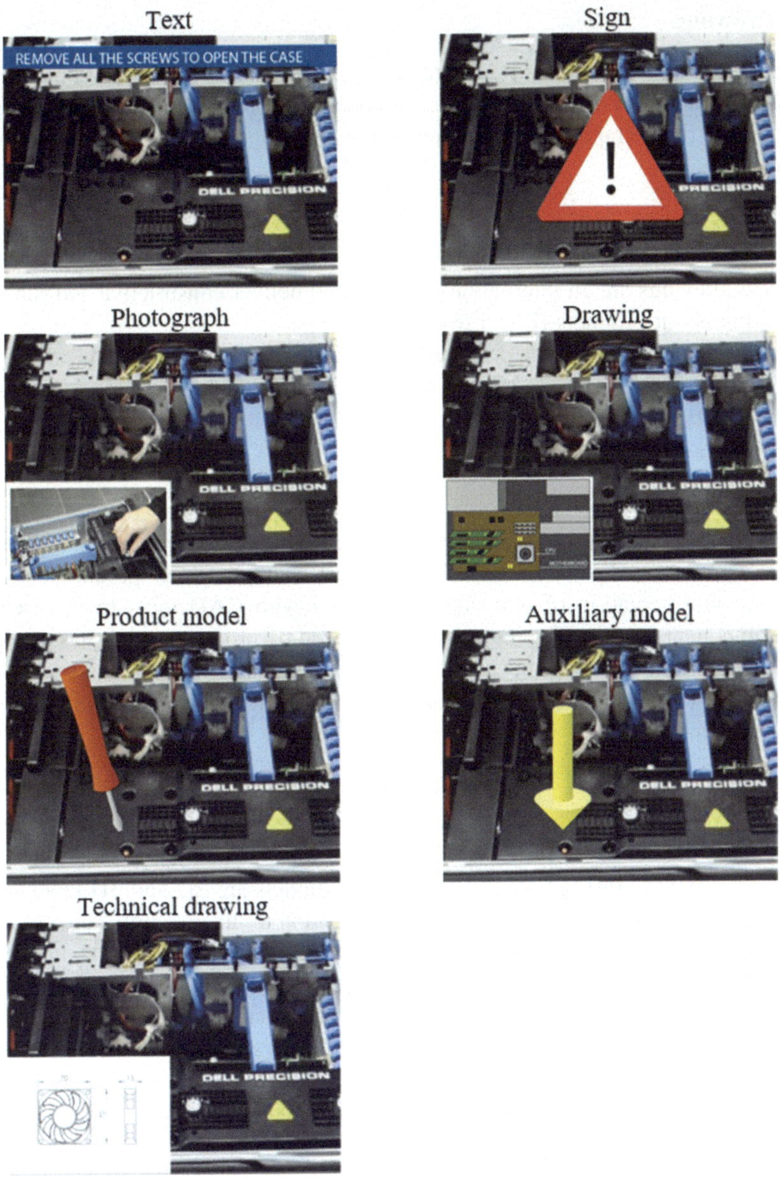

Fig. 1. Examples of using different visual assets in the same context.

2.4 Videos

In this class, we put just video recordings of real-world as acquired by a video camera or webcam. In the latest years, the number of video tutorials to convey technical information is widely increasing on the web since they are among the easiest and fastest way for "Do It Yourself" (DIY) authors to show technical instructions and for the end-users to learn them.

2.5 Drawings

In this class, we put digitized 2D drawings except for technical drawings on any support. Examples are freehand sketches, maps, charts. We also included in this group the annotated photographs, i.e. a combination of a photograph and 2D auxiliary models and/or text.

2.6 Technical Drawings

Technical drawings are an international standard to deliver constructive and functional information about products (ISO 128).

In this category, we included 2D representations in the form of technical drawings displayed as a static image on canvas, but also 3D graphical annotations according to ASME Y14.41 – 2003.

2.7 Product Models

We used the definition provided by Wang et al. [15]: "product models are 3D virtual models of product and parts". In most of the industrial applications, product models are the digital representation of real objects modeled with CAD tools (e.g., machinery parts, components, tools). The information conveyed by product models has an explicit meaning since it does not change as the feature it is anchored changes. For example, the 3D animation of a screw that fits in a hole has the same meaning for every hole in every real component.

2.8 Auxiliary Models

We also used the definition provided by Wang et al. [15]: "auxiliary models are virtual models for auxiliary instructions". The auxiliary models are 2D and 3D models, used by technical authors for delivering generic information to the operator. Some examples are arrows, circles, abstract sketches. Unlike the product models, the information conveyed by auxiliary models has not an explicit meaning, since it could change according to the feature it is anchored to: for example, an arrow can be used to locate objects in the real scene, to indicate a direction, to express a pushing action, and so on. We did not make a distinction between 2D and 3D elements since the same information can be conveyed by the 2D and 3D version of an auxiliary model.

3 Heuristic Evaluation of Visual Assets

We made heuristic evaluations to analyze the visual assets described in the previous section considering as heuristics the most common issues that we experienced in these years during the development of test cases for industrial companies [2, 5, 16]. A variable that could influence the choice of visual assets could be the display device used. However, we did not consider it in the heuristics list because our considerations are valid for all the devices commonly used in industrial AR (HWDs, handheld devices, desktop screen), except for Spatial AR that request specific attention.

- *Occlusion*. The use of AR to provide information for the operator implies that visual assets are merged with the real world. Then, visual assets can occlude a great portion of the real world. Occlusive visual assets may limit situation awareness and operator safety.
- *Registration*. As to registration, using the definition provided by Gabbard et al. [17], we can distinguish screen-fixed and world-fixed (or conformal) visual assets. The formers are rendered at a fixed location on the display and are generally not spatially anchored to any specific objects in the scene. These last are rendered such that they are perceived to exist at specific locations in the real world. Registration of world-fixed visual assets is affected by positioning and orientation errors due to tracking robustness. This misalignment influences the information correctness differently according to the visual asset used.
- *Communication time*. We defined it as the time that occurs between the moment since the visual asset appears until the moment the user comprehends the information associated. Different types of visual assets convey information in different ways then the communication time varies consequently. Some visual assets can provide a small piece of information in a more synthetic way, some others can provide more details about an operation. The information provided by a visual asset can be instantaneous or can be provided in a time cycle (e.g., an animation). In this second event, it is important that the user watches the augmentation for all the duration of the cycle. When it does not occur due, for example, to user distractions or tracking failures, the information provided by the visual asset can be compromised.
- *Authoring*. This is one of the most important issues for the development of AR interfaces. The creation of the virtual contents may require competencies that are not typically part of technical writers' background: 3D modeling, computer graphics/animation skills, programming.
- *Associated information*. The information provided by a visual asset can be either all included in the augmentation or in the combination of the augmentation with the real world. In this second case, the same visual asset can provide different information depending on the real object is associated with. If all the information relies on the visual asset, there could be more than one meaning conveyed by that visual asset.
- *Computational cost*. The hardware resources (e.g., CPU, GPU, RAM) employed during the rendering of the AR scene depend on the type of visual asset.

3.1 Text

Occlusion caused by text depends on the amount of text and visual style used. Descriptive instructions make use of many words and it is difficult to summarize them. The use of styles with large outlines and billboard is needed to improve the contrast between text and background and then legibility.

In most AR GUIs there is a dedicated area for text information, thus used screen-fixed. If text is world-fixed, information comprehension is not influenced by registration accuracy because a perfect matching between virtual text and the real scene is not needed.

The time spent by the user to acquire text information depends on text length. If some frames are lost due to tracking inaccuracy, users can miss some words, then communication time increases.

Authoring of text is very easy and does not request particular skills. Furthermore, text instructions can be imported by traditional manuals, if available in digital format. However, they need to be translated into all the languages requested by end-users.

A complete technical sentence must not have more than one meaning. Single words, often used in AR manuals, can lead to different interpretations and depend on the real object or event associated with (e.g., "right" could mean the direction of moving or a confirmation of a correct task execution).

Finally, rendering of text in AR interfaces require very low hardware resources.

3.2 Signs

Usually, a sign is monochromatic, and the background can be transparent. The dimension of the sign is the minimum to be distinguishable. Then, the occlusion caused by signs is very limited.

In an AR GUI, signs can be screen-fixed if coupled with other visual assets that indicate the real object the sign refers to. However, if they are world-fixed, they do not require a perfect registration with the real scene.

The memorization of the sign requires a little communication time, thus there is not the risk to miss information if some frames are lost.

The use of signs by AR designers is very simple since they would choose the desired one from a vocabulary of signs. Thus, authoring is limited to the definition of the vocabulary that should come from organizations for standardization. However, currently there are no standards for signs for all the industrial tasks (just for safety instructions), then the design of a vocabulary of signs is left to AR designers. Once that this vocabulary is defined, it can be used for all the AR industrial manuals designed.

The same sign could express different meanings: for example, arrows are used to indicate movements in a specific direction or forces to exert but are also used as pointers to focus the user's attention on an object or area. Ambiguity is easily avoided using common standards that are still not available for all the industrial tasks.

Rendering of signs in AR interfaces requires very low hardware resources.

3.3 Photographs

Opaque photographs occlude a large portion of the real world, hiding what is behind them. Use of transparent or reduced-size photographs could reduce occlusion, but information comprehension could be compromised.

Photographs are preferably used screen-fixed because AR designers can control the GUI area occluded by a photograph. However, if they are world-fixed, they do not require a perfect registration with the real scene.

Communication time depends on the amount of informative details present in the photograph acquired.

Authoring of photographs is accomplished by taking pictures from the real scene. It requires time for authorizations needed, to prepare the scene, to acquire pictures, to post-process them. However, in an industrial scenario not always it is possible to acquire this multimedia due to safety problems, patents problems, difficult to access to the components, and so on.

If in the photograph more than one object is present, this can lead to information ambiguity. This situation is very common in industrial facilities.

Computational cost depends on the original photograph resolution that can be reduced for the AR application.

3.4 Videos

As to occlusion and tracking issues, we can make the same considerations made on photographs.

Communication time depends on the video duration. If some frames are lost, there could be some missing information for users.

Authoring of videos is similar to photographs. However, post-processing requires more effort than a photograph because of scenes editing.

Most of the videos used as visual assets show a preview of the operation to accomplish, made by another operator. Then, users must repeat what they watch in the video, thus there is a very low risk of interpretations different from the one intended by the AR designer.

Computational cost depends on the original video resolution, that can be reduced for the AR application, and on the duration.

3.5 Drawings

As to occlusion and tracking issues, we can make the same considerations made on photographs and videos.

Communication time is related to the quantity of information included in the drawing. However, since a drawing is specifically designed by AR designers, the information presentation could be optimized to reduce communication time.

Authoring of drawing depends on the type of drawing. Annotated photographs require an authoring effort similar to photographs with, in addition, the annotation. Sketches or other user-generated drawings require a higher authoring effort.

A drawing is specifically designed to provide objective information, then the meaning associated is unique.

Computational cost depends on the picture resolution that can be optimized for the AR application.

3.6 Technical Drawings

Differently from photographs and drawings, technical drawings cannot be made transparent and reduced in size because some lines can be missed. Thus, occlusion of the real scene is hard to reduce.

Similarly, to photographs and drawings, technical drawings are preferably used screen-fixed. If they are world-fixed, a perfect registration with the real scene is not required.

Communication time depends on the complexity of the represented object and the quantity of information included in the technical drawing: shapes, dimensions, assembly information, bill of materials, and so on. The comprehension of the information requires skills in the engineering field and the knowledge of technical drawing standards.

Authoring of technical drawings is related to the complexity of the represented object. It requires skills in the engineering field since international standards must be followed. However, the use of 3D CAD software, compared to software for 2D drafting, helps in the creation of drawings starting from a 3D model of the object, but post-processing of the drawing is needed: views management, adding/removing lines, quotation alignment, and so on.

By definition, a technical drawing is a representation of an object aimed at transmitting technical and functional information objectively. Thus, technical drawings convey only the information specified by the designer.

The computational cost of technical drawing is usually lower than the one of photographs.

3.7 Product Models

World-fixed product models are used to create the impression of a scene in which the virtual and the real coexist. They can replace (e.g., in assembly tasks) or overlap to a real object (e.g., for locating tasks). However, when operators accomplish the task, large product models can hide the real scene where they are operating. In this case, the use of transparent models can help to reduce the occlusion of the real world.

Product models are highly sensitive to registration accuracy. A misalignment between the augmentation and the real object it refers to causes low user acceptance and can lead to interpretation errors.

For static models, communication time depends on its complexity. If the product model is animated, it depends also on the animation duration. If some frames are lost, there could be some missing information for users, increasing communication time.

Authoring of product models requires competence in 3D modeling. Authoring effort is related to the complexity of geometry and animation.

Product models have a unique meaning due to their resemblance to real objects. The use of animations provides a preview of the operation to accomplish, further improving the uniqueness of meaning.

Rendering of product models requires a high computational cost, especially if animated. It can be reduced with the decimation of the model mesh.

3.8 Auxiliary Models

Auxiliary models fill a small portion of the GUI, thus there is a little occlusion of the real world.

Auxiliary models are little sensitive to registration accuracy. This sensibility depends on the model size. A small misalignment between the augmentation and the real object it refers to does not cause interpretation errors.

The association of the information to this visual asset is very immediate and exploits the affordance of real objects. There is no risk of missing information if some frames are lost.

The authoring of auxiliary models is very simple. Designers can choose the desired one from a list of existing models or also draw a new one.

Auxiliary models have not a specific meaning, but it strongly depends on the real object they are associated with. For this reason, the use of these visual assets in an AR GUI makes sense only if they are world-fixed and coupled with other visual assets.

Rendering of auxiliary models in AR interfaces requires very low hardware resources.

4 Results of Heuristic Evaluation

As we can see from the analysis provided in the previous section, the reported issues are critical with some visual assets, while they can be overcome using other visual assets. We made heuristic evaluations to assess the usability of visual assets in Industrial AR. The six authors were the evaluators. For each visual asset, they reported if it is a good choice or not in terms of the considered heuristic: occlusion, registration, communication time, authoring, associated information, computational cost. We synthesized these evaluations in Table 1.

Table 1. Synthesis of visual asset analysis based on the considered issues: red dots indicate bad choices, green dots indicate good choices, blank cells indicate situations where the choice depends on multiple factors.

	occlusion	registration	communication time	authoring	associated information	computational cost
Text	🔴	🟢			🟢	🟢
Signs	🟢	🟢	🟢	🟢	🔴	🟢
Photographs	🔴	🟢		🟢	🔴	
Videos	🔴	🟢		🔴	🟢	🔴
Drawings	🔴	🟢	🟢		🟢	
Technical drawings	🔴	🟢	🔴	🔴	🟢	🟢
Product models	🔴	🔴	🟢	🔴	🟢	🔴
Auxiliary models	🟢		🟢	🟢	🔴	🟢

It is not by chance that occlusion, one of the issues most specifically related to AR visualization, is not a critical point for signs and auxiliary models, which are often specifically designed for AR applications, while the other visual assets are mainly adapted from previous documentation. Many of the latter also present other different issues: in particular, authoring is critical for product models, technical drawings and videos, since the development and update of this kind of documentation requires a significant amount time and efforts. For product models, registration accuracy is fundamental, since misalignments can lead to interpretation problems and errors. This adds difficulty to the implementation of AR applications, also depending on the AR technologies and lighting conditions, which in industrial environments may not be optimal.

Technical drawings demand a high communication time, meaning that the operator consulting the information needs time and effort to elaborate it and understanding it. Since one of the main advantages of using AR is the immediacy and simplicity with which information can be found and displayed, this criticality is particularly relevant. The use of photographs without additional explanation can instead lead to interpretation issues and ambiguity, especially if complex or several objects are depicted. Issues regarding the understanding of the related information are also a strong concern for signs and auxiliary models. In fact, as stated above, those are often newly designed for AR applications. For this reason, they are not subject to industry regulations and standards, differently from the other visual aspects, which inclusion in technical documentation has a long tradition.

5 Conclusion

From the results of the heuristic evaluation, it is possible to make the following insights for Industrial AR designers.

Signs and auxiliary models are the only visual assets that present only one critical issue; hence they are particularly suitable for AR communication of information and instructions. The development and adoption of new standards for their use seem to be the most sensible path to open the way to the employment of AR solutions supporting industrial tasks and procedures.

However, this is not trivial: the creation of international industrial standards requires a complex validation process, which should be implemented for all the designed assets until an exhaustive amount is approved to be used for AR communication, as it happens for ISO graphical symbols. For this process to be optimized and effective, an intense collaboration is needed between industries, academic researchers and developers.

Also, the text presents only one critical issue, i.e. occlusion. However, the authoring of text instructions is highly dependent on the translation effort. Thus, research works on text optimization for translations as [18], are of valuable interest for the community.

References

1. Paelke, V.: Augmented reality in the smart factory: supporting workers in an industry 4.0. environment. In: 19th IEEE International Conference on Emerging Technologies and Factory Automation, ETFA (2014)
2. Uva, A.E., Gattullo, M., Manghisi, V.M., Spagnulo, D., Cascella, G.L., Fiorentino, M.: Evaluating the effectiveness of spatial augmented reality in smart manufacturing: a solution for manual working stations. Int. J. Adv. Manuf. Technol. **94**, 509–521 (2018)
3. Henderson, S., Feiner, S.: Exploring the benefits of augmented reality documentation for maintenance and repair. IEEE Trans. Vis. Comput. Graph. **17**, 1355–1368 (2010)
4. Henderson, S.J., Feiner, S.: Evaluating the benefits of augmented reality for task localization in maintenance of an armored personnel carrier turret. In: Proceedings of the IEEE 2009 International Symposium on Mixed and Augmented Reality, ISMAR 2009, pp. 135–144 (2009)
5. Fiorentino, M., Uva, A.E., Gattullo, M., Debernardis, S., Monno, G.: Augmented reality on large screen for interactive maintenance instructions. Comput. Ind. **65**, 270–278 (2014)
6. Bottani, E., Vignali, G.: Augmented reality technology in the manufacturing industry: a review of the last decade. IISE Trans. **51**, 284–310 (2019)
7. Fite-Georgel, P.: Is there a reality in industrial augmented reality? In: 2011 10th IEEE International Symposium on Mixed and Augmented Reality, ISMAR 2011, pp. 201–210 (2011)
8. Lechner, M., Tripp, M.: ARML—an augmented reality standard. Coordinates **13**, 432–440 (2010)
9. Preda, M.A.M., Choi, B.S.: Augmented reality application format (2012)
10. Lechner, M.: ARML-augmented reality markup language. Ginzkeyplatz **11**, 5020 (2010)
11. IEEE-SA- Augmented Reality. https://standards.ieee.org/initiatives/ar/stds.html
12. Gattullo, M., Evangelista, A., Uva, A.E., Fiorentino, M., Gabbard, J.L.: What, how, and why are visual assets used in industrial augmented reality? A systematic review and classification in maintenance, assembly, and training. IEEE Trans. Vis. Comput. Graph. **49**, 215 (2019)
13. Peirce, C.S.: Collected Writings, vol. 8, p. 58. Harvard University Press, Cambridge (1931)
14. iFixit: The Free Repair Manual. https://www.ifixit.com
15. Wang, J., Feng, Y., Zeng, C., Li, S.: An augmented reality based system for remote collaborative maintenance instruction of complex products. In: 2014 IEEE International Conference on Automation Science and Engineering (CASE), pp. 309–314 (2014)
16. Boccaccio, A., Cascella, G.L., Fiorentino, M., Gattullo, M., Manghisi, V.M., Monno, G., Uva, A.E.: Exploiting augmented reality to display technical information on Industry 4.0 P&ID. In: Advances on Mechanics, Design Engineering and Manufacturing II, pp. 282–291. Springer (2019)
17. Gabbard, J.L., Fitch, G.M., Kim, H.: Behind the Glass: Driver challenges and opportunities for AR automotive applications. Proc. IEEE **102**, 124–136 (2014)
18. Gattullo, M., Uva, A.E., Fiorentino, M., Scurati, G.W., Ferrise, F.: From paper manual to AR manual: do we still need text? Procedia Manuf. **11**, 1303–1310 (2017)

Integrated Design Tools for Model-Based Development of Innovative Vehicle Chassis and Powertrain Systems

Emanuele Bonera, Marco Gadola, Daniel Chindamo, Stefano Morbioli, and Paolo Magri

Department of Industrial and Mechanical Engineering, Università degli Studi di Brescia, Via Branze 38, 25123 Brescia, Italy
{emanuele.bonera, marco.gadola}@unibs.it

Abstract. The paper provides a quick overview on system integration methodologies for advanced vehicle chassis and powertrains through an extension of the model-based design concept. Physics modeling and Driver-In-the-Loop (DIL) simulation techniques are described with an eye on the related hardware and software tools.

In greater detail, the design workflow pivots around a state-of-the-art device: the Driver-In-the-Loop simulator, where subjective, "human" factors can be taken into account together with objective, engineering factors. A framework of complementary tools with integrated functionalities supports the designer in order to merge traditional chassis engineering with advanced vehicle dynamics skills and modern control system design.

An interactive design framework is proposed in order to report the authors' experience. A self-developed utility for suspension and steering design, specifically conceived for intuitive integration in the design process, is also presented.

Keywords: Vehicle dynamics · Automotive chassis · Suspension · Steering · Driver-In-the-Loop simulator · Interactive design

1 Introduction

The advent of a new mobility based on hybrid-electric powertrains, and on full electric vehicles in perspective, raises new challenges for chassis design engineers, and new opportunities as well. Heavy limitations in terms of available energy stored on board require a massive effort directed at increasing efficiency. For example rolling resistance can be reduced with the adoption of narrower tyres with special tread design and compound, often at the expense of overall performance in terms of braking and road holding capabilities; cornering resistance can also be reduced by limiting the amount of built-in understeer, however cornering stability should not be put at risk. Moreover the need for bulky battery packs inevitably results in heavier passenger vehicles and can also make weight distribution more rear-biased than standard, once again requiring special care when determining vehicle dynamics behaviour and active safety characteristics.

C. Rizzi et al. (Eds.): ADM 2019, LNME, pp. 118–128, 2020.
https://doi.org/10.1007/978-3-030-31154-4_11

On top of that it is well-known that the dynamics of innovative powertrains is often perceived as not-so-fun-to-drive or even frustrating by car enthusiasts or even by the daily user. Emotional aspects should not be ignored as they can literally drive the potential customer in the complex process of choosing a brand new car, thus playing a major role in marketing. The impact of adverse effects on battery performance due to State-of-Charge (SoC) and environmental factors should also be addressed very carefully.

On the other side electric motors are quite suitable for accurate and fast motion control. Potentially active safety systems like Traction Control, ABS, ESC based on internal combustion engine management combined with braking can be replaced or complemented by purposely designed strategies based on motor control. Additional features like regenerative braking, active brake bias, torque vectoring, anti-rollover, etc. can be achieved in a relatively easy way depending on the vehicle architecture, and can be used as powerful tools to improve handling, stability, predictability, and perception, thus making the vehicle safer and more enjoyable to drive.

All the above calls for an innovative approach where traditional mechanical engineering skills for chassis design are integrated with digital knowledge, and the use of innovative tools makes for a research framework where subjective factors are also taken into account, thus making it also suitable for the development of Advanced Driver Assistance Systems (ADAS). On top of that, such an approach recalls what is usually defined as model-based design, where digital models are the key to product definition. In this case the models related to the vehicle components, subsystems and control logics entirely define the vehicle development stage and can be used for functional simulations at the same time.

2 The Design Process

2.1 Extension of the Model-Based Design Concept

It is well-known that Model-Based Design (MBD) is a modern methodology applied to complex system development, especially in the case of systems based on embedded software where multiple players take part in the development and validation process [1]. In the automotive case, given the complexity of safety requirements, the presence of extensive software embedded in various on-board units such as the VCU and active safety units (ABS, ESC etc.) and the interaction of multiple control systems to be designed and verified for compatibility and functionality, establishing a framework allowing to manage such a process is a necessity. Relevant case studies can be found in [2, 3].

2.2 Key Factors and Previous Work

It is well known that the design of both suspension and steering systems plays a major role in defining dynamics and active safety characteristics of an automobile. Suspension kinematics and elasto-kinematics allow tyres to interact with the road surface properly in order to develop consistent acceleration, braking and cornering forces.

Steering in particular is the key to a pleasant driving experience: the steering self-aligning properties are often considered a vital portion of the feedback required for safe and intuitive driving control [4, 5] because they have a primary impact on the tactile feel perceived by the driver through his hands acting on the steering wheel. This perception –often called "steering feel"- is considered to be vital "because steering is the driver's main line of communication with the car; distortion in this guidance channel makes every other perception more difficult to comprehend" [6]. According to [7], steering feel, or steering torque feedback, is widely regarded as an important aspect of the handling quality of a vehicle, as it is known to help the driver in reducing path-following errors. Some works even suggest that, apart from eyesight, the driving action is mainly based on feedback communication through the steering [4].

A basic dedicated tool can be an efficient alternative to multibody simulation when it comes to concept design of both suspension and steering systems. While various stand-alone software utilities were purposely created in the past [8–10], recent developments made it clear that integration is strongly required with the other tools involved in the design process, such as the utility described in [11], mainly conceived for educational purposes. An alternative solution is presented here: it has been conceived as a link between CAD and multibody simulation, to be then integrated with vehicle dynamics simulation and Driver-In-the-Loop virtual testing by means of a state-of-the-art driving simulator. Provision is made to involve system control design based on Matlab-Simulink in the process through co-simulation as well.

2.3 Features and Interface of the Tool for Steering and Suspension Design

The mathematics behind suspension kinematics computation is based on the Newton-Raphson algorithm applied to the Gough–Stewart platform. Examples are provided in [13–19] and again specific reference is made to [14]. The main features of the latest version were designed according to the following requirements. As a matter of fact the tool:

(a) incorporates the knowledge of experienced vehicle dynamics engineers
(b) features a very intuitive, self-explanatory, error-proof user interface (Fig. 1) allowing for real-time evaluation of design changes
(c) enables a typical, step-by-step design process based on a baseline geometry and subsequent iterations
(d) can overlap different geometries
(e) features output based on graphics and look-up tables suitable for post-processing with a spreadsheet or similar
(f) allows for kinematics evaluation of the complete axle in bump/rebound (Fig. 2), roll, and steer
(g) enables the design of the steering column geometry starting from the steering rack all the way up to the steering wheel, also taking into account the impact of the universal joints angular position on rack displacement vs steering wheel angle motion transfer (Fig. 3)

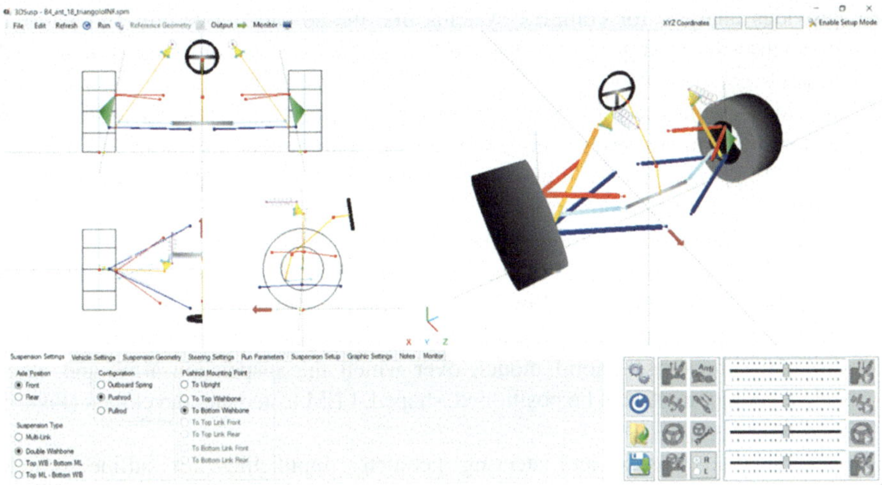

Fig. 1. Graphical interface for baseline suspension design.

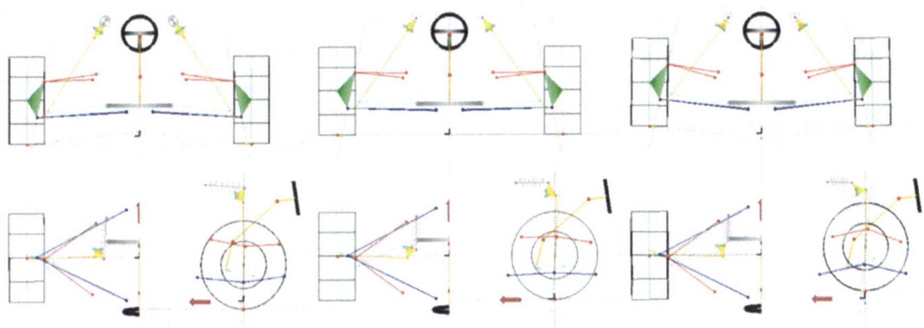

Fig. 2. Complete axle modeling and animation.

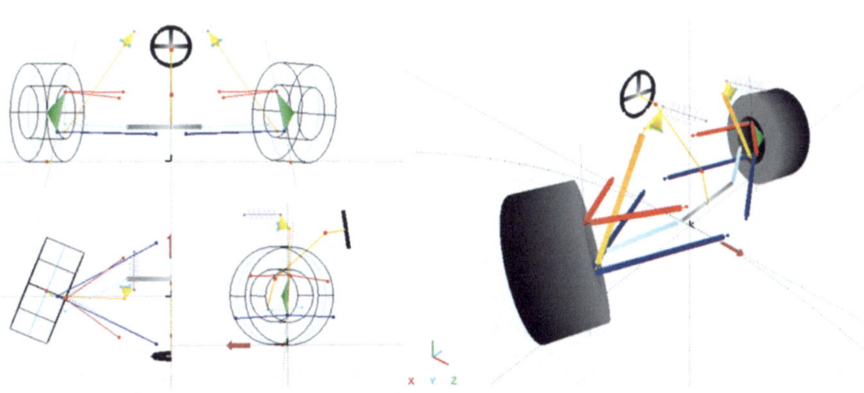

Fig. 3. Steering system and steering column design.

(h) provides analysis for complex systems like the so-called five-link and unusual geometries for racing cars [20] such as the upright-mounted push rod, as described in [21]
(i) provides simulation of built-in adjustments performed on real-world vehicles such as toe-in, camber, caster, roll center height, "anti" effect variations etc. (Set-up mode, Fig. 4) so that setup fine tuning can be designed for "special" vehicles like racecars and R&D prototypes
(j) provides calculation of fundamental properties related to handling and vertical dynamics like wheel rate, spring and damper motion ratio, wheel undamped frequency, roll stiffness, and main suspension derivatives
(k) generates direct input to CAD under the form of a parametric wireframe structure linked to the chassis solid model, over which the suspension arms and other related components can be positioned, shaped, FEM tested and developed (Figs. 5 and 6)
(l) generates suspension and steering geometry input files for offline vehicle dynamics simulation packages like VI-CarRealTime® and CarSim®
(m) generates suspension and steering geometry input files for Driver-In-the-Loop, real-time simulators

With regard to points (l) and (m), and unlike multibody-based systems, VI-CarRealTime for instance features the VI-SuspensionGen module that in turn requires just suspension coordinates to generate its own polynomial curves.

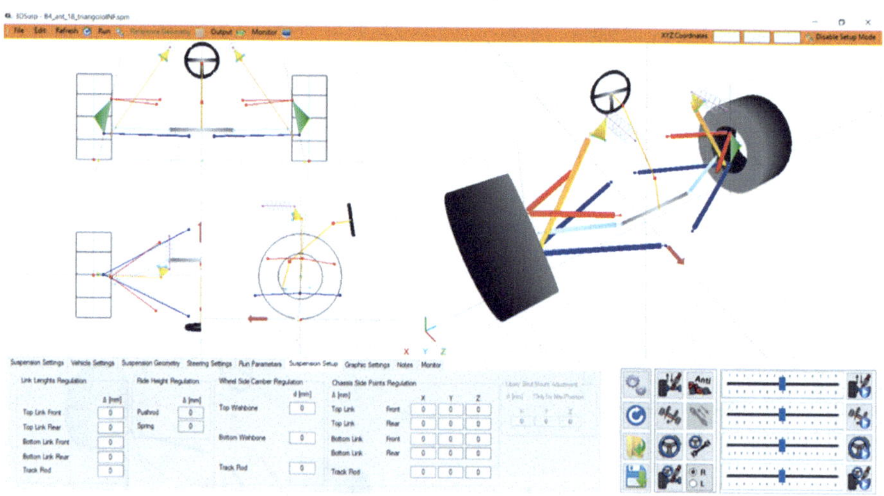

Fig. 4. Set-up mode: suspension fine-tuning can be designed or simulated.

2.4 Additional Tools for Driver-In-the-Loop Simulation

Professional driving simulators are the current state of the art in automotive simulations, enabling an immersive interaction between engineering models and virtual reality. The human can drive a vehicle physics model in any driving environment,

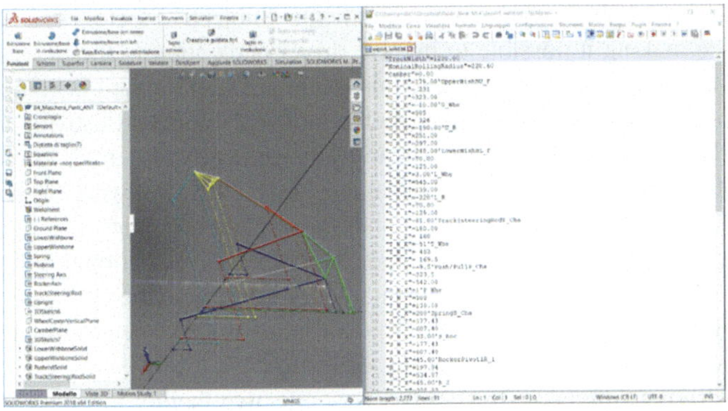

Fig. 5. Automatically generated CAD wireframe based on kinematics design data.

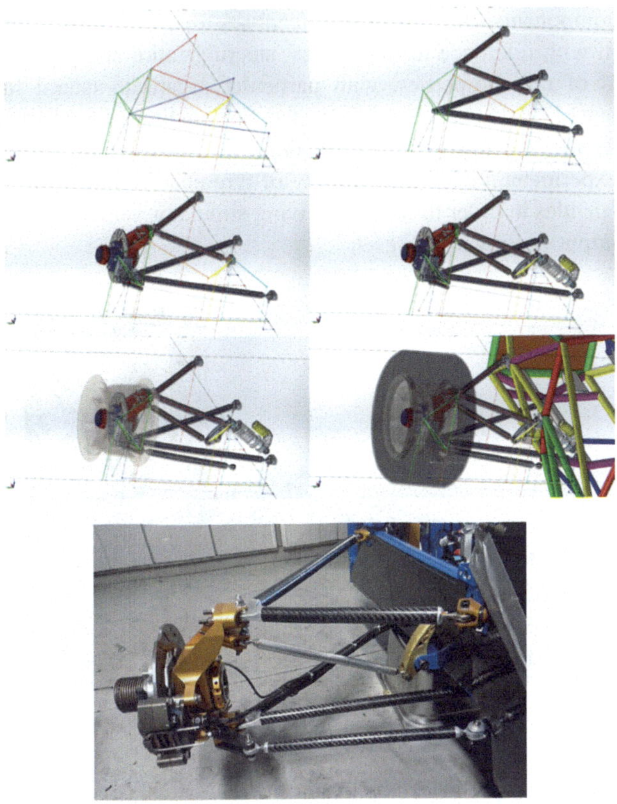

Fig. 6. From kinematics-based parametric wireframe to full suspension CAD to real-world design (from [11]).

thus delivering controlled, repeatable experiments. Subjective factors can be taken into account hence making it possible to involve driver perception in the process, although only to a certain extent [22]. In any case this is a major step forward in comparison with traditional offline simulations. Driving behaviour and training (also for drivers with special needs), driver physiology, human-traffic-infrastructure interaction (in urban environments for instance), Vehicle Dynamics, HMI, V2X, Software- & Hardware-In-the-loop, real-time applications, co-simulation of complex control systems, System Integration (for Hybrid-Electric vehicles for instance), and Functional Safety are some of the areas where simulators are most suitable for R&D activities [23], not to mention ADAS development (involving the well-known safety issues when testing is carried out in the real world) and finally engineering education.

A static driving simulator (Fig. 7) is a very effective facility for most of the above: it can even fit in a large office and does not involve complex motion cueing issues, as it is the case for expensive dynamic simulators (Fig. 8). According to the authors' experience however, the lack of cueing demands a supplementary modeling effort on systems and components affecting driver feedback directly. This is the case of the steering system with reference to kinematics, steering ratio, and power steering characteristics, and of tyre properties, with particular regard to lateral force and self-aligning moment. For this reason a dedicated "tyre editor" based on Pacejka's 5.2 Magic Formula was created in Matlab in order to provide intuitive processing and validation of experimental data, adjustability of tyre characteristics, and compatibility with standard *.tir* files in order to feed the driving simulator vehicle model. A sample of the graphics output is shown in Fig. 9.

Fig. 7. Static driving simulator.

Vehicle modelling is performed by means of VI-CarRealTime, currently considered the state of the art for vehicle dynamics simulations by the industry, e.g. see [24], while control logics and system integration can be built within the Matlab-Simulink environment and run in co-simulation. Traction Control, Torque Vectoring, and Rear-Wheel Steering systems are just examples of active strategies that can be successfully prototyped on Driver-In-the-Loop simulators even before real-world tests take place on a vehicle.

Fig. 8. Dynamic driving simulator.

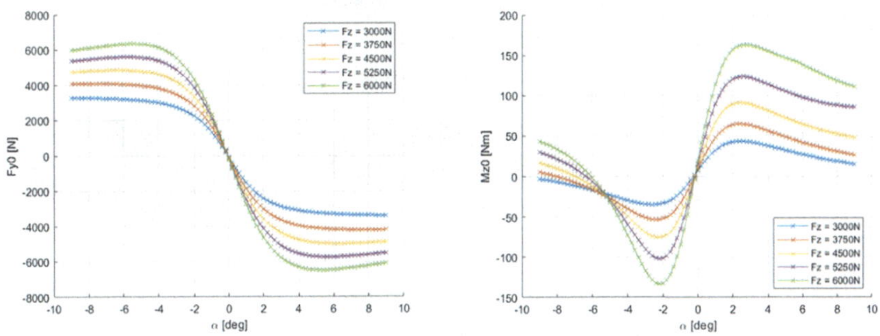

Fig. 9. Tyre data processing and modeling.

2.5 The Virtual Testing Process

The starting point is usually a benchmark vehicle, to be accurately tested with suitable instrumentation such as a VSA (Vehicle Sideslip Angle) sensor [25]. Specific, reference steady-state and transient maneuvers (e.g. steering pad and step steer), to be performed in a proving ground, are used to calibrate both the vehicle and tyre models.

Next a new vehicle model can be built on this validated baseline and driven on the simulator as a yardstick. The custom tools can then be used to edit tyre characteristics and steering/suspension kinematics, and related changes in vehicle behavior and steering feedback can be literally experienced first-hand. As a second step active controls and system integration logics, previously built and tested in Matlab-Simulink through offline simulation, can be driven on the real-time simulator.

As a third step the software code can be hardware-embedded on a Rapid Prototyping (RP) control unit and tested again on the simulator by means of the so-called Hardware-In-the-Loop (HIL) technique. Finally hardware and embedded code can be transferred on a prototype vehicle in order to start road testing. The process is outlined in Fig. 10, where a data analysis tool appears as well. It allows for real-world, offline and Driver-In-the-Loop data to be seamlessly overlapped and compared.

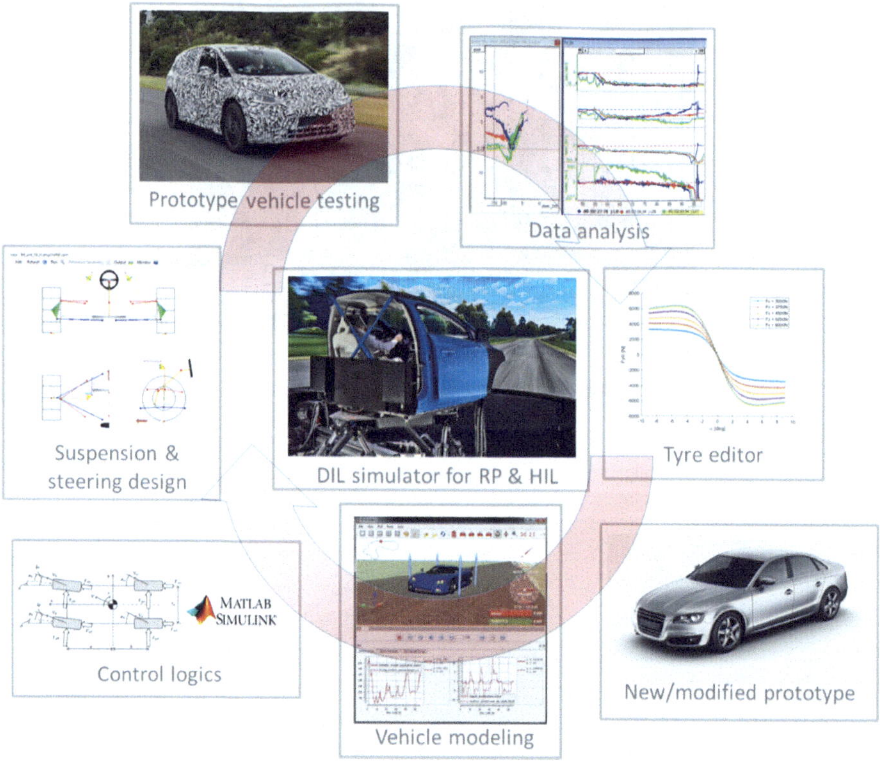

Fig. 10. Simulation and design framework based on Driver-In-the-Loop simulator.

2.6 Applications

An accurate description of relevant case studies is beyond the scope of this work. However some of the above cited papers [11, 12, 20, 21] can provide examples as an overview.

3 Conclusions

Innovative vehicles require peculiar techniques to make the most out of the opportunities offered by advanced control of innovative, electric motor-based powertrains. Driver-In-the-Loop simulators can foster R&D activities by merging traditional simulation tools and design techniques into a strongly interactive framework where human aspects and perception become the key to vehicle development. Such a framework basically forms the application of the Model-Based Design concept to the automotive engineering scenario, where the simulator plays the role of a pivot around which all the development and validation process activities grows.

The paper briefly describes the authors' experience in terms of integration between very different engineering tools and methods: from real-world testing to data analysis, from virtual vehicle modeling to suspension kinematics design with a custom tool, and finally to co-simulation techniques based on the driving simulator as well as on control algorithms developed for active safety and powertrain efficiency. Reference is made to some other works where these concept found application.

References

1. Wikipedia on the definition of Model-Based Design. https://en.wikipedia.org/wiki/Model-based_design
2. Alemanni, M., et al.: Model-based definition design in the product lifecycle management scenario. Int. J. Adv. Manuf. Technol. **52**(1–4), 1–14 (2011)
3. Valfrè, E., et al.: Integrating the Model Based Systems Engineering (MBSE) approach in a PLM platform for supporting data traceability in the aerospace domain. In: 1st IEEE International Symposium on Systems Engineering, ISSE 2015 (2015)
4. Sharp, R.S.: Vehicle dynamics and the judgement of quality. In: Pauwelussen, J.P. (ed.) Vehicle Performance – Understanding Human Monitoring and Assessment, pp. 87–96. Swets & Zeitlinger (1999)
5. Gritti, G., Peverada, F., Orlandi, S., Gadola, M., Uberti, S., Chindamo, D., Romano, M., Olivi, A.: Mechanical steering gear internal friction: effects on the drive feel and development of an analytic experimental model for its prediction. In: Proceedings of the International Joint Conference on Mechanics, Design Engineering & Advanced Manufacturing (JCM), Catania, Italy (2017)
6. Sherman, D.: Car & Driver magazine, December 2012
7. Kim, N., Cole, D.J.: A model of driver steering control incorporating the driver's sensing of steering torque. Veh. Syst. Dyn. **49**(10), 1575–1596 (2011)
8. Candelpergher, A., Gadola, M., Vetturi, D.: Development of a new software for racecar suspension kinematics. In: 2002 SAE Motorsport Engineering Conference (2002)
9. Cambiaghi, D., Gadola, M., Vetturi, D., Manzo, L.: Multi-link suspension analysis: an elasto-kinematic model. In: 29th ISATA, Florence, Italy (1996)
10. Martini, A., Ursella, A.: Elaborazione di un modello di cinematismo per sospensione automobilistica a bracci multipli. Degree thesis in Mechanical Engineering, University of Brescia (2002)
11. Gadola, M., Chindamo, D., Legnani, G., Comini, M.: Teaching automotive suspension design to engineering students: bridging the gap between CAD and CAE tools through an integrated approach. Int. J. Mech. Eng. Educ. **47**(1), 23–43 (2019)
12. Gadola, M., Chindamo, D.: Experiential learning in engineering education: the role of student design competitions and a case study. Int. J. Mech. Eng. Educ. **47**(1), 3–22 (2017)
13. Hiller, M., Woernle, C.: Elasto-kinematical analysis of a five-point wheel suspension. In: Revue de la SIA, Ingenieurs de l'Automobile, Paris, pp. 77–80 (1985)
14. Knapczyk, J., Dzierzek, S.: Displacement analysis of five-rod wheel-guiding mechanism by using vector method. In: Proceedings of the 25th ISATA, Mechatronics, paper 920893, Florence, pp. 687–695 (1992)
15. Knapczyk, J., Dzierzek, S.: Displacement and force analysis of five-rod suspension with flexible joints. Trans. ASME J. Mech. Des. **117**, 532–538 (1995)

16. Knapczyk, J., Dzierzek, S.: Position–force synthesis of five-rod suspension mechanism. In: 10th IFToMM World Congress on the Theory of Machines and Mechanisms, Oulu, Finland, 20–24 June 1999, vol. 2, pp. 671–676 (1999)
17. Totu, V., Alexandru, C.: Multi-criteria kinematic optimization of a front multi-link suspension mechanism using DOE screening and regression model. Appl. Mech. Mater. **332**, 351–356 (2013)
18. Angrosch, B., Ploechl, M., Reinalter, W.: Suspension design by means of numerical sensitivity analysis and optimization. Int. J. Veh. Des. **65**(1), 52–72 (2014)
19. Simionescu, P.A.: A unified approach to the kinematic synthesis of five-link, four-link, and double-wishbone suspension mechanisms with rack-and-pinion steering control. Proc. Inst. Mech. Eng. Part D J. Automob. Eng. **231**(10), 1374–1387 (2016)
20. Dal Bianco, N., Lot, R., Gadola, M.: Minimum time optimal control of a GP2 race car. Proc. Inst. Mech. Eng. Part D J. Automob. Eng. (2017, in press). https://doi.org/10.1177/0954407017728158
21. Marchesin, F.P., Barbosa, R.S., Alves, M.A.L., Gadola, M., Chindamo, D., Benini, C.: Upright mounted pushrod: the effects on racecar handling dynamics. The dynamics of vehicles on roads and tracks. In: Proceedings of the 24th Symposium of the International Association for Vehicle System Dynamics, IAVSD 2015, pp. 543–552 (2016)
22. Fisher, D.L., Rizzo, M., Caird, J.K., Lee, J.D.: Handbook of Driving Simulation for Engineering, Medicine, and Psychology. Taylor and Francis, Boca Raton (2011). ISBN 978-142006101-7
23. Minen, D., Martini, F., Scarsini, M., Valvason, E., Neto, A.C.: Human centric vehicle development using advanced real-time all-in-the-loop simulators. In: The dynamics of vehicles on roads and tracks, vol. 1, pp. 129–134 (2017)
24. Gómez, G.L., Lönnergård, A., Asher, M.H., Nybacka, M., Bakker, E., Drugge, L.: Analysis and optimisation of objective vehicle dynamics testing in winter conditions. Veh. Syst. Dyn. **55**(7), 945–969 (2017)
25. Chindamo, D., Lenzo, B., Gadola, M.: On the vehicle sideslip angle estimation: a literature review of methods, models and innovations. Appl. Sci. **8**(3), 355 (2018)

A Handheld Mobile Augmented Reality Tool for On-Site Piping Assembly Inspection

Fabio Bruno[1], Loris Barbieri[1(✉)], Emanuele Marino[1], Maurizio Muzzupappa[1], and Biagio Colacino[2]

[1] Department of Mechanical, Energy and Management Engineering (DIMEG), University of Calabria, Via Pietro Bucci, 87036 Arcavacata di Rende, CS, Italy
{f.bruno, loris.barbieri, emanuele.marino, muzzupappa}@unical.it

[2] BHGE Nuovo Pignone Srl, Contrada Le Marinate, 89900 Vibo Valentia, Italy
biagio.colacino@bhge.com

Abstract. In the context of Industry 4.0, Augmented Reality occupies an important role thanks to its unique capability to enhance the perception of the real world with virtual information.

Taking advantage of this capability, the paper presents a handheld mobile Augmented Reality tool that supports manufacturing and production workers and engineers to easily check on-site the ongoing operations carried out in the manufacturing environment for the tubing and piping assembly.

The tool runs on a modern tablet and performs an augmented reality visualization of the 3D models, as defined in the project plan, on the corresponding physical objects. In this manner the user can easily check the presence of assembly errors or detect design discrepancies. The level of acceptance of the proposed handheld mobile Augmented Reality tool has been assessed by means of a preliminary test carried out with representative users on a real case study. Results from the experiment are presented and discussed in the paper.

Keywords: Mobile augmented reality · Industrial augmented reality · Usability · Google ARCore · Industry 4.0

1 Introduction

In recent years, the latest advances in mobile device technology and mobile computing capacity is making mobile Augmented Reality (MAR) more popular and accessible. In fact, it does no longer require expensive hardware or dedicated equipment, such as head mounted displays (HMDs) and smart glasses, but it can now be performed with cost-effective commercial smartphones and tablets. In addition to the economic aspect, another important advantage offered by this mobile technology is its portability, especially for smartphones that are pocketable too, that facilitates movement of the device and allows users to adopt it at the workplace.

A sector that could benefit from MAR's capabilities is Oil & Gas industry and, in particular, the tubing and piping assembly. In fact, because of the complexity of the piping networks it is not uncommon to incur in human errors or in the need to intentionally make design changes and modifications. These assembly errors and

C. Rizzi et al. (Eds.): ADM 2019, LNME, pp. 129–139, 2020.
https://doi.org/10.1007/978-3-030-31154-4_12

design changes usually increase especially in the case of commissioned orders of custom products or small series production. In these cases, in fact, there is no existing knowledge about the specific product and it is therefore more likely that the operator in charge of the assembly operations makes errors or identifies possible improvements and optimizations compared to what was defined in the design phase by the technical office. In particular, about design changes, they consist in a series of adjustments and improvements performed by manufacturing workers to optimize the piping route by minimizing the number of bendings, reducing overall dimensions, and also by increasing the manoeuvrability space for easier the assembly of the components.

At the moment, the workers perform the piping assembly by means of paper-based solutions, such as technical drawings, paper files and written notes that indicate the sequence according to which the components must be mounted on the frame or baseplate. Then, at the moment, there is not a dedicated tool or instrument that supports workers and engineers, directly at the workplace, to check and inspect the ongoing operations carried out for the assembly of tubing and piping installations and networks.

On the basis of these considerations, and taking into due consideration end-users' needs and opinions, this paper proposes a handheld MAR tool that can be adopted on-site to facilitate the inspection of the ongoing piping assembly activities for the detection of design variations and assembly errors. The proposed tool, in fact, can be held with one hand and provides an augmented visualization of the 3D models, as defined in the project plan, on the actual design. In this manner, the user can easily check the progress of the assembly activities, assess if the operations are taking place correctly, identify and report any errors or design changes.

2 Mobile Computing Platforms for AR Applications

In the last years, AR technologies have been efficiently applied in the industrial field by numerous researchers especially to support the operator during training activities, assembly/disassembly, and maintenance operations [1–7]. On the basis of these researches, it is quite evident that AR is a very effective tool, compared to traditional paper-based solutions, when supporting and assisting the user in all those activities for which a considerable involvement of the cognitive and mnemonic capabilities' operator is required. In fact, the transfer of learning and cognitive facilitation when using AR-based systems and tools were confirmed by many studies [8].

Thanks to the technological progress of recent years, it is now possible to take advantage of the benefits of the AR technology by using mobile devices which can be easily transported and used directly at the workplace. Typical portable devices for mobile augmented reality applications are handheld devices, such as smartphones and tablets, and wearable smart glasses or HMDs, such as Microsoft HoloLens, Google Glass and Vuzix Blade.

In general, an AR-based application can be designed and deployed on a mobile device by means of dedicated software development kit, such as Vuforia, Wikitude, Tango, ARKit, ARCore. In particular, thanks to the massive installed hardware base of Google and Apple devices, ARCore and ARKit platforms will make AR accessible in the next years to hundreds of millions of Android and iOS users.

Compared to the HMDs, the main drawbacks of tablet- and smartphone-based AR applications are that users have to hold them in at least one hand and that the augmented scene is only visible on the display. But, to the detriment of these limitations, tablets and smartphones are very popular, independently by gender, age and education, therefore users are already accustomed to interact with such devices and, furthermore, do not require to be worn for their use.

The proposed tool has been developed through the ARCore SDKs [9] that has been preferred to the other abovementioned solutions because it is open source (Apache 2.0 license) and it does not need for specific and dedicated hardware components but it can run on any Android devices (version 7.0 and above).

3 ARCore

In the first half of 2018, Google's company has released the ARCore framework that allows to bring AR technology on a large number of Android consumer smartphones and tablets. At this moment there are almost a hundred of mobile devices that support ARCore SDKs, and more than one hundred of AR apps can be downloaded from the Google Play Store.

The main capabilities provided by ARCore SDKs are: motion tracking; environmental understanding and light estimation. The motion tracking algorithm allows estimating the device's pose in 3D space by means of a hybrid tracking technique that improves the accuracy of the estimation over time by combining visual and inertial tracking methods.

In particular, visual information, both marker-based and natural feature-based techniques, is combined with the inertial data that are provided by the inertial sensors (accelerometer, compass, gyroscope, etc.) mounted on the device itself. The visual tracking is performed by means of a COM (Concurrent Odometry and Mapping) process, patented by Google, that detects visual features of the real scene and use these data to perform pose estimation.

Environmental understanding is the capability to recognize planes by evaluating the size and location of the real environment. In this manner, it is possible to place and augment 3D assets in the physical real scene.

In addition to the above-mentioned capabilities, ARCore features also light estimation functionality that captures the information about the lighting of the real environment, in terms of color and light intensity, to increase the level of realism of the augmented virtual scenario.

4 Mobile AR Tool

As abovementioned, the proposed MAR tool has been developed by means of Google ARCore SDKs. The mobile hardware chosen is a Samsung Galaxy Tab S4. This tablet has been preferred to a smartphone in order to provide a better visualization experience and then overcome one of the main limitations related to the smartphone-based augmented reality that concerns the small devices screen. In particular, this is an Android device equipped with a Qualcomm Snapdragon 835 (Octa-core 4x2.45 GHz Kyro & 4x1.9 GHz Kryo), a 13 MP camera, and several sensors such as accelerometer, gyro, proximity and compass sensors. The tablet mounts a large screen of 10.5-in., and specifically a 16:10 ratios AMOLED display, for a total weight of the device of 0.493 kg.

The following picture (Fig. 1) depicts the software architecture of the AR tool which consists of different modules, implemented in Unity 3D, that communicate among them to generate the AR visualization. In particular, the modules responsible for managing the virtual scene, in terms of marker recognition, motion tracking, and AR scene generation, have been developed using the ARCore library.

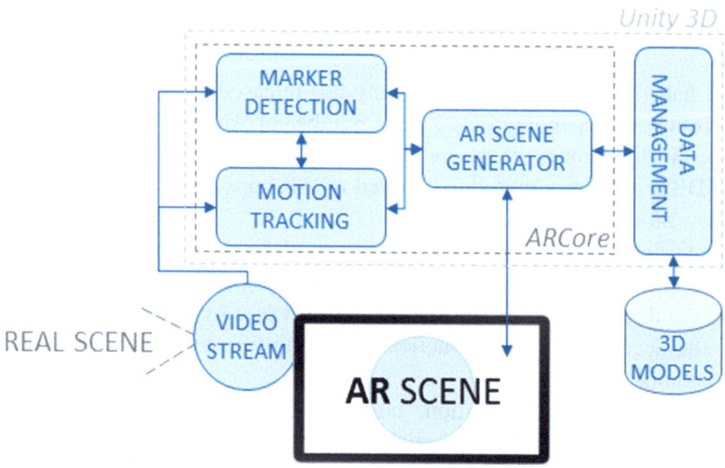

Fig. 1. MAR tool's hw/sw architecture

The rear camera of the device provides a real-world video stream to the marker detection module which adopts ARCore SDKs' routines to automatically recognize up to 20 markers. The choice to implement a multiple marker detection approach is due to the fact that, especially in the case of large products, a single marker is not sufficient to ensure a precise registration of virtual models on the real ones. This happens because, even if ARCore takes advantage of both marker- and feature-based methods and inertial sensors to perform hybrid tracking, there is a registration distortion caused by the sensor's drift that accumulates over time. Thanks to the multi-marker approach, the marker detection module assigns an ID and a local reference system to each detected marker in order to allow the motion tracking algorithm to calculate the relative device pose. In this manner, it is possible to improve the accuracy of the augmented visualization because every time a marker is recognized, the registration error is reset.

The 3D models adopted for the augmented visualization are stored in a local database and managed by the AR scene generator through the Data Management module. Once the 3D models are displayed the application enables the user to edit their graphic appearances, in terms of color and transparency levels, and add 3D notes and annotations. This last feature has been implemented by means of Raycast and Collision functions that recognize user touch and detect contacts between the ray originated by user touch and 3D models of the scene. For each annotation a set of data, consisting in the typology, position, and the textual contents introduced by the user, are automatically saved by the Data Management module.

4.1 User Interface

Mobile user interface (UI) design has been carried out according to the UCD (User Centered Design) principles [10, 11] by taking into account end-users' suggestions and opinions from the early design stage.

When the AR tool starts, the first screen offers to the user a set of options that allows him/her: to create a new work session; load a saved work session; get access to a control panel to choose the unit measurement system and set network connection parameters.

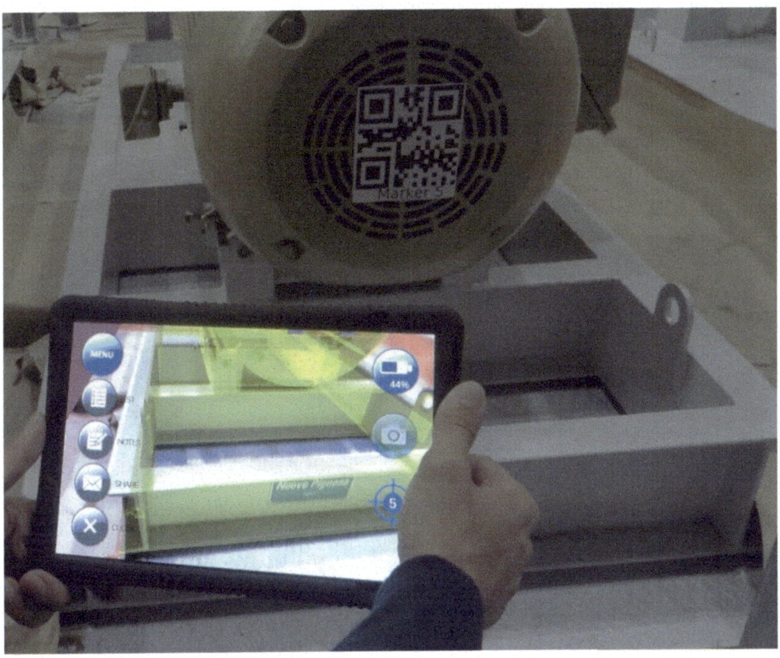

Fig. 2. UI of the handheld mobile AR tool © 2019 Baker Hughes, a GE company, LLC - All rights reserved

Figure 2 depicts a screenshot of the AR tool after the user selected the option for the creation of a new working session. Most of the screen is dedicated to the live video stream of the camera of the handheld device and the interface elements are placed on both sides of the screen to be easily accessible with the fingers of both hands. In particular, on the right side of the screen, two informational components provide information about the battery life and the actual marker detected by the tool. Furthermore, a camera icon allows the user to capture a screenshot of the entire screen, with both real and virtual components, that is automatically saved and associated with the current working session. On the top left side of the screen a menu command button, when activated, displays a drop-down list with four different options that allow the user respectively to: visualize a list of the 3D models, as single components or grouped in sub-assemblies, in order to easily select those ones that are augmented on the real scene; manage the 3D notes hiding or making visible all of them or only some categories; share the data, i.e., 3D notes and pictures, collected during the work session; quit the application.

Furthermore, when the user selects a specific 3D component, three command buttons appear on the left side of the screen and enable the user to: add a 3D note; customize color and transparency level of the 3D models; show or hide all the 3D notes (Fig. 3).

The capability to customize the visual style of the 3D models supports the user to better understand the different elements of assembly, what is inside them, identify assembly errors and detect design discrepancies.

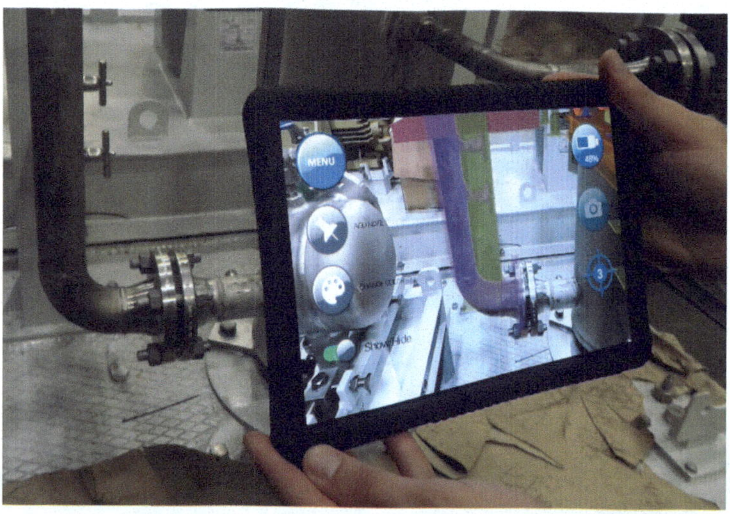

Fig. 3. UI of the handheld mobile AR tool © 2019 Baker Hughes, a GE company, LLC - All rights reserved

5 Experimentation

A preliminary experimentation has been carried out in the Baker Hughes (a General Electric company) plant (BHGE), located in Vibo Valentia (Italy), on a real case study with representative users in order to assess the level of acceptance of the proposed handheld MAR tool.

This experimentation has been undertaken following the positive results obtained in a laboratory exploratory study conducted to evaluate the potentialities offered by the proposed tool when compared to the traditional instruments used in this application field. In particular, the experimental findings showed that the MAR tool gains similar results with the other instruments, i.e., technical drawings and CAD systems, in term of effectiveness and very encouraging results about its efficiency [12].

The user study performed with presentative users has consisted in a simple experiment in which the proposed tool has been used to check the progress of the assembly activities carried out on an auxiliary baseplate system of 14 m long and 6 m wide. This system works as a start system for gas turbine, provides the lubrication oil for turbine bearings, and manage the oil pressure and temperature for turbine system. Such a system has been chosen not for its functions but for its dimensions that allow to highlight the benefits offered by the multi-marker tracking capabilities of the proposed tool.

As depicted in Fig. 4, the markers have been placed 1.5 m apart on the external perimeter of the baseplate system. Some of them have been fastened in the proximity of the motors and pumps sector, to ensure a correct visualization of these components, and on some vertical brackets with the aim to allows the user to visualize the piping networks inside the tank (the green component in Fig. 4).

Fig. 4. 3D model of the auxiliary baseplate system adopted for the user assessment of the AR tool

A total of 17 representative users have been recruited. The subjects involved in this preliminary user study are familiar with the paper-based solutions, such as technical drawings, and knowledge about the baseplate system adopted for the case study. Thanks to their skills and knowledge, the selected participants represent valid and appropriate witnesses to comprehend if the developed MAR tool could be a valid alternative to the traditional instruments to check on-site the progress of the assembly activities.

The participants have been divided into two groups:

- 10 factory workers (all males) with an average age of 39.7 (min 22, max 46, SD = 8,1);
- 7 engineers (4 males and 3 females) with an average age of 31.8 (min 26, max 45, SD = 7.8).

All the subjects are familiar with the use of smartphones and tablets. By contrast, only 30% of them used at least once augmented reality applications on their mobile devices. Nevertheless, all the participants never used AR-based applications to support industrial operations. This lack of knowledge and practical experience is functional for the user study because it reflects a characteristic of the user profile.

The test has consisted of four steps. In the first step, the participants began the test filling out a background questionnaire. In the second step, a tutor provided the participant a brief demo (usually lasting about 1 min) of the MAR tool. In the third step, each participant used the MAR tool without any limitation in time in order to visualize the 3D models on the actual design and then to verify the presence of assembly errors and variations, and the current state of progress of assembly activities. In the fourth step, each participant has been invited to fulfill a satisfaction questionnaire and to perform a one-on-one personal interview aimed to comprehend their personal opinions about the proposed MAR tool. The standard questionnaire adopted for the qualitative assessment of the developed tool is the Umux-lite [13, 14], it is a two-item questionnaire based on the usability metric for user experience [15] that is consistent with the SUS questionnaire but more closely conformed to the ISO definition of usability [16]. The two items are: "This tool's capabilities meet my requirements" and "This tool is easy to use". Furthermore, a third item has been added to the questionnaire to explicitly state their preference between the proposed MAR tool and the traditional paper-based instruments.

5.1 Results and Discussions

Descriptive statistics and *t*-tests have been used for statistically significant differences among the two groups. All analyses have been conducted using the statistical packages Microsoft Excel and IBM SPSS. The statistical significance level has been set at $p > 0.05$.

The following graph (Fig. 5) shows the result of the Umux-lite questionnaire fulfilled by participants after completing the experiment. Error bars represent 95% confidence interval. The histograms indicate high levels in subjective opinions for both groups.

In particular, the results of the first question, i.e., Q1, present similar values with an average score of respectively 5.70 with SD of 1.25 for the first group, and 5.71 with SD of 0.70 for the second group. These data have been compared by means of an independent sample t-test. The Levene's test for the homogeneity of the variances presents a value of 0.43 so it has been possible to confirm the null hypothesis. The t-test result $t(15) = -0.24$, $p = 0.98$ shows that there is not a statistically significant difference between the two groups.

Also, the outcomes of the second question, i.e., Q2, shows similar very high results with an average rate of respectively 6.60 with SD of 0.70 for the first group, and 6.71

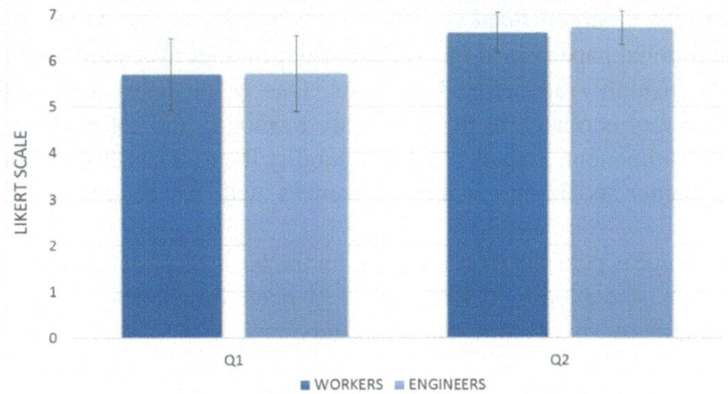

Fig. 5. Umux-lite results by groups

with SD of 0.49 for the second group. The Levene's test for the homogeneity of the variances presents a value of 0.34. The t-test result t(15) = −0.37, p = 0.71 shows that there is a significant difference between the two groups.

So, it is possible to confirm that both the operators and engineers have the same positive opinion about the proposed MAR tool, and specifically about its ease of use and the usefulness of its functionalities in accordance with the user's needs.

These outcomes have been confirmed also by the results of question three, i.e., Q3, in which the participants were asked to clearly express their preference for the AR-based or paper-based solutions. The findings, depicted in Fig. 6, reveal high values for both groups which resulted in a large consent for the proposed tool. Only 30% of the participants prefer to continue to adopt the traditional paper-based solutions.

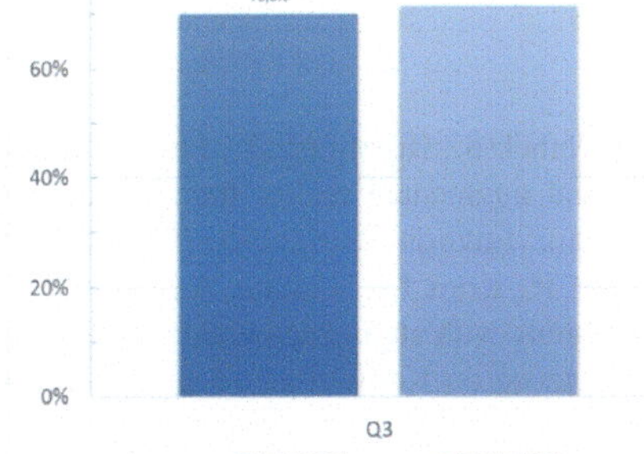

Fig. 6. Percentage of preference for the MAR tool in the groups

Furthermore, from the personal opinions gathered and the end of the test, very positive feedbacks emerged thanks to the functionalities of the MAR tool that can replace the traditional paper-based instruments and provides also additional capabilities which have been highly appreciated by users. One of the most appreciated feature provided by the tool is related to the possibility to modify the color and transparency level of the 3D components. Thanks to this capability the user can easily distinguish the various components and highlight any differences with respect to the real physical components.

In conclusion, it is possible to assess that, on the basis of the subjective metrics and personal opinions, the proposed MAR tool has been valued very positively by the workers and engineers involved in the user study.

6 Conclusions

The paper has presented a MAR tool, that runs on a consumer handheld tablet, to support on-site industrial operations and, in particular, the check of the assembly activities carried out by augmenting the 3D models, as designed by the technical office, on the real physical components. In this manner, the users can easily verify the progress of the assembly activities, the design change made on-site and the presence of assembly errors.

The preliminary test carried out in a real-world scenario with representative users has provided very encouraging results in terms of ease of use and correspondence with users' needs. Furthermore, from the opinions gathered at the end of the test, most users have shown a high level of acceptance and positive attitudes toward the proposed MAR tool. In addition to the subjective opinions of the participants to the study, it is important to highlight that thanks to the adoption of the ARCore SDKs this is a cost effective solution for deployment in an industrial environment.

Since this preliminary qualitative experimentation has shown encouraging results, in the next future further experimentations will be carried out with objective metrics in order to better understand the impact of mobile AR technologies on user performance.

References

1. Gavish, N., Gutiérrez, T., Webel, S., Rodríguez, J., Peveri, M., Bockholt, U., Tecchia, F.: Evaluating virtual reality and augmented reality training for industrial maintenance and assembly tasks. Interact. Learn. Environ. **23**(6), 778–798 (2015)
2. Funk, M., Kosch, T., Kettner, R., Korn, O., Schmidt, A.: motionEAP: an overview of 4 years of combining industrial assembly with augmented reality for Industry 4.0. In: Proceedings of the 16th International Conference on Knowledge Technologies and Data Driven Business, p. 4 (2016)
3. Gattullo, M., Scurati, G.W., Fiorentino, M., Uva, A.E., Ferrise, F., Bordegoni, M.: Towards augmented reality manuals for Industry 4.0: a methodology. Robot. Comput. Integr. Manuf. **56**, 276–286 (2019)

4. Ferraguti, F., Pini, F., Gale, T., Messmer, F., Storchi, C., Leali, F., Fantuzzi, C.: Augmented reality based approach for on-line quality assessment of polished surfaces. Robot. Comput. Integr. Manuf. **59**, 158–167 (2019)
5. Velosa, J.D., Cobo, L., Castillo, F., Castillo, C.: Methodological proposal for use of Virtual Reality VR and Augmented Reality AR in the formation of professional skills in industrial maintenance and industrial safety. In: Online Engineering & Internet of Things, pp. 987–1000. Springer, Cham (2018)
6. De Crescenzio, F., Fantini, M., Persiani, F., Di Stefano, L., Azzari, P., Salti, S.: Augmented reality for aircraft maintenance training and operations support. IEEE Comput. Graph. Appl. **31**(1), 96–101 (2010)
7. Morosi, F., Carli, I., Caruso, G., Cascini, G., Dhokia, V., Guefrache, F.B., O'Hare, J.A.: Analysis of co-design scenarios and activities for the development of a spatial-augmented reality design platform. In: DS92: Proceedings of the DESIGN 2018 15th International Design Conference, pp. 381–392 (2018)
8. Tang, A., Owen, C., Biocca, F., Mou, W.: Comparative effectiveness of augmented reality in object assembly. In: Proceedings of the SIGCHI Conference on Human Factors in Computing Systems, pp. 73–80. ACM (2003)
9. ARCore SDK. https://developers.google.com/ar/. Accessed 27 July 2019
10. ISO 9241-210:2019. Ergonomics of human-system interaction - Part 210: Human-centred design for interactive systems (2019)
11. Barbieri, L., Angilica, A., Bruno, F., Muzzupappa, M.: An interactive tool for the participatory design of product interface. In: ASME 2012 International Design Engineering Technical Conferences and Computers and Information in Engineering Conference, pp. 1437–1447. American Society of Mechanical Engineers (2012)
12. Barbieri, L., Marino, E.: An augmented reality tool to detect design discrepancies: a comparison test with traditional methods. In: De Paolis, L., Bourdot, P. (eds.) Augmented Reality, Virtual Reality, and Computer Graphics. AVR 2019. Lecture Notes in Computer Science, vol. 11614, pp. 99–110. Springer, Cham (2019)
13. Lewis, J.R., Utesch, B.S., Maher, D.E.: UMUX-LITE: when there's no time for the SUS. In: Proceedings of the SIGCHI Conference on Human Factors in Computing Systems, pp. 2099–2102. ACM (2013)
14. Lewis, J.R., Utesch, B.S., Maher, D.E.: Measuring perceived usability: the SUS, UMUX-LITE, and AltUsability. Int. J. Hum. Comput. Interact. **31**(8), 496–505 (2015)
15. Finstad, K.: The usability metric for user experience. Interact. Comput. **22**(5), 323–327 (2010)
16. ISO 9241-11:2018. Ergonomics of human-system interaction - Part 11: Usability: Definitions and concepts (2018)

Multisensory Augmented Reality Experiences for Cultural Heritage Exhibitions

Marina Carulli(✉) and Monica Bordegoni

Department of Mechanical Engineering, Politecnico di Milano, Milan, Italy
{marina.carulli,monica.bordegoni}@polimi.it

Abstract. Human beings interact with the external world through the perception that they get by touching, looking at, listening to, tasting and smelling it.

Even if this exploration is essential to identify opportunities and dangers, today it is also used to investigate, understand and enjoy objects that surround us and to use them to manage our life, have fun, increase our knowledge, relax, etc.

To date, interaction is based primarily on sight, on hearing and on touch. Very little interaction, however, is based on smell, a sense considered very difficult to manage and to use for creating more pleasant and effective experiences.

Yet, olfactory stimuli can make the interaction between users and objects more engaging and effective on sub-conscious levels and long-term memory. This can be particular relevant for museums, which are subjected to important changes in the way they use to involve visitors in their exhibitions.

This work presents two case studies of olfactory experiences integrated in applications for cultural heritage purposes, and effectively used to enhance the user interaction and experience.

Keywords: Multisensory experiences · Augmented Reality · Interactive exhibitions

1 Introduction

Museums have been subjected to important changes in the approach they use to involve visitors, also by using digital technologies. So, they are implementing a digital revolution with the aim to change the way people relate to and visit the museums.

Among the other trends, storytelling and interactive exhibitions are two of the most used approaches used to make the exhibition more and more interesting for the users who visit it. So, from the traditional concept in which museums and exhibitions were "show-and-tell" spaces [1], in this new scenario they are becoming spaces where it is more and more possible interacting with pieces of arts.

There are several approaches used to make museum exhibitions interactive. One of the most used and easy-to-implement approach used to carry out this process of innovation is the digital interaction. Specifically, Virtual Reality (VR) and Augmented Reality (AR) methods can be effectively used in the context of a museum exhibition to support both storytelling and interaction. In particular, by means of AR applications, visitors can enjoy immersive experiences in which the content, the history and the meaning of pieces of art are digitally added onto the real artifacts. Indeed, Augmented

C. Rizzi et al. (Eds.): ADM 2019, LNME, pp. 140–151, 2020.
https://doi.org/10.1007/978-3-030-31154-4_13

Reality is nowadays more and more used for museum exhibition, to add virtual contents to the real ones and "increase" the reality. So, pieces of art are able to self-explain their meaning, the story of their authors, the correlation with other pieces of art and so on. Conversely, in the case of VR methods, these are usually used to re-create historical sites with VR environments to "immerse" the users in ancient historical periods.

The primary objective of the use of these technologies is to make the visit of museums much more engaging, immersive, and suitable for different types of visitors. In particular, the common idea is to "extend" the target audience also to those who are not experts in art (or in the topics described in the exhibition) who, through the new co-communication methods, can better understand concepts often considered as very complex. In this sense, in fact, the possibility of viewing the history of the pieces of art and their meaning instead of reading or listening to the explanation, or rather of looking at images/animations instead of reading texts, reduces the need to have experienced and knowledgeable users, and increases the possibilities of learning and remembering information. In the "non-expert users" category, it is also possible to include children, who are often fascinated by science, nature and art, but hardly interested in static representations that are very difficult to understand.

New exhibitions aspire to raise acknowledge and curiosity among the visitors, especially children and teenagers. In fact, interactive exhibitions are becoming more and more appreciated and effective, involving people through different senses and calling them to action. They need to perform an action in order to see a corresponding reaction inside the environment or in the object shown.

Moreover, nowadays the user experience is one of the fundamental features of museums and exhibitions. Indeed, the user experience is strictly linked to the user interaction with pieces of arts and objects at physical, perceptual and cognitive levels. This means that it includes also the design of multisensory experiences that also satisfy sight, touch, hearing and smell senses in order to improve the exhibition understating and the user's engagement. In particular, the sense of smell plays an important role, since it recalls memories and deep emotions both at conscious and unconscious level. Thanks to these effects on humans, smells can help the customer to feel involved into and emotionally linked with pieces of arts.

The purpose of the research work presented in this paper is to design and develop interactive multisensory applications (based on sight, hearing, smell and touch senses), which can be used for improving the user experiences of cultural heritage exhibitions. Particular attention has been devoted to the integration of olfactory stimuli in the overall experience and their integration with the other senses. Specifically, the paper presents two case studies in which multisensory AR experiences of cultural exhibitions have been designed and developed. In both cases, the experiences involve the sight, hearing and smell senses and aim at improving the visitors' engagement in the exhibition and their understating of its content.

2 Literature Review

2.1 Multisensory Perception

Psychological and neuroscientific research has shown that perception is always multisensory: whenever we interact with the external world our brain integrates information from different sensory modalities as a function of certain binding principles [2].

The sense of sight has been thoroughly studied in the years [3]. Concerning the sense of hearing, several research studies have demonstrated that sounds can be used to elicit emotions in users [4]. The sense of touch is the first to develop in humans, well before vision [5]. In a general sense, touch plays an important role at both the emotional and perceptual level, even if the user is not aware of its use [6]. The sense of touch is important in the interaction with other people, but also when we interact with objects.

The sense of smell is one of the main means through which humans investigate the environment [7]. Odors can influence moods [8], emotions [9, 10], and memories [11]. Odors can induce activation or relaxation states in people [12], and can also stimulate cognitive abilities [13, 14]. Concerning the link between odors and memory, many studies have been carried out from different points of view, such as basic science, psychology [15] and marketing [16].

In modern cultures, the sense of vision is hegemonic [17]. So, Western society has been defined as "the society of olfactory silence" [18] and, up to a few years ago, there was the tendency to remove any unpleasant odor. It is only in recent years that the importance of smell on the rise, especially in regard to its potential to stimulate emotions and create positive product experiences combined with other sensory modalities. This property has been addressed especially in the marketing research field, in which many studies have been carried out on the use of odours for influencing customers' moods in shops, and for communicating information about products [19, 20].

2.2 Olfactory Displays for Virtual and Augmented Reality Environments

In the last years, several studies have focused on the integration of odors in Virtual and Augmented Reality environments by developing devices named Olfactory Displays.

Various technologies are used to develop an olfactory display. These are categorized by scent generation methods and scent delivery methods [7]. Several research works concerning olfactory displays have been presented in literature [21–24].

The authors have developed some prototypes of Olfactory Display and have evaluated their performances in several applications for products evaluation [25], for improving the drivers' attention level [26], for heritage studies [27], for improving the reading experience [28], and for improving virtual tours for the tourism sector [29].

Even if olfactory display devices are still evolving and some technical issues still remain unsolved, some companies have developed and sell olfactory displays [30]. Recently, some new devices have been launched on the market. For instance, Vaqso is a device that adds scents to the VR experience [31]. It can be customized by the user, who can feel the scent in conjunction with the scenes and actions on VR. The Olorama scents generator provides user a multisensory experience with wireless scent dispensers on VR/AR and home projection systems [32]. Feelreal simulates several scents to immerse the user into virtual worlds [33] by using a scent generator, which holds a replaceable cartridge containing nine individual aroma capsules.

2.3 Multisensory, Interactive and Digital Exhibitions

Nowadays interactive exhibitions in museums are becoming broadly common, thanks to their flexibility of application and effectiveness in engaging visitors in the experience. So, visitors are becoming active "consumers", and are able to define their own journey and experience at the museum. One emblematic example are interactive exhibitions for younger audiences, and specifically for children. In this case, museums frequently allow young visitors to "play" with exhibits, in order to make learning new knowledge an interesting and pleasant experience. One example is The Children's Museum of Manhattan, in New York. In 2016 the museum developed the "America to Zanzibar" exhibition [34], in which children were able to touch and experiment with artifacts of Muslim culture. The main idea was to create a multisensory experience to help children understand the Muslim culture using colors, images, objects and smells.

Concerning the use of digital technology, and specifically VR and AR applications, people are becoming more and more familiar with hybrid realities that make it easier to use this technology in public events and for cultural heritage purposes. Some museums present digital labels or touchscreen tables with texts and information as "side elements" of their exhibitions. Other museums, such as the MONA Museum in Hobart [35], effectively rely on the "O" device and application that replaces wall texts in the museum and represents the only way visitors can interact with pieces of art. Another interesting example is the Cooper Hewitt Pen at the Cooper Hewitt Smithsonian Design Museum [36]. In this case, the "Pen" is used by visitors to 'collect' objects around the gallery, which can be explored in further details at the end of the visit (by using interactive tables) and also online. At the Casa Batlló in Barcelona, visits are carried out by using an AR guide, which provides a dynamic and semi-immersive experience [37].

In many other cases, museums develop VR and AR experiences of some special exhibitions. Among the others, in 2010 the AR exhibition named "WeARinMoMA" by Sander Veenhof and Mark Skwarek has been held at the MoMA in New York [38]. The exhibition, performed within the context of another exhibition, allowed the visitors to choose what to see thanks to an application on their smartphones. In 2015, the British Museum's Samsung Digital Discovery Centre (SDDC) held a Virtual Reality Weekend in the Museum's Great Court, offering the visitors the possibility to enjoy VR environments via immersive and non-immersive devices [39]. Visitors were able to explore a virtual reality Bronze Age site, where three-dimensional scans of objects were placed. Moreover, interpretations of how these ancient objects might have been used in the past were presented. In 2017, the Natural History Museum in London featured in an interactive VR experience in which visitors were able to virtually handle fossils and other pieces from the Museum's collection in a one-on-one interactive experience [40].

Finally, concerning the use of stimuli from different senses, this approach can modify how visitors interact with and interpret the exhibition and its content. In this field, several experiences have been developed by using one or more senses at time. In 2012, the sense of touch has been embodied in the light installation of Takahiro Matsuo [41]. His interactive Aquatic Colors installation emulates a glowing, underwater sea filled with jellyfishes, which respond to the movements of visitors. The use of smells for cultural purposes is rare and still experimental, but some exhibitions that integrated olfactory displays have been presented in the last years. In "The Art of the Scent (1889-2012)" exhibition at the Museum of Art and Design in New York scents have been celebrated as a true artistic media [42]. The "Multisensory Met" exhibition at the Met Media Lab Intern Expo aimed at creating a multisensory environment to enhance the overall effect of pieces of art. So, they created more fulfilling experiences for museum visitors, which are satisfactory for multiple senses. Different multisensory products have been used: booklets, sculptures, clay replicas, scratch-and-sniff paintings, scented powder materials and sound paintings [43].

3 Case Studies

In this section two case studies of multisensory AR experiences of cultural exhibitions are presented. In both cases, the experiences involve sight, hearing and smell senses (in one case also touch) and aim at improving the visitors' engagement in the exhibition and their understating of the content.

3.1 The *Stati d'animo - A new AR olfactory experience* Case Study

The aim of *Stati d'animo – A new AR olfactory experience* case study has been to augment the experience of the user of an exhibition related to moods by using an AR application, tactile stimuli and olfactory displays for allowing him/her to be more immersed in the exhibition subject. The AR multisensory system intends to contribute in increasing the visitors' sense of presence in the environment and awareness of the meaning of each mood.

The case study consists of mixing Augmented Reality technology and simulated smells in the Palazzo dei Diamanti Museum in Ferrara (Italy), where the *Stati d'animo* exhibition took place in March-June 2018. The topic of the exhibition was a critical research of the origin and growth of moods and feelings in paintings in Italy in the 19th-20th century. The exhibition included many paintings focusing on different moods, one for each of the 12 areas of the museum. What was evident was the presence in every canvas of something related to a smell or to an olfactory sensation (Fig. 1). Specifically, in each room there was a link to a couple of smells that suggested a precise feeling.

So, the case study *Stati d'animo – A new AR olfactory experience* has the aim to give visitors a new and different way of comprehension of art in museums. Using AR technology and olfactory devices (developed by the authors and described in [29]) to solicit visual and olfactory stimuli creates immediately an exciting memory in the visitors' mind. Thanks to the specifically-developed AR application and the olfactory

Fig. 1. The second room of the exhibition – Involve the observer – gives a big suggestion to the viewer about different smells, in which many of them were used in the past from artists as drugs to augment the perception of reality.

experience, when visitors perceive the smells, their feelings and emotions are amplified. In fact, smells let them remember more the different feelings experienced when visiting the exhibition. In this way AR and smells are not used just as new experience means, but also as memories enablers.

Concerning the layout of the exhibition, the rooms tell stories are about the introduction of physiognomy study, i.e. the use of drugs to amplify experiences in life, melancholia and alienation, contemplation and empathy, the musical inspiration, communion and harmony, voluptuousness and feral instincts, fusion and ecstasy, happiness and enthusiasm, atmosphere and affection.

The AR multisensory experience consists of two main components, set up in all rooms.

The first main component consists of 3D printed objects – totems placed in big rooms, and frames placed in smaller rooms – accommodating inside olfactory displays that deliver odours if the visitor brings his/her hand close to an LDR (Light Dependent Resistor). The kind of odour released in the room has been carefully selected to create a link with the mood and augment the perception of the content of the painting. Each odour is a combination of two or three fragrances, according to a balanced creation of *"perfume notes"*. For example, the fragrance used in the room about *Fusion and Ecstasy* is a mix of rose and vanilla, a mix of tobacco and brandy for *Involve the observer*; lemon and cypress for *Melancholia and Alienation*; sandalwood and lavender for *Contemplation and Empathy*.

When triggered, the system also plays a music representing the mood of the room. Each 3D printed object used in the various rooms is modular and has a particular and suggestive shape, well related to the mood of the room. For example, in the room about *Contemplation and Meditation,* the pillar has the shape of an impoverished ionic column, representing the literal meaning of contemplation (from Latin, "*cum*" and "*templum*", delimited celestial space).

The second main component of the system consists of AR applications displayed on tablets or smartphones, which allow animating the most representative painting of each room, displayed on a wall. For example, in the room about *Contemplation and Meditation,* the most representative canvas of this mood is *Ave Maria a trasbordo* made by Giovanni Segantini in 1886. So, the authors developed an AR application in which the boat is moving (Fig. 2). A recorded voice accompanies the animation, telling the story of the piece of artwork and explaining the meaning of the room.

Fig. 2. The *Ave Maria a trasbordo canvas*, which has been animated in the AR application and the A scaled prototype of the totem about contemplation and empathy made with 3D printed material. In the background the Arduino Uno system connected with the spray diffuser in it.

The system consists of interactive devices, i.e. columns and frames displaying stories about the paintings on show in the room, and including olfactory displays delivering fragrances. The interactive devices are connected to Arduino Uno and AdaFruit boards, and an LDR sensor triggered when occluded (Fig. 3).

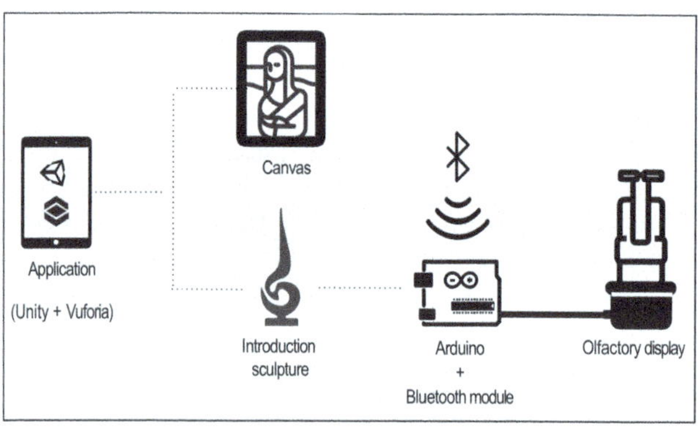

Fig. 3. The architecture of the *Stati d'animo - A new AR olfactory experience* system

The devices have particular shapes according to the mood of the room and are 3D printed with different materials. In order to suggest to the visitor a tactile interaction, some particular areas are printed with a softer material, similar to rubber, and highlighted with a different shade of the colour of the painting.

When the visitor touches the device, the fragrance is delivered in the environment and sounds related to the mood of the room is played back.

About the creation of the AR system, Unity3D and Vuforia software tools have been used. Specifically, the animations of the paintings have been developed in Unity3D, and then the Vuforia software has been used for the development of the AR application. In the AR application, one object per room is used as marker. So, in each room the AR application is able to recognize the frame (or the column), which triggers the animation related to the most representative painting of that room, displayed on a wall.

3.2 The *Senses of Religions* Case Study

Th *Senses of Religions* project aimed to develop a multisensory experience to improve knowledge and elicit curiosity about religions with an interactive approach that lead visitors inside the designed environment. Religion is one of the most important side of our history and origins; by analyzing it, it is possible to find the meaning of different behavioral patterns, usages, cultures and actions. So, the *Senses of Religions* case study has the aim to give visitors a new way of comprehension of religions. The authors decided to use a cultural and didactic approach, with the aim of describing the meaning, the history and the purpose of each religions, avoiding any subjective opinions and judgments. The use of odors in this contest is particularly important, thanks to the broad use of odors in religions (such as the use of incense in Christian religion). So, by using AR technology and olfactory devices (developed by the authors and described in [29]) it is possible to recall memories in the visitors' mind and create a cultural interest. The storytelling approach has been used to engage the user during the experience, and the possibility to interact with the application has been designed in order to make him/her live a full experience. In this case study, the storytelling concerns the main key elements and the history of religions.

The exhibition starts from a hall where the visitor finds five doors, one for each religion: Christian, Hinduism, Buddhism, Islam and Traditional Chinese. Above each door there is an illustration, representing a religious descriptive image that works as a marker for the AR application. Framing each marker with a tablet or a smartphone recalls the AR application with a representation of the world. For each marker (and door), a different area of the world is colored, which corresponds to the area of the world in which each religion is spread the most. Behind each door, visitors find a room dedicated to each religion. Moving from one room to another one is a very important moment in which the hall plays a key role to reset to a neutral point the nose of the visitor from the previous room.

One of the five represented religions is Buddhism. The storytelling of this religion is based on Siddhartha Gautama's life, the path he followed before becoming the Buddha.

Moving in the room, the first illustration to start is the Buddha itself, and the last one is the Bodhi tree, which is the place where Siddhartha became enlightened and received the supreme and final wisdom, hence becoming the Buddha (Fig. 4).

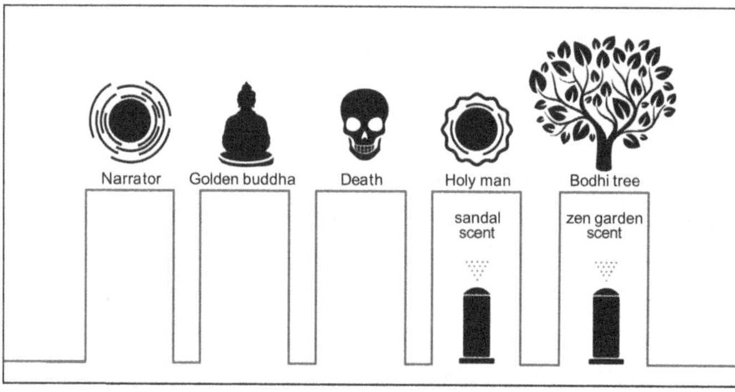

Fig. 4. The layout of the storytelling of the Buddhism room

Two olfactory displays are connected to the two Image Targets: "Holy man" and "Bodhi tree". Framing the Holy man marker, the olfactory display spays a smell of sandal, related to the 3D model, involving the spectator deeply. Lastly, when the Bodhi tree marker is framed, the olfactory display sprays a smell of Zen garden, which recalls in the visitor's mind the ideas of meditation, contemplation and peaceful. The olfactory displays are located inside the marker support (a totem) with a physical inclination of 90°, hidden from the visitors (Fig. 5).

Fig. 5. The Holy man marker (a conch shell) on the totem, which hides the olfactory display, and the AR application running.

For what concerns the AR application, the *Senses of Religions* case study has been developed by using Unity3D and Vuforia software tools. The interaction between visitors and the application has been created according to the storytelling. Specifically, for some objects in the path, visitors are required to interact with the digital device, listen to the story, touch the screen while immersed in the scent emanated from the olfactory display. In the AR application, five markers per room have been specifically developed. So, in the hall and in each room the AR application is able to recognize five markers, which trigger the animations related to the area of the world in which the religion is spread the most (for the hall) and related to the storytelling of the religion (for the other rooms). A recorded voice accompanies the animations, presenting the meaning of each key element. For what concerns the olfactory displays delivering fragrances, these are connected to the AR application by using an Arduino Uno and AdaFruit boards, and a Bluetooth module. Each time the visitor frames an image target linked to the scent, a script is triggered in order to send a signal to the Bluetooth module that activates the corresponding olfactory display (Fig. 6).

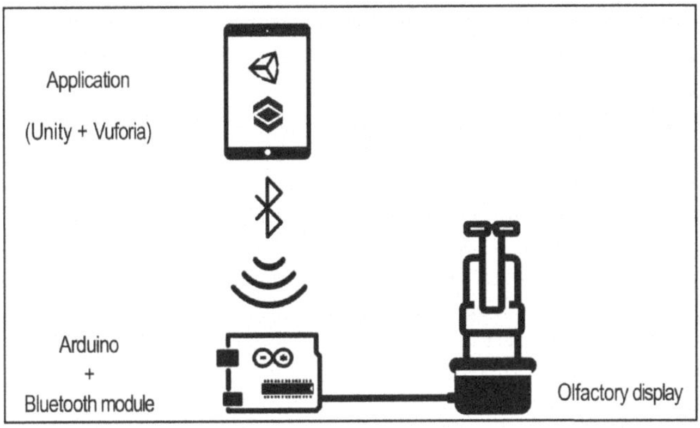

Fig. 6. The architecture of the *Senses of Religions* system

4 Conclusions

The research work presented in this paper focuses on the possibility to improve the user experience of cultural heritage exhibitions through multisensory AR applications (based on sight, hearing, smell and touch senses). Specifically, two case studies have been designed and developed with the aim of improving the visitors' engagement in the exhibition and their understating of the content. Particular attention has been devoted to the integration of olfactory stimuli and their integration with the other senses. Indeed, the olfactory stimuli have been designed with the aim of impacting on the subconscious and conscious levels of visitors to stimulate memories and emotions.

These case studies perfectly match with some of the most important trends used for innovating museum exhibitions, which are digital interaction and storytelling. More, while Augmented Reality is nowadays commonly used for museum exhibitions, the use of the sense of smell and/or multisensory approach is still a rare phenomenon. On the opposite, the integration of scents in these interactive AR applications can be effectively used to create an emotional link between the visitors and the pieces of art, to improve the users' experience and the users' memory of the exhibition. So, in an envisaged scenario, multisensory AR applications similar to those presented can be developed and implemented in a wide range of museums and exhibitions, for events related to cultural heritage and can be also remotely experienced by the user at home for virtual visits of museums.

References

1. https://blog.beaconmaker.com/5-ways-to-make-your-museum-exhibition-interactive-d302d300808c. Accessed 1 July 2019
2. Spence, C., Driver, J.: Crossmodal Space and Crossmodal Attention. Oxford University Press, Oxford (2004)
3. Marr, D.: Vision: a computational investigation into the human representation and processing of visual information. Henry Holt and Co., Inc., New York (1982)
4. Lu, W., Petiot, J.F.: A sound–based protocol to study product emotions. Int. J. Prod. Dev. **16** (3), 207–224 (2012)
5. Field, T.M.: Touch in Early Development. Lawrence Erlbaum Associates, Inc., Hillsdale (1995)
6. Gallace, A., Spence, C.: In Touch with the Future. Oxford University Press, Oxford (2013)
7. Nakamoto, T.: Human olfactory displays and interfaces: odor sensing and presentation, information science reference (2013)
8. Epple, G., Herz, R.S.: Ambient odors associated to failure influence cognitive performance in children. Dev. Psychobiol. **35**, 103–107 (1999)
9. Herz, R.S.: Are odors the best cues to memory?: A cross-modal comparison of associative memory stimuli. Ann. N. Y. Acad. Sci. **855**, 670–674 (1998)
10. Vernet-Maury, E., Alaoui-Ismaili, O., et al.: Basic emotions induced by odorants: a new approach based on autonomic pattern results. J. Aut. Nerv. Syst. **75**, 176–183 (1999)
11. Ehrlichman, H., Bastone, L.: The use of odor in the study of emotion. In: Van Toller, S., Dodd, G.H. (eds.) Fragrance: The Psychology and Biology of Perfume. Elsevier, New York (1992)
12. Bensafi, M., Rouby, C., et al.: Influence of affective and cognitive judgments on autonomic parameters during inhalation of pleasant and unpleasant odors in humans. Neurosci. Lett. **319**, 162–166 (2002)
13. Lorig, T.: Cognitive and non-cognitive effects of odour exposure: electrophysiological and behavioral evidence. In: Van Toller, S., Dodd, G.H. (eds.) Fragrance: The Psychology and Biology of Perfume. Elsevier, New York (1992)
14. Meamarbashi, A.: Instant effects of peppermint essential oil on the physiological parameters and exercise performance. Avicenna J. Phytomed. **4**(1), 72–78 (2014)
15. Chu, S., Downes, J.J.: Proust nose best: odors are better cues of autobiographical memory. Mem. Cogn. **30**(4), 511–518 (2002)
16. Krishna, A.: An integrative review of sensory marketing: engaging the senses to affect perception, judgment and behavior. J. Consum. Psychol. **22**, 332–351 (2012)
17. Classen, C., Howes, D., Synnott, A.: Aroma: The Cultural History of Smell. Routledge, London and New York (1994)
18. Corbin, A.: The Foul and the Fragrant: Odour and the French Social Imagination, London (1996)

19. Mitchell, D.J., Kahn, B.E.: There's something in the air: effects of congruent or incongruent ambient odor on consumer decision making. J. Consum. Res. **22**(2), 229–238 (1995)
20. Bosmans, A.: Scents and sensibility: when do (in)congruent ambient scents influence product evaluations? J. Mark. **70**(3), 32–43 (2006)
21. Hirota, K., Ito, Y., Amemiya, T., Ikei, Y.: Presentation of odor in multi-sensory theater. In: VAMR/HCII 2013 (2013)
22. Kim, D.W., Nishimoto, K., Kunifuji, S. et al.: Development of aroma-card based soundless olfactory display. In: Electronics, Circuits, and Systems, ICECS (2009)
23. Matsukura, H., Yoneda, T., Ishida, H.: Smelling screen: technique to present a virtual odor source at an arbitrary position on a screen. In: IEEE Virtual Reality 2012 (2012)
24. Narumi, T., Nishizaka, S., et al.: Meta Cookie+: an illusion-based gustatory display, virtual and mixed reality, pp. 260–269 (2011)
25. Bordegoni, M., Carulli, M.: Evaluating industrial products in an innovative visual-olfactory environment. J. Comput. Inf. Sci. Eng. **16**(3), 030904 (2016)
26. Bordegoni, M., Carulli, M., Shi, Y.: Investigating the use of smell in vehicle-driver interaction. In: Proceedings of the ASME Design Engineering Technical Conference, Smart Innovation, Systems and Technologies, vol. 65, pp. 513–523 (2016)
27. Terracciano, A., Dima, M., Carulli, M., Bordegoni, M.: Mapping memory routes: a multisensory interface for sensorial urbanism and critical heritage studies. In: Proceedings of Conference on Human Factors in Computing Systems, Part F127655, pp. 353–356 (2017)
28. Bordegoni, M., Carulli, M., Shi, Y., Ruscio, D.: Investigating the effects of odour integration in reading and learning experiences. In: Interaction Design and Architecture(s), vol. 32, pp. 104–125 (2017)
29. Bordegoni, M., Carulli, M., Ferrise, F.: Improving multisensory user experience through olfactory stimuli. In: Fukuda, S. (ed.) Emotional Engineering, vol. 7. Springer, Cham (2019)
30. Davide, F., Holmberg, M., Lundström, I.: Virtual olfactory interfaces: electronic noses and olfactory displays. In: Riva, G., Davide, F. (eds.) Communications Through Virtual Technology: Identity Community and Technology in the Internet Age. IOS Press (2001)
31. https://vaqso.com/#overview. Accessed 1 July 2019
32. https://www.olorama.com/en/. Accessed 1 July 2019
33. https://feelreal.com. Accessed 1 July 2019
34. https://www.pleasetouchmuseum.org/americatozanzibar/. Accessed 1 July 2019
35. https://mona.net.au/museum/the-o. Accessed 1 July 2019
36. https://www.cooperhewitt.org/new-experience/. Accessed 1 July 2019
37. https://www.casabatllo.es/en/visit/. Accessed 1 July 2019
38. http://www.sndrv.nl/moma/. Accessed 1 July 2019
39. Rae, J., Edwards, L.: Virtual reality at the British museum: what is the value of virtual reality environments for learning by children and young people, schools, and families? In: MW 2016: Museums and the Web 2016 (2016)
40. https://www.nhm.ac.uk/discover/news/2017/april/museum-to-star-in-sky-vr-experience-with-sir-david-attenborough.html. Accessed 1 July 2019
41. https://mymodernmet.com/takahiro-matsuo-aquatic-colors/. Accessed 1 July 2019
42. https://madmuseum.org/exhibition/art-scent. Accessed 1 July 2019
43. https://www.metmuseum.org/blogs/digital-underground/2015/multisensory-met. Accessed 1 July 2019

Reverse Engineering, Digital Acquisition and Inspection

Optical Stereo-System for Full-Field High-Frequency 3D Vibration Measurements Based on Low-Frame-Rate Cameras

Sandro Barone, Paolo Neri(✉), Alessandro Paoli,
Armando V. Razionale, Leonardo Bertini, and Ciro Santus

Department of Civil and Industrial Engineering, University of Pisa, Largo L. Lazzarino 1, 56122 Pisa, Italy
paolo.neri@dici.unipi.it

Abstract. Dynamic characterization of vibrating targets represents a critical issue for many industrial fields. In this paper, a stereo-camera system integrated with a Digital Image Correlation (DIC) algorithm is proposed with the aim at performing 3D full-field vibration measurements in the range of kHz. The system exploits two industrial low-speed cameras, and the Nyquist-Shannon frequency limitation is overcome by a down-sampling approach under the hypothesis that the vibration signal is characterized by a single known frequency component. Experimental results obtained from the measurement of vibrational responses of a cantilever plate excited at three high-frequency resonance values (1121 Hz, 2956 Hz and 4010 Hz) are provided. A comparison with numerical analyses evidences the effectiveness of the proposed approach.

Keywords: Vibration measurement · Digital Image Correlation · Low-frame-rate cameras · Stereo-camera setup

1 Introduction

The dynamic characterization of machine parts is essential in many industrial fields. In particular, rotating machinery always experience dynamic loading, which can result in fatigue failures if resonances are met. For this reason, experimental modal analysis and harmonic response analysis are widespread for critical components qualification [1–4].

Great interest is devoted to contact-less measurement techniques, which guarantee two main advantages: no need to transmit data from moving objects and no added mass. Single point measurement techniques, such as laser doppler vibrometry, are characterized by high sensitivity and accuracy [5], but require expensive setups and mainly provide punctual mono-directional measurements. Scanning 3D systems exist [6], but their cost and complexity does not allow for a widespread diffusion of their use. On the other hand, full-field measurement techniques were developed over the years, such as Digital Image Correlation (DIC) [7–9]. 2D and 3D displacement maps of the object surface can be obtained through different optical setups by exploiting image processing and stereo triangulation. The sensitivity of this method is generally related to the camera resolution and the field of view, and can be significantly high, in the order

C. Rizzi et al. (Eds.): ADM 2019, LNME, pp. 155–164, 2020.
https://doi.org/10.1007/978-3-030-31154-4_14

of microns. Nevertheless, this approach is commonly limited by the camera frame rate, since a tradeoff between resolution and sampling frequency must be applied. Thus, conventional approaches generally fail in many practical applications where small displacements need to be measured at high frequency values. An example is provided by compressors' and turbines' bladed wheels, where the frequency range of interest is in the range 1–10 kHz. For this reason, some approaches proposed in recent technical literature developed simplified systems based on low-frame-rate cameras integrated with down-sampling strategies. This allows to exploit high resolution camera and perform high-frequency vibration measurement without undergoing the Nyquist-Shannon theorem. In particular, the system proposed in [10] provided impressing measurement accuracy of a blade vibrating with an amplitude of 0.01 mm at about 3.5 kHz under a pure sinusoidal excitation, by using a 20 fps single camera and a bi-mirror setup.

The present paper proposes an approach which is similar to the one described in [10], but relies on a strongly improved hardware setup. In particular, a couple of synchronized stereo-cameras is used, instead of a single camera with two angled mirrors. The light source was drastically improved, providing intense and uniform illumination of the whole specimen. Also, a different stereo-matching algorithm is proposed to obtain 3D vibrations through a stereo camera system. The proposed measurement system is firstly described along with the adopted working principle. Experimental results are then presented by measuring the displacements of a target object moved by incremental small steps, in order to assess the sensitivity of the developed methodology. Finally, measurements of a cantilever beam vibrating at various resonance frequencies are presented with the aim at assessing the method capabilities in the detection of 3D high-frequency small-amplitude vibrations.

2 Measurement Strategy

2.1 Physical Principle

The 3D measurement of vibrating object is divided into two tasks: firstly, the stereo matching problem is solved, i.e. determining corresponding pixels on the left and right cameras. This allows to define a grid of points on the left image ($\text{grid}_{L,0}$), and to compute the corresponding grid point on the right image ($\text{grid}_{R,0}$) as shown in Fig. 1. Then, the displacement over time is measured on both left and right cameras separately, by tracking these grids through a 2D DIC algorithm, thus obtaining the deformed grids $\text{grid}_{L,i}$ and $\text{grid}_{R,i}$ on the i-th image. Since $\text{grid}_{L,i}$ represents the tracking of $\text{grid}_{L,0}$ on the left camera and $\text{grid}_{R,i}$ represents the tracking of $\text{grid}_{R,0}$ on the right camera, $\text{grid}_{L,i}$ and $\text{grid}_{R,i}$ can be triangulated without the need of repeating the stereo-matching task for each frame. This is schematically shown in Fig. 1.

The stereo-matching problem is commonly solved by applying a DIC algorithm between left and right undeformed images. Indeed, left and right images represent the same object viewed by two cameras placed in different positions. Thus, in principle, it is equivalent to acquire the object with left camera, then moving this camera to be aligned with the right camera and acquire the object again. The two different

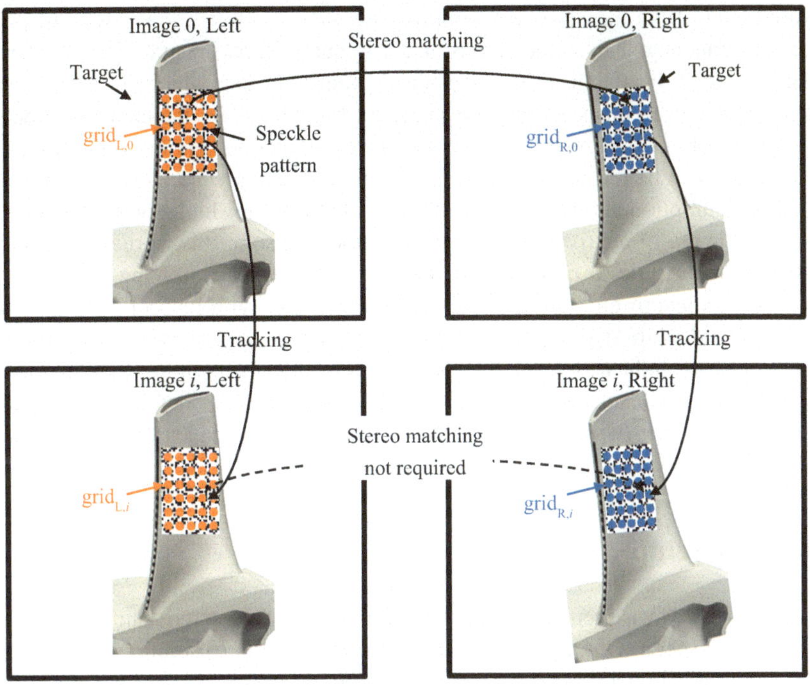

Fig. 1. Stereo-DIC approach used for the 3D shape reconstruction.

perspectives can be seen as a relative rigid motion between the object and a single camera, and can then be detected through the DIC algorithm. However, this strategy has the severe drawback that the pose of left and right cameras is significantly different, so that a virtual large displacement is to be measured. Even if this can be handled by modern DIC algorithms, it fails in the case of extremely different viewpoints, non-uniform illumination and high complex geometries. Thus, in this paper a slightly different approach was used. The stereo camera system is firstly calibrated [11]. Then, both left and right images (image 0) are rectified by exploiting the epipolar geometry [12]. This virtual image rotation is defined in order to have corresponding points of left and right images laying on the same image row. Thus, the *y*-direction of the stereo matching problem is automatically solved. Then, a disparity algorithm, which is faster and more robust with respect to common 2D DIC algorithms [13], is used to also determine the *x*-direction component.

2.2 Down-Sampling Approach

The Nyquist-Shannon theorem states that an unambiguous measurement of a sinusoidal signal can be achieved only if the sensor sampling frequency is at least double with respect to the signal frequency. Otherwise, aliasing phenomena may occur, and fictitious frequencies are detected. However, if the signal to be measured only contains one harmonic component of known frequency, the signal amplitude can be precisely measured even if the measurement sampling frequency is much lower with respect to the signal frequency. This is the case of many dynamic experimental setups, where the structure loading is obtained by an electrodynamic shaker which is driven with a known frequency. In this scenario, the sampling frequency (i.e. the camera frame rate) can be arbitrarily chosen respecting the hardware limitations, and the actual frequency and amplitude can be retrieved by various signal processing techniques, such as NHFA [14]. This assumption allows to drastically relax the need for high-speed cameras, thus opening to the use of relatively slow high-resolution sensors. It is worth noting that the exposure time becomes a limitation for the technique: since a moving object is to be imaged, the exposure time needs to be reasonably short with respect to the vibration period to avoid blurred images. This can be achieved by following two strategies: (1) using a strobe light with a small impulse width or (2) using industrial cameras with short exposure times. In the present paper, this latter solution has been adopted. An exposure time of 2 μs was selected, in order to be 50 times smaller than the period of a 10 kHz signal.

3 Optical Setup

The core of the measurement system is composed of a stereo camera pair. Two industrial low-speed Optomotive T-REX EVO cameras (resolution 2048 × 2048 pixels, minimum shutter time 2 μs) were used. The cameras allow for triggering, which is achieved by using the left camera as master and the right camera as slave. The exposure intervals of the two cameras thus result synchronized without using pulsed light source. Each camera is mounted on a spherical head, which allows to define a common working volume between the two fields of view. The spherical heads are then fixed on a rail, which establishes a proper baseline. In the assembled setup, the baseline was 260 mm, while the angle between cameras optical axis was about 25°.

The same lenses were mounted on both cameras, with a focus length of 16 mm. This setup determined a field of view of about 300 × 300 mm, placed at 500 mm (Fig. 2). Since a short exposure time was set, a high-power flicker-free light source was needed. A 100 W Stratus led V3 module was adopted (80 CRI, 140 lm/W, color temperature 5500 K), equipped with a parabolic reflector and mounted on the baseline rail. A light diffuser was used to ensure a more uniform target illumination. Some trial tests denoted a high level of light intensity stability over time. The system setup was completed by mounting the rail on a 2-axis manual translational stage, which is then assembled on a tripod through a 3-axis manual rotational stage. These degrees of freedom, along with the vertical regulation of the tripod, allows for a flexible and easy to use measurement system.

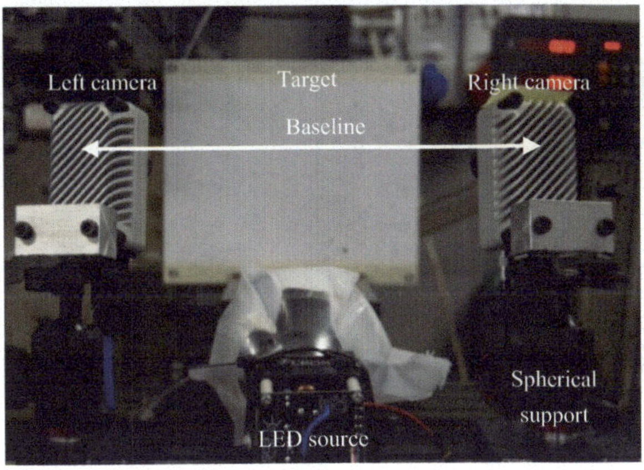

Fig. 2. Stereo camera setup.

4 Results and Discussion

4.1 Static Measurements

The described testing equipment was firstly tested through a static measurement, with the aim at assessing the system sensitivity to small displacements. Since dynamic measurement will be carried out with a short exposure time, also this static test was performed with the same acquisition parameters, thus illuminating the target with the LED module. The target object is a thick aluminium plate (250 $\times$ 200 $\times$ 20 mm). A white sheet of paper was stacked on the surface, and a speckle pattern was sprayed with black spray paint. The plate was then placed within the vise of a milling machine. The machine rails were used to apply a known displacement along one direction, roughly aligned with the x axis of the camera system. The linear encoder of the rails allowed to control the displacement with an accuracy of 0.01 mm. A known incremental displacement along the two directions was applied. Data elaboration is reported in Fig. 3: Fig. 3a shows the disparity map obtained by solving the stereo-matching task through the described disparity algorithm applied on the rectified images. Figure 3b shows the displacement values for each acquired image. It is worth noting that the imposed displacement is a rigid motion of the plate along one direction. Thus, in order to assess the system accuracy and reliability, the black solid line in Fig. 3b represents the mean value of the displacement as measured at 660 points of the target (22 $\times$ 30 sub-grid in the black box of Fig. 3a). The dashed black lines represent the maximum and minimum displacement values measured in the sub-grid for each frame. Finally, the red solid line represents the difference between maximum and minimum displacement values (Δ), as a metric of the measurement uncertainty. The standard deviation σ of each measurement stage was also computed, and the values are reported in μm in Fig. 3b. This static test demonstrated that the displacement uncertainty is in the order of 10 μm, giving a reference value for the method sensitivity.

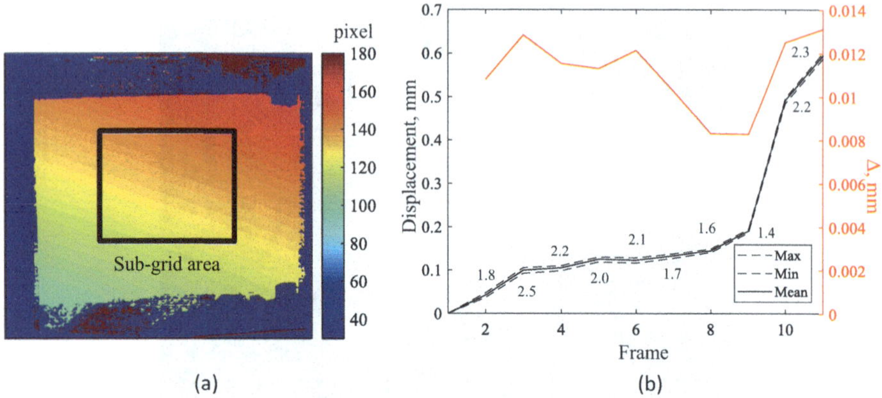

Fig. 3. Disparity map on the reference plate (a) and measured static displacement (b).

4.2 Dynamic Measurements

The aim of the described setup is to detect the displacement amplitude of a vibrating object. As pointed out in the previous section, the system sensitivity (considering the provided field of view) is in the order of 10 μm. Thus, a specimen was designed in order to have a vibration amplitude greater than 20 μm up to 4 kHz, at least at resonance frequencies. An aluminium 110 × 500 × 2 mm cantilever plate screwed to the moving part of the electrodynamic shaker TIRA Vib TV 51110 was used (Fig. 4). The shaker was equipped with a power amplifier (TIRA BAA 120, 100 N peak force, 2–7000 Hz).

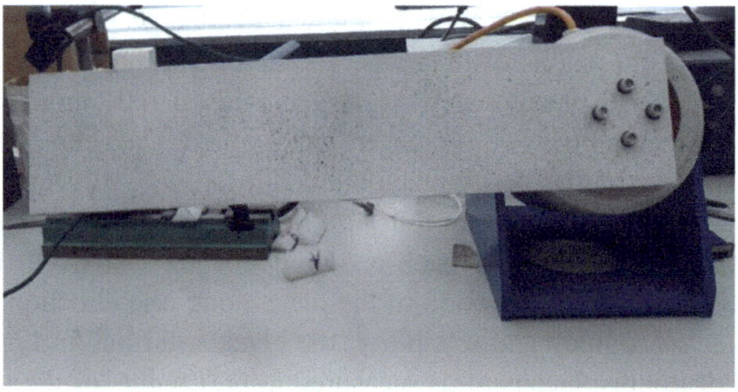

Fig. 4. Cantilever plate constrained to the electrodynamic shaker.

A simple finite element analysis was performed to assess the natural frequencies up to 6 kHz. Then, three modes were selected at increasing frequencies: 1121 Hz, 2956 Hz and 4010 Hz. It is worth noting that the first natural mode of the specimen was 121 Hz, thus the three selected modes are high order modes, which are characterized by complex deformed shapes and, hence, small displacements. An LMS-Siemens SCADAS was adopted to drive the shaker by imposing a sinusoidal excitation with the three selected frequency, separately. The plate vibration was then acquired through the described system and processed to obtain the full filed 3D displacement maps. Figure 5 shows the results obtained at 1121 Hz in terms of *x*, *y* and *z* displacement maps. These three coordinates are referred to the left camera reference frame, thus they are slightly misaligned with respect to the plate orientation. The direction normal to the plate surface roughly corresponds to the camera *z* axis. It is worth noting that the measured displacements maps are characterized by a low noise level in all the three directions, despite the complex mode shape. In particular, the displacement along *y* direction is in the range –60 μm/40 μm, and still is almost noise-free. The test was then repeated with the 2956 Hz frequency. This mode was characterized by smaller displacement values, in the range of about –25 μm/25 μm for both *x* and *y* directions and –65 μm/65 μm along the *z* direction. Despite this, the maps of Fig. 6 are still acceptable, and the noise level is reasonable. It is worth remarking how measurements along *z* direction are less sensitive to small displacements as evidenced by the comparison between Fig. 6c and Fig. 5b. The displacement map reported in Fig. 6c, indeed, results noisier with respect to the one shown in Fig. 5b, even if the displacement range is comparable.

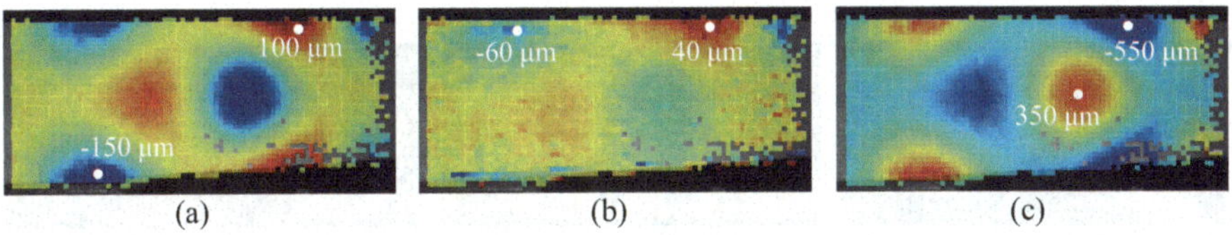

Fig. 5. Vibrational response at 1121 Hz. Displacement maps along *x* (a), *y* (b) and *z* (c) directions.

Finally, the test was repeated at the highest frequency (4010 Hz), which is a challenging value even by using more conventional experimental setups based on high speed cameras [15]. The results are reported in Fig. 7. The *z*-axis displacement map (in the range –60 μm/80 μm) has a low noise level. The *x*-axis displacement map (in the range 15 μm/–10 μm) still allows to assess the mode shape. On the other hand, this mode is characterized by an almost null *y*-axis displacement (which was also confirmed by the FEM analysis). Indeed, the corresponding displacement map reported in Fig. 7b is dominated by white noise, being the measured displacement below 10 μm, which is the limit of the presented setup (mainly depending on the large field of view).

The measured modes where also compared with the corresponding FEM analysis both in terms of shapes and frequencies. The shapes comparison is reported in Fig. 8 for the three measured resonance frequencies (the corresponding maps represent the displacement magnitude), showing a considerable agreement between experimental and numerical results, even for highly complex modes. Table 1 reports the frequencies comparison, evidencing errors in the order of 10% which are mainly to be ascribed to the applied boundary conditions: the FEM model was developed by using fixed

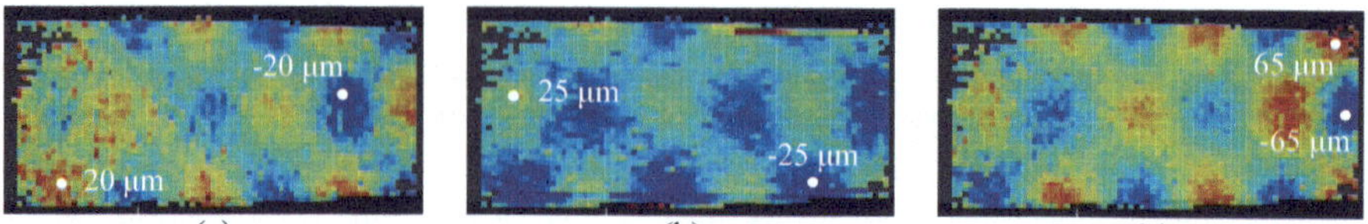

Fig. 6. Vibrational response at 2956 Hz. Displacement maps along x (a), y (b) and z (c) directions.

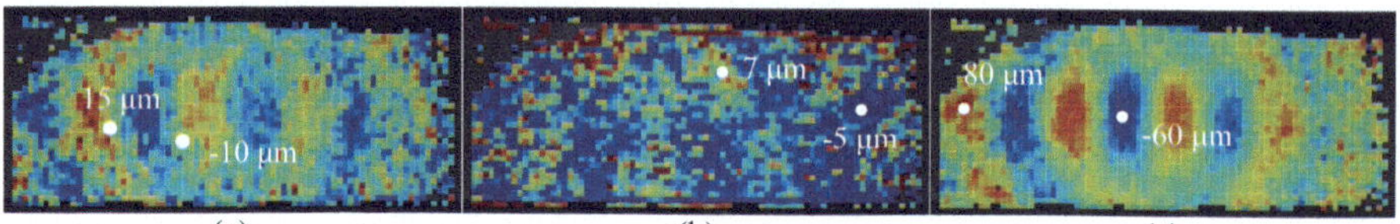

Fig. 7. Vibrational response at 4010 Hz. Displacement maps along x (a), y (b) and z (c) directions.

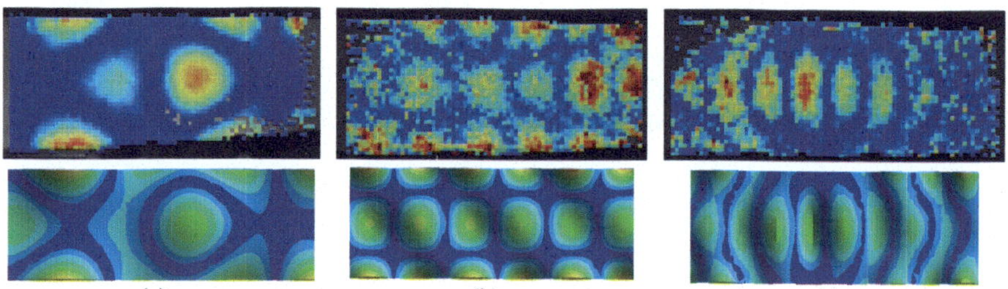

Fig. 8. Comparison between measured (top) and FEM (bottom) magnitude of displacement at 1121 Hz (a), 2956 Hz (b) and 4010 Hz (c).

constraint at the plate holes. This represents an infinite stiffness of the connection, which obviously does not represent physical constraints. Indeed, numerical frequencies are higher than experimental ones, confirming that the model is stiffer with respect to real conditions.

Table 1. Comparison between experimental and numerical frequencies

Mode	Exp. Freq. (Hz)	Num. Freq. (Hz)	Error (%)
1	1121	1220	8.2
2	2956	3115	5.4
3	4010	4435	10.6

5 Conclusions

This paper describes the use of a low-cost experimental setup for high-frequency full-field 3D vibration measurements. The main characteristic of the system is the use of conventional industrial cameras, instead of expensive high-speed cameras, to measure frequencies which are far beyond the camera maximum frame rate. The advantage is a cost reduction and, mainly, the possibility to exploit high resolution industrial sensors, while high-speed cameras always present lower resolutions and, thus, lower sensitivity to small displacements. The major drawback of the proposed approach relies on the constraint of having only a single sinusoidal component in the signal, which can be easily obtained by imposing the excitation with a shaker. The proposed measurement strategy was described, and the core assumptions of the developed software were presented. The system sensitivity to small displacements was firstly assessed through a static test. The system capabilities of measuring vibrational responses was finally validated on a cantilever plate excited at three high-frequency resonances values, i.e. 1121 Hz, 2956 Hz and 4010 Hz. Experimental shapes and frequencies were compared with numerical analysis, evidencing a good correlation. The displacement maps along the three main directions were also computed in order to evaluate the system limitations. It was shown that the frequency values do not represent a critical issue, since the adopted exposure time (2 μs) is much smaller than the signal period. On the other hand, the stronger limitation was found to be the detectable displacement amplitude since the system sensitivity, for the designed field of view (about 300 $\times$ 300 mm), resulted to be 10 μm for both static and dynamic tests.

Acknowledgments. The authors are grateful to the University of Pisa for supporting this research activity (Grant PRA_2018_80).

References

1. Kammerer, A., Abhari, R.S.: Experimental study on impeller blade vibration during resonance-Part I: blade vibration due to inlet flow distortion. J. Eng. Gas Turb. Power **131** (2009)
2. Bertini, L., Neri, P., Santus, C., Guglielmo, A.: One exciter per sector test bench for bladed wheels harmonic response analysis. In: Proceedings of the ASME Turbo Expo: Turbine Technical Conference and Exposition, vol. 7b (2017)
3. Berruti, T., Firrone, C.M., Gola, M.M.: A test rig for noncontact traveling wave excitation of a bladed disk with underplatform dampers. J. Eng. Gas Turb. Power **133** (2011)
4. Bertini, L., Neri, P., Santus, C., Guglielmo, A., Mariotti, G.: Analytical investigation of the SAFE diagram for bladed wheels, numerical and experimental validation. J. Sound Vib. **333**, 4771–4788 (2014)
5. Bertini, L., Neri, P., Santus, C., Guglielmo, A.: Automated experimental modal analysis of bladed wheels with an anthropomorphic robotic station. Exp. Mech. **57**, 273–285 (2016)
6. http://www.polytec.com/us/products/vibration-sensors/scanning-vibrometers/robovib-structural-test-station/

7. Beberniss, T.J., Ehrhardt, D.A.: High-speed 3D digital image correlation vibration measurement: recent advancements and noted limitations. Mech. Syst. Signal Process. **86**, 35–48 (2017)
8. Baqersad, J., Poozesh, P., Niezrecki, C., Avitabile, P.: Photogrammetry and optical methods in structural dynamics - a review. Mech. Syst. Signal Process. **86**, 17–34 (2017)
9. Helfrick, M.N., Niezrecki, C., Avitabile, P., Schmidt, T.: 3D digital image correlation methods for full-field vibration measurement. Mech. Syst. Signal Process. **25**, 917–927 (2011)
10. Barone, S., Neri, P., Paoli, A., Razionale, A.V.: Low-frame-rate single camera system for 3D full-field high-frequency vibration measurements. Mech. Syst. Signal Process. **123**, 143–152 (2019)
11. http://www.vision.caltech.edu/bouguetj/calib_doc/
12. Bradski, G., Kaehler, A.: Learning OpenCV: Computer Vision with the OpenCV Library. O'Reilly, Sebastopol (2008)
13. Hirschmuller, H.: Accurate and efficient stereo processing by semi-global matching and mutual information. In: Proceedings of CVPR, pp. 807–814. IEEE (2005)
14. Neri, P.: Bladed wheels damage detection through non-harmonic Fourier analysis improved algorithm. Mech. Syst. Signal Process. **88**, 1–8 (2017)
15. Molina-Viedma, A.J., Felipe-Sese, L., Lopez-Alba, E., Diaz, F.A.: 3D mode shapes characterisation using phase-based motion magnification in large structures using stereoscopic DIC. Mech. Syst. Signal Process. **108**, 140–155 (2018)

CAD Reconstruction: A Study on Reverse Modelling Strategies

Francesco Buonamici(✉), Monica Carfagni, Rocco Furferi, Lapo Governi, and Yary Volpe

Department of Industrial Engineering of Florence, Via di Santa Marta 3, 50139 Florence, Italy
francesco.buonamici@unifi.it

Abstract. Computer Aided reconstruction of mechanical parts nowadays heavily relies on dedicated RE software systems and highly skilled users to be carried out effectively. This paper investigates this scenario in order to identify most limiting factors in the current framework. With this respect, several considerations of general validity are drawn while presenting the problem from a theoretical perspective. A significant test case reconstruction, discussed in depth in the manuscript, is used to provide a practical point of view on real applications and help the reader acquire a hands-on comprehension of the current situation.

Keywords: CAD reconstruction · Reverse engineering · CAD

1 Introduction

Digital resources are nowadays fundamental in practically every aspect of human life. The possibility of easily sharing information, as well as the ability of executing complicated calculations, have made numerical models essential in all technical sciences. Computer Aided Design (CAD) models, in particular, provide a 3D digital description of an object's geometry. These are widely employed in practically every engineering field and they are among the most used tools by designers and engineers.

When the original CAD model of an object is not available, different techniques can be used to produce a valid substitute. In this context, Reverse Engineering (RE) aims at the reconstruction of a 3D digital model (i.e. CAD model) using information extracted from a physical object. This process, also called "CAD reconstruction", starts with measurements performed on the object. Depending on the object dimensions, the required accuracy and the type of hardware and software tools available to the reverse engineer, different techniques can be used to extract and process the required data.

This paper deals with the reconstruction of CAD models starting from 3D measured data and focuses, specifically, with the problem of retrieving CAD models of mechanical parts. Taking into consideration engineering applications, RE aims at reconstructing a parametric CAD model that is as close as possible to the original design of the part. The correct functioning of mechanical parts normally relies on a subset of functional geometric features; the geometry of these features is usually

C. Rizzi et al. (Eds.): ADM 2019, LNME, pp. 165–176, 2020.
https://doi.org/10.1007/978-3-030-31154-4_15

regulated by dimensions and geometric constraints. During the production phase, the precise manufacturing of functional features is essential to transfer the designer's intentions to the physical part; analogously, during the RE process, the functional surfaces must be accurately acquired and reconstructed to retrieve the original design intent. Specifically, it is important to achieve not only the correct retrieval of dimensions, but also of the geometric relations and of the modelling combinatorial structure defining the shape of the object (i.e. feature tree). The retrieval of a parametric and associative description is generally sought due to the high versatility and applicability of the produced result; the feature tree, in fact, allows for fast and simple modifications to the retrieved geometry.

Several RE methods offer solutions to produce non-parametric models because their information structure is easier to handle and they do not rely on established CAD software as medium. Specifically, research has been generally oriented into developing automatic and semi-automatic solutions to reach this kind of result. However, hardly any solutions have been proposed that produce a high-quality result expressed in a parametric associative format. To this day, the only practical solution to obtain a parametric CAD model remains a RE process carried out within commercial RE-CAD software systems starting from 3D scanned data. Considering this hypothesis, the typical workflow is depicted in Fig. 1. The modelling step, where the parametric surfaces are created using the scanned data as reference, results to be the most influencing phase for the achievement of a satisfying result.

Thanks to the development of commercial software systems, semi-automatic tools to perform every task depicted in Fig. 1 are nowadays available; the overall process, however, still remains tied to a human factor: a skilled user needs to guide the procedure across all phases. Indeed, the human intervention is essential to tune the procedure and, within the modelling step, to carefully select the reconstruction strategy (e.g. the order of operations to perform, which geometric relations to enforce, etc.). The importance of human contribution is further discussed in Sect. 2, in order to provide some perspective on the problems faced by reverse engineers.

Academic research, on the other hand, has been generally oriented towards the study of automatic procedures embracing the whole framework: these solutions, although interesting from a theoretical point of view, lack the required robustness, efficiency and reliability to be applied in real case scenarios. Moreover, they usually fail in producing parametric associative CAD models that can be easily handled and used in the CAD software of choice.

As a result, Fig. 1 represents the standard workflow that is adopted in practically every application. Although the industrial research has led to development of mature software tools, several bottlenecks and limitations could be improved. For the interested reader, a detailed analysis of the state of the art of RE software tools, focusing on the modelling step (Fig. 1) is provided in [1]. The present paper further expands the investigation of [1], providing an in-depth and hands-on description of the reverse modelling phase. Only through a detailed and practical description of the reverse modelling phase, presented in Sect. 3 following a reconstruction test case, it's possible to identify the aspects that need to be addressed by researchers in the next future. Moreover, this work provides fresh insight into the strategies adopted to retrieve design intent and impose meaningful geometric relations during the reconstruction.

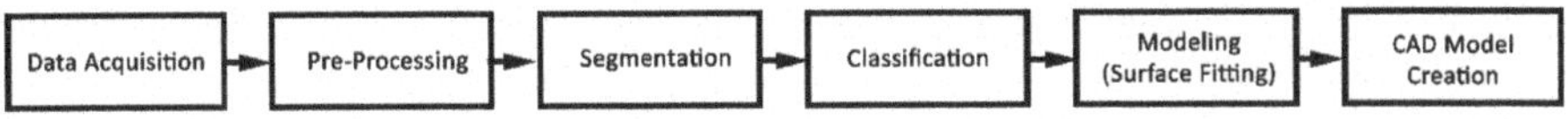

Fig. 1. RE workflow

It's noteworthy, with this respect, that many key-aspects and details need an in-depth description, enriched by practical examples, to be fully appreciated by the reader. Evidently, this kind of perspective is not usually offered by most research papers due to length limitations; accordingly, this area still needs to be filled. Moreover, the software landscape is a fast-changing environment, which is constantly updated and improved with novel features and tools. Summing up, a practical description of reverse modelling carried out within RE software systems is required to contribute to a deeper understanding of CAD reconstruction.

2 Design Intent: The Human Factor

The 3D data measured on the object is not the only source of information available to determine its topology. Indeed, reference data is essential to infer original dimensions (e.g. lengths, angles); nevertheless, considering the retrieval of a modelling feature tree, it can be integrated by other sources. As explained in [2], the shapes of common mechanical parts "follow standard conventions arising from tradition, utility or engineering design"; accordingly, it may be convenient to acquire some information to improve the RE process. Useful cues to understand how a part was modelled in the first place can be extracted from multiple sources: (i) functioning of the object – known information on the performances required to the object; (e.g. the external surface of a piston is a cylinder, due to its function); (ii) fitting with other parts; (e.g. the features and surfaces defining a linear guide/slide are parallel; hence, they can be described by a single extrusion); (iii) manufacturing process, which can impose geometric constraints to guarantee the producibility of the part; similarly, typical defects induced by production processes can be taken into account and filtered out of the scanned data (e.g. part produced on the lathe are usually axial-symmetric, presence of draft angles in die-casting); (iv) existing know-how and information on the original design process of the object. (e.g. it is known that two planes on the object are orthogonal).

As an example, if the measured data were not "filtered" by considering the described aspects, manufacturing and wear-induced defects visible in the scanned data would always be included in the final reconstructed geometry. Moreover, some degree of human interpretation is advisable in order to prevent negative effects caused by incomplete data. Summing up, human intervention is required to guide the reconstruction process. The user contribution is essential for the reconstruction of an effective modelling tree and for a correct interpretation of the scanned data. These inputs cannot be easily integrated in an algorithm, as they are influenced by all the aspects previously described, which depend directly on an extended skill set, as well as on a critical analysis of the specific reconstruction.

On a more general level, every aspect of the user' expertise can be useful for the retrieval of the original topology of a mechanical part. Specifically, the identification of the geometric constraints defining an object shape is a task that can be easily performed by a reverse engineer, while automatic algorithms can only provide a marginal contribution, since they lack direct knowledge of all the previously described aspects. This contribution is usually limited to the extraction and identification of geometric features within the segmented data [3] (i.e. classification step in Fig. 1). As Robert Fisher concludes in his discussion on knowledge-based reverse engineering [2], "Computers are good at data analysis and fitting; humans are good at recognizing and classifying patterns.". Summing up, the human factor is an essential element in the interpretation of geometries.

3 Reverse Modelling

The use of dedicated RE software is the typical solution adopted for the reconstruction of mechanical parts. The efficiency and reliability of the tools hosted within such systems positively influence their application on a day-to-day basis. In this section, most limiting factors and general issues affecting even high-quality software systems are highlighted in order to pinpoint areas to be improved.

The dissertation closely follows the reconstruction of a CAD model of an Electrical Socket Adapter (ESA - Fig. 2(a)) to guide the reader. The chosen object can be enclosed in a bounding box of 55 × 45 × 35 mm and has been acquired by a 3D optical scanner (i.e. a Romer RS1 mounted on a 7520-SI absolute arm by Hexagon Metrology) to produce the mesh depicted in Fig. 2(b). The RS1 scanner is characterized by a volumetric accuracy of 30 μm [4]. The acquired data has been processed using standard procedures (data integration, smoothing, decimation). The reconstruction is performed within Geomagic Design X (GDX) 2016, a dedicated RE software that offer the most advanced features for the reconstruction of mechanical parts.

Following the workflow of Fig. 1, the 3D data is first segmented into different regions; these are identified and isolated according to their geometrical properties. Depending on the specific tools offered by each software, segmentation usually leads to the direct identification of geometric features and primitives that are associated to the segmented regions (i.e. classification [3, 5]). The segmentation/classification result obtained in GDX can be tuned accordingly to a sensitivity and a smoothness parameter, which influence the size of the identified regions; the final result obtained for the ESA is depicted in Fig. 2(c).

Reverse modelling starts with a step that is usually not described in most dissertation on the topic: the information extracted in the classification step, together with other geometrical parameters obtained from the model, is used to orient the data in a convenient position to start the reconstruction. This choice is important, as it influences every downstream operation. Typical examples of geometric entities that are considered during this phase are symmetries, revolution axes, extrusion directions, plane faces, etc. GDX offers a dedicated tool able to compute a convenient data orientation: plausible alternatives to orient the data are suggested to the user, who performs the final choice. Alternatively, the user can extract a series of geometric elements (symmetry

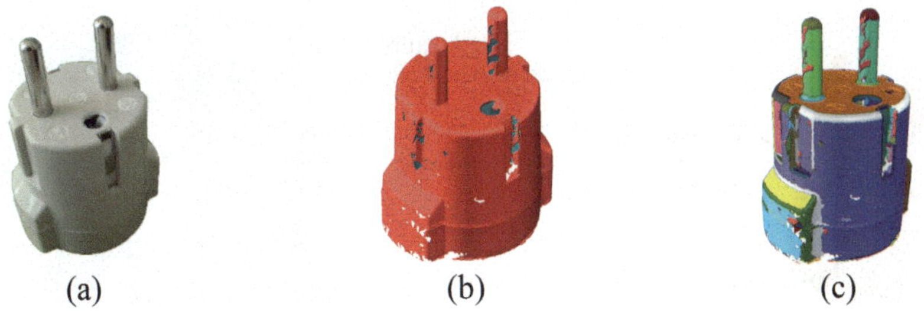

Fig. 2. (a) the ESA; (b) ESA 3D scanned data; (c) ESA classified regions.

axes, significant points, planes, etc.) associated with the classified regions in order to produce a manual orientation of the model. The result obtained in the ESA case are visible in Fig. 3. The selected configuration is characterized by the x-axis parallel to: (i) the normal of the support plane and other planar features (e.g. top surface of the ESA), (ii) the revolution axis of the main cylindrical surface, (iii) the extrusion direction of extruded features. Moreover, symmetries and other geometric features contribute also to the identification of the translational component: the origin is placed at the centre of the object base, located along the main axis of symmetry. It's important to note that, during the acquisition, the object was rigidly placed on a surface plate that has been acquired too; such planar surface, although not part of the object to be reconstructed, can be exploited as reference during the reconstruction.

After data alignment, the user needs to devise an effective modelling strategy for the reconstruction of the specific part. Practically, a main geometric feature is usually selected by the user in order to start the modelling process. Such feature will serve as reference in all subsequent steps of the reconstruction chain. The whole process is controlled by the designer that, using a series of CAD-like functions, manually controls the reconstruction process. The actual modeling of the reconstructed object is achieved by applying CAD/RE tools capable of adapting mathematical surfaces upon the previously classified mesh data. Different categories of tools can be used, depending on the application and the desired result. For mechanical parts, a reconstruction process based on regular features is commonly the most applied strategy whenever allowed by the software. Conversely, freeform reconstruction tools, although often available, are used only when explicitly required by the presence of complex non-regular shapes within the object. On a practical level, the user selects one or more of the regions previously identified, decides which type of feature is best suited to reconstruct the surface and generates the mathematical surface using pre-built tools. These tools, as previously mentioned, provide the capability of building a geometrical feature tailored on a segmented/classified region or extracting information on its position and dimensions from the 3D data. Depending on the tool used, different options may be available to the user (e.g. acceptable deviation, number of iterations to be performed computing the final CAD surface, etc.).

Dealing with the ESA test case, the most convenient strategy would be to start with the reconstruction of the main body (i.e. the big purple cylinder in Fig. 2c). The reconstruction of wide surfaces in the first steps of the process is beneficial for the

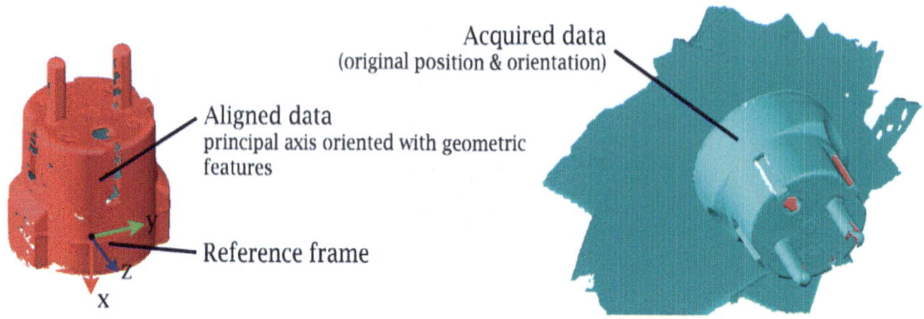

Fig. 3. Initial alignment of the scanned data.

quality of the whole procedure; such areas are described by a large set of points, which accordingly hold the highest share of information on the object shape. Moreover, from a statistical perspective, the reconstruction of a surface performed on a large sample of points offers more reliable results.

The list of CAD operation performed for the ESA reconstruction is described in Table 1; most surfaces are reconstructed considering them as the result of an extrusion feature (i.e. surfaces described by a 2D profile translated along a direction). Once that a feature is reconstructed, following steps can exploit it as reference surface. Accordingly, the definition of the order of operations is an important element of the whole process: its structure determines the way geometric relations are established between features. Moreover, it should be built prioritizing most important regions and performing the same kind of choices that were previously described for the selection of the first feature.

All the surfaces are defined throughout various subsequent modeling steps, exploiting previously built features (e.g. 2D sketches) to generate a tree of relations between the geometric entities and always using the scanned data as reference. In advanced software all these relations are explicitly stored in a proper feature tree, which allows for later access and retrieval. As showed in Table 1, depending on the complexity of the model, several modelling steps are generally required to reach the desired result (Fig. 4).

One of the most challenging aspects that the reverse engineer must face is the pursue of a correct balance between adherence to the scanned data and retrieval of original design intent. Acquired surfaces are affected by errors (acquisition, manufacturing). The original CAD model would have been flawless and characterized by dimensions that might be slightly different from those extrapolated by the scanned data.

Every time the engineer imposes a geometric relation (e.g. defining a surface as a cylinder, constraining the direction of a revolution axis, etc.) causes an increased deviation between CAD and reference surfaces. In other words, the "perfect" fit that would be obtained reconstructing a feature by pursuing only the minimization of point-to-point distances between the two objects is influenced by the introduction of additional aspects, hence changing an unconstrained minimization problem into a constrained one. As an example, let's consider the reconstruction of a cylinder from a single tessellated surface acquired by a 3D scanner (Fig. 5). The imposition of an

Table 1. ESA reverse modeling operations.

#Operation	CAD operation	Relations with previous steps	Dimensional/Geometrical Relations
1	Revolution Surface	Revolution axis obtained from the original alignment	
2	Cut-Extrusion	Extrusion direction parallel to the revolution axis in #1	The extrusion profile should be symmetric w.r.t. planes XZ and XY
3	Cut-Extrusion	Cut-Extrusion direction parallel to the revolution axis in #1 and #2	Extrusion profile symmetric w.r.t. planes XZ and XY
4	Filleting	–	Same diameter on all fillets
5	Extrusion	Extrusion direction parallel to the revolution axis in #1	Same diameter of the two pins. Position of two pins symmetric w.r.t planes XZ and XY
6	Revolution Surface	Revolution profile obtained from Extrusion #5 (seamless junction)	Same diameter of the two revolution profiles
7	Cut-Extrusion	Extrusion direction parallel to #1	Extrusion symmetric w.r.t. XZ plane

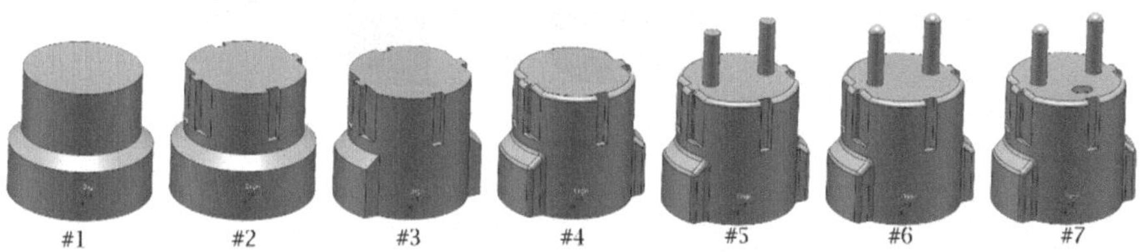

Fig. 4. Result associated with the steps described in Table 1.

orthogonality relationship between the revolution axis of the cylinder and a given plane heavily modifies the shape of the reconstructed surface, introducing significant differences w.r.t. the original shape (Fig. 5(b)).

Evidently, the reconstruction error w.r.t. the reference data (Fig. 5(f)) is made worse with the introduction of a geometric constraint. Although a constraint-free reconstruction could guarantee the minimum deviation error w.r.t. the scanned data, depending on the knowledge on the geometry that the reverse engineer has, this effect might be desired. Recalling the previous example, the orthogonality between the cylinder axis and the given plane could be known for sure a priori. This can happen also in real-case scenarios; the deviation from the reference data induced by the imposition of inferred or known constraints might be dimensionally minor, but nonetheless present. To reiterate, the progressive introduction of geometric relations guides last generated features away from the reference data.

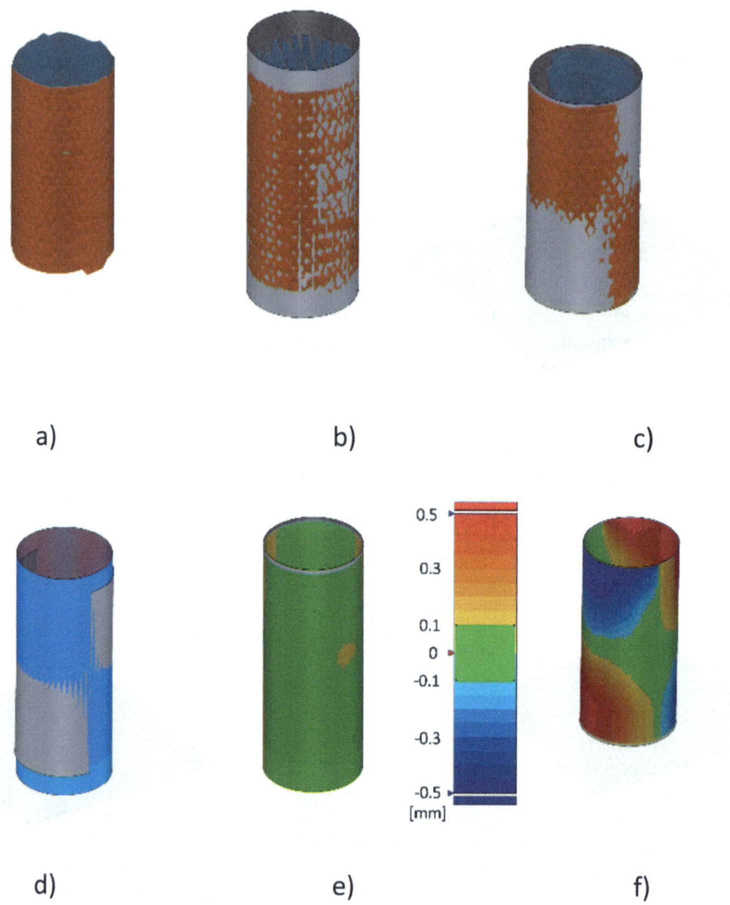

Fig. 5. Cylinder reconstruction performed in GDX; (a) original mesh; (b) cylinder reconstructed by minimizing the deviation error; (c) cylinder reconstructed as in (b) and imposing the orthogonality between the cylinder revolution axis and a given plane; (d) the two reconstructed cylinders: blue cylinder is (b) and grey is (c); (e) deviation map of cylinder (b); (f) deviation map of cylinder (c) (cylinder diameter: 25 mm, plane inclined by 1° w.r.t. the original extrusion)

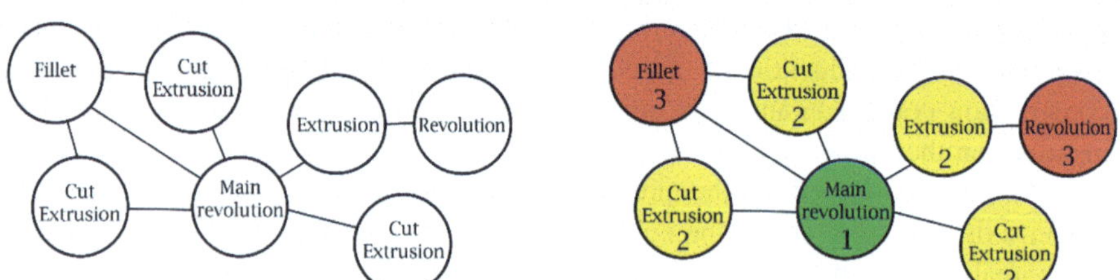

Fig. 6. ESA feature tree and Reverse Modeling tree

The same phenomenon can be observed in the ESA test case. Specifically, the deviation map in Fig. 7 shows an error that is not evenly distributed on the surface of the pins – i.e. a map of local errors – rather a macroscopic error gradient that develops on the vertical surface of the pins. The effect is accentuated by the fact that the features of Fig. 7 are the last created within the ESA. As a result, the errors observed in the deviation map are influenced by all the geometric constraints introduced throughout the entire reconstruction (not only by the constraints directly imposed on the geometry of the pins). To reiterate, all the reconstruction choices heavily influence the last reconstruction phases. It's usually at the end of the reconstruction, as a result, that the user becomes aware of possible mistakes made during the process.

The establishment of this error chain should be taken into consideration from the early stages of reconstruction, when drafting a reconstruction strategy and especially defining the order of operations. Once that the relations between the segmented and classified features are defined, a relation graph between features (e.g. Fig. 6) can be built. This type of graph, typical of many CAD applications (e.g. [6, 7]) is useful to plan the reconstruction, as the user should minimize the length of the relation chains. The first feature chosen to start the reconstruction, marked in green in Fig. 6, is reconstructed without any additional error ascribable to previous steps. Following steps are, on the other hand, affected by a chain of errors that depend on previous features. Accordingly, the features that depend directly on the first one (i.e. level 2, coloured in yellow in Fig. 6) are affected by the first level. The third level (coloured in red in Fig. 6) is even more affected by the chain of errors. It's important to note that the choice of the first feature determines the creation of the coloured graph of Fig. 6. In other words, a valid strategy is able to minimize the length of the chains and the effects caused by the error build-up.

In order to amend these observed defects of the procedure, some kind of global error evaluation is needed. At the moment, in fact, each CAD feature is generated and adapted to the scanned surface considering only the deviation data computed on the considered feature and the corresponding segmented region. Consequently, whenever geometric constraints are introduced in the scenario, the minimization of the deviation error for a feature could impose significant errors in other areas of the model. A possible reverse engineering approach, implementing this philosophy through the fitting of a CAD template to scanned data, is described in [8].

All the above described problems are accentuated by the fact that bad choices made during the process are usually not easily amendable in a RE software. Reconstruction mistakes become evident usually towards the end of the reconstruction process, modeling the last features (i.e. the most affected by the bad choices of the reverse engineer). Sadly, the typical RE modeling environment does not allow an efficient and convenient way of modifying the reconstructed geometry; this is one of the major differences w.r.t. traditional CAD software.

In CAD software, a set of parametric and associative functions allow for a fast and dynamic editing of the modelled part. The modeling tree defining the geometry of the part can be edited in every aspect (e.g. changing a dimension, enforcing a geometric relation, varying a parameter of a 3D modeling feature, etc.) and the geometry of the object is updated according to the user' input. As an example, the user would be able to change the diameter of a sphere freely, obtaining a sphere of different dimension.

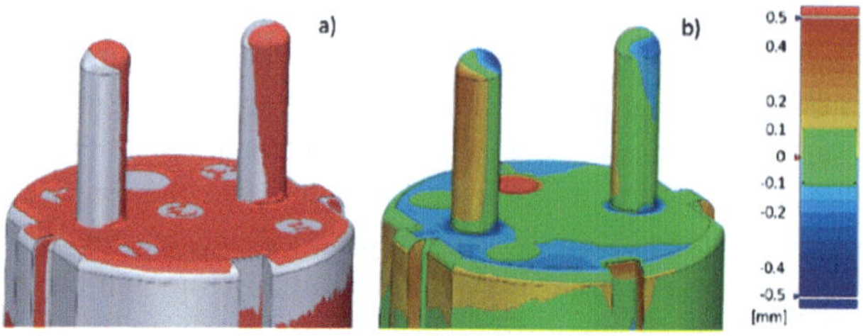

Fig. 7. Reconstruction errors on the two pins. (a) reconstructed solid in grey and reference mesh in red; (b) deviation map of the obtained result.

RE software systems, on the other hand, do not entirely guarantee this kind of parametric/associative relations, at least not w.r.t mesh-based operations. As previously discussed, RE tools generate a CAD surface by adapting it upon a segmented region of the reference data. To the best of the authors' knowledge, no RE software package guarantees an associativity relationship between CAD features and corresponding segmented mesh regions. As a result, modifications to the modeling history of the reconstructed CAD model do not involve a new evaluation of the mesh-derived feature parameters. Therefore, if users need to perform some changes to the modeling tree describing the shape of the object, they may obtain CAD features whose parameters (i.e. dimensions, position) are not correct with respect to the reference mesh. This concept is well described by the example in Fig. 8, where the reconstruction of a cylinder is depicted. In the example, the CAD feature is generated by means of an extrusion performed on a 2D sketch extracted from the original mesh. The 2D mesh-based sketch is obtained by cutting the mesh with a plane. Any change to the plane position/orientation is not reflected upon the sketch, resulting in a misinterpreted geometry within the following 3D modeling feature.

Therefore, it is evident that any change to the feature tree, especially when performed at the end of the reconstruction process, may result in a significant error. Depending on the modifications made to the feature tree, a CAD model that is completely unrelated to the scanned data might be generated. Whenever the errors are macroscopic and evident, the user can correct the geometry, even if not efficiently[1]. With this respect, the only applicable solution for the reverse engineer is represented by a complete remake of the whole procedure. In this case, the user can start the reconstruction again, learning from the mistakes previously made and trying to attain a better result.

[1] Whenever the generated effect is not as obvious, however, such errors can be neglected and may affect the final geometry generated by the user.

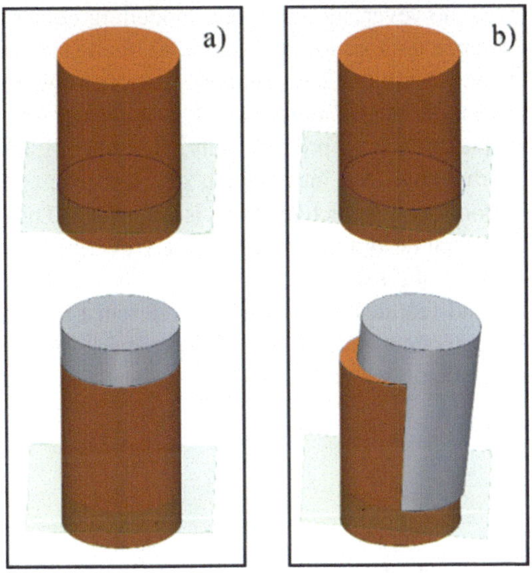

Fig. 8. Reconstruction of a cylinder: CAD cylinder (grey) obtained from a mesh-based 2D sketch that depends on a section plane (green). Position and orientation of the cutting plane is updated in (b). The 2D mesh sketch obtained (blue line) is not updated but only translated in (b).

4 Conclusions

Many negative aspects, described in the article and clearly highlighted in the discussed case study, can be identified within the typical reconstruction framework, even when it is hosted in a dedicated RE software package. Firstly, a competent user, highly skilled in the use of the software environment and capable of choosing the most appropriate "reverse modeling" strategy, is fundamental. The consolidation and honing of these skills can be achieved only through experience, at the cost of numerous hours of practice on RE software. Secondly, the overall process is time consuming and, being manually controlled by the user, requires their physical presence during the entire operation. Specifically, depending on the complexity of the model to be reconstructed and the tools available, several hours (occasionally days) can be required to obtain a valid result.

Finally, it is important to consider the effects caused by uncertainties in this specific framework. Obviously, the presence of reconstruction errors is physiological in every reconstruction process. Both the manufacturing and the 3D scanning phases introduce errors in the reference data. However, the human factor must be also taken into account: in the "traditional" framework, the user needs to make several choices throughout the procedure. Which CAD/RE tools to use, which reconstruction strategy to follow, the type of geometric features to use in the model, are all examples of variables controlled by the reverse engineer. In other words, the user needs to interpret the original design intent of the part and to try to use the tools available to produce a digital model that reflects his interpretation. Due to the temporal succession of the modelling steps (Fig. 2), however, all the wrong modelling choices, together with the noise affecting the reference data, cause the generation of an error that builds up during the reconstruction chain. Therefore, last features generated and modelled tend to be distanced from the corresponding reference data. Consequently, the correct retrieval of dimensions and other parameters is, progressively, significantly hindered. These negative effects are exacerbated by the introduction of geometric constraints during the reconstruction. The introduction of such relations is necessary to produce a CAD model that is faithful to the original design intent of the part. Nevertheless, the introduction of any additional constraint in the reconstruction process inevitably guides the generated mathematical surfaces and features away from the reference data.

Finally, the lack of associativity w.r.t. the reference mesh is often responsible for an inefficient reconstruction that hinders the generation of more accurate and high-quality reconstructed models.

References

1. Buonamici, F., Carfagni, M., Furferi, R., Governi, L., Lapini, A., Volpe, Y.: Reverse engineering modeling methods and tools: a survey. Comput. Aided Des. Appl. **15**, 443–464 (2018). https://doi.org/10.1080/16864360.2017.1397894
2. Fisher, R.B.: Applying knowledge to reverse engineering problems. In: Proceedings of the Geometric Modeling and Processing: Theory and Applications, GMP 2002, vol. 36, pp. 149–155 (2002). https://doi.org/10.1109/GMAP.2002.1027506
3. Agathos, A., Pratikakis, I., Perantonis, S., Sapidis, N., Azariadis, P.: 3D mesh segmentation methodologies for CAD applications. Comput. Aided Des. Appl. **4**, 827–841 (2007). https://doi.org/10.1080/16864360.2007.10738515
4. What are the different ROMER Scanner (RS1, RS2, RS3) specifications? https://support.hexagonmi.com/s/article/What-are-the-different-ROMER-Scanner-RS1-RS2-RS3-specifications-1527312292374
5. Di Angelo, L., Di Stefano, P.: Geometric segmentation of 3D scanned surfaces. Comput. Aided Des. **62**, 44–56 (2014). https://doi.org/10.1016/j.cad.2014.09.006
6. Di Angelo, L., Di Stefano, P., Morabito, A.E.: The *RGM* data structure: a nominal interpretation of an acquired high point density model for automatic tolerance inspection. Int. J. Prod. Res. **50**, 3416–3433 (2012). https://doi.org/10.1080/00207543.2011.579638
7. Mun, D., Kim, B.C.: Extended progressive simplification of feature-based CAD models. Int. J. Adv. Manuf. Technol. **93**, 915–932 (2017). https://doi.org/10.1007/s00170-017-0491-y
8. Buonamici, F., Carfagni, M., Furferi, R., Governi, L., Lapini, A., Volpe, Y.: Reverse engineering of mechanical parts: a template-based approach. J. Comput. Des. Eng. **5**, 145–159 (2018). https://doi.org/10.1016/j.jcde.2017.11.009

3D Scanning Procedure for the Evaluation of Lymphedema of Upper Limbs Using Low-Cost Technology: A Preliminary Study

Andrea Vitali[1(✉)], Daniele Regazzoni[1], Caterina Rizzi[1], and Guido Molinero[2]

[1] University of Bergamo, Bergamo, Italy
{andrea.vitali1,daniele.regazzoni, caterina.rizzi}@unibg.it
[2] Azienda Socio Sanitaria Territoriale (ASST) Papa Giovanni XXIII, Bergamo, Italy
gmolinero@asst-pg23.it

Abstract. This research work presents a preliminary study for the assessment of lymphedema using low-cost 3D scanning and modelling technology. The aim is to develop a methodology to measure the anatomical key features of the upper limbs lymphedema using a 3D scanning technology and an ad-hoc developed application, named Lym 3DLab. The application is able to automatically measure both perimeters of cross sections and volumes of arm segments for medical evaluation and design of compression stockings using the upper limb polygonal mesh. The scanning procedure is based on a Microsoft Kinect v1 as scanner and the low-cost Skanect tool, which creates the polygonal mesh of the 3D acquisition. Lym 3DLab has been developed using open-source Software Development Kits, such as Qt and Visualization Toolkit. The acquired volumes have been compared with the water displacement method, which is considered the gold standard for measuring volumes of limbs affected by lymphedema. A preliminary test has been performed to compare volumes measured using the developed procedure with the gold standard. Five volunteers have been involved who are not affected by lymphedema. The arm volume measured with water displacement have been compared with the volume computed using 3D model of arm in Lym 3DLab. The range of differences is between −6,75 cm^3 and 9,40 cm^3. Reached results are the base for planning further test with a large number of patients affected by lymphedema in collaboration with a hospital.

Keywords: Breast cancer · Lymphedema · 3D scanning · Microsoft Kinect · Upper limb modeling · Automatic measurement of arm volume

1 Introduction

Breast cancer is the most common malignancy that women can contract. To date, the survival rate at 5 years is over 90% [1]. Although oncologic research and prevention dramatically increased the probability of survival, some upper limb complications can arise after cancer therapy. In particular, removing the sentinel lymph node or all the

C. Rizzi et al. (Eds.): ADM 2019, LNME, pp. 177–188, 2020.
https://doi.org/10.1007/978-3-030-31154-4_16

axillary ones affects the lymph circulation in the limb. Therefore, lymph may accumulate in the tissues causing a visible swelling of the whole arm, or parts of it or of the breast. The lymph that stagnates originates the lymphatic edema, or lymphedema, which can appear gradually over a few days or even several years after surgery. It causes several ailments including pain along the arm, difficulty in performing simple movements such as adduction and abduction and an inclination for skin tissue infection. All these disorders lead also to a negative psychological impact and a worsening of patient's quality of life.

An early stage of lymphedema is assessed when the volume of the arm increases of 3–5% from baseline in swelling, which is determined by comparing the volume of the affected arm with the volume of the contralateral one [2]. The prevention and treatment of lymphedema is carried out in hospitals and rehabilitation centers through specific assessment methods based on the measure of the limb volume. These methods allow defining the severity of patient's condition and permit to gather the measures for standard or custom orthopedic stocking. Actually, elastic stocking is the main remedy for reducing the lymphedema by applying a constant pressure and, therefore, it must be done according to patient's shape and measures.

The aim of this research work is to introduce ICT technologies to support medical personnel to make the lymphedema assessment procedure more objective, reliable and repeatable adopting a low-cost philosophy. In particular, a low-cost 3D scanner is introduced for the acquisition of the limb and a SW application has been developed to automatically calculate, by means of the virtual model of the arm, the measures useful for the creation of the compression stocking and for the evaluation of the volume.

The following sections of the paper describe the state of the art related to the methods for the evaluation of lymphedema and to the use of 3D scanners for the acquisition of the virtual model of the limb. After that, the 3D acquisition procedure designed for this research work will be shown. The application developed is described by highlighting its software architecture based on open-source SDKs and its main functional characteristics. Finally, the first feedback from the involved medical personnel will be presented as well as the preliminary tests and the conclusions will be described.

2 Scientific Background

Current methods of measuring the volume of the limb affected by lymphedema have different levels of accuracy and different levels of usability by the medical personnel who perform them. The most precise method used is called Water Displacement (WD) [3, 4]. WD measures the upper limb volume by means of immersion into water. This method is the gold standard for the measurement of limb affected by lymphedema and is used in several scientific studies to verify the accuracy and precision of other methods for volume measurement. Anyway, WD, is often considered inconvenient for both physiotherapists and patients because it requires high and bulky vessels.

Circumference measurement (CM) is a way to indirectly measure the volume of a limb and is the most used in clinical practices [5]. It is assumed that each section of the arm, from the wrist to the shoulder, can be approximated to a circumference.

Different circumferences are measured with a tape measure at a predetermined distance (usually every 2–3 cm) and the volume of each conical section defined by each consecutive pair of circumferences along the limb is calculated. The sum of the volumes of the conic sections is considered the total volume. Although this method is inexpensive and simple for the physiotherapists, its accuracy is very low and the procedure lacks in inter-operator reproducibility.

There are some more advanced methods, which are the use of the perimeter (perometry) [6, 7], bioimpedance spectroscopy, [8] and Dual Energy X-ray Absorptiometry [9] that allow a precise measurement of the volume close to the WD gold standard. However, they are costly and, therefore, they are seldom adopted in clinical practices.

In recent years, 3D acquisition systems have been used in several different applications, such as for industrial and medical applications [10] adopting also virtual reality [11, 12] and additive manufacturing technology [13]. There are several categories of 3D scanners that are usually classified according to the level of accuracy and precision with which they are able to acquire an object [14, 15]. The level of accuracy and precision also defines the cost of the 3D scanner. For example, there are high-level 3D scanners with high accuracy like Artec Eva, Rodin 4D and Konica Minolta. Some other scanners have lower accuracy but are cheaper. Low-cost 3D scanners can be used easily and are based on sensors that can be connected to generic laptops, smartphones and tablets, such as Microsoft Kinect, Intel Real Sense or Structure sensor. Some research works have been done to measure the volume of limb affected by lymphedema through different 3D acquisition techniques. In particular, Lu et al. [16] have compared the use of low-cost scanners like the Microsoft Kinect v1 with the WD gold standard by defining an acquisition procedure that can be done by physiotherapists.

Hoevenaren et al. [17] have used stereo photogrammetry techniques to measure the difference in volume between the healthy hand and the hand affected by lymphedema. Armer et al. [3] have developed a web platform that allows the patient to acquire the 3D model of the arm with a smartphone in order to allow self-monitoring of the change in the volume of the limb affected by lymphedema.

Other works have defined specific 3D acquisition procedures analyzing accuracy and precision of the results obtained compared to the gold standard. Ohberg et al. [18] built a 3D acquisition tool based on the use of 3D Microsoft Kinect v1. The arm must be positioned horizontally in the acquisition volume common to the 3 devices. The 3D model generation is done using MatLab and it is simplified by using a set of cross section circumferences every 1 cm from the shoulder to the wrist and then calculating the volume of each conic section. This method was compared with both WD and CM [18]. The sample of volumes calculated shows an overestimation compared to the WD gold standard characterized by a standard error of 0.04 dm^3.

It is clear that no standalone application has been developed in order to automate the measurement of volumes of the entire limb or parts of it.

Cau et al. [19] used the Rodin4D scanner as an acquisition system. Rodin4D is a top-level scanner with high accuracy (up to 0.75 mm) which is used in orthopedics. The 3D acquisition is done after measuring the volume with the CM method. During the CM procedure, the measured circumferences were drawn down on the skin with a marker so that they were acquired during the 3D scan. Also in this case, the cross sections were simplified to circumferences and the sum of the volumes of the conic sections is used as the final volume of the limb. The volumes measured by both methods were compared and a difference of 0.05 dm^3 was calculated. The 3D acquisition procedure used can be difficult to replicate by doctors and physiotherapists who do not have advanced ICT skills.

Verhulst et al. [20] analyzed accuracy and reproducibility of limb measurement using 3D photogrammetry techniques. The 3dMD Cranial industrial system was used for the study. The system is based on 15 high definition cameras and has a very high cost. Furthermore, a tool was created in order to apply markers near the wrist in order to reconstruct the limb including the hand. The reconstruction of the 3D model was done with Autodesk 3D Max 2015. The study was done involving more operators and assessing intra-operator reproducibility. The final results show difficulties in reproducing the acquisition procedure mainly due to the calibration phase. Also in this case, no user-friendly application has been developed in order to make available a tool suitable by medical personnel for making the most appropriate measurements.

This research work aims at defining a procedure for the evaluation of limbs affected by lymphedema based on the use of technologies for 3D scanning and modeling. A specific 3D procedure was designed in order to provide a tool usable by physicians and physiotherapists. In particular, our aim is the definition of a 3D scanning procedure that is replicable in medical center and not time-consuming for medical personnel. Furthermore, there are no software applications in scientific literature that are specifically developed for extrapolating measurements useful for the medical assessment of lymphedema and the design of containing stocking. Therefore, we present a software application that is able to extrapolate useful measurements in automatic or semi-automatic way by starting from the 3D scanned model of the limb. In particular, the measurements of cross sections are not approximated to circumferences but, the application automatically compute the perimeter of each section. The volumes between consecutive cross sections are not simplified as conic sections but, specific algorithms have been exploited for calculating volumes of closed polygonal meshes. The acquisition procedure is based on the use of a low-cost solution; a Microsoft Kinect v1 is used as a 3D scanner and the Skanect application as a software for the automatic generation of the 3D limb model. The application has been developed using open source SDKs like VTK and Qt.

3 Method and Tools

The designed workflow is composed by two main steps: a procedure for 3D scanning of the upper limb and the use of an ad-hoc developed software application. The application uses the virtual arm obtained with the 3D scanning to automatically measure anatomical key features useful for both medical assessment and the design of the orthopedic compression stocking. In this section, the technological aspects of 3D scanning procedure are presented as well as the main features of the developed application.

3.1 3D Scanning Procedure

The 3D scanning procedure aims at obtaining the virtual upper limb through a clear list of steps that medical personnel can perform in an easy way. The procedure has also been thought to avoid fatigue and discomfort to the patients. The 3D scanning system is composed by a Microsoft Kinect v1. The acquired data (i.e., RGB images and depth data) are sent to a consumer laptop through a 10 meters long USB 2.0 cable. The acquired data are managed using the low-cost application Skanect. Skanect allows 3D scanning acquisition using a sequence of virtual tools, which simplify the generation of the final 3D model.

The developed procedure mainly exploits 3 virtual tools of Skanect. The first allows recording a real object using the Microsoft Kinect v1 device. After pushing the rec button, the user moves the Kinect devices around an object and Skanect automatically acquires the point clouds useful for creating the polygonal mesh. The second tool allows to clean the initial 3D acquisition by details which have been accidently acquired. The last tool automatically generates the polygonal mesh that can be exported in several standard file formats, such as STL, OBJ and PLY.

The patient has to stay upright during the whole acquisition with the arm in horizontal position. Therefore, a tripod is used to lay the hand until the acquisition is terminated. Besides preventing fatigue, this permits to reduce small movements of the arm, which could decrease acquisition quality. During the 3D acquisition of the arm, it is also important to acquire other parts of the upper human body that can be useful for lymphedema evaluation, i.e., the hand and the shoulder of the affected arm (Fig. 1(a)).

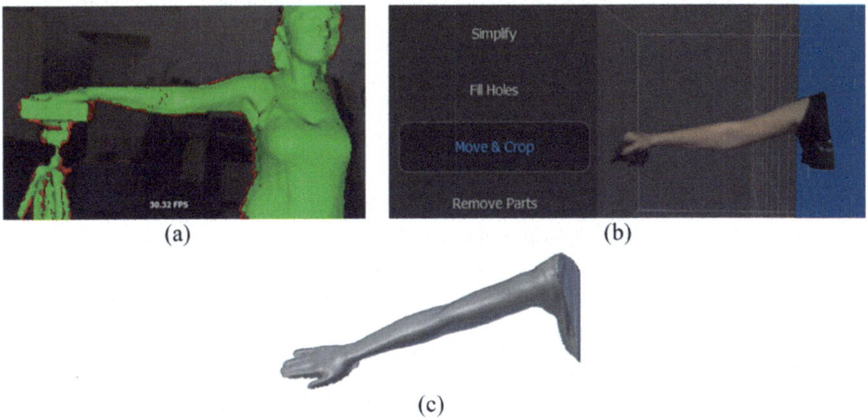

Fig. 1. Virtual tools of Skanect for recording (a) and modelling (b) the acquisition in order to obtain the final 3D model of the upper arm (c).

When the 3D acquisition is completed, Skanect automatically generates the polygonal mesh of the acquired points cloud by which the tripod and useless part of human body have to be removed by using the second tool of Skanect (Fig. 1(b)). Finally, the last tool allows exporting the polygonal mesh of the arm in both STL and

OBJ file format. The exported 3D model is ready to be used in the developed application for the medical assessment (Fig. 1(c)).

3.2 Developed Application: Lym 3DLab

The developed application, named Lym 3DLab (Lymphedema 3D Laboratory) allows easily measuring cross sections along the 3D model of the upper limb according to specific anatomical points, which are shown in Table 1. The evaluated measurements are necessary since they are an important part of the standard method traditionally used by physiotherapists to check the changes of the arm volume during the patient's rehabilitation. Furthermore, these measurements are also required by orthopedic technicians to create the orthopedic stockings.

Table 1. Anatomical key points where the cross sections have to be placed.

	Anatomical key points
1.	Level of armpit
2.	15 cm over the olecranon
3.	Elbow
4.	Midpoint between olecranon and the ulnar styloid
5.	Level of ulnar styloid
6.	Level of metacarpophalangeal joints with 4 finger
7.	Level of metacarpophalangeal joints with 5 finger

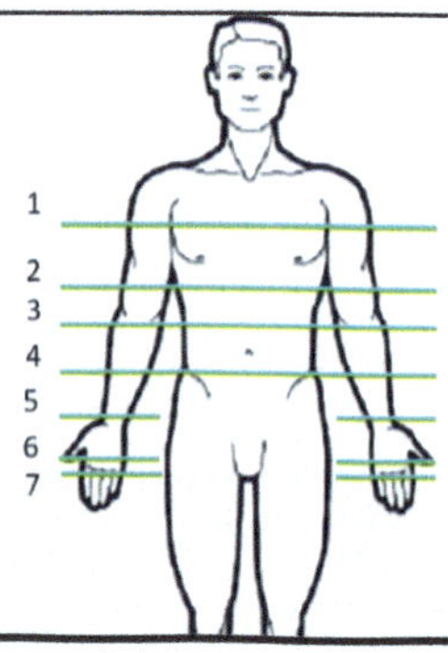

Lym 3DLab automatically computes the volume of the whole upper limb and the volume of each part of the 3D model between two consecutive cross sections. Figure 2 depicts the software architecture of Lym 3DLab based on two open-source software development kits (SDKs), which are Visualization ToolKit (VTK) and Qt. VTK is a software library that makes available many methods and functions for easily develop 3D modelling applications for several scientific research fields, such as medical and industrial sectors. Qt is a powerful SDK for developing innovative user interface. Both SDKs allows software development with several programming languages, such as C++ and Python. Lym 3DLab has been totally developed in C++ in order to guarantee high computing performances and an excellent 3D rendering.

Figure 3 presents the user interface of the developed software tool. First, the user has to fill in a digital form about patient's data. In particular, the user specifies if the ill arm is the left or the right arm. Then, the 3D model of the upper limb can be imported in both STL and OBJ file format. When the 3D model has been loaded Lym 3DLab automatically generates 7 cross sections and the relative planes evenly spaced along the arm by starting from the fingertips. To perform this task in an automatic way, the arm is

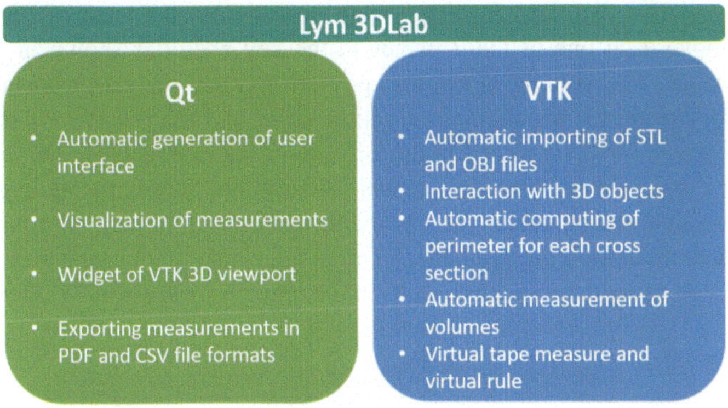

Fig. 2. Software architecture designed for developing Lym 3DLab. Several features of both QT and VTK have been exploited for creating the final application.

horizontally placed during the acquisition allowing in such a way Skanect to export the 3D model with the arm aligned along the x axis of the 3D environment. A standard algorithm for bounding box computation has been exploited to automatically find minimum and maximum x-y-z coordinates of the 3D model. If the acquisition concerns the horizontal right arm, the 3D point with minimum x coordinate belongs to the finger tip of the longest finger. If the 3D acquisition is relative to a left arm, the 3D point with maximum x coordinate belongs to the longest finger. The patient's data initially entered allow Lym 3DLab to automatically identify if the 3D model is a left or right arm. The 3D coordinates of the bounding box are also used to initially position the 7 cross sections equally distributed along the 3D model of the horizontal arm.

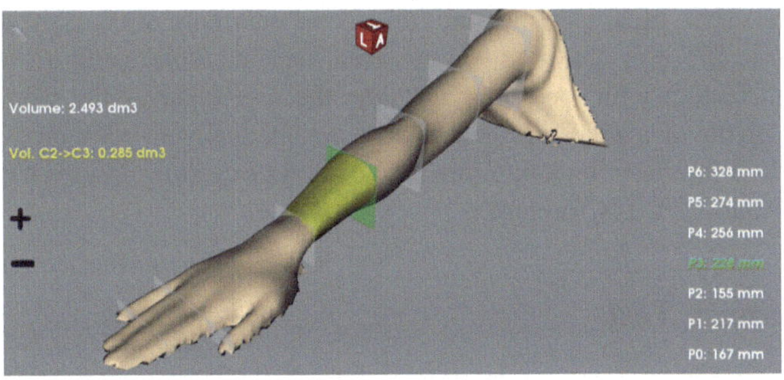

Fig. 3. User interface of Lym 3DLab.

The volume of the arm between the fingertips and the last cross section is automatically computed and shown in dm^3 in the left side of user interface. The perimeter of all cross sections is shown in millimeters on the right side of interface. The user is able to select a cross-section plane and move it along the arm and its perimeter is automatically updated. When the user selects a cross section plane, its color and the color of the relative perimeter become green. In this way, the medical personnel can select and move each one of the 7 planes to the key positions shown in Table 1 by using the mouse. The volume of the arm portion between the selected plane and the previous cross section is highlighted in yellow color and the relative volume is automatically computed and visualized in the left side of the interface. On the left side, the user interface has two buttons (i.e., a plus and a minus icons) for adding/removing further planes that can be useful for specific measurements in particular cases.

Two other virtual tools are available: a virtual ruler and a virtual tape measure, which can be activated using the respective icons in the upper left corner of 3D scene (Fig. 4).

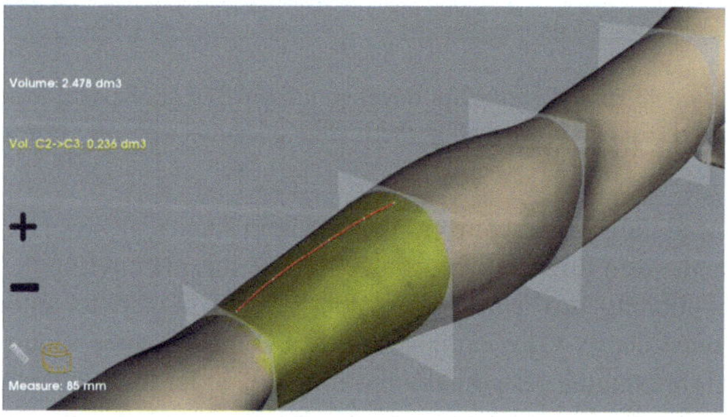

Fig. 4. The use of the red tape measure for measuring the distance along the surface of the upper limb between two cross sections.

The virtual ruler allows measuring the Euclidean distance between two points. The virtual tape measure is a 3D widget that measures a path along the polygonal surface of the 3D model. The virtual tape measure follows the curvature of the polygonal surface on which lies. The user can modify the virtual tape measure using a set of control points by interacting with mouse. Every time the user changes the path by control points, the length of virtual tape measure is automatically re-computed and shown through the user interface. Finally, Lym 3DLab allows exporting defined measurements in both PDF and CSV file formats.

4 Preliminary Test

The planned test aims to compare the volume of upper limb obtained with the standard gold WD and the volume measured with the presented procedure. Five engineering students, 4 males and 1 female, have been involved as volunteers. None of the testers is affected by lymphedema. Their average age is 26 years old.

Initially, each volunteer has to perform the water displacement test by immersing his/her upper left arm in a plastic box with partially filled with water. The height of the water is measured before and after the immersion of the arm.

The volume of arm is calculated as the difference between the volume with immersed arm and the initial volume of water. Before removing the arm, a line is marked on the skin of the upper limb with a pen just over the surface of the water. This is a reference line for placing the cross-section in Lym 3DLab to evaluate the same part of the virtual volume. The second step is relative to the 3D scanning acquisition of the arm.

The 3D acquisition is performed by another engineering student who previously learned the 3D scanning procedure based on Skanect and Microsoft Kinect v1 device. Each 3D acquisition of the arm has been approximately performed in 2 min. For each acquisition, the 3D model has been manipulated and repaired (e.g., an algorithm to fill holes is applied using Skanect) to prepare the initial polygonal mesh to be used in Lym 3DLab.

Afterward, the 3D model of the arm is loaded in Lym 3DLab, all the cross sections are positioned according to the anatomical key points and, finally, all measurements are automatically computed. Table 2 shows the volumes obtained with WD and those automatically computed by means of Lym 3DLab for all testers. The second column V_{WD} is the volume measured when the arm is immersed. V_{arm} is the arm volume measured as the difference between V_{WD} and the initial water volume. $V_{Lym3DLab}$ is the volume calculated using the 3D model of left arm using Lym 3DLab.

Table 2. Volumes obtained for each tester with WD and Lym 3DLab.

Tester	V_{WD} [dm^3]	V_{arm} [dm^3]	$V_{Lym3DLab}$ [dm^3]	Diff. [cm^3]
1	40,362	2,497	2,506	9,40
2	40,778	2,913	2,910	−2,70
3	40,986	3,121	3,115	−5,75
4	40,986	3,121	3,114	−6,75
5	40,362	2,497	2,493	−3,60

The differences between V_{arm} and $V_{Lymb3DLab}$ are reported in the last column. The range of differences is between −6,75 cm^3 and 9,40 cm^3. The differences are negligible for the sake of the medical assessment. These differences can be caused by a non-accurate 3D shape of the hand. When the scanned part of tripod is deleted for exporting the final 3D model, a big hole remains on the bottom part of hand and a filling hole algorithm is used in Skanect for creating a flat polygonal surface to close the final mesh and thus, the correct shape of palm is totally lost.

This problem could be avoided by leaning against the upper part of tripod only the tips of fingers. In this way, the surface of contact is minimized and only a small part of the 3D hand will be removed during the preparation of the final model of arm used in Lym 3DLab. Anyway, the preliminary test shows how the designed procedure is not time-consuming and easy for operators without specific technological skills. Lym 3DLab allows measurements of volumes in automatic way, which have a small difference compared with the gold standard WD and thus, this low-cost solution could be a very good alternative tool to be used in the medical centers for the assessment of lymphedema. Furthermore, the perimeters of cross sections can be easily exploited also for the design of orthopedic compression stockings instead of the approximated circumferences measured with CM method.

This preliminary research work has been evaluated also by physicians and physiotherapist, who considered Lym 3DLab a promising prototype that could be adopted as a powerful tool for the assessment of lymphedema. By starting from achieved results and medical feedback, further tests have been planned in collaboration with medical personnel for creating an observational research study with a large number of patients affected by lymphedema to improve and validate the presented preliminary research work.

5 Conclusion

The paper presents a low-cost procedure for measuring patient's upper limb volume, which is required for the assessment of lymphedema. The aim is the introduction of a 3D scanning technology to create the model of an upper limb. The model is processed in the ad-hoc developed application Lym 3DLab for automatically measuring key features useful for the evaluation of lymphedema. Moreover, gathered measures can be used to design personalized orthopedic compression stockings. Lym 3DLab is developed to be an alternative solution to both the gold standard WD and CM method. The developed procedure allows 3D scanning acquisition by using a Microsoft Kinect v1 and the low-cost commercial application Skanect. The 3D scanning procedure is highly usable and can be accomplished in no more than 2–3 min. Lym 3DLab is a software application developed by using open-source SDKs as VTK and Qt. The main features are the automatic measurements of both cross-sections perimeters and volumes of the arm, according to anatomical features for assessing the lymphedema.

The developed platform has been tested by involving 5 volunteers who did WD and 3D scanning acquisition in order to validate the measurement of arm volume. The preliminary results show a very good accuracy of the computed volume compared with volume measured with the standard gold WD. A further technological test should be done to compare perimeters of cross sections with CM method in collaboration with physiotherapists.

A future test has been planned in collaboration with a medical center to test the solution on a large sample of patients affected by lymphedema in order to validate the preliminary results presented in this research work.

References

1. Siegel, R.L., Miller, K.D., Jemal, A.: Cancer statistics, 2016. CA Cancer J. Clin. **66**(1), 7–30 (2016)
2. Executive Committee: The diagnosis and treatment of peripheral lymphedema: 2016 consensus document of the International Society of Lymphology. Lymphology **49**(4), 170–184 (2016)

3. Armer, J.M., Armer, N.C., DeSouza, G.N.: Mobile platform for assessment, early detection, and management of breast cancer-related lymphedema (2017)
4. Wanchai, A., Armer, J.M., Stewart, B.R., Lasinski, B.B.: Breast cancer-related lymphedema: a literature review for clinical practice. Int. J. Nurs. Sci. **3**(2), 202–207 (2016)
5. Cau, N., Galli, M., Cimolin, V., Grossi, A., Battarin, I., Puleo, G., Balzarini, A., Caraceni, A.: Quantitative comparison between the laser scanner three-dimensional method and the circumferential method for evaluation of arm volume in patients with lymphedema. J. Vasc. Surg. Venous Lymphat. Disord. **6**(1), 96–103 (2018)
6. Sharkey, A.R., King, S.W., Kuo, R.Y., Bickerton, S.B., Ramsden, A.J., Furniss, D.: Measuring limb volume: accuracy and reliability of tape measurement versus perometer measurement. Lymphat. Res. Biol. **16**(2), 182–186 (2018)
7. Sun, F., Hall, A., Tighe, M.P., Brunelle, C.L., Sayegh, H.E., Gillespie, T.C., Daniell, K.M., Taghian, A.G.: Perometry versus simulated circumferential tape measurement for the detection of breast cancer-related lymphedema. Breast Cancer Res. Treat. **172**(1), 83–91 (2018)
8. Xie, A.Z.T., Lee, M.J., Dylke, E.: Diagnosing male unilateral upper limb lymphedema: Determination of normatively determined thresholds using bioimpedance spectroscopy. Lymphat. Res. Biol. (2019)
9. Gjorup, C.A., Hendel, H.W., Klausen, T.W., Zerahn, B., Holmic, L.R.: Reference values for assessment of unilateral limb lymphedema with dual-energy x-ray absorptiometry. Lymphat. Res. Biol. **16**(1), 75–84 (2018)
10. Colombo, G., Regazzoni, D., Rizzi, C.: Markerless motion capture integrated with human modeling for virtual ergonomics. In: International Conference on Digital Human Modeling and Applications in Health, Safety, Ergonomics and Risk Management, pp. 314–323. Springer, Heidelberg (2013)
11. Colombo, G., Rizzi, C., Facoetti, G., Vitali, A.: A preliminary study of new interaction devices to enhance virtual socket design. In: ASME 2014 International Design Engineering Technical Conferences and Computers and Information in Engineering Conference. American Society of Mechanical Engineers (2014)
12. Rossoni, M., Bergonzi, L., Colombo, G.: Integration of virtual reality in a knowledge-based engineering system for preliminary configuration and quotation of assembly lines. Comput. Aided Des. Appl. **16**(2), 329–344 (2018)
13. Comotti, C., Regazzoni, D., Rizzi, C., Vitali, A.: Multi-material design and 3d printing method of lower limb prosthetic sockets. In: Proceedings of the 3rd 2015 Workshop on ICTs for improving Patients Rehabilitation Research Techniques, pp. 42–45. ACM (2015)
14. Carfagni, M., Furferi, R., Governi, L., Servi, M., Uccheddu, F., Volpe, Y.: On the performance of the intel sr300 depth camera: metrological and critical characterization. IEEE Sens. J. **17**(14), 4508–4519 (2017)
15. Redaelli, D., Barsanti, S.G., Fraschini, P., Biffi, E., Colombo, G.: Low-cost 3d devices and laser scanners comparison for the application in orthopedic centres. Int. Arch. Photogramm. Remote Sens. Spatial Inf. Sci. **42**(2), 953–960 (2018)
16. Lu, G., DeSouza, G.N., Armer, J., Shyu, C.R.: Comparing limb-volume measurement techniques: 3d models from an infrared depth sensor versus water displacement. In: 2013 IEEE 15th International Conference on e-Health Networking, Applications and Services (Healthcom 2013), pp. 685–691. IEEE (2013)
17. Hoevenaren, I.A., Verhulst, A., Hameeteman, M., Vreeken, R., Maal, T.J., Ulrich, D.: Three-dimensional stereophotogrammetry as an accurate tool for analyzing lymphedema of the hand. JPRAS Open **10**, 40–46 (2016)

18. Ohberg, F., Zachrisson, A., Holmner-Rocklöv, Å.: Three-dimensional camera system for measuring arm volume in women with lymphedema following breast cancer treatment. Lymphat. Res. Biol. **12**(4), 267–274 (2014)
19. Cau, N., Galli, M., Cimolin, V., Aranci, M., Caraceni, A., Balzarini, A.: Comparative study between circumferential method and laser scanner 3d method for the evaluation of arm volume in healthy subjects. J. Vasc. Surg. Venous Lymphat. Disord. **4**(1), 64–72 (2016)
20. Verhulst, A., Wesselius, T., Glas, H., Vreeken, R., Ulrich, D., Maal, T.: Accuracy and reproducibility of a newly developed tool for volume measurements of the arm using 3d stereophotogrammetry. J. Plast. Reconstr. Aesthetic Surg. **70**(12), 1753–1759 (2017)

Low Cost Device to Perform 3D Acquisitions Based on ChAruCo Markers

Luca Puggelli(✉), Rocco Furferi, and Lapo Governi

Department of Industrial Engineering, University of Florence, Florence, Italy
{luca.puggelli,rocco.furferi,lapo.governi}@unifi.it

Abstract. In the field of cultural heritage, operators make use of high resolution orthophotos of paintings both for purposes related to restoration and monitoring of art pieces and for realizing online documentations and exhibitions. Unfortunately, artworks to be restored and/or presented in digital museums are painted on canvas which are far to be perfectly planar. Therefore, technical documentation accompanying an artwork to be stored in digital archives or museums can be enriched by information related to the 3D shape of the canvas. In this paper, both the design of a portable low-cost device that allows the acquisition of the 3D geometry of the painting and a procedure to triangulate 3D data are proposed. Such a procedure, working using the principle of laser-camera triangulation, is based on the use of a set of fiducial markers to set and continuously control the reciprocal orientation of the laser source and of the camera.

Keywords: Reverse engineering · Cultural heritage · Marker detection · Pose estimation · 3D laser scanner

1 Introduction

High resolution pictures of paintings are widely used in cultural heritage field for several purposes, addressed both to the enjoyment of the general public and to professional use [1, 2].

While in the first case common undistorted digital photos are considered more than enough, professional uses are more demanding both in terms of resolution and in terms of dimensional accuracy. Especially in case of painting restoration and monitoring purposes, these two aspects assume a fundamental role. To obtain an image that is at the same time measurable (i.e. a measure taken in the image is equal to the corresponding measure value on the painting) and with a gigapixels resolution, the common workflow is based on the realisation of a master orthophoto, which is then used as primary reference for the successive stitching of multiple pictures. As the tiles of a mosaic, these pictures are undistorted and opportunely placed and stretched in order to fit exactly on the master orthophoto. The resulting image is an extremely defined picture, on which is possible to zoom in the finest detail of the art piece, that match exactly the real painting dimensions. Thus, the accuracy of the reference orthophoto is crucial for the success of the whole procedure. The orthophoto is obtained as a transformation of a undistort digital image by means of metrical references (points or planes), usually obtained by

C. Rizzi et al. (Eds.): ADM 2019, LNME, pp. 189–200, 2020.
https://doi.org/10.1007/978-3-030-31154-4_17

means of perspective-based operation. In particular, the cross-ratio method is one of the most adopted methods to evaluate metrics information from pictures [3, 4]. Unfortunately, artworks to be restored and/or presented in digital museums are painted on canvas which are far to be perfectly planar. Therefore, technical documentation accompanying an artwork to be stored in digital archives or museums can be enriched by information related to the 3D shape of the canvas. In other words, conservators, museum curators and practitioners, considers of relevant importance the availability of 3D data related to an artwork [5–7].

Common methods [8, 9] consists on projecting several convergent lines to the painting canvas; considering at least 3 reference points on each line, of which 3D coordinates with respect to a defined reference system are known, is possible to calculate the 3D coordinates of the points given by the intersection of the lines and the canvas. On a practical level, this can be obtained by projecting a laser line tangent to a reference table, on which lines and points are drawn. On a digital image, displaying both painting and table, is then possible to perform the operations described below. In order to have a univocal matching between the image to be transformed and the images on which metrical reference are evaluated, camera position must be unchanged during all the operations.

This procedure is very simple and for this reason is often carried out by hand. Unfortunately, this leads to three main drawbacks: (1) the manual selection of the reference points is subject to errors; (2) it is time consuming and requires the constant presence of an expert user; (3) it is limited to the retrieval of a selection of few points, due to timings.

With the aim of overcoming the above-mentioned drawbacks, the present paper proposes the design of a portable low-cost device to reliably acquire the 3D geometry of a canvas. Moreover, an appositely devised semi-automated procedure to obtain an accurate 3D acquisition is described. In fact, the main idea is to realize a cheap system, able to perform a 3D triangulation when coupled with the pre-calibrated user Single-lens reflex (SLR) camera, allowing the 3D measurement of thousands of points on the painting with high accuracy. At the same time, the camera used for triangulating the 3D scene together with the laser can be used to acquire 2D data for creating the orthophoto.

The remainder of paper is structured as follow: in Sect. 2 procedure and device are briefly introduced, in Sect. 3 procedure and its implementation are further discussed, in Sect. 4 physical prototype is described, in Sect. 5 tests and results are discussed.

2 Overall Description

As briefly mentioned, the proposed device aims at retrieving the 3D position of points on the surface of a painting by means of triangulation-based laser camera. Accordingly, the reciprocal positioning and orientation of laser and camera devices (i.e. the extrinsic calibration) needs to be maintained during the acquisition process. Consequently, laser position and orientation are required to be monitored constantly to avoid undesired reciprocal movements [10].

With the aim of keeping implementation costs low, thus avoiding the implementation of devices such as, for instance, encoders, this information is obtained by means of computer vision algorithms.

In particular, as shown in Fig. 1, a fiducial marker is placed tangential to the laser blade and framed by the camera. Since the marker is attached to be visible by the camera for different laser positions, by exploiting algorithms typical of computer vision and augmented reality, is it possible to retrieve the 3D positioning of the maker w.r.t. the camera system. This allow recovering the orientation of the laser beam, especially in case the fiducial marker is placed on a planar board, tangent to the laser itself.

Fig. 1. ChArUco fiducial marker.

By opportunely tuning the camera settings and lighting conditions, both laser projection on the painting canvas and marker are acquired. In this way, by using a single picture, is it possible to triangulate laser orientation and laser projection 3D coordinates.

Gradually changing laser orientation and repeating the same operation at each step, is it possible to obtain a 3D point cloud of points of the painting surface which are scanned by the laser beam. Such a point cloud will be more dense by reducing the displacement between consecutive laser positions.

From a procedural point of view, three main aspects have to be encompassed:

1. Detection and pose estimation of the fiducial marker
2. Laser line detection
3. Triangulation algorithm

For what concerns the designed acquisition device, except from an SLR camera, its lens and its tripod, only a few additional tools are needed:

1. Laser line generator;
2. Support for laser line generator;
3. Planar table with fiducial marker;
4. Tripod + panoramic head.

While laser generator, tripod and panoramic head are commercial products, the remaining ones have been appositely designed to meet the systems requirements in terms of portability (small dimensions and lightness) but offering at the same time sufficient stiffness and adequate performance in order to ensure a dimensional accuracy of 0.2 mm on the measurements.

Details on the design of the components and specifications of the commercial products will be discussed on Sect. 3.

3 Procedure Description

3.1 Marker Detection and Pose Estimation

In computer vision, marker detection is a widely discussed topic, which find application in several fields, from industry to gaming [11–13]. There are several different types of fiducial marker, each one with its own implemented function library, which allows various operation. Usually both markers and libraries are proprietary and are available under payment. In the current application, it has been used ChArUco fiducial marker [14], with its OpenCV implemented function libraries, which are open access and are usable in MATLAB environment (see Fig. 2).

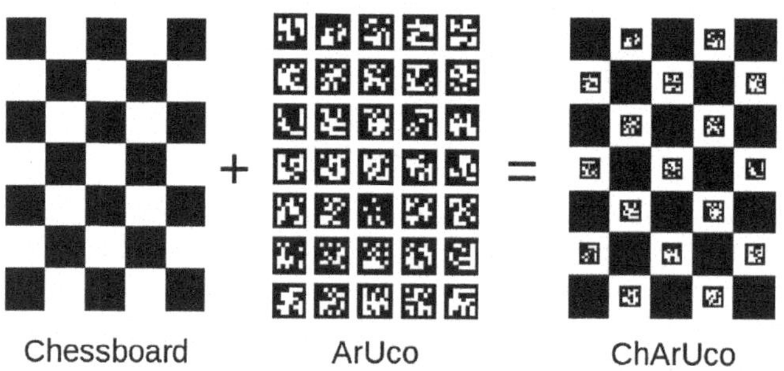

Fig. 2. ChArUco fiducial marker.

In detail, a ChArUco marker is the result of the combination of a black and white chessboard, with ArUco fiducial markers displaced inside its white boxes. ArUco markers are synthetic square markers composed by a black frame and an inner binary matrix which determines its id number, developed by Rafael Muñoz and Sergio Garrido [15]. The copresence of chessboard and ArUco markers allows to benefit of the accuracy in corner detection of the first and the extremely fast detection of the second.

Among the various functions included in the library, the following have been used:

1. Generate marker – to draw the desired ChArUco marker, by defining size of the chessboard grid and the ArUco markers shape;

2. Calibrate camera – to obtain camera calibration parameters by exploiting ChArUco marker;
3. Detect maker – to detect the ChArUco marker landmarks in the picture;
4. Estimate pose – to retrieve the 3D position and orientation of the ChArUco marker wrt camera reference system.

The general procedure, implemented in MATLAB, is completely based on these functions. A graphic user interface has been designed to guide the user during all the procedure. The accuracy of the procedure has been tested under many aspects. Different cameras and lens (see Fig. 3) have been used, as shown in Table 1.

Fig. 3. Adopted cameras: from left to right Canon EOS-M, Fujifilm X-T1, IDS uEye UI 1480 SE.

Table 1. Cameras and lens specifications.

	Canon EOS-M	Fujifilm X-T1	IDS uEye UI 1480 SE
Sensor	CMOS APS-C	CMOS APS-C	CMOS ½″
M. pixel	18.0	16.3	5.0
Lens	Canon EF-M 22 mm f/2 STM	Fujifilm XF 18-55 f/2.8-4R LM OIS	8 mm
Focal length [mm]	22 mm	18 mm	8 mm
Equivalent full frame focal length [mm]	35.2 mm	27 mm	48 mm

Each camera has been properly calibrated before each test session. Calibration pictures has been acquired following the procedure suggested in [16], see Fig. 4.

At least 32 pictures have been used for each performed calibration.

Fig. 4. Some photos of a calibration photo set (IDS uEye UI 1480 SE).

In order to evaluate the accuracy of the pose estimation phase, the displacement among different marker positions has been calculated by means of the implemented procedure and measured by means of a touch probe (3D *Romer Absolute Arm 7520-SI*).

Tests have been carried out placing the marker on a calibrated granite plate with a certified planarity error lower than 0.012 mm (grade 0), and on a plywood table with a planarity error of 0.6 mm (support characteristics in the following table) (Table 2).

Table 2. Supports for fiducial marker, used during tests.

	Calibrated plate	Plywood table
Size [mm]	300 × 200 mm	300 × 200
Planarity [mm]	<0.012 mm (certified)	0.6 mm (measured)
Material	Granite	Pine plywood

Since no noticeable difference has been detected by changing camera, only the first stages of the test campaign have been carried out using all the cameras. The following and main tests have been carried out adopting the Fujifilm X-T1 camera. In Tables 3 and 4, some of the obtained results are briefly reported.

Table 3. Test results, calibrated granite plate (Fujifilm X-T1).

Detected angle	Ground truth (Romer)	Error (degrees)	Error (mrad)
14.713	14.7045	0.0085	0.1484
12.104	12.0235	0.0805	1.4050
15.499	15.4951	0.0039	0.0681
14.289	14.1766	0.1124	1.9618
16.719	16.7047	0.0143	0.2489

Table 4. Test results, plywood table (Fujifilm X-T1).

Detected angle	Ground truth (Romer)	Error (degrees)	Error (mrad)
14.803	14.7641	0.0389	0.6789
13.978	13.8874	0.0906	1.5812
13.773	13.7495	0.0235	0.4101
18.033	17.9738	0.0592	1.0332
10.528	10.4573	0.0707	1.2339

Tables show only about 10% of all the obtained measurements. However, the results are significant. Two main observation can be made about them: 1. The overall accuracy of measurements can be considered enough and 2. There are no evidence of advantages in using particularly planar support. In fact, looking at Tables 3 and 4, the errors produced by using a calibrated plate and a plywood panel are approximately in the same range.

As shown in Tables 3 and 4, the results of these preliminary tests show a satisfactorily robustness of the procedure, which is able to obtain accurate measurements with a maximum error in the estimation of the angular position of approximately 1.5 mrad (0.08°).

Tests demonstrated also a slightly influence of support planarity. Thus, no strict requirement on the flatness of the support is required.

On the other hand, the procedure revealed (as predictable) sensible to errors in calibration parameters. Therefore, calibration phase must be carried out carefully and at least 32 pictures are required in order to get a reliable marker pose estimation. More in detail, calibration pictures have been shot by placing camera at a constant distance to the marker (to be maintained during scanning phase). The marker has been placed on the ground, upon a rotating support. In each picture, the marker must cover more than 80% of the frame. At least at three different angular position of the camera w.r.t. ground, the marker is rotated at list in to 8 different positions. Focal length and focus distance of the lens have been maintained during scanning test.

During tests it also emerged that brightness and contrast of the marker in the pictures also influences the accuracy of the detection phase. High contrast particularly benefits ChArUco landmarks detection. In case of low contrast images, corners are not clearly detected, with an unusable result. From this observation, it derives the necessity of provide an adequate lightning for the marker, when the scanning phase happens in dark rooms.

3.2 Laser Line Detection

Based on authors experience and scientific literature, a 4-step procedure has been implemented to detect the laser line in an image, briefly summed up in the following list:

1. Image colour filtering (enhancing laser colour);
2. Background removal;
3. Image filtering with a 2-D Gaussian smoothing kernel (to smooth laser's speckles);
4. Laser point detection (for each row/column of the image), calculated as the mean point of the Gaussian curve based on pixels laser intensity;

This sub-pixel precision procedure allows the detection of the laser line in the image row by row or column by column, depending on the main laser line orientation.

Colour filtering is applied in order to "enhance" laser colour. Since laser colour is green – only pixels on which green component is prevalent are considered, while the others are neglected. This filter is applied on all the image of the acquisition set (Fig. 5b).

$$\begin{aligned} I(x,y) &= I(x,y) \text{ } \mathit{if} \text{ } I_G(x,y) > I_R(x,y) \text{ } \mathit{and} \text{ } I_G(x,y) > I_B(x,y) \\ I(x,y) &= 0 \text{ } \mathit{in\ the\ other\ cases} \end{aligned}$$

Successively, background removal is obtained as a subtraction between the image under analysis and the reference image. In particular, the reference image is the last or the first of the entire image set if the considered image is respectively on the first or in the second half of the set.

$$\begin{aligned} I^{(k)} &= I^{(k)} - I^{(R)} \\ I^{(R)} &= I^{(1)} \text{ } \mathit{if} \text{ } k > N/2 \\ I^{(R)} &= I^{(N)} \text{ } \mathit{in\ the\ other\ cases} \end{aligned}$$

Fig. 5. (a) A test image. Detected reference system is plotted on the bottom-right corner; (b) a zoom of the image of the acquisition set.

The resulting image has an almost black background, on which laser line is clearly visible. However, brightness of the pixels belonging to the laser line is not smoothly distributed: speckles are clearly visible. To obtain a smooth laser line profile and remove noise, a gaussian filter is applied to the image (Fig. 6). In the implemented procedure, gaussian kernel is disk shaped, with a standard deviation equal to 5 (calculated on the basis of the expected laser line width, equal to 20 pixels).

According to [17], laser line is detected by fitting on each row (or each column, depending on main laser orientation) a gaussian curve, considering as x-value pixel coordinate and y-value pixel brightness. In this way, for each row (or column) only a single pixel is considered, and in particular the one that coincides with the centre (mean) of the fitted gaussian curve (see Fig. 6).

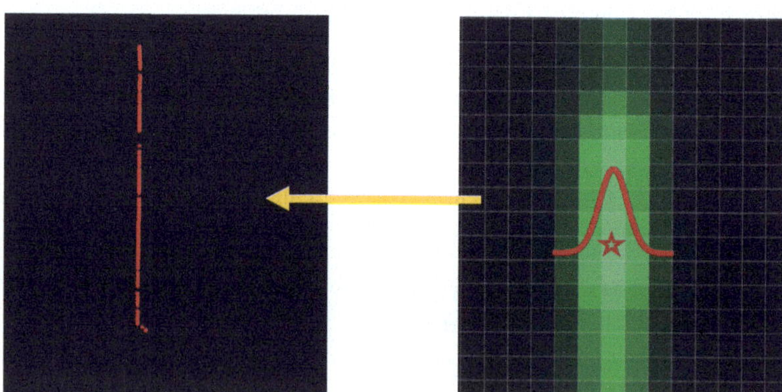

Fig. 6. Illustration of the plotted gaussian curve and the retrieved points.

3.3 Triangulation

Exploiting information on laser orientation and laser line coordinates in the image, it is possible to proceed with 3D coordinates retrieval by means of triangulation laser camera algorithms [18].

Generally speaking, knowing calibration parameters of the camera (camera matrix and distortion coefficients), the 3D position of each laser line pixel is retrieved as the intersection between the line passing through optical centre of the camera and the pixel and the plane identified by the laser line (i.e. the fiducial marker plane).

4 Hardware

Apart from the camera and its tripod, which are part of the standard equipment of a photographer, the auxiliary system for the acquisition of 3D points is composed by an Y-shaped component that support both laser line generator and orientable fiducial marker board. The support is mounted on a tripod by means of a panoramic head, which allows a horizontal rotation (or vertical, depending on head orientation).

The adopted laser line generator is focusable (i.e. allow to vary the laser focus distance) and emits a 532 nm light wave thus green. This choice is driven by the fact that green light is considered – in scientific literature - easier to detect within digital images. This is due to the fact that SLR camera sensors are generally built on the basis of green dominant RGB colour filters array, e.g. conventional Bayer filter has 2 photosites green out of 4 (50% green, 25% red, 25% blue) and X-Trans has 5 photosites green out of 9 (56% green, 22% red, 22% blue) – see Fig. 7.

Fig. 7. Conventional Bayer filter vs Fujifilm X-Trans.

The laser line generator is connected to the support by means of a specifically tailored 3D printed bracket [fig] and is powered by standard AAA batteries (3 V DC - <5 mW).

The planar board for the fiducial marker is jointed to the support by means of 3 countersunk head screws, secured to the planar board by means of bolts. A compression spring for each screw is placed between board and support, providing enough preload with the aim of assuring sufficient stiffness to the joint when board and support are placed a certain distance. Finally, each screw is secured to the support by means of a wing bolt and a spherical washer. Spherical washers have been used in order to assure a proper connection when support and board are not parallel each other (springs are perpendicular to the board, which is orientable w.r.t. the support) (Fig. 8).

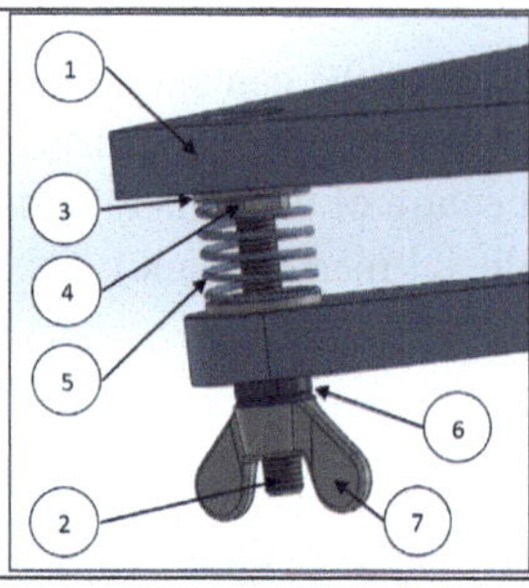

Components:

(1) Planar board
(2) Countersunk head screw
(3) Washer
(4) Bolt
(5) Compression spring
(6) Spherical washer
(7) Wing bolt

Fig. 8. Coupling between planar board and support.

The design of both board and support has been developed paying attention to lightness, portability and stiffness. On the light of the relatively low planarity influence of the fiducial marker to the final result, a planarity error of 0.6 mm can be considered acceptable for the board. For this reason, it was cut from a plywood panel. Other solutions have been previously taken into account, e.g. 3D printing by means of FDM technology or sandwich-like aluminium panels, but planarity requirement was not achieved on the first and costs was out of budget on the other. Moreover, FEM analysis demonstrated that the system entirely made from plywood panels (both board and support) is stiff enough to avoid exceeding deformation due to its own weight, in different position. In the worst scenario (horizontal board), a maximum displacement of 0.03 mm was calculated (see Fig. 9).

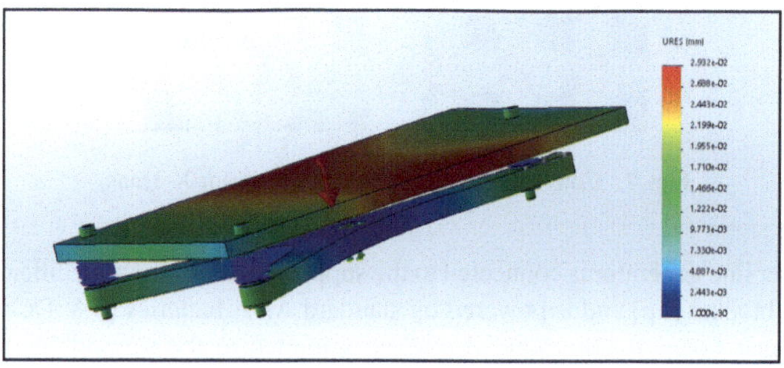

Fig. 9. FEM analysis on the device: resulting displacements.

5 Conclusions

The prototype has been finally tested for the acquisition of the geometry of a painting canvas, in the Confraternita Museum, in Arezzo (Italy), as shown in Fig. 10. The acquisition has been carried out by means of 112 pictures, one for each laser angular displacement. Approximately, laser lines were equally spaced on the plane of the painting, distant around 5–10 mm each other. The resulting point cloud is formed by approximately 200.000 points.

This final test verified the accuracy emerged during preliminary tests, carried out in laboratory.

The procedure and the device presented in this paper revealed to be a smart alternative to the cross-ratio based method for the metric acquisition of 3D points. The physical prototype is completely realised by means of inexpensive devices (general cost <200€) and it is easily portable, due to its reduced dimension (A3 format size) and its low weight (<1 kg). The procedure is fast, completely automatic and user friendly. 3D points are obtained directly from the set of pictures, without any user interaction, in a time of around 5–10 min for a set of 100 pictures by using a common notebook. Results have been judged by specialist (Culturanuova) as one step further with respect to the traditional method, both in terms of accuracy and usability.

Fig. 10. Test at the Confraternita Museum.

References

1. Vassilopoulou, S., Hurni, L., Dietrich, V., Baltsavias, E., Pateraki, M., Lagios, E., Parcharidis, I.: Orthophoto generation using IKONOS imagery and high-resolution DEM: a case study on volcanic hazard monitoring of Nisyros Island (Greece). ISPRS J. Photogramm. Remote Sens. **57**, 24–38 (2002). https://doi.org/10.1016/S0924-2716(02)00126-0
2. Jacobsen, K., Passini, R.: Accuracy of digital orthophotos from high resolution space imagery. In: Proceedings of the Workshop High Resolution Mapping from Space (2003)
3. Gros, P.: How to use the cross ratio to compute 3D invariants from two images how to use the cross ratio to compute projective invariants from two images? (1994). https://doi.org/10.1007/3-540-58240-1_6

4. Mohr, R., Gravir, B.T.: Projective geometry for image analysis (1996)
5. Remondino, F., Rizzi, A., Barazzetti, L., Scaioni, M., Fassi, F., Brumana, R., Pelagotti, A.: Review of geometric and radiometric analyses of paintings. Photogramm. Rec. **26**, 439–461 (2011). https://doi.org/10.1111/j.1477-9730.2011.00664.x
6. Liverani, A., Leali, F., Pellicciari, M.: Real-time 3D features reconstruction through monocular vision. Int. J. Interact. Des. Manuf. **4**, 103–112 (2010). https://doi.org/10.1007/s12008-010-0093-5
7. Volpe, Y., Furferi, R., Governi, L., Tennirelli, G.: Computer-based methodologies for semi-automatic 3D model generation from paintings. Int. J. Comput. Aided Eng. Technol. **6**, 88 (2014). https://doi.org/10.1504/IJCAET.2014.058012
8. Barazzetti, L., Remondino, F., Scaioni, M., Lo Brutto, M., Rizzi, A., Brumana, R.: Geometric and radiometric analysis of paintings, pp. 62–67 (2010)
9. Bimber, O., Coriand, F., Kleppe, A., Bruns, E., Zollmann, S., Langlotz, T.: Superimposing pictorial artwork with projected imagery. In: ACM SIGGRAPH 2006 Courses on - SIGGRAPH 2006, p. 10. ACM Press, New York (2006). https://doi.org/10.1145/1185657.1185805
10. Governi, L., Carfagni, M., Furferi, R., Puggelli, L., Volpe, Y.: Digital bas-relief design: a novel shape from shading-based method. Comput. Aided. Des. Appl. **11**, 153–164 (2014). https://doi.org/10.1080/16864360.2014.846073
11. Rosten, E., Drummond, T.: Machine learning for high-speed corner detection (2006). https://doi.org/10.1007/11744023_34
12. Santachiara, M., Gherardini, F., Leali, F.: An augmented reality application for the visualization and the pattern analysis of a Roman mosaic. IOP Conf. Ser. Mater. Sci. Eng. **364**, 012094 (2018). https://doi.org/10.1088/1757-899X/364/1/012094
13. Köhler, J., Pagani, A., Stricker, D.: Detection and identification techniques for markers used in computer vision. In: VLUDS 2010 - Visualization of Large and Unstructured Data Sets - Applications in Geospatial Planning, Modeling and Engineering (IRTG 1131 Workshop) (2010). https://doi.org/10.4230/OASIcs.VLUDS.2010.36
14. An, G., Lee, S., Seo, M.-W., Yun, K., Cheong, W.-S., Kang, S.-J., An, G.H., Lee, S., Seo, M.-W., Yun, K., Cheong, W.-S., Kang, S.-J.: Charuco board-based omnidirectional camera calibration method. Electronics **7**, 421 (2018). https://doi.org/10.3390/electronics7120421
15. Garrido-Jurado, S., Muñoz-Salinas, R., Madrid-Cuevas, F.J., Marín-Jiménez, M.J.: Automatic generation and detection of highly reliable fiducial markers under occlusion. Pattern Recogn. **47**, 2280–2292 (2014). https://doi.org/10.1016/j.patcog.2014.01.005
16. Furferi, R., Governi, L., Volpe, Y., Carfagni, M.: Design and assessment of a machine vision system for automatic vehicle wheel alignment. Int. J. Adv. Robot. Syst. **10**, 242 (2013). https://doi.org/10.5772/55928
17. Vosselman, G., Gorte, B.G.H., Sithole, G., Rabbani, T.: Recognising structure in laser scanning point clouds (2004). https://research.utwente.nl/en/publications/recognising-structure-in-laser-scanning-point-clouds
18. Scaramuzza, D., Harati, A., Siegwart, R.: Extrinsic self calibration of a camera and a 3D laser range finder from natural scenes. In: 2007 IEEE/RSJ International Conference on Intelligent Robots and Systems, pp. 4164–4169. IEEE (2007). https://doi.org/10.1109/IROS.2007.4399276

Automatic Segmentation of Constant Radius Secondary Features from Real Objects

Luca Di Angelo[1(✉)], Paolo Di Stefano[1], and Anna Eva Morabito[2]

[1] Department of Industrial Engineering, University of L'Aquila, L'Aquila, Italy
luca.diangelo@univaq.it

[2] Department of Engineering for Innovation, University of Salento, Lecce, Italy

Abstract. Secondary features, such as fillets, rounds, chamfers and grooves, are simple transitions between primary features, generally introduced in order to remove the sharp edges created by the intersection of primary features. Being able to distinguish secondary from primary features is important in various application contexts, such as reverse engineering, automatic geometric inspection of real scanned objects, and for preparation of models for FEM analysis and CNC tool-path generation.

The process for the recognition of secondary features from high-density tessellated models of real work-pieces is intrinsically complex for several reasons. This explains why, currently, there are no methodologies able to recognize automatically secondary features and the investigation on secondary features is mostly focused on B-Rep models. In a previous paper, the authors proposed a method for secondary features recognition from discrete geometric models synthetically generated. Here the methodology is extended to discrete geometric models experimentally acquired, for which the recognition is a very complex process, due to the object discretization, to its non-ideal geometry and to measurement errors.

Keywords: Feature recognition · Blending features · Point cloud segmentation · Fuzzy logic

1 Introduction

From a semantic point of view, the features of a mechanical component can be classified into primary and secondary features. Primary features represent the main geometric shapes of a mechanical component, those defining its engineering function. Secondary features are simple transitions between primary ones and they are generally introduced in order to remove the sharp edges created by the intersection of main geometric shapes (such as planes, cylinders or cones).

Typical secondary features are: fillets, rounds, chamfers and grooves. Fillets and rounds are rounded intersections of small radius, typically present in castings and forgings to increase the mechanical strength and to make the product shape more attractive. These features are also used to reduce stress concentration in corners improving the fatigue behavior of parts cyclically loaded. Grooves are features generally required to provide clearance for subsequent machining or assembly. The relief

C. Rizzi et al. (Eds.): ADM 2019, LNME, pp. 201–212, 2020.
https://doi.org/10.1007/978-3-030-31154-4_18

grooves, in particular, are typical for axially symmetric mechanical components and they are usually located on the edge where an axial-symmetric face (such as cylindrical or conical) matches a planar face normal to the axis of the component.

The importance of distinguishing secondary from primary features is evident especially in those specific application contexts, where a high-level semantic analysis of the component geometry is to be performed to automatically detect its most significant properties. For example, in the reverse engineering of a mechanical component from point clouds, the principal geometry should be differentiated from the secondary features [1]. In particular, the secondary features should be recognized, segmented, and assigned to the edges, in accordance with what is done in mechanical engineering practice. In other cases, such as in the automatic geometric inspection of scanned real objects, the secondary features should be differentiated since they are not subject to explicit geometric specifications but to indications of general tolerances [2]. Finally, secondary feature recognition plays an important role in downstream applications where CAD models are to be used for FEM analysis or for manufacturing analysis aimed at generating numerical codes ([3–6], and [7]).

Several automatic methods were proposed in the literature to recognize geometric features from unorganized point clouds or tessellated models ([8, 9] and [10]). The well-known RANSAC method iteratively extracts a set of feasible relationships (orientation, positioning, and equality) between the candidate primitives, by using a set of locally adapted primitives. These relations are then used to constrain and to optimize the local RANSAC-based primitives [8]. A severe drawback of this methodology is the inability to recognize complex-shaped features, such as those frequently describing the secondary features.

Currently, the development of methodologies for the automatic segmentation of secondary features from 3D high-density tessellated models experimentally acquired can be considered a relatively new topic. Most of the research on secondary features is still focused on B-Rep models [11–18], that is a relatively simple case of recognition problem since the surfaces, wrapping the model, are mathematically known. The recognition of secondary features from tessellated models is far more complex for various reasons. Firstly, secondary features have to be identified from a model composed of triangular facets approximating the original smooth surface of the object. Since the surface of the model is not fully differentiable, the geometric differential properties (such as principal curvatures), based on which the feature segmentation is carried out, must be recognized and cannot be directly evaluated [19]. Secondary features, moreover, due to the limited extension with respect to the sampling distance generally used during the discretization, are roughly described. This makes their recognition even more complex. In a previous paper, the authors proposed a method for secondary features recognition from discrete geometric models synthetically generated [20, 21]. In the case of mesh experimentally acquired from real objects, additional uncertainties, due to non-ideal geometry of the real object and measurement errors further complicate the recognition process. In [22] the authors proposed a new and original methodology for automatic segmentation of constant radius secondary features from tessellated models experimentally acquired during the 3D laser scanning of mechanical components. This methodology, based on the analysis of geometric-differential properties, also provided specific strategies in order to reduce its sensitivity to all of the above-mentioned uncertainties without affecting its selectivity.

In this paper, the recognition method is presented and tested for the case study of Fig. 1. Since it was manufactured by sand casting, the work piece is affected by a high value of the surface roughness and form errors. These issues make this case study critical for any methodologies aimed at the automatic segmentation of secondary features.

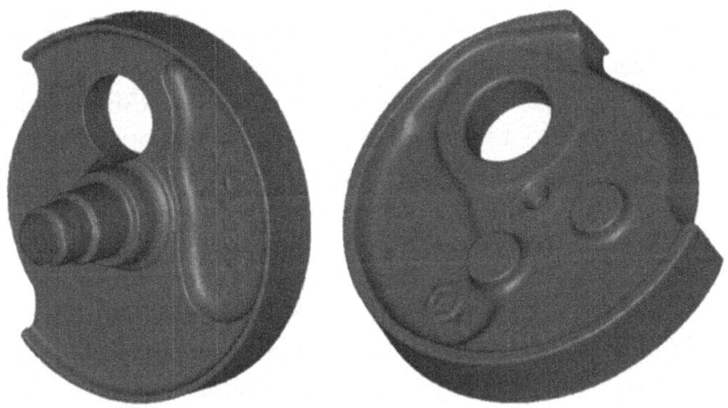

Fig. 1. The real work-piece used as case study for a critical testing of the methodology

2 The Concept of Constant Radius Secondary Feature (CRSF)

In order to develop a methodology aimed at the automatic recognition and segmentation of secondary features, a rule has to be established to identify univocally these features. This rule is founded on the concept of *Constant Radius Secondary Feature (CRSF)*.

A CRSF is a portion of the work-piece surface satisfying two main requirements. The first is of geometric type. CRSFs are features that develop along a trajectory, named sweep line, and whose section, carried out by a principal normal plane orthogonal to the sweep line, is approximately circular. This also explains why these features were called in this way. The radius of the circular section, indicated as R_1 in the following, is referred to as the characteristic radius of the CRSF and it remains almost constant along the feature. R_1 is associated with the inverse of the principal curvature maximum in absolute value (k_1).

The nodes pertaining to a CRSF, however, are characterized by values of k_1 usually affected by a significant variability. The amplitude of this variability depends on several factors, such as for example the method used to estimate the principal curvatures from the discrete model of the work-piece and the low resolution by which these detail features, of small sizes compared to the primary features, are usually acquired. In this paper, the principal curvatures were evaluated by means of the paraboloid approximation [23], which generally overestimates systematically the real value of the curvature radii. Further sources of uncertainty are the measurement errors and the non-ideal geometries of real objects, to which the method is applied here. This variability makes the CRSFs recognition a not trivial process. In order to deal with this variability, a fuzzy approach for CRSF segmentation was implemented.

The geometric requirement of k_1-values constancy at the nodes of a CRSF is, however, not sufficient to recognize univocally these features and, particularly, to differentiate them from primary features having the same geometric description; therefore, further requirements are necessary. An important additional prerequisite is of dimensional type. From this point of view, in fact, a CRSF is characterized by a value of the characteristic radius R_1 that is generally much lower than the radius R_2 of the sweep line of the feature (associated with the inverse of the principal curvature minimum in absolute value k_2). The typical R_1 values of a CRSF are relatively small ranging typically from one to a few millimeters. On the contrary, primary features are characterized by larger values. Secondary features generally do not have a predefined dimensional value for R_1, although in some cases and for specific kinds of secondary features, these values are established by ISO standards (for example the standard ISO 4755 unifying the grooves sizes for external threads).

Under the hypothesis that the R_1 values of the CRSFs are unknown, an approach was developed here in order to obtain a preliminary evaluation of the R_1 values. This approach, outlined as a flowchart in Fig. 1, is based on the analysis of the occurrences of the r_1-values (with $r_1 = 1/k_1$) at the mesh nodes.

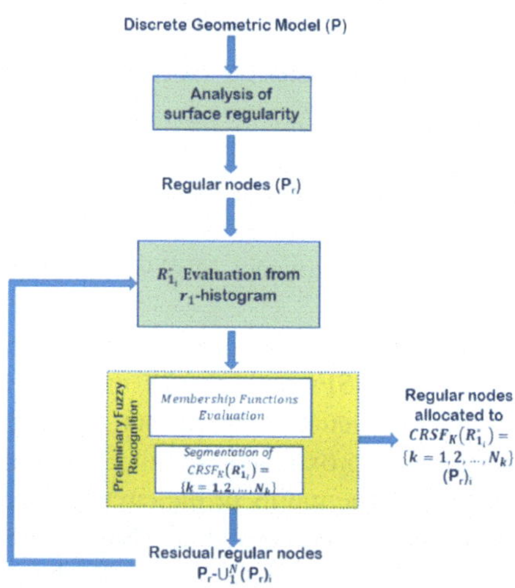

Fig. 2. Iterative methodology used for the preliminary evaluation of the R_1 values

If a CRSF exists, in fact, several nodes of the geometric model will pertain to it. Therefore, the CRSFs are associated with the most recurrent values of the radii of curvature r_1 and their characteristic radii can be preliminarily estimated from the peaks of the r_1-histogram, where the r_1–values are estimated at each regular point of the discrete model $\boldsymbol{P_r}$. In order to speed up the methodology, it is appropriate to limit this investigation to a given dimensional range ($r_1 \leq 10$ mm, for example, for the case study of Fig. 1). This range is established arbitrarily by specifying the radius value below which the characteristic radii of the CRSFs are expected to fall. These radii, denoted as R^*_{1i}, give only a preliminary estimation of the characteristic radii R_{1i} of the CRSFs. This difference between real and estimated values, however, is not a true problem because a relative, rather than absolute, evaluation of the CRSF radius is necessary for its recognition.

In the case of high-density synthetic discrete models, these peaks are generally well detectable [20, 21]. This is no longer true for mesh experimentally acquired where the effects of the experimental noise and of the manufacturing errors add to the intrinsic difficulties of secondary features recognition from tessellated modes. Due to the uncertainties affecting r_1, the peaks of the histogram tend to be wider and lower. This peak blur effect grows with the r_1 values and it can be so relevant to cover other potentially useful peaks. In other cases, some fictitious peaks, unrelated to real CRSFs, may appear [22].

To cope with this lack of well-defined peaks in the histogram, an iterative methodology was implemented. This methodology, outlined in Fig. 1, recognizes, one by one, the R^*_{1i} values. When a value of R^*_{1i} was identified, a two-step fuzzy algorithm, later described in this paper, is applied to recognize the CRSFs with that specific value of characteristic radius. During the first step, a suitable set of membership functions is built and evaluated for each node of the mesh. Then, a growing algorithm performs the identification of the regions of the surface, whose adjacent nodes pertain to CRSFs with characteristic radius equal to R^*_{1i} (i.e. $CRSF\left(R^*_{1i}\right)$). This algorithm is applied recursively as many times as the radius values R^*_{1i} identified. At the end of the generic iteration i, the points Q_{ri}, just assigned to $CRSF\left(R^*_{1i}\right)$, are removed from the evaluation of r_1-histogram of the following iteration i + 1. The iterative process in Fig. 2 is repeated until there are no more significant peaks in the histogram.

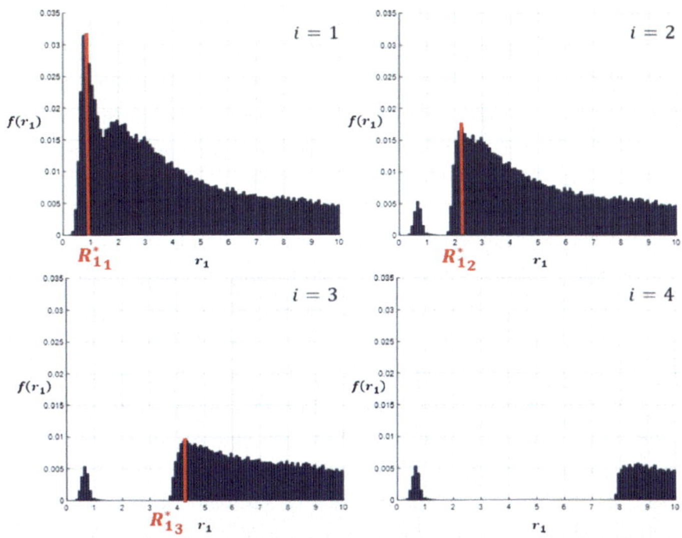

Fig. 3. The histograms for the evaluation of the characteristic radii R^*_{1i}

In Fig. 3, the r_1-histograms obtained by applying this recursive methodology to the case study of Fig. 1 are shown. The y-axis reports the frequencies of occurrence $f(r_1)$.

The values estimated for the corresponding characteristic radii are: R^*_{11} = 0.85 mm, R^*_{12} = 2.25 mm and R^*_{13} = 4.35 mm.

3 The Fuzzy Approach for CRSF Recognition

The fuzzy approach implemented here classifies unequivocally the points of the tessellated model according to three categories:

- regular points belonging to CRSFs with a characteristic radius equal to $R^*_{1_i}$;
- points belonging to sharp edges;
- regular points pertaining to primary features or to CRSFs with characteristic radius different from $R^*_{1_i}$.

The segmentation of the tessellated model according to these categories requires the definition of three membership functions, denoted respectively by $\mu_{r_1=R^*_{1_i}}$, μ_e and $\mu_{\neg r_1=R^*_{1_i}}$. Each of them measures the possibility that the generic node **p** of the object mesh belongs to one of the aforementioned categories. Since **p** can belong unequivocally only to those three categories, these functions must always satisfy the following equation:

$$\mu_{r_1=R^*_{1_i}} + \mu_e + \mu_{\neg r_1=R^*_{1_i}} = 1 \tag{1}$$

If **p** certainly belongs to a $CRSF(R^*_{1i})$, the membership functions, evaluated at that point, have the following values: $\mu_{r_1=R^*_{1_i}} = 1$, $\mu_e = 0$ and $\mu_{\neg r_1=R^*_{1_i}} = 0$.

To establish the membership of the generic node to one of the afore-mentioned categories, some node properties, able to affect this membership attribution, have to be considered. For the membership function μ_e, the property considered is the Sharpness Indicator (SHI), proposed in [23]. "This property compares the normal vector evaluated at the generic point **p** by a smooth function, which approximates the neighbourhood of **p** of order *h* (or *h-ring neighbourhood*), with the normal vectors of each planar facet belonging to the same neighbourhood of **p**." The ramp function, shown in Fig. 4(a), was selected to describe analytically the membership function $\widetilde{\mu}_e$.

The node property, able to affect the membership attribution of the node to the category of the $CRSF(R^*_{1i})$, is the quality of the surface tessellation. This property can be measured by the factor of curvature approximation γ. This factor, defined as the maximum value of the tangent of the dihedral angles between adjacent triangular facets incident at p, measures how well the geometry of a regular curved surface is described by the discrete model. A well-sampled surface has very small γ values. In particular, the γ factor is 0 for a flat surface. For curved surfaces, γ is different from 0, and its value increases as the surface is more coarsely sampled. The function selected for $\widetilde{\mu}_{r_1=R^*_{1_i}}$ is a trapezoidal function (Fig. 4(b)). A similar shaped function is used to evaluate the membership attribution of a node to a CRSF having a sweep radius equal to R_2^*. A detailed description of the meaning of the various parameters reported in Fig. 4 and of how evaluating them can be found in [22]. Among these parameters σ is particularly significant, since it measures the dispersion of the r_1-values evaluated for typical analytical shapes (such as tori, cylinders and spheres), whose mesh is obtained for a given maximum value of γ.

The parameters t_1 and t_2, important for the definition of the plateau of the membership functions $\widetilde{\mu}_{r_1=R^*_{1_i}}$ and $\widetilde{\mu}_{r_1=R^*_2}$, take into account of the dimensional variability due to the manufacturing errors of the workpiece analysed. t_1, in particular, is evaluated from the tolerance class specified by the ISO 2768 standard on linear sizes of rounds and fillets. t_2 depends, on the other hand, on the general quality characteristics of the analysed object, as prescribed by the table I of ISO 2768.

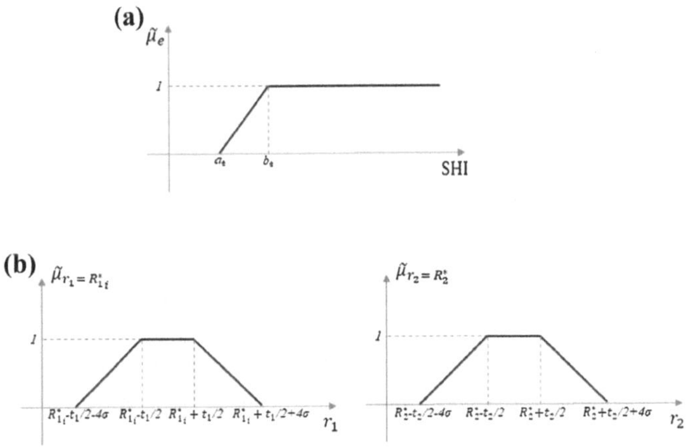

Fig. 4. (a) The membership function μ_e. (b) The membership functions $\widetilde{\mu}_{r_1=R^*_{1i}}$ and $\widetilde{\mu}_{r_2=R^*_2}$.

Since the possibility that a point belongs to a smooth surface is $(1-\widetilde{\mu}_e)$, the membership function, measuring the possibility that a point pertains to a $CRSF\left(R^*_{1i}\right)$ also depends by $\widetilde{\mu}_e$ by the following equation

$$\mu_{r_1=R^*_{1_i}} = \widetilde{\mu}_{r_1=R^*_{1_i}}(1-\widetilde{\mu}_e) \tag{2}$$

Analogously, the membership function for all the other regular features is:

$$\mu_{\neg r_1=R^*_{1_i}} = \left(1-\widetilde{\mu}_{r_1=R^*_{1_i}}\right)(1-\widetilde{\mu}_e) \tag{3}$$

In Fig. 5, the nodes for which a non-zero possibility of attribution of the radius R^*_{1i} are shown by a colour-map. These colour maps, however, do not allow distinguishing the single $CRSF\left(R^*_{1i}\right)$ from each other and from the other regular features. For this purpose, a further and last step, called region growing, is required.

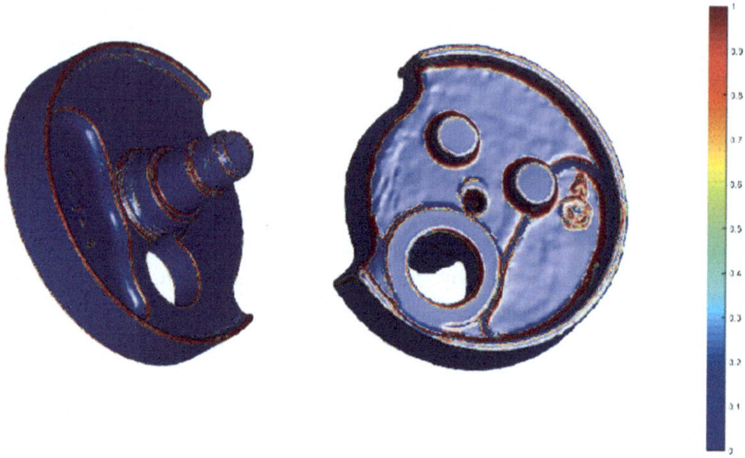

Fig. 5. Some examples of colour maps of the membership function $\mu_{r_1=R^*_{1_i}}$ for $R^*_{1_1}$ = 0.85 mm and $R^*_{1_2}$ = 2.25 mm.

4 Region Growing

$CRSF\left(R^*_{1i}\right)$ can be obtained as the result of a growing algorithm that aggregates the adjacent nodes, recognized to be similar to the growing region.

While in a single $CRSF\left(R^*_{1i}\right)$ the radius R^*_{1i} has a constant value, the radius R^*_2 may change gradually along the sweep line of the feature, depending on the geometric characteristics of the primary features encountering the secondary feature. For this reason, the dissimilarity of a node $\mathbf{p_j}$ adjacent to the growing region is evaluated with respect to the arithmetic mean of r_1-values and to the weighted average of the r_2-values, evaluated at the points $\mathbf{q}$ of the growing region. In particular, the generic weight to use for the R^*_2 estimation is selected to give more evidence to the r_2-values of the nodes $\mathbf{q}$ that are nearer to $\mathbf{p_j}$. The node aggregation to a CRSF, however, is not based on simple dimensional matching criteria. In order to assign a node $\mathbf{p_j}$ to a growing region, the algorithm looks at the similarity of $\mathbf{p_j}$ with respect to the growing region in terms of the values of the membership functions.

The region growing starts from a seed node of the mesh, selected among the nodes where the maximum membership degree $\mu_{r_1=R^*_{1_i}}$ is reached. The seed node is compared with every node $\mathbf{p_j}$ belonging to the 1-ring neighborhood to analyze similarity. The nodes recognized as similar are aggregated in the growing region. The growing algorithm stops when dissimilar nodes are met or all the mesh nodes have been analyzed.

In order to recognize better the boundary between a secondary and a primary feature, two different levels of the node neighborhood were used to evaluate the geometric-differential properties (r_1 and r_2) and the related membership functions. In the defuzzification process, this analysis simultaneously carried out on two amplitudes of the node neighborhood, allows satisfying two contrasting requirements: identifying selectively the boundary and eliminating the typical uncertainties due to noise and irregularities of a real surface. The first requirement needs that differential geometric properties are analyzed in a small level of neighborhood of the node. The second requires that the analysis is performed on a larger level of the node neighborhood.

5 Discussion and Conclusions

The CRSF recognition method, described in this paper, was applied to several critical cases highlighting interesting and satisfactory results. One of these is the case study of Fig. 1. It is a real object experimentally acquired by a 3D scanner FARO (Edge ScanArm) with a laser line probe (accuracy $4\sigma_s = 0.0688$ mm, sample rate $s_r = 0.1$ mm). The work-piece was manufactured by sand casting and some of its features are machined. For this reason, it is a case critical for the method, since it presents several recognition problems.

In Figs. 6 and 8, the CRSFs, recognized by the methodology respectively for the front and the back of the case study, were grouped based on the search radius value R^*_{1i} and displayed as colored regions. In particular, the light-blue color was selected to highlight the primary features and the magenta for the sharp edges. By comparing the images of Fig. 6 among them, it should be noted that, in some cases, the fillet zone is recognized as decomposed into two or more CRSFs. This result may seem to be a misinterpretation of the method, but this occurs because the fillet zone area is actually made up of not very large areas of different radius.

To highlight this aspect, the radial plane, shown on the left of Fig. 7, cut the piece so that the r_1 variability, along the resulting longitudinal profile, could be pointed out. Several colors, corresponding to those of the various CRSFs in Fig. 6, were used to identify the profile segments belonging to CRSFs, while the black and the magenta were used respectively for segments belonging to primary features and sharp points. In particular, as shown by the profile in Fig. 7, adjacent segments corresponding to different CRSFs are due to real differences in the radius of adjacent CRSFs. This confirms the great selectivity of the method, which also allows these details to be highlighted. It is important to note that, since this method is dedicated to the recognition of CRSFs rather than fillets, in the cases just highlighted it is convenient to expand the membership functions in order to identify the fillets. These considerations can also be extended to the CRSFs of Figs. 8 and 9.

The methodology proposed in this paper segments the various CRSFs by aggregating adjacent nodes, recognized as similar in terms of membership functions suitably defined. The fuzzy logic allows setting the recognition range of the CRSFs, so that the method can be tuned in order to segment features having a large variability of r_1-values. The methodology performs the CRSF segmentation, recognizing also shapes not generally identified by other methods that detect only analytical geometries (typically segments of tori and cylinders). For this reason, this method can be applied also to recognize complex swept features with constant radius transverse section [24].

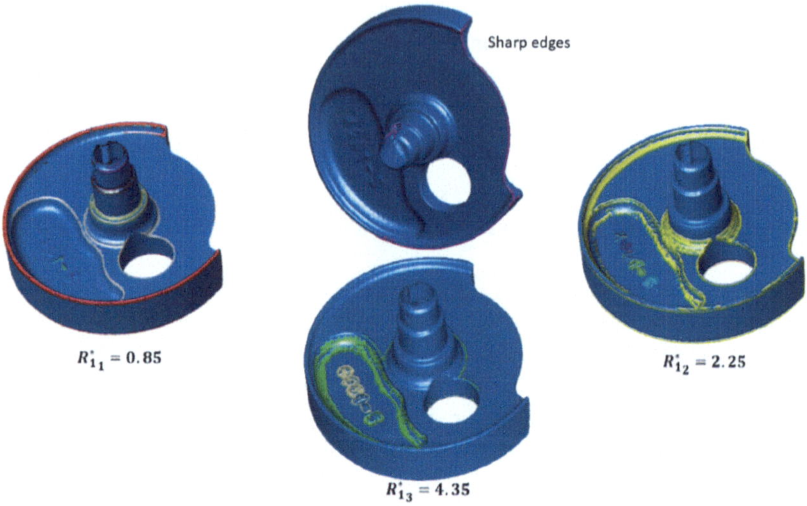

Fig. 6. The CRSF segmentation for the front of the case study of Fig. 1.

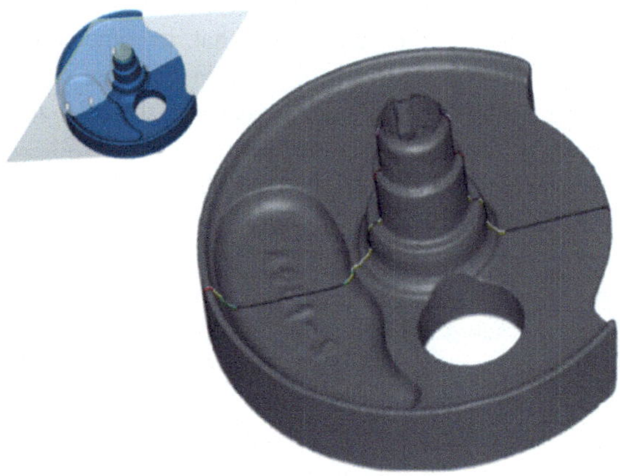

Fig. 7. The r_1-variability along the profile obtained by cutting the object with the radial plane on the left.

The methodology shows some limitations where a coarser or highly noisy acquisition of the surface has been performed. In these situations, the information contained in the tessellated model is unreliable or insufficient. At present, the methodology does not recognize the chamfers that are important secondary features substantially different from fillets, rounds and grooves. Chamfers are regions contained between two sharp edges, parallel and close to each other, so that they cannot be equated to CRSF. Since the methodology proposed is able to identify sharp edges efficiently, it is expected, in the near future, to expand it to include automatic recognition of these features as well.

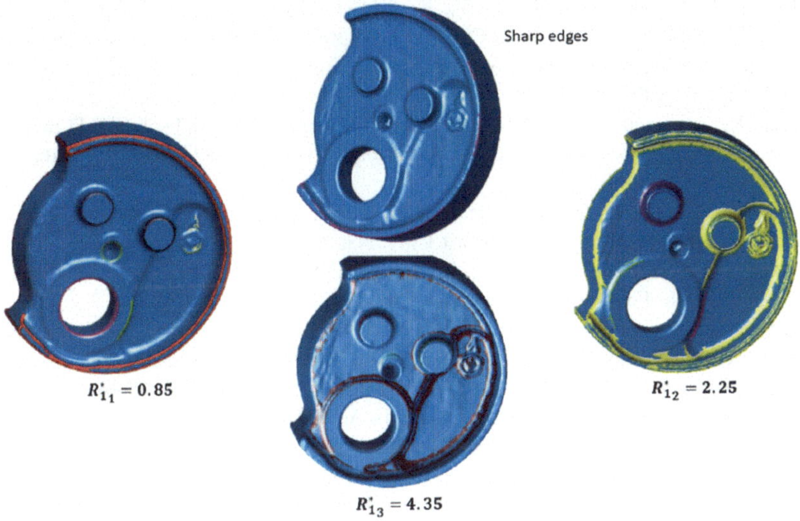

Fig. 8. The CRSF segmentation for the back of the case study of Fig. 1.

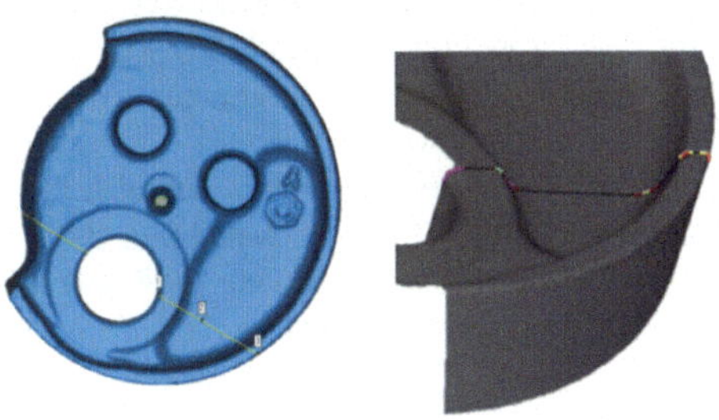

Fig. 9. The r_1 variability along the profile obtained by cutting the object with the plane on the left.

References

1. Buonamici, F., Carfagni, M.: Reverse engineering of mechanical parts: a brief overview of existing approaches and possible new strategies. In: Proceedings of the ASME Design Engineering Technical Conference, 1B-2016 (2016)
2. Di Angelo, L., Di Stefano, P., Morabito, A.E.: Automatic evaluation of form errors in high-density acquired surfaces. Int. J. Prod. Res. **49**(7), 2061–2082 (2011)
3. Sheen, D.P., Son, T.G., Myung, D.K., Ryu, C., Lee, S.H., Lee, K., et al.: Transformation of a thin-walled solid model into a surface model via solid deflation. Comput. Aided Des. **42**(8), 720–730 (2010)
4. Zhao, L., Tong, R., Dong, T., Dong, J.: B-rep model simplification for feature suppressing using local error evaluation. In: Proceedings of the Ninth International Conference on Computer Supported Cooperative Work in Design, Coventry, UK, 24–26 May 2005. IEEE (2005). http://dx.doi.org/10.1109/CSCWD.2005.194282

5. Hariya, M., Nonaka, N., Shimizu, Y., Konishi, K., Iwasaka, T.: Technique for checking design rules for three-dimensional CAD data. In: 3rd International Conference on Computer Science and Information Technology, Chengdu, China, 9–11 July 2010. IEEE (2010). http://dx.doi.org/10.1109/ICCSIT.2010.5565010
6. Otani, K.: Automatic FEM mesh generation of 3D mid-surface and solid CAD model with shape recognition technique. SAE Technical Paper 2016-01-1382 (2016). http://dx.doi.org/10.4271/2016-01-1382
7. Cheng, Y., Liu, X., Ni, Z.: Selectively inhibition and recognition treatment strategy of fillet oriented to manufacturing. J. Southeast Univ. Nat. Sci. **40**(4), 731–735 (2010)
8. Schnabel, R., Wahl, R., Klein, R.: Efficient RANSAC for point-cloud shape detection. Comput. Graph. Forum **26**(2), 214–226 (2007)
9. Lavoué, G., Dupont, F., Baskurt, A.: A new CAD mesh segmentation method, based on curvature tensor analysis. Comput. Aided Des. **37**(10), 975–987 (2005)
10. Li, Y., Wu, X., Chrysathou, Y., Sharf, A., Cohen-Or, D., Mitra, N.J.: GlobFit: consistently fitting primitives by discovering global relations. ACM Trans. Graph. **30**(4) (2011). Art. no. 52
11. Di Stefano, P., Bianconi, F., Di Angelo, L.: An approach for feature semantics recognition in geometric models. Comput. Aided Des. **36**(10), 993–1009 (2004)
12. Chen, Z., Yu, G., Zhang, Y.: Fillet recognition and suppression in CAD model based on feature chain(ring). China Mech. Eng. **22**(22), 2707–2711 (2010)
13. Li, B., Liu, J.: Detail feature recognition and decomposition in solid model. Comput. Aided Des. **34**(5), 405–414 (2002)
14. Cui, X., Gao, S., Zhou, G.: An efficient algorithm for recognizing and suppressing blend features. Comput.-Aided Des. Appl. **1**(1–4), 421–428 (2004)
15. Joshi, N., Dutta, D.: Feature simplification techniques for freeform surface models. J. Comput. Inf. Sci. Eng. **3**(3), 177–186 (2003)
16. Zhu, H., Menq, C.H.: B-rep model simplification by automatic fillet/round suppressing for efficient automatic feature recognition. Comput. Aided Des. **34**(2), 109–123 (2002)
17. Jiao, X., Alexander, P.J.: Parallel feature-preserving mesh smoothing. In: Proceedings of International Conference on Computational Science and Its Applications, pp. 1180–1189 (2005)
18. Bianconi, F., Di Stefano, P.: An intermediate level representation scheme for secondary features recognition and B-rep model simplification. In: Shape Modeling International, Seoul, South Korea, pp. 99–108. IEEE (2003)
19. Di Angelo, L., Di Stefano, P., Morabito, A.E.: The RGM data structure: a nominal interpretation of an acquired high point density model for automatic tolerance inspection. Int. J. Prod. Res. **50**(12), 3416–3433 (2012). https://doi.org/10.1080/00207543.2011.579638
20. Di Angelo, L., Di Stefano, P., Morabito, A.E.: Product model for dimensioning, tolerancing and inspection. In: Eynard, B., Nigrelli, V., Oliveri, S., Peris-Fajarnes, G., Rizzuti, S. (eds.) Advances on Mechanics, Design Engineering and Manufacturing. Lecture Notes in Mechanical Engineering, pp. 1033–1040. Springer, Cham (2017)
21. Di Angelo, L., Di Stefano, P., Morabito, A.E.: Secondary features segmentation from high-density tessellated surfaces. Int. J. Interact. Des. Manuf. **12**(3), 801–809 (2018)
22. Di Angelo, L., Di Stefano, P., Morabito, A.E.: Fillets, rounds, grooves and sharp edges segmentation from 3D scanned surfaces. CAD Comput. Aided Des. **110**, 78–91 (2019)
23. Di Angelo, L., Di Stefano, P.: Geometric segmentation of 3D scanned surfaces. CAD Comput. Aided Des. **62**, 44–56 (2015)
24. Di Angelo, L., Di Stefano, P., Morabito, A.E., Pane, C.: Measurement of constant radius geometric features in archaeological pottery. Meas. J. Int. Meas. Confederation **124**, 138–146 (2018)

Comparison of Algorithms for Recognition of Cylindrical Features in a Voxel-Based Approach for Tolerance Inspection

Michele Bici(✉) and Francesca Campana

DIMA, Sapienza Università di Roma, Via Eudossiana 18, 00184 Rome, Italy
michele.bici@uniroma1.it

Abstract. In injection molding production, automatic inspections are needed to control defects and evaluate the assigned functional tolerances of components and dies. With the "Smart Manufacturing" approach as a point of view, this paper resumes part of a wider research aiming the integration and the automation of a Reverse Engineering inspection process in components and die set-up. The paper compares two fitting approaches for recognition of portions of cylindrical surfaces. Therefore, they are evaluated in the respect of an automatic voxel-based feature recognition of 3D dense cloud of points for tolerance inspection of injection-molded parts. The first approach is a 2D Levenberg Marquardt algorithm coupled with a first guess evaluation made by the Kasa algebraic form. The second one is a 3D fitting based on the RANdom SAmple Consensus algorithm (RANSAC). The evaluation has been made according to the ability of the approaches of working on points associated to the voxel structure that locally divides the cloud to characterize planar and curved surfaces. After the presentation of the overall automatic recognition, the cylindrical surface algorithms are presented and compared trough test cases.

Keywords: Tolerance inspection · Injection molding · RANdom SAmple Consensus algorithm

1 Introduction

In fields like production of electromechanical components through injection molding in multicavity dies, automatic inspections, for die set-up and quality controls of components, are needed to avoid defects and to guarantee the assigned functional tolerances during production. Especially during die set-up, the high number of samples and tolerances to be detected makes the inspection process repetitive and time consuming, to that, many efforts have been spent to increase its automation during acquisition and in data post-processing. In fact, nowadays, automatized procedures are researched and developed not only during the productions or finishing phases [1] but, also, in measurement and control procedures. In previous works [2–5], authors presented procedures oriented to simplify and optimize standard Coordinate Measurement Machine (CMM) protocols, and to analyse how non-contact measurements may change the data-processing procedures. In the last decade, this kind of issues represented a difficult challenge for researchers in this field [6]. In particular, in this paper, we are going to

C. Rizzi et al. (Eds.): ADM 2019, LNME, pp. 213–225, 2020.
https://doi.org/10.1007/978-3-030-31154-4_19

focus on some new results achieved in a research carried out for integrating the Reverse Engineering (RE) inspection process in the die set-up, so that the final inspection protocol may be automatically filled, performing standard verifications, understanding and analysing dimensional and geometrical deviation of the manufactured surfaces.

The treated research is conducted with a non-contact RE system and original implementations that adopt MATLAB-based algorithms. The hardware system used for the acquisition is composed by a laser scanner (Nikon LC15Dx) mounted on a bridge CMM 3COORD Hera 12.9.7, which, coupled, allow an accuracy of 2,5 μm [2–4]. Then, algorithms, developed in MATLAB, are able to perform the data processing and segmentation of the obtained clouds. Finally, measurement algorithms are able to produce data that are useful in PDM. In the electromechanical field, injection molded components are mainly characterized by planar and cylindrical surfaces, with sharp angles, low thicknesses and many small ribs. Therefore, difficulties arise not from the characteristics of surfaces, but from the necessity of procedure automation on hundreds of samples. In [2], an original voxel-based implementation has been proposed to distinguish planar surfaces in electromechanical parts. Non-planar regions have been then investigated in [3, 4], so that, cylindrical surfaces may be recognized and measured according to a 2D fitting algorithm, developed starting from Kasa formulation. This paper compares this algorithm with another one, called Maximum Likelihood Estimate SAmple Consensus algorithm (MLESAC), and based on RANdom Sample Consensus algorithm (RANSAC). After a brief outline of the previous developments in Sect. 2, Sect. 3 presents the new approach and, then, in Sect. 4, some case studies, used to develop and test the procedure, are presented and discussed. Finally, in Sect. 5, major conclusions are highlighted.

2 Previous Developments

RE encompasses several tools and methods suitable to make digital models of real parts. As reported in [7, 8], generally, they can be divided in a sequence of four steps: (1) Data Acquisition; (2) Data Processing; (3) Segmentation and Surface (or feature) Fitting; (4) CAD Model Creation. The reported research is focused on the development and test of algorithms of data processing, feature segmentation and recognition (phases 2 and 3). The general workflow and phases of the developed procedure are reported in Fig. 1, [4] and explained in the following.

2.1 Voxel Structure

The adoption of voxel-based approach was defined considering the requirement of decreasing the post-processing duration. Although gradient analysis is one of the most adopted solutions [9, 10], it is clear that the number of points, their density and distribution influence the analysis accuracy. This problem is partially solved by filtering [11], but, in case of tolerance analysis and quality inspection, it may lead to a loss of data. Moreover, surface reconstruction of filtered acquisition may not represent a good solution, even if adaptive filtered, due to high computational time. For these reasons, supported also by [10], we decided to work with a voxel-based approach that localizes volumes according to a semantic significance, which in our case is a specific feature of the shape (planar/curved, planar/cylindrical). This kind of segmentation is related to a problem of feature recognition [12]. Doing so, we can analyze the surface recognition problem according to a scale larger than the tessellated mesh, without applying preliminary filters, or classical segmentation through curvature [13]. One of the advantages of the approach consists in avoiding mesh triangulation and point evaluation of local curvatures that are substituted with local surface reconstruction of the points inside voxels. Another peculiar point that has to be noticed is the fact that the usage of voxel segmentation, in this way, make the amount of segmentation elements not directly dependent from the point density and distribution.

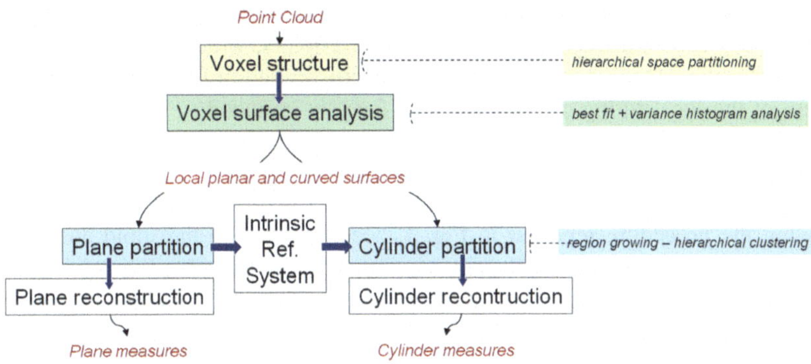

Fig. 1. The developed procedure for data processing, segmentation and surface fitting, [4].

The voxel-based approach is based on a grid method similar to an octree grid derived from hierarchical space partitioning [14]. In our case, a 3D voxel structure is superimposed on the acquired cloud of points. It is defined through an iterative procedure that starts from a single parallelepiped. It encompasses acquired cloud completely, then, along each direction (x, y and z), it is recursively split into smaller volumes, generating a voxel structure, until no more points are included into an element or if the assigned limit of a voxel size is reached. The final structure made with the smallest size voxel represents the voxel structure of the partition. In [2, 3], implementation details related to the voxel structure definition are described and discussed. From this part of the research, it resulted how the voxel ratio and dimensions influence the following results. Due to this, some procedures and tests to optimize the voxel lengths and ratio have been developed for each kind of component (prismatic, axisymmetrical and boxed-shape components).

2.2 Voxel Surface Analysis

Once the voxel discretization is performed, local surface recognition starts through the computation of geometric descriptors. Local surface recognition is carried out through best-fitting algorithms. Electromechanical parts made by injection molding are predominantly feature-based shapes, thus plane inspection is the starting part of our segmentation. For this reason, voxels with planar surfaces are firstly detected, through threshold analysis of the fitting parameters [10, 15]. The threshold works on the variance of the point distances from the local plane fit in the voxel. Thanks to the accuracy of the acquisition, planar surfaces populate always the first bin of the variance distribution on the voxels. The bin major value can be assumed as a threshold to exclude all V_{ijk}, which are filled with points that lay on curved surfaces. Optimal values for thresholds can be found with an iterative procedure, already developed, and explained in [4].

Local surface recognition is approached by fitting. In the first phase, for each "true" state voxel, a local plane is fit by least square minimization, through Single Value Decomposition, obtaining the geometrical descriptors parameters of each local plane. Voxel that do not present geometrical descriptors parameters inside the threshold are associated to "curved" local surfaces. They can be free-form or cylindrical or boundary voxels. In any case, finding their geometrical descriptors, which means recognizing a specific type of surface, means assuming a specific equation to be fit. In Sect. 3, the problem of the local fitting of cylindrical surfaces is provided.

2.3 Region Growing

Following the procedure reported in Fig. 1, according to the threshold analyses, the voxel structure is partitioned in planar, cylindrical and other curved voxels. From this distinction, the global recognition of the component planes or cylindrical surfaces may derive by means of a region growing algorithm [4, 16]. This kind of algorithm is able to aggregate together, according to an imposed threshold, contiguous voxels that have similar geometric descriptors (normal direction and distance from the origin, in the case of planes; axis and radius, in case of cylinders). Region growing algorithms define clusters starting from "seeds" able to describe a feature. In planar voxels case, the unit vectors associated to the normal of the best-fit planes inside the voxels, $(n_x\ n_y\ n_z)_{ijk}$, are verified to a specific unit direction, (a b c), by means of a threshold condition derived from the so-called L_2 orientation norm [2, 10] and to an adjacency matrix. This procedure, based on Single Value Decomposition, explained in [2], consent to perform the plane clustering according to every significant direction of the acquired components that are found in S.

Analogously, in cylindrical voxel, after the local fitting explained in the following section, regardless of the fitting procedure chosen, the clusterization is made according to seeds found from other suitable descriptors, as radius, considering that the aggregation by means of the axes is taken into account by the voxel connection and proximity. This conduct to obtain a cylindrical clustering according every significant radius of the component. Result of clusterization is the global surface recognition. By that, all the found global features, planar or cylindrical, can be submitted to the measurement procedure, explained in [5], able to perform tolerance control and to insert the result in a PDM file regarding the component, arriving to evaluations of dies and suppliers.

3 Fitting of Cylindrical Surfaces Inside Curved Voxels

In this section, the problem of finding cylindrical surfaces is faced. According to Fig. 2 it requires a proper local surface analysis inside the voxels associated to curved surfaces, then region growing based on eq. 2 must be run.

3.1 LM(Kasa) Approach

As first approach, local cylindrical surfaces have been recognized through a fitting strategy that is related to a common practice in CMM inspection: circle fitting of sections [4]. Circle fitting can be approached by geometric or algebraic fits. In the first case, the unknown parameters (position of the centre, C, and radius, R) are found through an iterative regression process, as, for example Levenberg-Marquardt (LM). Algebraic approaches like Taubin or Kasa [17, 18]. Kasa is the fastest method that works with good results in case of complete circles but it loses accuracy in case of points along arcs with small diameters. Regression approaches, like LM, are strictly related to the adopted first guess, moreover local minimum cannot be excluded. As reported in [4], our firstly developed application, LM(Kasa), couples together Kasa and LM, using Kasa to give a rough estimation of the curved voxel surface that is used as first guess of LM. It has been demonstrated that LM(Kasa) is, at least, performant as Taubin or even more, in some different conditions (high or scattering level, small and large arc lengths, small or high values of R).

Major problem related to LM(Kasa) approach is the necessity of finding proper projection directions, to find circular sections. It has to be associated with the concept of Intrinsic Reference System (IRS), treated in the following, and to the hypothesis of having pins and holes with axes along this reference system.

For satisfying this hypothesis, from the voxel structure over-imposed on the component, an IRS may be found, looking for the plane directions associated to the most populated sets of voxels. In our case, in fact, it resulted that it may be assumed as a consistent reasoning since these injection molded components are characterized by functional features on orthogonal planes or cylinders whose axes are parallel to three principal directions of the component. Thus, the most populated set of voxels are the ones that contain planar surfaces oriented along principal directions of the component. By this, The IRS computation derives, assigning as first reference axis the planar direction of the most populated set of voxels. Then the other two axes are found between the two subsequent more populated sets, mutually orthogonal, as tested in [3].

3.2 RANSAC Approach

To release cylinder fitting from IRS, point projection may be avoided. Thus, a second fitting strategy has been investigated. It works with a RANSAC-based algorithm, offering a more generalized approach to all the kind of surfaces, also planes [19, 20]. It is an iterative algorithm, able to fit the model parameters in case of data with noise. It starts with a preliminary evaluation of the model on a minimal data set, taken randomly from the point cloud. After this, the remaining points are checked concerning their significance with the adopted thresholds (statistical confidence of the parameters, inlier maximum distance from the evaluated surface). If they are not significant, they are evaluated as outliers. The evolution is iterative until the confidence is reached and, obviously, a proper model must be selected. In our case, to avoid the use of the IRS, the first application of RANSAC was made to fit cylinders through the points inside the curved voxels. In particular, we adopted the MLESAC algorithm, found in Matlab 2017. It starts from a sample of 6 points necessary to preliminary estimate the cylinder axis. Required thresholds are the maximum distance from the surface, to be an inlier point and the confidence (default 99%). Obviously, sometimes the algorithm may fail to evaluate an axis direction, thus a first guess must be provide. It means that, in a certain way, in some critical cases, MLESAC algorithms cannot work without a first guess information, connected to the axes. In general, it resulted that they are IRS-dependent, in some cases, as LM(Kasa) procedure is. In addition, a preliminary set-up of the proper thresholds for the maximum distance inlier-surface and confidence must be done.

Sensitivity Analysis for RANSAC Parameters. The first goal of the research activity is the evaluation of the sensitivity of the algorithm in the respect of the threshold analysis. We investigated if local fitting inside curved voxels is robust enough to provide a threshold for the next region growing. The sensitivity analysis has been carried out on the maximum distance, from the evaluated surface, used to define inliers and outliers. It has been made and compared with a least squares error fit, made in Catia V5. Two sets of points, one representing a whole cylindrical surface, the other only a portion, have been used (Fig. 2). Their parameters are resumed in Table 1. The following Figs. 3(a) and (b) resume the performed sensitivity analyses. Maximum distance (the abscissa) is correlated to the entity of "noise" present. A reduction of this value makes more stringent the evaluation of the "inlier condition", assignable to data points.

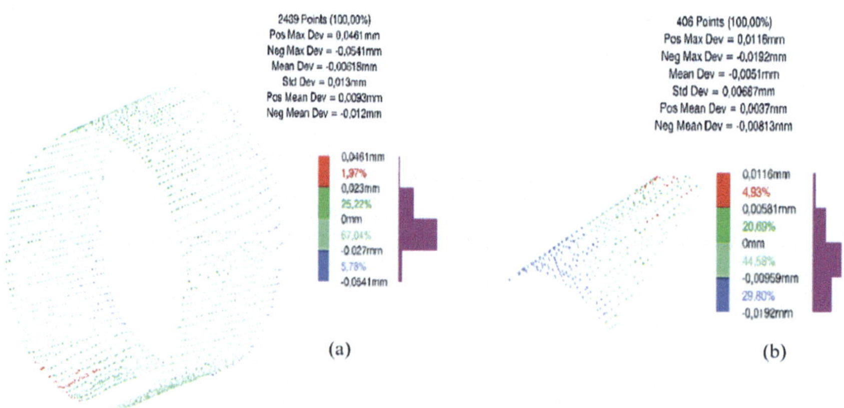

Fig. 2. Distance analysis from the cylinder fit made by CatiaV5R12: (a) Point cloud #1; (b) Point cloud #2.

Table 1. Fitting parameters found by CatiaV5R12.

	Point Cloud #1	Point Cloud #2
No. of points	2439	406
Estimated radius (mm)	0,9728	1.0103
Axis direction (mm)	[0.999, −0.007, −0.002]	[0.999, −0.360, −0.0165]
Mean error (mm)	0,0094	0.0055
Standard dev. error (mm)	0,0087	0.0033

To understand this effect, we have imposed a range for the distance from 0.5×10^{-3} mm to 0.06 mm with steps of 0.25×10^{-3} mm. These values, as said are reported as abscissa of the figures. The plot in the upper part of both show the fitting mean error in respect to MaxDistance. Smaller values of MaxDistance decrease the mean error, and obviously the number of inlier points, as reported in the graphs at the bottom of Fig. 3. Plots in the middle represent the trend of Radii of cylinders, varying the MaxDistance. For Fig. 3(a), a stable trend for Radius can be seen at about 0.97 mm, with a high scattering when MaxDistance increases. This is not suitable and consistent from a tolerance inspection point of view. It confirmed the necessity of a rule to define the Max Distance For this reason, through the comparison with Catia fittings, we investigate the Max Distance ranges nearby the maximum deviations (respectively 54.1×10^{-3} mm and 19.2×10^{-3} mm) of the distance analyses (Fig. 2). Doing so we are assuming that small probability of noisy data may occur and all the points can be inserted in the MLESAC computation, as in a standard fit. For first cloud, in a range of MaxDistance equal to $0.04 \div 0.06$ mm (in red in Figs. 3(a) and (b)), we found a fitting error of 0.017 ± 0.004 mm and a Radius of 0.9686 ± 0.0157 mm. Analogously, for the second set, in a range of $0.015 \div 0.025$ mm, we found a fitting error of 0.0053 ± 0.0017 mm and a Radius of 1.0117 ± 0.0644 mm

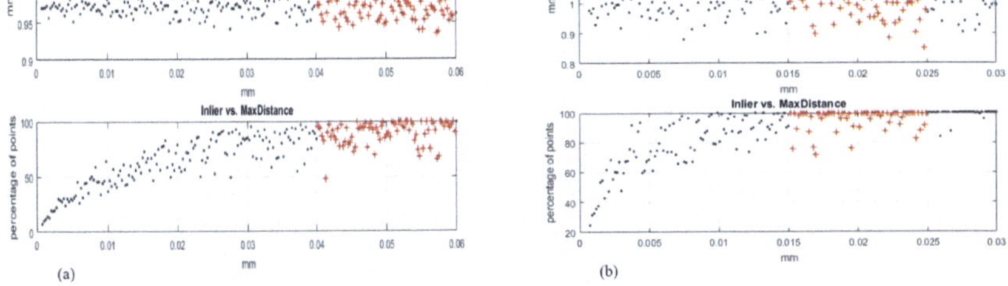

Fig. 3. Sensitivity Analysis according to the input parameter related to the maximum distance of the inliers from the evaluated surface (MaxDistance). Upper plot: Mean Error of the fit; Middle plot: Radius of the fitting; Lower plot: percentage of inliers in the respect of the total number of points: on the left Point Cloud #1, on the right Point Cloud #2.

4 Set-Up and Evaluation of the Algorithms Through Case Studies

Two different case-studies are reported: a flange and a latching lever. Both components have been used for set-up of the algorithms and their tests.

4.1 Flange for Aeronautical Applications

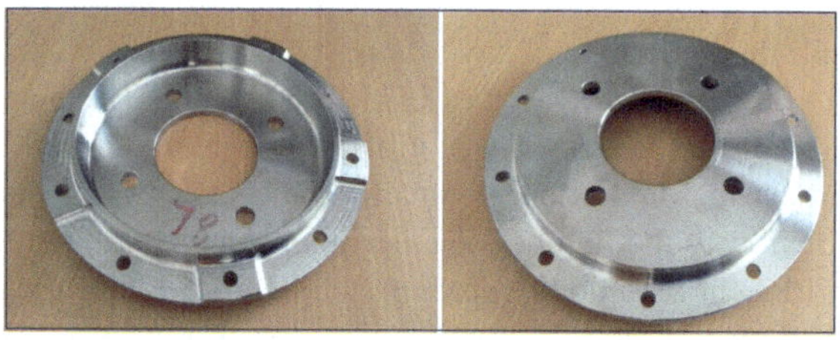

Fig. 4. Flange for aeronautical applications.

The flange is not an injection molded part (Fig. 4), but it has been selected, as testing case study, and already used in [3], to stress the algorithms with large shapes and curvatures, and details in a wide range of lengths. It is a component of a Boeing product, obtained through β-forging starting from Ti_6Al_4V Titanium alloy powders. Characteristic lengths are 208 × 208 × 30 mm, and after multiple view alignment, a point cloud of about 480000 has been obtained. Due to the fact that our focus is on cylindrical fitting, other parameters and parts of the entire procedure will be not treated here (as voxels subdivision and threshold analysis) and considered as optimized for LM (Kasa) (it is used a voxel structure of 64 × 64 × 16). As said, LM(Kasa) asks for a preliminary knowledge of the axes directions, MLESAC does not, with some exceptions. For LM(Kasa), a component IRS is defined through a specific elaboration of the most populated planar clusters that are derived from the voxel structure. In the flange case, the calculated IRS has the x-axis coincident with the longitudinal axis of the component. Therefore, LM(Kasa) assumes that axes directions of the founded cylinders are oriented as the x-axis of the IRS. This assumption is consistent with datum references. A good resolution of the proposed method has been confirmed also by the map of voxel osculating radii of the cylindrical features. The surface reconstruction of the cylinders has revealed 0.02 mm of tolerance from the nominal value and a good predictable ability of this kind of algorithm [3].

Regarding MLESAC algorithm, for the flange and the imposed voxel structure (64 × 64 × 16), it has been found that, in over the 30% of the voxels recognized as curved, the algorithm failed the surface fitting. This because of a low amount of points within each of these voxels (less than 6 points). This convinced us to use a less accurate voxel structure (32 × 32 × 16). The obtained result, reported in Fig. 5, demonstrated a not accurate fitting and recognition. Table 2 demonstrates this hypothesis according to a reduction of the number of voxels, thus, to an increase of the average number of points per voxel.

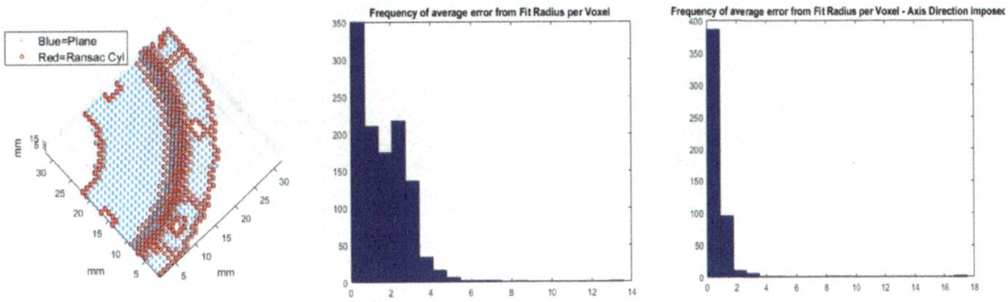

(a) (b) (c)

Fig. 5. (a) Final voxel partition in Cylindrical and Planar local surfaces; (b) Frequency distribution of the occurrence of the mean of the distance errors from the fit radius without axis imposition; (c) Frequency distribution of the occurrence of the mean of the distance errors from the fit radius with axis imposition.

More in detail, it shows a high scattering in terms of mean of the distance error among points in the cylindrical voxels and fit radius, the population of bins higher than first or second can be interpreted in this way. This represent the high difficulty of fitting cylinders with large curvature. Figure 5(c) shows the situation of the mean error distribution after the axis imposition. It seems to be improved for what concerns the possibility of finding a threshold able to divide inliers from outliers, performing a huge improvement for large curvature cylinders, but it resulted to have a higher number of failed curved voxels to be fitted as cylinders (up to the 60%). This makes the results obtained by MLESAC, in both forms, less accurate than the LM(Kasa) ones.

Table 2. Percentage of curved voxels that fail to find MLESAC solution – Maximum distance set to 0.1 mm, confidence = 99%.

Resolution	V_{ijk} length (mm)	% of curved V_{ijk} that fail MLESAC
16 × 16 × 16	6.5 × .6 × 1.9	1.50%
32 × 32 × 16	3.2 × 3.2 × 1.9	8.40%
64 × 64 × 16	1.6 * 16 * 1.8	35.80%

4.2 Latching Lever

In order to test the two strategies, another case study has been analyzed. It is an injection molded Latching Lever, used in assemblies of electromechanical switches of electric panels, made in PPS. Characteristic lengths are about 20 × 10 × 4 mm, more than one sample has been acquired obtaining, after multiple views alignment, point clouds of over 60000 points. We used this case study to evaluate performances of LM (Kasa) in comparison to the results obtainable through MLESAC, adopting, as imposed axis for the fitting, the same direction used in LM(Kasa). Using a 32 × 32 × 32 voxel structure, the local surface recognition of planar and curved featured is represented in the right side of Fig. 6.

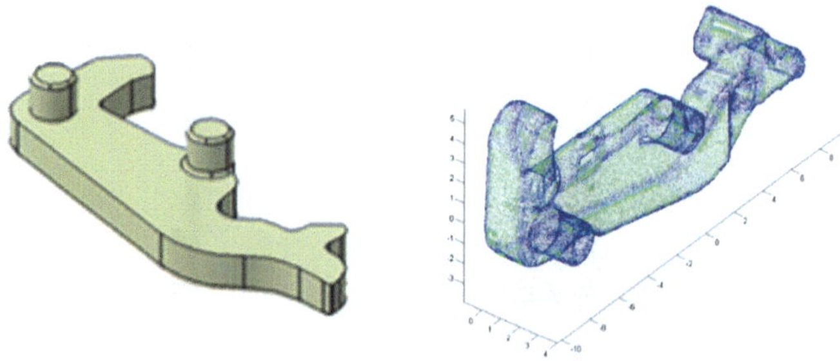

Fig. 6. Nominal CAD model of the Latching Lever (on the left); local surface recognition of planar feature, in green, and curved ones, in blue (on the right).

From the planar voxel clusterization, the IRS has been found. As already mentioned, the cylindrical features appear, also in this case, to have axis directed according to one of the principal direction, found through voxel population. In fact, in the lever, the x-axis resulted to be the one coincident with the directions of the two cylinders and the rounds present (on the left of Fig. 7). The cloud has been partitioned according to the found IRS, and the curved voxel are represented in the middle of Fig. 7. Then, cylindrical features, through LM(Kasa), have been obtained, as showed on the right of Fig. 7. To give an evaluation of this approach, cylinders A, B and C of Fig. 8 have been compared with interactive post-processing made by Catia. In this case, Cylinder A has an effective value of R = 0.974 mm, we also obtained R = 0.972 mm (for Cylinder B) and R = 0.999 mm (for Cylinder C). Standard deviation of the fittings are always less than 0.012 mm. Through the proposed algorithm, the radii of each feature are: 0.974 mm for cylinder A and B, 1.036 mm for cylinder C with a standard deviation less than 0.006 mm. Using MLESAC with imposed axis (1, 0, 0) as input parameter, similar results have been obtained, but improved in the detection of the common radii along the voxels on the width of the free form part (Fig. 9).

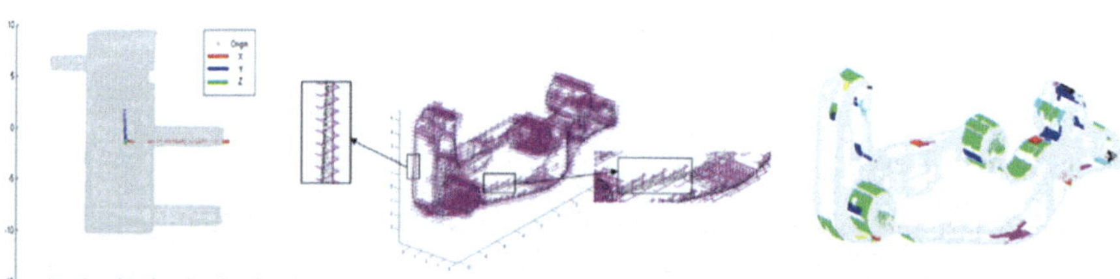

Fig. 7. The calculated IRS for the lever (on the left); curved voxel partition (in the middle); cylindrical feature recognition through LM(Kasa) (on the right).

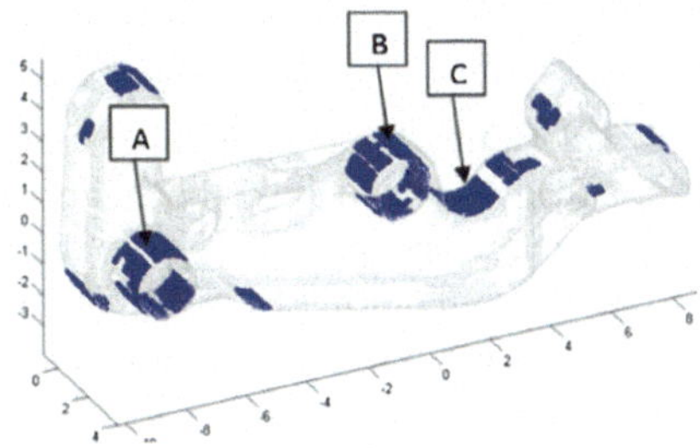

Fig. 8. Selected cylinders with R = 1 mm.

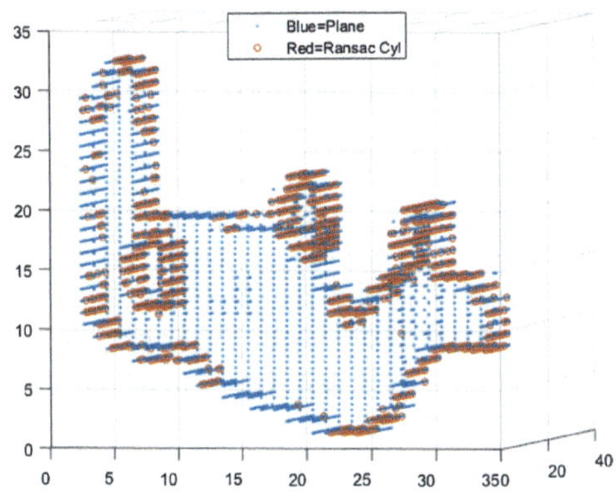

Fig. 9. MLESAC with imposed axis (1, 0, 0) applied on Latching Lever.

4.3 Discussion

For what concerns the case study of the flange, it can be said that LM(Kasa) showed good results despite the fact that asks for preliminary knowledge of the cylinder axis directions, taken out by the IRS calculation. MLESAC may avoid this request, not in totality of cases, but more problems arise to find a proper threshold for affirming that the voxel is cylindrical with any doubt. It is due to problems about finding large curvature cylinders through points on a small arc length, when no axis direction is given as first guess. This makes the fit model highly scattered also among contiguous curved voxels, ruining the region growing – hierarchical clustering of large radius areas. Imposing a first-guess direction, many curved voxels are not fit (60%), but radius evaluation highly improves. Concerning cylindrical fitting for the flange, MLESAC cannot overcome the problems of LM(Kasa) that results more reliable for large curvature cylinders through points distributed on a small arc length. On the contrary, in features with lower values of radii, as for latching lever, the two strategies can drive to similar results. It has to be noticed that also in this case, in order to perform a significant recognition, MLESAC needed the first guess, with the imposition of the cylinder axis direction, failing again to what we had thought could be a huge advantage of this method, proved instead to be not so disruptive, as depicted also in literature, in respect to that previously developed.

5 Conclusions

In this work, two kinds of algorithms for cylindrical feature recognition have been compared. One, LM(Kasa), already developed by the authors in the past, had already proved to be reliable and moderately accurate, presenting the disadvantage of being a 2D processing of points to be fit, so requiring the axis direction of the cylinder to be recognized. Due to this disadvantage, after a literature research, authors developed another procedure, based on RANSAC algorithms. The main advantages of this procedure, as depicted in the literature, were mainly connected to the possibility of avoiding a first-guess (6 points, so, a direction), thus, recognizing cylinders without the dependence from the definition of an IRS (Intrinsic Reference System). The comparison has been made through two different case studies: a flange for aeronautical application and a latching lever for electromechanical of electric panels. Due to the differences between the case studies, it has emerged that for what concerns cylinders with high radii, LM(KASA) performs better than MLESAC. In addition, this second approach revealed its needing of a first-guess direction exactly as LM(Kasa), but without arriving to the same level of accurate recognition. For what concerns cylinders with smaller radii, as the features present in the lever, the two developed algorithms resulted to be quite similar, depending both by the imposition of the axis of the cylinder to be recognized and performing results with the same level of accuracy. Further researches and tests are needed in order to obtain a more accurate evaluation of these algorithms, but this comparison showed that the already developed algorithm, LM (Kasa), remains currently the best and more robust choice for performing cylindrical feature recognition oriented to tolerance inspection and quality controls.

Acknowledgements. Author are pleased to thank Alessio Trifirò and the Quality Control Department of ABB Sace S.p.A., site of Santa Palomba (Rome), Italy, which have supported the research.

References

1. Hähnel, S., Pini, F., Leali, F., Dambon, O., Bergs, T., Bletek, T.: Reconfigurable robotic solution for effective finishing of complex surfaces. In: 2018 IEEE 23rd International Conference on Emerging Technologies and Factory Automation (ETFA), Turin, Italy, pp. 1285–1290 (2018)
2. Bici, M., Campana, F., Petriaggi, S., Tito, L.: Study of a point cloud segmentation with part type recognition for tolerance inspection of plastic components via reverse engineering. Comput.-Aided Des. Appl. **11**(6), 640–648 (2014)
3. Bici, M., Campana, F., Trifirò, A., Testani, C.: Development of automatic tolerance inspection through reverse engineering. In: 2014 IEEE Metrology for Aerospace (MetroAeroSpace), Benevento, Italy, pp. 107–112 (2014)
4. Bici, M., Campana, F., Trifirò, A.: Automatic post-processing for tolerance inspection of digitized parts made by injection moulding. Comput.-Aided Des. Appl. **13**(6), 835–844 (2016)

5. Bici, M., Broggiato, G.B., Campana, F., Dughiero, A.: Computer aided inspection procedures to support smart manufacturing of injection moulded components. Procedia Manuf. **11**, 1184–1192 (2017)
6. Germani, M., Mandorli, F., Mengoni, M., Raffaeli, R.: CAD-based environment to bridge the gap between product design and tolerance control. Precis. Eng. **34**(1), 7–15 (2010)
7. Varady, T., Martin, R.R., Cox, J.: Reverse engineering of geometric models - an introduction. Comput. Aided Des. **29**(4), 255–268 (1997)
8. Bradley, C., Currie, B.: Advances in the field of reverse engineering. Comput.-Aided Des. Appl. **2**(5), 697–706 (2005)
9. Attene, M., Falcidieno, B., Spagnuolo, M.: Hierarchical mesh segmentation based on fitting primitives. Vis. Comput. **22**(3), 181–193 (2006)
10. Shamir, A.: A survey on mesh segmentation techniques. Comput. Graph. Forum **27**(6), 1539–1556 (2008)
11. Ali, F., Chowdary, B.V., Imbert, C.A.: Part design and evaluation through reverse engineering approach. Int. J. Agile Manuf. **11**(1), 73–82 (2009)
12. Di Stefano, P., Bianconi, F., Di Angelo, L.: An approach for feature semantics recognition in geometric models. Comput.-Aided Des. **36**(10), 993–1009 (2004)
13. Attene, M., et al.: Mesh Segmentation - A Comparative Study. IEEE Computer Society, Washington, DC (2006)
14. Keller, P., Hering-Bertram, M., Hagen, H.: Reverse engineering with subdivision surfaces. Computing **79**(2–4), 119–129 (2007)
15. Courtial, A., Vezzetti, E.: New 3D segmentation approach for reverse engineering selective sampling acquisition. Int. J. Adv. Manuf. Technol. **35**(9), 900–907 (2008)
16. Lee, Y., Park, S., Jun, Y., Choi, W.C.: A robust approach to edge detection of scanned point data. Int. J. Adv. Manuf. Technol. **23**, 263–271 (2003)
17. Taubin, G.: Estimation of planar curves, surfaces, and nonplanar space curves defined by implicit equations with applications to edge and range image segmentation. IEEE Trans. Pattern Anal. Mach. Intell. **13**(11), 1115–1138 (1991)
18. Kasa, I.: A circle fitting procedure and its error analysis. IEEE Trans. Instrum. Measur. **IM-25**(1), 8–14 (1976)
19. Schnabel, R., Wahl, R., Klein, R.: Efficient RANSAC for point-cloud shape detection. Comput. Graph. Forum **26**(2), 214–226 (2007)
20. Torr, P.H.S., Zisserman, A.: MLESAC: a new robust estimator with application to estimating image geometry. Comput. Vis. Image Underst. **78**(1), 138–156 (2000)

Geometrical Product Specification and Tolerancing

Tolerance Prediction for Determinate Assembly Approach in Aeronautical Field

Rocco Mozzillo[1(✉)], Ferdinando Vitolo[1], Paola Iaccarino[2], and Pasquale Franciosa[3]

[1] Department of Industrial Engineering, University of Naples Federico II, Naples, Italy
rocco.mozzillo@unina.it
[2] Department of Research and Development, LAER Group S.p.a., Airola, Italy
[3] Digital Lifecycle Management, WMG, University of Warwick, Coventry, UK

Abstract. The optimization of the aeronautical assembly lines is one of the most challenging tasks in development of a new aircraft. To date the Aeronautical companies are founding project related to the optimization of the manufacturing processes, impacting on the reduction of the non-recurring costs. The work presented focuses on methods for the implementation of the determinate assembly approach in the case of manufacturing of aeronautical structures. The methods are based on the prediction of the coupling tolerances between airframe components in order to reduce the numbers of the features to be manufactured in assembly. All that to lead to a reduction of the non-recurring costs.

Keywords: CAD · Determinant Assembly · Determinate Assembly · Aeronautical products · Assembly process · Statistical distribution · Hole to hole · Tolerance prediction

1 Introduction

The aerospace assembly processes are characterized by long cycle time and consequently high costs due to reworking within the assembly jigs. Reliance on structures and the large number of different monolithic jigs has also meant this assembly approach has not resulted in flexibility since the jigs are highly inflexible, costly and have long lead times [1]. The high cost and long lead times of traditional tooling makes difficult for aircraft manufactures to efficiently meet ever-changing market demands. The large size, low relative stiffness and high positional tolerances required for aircraft components drive the requirement for rigid fixed tooling to maintain the precision part relationships over time. The use of heavy steel structures built on a concrete foundation for assembly tooling contributes to high capital costs and a lack of flexibility [2]. The assembly tooling which is used to control the form of assemblies is typically a heavy steel structure built on a concrete foundation. Assembly tooling accounts for approximately 10% of the total manufacturing cost of an airframe [3]. The design and management of the tooling during the program life depends on the management of the assembly processes. It has been reported that 67–70% of all design changes in the aerospace and

C. Rizzi et al. (Eds.): ADM 2019, LNME, pp. 229–240, 2020.
https://doi.org/10.1007/978-3-030-31154-4_20

automotive assembly processes are related to product-dimensional variations [4]. Since the assembly jigs cannot accommodate easily these variations, the assembly processes become critical and not oriented to adopt any changes in the aircraft project. Moreover, up to 40% of the total manufacturing cost of an airframe is incurred during assembly [5]. Use of today's advance 3-Dimensional CAD systems [6, 7], the cyber-physical system [8] coupled with advanced design method [9, 10] and the high accuracy of Computer Numerical Control (CNC) machines could enable the success of the implementation of Determinant or Determinate Assembly approach (DA). It could be said the two terms are more or less alternative in scientific literature. Indeed *Cadena* [11] defines the Determinant Assembly as an assembly process approach that allows for quicker assembly by using features of the parts, such as drilled holes, to quickly align components without the use of additional tooling to aid with alignment. *Hartmann* asserts in his work [12] Determinate Assembly is a term used to describe the practice of designing parts, which fit together at a predefined interface, and do not require setting gauges or other complex measurements and adjustments. In this optic DA could reduce the manufacturing costs and cycle time, increasing the quality of the products [12]. This will reduce the tooling costs and eliminates the need for local drilling, de-burring and re-assembly of the bracket to rib interface, reducing the cycle time of the operation. Ultimately, self-indexing components mean that the there is more flexibility as to what point in production the bracket can be attached to the rib. This allows for a reduction in cycle time and cost, whilst increasing production flexibility [13]. The present paper proposes two different methods for the tolerance prediction in the case of components assembled with DA approach. Since the aero - structures are characterized by components with flat and double curved coupling surfaces the authors presents for both kind of components a dedicated method for prediction of the manufacturing tolerances. The work has been carried out thanks to a close collaboration between LAER Spa, the department of industrial engineering of the University of Naples Federico II and the Digital Lifecycle Management of the University of Warwick.

2 State of Art: Traditional Assembly Issues

The production of large aerospace assemblies combines the inefficiency of craft production with the inflexibility of the early forms of mass production. Difficulties in maintaining very close tolerance; the large number of different operations for low production volumes; the interfaces often imprecise for composite components; the difficult to drill holes in different components which will match and lock the assembly into its correct form are some of the critical aspects related to the manufacturing and assembly of aeronautical structures. From the early days of aircraft manufacturing dedicated tooling has been used in the assembly process to ensure the attainment of strict assembly tolerances and product quality. Traditionally the jigs are used to support components and act as positioning gauges in order to guarantee the required quality level. The traditional solution consists on using monolithic jigs which holds flexible components to their correct final form, interface gaps are then measured in the jig so that shims can be fitted, and holes are drilled through the stack of components. Then the components are temporary disassembled in order to debur holes, clean and apply sealant before the final assembly takes place. Each aeronautical assembly has its own jig, which is designed to be suitable only to assemble that particular structure type. Jigs are critical in the development of new manufacturing techniques and largely dictate the level of flexibility that a manufacturing system can achieve: the large size, low relative stiffness and high positional tolerances required for aircraft components, drive the requirement for rigid fixed tooling to maintain the precision part relationships over time.

Some critical aspect related to the use of jigs for the assembly are the following: (i) difficulty in recovering tooling costs for short production runs; (ii) requirement for regular maintenance and calibration of the tooling possibly using 'master' tooling; (ii) need to store tooling and 'master' gauging when not in use.

Traditionally, the design of an aircraft is also used to produce the aircrafts assembly tools with which to manufacture that specific aircraft. Typically, precision tooling and assembly equipment are used to set the relationships between the myriad of parts, which make up the aircraft frame. In order to enable production rates to rise, it is necessary for the assembly equipment to be ready for production as soon as the aircraft components are ready for assembly; however, major assembly tools are typically larger than the aircraft parts themselves. There is a significant lead-time associated with the manufacture and calibration of tools, furthermore the use of heavy steel structures built on a concrete foundation for assembly tooling, contributes to high capital costs and a lack of flexibility. Any tooling realignment is time-consuming and can often be achieved by the trimming and shimming. In order to reduce cost and to increase affordability there is a growing requirement for tooling to be fulfilled:

- flexible, so that it can be quickly adapted to small variations within a single structure;
- reconfigurable, so that it can be changed within a structure family;
- reconstructable, so that it can be changed between different structure families.

The high cost and long lead times of traditional tooling makes it difficult for aircraft manufactures to efficiently meet ever-changing market demands: due to the long-lead time and effort required for design and manufacture, the signal-purpose of Dedicated Tooling is among the major cost drives in aircraft manufacturing, both in terms of non-recurring and recurring costs: traditionally, each different assembly process requires at least one dedicated rigid assembly tooling. When a manufacturing process is completed, or parts and operations are modified, manual set-up or modifications would be needed. For what has been said it is clear that the implementation of assembly approaches based on the simplification of assembly processes and consequently on the reduction of the monumental jigs plays a pivotal role in the optimization of the manufacturing process itself.

3 Tolerance Prediction and Determinant Approach

The assembly method here presented is based on the idea to use DA approach coupled with statistical prediction of the manufacturing tolerances. In particular the main objective consists on the prediction of the manufacturing tolerances that allows the assembly of the components without monumental jigs. It is clear that more costs are moved on parts rather than the assembly tools with consequent reduction of the assembly hours on the final line and reducing the non-recurring costs. The method is based on the statistical control of all processes involved in the production. Each manufacturing process is characterized by proper statistical distribution of the output parameters, the method here proposed manages and predicts this kind of parameters Since one of the most demanding assembly processes in aeronautical field is the drilling and the fastening of components with a large number of holes, our study investigates this process. All kind of coupling between aeronautical components can be classified in two different typologies: (i) components joined through flat shaped surfaces; (ii) components coupled with double curved surfaces. For each one, different strategies have been implemented to predict the tolerances.

In particular, let's assume the target clearance between hole and pin as the difference between the hole/pin diameter that allows for the assembly of the two component and the actual clearance the same values that the manufacturing processes can assure. In detail the target clearance can be defined starting from the technical specification of the typical fastener [16] and correspond to distribution of dimensional values of bolt and hole that guarantees the two component can be assembled. The actual clearance statistical distribution is the actual values distribution that the manufacturing processes can assure; it depends on the drilling processes and the characteristics of the tool. Once the statistical parameters of the two distributions (target and actual clearance) are calculated, a comparison between them is carried out, all that in order to define the optimized parameters of manufacturing processes, and consequently, to assure the assembly of the self-indexed components (Fig. 1). Moreover the method is based on the following assumptions: (i) the stiffness of each component shall not allow any changes in their shape due to own weight or positioning and assembly activities; (ii) the components to be assembled are drilled with finite diameters holes; (ii) the components are joined through the installation of the fasteners no dedicated drilling template are needed; (iii) no shimming activity is requested on the interface areas between the components; (iv) the components are machined in controlled temperature environment in order to guarantee their precision.

In this optic, starting from definition of holes diameters and bolt statistical distribution and assuming that both hole and bolt diameters are distributed following Normal Gaussian functions 2 sigma the characteristics of the Clearance (hole VS bolt) are defined (Fig. 2).

The achievable clearance and relevant connected probability are estimated through statistical prediction of the clearance between the bolt and hole. This approach has been implemented in the case of coupling between flat shaped components [17] and its feasibility has been demonstrated through trials on real prototypes [18].

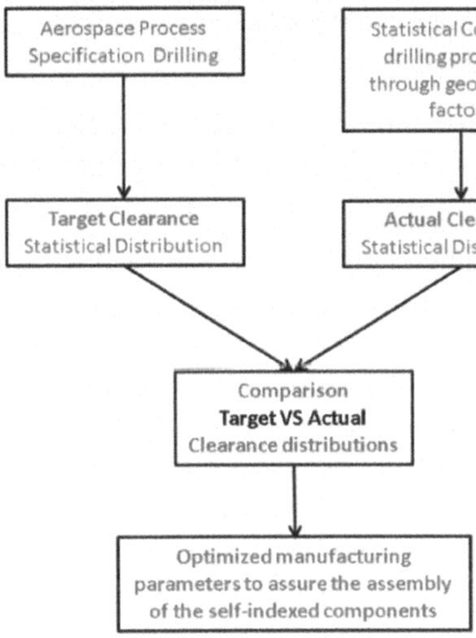

Fig. 1. Work flow tolerance prediction in the case of flat shaped surfaces

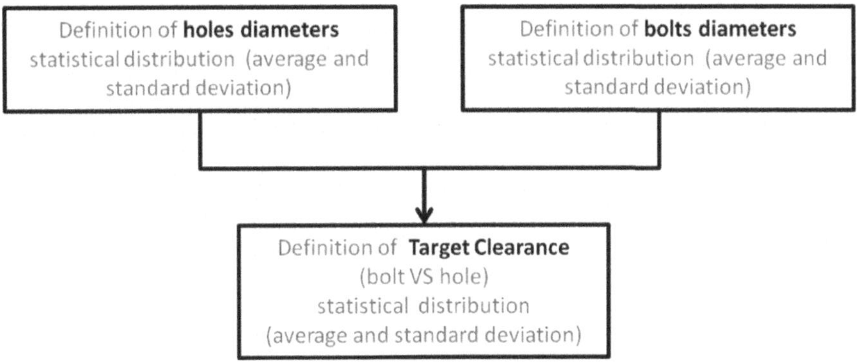

Fig. 2. Statistical prediction of target clearance (Hole VS bolt)

The proposed method is too restrictive in the case of double curved components since other degrees of freedom are added to the problem formulation. Due to parts variations, nominal clearances (difference between the maximum hole diameter and minimum pin diameter) between pin and holes could be insufficient to guarantee the assemblability of the components. In particular Let P1 and P2 two parts that have to be pinned and let assume P1 and P2 having a mating surface characterized by double curvature with large radius, thin thickness and rigid behavior. For each part a drilling process makes a set of n sequential holes. Hence mating condition is then obtained by a pin through two holes. For this reason a more generic assumption has been done: "*two components joined by fasteners can be assembled if each fastener passes through the coupled holes*".

The total variations of the holes are then decomposed in global and local variations. These variations are characterized as follow:

- global variation – it is computed as form variation related to the curvature;
- local variation – it is computed as hole variation in a planar zone nearby the hole.

Initially the manufacturing tolerances are those come from the design of the assembly itself. Deviation sets for the global variation are generated by using morphing while a random generation within tolerance specifications is used for holes and pins variations [6]. A statistical tolerance evaluation [14, 15] is carried out in order to compute the assemblability index which returns the percentage of the feasible assembly pins. Then a trade-off between the assemblability index and process parameters leads to an optimization of the manufacturing tolerances in order to allow the implementation of the DA approach (Fig. 3). The prediction tolerances of each component to be assembled are now defined taking into account not only the requirements come from the component design but also from its manufacturing processes.

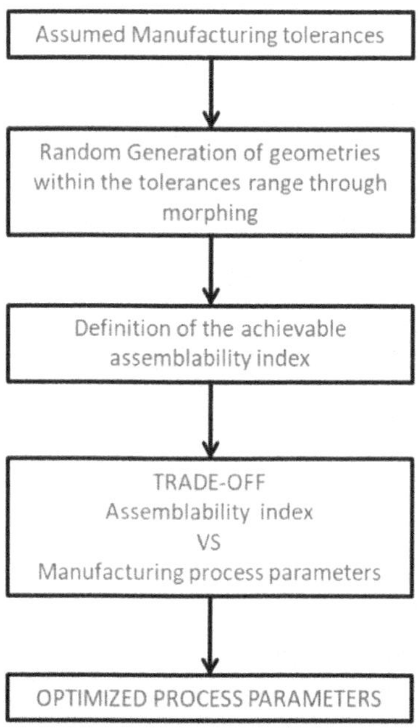

Fig. 3. Work flow method for double curved components

The approach is based on the following assumptions: rigid parts, drilling from the mating side through typical NC drilling process; big curvature radius, small hole diameters, parts assembled without jigs.

For the local variations there exist five main geometrical factors (Figs. 4 and 5):

- F1, dimensional tolerance of hole diameter;
- F2, positional tolerance of holes axis due to NC drilling machine;

- F3, orientation tolerance of holes axis due to NC drilling machine;
- F4, holes axis positional tolerance due to part supporting tool mis-position during drilling
- F5, holes axis orientation tolerance between normal to real surface and nominal hole axis due to part supporting tool mis-position during drilling

The F1 factor gives information about the dimension of the holes, the others concern their position and orientation. The aforementioned factors depend on drilling processes itself and on the tools used to fix the component on the NC machine working areas (Table 1).

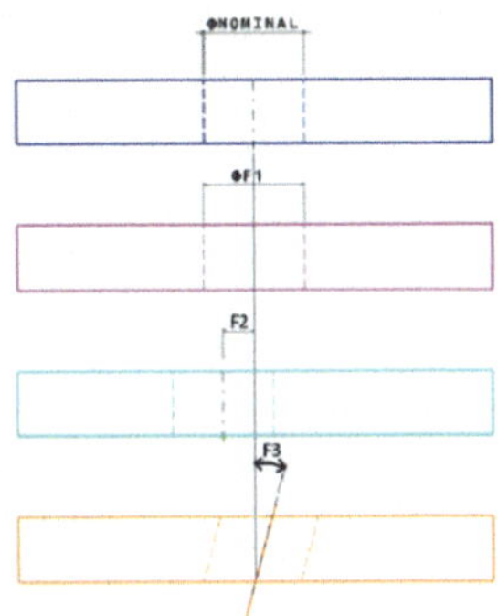

Fig. 4. Influence factors of Hole/Bolt tolerance due to drilling

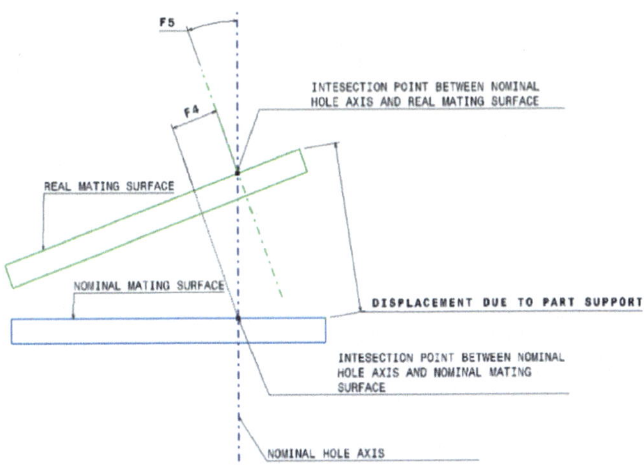

Fig. 5. Influence factors of Hole/Bolt tolerance due to part positioning and fixing on machining tools

Table 1. Geometrical factors for drilling tolerances estimation

Factor	Average	Sigma
F1	Statistical average of hole diameter distribution	Variability range of holes diameter/4
F2	Not assigned	Variability range of positional tolerance of the hole/4
F3	Not assigned	Variability range of orientation tolerance of the hole/4
F4	Not assigned	Variability range of positional tolerance of the hole/4
F5	Not assigned	Variability range of orientation tolerance of the hole/4

4 Assemblability Computation

The core of the method for the tolerance prediction of double curved component is the definition of an assemblability index. For this scope a dedicated algorithm has been developed. The general workflow of the algorithm to perform the variational analysis of compliant assemblies is depicted in Fig. 6. The analysis uses Morphing mesh: it is a technique used in computer graphic applications as a powerful tool for shape modelling and designing. The analysis algorithm is composed by the following steps:

- Modelling of nominal parts - generate a nominal CAD geometry;
- Tolerance specifications – define all critical features and its tolerances;
- FE model generation – a meshed model is generated starting from the CAD model;
- Surface form variation by using morphing mesh technique (global variation);
- Geometric variation of the holes (local variation);
- Assembly simulation;
- Assemblability index computation

Let the dimensional, positional and form tolerances for holes and pins are known; the assemblability between the two surfaces is guaranteed if the pin passes through both holes. According to Fig. 7, it can be defined a pin-holes local reference frame fixed to the hole center at the tool side.

Let Ω_1 the local reference, C_1^1 its origin which is the circumference center at tool side and D_1 the nominal value of the hole diameter. Due to geometrical factors F1-5, hole diameter, axis position and orientation varying. Let C_2^1 the circumference center at the tool outing side, the direction of hole axis can be defined as $\overline{C_1^1 C_2^1}$ segment. Generally, axis can rotate around the three reference axes. These rotations can be reduced to only one (α) by choosing a specific local reference frame orientation. Therefore, inclination can be reduced to a one-dimensional variation. Let C_1^1, C_1^2, w_1, w_2, α_1 and α_2 are known; it can be formally writing:

$$C_2^1 = C_1^1 \frac{w_1}{\cos(\alpha_1)} \tag{1}$$

$$C_2^2 = C_1^2 \frac{w_2}{\cos(\alpha_2)} \tag{2}$$

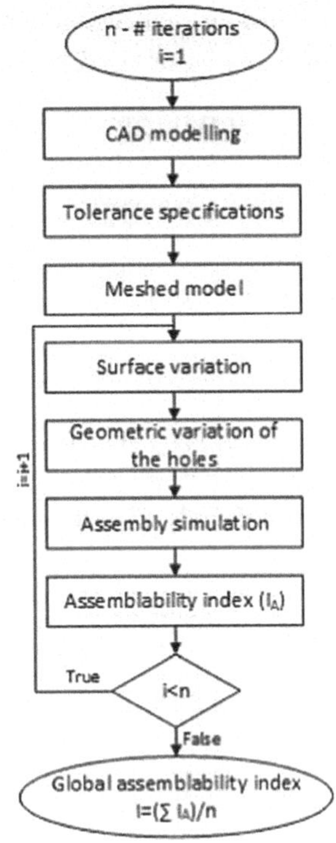

Fig. 6. Algorithm work flow

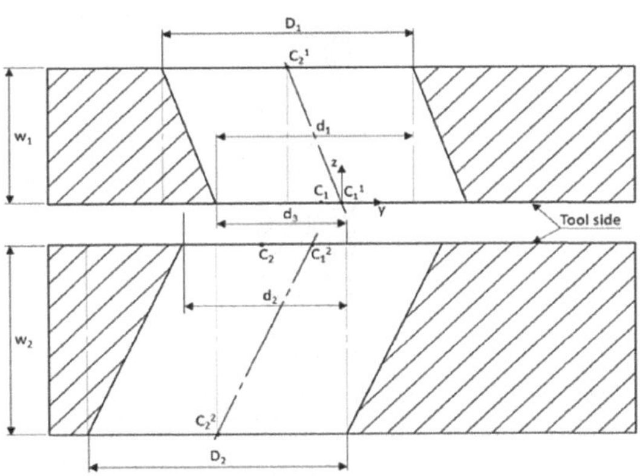

Fig. 7. Pin-holes scheme

For the varied hole can be defined an equivalent diameter d_1, centered in C_1, as the maximum diameter of the pin that can be inserted along nominal axis direction.

$$d_1 = D_1 - 2tg(\alpha_1) \tag{3}$$

$$d_2 = D_2 - 2tg(2) \tag{4}$$

At the mating side, it can be defined the mating diameter d_3, centered in C_3 which is the maximum diameter at tool side:

$$d_3 = R_1 + R_2 - \Delta \tag{5}$$

with $\Delta = \|C_2 - C_1\|$.

Due to mating condition and holes variations, the actual mating diameter d_{ef} centered in C, can be defined as the maximum diameter of the pin which can be inserted trough both holes

$$d_{ac} = \begin{cases} d_1 \text{ and } C = \mathrm{C}_1 \text{ if } \mathrm{d}_1 < d_2 \; and \; d_1 < d_3 \\ d_2 \text{ and } C = \mathrm{C}_2 \text{ if } \mathrm{d}_2 < d_1 \; and \; d_2 < d_3 \\ d_3 \text{ and } C = \mathrm{C}_3 \text{ if } \mathrm{d}_3 < d_1 \; and \; d_3 < d_2 \end{cases} \tag{6}$$

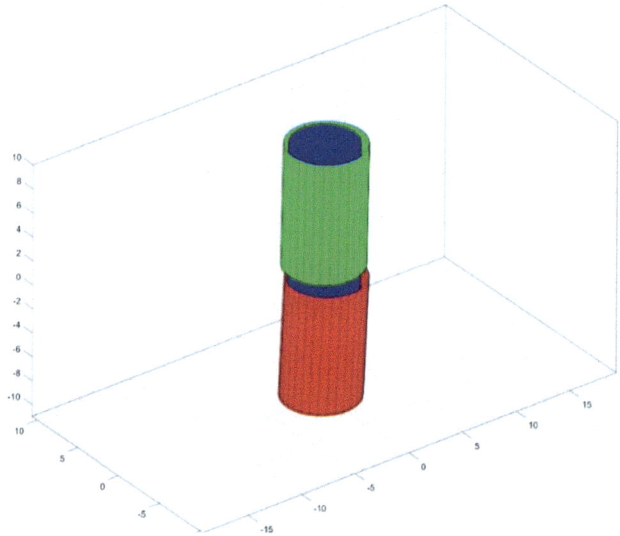

Fig. 8. Pin holes surfaces envelope

The assemblability index, I_A, defines the number of pins inserted between the two holes without any deformation; it can be defined as follow:

$$I_A = \frac{\sum_i^n \begin{cases} 1 & d_{p_i} \leq d_{ac,i} \\ 0 & d_{pi} > d_{ac,i} \end{cases}}{n} * 100 \tag{7}$$

where: i represent the ith mating, d_p the pin diameter within tolerance range and d_{ac} is the maximum effective diameter of the ideal cylinder that fit the two holes.

An example of pin holes statistical assembly simulation is shown in Fig. 8.

5 Conclusion

The work here proposed could be the winning key for the reduction of time and costs in the final assembly line. In this paper is reported part of the work that the authors are developing to achieve the implementation of the DA approach in assembly of aeronautical structures. Two methods for tolerance prediction are presented, one is dedicated to components with coupling surfaces flat shaped, the other deals with components with double curved coupling surface. The feasibility of the first has been proved trough trails on real prototypes, the second is still in development phase and needs its consistency has to be proven. In particular the algorithm for assemblability index definition has been provided and here described. Future development will be focused on the implementation and testing of the algorithm on real prototypes to check its consistency.

References

1. Muelaner, J.E., Kayani, A., Martin, O., Maropoulos, P.: Measurement assisted assembly and the roadmap to part-to-part assembly. In: 7th International Conference on Digital Enterprise Technology, pp. 11–19. University of Bath (2011)
2. Muelaner, J.E., Martin, O.C., Maropoulos, P.G.: Achieving low cost and high quality aero structure assembly through integrated digital metrology systems. ProcediaCirp **7**, 688–693 (2013)
3. Burley, G., Odi, R., Naing, S., Williamson, A., Corbett, J.: Jigless aerospace manufacture-the enabling technologies. SAE Technical Paper, No. 1999-01-2286 (1999)
4. Wang, H., Ceglarek, D.: Quality-driven sequence planning and line configuration selection for compliant structure assemblies. CIRP Ann. Manuf. Technol. **54**(1), 31–35 (2005)
5. Bullen, G.N.: The mechanization/automation of major aircraft assembly tools. Prod. Inventory Manage. J. **38**(3), 84 (1997)
6. Franciosa, P., Gerbino, S., Lanzotti, A., Patalano, S.: Automatic evaluation of variational parameters for tolerance analysis of rigid parts based on graphs. Int. J. Interact. Des. Manuf. (IJIDeM) **7**(4), 239–248 (2013)
7. Vitolo, F., et al.: Controlling form errors in 3D printed models associated to size and position on the working plane. Int. J. Interact. Des. Manuf. (IJIDeM) **12**(3), 969–977 (2018)

8. Tarallo, A., Mozzillo, R., Di Gironimo, G., De Amicis, R.: A cyber-physical system for production monitoring of manual manufacturing processes. Int. J. Interact. Des. Manuf. (IJIDeM) **12**(4), 1235–1241 (2018)
9. Marzullo, D., Bachmann, C., Coccorese, D., Di Gironimo, G., Mazzone, G., You, J.H.: Systems engineering approach for pre-conceptual design of DEMO divertor cassette. Fusion Eng. Des. **124**, 649–654 (2017)
10. Papa, S., Lanzotti, A., Di Gironimo, G., Balsamo, A., Ideas, F.J.: A new interactive railway virtual simulator for testing preventive safety. In: Computers in Railways XVI: Railway Engineering Design and Operation, vol. 181, p. 367 (2018)
11. Cadena, C.: Determined to find a better way. Boeing Front. 16–17, October 2006
12. Hartmann, J., Meeker, C., Weller, M., Izzard, N., Smith, A., Ferguson, A., Ellson, A.: Determinate assembly of tooling allows concurrent design of Airbus wings and major assembly fixtures. SAE Technical Paper, No. 2004-01-2832 (2004)
13. Irving, L., Ratchev, S., Popov, A., Rafla, M.: Implementing determinate assembly for the leading edge sub-assembly of aircraft wing manufacture. SAE Int. J. Aerosp. **7**(2), 246–254 (2014). https://doi.org/10.4271/2014-01-2252
14. de Laplace, P.S.: Analytical theory of probability (1812)
15. Erto, P.: Probabilità e statistica per le scienze e l'ingegneria. McGraw-Hill, Italy (2008)
16. Bombardier Aerospace Process Specification. Drilling of composite and Composite/Metallic Assemblies. BAPS 188-007_REVD
17. Mozzillo, R., Iaccarino, P., Vitolo, F., Franciosa, P.: Design and development of smart assembly process: the case of complex aeronautical systems. In: Proceeding of Metrology for Industry 4.0 and IoT, 4–6 June 2019 (in press)
18. Iaccarino, P., Mozzillo, R., Inserra, S., Cerreta, P.: Determinant assembly approach for composite one-shot cured structures. Int. J. Adv. Manuf. Technol. (in press)

Robust Parameter Analysis of Compliant Part Models for Computer Aided Tolerancing

Alberto Vergnano, Francesco Gherardini(✉), Andrea Petruccioli, Enrico Bonazzi, and Francesco Leali

"Enzo Ferrari" Department of Engineering, University of Modena and Reggio Emilia, Via Pietro Vivarelli 10, 41125 Modena, Italy
francesco.gherardini@unimore.it

Abstract. Sheet metal parts are widely used in automotive, aerospace, ship and consumer goods industries. The final dimensions of a sheet metal assembly result from the parts deformation, which in turn is affected by many variations in material, thickness and single parts dimensions. The tolerance analysis on sheet metal assemblies improve the knowledge about the process. Advanced simulations enable the optimization of product features, GD&T scheme and assembly process. Moreover, Variational Models of both the product and the assembly system enable to assess the sources of 3D error propagation from the different contributors. However, the simulation results are very affected by the modelling approach of critical components, such as the Fixture Systems. The present paper firstly introduces a strategy to model the Fixture System and the assembly process for compliant parts. Then, a robust analysis of the variations in the model with respect to the modelling factors is performed by a Design of Experiments. A case study on an automotive fender is discussed. The results demonstrate that the modelling strategy of the clamping operation have the main effects, while the modelling of locators scheme, spot joints and FEM meshing are less important.

Keywords: Computer Aided Tolerancing · Design method · 3D tolerances · Compliant parts · Robust analysis

1 Introduction

Nowadays, the integrated design of complex products together with their production process requires computer aided methods in order to reduce the time to market and the production costs. In particular, the simulation of the product assembly and the tolerance analysis enable the reduction of the variation of the key dimensions, thus optimizing performances and costs. In order for these simulations to be effective, all the factors of the production process affecting the dimensions must be modelled. Focusing on assemblies of compliant sheet metal parts, high final variations result from the stack-up of different assembly errors, other than single parts shape errors [1]. The part stiffness is affected by material and thickness variation [2]. However, at least the process dependent variations may be reduced by carefully designing the Fixture System and the assembly sequence [3] through Computer Aided Tolerancing tools.

C. Rizzi et al. (Eds.): ADM 2019, LNME, pp. 241–254, 2020.
https://doi.org/10.1007/978-3-030-31154-4_21

The tolerance stack-up analysis can follow worst-case or statistical approaches. The former is simpler but gives conservative results. The widespread use of 1D tolerance charts with a worst case approach to analyze the tolerance stack-up results uneffective in case of sheet metal assemblies [4]. In fact, the 3D error propagation through the assembly in multi-station manufacturing systems can be analyzed following the second approach, with a product and process Variational Model (VM) [5]. Assigned a dimension to be controlled, the model enables to study the contributors that cause its variation, its sensitivity on the contributors and the respective percentage contribution [4].

A Computer Aided Tolerancing (CAT) software enables the analysis of small errors from GD&T specifications [6] and leads to the development of 3D VMs. The software tools VisVSA® by UGS Co., eM-TolMate® by UGS Co., Mechanical Advantage® by Cognition Co., Sigmund® by Varatech Co, CETOL 6σ® by Sigmetrix LLC., and 3DCS® by Dimensional Systems Inc. can analyze assemblies of parts if they can be considered sufficiently rigid. However, in case of over constrained assemblies of compliant sheet metal parts, the deviation of the part points from their nominal positions requires a FEA software, such as VisVSA-FEA® by UGS Co., TAA® by Dassault Systemes, and 3DCS-FEA® by Dimensional Systems Inc. Many research studies arise, also with successful applications [7, 8]. However, the wide spreading in industry suffers the difficult practical application of the mathematical theories [9]. In fact, the models can be set up with different approaches, everyone involving a high number of factors that have to be finely tuned. Therefore, the operator experience is very important to achieve the results trustworthiness.

The Analysis of Variations can study how much the simulation results depend on the model parameters [10]. Other techniques introduce sensitivity indices [11]. These researches are remarkable but they do not aid the modeling and not reduce the gap from research to practice. The variations of different simulation models must be generally analyzed in order to validate their trustworthiness. In particular, FEA simulations are sensible on many factors, including material, constraints, loads, mesh and solver parameters. Up to date modelling guidelines are available in literature for rigid or even compliant parts [12–14]. However, no studies support guidelines for modelling sheet metal assemblies with high structural nonlinearities with the goal of optimizing the processes and computing a GD&T scheme. Moreover, they do not focus on critical components, such as the fixtures, and their modelling approach. Therefore, this work investigates the modelling strategies, the influence of techniques and parameters on the results in order to identify the critical and the practically negligible ones. In particular, we propose an approach to model the Fixture System and the assembly process for compliant parts. Due to the importance of tolerances in the automotive field [15–17], a case study on an automotive fender is discussed. With the aim of investigating the modelling techniques, the case study considers the product features, the GD&T schemes and the assembly process as given. Then a Design of Experiments (DoE) is defined to evaluate the variation effects due to different settings of the modelling factors. The experiments consist in simulations with the 3DCS-FEA® tool by Dimensional Control System Inc., as in [5, 18, 19].

The paper is organized as follows. In Sect. 2, we introduce the CAT modelling workflow and the involved parameters. In Sect. 3, a case study application on the assembly of an automotive fender is reported, together with the DoE plan results and discussion. Section 4 presents the conclusions and final remarks.

2 Method

The method requires modelling all the operations of the assembly process in a CAT environment, with the workflow here explained.

2.1 Computer-Aided Tolerancing Workflow

The CAT analysis workflow, shown in Fig. 1, follows the phases and subphases as:

A. Preparation:

- Data collection.
- CAD models import.

B. Functional analysis:

- Identification of the functional and technological requirements (e.g. position of feature, geometrical conditions, gaps, flushes, etc.).
- Definition of measurements consistent with the targets and requirements.
- Definition of the tolerance chains of functional features and the Datum Reference Frames (DRFs).

C. Tolerance set-up:

- Definition of tolerances types and values.
- Design of Experiments on tolerance stacks.

D. Assembly process modelling:

- Positioning of parts on the Fixture System.
- Clamping of the parts.
- Joining of the parts.
- Release of subassembly.

E. Analysis and results:

- Simulation run.
- Data analysis and comparison with the defined targets.

F. Eventual optimization:

- Optimization (synthesis) of tolerances, eliminating the criticalities.
- Final verification of the functional and technological requirements.

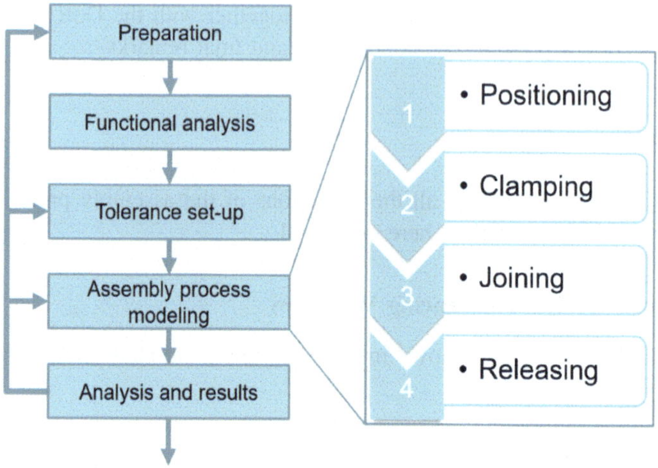

Fig. 1. CAT workflow, with a focus on assembly operations modelling.

2.2 The Variational Model

The VM expresses the possible variations on the single parts and their propagation through the assembly process by homogeneous transformation matrices [20]. A concept of the error propagation throughout the assembly sequence is shown in Fig. 2.

Assembly Sequence Modelling: The modelling starts with an analysis of all the operations in the assembly line. Considering a station k of the assembly line, $k = 1 \ldots N$, performing a specific operation on a specific sub-set of parts or sub-assemblies, the propagation of variation $\boldsymbol{X}(k)$ induced by the station is:

$$\boldsymbol{X}(k) = f_k(\boldsymbol{X}(k-1), \boldsymbol{V}(k)) \tag{1}$$

The variation $\boldsymbol{X}(k)$ on the output subassembly is function of the variation $\boldsymbol{X}(k-1)$ from the input subassembly and of the variation of the tool and Fixture System, expressed by the term $\boldsymbol{V}(k)$.

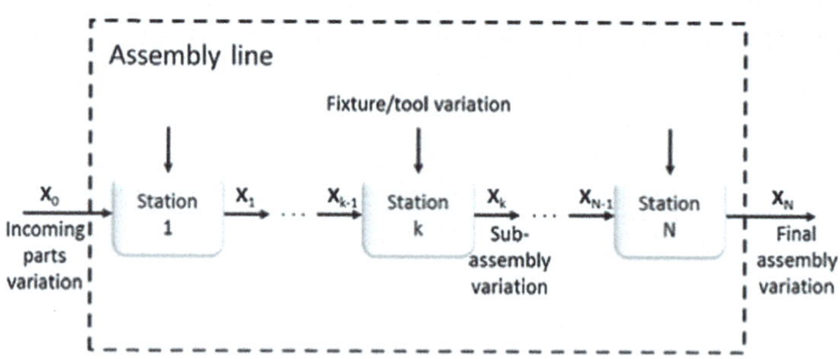

Fig. 2. General assembly sequence and propagation of variations through an assembly line.

The next step is the identification of the transfer function, defining the relation between the single parts variations and the final assembly ones (Fig. 3). It is important to consider that in a VM: each part is discretized as a mesh, the deformations and distortions are considered as linear elastic, the tools and the fixtures are considered rigid, the undeformed and deformed components have the same stiffness matrix, the thermal effects are not considered, the materials are isotropic [21].

Positioning: The first assembly operation is the Positioning, which considers the terms $\boldsymbol{U}_{u,i}$ (where the u subscript means unjoined) for locating each i part, i = 1..n, in the Fixture System. In particular, $\boldsymbol{U}_{u,i}$ defines the distance between the sheet metal part and the fixture element. $\boldsymbol{U}_{u,i}$ can be computed by simulations of forming process or by parts inspection.

Clamping: In this second operation, the action of the clamping devices on the part is considered. A force $\boldsymbol{F}_u$ is applied to contact and hold the parts against the Fixture System. The effect of this force is shown in Fig. 3 and described as:

$$\boldsymbol{F}_u = \sum\nolimits_{i=1}^{N} \boldsymbol{F}_i = \sum\nolimits_{i=1}^{N} \boldsymbol{K}_i \boldsymbol{U}_{u,i} \tag{2}$$

The force $\boldsymbol{F}_u$ results from the reaction forces $\boldsymbol{F}_i$ from the parts and subassemblies, in turn resulting from their stiffness matrix $\boldsymbol{K}_i$ and from the canceled positioning $\boldsymbol{U}_{u,i}$.

Joining: The parts or subassemblies are joined through spot welding, screwing or riveting. The result of this phase is the production of an output subassembly with a new stiffness matrix $\boldsymbol{K}_j$, resulting from the different $\boldsymbol{U}_{u,i}$ and $\boldsymbol{K}_i$.

Releasing: Finally, the clamps are opened and the subassembly returns from a clamping to a releasing position due to the spring back effect, described as

$$\boldsymbol{U}_j = \boldsymbol{F}_j \boldsymbol{K}_j^{-1} \tag{3}$$

in which the releasing force $\boldsymbol{F}_j = -\boldsymbol{F}_u$ is considered. Introducing (2) in (3), the final transfer function results

$$\boldsymbol{K}_j \boldsymbol{U}_j = -\sum\nolimits_{i=1}^{N} \boldsymbol{K}_i \boldsymbol{U}_{u,i} \tag{4}$$

which defines the direct relation between the variations of the single components and the final variations of the subassembly.

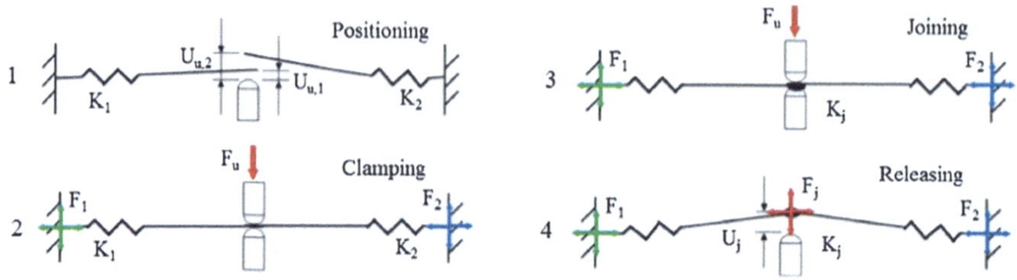

Fig. 3. The steps of a general assembly sequence.

2.3 CAT Modelling and Simulation Factors

This work focuses on modelling factors, not taking into account the effects of other variables, such as the GD&T schemes, the parts features and the assembly process. The evaluation of the effects of the modelling factors is supported by a full factorial DoE. In this plan, the modelling factors are subdivided in four categories, each one referring to a different assembly phase, as reported in Table 1.

Table 1. Modelling factors for CAT models.

Positioning	Locators scheme	F1
Clamping	DoF restrained by the clamp	F2
Joining	Number of nodes for the spot joints	F3
Releasing	FEM elements type	F4
	FEM element size	F5

Positioning: Different strategies affect the final result, due to choices in constraining all the Degrees of Freedom (DoFs) of the parts. In particular, although the *3-2-1* locating scheme would completely define the 6 DoFs of a rigid part in isostatic

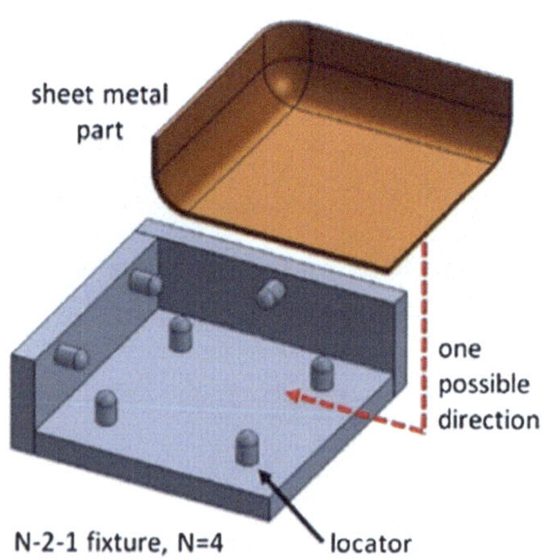

Fig. 4. Example of N-2-1 fixturing scheme for compliant sheet metal parts.

conditions, in case of compliant sheet metal parts a fixturing scheme as *N-2-1*, $N > 3$, is necessary, as shown in Fig. 4. The resulting overconstrained conditions make the order in the contact closure and the approach direction important variables of the model.

Clamping: The deformations induced on parts and subassemblies during the clamping operations are affected by the closing order of the clamps. Also for the clamps it is useful to underline the difference between ideal operating conditions and real ones. Ideally, a clamp constrains only one DoF of the component, along the pin axis direction, as shown in Fig. 5a. However, the actual equipment produce forces also in transversal directions, due to the static friction responsible of holding the parts. Such clamping reactions produce high stresses on parts and subassemblies that must be considered.

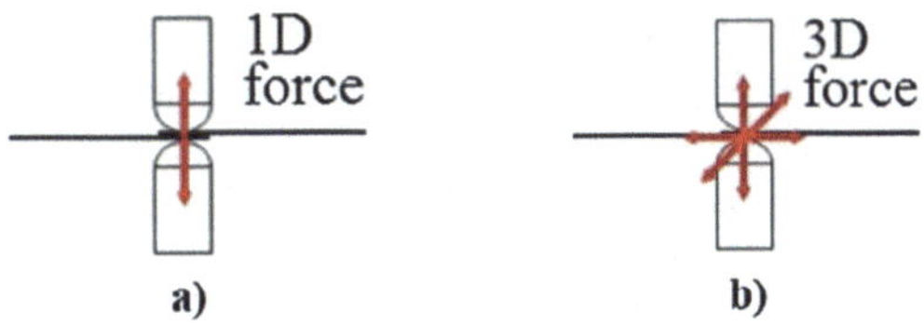

Fig. 5. Possible blocked or friction directions for a clamp device.

Joining: The main factor in modelling a spot joint is the choice of the number of constrained points. In fact, depending on the accepted level of simplification, a spot joint can be to modelled constraining a single or multiple points in the mesh. In the first case, only one node of the parts mesh is constrained (Fig. 6a). A joint model with three or more points gives more realistic results, representing a joint area and not just an ideal point, as shown in the Fig. 6b.

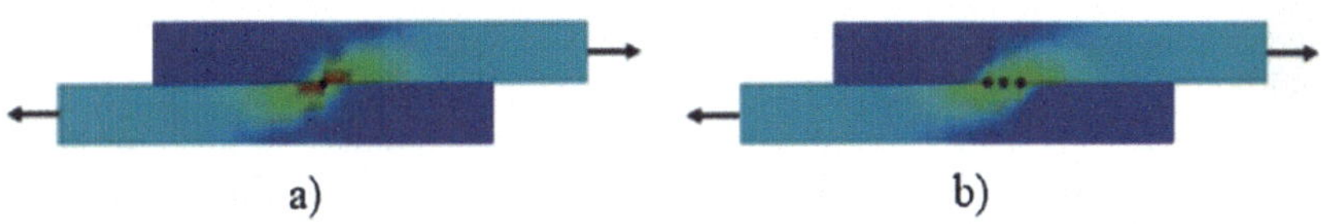

Fig. 6. Spot joint model with one or three constrained points.

Releasing: In the final releasing phase, the spring back of subassembly from the clamping to the releasing position is studied. In the model the $\boldsymbol{K}_j$ matrix is computed from the $\boldsymbol{K}_i$ ones. The main factors to be considered are the type and the size of the finite elements of the mesh, as shown in Fig. 7.

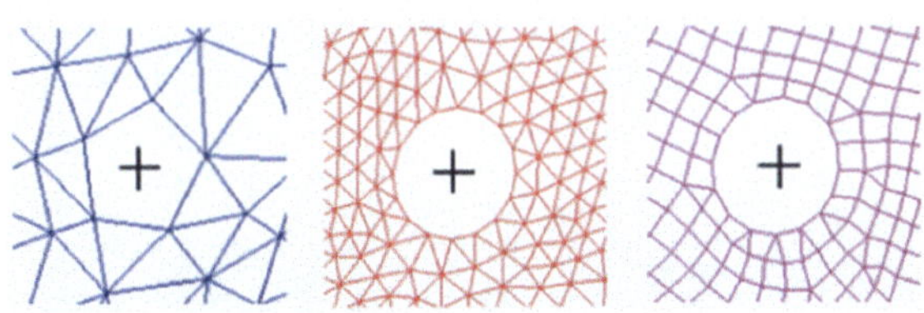

Fig. 7. Discretization strategy of the mesh for the stiffness matrix calculation.

3 Case Study on a Car Fender Assembly

The methodology defined in the Sect. 2 is applied on a front fender of a car in order to measure the variations of the simulation results depending on the factors of the CAT model.

3.1 Automotive Fender Characteristics

The fender is manufactured of steel sheet metal of 2 mm thickness. The stiffness matrices of each part are computed with mesh models within Abaqus CAE®, by Dassault Systèmes. With this CAE tool it is possible to directly introduce the properties of the parts in the 3DCS-FEA® CAT tool, by Dimensional Control System Inc. The tolerances are considered for the parts of the vehicle part and of the assembly station. Since the fixture devices can be considered much stiffer than the sheet metal parts, they are considered as rigid elements.

The assembly sequence consists of four different operations, as shown in Fig. 8:

1. The fender body panel is manually placed on the car Body-in-White (BIW), located under the panel in Fig. 8. Once the fender is positioned, the operator bolts the component on the structure below.
2. Then the fender panel, located by the fixture, is moved against the BIW. The fixture holds the panel only in the Car-Cross (CC) direction. Then, the fender brackets are positioned in the vertical direction by contact with the brace wheelhouse. The DoFs along the last direction, the Fore-Aft (FA) one, are constrained by a hand held fixture, moved against the door panel, in order to correctly locate the fender. After these fixturing operations, the fender is bolted on the two fixture devices and on the BIW, following the sequence 1 to 5. Finally, the clamps are opened.
3. Then, a different hand held fixture is moved against the rocker, in order to limit the flush of the lower part of the fender along the CC direction. So, the panel is clamped and bolted in the 6 to 9 sequence.
4. After clamps release, the last phase is the finesse operations. An operator locates a flush setter on the body near the A-pillar. Possible distortions of the body panel in CC direction are tuned by acting on the screw 10.

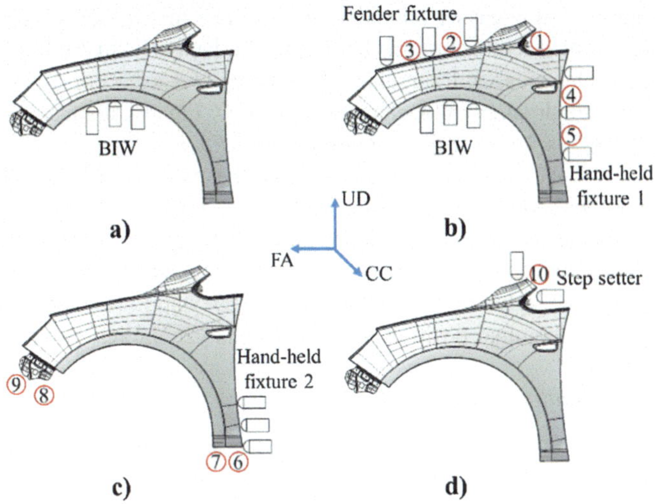

Fig. 8. Assembly process for the automotive fender.

3.2 Design of Experiments: Factors, Levels, and Responses

In order to evaluate the variation of the simulation results with respect to each factor and the first order interactions, a two level fractional factorial DoE $\left(2_{\mathrm{V}}^{5-1}\right)$, is implemented. The factors and their levels are reported in Table 2. In particular, the positioning operations of the parts on the fixtures in stations 1 and 2 can be modelled with different orders of contacts closure. The factor *F1* takes into account the effects of two different strategies (*F1* = −*1*/*1*) for locating the parts onto the same datums. The factor *F2* represents the clamp force in the clamping operation phase. In particular, (*F2* = −*1*) represents the case where the clamp holds the part only in the direction normal to the surface, constraining only one DoF, while in (*F2* = *1*) also the transversal reactions due to friction are considered, thus restraining three DoFs. The factor *F3* regards the spot joint modelling with a single point (*F3* = −*1*) or with three not aligned ones (*F3* = *1*).

Table 2. DoE factors and simulated levels.

Factors	Level	
F1	−1	First positioning model
	1	Second positioning model
F2	−1	1 DOF
	1	3 DOFs
F3	−1	1 point
	1	3 points
F4	−1	Tri
	1	Mixed
F5	−1	2 mm
	1	10 mm

Considering the last releasing operation, the stiffness matrix can be computed from different strategies of discretization of the mesh. The factor *F4* that takes into account the type of finite elements, as triangular S3 elements ($F4 = -1$) or a combination of triangular S3 and quadrilateral S4R elements ($F4 = 1$). Finally, the factor *F5* is related to the element size, as 2 mm ($F5 = -1$) or 10 mm ($F5 = 1$).

Figure 9 shows the measure points used for the evaluation. The 43 points are chosen on the border and the surface of the body panel, representing the target gaps (21 measures) and flushes (22 measures) defined by the functional requirements. In particular, the results of each *j* statistical simulations obtained with the Monte Carlo method give the Mean Value (MV) and the Standard Deviation (SD) of the variation of each *i* point from its nominal position. In order to analyze the results variability some criteria are adopted:

Variability evaluation criteria:

- The variation results are averaged for all the measure points for a single run.
- The variations are averaged only for the five points characterized by the five highest differences from highest to lowest MVs of position variation.
- The results are reported for the point with the highest difference from its highest to lowest MVs of position variation.

The variation responses, output of the DoE, are evaluated both for MV and SD as:

- R1. Average of the MVs of all the border points.
- R2. Average of the MVs of all the five points with the highest differences.
- R3. MV of the point with the highest difference.
- R4. Average of the SDs of all the border points.
- R5. Average of the SDs of all the five points with the highest differences.
- R6. SD of the point with the highest difference.

Considering Fig. 9, the five points of the border that have the highest difference of the MV among all the runs are Uf, Cf, Cg, Tf and Vf, listed in decreasing order. The results are shown in Figs. 10 and 11 as Main Effect plots, which enable the identification of the influence of each modelling factor on the overall effect. A Pareto chart is calculated, in order to determine the importance of each factor and to study the effects of the combinations of factors. Finally, an Interpolation plot is calculated, in order to show whether the response at factor level depends upon the level(s) of other factors.

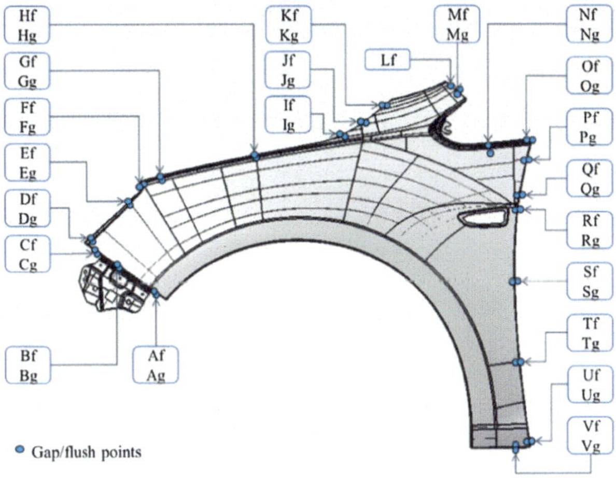

Fig. 9. Measure points on border and surface of the fender.

4 Results Discussion and Conclusions

Considering the results reported in Fig. 10, the Main Effects plots show that F2 is the main factor influencing the results. However, the effect of F2 for R1 is very small and, in particular, the level F2 = 1 leads to a reduction of the MV. Conversely, F2 presents a greater influence on both R2 and R3, as shown by their Main Effects plots. Regarding the contribution of the other factors, F4 and F5 slightly influence R2 and R3. In general, the MV increases for the level 1 of factors F3, F4 and F5 and for the level -1 of factors F1 and F2.

The analysis of the Pareto charts confirms the results of the Main Effects plots regarding the influence of single factors and identifies the influence of the combination of factors on the overall variability. In particular, the most important combination of factors is the F2 * F5 one. The result is also confirmed by the Interaction plots for each of the three responses R1, R2 and R3.

The analysis of the SD responses, namely R4, R5 and R6 (Fig. 11), reveals the same trend observed in the MV plots. Considering R4, the effects of the factors are negligible, except the factor F3, which has a little influence. Regarding R5 and R6, the results are highly influenced by the factor F2, which overtakes F3, becoming the main contributor. As for the MV plots, the Pareto charts and the Interaction plots confirm the trend shown in the Main Effects plots. In particular, the effect of the interaction F2 * F5 grows from R4 to R6, while the combination F2 * F3 shows an opposite trend, reducing its importance from R4 to R6.

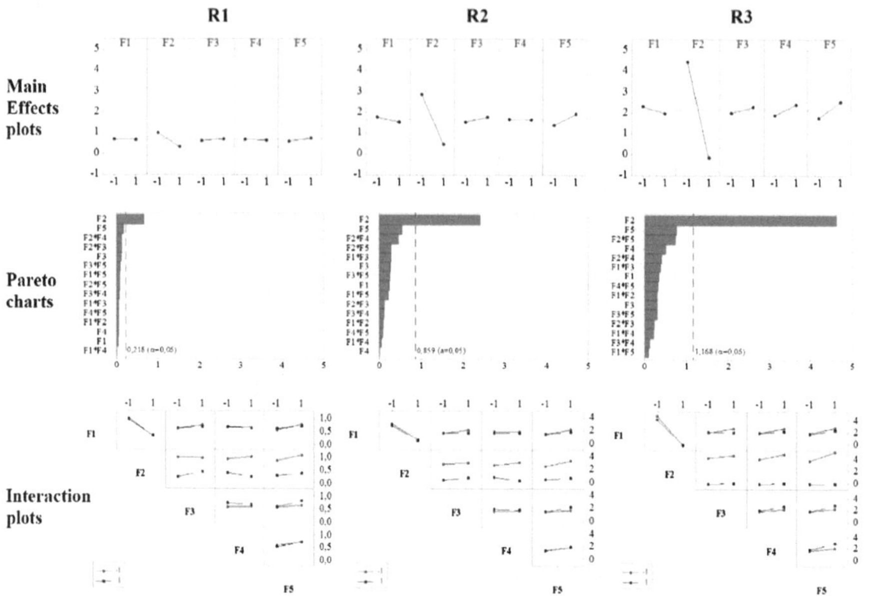

Fig. 10. Main Effects plots, Pareto charts and Interaction plots of MV for the responses R1, R2 and R3.

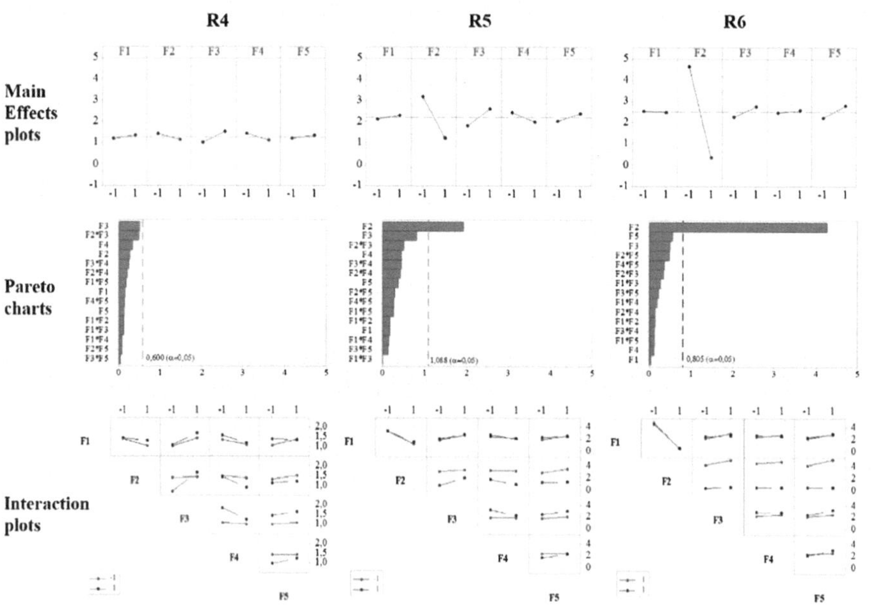

Fig. 11. Main Effects plots, Pareto charts and Interaction plots of SD for the responses R4, R5 and R6.

The simulation results underline that for sheet metal assemblies the correct modelling operations are very important, in order to reduce errors and provide a more robust virtual validation. In this sense, the DoE enables the analysis of the influence of the factors of the different assembly operations. In particular, the analysis of the results recommends to pay attention to the clamping operation, where the modelling strategy has the major influence on the responses variation. On the other hand, the models of positioning, spot joints and FEM mesh parameters, have a lower effect on the overall variations.

Future works will study the link between the modelling factors, the functional requirements, and also the measurements for dimensional verification.

References

1. Bici, M., Campana, F., Cimolin, F., Rizzo, L.: Robust die compensation in sheet metal design through the integration of dual response surface and shape function optimization. Math. Probl. Eng. **2019**, 7357135 (2019)
2. Yu, K., Jin, S., Lai, X., Xing, Y.: Modeling and analysis of compliant sheet metal assembly variation. Assem. Autom. **28**(3), 225–234 (2008)
3. Whitney, D.E.: Mechanical Assemblies. Oxford University Press, New York (2004)
4. Shen, Z., Ameta, G., Shah, J.J., Davidson, J.K.: A comparative study of tolerance analysis methods. J. Comput. Inf. Sci. Eng. **5**(3), 247–256 (2005)
5. Wang, H., Ceglarek, D.: Variation propagation modeling and analysis at preliminary design phase of multi-station assembly systems. Assem. Autom. **29**(2), 154–166 (2009)
6. Prisco, U., Giorleo, G.: Overview of current CAT systems: review article. Integr. Comput. Aided Eng. **9**(4), 373–387 (2002)
7. Franciosa, P., Gerbino, S., Patalano, S.: Simulation of variational compliant assemblies with shape errors based on morphing mesh approach. Int. J. Adv. Manuf. Technol. **53**(1–4), 47–61 (2011)
8. Yu, D.Z., Lv, W.P., Zhou, L.H., Xu, X.H.: Spatial error modeling and simulation on sheet metal assembly in single station, In: International Conference on Computer Information Systems and Industrial Applications, pp. 935–939. Atlantis Press (2015)
9. Qureshi, A.J., Dantana, J.-Y., Sabri, V., Beaucaire, P., Gaytond, N.: A statistical tolerance analysis approach for over-constrained mechanism based on optimization and Monte Carlo simulation. Comput. Aided Des. **44**, 132–142 (2012)
10. Saltelli, A., Annoni, P., Azzini, I., Campolongo, F., Ratto, M., Tarantola, S.: Variance based sensitivity analysis of model output. Design and estimator for the total sensitivity index. Comput. Phys. Commun. **181**(2), 259–270 (2010)
11. Homma, T., Saltelli, A.: Importance measures in global sensitivity analysis of nonlinear models. Reliab. Eng. Syst. Saf. **52**(1), 1–17 (1996)
12. Belytschko, T., Liu, W.K., Moran, B., Elkhodary, K.: Nonlinear Finite Elements for Continua and Structures. Wiley, Chichester (2013)
13. Haldar, A., Mahadevan, S.: Reliability Assessment Using Stochastic Finite Element Analysis. Wiley, Chichester (2000)
14. Ashraf, M., Gardner, L., Nethercot, D.A.: Finite element modelling of structural stainless steel cross-sections. Thin-Walled Struct. **44**(10), 1048–1062 (2006)
15. Bonazzi, E., Colombini, E., Panari, D., Vergnano, A., Leali, F., Veronesi, P.: Numerical simulation and experimental validation of MIG welding of T-joints of thin aluminum plates

for top class vehicles. Metall. Mater. Trans. A Phys. Metall. Mater. Sci. **48**(1), 379–388 (2017)
16. Panari, D., Renzi, C., Vergnano, A., Bonazzi, E., Leali, F.: Integrated design method for optimal tolerance stack evaluation for top class automotive chassis. Lecture Notes in Mechanical Engineering (2017)
17. Pini, F., Panari, D., Leali, F.: Computer-aided integrated method for effective development of robotic welding processes of car chassis assembly. DYNA **94**, 2 (2019)
18. Wei, C., Sun, J., Xin-min, L.: Tolerance optimization considerations applied to the sheet metal compliant assembly. Comput. Aided Des. Appl. **11**(1), 68–76 (2014)
19. Reich, J., Chiocchio, S., Cordier, J.J., Gallix, R., Guerin, O., Halcrow, T.: Three-dimensional tolerance investigation on main ITER components. Fusion Eng. Des. **84**(7), 1584–1587 (2009)
20. Colosimo, B.M., Senin, N.: Geometric Tolerances. Springer, London (2010)
21. Camelio, J.S., Hu, J., Ceglarek, D.: Modeling variation propagation of multi-station assembly systems with compliant parts. J. Mech. Des. **125**(4), 673–681 (2003)

Design for Manufacturing and Assembly

An Improved Design Method for Net-Shape Manufacturing in Powder Metallurgy

Marco Zago[1](✉), Mats Larsson[2], and Ilaria Cristofolini[1]

[1] Department of Industrial Engineering, University of Trento, Via Sommarive 9, 38123 Trento, Italy
marco.zago-1@unitn.it

[2] Höganäs AB, Bruksgatan 35, 263 83 Höganäs, Sweden

Abstract. The effectiveness of powder metallurgy as net-shape/near net-shape manufacturing technology is determined by the possibility of obtaining complex parts matching the required narrow tolerances. Sintering process determines change in volume of the green, and the related dimensional changes are significantly anisotropic. Anisotropy is affected by several variables, such as material, compaction and sintering parameters, geometry, whose influence is difficult to be distinguished and determined. Anisotropic dimensional change on sintering has been investigated in depth using an experimental approach, relating measurements results to the mechanisms responsible for the phenomenon. Main results concerning the influence of different variables are briefly presented in this work. Such results served as the basis for the development and further improvement of a design method, aimed at predicting anisotropic dimensional change. Main steps of the design method are presented and an example of application to a real part is described. Strong agreement between predicted and real dimensional changes has been observed, and compared to the attainable dimensional tolerances.

Keywords: Design for powder metallurgy · Precision engineering · Product development

1 Introduction

Net-shape technologies are significantly growing in importance as "green technologies", thanks to the possibility of avoiding (or strongly limiting) machining operations, which determine large amount of scrap, increased energy consumption and consequently higher cost of components [1]. Powder metallurgy (PM) is the most widespread net-shape/near net-shape technology, finding application in many different sectors, such as automotive, medical, consumer goods, construction etc.

Figure 1 compares powder metallurgy to traditional technologies, in terms of raw material utilization and energy consumption [2].

Depending on batch size, geometrical complexity, mechanical properties, dimensional and geometrical precision, and cost limitations, different alternatives can be considered, from press and sinter to metal injection molding to additive manufacturing. Among them, when large batches are needed, press and sinter is almost the unreplaceable

C. Rizzi et al. (Eds.): ADM 2019, LNME, pp. 257–267, 2020.
https://doi.org/10.1007/978-3-030-31154-4_22

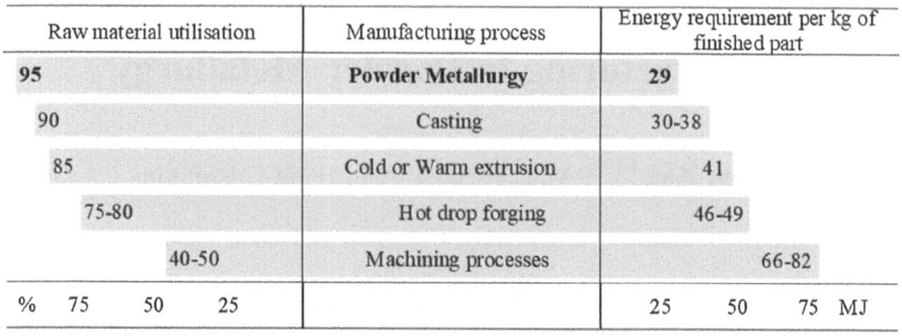

Fig. 1. Raw material utilization and energy consumption for different manufacturing processes.

technology. In the standard process, metal powders, added with a small amount of lubricant, are compacted in a die producing the so called "green part", having sufficient cohesion to be handled safely. Uniaxial cold compaction is commonly used. The green part is then heat treated in a protective atmosphere (sintering), so that particles weld together and confer sufficient strength to the material for the intended use [3].

Aiming at avoiding secondary operations, dimensional and geometrical precision is a crucial issue, which is strongly related to the change in volume (shrinkage or swelling) occurring on sintering [4].

In the design of PM parts, the change in volume has to be carefully considered, mostly in terms of linear dimensional changes. In fact, dimensional changes are the starting point of a design method aimed at obtaining net-shape/near net-shape parts, which only can be obtained from properly dimensioned green parts. Unfortunately, dimensional change on sintering is anisotropic, and it is differently affected by a large number of variables, related to the material, the process, and the geometry, whose influence is difficult to predict [5]. As a consequence, modeling is rarely used, mainly due to the large influence of geometry and to the lack of constitutive models for powder mixes. In most of the cases dimensional changes are quite small, but the required tolerances are very narrow, what is also limiting the reliability of modeling response, so that in the companies the approach is often "trial and error", until the required characteristics are obtained.

The anisotropy of dimensional change on sintering has been extensively studied by the authors, aiming at defining a design method to predict anisotropic dimensional change on sintering. The relationships used in the method are defined on the basis of a deep investigation of the mechanisms involved, starting from experimental data. Dimensions have been measured in the green and sintered state and dimensional changes have been derived for different parts, analyzing the influence of geometry, material, process parameters (such as compaction strategy, and sintering temperature and atmosphere). Main results are briefly recalled in the following paragraphs. An anisotropy parameter, aimed at quantifying the anisotropy extent, has been defined and used to develop the design method, which is described in the following and validated on a real part.

2 Dimensional Change on Sintering

Dimensional change has been derived measuring the same parts in the green and sintered state by a Coordinate Measuring Machine (CMM). The parts were measured by scanning mode and the data were processed by Matlab to calculate the features, from which dimensions were derived (diameters from cylindrical surfaces, distances from couples of parallel planes etc.). The procedure for data processing is explained in depth in [6]. In the last years, the attention was mainly focused on axi-symmetric, ring-shaped parts, which represent the most largely implemented geometry in press and sinter. Consequently, dimensional changes refer to entities, which can always be considered as external diameters (D_{ext}), internal diameters (D_{int}), and heights (h), as defined by Eqs. (1) to (3), where subscript g and subscript s refer to green and sintered state, respectively.

$$\varepsilon_{D_{ext}} = \frac{D_{ext_s} - D_{ext_g}}{D_{ext_g}} \tag{1}$$

$$\varepsilon_{D_{int}} = \frac{D_{int_s} - D_{int_g}}{D_{int_g}} \tag{2}$$

$$\varepsilon_h = \frac{h_s - h_g}{h_g} \tag{3}$$

Anisotropy is mainly related to the different dimensional change along the axis (ε_h) and in the plane orthogonal to the axis ($\varepsilon_{D_{ext}}$, $\varepsilon_{D_{int}}$), however dimensional change of external diameter may differ from dimensional change of the internal diameter too, as demonstrated in [7]. Dimensional changes and their anisotropy are strongly, and sometimes unexpectedly, affected by many different variables, whose influence has been investigated in depth in previous work. Main results are following briefly summarized. The aim is here highlighting the variety and complexity of variables involved, referring to the specific papers for explanation in depth.

2.1 Influence of Material

Metal powders used in press and sinter may differ for chemical composition, particle size and particle size distribution, and powder mix can be obtained by pre-alloyed or diffusion bonded powders. All of these parameters have a strong influence on dimensional change, from change in volume ΔV. Either swelling (increase in volume after sintering) or shrinking (decrease in volume after sintering) may occur, mainly depending on the chemical composition, as shown in Fig. 2. The considered materials are used for about 75% of press and sinter production.

Swelling is mainly related to the presence of copper, which determines liquid phase sintering due to its melting temperature, lower than sintering temperature. The larger the copper content, the higher the swelling amount [8]. In the shrinking powder mix, the increase in carbon content determines a lower shrinkage [9].

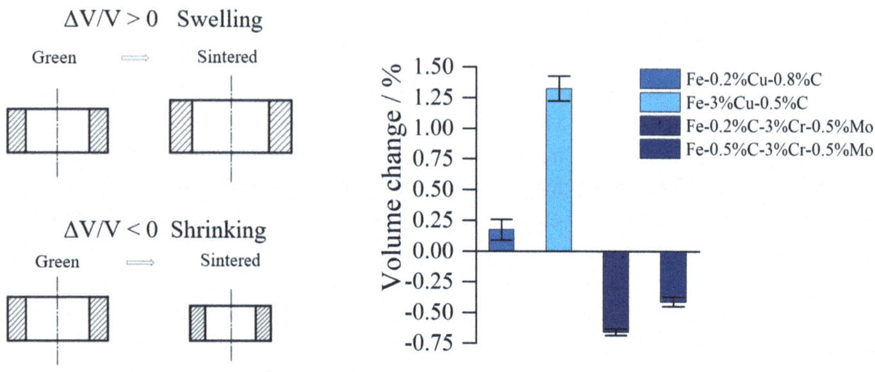

Fig. 2. Influence of chemical composition on the change in volume.

Figure 3 shows the influence of the type of powder mix and the particle size on the anisotropy of dimensional changes, left and right respectively.

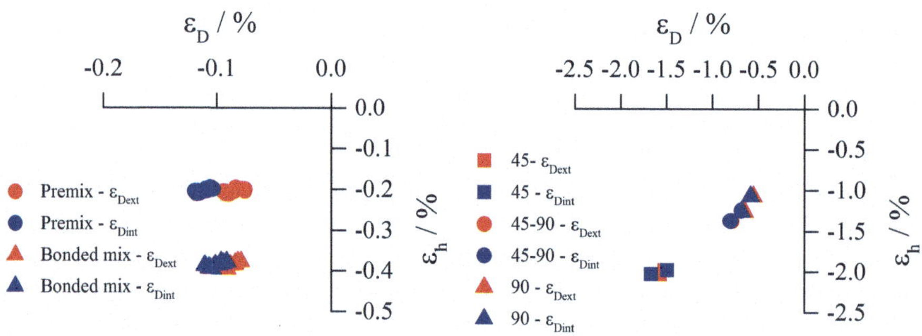

Fig. 3. Influence of alloying system (left) and particle size (right) on dimensional changes of height and diameters.

Anisotropy of dimensional changes is larger in bonded mix, dimensional change in height is about four time the dimensional change of diameters, while in premix it is about twice. However, a larger difference in the dimensional changes of external and internal diameters is shown in premix [9]. About the influence of particle size, a significant anisotropy is observed for finer particle size (45, meaning lower than 45 μm), while the coarser ones (90, meaning larger than 90 μm) show a tendency to a less anisotropic dimensional change [10].

2.2 Influence of Process Parameters

Uniaxial cold compaction determines green density and green density distribution, which in turn affect the density and density distribution of final part after sintering. Figure 4 shows the influence of green density and sintering temperature on dimensional changes. Dimensional changes are significantly anisotropic for any green density, but for green density up to 7.1 g/cm^3 dimensional changes of diameters are larger than dimensional change of height, while the opposite is observed for the highest green density. In all the cases a noticeable difference is also observed between dimensional change of internal and external diameters [11]. Concerning the influence of sintering temperature, it has to be pointed out that sintering at very high temperature determines an increase of dimensional changes, as expectable, but a tendency versus a less anisotropic behaviour, what might help the designers [12]. Work is in progress to highlight the influence of high sintering temperature on the stability of geometrical characteristics.

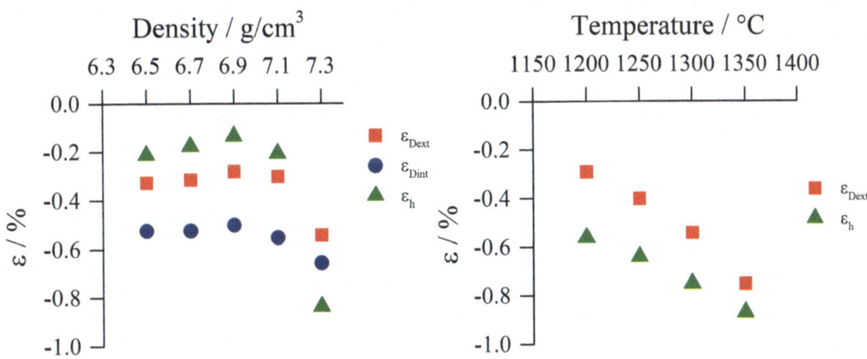

Fig. 4. Influence of green density (left) and sintering temperature (right) on dimensional changes of height and diameters.

2.3 Influence of Geometry

Designing anisotropic dimensional change of PM parts is further complicated by the influence of geometry [13, 14]. In fact, the influence of materials and process parameters above briefly described, also depends on geometry. Figure 5 shows the dimensional change of internal and external diameters along the axis, and the dimensional change of the height along the diameter. Dimensional change of internal and external diameters not only differ as mean values, but also as their trend along the axis. Moreover, dimensional change of the height may also differ, when measured in different positions along the diameter. This is mainly due to the density distribution and to the influence of compaction strategy [15].

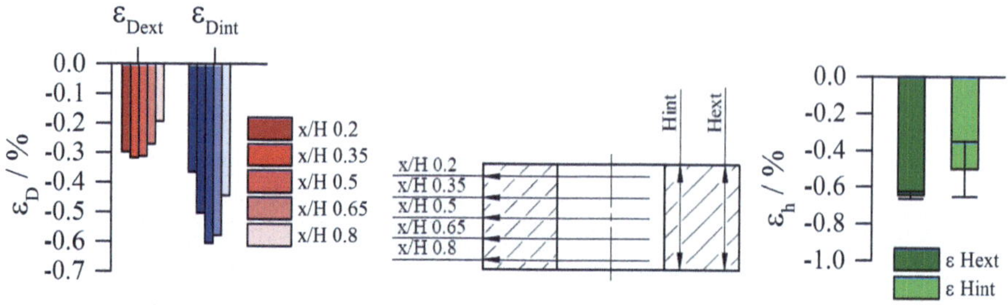

Fig. 5. Dimensional changes of diameters along the axis (left) and dimensional change of height in different positions (right).

3 Design Method to Predict Anisotropic Dimensional Changes

As briefly highlighted above, a wide experimental knowledge base is needed to improve the reliability of a model describing dimensional change on sintering. Main relationships used to develop the model are here briefly recalled, explanation in depth can be found in previous work [14, 16]. Starting from the main effect, dimensional changes can be related to the change in volume V, which is given by the difference between the volume of the sintered part V_s and the volume of the green part V_g. Under the hypothesis of isotropic dimensional change ε_{iso} the relationship is given by Eq. 4

$$1 + \frac{V_s - V_g}{V_g} = (1 + \varepsilon_{iso})^3 \tag{4}$$

As underlined above, dimensional change is actually anisotropic, so that the relationship between change in volume and dimensional changes turns from Eqs. (4) to (5)

$$1 + \frac{V_s - V_g}{V_g} = (1 + \varepsilon_h) \cdot \frac{(1 + \varepsilon_{D_{ext}})^2 - R^2(1 + \varepsilon_{D_{int}})^2}{1 - R^2} \tag{5}$$

Where $R = \frac{D_{int_g}}{D_{ext_g}}$. The relationship between isotropic and anisotropic dimensional changes has been defined through the change in volume, combining Eqs. (4) and (5), obtaining Eq. (6)

$$(1 + \varepsilon_{iso})^3 = (1 + \varepsilon_h) \cdot \frac{(1 + \varepsilon_{D_{ext}})^2 - R^2(1 + \varepsilon_{D_{int}})^2}{1 - R^2} \tag{6}$$

A relationship between dimensional change of external and internal diameter, expressed by Eq. (7), has been experimentally derived, considering a huge sampling differing for dimensions, materials and sintering conditions, which determine parameter α [9]

$$\frac{1+\varepsilon_{D_{int}}}{1+\varepsilon_{D_{ext}}} = \alpha = f(\varepsilon_{iso}, geometry) \tag{7}$$

Exploiting the equations above, anisotropic and isotropic dimensional changes have been represented as points on a graph, and an anisotropy parameter K has been defined as the distance between such points, according to Eq. (8), where γ is a geometrical parameter incorporating α [14].

$$K = \frac{(1+\varepsilon_{D_{ext}})\gamma - (1+\varepsilon_{iso})}{\varepsilon_{iso}} \tag{8}$$

A large experimental investigation allowed determining a database to be used in the design method, collecting the geometrical parameter $\gamma(\alpha)$, and the anisotropy parameter K, both of them defined as function of geometry and ε_{iso} [16]. The isotropic dimensional change ε_{iso}, in turn, depends on materials and process parameters. Depending on the several variables previously described, α ranges from 1 to 2 (and $\gamma(\alpha)$ accordingly), and K from −4 to 4. Work is in progress to enlarge and strengthen the existing database, in cooperation with the major European PM parts producers.

Using the above parameters and relationships, a design method aimed at predicting the anisotropic dimensional change on sintering has been developed, whose main steps are summarized in Fig. 6. Starting point is given by the isotropic dimensional change ε_{iso}, and the geometrical parameters α, $\gamma(\alpha)$, which allows determining the anisotropy parameter K. Consequently, dimensional changes of external diameter (Eq. 8), internal diameter (Eq. 7), and height (Eq. 6) can be predicted.

Material
Process
Geometry
⇨ ε_{iso}, $\alpha - \gamma(\alpha)$ ⇨ $K = f(\varepsilon_{iso}, \gamma(\alpha))$ ⇨ $\varepsilon_{D_{ext}} = \frac{\varepsilon_{iso}(K+1) - (\gamma - 1)}{\gamma}$

⇩

$$\varepsilon_{D_{int}} = \alpha(1 + \varepsilon_{D_{ext}}) - 1$$

⇩

$$\varepsilon_h = \frac{(1+\varepsilon_{iso})^3}{\gamma^2(1+\varepsilon_{D_{ext}})^2} - 1$$

Fig. 6. Schematic representation of the design method predicting anisotropic dimensional change on sintering.

4 Design Method Applied to a Real Part

The design method above refers to a ring-shaped part, consequently it can be applied to real axi-symmetric parts, providing that coaxial rings can be identified. Coaxial rings can be identified as by two different approaches: “column-based approach” (the powder columns, which can be identified during uniaxial compaction) or “maximum-section based approach” (the maximum amount of shrinking/swelling material, dragging after the other rings). The dimensional changes of diameters common to different rings are set congruent [17].

The design method has been applied to the gear shown in Fig. 7, split according to the two approaches above. As a first approximation, the teeth profile has been neglected. According to author’s experience, being the external diameter defining the envelope of the top of the teeth correctly predicted, the teeth profile and the diameter at the basis of the teeth derive accordingly.

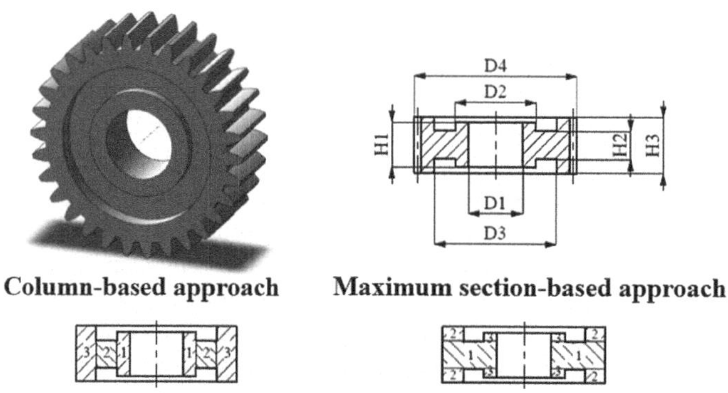

D1	20 mm
D2	30 mm
D3	45 mm
D4	60 mm
H1	16 mm
H2	10 mm
H3	20 mm

Fig. 7. Coaxial rings identified in a gear by “column-based approach” and by “maximum-section based approach”.

Diameters and heights shown in Fig. 7 have been measured in the green and sintered parts by CMM, deriving the real dimensional changes. The design method summarized in Fig. 6 has been applied to the coaxial rings identified in Fig. 6 and predicted dimensional changes have been calculated. Predicted dimensional changes are compared to the real ones in Fig. 8. Diameters are accurately predicted by both column-based and maximum section-based approaches, while the latter is more accurate in predicting heights. From the designer point of view, the significance of the result is further highlighted considering the IT tolerance classes related to the difference between predicted and measured dimensional changes.

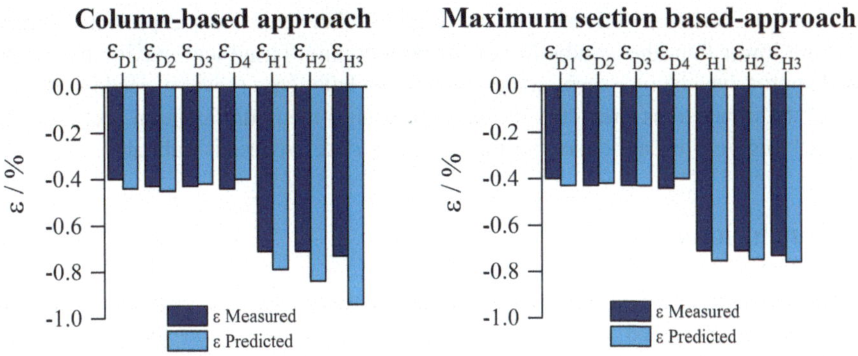

Fig. 8. Comparison between predicted and measured dimensional changes, over measured, by "column-based approach" and by "maximum section-based approach".

4.1 Tolerance Classes

Figure 9 shows the ISO IT classes derived comparing the measured dimensions of the real part with the predicted ones, as derived by the two different approaches. In the standard process, dimensional tolerances attainable in the compaction plane (diameters) are generally IT8-IT9, while in the axial direction (heights) are IT9-IT11. As by Fig. 9, the

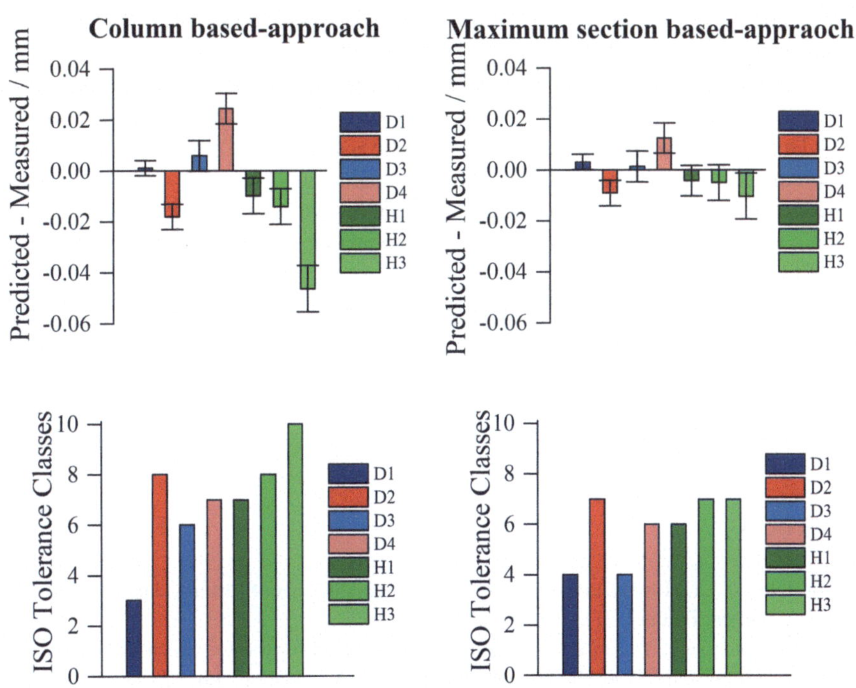

Fig. 9. ISO IT Tolerance classes derived form the difference between predicted and measured dimensions by "column-based approach" and by "maximum section-based approach".

dimensional tolerances derived from the difference between real and predicted dimensions are lower than the standard ones for both column-based and maximum section-based approaches, being exceptionally good in the latter case (never exceeding IT7).

All the dimensions predicted through the design method/maximum section-based approach are very accurate, so that the method's effectiveness is validated.

5 Conclusions

A design method aimed at improving powder metallurgy as a net-shape/near net-shape technology is presented in this work. The method is based on the possibility of accurately predicting anisotropic dimensional changes on sintering.

A large number of variables affect the anisotropy of dimensional changes, such as material (chemical composition determines either swelling or shrinking on sintering, anisotropy of dimensional changes is larger in diffusion bonded powder mix than in premix, and also in fine particle with respect to coarse particle size), process parameters (dimensional change is larger for height than for diameters for high green density; on increasing sintering temperature, dimensional changes tend to be less anisotropic), and geometry (dimensional change also depends on the position where features are measured).

Focusing on ring-shaped parts, an anisotropy parameter K and geometrical parameters α and γ have been defined on the basis of a large experimental campaign, as by the influence of the parameters above. Using K, α, and γ, a design method to predict anisotropic dimensional changes has been proposed.

The method has been applied to predict dimensional changes on a real part, using different approaches to split the part in order to apply the design method to coaxial rings. Through the approach called "maximum section-based" dimensional changes were predicted with very good accuracy. The dimensional tolerances derived from the difference between real and predicted dimensions were exceptionally good (never exceeding IT7) with respect to the standard dimensional tolerances in press and sinter (IT9-IT11).

The effectiveness of the design method has been demonstrated. The method can be used to predict the dimensional changes of any axi-symmetric part, leading to very accurate prediction. Work is in progress aiming at increasing the database to be used for reference, to further enlarge the field of application, mainly in the case of parts encountering liquid phase sintering.

References

1. Dornfeld, D.A.: Moving towards green and sustainable manufacturing. Int. J. Precis. Eng. Manuf.-Green Technol. **1**(1), 63–66 (2014)
2. EPMA Homepage: Vision 2025 – Future Developments for the European PM Industry ed. EPMA, Shrewsbury (UK). https://www.epma.com/. Accessed 17 June 2019
3. German, R.M.: Powder Metallurgy of Iron and Steel. Wiley, New York (1998)

4. Wanibe, Y., Yokoyama, H., Itoh, T.: Expansion during liquid phase sintering of iron-copper compacts. Powder Metall. **33**(1), 65–69 (1990)
5. Griffo, A., Ko, J., German, R.M.: Critical assessment of variables affecting the dimensional behavior in sintered iron-copper-carbon alloys. In: Advances in Powder Metallurgy and Particulate Materials, Metal Powders Industry Federation, Princeton, NJ, vol. 3, pp. 221–236 (1994)
6. Cristofolini, I., Menapace, C., Cazzolli, M., Rao, A., Pahl, W., Molinari, A.: The effect of anisotropic dimensional change on the precision of steel parts produced by powder metallurgy. J. Mater. Process. Technol. **7**(212), 1513–1519 (2012)
7. Cristofolini, I., Corsentino, N., Larsson, M., Molinari, A.: Analytical model of the anisotropic dimensional change on sintering of ferrous pm parts. Powder Metall. Prog. **16**(1), 27–39 (2016)
8. Cristofolini, I., Pilla, M., Molinari, A., Menapace, C., Larsson, M.: DOE investigation of anisotropic dimensional change during sintering of Iron- Copper-Carbon. Int. J. Powder Metall. **48**(4), 33–44 (2012)
9. Cristofolini, I., Pilla, M., Larsson, M., Molinari, A.: A DOE analysis of dimensional change on sintering of a 3%Cr-0.5%Mo-X%C steel and its effect on dimensional and geometric precision. Powder Metall. Prog. **12**(3), 127–143 (2012)
10. Zago, M., Cristofolini, I., Molinari, A.: New interpretation for the origin of the anisotropic sintering shrinkage of AISI 316L rings based on the anisotropic stress field occurred on uniaxial cold compaction. Powder Metall. **62**(2), 115–123 (2019)
11. Molinari, A., Zago, M., Amirabdollahian, S., Cristofolini, I., Larsson, M.: Anisotropic sintering shrinkage of ring shaped iron parts: effect of geometry and green density and correlation to the stress field during uniaxial cold compaction. In: EuroPM2017 Proceedings, vol. Sintering, 3686349, EPMA, Shrewsbury (2017)
12. Cristofolini, I., Pilla, M., Rao, A., Libardi, S., Molinari, A.: Dimensional and geometrical precision of powder metallurgy parts sintered and sinterhardened at high temperature. Int. J. Precis. Eng. Manuf. **14**(10), 1735–1742 (2013)
13. Cristofolini, I., Corsentino, N., Pilla, M., Molinari, A., Larsson, M.: Influence of the geometry on the anisotropic dimensional change on sintering of PM parts. In: Advances in Powder Metallurgy and Particulate Materials, Metal Powders Industry Federation, Princeton, NJ, pp. 1149–1161 (2013)
14. Cristofolini, I., Corsentino, N., Pilla, M., Molinari, A., Larsson, M.: Study of the influence of material and geometry on the anisotropy of dimensional change on sintering of Powder Metallurgy parts. Int. J. Precis. Eng. Manuf. **15**(9), 1865–1873 (2014)
15. Molinari, A., Amirabdollahian, S., Zago, M., Larsson, M., Cristofolini, I.: Effect of geometry and green density on the anisotropic sintering shrinkage of axisymmetric iron parts. Powder Metall. **61**(4), 267–275 (2018)
16. Cristofolini, I., Corsentino, N., Molinari, A., Larsson, M.: Influence of geometry and process variables on the anisotropy parameter K. In: Advances in Powder Metallurgy and Particulate Materials, Metal Powders Industry Federation, Princeton, NJ, vol. 01, pp. 31–40 (2015)
17. Cristofolini, I., Molinari, A., Zago, M., Amirabdollahian, S., Coube, O., Dougan, M.J., Larsson, M., Schneider, M., Valler, P., Voglhuber, J., Wimbert, L.: Design for powder metallurgy – predicting anisotropic dimensional change on sintering of real parts. Int. J. Precis. Eng. Manuf. **20**(4), 619–630 (2019)

Design for Assembly in the Conceptual Development of Aircraft Systems

Claudio Favi[1(✉)], Giovanni Formentini[1], Francois Bouissiere[2], Claude Cuiller[2], Pierre-Eric Dereux[2], and Corentin Malchair[2]

[1] Università degli studi di Parma, Parma, Italy
claudio.favi@unipr.it
[2] Airbus Operations S.A.S, Toulouse, France

Abstract. Conceptual design for assembly and installation is a key enabler for the improvement and development of an aircraft and related components. This work attempts to define a design for assembly methodology suitable for the evaluation and architecture design of aircraft systems in the preliminary phases of product development (conceptual design). Three main aspects are covered within this work: (i) the definition of a design framework, (ii) the characterization of suitable parameters driving the assessment and development of product architectures, and (iii) the formalization of internal knowledge for that purpose. The proposed approach has been tested in the assessment and development of an aircraft nose section with positive outcomes in terms of knowledge formalization and robustness of results in relation with the issues retrieved by the analysis of the assembly line. Future works will focus on the methodology optimization including automatic data input and mathematical models refinement.

Keywords: Conceptual design · Design for assembly · Knowledge formalization · Installation

1 Introduction and Context

Aircraft manufacturing is expected to dominate the market over the forecast period. Commercial aircraft led the market, which includes passenger and freighter aircraft. The global demand for commercial aircraft and cargo fleet is expected to grow in line with the annual number of passengers, which is expected to double over the next two decades. In this context, it is required to have an efficient production system which includes manufacturing, assembly and installation processes able to satisfy the market requests (e.g. time to market, lead time for delivery, etc.). Contrary to the automotive industry and other sectors, the development of new manufacturing/assembly facilities equipped with other technologies or a different arrangement of the already-existing ones are hardly achievable due to the extremely high-cost [17, 27]. In addition the aerospace industry has strong certification requirements, with consequences on development costs, technology solutions and product development lead time [4].

The first step to face the market challenge is to work at the product (aircraft) architectural level, improving the design concept with the aim to be more efficient during the assembly phase. Conceptual design allows engineers to have higher degrees

C. Rizzi et al. (Eds.): ADM 2019, LNME, pp. 268–278, 2020.
https://doi.org/10.1007/978-3-030-31154-4_23

of freedom designing alternative product architectures oriented to specific target (e.g. assembly) [15, 25]. In literature there are plenty of design methodologies oriented to the optimization of specific targets (Design for X - DFX) [5, 7, 8]. Design for Assembly (DFA) methodologies aiming at minimizing amount of components in complex assembly and at identifying critical tasks in assembly operations [3]. Among DFA methods, the most spread in academia and industry are: Hitachi method [24], Lucas method [11] and Boothroyd and Dewhurst method [2]. However, DFA techniques requires detailed design information and can be applied only during the detail design phase, where most of the choices have been already made [9] and costs engaged. Few attempts have been done to adopt DFA at the conceptual design stage [3, 23]. Those methods appear beneficial for this aim although few limitations have been identified: (i) method applicability considering specificities of aircraft systems, (ii) method feasibility considering the product level of complexity and, (iii) method efficiency in terms of quantitative indicators (numerical assessment vs. qualitative analysis).

In recent years, the majority of the work done in this sector focus on the definition of optimized aircraft architectures toward manufacturing and assembly [1, 12, 13]. In the work of [17] the overall manufacturing cost is condensed by reducing the lead time, using Resource Constrained Shortest Paths (RCSP) optimization applied to production assembly sequences linked to product architecture. Results obtained showed that a co-engineering approach presents many advantages, but it is also time-consuming, and it may require high-skilled operators to run all the necessary analysis. Another attempt has been done by [10] focusing on the relation between aircraft systems and final assembly stage. Unfortunately, the proposed method lacks design information at the right time for the early design and the available data are mostly qualitative than quantitatively. Thus, there is no generalized method for applying the DFA at the conceptual design level.

The goal of this work is to define a design for assembly methodology compliant with the preliminary phases of aircraft systems development (conceptual design). The overall methodology leads to tree main objectives: (i) the definition of a design framework, (ii) the characterization of suitable parameters driving the assessment and development of product architectures, and (iii) the formalization of the required internal knowledge. In particular, the characterization of design parameters together with the formalization of internal knowledge enable to define "scoring matrices" that can be used in the preliminary phases of product development without requiring detailed design information. The mentioned "scoring matrices" are used to translate assembly features in numerical values that are collected and combined through mathematical models defined within the design framework. The main novelty of the proposed approach is the knowledge formalization method for aircraft assembly that allows to characterize installation issues in quantitative manner early in the design phase (conceptual design). Knowledge from manufacturing department, coupled with functional and geometrical information allow to translate different data types (i.e. string, number) in scores that represent the base to apply a design for assembly method. Indeed, from the literature analysis knowledge formalization emerges as a research gap and the characterization of assembly/installation features in numerical scores arises as a potential improvement in this field. This aspect is also recognized as an important progress from the industrial perspective.

The paper is structured as follow: Sect. 2 presents the overall design for assembly framework, the definition of assembly attributes for conceptual product development and knowledge formalization method. As case study, Airbus A320 nose fuselage is presented in Sect. 3. Results are discussed in Sect. 4 and concluding remarks are highlighted in Sect. 5.

2 Materials and Methods

2.1 Conceptual Design for Assembly Framework

One of the main challenge of the conceptual design phase is the capability to elaborate on early architectural data. To overcome this drawback, the proposed methodology is built upon the use of industrial knowledge avoiding to elaborate specific design data. Therefore, a conceptual design for assembly framework is proposed (Fig. 1) using conceptual schemes (e.g. functional analysis and simplified mock-up) as input. Within the framework, a set of scoring matrices is used to translate the manufacturing knowledge into numerical values, and mathematical models allows to obtain two outcomes: (i) an absolute result, and (ii) a comparative analysis of the results.

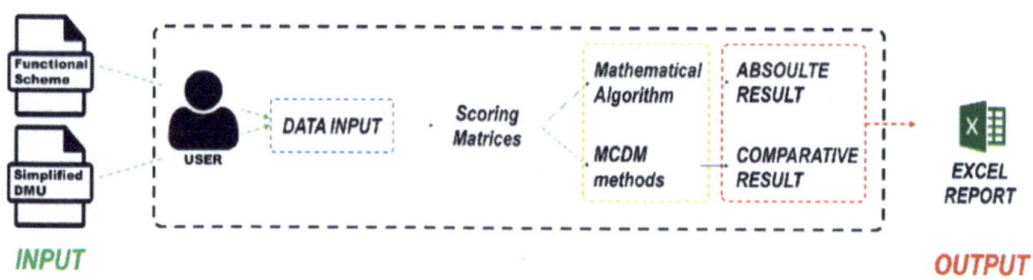

Fig. 1. Methodology framework

The proposed methodology starts with the classical functional analysis [14, 15, 20]. From the result of the functional analyses it is possible to derive *modules*. A *module* is a physical assembly that is designed according to the module's heuristics [23] applied on the functional analysis.

Once functional modules have been defined, it is necessary to identify *interfaces*. An *interface* describes a functional link between different functional modules. Several interfaces can be identified according to the product under study and using the fluxes defined through module's heuristics [23]. Concerning the aerospace industry, the following interfaces are considered as reference:

Air – Interface distributing air fluxes (i.e. air ducts)
Electrical – Interface distributing electrical fluxes (i.e. electrical harnesses)
Liquid – Interface distributing liquid fluxes (i.e. fluid pipes)
Mechanical – Interface providing mechanical connections (i.e. anchor points)

Starting from this functional derivation, a simplified geometrical representation of *Modules* and *Interfaces* is realized using a 3D CAD system. The overall result is a simplify geometry called "simplified Digital Mock-Up" (sDMU). In the sDMU, *Modules* are represented by simple geometries as parallelepipeds, while *Interfaces* are represented by cylinders (Fig. 2). *Modules* and *Interfaces* sizes are approximated considering the functions they need to perform. Moreover, the identification of *Modules* and *Interfaces* is coherent between the two representations (using the same color). Functional scheme and sDMU are used as input to start the design analysis.

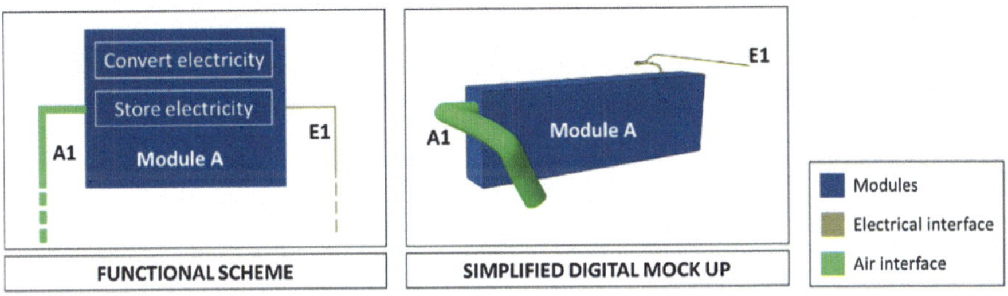

Fig. 2. Example of functional scheme and sDMU

Data from functional scheme and sDMU are allocated within the framework and translated into numerical parameters by the use of *scoring matrices*. Assembly attributes (i.e. working area, modules dimensions, and interface length) are characterized to describe the assembly process of modules/interfaces and they are clustered in specific *domains*. A *domain* is a group of attributes describing specific aspects of the assembly process for components of the product architecture. The concept of domain and attributes is further developed in Sect. 2.2.

Based on the attributes typology (i.e. string, numbers, percentage, etc.) a normalization process is required. Thus, the use of *scoring matrices* allows to have homogeneous data. A *scoring matrix* is defined as a table which translates assembly/installation features in numerical values (numbers in a range between 1 and 5). The matrix is composed of two columns: (i) range of a specific assembly feature, and (ii) score. Further explanations are provided in Sect. 2.3.

Finally, through the application of mathematical algorithms two different results are obtained: (i) an absolute result, and (ii) a comparative result. Absolute result is obtained by applying a Multi-Criteria-Decision-Making Method (MCDM) and it allows to identify criticalities within a specific architecture. Among all, the TOPSIS method has been chosen for its characteristics: (i) simple implementation and, (ii) ability to deal with more parameters without need of modifications [26]. Comparative result is obtained by applying simple mathematical operators (e.g. root mean square) and it allows to evaluate improvements among different product architectures. By doing so, it is possible to generate product architecture focusing on the reduction of installation issues highlighted in a given architecture.

2.2 Assembly Attributes and Domains

The definition of *domains* is a central aspect of this methodology, in fact they represent the starting point to perform the characterization of scoring matrices. A *domain* is a container of several attributes which are clustered based on common assembly features. Three *domains* have been identified considering the specificity of this sector: (i) Components domain – it describes physical aspects of interfaces that need to be brought inside the product and installed, (ii) Assembly domain – it describes the complexity of assembly operations referred to interfaces and, (iii) Ergonomic domain – it describes relation between the area in which operators perform the actions and human factors (Fig. 3).

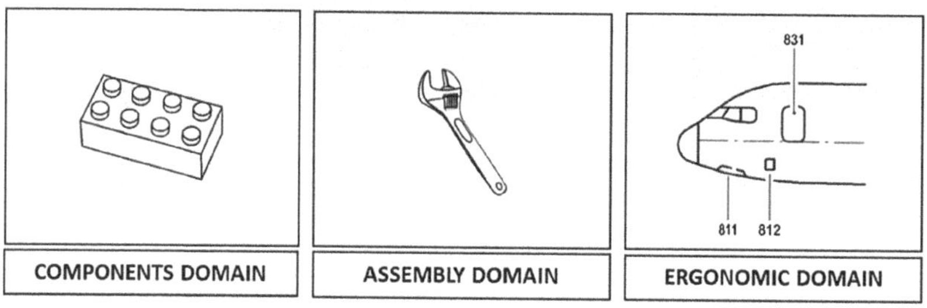

Fig. 3. Three domains defined for the collection of assembly attributes

As an example, an extract of attributes defined for the ergonomic domain is presented below (Table 1). The reason behind the definition of domains is the possibility to bundle attributes that are inter-related. In this way, specific mathematical models can be adopted to couple attributes inside a domain and other mathematical models can be adopted to combine results between domains. For each attribute a scoring matrix is defined, thus general information is translated in numerical value.

Table 1. Attributes of the ergonomic domain

Domain	Attribute	Description
Ergonomic	Working space size	Represents the available space to perform assembly operation
	Zones	Describe the zone where operation needs to be performed
	Working area	Represents the working area where operator needs to work
	Access	Describes the access to the working area

2.3 Knowledge Formalization

Nowadays, the formalization of the knowledge is becoming a great challenge for industries. Indeed, the translation of cases or lessons learned of previous product is necessary to keep internal knowledge [22]. This area of research is called Knowledge

Engineering (KE). In the design field, many methods have been proposed to face this issue such as (i) expert system [6], (ii) Case-based reasoning [16], (iii) Design for manufacturing with KE [18, 19, 21].

In the proposed methodology, the formalization of knowledge is achieved by defining *Scoring Matrices*. Scoring matrices are defined through a concurrent design process, indeed a close collaboration between manufacturing department and architectural department is required. Scoring matrices allow to normalize information inside each attribute (i.e. from string or number to a value), making possible the application of mathematical models (Fig. 4). Scoring matrices remain unchanged among different analyses of product architectures, in this way a comparison between *as-is* configuration and *to-be* configuration can be performed.

Range	Score
String 1	1
String 2	2
String 3	3

SCORING MATRIX ***String***

Range	Score
$A \leq X < B$	1
$B \leq X < C$	2
$X > C$	3

SCORING MATRIX ***Values***

Fig. 4. Scoring matrices

3 Case Study

Knowledge formalization technique for the application of the conceptual design for assembly methodology has been applied on the nose section of Airbus A320. This assembly is one of the most complex component of the aircraft due to the presence of several design constraints and confined areas to perform assembly operations. The aim of this study is to test the robustness of the overall framework by a real example. For this reason, scoring matrices have been defined with a *concurrent design* approach. In particular, four meetings between manufacturing and product architecture department have been scheduled: (i) an initial *face-to-face* meeting, (ii) two follow-up *web* meetings and, (iii) a *final review* meeting.

During the first meeting, the methodology framework has been presented and the necessity of defining scoring matrices has been explained. In the first follow-up *web* meeting, domains and attributes have been defined together with the company manufacturing and assembly department. The work has been carried out through a brainstorming approach. The brainstorming last, approximately, three hours. It is worth noting that brainstorming session allowed to define all and only attributes relevant for both manufacturing and architectural department. The identified attributes address the main features connected with assembly operations (valuable in the conceptual design phase) and then are clustered in the defined domains (Components, Assembly and Ergonomic domains). After this characterization, a *survey* with different questions for both manufacturing and architectural departments has been designed with the aim to define scoring matrices for each attribute of each domain. The survey consisted on a multiple-choice question where only one answer per question was allowed. Five levels of normalization have been chosen, thus the survey presented five multiple-choices, each one associated with a score ranging from one (1) to five (5). The obtained survey has been proposed during the second *web* meeting, in order to define scoring matrices. An extract of the provided survey for the definition of the Ergonomic domain scoring matrices is presented in Fig. 5.

1) Rate the accessibility of the Access IV	Very bad ☐	Bad ☐	Normal ☐	Good ☐	Very Good ☐
2) In what position the operator needs to work in the Working Area A?	Standing ☐	Kneeling ☐	Lying Down ☐	Stand on toes ☐	Curling Up ☐
3) ...	...				

Fig. 5. Extract of the survey

From the outcomes of the survey, scoring matrices have been defined and applied on the methodology. In the final in-loco meeting, the obtained scoring matrices and the methodology results were shown. Scoring matrices for attributes inside the ergonomic domain are shown in Fig. 6. The other scoring matrices have been defined as well, but for the sake of brevity, they have not been reported in this work.

Working Area	Score
A	1
E	2
H	3
B	4
C	5

SCORING MATRIX *Working Area*

Zone	Score
Upper	3
Middle	1
Lower	5

SCORING MATRIX *Zone*

Working Space Size	Score
Very Small	1
Small	2
Normal	3
Big	4
Very Big	5

SCORING MATRIX *Working Space Size*

Access	Score
I	1
IV	2
V	3
III	4
II	5

SCORING MATRIX *Access*

Fig. 6. Scoring matrices for ergonomic domain

Together with the scoring matrices, a document with explanation of each scoring matrix has been created (rationality). The document allows to storage and to transfer the internal knowledge created through scoring matrices.

4 Results

The definition of scoring matrices allowed the application of the methodology to assess the assembly time of product architecture at conceptual level. To check the goodness of the defined scoring matrices, the methodology has been tested over the already existing product architecture of A320 nose fuselage. The obtained results have been checked with the manufacturing and architectural departments. Among all, two (2) scoring matrices were updated (i.e. Working Area and Zone) based on the results derived from the application of the methodology. With the new scoring matrices (Fig. 7) results reflect the reality, highlighting modules that present difficulties from the assembly point of view. The definition of scoring matrices has been performed with four meetings, two face-to-face meetings and two (2) web meetings. The overall time invested in the activity was, approximately 8 h, with few more hours, approximately two (2) hours, necessary to analyze and create documents (i.e. survey and document for knowledge storage).

Working Area	Score
E	1
H	2
A	3
B	4
C	5

SCORING MATRIX *Working Area*

Zone	Score
Upper	3
Middle	2
Lower	5

SCORING MATRIX *Zone*

Fig. 7. Updated scoring matrices

During the first analysis, the five most critical modules and the five most critical interfaces, meaning modules and interfaces with the highest absolute score, have been identified. Results showed slight differences in relation with the real assembly and installation process. In particular regarding the modules assessment, among the most critical five modules, only one (i.e. Module-G) has not been highlighted by manufacturing department as critical. On the other side, regarding the interfaces assessment, the identified critical interfaces rebuilt the real assembly process (i.e. Interfaces E03, E29, E42 and A2). However, the list of critical interfaces is not fully reflected the ranking provided by the manufacturing department and scoring matrices required a fine-tuning. Based on the manufacturing feedbacks, scoring matrices have been re-worked and a new analysis has been done. It is worth noting that the scoring matrices have been re-worked considering the main differences highlighted by the analysis results and the manufacturing department feedback (gap in attributes allocation). Indeed, only specific scoring matrices have been modified. This aspect shows the potentiality of the proposed framework, in terms of: (i) attributes characterization and, (ii) attributes clustering in domains. The second assessment, with new scoring matrices, has shown no discrepancy with the real assembly process. The result can be considered as good starting point in the conceptual analysis of nose fuselage fit for assembly; firstly, as a tool for the assessment of a given configuration/architecture and, secondly for the implementation of architecture design improvements. Further developments need to be done in order to increase the capability of results in the representation of assembly complexity and to fit with the real observation of the assembly line. Indeed, the methodology results for modules and interfaces give a score from zero (0) to one (1) for both relative and absolute results. At the current state, even if the assessment of critical modules and interfaces is in line with the real assembly observation, the gap among final scores does not represent the real fit for assembly performance. For example, considering module-G with a relative final score of 0.8 and Module-T with a relative final score of 0.4, it looks that Module-G is two times longer to install than Module-T. However, this is not compliant with the feedback retrieved by the manufacturing department where a smaller difference is noticed. Evaluations are highly conditioned by the low number of attributes in a given domain, the number of classes characterizing a score (scores from 1 to 5), the personal sensitivity of respondents, etc. This issue can be solved by refining the mathematical models at the basis of the methodology or adopting ponderation (weights) that allow to make the results closer with the real process. In this way it would be possible to evaluate *how much* one module/interface is more/less critical than another from the assembly point of view.

5 Conclusions

This work describes a design for assembly methodology for the development and the assessment of aircraft architectures early in the conceptual design phase. In particular, it provides a design methodology structure (framework), key performance indicators (attributes and domains) and, knowledge elicitation example (scoring matrices). All the three aspects have been evaluated within a specific context for the assessment and development of an aircraft system: the A320 nose fuselage. In particular, knowledge formalization results are acceptable in this first example where the main driver was the possibility to represent with preliminary design information and simple mathematical models the same criticalities observed during the assembly phase of aircraft systems. With this framework, designers and engineers can use the formalized knowledge to compare existing architecture with new ones. Moreover, knowing critical aspects (domains and attributes) which are driving "bad" score for a specific module or interface, enables to set directives to design new architectures

However, by developing this approach some drawbacks have been noticed: (i) knowledge formalization is a time-consuming procedure and requires multiple working sessions with a deep cooperation between manufacturing and architectural departments, (ii) the involvement of senior resources are mandatory to define scoring matrices, and (iii) some iterative steps are mandatory to refine the model (including both knowledge formalization and mathematical models).

Additional work is therefore necessary: (i) automatize the knowledge collection by linking the retrieved information with current product items and systems within the enterprise design repositories, (ii) create a software tool able to support manual data input and information collection, and (iii) derive design guidelines starting from the analysis of critical modules and interfaces to help engineers and architects in the development of new architectures.

References

1. Barbosa, G.F., Carvalho, J.: Design for manufacturing and assembly methodology applied to aircrafts design and manufacturing. Proc. IFAC **46**(7), 116–121 (2013). https://doi.org/10.3182/20130522-3-BR-4036.00044
2. Boothroyd, G., Dewhurst, P., Knight, W.: Product Design for Manufacture and Assembly. Marcel Dekker, New York (2002)
3. Favi, C., Germani, M.: A method to optimize assemblability of industrial product in early design phase: from product architecture to assembly sequence. Int. J. Interact. Des. Manuf. **6** (3), 155–169 (2012). https://doi.org/10.1007/s12008-012-0147-y
4. Hickie, D.: Knowledge and competitiveness in the aerospace industry: the cases of Toulouse, Seattle and north-west England. Eur. Plan. Stud. **14**(5), 697–716 (2006). https://doi.org/10.1080/09654310500500254
5. Holt, R., Barnes, C.: Towards an integrated approach to "Design for X": an agenda for decision-based DFX research. Res. Eng. Design **21**(2), 123–136 (2010). https://doi.org/10.1007/s00163-009-0081-6
6. Hopgood, A.A.: Intelligent Systems for Engineers and Scientists. CRC Press, Boca Raton (2001)
7. Huang, G.Q.: Design for X: Concurrent Engineering Imperatives. Springer, Dordrecht (1996). https://doi.org/10.1007/978-94-011-3985-4
8. Kuo, T.-C., Huang, S.H., Zhang, H.-C.: Design for manufacture and design for 'X': concepts, applications, and perspectives. Comput. Ind. Eng. **41**(3), 241–260 (2001). https://doi.org/10.1016/S0360-8352(01)00045-6
9. Lefever, D.D., Wood, K.L.: Design for assembly techniques in reverse engineering and redesign. In: The Proceedings of ASME Design Theory and Methodology Conference (1996)
10. Li, T., Lockett, H.: An investigation into the interrelationship between aircraft systems and final assembly process design. Procedia CIRP **60**, 62–67 (2017). https://doi.org/10.1016/j.procir.2017.01.056
11. Lucas Engineering Systems Ltd., University of Hull, Design for Manufacture and Assembly Practitioners Manual. Version 10 (1993)
12. Mas, F., Ríos, J., Gómez, A., Hernández, J.C.: Knowledge-based application to define aircraft final assembly lines at the industrialization conceptual design phase. Int. J. Comput. Integr. Manuf. **29**(6), 677–691 (2016)
13. Mas, F., Ríos, J., Menéndez, J.L., Gómez, A.: A process-oriented approach to modeling the conceptual design of aircraft assembly lines. Int. J. Adv. Manuf. Technol. **67**, 771–784 (2013)
14. Otto, K.N., Wood, K.L.: Product evolution: a reverse engineering and redesign methodology. Res. Eng. Design **10**(4), 226–243 (1998). https://doi.org/10.1007/s001639870003
15. Pahl, G., Beitz, W., Feldhusen, J., Grote, K.-H.: Engineering Design. Wallace, K. (ed.) London (2007). https://doi.org/10.1007/978-1-4471-3581-4
16. Pal, S.K., Shiu, S.C.K.: Foundations of Soft Case-Based Reasoning. Wiley, Hoboken (2004)
17. Polacsek, T., et al.: Towards thinking manufacturing and design together: an aeronautical case study. Lecture Notes in Computer Science, pp. 340–353 (2017). https://doi.org/10.1007/978-3-319-69904-2_27
18. Ramana, K.V., Rao, P.V.M.: Automated manufacturability evaluation system for sheet metal components in mass production. Int. J. Prod. Res. **43**, 3889–3913 (2005)

19. Sandberg, S., Kalhori, V., Larsson, T.: FEM simulation supported KEE in high strength steel car body design. In: Proceedings of IMECE04 2004 ASME International Mechanical Engineering Congress, Anaheim, California, USA (2004)
20. Schuh, G., Rudolf, S., Vogels, T.: Development of modular product architectures. Procedia CIRP **20**, 120–125 (2014). https://doi.org/10.1016/j.procir.2014.05.042
21. Shehab, E.M., Abdalla, H.S.: Manufacturing cost modelling for concurrent (2001)
22. Stokes, M.: Managing Engineering Knowledge - MOKA: Methodology for Knowledge Based Engineering. ASME Press, New York (2001)
23. Stone, R.B., McAdams, D.A., Kayyalethekkel, V.J.: A product architecture-based conceptual DFA technique. Des. Stud. **25**(3), 301–325 (2004). https://doi.org/10.1016/j.destud.2003.09.001
24. Suzuki, T., et al.: Assembly reliability evaluation method (AREM). In: Proceedings of the IEEE International Symposium on Assembly and Task Planning (2001)
25. Ulrich, K., Eppinger, S.A.: Product Design and Development. McGraw-Hill Education, New York (2011)
26. Velasquez, M., Hester, P.: An analysis of multi-criteria decision making methods. Int. J. Oper. Res. **10**, 56–66 (2013)
27. Yancey, R.N.: Challenges, opportunities, and perspectives on lightweight composite structures: aerospace versus automotive. In: Lightweight Composite Structures in Transport, pp. 35–52 (2016). https://doi.org/10.1016/B978-1-78242-325-6.00002-5

A Knowledge Formalization Approach for Manufacturing Cost Estimation

Marco Mandolini[1(✉)], Claudio Favi[2], Federico Campi[1], and Roberto Raffaeli[3]

[1] Università Politecnica delle Marche, Via Brecce Bianche 12, 60131 Ancona, Italy
m.mandolini@univpm.it

[2] Università degli Studi di Parma, Parco Area delle Scienze 181/A, 43124 Parma, Italy

[3] Università degli Studi eCampus, Via Isimbardi 10, 22060 Novedrate, Como, Italy

Abstract. Target pricing is a methodology to develop competitive products by determining the target price from market analyses. To guarantee the right profit margin, target cost is a direct consequence of target price. In this situation, the manufacturing cost estimation at the design phase becomes an essential task.

The paper presents a framework for collecting knowledge required for estimating manufacturing cost of components. The framework consists of: (i) a cost breakdown structure used for splitting manufacturing costs, (ii) a data model for collecting that knowledge required for defining manufacturing processes, (iii) a data model for the collecting that knowledge required for computing the manufacturing cost of each operation within a manufacturing process and (iv) a workflow for analytically estimating cost of components. The framework has been mainly conceived for managing components realized through forming and shaping processes. The result presented in this paper guarantee the following benefits: (i) knowledge elicitation on product manufacturing cost, (ii) knowledge sharing among design/engineering departments, and (iii) knowledge capitalization for decision-making process.

Keywords: Design for manufacturing · Design to Cost · Manufacturing cost estimation · Knowledge management

1 Introduction and State of the Art

Target pricing is a methodology for determining the target price from market analyses, with the aim to develop competitive products [1]. To guarantee the right profit margin, the target cost is a direct consequence of the target price. Hence, the cost becomes a constraint that designers must consider during the product design process [2, 3]. Manufacturing and assembly costs are decided during the design stage and their definition tend to affect the selection of materials, machines and human resources that are being used in the production process [4]. In this situation, the manufacturing cost estimation at the design phase becomes an essential task.

C. Rizzi et al. (Eds.): ADM 2019, LNME, pp. 279–290, 2020.
https://doi.org/10.1007/978-3-030-31154-4_24

Manufacturing cost estimation at the design stage, is currently managed through customized and very complex electronic spreadsheets, developed by cost engineers that collected their experience. However, the process used for analysing and developing such spreadsheets does not follow a specific framework, so that, the results (i.e. cost models) are commonly not structured, they depend by the engineer that analysed the process and they are not homogeneous. As consequence, such results cannot be easily used by designers. Manufacturing companies that want standardizing and wide spreading the method for manufacturing cost estimation, by transforming experience of their technicians (i.e. cost and production engineers and skilled operators), as well as of their certified suppliers, in knowledge, are looking for proper frameworks.

So far, qualitative and quantitative techniques can be employed for the product cost estimation [5]. The first ones are based on a comparative analysis between the new product and products previously manufactured, to identify similarities. Qualitative techniques are more appropriate when the cost estimation time is limited, past data or expert's knowledge is available, and the estimating accuracy requirement is not mandatory. Quantitative techniques are preferable when relationships among the cost variables are identifiable, and when more accurate cost analyses are needed. Quantitative techniques are based on a detailed analysis of the product to develop and can be further categorized into parametric and analytical methods. Using the first ones, cost is expressed as a function of its constituent variables. Parametric techniques could be effective when parameters, named also cost drivers, could be easily identified. With analytical techniques, the product cost is decomposed into elementary units, operations, and activities that represent different resources consumed during the production cycle.

The manufacturing knowledge represents the groundwork for a proper implementation of analytical cost estimation methods [6]. Knowledge can be divided in tacit and explicit [7]. Tacit knowledge refers to that expertise people carry in their minds. Hence, this is unformalized and cannot be widely used by an organization. Explicit knowledge, instead, refers to a set of information that can be articulated, codified, and stored in certain media. To make knowledge usable, a representation framework is needed to deposit and then make it accessible to everyone involved within an enterprise [8]. Few attempts are available on this aim. Streppel [9] developed a framework for cost estimation and cost control where the product is divided into different levels: assembly, component and feature. Each of these levels has its own cost attributes, such as geometry, material, production process, and product planning. Zhang [10] analysed knowledge representation for unit manufacturing processes using an ontology model. Unitary process or operation is represented as an element (block) connected with others. These blocks are: incoming and outgoing flows (energy and materials), machines (equipment), tools, process capacity. Kang [11] used an ontology knowledge model for sequencing a machining process. The knowledge model incorporates information on process characteristics, the relationship between machining characteristics and machining processes, and the process capability to meet production requirements. Kulon [12] developed a knowledge-based engineering (KBE) system for the integration of hot forging design process into a single framework to capture the experience and knowledge of the designers. In this framework, the forged part is classified and defined according to its characteristics, such as material and features (holes, tolerances, etc.). On the same aim, Shehab [13] presented an intelligent KBE system for product cost modelling of machining and injection moulded products at the design stage of the product life cycle. The main feature of those systems is their link with feature-based CAD systems which allow the definition of raw material, machining processes and related parameters based on a set of design and production parameters. Even if these systems estimate the product development cost (including the assembly phase), they show some limitations concerning the covered technologies (i.e. injection moulding, machining processes) and there is no evidence that it can be extended to other processes.

To summarize, the state-of-art shows how dedicated cost models were developed to address specificity of each manufacturing process. However, when multiple technologies are adopted for producing complex products, the manufacturing cost estimation activity requires a concurrent assessment of several processes as well as the definition of a precise cost breakdown structure (material, setup, equipment, consumable, etc.). To date, methods available for the elicitation of manufacturing cost related knowledge, for different technologies, have not developed yet.

Following the limitations highlighted by the literature review, this research work aims at defining a *framework* for formalizing knowledge required for analytically estimating the manufacturing cost of mechanical components, to be employed for shaping processes.

The *framework* consists of four main constructs:

- a *cost breakdown structure* used for splitting the manufacturing cost of a component;
- a *procedure* for defining a manufacturing process;
- a data model (*cost routing*) for collecting knowledge to be used for estimating a manufacturing process;
- a data model (*cost model*) for collecting the knowledge required for computing the manufacturing cost of each operation within a manufacturing process.

This framework is a sort of pre-configured structure that can be used by cost engineers for developing cost models and cost routings. In this manner, it is possible to formalized, in a proper way, the knowledge required for analytically estimating the manufacturing cost of products. Such approach guarantees that each cost model and cost routing is formalized according to a specific standard and that the manufacturing cost is structured according to a specific breakdown. Cost models and routings developed with such an approach, can be coupled with a procedure for automatically estimating a manufacturing process from a 3D CAD model. The combination of such constructs (namely, the proposed framework) will support design engineers during cost estimation and Design to Cost.

2 The Knowledge Formalization Approach

2.1 Manufacturing Cost Breakdown Data Structure

A cost breakdown data structure (Fig. 1) is necessary to collect cost-related information of each production phase with the aim to standardize the output of a cost estimation activity. Manufacturing costs can be divided into six macro categories: (i) material, (ii) machine, (iii) labour, (iv) equipment, (v) consumables and (vi) energy. This organization is a result of literature analysis, combining and generalizing the most common classification methods (different manufacturing technologies were analysed) [14–17].

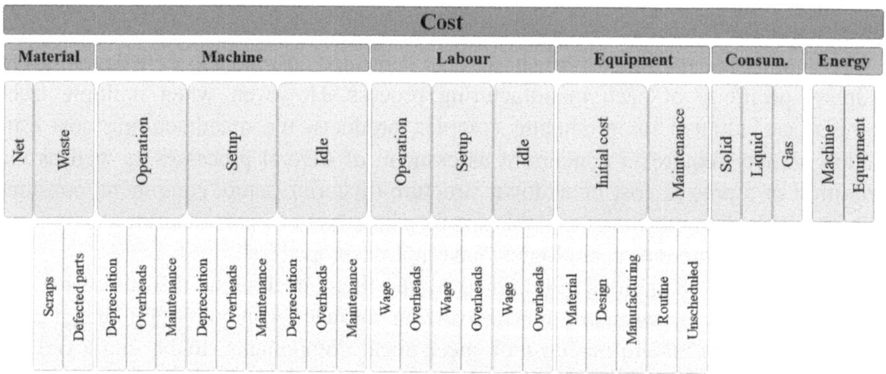

Fig. 1. Cost breakdown structure

The material category refers to the costs of raw material necessary to produce a specific part/component. The raw material cost is the sum of the part net cost and waste cost. Material waste is divided in two categories: (i) scraps and (ii) defected parts. For scraps, authors intend the excess of material necessary for processing (e.g. flash in the forging process). Defected parts refer instead to non-compliant components realized during the initial process start-up or during production.

Machine and *labour* categories refer to the cost-centres used for performing an operation. These costs are further classified in *operation*, *setup* and *idle*. For each process operation, according to the degree of automation, one/none machines and/or one/multiple workers can be employed. The hourly cost rate of a machine considers its maintenance, overhead and depreciation, whereas, the rate for an operator considers its wage and overhead. The *operation* sub-category refers to the manufacturing operations (e.g. chip removal, plastic deformation, etc.) that directly contribute toward the realization of the final component. This item is considered a product direct cost. The *idle* sub-category refers to a passive manufacturing phase when, for example, one operation has been completed, and tooling or materials for the next one is not yet completed or available. The *setup* sub-category refers to those operations, such as tool setting and machine cleaning, required before beginning the production. These operations are independent by the batch dimension; hence, the related cost must be split for the batch quantity for calculating the setup cost for each component.

The *equipment* category refers to those tools, such as mould jigs and fixtures, required for performing a specific process operation. The cost is the sum of its initial expenditure and maintenance cost during its usage. The initial expenditure considers the cost for its design and manufacturing, plus the material cost. This cost is independent by the production volume; hence, the related cost must be split for the production volume for calculating the equipment cost for each component.

The *consumables* category refers to those materials that enable the process itself (e.g. lubricants used for forging, gas cutting assistance for laser cutting, etc.). This is a direct and accessory cost directly allocated to the cost of each component.

The *energy* category refers to the energy vectors (e.g. electricity, water, steam, etc.) that guarantee the process working. Energy may be requested by machines and/or equipment and the related cost is function of their power and working time.

2.2 Procedure for Manufacturing Process Estimation

Manufacturing cost can be analytically calculated only once a manufacturing process has been defined. As stated in the introduction, since cost estimation should be carried out at the design stage, is rather important to use/define a method through that it is possible to establish the best or all the feasible manufacturing processes able to transform a raw material in a finished component. This section presents a step-by-step approach, that starting from the virtual prototype of a component, it can define a manufacturing process plan.

The workflow consists of five decision steps (Fig. 2), each one supported by a specific database and a set of knowledge-based rules (cost routing). Most of the rules described in this workflow can be classified in two types. Validity rules are used for removing manufacturing-related solutions that are not applicable in certain conditions. Priority rules are used for sorting the valid solutions. The cost estimation process is based on a set of product and process related information, which are: 3D CAD model, geometrical and non-geometrical attributes, Product Manufacturing Information (PMI) and process attributes.

The *first step* for defining a manufacturing process consists in establishing the overall production scenario. Indeed, the manufacturing process and the related costs firstly depend by the production environment (production facility, raw material warehouse and sourcing strategy).

Once defined the production environment, the *second step* aims to define the production strategy, consisting in the selection of the raw material and manufacturing process. The material selection (e.g. commercial semi-finished product vs custom stock) and the manufacturing process (e.g. die casting vs chip forming) should be carried out at the same time since the last one depends by the raw material and vice-versa. For example, the injection moulding process is valid only for thermoplastic polymers (validity rule), whereas it is convenient only for producing, at least, thousands of components (priority rule).

The *third step* aims to define raw material. The information to be calculated are the type of material (e.g. commercial bar, sheet metal), shape (e.g. circular, rectangular, solid/hollow), dimensions (e.g. thickness, length, width), supply status (e.g. hot rolled, extruded, grinded), volume, weight and unitary cost. For the injection moulding, only thermoplastic polymers granules can be used (validity rule), while the raw material volume is computed considering the part volume plus the volume of runners (selection/calculation rules). For the calculation of such information, feature recognition algorithms should be employed for analysing the 3D CAD model with the aim to compute a specific raw material feature, consisting of a set of geometrical information required for selecting the stock. The raw material cost is computed by multiplying the amount of requested material by the unitary cost.

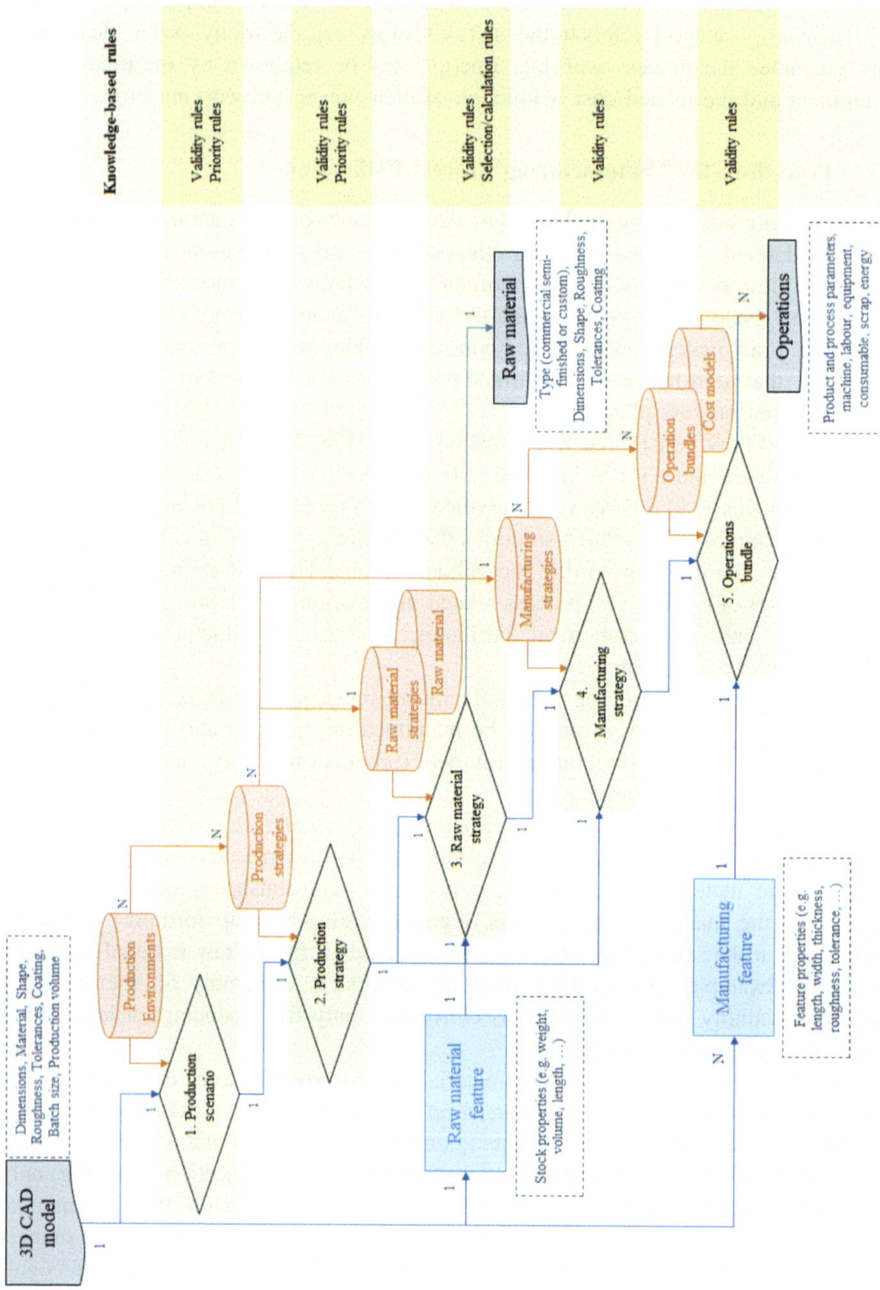

Fig. 2. Workflow for defining a manufacturing process

The *fourth step* aims to define the manufacturing strategy to be employed for making a component/product. A manufacturing strategy consists of a list of operations bundles, described in the next step. Each one has a list of validity and priority rules. A manufacturing strategy is triggered by validity rules. Considering the injection moulding as example, this is valid only for thermoplastic materials.

The *fifth step* is an intermediate phase before the calculation of the operations sequence. Indeed, a product can be realized employing multiple and different operations. For example, a hole, according to its shape, diameter, depth, roughness, tolerance, product material, production volume can be realized adopting different operations despite already a specific manufacturing strategy was yet defined. For simplifying the definition of an operations sequence, the concept of operations bundle has been defined. This one consists of a group of operations required to produce a specific product manufacturing feature (PMF). A PMF is an object consisting of a list of faces and properties (e.g. hole depth, hole diameter, hole shape, minimum tolerance, minimum roughness). The latter are bundle dependant and are used for establishing the operations bundle to be used. PMFs (e.g. hole, cutout, chamfer, fillet, turning, welding) can computed by feature recognition algorithms (their description is beyond the scope of this paper). The bundle is also responsible for transferring the PMF properties to the valid operations defined in its inside (the bundle contains one or multiple validity rules for each operation). For each bundle, a list of validity rules is also defined.

By assembling all the valid operations within the valid bundles, for a specific manufacturing strategy, it is possible to define the operations list that represent the manufacturing process of a product.

2.3 Manufacturing Cost Model

Cost model is a data model containing that knowledge required for estimating the production time and cost for each operation. Cost model can be considered as a structured object of information as illustrated in Fig. 3. It consists of a list of product and process parameters. The first ones define the shape and dimension of the operation to be carried out, while the second ones characterize the manufacturing operation from the technological standpoint. These parameters (e.g. injection temperature and pressure for injection moulding), computed using specific calculation rules, are based on the product parameters, other information available from a database and analytical/empiric calculation rules. The latter could be retrieved from industrial and scientific literature.

A cost model contains also several validity and calculation rules. The first ones are used for limiting the possible cost centres (machine and operator) energy vectors, consumables, equipment and wastes applicable for a specific operation. The second ones are used for calculating the consumption of energy vector, consumables, equipment and the generation of waste. At last, an operation contains rules for computing the manufacturing time and cost, respecting the cost breakdown presented in Fig. 1.

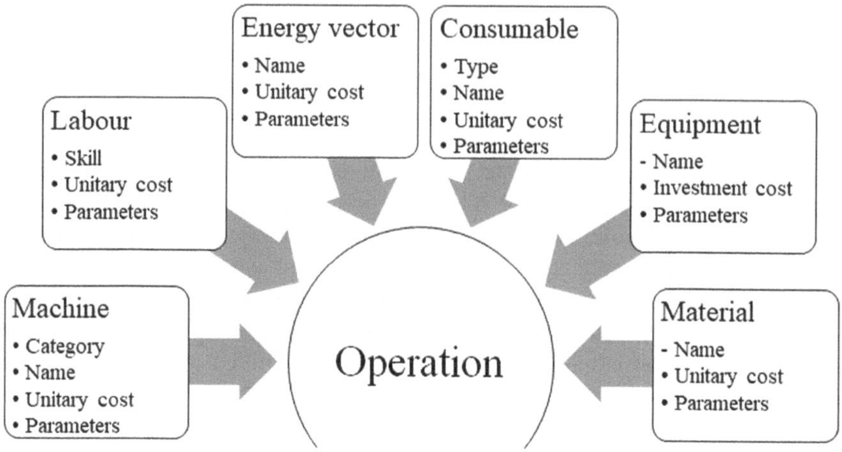

Fig. 3. Manufacturing cost model structure

To compute the process parameters, it is necessary to establish the following information (the examples refer to the injection moulding process):

- *Machine*: this is the cost centre used for realizing the operation. Each operation has a list of available machines, restricted by a list of validity rules. It is worth noting that process parameters are influenced by the machine (e.g. injection time depends by the press power);
- *Labour*: this is another cost centre that can be used for realizing the operation. Its behaviour is the same one of *Machine*;
- *Energy*: each operation may use one, multiple or none energy vector, such as electricity. The energy consumption mainly depends by the machine, product and process parameters (e.g. electricity consumption depends by the machine power and time of usage);
- *Consumable*: each operation may use one, multiple or none consumable, such as lubricant, cutting tools, cutting assistance gas, etc. The consumable consumption mainly depends by the machine, product and process parameters;
- *Equipment*: each operation may use one, multiple or none equipment, such as jigs, fixture and moulds. The equipment depends by the machine and some process parameters, while influences other process parameters;
- *Waste*: each operation may generate scraps or defected parts during the process start-up or standard production. Waste depends by both product and process parameters and by the maturity of a process.

All the previous information contributes to the calculation of the operation cost.

3 Achievements and Benefits

The proposed method has been applied for modelling the manufacturing knowledge related to open-die forging and plastic injection processes. For clarifying the usage of the method proposed in the previous section, Table 1 and Appendix 1 present respectively a production strategy and a cost model for an open die forging process.

Table 1. Production strategies for the open-die forging

Production strategy	Raw material strategy	Manufacturing strategy	Validity rules	Priority rules
Disc Open Die Forging From Bar	Bar	Disc Open Die Forging	–	IF (Production. BatchQuantity > 10) THEN Score = 0 ELSE Score = 10
Disc Open Die Forging From Billet	Billet	Disc Open Die Forging	Piece.Material.Category = "Metal" Piece.Volume > 8dm3 NOT (Piece.Shape = "Hollow") NOT (Piece.Shape = "SheetMetal")	IF (Production. BatchQuantity > 10) THEN Score = 10 ELSE Score = 0 IF (Piece.Volume > 25dm3) THEN Score = 10 ELSE Score = 0

The other constructs previously presented (i.e. production logic, raw material strategy, manufacturing strategy and operations bundle) are not reported for confidentiality reasons.

For injection moulding and open-die forging, the constructs of the proposed framework (cost breakdown, cost routing, procedure for manufacturing process definition and cost model) have been evaluated considering a set of requirements defined within the literature analysis and the findings of the specific case study. For each requirement, Table 2 presents the results achieved in this research work and relative comments. Two outcomes for each cost item have been identified:

- *Improvement addressed considering existing barriers of the state-of-art*: the requirement, emerged during the literature review, was satisfied, hence, the proposed framework is completer and more comprehensive than the ones proposed in the literature;
- *Improvement to be addressed considering existing barriers of the state-of-art*: the requirements, emerged during the literature review and the analyses of the results, represent interesting improvement to be managed in a future research.

Very positive outcomes are highlighted in relation to the cost breakdown structure and cost model, where the most important requirements were addressed. Some future improvements are required for the cost routing and for the management of different objectives, variables and constraints of an optimization problem as well as for the management of rules to be used for sorting operations. Looking at the workflow, a positive outcome is the possible extension to other forming processes, while a workflow for assemblies is required as improvement.

Table 2. Outcomes of the framework implementation: (TbA: To be addressed, A: Addressed)

Requirement	Construct	Outcome
Detailed cost breakdown structure to be used for in-depth cost analyses	Cost breakdown	A
General cost breakdown structure to be used for forming processes	Cost breakdown	A
Workflow for defining manufacturing processes from 3D virtual prototypes of components	Procedure	A
Workflow for defining manufacturing processes from 3D virtual prototypes of assemblies	Procedure	TbA
General structure for collecting knowledge-based rules for defining a manufacturing process	Cost routing	A
Optimization methods at cost routing	Cost routing	TbA
Process yield	Cost routing	TbA
Sorting of process operations	Cost routing	TbA
Cost routings for joining processes	Cost routing	TbA
General cost model to be used for forming processes	Cost model	A
Cost model to provide cost breakdown according to the structure proposed in section	Cost model	A
Optimization methods at cost model	Cost model	TbA

4 Conclusions

The paper originated from the need to support enterprises in formalizing the manufacturing knowledge to be used for the manufacturing cost estimation of products (analytical approach) during the design process. The paper attempts to close the research gap between detailed cost models of single manufacturing processes (available in literature for most of the knowing technologies) and the need to have a suitable framework for cost estimation that can be representative of each manufacturing technology used to realize mechanical components. The research work presented a knowledge-based procedure for defining a manufacturing processes starting from the product virtual prototype. This procedure is based on a set of repositories and cost routings, properly defined for collecting the knowledge required for estimating a manufacturing process. Once defined the process, the knowledge behind each manufacturing operation (cost model) allows calculating the manufacturing cost. Economic information is obtained according to a precise breakdown that allows designers and production technologists to evaluate in detail product and process criticalities. The framework presented in this paper allows production companies to capitalize their manufacturing best practices (often in the minds of a few qualified engineers) and make them available to all stakeholders involved in the product development and design to cost actions.

Future research will be focused to further improve the proposed framework and increase its boundaries of application. Cost routings and cost models should include rules for optimizing the manufacturing cost of a single operation as well as of the whole process.

Appendix 1: Open-Die Forging Cost Model

Material	Machine	Labour	Equipment	Consumable	Energy
Validity rules N/A	**Validity rules** Machine.Category = ("HydraulicPress" OR "*Hammer*") Machine.Tonnage > RequestedTonnage Machine.Energy * BlowNumbers > RequestedEnergy Machine.MaxSizeX > Piece.Width Machine.MaxSizeY > Piece.Width Machine.MaxSizeZ > Billet.Lenght	**Validity rules** Labour.Skill = "*Hot Forging Operator*"	**Validity rules** N/A	**Validity rules** **Solid** IF RawMaterial.Material = "*Aluminium*" THEN Consumable.Solid.Type = "*Solids*" ELSEIF RawMaterial. Material = "*Steel*" OR RawMaterial.Material = "*StainlessSteel*" THEN Consumable. Solid.Type = Graphite **Liquid** Consumable.Liquid.Type = NA **Gas** Consumable.Gas.Type = NA	**Validity rules** Energy.Vector = "*Electricity*"
Priority rules N/A	**Priority rules** Minimum hourly rate	**Priority rules** N/A	**Priority rules** N/A	**Priority rules** N/A	**Priority rules** N/A
Calculation rules N/A	**Calculation rules** **Operation** ProcessTime = BlowNumbers/Machine. BlowRate MachineProcessCost = ProcessTime * Machine.HourlyRate **Setup** SetupTime = Machine. SetupTime/Production. BatchQuantity MachineSetupCost = SetupTime * Machine. HourlyRate **Idle** IdleTime = 0 MachineIdleCost = IdleTime * Machine. HourlyRate	**Calculation rules** **Operation** LabourProcessCost = ProcessTime * Labour. HourlyRate * OperatorsQuantity **Setup** LabourSetupCost = SetupTime * Labour. HourlyRate * OperatorsQuantity **Idle** LabourIdleCost = IdleTime * Labour. HourlyRate * OperatorsQuantity	**Calculation rules** N/A	**Calculation rules** **Solid** Consumable.Quantity = Consumable. Hourly. Quantity* Operation. ProcessTime **Liquid** N/A **Gas** N/A	**Calculation rules** **Machine** EnergyCost.Machine = (Machine. Power * Machine.LoadFactor * Operation.ProcessTime) * Energy.UnitaryCost **Equipment** EnergyCost.Equipment = 0

References

1. Cooper, R., Slagmulder, R.: Target Costing and Value Engineering, 1st edn. CRC Press, New York (1997)
2. Pahl, G., Beitz, W.: Engineering Design: A Systematic Approach, 3rd edn. Springer, London (2007)
3. Ulrich, K.T., Eppinger, S.D.: Product Design and Development, 5th edn. McGraw-Hill Inc., New York (2011)
4. Nitesh-Prakash, W., Sridhar, V.G., Annamalai, K.: New product development by DfMA and rapid prototyping. J. Eng. Appl. Sci. **9**(3), 274–279 (2014)
5. Niazi, A., Dai, J.S., Balabani, S., Seneviratne, L.: Product cost estimation: technique classification and methodology review. J. Manuf. Sci. Eng. **128**, 563–575 (2006)
6. Hoque, A.S.M., Halder, P.K., Parvez, M.S., Szecsi, T.: Integrated manufacturing features and design-for-manufacture guidelines for reducing product cost under CAD CAM environment. Comput. Ind. Eng. **66**, 988–1003 (2013)
7. Darai, D.S., Singh, S., Biswas, S.: Knowledge engineering-an overview. Int. J. Comput. Sci. Inf. Technol. **1**(4), 230–234 (2010)
8. Jiang, Y., Peng, G., Liu, W.: Research on ontology-based integration of product knowledge for collaborative manufacturing. Int. J. Adv. Manuf. Technol. **49**, 1209–1221 (2010)
9. Streppel, A.H., Weustink, I.F., Brinke, E., Kals, H.J.J.: A generic framework for cost estimation and cost control in product design. J. Mater. Process. Technol. **103**, 141–148 (2003)
10. Zhang, Y., Luo, X., Zhang, H., Sutherland, J.W.: Knowledge representation for unit manufacturing processes. Int. J. Adv. Manuf. Technol. **73**, 1011–1031 (2014)
11. Kang, M., Kim, G., Lee, T., Jung, C.H., Eum, K., Park, M.W., Kim, J.K.: Selection and sequencing of machining processes for prismatic parts using process ontology model. Int. J. Precis. Eng. Manuf. **17**(3), 387–394 (2016)
12. Kulon, J., Broomhead, P., Mynors, D.J.: Applying knowledge-based engineering to traditional manufacturing design. Int. J. Adv. Manuf. Technol. **30**, 945–951 (2006)
13. Shehab, E.M., Abdalla, H.S.: An intelligent knowledge-based system for product cost modelling. Int. J. Adv. Manuf. Technol. **19**, 49–65 (2002)
14. Chiadamrong, N.: The development of an economic quality cost model. Total Qual. Manage. Bus. Excell. **14**(9), 999–1014 (2003)
15. Wang, H.S.: Application of BPN with feature-based models on cost estimation of plastic injection products. Comput. Ind. Eng. **53**(1), 79–94 (2007)
16. Hmida, F., Martin, P., Vernadat, F.: Cost estimation in mechanical production: the cost entity approach applied to integrated product engineering. Int. J. Prod. Econ. **103**(1), 17–35 (2006)
17. Xu, Y., Elgh, F., Erkoyuncu, J.A., Bankole, O., Goh, Y., Cheung, W.M., Baguley, P., Wang, Q., Arundachawat, P., Shehab, E., Newnes, L., Roy, R.: Cost engineering for manufacturing: current and future research. Int. J. Comput. Integr. Manuf. **25**(4–5), 300–314 (2012)

Vibration-Assisted Face Grinding of Mould Steel

Sebastian Hähnel[1(✉)], Tim Grunwald[1], Thomas Bergs[1], Fabio Pini[2], and Francesco Leali[2]

[1] Fraunhofer Institute for Production Technology IPT, Steinbachstr. 17, 52074 Aachen, Germany
{sebastian.haehnel, tim.grunwald, thomas.bergs}@ipt.fraunhofer.de

[2] Department of Engineering "Enzo Ferrari", University of Modena and Reggio Emilia, Via Vivarelli 10, Modena 41125, Italy
{fabio.pini, francesco.leali}@unimore.it

Abstract. This work investigates vibration-supported, force-controlled fine machining with elastic bonded mounted points for automated fine processing of mould steel samples. The aim is to compare conventional robot- or machine-tool-based face grinding with a vibration-supported grinding process. The influence of vibration support on the surface topography is investigated primarily to minimize kinematically caused grinding traces. First, the state of the art for the production of tool moulds and vibration-supported fine machining is explained. On this basis, the potentials for the reduction of grinding marks through vibration support for an increase in the degree of automation are derived and the experimental procedure is introduced. Subsequently, robot-based grinding tests with vibration support are carried out and compared with conventional grinding tests. After the tests carried out, the results are evaluated using tactile and optical measuring methods.

Keywords: Vibration-assisted · Grinding · Finishing · Moulds

1 Introduction

In the tool and mould making sector, fine machining is the only step in the entire manufacturing process chain that is largely carried out manually and is not automated. The fine machining step can take up approximately 50% of the total production time with a cost share of up to 30%. In particular, large-area mould steel for the production of car body plastics components from the automotive industry most often require time-consuming manually executed finishing works such as grinding and polishing. In addition, manual fine machining is characterised by monotonous, strenuous work carried out under adverse conditions (grinding dust, noise, etc.). Above there is a rising shortage of skilled workers. Due to the manual and mostly individual processing strategy, the work result is subject to quality fluctuations and has limited reproducibility. This has an overall negative effect on the machining result and is contrary to a trend of constantly increasing component requirements with regard to dimensional

C. Rizzi et al. (Eds.): ADM 2019, LNME, pp. 291–303, 2020.
https://doi.org/10.1007/978-3-030-31154-4_25

accuracy and surface quality. For these reasons, there is significant potential for optimisation through the introduction of partial or full automation of fine machining in tool and mould making.

The process chain for the production of large injection moulds and thermoforming tools for car body parts (e.g. bumper, rocker panel, etc.) is shown schematically in Fig. 1, which also provides a comparison between the manual process chain and the potential of a partly automated process [1]. After component and process development by means of CAD/CAM-supported systems, the component is usually machined using a milling (or eroding) process. The finishing process then removes any remaining milling grooves and smoothes the surface to a desired level of surface quality up to high gloss. With respect to the actual process, this is most often done by manual finishing operations with honing stones, grinding paper, polishing felt/paste in several iterative steps whereby a change in the active surface geometry is strictly avoided. For example, undesirable edge rounding can lead to time-consuming reworking.

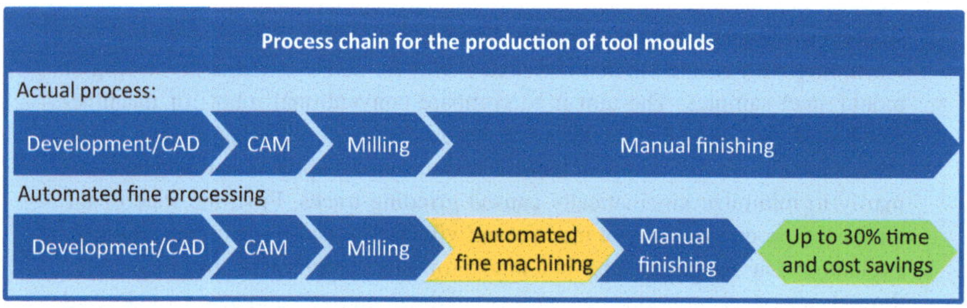

Fig. 1. Comparison of manual and (partly-) automated manufacturing process chain of moulds.

Following the finishing process, so-called "spotting" is carried out, in which the moulds are adjusted on the upper and lower sides as well as on the press by means of local reworking [1]. Due to the machine- or robot-based process kinematics, traces of machining occur on the surfaces of workpiece, which have to be removed by manual reworking, e.g. cross grinding, depending on the customer's requirements.

The aim of automated fine machining is to partially substitute the final manual processing for the removal of kinematically caused grinding marks and to maximise the degree of automation. As the basis for the automated approach, the Fraunhofer IPT developed the system CTHone, an axially compliant tool holder for both machine-integrated and robot-based fine machining. The approach feasibility has been successfully proved in human-robot reconfigurable system where manual finishing occurs after automated polishing and quality assessment sessions [2, 3].

To further improve the surface quality, subsequent studies were carried out on a hybrid solutions based on vibration-supported grinding with rotating grinding tools acting on the surface at a specific vibration frequency. The aim of vibration-supported grinding is to superimpose the conventional grinding kinematics with an additional effective movement, which was carried out uniaxial and workpiece-based. The hybrid process solution is intended to improve surface quality, material removal rate and tool life. In particular, the homogeneity of the surface is to be improved. With regard to component quality, kinematically induced machining traces are to be minimised and a unidirectional or specifically adjustable surface texture is to be achieved.

Preliminary work for vibration-supported machining with geometrically indeterminate cutting edges mainly deals with high-frequency ultrasonic-supported surface grinding/lapping, circumferential longitudinal grinding, transverse circumferential grinding, plunge grinding or cylindrical grinding with grinding wheels or grinding pins as well as dressing processes of grinding wheels. Brittle hard materials such as ceramics and tungsten carbide as well as occasional steels and hard turning of hardened steels are considered in most cases, with the majority of scientific work concentrating on an excitation frequency in the range of 20 kHz [4–16].

In contrast to high-frequency oscillation units (e.g. ultrasonic systems), oscillation systems for the low-frequency range are not necessarily designed electrically and can be realised with less apparatus and correspondingly reduced costs. In addition, significantly higher vibration amplitudes up to several millimetres can be achieved.

This means that vibration-induced superimpositions of grinding paths can also be achieved at path distances in the millimetre range, which occur during automated fine machining. Studies have already successfully shown that an increasing and thus relatively large vibration amplitude correlates positively with a reduction of the grinding forces and improvement of the surface of hard brittle materials, so that the choice of low frequencies appears to be sensible. Moreover, investigations on a vibration-supported grinding process with galvanically bonded diamond face grinding tools and grey cast iron have shown that an improved homogeneity of the grinding profile can be achieved by reducing the grinding traces and increasing the surface quality [16].

The present work evaluates the effectiveness to implement the hybrid face grinding low-frequency vibration solutions for automated finishing of steel mould. Consequently, the subsequent Sect. 2 details the reference model for the process and the related mathematical definition, as well as it describes the experimental set-up. The results are collected and compared in Sect. 3. Conclusive remarks and future developments close the paper in Sect. 4.

2 Experimental Methodology

A reference model with workpiece-side excited vibration was developed for the evaluation of a low-frequency vibration-supported fine machining of plastic mould steel. Vibration support on the workpiece side is limited in terms of component weight and conceptually designed for fine machining of plane samples. For preliminary investigations and an evaluation of the potential of the hybrid process for minimising kinematically induced machining traces, the limitations mentioned are initially not relevant.

Figure 2(a) describes the process kinematics by schematically depicting the test set-up and knitting movement of the tool/workpiece interaction with vibration-assisted grinding kinematics. A "zig-zag" path strategy with a constant lead/tilt angle α/β was selected as the path strategy. In this way, the tool performs a leading movement in the direction of the feed arrow and a trailing movement in the opposite direction to the arrow with the lead α. During the test, the direction of oscillation in relation to the machining direction was carried out with a varying superposition angle γ. Cutting speed v_c, contact pressure F_A, lead/tilt angle α/β, feed rate v_f and path distance a_p were set and kept constant based on preliminary investigations.

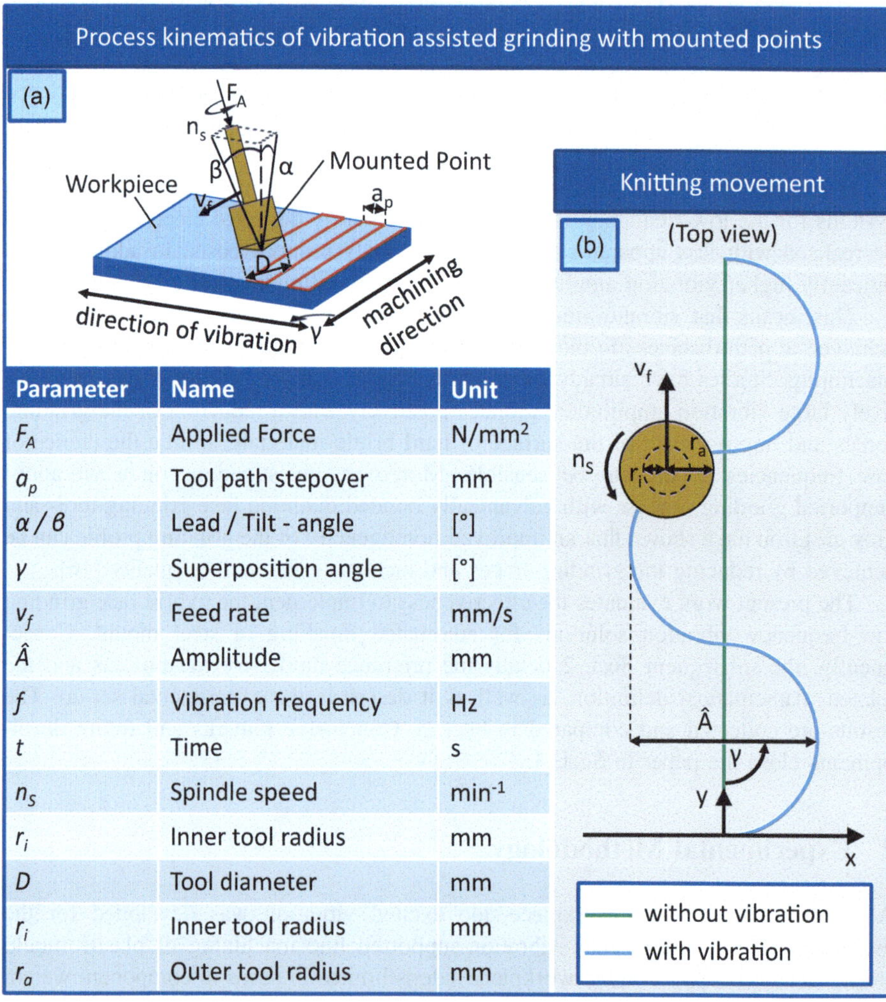

Parameter	Name	Unit
F_A	Applied Force	N/mm²
a_p	Tool path stepover	mm
α / β	Lead / Tilt - angle	[°]
γ	Superposition angle	[°]
v_f	Feed rate	mm/s
$\hat{A}$	Amplitude	mm
f	Vibration frequency	Hz
t	Time	s
n_S	Spindle speed	min⁻¹
r_i	Inner tool radius	mm
D	Tool diameter	mm
r_i	Inner tool radius	mm
r_a	Outer tool radius	mm

Fig. 2. Process kinematics (a) and knitting movement (b) of vibration-assisted grinding with mounted points.

The force-controlled process kinematics used have already been successfully tested in initial investigations for the partial automation of the fine machining of plastic mould steels [2]. In this thesis, the tool kinematic is supplemented by a low-frequency work piece-based vibration support to minimize kinematic induced grinding traces.

The knitting movement of a vibration-assisted grinding process with mounted points can be basically described as follows: The induced oscillatory motion of the frequency f and amplitude A is sinusoidal in shape. The introduction of a superimposed oscillation leads to a superposition of the effective movement between tool and workpiece. The knitting movements in tool engagement are shown in Fig. 2(b).

The following model was derived for a face grinding process with vibration support with variable superposition angle γ and is based on already developed models for ultrasonic-assisted cross-side grinding kinematics [17]. The model accounts the overall effective knitting movement as composed by the individual effective movements of cutting, feed rate and vibration. Equations (1–3) respectively provide the three movements with respect the kinematic model (the influence of lead angle α and tilt angle β being neglected in this work).

$$\vec{l}_{cutting} = \begin{pmatrix} l_{cutting,x} \\ l_{cutting,y} \\ l_{cutting,z} \end{pmatrix} = \begin{pmatrix} \frac{r_i + r_a}{2} \cdot cos\left(2\pi \cdot \frac{n_s}{60} \cdot t\right) \\ \frac{r_i + r_a}{2} \cdot sin\left(2\pi \cdot \frac{n_s}{60} \cdot t\right) \\ 0 \end{pmatrix} \tag{1}$$

$$\vec{l}_{feed\ rate} = \begin{pmatrix} l_{feed\ rate,x} \\ l_{feed\ rate,y} \\ l_{feed\ rate,z} \end{pmatrix} = \begin{pmatrix} 0 \\ v_f \cdot t \\ 0 \end{pmatrix} \tag{2}$$

$$\vec{l}_{vibration} = \begin{pmatrix} l_{vibration,x} \\ l_{vibration,y} \\ l_{vibration,z} \end{pmatrix} = \begin{pmatrix} \frac{\gamma}{90^\circ} \cdot \frac{\hat{A}}{2} \cdot sin(2\pi \cdot f \cdot t) \\ \left(1 - \frac{\gamma}{90^\circ}\right) \cdot \frac{\hat{A}}{2} \cdot sin(2\pi \cdot f \cdot t) \\ 0 \end{pmatrix} \tag{3}$$

Finally, the total knitting movement $\vec{l}_{total} = \vec{l}_{vibration} + \vec{l}_{feed\ rate} + \vec{l}_{cutting}$ can be described by Eq. (4):

$$\vec{l}_{total} = \begin{pmatrix} l_{total,x} \\ l_{total,y} \\ l_{total,z} \end{pmatrix} = \begin{pmatrix} \frac{\gamma}{90^\circ} \cdot \frac{\hat{A}}{2} \cdot sin(2\pi \cdot f \cdot t) + \frac{r_i + r_a}{2} \cdot cos\left(2\pi \cdot \frac{n_s}{60} \cdot t\right) \\ \left(1 - \frac{\gamma}{90^\circ}\right) \cdot \frac{\hat{A}}{2} \cdot sin(2\pi \cdot f \cdot t) + v_f \cdot t + \frac{r_i + r_a}{2} \cdot sin\left(2\pi \cdot \frac{n_s}{60} \cdot t\right) \\ 0 \end{pmatrix} \tag{4}$$

As test samples, workpiece plan samples made of plastic mould steel with the specification 1.2738 and dimension (L × W × D) 100 × 100 × 10 mm were selected. The sample selection is based on the typical use of this material in the field of plastic injection moulds for e.g. the automotive industry. The samples were pre-machined with a finish milling process analogous to the described process chain before grinding (cf. Fig. 1).

The experimental set-up is based on a 6-axis ABB IRB4400/60 articulated-arm robot. The end effector of the robot is equipped with a machine spindle, on whose HSK-supporter the axial-compliant tool holder CTHone with clamped grinding tool was mounted. Face grinding tools in the form of elastic bonded, chamfered mounted points of the company Novapax GmbH of the type K320 were used (D = 15 mm).

The tool holder makes it possible to apply a defined, controlled, constant contact force to the workpiece, F_A as reported Fig. 2(a). In combination with the feed movement, the tool performs a screw movement on the surface in the y-direction, which is superimposed by a low-frequency oscillating movement dependent on the superimposition angle γ. Figure 3(a) shows the experimental design for the performance of the test. The robot feed rate was set to v_f = 15 mm/s, the contact force to F_A = 7 N and the cutting speed to v_c = 10 m/s. The lead angle was set to α = 15° and the tilt angle to β = 0°. The workpiece oscillation was set in longitudinal, diagonal and lateral orientation to the machining direction (γ = 0°, 45°, 90°).

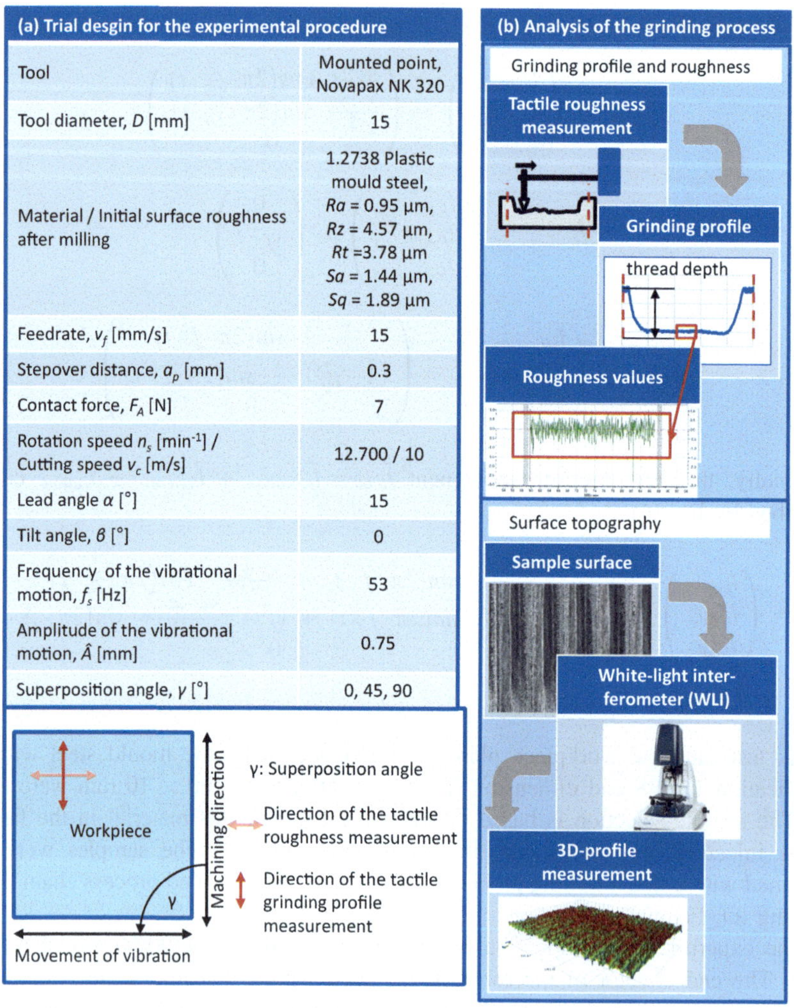

Tool	Mounted point, Novapax NK 320
Tool diameter, D [mm]	15
Material / Initial surface roughness after milling	1.2738 Plastic mould steel, Ra = 0.95 µm, Rz = 4.57 µm, Rt =3.78 µm Sa = 1.44 µm, Sq = 1.89 µm
Feedrate, v_f [mm/s]	15
Stepover distance, a_p [mm]	0.3
Contact force, F_A [N]	7
Rotation speed n_s [min^{-1}] / Cutting speed v_c [m/s]	12.700 / 10
Lead angle α [°]	15
Tilt angle, β [°]	0
Frequency of the vibrational motion, f_s [Hz]	53
Amplitude of the vibrational motion, $\hat{A}$ [mm]	0.75
Superposition angle, γ [°]	0, 45, 90

Fig. 3. Process parameters set-up and experimental execution (a) and measurement strategy for analysis of the grinding process (b).

The frequency of the vibration movement excited on the workpiece side was set to *53* Hz, whereby a vibration amplitude of $\hat{A} = 0.75$ mm could be achieved. The oscillation movement is sinusoidal and both the shape and amplitude of the oscillation were measured in situ using a *20* kHz laser measuring device. For comparison, a reference test with the same process parameters was carried out without vibration-supported processing.

After the tests, the surface topography was optically and tactilely measured, analysed and subsequently evaluated. The surface analysis strategy is depicted in Fig. 3(b).

The surface quality in terms of roughness and the grinding profile was determined and evaluated using a stylus method with the aid of a Form Talysurf Series 2 stylus-cutting device from Taylor Hobson. The arithmetic mean roughness value *Ra*, the average roughness depth *Rz* and the maximum roughness depth *Rt* were evaluated [18]. Measurements were carried out both transversely and longitudinally in orientation to the machining direction. Above, measurements of the initial surface after milling were performed.

The surface topography was measured and evaluated using an optical white light interferometer type Bruker Vision ContourGT. The mean arithmetic height *Sa* and the mean square height *Sq* of the surfaces confirm the visual impressions of the topography created on the basis of the macrographs and back up the tactile roughness measurements. Both *Sa* and *Sq* are parameters for describing surface roughness [19]. *Sa* can be interpreted as an addition to the *Ra* value in order to be able to carry out a surface-related description, whereas the *Ra* value only describes tactilely measured, linear sections of a surface. The *Sa* value gives the magnitude of the height difference of each point compared to the arithmetic mean of the surface. The *Sq* value is the quadratic mean of the ordinate values within the definition range and corresponds to the standard deviation of the heights on a surface.

Thus, the surface characterization was carried out with both one- and two-dimensional surface parameters by means of tactile probe cuts and optically recorded measuring fields.

3 Results

Figure 4(a) shows the roughness values before and after machining determined with the stylus method as a function of the superposition angle and in comparison to the reference surface without oscillating motion. The evaluation across in the orientation of the machining direction has consistently resulted in higher roughness values than the evaluation along in the orientation of the machining direction. Therefore, measurements crosswise orientated to the machining direction were used as comparison values.

The evaluation with the stylus method of the measured roughness values showed that all vibration-supported machined surfaces have no significant deviations of roughness in comparison to the conventionally machined reference surface. In addition to the roughness values determined, the roughness of the milling line was also measured locally before machining and averaged on the basis of two measurement repetitions.

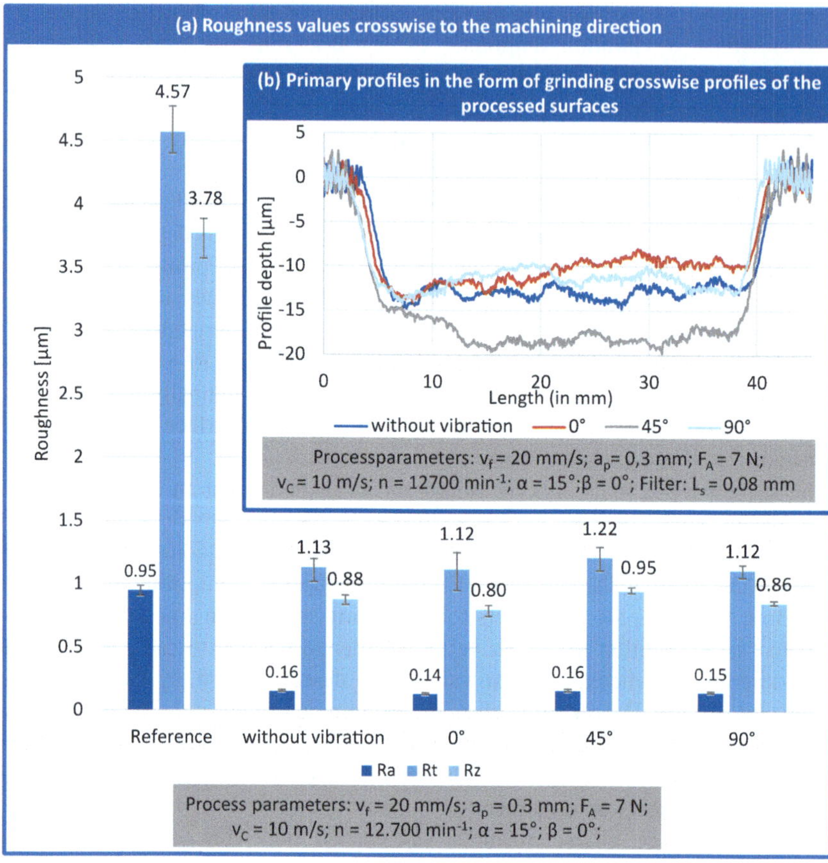

Fig. 4. Measured roughness in crosswise relation to the superposition angle compared to a conventional, non-vibration-supported grinding operation (a). Depth of the crosswise grinding profile in relation to the superposition angle in comparison to a non-vibration-supported finishing operation (b).

After two repetitions of the initial roughness measurements with distribution of the measurement series over the entire area of the face sample, it was found that the initial roughness variance was Ra = 0.90–0.98 µm, Rz = 3.5–3.8 µm and Rt = 4.4–4.7 µm. The scattering of the initial roughness, however, had no correlative effect on the final roughness produced. The bandwidth of the final roughness was Ra = 0.13–0.17 µm, Rz = 0.75–0.98 µm and Rt = 0.95–1.2 µm.

In addition to the surface roughness, the grinding profile was also considered. The depth of the grinding profile was analysed with regard to its characteristics with regard to a uniform profile depth and the associated homogeneity of the grinding profile.

Figure 4(b) shows the recorded primary profiles in the form of longitudinal grinding profile depths as a function of the superposition angle in comparison to the reference surface without oscillating motion. Depending on the superposition angle, there are significant differences in the profile depth and the profile course.

The profile progressions of the surfaces with superposition angles $\gamma = 45°$ and $\gamma = 90°$ are more uniform and constant than the progressions of the reference surface and the surface with superposition angle $\gamma = 0°$. The surface with a superposition angle of $\gamma = 45°$ has a grinding profile depth of $\sim$18–20 µm, which is significantly higher than the reference surface (without vibration). The profile depth of the surfaces with superposition angle $\gamma = 0°$ and $\gamma = 90°$, is not significantly higher than the grinding profile depth of the conventionally grinded surface. The grinding profile depth is proportional to the material removal rate.

Since the aim is to achieve as uniform a profile depth as possible, a distinction must be made between the individual superposition angles. The angle $\gamma = 45°$ produced the best results with regard to a constantly pronounced profile depth of the primary profiles measured.

In addition to roughness and profile depth gradients, the surface topography of the machined surfaces was also considered. Figure 5 shows sections of the machined surfaces, which were created with a Canon EOS 700 D camera with macro lens and evaluated with the Emagic software. Figure 5(a) shows the conventionally grinded reference surface. The horizontal kinematically caused grinding tracks are clearly visible. Figure 5(b) shows the sample surface with vibration-supported processing with orientation of the superposition angle in longitudinal direction to the machine feed movement $\gamma = 0°$. The characteristic machining trace can no longer be clearly identified in Fig. 5(b) and the grooves generated by the grinding process appear unidirectional. At a superimposition angle of $\gamma = 45°$, Fig. 5(c), the texture can be seen in an attenuated form. If the superposition angle is laterally aligned in orientation to the machining direction on $\gamma = 90°$, the texture appears to be similarly pronounced, without visible grinding traces, see Fig. 6(d).

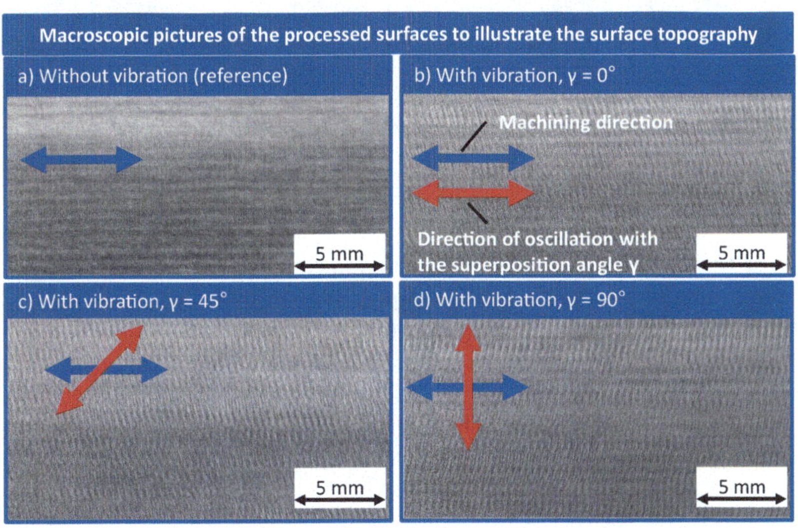

Fig. 5. Topographies of the machined surfaces from (a): without vibration; (b): superposition angle $\gamma = 0°$; (c): superposition angle $\gamma = 45°$ and (d): superposition angle $\gamma = 90°$.

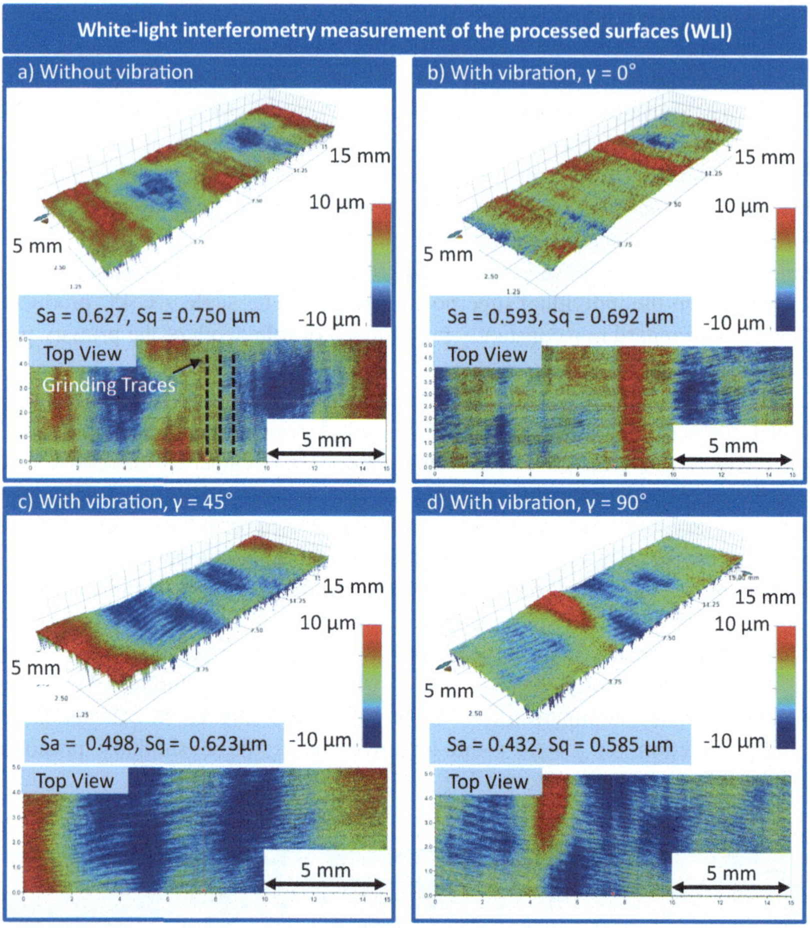

Fig. 6. White light interferometry measurements of the plastic mould steel surfaces of (a): without vibration; (b): superposition angle $\gamma = 0°$; (c): superposition angle $\gamma = 45°$; (d): superposition angle $\gamma = 90°$.

Figure 6 shows area-like white light interferometric images. A 5-fold magnification with a back scan of 50 µm was set. The topography of the reference surface (without oscillation movement) shows a periodic formation of grinding traces in the machining direction, which are repeated almost periodically (see Fig. 6(a)). The traces represent grooves which form peaks and valleys which can be identified as kinematically conditioned grinding traces. In comparison to the conventionally ground specimen and longitudinal vibration-assisted grinded specimen, no clear assignment of kinematic grinding marks could be made on the diagonally and lateral vibration-supported ground specimens, see Fig. 6(a)–(b).

The scores appear more homogeneous and show no significant repeating features which can be assigned to grinding marks, Fig. 6(c)–(d). Furthermore, there is a correlation between the texture and the superposition angle. The surface with $\gamma = 45°$, Fig. 6(c), shows the lowest expression of the grinding topography of all comparable surfaces.

It can be seen that the vibration-supported processing reduces the texture on the surface depending on the superposition angle.

There is a contradiction between the roughness values of the tactile line measurements with the stylus method and the surface evaluation. Local line measurements show no qualitative improvement of the surface, whereas the surface evaluation shows significantly lower roughness values. The test surfaces with $\gamma = 45°$ and $\gamma = 90°$ with $Sa = 0.4–0.5$ µm and $Sq = 0.56–0.62$ µm show significantly better surface characteristics, Fig. 6(c)–(d), and confirm the literature-based experiences of a reduction of surface roughness based on ultrasonic investigations.

The basic explanation for vibration-supported grinding is explained in the following paragraph. Under the condition that there is continuous contact between the tool and the workpiece, there is an increase in the kinematic contact length between the grinding wheel and the workpiece due to the vibration support. At constant grain concentrations and material removal rates, the low-frequency vibration-assisted grinding results in a lower single grain thickness, which results in lower roughness values [20]. Furthermore, it was observed that different superposition angles also have an influence on the surface quality achieved. The samples measured with $\gamma = 90°$ superposition angle had the lowest *Sa* roughness, see Fig. 6(d). In order to make a valid statement about the influence of the low-frequency vibration support on the surface roughness, further tests must be carried out.

4 Summary and Outlook

Plastic mould steel samples were prepared with a fine-finished surface usual for automotive car parts (bumper, rocker panel etc.). After milling, the surfaces were automatically grinded with an elastic mounted point. The fine machining process was carried out with force control using an axially compliant CTHone tool holder. Subsequently the surface quality was measured and evaluated both tactilely and optically. For comparison, grinding tests were carried out with a workpiece-based, low-frequency vibration support lateral, diagonal and longitudinal to the feed direction and evaluated analogously to the reference test.

The results of the investigations presented show that improvements in surface quality can be achieved with the aid of a low-frequency, vibration-supported face grinding process, accompanied by a reduction or elimination of kinematically induced grinding marks. Above there is a potential to increase in the metal removal rate. However, there is a contradiction between the evaluation of the linear measurements and the surface evaluation. While the tactile measurements show no significant improvement of the surface quality, the surface evaluation using WLI interferometry shows an overall significant improvement of the surface quality.

The focus of this contribution was the analysis of the surface topography in the form of kinematically determined grinding tracks characteristic for an automation process and the surface roughness as well as homogeneity of the grinding profile depth.

The samples processed with lateral and diagonal vibration excitation showed significantly lower kinematically induced grinding traces. When the vibration support was performed diagonally and lateral to the grinding direction, no original direction based on the path guidance could be determined. The alignment of the pattern continues to be unidirectional.

This contrasting results of the surface roughness can be due to the oscillation frequency used in the test set-up, which reaches peak cutting speeds transverse to the tool cutting speed of only approximately 20% to the maximum cutting speed. It can thus be concluded that due to the lower cross section speeds, transverse or diagonally aligned grooves, which can occur kinematically at superposition angles of $\gamma = 0°, 45°$, are far less pronounced in the grinding pattern. Furthermore, the samples tested with $\gamma = 45°$ vibration support showed a higher grinding profile depth, c.f. Fig. 4(b). This leads to the reverse conclusion that the same metal removal rates or grinding profile depths can be achieved with lower forces as with conventional face grinding. This in turn can have a positive effect on tool life and surface quality, since a lower contact pressure reduces tool wear and improves surface roughness.

Based on the presented investigations, higher vibration frequencies at vibration amplitudes in the millimetre range are to be tested in the future in order to investigate the influence of higher cutting speeds on the development of a bidirectional texture. In addition, the test room is to be expanded to a set-up for vibration-supported fine machining of freeform surfaces. For future works, a vibration-assistance device will be developed to access free form areas on surfaces. The key-technology to enable a vibration-assisted movement on free form surfaces, is to enable tool-based vibration assistance at e.g. the end effector of the robot.

References

1. Timmermann, S.: Automatisierung der Feinbearbeitung in der Fertigung von Hohlformwerkzeugen. Dissertation, Universität Karlsruhe, pp. 1–10 (1990)
2. Hähnel, S., Pini, F., Leali, F., Dambon, O., Bergs, T., Bletek, T.: Reconfigurable robotic solution for effective finishing of complex surfaces. In: 23rd International Conference on Emerging Technologies and Factory Automation (ETFA 2018), pp. 1285–1290. IEEE, Turin (2018)
3. Pini, F., Leali, F.: Human-robot collaborative reconfigurable platform for surface finishing processes. In: 29th International Conference on Flexible Automation and Intelligent Manufacturing (FAIM 2019). Limerick (2019)
4. Bleicher, F.; Brier, J.: Schwingungsunterstützte Schleifbearbeitung. In: Schweizer Schleifsymposium 2016, Zurich (2016)
5. Uhlmann, E., Hübert, C.: Ultrasonic assisted grinding of advanced ceramics. Institute for Machine Tools and Factory Management, Technische Universtität Berlin, Berlin (1996)
6. Nik, M.G., Movahhedy, M.R., Akbari, J.: Ultrasonic-assisted grinding of Ti6A14 V alloy. In: 5th CIRP Conference on High Performance Cutting 2012 (HPC 2012). Procedia CIRP, vol. 1, no. 1, pp. 353–358. Elsevier B.V., Zurich (2012)
7. Tawakoli, T., Azarhoushang, B., Rabiey, M.: Ultrasonic assisted dry grinding of 42CrMo4. Int. J. Adv. Manuf. Technol. **42**(9–10), 883–891 (2009)

8. Yanyan, Y., Bo, Z., Junli, L.: Ultraprecision surface finishing of nano-ZrO_2 ceramics using two-dimensional ultrasonic assisted grinding. Int. J. Adv. Manuf. Technol. **43**(5–6), 462–467 (2009)
9. Ding, K., Fu, Y., Su, H., Gong, X., Wu, K.: Wear of diamond grinding wheel in ultrasonic vibration-assisted grinding of silicon carbide. Int. J. Adv. Manuf. Technol. **71**(9–12), 1929–1938 (2014)
10. Klocke, F., Bresseler, B., Weber, A.: Ultraschallunterstütztes Schleifen – Chancen und Grenzen der Technologie am Beispiel der Zerspanung von Glas und Keramik. wt Werkstattstechnik online 94, pp. 258–262 (2004)
11. Brecher, C., Klocke, F., Winterschladen, M.; Heselhaus, M.: Ultraschallunterstütztes Hartdrehen für die Fertigung von gehärteten Präzisionsstahlbauteilen. wt Werkstattstechnik online 6, pp. 396–401 (2006)
12. Tawakoli T., Lohner R.D.: Schleifen von Hartmetallen und Keramik mit magnetgelagerten Spindeln - Grinding of Carbide and Ceramics with Magnetic Bearing Spindles. In: Moderne Schleiftechnologie und Feinstbearbeitung, Stuttgart, pp. 1–11 (2014)
13. Guo, B., Zhao, Q.-L., Jackson, M.-J.: Ultrasonic vibration-assisted grinding of micro-structured surfaces in silicon carbide ceramic materials. Proc. Inst. Mech. Eng. Part B: J. Eng. Manuf. **226**(3), 553–559 (2011)
14. Klocke, F., König, W.: Manufacturing Processes 2 – Grinding, Honing, Lapping, pp. 338–346. Springer, Heidelberg (2009)
15. Chen, H., Tang, J., Zhou, W.: An experimental study of effects of ultrasonic vibration on grinding surface roughness of C45 carbon steel. Int. J. Adv. Manuf. Technol. **68**, 2095–2098 (2013)
16. Hähnel, S., Hoff, S., Grunwald T., Bergs, T.: Vibration-assisted grinding – Vibration assisted grinding of cast iron to reduce kinematically induced grinding traces, wt Werkstattstechnik online 6, pp. 48–488 (2019)
17. Daus, N.: Ultraschallunterstütztes Quer-Seiten-Schleifen, Dissertation, Berichte aus dem Produktionstechnischen Zentrum, Berlin, pp. 27–32 (2004)
18. DIN EN ISO 4287: Geometrical Product Specifications (GPS) – Surface texture: Profile method – Terms, definitions and surface texture parameters (2010)
19. DIN EN ISO 25178-2: Geometrical product specifications (GPS) - Surface texture: Areal – Part 2: Terms, definitions and surface texture parameters (2012)
20. Denkena, B.; Hahmann, D.: Ultraschallschleifen dünnwandiger Bauteile, wt Werkstattstechnik online 97, pp. 862–867 (2007)

Virtual Design for Assembly Improving the Product Design of a Two-Way Relief Valve

Daniela Francia(✉), Davide Seminerio, Gianni Caligiana, Leonardo Frizziero, Alfredo Liverani, and Giampiero Donnici

University of Bologna, Bologna, Italy
d.francia@unibo.it

Abstract. Design for Assembly is a strategy of design aimed at minimising product cost through design and process improvements. It led to a revolution in the manufacturing industry, resulting in reduced product costs, improved quality, shorter time to market, lower inventory, fewer suppliers and many other improvements. DFA is the method of design of the product for ease of assembly and in this context this paper presents an application of the strategy to a virtual case study represented by a two-way relief valve modelled by CAD. The aim of this paper was to test some DFA methods proposed by the literature and search for an efficient assembly of the virtual two-way relief valve reproduced by CAD, then verifying its assembly effectiveness. Some evaluation methods proposed by literature about the optimisation of design efficiency have been applied to the case study and this led to reconsider the design of some parts of the product. A final new design for the valve is proposed and the evaluation methods applied have been tested again on the new solution, in order to validate the results. A comparison between the original valve and the new version proposed by the authors has been made, in order both to check the feasibility of the new valve and in order to check the evaluation methods proposed.

Keywords: DfA · CAD · Optimization

1 Introduction

Design is the first step in the production process, and it is where the most important decisions are made, which will influence the final cost of a product.

Since 1980, analysis techniques have been implemented that guide designers towards the creation of products, which are easy to produce and assemble.

The availability of these techniques has created a revolution in the manufacturing industry, resulting in reduced product costs, improved quality, shorter time to market, lower inventory, fewer suppliers and many other improvements.

This wants to underline, first of all, the importance of paying attention to manufacturing and assembly problems, already in the initial stages of product design [1].

For the assembly to be done as effectively as possible, guidelines are needed to direct the designer, and assessment methods that estimate the effectiveness of a product in its assembly.

C. Rizzi et al. (Eds.): ADM 2019, LNME, pp. 304–314, 2020.
https://doi.org/10.1007/978-3-030-31154-4_26

This led to the creation of Design for Assembly (DFA), a design aimed at simplifying product design and at improving assembly, with consequent savings in terms of time and money.

The guidelines to be followed in an initial design phase or, subsequently, to optimize an already existing piece will then be explained [2].

This paper has the purpose of identifying the optimal assembly sequence of a product, studying its assembly effectiveness; an alternative model will then be proposed, following the guidelines, with a different design of the object, with the aim of improving the assembly phase.

The product has been virtually reproduced by means of CAD (Creo Parametric CAD software, 2017 version). CAD modeling reveals very useful in these cases of study because of the possibility to extract features directly from the system database [3]. Later, three different evaluation methods based Design for Assembly will be analyzed, to be applied to the case study of the limiting valve, in order to find the optimal approach to the design of the product: they are Hitachi's technique (Japan), Boothroyd-Dewhurst's (USA) and Lucas' technique (United Kingdom).

2 DFA Guidelines

In this section, some general guidelines are described because the assembly design method must implement some important objectives:

1. Provide a tool for the designer, or design team, which ensures considerations on the complexity of product assembly, which must take place in the early stages of the project. This eliminates the danger of focusing exclusively on the functionality of the product, without adequate consideration for the cost and competitiveness of the piece.
2. Guide the designer, or the design team, to simplify the product so as to save both on assembly and on the production of the parts.
3. Collect the information, normally owned by the expert designer, to arrange it conveniently for use by less experienced engineers.
4. Establish a database that includes assembly times and cost factors for various design situations and production conditions.

From the experience in the application of the DFA it was thus possible to develop design guidelines that attempt to consolidate production knowledge, presenting themselves to the designer in the form of simple rules to follow.

Generally, he manual assembly process can be divided into two separate parts:

- Management (acquisition, orientation and movement of parts)
- Insertion and fixing (coupling of one part to another or to a group of parts)

Guidelines for parts management suggest that in general, to facilitate parts management, a designer should attempt to:

1. Design parts that have end-to-end symmetry (i.e. from one end to the other) and rotational symmetry with respect to the insertion axis. If this cannot be achieved, try to design parts with the maximum possible symmetry.
2. Design parts that, in those cases where the part cannot be symmetrical, must obviously be asymmetrical.
3. Provide functionalities that prevent the jamming of parts that tend to tangle if inserted in large quantities.
4. Avoid the features that will allow the tangling of parts if stored loose.
5. Avoid parts that stick or are slippery, delicate, flexible, very small or very large or dangerous for the conductor (i.e. parts that are sharp, easily splintered, etc.).

Guidelines for the insertion and the fixing of parts, instead, suggest that a designer should attempt to:

1. Draw so that there is little or no resistance to insertion and provide chamfers to guide the insertion of two mating parts. Generous play should be guaranteed, but care should be taken to avoid results that would result in the parties having a tendency to jam or hang up during insertion.
2. Standardize using parts, processes and methods common to all the models and also through the production lines, to allow the use of a higher volume of processes, which normally involve a lower cost.
3. Use the "pyramid" assembly: that is, provide a progressive assembly around a reference axis. In general, it is better to assemble from above.
4. Avoid, where possible, the need to keep the parts lowered to maintain orientation when handling the subassembly or when positioning another part. If it is necessary to hold down, try to design the part so that it is secured as soon as it is inserted.
5. Design so that a part fits before it is released. A potential source of problems comes from a positioned part that, due to design constraints, must be released before being placed in the assembly. Under these circumstances, the interest is placed on the trajectory of the part, which must be sufficiently repeatable to identify it coherently.
6. Avoid the need to reposition the partially completed unit in the equipment.
7. When using common mechanical fasteners, follow the following sequence that indicates the relative value of the different fixing processes, listed in ascending order of manual assembly costs:
    a. Snap mounting
    b. Plastic bending
    c. Riveting
    d. Screw closure

These guidelines, therefore, are simply a set of rules which, taken together, provide the designer with the appropriate basic information to be used for the development of a design that will be more easily assembled.

It is therefore necessary to use an approach that provides the designer with an organized method, which involves designing a product that is easy to assemble. The method must also provide an estimate of how much easier it is to assemble a design, with certain characteristics, rather than assemble another with different characteristics.

For this reason, several design methods for assembly have been formulated to rationally evaluate one assembly over another.

3 DFA's Methods Analysis

The idea behind most of DFA's systematic methods is to consider each part when it is inserted into the product, to evaluate the difficulty of the assembly process, then to sum the results to obtain a numerical evaluation of it.

Of all the quantitative assessment techniques, the best known are those of Hitachi (Japan) [4, 5], Boothroyd-Dewhurst (USA) [6–8] and Lucas (United Kingdom) [9]. All three have been used in the industrial world.

The method, so to speak "Japanese", takes the name of Assembly Evaluation Method (AEM) and was developed by Hitachi in the late 1970s; later, around 1980, the Design for Assembly (DFA) was introduced which reflects the work of Professor Geoffrey Boothroyd at the University of Massachusetts who, with the help of Dr. Peter Dewhurst, have achieved a further analysis also on manufacturing, creating the Design for Manufacture and Assembly (DFMA); as regards the last method examined, it is a review of the first two, made in the 1990s by Lucas Corp. in the United Kingdom.

These techniques are evaluative methods that analyse the assembly of the products, at an early stage of the design process; although they use the same philosophy, they differ in the way they quantitatively evaluate the project.

In general, there are two ways:

1. in the first one we compare the number of parts in an assembly and how each part is assembled, with an ideal, perceived, that will have the least number of possible parts and each part will be optimised for assembly; the recommendations for the redesign will take into account two factors: the number of parts, to be as close as possible to the ideal one, and the assembly process as similar as possible to the ideal one.
2. The other way uses a classification system to compare the number of parts and the assembly of alternative projects; the evaluation system can be based on assembly costs, effort or time, and the scores will be based on the number of parts and on the ease with which it is possible to assemble them and, the best design, is the product with the highest or lowest score in based on the method.

4 Methods Application to the Two-Way Relief Valve

The case study presented is a two-way relief valve and in this section all three methods of evaluation of the assembly previously treated are applied, in order to examine the piece as it is on the market.

Once this is done, an alternative model of the valve will be proposed, following the guidelines to improve the design for better assembly, subsequently checking whether or not there will be an improvement in the assembly indexes.

First of all, however, by providing an exploded view of the piece reproduced at the CAD and the BOM (Fig. 1 and Table 1), the optimal assembly sequence of the valve components is determined.

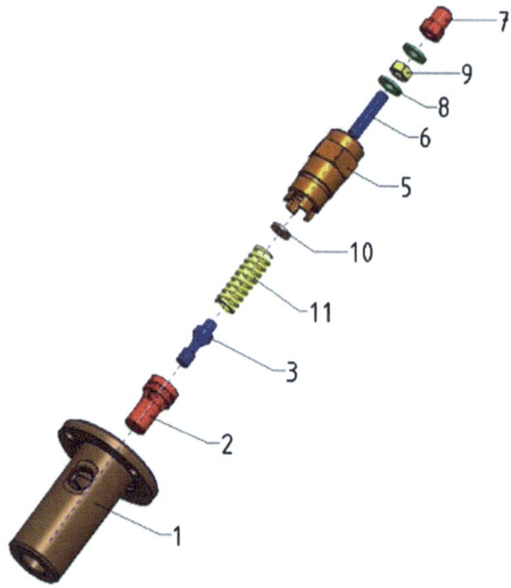

Fig. 1. Exploded view of the assembly

The sequence will then be listed by points, as follows, assuming that the OR rings were already assembled in the special seats, in the valve parts:

1. The valve body is positioned on the base.
2. The limiting body is inserted into the valve body.
3. Insert the valve piston inside the limiting body, already present in the valve body.
4. Position the spring inside the valve body which will rest on the valve piston.
5. The pressure plate is placed at the end of the spring.
6. Screw the register holder to the valve body, taking care to contain the spring and the pressure plate.
7. Screw the screw into the hole on the top of the register holder, which will rest and push the pusher plate. The screwing is proportional to the spring preload value.
8. Insert the washer into the screw.
9. Screw the M8 nut to the screw.
10. Insert the second washer into the screw that will rest on the bolt.
11. Screw the nut onto the screw, acting as a cover, and together with the M8 nut, form a nut-lock nut anti-loosening system.

Table 1. Bill of materials.

Position	Description	N. Parts	Materials and notes
1	Valve body	1	
2	Limiting body	1	
3	Valve piston	1	
4	Register holder	1	
5	Screw	1	
6	Nut	1	
7	Washer	2	Washer 8,4x17 UNI 6592-R80
8	Nut	1	
9	Nut	1	M8 UNI 5588-A-6S
10	Spring pressure plate	1	
11	Spring	1	

5 Valve Optimization: Results

The studies carried out on the assembly of the valve according to the three methods have led, in any case, an estimate of a very good assembly efficiency; for this reason, the piece is difficult to improve from a purely assembly point of view.

But from the application of the theories of Boothroyd-Dewhurst and Lucas it can be noted that the valve can be optimized, working on the parts that are not indispensable to the function of the valve itself, which are those components that concern the regulation.

For this reason, the study focused on the following components:

- pusher plate;
- lives;
- nut M8;
- washer;
- nut.

The model proposed foresees an overall reduction of the pieces, passing from eleven to eight components: the changes were made to the register screw (Fig. 2), in which, instead of an excavation on the top, a quarry was obtained hexagonal shape for inserting hexagonal keys; in addition, the M8 nut, the washers and the upper nut have been eliminated, giving way to a single M8 nut with a built-in washer-proof system (Fig. 3).

By providing the new exploded view and the new bill of materials (Fig. 4 and Table 2), we are going to define the new assembly sequence:

1. The valve body is positioned on the base.
2. The limiting body is inserted into the valve body.
3. The valve piston is inserted into the limiter body, already present in the valve body.
4. Position the spring inside the valve body which will rest on the valve piston.
5. Place the spring plate on top of the spring.

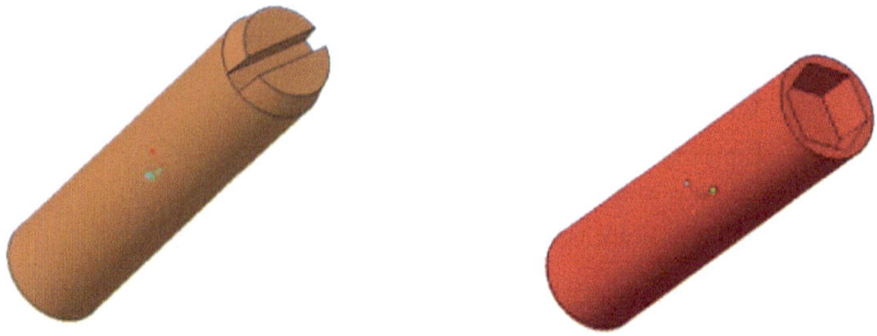

Fig. 2. Comparison between old adjusting screw (left) and new adjusting screw (right)

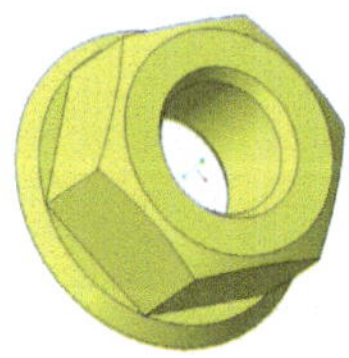

Fig. 3. New M8 nut with incorporated washer anti-loosening system

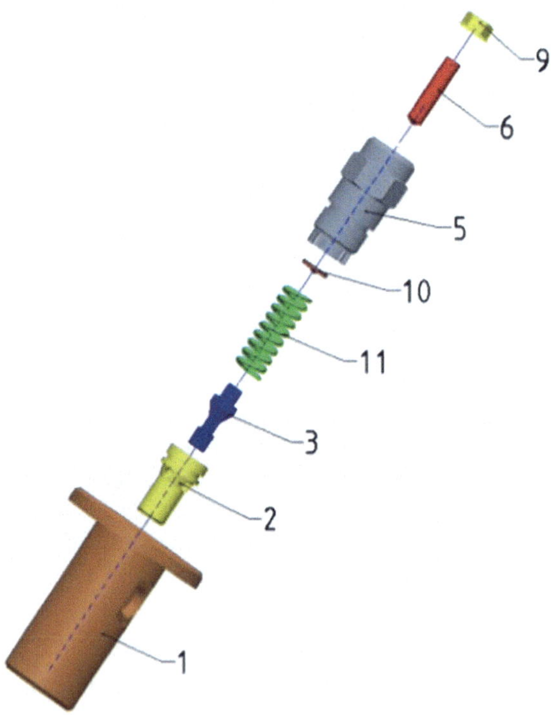

Fig. 4. Exploded view of the new two-way relief valve

6. Screw the register holder to the valve body, taking care to contain the spring and the pressure plate.
7. Screw the new screw into the hole on the top of the register holder, which will rest and push the spring push plate. The screwing is proportional to the spring preload value.
8. Screw the nut with a built-in washer to prevent screw loosening.

Table 2. Bill of materials for the optimized valve.

Position	Description	N. Parts	Materials and notes
1	Valve body	1	
2	Limiting body	1	
3	Valve piston	1	
5	Register holder	1	
6	Screw	1	
9	Nut	1	
10	Spring pressure plate	1	
11	Spring	1	

It is immediately evident that, in addition to the decrease in components, a decrease in assembly actions can be observed, thus making the overall piece simpler.

Now, in order to verify the effectiveness of the model just proposed, the three assembly evaluation methods are applied again

5.1 AEM Method

Applying the Hitachi method, the original and the new valve are compared and, although the number of components and the number of assembly operations in the new solution have been reduced, it can be registered a 1% reduction in assembly efficiency (Figs. 5 and 6).

# operazione	symbol	punti penalità	part assemblability evaluation score	E assemblability evalutaion score
1	↓-	0	100	
2	↓	0	100	
3	↓	0	100	
4	↓	0	100	
5	↓	0	100	
6	↓ nc	30	70	
7	↓ nc	30	70	
8	↓	0	100	
9	↓ nc	30	70	
10	↓	0	100	
11	↓ nc	30	70	
			980	89,09090909
			Ei	E

Fig. 5. Application of the AEM method to the original valve

# operazione	symbol	punti penalità	part assemblability evaluation score	E assemblability evalutaion score
1	↓-	0	100	
2	↓	0	100	
3	↓	0	100	
4	↓	0	100	
5	↓	0	100	
6	↓ ∩C	30	70	
7	↓ ∩C	30	70	
8	↓ ∩C	30	70	
			710	88,75
			Ei	E

Fig. 6. Application of the AEM method to the new valve

This means that the method does not present a systematic method for identifying the parts to be eliminated, so that the final result is not improved.

5.2 DFA Method

The application of Boothroyd-Dewhurst method instead, provided different results (Figs. 7 and 8).

pezzo	# of items	total angle simmetry (deg)	handling code	handling time	insertion code	insertion time	operation time	operation cost	estimation of theoretical minimum parts
1	1	360	10	1,5	00	1,5	3	1,2	1
2	1	360	10	1,5	20	5,9	7,4	2,96	1
3	1	360	10	1,5	10	3,7	5,2	2,08	1
11	1	180	03	1,84	00	1,5	3,34	1,34	1
10	1	360	12	2,06	00	1,5	3,56	1,42	1
5	1	360	10	1,5	23	9,6	11,1	4,44	1
6	1	360	10	1,5	01	3	4,5	1,8	1
8	2	180	02	1,69	01	3	4,69	1,88	0
9	1	180	01	1,43	01	3	4,43	1,77	0
7	1	360	10	1,5	01	3	4,5	1,8	0
							51,72	20,69	7
							TM	CM	NM

$$DESIGN\ EFFICENCY = \frac{3 \cdot NM}{TM} = 0,40$$

Fig. 7. Application of the DFA method to the original valve

Going to work on those components that the method emphasised as unnecessary, there is a 1% improvement in design efficiency; this underlines the best effectiveness of the method which, in addition to an improvement in the assembly, leads to an economic saving in the assembly of the object, of 0.8%.

5.3 LUCAS Method

Applying the last assembly evaluation method (Figs. 9 and 10), we have a result similar to that previously found with the DFA method.

By reducing the number of components that are part of group B, i.e. the set of parts not strictly necessary for the functionality of the valve, we immediately notice a great improvement in the design efficiency, which reaches 87% well against 58% of the

pezzo	# of items	total angle simmetry (deg)	handling code	handling time	insertion code	insertion time	operation time	operation cost	estimation of theoretical minimum parts
1	1	360	10	1,5*	00	1,5	3	1,2	1
2	1	360	10	1,5*	20	5,9	7,4	2,96	1
3	1	360	10	1,5*	10	3,7	5,2	2,08	1
11	1	180*	03	1,84*	00	1,5	3,34	1,34	1
10	1	360	12	2,06*	00	1,5	3,56	1,42	1
5	1	360	10	1,5*	23	9,6	11,1	4,44	1
6	1	360	10	1,5*	01	3	4,5	1,8	1
9	1	180*	01	1,43*	01	3	4,43	1,77	0
							42,53	17,01	7
							TM	CM	NM

$$DESIGN\ EFFICENCY = \frac{3 \cdot NM}{TM} = 0{,}49$$

Fig. 8. Application of the DFA method to the new valve

parte	functional analysis	feeding analysis					fitting analysis						
	A	B	C	D	TOT	A	B	C	D	E	F	TOT	
1	A	1	0,1	0	0	1,1	1	0	0	0	0	0	1
2	A	1	0,1	0	0	1,1	1	0	0	1,5	0	0	2,5
3	A	1	0,1	0	0	1,1	1	0	0	1,5	0	0	2,5
11	A	1	0	0	0	1	1	0	0	0	0	0	1
10	A	1	0	0,4	0	1,4	1	0	0	0	0	0	1
5	A	1	0,1	0	0	1,1	4	0	0	0	0,7	0	4,7
6	A	1	0,1	0	0	1,1	4	0	0	0	0,7	0	4,7
8	B	1	0	0	0	1	1	0	0	0	0,7	0	1,7
9	B	1	0	0	0	1	4	0	0	0	0,7	0	4,7
7	B	1	0,1	0	0	1,1	4	0	0	0	0,7	0	4,7
						11							28,5
	#A=7					HI							FI

$$DESIGN\ EFFICENCY = \frac{\#A}{\#TOTALE\ PEZZI} * 100 = 58\% \qquad FEEDING\ HANDLING\ RATIO = \frac{HI}{\#A} = 1{,}57 \qquad FITTING\ RATIO = \frac{FI}{\#A} = 4{,}07$$

Fig. 9. Application of the Lucas method to the original valve

parte	functional	feeding analysis					fitting analysis						
	A	B	C	D	TOT	A	B	C	D	E	F	TOT	
1	A	1	0,1	0	0	1,1	1	0	0	0	0	0	1
2	A	1	0,1	0	0	1,1	1	0	0	1,5	0	0	2,5
3	A	1	0,1	0	0	1,1	1	0	0	1,5	0	0	2,5
11	A	1	0	0	0	1	1	0	0	0	0	0	1
10	A	1	0	0,4	0	1,4	1	0	0	0	0	0	1
5	A	1	0,1	0	0	1,1	4	0	0	0	0,7	0	4,7
6	A	1	0,1	0	0	1,1	4	0	0	0	0,7	0	4,7
9	B	1	0	0	0	1	4	0	0	0	0,7	0	4,7
						8,9							22,1
	#A=7					HI							FI

$$DESIGN\ EFFICENCY = \frac{\#A}{\#TOTALE\ PEZZI} * 100 = 87\% \qquad FEEDING\ HANDLING\ RATIO = \frac{HI}{\#A} = 1.27 \qquad FITTING\ RATIO = \frac{FI}{\#A} = 3{,}16$$

Fig. 10. Application of the Lucas method to the new valve

previous valve; this brings with it the improvement of the remaining indexes, both in the feed ratio, which improves by 0.8%, and in the insertion ratio, which improves by 0.8%.

6 Conclusion

The aim of this paper was to test some DFA methods proposed by the literature and search for an efficient assembly of a virtual two-way relief valve reproduced with the CAD, then verifying its assembly effectiveness. Each method proposed some step of evaluation that have been rigorously tested on the case study.

Some product improvement has been foreseen and an alternative design of the valve was then proposed, always reproducing it by CAD, in order to test its assembly efficiency. A comparison between the original valve and the new version proposed by the authors has been in order to check the feasibility of the new valve and in order to check the evaluation methods proposed.

During the study, it became clear that the subject is still very immature; in spite of the fact that the information in this regard is high and the proposed evaluation methods have been many, in none of these has it yet been possible to obtain the best model of estimation, so much so that there have been conflicting results.

The documents cited in this paper underlined how designers and companies have developed a greater interest in this subject, starting to exploit their potential in today's productions.

CAD reproduction of the two-way relief valve made it possible, simply by the use of 3D modelling, to check and test a product, without having to resort to its immediate prototyping.

This aspect is favourable to the Design for Assembly, because it facilitates the study in the initial phase of the design, without having to propose, materially, different models to be subjected to verification.

References

1. Warnecke, H.J., Bassler, R.: Design for assembly - part of the design process. Annals of CIRP **37**(1), 1 (1988)
2. Kroll, E., LenZ, E., Wolberg, J.R.: A knowledge-based solution to the design for assembly problem. Manuf. Rev. **1**(2), 104–108 (1988)
3. Rosario, L.M., Knight, W.A.: Design for assembly analysis: extraction of features from a CAD system database. Ann. CIRP **38**(1), 13 (1989)
4. Miyakawa, S., Ohashi, T.: The Hitachi Assemblability Evaluation Method (AEM). In: Proceedings of the International Conference on Product Design for Assembly, Newport, RI, 15–17 April 1986
5. Miyakawa, S., Ohashi, T., Iwata, M.: The Hitachi New Assemblability Evaluation Method (AEM). Trans. North Am. Manuf. Res. Inst. SME (1990)
6. Boothroyd, G.: Product design for assembly. Department of Mechanical Engineering, University of Massachusetts, Amherst (1980)
7. Yoosufani, Z., Boothroyd, G.: Design of parts for ease of handling. Report No. 2, Department of Mechanical Engineering, University of Massachusetts, September 1978
8. Boothroyd, G., Dewhurst, P.: "Product Design for Assembly," Designers Handbook, Boothroyd Dewhurst, Inc., 138 Main Street, Wakefield, RI 02879 (1979)
9. Leaney, P., Wittenberg, G.: Design for assembling: the evaluation methods of Hitachi, Boothroyd and Lucas. Assembly Autom. (1992). https://doi.org/10.1108/eb004359

Integrated Product and Process Design

Improving the Shoes Customization Process Through a Digitally-Enabled Framework

Marco Marconi[1](✉), Alessandra Papetti[2], Marta Rossi[2], and Giulia Di Domizio[2]

[1] Università degli Studi della Tuscia, Largo dell'Università, 01100 Viterbo, Italy
marco.marconi@unitus.it

[2] Università Politecnica delle Marche, Via Brecce Bianche, 60131 Ancona, Italy

Abstract. Product customization is largely considered one of the most important competitive factors in several industrial sectors, including high-end footwear. However, in this sector, products are generally manufactured through manual and artisanal operations, processes are not standardized and repeatable, and the personal skills of expert operators play a leading role. The objective of the present paper is to propose a workflow enabled by a framework including several technologies (computer aided design tool, augmented reality systems, traceability infrastructure), to support industrial companies of the high-end footwear sector during the different phases of shoes customization (from configuration to delivery). All these technologies jointly contribute to innovate the shoes customization process by increasing the flexibility of internal processes, improving the ability of companies to answer to specific requirements thanks to the direct involvement of customers, maximizing the efficiency of data sharing, making the organizational, design, production and management processes more efficient and repeatable, and reducing the customer response time.

Keywords: Shoes configuration · Customization process · Customization framework

1 Introduction

Product customization is largely considered an effective way to take into account individual customer preferences during the design and development of products. By enlarging the variety of the offered products, industrial companies can answer the individual requirements, maximize the satisfaction of customers, differentiate their production against competitors and maintain competitiveness in the market [1]. This is the reason why product customization is currently one of the most important competitive factors in several industrial sectors (e.g. fashion goods, automotive), and particularly in developed countries where the very high cost of labor makes the mass production no longer sustainable from the economic point of view [2, 3].

In this context, manufacturing companies need to increase their flexibility to be able to produce small and heterogeneous lots, and to be strongly oriented to customers in order to maximize their satisfaction and the value that they derive from buying products [4]. A radical change in the organizational, design, production and management

C. Rizzi et al. (Eds.): ADM 2019, LNME, pp. 317–328, 2020.
https://doi.org/10.1007/978-3-030-31154-4_27

processes is therefore required. However, the implementation of this paradigm requires to consider a very complex environment that includes production plants, human resources and external stakeholders, other than to manage and exploit a very huge amount of data [5].

Concerning the high-end footwear industrial sector, products are generally manufactured through manual and artisanal operations, processes are not standardized and repeatable, and the personal skills of expert operators play a leading role [6]. Also in this sector, customization is assuming a key role to guarantee customer satisfaction, as well as foot comfort [7]. Several methods and tools to support the processes affected by shoes customization have been developed at both academic and industrial levels. They mainly aim to support the design of shoes or essential shoe components (e.g. shoe last, insole), also in case of customers with special needs [8–10]. However, such studies only focus on a specific phase of the shoe customization (i.e. design), and do not allow to jointly consider all the other aspects (e.g. production phase, exchange of data, supply of components).

In order to support the different processes affected to product customization, several research studies focused on the development of customization frameworks to represent and manage data in an effective way [11, 12]. These studies are often general-purpose and do not take into account the specific features of the footwear industrial sector. Only one research study [13] investigated the possibility to develop a reference model for mass customization of fashion products by using two different representation techniques (i.e. IDEF0 and GRAI). However, such framework is only theoretically defined, while the description of enabling technologies and supporting tools to implement in real applications is lacking. Another correlated study [6] focused on the development of a customization framework in the footwear industrial sector, but it is only dedicated to improve the internal production and management processes of last manufacturers, while the customer is not directly involved and the other phases of the shoe production are not considered (e.g. design of the shoe, choice of the leather, seam of the shoe upper, customization of accessories, assembly of the shoe).

This paper aims to propose a framework to support industrial companies of the high-end footwear sector during the different phases of shoes customization. The framework considers all the relevant phases, from the definition of customer requirements to final product manufacturing and delivery, including also the management of product and customer-related data. The proposed workflow that supports both internal (e.g. designers, operators of the production line) and external (e.g. suppliers) stakeholders is enabled by several technologies: computer aided design (CAD)-based tools, virtual/augmented reality (VR/AR) systems, product lifecycle/database management (PLM/PDM) systems, traceability infrastructures, additive manufacturing (AM) equipment. All these technologies strongly contribute to innovate the shoes customization process by increasing the flexibility of internal processes, improving the ability of companies to answer to specific requirements thanks to the direct involvement of customers, making the processes more efficient and repeatable and maximizing the efficiency of data sharing. The final objective is to support the transition of artisanal companies involved in the manufacturing of high-end footwear products from the traditional mass production to the mass customization model, where products are co-ideated/designed by producers and customers, and finally produced in a reduced customer response time.

The rest of the paper is organized as follows. Section 2 describes the "traditional" customization process for footwear products, highlighting the main limitations and weaknesses. Section 3 introduces the proposed framework and workflow to support high-end shoes manufacturers in improving the processes related to personalized products. Section 4 describes the preliminary implementation of some modules of the proposed framework in the context of an Italian company and discusses potential benefits and limitations. Finally, Sect. 5 reports conclusions and some proposals for future work.

2 The AS-IS Shoes Customization Process

Analyzing the process of several companies involved in the high-end footwear sector, a general AS-IS customization workflow has been defined (Fig. 1).

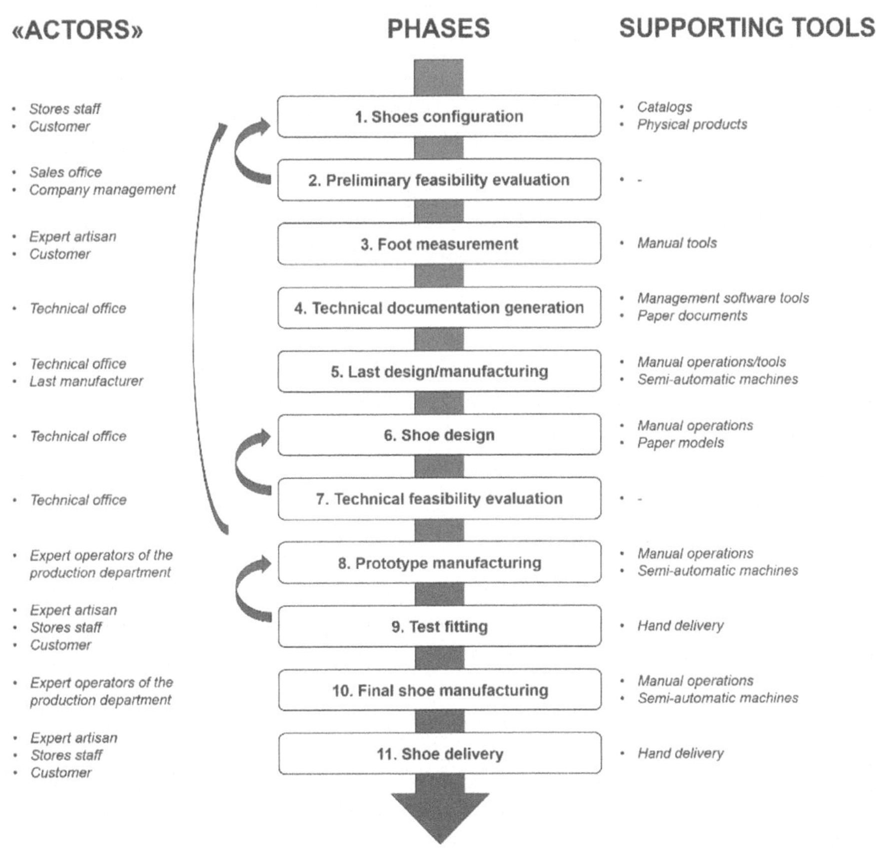

Fig. 1. The AS-IS shoes customization process

The process starts at stores where customers interact with store operators to set the features of customized shoes (*1. Shoe configuration*). During this phase, only catalogs, booklets and standard products are generally used to support customers.

The preliminary order, generally collected in non-structured paper or electronic documents, is sent to the company sales office to evaluate the feasibility of the solution (*2. Preliminary feasibility evaluation*). If the order is not feasible, the customer has to be call back to the store to change some chosen features.

In case of positive feasibility check, the feet need to be measured by a specialized artisan who reach the customer in stores potentially located all over the world (*3. Foot measurements*). Manual tools (e.g. meter sticks) are used to take the measurements in the right way. On the basis of his/her knowledge, the artisan suggests to the customer the shoe and last models that best fit with the shape of his/her feet.

At this stage the needed documentation to start with the design and production phases is generated (*4. Technical documentation generation*). It usually consists of paper documents, organized according to internal codifications, with the addition of notes to distinguish customized products from the standard and seasonal collections.

During the design phase, at first, the customized last is designed and manufactured in collaboration with a last manufacturer (*5. Last design/manufacturing*). Then, the technical office is in charge to develop a shoe model (*6. Shoe design*), made of paper, to simulate the manufacturing and assembly of the different shoe components (e.g. shoe upper, toe). Only during these phases, the specifications are deeply analysed from the technical and stylistic points of view and the real technical feasibility of the configured customized shoe can be finally verified (*7. Technical feasibility evaluation*). In case of issues the shoe design must be iterated to modify the project. If the issues are relative to specific features of the shoe configuration (e.g. not feasible stylistic combinations), a further long and complex negotiation phase with the customer is generally needed to try to reach a feasible trade-off solution.

When a feasible configuration has been defined the manufacturing phase can start with the production of the first prototype (generally made of less valuable materials), under the responsibility of the expert artisan, who leads the customization activities, and the most expert operators of the production department (*8. Prototype manufacturing*). In some unfortunate cases, the unfeasibility of a customized shoe can emerge during this very late stage of the process.

The first prototype is then hand-delivered to the customer to verify the comfort level and the fulfillment of the stylistic requirements (*9. Test fitting*). Also in this case a long iterative process of design changes-prototype manufacturing-test fitting can start until an acceptable solution is reached (on average more than two prototypes are manufactured before reaching the final shoe version).

After the final approval, the customized shoes are manufactured by expert operators and by using the chosen materials and accessories (*10. Final shoe manufacturing*). Finally, the shoes are packed in customized boxes (e.g. box with the customer initials), together with accompanying goods (e.g. polishing products, letter of thanks) and hand-delivered by the expert artisan to the customer at the store where the configuration started (*11. Shoe delivery*).

Summarizing, the AS-IS process allows to produce very high-quality products, but several inefficiencies can be observed:

- The choices made during the first configuration are influenced by the existing shoes, thus customers are not completely free to create his/her favorite solution;
- Technical constraints are not verified during the first stages of the customization process;
- The process is strongly influenced by the knowledge of expert operators (e.g. expert artisan for configuration suggestions and foot measurements, expert technicians for shoe and last design, expert operators of the production line for shoe manufacturing), thus it is not repeatable;
- Data about configuration are not stored and shared in a standardized way (e.g. customer requirements are annotated in paper or electronic documents and then generally transmitted via email, the technical and production documentation is composed by paper documents with the addition of non- standardized notes);
- The production of customized shoes is only supervised by expert operators, while the customer is not aware about the state of his/her order;
- The iterative phases of the customization process (e.g. configuration and preliminarily feasibility evaluation, shoe design and technical feasibility evaluation, test fitting and prototype manufacturing) could cause a delay in the final shoe delivery and customer response time.

3 The Digitally-Enabled Shoes Customization Process

The process described in the previous section has a series of potential technical and economic issues that contribute to reduce the overall process efficiency and the customer experience. This is the reason why a customization framework is proposed (Sect. 3.1) and a customization workflow is defined (Sect. 3.2) to support footwear companies in improving all the activities related to product customization.

3.1 The Proposed Customization Framework

The proposed customization framework (Fig. 2) is composed by a combination of existing and customized technologies, able to support the configuration, design and production phases, other than an effective data management and exploitation. This framework enables the "see, design, buy" paradigm which allows customers to co-ideate his/her ideal customized shoes directly at stores that become places of collaboration among customers and company employees.

As shown in Fig. 2, the configuration can be supported by the following technologies:

- *Scanner*: this equipment, based on the laser technology, allows scanning the customer's feet at stores, to automatically extract the characteristic measurements;
- *CAD-based configurator*: this tool supports the store staff during the shoes/accessories configuration, according to customer requirements, historical data (if available) and/or feet characteristics (measured through the scanner). These parameters are used by specific machine learning-based searching algorithm that allow finding the last and shoe models that best fit with the customer's feet and

stylistic preferences. In addition, this tool implements specific configuration rules to automatically verify the technical feasibility of the configured solutions;

- *VR/AR fitting system*: these technologies allow the virtual fitting of the configured shoe, offering the opportunity to customers to evaluate in real time if the solution fully reflects his/her preferences.

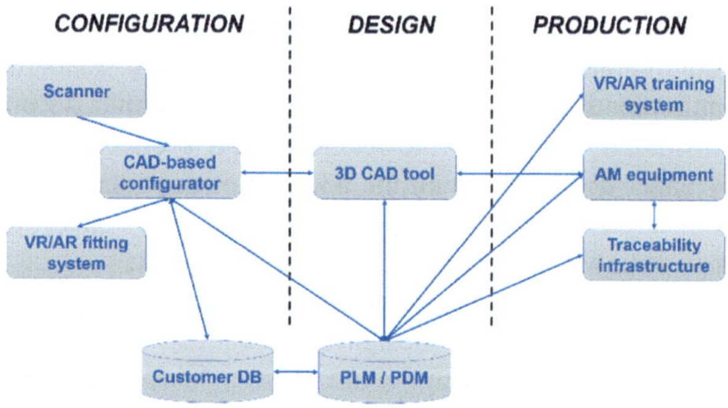

Fig. 2. The proposed customization framework

Concerning the design phase, the key enabling technology adopted in the proposed framework is A *3D CAD software tool* dedicated to shoe design, which allows the technical department to model the new configured shoes and develop the technical documentation needed for the successive phases.

During the production, instead, the following technologies should be adopted to improve the customization process:

- *Traceability infrastructure*: these systems, based, for instance, on the RFID technology, are able to monitor in real time the status of each order, with the aim to: (i) keep customer response time under control, (ii) control the efficiency of production processes preventing bottlenecks, and (iii) make the customer aware about the work progress of his/her order;
- *AM equipment*: this technology can be implemented to realize prototypes of configured accessories to provide to customers the opportunity to evaluate their aesthetic performance during the different test fittings;
- *VR/AR training system*: in the context of the production phase, these technologies can support the training of non-expert operators. For instance, a dedicated off-line workstation can be implemented to allows operators to learn how to carry out specific manufacturing operations, in order to produce high-quality products.

The abovementioned technologies must be enabled by a centralized data management, composed by the following items:

- *Customer DB*: this database allows to store large quantity of statistical data related to customers. Data contained in this repository are essential to feed the CAD-based configurator in order to suggest the solutions that best fit with the individual characteristics of each customer;

- *PLM/PDM*: these repositories are essential to guarantee an efficient sharing of information along the whole flow of configuration, design, production and sale of customized shoes.

3.2 The Proposed Shoes Customization Workflow

The implementation of the proposed framework allows the definition of an improved customization process for high-end footwear products. The steps of the proposed process are represented in Fig. 3.

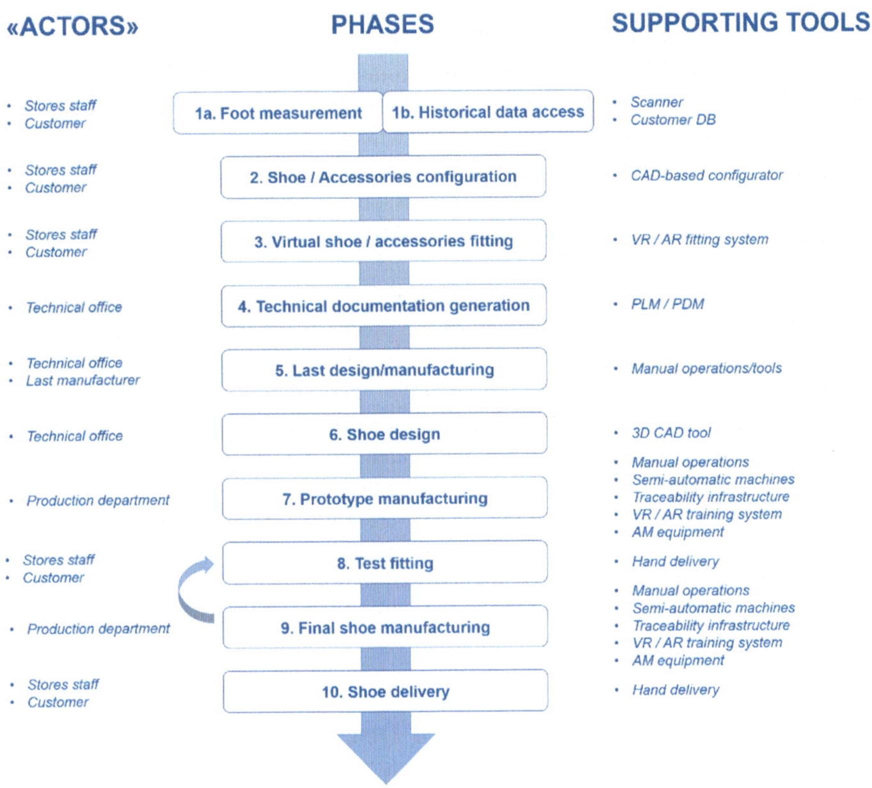

Fig. 3. The proposed shoes customization workflow

Starting from the first step, the differences among the AS-IS and the proposed workflows are relevant. In the proposed workflow, the configuration phase is based on the measurement and/or recovery of customer's biometric characteristics. If the customer is new the store staff registers the customer data in the Customer DB for future configurations and can directly use the scanner to measure his/her foot, by following a standardized procedure (*1a. Foot measurement*). In case of loyal customers, instead, historical data about biometric characteristics and stylistic preferences can be automatically retrieved (*1b. Historical data access*).

Successively, the CAD-based configurator can be used by the store staff, together with the customer, to configure a shoe that fit with the characteristic of the customer foot and is compliant with his/her preferences (*2. Shoe/Accessories configuration*). In addition, also customized shoe accessories can be co-designed (e.g. buckles, eyelets). The implemented configuration rules allow configuring feasible solutions, preliminarily avoiding the definition of shoes that cannot be produced.

After the configuration, the VR/AR fitting system enables a first virtual test fitting (*3. Virtual shoe/accessories fitting*). This stage is essential to give to the customer the sensation of wearing the configured shoes and accessories, and to check if they are compliant with all his/her stylistic preferences.

Once a satisfying configuration has been found, the shoe order is formalized and stored in the centralized PLM/PDM database. In this way, the technical/production documentation can be automatically generated without information losses (*4. Technical documentation generation*).

The last is developed as in the AS-IS process, but the availability of the virtual foot model, acquired through the scanner, facilitates the work of the last manufacturer (*5. Last design/manufacturing*)

Regarding the design and development of the shoe components, this phase is enabled by the 3D CAD tool, included in the proposed framework, which allows to work with virtual models, reducing the time to implement design modifications (*6. Shoe design*).

The steps concerning the production of prototypes and the final product, as well as delivery to the customer (*7. Prototype manufacturing*, *8. Test fitting*, *9. Final shoe manufacturing* and *10. Shoe delivery*), are essentially the same described for the AS-IS process. However, the manufacturing phases are supported by three different technologies: (i) the traceability infrastructure to monitor the status of each individual order, (ii) the AM equipment for the production of accessories prototypes, and (iii) the VR/AR training system to support the learning process of non-expert operators of the production line.

4 Framework Implementation and Discussion of Potential Benefits and Limitations

4.1 Preliminary Implementation in an Italian Shoe Manufacturer

Currently, the implementation of the customization framework started in collaboration with an Italian company involved in the production of high-end footwear products. The involved company, specialized in the production of classic leather shoes, mainly for men, registered a relevant growth of the market for highly customized shoes (e.g. personalized last) during the last years. The management of these additional orders with the standard organization and processes is very challenging, and, as a consequence, require the adoption of innovative tools to support the different phases affected by the product customization.

Only some modules related to the configuration phase have been developed and a preliminary implementation started:

- A commercial laser scanner (Ortho FootScan 2 by Duna) has been chosen and implemented in a single-brand store of the company located in Italy. This system allows the easy scanning of customer's feet and the generation of mesh models in standard STL format;
- Regarding the CAD-based configurator, at first, a specific product archetype has been defined to represent the shoe product structure, to identify the items which can be personalized (e.g. leather typology/color, heel height, surface treatments) and to define the configuration rules and constraints. Successively a prototype of a JavaScript-based software application has been developed to provide to the store staff all the functionalities needed for the shoes configuration;
- The configurator is able to interface with the developed database, dedicated to the management of customer's characteristics;
- Finally, a preliminary version of an AR application for mobile devices has been developed to project the configured shoe under the customer foot.

The following Figs. 4 and 5 shows some details about the developed/implemented modules and the product archetype related to the upper part of the shoe, respectively.

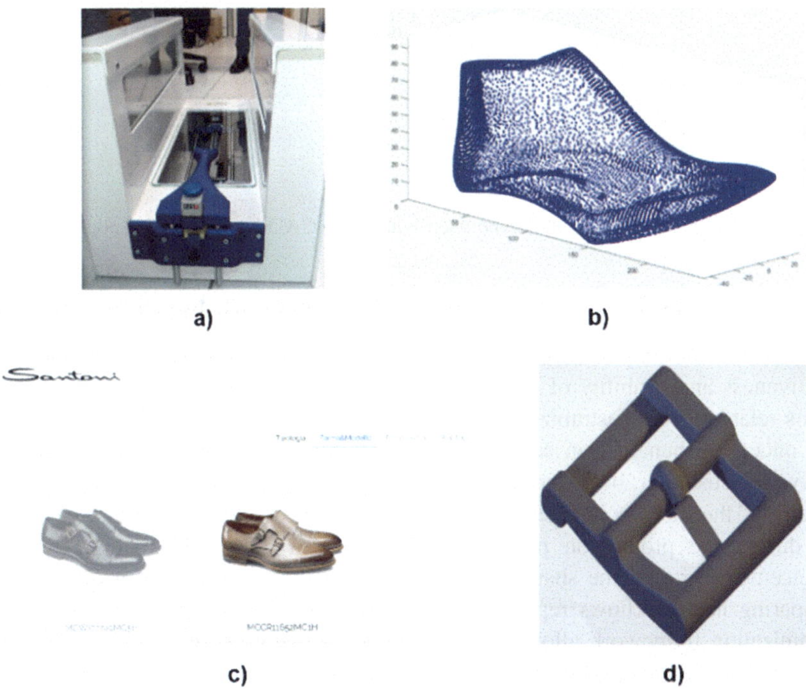

Fig. 4. (a) Foot scanner, (b) last cloud points, (c) configurator interface, (d) buckle virtual model

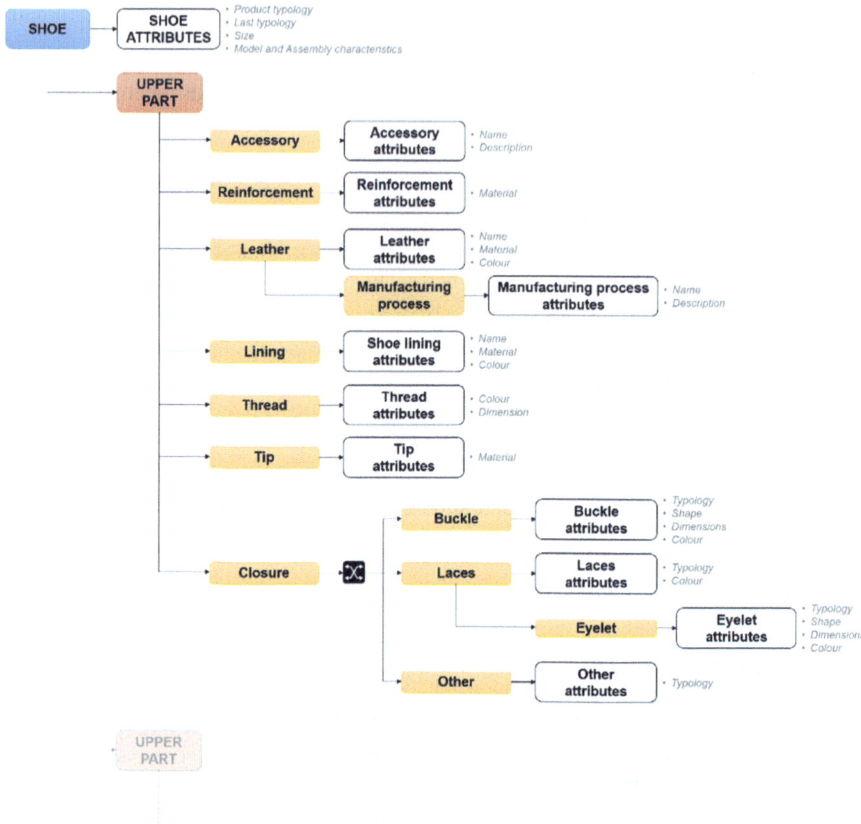

Fig. 5. Product archetype used within the CAD-based configurator

4.2 Benefits and Limitations Related to the Framework Implementation

The experimentation phase has not yet started, therefore no quantitative results on the effectiveness and usability of systems are available. Only potential benefits and limitations related to the customization framework implementation can be discussed.

Concerning benefits, an essential added value of the framework is the ability to involve the customer, both during the configuration phase at the store when he/she can "co-create" the shoe/accessories and virtually test the aesthetic performance of the shoe, and during the production phase when he/she can constantly monitor the order advancement through the sharing of data collected by the traceability infrastructure. Comparing the workflows represented in Figs. 1 and 3 it appears that the proposed customization framework allows also to reduce the iterative phases of configuration-design-prototype-test-modification, thanks to the possibility to monitor the technical feasibility of the configured shoe already during the first stages of the customization process. The whole process is more repeatable and generally less dependent from the knowledge of expert operators: (i) the customer's feet can be directly measured in a standardized way by the store staff through the scanner, and (ii) operators of the production line can be trained in performing specific operations through the AR-based off-line workstation. In addition, operators' work is supported with the aim to minimize errors and waste of time and resources. Finally, the centralized data management systems (Customer DB and PLM/PDM) allow to make better use of available data (e.g. historical data about customers can be reused for supporting future configurations), as well as to optimize the information flow among the different phases and involved "actors" and to reduce information losses (e.g. non-standardized annotations in paper documents are substituted by electronic notes stored and shared through the centralized DBs).

It must also be said that different barriers against the implementation of the whole customization framework exist. First of all, the technology acceptance at both internal (i.e. company operators) and external (i.e. customers) levels need to be monitored to maximize the usability and effectiveness of the adopted technologies, reducing the time needed to run the entire system at full capacity. Regarding the initial costs for the adoption of the proposed technologies, the economic investment could be not negligible, since software tools and technologies need to be customized for the implementation in each specific context. It is therefore essential to accurately evaluate the implementation costs in a preliminary phase, against the potential revenues and the market trend for customized products. In addition, the introduction of innovative tools requires the careful management of the additional complexity and activities and the mitigation of possible negative impacts on internal company processes. Generally speaking, the adoption of the proposed customization framework requires to set a long-term strategy and a "change of mentality", not a common situation especially in case of very small companies that are usually busy with day-by-day issues.

5 Conclusions

The present paper aims to propose an innovative framework for the management of the whole customization process of footwear products. The combination of different specific and existing technologies allows to manage all the phases and processes (i.e. configuration, design, production) affected by the customization of shoes, as well as to involve all the "actors" who play an active role (e.g. customers, store staff, technical office, production department). The final objective is to improve the efficiency of the shoes customization process by overcoming well-known limitations of the traditional approach: high number of iterative phases, low repeatability of the process, influence of personal skills of expert operators, low involvement of customers, inefficient management and exploitation of data.

Currently the framework and the key enabling technologies have been defined, but only few modules have been developed and implemented in collaboration with an Italian company. Therefore, the first future work will be related to the experimentation of the whole framework, in in order to quantify the benefits in terms of system usability and effectiveness, customer involvement, user acceptance, reduction of customer response time for new products, economic savings, etc. Successively, other heterogeneous industrial contexts (e.g. different companies, different countries, different products) could be considered to evaluate the replicability of the proposed approach and infrastructure.

References

1. Ellena, T., Mustafa, H., Subic, A., Pang, T.Y.: A design framework for the mass customisation of custom-fit bicycle helmet models. Int. J. Ind. Ergon. **64**, 122–133 (2018)
2. Kumar, A.: Mass Customization: Metrics and Modularity. Int. J. Flex. Manuf. Syst. **16**, 287–311 (2004)
3. Wong, H., Lesmono, D.: On the evaluation of product customization strategies in a vertically differentiated market. Int. J. Prod. Econ. **144**, 105–117 (2013)
4. Vickery, S., Dröge, C., Germain, R.: The relationship between product customization and organizational structure. J. Oper. Manag. **17**, 377–391 (1999)
5. Zezulka, S., Marcon, P., Vesely, I., Sajdi, O.: Industry 4.0 – An Introduction in the phenomenon. IFAC-PapersOnLine 49, pp. 8–12 (2016)
6. Marconi, M., Manieri, S., Germani, M., Raffaeli, R.: A Digitally-enabled Integrated Approach to Design and Manufacture Shoe Lasts. Comput. Aided Des. Appl. **16**, 593–610 (2019)
7. Jimeno-Morenilla, A., Garcia-Rodriguez, J., Orts, S., Davia-Aracil, M.: GNG based foot reconstruction for custom footwear manufacturing. Comput. Ind. **75**, 116–126 (2016)
8. Luximon, A., Luximon, Y.: Shoe-last design innovation for better shoe fitting. Comput. Ind. **60**, 621–628 (2009)
9. Weerasinghe, T.W., Goonetilleke, R.S.: Getting to the bottom of footwear customization. J. Syst. Sci. Syst. Eng. **20**, 310–322 (2011)
10. Bernabéu, J.A., Germani, M., Mandolini, M., Mengoni, M., Nester, C., Preece, S., Raffaeli, R.: CAD tools for designing shoe lasts for people with diabetes. Comput. Aided Des. **45**, 977–990 (2013)
11. Zhang, L.L., Lee, C.K.M., Xu, Q.: Towards product customization: An integrated order fulfillment system. Comput. Ind. **61**, 213–222 (2010)
12. Klein, P., Pugliese, D., Lützenberger, J., Colombo, G., Thoben, K.-D.: Exchange Of Knowledge In Customized Product Development Processes. Procedia CIRP **21**, 99–104 (2014)
13. Ducq, Y., Vallespir, B.: Reference Models for Mass Customisation Production of High Fashionable Products: Application to the Shoe Manufacturing Domain. In: Koch, T. (ed.) Lean Business Systems and Beyond. IFIP – The International Federation for Information Processing, vol. 257, pp. 165–173. Springer, Boston (2008)

Conceptual Design of a Functional Orthodontic Appliance for the Correction of Skeletal Class II Malocclusion

Luca Grigolato[1(✉)], Stefano Filippi[2], Daniela Barattin[2], Daniele Cantarella[3], Won Moon[4], Roberto Meneghello[5], Gianmaria Concheri[1], and Gianpaolo Savio[1]

[1] ICEA Department, University of Padova, Via Venezia, 1, 35131 Padua, Italy
luca.grigolato@unipd.it
[2] DPIA Department, University of Udine, Viale delle Scienze, 206, 33100 Udine, Italy
[3] SBCO Department, University of Milano, Via Commenda, 10, 20122 Milan, Italy
[4] UCLA - School of Dentistry, 10833 Le Conte Ave, Los Angeles, CA 90095-1668, USA
[5] DTG Department, University of Padova, Stradella S. Nicola, 3, 36100 Vicenza, Italy

Abstract. The current requests for continuous innovation represent a challenge in every industry as well as in the field of orthodontics. Aim of this work was to develop new concepts of a functional appliance for the correction of class II skeletal malocclusion through a systematic design methodology. Staring at the existing devices in this field, taking into account the literature and the patient's needs, the customers' requirements were identified by Quality Functional Deployment. Systematic methods such as morphological method, theory of inventive problem solving and other creative methods were used for generating concepts some of which are presented at the end of the paper.

Keywords: Conceptual design · Functional appliance · Quality function deployment · Morphological method · TRIZ

1 Introduction

Class II malocclusion represents one of the most frequent problems in the orthodontic practice, affecting 37% of school children in Europe and occurring in 33% of all orthodontic patients in the USA [1]. Class II malocclusion is characterized by a functional and aesthetic deviation from ideal occlusion, in which the mandible is positioned too far backward in relation to the maxilla. There are different solutions for this issue but nowadays it is frequently corrected with special orthodontic devices called functional appliances that act on both the maxilla and the mandible in order to advance the lower jaw into the correct forward position, to obtain a neutral occlusion with the right relationship of the jaws [1]. Also, several solutions for functional

C. Rizzi et al. (Eds.): ADM 2019, LNME, pp. 329–341, 2020.
https://doi.org/10.1007/978-3-030-31154-4_28

appliances exist in the market or are described by literature, typically divided into fixed and removable [1–4]. The most popular and studied device is the Herbst appliance (Dentaurum, Ispringen, Germany), also called "bite-jumping" device [5].

An overview of the design and the orthodontic workflow for this type of appliances identified a well-established approach. These devices almost always need to be constructed by an orthodontic technician based on the clinician instructions, so that they are conditioned by the manufacturing process which influences their reliability [6]. This is called by Proffit a 'components' approach' to design, which means that there is no ideal appliance to be used in every clinical situation and so there is not one single design of appliance to treat each type of malocclusion.

Today several types of functional appliances are produced by many orthodontic companies and laboratories imitating the Herbst appliance in order to solve the complications acknowledged by the clinicians and widely described in the literature as problems of breakages, decalcifications, excessive mesial movement of lower dentition and others [6]. It appears that the evolution of functional devices is like a trial-and-error process in order to solve the design issues of the Herbst appliance. For this reason, it is observed that there is inertia for this type of devices into a funnel that causes no real improvement and no very innovative solutions. In particular, design efforts, also described in patents, concern the types of stabilization or the anchorages to the teeth and the connections of the mechanism.

At this stage, two main problems arise: the first is the design approach methodology for orthodontic functional appliances, the second is the manufacturing processes, strongly correlated to the product design solutions implemented until now [6].

To overcome this inertia and the lack of structured process in functional appliance design, in this work, a systematic design method is proposed for the conceptual stage. This should lead to a new and different approach in this medical area, trying to go beyond the traditional way of thinking and improving these types of products. As a result, several conceptualized solutions for a functional appliance are proposed, using a systematic approach that combines valuable methodologies of the scientific literature.

2 Background

2.1 Malocclusion and Functional Appliances

Malocclusion is defined as an abnormal deviation either aesthetic, functional or both, from the ideal occlusion and it is the result of a combination of both genetic and environmental factors. It reflects an abnormal deviation of mandible or maxilla growth and dimensions. As defined by E. Angle, three different types of malocclusion exist: class I, class II and class III [4]. To solve this issue, orthodontic treatment can focus on dental movements only, or deal with the control and modification of facial bones growth. In the latter case, it is better defined as "dentofacial orthopedics". The main current approach to correct class II malocclusion, in which the mandible is in a retruded position with respect to the maxilla, is using functional appliances [4].

It is acknowledged by the orthodontics experts that the main issue for a good medical result is related to the patient's compliance [1, 4], especially because the treatment takes place in the adolescents during the growing age. For this reason, functional appliances are usually divided into non-compliance and compliance devices [1]. The first ones refer directly to fixed appliances that are usually cemented on patient's teeth, which implies full-time wear, while the second ones refer to removable appliances, that can be removed by the patient, *e.g.* while eating. Due to historical reasons, fixed functional appliances are preferred by U.S. clinicians because of their effectiveness and shorter therapy duration, while removable appliances are mostly used by the European clinicians, due to the reduced invasiveness for the patients [7].

There are several fixed functional appliances, but every system can be seen as derived from the Herbst functional appliance (Fig. 1.a). The Herbst appliance is an artificial bilateral joint between the maxilla and the mandible. The original design had a telescopic mechanism attached to orthodontic bands on the maxillary first permanent molars and on mandibular first permanent premolars [1]. Also, several types of removable functional appliances exist, some of the most common are the Twin Block, Bionator and Frankel (Fig. 1b) [1, 4]. The action mechanism of all these appliances is based on two main theories: the pressure-tension theory and the functional matrix theory [4]. Furthermore, other devices have been developed for the therapy of Obstructive Sleep Apnea Syndrome (OSAS), with the rising interest for innovative manufacturing technologies, such as additive manufacturing, taking advantage from the same functional principles of malocclusion class II appliances (Fig. 1c).

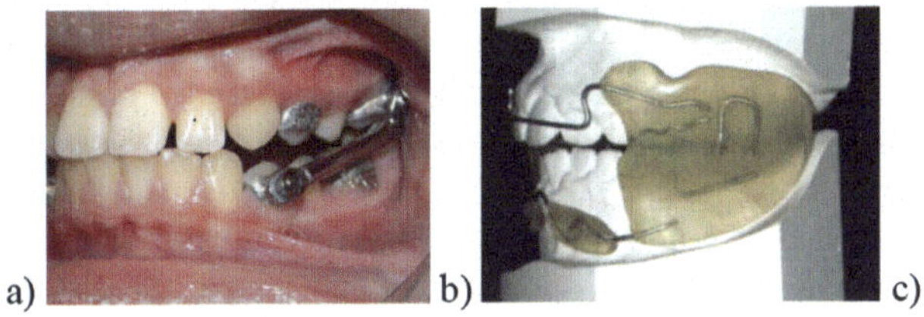

Fig. 1. Main types of functional appliances and an additively manufactured OSAS device: (a) Fixed functional device, Herbst appliance [1]; (b) Removable functional device, Frankel appliance [4]; (c) OSAS device, Narval CC [8].

To avoid patient's compliance problems, the focus has been addressed to fixed devices.

2.2 Systematic Approaches for Conceptual Design

In order to make the creation process more formalized several design methodologies have been developed and accepted [9–12]. These methodologies, also called rational design methods, try to explain a guided pattern to approach the design activities. Newbies, as well as design experts, can take advantage of this type of approach, in order to improve quality design decisions [12].

Even if there are several methodologies, a common logical path is acknowledged [11]: clarification of the task phase, conceptual design phase, embodiment design phase, and detail design phase. In the first phase, the design process is organized, and every useful data is collected. In conceptual design, the fundamental traits of the product are defined. In embodiment design, the layout of the product is developed and in detail design, the product is completely described.

For the clarification of the task phase, there are several systematic tools that bridge customer's expectations with product engineering parameters that represent the main attended output. Quality Function Deployment (QFD) [13] seems to be the most popular method for quality improvement [14]. It helps not only in understanding the design problem, but it also supports setting the foundation for the concept generation phase [9]. The method consists of a series of steps and procedures in order to translate qualitative information into quantitative parameters. All the data obtained is usually filled and visualized in a combined matrix called House of Quality (HOQ) to be more accessible in the further design phases.

One of the most structured techniques in conceptual design is the morphological method [9]. The morphological method is a technique for systematically structuring and exploring the set of relationships contained in a multi-dimensional complex problem. Defined as the "study of form and structure", it is usually visualized as a 2D matrix. This method consists of 3 sub-steps: generating the functions that the system must accomplish, generating as many as possible concepts for each function, and combining the individual concepts into macro concepts [9].

The generation of concepts is the less rational activity and one of the most important of the design process. In order to develop concepts, several methods for supporting the generation of creative solutions are proposed by the literature [9]: creative methods such as Brainstorming, 6-3-5, Synectics or conventional tools such as patents analysis, experts' knowledge or personal knowledge. While these are based on psychology and analytical techniques, a more systematic and complete method is found in "theory of inventive problem solving" (TRIZ) [15]. It can be interpreted and applied as a philosophy or a method or a set of problem definition and solving tools [16, 17]. Contradictions matrix and the 40 Inventive Principles are the simplest tools for problem-solving and, for this reason, the most used TRIZ techniques [18–20].

Also, Biomimicry [21] is another tool in the hand of the designer to enlarge his point of view, that is having now an increasing interest. Biomimicry consists of linking analogies from biology to technology, emulating the models, systems, and elements of nature for the purpose of solving human challenges [21]. AskNature database [22] is a helpful instrument to conceptual design, having the biological solutions organized in a functional way, so it matches with a functional approach.

These methods are based on functions knowledge and can be used together [9]. While TRIZ theory is based on human technology, on the other hand, nature can be another source of inspiration. In addition, TRIZ and Biomimicry find a possible integration as described by Vincent et al. [23].

Several studies suggesting interactions and integration between these tools can be found in the literature, *e.g.* between QFD, functional modeling and TRIZ [9, 10, 18, 24–29]. The most valuable connection is that QFD reveals negative correlations in couples of quality parameters, which can be seen as TRIZ contradictions [14]. QFD can also help the generations of functions as proposed by Ullman [9].

3 The Proposed Design Methodology and the Case Study

The systematic design process as described by Ullman [9] is taken as a base for the proposed conceptual design of an orthodontic functional appliance. The attempt is to integrate QFD, morphological method, TRIZ tools, and biomimicry into a systematic approach, because of the synergic potentialities in each method. The overall process in this work is represented by the activity "conceptualize an orthodontic functional appliance". Figure 2 represents the proposed approach showed in an IDEF0 like diagram. The process contains 3 tasks defined as:

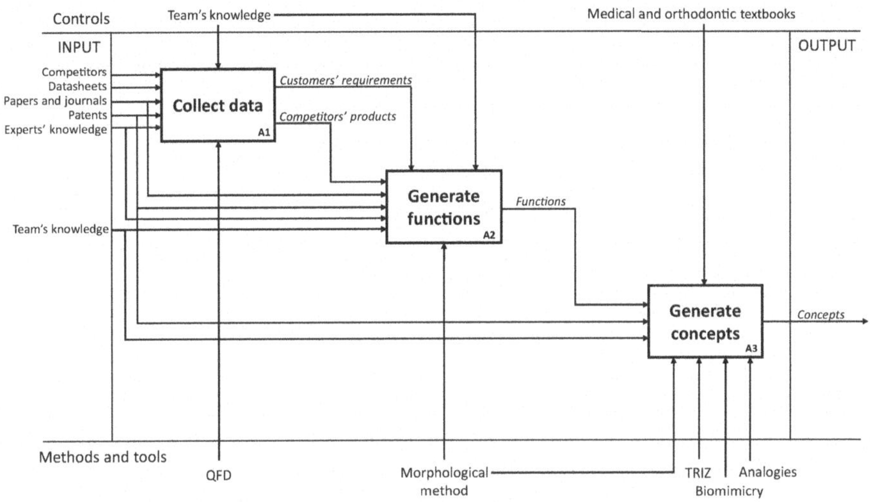

Fig. 2. Description of the proposed design process using IDEF0 formalism.

1. Collect data;
2. Generate functions;
3. Generate concepts.

The input data, on the left side, used for this work are papers, patents, experts' knowledge, competitors, datasheets and team's knowledge, the team being composed by 3 engineers, 2 orthodontists and 1 designer. In the upper side, the external controls are the team's knowledge and medical and orthodontic reference textbooks. On the right side, the main attended outputs are the non-evaluated concepts. In the downer side, the methods used to accomplish the result are reported: QFD, morphological method, TRIZ and other creativity tools such as biomimicry and analogies.

Each activity contains sub-activities that are explained in the following, accomplished with the use of the aforementioned methods.

3.1 Collect Data

The collected data comes from the QFD application, following the guidelines proposed in the literature [9]. A rough overview of the product life cycle for a functional appliance helps the identification of the main customers, the typical way of the product use and the main phases during its life. It is found that a fixed appliance product life can be divided into 3 main and sequential phases as shown in Table 1: production, when the product is manufactured; utilization, when the product is inserted in the patient mouth for a certain period; disposal, when the product is dismantled and wasted. The utilization phase can be also split into three sub-steps: positioning, use/therapy, removal. The principal customers are identified in orthodontist, patient, and dental technician.

Table 1. Identification of the customers of an orthodontic fixed functional appliance.

Main life cycle phases	Customers		
1) Production	Dental technician	Orthodontist	Patient
2) Utilization	Orthodontist	Patient	
– Positioning	Orthodontist		
– Use/Therapy	Patient	Orthodontist	
– Removal	Orthodontist		
3) Disposal	Orthodontist		

The orthodontist represents the main customer, because he is present in every phase, while the patient becomes more important during the use/therapy phase. The choice for the specific appliance solution is strongly influenced by the orthodontist, mediated by his experience and knowledge that guides the patient's final decision.

The individuation of customers' needs is performed with the help of experts' knowledge and literature review. After the translation of needs into more operational objects, usually defined as 'customers' requirements', they are grouped together into domains that reflect different aspects of the appliance design. Six domains have been identified (Table 2): mounting, comfort, hygiene, aesthetics, reliability and unmounting. A total of sixteen independent requirements are recognized. The customer requirements are divided into two groups usually referred to as attributes (vertical character set in Table 2) and functional requirements (sloped character set in Table 2) [9]. The attributes group together non-functional requirements, like aesthetic requirements. Functional requirements are used as guidelines for generating functions of the technical system, as shown in Fig. 2.

The relative importance of the customers' requirements is evaluated with the help of orthodontic experts. The tool used here is a survey, where it has been asked to assign a value between 1 (undesirable) to 5 (fundamental) that represents the perceived importance of each requirement.

The engineering requirements and the direction of improvement of each requirement are then derived from the customers' requirements. A total of 29 engineering requirements are generated.

Table 2. Customers' requirements grouped into domains.

	Domains					
	Mounting	Comfort	Hygiene	Aesthetics	Reliability	Unmounting
Customers' Requirements (CR)	*Easy to mount* Small number of parts	*Allow adjustability,* *Assure physiological opening movements of mandible* *Assure physiological lateral movements of mandible* Should be small Should be light *Not create injuries and harm*	*Low cleaning time required* *Holds little amount of plaque*	It is hardly visible	*It is resistant* *It is reliable* *Not cause dentoalveolar movements*	*Easy to unmount* *Not painful during removal*

A matrix, called correlation matrix, filled by the team, expresses which customers' requirements are directly related to engineering requirements and how much the relationship is strong, assigning the values 1-3-9 [9]. In this way, evaluating column by column the weighted sum of the values, the relative importance of the engineering requirements is obtained. In our case, the more important are 'functional principle', 'number of pieces' and 'stiffness of the connections'. A portion of the matrix is reported in Table 3.

A correlation between the engineering requirements is analyzed in another matrix, usually called the HOQ roof, part of the QFD. It highlights the bad and good relationships among the engineering requirements, to find possible contradictory parameters that can be resolved by TRIZ. For instance, an improvement of the connection strength suggests an increase in the total volume of the device. This contradiction can be resolved by the use of TRIZ contradiction matrix that suggests the following principles: prior counteraction, spheroidality, another dimension, and dynamics.

At the same time, competitors' products benchmarking is performed. The problems of actual systems and the working principles of these systems are also highlighted. From the market, the main products are identified with the help of orthodontics experts. The products chosen for the competitors' evaluation are the most commonly used fixed functional appliances, as defined by Papadopoulos [1]: Herbst, MARA (Allesee Orthodontic Appliances, Sturtevant, WI), Jasper Jumper (American Orthodontics, Sheboygan, WI) and Forsus (3 M Unitek, Monrovia, CA). These products are evaluated on the basis of the requirements' satisfaction. The evaluation process is made by

Table 3. A portion of the correlation matrix customers' requirements/engineering requirements.

		Engineering Requirements (ER)					
		Average time needed for mounting	Number of steps needed for mounting	Number of pieces	Number of regulations	…	CR relative importance
Customers' Requirements (CR)	Easy to mount	**9**	**9**	**1**			3.54
	Small number of parts	**3**	**3**	**9**			3.77
	Allow adjustability		**1**		**9**		3.69
	Assure physiological opening						5.00
	…						
	ER relative importance percentage	3.45%	2.06%	6.54%	2.61%		

orthodontics experts. It is found that every system doesn't completely satisfy the customers' requirements and for this reason, possible space for improvement exists. This elaboration, together with the importance of the requirements, gives the first suggestions to designer for product development. From this analysis, technical data of the competitors' products are also extracted, mainly from datasheets and literature reviews [1].

Lack of technical data for these type of appliances is one issue that was encountered and it was quite impossible to find overall technical values for the engineering requirements identified. This confirms the considerations about manufacturing mentioned in the introduction.

Among the QFD outputs, only customers' requirements and competitors' products needed for the conceptual phase are reported.

3.2 Generate Functions

Once the QFD has been filled the next activity is to generate the functions of the system, which is the first step of the morphological method.

First of all, the overall function that the system has to perform was defined. From QFD competitor's products and team's knowledge elaboration, the system to be investigated and its overall task have been identified. It is also clear that the overall function of the system cannot be explicit during the requirements individuation phase, because customers are expecting this from the product, like a must-be attribute. From experts and literature sources, a general overall function for the functional appliance has been recognized as: "Positioning the mandible in the correct relationship respect with the maxilla". To restrict the exploration area, two boundary conditions were applied: the system has to be not removable by the patient, *i.e.* a non-compliance device, and has to be intraoral.

Every competitors' product analyzed in the collect data phase, is composed by 2 main sub-systems: the principal system, the one that performs the correction and the stabilization system, which connects and stabilizes the appliance to the maxillary and mandible teeth.

Analyzing the competitors' products, 6 main functions were established (Table 4), following the Energy-Material-Signal (EMS) like approach [9–11]. This approach permits to model an engineering system as a black-box using the flows of energy, material, and signal, to define the system functions.

Table 4. Main functions for the orthodontic functional appliance.

Main functions for a functional appliance	Approach	Principal system	Stabilization system
Influence the mandible motion	EMS-Energy	X	
Relate mandible and maxilla	EMS-Energy		X
Guarantee correct hygiene maintenance	EMS-Signal	X	X
Generate feedback	EMS-Signal	X	X
Maintain efficiency	EMS-Signal	X	X
Provide effectiveness	EMS-Signal	X	X
Assure physiological movements of mandible	Team discussion	X	
Allow therapy customization and adjustment	Team discussion	X	
Ensure protection to human tissues	Team discussion	X	

The energy flow directly defines two main functions: influence the mandible motion and relate mandible and maxilla.

These 2 functions identify, at an abstract level, the 2 sub-systems recognized during competitors' products analysis: 'influence the mandible motion' is the high-level function of the principal system and 'relate mandible and maxilla' represents the high-level function of the stabilization system.

Following the signal flow, 4 functions are defined: 'guarantee the correct hygiene maintenance', 'generate feedback', 'maintain efficiency' and 'provide efficacy'. Another group of functions doesn't follow any flow, but it was identified by the team through discussion. It identifies the comfort/ergonomic functions: 'allow therapy customization and adjustment', 'assure physiological movements of mandible', 'ensure protection to human tissues'.

The main functions are then organized into three main groups in order to better perform the next concepts generation phase. As shown in Table 4, the first group contains the 4 main functions of the principal system, marked in light grey; the second group collects the main function related to the stabilization system, marked in dark grey and the last group summarizes the main common functions of both systems, in the white cells.

The simplified morphological matrix is filled in the first column with the 4 functions of the principal system for each row.

3.3 Generate Concepts

Several tools can be used consciously or unconsciously to generate concepts, *i.e.* how to perform the functions established. Mainly TRIZ tools have been used to direct the ideas generation phase. The principal TRIZ tools used in this phase are trends, 40 inventive principles, and contradictions.

For the 4 main functions of the principal system highlighted in Table 4, different concepts were generated and summarized in the simplified morphological matrix reported in Table 5. In particular, two possible combinations for each concept's solution relating to the function "assure physiological movements of mandible" have been reported. Each combination interconnects a cell for each function.

Table 5. A portion of the Morphological Matrix. 4 concepts combinations identified for the functional appliance.

Principal System Functions	Concept 1	Concept 2	Concept 3	Concept 4	Concept 5	Concept 6
1) Influence mandible motion	Obstacle ■	Guide ▲●	Force	Momentum	Magnetic attraction	Magnetic repulsion ★
2) Allow regulation	Continuous ■	Discretized ●	Inter-changeable structures ▲	No regulation ★		
3) Ensure protection to human tissue	Surface and Shape study	Cover tissues	Cover the appliance ●▲★	Release healing substances	Position change ■	
4) Assure physiological movements	Flexible linkage ■●	No linkages ▲★				

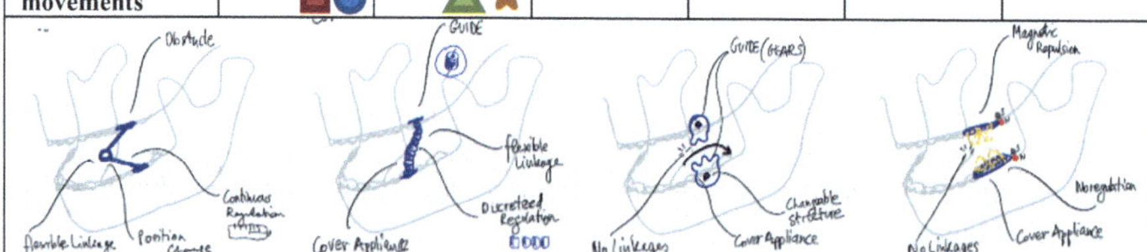

■ Combination 1 ● Combination 2 ▲ Combination 3 ★ Combination 4

To fill the alternative concepts for the function 'influence mandible motion', the dynamization trend of TRIZ was used starting from the ideas of mechanical solutions, in the first 2 cells, and finishing with the use of magnetism. The remaining alternatives use the trend of system completeness to suggest the addition of an internal linear force or a momentum to the system. Also the alternatives for the function 'allow regulation' has been generated following the dynamization trend. For the 'ensure the protection of human tissue' function the first cell contains the suggestion from the 14th inventive principle of TRIZ, *i.e.* curvature-spheroidality. Covering concepts come from analogies with solutions from other technological areas and they refer to thin films or shields; release healing substances comes from biomimicry analogies with some types of plants and animals that use the mechanism of secretion to protect themselves, for example, latex secretion in the trees. Position change comes from the 2nd inventive principle of TRIZ. The last function can be accomplished with 2 concept alternatives, the first one suggested by the 1st inventive principle, *i.e.* segmentation, while no linkages come from ideality trend suggestion.

Combination 1 originates from the possible mechanical solution. An obstacle with a continuous regulation in one rod, where the hinge is in a different position regarding current competitors' solutions, as Herbst, and to add flexibility, so in the sketch, the number of rods is two. Following the dynamization trend of TRIZ, combination 2 is also generated. In particular, this solution found analogies in some biological structure in nature like some arthropods legs, as suggested by the function 'modify position' in the AskNature database. It acts like a discretized guide, so the single pieces can be disassembled for regulation purpose. The tessellated structure of the mechanism has to be covered.

Combination 3 comes from mechanical knowledge analogies. Usually, gears are adopted to transmit motion. In this way, the gears can be thought of as guides, and interchangeable structures, substituting one gear with a larger diameter to increase the correction. The mechanism is covered and there are no linkages in between.

One of the main contradictions is linked to force transmission with physical contact, because it creates all of the issues described by the literature, such as breakages and dentoalveolar modifications. Following the ideality trend, combination 4 is generated. Combination 4 consists in a system that uses magnets in a repulsing configuration in order to act like physical obstacles, and there are no mechanical linkages between maxilla and mandible. In this way, the breakages problems related to the mechanical parts can be resolved or reduced. The system can be covered and with no regulation because the magnets are supposed to be permanent and fixed.

The evaluation step inside the conceptual phase is part of future works, and it is going to be performed by the decision matrix method [9].

4 Conclusions

This work represents a systematic approach applied to the conceptual design case of a functional orthodontic appliance. The attempt to use a systematic approach leads to similar existing solutions but also to different concepts. The method has been briefly described and few conceptualizations have been reported. This methodology, especially due to the functions-based approach, aids the integrated consideration of the functional appliances' parts that until now have been developed asynchronously.

In further works, the transition from the conceptual phase into the embodiment phase with the best and promising concepts are going to be described. Moreover, future works may be addressed to the implementation of the proposed approach in a

Computer-Aided Innovation (CAI) tool, reducing the external inputs from customers and disengaging from other evaluations that influence, for better or for worse, the process of design in every stage.

Acknowledgments. This project was partially funded by the grant "FSE 2105-51-11-2018" by Regione Veneto.

References

1. Papadopoulos, M.A.: Non-compliance approaches for management of Class II malocclusion. In: Skeletal Anchorage in Orthodontic Treatment of Class II Malocclusion, pp. 6–21. Elsevier (2015). https://doi.org/10.1016/B978-0-7234-3649-2.00002-6
2. Bishara, S.E., Ziaja, R.R.: Functional appliances: a review. Am. J. Orthod. Dentofac. Orthop. **95**, 250–258 (1989). https://doi.org/10.1016/0889-5406(89)90055-3
3. Moro, A., Borges, S.W., Spada, P.P., Morais, N.D., Correr, G.M., Chaves Jr., C.M., Cevidanes, L.H.S.: Twenty-year clinical experience with fixed functional appliances. Dental Press J. Orthod. **23**, 87–109 (2018). https://doi.org/10.1590/2177-6709.23.2.087-109.sar
4. Proffit, W.R., Fields Jr, H.W., Sarver, D.M.: Contemporary orthodontics. Elsevier Health Sciences, Amsterdam (2006)
5. Pancherz, H.: Treatment of Class II malocclusions by jumping the bite with the Herbst appliance A cephalometric investigation. Am. J. Orthod. **76**, 423–442 (1979). https://doi.org/10.1016/0002-9416(79)90227-6
6. Schiavoni, R.: The Herbst appliance updated. Prog. Orthod. **12**, 149–160 (2011). https://doi.org/10.1016/j.pio.2011.06.004
7. Bishara, S.E., Saunders, W.B.: Textbook of Orthodontics. Saunders Book Company, Collingwood (2001)
8. Redazione DentalAcademy: Narval CC. http://www.dentaljournal.it/narval-cc/
9. Ullman, D.G.: The Mechanical Design Process. McGraw-Hill, New York (2010)
10. Ulrich, K.T., Eppinger, S.D.: Product Design and Development. McGraw-Hill, New York (2012). https://doi.org/10.1016/B978-0-7506-8985-4.00002-4
11. Pahl, G., Beitz, W., Feldhusen, J., Grote, K.-H.: Engineering Design. Springer, London (2007). https://doi.org/10.1007/978-1-84628-319-2
12. Cross, N.: Engineering design methods - Strategies for Product Design (2000)
13. Akao, Y.: QFD: Quality Function Deployment - Integrating Customer Requirements into Product Design (1990). ISBN: 9781563273131
14. Chechurin, L., Borgianni, Y.: Understanding TRIZ through the review of top cited publications. Comput. Ind. **82**, 119–134 (2016). https://doi.org/10.1016/j.compind.2016.06.002
15. Altshuller, G.S.: The Innovation Algorithm: TRIZ Systematic Innovation and Technical Creativity. Technical Innovation Center, Worcester (2000)
16. Cascini, G.: TRIZ-based anticipatory design of future products and processes. J. Integr. Des. Process Sci. **16**, 29–63 (2012). https://doi.org/10.3233/jid-2012-0005
17. Mann, D.: An introduction to triz: the theory of inventive problem solving. Creat. Innov. Manag. **10**, 123–125 (2001). https://doi.org/10.1111/1467-8691.00212
18. Spreafico, C., Russo, D.: TRIZ industrial case studies: a critical survey. Procedia CIRP. **39**, 51–56 (2016). https://doi.org/10.1016/j.procir.2016.01.165

19. Ilevbare, I.M., Probert, D., Phaal, R.: A review of TRIZ, and its benefits and challenges in practice. Technovation **33**, 30–37 (2013). https://doi.org/10.1016/j.technovation.2012.11.003
20. Chechurin, L.: TRIZ in Science Reviewing Indexed Publications. Procedia CIRP **39**, 156–165 (2016). https://doi.org/10.1016/j.procir.2016.01.182
21. Benyus, J.M.: Biomimicry: Innovation Inspired by Nature. Morrow, New York (1997)
22. Institute, B.: AskNature. https://asknature.org/
23. Vincent, J.F., Bogatyreva, O.A., Bogatyrev, N.R., Bowyer, A., Pahl, A.-K.: Biomimetics: its practice and theory. J. R. Soc. Interface **3**, 471–482 (2006). https://doi.org/10.1098/rsif.2006.0127
24. Hua, Z., Yang, J., Coulibaly, S., Zhang, B.: Integration TRIZ with problem-solving tools: a literature review from 1995 to 2006. Int. J. Bus. Innov. Res. **1**, 111 (2006). https://doi.org/10.1504/IJBIR.2006.011091
25. Frillici, F.S., Fiorineschi, L., Cascini, G.: Linking TRIZ to conceptual design engineering approaches. Procedia Eng. **131**, 1031–1040 (2015). https://doi.org/10.1016/j.proeng.2015.12.421
26. Donnici, G., Frizziero, L., Francia, D., Liverani, A., Caligiana, G.: TRIZ method for innovation applied to an hoverboard. Cogent Eng. **5**, 1–24 (2018). https://doi.org/10.1080/23311916.2018.1524537
27. Yamashina, H., Ito, T., Kawada, H.: Innovative product development process by integrating QFD and TRIZ. Int. J. Prod. Res. **40**, 1031–1050 (2002). https://doi.org/10.1080/00207540110098490
28. Fiorineschi, L., Frillici, F.S., Rotini, F.: Enhancing functional decomposition and morphology with TRIZ: literature review. Comput. Ind. **94**, 1–15 (2018). https://doi.org/10.1016/j.compind.2017.09.004
29. Mann, D.: Common Ground – Integrating The World's Most Effective Creative Design Strategies. http://120.55.91.217/wp-content/uploads/soft/100921/6-100921130335.pdf

ANOVA Applied to the Taguchi Method: A New Interpretation

Sergio Rizzuti and Luigi De Napoli(✉)

Department of Mechanical, Energy and Management Engineering (DIMEG), University of Calabria, Ponte Pietro Bucci 46/C, 87036 Rende, CS, Italy
{sergio.rizzuti,luigi.denapoli}@unical.it

Abstract. ANOVA is currently employed in association with the Robust Design in order to discover the parameters most influencing a particular performance of a device in the phase of development. Such relation has been studied for many times and literature is abundant. The peculiar nature of the Taguchi Method with the employment of orthogonal arrays introduces new elements in the investigation of the most influencing parameters. Considering that the plan of the experiment is organized on the base of level parameters combination orthogonal and balanced, ANOVA can be applied to each parameter individually and this gives new insight on the occurrence of noises in correspondence of some levels parameter. The paper analyses this kind of relation, reproducing an example from literature and discussing how ANOVA can be employed in such a study. Robust Design is an extraordinary method to be used in product development with the experiments simulated from CAE analyses. In this context, more conscious employment of such methodologies can aid in managing and organizing the study and experimentation on new products or processes. In another section a second example from the authors is re-proposed and some new insights are shown.

Keywords: ANOVA · Robust Design · Taguchi Method

1 Introduction

It is almost frequent to study a problem, even experimental or designerly, organizing a set of alternatives, in which a certain number of parameters are fixed to different values, on which an answer or performance is evaluated. The Robust Design philosophy, proposed by Taguchi [1] is generally employed, considering its friendliness in performing study and managing results. Robust Design derived from the Design of Experiment (DOE) invented by Ronald Fisher is considered a valid method to investigate the effect of some parameters on its performance [2].

In conjunction to Robust Design, the ANOVA approach, also invented by Fisher, gives the possibility to identify importance and influence that the parameters involved in the experiment have on the performance. It is a great tool that allows designer to investigate about the performance of the device in course of development.

Just to verify the interest of the scientific community and industry to such methods, Scopus was accessed on July 2019 and a query was submitted to search the occurrence

C. Rizzi et al. (Eds.): ADM 2019, LNME, pp. 342–351, 2020.
https://doi.org/10.1007/978-3-030-31154-4_29

to ‹‹ANOVA and "Taguchi Method"›› from 2010 till 2019 in ‹‹Title, Abstract and Keywords››. The report registered 1830 products with a steady growing trend of near 70 papers on 2010, 200 papers on 2015 and 300 papers on 2018. The papers have a wide range of topics and this shows how the interest is really great on these conjoint methodologies. Some examples of this scientific production, two of the most cited and one of the most recent, are reported in [3–5].

As discussed in the vast literature about Robust Design, it is interesting examining in deep the effects of the most important parameters. ANOVA is the method for pruning the right parameters. Starting from the results of the plan of experiment, designers can select the most influencing parameters employing the "basic relation" between of the sum of squares of the experiments and the sum of the squared average effects of each parameter.

The paper presents the application of ANOVA to an easy experiment taken from in literature [6] and reasons about the aggregation of data. The "basic relation" emerges as a natural consequence of the influence of the parameters average values.

Phadke [7] based the "basic relation" on the prerequisite that an additive model is present on the effects of each parameter. Such condition that could guarantee the applicability of the effects superposition principle cannot be assured at the beginning of design of the experiment. The authors believe that such assumption is too heavy, considering that any kind of interactions among parameters should be excluded, when such occurrences can be verified only at the end of the study.

The paper suggests the "basic relation" of ANOVA applied to the Taguchi Method after the application of the ANOVA identity to each design parameter. The assessment of the sum of squares among mean level values and the sum of squares inside the levels can aid to discern the importance of the parameters on the performance. A case study proposed by the authors [8] has been re-analyzed enhancing the relation between ANOVA and Robust Design. Further insights are given on such case.

2 Definition of the Problem

The example re-proposed in the paper is extracted by the book written by Roy [6]. It is related to an experiment in the process of popping corn and it is present in Chap. 7.

Table 1. Experimental results.

Expt. No.	Column Number and factor assigned			Observation $\mathbf{y}_i$
	Power level (A)	Type of oil (B)	Mixing/Stirring (C)	
1	1	1	1	3
2	1	2	2	6
3	2	1	2	4
4	2	2	1	2

Avoiding to report the context, that could be read in the book, here the plan of the experiment is presented of a L4 Orthogonal Array, where three parameters (A – power level, B – type of oil, C – mixing/stirring) vary between two levels each one. Table 1 reproduces the results of the experiment $\mathbf{y}_i$ that represents the amount of unpopped kernel, weighted and recorded in ounces.

2.1 The ANOVA Identity

As reported in literature, the identity associated to an empirical experiment in which the influence of one parameter considered at several n levels is evaluated in relation to the answer of the system. The experiment must be performed m times. The symbols employed are clearly explained: y_{ij} is one answer of the system; $\bar{y}$ is the grand average of the answers; $\bar{y}_i$ is the average of the system answer to the i-th level of the parameter.

$$\sum_{i=1}^{n}\sum_{j=1}^{m}(y_{ij}-\bar{y})^2 = m\sum_{i=1}^{n}(\bar{y}_i-\bar{y})^2 + \sum_{i=1}^{n}\sum_{j=1}^{m}(y_{ij}-\bar{y}_i)^2 \quad (1)$$

In Eq. (1) the left part of the identity is the so called "Total Sum of Squares of the experiment answers", the first term of the right part is the "SS_{BL} Sum of Squares of the difference of answer average values *between* the levels and the grand average" and the second term is the "SS_{IL} Sum of Squares of the difference of answers *inside* the levels and the level average". This relation can be applied to each parameter of the plan of experiment reported in Table 1. It is important to clarify that each parameter is here employed at two levels and that 2 repetitions are present in the plan of experiment. In this experiment $n = 3$ and $m = 2$.

The grand average $\bar{y}$, valid for each parameter A, B and C, is

$$\bar{y} = \frac{1}{n}\sum_{i=1}^{n} y_i = \frac{1}{4}(3+6+4+2) = \frac{15}{4} = 3.75$$

Also, each parameter has the same "Total Sum of Squares of the experiment answers". This is true every time an experiment is organized by means of an orthogonal array, or better when the fractional factorial plan employed is orthogonal and balanced. Every parameter has the following Total Sum of Squares, SS_{T}:

$$\sum_{i=1}^{n}\sum_{j=1}^{m}(y_{ij}-\bar{y})^2 = (3-3.75)^2 + (6-3.75)^2 + (4-3.75)^2 + (2-3.75)^2$$
$$= 8.75$$

2.2 The ANOVA Identity Applied to Each Parameter

In order to consider the effects of each parameter on the system answer it is necessary to evaluate the average effects of each level.

For parameter A, the average effects are signed as $\overline{A_1}$ and $\overline{A_2}$

$$\overline{A_1} = \frac{3+6}{2} = 4.5 \qquad \overline{A_2} = \frac{4+2}{2} = 3$$

and SS_{BL} and SS_{IL} are

$$SS_{BL} = m\sum_{i=1}^{n}(\bar{A}_i - \bar{y})^2 = 2\left[(4.5 - 3.75)^2 + (3 - 3.75)^2\right] = 2.25$$

$$SS_{IL} = \sum_{i=1}^{n}\sum_{j=1}^{m}(A_{ij} - \overline{A_i})^2 = \left[(3 - 4.5)^2 + (6 - 4.5)^2\right] + \left[(2 - 3)^2 + (4 - 3)^2\right]$$
$$= 6.5$$

For parameter B, the average effects are signed as $\overline{B_1}$ and $\overline{B_2}$

$$\overline{B_1} = \frac{3+4}{2} = 3.5 \qquad \overline{B_2} = \frac{2+6}{2} = 4$$

and SS_{BL} and SS_{IL} are

$$SS_{BL} = m\sum_{i=1}^{n}(\bar{B}_i - \bar{y})^2 = 2\left[(3.5 - 3.75)^2 + (4 - 3.75)^2\right] = 0.25$$

$$SS_{IL} = \sum_{i=1}^{n}\sum_{j=1}^{m}(B_{ij} - \bar{B}_i)^2 = \left[(3 - 3.5)^2 + (4 - 3.5)^2\right] + \left[(6 - 4)^2 + (2 - 4)^2\right]$$
$$= 8.5$$

For parameter C, the average effects are signed as $\overline{C_1}$ and $\overline{C_2}$

$$\overline{C_1} = \frac{3+2}{2} = 2.5 \qquad \overline{C_2} = \frac{6+4}{2} = 5$$

and SS_{BL} and SS_{IL} are

$$SS_{BL} = m\sum_{i=1}^{n}(\bar{C}_i - \bar{y})^2 = 2\left[(2.5 - 3.75)^2 + (5 - 3.75)^2\right] = 6.25$$

$$SS_{IL} = \sum_{i=1}^{n}\sum_{j=1}^{m}(C_{ij} - \bar{C}_i)^2 = \left[(3 - 2.5)^2 + (2 - 2.5)^2\right] + \left[(6 - 5)^2 + (4 - 5)^2\right]$$
$$= 2.5$$

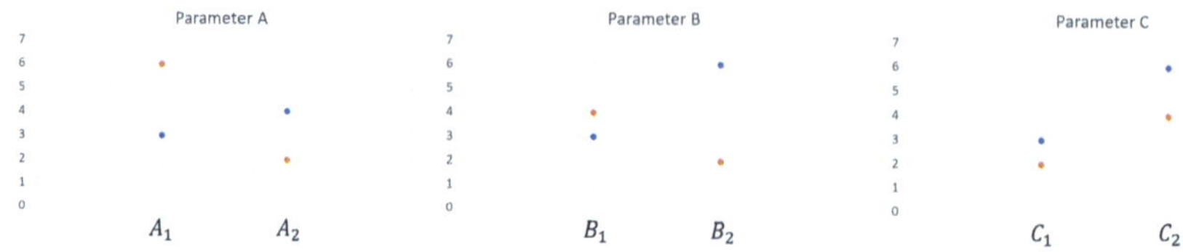

Fig. 1. A graphic representation of the performance reached by each level parameters in the experiment.

Figure 1 gives an immediate visualization of the parameters influence on system performance. Generally, SS_{IL} is considered an estimate of the error or noise present in the experiment, named SS_E.

The ANOVA identity generally appears as

$$SS_T = SS_{Q_i} + SS_E \tag{2}$$

where subscript Q_i refers to each parameter involved in the analysis.

In Fig. 1 it can be seen that parameter B is greatly affected by noise, mainly at level 2 (B_2). Strictly related to this, SS_{BL} has the lowest value associated to the parameter average values (see Fig. 2). The ANOVA suggests how to take into account the effect of the parameters average values and the behavior of the noise.

2.3 The Basic Relation of ANOVA Applied to the Taguchi Method

A great conceptual difference exists when ANOVA is associated to Robust Design, or more in general to a Design of Experiment. Mainly during a design activity, the employment of Robust Design has a big influence on designers to allow them to identify the right "parameter levels". After performing the experiments, pruning parameters the least influence is a prerequisite to focusing the attention on those more promising. The investigation of the parameter effects is based on the average effects of each parameter levels. Referring to Table 1, the study moves from the analysis of the parameters to the average effects. This introduces a change of perspective in the subject named ANOVA, because starting from this point of view it is not important to study the effect *inside* the levels. Furthermore, considering that the "grand average", coincides with the mean value of the average of the levels effects, the problem must be seen in a different point of view. Figure 2 presents a new situation that describes the effects

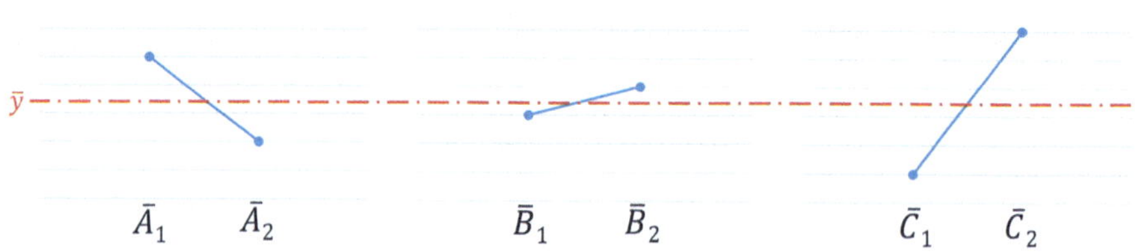

Fig. 2. The average values of each levels in relation to the grand average (red dashed).

of the averages levels on the system performance. In the figure the average values of the levels of the three parameters are reported, that have been computed previously.

The ANOVA Eq. (2), when applied in a Robust Design experiment, changes in a new relation

$$SS_T = SS_{Q_i} \tag{3}$$

Where the term SS_E is no longer considered and subscript Q_i now represents the average values of the levels of the parameters involved in the experiment.

Before starting the computation of Eq. 3 a new table with the average value must be filled (Table 2).

Table 2. Parameters average effects.

Levels	Parameters average effects on performance		
	$(\bar{A})$	$(\bar{B})$	$(\bar{C})$
1	4.5	3.5	2.5
2	3	4	5

The identity (3) can be verified *a posteriori* considering that SS_{Q_i} is obtained as

$$SS_{Q_i} = SS_{\bar{A}} + SS_{\bar{B}} + SS_{\bar{C}} = 2.25 + 0.25 + 6.25 = 8.75$$

And that SS_T now must be seen as associated to the experiments performed

$$SS_T = \sum_{i=1}^{num.\,experiments=4} (y_i - \bar{y})^2 = 8.75 \tag{4}$$

Relating each SS_{Q_i} to the Total Sum of Squares allow to quantify the relative importance of each parameter on the system performance and gives designers the opportunity to compare their effects and decide to prune some design parameter from a finer investigation, in strictly relation to the classical graphics of mean values and S/R ratios, from the Taguchi analysis.

2.4 Discussion on the Such Basic Relations

Following the elaboration and reflections by the authors on the subject Robust Design [8], it remained to investigate some aspects of the ANOVA associated to the Taguchi approach. Even the presentation of such relation, reported in the paper, might be seem too "simple", the authors assure that this has been done only to guarantee an easy tracking of arithmetic. In all context in which it has been applied the arithmetic worked well.

Just to point out some prerequisites that this conjoint application of ANOVA and Taguchi Method must follow, can be bulleted:

- The plan of experiment that must be designed MUST be orthogonal and balanced. This assures that Eq. (3) is satisfied. Any other experiment, designed combining parameters with different levels, or with parameters omitted in some experiments of the plan introduces Sum of Squares of Errors, that can be unavoidable.
- Performing the ANOVA on all single parameters can reveal the influence of some levels chosen for this on system performance. As in the case of parameter B, the level 2 introduces a high discrepancy on system performance. The designer can now decide if such level of the parameter range must be changed or not. Surely the global effect of B is not relevant for the performance.
- It is interesting to note the different meaning in the application of ANOVA identity in a plan of experiment, and the different kind of information that can be gathered if this is applied to a single parameter or on the parameters average levels.

3 Re Elaboration of the Results on a Test Case

In Reference [8] was proposed by the authors the study of an E-byke frame (see Fig. 3) under three different operation conditions. Here the plan of the experiment and the mean values of the objective function are reported, where the factor of safety of the frame has been considered as objective function. The discussion on the relation ANOVA-Robust Design is now given on the basis of the relations of Sect. 2.

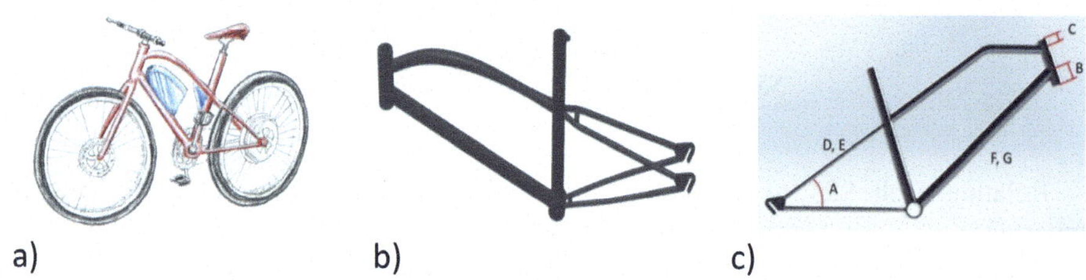

Fig. 3. (a) The first drawing of the E-bike; (b) The geometric model of the frame; (c) The set of design parameters employed during the investigation.

In Table 3 the results of the experiment are reported. The reader is invited to read reference [8] for more detail on the case. The plan of experiments was carried out following an L8 orthogonal array, with the loss function that try to minimize the objective function.

Table 3. Plan of experiment and mean values of the test case.

	A [deg]	B [mm]	C [mm]	D [mm]	E [mm]	F [mm]	G [mm]	y_i
E1	1	1	1	1	1	1	1	1.122
E2	1	1	1	2	2	2	2	2.227
E3	1	2	2	1	1	2	2	1.808
E4	1	2	2	2	2	1	1	1.729
E5	2	1	2	1	2	1	2	0.927
E6	2	1	2	2	1	2	1	1.162
E7	2	2	1	1	2	2	1	0.965
E8	2	2	1	2	1	1	2	0.873

The mean value and the Total Sum of Squares are

$$\bar{y} = 1.352$$

$$SS_T = \sum_{i=1}^{Num.Experiments=8} (y_i - \bar{y})^2 = 1.765 \tag{5}$$

In Table 4 the average values of each parameter are reported and in Fig. 4 the graphs of each parameter gives an easy interpretation of the parameter importance.

Table 4. Parameters average effects on the test case.

Levels	Parameters average effects on performance						
	$(\bar{A})$	$(\bar{B})$	$(\bar{C})$	$(\bar{D})$	$(\bar{E})$	$(\bar{F})$	$(\bar{G})$
1	1.721	1.359	1.297	1.206	1.241	1.163	1.244
2	0.982	1.344	1.406	1.498	1.462	1.540	1.459

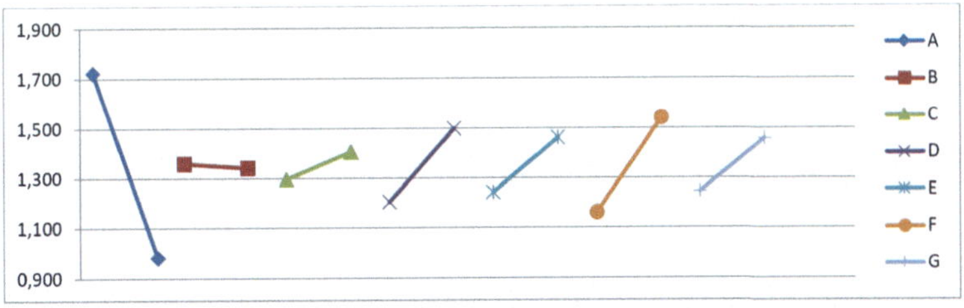

Fig. 4. The average values of each levels.

The nature of the influence of each parameter on the system performance can be analyzed from the comparison of the Sum of Squares "Between Levels" and "Inside Levels".

Table 5. Parameters average effects on the test case.

	SS_{BL}	SS_{IL}	SS_T
A	1.0922	0.6701	1.7623
B	0.0005	1.7639	1.7644
C	0.0238	1.7403	1.7641
D	0.1705	1.5937	1.7642
E	0.0977	1.6670	1.7647
F	0.2843	1.4792	1.7635
G	0.0925	1.6726	1.7651

In Table 5 all the terms of the ANOVA identity are reported, derived from the associated data. For each parameter the identity is demonstrated, with a maximum error of 1/1000.

Really more interesting is the analysis of the Sum of Squares "Inside Levels". This describes how the random results may be a consequence of parameter levels set at values that non reduce the spread of results, conflicting with the main assumption of Robust Design: try to work with value that influence in a weaker sense the system performance.

In Fig. 5 the spread of results is reported. All parameters, except A, have great spread of the results for each level. Parameter A has the lower level of SS_{IL}, even only the value set at A_2 has very close results. The analysis of such graphs can suggest better where address the range of parameter must be moved. Obviously this kind of graphs must be analyzed in conjunction with the behavior of the mean value and the "reduce of noise", as suggested in [8]. From Fig. 5 in fact only parameter A has a great influence

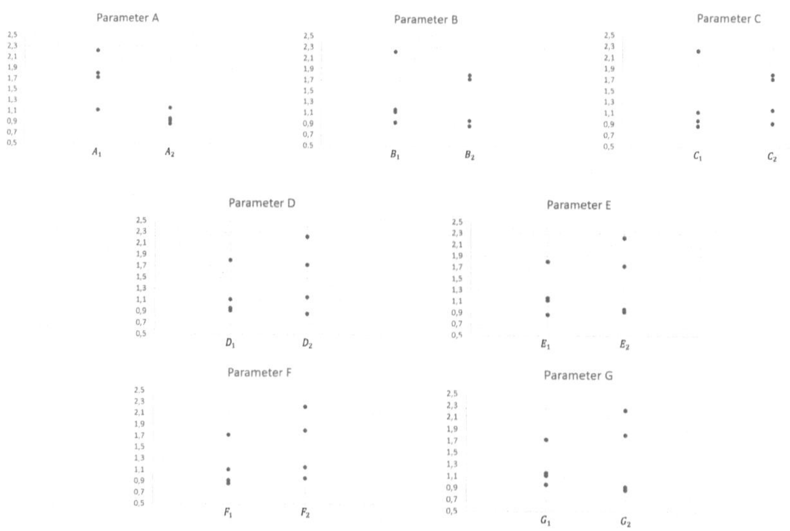

Fig. 5. Representation of the performance reached by each level parameters in the experiment.

on the system performance. Furthermore, level A_2 goes in the direction to minimize the factor of safety of the frame.

From Table 5, comparing the values of SS_{BL} and SS_{IL} designers can be more conscious during selection of the parameters that must be better analyzed with greater detail. A second step of analysis, can be limited only to parameters A, D and F.

The figures of the SS_{BL} column in Table 5 added up amounts to 1.7615, that satisfy Eq. 3, with an error of 2/1000.

4 Conclusion

In conclusion it can be said that Robust Design and ANOVA can validly support the design choices that must be done during product development. A set of information can aid designers to perform the "parameter design" since the early phase of design. Furthermore, it is sufficient to organize the data and elaborate them employing a friendly data sheet, instead to follow the blind answers of a software package.

Acknowledgements. The authors would like to thank Francesco Careri and Francesco Lago for the numerical and graphic elaborations.

References

1. Taguchi, G., Chowdhury, S., Wu, Y.: Taguchi's Quality Engineering Handbook. Wiley, Hoboken (2005)
2. Montgomery, D.C.: Design and Analysis of Experiment, 8th edn. Wiley, New York (2012)
3. Asilturk, I., Akkus, H.: Determining the effect of cutting parameters on surface roughness in hard turning using the Taguchi method. Measur. J. Int. Measur. Confederation **44**(9), 1697–1704 (2011)
4. Altan, M.: Reduction shrinkage in injection molding via the Taguchi, anova and neural network methods. Mater. Des. **31**(1), 599–604 (2010)
5. Suresh, A., Diwakar, G.: Optimization of process parameters in turning operation of austenitic stainless steel rod using Taguchi method and anova. Int. J. Mech. Prod. Eng. Res. Dev. **9**(3), 379–386 (2019)
6. Roy, R.K.: Design of Experiments Using the Taguchi Approach. Wiley, New York (2001)
7. Phadke, M.S.: Quality Engineering using Robust Design. Prentice-Hall International, Englewood Cliffs (1989)
8. Rizzuti, S., De Napoli, L.: An interactive robust design approach in product development. Int. J. Interact. Des. Manuf. (IJIDeM) **12**(2), 677–688 (2018)

Proposal of a Framework Based on Continuous Brainwriting to Expand Mindfulness in Concept Generation

Sergio Rizzuti and Luigi De Napoli(✉)

Department of Mechanical, Energy and Management Engineering (DIMEG), University of Calabria, Ponte Pietro Bucci 46/C, 87036 Rende, CS, Italy
{sergio.rizzuti, luigi.denapoli}@unical.it

Abstract. The first phases of designing an industrial product are those in which creativity has the predominant weight in all the design process. Teaching to be creative is extremely complex and ambiguous, given the elusiveness of the mechanisms that guide it. In this work, the process of generating concepts has been observed during the laboratory of a dyad workgroup of master's degree students in mechanical engineering in a course of Product Design and Development. Starting from a standard session of the 635 method (Brainwriting) it has been observed how the concepts have evolved in a continuous Brainwriting developed in five steps, between feared and almost manifested moments of fixation and exhaustion, towards increasingly mature and conscious solutions, despite the inexperience of the team members. Some aspects have been highlighted on the rotation mechanism of the different Brainwriting sessions and how this has contributed to concepts development; therefore a framework of a three-step Brainwriting session is proposed, which takes inspiration from this experience and those performed of past years.

Keywords: Concept generation · Brainwriting · 635 method · Continuous Brainwriting

1 Introduction

There are many methods to guide the creative process during the early phases of designing an industrial product [1]. The most accredited and established methodologies over the years have evolved and undergone changes and adaptations depending on the domain in which they were applied [2, 3]. Particular attention, also in the literature has been paid to those people who do not have great experience of designing; among these, we can certainly count the students of master degree in mechanical engineering. In [4], the authors observe in deep the concept generation phase in students at different levels of the learning process, during a continuous experiencing between learning blocks and ideation techniques.

Much attention is placed by researchers on how to teach [5] and investigate [6] the design methods and fostering methods of creativity in the early stages of design (in [7] by mean of humorous mood), most of which are based on PBL (Project Based

C. Rizzi et al. (Eds.): ADM 2019, LNME, pp. 352–360, 2020.
https://doi.org/10.1007/978-3-030-31154-4_30

Learning). In [8], the authors underline how PBL is one of the more effective ways for students to learn design by experiencing design as active participants.

On another side, one of the problems in concept generation is how to overcome the problem of fixation [9] and exhaustion [10]. In [9] the authors, studying the work of dyads of students, demonstrate that task alternation facilitates group idea generation, most likely via overcoming fixation. In [10] the authors propose a tool to overcome exhaustion and demonstrate how the new concepts after the first exhaustion present a higher level of novelty.

In the present article, the evolution of concepts generation, performed by students, has been observed during a workgroup for the design of an industrial product.

The concept generation activity has been carefully observed over the years in the course of "Product Design and Development" of the Master Degree in mechanical engineering at the University of Calabria.

We realized the need to support the creative phase for students with particular care.

The Brainwriting method has been constantly proposed in the practice of the 635 method.

The self-governing work of the students, who repeated Brainwriting sessions after the initial one, developed in the classroom, has been observed. The experience of the 2018/19 academic year is reported in detail.

Once the task of the design was established, having heard the voice of the customers, having compiled a list of technical requirements and then, on all this, having clarified the ideas together with the teacher, the group was invited to generate a series of concepts by applying the 635 method (Brainwriting) [11]. This was only the beginning of the process which lasted about 3 weeks and which, in several steps, led to the generation of a consolidated concept. The subject of the design was a leisure bottle.

From this last one and mindful of the past experiences, a framework is proposed in order to structure an operational modality that allows familiarizing with the creativity and the generation of original concepts.

2 Conceptual Design Experience

Over the years, numerous topics have been proposed to the students in really spread areas of interest and with a level of difficulty that is substantially similar, even distributed with different task levels. The organization of the teams included groups ranging from 2 to 5 components and therefore the 635 method was adapted to groups number.

Figure 1 shows some examples of the results of Brainwriting session in past projects. Figure 1a shows the concept of a smart bench with energy harvesting systems; Fig. 1b shows a system to allow people with walking difficulties to access swimming pools; Fig. 1c shows the concept of a domestic composter; Fig. 1d shows an innovative window; Fig. 1e shows a swing for wheelchair users.

The aim of this work is to observe and investigate the concept generation and growth in a workgroup of a team composed of two students, a dyad, their initial endeavor and how they developed their concept.

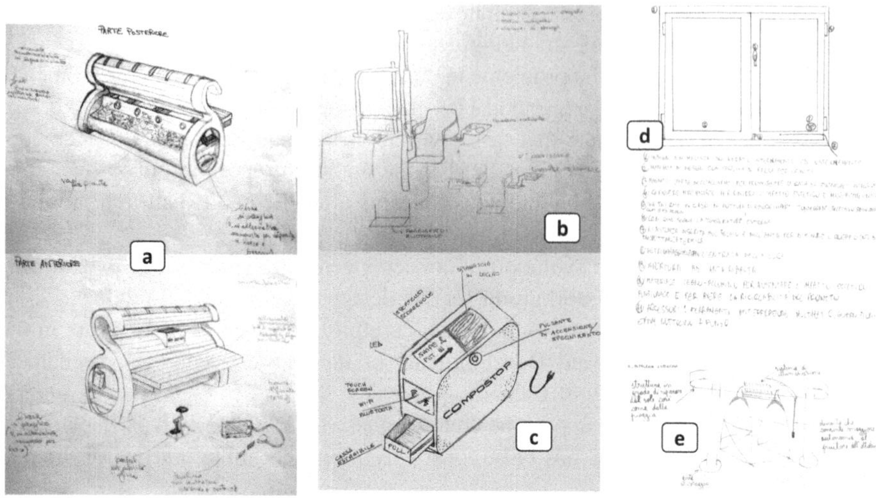

Fig. 1. Concepts generated in Brainwriting session over past academic years.

The subject of the design activity was the development of a new leisure bottle. This type of product is somewhat subtle and insidious since, representing an object of very common use, it presents some design drawbacks. First of all, the designer runs the risk of starting his work already with the solution in mind, so, due to this fixation, it is difficult for him to find important innovations. Just for this reason, starting from a classic Brainwriting session, adapted to the particular team of only two components (dyad), we proceeded to an evolution of the initial concepts in successive steps.

The beginnings of the multistep Brainwriting activity consisted in identifying the functional aspects of the product, with a sort of Brainstorming session led by the teacher. The first session of Brainwriting, which could be called "2-3-2 method" in this case, consists in the generation, by each member of the dyad of concepts in which three particular aspects of the product are highlighted (cap/spout, thermal insulation, and type of connection), those that from the analysis of the questionnaires seemed more important for the interviewees. Since the team consists of only two people, it was decided to perform two rotations, to have a better definition in the generation of the concept. Furthermore, the teacher motivated the team members before starting to advise them to leave their imagination as free as possible regarding the shape of the product and the inclusion of all the options resulting from the survey carried out with prospect customers.

During the first step of the methodology, and also during the following ones, the concepts were generated individually, on opposite sides of the classroom, to not see each other's work and be influenced by it. The first rotation lasted 20 min, then the sheets were exchanged and in 10 min each component understood the concept expressed by the colleague and made the most appropriate upgrades. Subsequently, on the second rotation, the same procedure was repeated. At the end of the step, together with the teacher, the concepts generated were evaluated, explaining the reasons for the changes made to the work of the colleague, and trying to get suggestions for new ideas, to generate further concepts in the subsequent steps. Furthermore, with the teacher and in the same session, existing concepts have been investigated through research on patented products, and on the technologies used, to broaden the knowledge especially towards the needs and technologies. Figure 2 shows the result of this first step.

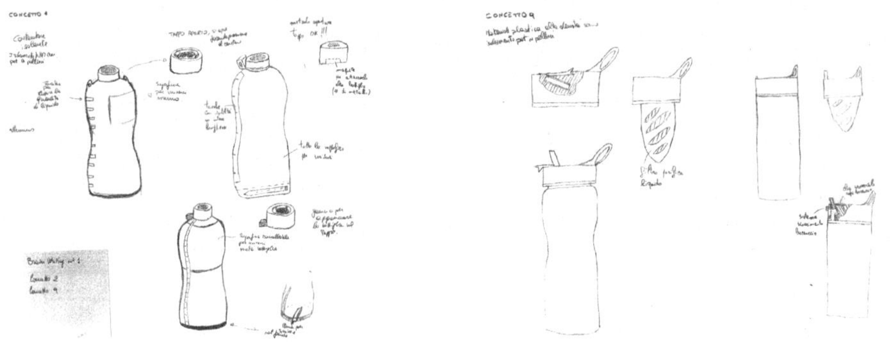

Fig. 2. Concepts generated during the first step.

In order to free the mind from fixation and to prevent the exhaustion, before the second step of Brainwriting each of the members of the dyad has reflected independently for 2–3 days. Given the diffusion of the typology of the product in question, everyone looked around and also took inspiration from simple objects not directly related to the case study, whose functionalities could give inspiration to the evolution of the generated concepts, even to the generation of completely new concepts.

The second step, as well as the following ones, were carried out by the dyad independently, that is not in the presence of the teacher. Since the concepts generated in the first step seemed too similar to existing products, the dyad tried to completely change its appearance concerning the concepts generated in the first application. Here is the reflection of the dyad after the second step: "At the end of the application we were quite satisfied with the work done, but the problems were just around the corner!". Figure 3 shows the result of this second step.

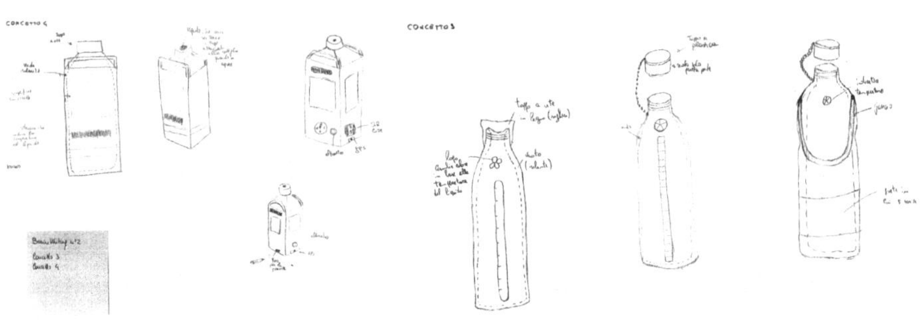

Fig. 3. Concepts generated during the second step.

Between the second and third steps, a drop in creativity was detected, both for fixation and exhaustion: ideas, when they came, were trivial. A period of suspension was therefore tried between the second and the third step of Brainwriting, which lasted about a week, not thinking at all about the case study. This gave way to grasp different aspects, which otherwise would never have been noticed if the dyad had been continually projected to the problem and its solution. After a week, the third step was taken which was the trigger point of the work: the concepts generated were already more particular, different from the usual bottles. From that moment on, new ideas were the order of the day, even a simple pencil or coffee maker suggested new solutions. Many were not applicable solutions, and they were discarded a priori, but this nevertheless showed that there was the right inspiration to be able to create something inventive, applicable and maybe winning.

Figure 4 shows the result of the third step.

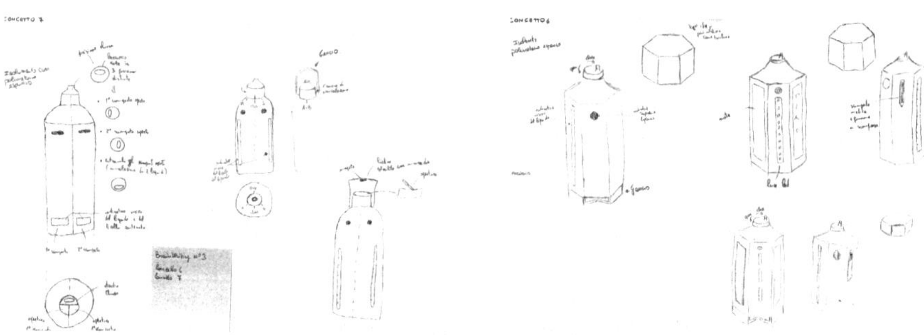

Fig. 4. Concepts generated during the third step.

Once the trigger point was passed with the previous step, the last two steps, completed a few days apart, just the time to rearrange some ideas, has been fluent and generative. Figure 5 shows the results of the fourth and fifth steps.

After the last step the new ideas did not seem interesting anymore, and for fear of falling back into the fixation, the process of generating concepts was stopped. The ten concepts generated in this way, after a light make-up followed by a summary Brainstorming session with the teacher, then moved on to the screening and scoring phases to select the best one according to some criteria and according to consolidated techniques. The winning concept, as expected and hoped for, was one of those generated in the last session.

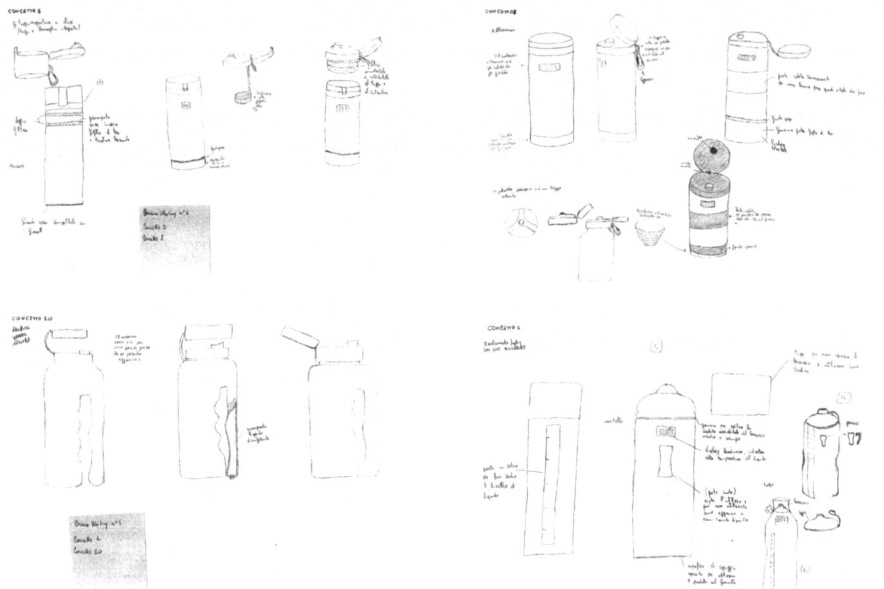

Fig. 5. Concepts generated during the fourth and fifth steps.

3 Proposal of a Framework to Support Creativity

In literature, much is said about the mechanisms that guide creative processes. Some have treated creative processes by proposing biological bases according to which creative and innovative ideas are developed [12]. For more than a century many others have suggested explanations arising from the observation of creative processes and among the first were the reflections of Poincaré in the early twentieth century [13] and Wallas, with his pioneering model of the four stages of creativity [14].

For some time, studies have been carried out that highlight the time variable in creative processes. In [15] Paulus and Brown, among the schemes of enhancing ideational creativity in groups, introduce, for brainstorming, a variant with breaks and Brainwriting sessions. In [16], Gilhooly deals with incubation in-depth for creative problem solving, studying in particular either delayed and immediate incubation effects.

Following the observations made on the workgroup projects of the students and comforted by what has been found in the literature, a framework that summarizes what naturally emerged from field experience is proposed.

After assigning the task, having defined the main function that the device to be designed must achieve, a series of customer and technical requirements are established in the classroom.

At this point, the first Brainwriting session can begin, carried out with the 635 method in the classroom. The final meeting of the group at the end of the first Brainwriting session begins to instill the first doubts and to arouse the first thoughts that everyone will freely begin to elaborate independently once they leave the classroom.

In the following days, the teacher stimulates the class by proposing some observations and suggesting stimuli from external research, for example from the market and the patent databases.

After about a week from the first Brainwriting session, each team meets in a second session, again in the same way as the first, without the presence of the teacher or an instructor.

At this point, a new break is made, during which every component with much less continuity pursue to think about the theme, occasionally finding stimuli from the outside, possibly involuntary.

After a week or two, the third and final Brainwriting session is held, which will conclude with the latest generation of concepts, in which there is always more mindfulness of the growing product and its functions, towards the exhaustion of new ideas.

Finally, all the concepts generated in the three sessions will be screened to decide which should then be developed in detail.

In Fig. 6 the proposed method is summarized schematically.

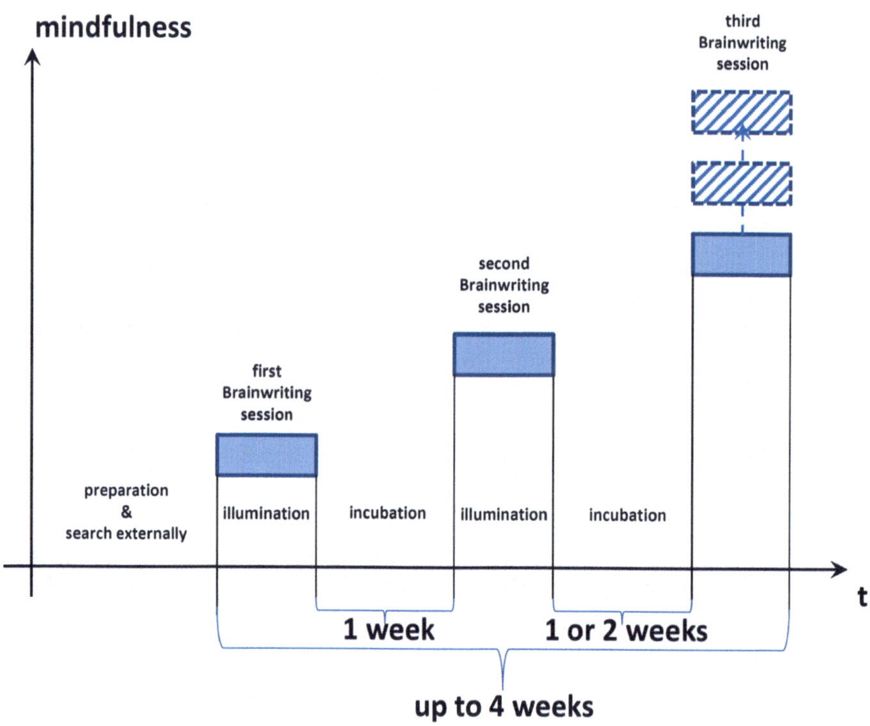

Fig. 6. Framework for three-step concept generation based on 635 method.

The proposed scheme will be experienced in the next academic year in the courses of "Product Design and Development" of the master degree in mechanical engineering and the course of "Sustainable Product Design" of the master degree in management engineering.

4 Discussion and Conclusions

Observing the activity of the dyad of students of master degree in mechanical engineering on which has been focused this work, some facts should be pointed out.

Given the fear of incurring fixation and exhaustion, this conceptual design activity has been developed in five successive steps in a sort of continuous Brainwriting. The 635 method has been chosen because over the years, in our "Product Design and Development" courses, it has been shown to be the one that best allowed to highlight, in group projects, the most creative ideas, giving the possibility, for each element of the group, to be able to express themselves freely. Dealing with teamwork with a group of minimal consistency could have led to the risk of creative stagnation. The risk has been overcome thanks to the use of a particular slowness (something similar to the slow thinking of the Nobel prize Daniel Kahneman?) and the use of right pauses, which has allowed both the sedimentation of some robust ideas about the concepts, and the awareness and consciousness that the innovation could nest in the most unexpected aspects, and this has triggered numerous "aha moments" [17].

Such a mechanism activates, for example, when a musician tries and tries a melody for a day, stops for a few days, also taken by dejection because it does not improve the performance and the melody does not come out, then, after the pause, regains vigor almost by magic (still aha moments?), with the hands, for example, of the pianist who play automatically, without even the score.

Quoting Søren Kierkegaard: "Far from idleness being the root of all evil, it is rather the only true good."

Acknowledgements. The authors would like to thank Francesco Careri and Francesco Lago for the contribution to the survey.

References

1. Otto, K.N., Wood, K.L.: Product Design: Techniques in Reverse Engineering and New Product Development. Prentice-Hall, Upper Saddle River (2001)
2. Conradie, P.D., Nafzger, R., Vanneste, C., De Marez, L., Saldien, J.: Methods for ideation: reviewing early phase concept generation among industrial design engineer students. In International Conference On Engineering And Product Design Education (EPDE 2015). Design Society, Loughboroug (2015)
3. Jensen, D.J., Weaver, J., Wood, K.L., Linsey, J., Wood, J.: Techniques to enhance concept generation and develop creativity. In: ASEE Annual Conference, AC 2009-2369, Austin, Texas, USA, pp. 1–25 (2009)
4. Lee, J.W., Daly, S.R., Vadakumcherry, V.: Exploring students' product design concept generation and development practices. In: Proceedings of the 2018 ASEE Annual Conference & Exposition, Salt Lake City, UT, USA (2018). https://peer.asee.org/30494
5. Gero, J.G., Jiang, H., Williams, C.B.: Does using different concept generation techniques change the design cognition of design students? In: The ASME 2012 International Design Engineering Technical Conferences & Computers and Information in Engineering Conference, Chicago, Illinois, USA (2012)

6. Linsey, J.S., Clauss, E.F., Kurtoglu, T., Murphy, J.T., Wood, K.L., Markman, A.B.: An experimental study of group idea generation techniques: understanding the roles of idea representation and viewing methods. J. Mech. Des. **133**(3), 031008-031008-15 (2011)
7. De Napoli, L., Rizzuti, S., Raco, A.: How mood fosters creativity in product design? Experimental evidences on humour impact during a conceptual design session in a master degree class, In 20th International Conference on Engineering and Product Design Education, E&PDE 2018, Dyson School of Engineering, Imperial College London, United Kingdom (2018)
8. Dym, C.L., Agogino, A.M., Eris, O., Frey, D.D., Leifer, L.J.: Engineering design thinking, teaching, and learning. J. Eng. Educ. **94**(1), 103–120 (2005)
9. Sio, U.N., Kotovsky, K., Cagan, J.: The facilitating role of task alternation on group idea generation. J. Appl. Res. Mem. Cogn. **6**(4), 486–495 (2017)
10. Gray, C.M., McKilligan, S., Daly, S.R., Seifert, C.M., Gonzalez, R.: Using creative exhaustion to foster idea generation. Int. J. Technol. Des. Educ. **29**(1), 1–19 (2017)
11. Rohrbach, B.: Kreativ Nach Regeln - Methode 635, Eine Neue Technik Zum Lösen Von Problemen. Absatzwirtschaft **12**(19), 73–76 (1969)
12. Martindale, C.: Biological bases of creativity. In: Sternberg, R. (ed.) Handbook of Creativity, pp. 137–152. Cambridge University Press, New York (1999)
13. Poincaré, H.: Mathematical creation. The Monist **20**(3), 321–335 (1910)
14. Wallas, G.: The Art of Thought. J. Cape, London (1926)
15. Paulus, P.B., Brown, V.: Enhancing ideational creativity in groups: lessons from research on brainstorming. In: Paulus, P.B., Nijstad, B.A. (eds.) Group Creativity, pp. 110–136. Oxford University Press, New York (2003)
16. Gilhooly, K.J.: Incubation and intuition in creative problem solving. Front. Psychol. **7**(1076) (2016)
17. Goleman, D.: New insights on the creative brain. https://www.psychologytoday.com/us/blog/the-brain-and-emotional-intelligence/201108/new-insights-the-creative-brain. Accessed 04 Aug 2019

Morphological and Mechanical Characterization of P-Scaffolds with Different Porosity

Marta De Giorgi(✉), Nunzia Gallo, Marta Madaghiele, and Anna Eva Morabito

University of Salento, Lecce 73100, Italy
marta.degiorgi@unisalento.it

Abstract. The aim of this paper is to model and to compare the results of the mechanical characterization, carried out on numerical models and real specimens, of uniform P-scaffolds with different porosity values. The analysis includes the morphological characterization of 3D printed specimens and the implementation of a FEM shell model to reproduce a compressive test suitable for mechanical properties evaluation of PLA scaffolds. Young modulus and yield strength were also obtained, in order to verify the numerical model accuracy, by experimental tests on 3D printed PLA scaffolds. Numerical results showed that the shell model was able to reproduce, more efficiently compared to a solid model proposed in a previous work, both elastic and plastic behavior of the scaffolds, providing elastic modulus values very close to the experimental ones. On the other hand, the not very high quality of the 3D printing, detected by MicroCT analysis, caused a significant dispersion in the yield strength numerical values respect to the real data. Anyway, an inverse correlation between mechanical properties and porosity was found as expected. The elastic modulus values were similar to the typical values of the trabecular bone for whose regeneration this kind of scaffolds is usually employed.

Keywords: TPMS scaffold · MicroCT · FEM analysis · Experimental tests

1 Introduction

In recent years, tissue engineering has become a topic of great interest, emerging as suitable approach to replace and regenerate damaged and/or diseased tissue. In the case of tissues designed for making bone substitutes with predominantly biomechanical functions, however, several open issues remain. The load-bearing scaffolds do not often promote a successful tissue ingrowth or the complete scaffold integration in the host tissue. Sometimes, the mechanical properties of the scaffold are not similar to those of the surrounding bone and these mismatches can affect the operation or longevity of the implant, because of the effect known as stress shielding [1].

In the design of load-bearing structures for the bone tissue regeneration, the aspects to be considered are numerous and concern the shape of the bone defect and the performances required to the scaffold in terms of mechanical, fluid-dynamic, and cellular proliferation properties. Thanks to the Additive Manufacturing (AM), it is now

C. Rizzi et al. (Eds.): ADM 2019, LNME, pp. 361–372, 2020.
https://doi.org/10.1007/978-3-030-31154-4_31

possible to manufacture porous bone substitutes with complex geometry able to reproduce perfectly bone defect shapes [2]. In order to promote the cellular proliferation and transport of nutrients, an appropriate porous structure with interconnected porosity has to be designed. Given the pivotal role that nutrients flow plays in promoting cellular proliferation and differentiation, the scaffold surface must be characterized by high values of permeability that is a parameter related to geometrical characteristics such as pore shape and size. High-values of surface to volume ratio are also beneficial in order to allow large number of cells to attach and migrate into porous scaffolds [3]. Furthermore, it was shown that curved surfaces can promote cell proliferation more than lattice-based architectures and, recently, a correlation between support curvature and tissue regeneration was found in [4], which points out that concave surfaces are preferable to convex and flat surfaces.

Several are the methodologies available to model these structures; some examples are: CAD-based methods [5, 6], Image-based methods (MRI/CT) [7] and topology optimization [8]. Recently, methods for the optimization of scaffolds microstructure geometry, based on mechano-biological criteria, have been also proposed [9].

Triply periodic minimal surfaces (TPMS) are very attractive candidates for the design of biomorphic porous scaffolds. This interest is justified by the several natural manifestations that these surfaces show, such as biological structures, block copolymers and equipotential surfaces in crystals [10]. Moreover, most TPMS forms exist as an interface between two phases [11]. From a mathematical point of view, TPMS are surfaces periodic along three directions mutually perpendicular between each other and with zero mean curvature at each point. Different are the surfaces attributable to this class that are often considered for scaffold design [12]. For the TPMS modelling, the approach based on a single implicit function (called implicit surface modelling, ISM) is very suitable [13].

Scaffolds, based on these surfaces, have shown to provide high values of permeability [14] and their curved configuration can promote cellular proliferation more than scaffold architectures based on lattice cells [15]. While, on the one hand, high values of porosity and large pore sizes favor nutrients flow and cell ingrowth, on the other they can limit the load bearing capability of the scaffold. Therefore, when designing these structures, a compromise has to be found between biological and mechanical requirements, which means that the ideal material for scaffold fabrication should satisfy specific criteria. It must be biocompatible and capable to reproduce the network architecture of the natural bone in order to create a microenvironment stimulating cell attachment. The design of temporary scaffolds, as polymeric scaffolds, moreover, must ensure the compatibility of degradation products, the rate of the degradation process and how each of these affects the regenerative process. The degradation phenomenon affects the mechanical properties of the scaffold. On the other hand, the use of degradable polymers is required if the surgical removal must be avoided. A slow degradation process allows maintaining the scaffold mechanical properties up to the regenerative process is in an advanced stage. On the basis of these premises, the most usually used degradable synthetic polymer is the FDA-approved poly(lactic acid) (PLA) because the rate of degradation can be tuned by composition obtaining time of degradation from months to several years [16–18]. Due to the availability, biocompatibility, and processibility, PLA is used for 3D culture and transplantation of articular, auricular, nasal, costochondral, tracheal, and intervertebral disk cartilage cells [19]. PLA is also recommended for bone fixing after a cancellous bone fracture, osteotomy, or arthrodesis/joint fusion. PLA is also a material suitable for 3D printing that allows the fabrication of complex, multiscale structures through computer-aided design (CAD) [20–24].

In the adult human skeleton, the cortical bone represents the 80% of the skeletal mass and performs the supportive and protective function. The cortical bone constitutes the external wall of all bones. It shows a solid mass with microscopic channels (Fig. 1). The internal portion of bones, i.e. the remaining 20% of the bone mass, is formed by trabecula and it is called cancellous bone (Fig. 1). Large difference exists in the cortical and cancellous distribution depending on bone type. The ulna, for example, consists of 92% cortical and 8% cancellous bone, while a vertebra presents a 62% cortical and 38% cancellous bone [25].

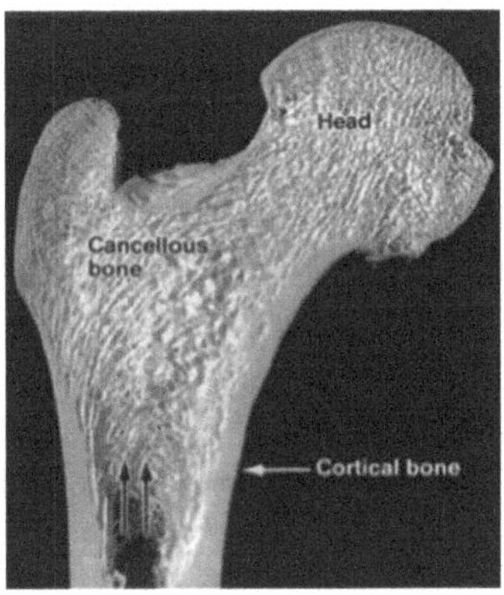

Fig. 1. Example of bone composition: cortical and cancellous region [25].

The more suitable test to determine the mechanical properties of a bone mainly depends on the kind of bone being tested (cortical or trabecular bone), the age of the bone and its location in the skeleton. In the case of cortical bones, the bending test should be performed, while in the case of the trabecular bone, the compression test is suggested. The compression test is usually performed to evaluate the biomechanical properties of the trabecular bone, which is present in large quantities in the vertebrae [26]. In the vertebrae, in fact, compressive tests reproduce the in vivo loading conditions to which the bone is subjected. The structural properties depend on the cancellous bone density and trabecular orientation. As reported in literature [19, 26], the Young's modulus in the cancellous bone can vary from 0.1 to 4.5 GPa.

In this paper, PLA scaffolds having different porosity were characterized, in terms of compressive mechanical properties, from a numerical and experimental point of view. The surface model of a cubic scaffold, generated by an original MATLAB routine, was used as input in order to implement the FEM model as shell model. This solution allowed overcoming the significant difficulties encountered in a previous work [27] where a solid FEM model was considered. Several problems associated to the data exchange between different s/w were avoided and laborious manual operations for mesh correction were eliminated using a shell model. By adopting this modelling strategy, several scaffold models having different morphological properties can be evaluated fast and efficiently. The above-mentioned geometric model was also used for 3D printing of PLA specimens to be subjected to the compressive test. Moreover, a deep morphological characterization was performed through a micro-tomographic analysis on the printed specimen, in order to evaluate, by the comparison of a set of specific cell properties, its deviations from the nominal geometry.

2 Materials and Methods

2.1 P-Scaffold Modelling

Scaffolds for bone tissue regeneration can be obtained by replicating a single TPMS cell along the three orthogonal directions or by combining TPMS cells with different geometrical configuration in order to obtain graded porosity scaffolds. Among the various TPMS, the Schwarz's Primitive minimal surface (hereinafter referred as P-surface) has been considered in this study. The P-surface can be described, to the first order of approximation, by the following nodal equation:

$$f(x, y, z, s, k) = \cos(x/s) + \cos(y/s) + \cos(z/s) - k \tag{1}$$

under the boundary conditions $x = [-\pi, \pi]$, $y = [-\pi, \pi]$ and $z = [-\pi, \pi]$. A detailed description of the P-cell can be found in a previous work [28].

In this paper, the implicit function for a P-surface with $s = 1$ and $k = 0$ was considered to model, by an original MATLAB routine, a cubic scaffold with a number N of cells along the side equal to 7. The unit cell presented a length of side L equal to 2π mm. This model is given in the form of a tessellated model described by a .stl file and presents 117992 faces and 59976 vertices (Fig. 2).

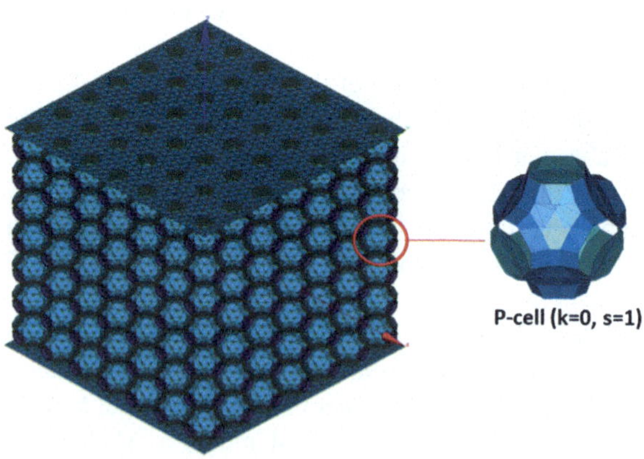

Fig. 2. Tessellated cubic scaffold and detail of the unit cell with $L = 2\pi$ and $N = 7$.

2.2 Experimental Characterization of P-Scaffold

In this work, experimental data used for comparison with numerical models were obtained through compressive tests. The uniaxial compression test is the easiest and most efficacious way to determine the mechanical behaviour of porous structures under static conditions and it is the recommended test for cancellous bone characterization [26, 29, 30]. Therefore, the stress-strain compressive curve is assumed as the term of comparison to evaluate the numerical model reliability.

Nine specimens having three different porosity values of the cubic scaffold were manufactured by the 3D printer Ultimaker 2.0. In particular, scaffolds having nominal porosity equal to 88, 82 and 76% were modeled and realized assigning a thickness t of 0.4, 0.6 and 0.8 mm to the cell wall respectively. The nominal values of porosity were calculated on the solid cell obtained in Meshlab through an offset operation on the tessellated surface of Fig. 2. The PLA filament with 2.85 mm diameter was extruded at an optimal temperature of 210 °C using 0.4, 0.6 and 0.8 mm nozzle for 0.4, 0.6 and 0.8 mm cell wall thickness respectively. The print speed and the layer thickness were set to 40 mm/s and 0.1 mm respectively using a flow equal to 100%.

The 3D structure of the printed scaffolds was assessed by microcomputed tomography (Bruker Skyscan 1172, Kontich, Belgium). Three samples, one for each porosity value, were scanned at a nominal resolution of 26.8 μm, with a camera pixel binning of 4 × 4. The X-ray source was set at 50 kV and 200 μA, and no filter was used. The scan orbit was 180°, with a rotation step of 0.3°. Camera offset was applied to scan the entire cross-section of the samples having 5 cells for side (31.42 × 31.42 mm), while the scanned height was about 16 mm. After reconstruction with NRecon software, DataViewer software was used to visualize the sections of the scaffolds in the XY, XZ and YZ planes. The reconstructed grey scale images of the XY sections were then analyzed with CTAn software to quantify the porosity and the average thickness of the pore walls, after selection of the volume of interest (VOI) and appropriate thresholding. Two VOIs were analyzed to compare the structure of the printed scaffolds with that of the original .stl models, respectively a unit cube (6.28 × 6.28 × 6.28 mm) corresponding to the single P-cell (taken at the center of the scanned sample) and a square prism (31.5 × 31.5 × 16 mm) enclosing the overall cross-section of the scaffolds.

Compression tests were carried out on a servo-hydraulic MTS 810 machine having 100 kN load capacity equipped with two planar plates (Fig. 3), according to the ASTM D 1621-00 Standard [31]. The crosshead speed was fixed at 4 mm/min. Strain was calculated dividing the cross-head displacement by the cubic specimen side approximately equal to 44 mm. Three specimens for each cell wall thickness t were tested. Moreover, in order to characterize the mechanical behavior of PLA base material, three bulk specimens 20 × 20 × 20 mm were tested.

2.3 FEM Analysis

Numerical analysis was performed using the open source FEM software Code-Aster R13.6. A discretization in triangular shell elements, named DKT and having a linear formulation, was in particular adopted. The models consisted of 75308 elements and 36668 nodes and they differed for the value of the cell wall thickness t (0.4, 0.6 and 0.8 mm). The study was performed in the large displacements range. An isotropic elastic-perfectly plastic stress-strain behaviour of the material was, in particular, hypothesized. The material properties, required by the numerical analysis, were obtained as the average of the mechanical properties resulting from the compressive tests carried out on the bulk specimens. Boundary conditions were applied on the lower face, imposing a zero value to z-direction displacement and rigid body boundaries; the load was applied to the upper surface as a 10 MPa pressure load distributed in 100 substeps to overcome convergence troubles. Figure 2 shows the mesh of the FEM model.

Fig. 3. Example of specimen under compression.

3 Results and Discussions

3.1 MicroComputed Tomography (MicroCT)

As shown in Fig. 4, MicroCT scanning of printed samples confirmed the highly interconnected porosity of the scaffolds and the different thickness of their pore walls, the latter in rough agreement with the values set for the printing. Quantitative 3D analysis evidenced an overall porosity of 87.4%, 84.2% and 79.9% for PLA04, PLA06 and PLA08 samples, respectively, which is consistent with the increasing pore wall thickness and with the values of porosity evaluated from the tessellated nominal models obtained applying the thickness t to the scaffold surface of Fig. 2.

Porosity was mostly open, as expected, due to the high interconnectivity of the P-cells. However, a certain amount of closed porosity was also detected, which was found to increase when the pore wall thickness was increased (0.013%, 0.108% and 0.331% of closed porosity for PLA04, PLA06 and PLA08, respectively). Closed porosity indicated the presence of a certain number of voids (air bubbles) in the cell

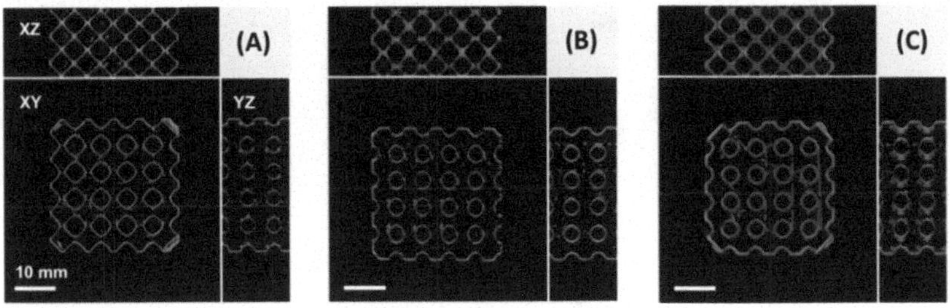

Fig. 4. Representative cross-sections of the PLA scaffolds, obtained from MicroCT analysis: (A) PLA04, (B) PLA06 and (C) PLA08.

walls. Furthermore, the average structure thickness of the scaffolds was found to be lower than the theoretical set values (0.4, 0.6 and 0.8 mm), and in particular was found the values in Table 1. Overall, these observations suggested that the printing accuracy was greatly reduced when increasing the theoretical thickness of the deposited material. This was also confirmed by the presence of a large amount of undesired material within the porous cells, for PLA06 and PLA08 samples, in the form of filaments that the

Table 1. Characteristics of the porous specimens and experimental mechanical data (nom stands for nominal and act for actual).

Specimen	t_{nom} [mm]	t_{act} [mm]	P_{nom} [%]	P_{act} [%]	E_{exp} [MPa]	σ_{y_exp} [MPa]	E_{num} [MPa]	σ_{y_num} [MPa]	$Error_E$ [%]	$Error_{\sigma y}$ [%]
P1_0.4		0.35		87	34	0.9				
P2_0.4	0.4	± 0.125	88		38	1.0	39	2.7	3	170
P3_0.4					43	1.0				
P1_0.6		0.45		84	75	2.2				
P2_0.6	0.6	± 0.167	82		80	2.3	69	4.0	13	82
P3_0.6					83	2.2				
P1_0.8		0.50		80	131	3.9				
P2_0.8	0.8	± 0.229	76		108	2.8	108	5.8	16	61
P3_0.8					147	4.1				

nozzle leaves during its translation (Fig. 4). When analyzing the cubic VOI of a single P-cell and visualizing the corresponding .stl model, unexpected holes were also found through the cell wall, and these irregularities were more evident for PLA06 and PLA08 samples with respect to PLA04 one (Fig. 5).

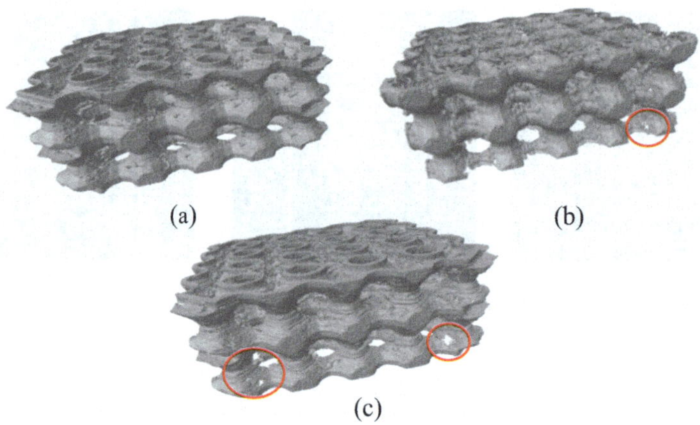

Fig. 5. 3D visualization from μ-CT analysis for 0.4 (a), 0.6 (b) and 0.8 (c) mm cell wall thickness.

3.2 Numerical and Experimental Mechanical Characterization

Three preliminary compression tests were performed on bulk specimens in order to evaluate the mechanical properties of PLA printed material. Compression curve is reported as example in Fig. 6a. The resulting Young modulus and yield strength, used as inputs in the FEM analysis, were 1280 MPA and 67 MPa respectively. Stress-strain curves, deduced from the experimental tests on P-scaffolds, were similar to those reported in literature for bone (Fig. 6b) [26, 32]. The yield point is an imaginary boundary, above which stresses cause permanent damage (plastic deformation) to the bone structure. The last phase, represented by the rapid load increase, is due to the specimen densification.

The remaining compressive tests were carried out on three scaffolds for each cell wall thickness. Table 1 reports the characteristics of the specimens and the resulting mechanical data. These values were used to evaluate the reliability of the FEM models, whose results, in term of displacement, are shown in Fig. 7 for the model with the highest porosity and at a load corresponding to 1, 2 and 3 MPa. Note that 3 MPa represented, in this case, a stress level slightly higher than yield strength estimated for the numerical scaffold.

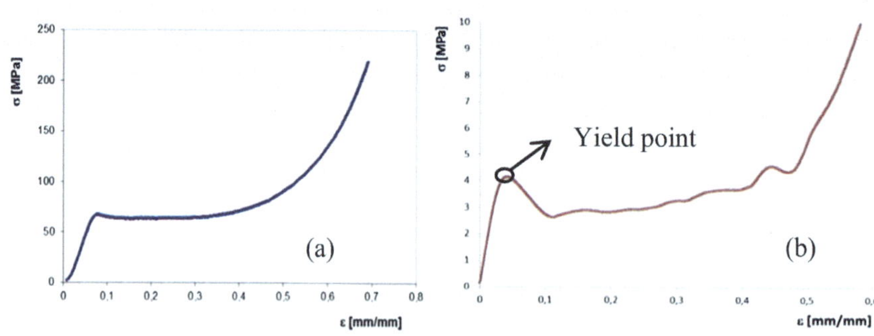

Fig. 6. Compression curves of the bulk specimen (a) and 3D printed scaffold (b).

Fig. 7. Maps of displacements at the load of 1, 2 and 3 MPa for the most porous P-scaffold.

The ratio of upper surface displacement to cubic model side allowed calculating the global strain of the scaffold. The stress-strain curves are reported in Fig. 8 in comparison with experimental curves.

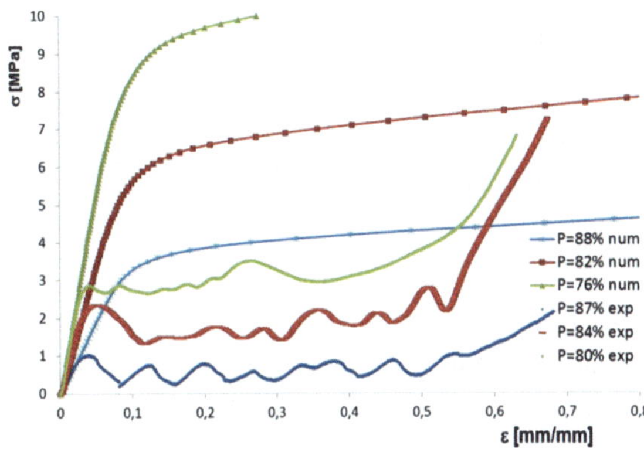

Fig. 8. Experimental and numerical stress-strain curves for different porosity values.

Figure 9 shows the detail of the linear portion of the experimental and numerical compression curves.

The first evident result is that the numerical model is able to predict the elastic behavior of the scaffold with high accuracy. The error in the Young modulus prediction is significantly lower than 20% as reported in Table 1. As regards to the plastic behavior, the model is also adequate to reproduce the plastic phase but with a higher error in the yield strength prediction. This is due to the microplasticizations that actually occur at the interface between two cells and to the buckling of the cell walls. In fact, it can be seen that the error increases for lower cell wall thickness that is responsible of higher plasticization and lower buckling strength. These phenomena are not considered in the present analysis. Moreover, the real foam structure obtained from the 3D printing is not defects free, as demonstrated from the MicroCT analysis and visible in Fig. 5. This causes lower strength at loads over yield point. A significant improvement on numerical model can be achieved also considering for the base material a damage model suitable for porous structures. In particular, the Gurson-Tvergaard-Needleman (GTN) damage law is able to explain the local strength decrease [33, 34]. The GTN damage model has been implemented in the last version of the Code-Aster FEM software and it will be taken in account in the future numerical analysis.

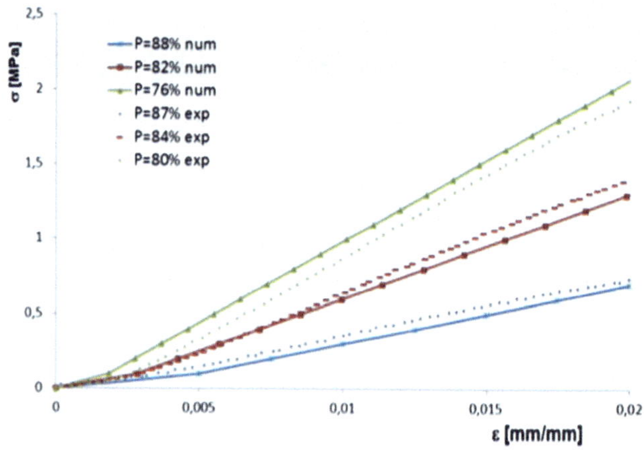

Fig. 9. Linear portion of compression curves: comparison between numerical and experimental results for different porosity values.

4 Conclusions

The shell model of the P-scaffold implemented here allowed overcoming some significant difficulties encountered during the data exchange pertaining to the solid model proposed in a previous work [27]. Several laborious manual operations for mesh correction were eliminated. A morphological and mechanical characterization on uniform P-scaffolds with different porosity has been performed. MicroCT analysis revealed a real structure of the scaffold coherent with the geometrical model but not defect free. Closed micro-porosities and holes through the cell walls were the principal cause of the different compressive behaviour in the plastic region between FEM models and real specimens. The micro-plasticizations that actually occur at the interface between two cells and the buckling of the cell walls were amplified because of the defects in the cell walls, particularly in the scaffold with lowest cell wall thickness. On the other hand, the shell FEM model simulated the elastic behaviour of this complex structure very well providing elastic modulus values comparable with those of cancellous bones. Moreover, an inverse correlation between mechanical properties and porosity was found as expected. A significant improvement of the numerical model can be achieved using as input the real geometric model obtained by MicroCT and considering a damage model suitable for porous structures, as the Gurson-Tvergaard-Needleman (GTN) damage law. Results obtained on the shell model will be compared with data resulting from the solid model implemented in [27], assigning to this last one the PLA mechanical properties. Moreover, graded porosity scaffolds, i.e. structures where pore architecture varies throughout the scaffold, will be analyzed too. These scaffolds are of particular interest, since they are more similar to the stochastic configuration of real bone [35].

References

1. Gao, W., Zhang, Y., Ramanujana, D., Ramani, K., Chen, Y., Williams, C.B., Wang, C.C., Shin, Y.C., Zhang, S., Zavattieri, P.D.: The status, challenges, and future of additive manufacturing in Engineering. Comput.-Aided Des. **69**, 65–89 (2015)
2. Do, A.V., Smith, R., Acri, T.M., Geary, S.M., Salem, A.K.: 9 − 3D printing technologies for 3D scaffold engineering. Funct. 3D Tissue Eng. Scaffolds, 203–234(2018)
3. Lanza, R., Langer, R., Vacanti, J.P.: Principles of Tissue Engineering, 1344 p. Academic Press, Cambridge (2011)
4. Park, J.Y., Lee, D.H., Lee, E.J., Lee, S.-H.: Study of cellular behaviors on concave and convex microstructures fabricated from elastic PDMS membranes. Lab Chip **9**(14), 2043–2049 (2009)
5. Chantarapanich, N., Puttawibul, P., Sucharitpwatskul, S., Jeamwatthanachai, P., Inglam, S., Sitthiseripratip, K.: Scaffold Library for Tissue Engineering: A Geometric Evaluation. Comput. Math. Methods Med. vol. 2012, 14 p (2012). Article ID07805
6. Carofalo, A., De Giorgi, M., Morabito, A.E.: Geometric modelling of metallic foams. Eng. Comput. **30**(7), 924–935 (2013)
7. Hollister, S.J., Maddox, R.D., Taboas, J.M.: Optimal design and fabrication of scaffolds to mimic tissue properties and satisfy biological constraints. Biomaterials **23**, 4095–4103 (2002)
8. Wang, X., Xu, S., Zhou, S., Xu, W., Leary, M., Choong, P., Qian, M., Brandt, M., Min, Xie Y.: Topological design and additive manufacturing of porous metals for bone scaffolds and orthopaedic implants: a review. Biomaterials **83**, 127–141 (2016)
9. Boccaccio, A., Uva, A.E., Fiorentino, M., Lamberti, L., Monno, G.: A mechanobiology-based algorithm to optimize the microstructure geometry of bone tissue scaffolds. Int. J. Biol. Sci. **12**, 1–17 (2016)
10. Lord, E.A., Mackay, A.L.: Periodic minimal surfaces of cubic symmetry. Curr. Sci. **85**(3), 346–362 (2003)
11. Han, S.C., Choi, J.M., Liu, G., Kang, K.: A microscopic shell structure with Schwarz's D-Surface. Sci. Rep. **7**, 13405 (2017)
12. Yan, C., Hao, L., Hussein, A., Young, P.: Ti-6Al-4 V triply periodic minimal surface structures for bone implants fabricated via selective laser melting. J. Mech. Behav. Biomed. Mater. **51**, 61–73 (2015)
13. Gabbrielli R., Turner I.G., Bowen C.R.: Development of modelling methods for materials to be used as bone substitutes. In: Daculsi, G., Layrolle, P (Eds) Key Engineering Materials, vol. 361–363, pp. 903–906. Scientific.net Zurich (2008)
14. Zhang, Z., Jones, D., Yue, S., Lee, P.D., Jones, J.R., Sutcliffe, C.J., Jones, E.: Hierarchical tailoring of strut architecture to control permeability of additive manufactured titanium implants. Mater. Sci. Eng. C **33**, 4055–4062 (2013)
15. Rajagopalan, S., Robb, R.A.: Schwarz meets Schwann: Design and fabrication of biomorphic and durataxic tissue engineering scaffolds. Med. Image Anal. **10**, 693–712 (2006)

16. Shoichet, M.S.: Polymer scaffolds for biomaterials applications. Macromolecules **43**(5), 581–591 (2010)
17. Sultana, N., Wang, M.: Fabrication of HA/PHBV composite scaffolds through the emulsion freezing/freeze-drying process and characterisation of the scaffolds. J. Mater. Sci. Mater. Med. **19**, 2555–2561 (2008)
18. Zhang, T., Zhou, S., Gao, X., Yang, Z., Sun, L., Zhang, D.: A multi-scale method for modelling degradation of bioresorbable polyesters. Acta Biomater. **50**, 462–475 (2017)
19. Moran, J.M., Pazzano, D., Bonassar, L.J.: Characterization of polylactic acid polyglycolic acid composites for cartilage tissue engineering. Tissue Eng. **9**, 63–70 (2003)
20. Kucharska, M., Butruk, B., Walenko, K., Brynk, T., Ciach, T.: Fabrication of in-situ foamed chitosan/β-TCP scaffolds for bone tissue engineering application. Mater. Lett. **85**, 124–127 (2012)
21. Fayyazbakhsh, F., Solati-Hashjin, M., Keshtkar, A., Shokrgoza, M.A., Dehghan, M.M., Larijani, B.: Novel layered double hydroxideshydroxyapatite/gelatin bone tissue engineering scaffolds: fabrication, characterization, and in vivo study. Mater. Sci. Eng., C **76**, 701–714 (2017)
22. Cao, H., Kuboyama, N.: A biodegradable porous composite scaffold of PGA/β-TCP for bone tissue engineering. Bone **46**(2), 386–395 (2010)
23. Cui, L., Zhang, N., Cui, W., Zhang, P., Chen, X.: A novel nano/micro-fibrous scaffold by melt-spinning method for bone tissue engineering. J. Bionic Eng. **12**(1), 117–128 (2015)
24. Sanatgar, R.H., Campagne, C., Nierstrasz, V.: Investigation of the adhesion properties of direct 3D printing of polymers and nanocomposites on textiles: effect of FDM printing process parameters. Appl. Surf. Sci. **403**, 551–563 (2017)
25. Cowin, S.C.: Bone Mechanics, 2nd edn. CRC Press, Boca Raton (2001)
26. Disease, K., Oksztulska-Kolanek, E., Znorko, Małgorzata B., Pawlak, M.K.: Nephron **132**, 51–58 (2016)
27 Ambu, R., Morabito, A.E.: Design and analysis of tissue engineering scaffolds based on open porous non-stochastic cells. Lecture Notes in Mechanical Engineering, pp. 777–787 (2017). Springer, Cham
28. Ambu, R., Morabito, A.E.: Modeling, assessment, and design of porous cells based on schwartz primitive surface for bone scaffolds. Sci. World J. 2019, Article ID 7060847, 16 p (2019). https://doi.org/10.1155/2019/7060847
29. Turner, C.H., Burr, D.B.: Basic biomechanical measurements of bone a tutorial. Bone **14**(4), 595–608 (1993)
30. Turner, C.H., Cowin, S.C., Rho, J.Y., Ashman, R.B., Rice, J.C.: The fabric dependence of the orthotropic elastic constants of cancellous bone. J. Biomech. **23**, 549–561 (1990)
31. ASTM D1621-16, Standard Test Method for Compressive Properties of Rigid Cellular Plastics, ASTM International, West Conshohocken, PA (2016)
32. Mercer, C., He, M.Y., Wang, R., Evans, A.G.: Mechanisms governing the inelastic deformation of cortical bone and application to trabecular bone. Acta Biomater. **2**, 59–68 (2006)
33. Cricrì, G.: Consistent use of the Gurson-Tvergaard damage model for the R-curve calculation. In: Proceedings of IGF XX, Torino, pp. 138-150, 24–26 June 2009
34. Gurson, A.L.: Continuum Theory of ductile rupture by void nucleation and growth – Part I. Yield criteria and flow rules for porous ductile media. J. Engrg. Mat. Tech. **99**, 2–15 (1977)
35. Ambu, R., Morabito, A.E.: Porous scaffold design based on minimal surfaces: Development and assessment of variable architectures. Symmetry **10**(9), 361 (2018)

Automotive Design Engineering: Material and Processes Selection Problems

Cristina Renzi[1(✉)], Luca Di Angelo[2], and Francesco Leali[1]

[1] Dipartimento di Ingegneria Enzo Ferrari, Università di Modena e Reggio Emilia, Modena, Italy
{cristina.renzi,francesco.leali}@unimore.it

[2] Dipartimento di Ingegneria industriale e dell'informazione e di economia, Università degli Studi dell'Aquila, L'Aquila, Italy
luca.diangelo@univaq.it

Abstract. Multicrieria Decision-making methods (MCDM) are efficiently used as a support to engineering design, even if tested procedures and experience-based approaches are often preferred in SMEs and several industrial contexts.

An extended critical review investigates how decision-making methods can help solving engineering design problems in automotive design. In particular, in this work, a classification of decision-making methods related to engineering design, is proposed, in which decisional techniques are matched with the design phases, and to corresponding automotive industry design problems. In literature a large amount of paper is dedicated to the selection of materials and manufacturing processes in the detail design phase. Also, hybrid methods, namely a combination of decision-making methods with other mathematical methods will be investigated to overcome some methodological drawbacks in MCDM methods. Due to the high impact of the problem in both industrial and research field, this work could help increasing production rate by reducing redesign errors related to material/processes selection problems, and concurrently transferring the knowledge of decision-making methods within the industrial context.

Keywords: Automotive design · Multicriteria decision making · Detail design · Material selection

1 Introduction

Engineering design can be conceived as a complex process made of a series of decisions (i.e. 'either–or') and compromises (a trade-offs) [2, 47]. Even if decision-making methods could be very useful in engineering design, automotive engineering designers seem to prefer tested procedures and experience-based approaches [9, 24]. On the one hand, in several automotive industry contexts, high-quality parts must be designed and produced avoiding redesign errors and reducing time to market operations. On the other hand, industries are oriented to adopt multicriteria decision making (MCDM) techniques in the industry standards, only after they know how this will impact on the final product success rate. The importance of the use of decision making for engineering design in automotive is described in [48].

C. Rizzi et al. (Eds.): ADM 2019, LNME, pp. 373–384, 2020.
https://doi.org/10.1007/978-3-030-31154-4_32

This work is aimed at investigating the most suitable decision-making methods for solving automotive industry design problems. In this work, problems related to the selection of the most suitable material and manufacturing process in automotive design. Design techniques have been matched with typical design problems for each step of the design process. Engineering design problems have been defined in the review method, as well as decision-making methods used in the design process. Results and discussion have been presented in Sect. 3.

2 Analysis of Decision-Based Applications in Automotive Engineering Design: Selection Problems in the Detail Design

2.1 The Review Method

A review method is proposed, following the flowchart depicted in Fig. 1. First, a state-of-the-art on decision making methods is proposed, as well as a classification of the engineering design process keywords. A literature research has been performed on the Scopus database, as wide as possible at first, in order to point out the occurrence of decision-making methods in industrial design applications.

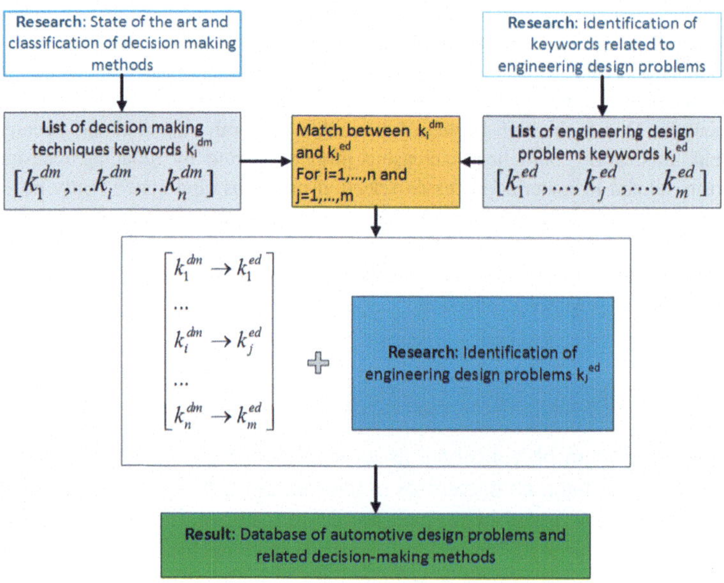

Fig. 1. The flow chart of the proposed method

Decision-making methods dedicated to the solution of engineering design problems involve three main groups: Multi-Criteria Decision-Making (MCDM) methods [6], Problem Structuring Methods (PSMs) [6], and Decision-making Problem-Solving (DPS) methods. The MCDM methods are grouped into two additional sub-areas:

Multi-Attribute (MADM) and Multi-Objective (MODM) decision-making methods. The former deals with a finite number of alternatives and a discrete solution space, while the latter searches within an infinite (and unknown) set of solutions in a continuous space. Among MADM are the Multi-Attribute Utility Theory (MAUT) and the Outranking approaches such as the Elimination Et Choix Traduisant la REalité (ELECTRE), the Analytic Hierarchy Process (AHP), the Analytic Network Process (ANP), the Preference Ranking Organisation METHod for Enrichment Evaluations (PROMETHEE), and the VIKOR method. Goal Programming (GP) and Technique for Order Preference by Similarity to Ideal Solution (TOPSIS) belong to the MODM techniques. The PSM focus on the solution of decision problems that deal with ill-posed problems and uncertain contexts [6].

The DPS methods have been defined in this paper as a group of decision-making methods addressing engineering design problems, related to the early design stages. They are the Quality Function Deployment (QFD), as well as the Axiomatic Design (AD) method [54], which are aimed at transforming customer requirements (CRs) into product specifications; the TRIZ method is used either as a problem-solving method or for generating design alternatives [4]; the Pugh's Controlled Convergence (PuCC) [45] is used for the generation and selection of design alternatives.

2.2 Identification of Engineering Design Key Words

The engineering design keywords classification related to the phases of the design process (as in [40] have been exploited in Table 1. For each phase, design activities, design problems and design keywords have been found. In the following the detail design related problems are explicated, with a view on the selection problems for material and manufacturing processes. Collected papers have been grouped according to the design problems and decision-making methods. Accordingly, material selection and manufacturing process selection problems in automotive design engineering are investigated in the following.

2.3 Detail Design: Material Selection

In the detail design, the applications of AHP, ANP, PROMETHEE, and TOPSIS, alone or combined, address the selection of materials in automotive industry applications. Occurrence of contributions are reported in Fig. 2 and analysed in the following.

Several review works regard the MCDM methods for selecting natural fibre composites for automotive components (e.g. [3, 38], as well as biocomposites (e.g. [50]. In [38], a review including literature from 2000 to 2016 regarding decisional criteria for the selection of natural fibre composite for automotive components. Most used criteria have been found as cost, density, tensile strength, Young's modulus and elongation at break. Another work [39] provide an interesting work related to the occurrence of MCDM methods for solving material selection problems for natural fibres in the automotive field. They list decisional techniques for selecting material, finding that AHP is mostly applied, followed by TOPSIS and ANP. The least applied is MAUT. **AHP**. An AHP-based systematic method is proposed in [16], with the aim of selecting the most suitable polymeric composite material for an automotive bumper

Table 1. Design activities related to the design process (first two columns adapted from [40])

Design phases	Design activities	Design problems	Design keywords
Planning	Clarify the task and build out the specification and requirements	- Collection and management of data and information in uncertain contexts - Specification of product technical requirements - Translation of customers' needs into specifications for the product	- Uncertain data in planning - Product requirements - Customers' needs (translation)
Conceptual	- Abstract the task to identify the essential problem - Establish the functional structures - Search for solution principles - Combine the solution principles with concept variants - Evaluate concept variants using technical and economic criteria	- Description of the decisionmaker preferences - Generation of design alternatives - Selection of design alternatives	- Decision-maker preferences - Design alternatives generation - Design alternatives selection
Embodiment	- Develop a definitive layout and check that the requirements are met - Evaluation of product performance - Preparation of product design documentation	- Definition of the complete architecture of product assembly	- Product performance
Detail design	- Select alternatives according to the physical realization of the product - Selection of industrial process and equipment - Final optimization of components and assembly - Release of the design documentation	- Selection of material - Industrial process or equipment selection	- Material selection - Manufacturing process selection

beam with six materials and twelve sub-criteria. In this context, AHP is used as a structured approach to solve problems related to the definition of uncertain data result from early investigations on CS. Since the Pugh method does not require a measure of the importance of each criterion and does not allow for a coupled decision, [15] the final result could be imprecise. AHP is used for selecting the best composite material

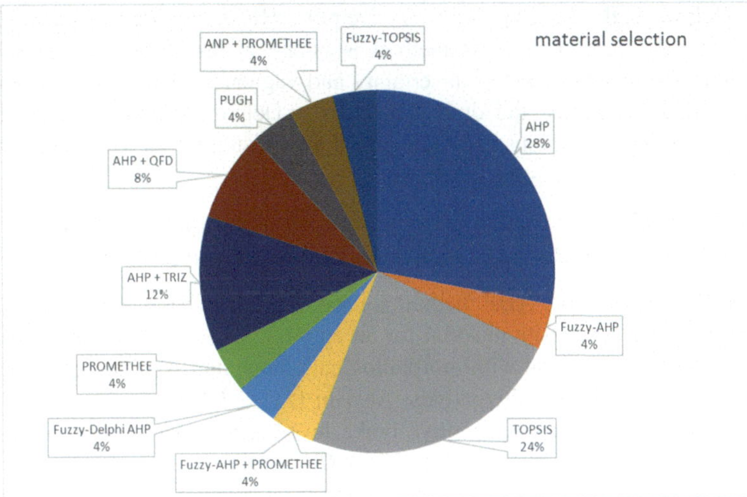

Fig. 2. Percentage of occurrence of MCDM methods in material selection problem in automotive engineering design

for automotive components (e.g. [1] as well as for selecting the most suitable natural fibres for automotive applications [3]. Also in [53], AHP is used to include quality indicators in the evaluation of the most suitable material for dry fiber tapes placement for manufacturing a L-section composite component. Green design is a fundamental approach in the nowadays automotive industry, so that natural fiber composites are substituting metal parts. In [48], AHP method is used to select the most appropriate thermoset matrix for natural fiber composites automotive armrest. Paper [33] proposes a combination of entropy and analytic hierarchy process (AHP) for evaluating the weights, while a comparison between MOORA, TOPSIS and VIKOR methods are used as material selection methods, for braking booster valve body.

FUZZY-AHP. FUZZY logic is adopted for including uncertainty due to the lack of material data. This is the case of [10], in which AHP is used to select the most suitable material for an automobile torsion bar.

TOPSIS. TOPSIS is used for the selection of a composite materials in automotive (e.g. selection of a biocomposite material for a car bumper beam in [12] It is used also to rank different resins from a given database [58]. In [7], Fuzzy TOPSIS is used to select the best aluminum alloy for automotive industries. In [62], the application of Fuzzy TOPSIS to the selection of materials for remanufacturing an engine block and an intake manifold is investigated. In [44], Grey TOPSIS is adopted for optimal lightweight material selection for automobile applications. In [25], TOPSIS and Information Entropy Method are used for the selection of the most suitable material for an automotive door outer panel

FUZZY-AHP + PROMETHEE. The selection of the most suitable material for an automobile damper is carried out by means of PROMETHEE and Fuzzy AHP [19].

FUZZY-Delphi AHP. In [21], the Fuzzy-Delphi AHP is proposed to select material for a car interior dashboard. This method is aimed at taking account of decision-makers' judgements and reducing uncertainty and vagueness. Fuzzy methodology has been adopted to overcome the drawbacks of the Delphi method, such as the low convergence of expert opinions, the high cost of administration, as well as the probable elimination of opinions of some customers.

PROMETHEE II. The selection of the most suitable material for engine bearings is carried out by means of PROMETHEE II ranking method in [26].

AHP + TRIZ. In [27], the AHP method is used for the selection of the most suitable reinforced polymer composite material for an automotive brake lever. In a successive work [28] TRIZ is combined with morphological charts, and AHP for generating and selecting conceptual design alternatives. Also in [51], AHP is used for proper composite fibre selection for an automotive brake lever. The generation of design alternatives is carried out by means of TRIZ. TRIZ is also used in [20] to eliminate contradictions in the design and evaluation of the most suitable fibre metal laminate for the car front hood.

TOPSIS. Mansor et al. [29] used the TOPSIS method for selecting optimal materials for an automotive bumper beam. The same case study is proposed in [17], in which TOPSIS is used together with multi-objective optimization based on ratio analysis (MOORA) to select the most suitable polymeric nanocomposite. In [25], TOPSIS and Information Entropy Method are used for the selection of the most suitable material for an automotive door outer panel. **FUZZY-TOPSIS.** In [62], Fuzzy-TOPSIS is adopted for the performance evaluation the of the candidate materials for the design of an engine block and an intake manifold, according to re-manufacturability criteria.

AHP + QFD. AHP and QFD have been compared in [30] for the selection of materials for the body in white. QFD seems to better include the customer needs in the analysis. As an objective scaling method for the judgment of criteria and alternatives a (1–10) scale can be adopted as in the AHP method. In particular, a combination of AHP and QFD is used to select fibres for a hybrid bio-composite material. The case study is the design for an automotive anti-roll bar. QFD is implemented in a software, called QFD4Mat which is largely applied for the evaluation of the most suitable material in the automotive sector. **PUGH**. In [34] the Pugh's controlled convergence (PuCC) is used for the selection of the most suitable eco-aware friction material for car components (brake pads). **ANP + PROMETHEE.** In [42], a combined ANP and PROMETHEE is used for the material selection for journal bearings. ANP is used to eliminate interdependencies among the attributes, and to determine the attribute weights., PROMETHEE method is chosen in place of ELECTRE for selecting alternatives, thanks to a minor computational requirement and because it offers more preference functions to select.

2.4 Detail Design: Manufacturing Process Selection

Often the process selection is strictly correlated to the material selection so that they should be performed concurrently [39]. Nevertheless, several applications are strictly

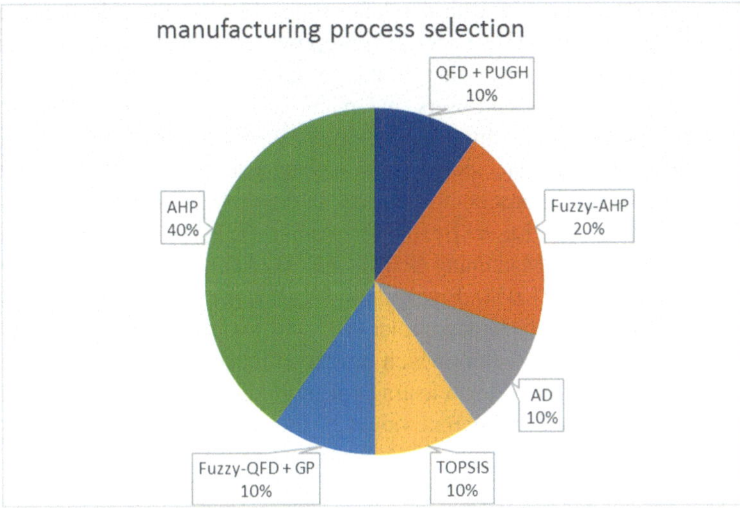

Fig. 3. Percentage of occurrence of MCDM methods in the manufacturing process selection problem in automotive engineering design

related to the process selection problem in the detail design phase. Occurrences are reported in Fig. 3 and problems are detailed in the following.

QFD + PuCC. A combined QFD and Pugh method is proposed [14], QFD is adopted for the translation of customer needs into requirements. Pugh matrix is used as a selection method for the most suitable technology to join metal pieces in a car manufacturing company.

FUZZY-AHP. The most suitable automated manufacturing system selection for an automotive gear lever is carried out in [32], by means of FUZZY-AHP. uncertainty is handles by means of Fuzzy set theory. In [15], Fuzzy-AHP is used for selecting the most suitable composite manufacturing process for manufacturing automotive bumper beams. The Expert Choice software is used for applying AHP.

AD. Complex design contexts seem to be efficiently handled by means of Axiomatic design [18]. An integrated assembly and welding system for producing the handle of a motorcycle has been proposed with an efficient increase of system quality and efficiency.

TOPSIS. In [41], an optimal assembly laser welding system for an automobile side panel is handled by means of TOPSIS, which is used as an efficient comparison method.

QFD, Fuzzy Linear Regression, and Zero-One GP. In [43], a combined QFD, fuzzy linear regression, and Zero-one GP is used for machine selection purposes, in a Turkish steel automobile parts manufacturer. While QFD addresses the problems related to the inclusion of customer needs and technical requirements, fuzzy linear regression is used to handle imprecise or unclear causal relationships. Finally, the selection technique adopted is a Zero-one GP. **AHP.** Paper [55] proposes a modular approach to integrate additive manufacturing and machining, in the design and manufacturing domains. The aim is to increase value for end-users and manufacturers. AHP is applied to select between fused deposition modelling and computer numerical control (CNC) machining for manufacturing a V6 engine section sand casting pattern. In [53] AHP is used to select a suitable machining process to perform an engraving operation on a gear face. Alternative processes are Laser Beam Machining (LBM), Ultrasonic Machining (USM), Electric Discharge Machining (EDM), Electrochemical Machining (ECM), and Electron Beam Machining (EBM). Five criteria are used, namely: material removal rate (MRR), surface finish, depth damage, tolerance, and toxicity. In [46], an AHP-based framework for assessing the factors which mostly influence additive manufacturing is proposed. In [21], AHP is used for selecting optimum die semi angle by estimating the overall performance score for different values of die semi angles.

3 Discussion and Conclusions

Matches have been built between decision-making methods and automotive design problems. As a first result, a database has been built for each design phase for handling specific automotive design problems in each design phase.

Hybrid methods have been adopted in several contexts. As for the selection of conceptual design alternatives, they are carried out mainly by means of AHP and QFD. Also. TRIZ or AD is used to generate design alternatives or to overcome design conflicts, QFD for evaluating purposes, and AHP or TOPSIS for selecting them.

AHP is either used as a selection method, or for generating alternatives when combined with TRIZ. AHP and QFD have been used combined to overcome the drawback of a subjective scale in QFD. FUZZY-AHP (or sometimes ANP) has been combined to PROMETHEE II for ranking materials. As for *material selection*, in [42], a comparison between MADM methods and the Ashby charts for material selection is proposed. The use of MADM seem to overcome the Ashby charts [5] due to the fact that no physical relations of material properties must be known for their application. Besides, combination of fuzzy techniques could reduce uncertainty in the translation of verbal judgements. As for *manufacturing process selection*, QFD is used to address problems related to the inclusion of customer needs and technical requisites again. AHP is mostly used, alone or combined with fuzzy sets in the selection phase and less TOPSIS and GP.

References

1. Ahmed Ali, B.A., Sapuan, S.M., Zainudin, E.S., Othman, M.: Implementation of the expert decision system for environmental assessment in composite materials selection for automotive components. J. Clean. Prod. **107**, 557–567 (2015)
2. Allen, J.K., Mistree, F.: Decision-based design: where should we go from here?. In: Notes of 1997 Decision-Based Design Workshop (1997)

3. Al-Oqla, F.M., Sapuan, S.M., Ishak, M.R., Nuraini, A.A.: A decision-making model for selecting the most appropriate natural fiber–polypropylene-based composites for automotive applications. J. Compos. Mater. **50**(4), 543–556 (2016)
4. Altshuller, G.S.: TRIZ The Innovation Algorithm; Systematic Innovation and Technical Creativity. Technical Innovation Center, Worcester, MA (1999)
5. Ashby, M.F., Cebon, D.: Materials selection in mechanical design. Le Journal de Physique IV **3**(C7), C7-1–C7-9 (1993)
6. Belton, V., Stewart, T.J.: Multiple Criteria Decision Analysis: An Integrated Approach. Kluwer Academic, Boston (2002)
7. Bhattacharjee, P., Debnath, A., Chakraborty, S., Mandal, U.K.: Selection of optimal aluminum alloy using TOPSIS method under fuzzy environment. J. Intell. Fuzzy Syst. **32**(1), 871–876 (2017)
8. Borjesson, F., Fancher, R., Sellgren, U.: On a methodology for component selection in modular branding: an industrial pilot study. Concur. Eng. Res. Appl. **22**(2), 93–105 (2014)
9. Cagno, E., Trucco, P.: Integrated green and quality function deployment. Int. J. Prod. Lifecycle Manag. **2**(1), 64–83 (2007)
10. Chandrasekar, V.S., Raja, K., Marimuthu, P.: Composite material selection using multi criteria decision making for automobile torsion bar. Appl. Mech. Mater. **573**, 649–654 (2014)
11. Chen, W., Hoyle, C., Wassenaar, H.J.: Decision-Based Design: Integrating Consumer Preferences into Engineering Design. Springer-Verlag, London (2013)
12. Davoodi, M.M., Sapuan, S.M., Ahmad, D., Aidy, A., Khalina, A., Jonoobi, M.: Concept selection of car bumper beam with developed hybrid bio-composite material. Mater. Des. **32** (10), 4857–4865 (2011)
13. Earl, C., Johnson, J., Eckert, C.: Complexity. In: Clarkson, J., Eckert, C. (eds.) Design Process Improvement: A Review of Current Practice, pp. 174–197. Springer, London (2005)
14. Fernandes, M.M., Rosati, A.C., Neto, D.G., Goto, F.K., Maciel Jr., H., Mologni, J.F.: Quality function deployment (qfd) and pugh matrix on innovative concept selection: an application in automotive sector. SAE technical papers (2008)
15. Hambali, A., Sapuan, S.M., Ismail, N., Nukman, Y.: Application of analytical hierarchy process in the design concept selection of automotive composite bumper beam during the conceptual design stage. Sci. Res. Essays **4**(4), 198–211 (2009)
16. Hambali, A., Sapuan, S.M., Ismail, N., Nukman, Y.: Material selection of polymeric composite automotive bumper beam using analytical hierarchy process. J. Central South Univ. Technol. (English Edition) **17**(2), 244–256 (2010)
17. Hasanzadeh, R., Azdast, T., Lee, R.E., Ghazi, A.A.: Experimental polymeric nanocomposite material selection for automotive bumper beam using multi-criteria decision making methods. Iran. J. Mater. Sci. Eng. **14**(3), 1–10 (2017)
18. Hou, L., Wen, Z., Han, D.: Axiomatic design of an integrated automatic welding system for the handle of motorcycle. In: Paper Presented at the 2007 IEEE International Conference on Control and Automation, ICCA, pp. 1245–1249 (2008)
19. Ilangkumaran, M., Avenash, A., Balakrishnan, V., Barath Kumar, S., Boopathi Raja, M.: Material selection using hybrid MCDM approach for automobile bumper. Int. J. Ind. Syst. Eng. **14**(1), 20–39 (2013)
20. Ishak, N.M., Sivakumar, D., Mansor, M.R.: The application of TRIZ on natural fibre metal laminate to reduce the weight of the car front hood. J. Brazilian Soc. Mech. Sci. Eng. **40**(2), art. no. 105 (2018)
21. Karanjule, D.B., Bhamare, S.S., Rao, T.H.: Die semi angle optimization for cold drawing of seamless tubes based on multiple attribute decision making methods. Int. Rev. Mech. Eng. **11**(7), 486–493 (2017)

22. Kazemi, S., Homayouni, S.M., Jahangiri, J.: A Fuzzy Delphi-Analytical Hierarchy Process Approach for Ranking of Effective Material Selection Criteria. Adv. Mater. Sci. Eng. **2015**, 12 p. (2015). art. no. 845346
23. Koganti, R., Zaluzec, M., Chen, M., Defersha, F.M.: Design for Assembly: An AHP Approach for Automotive Front end Component Design Evaluation (No. 2007-01-0522). SAE technical paper (2007)
24. Krishnamurty, S.: Normative decision analysis in engineering design. Decis. Making Eng. Des. **4**(4), 21–33 (2006)
25. Ma, F., Zhao, Y., Pu, Y., Li, J.: A comprehensive multi-criteria decision making model for sustainable material selection considering life cycle assessment method. IEEE Access **6**, 58338–58354 (2018)
26. Maity, S.R., Chakraborty, S.: A decision-making approach for bearing material selection. Int. J. Comput. Mater. Sci. Surf. Eng. **5**(4), 289–303 (2013)
27. Mansor, M.R., Sapuan, S.M., Zainudin, E.S., Nuraini, A.A., Hambali, A.: Hybrid natural and glass fibers reinforced polymer composites material selection using analytical hierarchy process for automotive brake lever design. Mater. Des. **51**, 484–492 (2013)
28. Mansor, M.R., Sapuan, S.M., Zainudin, E.S., Nuraini, A.A., Hambali, A.: Conceptual design of kenaf fiber polymer composite automotive parking brake lever using integrated TRIZ-morphological chart-analytic hierarchy process method. Mater. Des. **54**, 473–482 (2014)
29. Mansor, M.R., Sapuan, S.M., Hambali, A., Zainudin, E.S., Nuraini, A.A.: Materials selection of hybrid bio-composites thermoset matrix for automotive bumper beam applicationusing topsis method. Adv. Environ. Biol. **8**(8), 3138–3442 (2014)
30. Mayyas, A.T., Omar, M.A.: Eco-material selection assisted with decision-making tools, guided by product's attributes; functionality and manufacturability. Int. J. Mater. Struct. Integrity **6**(2–4), 190–219 (2012)
31. Mayyas, A., Shen, Q., Mayyas, A., Abdelhamid, M., Shan, D., Qattawi, A., Omar, M.: Using quality function deployment and analytical hierarchy process for material selection of body-in-white. Mater. Des. **32**(5), 2771–2782 (2011)
32. Monitto, M., Pappalardo, P., Tolio, T.: A new Fuzzy AHP method for the evaluation of automated manufacturing systems. CIRP Ann. Manuf. Technol. **51**(1), 395–398 (2002)
33. Moradian, M., Modanloo, V., Aghaiee, S.: Comparative analysis of multi criteria decision making techniques for material selection of brake booster valve body. J. Traffic Transp. Eng. (English Edition) (2018)
34. Mustafa, A., Abdollah, M.F.B., Ismail, N., Amiruddin, H., Umehara, N.: Materials selection for eco-aware lightweight friction material. Mech. Ind. **15**(4), 279–285 (2014)
35. Nahm, Y.: New competitive priority rating method of customer requirements for customer oriented product design. Int. J. Precis. Eng. Manuf. **14**(8), 1377–1385 (2013)
36. Nahm, Y.-E.: A novel approach to prioritize customer requirements in QFD based on customer satisfaction function for customer-oriented product design. J. Mech. Sci. Technol. **27**(12), 3765–3777 (2013)
37. Nahm, Y.-E., Ishikawa, H., Inoue, M.: New rating methods to prioritize customer requirements in QFD with incomplete customer preferences. Int. J. Adv. Manuf. Technol. **65** (9–12), 1587–1604 (2013)
38. Noryani, M., Sapuan, S.M., Mastura, M.T.: Multi-criteria decision-making tools for material selection of natural fibre composites: a review. J. Mech. Eng. Sci. **12**(1), 3330–3353 (2018)
39. Noryani, M., Sapuan, S.M., Mastura, M.T., Zuhri, M.Y.M., Zainudin, E.S. Material selection criteria for natural fibre composite in automotive component: A review (2018) IOP Conference Series: Materials Science and Engineering, 368 (1), art. no. 012002
40. Pahl, G., Beitz, W.: Engineering design: a systematic approach, 2nd ed. In: Wallace, K., Blessing, L., Baurt, F. (eds.) Trans, pp. 125–143. Springer, London (1996)

41. Park, H.S., Lee, G.B.: Development of digital laser welding system for automobile side panels. Int. J. Automot. Technol. **8**(1), 83–91 (2007)
42. Peng, A.-H., Xiao, X.-M.: Material selection using PROMETHEE combined with analytic network process under hybrid environment. Mater. Des. **47**, 643–652 (2013)
43. Perçin, S., Min, H.: Optimal machine tools selection using quality function deployment and fuzzy multiple objective decision making approach. J. Intell. Fuzzy Syst. **24**(1), 163–174 (2013)
44. Pu, Y., Ma, F., Zhang, J., Yang, M.: Optimal lightweight material selection for automobile applications considering multi-perspective indices. IEEE Access **6**, 8591–8598 (2018)
45. Pugh, S.: Concept selection: a method that works. In: Proceedings of the International Conference on Engineering Design ICED, Rome, Italy (1981)
46. Rahim, S.L., Maidin, S.: Feasibility study of additive manufacturing technology implementation in malaysian automotive industry using analytic hierarchy process. Appl. Mech. Mater. **465**, 715–719 (2014)
47. Rajan, V.N.: A framework for concurrent assembly design and planning. In: Proceedings of the First Annual International Conference on Industrial Engineering: Applications and Practice, Houston, TX, 4–7 December 1996
48. Renzi, C., Leali, F., Di Angelo, L.: A review on decision-making methods in engineering design for the automotive industry. J. Eng. Des. **28**(2), 118–143 (2017)
49. Rosli, M.U., Jamalludin, M.R., Khor, C.Y., Ishak, M.I., Jahidi, H., Wasir, N. Y., Faizal, W. M., Draman, W.N.A.T.W., Lailina, N.M., Ismail, R.I.: Analytical hierarchy process for natural fiber composites automotive armrest thermoset matrix selection. In: MATEC Web of Conferences, vol. 97, p. 01039. EDP Sciences (2017)
50. Sanyang, M.L., Mansor, M.R., Sapuan, S.M., Ahmed Ali, B.A.: Conceptual design of biocomposites for automotive components. Green Energy and Technology, pp. 101–126 (2017)
51. Sapuan, S.M., Kho, J.Y., Zainudin, E.S., Leman, Z., Ali, B.A., Hambali, A.: Materials selection for natural fiber reinforced polymer composites using analytical hierarchy process. Indian J. Eng. Mater. Sci. **18**(4), 255–267 (2011)
52. Shi, Z., Mocko, G.: Knowledge base representation for axiomatic design through ontologies. In: ASME 2013 International Design Engineering Technical Conferences and Computers and Information in Engineering Conference, V02BT02A016–V02BT02A016. American Society of Mechanical Engineers, New York (2013)
53. Srinivasan, S., Srivatsan, L., Sathyanarayan, R., Vijaya Ramnath, B.: MCDM model for selection of optimum machining process. Mater. Sci. Forum **773–774**, 348–354 (2014)
54. Suh, N.P.: Axiomatic design theory for systems. Res. Eng. Des. **10**(4), 189–209 (1998)
55. Townsend, V., Urbanic, J.: Relating additive and subtractive processes in a teleological and modular approach. Rapid Prototyping J. **18**(4), 324–338 (2012)
56. Veldenz, L., Di Francesco, M., Giddings, P., Kim, B.C., Potter, K.: Material selection for automated dry fiber placement using the analytical hierarchy process. Adv. Manuf. Polym. Compos. Sci. **4**(4), 83–96 (2018)
57. Vinodh, S., Kamala, V., Jayakrishna, K.: Integration of ECQFD, TRIZ, and AHP for innovative and sustainable product development. Appl. Math. Model. **38**(11–12), 2758–2770 (2014)
58. Waigaonkar, S., Babu, B.J.C., Prabhakaran, R.T.D.: A new approach for resin selection in rotational molding. J. Reinf. Plast. Compos. **27**(10), 1021–1037 (2008)
59. Xu, T., Moon, D.H., Baek, S.G.: A simulatstudy integrated with analytic hierarchy process (AHP) in an automotive manufacturing system. Simulation **88**(4), 450–463 (2012)

60. Yang, S., Ong, S.K., Nee, A.Y.C.: Design for Remanufacturing-A Fuzzy-QFD Approach. In: Seliger, Günther, Kohl, Holger, Mallon, Jürgen (eds.) Re-engineering Manufacturing for Sustainability, pp. 655–661. Springer, Singapore (2013)
61. Yang, S.S., Ong, S.K., Nee, A.Y.C.: A decision support tool for product design for remanufacturing. Procedia CIRP **40**, 144–149 (2016)
62. Zahedi, F.: The analytic hierarchy process – a survey of the method and its applications. Interfaces **16**(4), 96–108 (1986)
63. Yang, S.S., Nasr, N., Ong, S.K., Nee, A.Y.C.: Designing automotive products for remanufacturing from material selection perspective. J. Clean. Prod. **153**, 570–579 (2017)

Integrated Methods for System Design, Simulation, Analysis and Optimization

Development of an Exhaust System for Agricultural Tractors

Stefano Uberti(✉) and Alessandro Copeta

DIMI, Università degli Studi di Brescia, via Branze 38, 25123 Brescia, Italy
stefano.uberti@unibs.it

Abstract. The main sources of sound emitted by vehicles are the engine and the rolling noise produced by the tires. In case agricultural tractors are considered, the engine and the transmission are responsible of the main emissions since they often work in the fields on soft ground at low speeds. This work focuses on the combined acoustic and mechanical development of a new exhaust system for an existing agricultural tractor, with the aim of providing a product with improved acoustic performances and a neater design. The host vehicle is equipped with a 300 hp diesel engine which can be particularly noisy at low rpm, with a further constraint being the "under the bonnet" available volume to fit the system into. The existing exhaust system is the baseline for the engineering process. The acoustic design is carried out by a 1D simulation software based on an electro-acoustic analogy, while the design modelling is performed by using Solidworks® 3D CAD. After the design stage a prototype has been manufactured and tested at the MWL/KTH laboratory. The engineering process gave the product a cleaner design and allowed to identify feasible solutions ensuring increased sound attenuation performances.

1 Introduction

This work starts from the desire of one of the world's leading manufacturers of agricultural equipment, based in the North of Italy, to incorporate new exhaust systems in its tractors in order to provide the operator with more comfortable working conditions. Among the several sources of farm noise, agricultural tractors are indeed not the noisiest, but, being used continuously for long periods, they become potentially more hazardous than other noisier machines, especially when they are old or even uncabbed. In particular, the exhaust system is responsible for most of the emitted noise, which exits the vehicle with the gases expelled by the engine. The reference legislative framework in terms of noise disturbance is complex [1–3] because standards and regulations vary from country to country and are often updated, which makes it more difficult for the manufacturer to offer always compliant products; in this particular case, the vehicle has to be marketed in a country with an emission legislation equivalent to "Euro 1". Any time a new part must be designed or, even more so, an existing part must be modified, integrating acoustic and mechanical constraints can be a challenging task, and the need arises to adopt an agile selection process which allows to rapidly identify a small number of solutions to be tested and evaluated thoroughly. On the one hand, trial-and-error method alone can be time and money consuming and often

C. Rizzi et al. (Eds.): ADM 2019, LNME, pp. 387–399, 2020.
https://doi.org/10.1007/978-3-030-31154-4_33

provides a limited amount of data or even misleading conclusions; on the other hand, simulations with computed aided engineering tools nowadays are a standard option, but the best tool among CFD [4], FEM [5, 6] or 1D models [7] must be selected to isolate the contributions of the noise sources under investigation with the desired degree of detail.

In this work, a 1D model applicable to any duct system, from exhaust systems to HVAC ducts [8], is used during the acoustic design phase to identify the key features that the new exhaust system must possess, then they are passed to the mechanical engineers to check the feasibility of the solution. After several iterations, the final 3D model has been designed, according to the procedure outlined in [7]. The structure of this iterative methodology can be applied in similar fashion to other design problems, in particular to mechanical systems and kinematic design, as outlined in [9–12]. The paper is organised as follows: Sect. 3 presents the materials and methods used in the work, including a description of the tractor characteristics, the relevant acoustic theory and testing methods and the employed software tools; in Sect. 3, the results of the acoustic simulation and a comparison with tests performed at Marcus Wallenberg Laboratory for Sound and Vibration Research, the Royal Institute of Technology (MWL) are reported. In Sect. 4, conclusions are drawn.

2 Materials and Methods

The aim of the study is to design an exhaust system for the tractor shown in Fig. 1.

- The tractor features are:
- 7.8 l six-cylinder engine
- 300 hp with two-stage turbocharger and intercooling
- 90% of torque at 1000 rpm, with max torque constant over 1200–1600 rpm.

Fig. 1. The agricultural tractor undergoing exhaust system re-design.

Fig. 2. The original exhaust system.

The acoustic optimisation has been carried out with 1D frequency domain simulation software SIDLAB® [13], which models each segment of the exhaust system as a two-port or multi-port element, whose characteristic parameters can be tuned to obtain the desired sound attenuation spectrum. The result of the simulation is the transmission loss of the system, defined as the reduction in sound power level between the sound power entering the component and the sound power exiting it (Fig. 7).

Several elements are available to compose the exhaust systems acoustic attenuation, such as expansions or contractions, quarter-wave resonators, orifices, perforated ducts or lined ducts. In particular, area discontinuities and quarter-wave resonators are reactive elements used to attenuate noise below the so-called cut-on frequency, $f_{cut\text{-}on}$, which can be estimated ad $f_{cut\text{-}on} = 0.5\ c/D$, where c is the speed of sound in air and D the internal diameter of the duct. At high frequencies resistive elements, such as absorbing materials, perforated ducts and orifices, can provide high transmission loss over a wide spectrum. The sequence and characteristic parameters of these element can be adjusted in the software until the desired acoustic performances are reached. The original exhaust system was the result of several functional and design changes applied to the component over time. For this reason, it was chosen to develop the new exhaust system from scratch, on the basis of the information collected by the manufacturer in terms of engineering process and space constraints. In particular, the available volume to fit the exhaust system was the major issue since it had to fit perfectly in a very packed under-the-bonnet space (Figs. 3 and 4). Finally, a cost target of 25% of original system was given.

Fig. 3. The tractor undergoing re-design.

Fig. 4. The original under the bonnet space.

The iterative-operative method followed during this stage of the work is well illustrated in reference [7], where it is also illustrated with a clear graphic.

The prototype has been tested according to the procedure of ASTM E2611, which describes the measurement of normal incidence sound transmission in a four-microphone test rig. This method can be used to characterise samples of materials but even finished products such as complete exhaust systems. As shown in Fig. 5, the part to be tested is placed in a testing rig featuring a loudspeaker and an anechoic or reflective termination at the endpoints. The part is mounted between microphones that simultaneously record signals. As shown in [14], this allows to decompose the sound wave among incident and reflected fractions on either side of the component, and to calculate the transfer matrix between the inlet and the outlet state variables. The transmission loss TL of the tested part can therefore be obtained from the matrix elements T_{ij}, according to the following equation:

$$TL(f) = 10\log_{10}\left(\frac{1}{4}\left|T_{11} + \frac{T_{12}}{\rho_0 c} + \rho_0 c T_{21} + T_{22}\right|^2\right) \tag{1}$$

where ρ_0 and c are the density of air and the speed of sound in air, respectively. In particular, the prototype has been mounted into the experimental rig in the flow acoustic facility at the MWL. The flow velocity has been measured by means of a Pitot tube installed at the centre of the pipe.

The test bench is a standing wave tube, and the measurement of the transmission loss is based on the evaluation of the sound pressure at the walls of the tube under the assumption of plane-wave propagation. The results can thus be considered reliable up to the so-called cut-on frequency, which is the frequency delimiting the transition from plane-wave to non-plane-wave region. The cut-on frequency is determined as:

$$f_{cut-on} = \frac{c}{2D} \tag{2}$$

where c is the speed of sound in the medium and D is the diameter of the tube. For the specific case at hand the inlet duct diameter is 0.123 m, thus the limit of validity of 1D modelling is 1400 Hz.

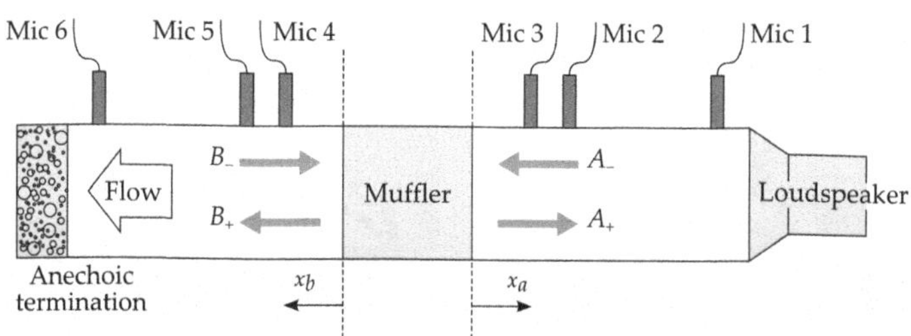

Fig. 5. Schematic of the flow test facility at MWL.

3 Results and Discussion

3.1 Acoustic and Mechanical Design of the New Exhaust System

The details of the acoustic simulations are reported in [15]. The acoustic model of the original exhaust system is shown in Fig. 6. The acoustic behaviour of the existing exhaust system has been carried out with zero flow. Figure 7 shows the calculated transmission loss, which is good in the plane-wave frequency range except in the region around 500 Hz. This result helped the manufacturer realise that the development of a quieter exhaust system was indeed an interesting improvement.

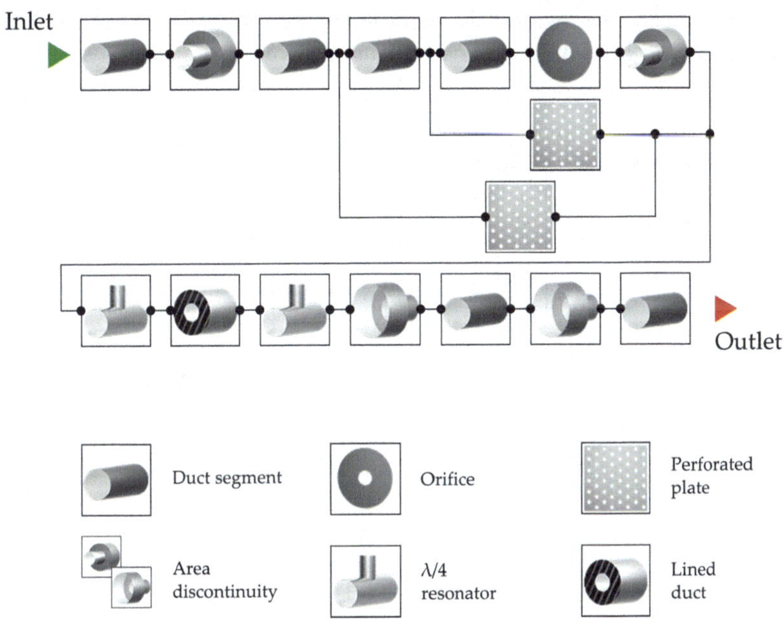

Fig. 6. 1D model of the original exhaust system.

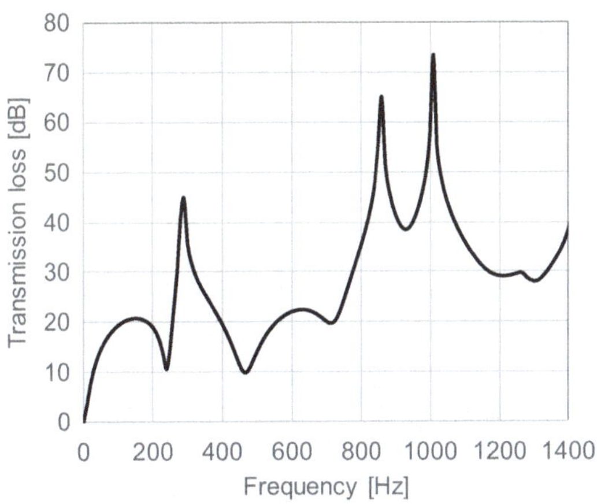

Fig. 7. Transmission loss of the original exhaust system (SIDLAB® simulation).

The acoustic study led to the model shown in Fig. 8. Figure 9 shows the comparison between calculated transmission losses for the new and original exhaust systems, where it can be noted that the transmission loss has been improved in most of the frequency range, particularly between 400 and 800 Hz. As a further possible improvement, Fig. 10 highlights the effect of internally lining the inner chamber of the 4-port element with steel wool: it is worth noting that the addition of the absorbing layer is essential to exceed the original exhaust system performances.

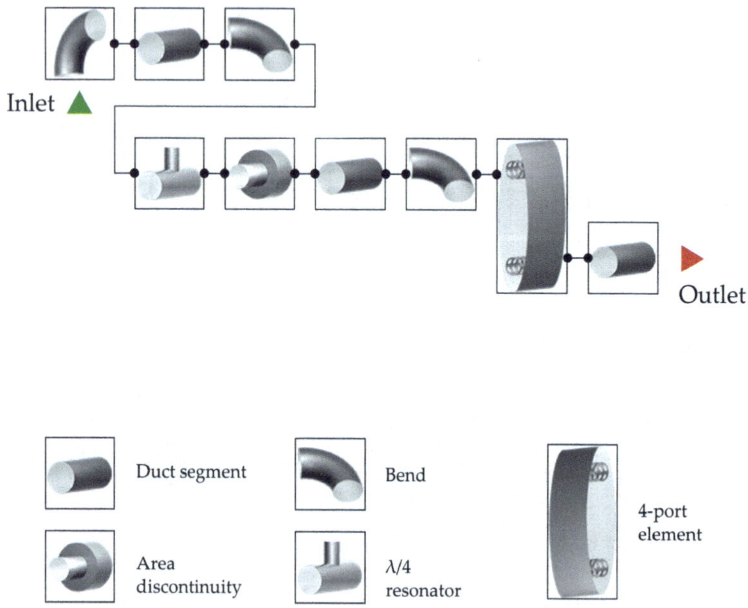

Fig. 8. 1D model of the new exhaust system.

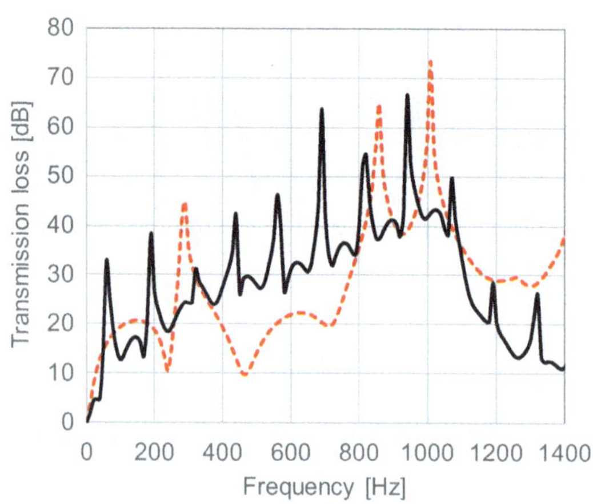

Fig. 9. Transmission loss associated to the new exhaust system design (solid line), compared to the original one (dashed line). SIDLAB®

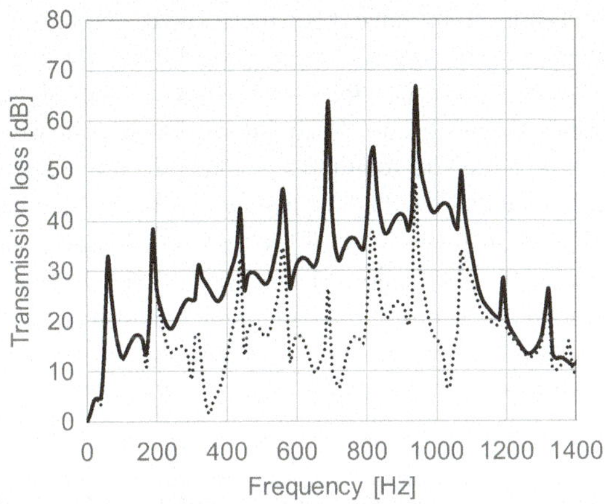

Fig. 10. TL associated to the new exhaust system design with (solid line) and without (dotted line) internal lining of the chamber. SIDLAB® simulations.

The acoustic optimisation and the mechanical design have been performed interactively. At each step of the acoustic optimization process, the new muffler has been modeled by using Solidworks® parametric CAD software and virtually tested using a finite element engine to double check the pressure drop, flow rate and gas velocity field (in Figs. 11 and 12 a couple of examples extracted from early tests are given).

Simulations have been performed with the following parameters checking the flow trajectories for avoiding stagnation zones: temperature = 350 °C, mass flow rate = 0.28 kg/s. Particular attention has been paid to estimate the total backpressure, because one of the engine specifications was to avoid a total backpressure on the turbo outlet bigger than 50 mbar.

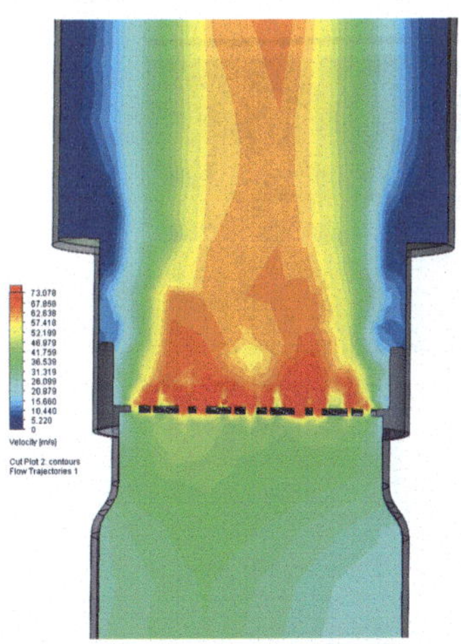

Fig. 11. Flow velocity at the 4 port exit element, revealing no sonic blocks.

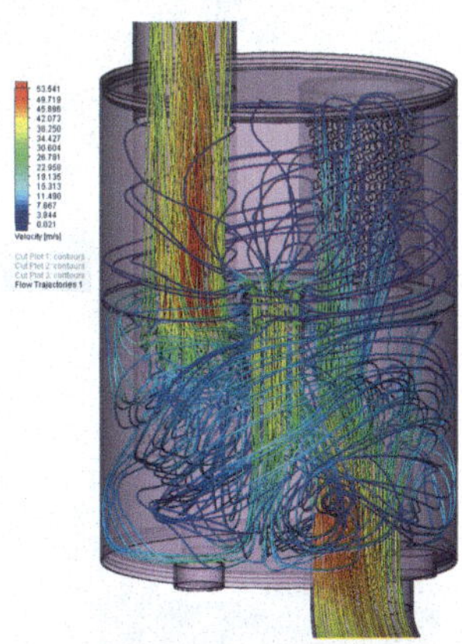

Fig. 12. Stream lines (colours: velocity magnitude), showing flow distribution.

A total of 29 iteration steps have been performed, exploring a lot of possible geometries and solutions. The typical approach adopted for the iteration steps is described in [7] and is based on the close interaction between the mechanical design group and the acoustics design group. Such interaction is primarily needed to find a compromise between the large size required to control the noise propagation in the low frequency region and the actual available space, which is evaluated at each design step through the 3D modelling software. One of the most challenging problems during this phase was the choice of the dimension of the mesh element for the simulations. An approach with an equivalent porous surface element was tested, but it proved not to be satisfactory. For this reason, an adaptive mesh approach with a smaller element size of (1/6 the hole dimension) was implemented (Figs. 13 and 14).

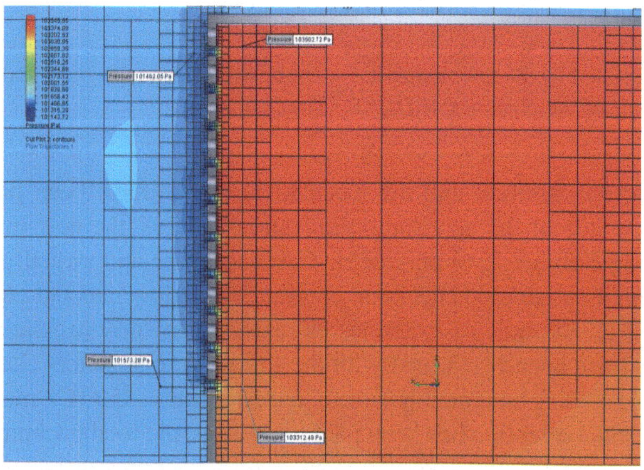

Fig. 13. Pressure simulation mesh.

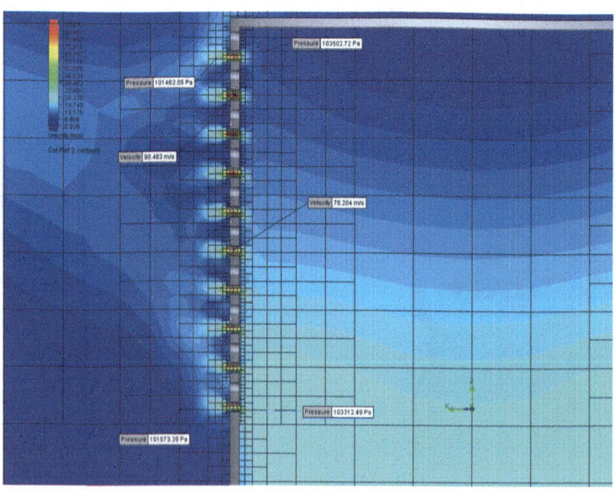

Fig. 14. Velocity simulation mesh.

A peculiar problem to be managed was the optimisation of the holes dimension and pattern that had to be defined according to the available commercial formats. In fact the holes size and the spacing between holes had to be chosen among what was available "off-the-shelf"; it is to be considered that perforated sheet metal is manufactured by the use of special tools and plants and it is not economically manageable to have a custom product in a small batch. Provision for manual assembly had also to be considered due to possible manufacturing issues.

Finally, the manufacturer was provided with the necessary suggestions to build a prototype. One of the key elements in the new solution is a 4-pole element with two perforated inlet and outlet pipes, running through the chamber so that the flow is forced to pass through the perforated wall twice. In this system, the 4-port block is the only element with a certain construction complexity, the other parts being mere pipes, bends and adapters. Another key element is a one fourth wavelength resonator, tuned to cut a particular frequency in the lower frequency span (Figs. 15 and 16). In this case the problem was to find the correct insertion point of this element and the space to fit it.

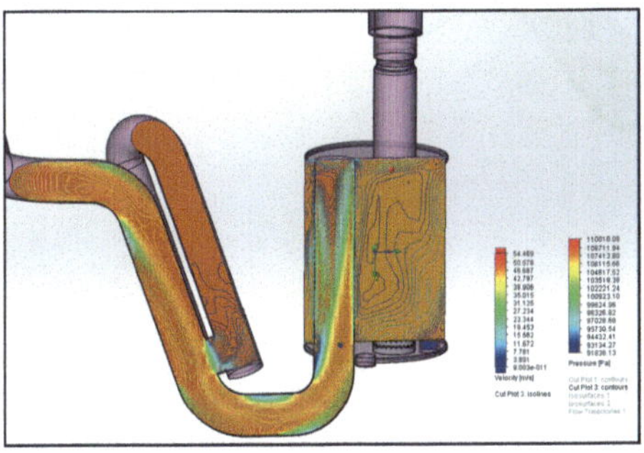

Fig. 15. Final solution pressure and velocity simulation.

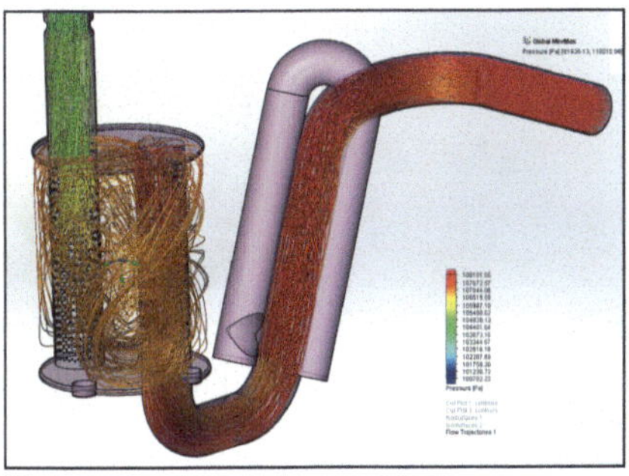

Fig. 16. Final solution flow trajectories and pressure drop

The evolution led to a configuration corresponding to the model of Fig. 8 as depicted also in Fig. 17 At the end of this process, the detailed new exhaust system modelled by using Solidworks® parametric CAD software has been produced (Figs. 17 and 18).

When the design has been finalised, the manufacturer has been provided with the necessary indications on how to build a prototype. The first prototype design was provided with a regulation system in the resonating chamber, to test some different length of it, if tuning of the length itself was needed (Fig. 19). After testing, deliberated prototype has been installed on vehicle, according layout depicted in Fig. 20.

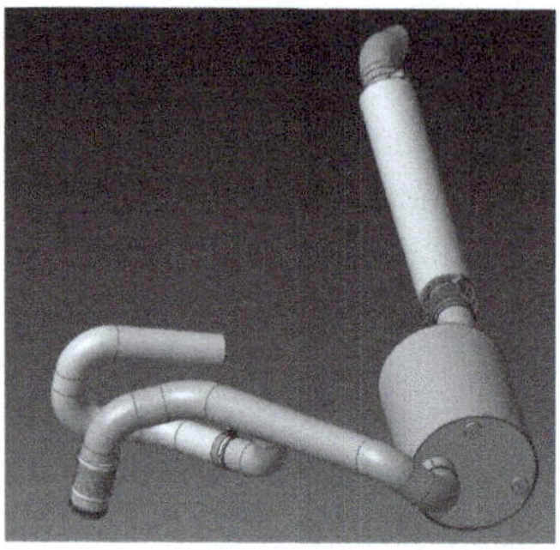

Fig. 17. Mechanical design of new system.

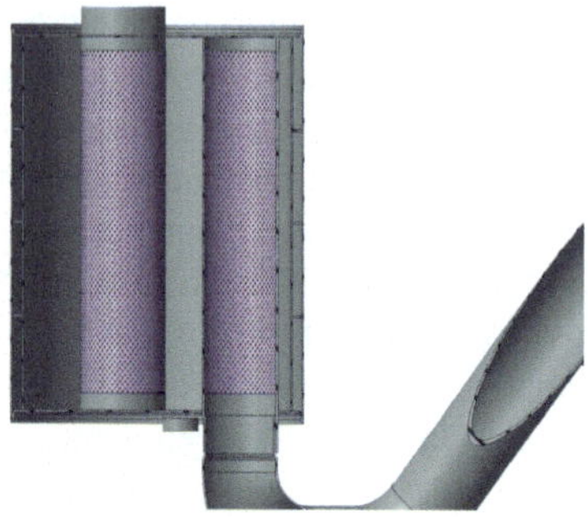

Fig. 18. Detail of 4-port element 3D model.

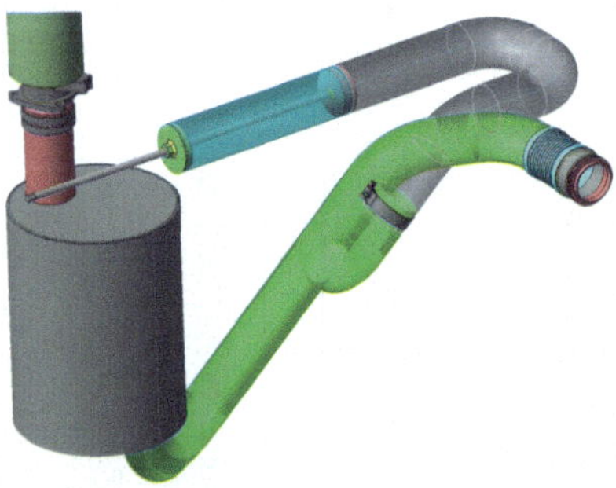

Fig. 19. Final prototype design with tuneable resonator.

Fig. 20. Layout of the installation of the muffler on the tractor.

3.2 Experimental Validation

The prototype manufactured based on the specifications described in the previous section and featuring no lining inside the 4-port element, has been tested in the experimental rig at MWL (Fig. 21).

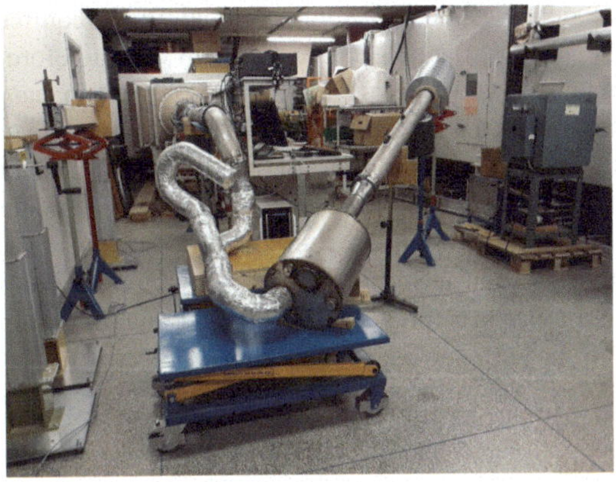

Fig. 21. New exhaust system under test at MWL facility.

Figure 22 shows a comparison between simulated and measured data, with flow rate 0.170 l/s, in the assumption of airflow at room temperature. The agreement between the curves looks good, except around 700 Hz and 1050 Hz, the reason being the effect of possible variations to the simulated prototype introduced during the manufacturing process.

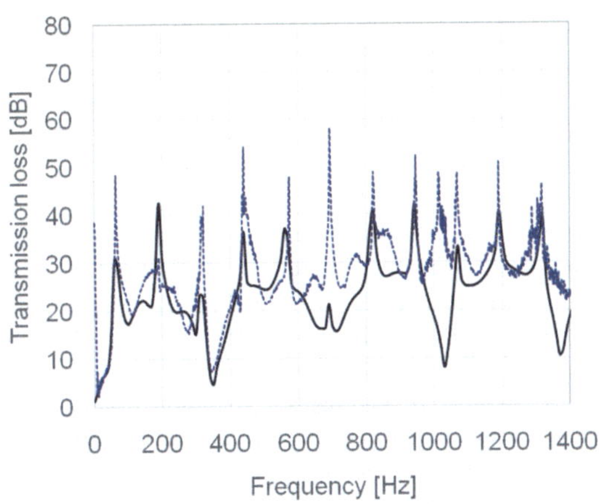

Fig. 22. Comparison between simulated (solid line) and experimental (dashed line) transmission loss of the new exhausted system.

4 Conclusions

In this work, a case study describing the complete re-design of an exhaust system for a commercial agricultural tractor has been presented.

The acoustic performances of the existing system and of new solutions have been studied by means of 1D simulations, whose main benefits are the quick set-up and adjustment of the models and the low computational effort required to run the calculation. Based on this analysis, the new exhaust system was therefore engineered from a mechanical viewpoint, and a prototype has been built which has been successfully tested in the flow facility at MWL. This combined approach led to an increase in noise attenuation capability between 400 and 600 Hz and to a far cleaner design, thus providing the manufacturer with a dually improved product. In particular, the 4-port block included in the new exhaust system realises most of the noise reduction when it is internally lined with a layer of absorbing material.

Final cost of the muffler is estimated less than 1k € in comparison with the original equipment, valued to 4k €.

References

1. European Parliament: Regulation (EU) No 167/2013 of the European Parliament and of the Council of 5 February 2013 on the approval and market surveillance of agricultural and forestry vehicles (2013)
2. Commission Delegated Regulation (EU) 2015/96 of 1 October 2014 supplementing Regulation (EU) No 167/2013 of the European Parliament and of the Council as regards environmental and propulsion unit performance requirements of agricultural and forestry vehicles (2015)
3. Directive 2009/76/EC of the European Parliament and of the Council of 13 July 2009 relating to the driver-perceived noise level of wheeled agricultural or forestry tractors (2009)
4. Zhang, H., Fan, W., Guo, L.-X.: A CFD results-based approach to investigating acoustic attenuation performance and pressure loss of car perforated tube silencers. Appl. Sci. Basel **8**, 545 (2018). https://doi.org/10.3390/app8040545
5. Yadav, P.S., Gaikwad, A.A., Badgujar, S.Y., Surkutwar, Y.V., Karanth, N.V.: Noise reduction on agricultural tractor. SAE Technical Papers, 5 (2013) https://doi.org/10.4271/2013-26-0103
6. Neihguk, D., Fulkar, S.: Acoustic analysis of a tractor muffler. SAE Technical Papers, June 2017. https://doi.org/10.4271/2017-01-1791
7. Piana, E.A., Uberti, S., Copeta, A., Motyl, B., Baronio, G.: An integrated acoustic–mechanical development method for off-road motorcycle silencers: from design to sound quality test. Int. J. Interact. Des. Manuf. **12**, 1139–1153 (2018). https://doi.org/10.1007/s12008-018-0464-x
8. Piana, E.A., Grassi, B., Bianchi, F., Pedrotti, C.: Hydraulic balancing strategies: a case study of radiator-based central heating system. Build. Serv. Eng. Res. T. **39**, 249–262 (2018). https://doi.org/10.1177/0143624417752830
9. Uberti, S., Copeta, A., Baronio, G., Motyl, B.: An eco-innovation and technical contaminated approach for designing a low environmental impact off-road motorcycle. Int. J. Interact. Des. Manuf. **12**, 281–295 (2018). https://doi.org/10.1007/s12008-017-0382-3

10. Baronio, G., Copeta, A., Motyl, B., Uberti, S.: Gölem project: concept and design of a trekking/hiking wheelchair. Adv. Mech. Eng. **9**, 1–11 (2017). https://doi.org/10.1177/1687814017730541
11. Uberti, S., Gadola, M.: Design of a new high-end street bike. In: 11th International Design Conference, DESIGN 2010, Dubrovnik, Croatia, pp. 1741–1752 (2010)
12. Uberti, S., Gadola, M., Chindamo, D., Romano, M., Galli, F.: Design of a double wishbone front suspension for an orchard-vineyard tractor: kinematic analysis. J. Terramechanics **57**, 23–39 (2015). https://doi.org/10.1016/j.jterra.2014.11.001
13. SIDLAB Acoustics. http://sidlab.se/products/sidlab-acoustics/. Accessed 10 June 2019
14. Piana, E.A., Roozen, N.B., Scrosati, C.: Impedance tube measurements on the DENORMS round robin test material samples. In: Proceedings of the 26th International Congress on Sound and Vibration, ICSV 2019, Montréal, Canada. IIAV (2019)
15. Piana, E.A., Uberti, S., Copeta, A., Carlsson, U.E.: Design and testing of a muffler for agricultural tractors. In: Proceedings of Euronoise 2018, EAA, Heraklion, Greece, pp. 2277–2284 (2018)

A Topology Optimization of a Motorsport Safety Device

Filippo Cucinotta(✉), Marcello Raffaele, and Fabio Salmeri

Department of Engineering, University of Messina, Contrada Di Dio (S. Agata), 98166 Messina, Italy
{filippo.cucinotta,marcello.raffaele, fabio.salmeri}@unime.it

Abstract. The work proposes a method of Topological Optimization of 3D surfaces using an algorithm that works by considering the distribution and intensity of stress on the studied component. Rather than acting on the mesh, this algorithm modifies directly the CAD, allowing its direct use avoiding any subsequent intervention. The algorithm has been developed using Rhino – Grasshopper. The FEM analyses have been performed using the Nastran solver within the Siemens NX environment. With this method the original non-optimized model is hollowed through a Voronoi tessellation that is managed through several parameters. Through an iterative process, the algorithm performs the hollowing on the original CAD, varying the size and distribution of the holes in function of the stresses. As case study, the authors considered a safety device that helps to prevent injuries to the necks of pilots of various high-speed motorsports (Head and Neck Support, HANS). The results of this work show the potentiality of this methodology, with which it is possible to obtain a much lighter device with the same mechanical performance.

Keywords: Topology Optimization · Additive Manufacturing · Selective Laser Sintering · Mechanical design

1 Introduction

Topology Optimization (TO) is a quite new and very powerful tool [1]. It allows to obtain very performing structures, optimizing one or more parameters such as weight, stresses, costs and environmental impact [2]. Recently, thanks to the rapid development of the Additive Manufacturing (AM) technologies, also TO has had a rapid increase in interest and application [3]. The fields of application of TO range from biomedical [4, 5] to sports [6], automotive [7] and aerospace [8].

Cucinotta et al. proposed a TO method based on a new algorithm developed in the Rhinoceros-Grasshopper environment (grasshopper3d.com). It basically consists in a Voronoi tessellation of the surface guided by the stress concentration coming from a Finite Element Analysis (FEA) [9]. The great potentiality of the method consists in the fact that it acts directly on a CAD level, and not on the mesh of the FEA. In this way it is possible to obtain geometries ready to use, with the desired dimensions and tolerances.

C. Rizzi et al. (Eds.): ADM 2019, LNME, pp. 400–409, 2020.
https://doi.org/10.1007/978-3-030-31154-4_34

The method is oriented for a Selective Laser Sintering (SLS) AM technology, or similar. These techniques consist in the sintering of metal powders with the energy supply coming from a laser light [6, 10].

As case study, in order to show, the potentiality of the proposed algorithm, a Head and Neck Support (HANS) has been optimized [11], starting from a commercial one made of carbon of 750 g of weight. This kind of device is largely used in motorsport, automotive and powerboats [12], in order to reduce craniovertebral junction injuries due to sudden braking and collisions. The choice of a HANS device as a case study depends on the fact that, to test the competitiveness of the algorithm, in terms of weight reduction, in the manufacturing scenarios that can be expected for the near future, it is necessary to evaluate the optimization of already highly performing objects.

The case study has been chosen to show the potentiality of the proposed algorithm. To make the object usable in the sports field, the homologation process must still be carried out.

2 Materials and Methods

2.1 The Topology Optimization Method

The proposed method consists in a reconstruction of the surfaces by means of, firstly, a regular hollowing of the surfaces and, subsequently of a reduction of the holes size and distribution in function of the stresses evaluated by a FE analysis. The pattern for the hollowing is based on a Voronoi tessellation. The FE analysis is used for reducing the holes diameters where the stress overcomes an imposed threshold, depending on the maximum allowable stress of the material. Where the size of the holes is too little, for technological feasibility and to avoid stress concentrations, the algorithm closes the holes. After the hollowing, each hole is filleted with a set radius by the user.

The algorithm has been developed in the Rhinoceros – Grasshopper environment. The Grasshopper is a visual programming language that allows to parametrize the drawing tools of Rhinoceros. An image of the canvas referring to the algorithm is reported in Fig. 1.

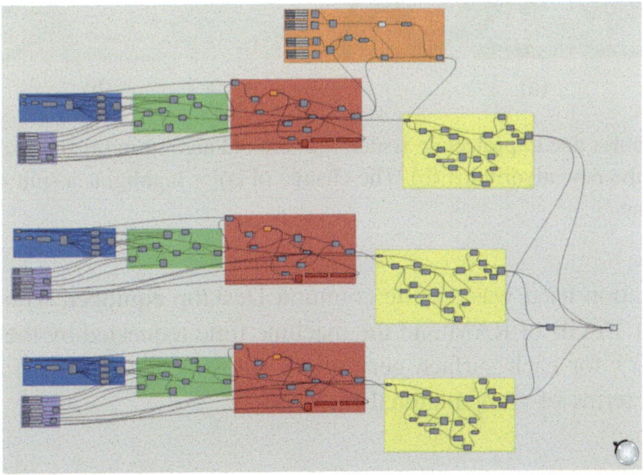

Fig. 1. The flowchart of the algorithm on the Grasshopper canvas

In the canvas, the main blocks, each one of them contains several functions, are identified by a different colour. The colours mean, respectively:

- Blue: Point and surface reading
- Violet: Parameter setting
- Green: Voronoi tessellation
- Red: Curve generation
- Orange: Constraint imposition
- Yellow: Surface generation

Each function inside the main blocks, is represented by a grey palette and means a different command. The relation between each function is represented by a grey connection line.

The used algorithm is an evolution of that presented by Cucinotta et al. [9]. In this evolution, the authors improved some characteristics of the algorithm, in particular, the management of surfaces that significantly change the orientation in space. Indeed, when the surface to be optimized has an abrupt spatial variation in terms of curvature and orientation, the holes may risk self-intersecting or intersecting with one another, leading to the construction of impossible geometries, angular points and broken holes (Fig. 2a). The new algorithm instead projects the Voronoi pattern with the normal of the surface point by point, instead of using a unique fixed normal. Where even this procedure is not possible, it splits the surface into two sub-surfaces, so as to always reach a good quality hollowing (Fig. 2b).

Fig. 2. Broken holes due to projecting issues with the old algorithm (a) and fixed holes with the improvement of the new algorithms (b). The change of color highlights a split into sub-surfaces of the original one.

The workstation used was a quite common Desktop equipped with Intel Core I7-8700 CPU with 16 GB of RAM and the machine time requested by the algorithm was about of 20–30 s for each surface generation, showing the high performance of the method also in terms of computational resources.

2.2 Geometry Acquisition

A 3D structured light optical scanner with two stereo cameras was used, in particular it was an ATOS Compact Scan 500-2M by GOM. The scanner, using the triangulation method, allows to obtain point clouds from the acquired surfaces guaranteeing a resolution of 0.02 mm [13]. An STL polygonal mesh was obtained from the scanning process. From the mesh through a reverse engineering process a CAD model was obtained using the loft function of the NURBS that make up the model Fig. 3.

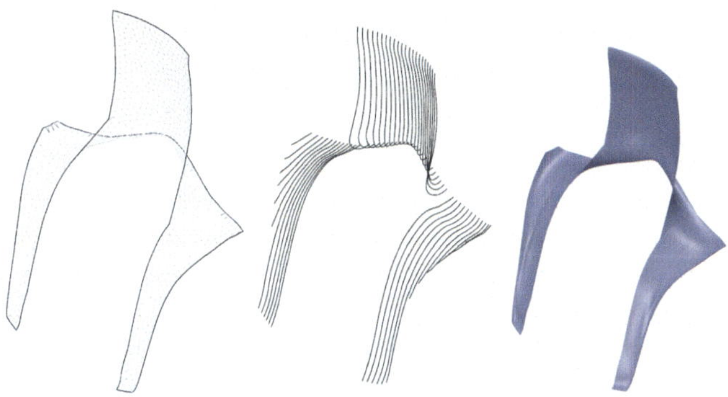

Fig. 3. The HANS, respectively in StL triangulation, sections and reconstructed CAD surface

2.3 The Material

For this work a Titanium alloy has been used for the AM process in SLS technology. In particular, the alloy takes the name of EOS Titanium Ti64, made by EOS GmbH. This type of material is used in various industrial sectors, in relation to its good mechanical properties and in the biomedical field in consequence of its biocompatibility [14].

Material properties are reported in Tables 1 and 2.

Table 1. Physical and chemical properties

Material composition	Ti (balance) Al (5.5–6.75 weight %) V (3.5–4.5 weight %) O (<2000 ppm) N (<500 ppm) C (<800 ppm) H (<150 ppm) Fe (<3000 ppm)
Density	4410 kg/m^3

Table 2. Mechanical properties of parts

	As built
Horizontal tensile strength	1230 ± 50 MPa
Vertical tensile strength	1200 ± 50 MPa
Horizontal yield strength	1060 ± 50 MPa
Vertical yield strength	1070 ± 50 MPa
Horizontal elongation at break	10 ± 2%
Vertical elongation at break	11 ± 3%
Horizontal Modulus of elasticity	110 ± 10 GPa
Vertical Modulus of elasticity	110 ± 10 GPa

2.4 The FE Model

FEM simulations were performed using the Nastran solver within Siemens NX 1859. The choice of mesh fell on triangular elements of 1 mm, in order to faithfully follow the topology of the HANS device, which, once optimized, becomes increasingly complex. The device has a constant thickness of 4 mm. A linear static analysis was performed using a Nastran SOL 101. To be sure that the boundary conditions were always the same in each simulation, fixed selection boxes have been created.

The system was constrained along the lateral shoulders of the HANS device with fixed constraints. That constraint simulated the action of the safety belts which are able to block all translations and rotations (Fig. 4).

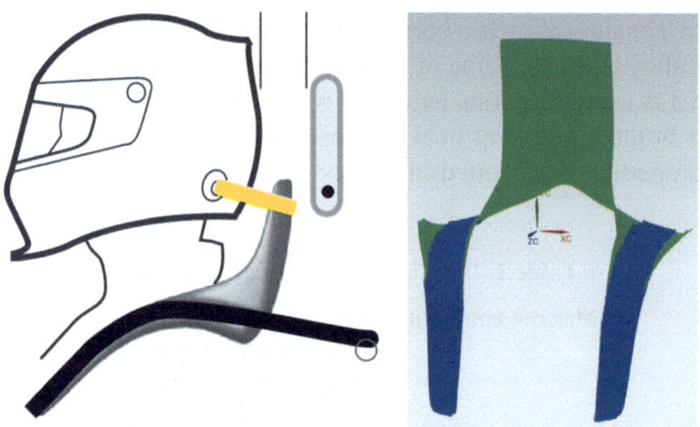

Fig. 4. A scheme of the HANS and safety belts configuration (on left) and the fixed constraints (on right)

The applied load consists of a force that simulates the action of the weight of the driver's head (5 kg) plus the weight of the helmet (1 kg) with a negative acceleration of 45 g (about 441.3 m/s^2). This load is distributed on the little belt the connects the helmet with the HANS device (Fig. 5). The HANS device is perfectly symmetric.

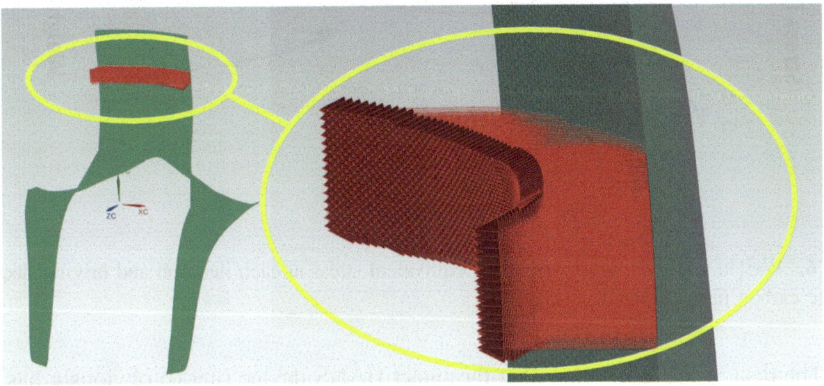

Fig. 5. The applied load

3 Results and Discussion

3.1 The Topology Optimization

A TO study was carried out on a HANS device made of SLS Titanium alloy. The first iteration, called Iteration 0, was carried out on the original surface. After that, the iterations, from 1 to 3, allowed to reach a right balance between the collar weight and an allowable stress field. Table 3 shows the weight and maximum stresses obtained at each iteration.

Table 3. Titanium alloy stress and weight

Iteration	Max stress [MPa]	Weight [kg]
0	364.99	0.828
1	2122.59	0.472
2	982.48	0.514
3	661.83	0.517

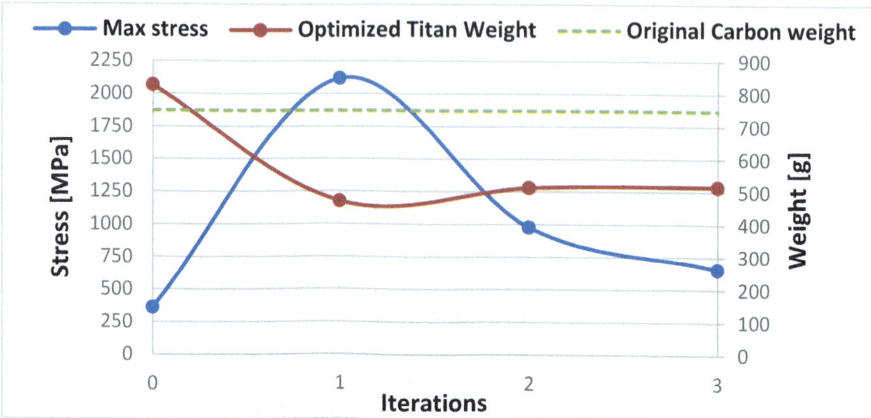

Fig. 6. Weight and maximum von Mises equivalent stress at each iteration and original weight of the carbon fiber device

The data in the table show that the intact HANS device (iter 0) obviously has the lowest stress and the highest weight. The iteration 1 is characterized by a too high maximum stress, greater than the yield strength of the material, and a weight reduced by 43% compared to the original HANS. At the iteration 2, there is a small increase in weight with respect to the previous iter, but the maximum stress decreases to a value similar to the yield stress. In the last iteration, it is possible to observe a very small increase in weight with respect to iter 2 and a furthermore decrease in stress that lead to an allowable condition. The total weight reduction of the iteration 3 compared to the initial conditions is of 37.5%. Figure 6 shows the trend of the HANS weight and the maximum stress at each iteration.

In Fig. 7 it is possible to evaluate the von Mises equivalent stress distribution at each iteration.

At the iteration 1 it is possible to evaluate a regular distribution of the holes, both in shape and in position, with consequent great reduction in weight but also in the resistance of the device. At the last iteration, a local condensation in the most stressed points allows to obtain globally a HANS that resists the external stresses for which it was designed with a drastic reduction of weight.

Iter	Global view	Particular
0	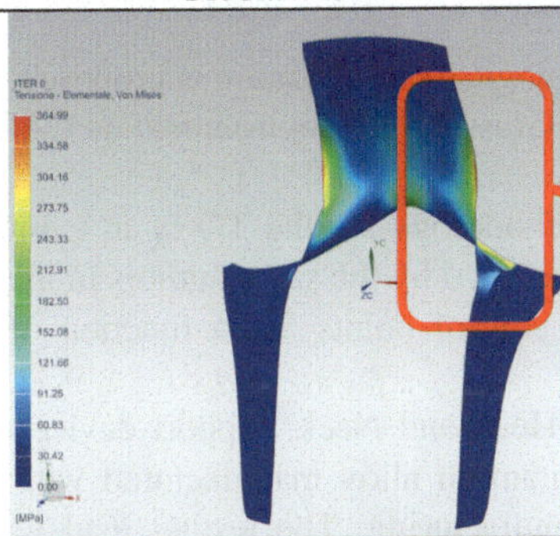	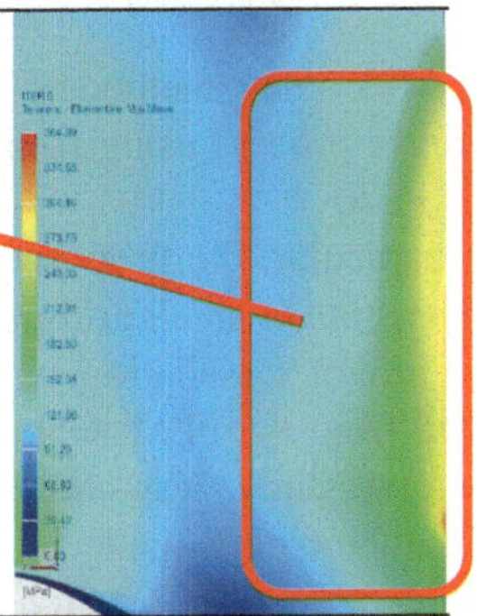
1	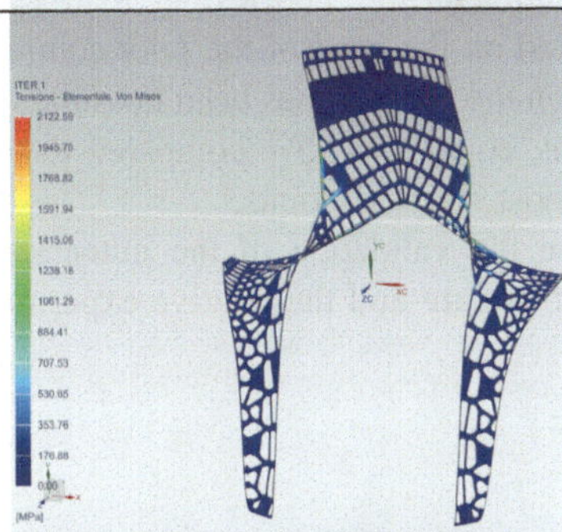	
2	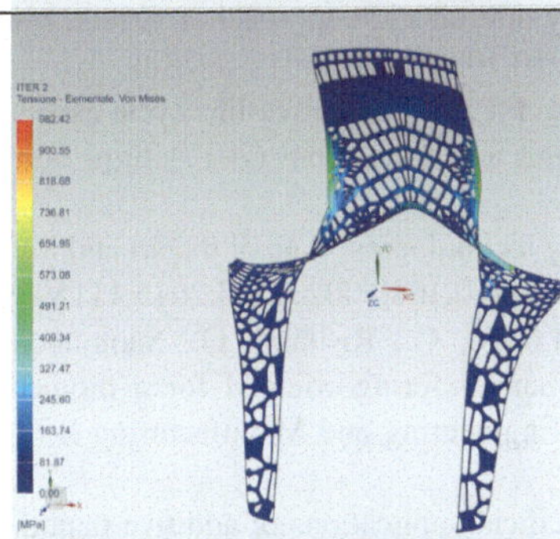	
3	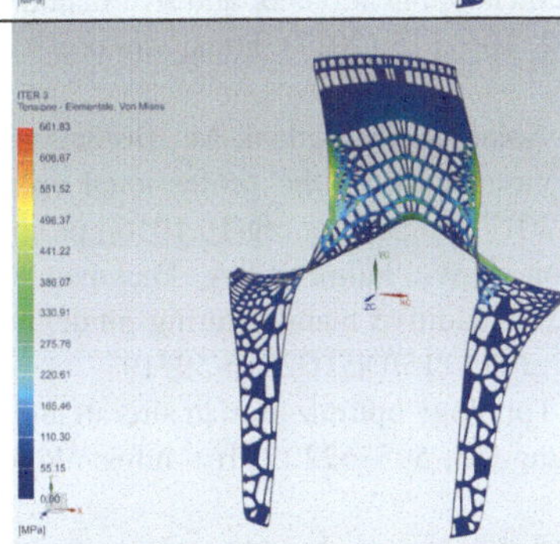	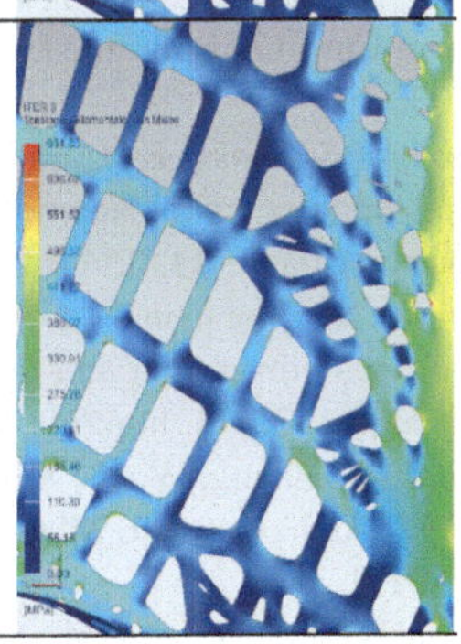

Fig. 7. The stress map at each Iterations

4 Conclusions

In this work a new method of Topological Optimization is proposed, which is based on an iterative process using Rhino - Grasshopper environment as CAD and Nastran NX for the FEA.

In particular, the proposed methodology for the TO is an evolution, with several improvements of an algorithm, developed by the same authors in the past, that is based on a Voronoi tessellation whose control points are a function of the stress values obtained from the FEM analysis.

As case study, was chosen a Head and Neck Support device designed for competitive motorsport, made by a Titanium alloy manufactured with a Selective Laser Sintering process in Additive Manufacturing. The testing load was the action of a negative acceleration of 45 g applied on the head of the pilot equipped with a helmet.

Through four iterations acting on the CAD, it has been possible to achieve a drastic weight reduction of the component, equal to 37.5% compared to the original model, maximizing the resistance in the most stressed points.

Further developments could be the validation of the numerical results with the application of the homologation procedure and the relative experimental tests.

References

1. Sigmund, O., Maute, K.: Topology optimization approaches. Struct. Multidiscip. Optim. **48**, 1031–1055 (2013). https://doi.org/10.1007/s00158-013-0978-6
2. Barone, S., Cucinotta, F., Sfravara, F.: A comparative life cycle assessment of utility poles manufactured with different materials and dimensions (2017). https://doi.org/10.1007/978-3-319-45781-9_10
3. Gardan, J.: Additive manufacturing technologies: state of the art and trends. Int. J. Prod. Res. **54**, 3118–3132 (2016). https://doi.org/10.1080/00207543.2015.1115909
4. Cucinotta, F., Guglielmino, E., Longo, G., Risitano, G., Santonocito, D., Sfravara, F.: Topology optimization additive manufacturing-oriented for a biomedical application. In: Advances on Mechanics, Design Engineering and Manufacturing II (2019). https://doi.org/10.1007/978-3-030-12346-8_18
5. Singh, S., Ramakrishna, S.: Biomedical applications of additive manufacturing: present and future. Curr. Opin. Biomed. Eng. **2**, 105–115 (2017). https://doi.org/10.1016/j.cobme.2017.05.006
6. Graziosi, S., Rosa, F., Casati, R., Solarino, P., Vedani, M., Bordegoni, M.: Designing for metal additive manufacturing: a case study in the professional sports equipment field. Procedia Manuf. **11**, 1544–1551 (2017). https://doi.org/10.1016/j.promfg.2017.07.288
7. Reddy, N., Maranan, V., Simpson, T.W., Palmer, T.A, Dickman, C.J.: Application of topology optimization and design for additive manufacturing guidelines on an automotive component (2016). http://dx.doi.org/10.1115/DETC2016-59719
8. Zhu, J.-H., Zhang, W.-H., Xia, L.: Topology optimization in aircraft and aerospace structures design. Arch. Comput. Methods Eng. **23**, 595–622 (2016). https://doi.org/10.1007/s11831-015-9151-2
9. Cucinotta, F., Raffaele, M., Salmeri, F.: A stress-based topology optimization method by a Voronoi tessellation Additive Manufacturing oriented. Int. J. Adv. Manuf. Technol. **102** (2019). https://doi.org/10.1007/s00170-019-03676-4

10. Fotovvati, B., Namdari, N., Dehghanghadikolaei, A.: Fatigue performance of selective laser melted Ti6Al4V components: state of the art. Mater. Res. Express. **6**, 14 (2019). https://doi.org/10.1088/2053-1591/aae10e
11. Kaul, A., Abbas, A., Smith, G., Manjila, S., Pace, J., Steinmetz, M.: A revolution in preventing fatal craniovertebral junction injuries: lessons learned from the Head and Neck Support device in professional auto racing **25**, 756–761 (2016). https://doi.org/10.3171/2015.10.SPINE15337.756
12. Abrami, S., Cucinotta, F., Guglielmino, E., Sfravara, F.: Human factors assessment for comfort and safety in the XCAT powerboats rules. Springer, Cham (2019). https://doi.org/10.1007/978-3-030 12346-8
13. Cucinotta, F., Paoli, A., Risitano, G., Sfravara, F.: Optical measurements and experimental investigations in repeated low-energy impacts in powerboat sandwich composites. Proc. Inst. Mech. Eng. Part M J. Eng. Marit. Environ. (2017). https://doi.org/10.1177/1475090217720619
14. Kumari, R., Scharnweber, T., Pfleging, W., Besser, H., Dutta, J.: Applied surface science laser surface textured titanium alloy (Ti – 6Al – 4V) – Part II – Studies on bio-compatibility, vol. 357, pp. 750–758 (2015)

A Cooperative Monitoring System for Diver Global Localization and Operation Support

Fabio Bruno[1,2], Loris Barbieri[1(✉)], Antonio Lagudi[2], Marino Mangeruga[2,3], Francesco Pupo[3], and Alessandro Casavola[3]

[1] Department of Mechanical, Energy and Management Engineering (DIMEG), University of Calabria, via P.Bucci 46c, 87036 Rende, CS, Italy
{f.bruno,loris.barbieri}@unical.it

[2] 3DResearch S.r.l., via P.Bucci 45c, 87036 Rende, CS, Italy
{a.lagudi,m.mangeruga}@3dresearch.it

[3] Department of Electronics, Computer Science and System Sciences (DIMES), University of Calabria, via P.Bucci 42c, 87036 Rende, CS, Italy
{pupo,a.casavola}@dimes.unical.it

Abstract. The paper presents the MOLUX project which aims to design and develop a system capable of supporting the underwater operators by means of innovative technologies for the diagnosis, recovery, management, conservation, and valorization of the archaeological and naturalistic assets in the marine environment. This system allows to increase the efficiency and the safety of the operations carried out in the underwater environment thanks to its capabilities: to make the diver aware of his/her position; to acquire geolocated data (images, videos, notes, environmental parameters); to optimize and monitor the planned dive route.

Keywords: Underwater localization · Autonomous surface vehicle · Diver operation support · Monitoring · Underwater tablet

1 Introduction

The activities carried out in the marine environment, specifically in archaeological and biological underwater sites, present many problems and limitations compared to those performed in dry conditions.

Over the past few years, these activities have benefited from the advent of unmanned marine vehicles, both for the surface (ASV, Autonomous Surface Vehicle) and the submerged (UUV, Unmanned Underwater Vehicle) environment, that support and facilitate the exploration and survey of the marine ecosystem thanks to their reduced overall dimensions and reduced costs, when compared to research vessels, and the overcoming of all those limitations due to the presence of human operators [1–6]. Although these technologies have given a substantial boost to the marine scientific research, there are still many scenarios in which the human presence cannot be replaced by autonomous and robotic systems. In particular, where the depths allow it, human intervention is often preferred. However, this preference is made critical by the lack of technological tools on the market that can support divers in their work. In fact, the gap

C. Rizzi et al. (Eds.): ADM 2019, LNME, pp. 410–421, 2020.
https://doi.org/10.1007/978-3-030-31154-4_35

between operational needs and the support provided by the technological devices on the market is still very large.

For example, in the archaeology field, great efforts have been focused on the documentation, monitoring and conservation of the underwater cultural heritage. These activities could greatly benefit from the introduction of new devices and technological tools capable of supporting divers in the gathering of data related to the archaeological remains and to the marine environment in which they are located. Such data would be then of great help for a correct and efficient planning of the restoration and conservation activities to perform on site by the underwater archaeologists. The cooperation between man and technology can therefore represent an efficient solution to overcome the difficulties in working in a difficult and unpredictable context like the marine environment.

Another example that could benefit from the development of appropriate technological tools is represented by the inspection and recovery activities undertaken in the submerged environment by the enforcement authorities. In fact, due to the lack of a precise and reliable positioning system for the underwater environment, these activities are usually carried out by exploring the search area through systematic movements, e.g., starting from the center of the area and moving in a spiral direction, or starting from one end and moving along parallel and equidistant lines. The search and exploration activities carried out in this manner could be subject to many difficulties and problems with orientation due to the water turbidity. Therefore, a navigation and geolocation system would be of great support to efficiently and safely perform this kind of activities.

Furthermore, it is very difficult for a diver to manually calculate the safest and shortest route, that passes through predefined targets, taking into account all the different variables of the case: travel distance, the amount of air remaining in the tank, decompression stops, etc. At present, the dive computers that are on the market allow to measure time and depth of a dive and calculate a safe ascent profile, but they are not able to provide information neither reliable estimates of the route ahead.

On the basis of these considerations, the paper presents the MOLUX (MObile Lab for Underwater Exploration) project that has been founded by the Structural Funds (ERDF and ESF) of the European Commission by means of the POR (Operative Regional Programme) Calabria 2014-2020 "Priority 1 - research and innovation" for proposals in Action 1.2.2 "Support for the implementation of complex projects of research and development activities on a few important thematic areas and for the application of technological solutions in accordance with S3 strategies".

The aim of the research project is the design, development, and testing of an innovative system that supports underwater operators for the diagnosis, recovery, management, conservation and valorization of the archaeological and naturalistic assets in the marine environment. A key capability provided by the proposed system consists of a 3D navigation feature, provided by means of an underwater tablet, that allows the diver to know his/her position within the underwater environment and perform a real-time calculation of the shortest path through which it is possible to safely reach the desired targets.

2 Competitive and Technological Reference Context

At the moment there are no integrated systems or prototypes, both on the market and in the scientific research field, that adopt autonomous marine vehicles to support the operations carried out by the diver in the underwater environment and provide his/her global position. Furthermore, also the features of the MOLUX system, detailed in the next sections, their integration level represents a significant innovation in the field of the research applied to the underwater environment.

If instead consideration is given to the individual main components or functionalities that characterize the proposed system, then the market offers some potentially competing products.

About the underwater navigation systems, most of the existing commercial solutions have been specifically developed for the military field in order to perform operations in an unknown underwater environment without the presence of a support vessel. These systems cannot be efficiently adapted to other fields, in fact, in the archaeological and scientific fields the activities are usually undertaken with the adoption of the bathymetric map of the seafloor and the supervision of the operations from the shore or from a support vessel. Based on this premise, a first example of underwater navigation system is DiNIS (Diver Navigation and Imaging System) [7], developed by Kenautics, that has been designed primarily to meet the needs of the military special forces. It consists of a display and a series of commercial sensors integrated into a single housing without other external components. The main features of this system are the navigation between waypoints and the acquisition of sonar images and video. The navigation is performed by means of underwater sensors, such as Doppler Velocity Log (DVL), Inertial Motion Unit and Attitude, Heading, and Roll Sensor (IMU/AHRS), pressure sensor, and GPS which can provide only data about the positioning on the sea surface and not underwater. Moreover, the system adopts a proprietary algorithm based on Kalman filter to provide to the diver accurate information about position and velocity.

Other commercial solutions come from the product line offered by Mistral, among which the EDGE [8] system is the one with more functionalities that have been designed specifically to perform secret military operations in the underwater environment. This system creates a connection network between different nodes, with or without the visual horizon, to assure a long-range data transfer through acoustic techniques. The system consists of an EDGE Commander, placed on the support vessel, and an EDGE Nav, used by the divers, for the underwater navigation and communication. In particular, the EDGE Nav consists of a display and a number of sensors integrated into a single compact package, and the navigation system adopts GPS, IMU and DVL sensors to calculate the position of the diver.

The Blueprint Subsea company also produces underwater navigation systems. In particular, this company has developed an underwater laptop, Artemisis [9] that integrated a sonar for the identification of a target, DVL, GPS, and S57/S63 electronic navigation maps. This system has been designed for military purposes and for supporting the divers in the navigation and localization of objects placed on the seabed. The system is equipped with a display and five buttons that allows the diver to interact with it. Before the diving, it is possible to upload on the device files related to the mission, which contain the targets that the diver can decide to visit.

With regards to non-commercial solutions, the CADDY (Cognitive Autonomous Diving Buddy) [10] project proposes an autonomous surface vehicle to improve the monitoring, the assistance and the safety of the diver. It features cognitive functions to interpret, with a camera, the diver's actions and his/her health conditions. Nevertheless, the system does not support the diver for the exploration activities and for collecting geolocated data.

A patent search on specialized databases showed some patents on underwater virtual reality systems [11] and tools for underwater navigation and communication [12–14], which however provide only specific and limited functionalities.

About surface vehicle, the market offers several solutions for environmental monitoring and hydrographic surveys. The Teledyne Marine offers the Z-Boat 1800 [15], an autonomous surface vehicle available in five configurations that can transport up to 30 kg of equipment, such as Side Scan Sonar, Echosounder, and cameras. The vehicle's structure consists of an ABS hull made of 1,8 m in length, equipped with brushless motors, but does not provide instruments to support the diving activities.

The American ASV Global company [16] proposes three lightweight vehicles equipped for water quality monitoring, sample acquisition, wave energy monitoring, passive acoustic monitoring (PAM), and monitoring of aquatic fauna.

In particular, C-Enduro, C-Cat 2, and C-Cat3 vehicles have length and weight ranged respectively, from 2.4 to 4.2 m, and from 80 up to 350 kg. Furthermore, Yunzhous Tech company [17] offers small-size autonomous vehicles, with length less than 4 m, equipped with tools for environmental monitoring and seabed mapping.

In conclusion, differently from the abovementioned researches and products, the MOLUX project aims to develop an innovative system able support the underwater operator during the exploration and monitoring of the marine environment and make him/her operate safely thanks to its capability to know the position of the diver. The MOLUX system represents, in fact, a clear novelty in the reference market, with an original architecture and new functionalities, that can satisfy the needs of archaeologists, biologists, geologists, and also enforcement authorities engaged in inspection, documentation, protection, and monitoring of the underwater cultural and naturalistic assets.

3 Project Overview

The project partnership is composed of private companies and public research centers and in particular by: University of Calabria, which participates through the Department of Mechanical, Energy and Management Engineering (DIMEG); 3D Research S.r.l., a private firm specializing in computer graphics and software engineering that coordinate the research activities and is in charge of the development of the different software devoted to the control and communication among the various systems and tools adopted in the MOLUX project; Tech4Sea S.r.l. is an innovative start-up engineering company that provides the hardware and technological solutions for the underwater environment; and Italia Compositi S.r.l. that is in charge of the prototyping of the ASV's hulls.

The MOLUX project has been conceived on the basis of the know-how possessed by the proponents on the underwater technologies field. In particular, it is an evolution of the Lab4Dive european project [18, 19] in which Atlantis Consulting, 3D Research, Università Politecnica delle Marche and the Hellenic Institute of Marine Archaeology have collaborated to develop an innovative product for supporting underwater archaeologists in surveying, documenting and preserving the underwater cultural heritage. Furthermore, DIMEG and 3DResearch have carried out the COMAS [20] and VISAS [21, 22] projects financed by the Italian MIUR, and they are participating to four European research projects (iMARECulture [23], BLUEMED, Underwater Cultural Route in Classical Antiquity [24]) all devoted to the development of new technologies for the underwater cultural heritage.

As above mentioned in Sect. 1, the general purpose of the MOLUX project is the design, development, and testing of an innovative system for supporting divers engaged in environmental monitoring activities, biological or archaeological documentation, bathymetric survey, search and recovery of objects or in judicial police investigations. The proposed system aims to increase the efficiency and safety of underwater operations by allowing the diver to know, at any time, his/her position within the underwater environment, to acquire geolocated data (such as images, videos, notes, environmental parameters), and to optimize and monitor the path to follow for the execution of the mission. Figure 1 depicts the main components of the MOLUX system:

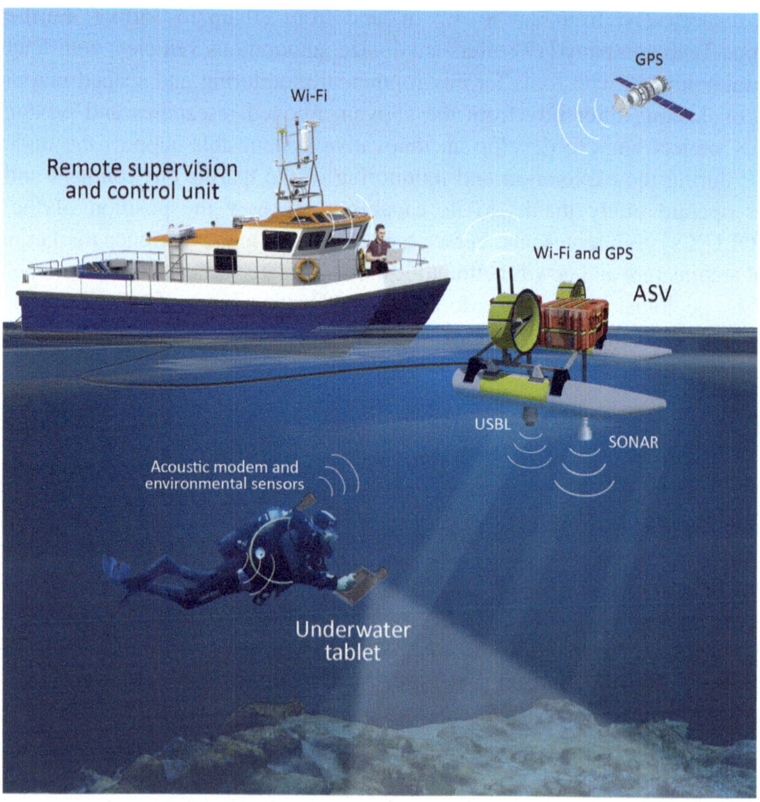

Fig. 1. Main components of the MOLUX system

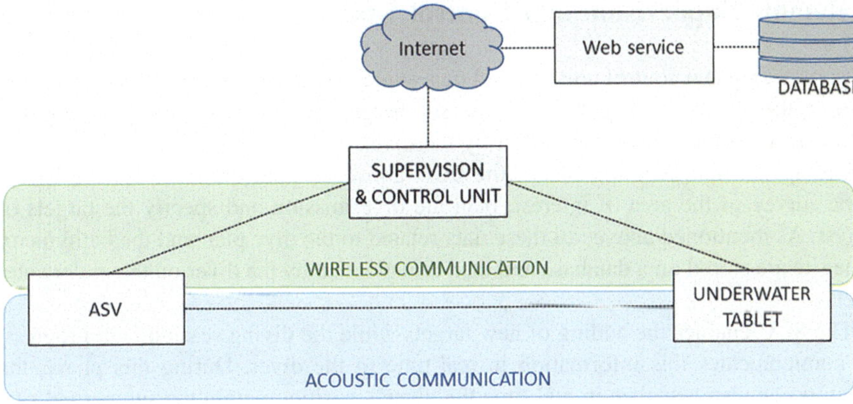

Fig. 2. Network infrastructure of the MOLUX system

- a supervision and control unit (SCU), placed on a support vessel or on the shoreline, that allows the operators to plan and monitor the mission;
- an underwater tablet that performs an acoustic geolocalization of the diver's position and it is equipped with different sensors to provides diverse functionalities that support and facilitate the research and scientific activities carried out in the submerged environment;
- an autonomous surface vehicle (ASV) that allows to perform bathymetric surveys and follow the diver during his/her movements in order to optimize the reliability and accuracy of the geolocation. Furthermore, it acts as the communication channel between the SCU and the underwater tablet.

The communication among the different devices occurs according to the network infrastructure depicted in the following image (Fig. 2). In particular, within the underwater environment, the communication between the SCU and the underwater tablet occurs thanks to the ASV that serves as a bridge between the two devices and performs the data exchange through acoustic signals. Instead, outside of the marine environment, the SCU communicates with the ASV through dedicated wireless communication channels consisting of long-range telemetry radio and WiFi transmissions. The direct communication between the SCU and the underwater tablet can take place only in the dry environment by means a WiFi communication channel.

Furthermore, for saving, downloading and synchronizing information (such as bathymetries, planned missions, mission logs, etc.) among the different devices, the SCU operates on the database by means a Web service implemented on a local network or internet.

4 Remote Supervision and Control Unit

The supervision and control unit, located on a support vessel or on the shoreline, allows the operators to perform, outside of the submerged environment, the planning and control of the diving missions carried by the diver.

During the first planning phase, the SCU allows to plan the route for the bathymetric survey of the area of interest, plan the dive mission and specify the targets of interest. As mentioned above, all these data related to the dive plan and the bathymetry of the site are stored on a database and then downloaded by the diver on the underwater tablet.

The SCU enables the adding of new targets while the diving session is in progress, and communicates this information in real-time to the diver. During this phase, the operators can also visualize in real-time the diver's position within the submerged site, the path traveled, the parameters measured by the sensors equipped on the underwater tablet and send and receive predefined short messages with the diver. When the mission is ended, the SCU is responsible for the download of the data (geolocated pictures, videos, and log files) acquired with the underwater tablet.

Furthermore, this device allows to remotely manage the ASV and if necessary perform a manual control of the vehicle.

5 Underwater Tablet

The underwater tablet (Fig. 3) consists of a watertight house equipped with an acoustic communication and localization system, and a number of sensors. The underwater housing, manufactured and commercialized by Easydive, is made from a solid block of anticorodal aluminum, subsequently anodized, and a frontal cover made with a curved transparent plexiglass that allows to operate up to 60 m deep. Inside the housing there is a double adjustable caliper support that allows to fix the tablet firmly.

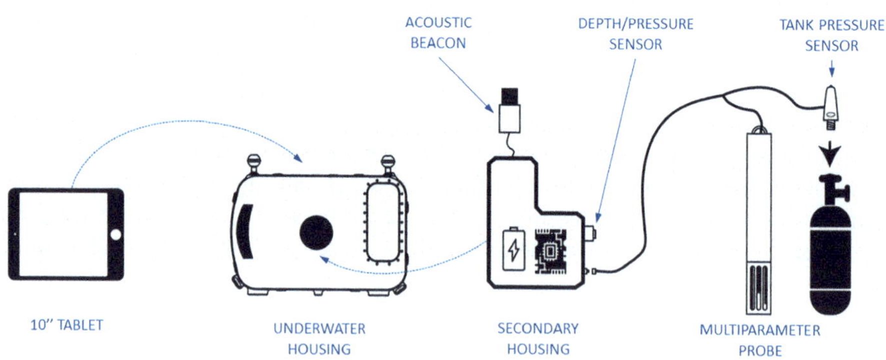

Fig. 3. Underwater tablet's architecture

The user interaction occurs by means of five command buttons, placed on the right side of the case, that can be activated through optical sensors and communicate with the tablet via a Bluetooth connection. This housing has been properly modified in order to assemble to its back a secondary case that house the acoustic localization system, an electronic control board, and a depth/pressure sensor. A multiparametric probe with water quality sensors is placed on the diver's tank to allow a real-time monitoring of the portion of the sea in which the diver is operating. In particular, the environmental parameters measured are: temperature; conductivity; pH and turbidity. Furthermore, a tank pressure sensor is mounted on the high pressure outlet on the first stage of the regulator in order to acquire the cylinder pressure data which are of fundamental importance to assess or optimize the planned diving session and ensure that it is carried out safely. All the sensors are managed by the control board that collects the various data and transmits them to the tablet via a WiFi communication channel.

On the tablet runs a software, with a graphic user interface (Fig. 4) and functionalities specifically developed to be used with the abovementioned hardware components, that allow the divers to:

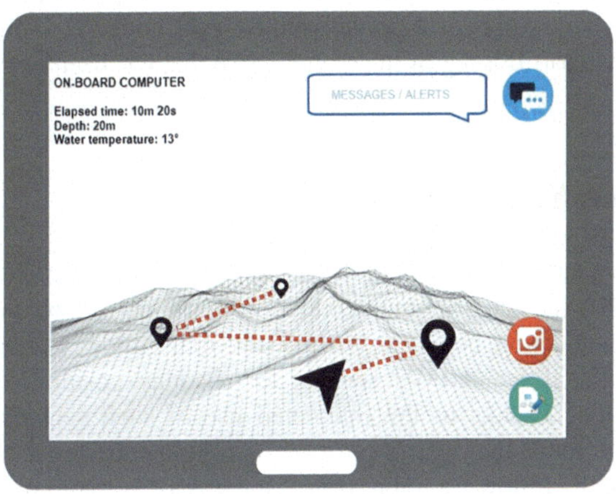

Fig. 4. GUI mockup of the underwater tablet

- download the dive plan and the bathymetry of the area of interest before starting the diving session;
- display a 3D map of the site, based on the bathymetry, and the mission targets;
- visualize a 3D map of the underwater site that shows his/her position in the submerged environment;
- view the current position and the path travelled and the target to reach and areas to explore;
- use 3D navigation features to calculate the optimal route to visit the highest number of targets;
- add new targets and communicate them to the control station located on the shore or on a support vessel;

- acquire geolocated picture with the HD camera for reporting areas of interest or for the subsequent 3D reconstruction of the explored area;
- acquire geolocated textual notes to report the location where samples were taken;
- exchange predefined short messages with the control unit.

In particular, the path optimization feature has been implemented developing a pathfinding algorithm that: recognizes obstacles within the 3D environment; calculates the shortest paths through the remotely prescribed waypoints taking into account obstacles and minimizing the pressure at which the diver is subjected; verifies the generated paths' admissibility according to constraints imposed by the underwater environment. In particular, two constraints have been taken into account: the air in the diving tank and the decompression stops needed to complete the scuba diving session safely. Given the prescribed waypoints, if there is no path that satisfies the constraints, the diver cannot visit all the waypoints in the remaining time. In this case, it is necessary to discard a waypoint at a time until a path that satisfies the constraints exists. In any case the priorities between the waypoints are respected. The pathfinding feature has been developed on the basis of the A* searching algorithm [25] that constructs a tree of weighted paths starting from a specific node and expands the paths, one step a time, until one of them ends at a predetermined goal node. To achieve the goal the algorithm requires a partitioning of the search space that has been performed with a 3D grid.

6 Autonomous Surface Vehicle

In order to improve the performance of the positioning system, the acoustic localization device, i.e., a USBL sensor, is mounted on the ASV which follows the diver to ensure an optimal communication between the acoustic transponder beacons. The ASV is equipped also with GPS and IMU sensors for absolute position, accelerations and rotations calculation. Thanks to the integrated use of such sensors, it will be then possible to geo-reference the diver's position on a global reference system.

As depicted in the following figure (Fig. 5), the ASV is based on a catamaran type structure, 1.5 m long and an overall width of about 1 m. ASV's hulls have been designed on the basis of CFD analysis and experimental test carried out with a scaled model in a recirculating water channel taking into account the following requirements: weight of approximately 50 kg; low draft; two electric powered jet-drive propellers; cruise speed of 1.5 knots [26]. It has also a central fin equipped with the USBL beacon and an electrical thruster that facilitates maneuvering operations.

The hulls are made of composite material and house the propulsion system and batteries. While, the support structure, which connects the hulls, is made of stainless steel. The vehicle also carries an onboard computer responsible for the execution of autonomous or remotely controlled missions and the management of the various sensors mounted and communication channels with the underwater tablet and the control unit.

Fig. 5. ASV's architecture

7 Conclusions

The paper has presented an innovative system, developed in the context of the MOLUX project, that supports underwater operators while performing activities for the environmental monitoring, biological and archaeological documentation, bathymetric surveys, search and recovery of objects and police investigations.

In particular, the MOLUX system increases the efficiency and safety of underwater operations by allowing the diver to know at any time his/her position within the submerged environment, to acquire geolocated data (images, videos, notes, environmental parameters), to optimize and monitor the path for the execution of the mission.

The most innovative aspect of the proposed system concerns the 3D navigation capability that allows to calculate, optimize and carry out safely the diving route thanks to a real-time monitoring of the diver's position, depth from the sea surface, and amount of residual air contained in the diving tank. Thanks to this capability the diver has a much better perception of the submerged environment. The dive area is represented through a 3D bathymetry map which can be acquired, before the diving mission, through an autonomous marine surface vehicle (ASV).

Another innovative aspect concerns the adoption of the ASV as a bridge for the communication between the diver and the operators who monitor and control all the activities from the shore or a support vessel.

The proposed system can be used also without the bathymetry map of the area, losing then the functionalities for the automatic calculation of the optimal path, but maintaining all the others, such as the geolocation of the diver and the acquisition of geolocated photos, environmental data, and notes.

Acknowledgements. This research is funded by the Calabria Region in the framework of the project POR CALABRIA FESR-FSE 2014-2020. ASSE I - PROMOZIONE DELLA RICERCA E DELL'INNOVAZIONE Obiettivo specifico 1.2 "Rafforzamento del sistema innovativo regionale e nazionale" Azione 1.2.2 "Supporto alla realizzazione di progetti complessi di attività di ricerca e sviluppo su poche aree tematiche di rilievo e all'applicazione di soluzioni tecnologiche funzionali alla realizzazione delle strategie di S3".

References

1. Steimle, E.T., Hall, M.L.: Unmanned surface vehicles as environmental monitoring and assessment tools. In: OCEANS 2006, IEEE, pp. 1–5 (2006)
2. Dunbabin, M., Grinham, A., Udy, J.: An autonomous surface vehicle for water quality monitoring. In: Australasian Conference on Robotics and Automation (ACRA), pp. 2–4 (2009)
3. Scaradozzi, D., Sorbi, L., Zoppini, F., Conte, G., Ramirez, P.A.Z., Cerrano, C.: Innovative strategy and process for underwater data gathering and results elaboration. In: 22nd Mediterranean Conference on Control and Automation, IEEE, pp. 652–657 (2014)
4. Vasilijević, A., Nađ, Đ., Mandić, F., Mišković, N., Vukić, Z.: Coordinated navigation of surface and underwater marine robotic vehicles for ocean sampling and environmental monitoring. IEEE/ASME Trans. Mechatron. **22**(3), 1174–1184 (2017)
5. Scaradozzi, D., Zingaretti, S., Panebianco, L., Altepe, C., Egi, S.M., Palma, M., Micheli, F.: DocuScooter: a novel robotics platform for marine citizen science. In: The 27th International Ocean and Polar Engineering Conference. International Society of Offshore and Polar Engineers (2017)
6. Jorge, V.A., Granada, R., Maidana, R.G., Jurak, D.A., Heck, G., Negreiros, A.P., Amory, A. M.: A survey on unmanned surface vehicles for disaster robotics: main challenges and directions. Sensors **19**(3), 702 (2019)
7. Kenautics. http://kenautics.com. Accessed 01 July 2019
8. MISTRAL. http://www.mistralinc.com/Technologies/Underwater-Communications-and-Navigation/EDGE. Accessed 01 July 2019
9. BLUEPRINT SUBSEA. http://www.blueprintsubsea.com/pages/product.php?PN=BP00979. Accessed 01 July 2019
10. Mišković, N., Bibuli, M., Birk, A., Caccia, M., Egi, M., Grammer, K., Marroni, A., Neasham, J., Pascoal, A., Vasilijevic, A., Vukić, Z.: Overview of the FP7 project "CADDY - Cognitive Autonomous Diving Buddy". In: OCEANS 2015-Genova, pp. 1–5. IEEE (2015)
11. Virtual Diving System and Method, US 20100302233 A1
12. Scuba diving device providing underwater navigation and communication capability, US 20110055746 A1
13. Underwater navigation system, US 8700323 B1
14. Autonomous underwater navigation, US 8346415 B1
15. Teledyne Oceanscience. http://www.teledynemarine.com/z-boat-1800. Accessed 01 July 2019
16. American ASV Global company. https://www.asvglobal.com/marine-science/. Accessed 01 July 2019
17. Yunzhou Tech company. http://www.yunzhou-tech.com/. Accessed 01 July 2019
18. Scaradozzi, D., Bruno, F., Zingaretti, S., Panebianco, L., Ciuccoli, N., Screpanti, L., Ritacco, G.: The Lab4Dive Project: a mobile smart lab for augmented archaeological dives-overview and first results. In: IMEKO International Conference on Metrology for Archaeology and Cultural Heritage Lecce, Italy (2017)
19. Scaradozzi, D., Zingaretti, S., Ciuccoli, N., Costa, D., Palmieri, G., Bruno, F., Tzifopanopoulos, A.: Lab4Dive mobile smart lab for augmented archaeological dives. In: IOP Conference Series: Materials Science and Engineering, vol. 364, no. 1, p. 012054. IOP Publishing (2018)

20. Bruno, F., Muzzupappa, M., Barbieri, L., Gallo, A., Ritacco, G., Lagudi, A., Comite, V.: The CoMAS project: new materials and tools for improving the in situ documentation, restoration, and conservation of underwater archaeological remains. Mar. Technol. Soc. J. **50** (4), 108–118 (2016)
21. Bruno, F., Lagudi, A., Muzzupappa, M., Lupia, M., Cario, G., Barbieri, L., Saggiomo, R.: Project VISAS: virtual and augmented exploitation of submerged archaeological sites-overview and first results. Mar. Technol. Soc. J. **50**(4), 119–129 (2016)
22. Bruno, F., Barbieri, L., Lagudi, A., Cozza, M., Cozza, A., Peluso, R., Muzzupappa, M.: Virtual dives into the underwater archaeological treasures of South Italy. Virtual Reality **22** (2), 91–102 (2018)
23. Bruno, F., Lagudi, A., Ritacco, G., Agrafiotis, P., Skarlatos, D., Čejka, J., Mudur, S.: Development and integration of digital technologies addressed to raise awareness and access to European underwater cultural heritage. An overview of the H2020 i-MARECULTURE project. In: OCEANS 2017-Aberdeen, pp. 1–10, IEEE (2017)
24. Petriaggi, B.D., Bruno, F., Puglisi, V., Medaglia, S., Lagudi, A.: UCRCA: underwater cultural routes in classical antiquity: un progetto europeo per la promozione del turismo archeologico subacqueo. Archaeologia maritima mediterranea **15**(15), 103–137 (2018)
25. Hart, P.E., Nilsson, N.J., Raphael, B.: A formal basis for the heuristic determination of minimum cost paths. IEEE Trans. Syst. Sci. Cybern. **4**(2), 100–107 (1968)
26. Barbieri, L., Cucinotta, F., Gallo, A., Bruno, F., Penna, N., Gaudio, R.: Design and simulation of the hull of a small-sized Autonomous surface vehicle for seabed mapping. In: Proceedings of the ADM 2019 International Conference, Springer (2019)

Design and Simulation of the Hull of a Small-Sized Autonomous Surface Vehicle for Seabed Mapping

Loris Barbieri[1(✉)], Filippo Cucinotta[2], Alessandro Gallo[3], Fabio Bruno[1,3], Maurizio Muzzupappa[1], Nadia Penna[4], and Roberto Gaudio[4]

[1] Department of Mechanical, Energy and Management Engineering (DIMEG), University of Calabria, via Bucci 46C, 87036 Rende, CS, Italy
{loris.barbieri,fabio.bruno,muzzupappa}@unical.it
[2] University of Messina, C.da di Dio, 98166 Messina, Italy
filippo.cucinotta@unime.it
[3] Tech4Sea S.r.l., via Bucci, 87036 Rende, CS, Italy
alessandro.gallo@tech4sea.it
[4] Department of Civil Engineering (DINCI), University of Calabria, Via Bucci 42b, 87036 Rende, CS, Italy
{nadia.penna,roberto.gaudio}@unical.it

Abstract. Autonomous Surface Vehicles are versatile marine vehicles that allow to fulfill a variety of offshore activities. Their versatility has been appreciated by the marine and aquatic science community, in fact, in the last years, a large number of ASVs have been developed in research projects and introduced in the market.

In this paper, the design and simulation of a small-sized ASV for seabed mapping of shallow waters are described. The vehicle is characterized by catamaran shape, low draft, jet-drive propellers that allow its deployment from the shore, and a payload of 20 kg. The design process has been carried out with the aim to realize a vehicle characterized by ease of transportability and deployment, available payload and performance in terms of speed and endurance.

Three hull types have been modelled in a computer-aided design environment and then optimized through fluid dynamics analysis for a cruise speed of 1.5 kN. The results of these simulations have been used to choose the best hull shape in terms of resistance, in order to comply with the constraints of autonomy and available payload.

Finally, a scaled model of the best hull shape has been then tested in a circulating water channel to validate simulation data.

Keywords: Shape design · Modeling and simulation · Autonomous Surface Vehicle · Additive manufacturing

C. Rizzi et al. (Eds.): ADM 2019, LNME, pp. 422–431, 2020.
https://doi.org/10.1007/978-3-030-31154-4_36

1 Introduction

Over the last three decades, a large variety of Autonomous Surface Vehicles (ASVs) have been developed and put in the market [1–3]. Their widespread use is due to their versatility and variety of sizes and prices that make them be adopted in many operating contexts, such as oceans, lakes, and rivers, and for a large number of missions such as scientific studies, marine environment monitoring, target object searching, seabed mapping, hydrologic survey and sampling [4–6]. Furthermore, compared to AUVs (Autonomous Underwater Vehicles), ASVs allow larger payloads and battery capacity, they can employ solar or wind power, and use various methods of propulsion.

In this paper, the focus will be given on the design and simulation of a small-sized ASV for the seabed mapping of shallow waters. The design is carried out on the basis of hull series for catamaran vehicles and the assessment of the geometry is performed by means of CFD simulations and experimentation in a recirculating water channel of a scaled model prototyped through additive manufacturing technology.

This research has been carried out in the context of the MOLUX (mobile lab for underwater exploration) project that has been founded by the Structural Funds (ERDF and ESF) of the European Commission through the POR (Operative Regional Programme) Calabria 2014–2020 "Priority 1 - research and innovation". In particular, this project aims to develop an innovative system able to support underwater operators for the diagnosis, recovery, management, conservation, and valorization of the archaeological and naturalistic cultural assets in the marine environment.

2 Hull Design

The first step in the hull design process is represented by the definition of vehicle constraints. In particular, the main requirements for the ASV, and consequently for the hull, are:

- ease of transport. The ASV should fit into the trunk of a medium-size car, then the hull dimensions must not exceed the 1500 mm of length;
- the overall weight of the ASV, that consists of hulls, motors, and batteries, must not exceed 30 kg;
- payload of 20 kg;
- cruise speed of 1.5 kN;
- the vehicle must be deployed from the shore and used in shallow waters, for this reason traditional propellers or thruster cannot be used. As a consequence, a jet-drive propeller has been chosen to ensure a very low draft.

As it is common in naval architecture, in order to obtain the best performance with as little effort as possible, hull series have been used. The catamaran series, among the best known, for which reference was made in this study are the following:

- Muller-Graf series (1989) [7]
- NPL series, as modified by Molland (1994) [8];
- Sahoo series (2004) [9].

The modelling of the parametric geometry has been carried out by 20 transversal sections of the hull, by imposing the same length of the hull of 1·500 mm. The sections have been drawn for points, according to the series recommendations.

Since the models have different shapes, they have also different weight and draught. The tests have been carried out in the two standard load conditions: Light Ship Weight (LSW) and Dead Weight Tonnage (DWT). The difference consists of a payload estimated in 10 kg. The weights on board, expressed in kilograms, are summarized in the following table (Table 1):

Table 1. LSW and DWT loading conditions for each hull.

Models	Hull	Structures	Engines and electronics	Battery	LSW	Payload
Muller-Graf	3.8	3.0	2.0	5.0	13.8	10.0
NPL Molland	3.8	3.0	2.0	5.0	13.8	10.0
Sahoo	5.5	3.0	2.0	5.0	15.5	10.0

The models' main characteristics are reported in Table 2, in which D is the construction height and T the draught in the various displacement conditions.

Table 2. Model's main characteristics.

Models	D [mm]	T_{LSW} [mm]	T_{DWT} [mm]	LSW [kg]	DWT [kg]
Muller-Graf	280	128	172	13.8	23.8
NPL Molland	225	120	160	13.8	23.8
Sahoo	300	100	160	15.5	25.5

As above mentioned, the sections extracted from each series recommendation, have been imported into a CAD system and adopted as reference for the surface modelling of the three different hull's geometries (Fig. 1). Furthermore, curvature analysis has been conducted on each section in order to obtain a smooth surface.

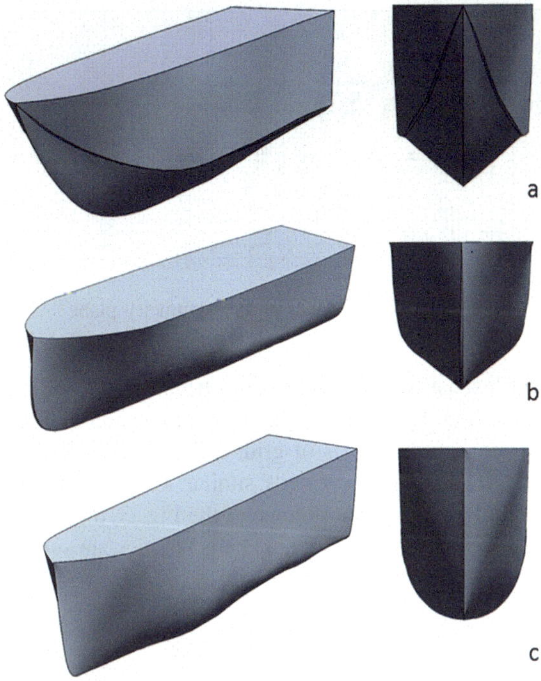

Fig. 1. Hull's geometries modelled according to the Muller-Graf (a), NPL Molland (b) and Sahoo (c) series.

3 CFD Analysis

The CFD analysis has been carried out through a commercial parametric tool that integrates a package for the Reynolds-Averaged Navier-Stokes equations (RANSe) solution. The mesh has been determined by means of an immersed-body approach that allows, despite the traditional body-fitted method, to obtain an unstructured grid with characteristics very similar to a structured one, in terms of robustness of the differential scheme and discretization of the equations, but with much higher simplicity of realization [10].

The solution is a transient Volume of Fluid (VOF) [11], with the use of the Free Surface modulus and the k-epsilon turbulence model, as modified by Lam and Bremhorst [12].

Taking advantage of the catamaran's symmetry plan, only one hull has been simulated, to reduce the grid size. Considering the fact that even in the recirculating water channel only one hull was tested, the inevitable interference effect between the two hulls in this stage was not considered. The main dimensions of the volume of control have been chosen according to the work of Ahmed [13] that parametrizes each dimension with the ship length and in particular: 1.5 L fore and 2.5 aft of the ship (globally 5 L); 1.5 L towards the side; 1.2 L under the free surface and 0.125 L above.

The mesh has been generated with cuboid elements and a refinement near the walls (Fig. 2).

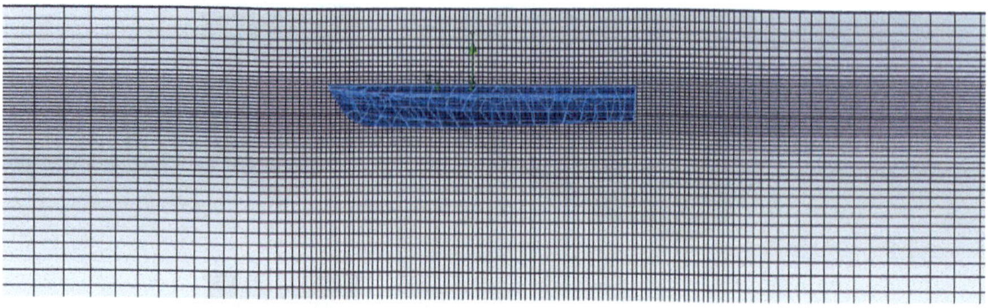

Fig. 2. The domain on the symmetry plane

A simplified mesh sensitiveness analysis has been carried out on the Sahoo model at 3 kN in DWT condition, by changing the mesh dimension and the refinement ratio, and using four different combinations of grid: coarse, medium, fine and refined. The refined mesh has a global number of cells similar to the medium mesh but a local refinement near the hull, on the green volume of the Fig. 3, that leads a local mesh size similar to the fine mesh one. The mesh main characteristics are summarized in Table 3.

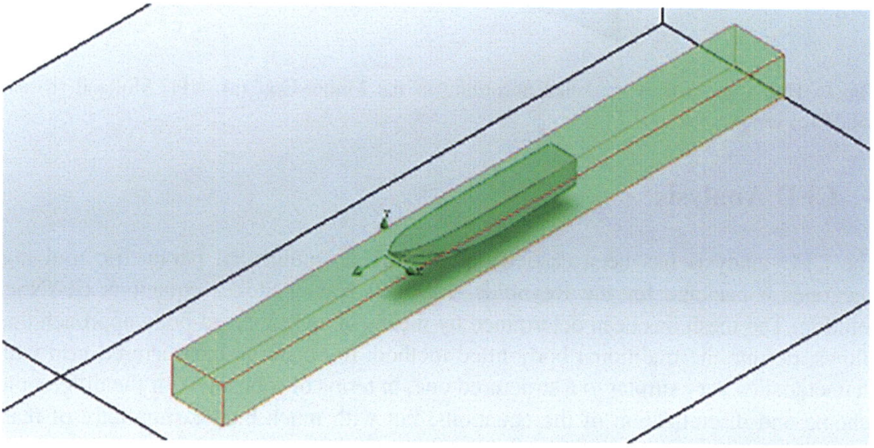

Fig. 3. The local refinement volume.

Table 3. Main characteristics of the tested meshes for sensitiveness analysis.

	Refinement ratio	Cells in the volume	Cells on the hull
Coarse	2	134141	962
Medium	3	280292	2685
Fine	4	532763	6059
Refined	4 - locally	286545	46657

An additional friction coefficient, in order to take in account, the surface roughness of the model, has been adopted according to the ITTC1978 Performance Prediction Method:

$$\Delta C_F = \left[105\left(\frac{k_s}{L}\right)^{\frac{1}{3}} - 0.64\right] \times 10^{-3} = 0.0004$$

in which k_s is a roughness coefficient assumed equal to 0.017.

The analysis was stopped when a convergence criterion was satisfied i.e., in particular, when the current oscillation of the force was less than the 3% of the average fluctuation in the interval analysis. In this way, at the achievement of the convergence, the results were those reported in Table 4.

Table 4. Main output in the tested meshes for sensitiveness analysis.

	Number of iterations	CPU Time [h:min]	Resistance [N]
Coarse	719	1:04	8787
Medium	485	1:29	8326
Fine	662	4:58	9953
Refined	708	1:44	9547

The mesh that was chosen is the Refined one because it allowed to have similar results to the Fine one, but with calculation times comparable to the Medium one.

Each one of the three models (Muller-Graf, NPL Molland and Sahoo) has been tested in unload (LSW) and load (DWT) conditions at 5 different speeds (1.5, 2, 3, 5 and 10 kN) for a total of 30 simulations.

The following figure (Fig. 4) shows the Drag curves for the LWS and DWT load conditions, for the three models. As expected, the Muller-Graf design, that is designed to be a high-speed semi-planing hull, guarantee the best performance for the high velocities at full load. However, the hull that performs the highest drag reduction in the low speed range is the Sahoo model. For these reasons the Sahoo catamaran has been chosen for the realization of the model for the experimental tests.

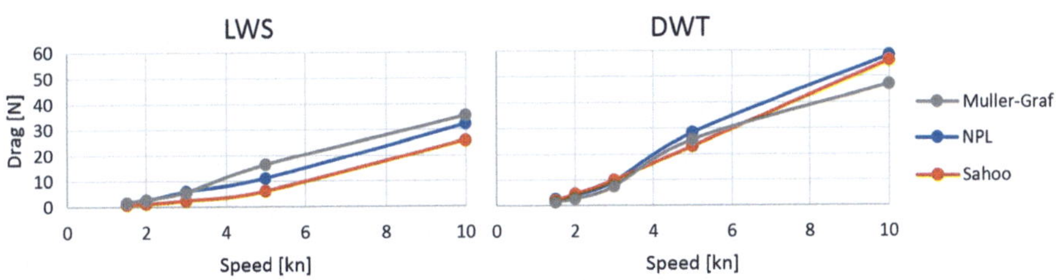

Fig. 4. Drag curves for the LWS (left) and DWT (right) conditions.

4 Model Validation in a Recirculating Water Channel

For the validation of the results, experimental tests were carried out at the *Grandi Modelli Idraulici* laboratory of the Department of Civil Engineering of the University of Calabria in a recirculating tilting flume. As depicted in the following table (Table 5), the tilting flute has a working section length 10.0 m, width 1.0 m, and depth 0.50 m. The flow was supplied with an electric pump, capable of a discharge of 60 l/s, connected to a 250 mm pipe with a diffuser, which fed an inlet tank. Further, a honeycomb filter was placed at the flume inlet; it was made of ducts with a diameter of 10 mm, to minimize the instabilities in flow at the entry. In order to regulate the flow depth, an adjustable tailgate was operated at the flume outlet. The main characteristics of the channel are summarized in the following table:

Table 5. Recirculating water channel's characteristics.

Test section length	18 m
Test section width	1 m
Pump capacity	60 l/s
Maximum speed	0.5 m/s (0.97 kN)

Due to the size and speed limits of the channel, the tests have been performed according to the Froude similitude, which involves a scale factor proportional to the ratio of the squares of the velocities, as described by Cucinotta et al. [14]. Consequently, with an operating speed of 1.5 kN (0.77 m/s), and wanting to exploit the maximum velocity allowed by the channel, a scale factor of $\lambda = V_M^2/V_S^2 = (0.5)^2/(0.77)^2 = 0.42$ has been calculated, where V_M is the speed of the scaled model and V_S is the speed of the full-scale boat. On the basis of the scale factor calculated according to the Froude similitude, the main dimensions of the scaled model of the Sahoo hull have been therefore computed (Table 6):

Table 6. Boat's and scaled model's characteristics.

	Boat	Scaled model
Length between perpendicular [mm]	1369	575
Beam width [mm]	469	197
Construction height [mm]	186	78
Draught [mm]	73	31
Wet surface [m^2]	0.480	0.086
Displacement [m^3]	0.013	0.0009
Block coefficient	0.26	0.26

4.1 Prototyping

The scaled model, with the dimensions depicted in Table 5, has been prototyped by means of additive manufacturing technologies and in particular through an SLS (Selective Laser Sintering) process performed with an EOS Formiga P110 printer. The model has been obtained by fusing a nylon powder bed by a laser. The adoption of additive manufacturing technologies required some design changes in order to take into account the technological constraints [15]. In particular, since the building volume of this printer measures 250 × 250 × 320 mm, it has been necessary to split the hull in four sections. In order to assemble the four parts, proper joints have been modelled on interface sections. The parts have been oriented with hull sections orthogonal to z-axis to reduce the staircase effect, and the printing volume has been optimized to reduce printing time to 15 h (Fig. 5a).

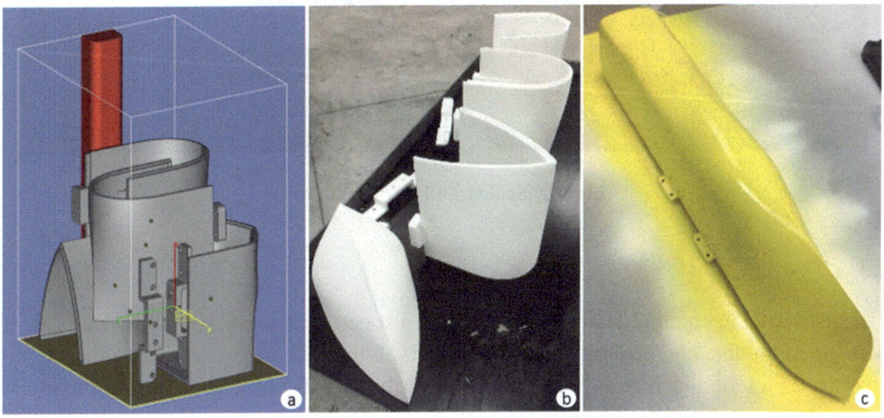

Fig. 5. Printing volume optimization and orientation of the parts (a); prototyped sections (b) and post-treatment of the assembled hull (c)

At the end of the printing process, the parts have been cleaned by sandblasting and subsequently assembled together with cyanoacrylate (Fig. 5b). The hull has been refined with different stages of painting in order to obtain a smooth surface (Fig. 5c), similar to the one used in the numerical tests.

4.2 Experimental Setup

The experimental setup consists of a modular aluminum structure and a rail with a linear recirculating bearing that support the hull allowing it to slide without friction in the longitudinal direction (Fig. 6). The thrust is measured by a load cell fixed at the end of the rail. The load cell is characterized by a capacity of 780 g and is equipped with an amplifier based on the HX711 24-bit analog-to-digital converter (ADC). The ADC signal is managed by an Arduino Mega board that sends it to the PC via USB interface.

Fig. 6. Scaled hull during the test

4.3 Test

The tests have been conducted at the maximum speed of 0.5 m/s (0.97 kN) that can be performed in the channel on the scaled hull of the Sahoo model. Considering the characteristics of the recirculating flume described in Table 5 and, in particular, the flow rate capacity of 60 l/s, implies a water depth of 120 mm. The results show a resistance of 0.42 N. Taking into account the particular test conditions, due to the limitations on the maximum flow velocity and water depth, a 25% difference between the results of CFD simulations and experimental test in the recirculating water channel will be acceptable.

5 Conclusions

This work has presented the design and simulation of the hull's geometry of a small-sized ASV that will operate on shallow waters for the mapping of the seabed.

The hull' dimensions have been defined on the basis of design requirements. The most important of them concerns ease of transport, and its geometry has been designed in accordance with three catamaran series. The final geometry has been chosen on the basis of CFD simulations and assessed by means of an experimental study conducted on a scaled model in a recirculating tilting flume. Future works are still necessary in order to map a complete range of speeds with the experimental tests and, finally, to carry out full-scale sea trials.

References

1. Manley, J.E.: Unmanned surface vehicles, 15 years of development. In: OCEANS 2008, pp. 1–4. IEEE (2008)
2. Caccia, M.: Autonomous surface craft: prototypes and basic research issues. In: 2006 14th Mediterranean Conference on Control and Automation, pp. 1–6 (2006)
3. Bertram, V.: Unmanned surface vehicles-a survey. Skibsteknisk Selskab, Copenhagen, Denmark **1**, 1–14 (2008)
4. Bayat, B., Crasta, N., Crespi, A., Pascoal, A.M., Ijspeert, A.: Environmental monitoring using autonomous vehicles: a survey of recent searching techniques. Curr. Opin. Biotechnol. **45**, 76–84 (2017)
5. Dunbabin, M., Grinham, A., Udy, J.: An autonomous surface vehicle for water quality monitoring. In: Australasian Conference on Robotics and Automation (ACRA), pp. 2–4 (2009)
6. Jorge, V.A., Granada, R., Maidana, R.G., Jurak, D.A., Heck, G., Negreiros, A.P., Amory, A. M.: A survey on unmanned surface vehicles for disaster robotics: main challenges and directions. Sensors **19**(3), 702 (2019)
7. Muller-Graf, B., Radojcic, D., Simic, A.: Resistance and propulsion characteristics of the VWS hard chine catamaran hull series' 89. SNAME Trans. **110**, 1–29 (2002)
8. Molland, A.F., Wellicome, J.F., Couser, P.R.: Resistance experiments on a systematic series of high speed displacement catamaran forms: variation of length-displacement ratio and breadth-draught ratio. Ship Science Report, University of Southampton (1994)
9. Sahoo, P.K., Browne, N.A., Salas, M.: Experimental and CFD study of wave resistance of high-speed round bilge catamaran hull forms. In: Proceedings of 4th International Conference on High Performance Marine Vehicles, Rome, Italy (2004)
10. Cucinotta, F., Nigrelli, V., Sfravara, F.: A preliminary method for the numerical prediction of the behavior of air bubbles in the design of Air Cavity Ships. In: Advances on Mechanics, Design Engineering and Manufacturing, pp. 509–516. Springer, Cham (2017)
11. Hirt, C.W., Nichols, B.D.: Volume of fluid (VOF) method for the dynamics of free boundaries. J. Comput. Phys. **39**(1), 201–225 (1981)
12. Lam, C.K.G., Bremhorst, K.: A modified form of the k-ε model for predicting wall turbulence. J. Fluids Eng. **103**(3), 456–460 (1981)
13. Ahmed, Y.M.: Numerical simulation for the free surface flow around a complex ship hull form at different Froude numbers. Alex. Eng. J. **50**(3), 229–235 (2011)
14. Cucinotta, F., Guglielmino, E., Sfravara, F.: An experimental comparison between different artificial air cavity designs for a planing hull. Ocean Eng. **140**, 233–243 (2017)
15. Barbieri, L., Calzone, F., Muzzupappa, M.: Form and function: functional optimization and Additive Manufacturing. In: Advances on Mechanics, Design Engineering and Manufacturing II, pp. 649–658. Springer, Cham (2019)

Machine Health State Recognition Through Images Classification with Neural Network for Condition-Based Maintenance

Marco Rossoni(✉), Andrea Fumagalli, and Giorgio Colombo

Politecnico di Milano, Via La Masa 1, 20156 Milan, Italy
marco.rossoni@polimi.it

Abstract. A robust machine health state recognition tool is a pillar for condition-based maintenance and Deep Learning approach finds its natural application in such a context. This paper investigates the recognition of machine failures by image classification through a convolutional neural network in a condition-based maintenance environment. The case study involves a refrigerator for large retail establishments. Experimental measures, while the machine is approaching failure, are difficult to be collected, especially in the quantity needed for training and testing the neural network. For this reason, a digital twin of the asset has been created to simulate the behavior of the machine and generate as many data as needed: physically-based models of the machine and failure modes have been included and the simulated behavior has been tuned by using experimental data. Finally, it has been employed to generate signals that, translated into images, test the suitability of the neural network.

Keywords: Digital twin · Condition based maintenance · Image classification · Neural network

1 Introduction

Manufacturing companies are continuously looking for gaining more and more revenues. In the last past years, the trend seems to be the exploitation of those activities that historically have been considered as a source of losses. Maintenance is one of those "necessary evils" which no manufacturer would like to care about. Betting on the huge amount of data swamping companies today is a great opportunity to pursue improved maintenance approaches. This scenario enables the integration of prognostics/diagnostic techniques and maintenance policies leading to innovative approaches, like condition-based maintenance (CBM). A whole system implementing such strategies grounds on the capabilities to recognize whether a machine is healthy or not. In this view, the ingredients for the effective implementation of a health state recognition tool are (i) data and (ii) digital tools able to extract and process information. The route is straightforwardly marked: deep learning approaches find their natural application in such

C. Rizzi et al. (Eds.): ADM 2019, LNME, pp. 432–443, 2020.
https://doi.org/10.1007/978-3-030-31154-4_37

contexts. Since artificial intelligence aims to mimic the behavior of the human brain, the question is in turn: how do people recognize illness state? Research studies have highlighted a certain diagnostic capability of humans when looking to a person face [2]. Whether it is not possible to identify the disease, it is still possible to guess unhealthy conditions affecting the analyzed person. Given the outstanding performance of Convolutional Neural Network (CNN) in image classification, it is worthy to investigate whether it is feasible to perform evaluations of the health of a machine relying on the same mechanism.

This paper aims at investigating an approach to recognize and classify the failures of a machine through a CNN, converting sensor signals to images. The approach has been tested on a refrigerator for large trade establishments, such as horizontal freezers or upright display refrigerators. Such a machine is a good benchmark to think of maintenance issues from an Industry 4.0 perspective, given also the typical hierarchical structure of large trade establishment network. A proper quantity of data for training and testing the approach would require a huge amount of monitoring time and resources. In order to shrink this phase, a digital twin of the equipment is developed to simulate the behavior of the refrigerators, considering also a subset of failures that can occur in the real machine. The digital twin has been tuned using experimental pressure and temperature signals. The data for feeding and training the CNN have been then generated digitally through the digital twin. The time series have been then translated into images exploiting Gramian Angular Field algorithm and creating a three layers matrix interpreted as an RGB image. Given their outstanding performance in image recognition and classification, the convolutional architecture has been selected for the artificial neural network. The set of maps depicting the health status of the equipment is the input for the artificial net which aims at classifying the type of failure.

2 Background

According to [14], maintenance can be defined as "all technical and managerial actions taken during usage period to maintain or restore the required functionality of a product or an asset". Although the digital revolution is deeply affecting the way maintenance is performed, several models have been already developed. One of the most recent and investigated is the CBM. Following this approach, the maintenance intervention is performed at the exact moment one of the monitored parameters exceeds a threshold previously defined. In other words, it aims at linking the maintenance intervention to the actual condition of the asset (diagnostic) and to the estimation of its residual life (prognostic). Examples of successful case studies implementing the CBM approach are provided in [1,6,16]. It is clear how the successful implementation of CBM approach relies on (i) a reliable sensor network that monitors all the relevant working parameters of the asset; (ii) a suitable signal analysis system for interpreting the eventual approaching failure and (iii) an effective maintenance policy strategy. This paper focuses on the second point and an extensive and clear literature review on this topic can be found in [8].

Table 1. List of available measures and related symbols

Items	Measured parameters	Symbols
Condenser	Temperature at inlet	T_{inC}
	Temperature at outlet	T_{outC}
Compressor	Pressure at outlet	$P_{outComp}$
Expansion valve	–	–
Evaporator	Temperature at inlet	T_{inE}
	Temperature at outlet	T_{outE}

Since our objective is to classify an image representing a machine state, the selection of CNNs seems a reasonable choice: the outstanding performances of such nets in image classification is well known in literature and supported by several case studies [2,7]. In this work, a "ready-to-use" CNN has been used and trained with transfer learning approach: AlexNet [12] is the CNN selected. It has been applied in [5] for object recognition in heterogeneous formats belonging to the manufacturing industry. Other examples can be found in medicine [9–11] and rehabilitation [4]. In the case networks that have been already developed and trained, transfer learning is convenient because (i) it permits to save time and (ii) it improves feature extraction capability of the net [13].

3 Introduction to the Case Study: Retail Refrigerators

3.1 Refrigerators and Their Working Principle

Refrigerators can be very complex items with thousands of parts and, following the "Internet of Things" technology, they are now equipped with sensors and intelligent board gathering information both from the refrigerator itself and from the surroundings. Refrigerators for retail stores are the types of machine under investigation, such as upright display cabinets or horizontal freezers. A complete description of the refrigerator's components goes beyond the scope of this paper, hence the following is just the subset of items needed for our discussion. The core of every refrigerator is the refrigeration circuit: it has a compressor, an evaporator, an expansion valve and a condenser. For each of these components, a set of physical quantity has been monitored and measured, as reported in Table 1. The setup of the measuring system was already available on the machine, hence no additional sensors were added.

The proposed solution grounds on the development of a digital twin able to mirror the working principle of the machine and to simulate failures. This need originates from the lack of experimental data which clash with the NN approach pursued in this work. Two concerns can be highlighted about these experimental data: (i) they are not enough for training the NN and only a few types of failures have been recorded; (ii) a measuring campaign in order to catch every type of

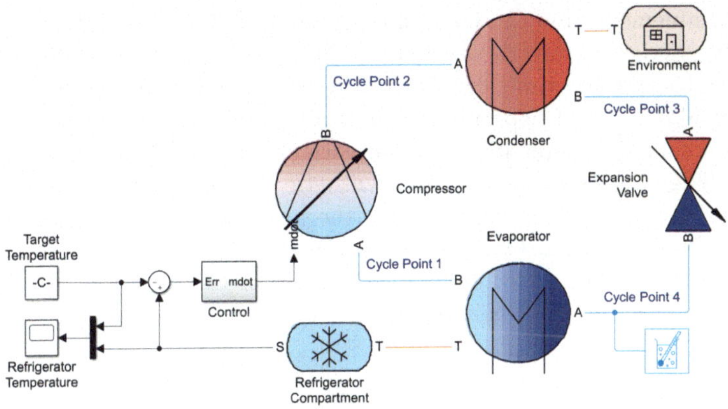

Fig. 1. Simulink model of the refrigerator circuit

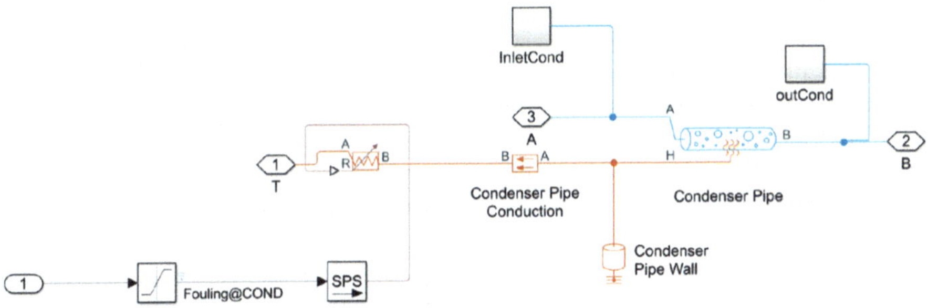

Fig. 2. Exploded view of the condenser

possible failures (with statistical meaning) could last for several years. This time span is not compatible with the industrial rhythms.

The virtual model offers the following features: (i) it is parametric in the sense that it can be adapted to any refrigerators (of the same kind) just changing the value of the parameters (e.g. diameter of tube, power, mass flow rate, etc.); (ii) it is physically consistent; (iii) the user can decide the specific faulty condition under which to generate data. The digital twin is a virtual replica of the physical refrigerator. It is desired to mirror just relevant aspects of the issue under consideration. In the case under analysis, the refrigeration circuit cycle has been modeled and implemented in a graphical programming environment for modeling and analyzing multi-domain dynamic systems (i.e. Mathworks Simulink). Figure 1 shows the visual model of the digital twin which has been modified and improved starting from a model already available. The model is able to simulate the thermodynamic cycle of the refrigerator in the normal operating conditions (i.e. when all the components work exactly as they should). Furthermore, it also has the ability to simulate some off-design conditions (i.e. failures is approaching and/or is occurred). More details on the types of failures and their implementation can be found in Sect. 3.2. Figure 2 shows the exploded view of the condenser with the detail of fouling failure blocks. The digital twin is finally able to generate time signals tracking the variables presented in Table 1.

The available set of experimental data available lasts for 16 h of monitoring, with a sampling time 45 s, and the variables reported in Table 1 has been recorded. The alignment between the synthetic signals generated by the digital twin and the experimental data ensures the physical consistency and the suitability of the simulation. Once the model has been calibrated, a (potentially) infinite source for data is available, in the form of time series. These data are employed for training the NN and validate the final results as well.

3.2 Failures Modeling

Given that the virtual model of the refrigerator aims at generating data even under faulty conditions, it has been necessary to define which faults to simulate and how to implement them. It is not always possible to define a priori all the faults that may occur during the working life of the system. The adoption of techniques like the Failure Modes and Effects Analysis (FMEA) [15] or the fault trees would have been the systematic ways to approach this problem. However, according to a previous author's work [3], it has been preferred to implement the list of faults reported in Table 2. This limited set of failure modes comes from the suggestions of experts within this field and can be regarded as exhaustive for refrigerators operating above the 0 °C, which is the same class of machines the refrigerator of reference belong to. As can be noticed from Table 2, the majority of the selected faults, e.g. the fouling and the fan breakage issues, generates a deterioration of the heat exchange capability, increasing the conductive and convective resistance between the refrigeration fluid and the environment. It follows that a physics consistent way to represents this behavior is by adding a variable thermal resistance. The other conditions cannot be modeled with a common strategy, so they require specific modeling approaches. The sensor drift is simply identified with a steep increment of the measured values and it is easily obtained by adding a constant value of one orders of magnitude above the healthy level. The clogging is instead interpreted as a decreasing cross-section of

Table 2. List of faults taken into consideration and grouped

ID	Modeling technique	Group
Mass Leakage	Reduced mass flow rate imposed	1
Fouling at Evaporator	Variable thermal resistance	2
Fouling at Condenser		
Evaporator Fan Failure		
Condenser Fan Failure		
Clogging	Additional thermal expansion valve	3
Sensor Drift	Sensor's value steep increment	4

the tubes in a specific point of the refrigerator circuit. Physically, it is identical to the thermal expansion valve working principle with the only difference in the expansion behavior: the expansion valve presents a periodic opening motion while the clogging phenomenon a progressive closure. Finally, the mass leakage has been modeled by imposing along the refrigerator circuit a lower mass flow rate value with respect to the nominal one.

4 Image Generation and Setup of the Virtual Experiments

Once the digital twin has been built, the classifier has been selected. AlexNet has been the choice since it has already shown great effectiveness in the recognition of real objects. The motivation for such a choice stands in the spirit of this investigation, in which two images representing two different faulty conditions are expected to yield two different objects, with their own colors and patterns, as it happens for two distinct physical objects.

The image of the health condition of the machine is the core objective of this work and deserves a detailed explanation. It is the result of the encoding of the data measured on the machine. The encoding procedure adopted has been developed in order to make the health evaluation performed by the NN fully compliant with the analogy with the human eye diagnostic capabilities. The method developed guarantees that the created object embedding the information of the sensor can be evaluated by the human eye as well. The developed approach grounds on a modified Gramian Angular Field (GAF) for single time series [17] and Fig. 3 shows the steps for the conversion of the time series into images. This algorithm translates three signals in a single image, as shown in Fig. 4.

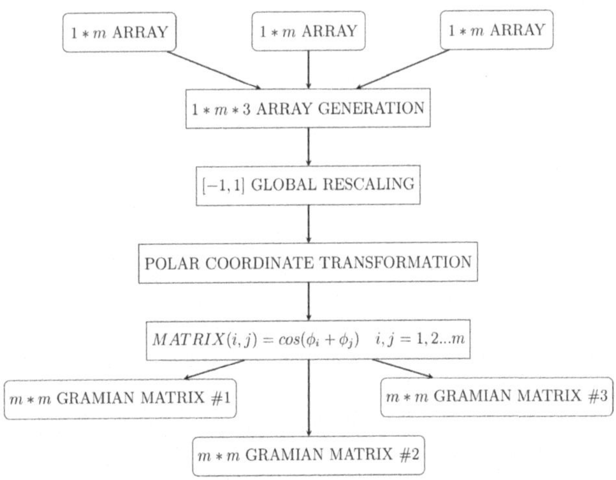

Fig. 3. Gramian Angular Field algorithm schema

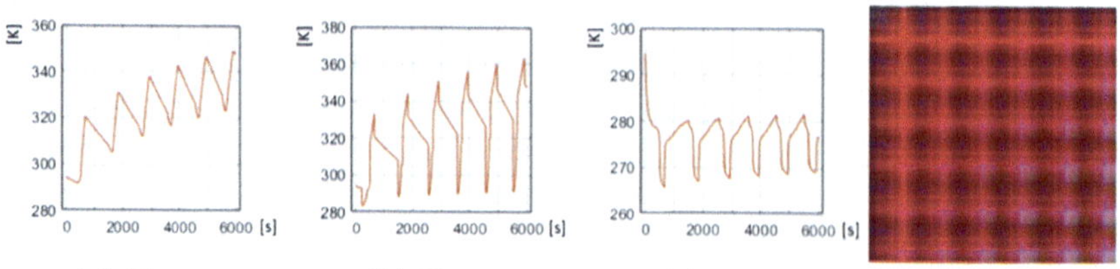

(a) T_{outC} (b) T_{inC} (c) T_{inE} (d) Related image

Fig. 4. A sample of three time histories extracted in the condenser fan failure condition and the respective image

The first step is totally demanded to the virtual refrigerator. At this very first stage it is necessary to define: (i) which fault types to simulate; (ii) the desired time length of the simulation; (iii) the three physical signals to be measured. The number of fault modes to be simulated varies from 1 to the whole set of faulty conditions the virtual model has been designed to represent. The direct impact of this first choice lays in the duration of the training phase of the NN since it implicitly means a larger amount of data available to train the NN. The selection of the three physical signals to measure is a crucial part of the overall process. Referring once again to the biological analogy, the question is: which is the set of vital parameters that is effective to detect the larger amount of diseases? In machinery, this means that it is necessary to find that triplet of physical quantities measured on the machine that would produce images that will be accurately labeled by the trained NN. For what concerns these virtual experiments, an intuitive approach has been employed. For example, when thermal faults at the condenser and at the evaporator are simulated, the selected signals are the temperatures laying at the extremities of these two devices. The next step is the translation of the three selected time series into a database of images with the developed mapping algorithm, generating as many images as it is necessary. It is now possible to execute the experiments: a part of the database has been devoted for training while the remaining part will be employed to determine network effectiveness in the classification process.

Among the different experiments conducted, the most significant are reported hereafter. In the first experiment, a triplet consisting in $P_{outComp}$, T_{outC} and T_{outE} are selected (Table 1). Starting with the description of the random images sample of each of the health classes simulated in Fig. 5, the inspection by human eye reveals a high resemblance among all the classes. The status that can be detected at the first glance is the sensor drift, characterized by uniform color, that just at a deeper analysis reveals a very light pattern. Because of this, it is expected that the chosen combination will yield a worse classification performance and consequently an unsuccessful experiment. Despite presenting a notable accuracy on the global scale, the first experiment presented a very bad local accuracy on the healthy and clogged state classification, as can be noticed in the confusion matrix in Fig. 6. Each row of the matrix represents the outcomes of the prediction while each column represents the actual values.

(a) Healthy

(b) Clogging

(c) Cond. fan

(d) Cond. fouling

(e) Eva. fan

(f) Eva. fouling

(g) Sensor drift

Fig. 5. Random samples of the simulated health conditions in experiment 1

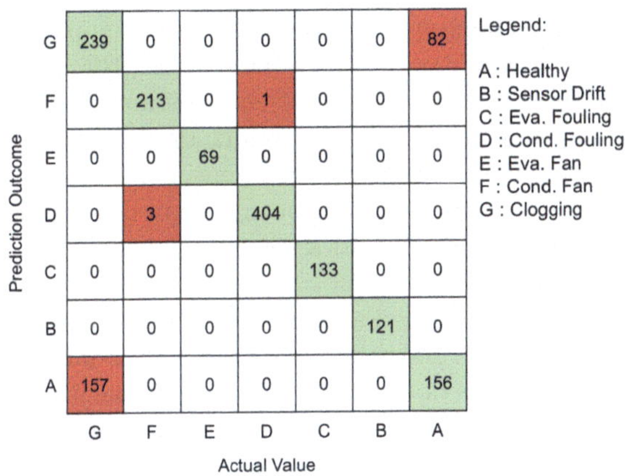

Fig. 6. Experiment 1 confusion matrix, the row index is the classification while the columns stand for the real class

The causes of this poor performance have been investigated, leading to the following hypothesis: when channels with values of a different order of magnitude are put together, the mapping algorithm generates images that don't present evident patterns, leading the classifier to wrong classifications. The problem has been identified in the global rescaling step. Since a pressure signal is combined with two other temperature signals, the global rescaling causes the normalized value of the pressure to be "hidden" by the temperature signals. In fact, the rescaled pressures are two orders of magnitudes less than the temperature signals. Therefore, a modification of the rescaling has been undertaken, turning it from a global approach to a local one. In this second version of the algorithm (Fig. 7),

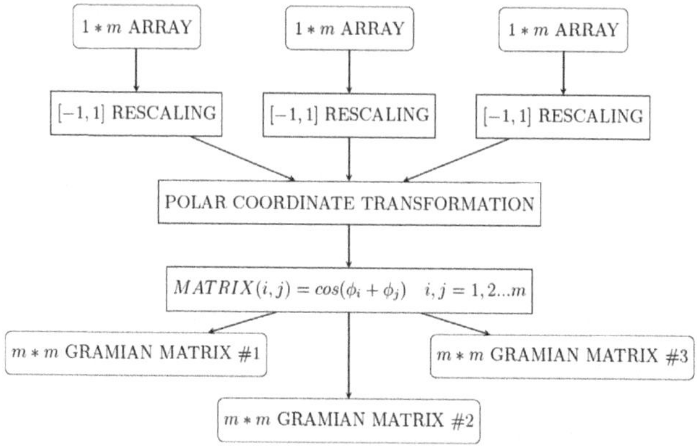

Fig. 7. Gramian Angular Field algorithm schema modified with local rescaling

the normalization has been moved forward and each of the signals has been normalized separately. With the new mapping algorithm, a new set of experiments have been carried out. The employment of the new mapping algorithm has a cost, which is the impossibility to detect the sensor drift. Since this fault has modeled through a constant value added at each instant of the signal, this information is lost during the single-channel normalization procedure, therefore there is no chance that an image generated from a sensor drift condition would differ from a healthy one. This compromise has been accepted since this failure doesn't belong to the refrigerator itself but to a sensor (more effective approaches are available to detect signal offset).

Experiment 2 has been designed also to understand the cause of such deterioration of accuracy. The local re-scaling has produced a database with a totally different aspect from the previous method. A preliminary visual inspection of Fig. 8 suggests a confusing scenario as in experiment 1, with the same poor "at glance" classification capability.

The confusion matrix for experiment 2 is shown in Fig. 9. The local re-scaling algorithm has shown to solve the issue encountered in experiment 1. The local accuracy is comparable for all the fault classes, mass leakage included.

5 Discussion of the Results

The described virtual experimental campaign has reached its goal, having performed diagnostic through images generated from the merging of data coming from three different signals with a high level of accuracy. In the case of experiment 1 the accuracy was 85% while, after slight editing of the mapping strategy, the accuracy has increased up to 99% for experiment 2. However, many are the improvements either in the approach itself or needed to get ready real industrial application.

For example, the *mapping algorithm* adopted has been effective but it is necessary to verify how the adoption of different mapping strategies would affect

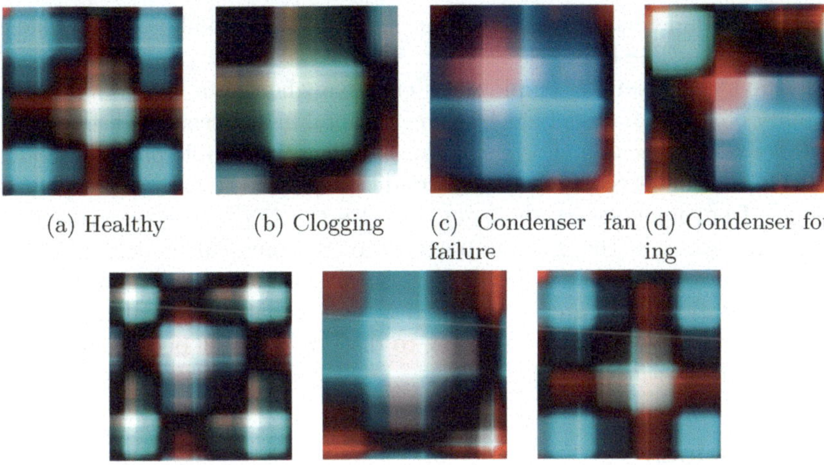

(a) Healthy (b) Clogging (c) Condenser fan failure (d) Condenser fouling

(e) Evaporator fan failure (f) Evaporator fouling (g) Mass Leakage

Fig. 8. Random samples of the simulated health conditions in experiment 2

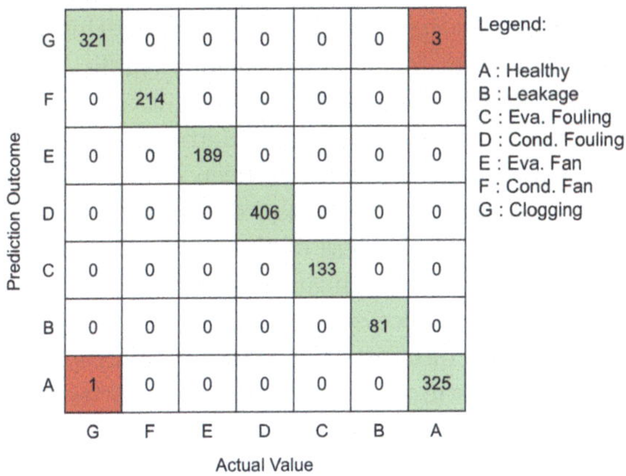

Fig. 9. Experiment 4 confusion matrix, the row index is the classification while the columns stand for the real class

the final results. Furthermore, the images are an effective mean for explaining the overall method. However, the image is nothing but a human interpretable way to see the three data layers. The fact that the CNNs are outstanding in image recognition is equivalent to say that CNNs are great in operating on structured data. The three channels limitation exists just if an image wants to be generated. In fact, the different layers generated from different signals can be easily piled up, increasing the number of channels mixed. An effective monitoring system relying on the proposed approach should be able to process data structures embedding all relevant parameters belonging to the machine that, combined, are able to identify every possible fault.

The supervised training employed is effective when data are available. The availability of a virtual model has been very helpful in this sense and stresses the importance of the research efforts spent on building up reliable physical models. Being able to simulate certain faulty condition permits to spare the physical equipment from unnecessary damages. Despite the CNN based approach is a data-driven approach, the great importance of the physical models makes the proposed approach actually hybrid. The use of synthetic signals raises embed a strong limitation: they are not affected by noise that usually alters physical signals. The sensitivity of this workflow to disturbs has not been tested and it will be matter of future investigations.

6 Conclusions

Given the outstanding performances encountered on the field by the CNN in the traditional computer vision challenges, a virtual experiment campaign has been set up and executed in order to verify whether this capability can be useful also in the classification of the health state of the machine, transforming the acquired time series into images. Before to verify this guess, it has been necessary to define what is an image of the health condition of a machine. An arithmetic scale RGB image, yielded by three arbitrary selected signals measured on the virtual machine, transformed in matrices with the GAF and piled up all together in a three layers matrix has been employed. The time histories database has been translated into an image database for training the selected CNN as well as testing the effectiveness of the NN in discriminating among the different health conditions. The classification accuracy encountered to make the virtual experiment campaign successful, justifying further efforts in this direction. The final goal will be to integrate this classification tool withing a complete CMB tool, providing diagnostic and prognostic capabilities as well.

References

1. Accorsi, R., Manzini, R., Pascarella, P., Patella, M., Sassi, S.: Data mining and machine learning for condition-based maintenance. Procedia Manuf. **11**, 1153–1161 (2017)
2. Axelsson, J., Sundelin, T., Olsson, M.J., Sorjonen, K., Axelsson, C., Lasselin, J., Lekander, M.: Identification of acutely sick people and facial cues of sickness. Proc. R. Soc. B Biol. Sci. **285**(1870), 20172430 (2018)
3. Bergonzi, L., Colombo, G., Rossoni, M., Furini, F.: Data and knowledge in IIoT-based maintenance application. In: ASME 2017 International Mechanical Engineering Congress and Exposition, pp. V011T15A029–V011T15A029. American Society of Mechanical Engineers (2017)
4. Cohen, M.W., Voldman, I., Regazzoni, D., Vitali, A.: Hand rehabilitation via gesture recognition using leap motion controller. In: 11th International Conference on Human System Interaction (HSI). IEEE (2018)

5. Dekhtiar, J., Durupt, A., Bricogne, M., Eynard, B., Rowson, H., Kiritsis, D.: Deep learning for big data applications in CAD and PLM – research review, opportunities and case study. Comput. Ind. **100**, 227–243 (2018)
6. Diaz-Rozo, J., Bielza, C., Larrañaga, P.: Machine learning-based CPS for clustering high throughput machining cycle conditions. Procedia Manuf. **10**, 997–1008 (2017)
7. Hansen, M.F., Smith, M.L., Smith, L.N., Salter, M.G., Baxter, E.M., Farish, M., Grieve, B.: Towards on-farm pig face recognition using convolutional neural networks. Comput. Ind. **98**, 145–152 (2018)
8. Jardine, A.K., Lin, D., Banjevic, D.: A review on machinery diagnostics and prognostics implementing condition-based maintenance. Mech. Syst. Signal Process. **20**(7), 1483–1510 (2006)
9. Lakhani, P., Sundaram, B.: Deep learning at chest radiography: automated classification of pulmonary tuberculosis by using convolutional neural networks. Radiology **284**(2), 574–582 (2017)
10. Lanitis, A., Draganova, C., Christodoulou, C.: Comparing different classifiers for automatic age estimation. IEEE Trans. Syst. Man Cybern. Part B (Cybern.) **34**(1), 621–628 (2004)
11. Li, Q., Cai, W., Wang, X., Zhou, Y., Feng, D.D., Chen, M.: Medical image classification with convolutional neural network. In: 13th International Conference on Control Automation Robotics & Vision (ICARCV). IEEE (2014)
12. Russakovsky, O., Deng, J., Su, H., Krause, J., Satheesh, S., Ma, S., Huang, Z., Karpathy, A., Khosla, A., Bernstein, M., Berg, A.C., Fei-Fei, L.: ImageNet large scale visual recognition challenge. Int. J. Comput. Vision (IJCV) **115**(3), 211–252 (2012)
13. Shin, H.C., Roth, H.R., Gao, M., Lu, L., Xu, Z., Nogues, I., Yao, J., Mollura, D., Summers, R.M.: Deep convolutional neural networks for computer-aided detection: CNN architectures, dataset characteristics and transfer learning. IEEE Trans. Med. Imaging **35**(5), 1285–1298 (2016)
14. Shin, J.H., Jun, H.B.: On condition based maintenance policy. J. Comput. Des. Eng. **2**(2), 119–127 (2015)
15. Stamatis, D.H.: Failure mode and effect analysis: FMEA from theory to execution. ASQ Quality Press (2003)
16. Walker, R., Perinpanayagam, S., Jennions, I.K.: Rotordynamic faults: recent advances in diagnosis and prognosis. Int. J. Rotating Mach. **2013**, 1–12 (2013)
17. Wang, Z., Oates, T.: Encoding time series as images for visual inspection and classification using tiled convolutional neural networks. In: Workshops at the Twenty-Ninth AAAI Conference on Artificial Intelligence, pp. 40–46 (2015)

Mechanics–Based Virtual Prototyping of Robots with Deformable Bodies and Flexible Joints

Stanislao Grazioso(✉), Giuseppe Di Gironimo, and Antonio Lanzotti

Fraunhofer Joint Lab IDEAS, Dipartimento di Ingegneria Industriale,
Università degli Studi di Napoli Federico II,
P.le Tecchio 80, 80125 Naples, Italy
{stanislao.grazioso,giuseppe.digironimo,antonio.lanzotti}@unina.it

Abstract. This paper describes a mechanics–based framework for virtual prototyping of soft robots, i.e. robots with deformable bodies and flexible joints. The framework builds on top of the screw theory, and uses geometrically exact nonlinear beam models for describing the behavior of deformable bodies, as well as the finite element method for space discretization. The computer implementation of this framework results in SimSOFT, a physics engine for soft robots. The capabilities of the framework are illustrated with one general example, an articulated chain of rigid and soft links connected through rigid and flexible joints. Furthermore, several case studies are shown for industrial and medical applications.

Keywords: Virtual prototyping · Design methods · Soft robotics · Continuum mechanics · Multibody dynamics

1 Introduction

The standard hypothesis underlying robot modeling, simulation and control is that manipulators consist of a discrete set of rigid bodies connected by joints; this is valid for most of industrial robots working at slow motion and with small interaction forces [1]. However, in real conditions, mechanical flexibility can arise due to the particular geometric configuration and external loads that the robots undergone during operations, or it is introduced on purpose to let the robot having a more compliant behavior. With this respect, the need for developing human–friendly and collaborative robots has led to the introduction of flexible elements inside the joints: the literature refers to these systems as soft articulated robots [2]. A different strategy is to move the mechanical flexibility along the manipulator's structure, through slender concept designs and lightweight materials applied to the links. This design approach generates internal deformations, which in most of cases are limited to the linear domain. Since these deformations in the past were seen as undesired, the most diffused paradigm was to design

C. Rizzi et al. (Eds.): ADM 2019, LNME, pp. 444–457, 2020.
https://doi.org/10.1007/978-3-030-31154-4_38

control algorithms to avoid and suppress them. Conversely, a recent trend is to exploit (large) internal deformations to provide robotic systems novel capabilities with respect to traditional robots, such as increased safety in cooperating with human beings and morphological adaptation to complex environments. Robots deliberately made with a continuously deformable body (but without joints) are referred to as soft continuum robots or soft–bodied robots [3,4]. In the literature, soft articulated robots and soft continuum robots are generically named soft robots.

Robotic systems made of deformable bodies and/or flexible joints are becoming pervasive in several applications, including medicine, rehabilitation, manufacturing, inspection and maintenance, remote explorations. However, despite the recent growing interest in soft robotics, most of existing applications are limited to laboratories.

In order to further advance the development of soft robots towards commercially available solutions, it is important to put a great effort in their design phase; in this respect, their effective development should be based on robust and efficient virtual prototyping techniques [5]. Due to the mechanical structure of soft robots, it is clear that such virtual prototyping techniques should be based on continuum mechanics and multibody dynamics theories. Therefore, computational tools with the ability to handle finite deformations, multi–joint systems, linear and nonlinear materials are highly desired in this context.

Virtual prototyping tools for soft robots would have impact in various domains: (i) aiding the first phases of concept design; (ii) analysis of mechanisms; (iii) testing and verification of geometry and materials; (iv) simulation of complex motions; (v) development of model–based controllers; (vi) planning of input trajectories.

Currently, the design and development of soft robots relies on the subsequent use of computer aided design (CAD) software for design of the undeformed configuration of the system, and computer aided engineering (CAE) software for multibody and structural analyses. With this respect, a major problem is in the data exchange process between different software tools. Generically, the CAE phase involves the combined use of multibody systems (MBS) packages with finite element (FE) analysis tools. First, a FE model is created from the CAD model; then, the FE solver generates an input model for the MBS software, and in the latter the multibody analyses are performed. To implement this process, commercial available software solutions are: ABAQUS and SIMPACK from Dassault Systèmes[1]; NASTRAN/PATRAN and ADAMS from MSC Software.[2] This approach presents two major limitations: (1) the overall motion of a flexible systems is usually seen as a superposition of a local motion (available from a modal analysis) to a mean rigid body motion; thus, the geometrically nonlinear effects of deformations are not captured [6]; (2) it is time expensive, since a FE expert and a MBS expert are required for developing two different models, and since the solutions of these two models are usually computationally expensive.

[1] https://www.3ds.com/.
[2] https://www.mscsoftware.com/.

In order to overcome the second issue, RecurDyn from Functionbay[3] integrates multibody dynamics analysis with non–linear finite element analysis. Examples of using RecurDyn for compliant mechanisms can be found in [7], where the authors modeled the flexible members via a series of rigid links connected by spring–loaded joints. However, the models developed using commercially available solutions are still heavy from the computational point of view, and thus they can be used only for simulation purposes (not for real–time control applications). Furthermore, as a manageable mathematical model is not available in this case, the development of model–based controllers is not possible.

An ideal solution for rapid and effective product development of soft robots would foresee a unique software environment where it is possible to integrate structural and multibody analyses, together with planning and control algorithms. With this respect, it has also been pointed out that integrating the geometry of soft mechanisms with their analysis is a major challenge in the soft robotics community [8]. Such environment should be based on models which, from one side, are able to capturing all the nonlinearities of the problem, and from the other side, are computationally efficient in order to get simulations closer to real–time. Furthermore, in order to use this environment for control purposes, the models should be control–oriented.

Existing mathematical formulations are well–known for soft articulated robots [9]. The literature on modeling of soft continuum robots is more recent, and the most promising approaches are those based on geometrically exact theories from continuum mechanics [10–13]. To date, there is not yet an established mathematical framework which can accommodate the simulation and control of robots made of both deformable bodies and flexible joints, i.e. multi–link soft–bodied robots. One of the first attempt in developing such kind of mathematical framework can be found in our past works [14]. In the rest of this paper, we show how this framework can be used for mechanics–based virtual prototyping of generically complex soft robots.

2 The Mechanics–Based Method for Virtual Prototyping of Soft Robots

The virtual simulation workflow for mechanical systems used in this work is illustrated in Fig. 1. First, the system is designed using classic CAD software. Then, the CAD model is translated in a simulation model for performing both structural and multibody analyses. In this phase (called also pre–processing phase), the simulation engineer has to: (1) extract the geometric topology of the system from the CAD model; (2) assign the physical behavior to the elements involved in the model (i.e. rigid and/or soft bodies); (3) define the kinematic joints; (4) define the boundary conditions. Hence, the model is ready for the analyses, made through a proper solver engine, which has to solve the equations of motion of the system according to the input trajectories and control algorithms. At the end,

[3] https://functionbay.com/.

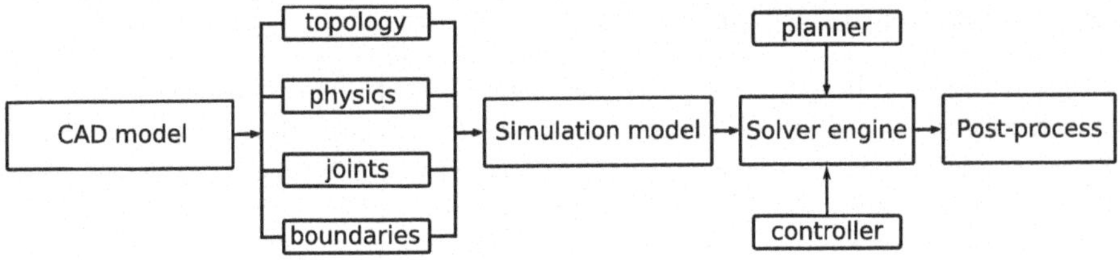

Fig. 1. Mechanics–based virtual simulation workflow for mechanical systems.

the solver generates output files for post–processing purposes (positions, velocities and accelerations of joints and bodies; forces and torques at boundaries and joint location; stresses and strains of the deformable elements; animations).

To make this workflow effective, solver engines with the following capabilities are required: (i) simulation of rigid and deformable bodies; (ii) modeling of all kind of joints and external boundary conditions; (iii) modeling of serial and parallel kinematic chains; (iv) modeling of multiple actuation sources, in terms of motion and forces; (v) planning of input trajectories from a set of motion primitives; (vi) providing of control algorithms. It is important that such solver engine would balance the trade–off between accuracy of the solution and computational efficiency, in view of possible real–time applications related to simulation and control of deformable–based robots.

In the following we describe the basis of a mechanics framework whose computer implementation might result in a solver engine with the desirable features reported above.

2.1 The Geometric Finite Element Approach for Modeling of Soft Articulated and Soft–Bodied Robots

The mechanics–based framework for soft articulated and soft–bodied robots is based on three main pillars:

- Differential geometry of Lie groups and Lie algebras. The use of geometric techniques allow to capture in an elegant way the most salient physical features of a robot [15].
- Cosserat rod theory. In robotics, we usually deal with solids in which one dimension is predominant over the two others. Therefore, we need to refer to beam theories from computational mechanics. Since soft–bodied robots are usually subject to finite deformations, we need to refer to nonlinear geometrically exact beam theories. From those, one of the most appealing for robotics applications in the Cosserat rod theory [16,17].
- Finite element method. This method is a spatial discretization technique for solving partial differential equations (PDE) – which in this case describe the behavior of deformable bodies. Furthermore, due to its assembly process, it allows to simulate both serial and/or parallel chains of soft bodies.

In the following, we report the equations of motion of a generic soft robot.

2.2 Equations of Motion

A generic soft robot is composed by rigid/soft bodies connected through rigid/flexible joints. The effects of the joints connecting the bodies are taken into account by imposing a set of algebraic constraints, which prevent the non–allowed motion as imposed by the joint. A single rigid body is described by the motion of one node in its center of gravity, to which a frame $\boldsymbol{H}_{CoG} \in SE(3)$ (special Euclidean group) is attached; a soft body is described by the motion of two extreme nodes, to which two frames $\boldsymbol{H}_A \in SE(3)$ and $\boldsymbol{H}_B \in SE(3)$ are attached and connected through an helical shape function [18]. Indeed, the relative motion between two nodes 1 and 2, belonging to two different bodies, can be described by a relative frame $\boldsymbol{H}_{J,I} \in$ Lie subgroup of SE(3) as $\boldsymbol{H}_2 = \boldsymbol{H}_1 \boldsymbol{H}_{J,I}$ [19,20]. Collecting the motion variables in the matrix $\boldsymbol{H} = \text{diag}(\boldsymbol{H}_1, \ldots, \boldsymbol{H}_n, \boldsymbol{H}_{J,1}, \ldots, \boldsymbol{H}_{J,k})$, with n the number of nodes and k the number of joints of the system, the strong form for the global dynamic equilibrium of the system take the form

$$\boldsymbol{M}(\boldsymbol{H})\dot{\boldsymbol{\eta}} - \boldsymbol{C}(\boldsymbol{H})\boldsymbol{\eta} + \boldsymbol{f}_{int}(\boldsymbol{H}) + \boldsymbol{f}_{\varphi}(\boldsymbol{H}, \boldsymbol{\lambda}) = \boldsymbol{f}_{ext}(\boldsymbol{H}) \tag{1}$$

where $\boldsymbol{\eta}$ contains the absolute and relative velocities of the nodes and joints of the system, $\boldsymbol{M}$ and $\boldsymbol{C}$ are the global discretized mass and velocity matrices; $\boldsymbol{f}_{int}$ are the discretized global internal forces, including elastic forces of the soft bodies as well as elastic and dissipative forces of the flexible joints; $\boldsymbol{f}_{\varphi}$ are the discretized constraint forces, with $\boldsymbol{\lambda}$ the Lagrange multiplier vector; $\boldsymbol{f}_{ext}$ are the discretized global external forces, including also gravity. An algebraic constraint equation $\boldsymbol{\varphi}(\boldsymbol{H}) = \boldsymbol{0}$ must be append to the differential system (1) to define a differential-algebraic equation (DAE) system that must be solved for $(\boldsymbol{H}, \boldsymbol{\lambda})$. Finally, a geometric implicit integration scheme and a Newton-Raphson iterative method are used to numerically solve the DAE system [21].

3 Example

As an illustrative example of the capabilities of the method, in this section we show the dynamic analysis of a generic robotic mechanism (GRM). The GRM is composed by rigid and soft bodies, articulated in a kinematic chain with one branched tree and one closed loop, according to the topology shown in Fig. 2.

The GRM comprises six rigid bodies, each one with the following mass and rotation inertia properties: $m = 0.15\,\text{kg}$; $J_{xx} = J_{yy} = J_{zz} = 1 \times 10^{-4}\,\text{kgm}^2$. Indeed, the cross sections of the soft bodies have the mass and stiffness matrices

$$\mathbf{M} = \text{diag}(0.1\,\text{kgm}^{-1}, 0.1\,\text{kgm}^{-1}, 0.1\,\text{kgm}^{-1}, 0.5\,\text{kg m}, 0.5\,\text{kg m}, 0.5\,\text{kg m}) \tag{2}$$
$$\mathbf{K} = \text{diag}(1 \times 10^6\,\text{N}, 1 \times 10^6\,\text{N}, 1 \times 10^6\,\text{N}, 1\,\text{Nm}^2, 1\,\text{Nm}^2, 1\,\text{Nm}^2) \tag{3}$$

The initial configuration of the GRM is given in Table 1, while the joints are defined in Table 2. There are a total of eight joints: four passive and four actuated. All joints are revolute about the z–axis, except joint 5 which rotates about x–axis and joint 3 which is prismatic.

Table 1. Initial configuration of the GRM.

Point	x [m]	y [m]	z [m]	Point	x [m]	y [m]	z [m]
0	0	0	0	4	0.5	1.3	0.2
0	0.2	0.6	0	5	0.8	1.6	0.2
$2'$	1	1	0	6	1	1.8	0.3
$3'$	1.2	0	0	7	1.2	2	0.4
2	0	1	0	8	1.4	2.2	0.5
3	0.2	1.5	0				

Table 2. Kinematic joint definition of the GRM. A = actuated; P = passive. $c = \cos$; $s = \sin$. $\boldsymbol{a}_u, \boldsymbol{a}_\omega$ = displacement and rotation parts of the relative motion.

Joint	a_u	a_ω	A/P	Joint	a_u	a_ω	A/P
0	$\mathbf{0}_{3\times1}$	$[0\ \ 0\ \ 1]^T$	A	2	$\mathbf{0}_{3\times1}$	$[0\ \ 0\ \ 1]^T$	A
1	$\mathbf{0}_{3\times1}$	$[0\ \ 0\ \ 1]^T$	P	3	$[c(1.1903)\ \ s(1.1903)\ \ 0]^T$	$\mathbf{0}_{3\times1}$	A
$2'$	$\mathbf{0}_{3\times1}$	$[0\ \ 0\ \ 1]^T$	P	4	$\mathbf{0}_{3\times1}$	$[0\ \ 0\ \ 1]^T$	P
$3'$	$\mathbf{0}_{3\times1}$	$[0\ \ 0\ \ 1]^T$	P	5	$\mathbf{0}_{3\times1}$	$[1\ \ 0\ \ 0]^T$	A

Table 3. (Left) Point-to-point motion of the actuated joint with bang-bang acceleration profiles. q_i, q_f = initial, final values. (Right) Point-to-point actuation forces with S–shaped force profiles. f_i, f_f = initial, final values.

Joint	q_0	q_2	q_3	q_5	Act. force	f_{1x}	f_{1y}	f_{2x}	f_{2y}	f_{3x}	f_{3y}
q_i [m] or [rad]	0	0	0	0	f_i [N]	0	0	0	0	0	0
q_f [m] or [rad]	$\pi/6$	$\pi/6$	0.75	$\pi/3$	f_f [N]	100	-100	-100	100	100	-100
$\ddot{q}$ [ms^{-2}] or [rads^{-2}]	$2/3\pi$	$2/3\pi$	3	$4\pi/3$	$\ddot{f}$ [Ns^{-2}]	400	-400	-400	400	400	-400

The GRM is subject to: (1) prescribed joint motion to the four actuated joints and actuation forces on three points on the system. The active joints are actuated with a bang–bang acceleration profile (triangular velocity profile) for 1 s, according to the data given in Table 3. Three actuation forces are applied on the system at points 6, 7 and 8. These forces follow a S–shaped profile for 1 s (see Table 3). The system is subject to gravity downward z–direction.

The dynamic simulations are performed using SimSOFT, our physics engine for soft robots. Snapshots of the simulation are shown in Fig. 3a–d. As output, we plot in Fig. 4 the 3D displacements, velocities and accelerations of the tip of the GRM (point 8) as well as of the free joint (4). A total of three simulations are performed, each one with the assumption of: (i) rigid joints; (ii) soft joints with internal stiffness $K = 10\,\text{Nm}^{-1}$ for the prismatic joint and $K = 10\,\text{Nmrad}^{-1}$ for the revolute joints; (iii) soft joints with internal stiffness (same as before) and damping $D = 5\,\text{Nsm}^{-1}$ for the prismatic joint and $D = 5\,\text{Nmsrad}^{-1}$ for the revolute joints.

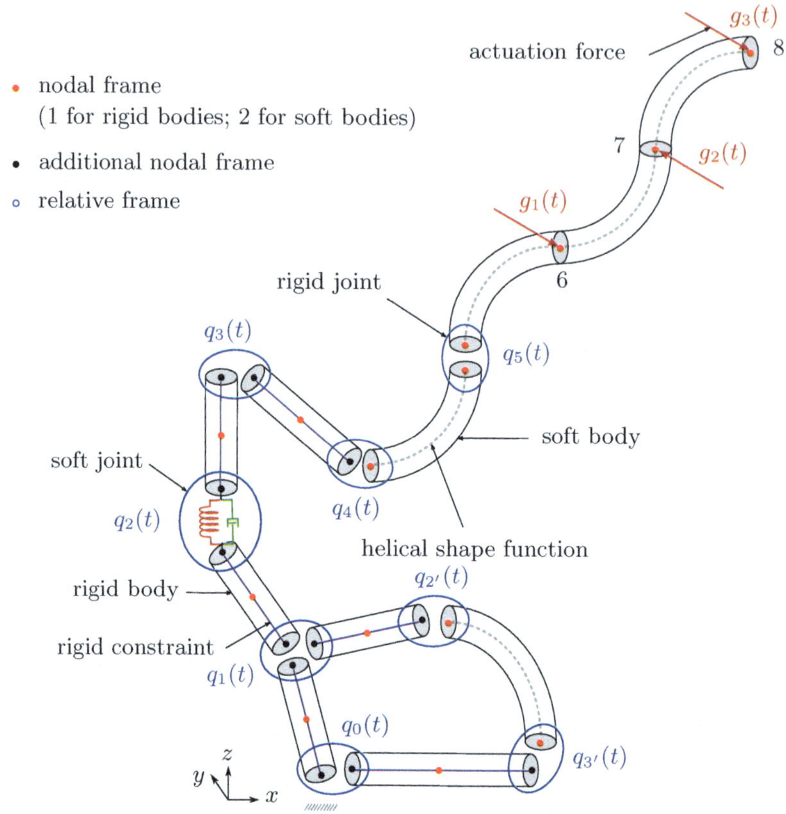

Fig. 2. Geometric description of the GRM.

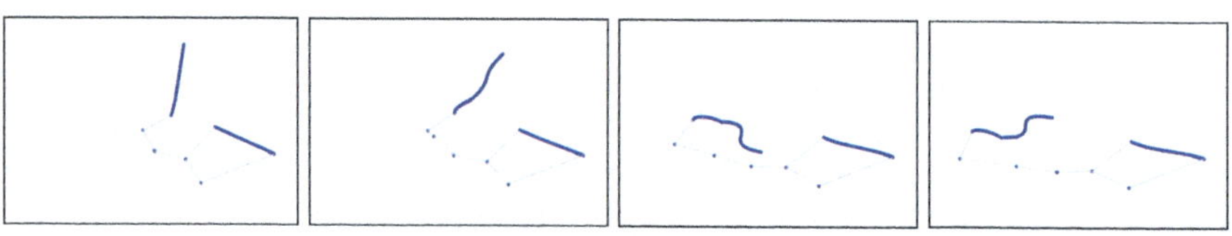

(a) t = 0 s (b) t = 0.33 s (c) t = 0.66 s (d) t = 1 s

Fig. 3. Snapshots of the GRM in SimSOFT

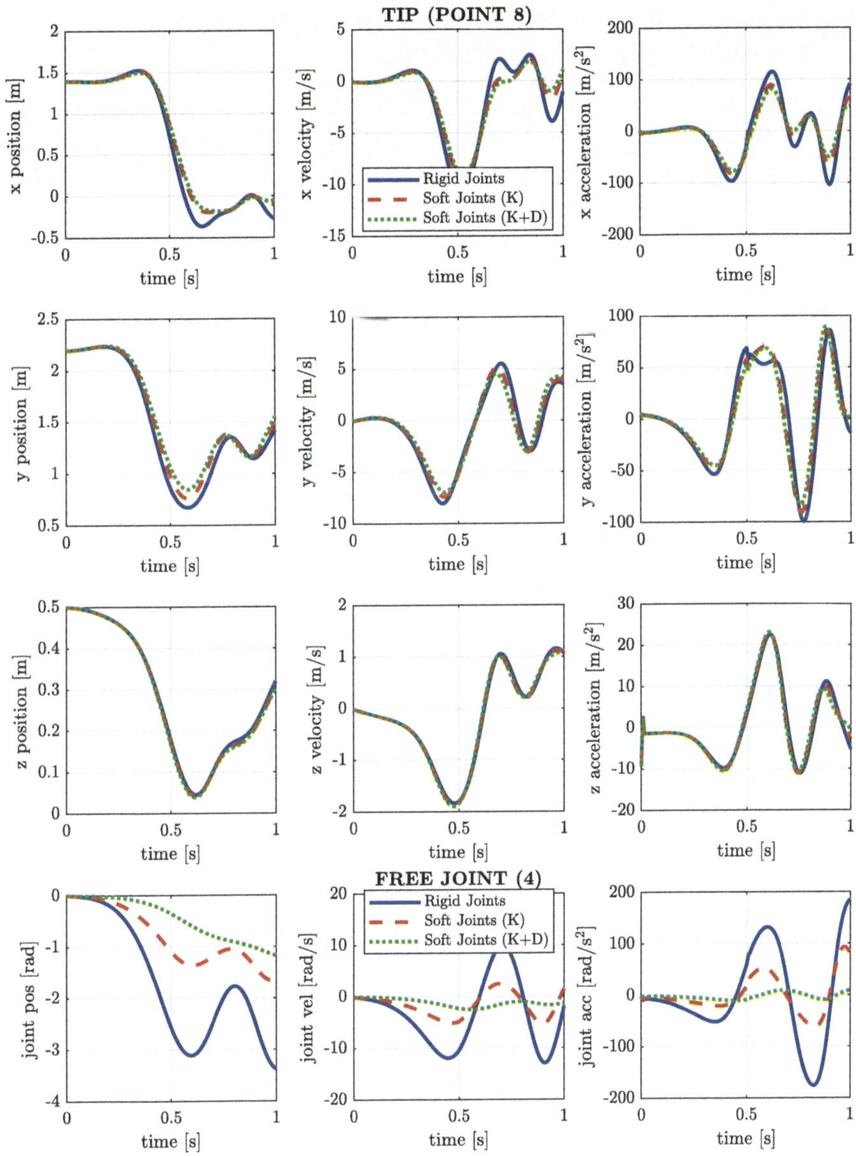

Fig. 4. Displacements, velocities and accelerations of the tip (8) and of the free joint (4) of the GRM.

4 Applications

We present here some models and simulations developed using the mechanics-based virtual prototyping framework described in this paper. These case studies illustrate the diversity and the flexibility of this framework to handle different kind of mechanisms.

4.1 Serial/Parallel Soft Articulated Robot for Remote Transportation of Large Payloads

The Hybrid Kinematic Mechanism (HKM) is the current proposal for remote transportation of breeder blanket segments for the DEMO reactor [22]. It includes a parallel and serial kinematic structure. The parallel section comprises three elastic prismatic joints which position the mechanism in space. Then, a serial section composed by three revolute joints orientate the mechanism. All the links of the HKM are modeled as rigid bodies, while the joints as flexible elements with internal stiffness. A rigid payload is attached to the tip of the HKM. Figure 5a shows the CAD model of the HKM, while Fig. 5b the simulation model in our simulation environment. For the complete description of this case study, the reader can refer to the work in [23].

(a) CAD model (b) Simulation model

Fig. 5. HKM.

4.2 Hyper–redundant Soft Articulated Robot for Remote Inspection and Maintenance

The Telescopic Articulated Remote Manipulator (TARM) is a 9 joints hyper–redundant robot, used for testing remote inspection and maintenance operations inside nuclear reactors. In order to test how a payload can eventually deform during the maintenance procedures, a special end–effector has been developed and here it is attached to the TARM (see Fig. 6a). The simulation model in Fig. 6b considers rigid links, elastic joints and a flexible payload modeled as a nonlinear beam element. This case study is described in our previous works [24,25].

4.3 Soft Continuum Robot for Intravascular Shaping Operations

Soft continuum robots are used in minimally invasive surgery, as they can exploit their internal deformations to access in anatomical sites through one single incision on the patient's body [26]. In this context, it is important to predict the real shapes that these systems undergone when subject to actuation and external loads. One commercial surgical continuum instrument is the Magellan®10Fr Robotic Catheter from Hansen Medical (Auris Health Inc., Redwood City, CA).

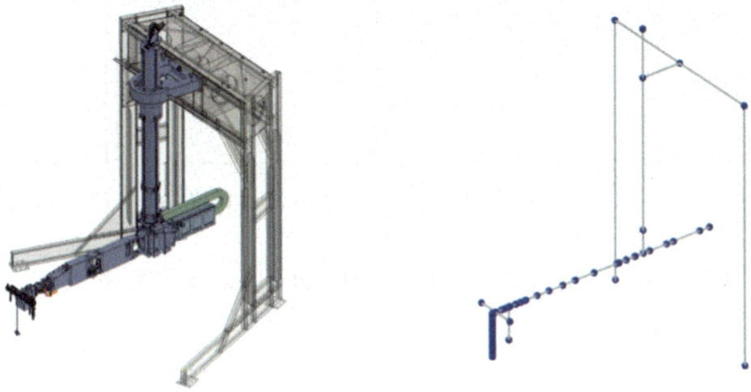

(a) CAD model (b) Simulation model

Fig. 6. TARM with flexible payload.

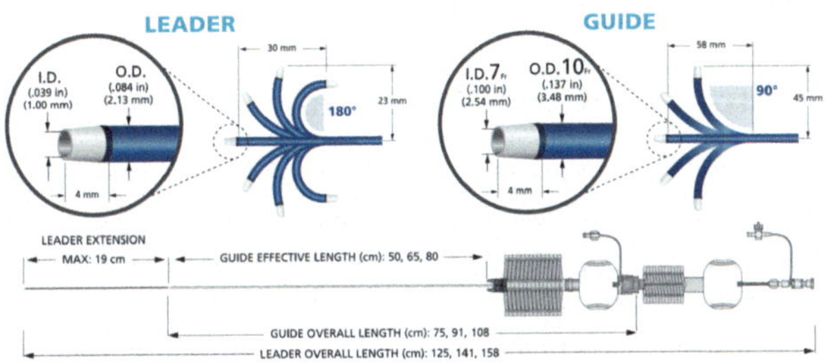

Fig. 7. Magellan®10Fr Robotic Catheter (https://www.aurishealth.com/hansen-medical)

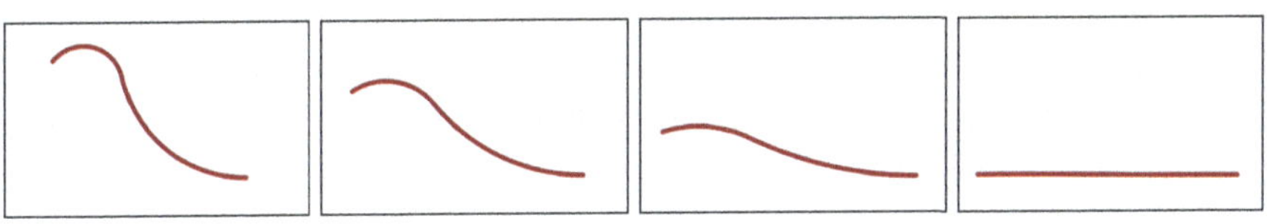

(a) t = 1 s (b) t = 0.66 s (c) t = 0.33 s (d) t = 0 s

Fig. 8. Snapshots of the soft continuum robot in SimSOFT

Fig. 9. Soft bending actuator (https://softroboticstoolkit.com/)

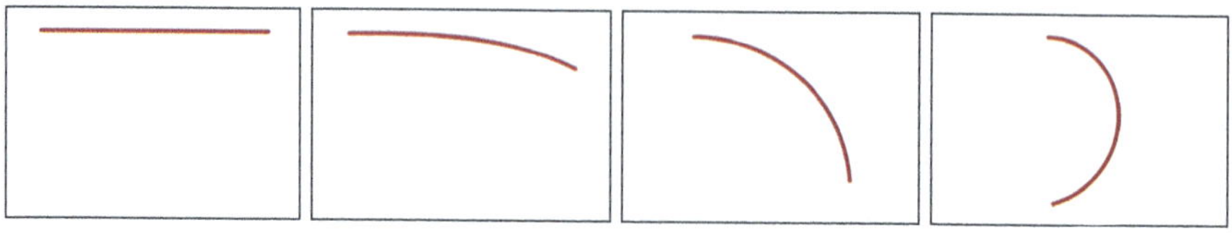

(a) t = 0 s (b) t = 0.33 s (c) t = 0.66 s (d) t = 1 s

Fig. 10. Snapshots of the soft actuator in SimSOFT

This instrument is composed by a guide and a robotically steerable inner leader; both the guide and leader have the possibility to bend (see, e.g. Fig. 7). When the leader is in its minimal extension, the instrument is composed of two consecutive bendable elements, which we have modeled as active nonlinear beam elements. The time evolution of the shape of these two elements, subject to specific actuation loads, is given in Fig. 8a–d.

4.4 Soft Actuators for Rehabilitation Robots

Soft actuators are used in the development of soft othoses and/or prostheses [27]. In this context, dynamic analyses using a fast finite element solver which uses mono–dimensional elements could play an essential role in the first design phases of the robotic system, when we are interested in the simulation of the overall motion of the actuator. A typical circumstance could be the one involved in robust design optimization, i.e. how changing the design parameters regarding the topology and the geometry of the actuator, could affect the motion performance of the system. In this case, a fast physics engine could predict the resulting motion in short time, if compared to classic finite element software as ABAQUS. After, once defined the overall parameters of the actuator, one could, eventually, optimize the internal design of the actuator's chambre by using finite element meshes with three–dimensional solids, available in commercial software. A typical example of a soft actuator is shown in Fig. 9, with some snapshots of its motion shown in Fig. 10a–d.

5 Three Major Challenges

Research on virtual prototyping techniques for soft robotics is still in its early stage. In order to facilitate the design and fabrication of soft robotic systems, three grand challenges are in the development of: (1) Conceptual design tools; (2) Integrated software tools for design, analysis and control; (3) Interactive and real–time virtual simulation tools.

5.1 Conceptual Design Tools

Virtual prototyping tools usually deal with analysis and verification of systems, when a detailed CAD model is already available. This strategy can result in wasted time and effort for detailed designs which may not work. A special effort is worth to be made towards the development of computer aided conceptual design tools specifics for soft robots. Indeed, model–based conceptual design tools might speed up the development of effective soft robots, by providing non–experts key design parameters of the systems according to their specific application.

5.2 Integrated Software Tools for Design, Analysis and Control

Virtual prototyping tools for soft robots should foresee a unique software environment where the integration of geometry and analysis of soft mechanisms is natural. As a soft robot usually works in its deformed configuration, it is necessary to have an environment where the CAD geometry follows the real physics behavior of the manipulator during the working trajectories. Such integrated environment should provide multiple material models, multibody dynamics, computational mechanics and (eventually) computational fluid dynamics capabilities (the latter is required for flying and underwater soft robots). Furthermore, effective virtual prototyping tools for soft robots should foresee the possibility to test control algorithms and different input trajectories. To this end, models specifically developed for control purposes and not only for simulation purposes are required. However, having an integrated software environment for design, analysis and control of soft robots is still hard to achieve. A first effort could be done in developing software tools for automatically generating input files from the CAD model to an existing solver engine for analysis and control purposes. Another aspect which is worth to be investigated is in the development of a unique product design representation of soft robots.

5.3 Interactive and Real–Time Virtual Simulation Tools

Interactive and real–time virtual simulation tools [28], together with advanced interface devices might allow engineers and users to moving closer to a better model viewing, manipulation and feedback on the design for soft robotics. To do that, the following features are needed: (i) rapid computational mechanics tools based on efficient dynamic models; (ii) integration of graphics and mechanics; (iii) interface devices with mapping algorithms able to move a distributed system as a soft robot with few inputs.

6 Conclusions

Mechanics–based virtual prototyping techniques can play a relevant role in the development of robust and effective designs for soft robots. As a matter of fact, currently, soft robots are more widespread in laboratories and only few examples of commercial solutions exist. In this work we have presented a modeling framework which can be eventually used for different virtual prototyping applications related to soft articulated and soft–bodied robots: design optimization, design analysis and verification, simulation of motion. Due to its good balance between accuracy (below 5% according to experimental trials in previous papers) and computation time (average of 2 s of computer time for simulating 1 s of motion) for simulation of such kind of complex systems, this framework can be the basis for the development of conceptual design tools, integrated software tools for design, analysis and control, as well as interactive and real–time virtual simulation tools, tailored for soft robots. However, the integration of multiple tools to create a robust virtual prototyping simulation software for soft robots still remains an open topic.

References

1. De Luca, A., Book, W.J.: Robots with flexible elements. In: Springer Handbook of Robotics, pp. 243–282. Springer (2016)
2. Della Santina, C., Bianchi, M., Grioli, G., Angelini, F., Catalano, M., Garabini, M., Bicchi, A.: Controlling soft robots: balancing feedback and feedforward elements. IEEE Robot. Autom. Mag. **24**(3), 75–83 (2017)
3. Trivedi, D., Rahn, C.D., Kier, W.M., Walker, I.D.: Soft robotics: biological inspiration, state of the art, and future research. Appl. Bionics Biomech. **5**(3), 99–117 (2008)
4. Majidi, C.: Soft robotics: a perspective-current trends and prospects for the future. Soft Robot. **1**(1), 5–11 (2014)
5. Zorriassatine, F., Wykes, C., Parkin, R., Gindy, N.: A survey of virtual prototyping techniques for mechanical product development. Proc. Inst. Mech. Eng., Part B: J. Eng. Manuf. **217**(4), 513–530 (2003)
6. Wasfy, T.M., Noor, A.K.: Computational strategies for flexible multibody systems. Appl. Mech. Rev. **56**(6), 553–613 (2003)
7. Bilancia, P., Berselli, G., Bruzzone, L., Fanghella, P.: A CAD/CAE integration framework for analyzing and designing spatial compliant mechanisms via pseudo-rigid-body methods. Robot. Comput.-Integr. Manuf. **56**, 287–302 (2019)
8. Shabana, A.A.: Continuum-based geometry/analysis approach for flexible and soft robotic systems. Soft Robot. **5**(5), 613–621 (2018)
9. Albu-Schaffer, A., Eiberger, O., Grebenstein, M., Haddadin, S., Ott, C., Wimbock, T., Wolf, S., Hirzinger, G.: Soft robotics. IEEE Robot. Autom. Mag. **15**(3), 20–30 (2008)
10. Trivedi, D., Lotfi, A., Rahn, C.D.: Geometrically exact models for soft robotic manipulators. IEEE Trans. Robot. **24**(4), 773–780 (2008)
11. Rucker, D.C., Jones, B.A., Webster III, R.J.: A geometrically exact model for externally loaded concentric-tube continuum robots. IEEE Trans. Robot. **26**(5), 769 (2010). A Publication of the IEEE Robotics and Automation Society
12. Renda, F., Boyer, F., Dias, J., Seneviratne, L.: Discrete cosserat approach for multisection soft manipulator dynamics. IEEE Trans. Robot. **34**(6), 1518–1533 (2018)
13. Grazioso, S., Di Gironimo, G., Siciliano, B.: A geometrically exact model for soft continuum robots: the finite element deformation space formulation. Soft Robot. (2018)
14. Grazioso, S.: Geometric soft robotics: a finite element approach. Ph.D. thesis, University of Naples Federico II (2018)
15. Lynch, K.M., Park, F.C.: Modern Robotics. Cambridge University Press, New York (2017)

16. Simo, J.C., Vu-Quoc, L.: A three-dimensional finite-strain rod model. Part II: computational aspects. Comput. Methods Appl. Mech. Eng. **58**(1), 79–116 (1986)
17. Sonneville, V., Cardona, A., Brüls, O.: Geometrically exact beam finite element formulated on the special euclidean group SE (3). Comput. Methods Appl. Mech. Eng. **268**, 451–474 (2014)
18. Grazioso, S., Di Gironimo, G., Siciliano, B.: From differential geometry of curves to helical kinematics of continuum robots using exponential mapping. In: International Symposium on Advances in Robot Kinematics, pp. 319–326. Springer (2018)
19. Sonneville, V., Brüls, O.: A formulation on the special Euclidean group for dynamic analysis of multibody systems. J. Comput. Nonlinear Dyn. **9**(4), 041002 (2014)
20. Grazioso, S., Sonneville, V., Di Gironimo, G., Bauchau, O., Siciliano, B.: A nonlinear finite element formalism for modelling flexible and soft manipulators. In: 2016 IEEE International Conference on Simulation, Modeling, and Programming for Autonomous Robots, pp. 185–190. IEEE (2016)
21. Brüls, O., Cardona, A., Arnold, M.: Lie group generalized-α time integration of constrained flexible multibody systems. Mech. Mach. Theory **48**, 121–137 (2012)
22. Keep, J., Wood, S., Gupta, N., Coleman, M., Loving, A.: Remote handling of demo breeder blanket segments: blanket transporter conceptual studies. Fus. Eng. Des. **124**, 420–425 (2017)
23. Grazioso, S., Di Gironimo, G., Iglesias, D., Siciliano, B.: Screw-based dynamics of a serial/parallel flexible manipulator for demo blanket remote handling. Fus. Eng. Des. **139**, 39–46 (2019)
24. Grazioso, S., Di Gironimo, G., Siciliano, B.: Modeling and vibration control of flexible mechanical systems for demo remote maintenance: results from the flexARM project. Fus. Eng. Des. (2019)
25. Grazioso, S., Powell, R., Skilton, R., Di Gironimo, G., Siciliano, B.: Multibody simulations of the telescopic articulated remote manipulator with flexible payload for demo studies on remote handling. Fus. Eng. Des. (2019)
26. Burgner-Kahrs, J., Rucker, D.C., Choset, H.: Continuum robots for medical applications: a survey. IEEE Trans. Robot. **31**(6), 1261–1280 (2015)
27. Polygerinos, P., Correll, N., Morin, S.A., Mosadegh, B., Onal, C.D., Petersen, K., Cianchetti, M., Tolley, M.T., Shepherd, R.F.: Soft robotics: review of fluid-driven intrinsically soft devices; manufacturing, sensing, control, and applications in human-robot interaction. Adv. Eng. Mater. **19**(12), 1700016 (2017)
28. Di Gironimo, G., Lanzotti, A.: Designing in VR. Int. J. Interact. Des. Manuf. **3**(2), 51–53 (2009)

Virtual Prototyping Design Method to Optimize Mechanical Spring Devices for MV Switch Disconnector

Michele Calì[1](✉), Salvatore Massimo Oliveri[1], and Sebastiano Zuccarello[2]

[1] Department of Electric, Electronics and Computer Engineering, University of Catania, V.le A. Doria, 6, 95125 Catania, Italy
mcali@dii.unict.it

[2] NT&T, S.P. 14 Zona Industriale Piano Tavola, 95040 Belpasso, Catania, Italy

Abstract. This paper proposes a replicable methodology based on virtual prototyping design in multibody environment to optimize the functionality of Mechanical Spring Devices (MSD). These devises are assembled to control the shaft angular velocity in medium-voltage (MV) switch disconnector. The angular velocity of switch disconnector, moving the contact fingers, is directly linked to arcing time, which is the parameter that mainly influences accuracy, safety drives and a longer service life of device. Design of experiment (DoE) techniques, integrated with tridimensional geometric parametrization, were used in multibody environment to optimize the displacement of switch disconnector shaft. The best values of shape, stiffness and preload of the main cylindrical helical spring of MSD were obtained in every functional condition. Optimization results were compared with the limits values measured in homologation and with the acceptance limits values released by ENEL technical specifications for the MSD studied proving the effective methodology and the improvement obtained in terms of the safety of the system.

Keywords: Parameterization · Multibody model · DoE · Arcing time · MV electromechanical compartments

1 Introduction

Many electrical devices are designed to conform the relative requirements, such as design, performance and test features, to the switching and protection tasks. The switch disconnector is commonly used in the medium-voltage (MV) distribution system to guarantee the isolation from pole to pole, to break up load currents and the functions of a switch, realizing a suitable design [1–3]. In the modern electric power system, it ensures the reliability of the electricity supply, providing protection for expensive equipment with preventing downtime and darkness and verifying breaker performance, efficient test instruments and methodology. It is clear that the correct functioning of breaker circuits or switch relies on the number of the individual components as implemented and tested. Therefore, modelling the properties of these electromechanical devices has attracted some considerable interest and attention in the last years [4, 5].

C. Rizzi et al. (Eds.): ADM 2019, LNME, pp. 458–469, 2020.
https://doi.org/10.1007/978-3-030-31154-4_39

The geometrical design and optimization of the mechanical parts of electronic device done entirely on the computer can provide the advantages of testing and refining the virtual, accurate physical models prototype, when forces are applied to parts with mass and inertia properties [6–8].

The goal of this research is to optimize the functionality of Mechanical Spring Device (MSD) of switch disconnector for MV substations of electromechanical compartments using tridimensional geometric parametrization and multibody optimization techniques. In particular, the shape of the main cylindrical helical spring of MSD was geometrically optimized using Design of experiment (DoE) technique implemented in the commercial multibody software MSC Adams® ver. 2018 and the tridimensional geometric parametrization realized by the commercial parametric software Solid-works® ver.2018.

In the contest of virtual prototyping design, assembled systems with flexible components (such as that analyzed in this study) are widely studied by researchers. The functionality optimization of such assemblies, however, often requires a long design phase characterized by many expensive experimental testings. In this paper the virtual prototyping methodology based on the geometric optimization with constraints of shape, stiffness and preload of a cylindrical helical spring is described. The proposed method allows limiting (or even avoiding) prototypes manufacture and the execution of experimental testing during the designing phase.

The results obtained were compared with the limits values measured in homologation and with the acceptance limits values released by ENEL technical specifications.

2 Mechanical Spring Device

The electromechanical compartments in secondary cabins, designed to convert electrical energy from medium to low voltage, are composed by self-supporting metal subdivided into two parts: one above (busbars side) and one below (side cables); with a compartment door in order to allow the maintenance (Fig. 1a). The metal casing of the equipment is divided into two sides, right and left, and a rear panel made of sheet metal. Positioned above the compartment door is the carter for the controls and the operational phases thereof. The MSD and a metal container made of stainless steel, sealed for life, are contained inside the electrical switch disconnector isolated by pressurized sulfur hexafluoride gas SF6 [9]. Into the disconnector, the movable and fixed contacts are divided into three cells. The fixed contacts are represented by copper and silver components bound together and fixed to the cylindrical surface of the switchgear. The moving contacts are fixed, instead, to a rotary axis moved by the external mechanical control. The switch has three positions: closed, open, and electrically grounded (Fig. 1b). The three positions are controlled by MSD with handled manually, or electrically through a geared motor (Fig. 2). The main functions performed by the switch disconnector are the starting up or safety of the electromechanical compartments.

The opening takes place when the command is located in the idle position and the main lever is perpendicular to the clamping plane; namely, the switch disconnector is in

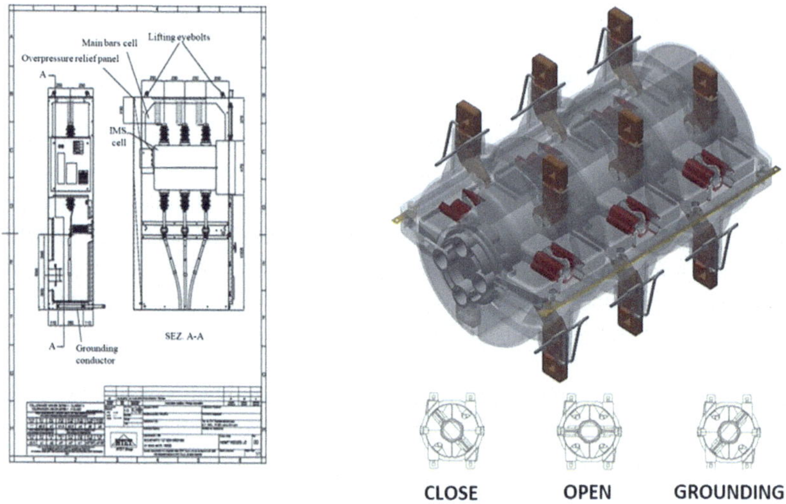

(a) (b)

Fig. 1. (a) Electromechanical compartments; (b) Parametric model of switch disconnector.

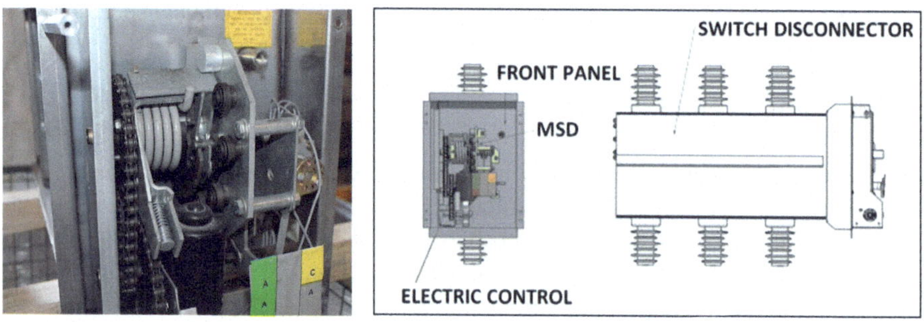

(a) (b)

Fig. 2. (a) MSD; (b) MSC-switch disconnector group: frontal and longitudinal view.

horizontal position without contact between the fixed and mobile parts. There is not any current flow in the compartment but security is not guaranteed.

The closure occurs during the current, the main lever is upwards. Turning to right, the switch-disconnector connects the fixed and mobile parts. The circuit is closed and the compartment is activated (Fig. 3).

The grounding occurs by moving the main lever towards down, the switch disconnector rotates to the left, the contacts close the circuit of the ground and lock the system.

The MSD is a set of fixed and movable parts connected to the front panel. It entitles the holder to engage the three control shunting under the action of an external force. According to the maneuver, the system can be in manual or electric control mode.

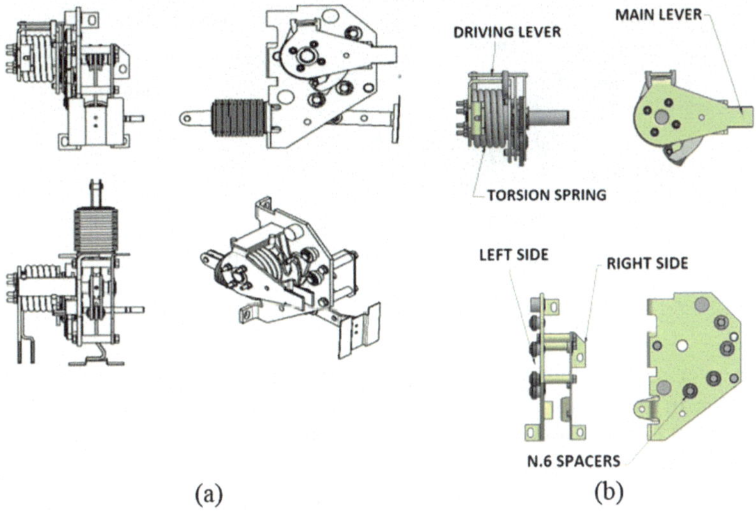

Fig. 3. (a) Manual control in MSD; (b) Main lever, side panels and spacers.

The manual control of the switch disconnector operates in rotary type at the overcoming of the dead point during the opening and closing processes all the time. A slight force capable of overcoming the resistance of the Main Torsion spring is applied to special operating lever, main control lever, preloaded during the control (Fig. 3a).

The switch disconnector electrical control takes place in overcoming dead point by the gear motor, allowing the contacts stability. If, during an electrical operation, the power failure is verified, the system allows to complete the manoeuvre manually. The MSD is connected to the geared motor through two lower and superior chains. The opening and closing stages are controlled via push-button positioned on the left side of carter.

The system is made up of three main parts: the main lever, the driving lever and torsion springs (Fig. 3b). The three parts are connected to each other through the preloaded springs. The preload enables the rectilinear parts of the spring to support the two levers positioned in the middle. The main lever with spring allows to diffuse movement to the system.

The backbone of the mechanism is represented by sides. They are rigidly connected to the front panel through bolts, and have a support function for the moving parts. The right and left side panels are connected rigidly to each other with three bolts. On the left side, six spacers are fixed in order to block the drive lever movement after the manoeuvre has been performed. The other important components in stainless steel are the bellows jack gear (Fig. 4a) and the curved lever (Fig. 4b).

The bellows jack gear is responsible for connecting the internal switch with external MSD and isolating the gas inside the armored steel with a gas burner made of rubber. The secondary lever is made up of two parts hinged together and connected to

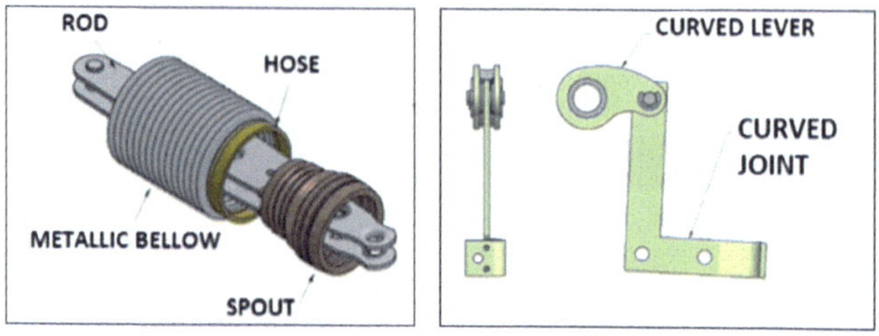

(a) (b)

Fig. 4. (a) Bellows jack gear; (b) Curved lever.

the main lever shaft. It has the task of transferring the motion from MSD to switch-disconnector. In the system, there are three torsional springs, two of these are equal and small sized and the largest one (main spring) has the task to moving to the system [10, 11]. All the springs are subject to pre-load ensuring that the system starts up. The specifications of spiral torsion spring and the specified parameters are reported in Table 1.

Table 1. Springs data specifications.

	Small size spring	Main torsional spring
Number of coils	1.402	5.0
Flexure force [N/mm^2]	1190/1260	980
Angular deflection [°]	22°	34°
Preload [N · mm]	4250	45000
Rigidity [N · mm/degree]	223	1030

The geometric parameters and momentum and the stiffness characteristic of the main cylindrical spring are reported in Fig. 5a and Fig. 5b respectively.

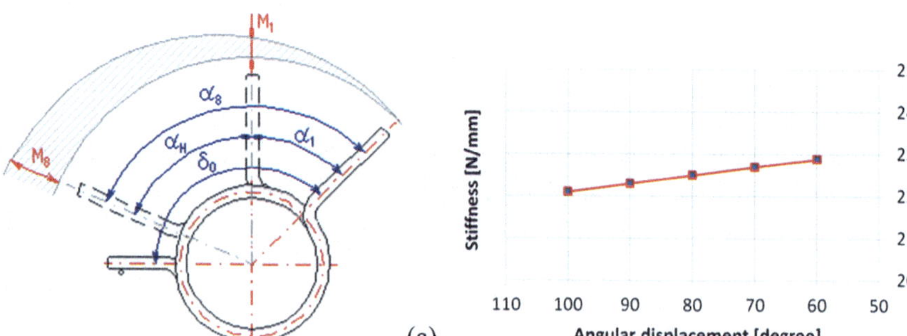

(a) (b)

Fig. 5. Main spring: (a) geometric parameters and momentum; (b) stiffness.

3 Parametric Modelling and Optimization

In order to optimize the angular velocity of switch disconnector shaft and obtain enhanced arcing time, a complete parametric multibody model was generated. The 41 rigid and flexible parts and the 3 torsion springs of the system (MSD and switch disconnector) were modelled by the commercial parametric software Solidworks® ver.2018. Then they were imported as 3D parametric parts into the commercial multibody software MSC Adams® ver. 2018 (Fig. 6).

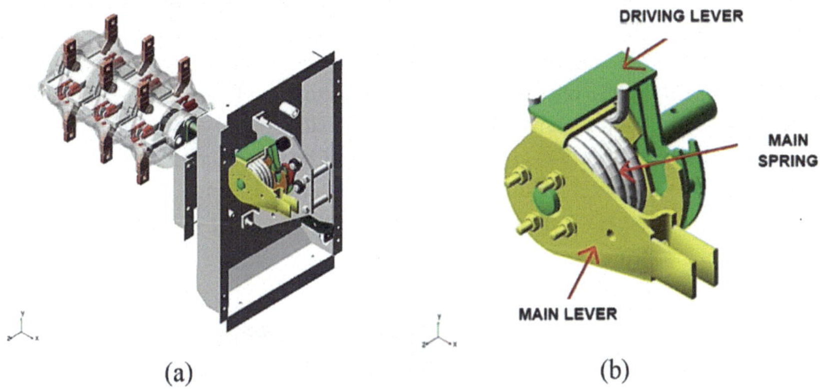

Fig. 6. (a) MSD and switch disconnector multibody model; (b) Particular of MSD.

All parts were assembled with each other through contact forces; spring forces (with their elastic curve and preload) and external forces applied to the main levers to start the movement of the system. The contact forces were embedded to the spacers, the latches and the drive lever. The spacers stopped the motion of the drive lever when it rotated on its axis. The torsion springs were applied on latches keeping the contact with the drive lever.

The main torsion spring was placed between the two levers: main and driving lever. Subsequently, the contact forces were applied in the front support of the switchgear. Finally, the external force applied to the main lever was inserted to start the mechanism. This force passed the buried suture and overcame the resistance of the main spring torsion, cancelled its intensity and the mechanism automatically completed its movement. The shaft angular velocity of MSD depends on values of shape, stiffness and preload of the main cylindrical helical spring of MSD and on the moments of inertia of the other moving parts connected to the shaft. No external force intervenes on the movement of the shaft.

3.1 DOE Optimization

Keeping the volume and dimensions of the device constant, the best geometry of the main cylindrical spring was determined and the best value of his preload was found so to have the maximum angular velocity of switch disconnector. The design spring preload was 45000 N • mm. Assuming the variation of the preload from a minimum value of 40000 N • mm to a maximum value of 58000 N • mm, the best value of spring stiffness combination of spring geometry and preload was calculated with DoE multibody analysis.

Varying coils diameter ϕ of the main spring, the preload is increased or decreased. At the same internal coils diameter (ϕ_i) and axial length (l = 55 mm) of the spring, the preload and stiffness of spring can be increased enlarging coil section and decreasing their total number.

The DoE multibody analysis was carried out varying design coils diameter (ϕ = 9.2 mm) by 30%, design axial length (l = 55 mm) by 20% and external diameter (ϕ = 75 mm) by 30%. The multibody analysis as mentioned allowed to take into account the moments of inertia of the other moving parts connected to the shaft.

The DoE analysis has underlined that shaft angular velocity has always different trends in the opening and closing phases. The optimal geometry founded is constituted by a spring with one less coil (4 coils), with a largest by 20% diameter of the coil cross-section and with a greater than 13% outer diameter. In Fig. 7 the design initial geometry of the spring is compared with the optimized spring geometry.

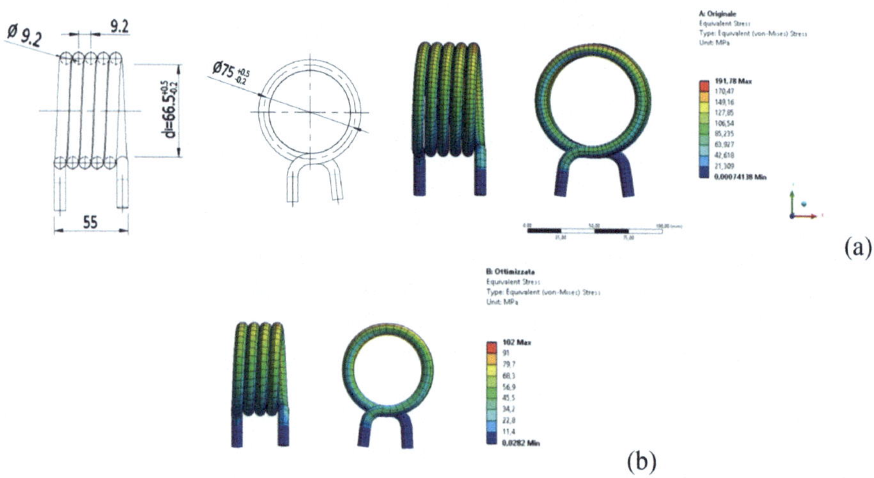

Fig. 7. Main cylindrical spring: (a) Initial geometry; (b) Optimized geometry.

Verifying the optimized model with FE method, it results that the final optimized geometry presents inferior values of von Mises max tension (σ_{lim}) with respect to the initial project design and an increased safety coefficient $\gamma = \frac{\sigma_s}{\sigma_{lim}} = 3.8$ (σ_s = 410 N/mm^2 value of yield strength tension for steel AISI 316).

To estimate model parameters, the DoE optimization is formulated as a computational problem of non linear systems equations such that assuming $f\colon \mathbb{R}^n \to (-\infty, \infty)$ as smooth function, it is necessary finding the minimum (or maximum).

$$\min_{x \in \mathbb{R}^n}\{f(x)\} \text{ and } x_* \text{ s.t. } f(x_*) = \min_{x \in \mathbb{R}^n}\{f(x)\} \quad (1)$$

The graph in Fig. 8 shows the angular velocity of switch disconnector during the closing phase at four different preloads between 40000 N · mm and 58000 N · mm. The increase in preload generated an increase of the switch-disconnector rotation speed maximum value.

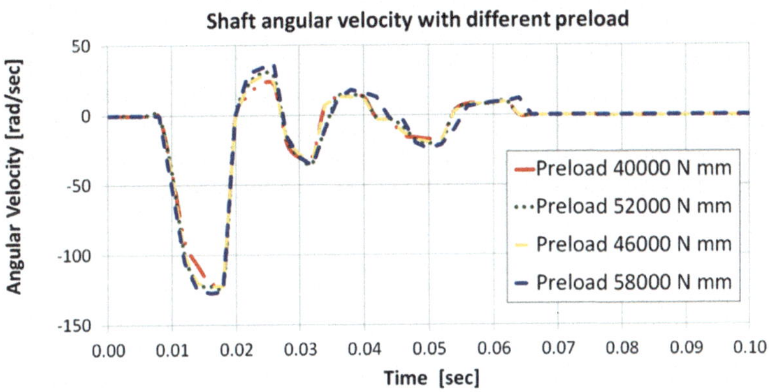

Fig. 8. (a) Angular velocity of switch disconnector shaft during the closing phase at 4 preload values.

3.2 Optimality Criteria

Minimize (x), the objective function, is called as optimization problem, usually referred to as a nonlinear problem. Assuming that f is continuously differentiable from $D \subseteq \mathbb{R}^n$ to $\mathbb{R}^n$ and setting the partial derivatives of f equal to zero at a point x_* at which f is minimum leads to a system of nonlinear equations, we can also define the Jacobian matrix associated with f and M the information matrix for the parameters estimation problem.

$$M = \nabla f(x_*)^T \nabla f(x_*) \quad (2)$$

The optimal experimental design (OED) problem can now be formulated as a nonlinear constrained optimal control problem, starting from a model structure, selecting a set of dynamic or static experiments, estimating the parameters under varying realistic conditions and minimizing a scalar function of the matrix by an optimal choice of the experimental input. The design optimality criteria often used are:

The A-criterion with $\min(\mathrm{trace}(M^{-1}))$ minimizes the arithmetic mean of the canonical variances in the optimal design.

The D-criterion with $\max(\det(M))$ minimizes the geometric mean of the parameter estimation errors with the variance covariance.

The E-criterion (maximum eigenvalue criterion) is focused on the minimization of the largest parameters error such as the smallest eigenvalues, corresponding to minimizing the longest axis of the confidence region. The simulation time is set up to 0.15 s and each processing step time takes 250 since the mechanism of the displacement occurs very quickly and almost instantaneously. The virtual mechanism movement starts with an external force applied to the main control lever, positive or negative, respectively the lever moves up or down. Overtaken the dead center regulated by the two torsion springs via the two latches, the main spring torsion pulls on the control lever which is connected to a shaft, allowing the mechanism to move. This movement, through two levers, is transmitted to the sliding support of switchgear. On the sliding support a fixed draw bolt pulls with it a lever (sliding lever) that rotates around its axis; in turn, by contact it trails the switch disconnector, making it perform a rotation likely to close the contacts. In the rotation the speed of switch disconnector is obtained as a function of time. From the graph in Fig. 9 it is clear that the absolute minimum point of the curve, corresponding to 120.6 rad/s indicates the maximum angular speed which reaches the switch disconnector at the putting into service.

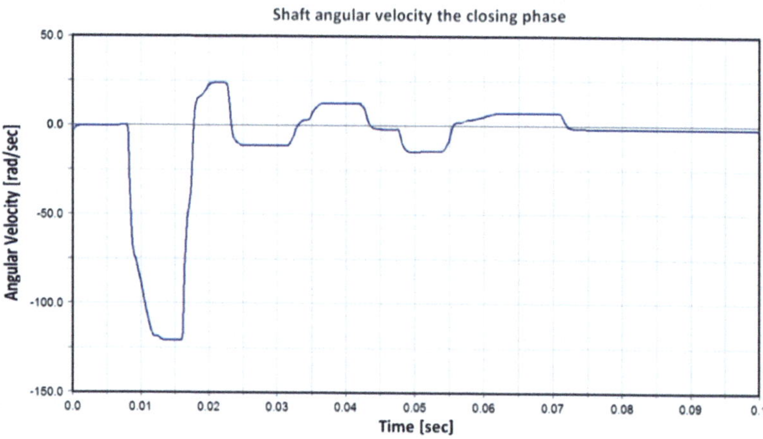

(a)

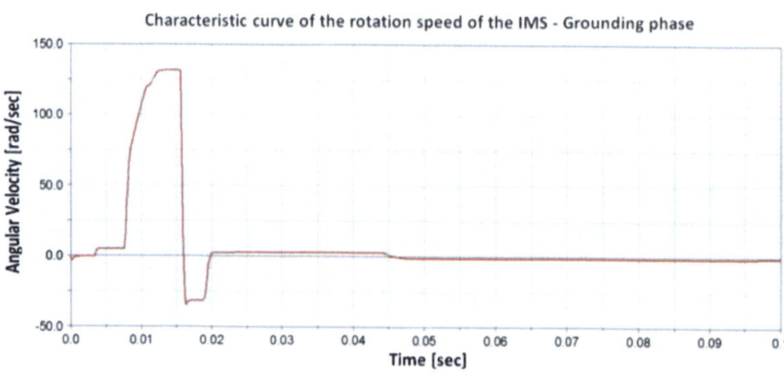

(b)

Fig. 9. Shaft angular velocity: (a) during opening phase; (b) during closing phase.

In the second test for the grounding, the main lever moves downwards. From the graph, the absolute maximum point of the curve corresponds to 131.6 rad/s, indicates the maximum speed reached by closing the switch disconnector at the grounding. In the simulation tests, the minimum and maximum points are recorded, representing the oscillations of the swing disconnector to come back to equilibrium, this oscillating phenomenon tends to reduce considerably in reality due to friction between the fixed and movable contacts. From the simulations we have shown a comparison between the speed curve of the swing disconnector (in the closing phase) and the preload trends, as a function of time (Fig. 10).

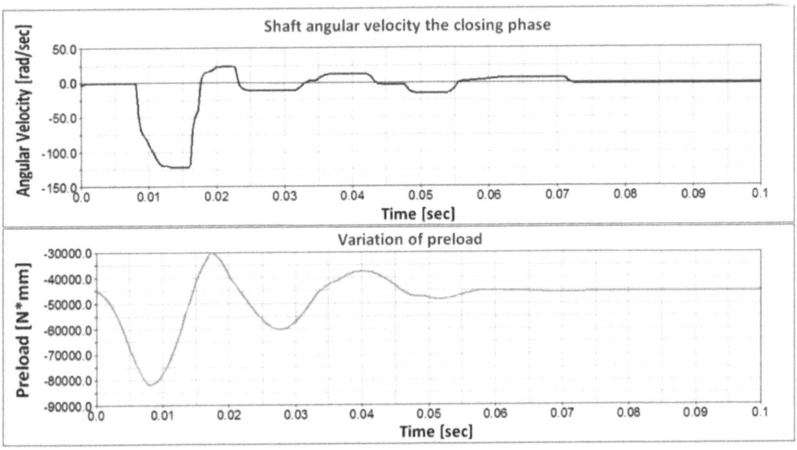

Fig. 10. Shaft speed in the closing phase and preload of disconnector.

The two graphs exhibit that at the activation systems, the main torsion spring undergoes a further increase of the preload until it exceeds the dead center reaching a value of 81 000 N • mm.

After overcoming the dead center, the stored energy is transferred to the mechanism reaching the maximum speed measured previously. The main spring, after a short time between 0.01 s and 0.06 s, loses its preload and gets the system in equilibrium state, ceasing at the same time the disconnector rotations.

4 Result

The simulation data so obtained were compared with measured approval and acceptance limit values issued by ENEL. During the approval procedure a test, measuring the closing speed of the switch disconnector, was carried out during the putting into service via a bidirectional encoder. The encoder was fixed on a support in the back of the switch disconnector united to the rotation axis by a flexible coupling. The average time of "n" closing and average speeds during the encoder rotation were calculated by the embedded dedicated software. The times and the relative speeds were referred to the

25% of the sectioning process. The performed simulation results closely approximated to the value obtained in the experimental test in accordance with range of acceptance limits values, and the percentage of error between simulation and test of 4.9%. Figure 11 shows the improvement of arcing time for 500 V of nominal tension and current variable between 80 A and 340 A.

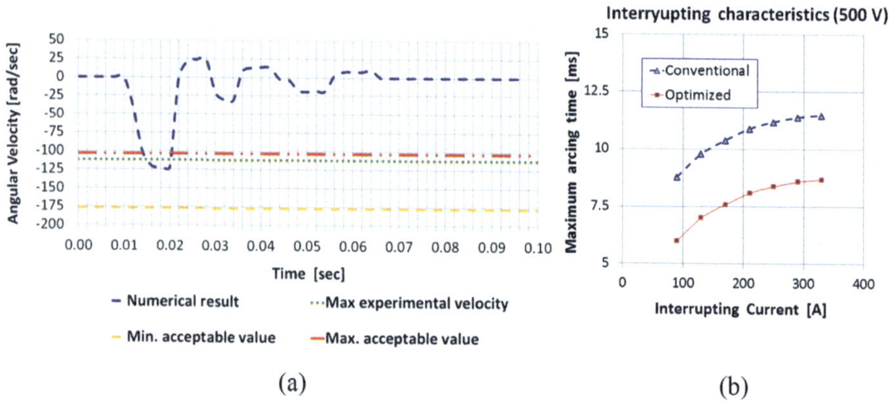

Fig. 11. (a) Comparison with experimental data; (b) Arcing time improvement.

5 Conclusions

A methodology to optimize Mechanical Spring Device (MSD) of a switch disconnector in medium-voltage substations of electromechanical compartments was presented. DoE optimization techniques in multibody environment were used integrated with tridimensional geometric parametrization of the parts of the system (springs included). It was found that the right choice of stiffness and preload in the main cylindrical spring of MSD improved considerably angular velocity of switch disconnector shaft with a remarkable advantage on the arcing time of the device. Optimization of these characteristics was obtained deriving the optimal geometric shape of the main cylindrical spring. The optimal geometric shape of the spring was obtained complying with the geometric and functional constraints of the device.

The experimental data provided by the Company NTET enabled to make an accurate analysis comparing the numeric data with the data obtained in the approval test reported in the technical ENEL specifications.

References

1. Peng, C., Song, X., Rezaei, M.A., Huang, X., Widener, C., Huang, A.Q., Steurer, M.: Development of medium voltage solid-state fault isolation devices for ultra-fast protection of distribution systems. In: IECON 2014-40th Annual Conference of the IEEE Industrial Electronics Society, pp. 5169–5176. IEEE, October 2014

2. Schellekens, H., Moesch, G., Biasse, J.M., Hazel, T., Cabaret, B.: Impact of vacuum switch-disconnectors on reliability of MV distribution networks. In: CICED 2010 Proceedings IEEE, pp. 1–8, September 2010
3. Rochette, D., Rachid, T., William, B.: Numerical study of the short pre-arcing time in high breaking capacity fuses via an enthalpy formulation. J. Phys. D: Appl. Phys. (2007)
4. Stosur, M., Szewczyk, M., Piasecki, W., Florkowski, M., Fulczyk, M.: GIS disconnector switching operation—VFTO study. In: Modern Electric Power Systems, pp. 1–5. IEEE, September 2010
5. Yinbiao, S., Bin, H., Ji-Ming, L., Weijiang, C., Liangeng, B., Zutao, X., Guoqiang, C.: Influence of the switching speed of the disconnector on very fast transient overvoltage. Trans. Power Delivery 2080–2084 (2013)
6. Sequenzia, G., Oliveri, S., Calabretta, M., Fatuzzo, G., Cali, M.: A new methodology for calculating and modelling non-linear springs in the valve train of internal combustion engines. SAE Technical Paper (2011)
7. Calì, M., Oliveri, S.M., Ambu, R., Fichera, G.: An integrated approach to characterize the dynamic behaviour of a mechanical chain tensioner by functional tolerancing. Strojniski Vestnik/J. Mech. Eng. (2018)
8. Calì, M., Oliveri, S.M., Cella, U., Martorelli, M., Gloria, A., Speranza, D.: Mechanical characterization and modeling of downwind sailcloth in fluid-structure interaction analysis. Ocean Eng. **165**, 488–504 (2018)
9. Okabe, S., Yuasa, S., Kaneko, S.: Evaluation of breakdown characteristics of gas insulated switchgears for non-standard lightning impulse waveforms-breakdown characteristics for non-standard lightning impulse waveforms associated with disconnector switching surges. Trans. Dielectr. Electr. Insul. IEEE 721–729 (2008)
10. Carsimamovic, S., Bajramovic, Z., Ljevak, M., Veledar, M., Halilhodzic, N.: Current switching with high voltage air disconnector. In: International Conference on Power Systems Transients IPST, pp. 19–23, June 2005
11. Carsimamovic, S., Bajramovic, Z., Ljevak, M., Veledar, M.: Very fast electromagnetic transients in air insulated substations and gas insulated substations due to disconnector switching. In: International Symposium on Electromagnetic Compatibility, pp. 382–387. IEEE, August 2005

Design and Process Optimization of a Sintered Joint for Power Electronics Automotive Applications

Michele Calabretta[1(✉)], Alessandro Sitta[1,2], Salvatore Massimo Oliveri[2], and Gaetano Sequenzia[2]

[1] STMicroelectronics S.r.l., Stradale Primosole 50, Catania, Italy
michele.calabretta@st.com

[2] Dipartimento di Ingegneria Elettrica, Elettronica ed Informatica (DIEEI), Università degli Studi Catania, Viale Andrea Doria 5, Catania, Italy

Abstract. The massive development of Hybrid and Electrical Vehicles (HEV) is strongly impacting the semiconductor industry demanding for highly reliable Power Electronic components. Within the engine compartment installation space is of major concern, therefore small size and high integration level of the modules are needed. Conventionally devices are typically soldered to ceramics substrates that are vacuum soldered to water-cooled base plates. The known reliability limitations of traditional solder joints are significantly limiting the power density increase, limiting the maximum operative temperature and representing a strong constrain for using high performances devices such as wide bandgap compound like Silicon Carbide (SiC). Silver sintering today has started to replace the solder joint from chips to carrier substrates, leaving one major reliability bottleneck. Combining properly temperature, time and pressure, a strong, highly electrically and thermally conductive bond is formed. The aim of this work is to develop an integrated methodology, numerical and experimental, to assess the Ag sintering die attach process for a SiC power MOSFET. Different process parameters have been benchmarked by means of physical analyses, performed at time zero and also after liquid-to-liquid thermal shock aging test. The sintering flakes densification process has been reproduced by Finite Element Analysis and the obtained morphological texture has been used for extracting the mechanical properties of the layer as a function of the thermo-compression process itself. A simulation method, based on the evaluation of the inelastic strain accounted per cycle has been used for matching the experimental results according to an aging model.

Keywords: Die attach · Thermal shocks · Finite Element Model · Silicon Carbide · Reliability

1 Introduction

The constant increase of mobility electrification trends with the massive diffusion on the Market of HEV is demanding to the semiconductor industry for Power

C. Rizzi et al. (Eds.): ADM 2019, LNME, pp. 470–480, 2020.
https://doi.org/10.1007/978-3-030-31154-4_40

Electronics components with challenging electrical and reliability performances. In particular power devices for automotive traction applications may allow an usage at higher temperatures and power density. Currently, wide bandgap semiconductors (WBG) like silicon carbide (SiC) are substituting conventional silicon (Si) devices due to their excellent electrical and physical properties. Furthermore, they feature high switching speeds and high junction operating temperature (above 200 °C) reducing the power losses and also allowing more compact cooling systems. Due to this harsh environment, materials and interfaces have to be developed enabling Silicon Carbide device to benefit of its superior properties without constrains by the surrounding parts. Therefore, reliability is one of the main challenge for SiC device at high temperature operative conditions and the main bottleneck is represented by die attach degradation [10]. In fact, actual die attach materials (typically Sn-rich solder) are not suitable to reliably endure temperatures higher than 150 °C due to their low melting temperature [26]. For these reasons, it is important to explore alternative die-attach materials and processes that can provide significantly improved performance and reliability metrics over the current Sn-solder brazing process.

One approach is the sintering technology [22], in which chips and substrates are joined together using a metal powder, such as silver or copper, sintered by thermo-compression [9,24,25]. These metals have high electrical and thermal conductivity and do not have the creep fatigue-failure issue typical of high homologous temperature material like solder [17]. In fact, their melting points are high enough to suit high-temperature packaging solutions, avoiding the creep degradation that affects solder. Some authors have shown that use of pressure may reduce the process needed temperature for sintering the silver powder to attach the power semiconductor devices to their substrate or frame, reducing at the same time the induced thermo-mechanical stress [15].

It is crucial to experimentally reproduce the mission profile aging, in order to evaluate the reliability benefits compared to current die-attach solutions. It has been proven the achievement of significant improvements in performance and reliability for sintered power devices over the soldered solutions, as well in passive temperature cycling capability as in active temperature cycling lifetime [16,27].

Some authors [18,28] reproduced the mission profile lifetime by means of liquid-to-liquid thermal shock, demonstrating that it is possible to replicate some mission profile stress, such as the sharp temperature gradient during the switch-on, which can reach 2500 °C/s [14]. Compared with thermal cycle, liquid shocks can achieve a much greater temperature rate and, as consequence, they can induce a greater strain rate to tested devices. As shown by Herboth, the stress-strain relation of sintered silver is sensible to strain rate, both for low and medium porous samples: higher strain rate implies a stiffer mechanical behavior of sinter layer [11,12]. For innovative and high reliable packaging solution not affected by conventional failure mechanism (Sn-based solder degradation or wire-bonding lift-off), such as sintering die attach with source clip connection, thermal shocks represents an effective approach to exploit fatigue behavior in reasonable time, much faster than other reliability techniques such as passive or active cycling [13,21].

The purpose of this article is the development of an experimental technique to analyze the reliability of pressure-assisted silver sintering at die attach level.

2 Methods

Different sets of the parameters pressure, time and temperature have been considered. In order to benchmark the reliability effect of process parameters during thermal shocks, the samples degradation have been monitored in order to analyze the evolution of delamination between die and Active Metal Brazed (AMB) substrate, quantifying the delaminated area. Devices have been assembled with a commercial sintering equipment, considering different process parameters as later explained in Sect. 3. The formed silver sintering joints have been evaluated just after assembly by means of optical microscopy, evaluating the porosity with a dedicated software routine. Sintered layers have been aged with liquid-to-liquid thermal shocks, described in Sect. 4. The quality of interconnections after thermal shocks have been investigated with Scanning Acoustic Microscopy (SAM) on device backside in order to analyze the evolution of delamination between die and Active Metal Brazed (AMB) substrate, quantifying the delaminated area. Furthermore, destructive examination methods such as Scanning Electron Microscopy (SEM) and metallographic analyses of cross sections have been performed. A simulation methodology, able to estimate the mechanical behaviour of the sintered joint and therefore to reproduce the accounted inelastic strain per cycle, has been developed for creating an aging model useful to convert the Mission Profile Application in an equivalent accelerated Thermal Shocks test.

3 Sample Preparation

The considered test vehicles are Silicon Carbide power MOSFET devices, sintered on Ag-plated AMB substrate according to the process schematic shown in Fig. 1.

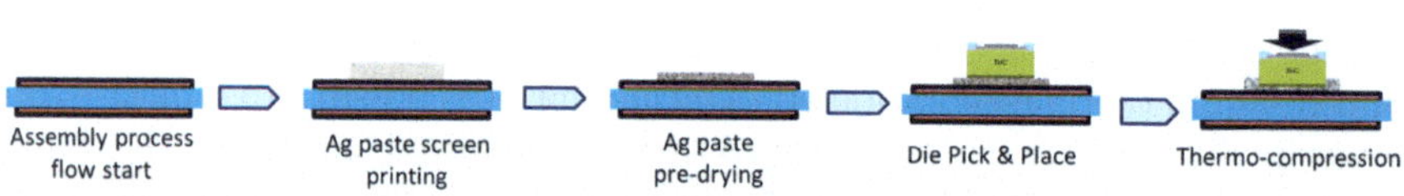

Fig. 1. Flow-chart for silver sintering pressure assisted die attach.

The sintering process has been performed considering three different value of pressure ($pA < pB < pC$), between 1 and 50 MPa, applied for 120 s. Process temperature has been set between 200 and 250 °C.

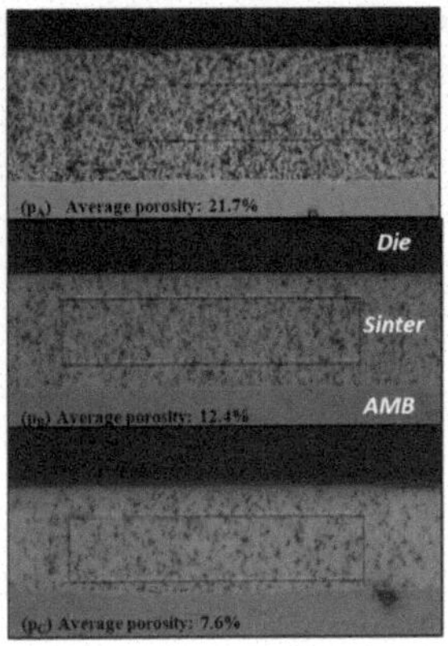

Fig. 2. Porosity measurements by microscopic pictures for the three sintering pressure.

The physical parameter that can allow a direct link to applied pressure during the sintering process is the porosity of the sintering joint that can be experimentally measured for each pressure leg, at cross section level, using a software routing on the Scanning Electron Microscope (SEM) as reported in Fig. 2.

The state of aggregation of silver layer, just after sintering process at the three different pressure value, has been evaluated performing a microscopy analysis on sample cross-sections. Microscopic pictures have been processed by software to quantify the porosity inside the sintered layer (Fig. 2).

Porosity decreases with higher pressure as expected [12]. As consequence, the Ag sintering mechanical properties (in terms of Young modulus, ductility, ultimate strength) changes between the three scenarios. Higher pressure leads to higher tensile strength, so to improve the formed joint. On the other side, Young's modulus increases, implying a less reduction of thermo-mechanical stress [13]. Therefore, thermal shocks have been aimed to benchmark the reliability behavior of different sintering parameters.

4 Experimental Characterization

The mechanical aging phenomena has been reproduced and accelerated by means of thermal shocks test performed in a liquid-to-liquid exchange between 150 °C and −65 °C, using a perfluoropolyether (PFPE) fluid, with a cycle time of six minutes (three minutes at each temperature). During and after reliability tests, metallographic samples have been investigated with optical microscopy and SEM (equipped with backscattered detector), in order to benchmark the different sintering condition at different stress application conditions.

Metallographic and Ag grain characterization has been performed by SEM analysis on specimens cross-sections at time 0 (t0) before the cycling and after 3000 thermal shocks (Fig. 3). SEM investigations clearly show the difference in terms of crack evolution after 3000 cycles for the three considered pressure.

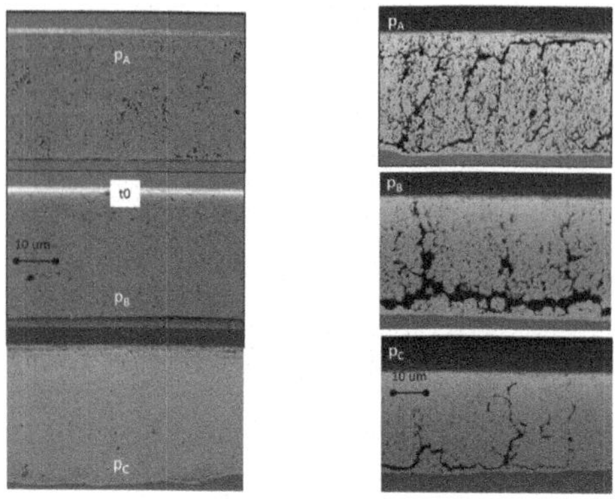

(a) (b)

Fig. 3. SEM analyses at time 0 (a) and after 3000 thermal shocks (b).

It has been observed that cracks are shorter in the condition with the highest pressure (pC). This means that higher process pressure produces a stronger sintering layer with an increased reliability.

The evolution of degradation during thermal shocks has been studied by SAM analysis considering different test time with a readout every 250 thermal shocks (Fig. 4) in which the progressive delamination has been highlighted by the contrast reduction between dies and substrate.

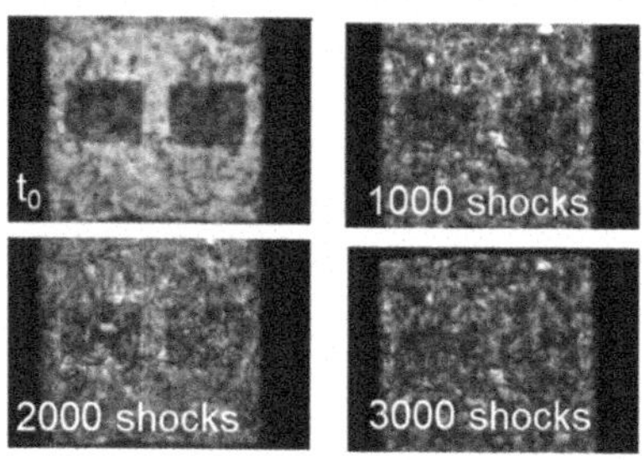

Fig. 4. Delamination of sinter layer at die-AMB interface, as function of number of cycles. The delamination occurrence has been confirmed due to the fact that the contrast between dies (in black) and substrate (white) decreases during the cycling.

The used method consist in a noninvasive ultrasound technique to determine the integrity of the sintering joint using the Scanning Acoustic Microscope in Reflection Mode. In this case the transducer send and receive the ultrasound waves as shown in Fig. 5.

The transducer transmits pulses of acoustic waves through the sample. At the interface of different materials inside the specimen, a portion of the pulse is reflected back to the transducer and the remainder propagates through the

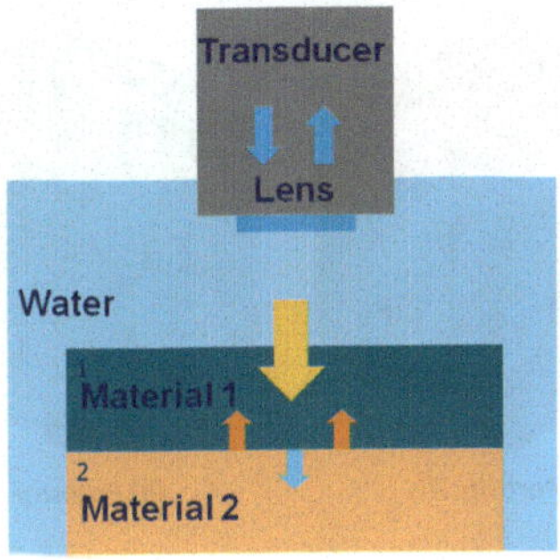

Fig. 5. The higher the impedance mismatch at the interface, for instance due to material degradation, the higher the intensity of the reflected signal (in orange).

interface. The degree of ultrasonic reflection from that interface is governed by the following acoustic impedance formula:

$$R = \frac{(z_2 - z_1)}{(z_2 + z_1)} \tag{1}$$

where R is the amplitude of the reflected pulse, and z_1 and z_2 are the acoustic impedances (density times sound speed) of the two materials [1,2].

5 Modelling

5.1 Sintering Unit Cell: CAD and FE Model

Starting from FIB/SEM morphological analysis (Fig. 6) of microstructure texture of sinter layers, a micro-mechanical cell models has been developed for theoretical modeling of effective properties of Ag sintering materials.

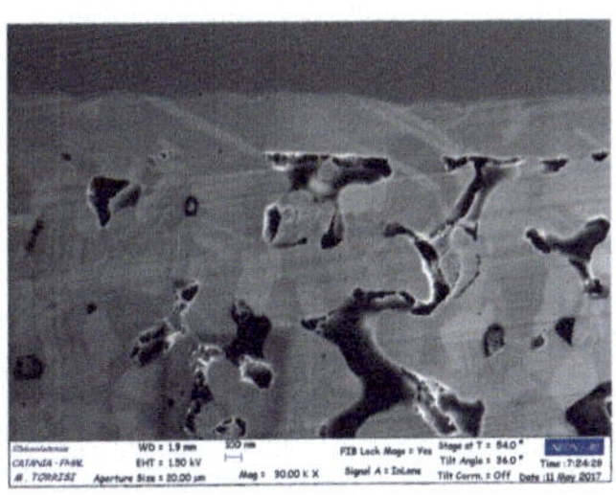

Fig. 6. SEM picture of a FIB cut showing the morphological details of the sintering layer's microstructural texture.

In the unit cell model, reported in Fig. 7, different sintering void fractions have been considered and reproduced by a CAD model taking into account the different arrangements of Ag-spheres [7,8].

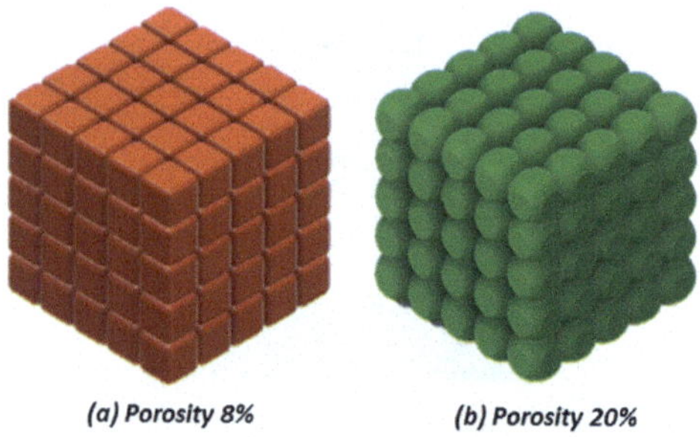

Fig. 7. CAD models representing the silver sintering grain structure for two porosity value: 8% (a) and 20% (b).

The effective elastic-plastic properties have been derived from the FE calculation results for the different porosities depicted in Fig. 2. In Fig. 8 it has shown the elastoplastic relation, as function of densification level.

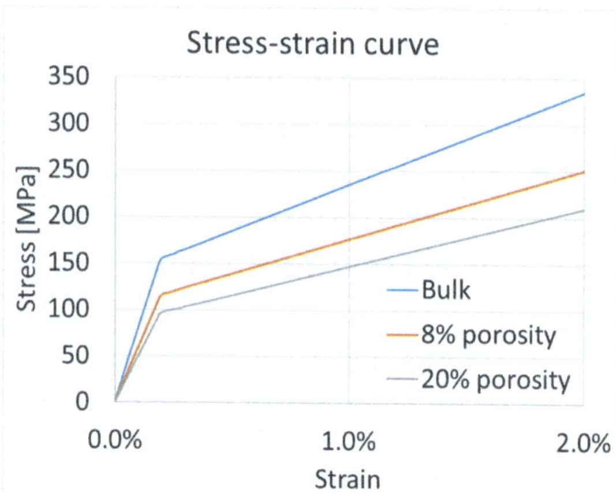

Fig. 8. Stress-strain curves at Room Temperature for silver bulk and for different Ag sintering layer porosity.

For pure Ag an initial yield strength of 160 MPa at RT was taken from literature published data [7] and true-stress/true strain measurements done in house on dogbone Ag sintering samples. Simulations have quantified the overall silver sintering mechanical softening due to porosity percentage. This effect has been also proven by mechanical Ag layer characterization by means of nanoindentation technique [4,5]. The temperature dependence has been also considered in the elastic-plastic model in accordance with [19]. Therefore, mechanical thermal softening has been also taken into account by simulation in accordance to literature data, even if, n this study, it has not been proven by experimental tensile test.

The calculated material properties at Room Temperature shown in Fig. 8 and the related final layer thickness (higher applied pressure produces a lower layer thickness) have been used as input for reproducing by simulation the thermal shock test. Even if the assumption of using the material mechanical properties from quasi-static experimental characterization is, due to the strain rate sensitivity of sintering layer, physically not correct [23], it can be acceptable if considering the simulation results as comparative benchmark at a fixed strain rate conditions.

5.2 Thermal Shocks Simulation

The experimental setup and specimens configuration used for Thermal Shocks Tests has been reproduced by Finite Element Modeling using Comsol Multiphysics with customized isoparametric tetrahedral finite elements mesh based on the absolute nodal coordinate formulation (ANCF) [6,20]. The silver sintering layer has been simulated as a bulk structure considering the mechanical properties calculated in the sintering unit cell, according to the methodology explained in Sect. 5.1. The considered elasto-plastic behaviors for sintering material have shown in Fig. 8, at different densification rates. Simulated geometry has been depicted in Fig. 9(a). It is made by a copper-ceramic substrate (respectively in brown and gray), silver sinter connections (in blue, between silicon device and substrate) and two silicon carbide devices (in orange).

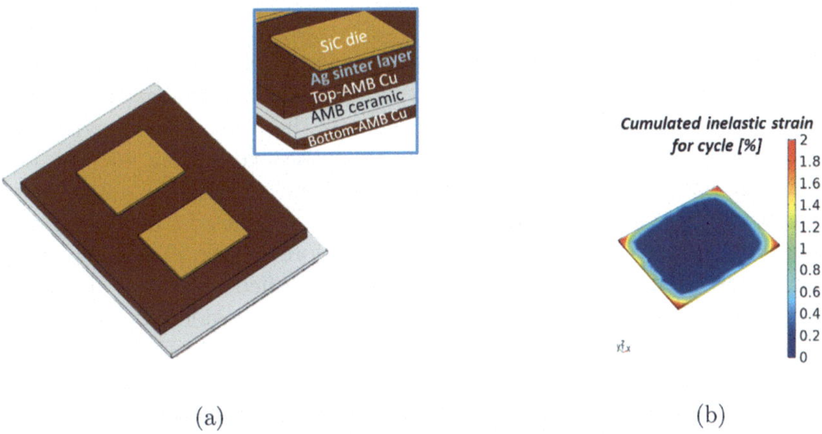

Fig. 9. Considered geometry for Finite Element Analysis (a) and plastic strain per cycle on 20% porosity sintering layer (b).

The obtained results in terms of silver layer cumulated plastic strain per cycle, as plotted in Fig. 9(b), have been used as reference parameter for building up a reliability model. More in details, the calculated values have been coupled with the experimental number of cycle to failure for the two simulated layer density, as resumed in Table 1. The used End of Life Criteria (EOL) has been the maximum degradation area, detected by SAM analysis over 20% of total sintering area. Device-substrate interface is a bottleneck for package reliability because the heat dissipation occurs through this surface during the operative life. A critical decrease of interconnections area induces a temperature increase, with consequent device failure. Then, the proposed approach could be useful to predict the long-term behavior of semiconductor devices.

Table 1. Calculated cumulated plastic strain and cycles to failure.

Porosity %	Cumulated plastic strain $\Delta\varepsilon_{cycle}$	Cycles to failures N_F
8	1.27%	5000
20	1.75%	3000

Starting from data reported in table, it is possible to derive coefficients a and m for Coffin-Manson equation [26]:

$$N_F = a \cdot \Delta\varepsilon_{cycle}{}^{m} \tag{2}$$

whereas a and m are equal respectively to 3.34 and −1.67.

6 Conclusion

In the presented work, the setup of a pressure-assisted silver sintering process has been performed by means of an experimental methodology, made by different microscopic analyses. In order to evaluate the reliability behavior, device aging has been enhanced by high strain rate stress condition, such as liquid-to-liquid thermal shocks. In particular it has been shown how sintering pressure (considering constant time and temperature parameters) have a significant influence on porosity, microstructure and therefore on the mechanical properties and consequently on reliability. It has been verified by means of cyclic fast transient thermomechanical stress that the highest considered sintering pressure (pC) implies, at the same time, higher sinter layer density and improved reliability performances. This experimental evidence has been also linked with the simulated cumulative inelastic strain per cycle in order to generate a reliability model based on Coffin Manson formulation. The proposed methodology, integrating experimental and simulation results, allows a reliability prediction of the sintering layer as function of the process parameters mechanically well represented by the layer porosity.

Acknowledgment. The research activity done in STMicroelectronics leading to the results shown in this paper was partially funded from the H2020-ECSEL Joint Undertaking under grant agreement no. 783158, REACTION (first and euRopEAn siC eigTh Inches pilOt liNe) Project [3].

References

1. Bai, J., Zhang, Z., Calata, J., Lu, G.: Characterization of low-temperature sintered nanoscale silver paste for attaching semiconductor devices. In: Conference on High Density Microsystem Design and Packaging and Component Failure Analysis, pp. 1–5. IEEE (2005)
2. Balk, L.J.: Scanning acoustic microscopy. Surf. Interface Anal. **9**(1), 47–54 (1986)
3. Bieniek, T., Janczyk, G., Sitnik, A., Messina, A.A.: The first and euRopEAn siC eigTh inches pilOt line REACTION project as a driver for key European SiC technologies focused on power electronics development. In: TechConnect Briefs, pp. 256–259 (2019)
4. D'Arrigo, G., Christian, M., Morandi, V., Favaro, G., Bongiorno, C., Mio, A., Russo, M., Sitta, A., Calabretta, M., Sciuto, A., et al.: Mechanical and electrical characterization of CVD-grown graphene transferred on chalcogenide Ge2Sb2Te5 layers. Carbon **132**, 141–151 (2018)
5. D'Arrigo, G., Mio, A., Favaro, G., Calabretta, M., Sitta, A., Sciuto, A., Russo, M., Calì, M., Oliveri, M., Rimini, E.: Mechanical properties of amorphous Ge2Sb2Te5 thin layers. Surf. Coat. Technol. **355**, 227–233 (2018)
6. De Simone, M.C., Rivera, Z., Guida, D.: Finite element analysis on squeal-noise in railway applications. FME Trans. **46**(1), 93–100 (2018)
7. Dudek, R., Döring, R., Sommer, P., Seiler, B., Kreyssig, K., Walter, H., Becker, M., Günther, M.: Combined experimental-and FE-studies on sinter-Ag behaviour and effects on IGBT-module reliability. In: 15th International Conference on Thermal, Mechanical and Multi-Physics Simulation and Experiments in Microelectronics and Microsystems (EuroSimE), pp. 1–9. IEEE (2014)
8. Dudek, R., Sommer, P., Fix, A., Rzepka, S., Michel, B.: Reliability issues for high temperature interconnections based on transient liquid phase soldering. In: 14th International Conference on Thermal, Mechanical and Multi-Physics Simulation and Experiments in Microelectronics and Microsystems (EuroSimE). pp. 1–8. IEEE (2013)
9. Göbl, C., Faltenbacher, J.: Low temperature sinter technology die attachment for power electronic applications. In: 6th International Conference on Integrated Power Electronics Systems, pp. 1–5. IEEE (2010)
10. Goehre, J., Schneider-Ramelow, M., Geißler, U., Lang, K.D.: Interface degradation of al heavy wire bonds on power semiconductors during active power cycling measured by the shear test. In: 6th International Conference on Integrated Power Electronics Systems (CIPS), pp. 1–6. IEEE (2010)
11. Herboth, T., Früh, C., Günther, M., Wilde, J.: Assessment of thermo-mechanical stresses in low temperature joining technology. In: 13th International Thermal, Mechanical and Multi-Physics Simulation and Experiments in Microelectronics and Microsystems, pp. 1–7. IEEE (2012)
12. Herboth, T., Guenther, M., Fix, A., Wilde, J.: Failure mechanisms of sintered silver interconnections for power electronic applications. In: IEEE 63rd Electronic Components and Technology Conference, pp. 1621–1627. IEEE (2013)
13. Herold, C., Sun, J., Seidel, P., Tinschert, L., Lutz, J.: Power cycling methods for SiC MOSFETs. In: 29th International Symposium on Power Semiconductor Devices and IC's (ISPSD), pp. 367–370. IEEE (2017)
14. Huang, Y., Luo, Y., Xiao, F., Liu, B.: Failure mechanism of die-attach solder joints in IGBT modules under pulse high-current power cycling. IEEE J. Emerg. Sel. Top. Power Electron. **7**(1), 99–107 (2019). https://doi.org/10.1109/JESTPE.2018.2871474

15. Klaka, S., Sittig, R.: Reduction of thermomechanical stress by applying a low temperature joining technique. In: Proceedings of the 6th International Symposium on Power Semiconductor Devices and Ics, pp. 259–264 (1994). https://doi.org/10.1109/ISPSD.1994.583736
16. Knoerr, M., Kraft, S., Schletz, A.: Reliability assessment of sintered nano-silver die attachment for power semiconductors. In: 12th Electronics Packaging Technology Conference, pp. 56–61. IEEE (2010). https://doi.org/10.1109/EPTC.2010.5702605, http://ieeexplore.ieee.org/document/5702605/
17. Knoerr, M., Schletz, A.: Power semiconductor joining through sintering of silver nanoparticles: evaluation of influence of parameters time, temperature and pressure on density, strength and reliability. In: 6th International Conference on Integrated Power Electronics Systems, pp. 1–6. IEEE (2010)
18. Liu, C., Brem, F., Riedel, G., Eichelberger, E., Hofmann, N.: The influence of thermal cycling methods on the interconnection reliability evaluation within IGBT modules. In: 4th Electronic System-Integration Technology Conference, pp. 1–5 (2012). https://doi.org/10.1109/ESTC.2012.6542180
19. Mertens, C.: Die niedertemperatur-verbindungstechnik der leistungselektronik. VDI Fortschritt-Bericht, Dissertation (2000)
20. Pappalardo, C.M., Wang, T., Shabana, A.A.: Development of ancf tetrahedral finite elements for the nonlinear dynamics of flexible structures. Nonlinear Dyn. **89**(4), 2905–2932 (2017)
21. Russo, S., Bazzano, G., Cavallaro, D., Sitta, A., Calabretta, M.: Thermal analysis approach for predicting power device lifetime. IEEE Trans. Device Mater. Reliab. **PP**, 1 (2019). https://doi.org/10.1109/TDMR.2019.2892185
22. Santopà, M., Russo, S., Torrisi, M., Renna, M., Sitta, A., Calabretta, M.: Fast transient thermomechanical stress to set a pressure-assisted sintering process. In: Proceedings of PRIME 2019. IEEE (2019, in press)
23. Schaal, M., Klingler, M., Wunderle, B.: Silver sintering in power electronics: the state of the art in material characterization and reliability testing. In: 7th Electronic System-Integration Technology Conference (ESTC), pp. 1–18. IEEE (2018)
24. Schmitt, W., Chew, L.M.: Silver sinter paste for sic bonding with improved mechanical properties. In: IEEE 67th Electronic Components and Technology Conference (ECTC), pp. 1560–1565. IEEE (2017)
25. Schwarzbauer, H., Kuhnert, R.: Novel large area joining technique for improved power device performance. IEEE Trans. Ind. Appl. **27**(1), 93–95 (1991). https://doi.org/10.1109/28.67536
26. Sitta, A., Calabretta, M., Renna, M., Cavallaro, D.: Solder joint reliability: thermo-mechanical analysis on power flat packages. In: Advances on Mechanics, Design Engineering and Manufacturing, pp. 709–716. Springer (2017)
27. Sitta, A., Russo, S., Bazzano, G., Cavallaro, D., Greco, G., Calabretta, M.: Numerical approach to predict power device reliability. In: 13th International Conference on Design & Technology of Integrated Systems in Nanoscale Era (DTIS), pp. 1–5. IEEE (2018)
28. Ying, L.C., Fei, C.C.: Temperature cycling and thermal shock correlation in DPAK & DSO packages. In: 10th IEEE International Conference on Semiconductor Electronics (ICSE), pp. 458–461 (2012). https://doi.org/10.1109/SMElec.2012.6417185

An Integrated Approach to Optimize Power Device Performances by Means of Stress Engineering

Michele Calabretta[1(✉)], Alessandro Sitta[1,2], Salvatore Massimo Oliveri[2], and Gaetano Sequenzia[2]

[1] STMicroelectronics S.r.l., Stradale Primosole 50, Catania, Italy
michele.calabretta@st.com

[2] Dipartimento di Ingegneria Elettrica, Elettronica ed Informatica (DIEEI), Università degli Studi di Catania, Viale Andrea Doria 5, Catania, Italy

Abstract. In the present work it is shown how stress engineering can be used in semiconductor industry to improve Power MOSFET transistor's performance beyond simple geometrical downscaling. The aim of this paper is to present an integrated methodology, coupling modelling and experimental results, focused on the structural optimization of a power device by means of final passivation mechanical stress tuning. The proposed approach is based on a Finite Element Model that describes and predicts the mechanical strain of a singulated power device (MOSFET) validated by the correlation with interferometric experimental warpage measurements (Topography and Deformation Measurements). Scope of the activity is to engineer Power Devices with reduced intrinsic stresses in order to optimize the reliability performances. Controlled stress into a single semiconductor crystal oriented substrate can be managed at manufacturing level by several methods, including the introduction of a layer on the top of the substrate or around the gate region. From the knowledge of the mechanical boundaries, as a function of temperature, it is possible to predict the stress conditions impacting on device fabrication and on reliability performances. Moreover, according to the piezoresistive model, it has been evaluated the electrical characteristics (on-resistance) in the operative working condition range. According to the proposed approach an optimized passivation layer solution has been proposed, simulated by Finite Element model and validated by experiments.

Keywords: Silicon MOSFET · Warpage · FEM · Moiré topography · Reliability

1 Introduction

Stress engineering is currently one of the main approach for performance and reliability enhancement of device. For vertical n-channel power MOSFET, which are typically tensile stressed on the device front, the stress management and its

C. Rizzi et al. (Eds.): ADM 2019, LNME, pp. 481–491, 2020.
https://doi.org/10.1007/978-3-030-31154-4_41

optimization concurrently act as reliability improvement, by means of reduction of overall stress and warpage, and also as piezoresistive characteristics improvement. It is based on the fact that silicon resistivity improves or degrades when the material is strained, both in tensile and compressive way. The coupling between electrical resistivity and mechanical strain is the so-called piezoresistivity.

Considering the silicon substrate used for device manufacturing, piezoresistivity depends on considered crystallographic direction, being material anisotropic, and on dopant type (holes or electrons) [8]. This means that tensile strain enhances silicon resistivity in a specific crystal direction, but it can degrade resistivity for different crystal orientation or changing the dopant type. State-of-the-art strained devices are manufactured by substrate and process mechanical strain techniques.

Techniques based on substrate-level generates compressive or tensile stresses over the entire semiconductor device, exploiting the differences in the crystal-lattice constant of different semiconductor materials, typically silicon and germanium. The known solutions envisage, for example, growth of a thin film of a silicon-germanium (SiGe) crystal on a silicon (Si) structural layer, for generation of biaxial compressive stresses, or formation of a thin silicon (Si) film on a structural layer constituted by a silicon-germanium (SiGe) crystal, for generation of biaxial tensile stresses [23]. The main disadvantages of stress engineering methods based on substrate are the costs and the manufacturing complexity.

Process-based techniques manage stress in the MOSFET channel modifying Front-End manufacturing processes [10]. Some methods are the gate stack scaling [11], the deposition of intrinsically stressed contact etch stop layers (CESL) [12] and the creation of a nitride capping layer, which can generate stress up to 700 MPa [20]. Varying the SiN capping layer deposition method, stress could be tensile (in case of chemical vapor deposition) or compressive (in case of plasma enhanced deposition). An immediate effect of piezoresistivity is the modification of device drain-source on-state resistance (R_{DSon}). An adequate strain is able to reduce the R_{DSon}, which means to limit the dissipated power and temperature swing during operation life.

Furthermore, stress engineering permits to reduce warpage already at wafer level, achieving a more reliable manufacturing during Front End processes, such as wafer handle and the adsorption of vacuum suction, but also more accuracy of photolitographic processes [1]. Less warped wafer makes more robust Back End processes, such as die attach and wire bonding [2,14]. Warpage reduction means a stress reduction at the interfaces, which improves the device reliability performances, both in passive and active cycles [15,17,18]. These aspects, affecting die attach and wire bonding, are the typical bottlenecks for device durability. Combining these described improvements with advanced die attach technique, such as pressure silver sintering [16,22], it is possible to achieve the challenging reliability requirements of novel high-power application for automotive market. At the state-of-art, it seems that to be missing a calculation technique able to predict, at the same time, warpage and electric performances shifts induced by device modifications.

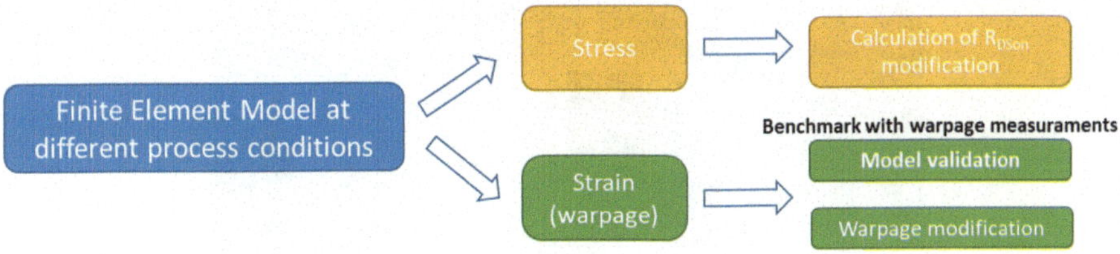

Fig. 1. Schematic of proposed approach.

In this work, it has been presented a novel methodology aimed to optimize the stress conditions of power device by means of final passivation mechanical stress tuning. The proposed method combines experimental measurements for warpage and R_{DSon} with theoretical models. Mechanical Finite Element Model has been developed to calculate stress and strain induced by final passivation process as function of temperature, considering the standard layer and a modified version with reduced thickness. In order to validate the calculation, FEM warpage values have been benchmarked with dedicated warpage measurements. Stress coming from Finite Element Analysis have been considered to estimate the piezoresistivity effect on R_{DSon} reduction according to a developed theoretical model. The proposed approach has been resumed in Fig. 1. The final purpose of the presented activity is to verify if and also how passivation thickness reduction improves the reliability (warpage) and electric (R_{DSon}) performances.

2 Device Description

Considered device is a low voltage n-channel Power MOSFET for automotive costumers, with die size of 5 mm^2 (2.85 × 1.85 mm, silicon thickness of 0.28 mm). A wafer of standard device has been reported in Fig. 2. According to experimental setup later explained, wafer has been sliced in "die matrix". Due to its logical function, several layers of different materials, both conductive and insulator, have been manufactured above silicon. Due to industrial policy about data disclosure, simplified stacks have been reported in Fig. 3.

Considering that these layers have been formed at higher temperature than operative ones, the CTE mismatch generates stress and deformation ("thermal stress"). In addition, layer manufacturing process implies structural modification in material, therefore an additional source of stress ("intrinsic stress") has to be considered. Device electric performance has been optimized decreasing the thickness of top passivation layer, as depicted in Fig. 3. In order to perform electric measurements, devices have been assembled in typical package for LV application (PSSO16).

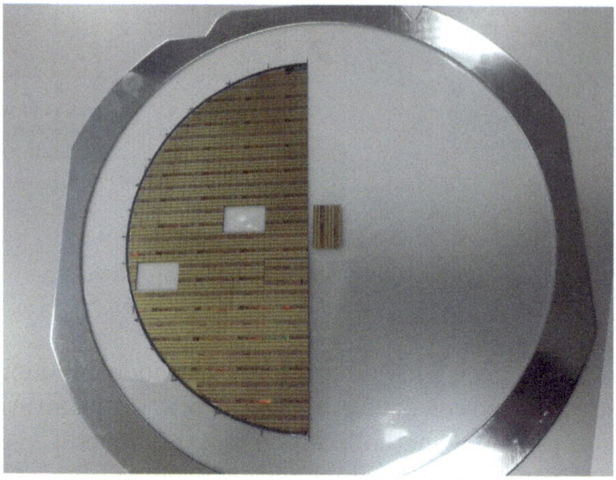

Fig. 2. Considered wafer for measuring standard devices.

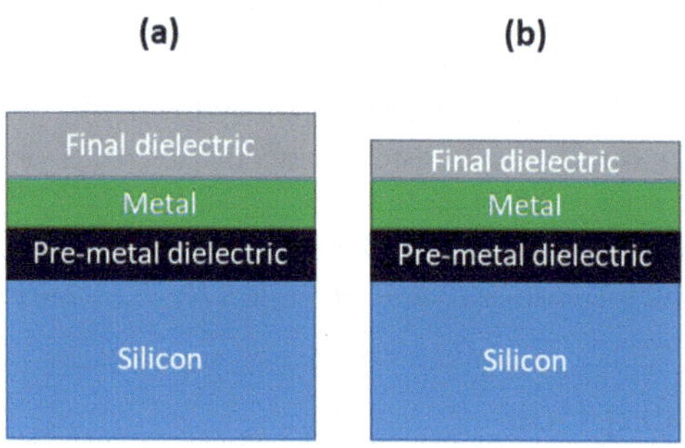

Fig. 3. Simplified cross sections for standard (a) and optimized (b) options.

3 Warpage Measurement

3.1 Mathematical Model

A method based on Moirè Phase Shift has been considered to experimentally quantify the warpage [24]. The sample to be analysed is illuminated by a structured light made by stripe pattern. The stripe pattern is deformed by the sample's surface structure and the resulting image is captured by a CCD camera, that correlates the out-of-plane deformation with the xy coordinates. Equipment, TDM Compact by Insidix, has been shown in Fig. 4. The principle of working has been depicted in Fig. 5.

Considered equipment is mainly made by two sub-parts: the projection (*composite light source*) and the viewing (*CCD camera + optics*) sub-systems. The projection sub-system is a commercial projector, regulated by computer in order to project strip patterns. The viewing sub-system includes a CCD camera and viewing lens.

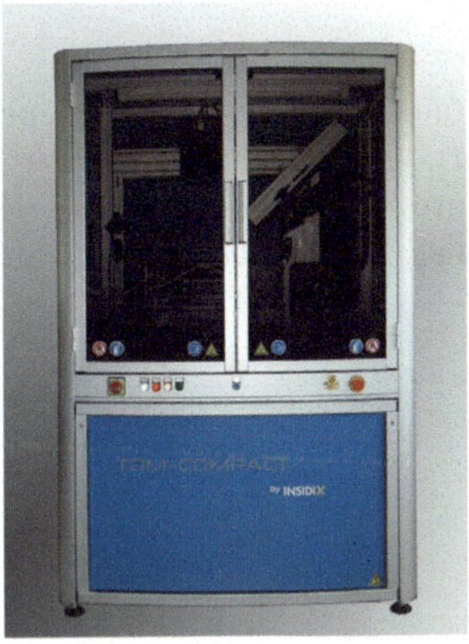

Fig. 4. TDM Compact equipment.

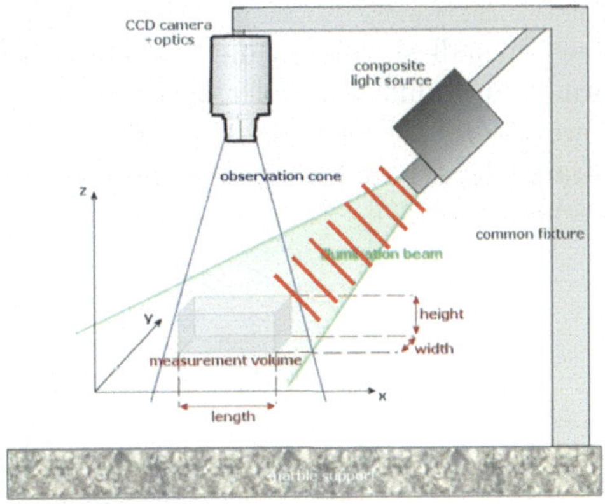

Fig. 5. Schematic of experimental apparatus.

The transmission of the square wave gratings through the optical system can be approximated as a cosine function of the profile number. Thus the intensity pattern projected onto the object surface through the projection lens (I_p) is described as:

$$I_p(x,y) = I_s R(x,y) A\left[1 + cos(2\pi n_p(x,y))\right] \tag{1}$$

where I_s is the intensity of the light source, $R(x,y)$ is the reflectance, A and n_p are respectively amplitude and profile number of the projection grating. Meanwhile, the virtual intensity pattern on the sample (I_v), coming from viewing system, could be described by:

$$I_v(x,y) = A\left[1 + cos(2\pi n_v(x,y))\right] \tag{2}$$

where n_v is the profile number of the viewing grating line that virtually illuminates the point (x, y). This described overlap generates an interference pattern, which can be expressed [7]:

$$I_k(x,y) = A(x,y) + B(x,y) \cdot cos\left(\frac{2\pi d\,(z+\delta_k)}{p\,(h+z+\delta_k)}\right) \qquad \delta_k = \frac{k-1}{m}\frac{ph}{d} \tag{3}$$

in which I_k is the recorded fringe interference pattern for the k-th shift ($k = 1, 2, ..., m$), d is the distance between the light source and the observer, h is the distance from the grating to the light source and observer, z is the out-of-plane deformation of measured surface, p is the pitch of the Moirè grating and $A(x,y)$ and $B(x,y)$ are respectively the background and local contrast. In order to find out z, deleting experimental noise, measurements have been performed with different m grating pitches, as explicated in Eq. 3 by terms δ_k. Considering a Gaussian least squares approach, it is possible to solve iteratively the m equations (one for each considered phase-shift), non-linear in the coordinate z. The described approach has been practically implemented by the used equipment considering a "virtual" grating, according to [7].

3.2 Experimental Procedure

Due to small singulated die dimension, warpage measurement have been performed on 3 different wafer portions (22 × 13 mm), which includes 56 (8 × 7) Power MOSFET devices. Measurement temperature can be controlled by the infrared heater and air cooler and it is continuously monitored by dedicated thermocouple [24].

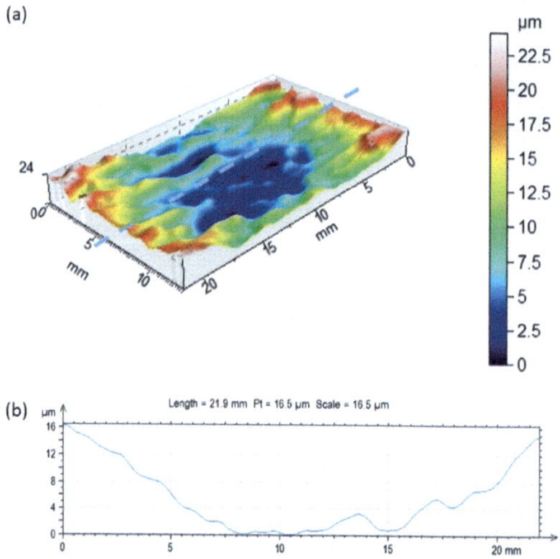

Fig. 6. Measured 3D map for standard solution a 150 °C (a) and warpage along its cutline (b).

For the measurements, a camera with a field of view of 80 × 60 mm has been used, setting an exposure time of 20 ms. Sample has been heated from room temperature to 150 °C, with temperature rate of 0.5 °C/s, representative of a quasi-steady heating. Figure 6 depicts the warpage map and out-of-plane deformation along a specific path at the maximum temperature (150 °C).

Maximum warpage values have been reported in Fig. 9 at 6 different temperature (25, 50, 75, 100, 125 and 150 °C). Plotted values are the average between 5 identical samples.

4 Modeling and Experimental Correlation

4.1 Finite Element Model

Finite Element Models are widely used to optimize device and assembly technologies, calculating deformation and force inside the structure. Considering electric device, mechanical stresses could be distinguished in thermal and intrinsic stress [9]. Thermal stress are induced due to the fact that device is composed by different materials, as reported in Fig. 3. Each material is characterized by own coefficient of thermal expansion (CTE), therefore thermo-mechanical stress is generated due to temperature variation. Considering a thin film ("f") deposited on a much thicker substrate ("s"), induced stress in film due to CTE mismatch is:

$$\sigma_f = \frac{E_s}{1 - \nu_s} (\alpha_f - \alpha_s) \Delta T \tag{4}$$

in which E_s and ν_s are respectively the Young modulus and the Poisson ratio of substrate, α is the coefficient of thermal expansion and ΔT the temperature variation. Some authors have been developed analytic models for calculate thermal stress in multi-layer structure [19,21]. However, developed approaches are limited to elastic material and 2D geometry. The intrinsic stress reflects, instead, the internal structure of a material during its deposition, depending on some process parameters, for example the deposition rate and temperature.

The followed approach for stress/strain calculation has replicated the generated thermal stress, according to the material characteristics reported in Table 1. Considered elastic modulus have been experimental characterized by means of nanoindentation [4,5]. AlCu metal has been considered as elasto-plastic material according to internal uni-axial tensile characterization, introducing a nonlinearity in the model.

Table 1. Material data considered for simulation.

Material	Young modulus GPa	CTE ppm/K	Poisson ratio
Silicon	169	2.6	0.2
Dielectric	80	4	0.17
Metal	70	23	0.33
Final passivation	300	0.8	0.26

In order to properly estimate the thermal stresses of device, thermal assembly process has been replicated so that the appropriate layer is activated at its respective stress-free temperature. This can be accomplished considering layer with a fictive very low elastic modulus when temperature is greater than stress-free temperature. This low modulus ensures that these "low stiff" elements do not influence the bending stiffness of the stack at the first simulation step but strain will result in these elements [3]. Intrinsic stresses has been modeled adding an equivalent thermal coupling effect for layers depicted in Fig. 3. Model mesh has been shown in Fig. 7 [6,13]. Body forces such as gravity are neglected such that the resulting curvature is purely a function of the CTE mismatch and intrinsic stresses. Used software has been COMSOL Multiphysics.

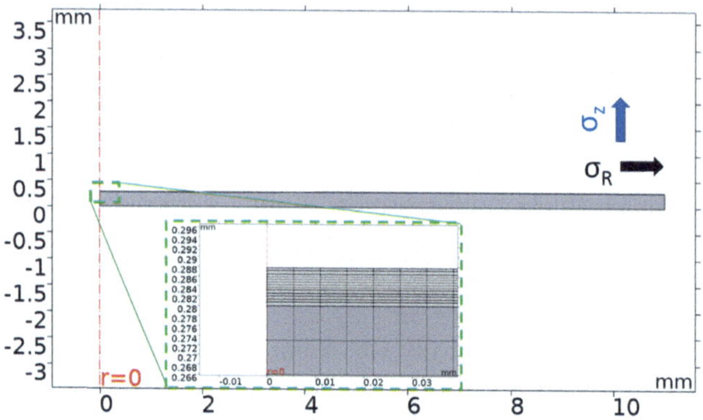

Fig. 7. 2D FE model, with mesh and stress orientation.

Figure 8 shows the correlation between experimental and calculated warpage for standard and optimized solutions. Furthermore, stress in radial direction r on top silicon surface (interface between silicon and the pre-metal dielectric) is plotted as function of temperature both for standard and optimized solution in Fig. 9. Due to very low thickness, stress in z-direction (σ_z) is quite negligible. New passivation has reduced both stress and warpage of about 10%. Calculated stress serves also to predict the entity of piezoresistivity effect, as shown in the next paragraph.

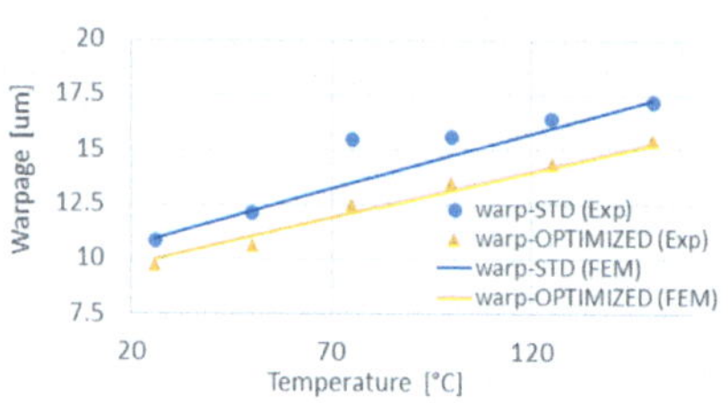

Fig. 8. Maximum warpage at different temperature (both experimental and FEM).

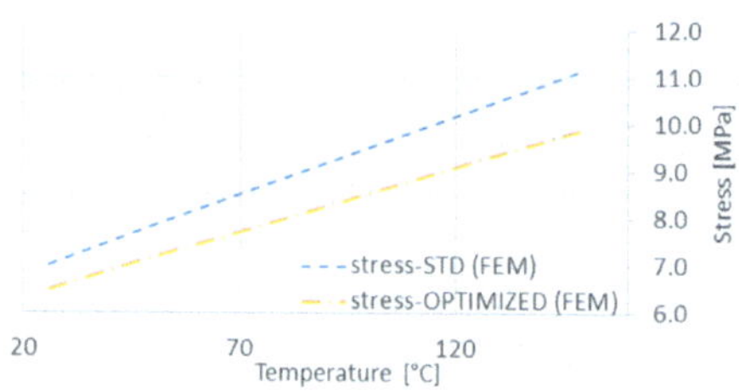

Fig. 9. FEM calculated stress in radial direction, as function of temperature.

4.2 Piezoresistive Effect

Analyzed device is n-type vertical channel Power MOSFET. Considering vertical channel power, piezoresistive effect on R_{DSon} is given by:

$$\Delta R_{DSon}/R_{DSon} = P(N,T) \cdot \pi_t \cdot (\sigma_x + \sigma_y) \tag{5}$$

where π_t is piezoresistive coefficient expressed in m^2/N, $\sigma_x + \sigma_y$ is the planar (perpendicular to channel) mechanical stress: positive if tensile, negative if compressive [8]. For n-type Si, $\pi_t > 0$ and it depends both on doping level (N) and temperature (T), as considered by terms $P(N,T)$. For considered conditions, $\pi_t = 53.4 \cdot 10^{-11}$ m^2/N. This means that, if resultant planar stress is compressive, ΔR_{DSon} is negative and R_{DSon} decreases. Otherwise, R_{DSon} increases in case of tensile stress.

A theoretical model for on-state resistance has been set to distinguish the contribute on overall resistance of each part of system (active area, epitaxy, silicon substrate, package connections) and the resistivity degradation with temperature. According to this model, the major contribution (about 80%) for R_{DSon} is in the active part of devices, physically placed on the topside of silicon. Therefore, stress on silicon topside (plotted in Fig. 9) is the main contribution for piezoresistivity. Due to the low stress difference between analyzed configurations (around 1 MPa), the predicted relative R_{DSon} reduction is negligible (below 0.1%).

5 Conclusion

A novel numerical technique has been proposed to calculate the differences in terms of stress and warpage on a semiconductor device induced by process modifications. In the specific work, the performed optimization has been the thickness reduction of top passivation layer. The developed Finite Element Models have been predicted that the optimized device, with reduced top passivation thickness, has been reduced warpage and stress of about 10%, as shown respectively in Figs. 8 and 9. Calculated warpage values have been in good agreement with measurements (Fig. 8). This means that device modification has improved the whole device workability and reliability. Due to the low stress entity, a piezoresistive improvement is present, even if the absolute benefit is quite marginal.

Nevertheless, this method allows to explore different solutions at Front End level, e.g. in terms of material stack (as in the specific work), mask layout and deposition temperature. According to this approach, it is possible to evaluate both the impact on mechanical characteristics (warpage, stress) and the effect on electric resistance (R_{DSon}). Considering the incoming device technologies and the challenging applications, the proposed innovative technique seems to be crucial to design reliable devices, without electric performance losses.

References

1. Cheng, G., Xu, G., Gai, W., Luo, L.: Deep understanding the role of Cu in RDL to warpage by exploring the warpage evolution with microstructural changes. In: IEEE 68th Electronic Components and Technology Conference (ECTC), pp. 2416–2421. IEEE (2018)
2. Cheng, G., Xu, G., Gai, W., Luo, L.: Influence of observed anelasticity of Cu on the wafer warpage evolution during thermal processes. In: 19th International Conference on Electronic Packaging Technology (ICEPT), pp. 1119–1122. IEEE (2018)

3. Conrad, H., Klose, T., Sandner, T., Jung, D., Schenk, H., Lakner, H.: Modeling the thermally induced curvature of multilayer coatings with COMSOL multiphysicstm. In: Proceedings of COMSOL Conference (2008)
4. D'Arrigo, G., Christian, M., Morandi, V., Favaro, G., Bongiorno, C., Mio, A., Russo, M., Sitta, A., Calabretta, M., Sciuto, A., et al.: Mechanical and electrical characterization of CVD-grown graphene transferred on chalcogenide Ge2Sb2Te5 layers. Carbon **132**, 141–151 (2018)
5. D'Arrigo, G., Mio, A., Favaro, G., Calabretta, M., Sitta, A., Sciuto, A., Russo, M., Calì, M., Oliveri, M., Rimini, E.: Mechanical properties of amorphous Ge2Sb2Te5 thin layers. Surf. Coat. Technol. **355**, 227–233 (2018)
6. De Simone, M.C., Rivera, Z., Guida, D.: Finite element analysis on squeal-noise in railway applications. FME Trans. **46**(1), 93–100 (2018)
7. Degrieck, J., Van Paepegem, W., Boone, P.: Application of digital phase-shift shadow Moiré to micro deformation measurements of curved surfaces. Opt. Lasers Eng. **36**(1), 29–40 (2001)
8. Kanda, Y.: A graphical representation of the piezoresistance coefficients in silicon. IEEE Trans. Electron Devices **29**(1), 64–70 (1982)
9. Mallik, A., Stout, R.: Simulation of process-stress induced warpage of silicon wafers using ANSYS® finite element analysis. In: International Symposium on Microelectronics, vol. 2010, pp. 000364–000371. International Microelectronics Assembly and Packaging Society (2010)
10. Ngai, T., Wang, Q., Sharp, J.: Power MOSFET having a strained channel in a semiconductor heterostructure on metal substrate. US Patent 8,237,195, 7 August 2012
11. Oishi, A., Fujii, O., Yokoyama, T., Ota, K., Sanuki, T., Inokuma, H., Eda, K., Idaka, T., Miyajima, H., Iwasa, S., et al.: High performance CMOSFET technology for 45 nm generation and scalability of stress-induced mobility enhancement technique. In: IEEE International Electron Devices Meeting, 2005. IEDM Technical Digest, pp. 229–232. IEEE (2005)
12. O'Neill, A., Tsang, Y., Gallacher, B., Olsen, S.: Piezomobility description of strain-induced mobility. In: 9th International Conference on Solid-State and Integrated-Circuit Technology, pp. 464–467. IEEE (2008)
13. Pappalardo, C.M., Wallin, M., Shabana, A.A.: A new ANCF/CRBF fully parameterized plate finite element. J. Comput. Nonlinear Dyn. **12**(3), 031008 (2017)
14. Reigosa, P.D., Wang, H., Yang, Y., Blaabjerg, F.: Prediction of bond wire fatigue of IGBTs in a PV inverter under a long-term operation. IEEE Trans. Power Electron. **31**(10), 7171–7182 (2015)
15. Russo, S., Bazzano, G., Cavallaro, D., Sitta, A., Calabretta, M.: Thermal analysis approach for predicting power device lifetime. IEEE Trans. Device Mater. Reliab. **19**(1), 159–163 (2019)
16. Santopà, M., Russo, S., Torrisi, M., Renna, M., Sitta, A., Calabretta, M.: Fast transient thermomechanical stress to set a pressure-assisted sintering process. In: Proceedings of PRIME 2019. IEEE (2019, in press)
17. Sitta, A., Calabretta, M., Renna, M., Cavallaro, D.: Solder joint reliability: thermo-mechanical analysis on power flat packages. In: Advances on Mechanics, Design Engineering and Manufacturing, pp. 709–716. Springer (2017)
18. Sitta, A., Russo, S., Bazzano, G., Cavallaro, D., Greco, G., Calabretta, M.: Numerical approach to predict power device reliability. In: 13th International Conference on Design & Technology of Integrated Systems in Nanoscale Era (DTIS), pp. 1–5. IEEE (2018)

19. Suhir, E.: Analysis of interfacial thermal stresses in a trimaterial assembly. J. Appl. Phys. **89**(7), 3685–3694 (2001)
20. Suthram, S., Ziegert, J., Nishida, T., Thompson, S.: Piezoresistance coefficients of (100) silicon nmosfets measured at low and high (∼1.5 GPa) channel stress. IEEE Electron Device Lett. **28**(1), 58–61 (2006)
21. Vasudevan, M., Johnson, W.: On multi-metal thermostats. Appl. Sci. Res. Sect. B **9**(6), 420–430 (1961)
22. Weber, C., Hutter, M., Schmitz, S., Lang, K.D.: Dependency of the porosity and the layer thickness on the reliability of Ag sintered joints during active power cycling. In: IEEE 65th Electronic Components and Technology Conference (ECTC), pp. 1866–1873. IEEE (2015)
23. Wee, C., Maikop, S., Yu, C.Y.: Mobility-enhancement technologies. IEEE Circuits Devices Mag. **21**(3), 21–36 (2005)
24. Weidmann, D., Dubois, G., Hertl, M., Chauffleur, X.: Determination of strength of interface in packages based on an approach using coupling of experimental and modeling results. In: 12th International Conference on Thermal, Mechanical & Multi-Physics Simulation and Experiments in Microelectronics and Microsystems, pp. 1–6. IEEE (2011). https://doi.org/10.1109/ESIME.2011.5765836

Iterative and Participative Axiomatic Design Process to Improve Conceptual Design of Large-Scale Engineering Systems

Domenico Marzullo[1(✉)], Giuseppe Di Gironimo[1], Danilo Nicola Dongiovanni[2], Antonio Lanzotti[1], Rocco Mozzillo[1], and Andrea Tarallo[1]

[1] Department of Industrial Engineering (DII), University of Naples Federico II, Piazzale Tecchio 80, 80125 Naples, Italy
{domenico.marzullo,giuseppe.digironimo, antonio.lanzotti,rocco.mozzillo, andrea.tarallo}@unina.it

[2] Nuclear Fusion and Safety Technologies Department, ENEA, via Enrico Fermi 45, 00040 Frascati, Italy
danilo.dongiovanni@enea.it

Abstract. This research discusses the use of a systematic design method, the Iterative and Participative Axiomatic Design Process (IPADeP), for the early conceptual design stage of large-scale engineering systems. The involvement of multiple and competing requirements has imposed high challenges for achieving an affordable design of complex systems in a reasonable lead time. Systems Engineering (SE) focuses on how to design and manage complex systems over their life cycles. Both must begin by discovering the real problems that need to be resolved and identifying from the early stage of the design the main stakeholder requirements and customer needs. The Axiomatic Design (AD) methodology is widely recognized in the literature to efficiently support the design of complex systems from the early conceptual stage. IPADeP provides a systematic methodology for applying AD theory in the conceptual design of large-scale engineering systems, aiming to minimize the risks related to the uncertainty and incompleteness of requirements and to improve the collaboration of multidisciplinary design teams.

IPADeP has been adopted as design methodology in the pre-conceptual design stage of a subsystem of the DEMOnstration fusion power plant (DEMO): the divertor cassette body-to-vacuum vessel locking system. In this paper improvements in IPADeP are presented and its validity is discussed by presenting the application to the divertor system design.

Keywords: Axiomatic Design · Systems Engineering · Design methods · Tokamak design

C. Rizzi et al. (Eds.): ADM 2019, LNME, pp. 492–505, 2020.
https://doi.org/10.1007/978-3-030-31154-4_42

1 Introduction

The large and complex engineering system design is involving increasingly geographically dispersed and multi-disciplinary working groups, dealing with multiple and competing design objectives, so more and more attention is paid to global cooperation, especially during the conceptual design stage. Currently, most design groups use local and segmented approaches, which cannot provide a shared evaluation of competing design alternatives among the involved stakeholders and partners [1]. Moreover, due to the long lead time of the implementation process for the large system design, the implementation tasks are usually determined based upon incomplete design information [2]. Consequently, the information and changes coming in the project during the design process usually require several iterations to search for a proper result, having significant impact on the cost, quality and schedule of projects.

In this context, a design methodology would provide a systematic approach to design from the early conceptual design stage, aiming to avoid traditional design-build-test-redesign cycles and to allow for the evaluation of the design alternatives on a common basis with different stakeholders and across different design phases. The early conceptual design stage, dealing with an high level of abstraction, have recently attracted increasing attention from the academia [3, 4]. The conceptual design stage is responsible for more than 75% of the cost of a product and it is the top cause of troubled projects since the requirements sometimes are unclear, imprecise, with lack of agreement and priority [5].

The traditional practice of systems engineering management [6] involves the determination of requirements at or near the beginning of a system development project. All subsequent steps are dependent upon the completeness, accuracy and specificity of these requirements. Consistently with Systems Engineering principles, a systematic and efficient design methodology is needed to deal with the early conceptual design stage of large and complex system. In this methodology, the design should start from abstract ideas and proceed to detailed design with an incremental process. It should ensure the traceability and documentation of the design and should provide an efficient method to evaluate competing conceptual design alternatives.

Several design methods and theories are available in literature, some focusing on concept generation and selection, others helping the requirement management and quality development, others highlighting the steps to be performed during the design development.

The Axiomatic Design (AD) methodology [7] is recognized to provide designers with a tool to structure their thought processes in the early design stage and for optimization later in the design process. AD provides design parameters specification from the higher qualitative level to the lower quantitative level. Moreover, the design matrix and decomposition process facilitate the design documentation, the information traceability, the identification of changes impact and the achievement of multiple competing design objective.

According with these considerations in [8] we proposed an axiomatic design methodology, the Iterative and Participative Axiomatic Design Process (IPADeP), which provides a systematic approach to the conceptual design activities, based on the theory of AD. It aims to optimize the collaboration among the parties involved in complex interdisciplinary projects during the early stage of design and to minimize the risks related to the uncertainty and incompleteness of requirements. The conceptual stage is characterized by fuzzy and incomplete information, making the design process quite difficult and challenging. For this, IPADeP proposes an iterative process focused on the experience of the people involved and deals with the decision-making phase using a multicriteria decision making technique (MCDM).

This paper presents recent improvement implemented in the IPADeP process, mainly focusing on the requirements engineering aspects and on the management of the design changes. A case study addressing the early conceptual design stage of tokamak fusion reactor components is discussed to highlight the benefits of the IPADeP application.

2 An Innovative AD Process: IPADeP

2.1 Motivation

Systems engineering theory [6] defines six life cycle stages, with predefined levels of developemnt, in order to establish a framework for meeting the stakeholders' needs in an orderly and efficient manner.

The product development through the six stages is supported by the technical processes, which are usually represented by the V model, which highlights the need to define verification plans during requirements development, the need for continuous validation with the customers and the importance of continuous risk and opportunity assessment [9]. The assumption is that the elicited requirements provide all necessary information needed to move forward. However, this does not usually happen in real-world design of large-scale systems, in particular as regards the interfaced sub-system, the development of which proceeds in parallel and involves the continuous updating and refinement of the technical interface requirements.

The main motivation that leads to the development of IPADeP comes from the finding that, as discussed, in many projects regarding large and complex systems, there is a need to have a process that provides a robust structure and systematic thinking to support design activities in the early conceptual design stage. The necessity of reducing lead-time commonly imposes to start design process at a stage suffering from lack of information and incomplete set of requirements which is generally integrated during the project from the other actors involved in the design activities (i.e. interface requirements).

A suitable method to support the design activities in this environment must first have an incremental and iterative nature that provides for continuous updating and refinement of requirements and the continuous improvement of the conceptual solution. During all process activities the experience of designers is fundamental, from the "Customer need identification" (especially in the first iteration of the process) to the generation of the conceptual alternatives and the selection of best alternative. Continuous design documentation throughout the process and dynamic requirements traceability play a central role providing the possibility to evaluate how each new requirement completed during the design activities affects higher level decisions.

The AD and the APDL methodologies address the problem of requirements traceability and generating design solutions but, in some aspects, they miss a clear and systematic approach to design activity in the early conceptual design phase. Moreover, the new methodology has to provide a quantitative technique able to deal with the selection of the best conceptual solution considering the "fuzzy" nature of the information at this stage.

First applications of IPADeP, presented in [10] and [11] suggest some aspects to be better evaluated and discussed. A main point is that during the design process of a large-scale system, the first source of complexity lies in the identification customer and stakeholders and their distinction. For a technical complex system the customers define, through statements, the system functions and its expected behavior, leading to the main design drivers. In parallel, there are several stakeholders (technical partners, regulators, etc.) which provides constraints, interfaces and functional requirements.

Both customer needs and stakeholders needs are better being captured from the beginning or as soon as they become available during the design process, since (i) they represent the initial set of guidelines for the design of the system structure and the development of alternatives, and (ii) the selection of a balanced solution depends on how they are clear and complete. Furthermore, depending on the nature of the system being design, the relative contribution of these different sources of needs may vary depending on the level of complexity and/or technical readiness of the system as well as the applicable regulation, etc.

For instance, the initial design phase for systems that provide for a broader range of users, e.g. a cruise ship, is mainly driven by the customers' needs and expectations and it is devoted to the elicitation and focalization of such needs.

Alternatively, as for the chosen study on tokamak fusion reactor discussed in Sect. 3, design activities are mainly driven by the technical requirements coming from the stakeholder needs. By the way, in both cases, the transformation of the needs into a set of clear and technically usable requirements is needed to proceed with the design development.

Another main characteristic of complex systems is that a prospective system element may itself need to be considered as a system (that in turn is comprised of system elements) before a complete set of system elements can be defined with confidence (ISO/IEC15288, 2008), as depicted in Fig. 1. As the system is decomposed, the requirements are also decomposed into more specific requirements that are allocated to the system components.

This implies the design process to be hierarchically structured and it allows for the easy understanding of the cross impact between system elements, sub-systems and system of interest. In other words, there is the need for a tool to check how the requirements and constraints on each element hierarchically impact on the system structure.

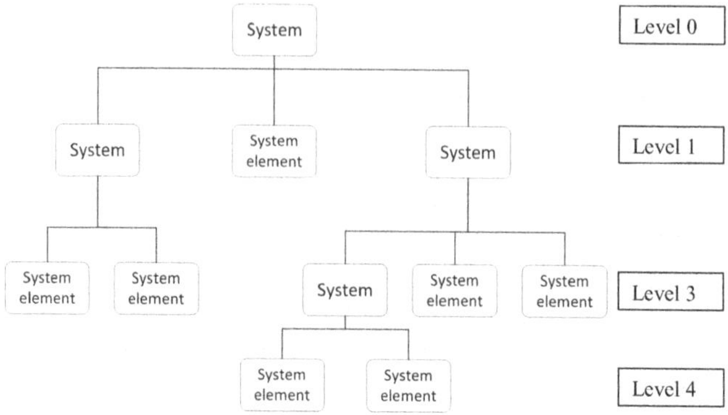

Fig. 1. System structure

2.2 Iterative and Participative Axiomatic Design Process

The IPADeP flowchart is presented in Fig. 2. Based on the APDL it was developed according to the design process roadmap proposed by Tate and Norlund [12] to propose a systematic thinking to support design activities in the early conceptual design stage. It is an iterative incremental design process, participative and requirements driven. The process highlights the iterative nature of the design activities; for each level of decomposition iteration is performed, and from the 2nd iteration also new information could come in the process from the stakeholders. IPADeP aims to drive the conceptual design activities avoiding traditional design-build-test-redesign cycle. It integrates brainstorming sessions, MCDM techniques and the AD method, taking advantages of its systematic and logic approach for design derivation, documentation and optimization. Furthermore, it proposes the use of CAD and simulation software from the early stage to improve idea generation and communication among stakeholders and takes advantages of documentation templates and of the Master design matrix to document the design and evaluate the impact of requirement changes during the project.

IPADeP highlights the iterative nature of the design activities and the central role of the "human factor", with the involvement of experts' panel during the requirements elicitation and concept evaluation. The smooth evolution from uncertain information during the early stages towards more detailed solutions emerging across subsequent design iterations is dealt with using Fuzzy- AHP during decision making steps. Compared to the previous version [8] the process presented in this paper has been improved in particular as regards the requirements definition and change management. Furthermore in this work it is highlighted the hierachical structure of the design process as a main point to avoid re-design cycles and minimize the impact of requirements changes during the design activties. As discussed above, during the design process of a large-scale complex system the informations are continuously updated and improved and new requirements needs to be managed and integrated (e.g. interface requirements). Differing from the requirements coming from the selection of a higher-level concept solution, these requirements are not well handled by AD, and require the looping provided by IPADeP. The process is exemplified in following sub-sections.

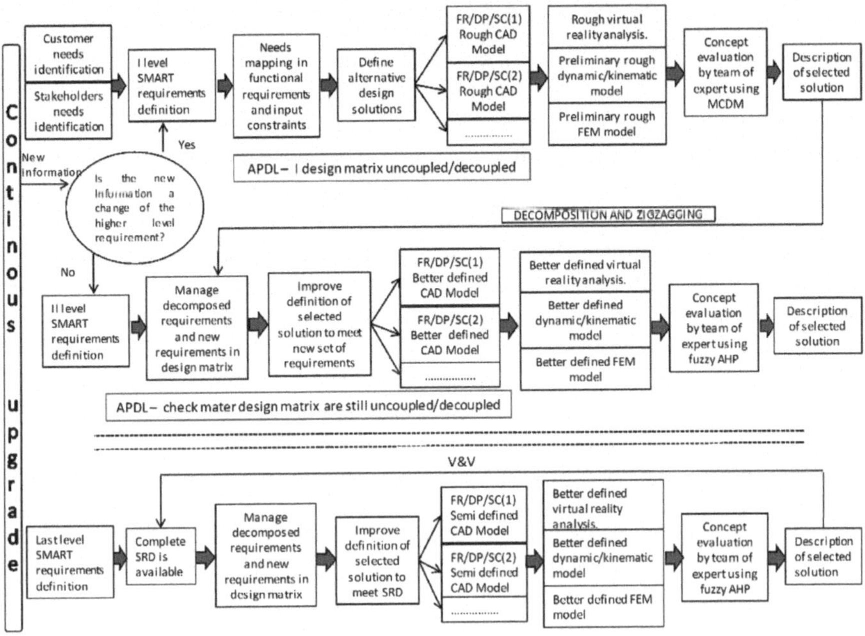

Fig. 2. New IPADeP flowchart. (FR: Functional requirements, DP: Design Parameters, SC: System Components, MCDM: Multi-criteria Decision Making, SRD: System requirements document, AHP: Analytic Hierarchy Process)

2.2.1 First Iteration

The process starts with first iteration corresponding to the first level of decomposition. Each iteration of the process can be divided in three macro area:

<u>(1) Requirements identification and analysis</u>

At this level the systems functions are known but there is not yet a set of defined requirements. To start the process, a brainstorming session between sector experts, customers and stakeholders is performed in order to collect few generic high-level needs. At this point, to start with design activities the transformation of the needs into a set of technically usable functional requirements is needed. These requirements should be also SMART requirements, where SMART is a mnemonic acronym giving criteria to write good requirements. Letters S and M usually indicate specific and measurable respectively, while the letters A, R and T indicate respectively achievable, relevant and traceable.

The Customer and stakeholders' needs are then mapped in the initial functional requirements (FRs) and input constraints (ICs). This mapping process is done according to the APDL method and using Requirement Matrix and Constraint Matrix to document and trace the process.

The mapping between the CNs and the initial FRs and ICs is captured by the equations:

$$\{CN\} = [R]\{FR\} \tag{1}$$

$$\{CN\} = [C]\{IC\} \tag{2}$$

(2) Design solutions development

Once CNs are mapped to FRs and ICs, the top-level design parameter (DP) and the top-level physical system components (SC), are proposed in order to start the decomposition and zigzagging process. Generally speaking, from the first brainstorming session enough information for a first level of decomposition is available. Several different DPs could satisfy a single FR and several SCs could be used to apply a DP. So, several design solutions should be developed and modelled in a CAD system to show and clarify DPs and SCs.

For each solution, templates for design activities documentation are used and a design matrix to map FRs onto DPs is developed. For each solution the design matrix has to be diagonal (uncoupled design) or triangular (decoupled design) to satisfy the Independence Axiom (Eqs. (3) and (4)). Also, system Structure matrix to DP-SC mapping is developed.

$$\begin{Bmatrix} FR1 \\ FR2 \\ FR3 \end{Bmatrix} = \begin{bmatrix} X & 0 & 0 \\ 0 & X & 0 \\ 0 & 0 & X \end{bmatrix} \begin{Bmatrix} DP1 \\ DP2 \\ DP3 \end{Bmatrix} \tag{3}$$

$$\begin{Bmatrix} FR1 \\ FR2 \\ FR3 \end{Bmatrix} = \begin{bmatrix} X & 0 & 0 \\ 0 & X & 0 \\ X & 0 & X \end{bmatrix} \begin{Bmatrix} DP1 \\ DP2 \\ DP3 \end{Bmatrix} \tag{4}$$

(3) Concepts evaluation and first level solution selection

The comparison of concepts, their evaluation and the choice of the best solution is performed using a multiple-criteria decision analysis (MCDA). Concept selection is a complex task for engineering designers as it can be considered as the most critical decision-making step in the product development process [13]. During this phase, erroneous solutions need to be minimized, which means that several facets of the problem have to be considered concurrently. Analytic Hierarchy Process (AHP) has been proposed in literature as a methodology to large, dynamic and complex real-world MCDA problems [14]. Since decision maker's requirements may contain ambiguity and the human judgment on quality attributes may be imprecise [15], the crisp aspect of the conventional AHP seems inappropriate in depicting the uncertain nature of this decision phase. To consider uncertainties during the early stages of design and deal with the variables in verbal judgments, in this research AHP is used with a fuzzy approach, using triangular fuzzy numbers [16].

Fuzzy AHP allows dealing with the multicriteria decision making stage considering uncertainties related to the early stages of design and to the judgements of the decision makers. The application of the fuzzy-AHP in the concept evaluations stage of the IPADeP process is detailed discussed in [8] and [11].

2.2.2 Subsequent Iterations

Proceeding with the iterations, when enough information are available to decompose the solution to the subsequent level, according to zigzagging and decomposition, the solution selected in the previous iteration is improved to meet the new requirements and constraints. One of the main improvements of IPADeP with respect to classical AD application is that a new iteration could start also if new information is made available from other stakeholders, and the needs are accordingly updated. New information could invalidate a precedent assumption, therefore requiring the process to restart, or can introduce a new FR or IC. In the latter case, one or more DP must be developed to meet the new FRs; so the master design matrix (Table 1) is exploited to check whether the design still respects the independence axiom, or the early design decision is violated.

Table 1. Master design matrix

	DP1.1	DP1.2	DP2.1	DP2.2	DP2.3	DP3
FR1.1	X	O	O	O	O	O
FR1.2	X	X	O	O	O	O
FR2.1	O	X	X	O	O	O
FR2.2	O	X	O	X	O	O
New FRs → FR3.1	O	X	O	O	O	X

If lower levels DPs violate the higher-level design, the design issue can be addressed by modifying the lower level DPs, revising the higher-level design matrix or imposing constraints to prevent DPs unwanted effects.

During the decomposition and iterations, the SMART requirements are collected in a System Requirements Document (SRD). The iterations concerning the conceptual phase stop when this document is completed, all functional requirements and input constraints are well defined and no further decomposition is needed. At this point all requirements are verifiable, attainable and approved by stakeholders, so Verification and Validation activities can be performed to arrive at the first lifecycle decision gate: Conceptual Design Review.

3 Design Progress of DEMO Divertor Cassette to Vacuum Vessel Locking System

IPADeP was used to deal with the conceptual design of the DEMO fusion reactor mechanical components. In particular, the driving case study was focused on the design of an internal reactor component: the divertor cassette body to vacuum vessel locking system. The main aim of the divertor locking system within a fusion reactor is to keep locked the divertor in its relative position to the vacuum chamber (named Vacuum Vessel, green in Fig. 4), withstanding the electromagnetic, seismic, thermal loads avoiding vibrations.

First design iterations were performed and presented by authors in [10] and [11].

According to IPADeP, the design process started from the main high-level customer needs and stakeholder needs, collected and document in the traceability tables as shown in Table 2. The needs have been than translated in SMART requirements for the first design iteration. Keeping good documentation and traceability, this kind of approach supports the requirements elicitation and the change management, avoiding redesign cycle. The complete decomposition levels 1 and 2, including system components, CAD models, FEM analysis and design matrices are reported in [10] and [11]. Following the IPADeP flowchart (Fig. 2), Sect. 3.1 reports the third level design iteration developed because of significant changes occurred in the high-level requirements.

3.1 Third Iteration

The third iteration of the locking system conceptual design started from three main updates in the available information, regarding the configuration model and a required function of the locking system:

- Divertor locking system shall be compatible with the divertor configuration model 2017. Differences with divertor 2014 are shown in Fig. 3.
- The locking system shall ensure the electrical connection to the vessel and shall be able to carry the maximum current during plasma disruption
- Avoid sliding surfaces in vacuum environment

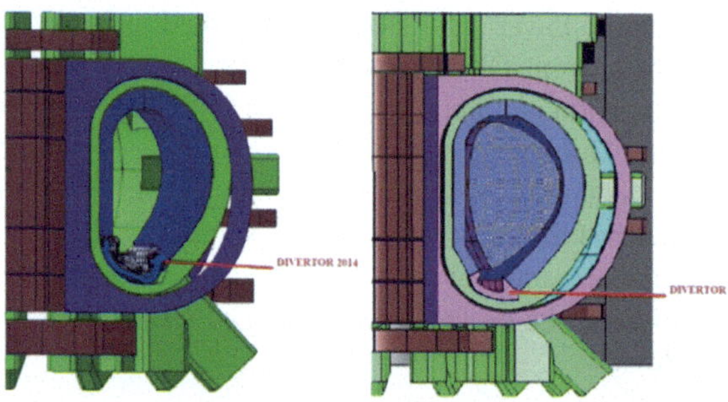

Fig. 3. New divertor configuration model

The first information represents a higher-level change. From the mapping tables it was easy to check that this change affects a higher-level input constraint and all the design parameters developed during the first iteration can be adopted also with the new constraint. This implies that, since according to IPADeP the project started from a high level of abstraction, all the second level DPs can be adapted to respect the new geometrical boundaries. The second information adds new SNs (SN3 and SN4) as shown in Table 2. The updates FRs and ICs are reported in Table 3.

Table 2. Third iteration: Customer and stakeholder needs

CN ID	Statement
CN1	Lock divertor in place after placement operations, avoid displacement in any load conditions
CN2	Maximize reactor availability using systems with short maintenance time and avoid unplanned stop
SN ID	Statement
SN1	Avoid "shaking" due to sudden change of magnetic field
SN2	*Accommodate distortions*
SN3	*Provide electrical connection between divertor cassette and Vacuum Vessel during operations*
SN4	*Avoid sliding surfaces*

Table 3. Third iteration: FRs and ICs

FRi ID	FRi description	CN/SN					
		CN1	CN2	SN1	SN2	SN3	SN4
FRil	Remove clearances to avoid vibrations – *clearances of maximum 5 mm*	0	X	0	0	X	0
FRi2	Provide an outer locking system able to take force in any direction – *ITER-like loads to be considered*	X	0	0	0	X	0
FRi3	*Provide a system to accommodate thermal distortion for a total displacement of 10 mm*	*0*	*0*	*0*	*X*	*0*	*X*
Fri4	*Provide a system to ensure electrical connection during sudden change of magnetic field*	*0*	*X*	*0*	*X*	*X*	*0*
ICi ID	IC description						
ICi1.1	Locking System shall be compatible with remote installation and disassembly during divertor maintenance – *take as reference ITER RH tools*	X	X	X	0	0	X
ICi1.2	*As simple as possible* mechanism to lock and preload in order to reduce operational time	X	X	X	0	0	X
ICi1.3	Locking System shall be the same for all standard cassette (left and right)	X	X	X	X	0	0
ICi1.4	Structural robust looking system – *withstand ITER-like extraordinary events*	X	X	X	X	0	0
ICi1.5	*Geometry and interface consistent with Divertor CAD model 2015*	*X*	*X*	*X*	*X*	*0*	*X*
ICi1.6	*Deadweight 4 ton*	*X*	*X*	*0*	*0*	*0*	*0*
ICi1.7	*Avoid sliding surfaces*	*0*	*0*	*0*	*X*	*0*	*X*

The DP meeting these functional requirements added in the third iteration are reported in Table 4. They has to be considered as added to the ones reported in the first and second iteration tables, available in [10] and [11].

Table 4. Third iteration: FRs and DPs

Level	ID	FR	DP
0	1.4	Ensure electrical connection between cassette and vacuum vessel	Avoid relative displacement between cassette and Vacuum Vessel under ITER-like load conditions
I	1.4.1	Avoid relative displacement between cassette and vessel under ITER-like load conditions	(a) Preload cassette to ensure the connection (b) Provide electrical strap between cassette and vacuum vessel (c) Provide elastic elements in the outboard area to ensure connection in any condition
II	1.4.1.1	(a) Insert tool to preload cassette (b) Provide electrical strap (c) Provide elastic elements	(I) Transports the divertor on a tilted rail slightly raised from the rest position. Releasing the divertor it moves forward due to the inclination of the rail, preloading the cassette. The surface of the divertor should have a spherical shape to ease the preload. Insert a removable hydraulic jack to help the preload. (II) Bolted electrical strap (III) Disc spring in the outboard area to preload a Stainless Steel component against the Vacuum Vessel

The new DPs have been added to the master design matrix for the option selected during second iteration. As example, one of the master design matrices updated with the new DPs is reported (5). It shows that the design solution is still uncoupled.

$$\begin{Bmatrix} FR1.1.1 \\ FR1.2.1 \\ FR1.2.2 \\ FR1.3.1 \\ FR1.4.1 \end{Bmatrix} = \begin{bmatrix} X & 0 & 0 & 0 \\ 0 & X & 0 & 0 \\ X & 0 & X & 0 \\ 0 & 0 & 0 & X \\ X & 0 & X & X \end{bmatrix} \begin{Bmatrix} DP1.1.1(a) \\ DP1.2.1(a)(b) \\ DP1.2.2 \\ DP1.3.1(a)(b) \\ DP1.4.1(a)(b) \end{Bmatrix} \tag{5}$$

Basing on the new DPs and ICs, the design of higher-level solutions has been improved and other solutions have been proposed, given the new geometric constraints, the lower weight and the new FR4 and IC7. The solutions are showed in Fig. 4. Also, in this case the application of IPADeP allowed for avoiding re-design cycle thanks to the hierarchical development from higher level solution towards more detailed solutions and the use of traceability matrix to easily check the FRs affected from each DPs modified.

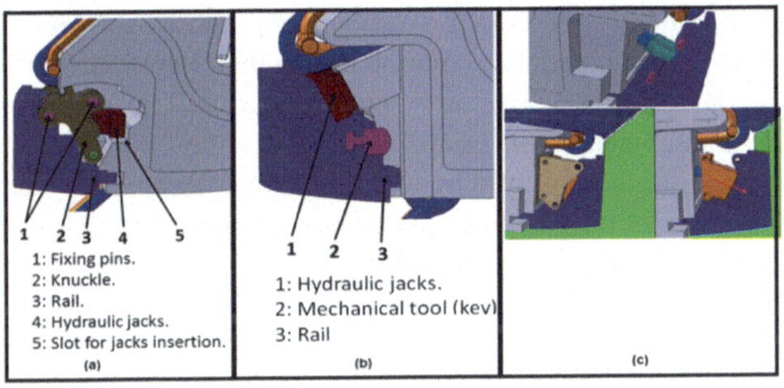

Fig. 4. Third iteration design solutions

According to IPADeP, as for second iteration, solutions were compared by means of Fuzzy-AHP, leading at the preferred solution, identified in the “Knuckle system” (Fig. 4a) emerging as an improvement of the second level solution considering the new geometric constraints.

4 Conclusions

The work presents IPADeP improvements in requirements engineering aspects and provides systematic procedure for the definition of SMART requirements. IPADeP seems to be well suited for drafting conceptual solution of large and complex systems. Basing on the AD theory, it provides a systematic approach to address the early stage of the design, dealing with the uncertainty of the available information. Moreover, proceeding iteratively layer by layer it allows an easy integration of the new requirements and subsequent design parameters, avoiding redesign cycles. The main characteristics of the new IPADeP version here presented can be summarized in: (i) IPADeP supports the management of new information coming late in the design process due to parallel development of high technical complex sub-systems; (ii) using APDL templates and design matrix it aims to provide good traceability of the design activities, improve design documentation and communication; (iii) the definition of SMART requirements allows for improved requirements statement. The writing of “good” requirements from the beginning is fundamental to correctly evaluate the alternative solutions and avoid re-design cycles; (iv) the design process is hierarchically structured and this allow for the integration of sub-systems and system elements.

IPADeP has been adopted for the conceptual design activities of DEMO divertor locking system. The design started from few high-level requirements, which led to some "high level" conceptual solutions.

In this work the design progress of locking system is re-discussed according to the improvements in IPADeP and the second and third design iterations are presented. The improvements in the method allowed for better definition of requirements and design solutions and for a more robust decision-making stage. New solutions for divertor cassette body-to-vacuum vessel locking system were proposed and pairwise compared, leading to the selection of a reference solution for the third iteration. Starting from these results a new iteration for the divertor locking system will be performed when new information and requirements will be available. Moreover, further studies about the IPADeP application in design of large-scale system, technically challenging and involving a wide range of physics (e.g. new aerospace systems), shall be performed, in order to enhance its characteristics and validity.

References

1. Thielman, J., Ge, P.: Applying axiomatic design theory to the evaluation and optimization of large-scale engineering systems. J. Eng. Des. **17**(1), 1–16 (2006)
2. Xue, D., Cheing, S., Gu, P.: Configuration design considering the impact of design changes on downstream processes based upon the axiomatic design approach. J. Eng. Des. **17**(6), 487–508 (2006)
3. Chen, W., Luo, X., Su, H., Wardle, F.: An integrated system for ultra-precision machine tool design in conceptual and fundamental design stage. Int. J. Adv. Manuf. Technol. **84**(5–8), 1177–1183 (2016)
4. Mozzillo, R., Marzullo, D., Tarallo, A., Bachmann, C., Di Gironimo, G.: Development of a master model concept for demo vacuum vessel. Fusion Eng. Des. **112**, 497–504 (2016)
5. Pm Solutions. Strategies for project recovery (2011)
6. Haskins, C., Forsberg, K., Krueger, M., Walden, D., Hamelin, D.: Systems Engineering Handbook: INCOSE (2006)
7. Suh, N.P.: The Principles of Design. Oxford University Press, New York (1990)
8. Di Gironimo, G., Lanzotti, A., Marzullo, D., Esposito, G., Carfora, D., Siuko, M.: Iterative and participative axiomatic design process in complex mechanical assemblies: case study on fusion engineering. Int. J. Interact. Des. Manuf. (IJIDeM) **9**, 1–14 (2015)
9. Vitolo, F., Patalano, S., Lanzotti, A., Timpone, F., De Martino, M.: Window shape effect in a single bowden power window system. In 2017 IEEE International Systems Engineering Symposium (ISSE), pp. 1–5. IEEE, October 2017
10. Di Gironimo, G., Carfora, D., Esposito, G., Lanzotti, A., Marzullo, D., Siuko, M.: Concept design of the demo divertor cassette-to-vacuum vessel locking system adopting a systems engineering approach. Fusion Eng. Des. **94**, 72–81 (2015)
11. Marzullo, D., et al.: Design progress of the DEMO divertor locking system according to IPADeP methodology. Procedia CIRP **34**, 56–63 (2015)
12. Tate, D., Nordlund, M.: A design process roadmap as a general tool for structuring and supporting design activitiesed. In: Proceedings of the Second World Conference on Integrated Design and Process Technology (IDPT), vol. 3, pp. 1–4. Society for Design and Process Science, Austin (1996)

13. Sebastian, P., Ledoux, Y.: Decision support systems in preliminary design. Int. J. Interact. Des. Manuf. **3**(4), 223–226 (2009)
14. Murat Albayrakoglu, M.: Justification of new manufacturing technology: A strategic approach using the analytical hierarchy process. Prod. Inventory Manage. J. **37**, 71–76 (1996)
15. Renzi, C., Leali, F., Pellicciari, M., Andrisano, A., Berselli, G.: Selecting alternatives in the conceptual design phase: An application of fuzzy-AHP and pugh's controlled convergence. Int. J. Interact. Des. Manuf. (IJIDeM) **9**, 1 (2013)
16. Chang, D.-Y.: Applications of the extent analysis method on fuzzy AHP. Eur. J. Oper. Res. **95**(3), 649–655 (1996)

Industrial Noise Modelling and Control: The Case of Natural Gas Distribution Systems

Chiara Bartalucci(✉), Francesco Borchi, and Monica Carfagni

Department of Industrial Engineering, University of Florence, via di S. Marta 3, 50139 Florence, Italy
chiara.bartalucci@unifi.it

Abstract. It is widely known that exposure to noise may have negative impacts on human health and well-being. Among the several sources of noise pollution, pressure monitoring stations associated with natural gas distribution networks can have a significant noise impact on the surrounding areas. A collaboration between Toscana Energia S.p.A. and the Department of Industrial Engineering of the University of Florence has allowed to characterize the emission and sound propagation of these stations, through both theoretical and experimental analysis.

In the current manuscript, results obtained from the validation of the algorithm for the estimation of the sound pressure level due to the gas distribution systems proposed by Tartarini manufacturer according to noise measurements campaigns are illustrated. Moreover, the design of a semi-automatic tool for the noise evaluation impact of natural gas distribution systems at the receivers is presented.

Keywords: Environmental noise · Noise control · Gas distribution system

1 Introduction

Environmental noise is currently considered as one of the most important public issues, since an exposition to the latter can lead to health implications such as annoyance, stress, hypertension, coronary heart disease and stroke [1–4]. Many projects and researches have been carried out until now also by the authors of the current manuscript [5–9] in order to contribute to provide methodologies and strategies to solve these issues.

In natural gas distribution networks, due to the turbulence of the fluid that is generated at the pressure reductions in the control stations/plants in correspondence to the distribution networks, there may be scenarios of variable noise emission over time [10], depending on the load conditions. In order to ensure compliance with current regulations, the gas distribution companies carry out acoustic measurements on the individual control system, in response to complaints from citizens or on the advice of maintenance personnel [11, 12]. However, the single noise measurement characterizes the noise emissions of the measurement period, but, given the variability of the latter over time, it is generally not able to correctly represent the maximum acoustic impact of the system. At the same time, the maximum noise emission data defined by the manufacturers are often overestimated to represent the actual noise conditions, since the

C. Rizzi et al. (Eds.): ADM 2019, LNME, pp. 506–515, 2020.
https://doi.org/10.1007/978-3-030-31154-4_43

systems generally work in intermediate flow conditions. For these reasons, given the territorial diffusion of sound sources in the Tuscany region (38 first-stage stations/plants - REMI - and about 2000 second-stage stations/plants - Final Reduction Groups, GRF), the company Toscana Energia S.p.A. (hereinafter TE) for years is developing systemic procedures for the management of the acoustic impact generated by pressure control stations/plants.

In 2011 a collaboration between TE and the Department of Industrial Engineering of the University of Florence (hereinafter DIEF) started and it is still ongoing, aimed at verifying the reliability of algorithms proposed by manufacturers to estimate the sound pressure level due to the gas distribution systems. A second prior objective of the collaboration is to analyze and integrate the company's territorial information system for its use as a support tool for the assessment of the acoustic impact of the stations/plants of the managed distribution networks.

To the best of authors knowledge, no similar studies have been carried out yet in order to design and test a procedure to evaluate the acoustic impact of pressure control plants. In addition, beyond its practical application, this procedure can provide an important scientific contribution in the sector, making it possible to assess the noise impact of pressure reduction plants but also being able to be extended, with the necessary adjustments, to other similar types of plants.

2 The Tartarini Algorithm

With reference to the noise emission of pressure control stations/plants, as indicated in the product standard UNI EN 334 [13], the manufacturers of pressure regulators are required to give information about the noise emissions generated with an accuracy of ±5 dB, with reference to the conditions of maximum noise related to the configurations of maximum flow rate. However, as already mentioned in the Introduction, the service conditions at maximum flow rate are often unrepresentative, even at low temperatures. At the same time, therefore, the need to analyze emissions under intermediate flow conditions arose. The latter are usually unknown or foreseeable and are not currently required from suppliers, although this possibility is envisaged by industry standards. At the moment only some manufacturers, in particular Tartarini manufacturer, provide the useful equation to be applied for the calculation of the noise emission of the plant in intermediate conditions of flow rate, with declared uncertainty of ±3 dB. According to the Tartarini manual [14], the algorithm is reported in (1):

$$\mathrm{SPL_A} = \mathrm{SPL_{\Delta P}} + \mathrm{SPL_{CGf}} + \mathrm{SPL_{\frac{\Delta P}{P1}}} + \mathrm{SPL_K} - \mathrm{SPL_S} - \mathrm{SPL_T} \quad (1)$$

where

SPL_A is the overall noise pressure level, in dB(A), due to the distribution system and calculated at 1 m-distance from the latter;

$SPL_{\Delta P}$ is the noise pressure level, in dB(A), due to the pressure loss between the inlet and the outlet;

$SPL_{\frac{\Delta P}{P1}}$ is a correction, in dB, depending on the pressure regulator type and on the ratio between the pressure variation and the inlet's one;
SPL_K is a correction, in dB, related to the piping thickness;
SPL_S, SPL_T are corrections, in dB, due to the presence of different types of attenuators;
SPL_{CGf} is a correction related to the outflow coefficient Cg_f

In this frame, it is necessary to go deeper in the description of the last term of the equation SPL_{CGf}, while it is possible to refer to the Tartarini manual [14] for the description of the evaluation modalities of all the other terms. The SPL_{CGf} term can be evaluated by using the equation reported in Fig. 1 according to the coefficient Cg_f depending on the flow rate and the inlet pressure. The equation reported in the Fig. 1 is determined by the authors starting from the abacus provided into the Tartarini manual [14]. However, due to the difficulties in measuring the flow rate, it is usually foreseen by using a forecasting software or calculated in an indirect way starting from other parameters such as the height of the pin pressure regulator. In particular, Cg_f can be also evaluated as in (2):

$$\mathrm{Cg_f} = \frac{\mathrm{Cg_{hpin}}}{\mathrm{hmax_{pin}}} \tag{2}$$

Where
Cg_{hpin} is an implicit asset of the distribution system;
$hmax_{pin}$ can be measured on each distribution system.

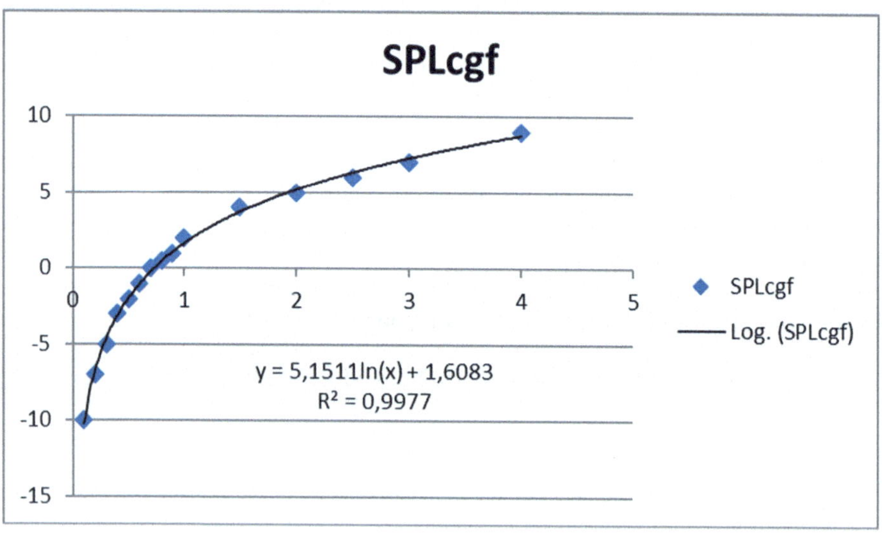

Fig. 1. Evaluation of SPL_{CGf} from $\mathrm{Cg_f}$ term.

The algorithm (1) allows the evaluation of the noise emissions of the plants in a free field condition, in an external environment and in the absence of a containment box.

3 Attenuations Due to the Containment Box

The Tartarini algorithm described in Paragraph 1 does not take into account the possible positive effect due to the presence of pressure reduction plants' containment boxes.

In fact, there are various acoustic attenuations due to the containment box of the reduction plants (in general referred to by the term *Att*):

- attenuation due to insulation of the containment box;
- attenuation due to the introduction of sound-absorbing material;
- attenuation due to the introduction of soundproofing grids.

As part of the study, the attenuations due to both metal cabinets and masonry buildings were determined by DIEF both for plants located in metal cabinets (Fig. 2) and in masonry buildings (Fig. 3).

Fig. 2. Example of gas natural distribution system in a metal cabinet.

Fig. 3. Example of gas natural distribution system in a masonry building.

With specific reference to the metal cabinets, according to the results of the present study, the *Att* term has been determined as it follows:

- attenuation due to insulation of the containment box equal to 9 dB in case of absence of inner lining and in which an airtight seal on the doors is present and equal to 5 dB if the latter is not present;
- attenuation due to the presence of sound-absorbing material (thickness = 4.5 cm) equal to 10 dBA;
- attenuation due to the presence of soundproofing grids equal to 2 dBA.

On the other hand, with reference to masonry buildings, where in general there is no internal sound-absorbing coating or soundproof grid, the relationship to be applied is based on the assumption that the sound energy produced by the system flows entirely from the openings of the same.

As a consequence, the broadband sound power level expressed in dB(A), associated with the windowed surface of each façade, can be assessed as in (3):

$$Lw_{wi} = L_w - 10 \cdot \log_{10}(\frac{S_{wi}}{S_w}) \tag{3}$$

while the overall sound power level Lw is determined as described later in relation (5).

4 Algorithm's Validation Procedure

In order to verify the applicability and the coherence of the Tartarini algorithm to the specific natural gas distribution systems at partial loads, several noise measurement campaigns have been carried out by the authors.

The first one was conducted in 2015 on five natural gas distribution systems, running at almost full load conditions, located in metal cabinets. An overestimation of the Tartarini algorithm up to 2 dBA turned out for four of the considered systems and up to 5 dB(A) for all the considered systems [15].

Then, in 2017 it has decided to carry out further noise measurements campaigns with systems running at partial load conditions, to corroborate the previously obtained results.

In fact, in order to validate the results of the calculation reported in Paragraph 1, a verification procedure based on field measurements has been elaborated. In particular, with regard to systems located inside metal cabinets, noise measurements have been carried out both in the open and closed cabinet condition. In particular, the measurements performed in the open cabinet conditions, possibly corrected considering the noise reflections due to the cabinet itself, are directly comparable to the Tartarini's algorithm. For the masonry systems, the measurements were made at each opening (window).

All the carried-out measurements have had a general duration of 60 s, however sufficient to characterize the acoustic phenomenon. The noise levels reported for comparison both with the simulated levels in terms of LAeq, are relative to the measurement period obtained excluding recognizable anomalous noisy events due to other sources (e.g. car passages). In addition, measurements of the height of the membrane pin have also been carried out on each plant to use that data as input for the determination of the coefficient Cg_f.

In Fig. 4 an example of a time history of measured noise levels is shown.

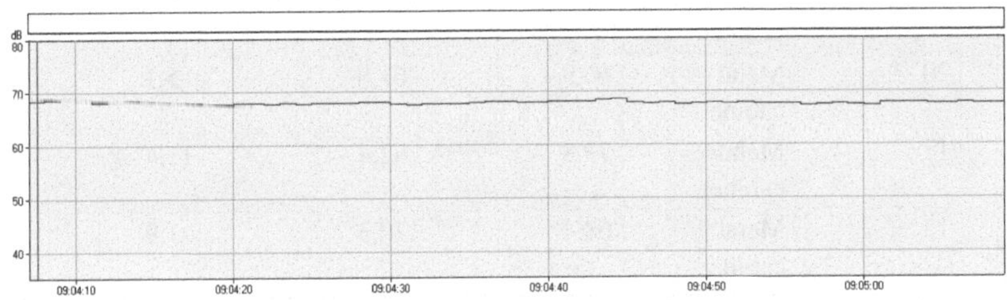

Fig. 4. Time history of measured noise levels.

5 Comparison Between Measured and Simulated Results

In Table 1 the comparison between noise levels evaluated with the Tartarini algorithm ($Lp_{Tartarini}$), considering also the additional attenuation due to the presence of the box, and the measured ones ($Lp_{measured}$) is shown referring to plants located both inside metal cabinets and in masonry buildings.

All noise levels are evaluated at 1 m from the plants, in the open cabinet conditions (in the case of masonry buildings the measurements are performed at windows), in a condition of flow rate correspondent approximately to the 20% of the maximum one. In addition, on two plants located inside metal cabinets it was also possible to evaluate the configuration characterized by a twice gas flow rate with respect to the current one. In fact, the pressure demand at the two plants was increased for a few minutes and new acoustic measurements were carried out (6* and 9* in Table 1), keeping the cabinet open.

The obtained results show that, with regard to situations in which the required flow rate to the system is less than half the design flow rate (of the order of 20% of the nominal flow rate), the Tartarini algorithm provides results that generally overestimate by at least 8 dBA the sound pressure level. Instead, it is possible to deduce an excellent adaptability of the Tartarini algorithm to situations in which the required flow rate of the systems is equal to or higher than 40–50% of the nominal flow rate.

Table 1. Comparison between noise levels evaluated with the Tartarini algorithm and the measured ones.

Plant	% Max flow rate	Typology	Lp_Tartarini [dB]	Lp_measured [dB]	Lp_Tartarini–Lp_measured [dB]
1	13	Metal cabinet	72,8	64,9	7,9
2	13	Metal cabinet	73,0	61,2	11,8
3	20	Metal cabinet	66,9	61,8	5,1
4	17	Metal cabinet	73,8	62,4	11,4
5	13	Metal cabinet	68,4	70,3	−1,9
6	23	Metal cabinet	75,4	60,7	14,7
7	10	Metal cabinet	70,5	59,2	11,3
8	16	Metal cabinet	74,2	59,4	14,8
9	16	Metal cabinet	71,6	60,7	10,9
6*	40	Metal cabinet	78,9	80,6	−1,7
9*	40	Metal cabinet	75,2	77,0	−1,8
10	20	Masonry building	78,0	65,0	13

6 A New Simplified Acoustic Impact Assessment Procedure

A simplified acoustic impact assessment procedure has been developed which, starting from the emission level of the system and the actual attenuation of the metal cabinet, aims to determine the sound pressure level produced at the façade of the receiving buildings closest to the system itself (L_p).

The procedure involves the use of different relationships depending on the distance between the source and receiver. In particular, for source-receiver distances less than 2 times the maximum size of the cabinet, it is considered to use the Eq. 4 which provides the level expected at 1 m from the closed cabinet:

$$L_p = L_{Tartarini} - Att + f \qquad [dB(A)] \tag{4}$$

where:

$L_{Tartarini}$ is the sound pressure level of the system evaluated at one meter from the same according to the relation defined in [6] and relative to the effective conditions of flow rate;
Att is the attenuation guaranteed by the cabinet;
f is the contribution of the façade reflection supposed equal to 3 dB.

For source-receptor distances greater than or equal to 2 times the maximum size of the cabinet, or for distances at which the cabinet can be considered equivalent to a point source [16], the sound power level is dcfined in Eq. 5 and the sound pressure level at the receiver in Eq. 6 according to [17]:

$$L_W = L_{Tartarini} + 10 \cdot \log_{10} S \quad [dB(A)] \tag{5}$$

$$L_{p\_receiver} = L_W - 20 \cdot \log_{10} d - 8 + f - Att \quad [dB(A)] \tag{6}$$

where:

S is the measurement area defined at 1 m from the plant (equal to 20 m^2 according to the measurement scheme defined in [13]);
d is the distance between the source and the receiver.

A similar procedure is developing for the systems located in technical masonry buildings.

6.1 Design and Implementation of an Automated Tool for Simulation

Regarding systems located inside metal cabinets, also a schematic procedure has been defined in order to estimate noise levels at the receivers due to the natural gas distribution systems and a semi-automatic tool has been implemented in ArcGis environment. Moreover, the verification of the noise limits according to the Italian legislation [18] has been implemented.

The algorithm is structured according to the following steps:

STEP 1 – Input data: cabinet's dimensions, presence of coatings and/or grates, attenuations of the latter, noise limits, etc.;
STEP 2 – Calculations: noise pressure level at 1-m distance from the plant according to the Tartarini algorithm, sound power level of the plant considering the attenuation due to the cabinet, noise pressure level at the receiver;
STEP 3 – Verification of noise limits according to the National legislation.

The developed automated tool can support the design of new gas distribution networks in the near future.

7 Conclusions

With reference to the assessment of the acoustic impact of the stations/plants associated with the natural gas distribution networks, the collaboration between Toscana Energia (TE) and the Department of Industrial Engineering of the University of Florence (DIEF) made it possible to verify the reliability of the Tartarini algorithm for calculating the noise emissions of the stations/plants under intermediate flow conditions. In fact, from the comparison between noise pressure levels estimated with the Tartarini algorithm and the measured ones it turned out that with regard to situations in which the required flow rate to the system is less than half the design flow rate (of the order of 20% of the nominal flow rate), the Tartarini algorithm provides results that generally overestimate by at least 8 dBA the sound pressure level produced at a distance of one meter from the system for both typologies of plants.

Instead, it is possible to deduce an excellent adaptability of the Tartarini algorithm to situations in which the required flow rate of the systems is equal to or higher than 40–50% of the nominal flow rate.

In addition, simplified and semi-automatic procedures for assessing the noise impact of stations/pressure located inside metal cabinets control systems have been defined.

The study for the definition of a similar procedure and of a semi-automatic tool for plants located inside masonry buildings is on-going and it will be completed in the next months.

Acknowledgment. The authors would like to thank the company "Toscana Energia S.P.A." for the support provided in preparing this work.

References

1. World Health Organization. Environmental Noise Guidelines for the European Region (2018)
2. European Environment Agency: Noise in Europe 2014, ISSN 1977-8449 (2014)
3. Kim, M., Chang, S.I., et al.: Road traffic noise: annoyance, sleep disturbance, and public health implications. Am. J. Prev. Med. **43**(4), 353–360 (2012)
4. Houthuijs, D.J.M., Van Beek, J., Swart, W.J.R., Van Kempen, E.E.M.M.: Health implication of road, railway and aircraft noise in the European Union, RIVM Report 2014-0130 (2004)
5. Bartalucci, C., Borchi, F., Carfagni, M., Governi, L., Zonfrillo, G., Bellomini, R., Wolfert, H., Aspuru, I., Gaudibert, P.: Life+2010 QUADMAP project (QUiet areas definition and management in action plans): results of post operam data analysis and the optimized methodology. In: ICSV22, Florence, Italy (2015)
6. Aspuru, I., Garcia, I., Bartalucci, C., et al.: LIFE+2010 QUADMAP project: a new methodology to select, analyze and manage quiet urban areas defined by the European Directive 2002/49/EC. Noise Mapp. J. **3**(1), 120–129 (2016)
7. Lapini, A., Bartalucci, C., Borchi, F., Argenti, F., Carfagni, M.: Audio virtualization of façade acoustic insulation by convex optimization. In: European Signal Processing Conference, 29 November 2018, vol. 2018-September, pp. 2185–2189 (2018). Article number 8553035

8. Bartalucci, C., Borchi, F., Carfagni, M., Furferi, R., Governi, L., Silvaggio, R., Curcuruto, S., Nencini, L.: Design of a prototype of a smart noise monitoring system. In: 24th International Congress on Sound and Vibration. In: ICSV24, London, UK (2017)
9. Bartalucci, C., Borchi, F., Carfagni, M., Governi, L., Bellomini, R., Luzzi, S., Asdrubali, F., D'Alessandro, F., Schiavoni, S.: Contributions to end interpretation and implementation from the Italian case studies of EU funded projects HUSH, NADIA and QUADMAP. In: ICSV23, Athens, Greece (2016)
10. Tupov, V.B., Skvortsov, V.S.: Features of noise emission from gas distribution stations and gas pipelines. Power Technol. Eng. **52**(4), 448–450 (2018)
11. Halim, H.A., Hayder, G.: Operational noise assessment in an open cycle power plant: study on the existing acoustical performance. Int. J. Eng. Technol. (UAE) **7**(4), 11–14 (2018)
12. Kisku, G.C., Bhargava, S.K.: Assessment of noise level of a medium scale thermal power plant. Indian J. Occup. Environ. Med. **10**(3), 133–139 (2006)
13. UNI EN 334:2005+A1:2009: Regolatori di pressione del gas per pressioni di entrata fino a 100 bar
14. Tartarini: Calcolo del livello di rumore regolatori di pressione gruppi ed impianti, Tab. N° 2863-I, 28/08/1997
15. Bartalucci, C., Borchi, F., Carfagni, M.: Metodologia per la valutazione previsionale di impatto acustico delle reti di distribuzione del gas naturale, Atti del 44° Convegno Nazionale dell'Associazione Italiana di Acustica (2017)
16. ISO 9613-2:1996. Acoustics - Attenuation of sound during propagation outdoors - Part 2: General method of calculation
17. UNI EN ISO 3744:2010. Acustica - Determinazione dei livelli di potenza sonora e dei livelli di energia sonora delle sorgenti di rumore mediante misurazione della pressione sonora - Metodo tecnico progettuale in un campo essenzialmente libero su un piano riflettente
18. D.P.C.M 14/11/1997. Determinazione dei valori limite delle sorgenti sonore

Design Optimization: Tools and Methods for ETO Products

Paolo Cicconi[1(✉)], Marco Mandolini[2], Miriam Nardelli[2], and Roberto Raffaeli[1]

[1] Università degli Studi eCampus, 22060 Novedrate, CO, Italy
p.cicconi@staff.univpm.it
[2] Università Politecnica delle Marche, 60131 Ancona, Italy

Abstract. The design of Engineer-To-Order products needs tools and methods for reducing time and cost during the phase of the quotation preparation. Modularization is one of the more applied design methods for ETO systems; however, it is necessary to integrate traditional tools with practices of design optimization to improve the development of a proposal. Even if commercial design tools for modeling specific types of engineering systems are available, the application of design optimization is still based on the use of tools not integrated with each level of the design phases. Moreover, these tools often require software customization. The integration of geometrical modeling, simulations, analysis, and optimization concerns the interaction between different tools. This paper describes an approach to support the Multi-Object Optimization related to the design of complex ETO systems with a focus on the oil & gas context. In this context, Genetic Algorithms and Constraint Satisfaction Problems are introduced as tools to support the design optimization of steel structures. The approach includes the employment of different and integrated tools throughout the design phases. This paper also shows a collection of tools to support the different levels for the design of different ETO products during the preparation of an offer related to proposal submission.

Keywords: Design tools and methods · Design optimization · Engineered-To-Order · Oil & gas

1 Introduction

Focusing on the context of industrial engineering, two types of approach can be compared to achieve specific customer's needs. These different approaches are Configure-To-Order (CTO) and Engineer-To-Order (ETO). Both approaches require a customer's order. While a CTO production is based on already existing configurations [1], an ETO approach requires the development of engineering solutions before product fabrication [2]. ETO applications are necessary when customer's requirements cannot be fulfilled through a catalog of pre-configurated solutions. ETO solutions can be classified as "light" if concern minor customizations or as "heavy" when the final solution must be completely engineered with a major level of customization [3].

C. Rizzi et al. (Eds.): ADM 2019, LNME, pp. 516–527, 2020.
https://doi.org/10.1007/978-3-030-31154-4_44

Generally, configurations tools are very used in CTO production because these tools support customers and designers in the selection of already defined solutions. Configurators are tools that implement rules and a knowledge base for a rapid selection of the most suitable product configurations [4]. The main advantage of this approach is the reusing of past solutions avoiding new design efforts related to simulations, engineering design, cost estimations, etc. In this context, products can be simple machines or even types of machinery. Otherwise, the design of more complex systems often requires an ETO approach with customization. The customization can be related to specific normative, boundary conditions, sizing, special requirements or other. Therefore, ETO products are often complex structures, types of machinery, and even plants. In this case, configurations tools are limited by the impossibility to reuse past solutions that are designed and optimized for different cases.

Configurations are often related to the product modularization. Modularization is often applied in the context of the product platforms which enable the development of product families using shared assets with a cost-effectiveness advantage [5]. Modularization brings cost savings and other gains in terms of sourcing, manufacturing, and quality control. Modularized products can be modeled in 3D using product families in CAD tools. The use of a product family in the 3D modeling allows a wide range of variants to be generated by the modeling of one parametric CAD document [6]. However, the definition of a product structure is still a challenging task in the modular design, as highlighted by Windheim et al. [6]. Modularization is also applied in ETO systems such as oil & gas plants [7]. A modular plant consists of many modules which include pre-fabricated units [8]. A dedicated design strategy is needed to exploit all the advantages offered by modularization in terms of cost and project risk reduction.

Regarding the ETO design, the receipt of a specific order is a phase called Request-For-Proposal (RFP). The RFP phase gives the beginning to the early design phase. This phase is a preliminary analysis that has to take into consideration the draft project including geometry, layout, simulations, and cost estimation. An important issue in ETO companies is the time necessary to provide a bid for the proposal submission [9]. ETO companies must be able to prepare a competitive bid as soon as possible. Therefore, reliable solutions are necessary to be evaluated during the early design phase, to avoid further risks related to a major design effort and final production cost.

Long engineering lead-times significantly affect the development of ETO products. Even if agile, rapid and flexible design approaches are ever searched in this filed, the analysis carried out on different application cases highlights the lack of a design platform integrating every design tool and focusing on the design optimization. The authors have observed different ETO contexts from the design of steel structures [10] to the development of customized and modularized plants [11].

Nowadays, the most used design strategies in ETO companies are Design Automation [12], Modularization [13], and Configurations [2]. To reduce time and cost in design phases and fabrication, configurations tools have been also implementing in ETO companies [3]. However, these tools are very difficult to be implemented if they have to provide new product solutions. In this case, Multi-Objective Optimization (MOO) is necessary. The ever-increasing competitiveness, due to market globalization, has been forcing the industries to make design optimization of their products [14]. ETO products are more difficult to be optimized because they often are complex systems.

Generally, the optimization in ETO solutions is parameter-based and parameters are related to normative and geometry. A modular approach can be extended to a multi-step optimization to reduce computing time and complexity [10].

Regarding past research on ETO approaches, other topics have been investigated in the last 20 years such as Design for Manufacturing and Assembly [15], Feature-Based Costing [16] and Design-To-Cost [17]. However, these applications for the ETO industry requires the definition of a Knowledge Base (KB) to formalize the implicit, explicit and tacit knowledge [18]. These resulting software tools are often described as Knowledge-Based Engineering (KBE) systems. While the implicit knowledge is embedded from product documents, drawings, and engineering dimensioning algorithms, the tacit knowledge comes from the experience of people with technical expertise. The knowledge formalization is a critical issue which involves capturing, representing, and reusing of past design cases [19]. Moreover, the knowledge about design requirements and constraints available during the conceptual design phase is often imprecise, approximate or incomplete [20]. In this context, the modeling of a digital product structure related to ETO products is often represented by an Objected-Oriented approach [21]. The development of a KBE tool allows repetitive design tasks to be automated with an increase of efficiency and a reduction of time. This achievement is searched because long engineering lead-times significantly affect the development of ETO products [22]. Time-related problems are mainly attributable to the required customization studies which can produce delays in the project development. In this context, the integration between optimization tools and model-based simulations is necessary to manage the complexity [23].

The paper proposes a methodological approach to support the design workflow of ETO products with optimization throughout the design phases. Here a design platform is described including optimization tools such as Genetic Algorithms, Constraint Satisfaction Problem [24], analytical calculations, configurations, design automation, and FEM simulations. The approach is based on a multi-objective optimization analysis to reduce cost and increase the mechanical performance.

2 Materials and Methods

This section describes the proposed methodological approach defined to support the design of ETO products (Fig. 1). The design study is focused on two actions: Request-For-Proposal and Proposal Submission. The approach aims at supporting the designers in a rapid and reliable definition of an embodiment project in order to participate in a Proposal Submission. This design workflow presents a set of stand-alone tools that can be integrated inside the life cycle using a Knowledge Base which consists of a repository of rules, formulas, documents, past projects, etc.

The design input is data and synthesis related to the RFP receipt from the customer. Since the development of ETO products is characterized by a quotation process, the proposed design workflow has been defined to reduce time integrating different types of design tools and practices. A company operating in the business of ETO solutions is commonly involved in many quotation processes. Even if objectives and targets are

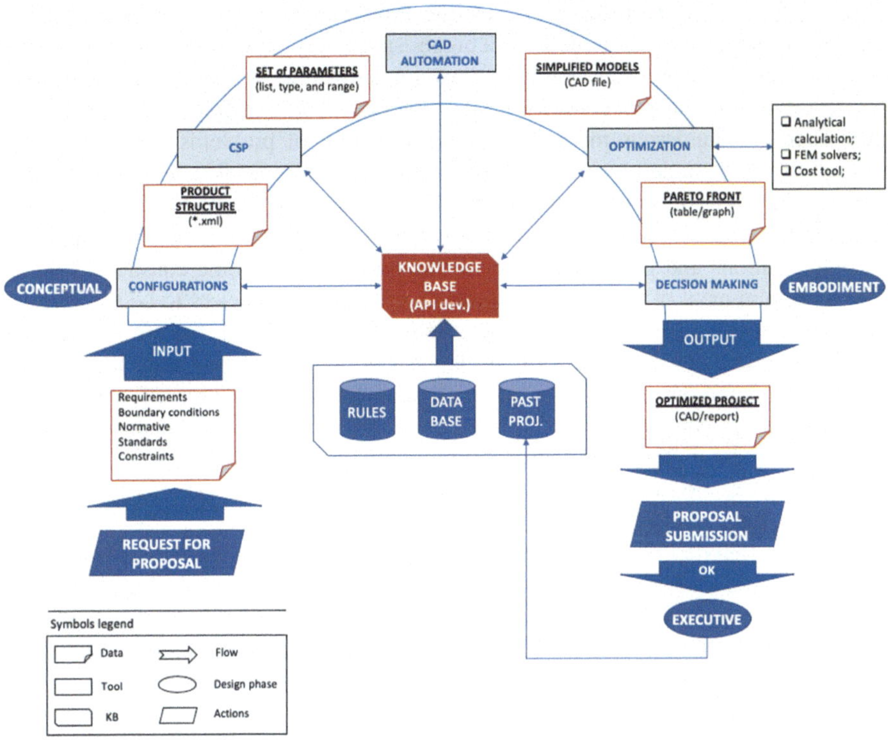

Fig. 1. The methodological approach to support the design of ETO product throughout the design phases including design optimization practices.

defined with the specific customer, an optimization activity is necessary to achieve a high product quality at a reduced cost.

The first design phase described in Fig. 1 is the Conceptual level which begins with the early elaboration of the input data to define a first scheme of the Product Structure. Since the study is focused on the design of ETO systems, already configured solutions are not available or reusable for new engineering design. Therefore, Configuration tools can only release a draft of a Product Structure with the main functional groups. The output of this phase is an XML file which includes the description of the Product Structure where the relations between each component is represented as a tree graph. A configurations tool can be implemented using commercial software or developing a customized software solution. The outcome is a Product Structure and the values of each parameter are analyzed in the next step of the methodological approach, using a constraints analysis.

A Knowledge Base is proposed to collect each rule, formula, and data related to the product design. This repository of knowledge can interact with commercial software using the Application Programming Interface (API) tools. The Knowledge Base is connected to each stage of the proposed design, to integrate each input and output into a common design platform.

Tools for solving Constraint Satisfaction Problems (CSP) are here introduced to select the feasible configurations from a bigger space of possible combinations. A CSP model solves the parameter configurations implementing mathematical constraints. Regarding the solving processes for CSP problems, it can be depicted in a tree graph [24]. There are numerous methods to solve this kind of problems and many other methods are also studied to improve calculation time and precision in CSP problems. Backtracking, branch and bound, and depth-first search are the more common algorithms in the CSP solutions. Backtracking is a technique that consists to enumerate all possible solutions and prunes all the ones that do not satisfy some constraint [25].

The employment of a CSP tool allows the setting of parameters to be investigated and defined in terms of range and values. Before to proceed with further optimization and design embodiment phases, a CSP analysis is defined to limit the range of not-possible combinations from a set of configurations. The selection of feasible and unfeasible combinations of parameters is based on the constraint's definition. While a part of these constraints is related to the proposal specifications, the rest of the constraints are related to rules and formulas which have to be formalized into the proposed Knowledge Base.

The next step described in the proposed Knowledge Base is the CAD Automation level. Designers can use tools to generate parametric CAD models from the define Product Structure. Each geometrical parameter can be edited considering the set of values defined by the previous CSP analysis. The levels of detail are related to the set of CAD templates used for CAD automation. However, a simplified level of detail is suitable to achieve fast simulations.

Simulations and other calculations are presented within the Optimization level. Here the optimization level is presented as a multi-object and multi-level optimization analysis. The designer can use analytical tools or FEM solvers for an optimization study. In the case of a two-level optimization, a first study can involve an analytical tool and a second one a FEM solver. Genetic algorithms or similar can be used into the design workflow to optimize the geometrical parameters of the analyzed ETO systems.

The approach considers a multi-objective optimization including the cost reduction as an objective together with the structural behavior. Regarding the employed tools, while the optimization tool can be a commercial tool or a customized one, a FEM solver can be selected from a list of available commercial tools. The tool for the evaluation of the product cost should be customized on the specific product. Therefore, this tool can be parametric and developed by the design team of specific ETO products, or a commercial software solution with customized functions. The difference between these two solutions depends on the level of accuracy to be achieved.

The proposed Optimization level also presents an integration with the Knowledge Base. The interaction concerns design constraints to be applied to the optimization workflow and the formulas used for cost analysis and analytical calculations. Finally, decision-making tools are necessary to complete such design workflow because the result of the multi-optimization study is not a final configuration of parameters but a Pareto front which represents a set of optimized solutions. Every point of the Pareto front are optimum solutions, but the decision-making process has to select only one optimum case. A strategy is to change second objectives in design constraints, in order to obtain a one-level optimization and reduce the complexity of such analysis. As a second strategy, threshold values can be applied to each objective in order to reduce the points of the Pareto front. The connection between decision-making tools and Knowledge Base is related to the definition of threshold values and constraints to be considered. The decision-making process can be also supported by the analysis of past-projects stored into the database of the design platform.

3 Test Case

According to the proposed methodological approach, this section presents a collection of tools developed to support the different design levels for proposal submission. These tools are related to four different researches focused on the design optimization and configurations of ETO products for oil & gas applications. In particular, these tools are software prototypes used in research activity to compare effort and time related to traditional design practices with the use of KBE tools. These tools implement algorithms and knowledge to avoid repetitive and time-consuming tasks. They are described following the order presented in Fig. 1: Configurations, CSP, Design Automation, and Optimization.

3.1 Configurations

A configurator tool was developed to support the early definition of steel structures for oil & gas applications (Fig. 2). This software tool implements the required Knowledge Base to define the steel structure of a 700-ton plant used as a single module of a bigger modular power plant [11]. Oil & gas plants are modular and consist of several similar units to reduce the installation cost and time. Therefore, the same steel structure is used in a different number of modules in the same plant.

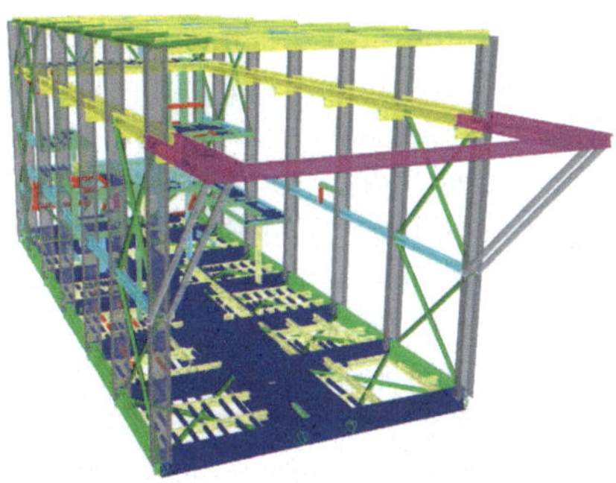

Fig. 2. An example of a steel structure for Oil & Gas applications.

Figure 3 describes the GUI interface of the configurator tool which implements design rules and metrics formalized from the analysis of past solutions. The configuration of the steel structures also requires the definition of specifications in terms of machinery employed, installation type, location, transportation phases, etc. All this information is evaluated by the software in order to define the maximum and minimum dimensions, the number of decks and columns, materials, list of normative, and the loading conditions to be considered. The outcome of this software tool is the definition of an early Product Structure. An XML format has been chosen for the data export. Any additional information relative to normative and boundary conditions are included into this file within the scheme of the Product Structure.

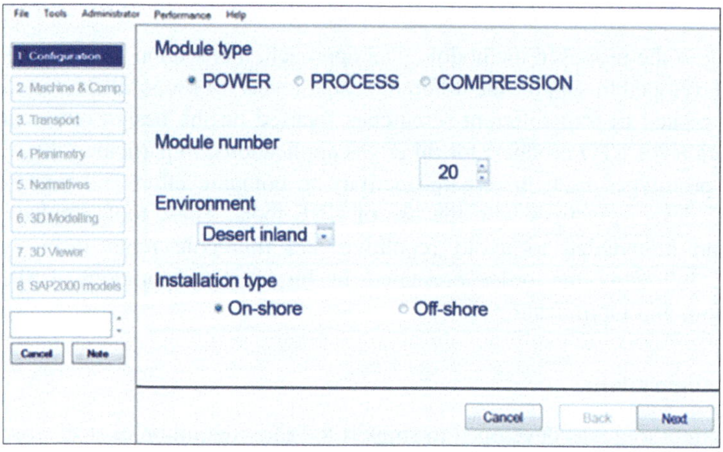

Fig. 3. The GUI interface of the configurations tool.

3.2 CSP

This tool has been developed to support the design of the ETO structures configured in the previous section. The target is the study of a parameters configuration to achieve the weight and cost reduction for 700-ton steel structures [26]. The CSP model implements a set of variables, values, and constraints. In particular, these variables concern: external diameters, thickness, dimensions of standard beams, etc. The constraints regard the geometrical limits for each dimension, and the ratio between length, height, and thickness. Constraints are described by mathematical functions. The definition of each constraint is related to the formalized knowledge base. This approach is fast and allows hundreds of configurations to be solved in a few minutes.

In this case, the CSP study also includes some design constraints related to the structure's weight and frame sizing. The objective of the test case is to achieve an optimum configuration by reducing the design time and the number of simulations. In this context, the CSP analysis is able to calculate all feasible solutions from a bigger space of possible solutions. The search engine of the proposed CSP tool uses the backtracking algorithm combined with the constraint propagation [24]. In particular, this tool uses a simple backtracking algorithm, combined with the constraint propagation method due to the dimension of the space of solutions to be investigated.

The prototypical tool has been developed using VB.NET. This tool has a GUI interface (Fig. 4) and a solution module that interacts with other two tools called

MiniZinc and Gecode. In particular, Gecode is an open tool that implements algorithms for solving CSP problems [25]. MiniZinc is a medium-level constraint modeling platform based on Zinc language to run problems using the Gecode developing framework [27].

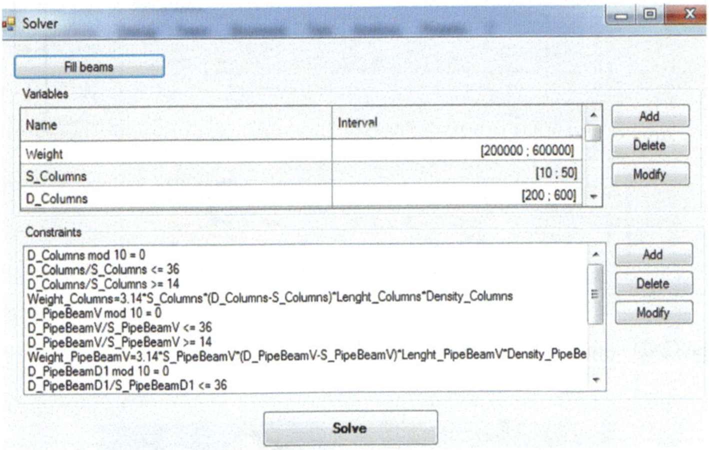

Fig. 4. The GUI interface implementing for the CSP tool.

The results of using such a CSP tool confirmed the possibility of reducing the space of solutions without considering not feasible or not optimized solutions. The experiment activity showed the possibility to select 5% of all possible combinations for further analysis. The limits of this CSP approach are the impossibility to add numeric analysis inside the definition a mathematical constraint, and the difficulty to run an analysis with millions of combinations. Even if this approach is fast, a great number of combinations can considerably increase the cost of computing.

3.3 CAD Automation

Here a CAD automation tool is proposed for generating 3D models of exhaust ducts employed in Gas Turbine (GT) plants. This tool, called Duct Designer, requires an input of geometrical and non-geometrical parameters to generate the assembly and each duct items of an exhaust manifold system for GT (Fig. 5).

This CAD automation tool supports the designers to configure desired solutions of exhaust ducts by simplified layout [23]. This tool implements an objected-oriented product structure based on product architecture. Different design alternatives can be configured and automatically generated in CAD documents based on the product parameterization (Fig. 6).

Involving such a tool inside the proposed methodological workflow it is possible to obtain a rapid generation of CAD models to be used for further analysis such as simulations, optimization and cost evaluation. The limit of such tools is related to the developing effort because they have to be specialized for a specific product.

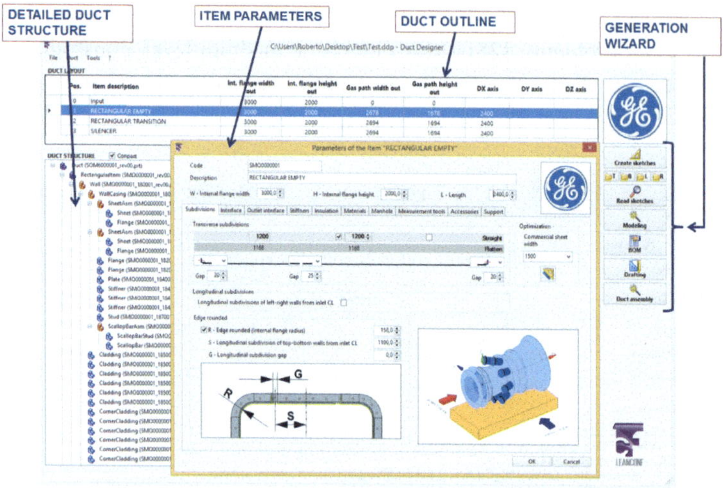

Fig. 5. The CAD automation tool implemented to generate the 3D models of gas turbine ducts.

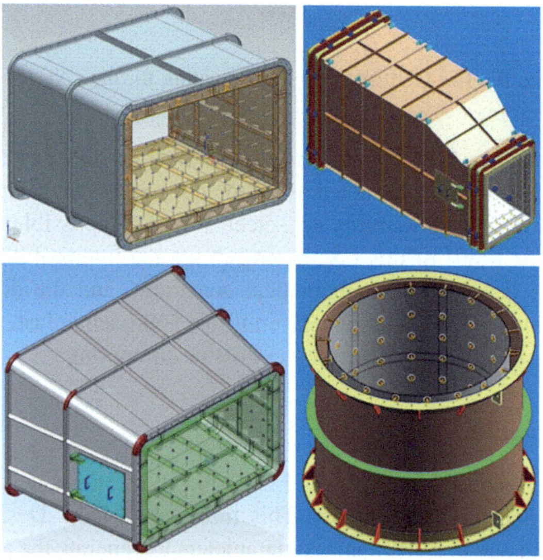

Fig. 6. Example of duct items generated by the CAD automation tool.

The designers can substitute this tool with the use of pre-defined and parametric templates. These tools can offer the opportunity to achieve and a higher level of details which can be re-used for the further engineering design phases.

3.4 Optimization

This section describes the optimization approach based on a platform of tools to support the multi-level and multi-object design optimization. The context of this fourth test case is the design of GT chimneys, which are the final vertical items of an exhaust duct [10]. The multi-objective optimization (MOO) approach involves the minimization of the product weight and the cost related to the manufacturing and assembly phases.

Figure 7 describes the proposed approach. Three levels of optimization are described. The first level considers a calculation based on 1-D simplified models. Here the analytical calculation is solved by applying equations of a pole fixed in the ground, and the cost evaluation is based on an easy parametric approach. This level of MOO was implemented inside a MATLAB® application.

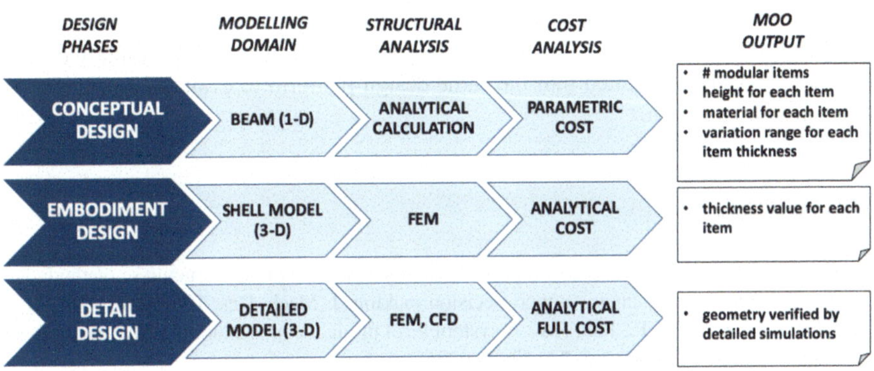

Fig. 7. Description of the multi-level and MOO approach for the design of GT chimneys.

The second level of MOO considers the employment of SAP2000®, a commercial FEM tool, for solving the stress-strain behavior of the configured structure under the related boundary conditions. At this level, the manufacturing cost estimation is related to an analytical analysis using a tool that implements the required Knowledge Base for assigning cost values to the 3D geometrical dimensions.

The third MOO level is focused on the detailed design; therefore, it is related to the phase of the order execution. This level is based on the use of full 3-D models. The engineers simulate the overall product using a detailed FEM analysis. The cost analysis also includes the definition of manufacturing parameters related to machine tools, setups, and raw materials.

All these optimization levels are based on MOGA-II, which is a Genetic Algorithm. This algorithm is widely used in the literature [28] because it uses a smart multi-search elitism approach that preserves excellent solutions without spurring premature convergence to local-optimal frontiers. The main difference between each level is related to the number of parameters involved. While the first stage considers a small number of geometric parameters, the second stage focuses on a larger number of variables related to the 3-D geometries involved. Finally, the third stage of optimization considers an increased number of parameters related to manufacturing and assembly.

4 Conclusions

A design methodology has been described to support a fast proposal definition in the context of ETO products. The approach is focused on the definition of tools and methods to support designers from the RPF phase to the proposal submission. The scope of the research is to increase the competitiveness of ETO companies analyzing the design methods and tools to support a multi-objective optimization. Therefore, the approach aims at reducing the time-consuming iteration by integrating different software tools.

Four tools have been described as an example of applications to be integrated inside the proposed methodological workflow. These tools support the designers in configurations, CPS analysis, CAD automation, and design optimization. As future work, all these tools will be integrated into the same design platform to evaluate the validity of the proposed approach.

References

1. Ripperda, S., Krause, D.: Cost effects of modular product family structures: methods and quantification of impacts to support decision making. J. Mech. Des. **139**(2), 021103 (2017)
2. Kristianto, Y., Helo, P., Jiao, R.J.: A system level product configurator for engineer-to-order supply chains. Comput. Ind. **72**, 82–91 (2015)
3. Sylla, A., Guillon, D., Vareilles, E., Aldanondo, M., Coudert, T., Geneste, L.: Configuration knowledge modeling: how to extend configuration from assemble/make to order towards engineer to order for the bidding process. Comput. Ind. **99**, 29–41 (2018)
4. Haug, A., Shafiee, S., Hvam, L.: The costs and benefits of product configuration projects in engineer-to-order companies. Comput. Ind. **105**, 133–142 (2019)
5. Otto, K., et al.: Global views on modular design research: linking alternative methods to support modular product family concept development. J. Mech. Des. **138**(7), 071101 (2016)
6. Windheim, M., Gebhardt, N., Krause, D.: Towards a decision-making framework for multi-criteria product modularization in cooperative environments. Procedia CIRP **70**, 380–385 (2018)
7. Tanabe, M., Miyake, A.: Safety design approach for onshore modularized LNG liquefaction plant. J. Loss Prev. Process Ind. **23**(4), 507–514 (2010)
8. Gibb, A.G.: Standardization and pre-assembly-distinguishing myth from reality using case study research. Constr. Manag. Econ. **19**(3), 307–315 (2001)
9. Willner, O., Gosling, J., Schönsleben, P.: Establishing a maturity model for design automation in sales-delivery processes of ETO products. Comput. Ind. **82**, 57–68 (2016)
10. Cicconi, P., Castorani, V., Germani, M., Mandolini, M., Vita, A.: A multi-objective sequential method for manufacturing cost and structural optimization of modular steel towers. In: Engineering with Computers (2019)
11. Castorani, V., Cicconi, P., Germani, M., Bondi, S., Marronaro, M.G., Coppolaro, S.: A framework to support the optimization of modularized oil and gas structures. In: Volume 2B: 44th Design Automation Conference (2018)

12. Rigger, E., Vosgien, T.: Design automation state of practice - potential and opportunities. In: Proceedings of the DESIGN 2018 15th International Design Conference (2018)
13. Sievers, S., Seifert, T., Franzen, M., Schembecker, G., Bramsiepe, C.: Fixed capital investment estimation for modular production plants. Chem. Eng. Sci. **158**, 395–410 (2017)
14. Gholizadeh, S., Baghchevan, A.: Multi-objective seismic design optimization of steel frames by a chaotic meta-heuristic algorithm. Eng. Comput. **33**, 1045–1060 (2017)
15. Lewis, G.: Product design for manufacturing and assembly. In: Mechanical Engineers' Handbook, pp. 1–19 (2015)
16. Opiyo, E.Z.: A feature-based approach to conceptualisation, upfront modeling, and planning for the future of complex systems. Int. J. Inf. Technol. Manag. **16**(1), 91 (2017)
17. Ehrlenspiel, K., Kiewert, A., Lindemann, U.: Factors that influence manufacturing costs and procedures for cost reduction. In: Cost-Efficient Design, pp. 143–384 (2007)
18. McMahon, C., Lowe, A., Culley, S.: Knowledge management in engineering design: personalization and codification. J. Eng. Des. **15**(4), 307–325 (2004)
19. Verhagen, W.J.C., et al.: A critical review of knowledge-based engineering: an identification of research challenges. Adv. Eng. Inform. **26**(1), 5–15 (2012)
20. Wang, L., et al.: Collaborative conceptual design—state of the art and future trends. Comput. Aided Des. **34**(13), 981–996 (2002)
21. La Rocca, G.: Knowledge based engineering: between AI and CAD. Review of a language based technology to support engineering design. Adv. Eng. Inform. **26**(2), 159–179 (2012)
22. Zhu, Z., La Rocca, G., van Tooren, M.J.L.: A methodology to enable automatic 3D routing of aircraft electrical wiring interconnection system. CEAS Aeronaut. J. **8**(2), 287–302 (2017)
23. Cicconi, P., Raffaeli, R., Germani, M.: An approach to support model based definition by PMI annotations. Comput. Aided Des. Appl. **14**(4), 526–534 (2017)
24. Kumar, V., Lin, Y.-J.: A framework for intelligent backtracking in logic programs. In: Foundations of Software Technology and Theoretical Computer Science, pp. 108–123 (1986)
25. Tsang, E.: Foundations of Constraint Satisfaction. Academic Press, London (1993)
26. Nardelli, M., Cicconi, P., Raffaeli, R., Germani, M.: Supporting design tasks through constraint satisfaction tools. In: Proceedings of the DESIGN 2018 15th International Design Conference (2018)
27. Nethercote, N., Stuckey, P.J., Becket, R., Brand, S., Duck, G.J., Tack, G.: MiniZinc: towards a standard CP modeling language. Lecture Notes in Computer Science, pp. 529–543 (2007)
28. Karpat, F.: A virtual tool for minimum cost design of a wind turbine tower with ring stiffeners. Energies **6**, 3822–3840 (2013)

Design and Optimization of the Thermo-Mechanical Behavior in Glass Reinforced Polyamide 6 for Automotive Application

Silvia Barbi[1(✉)], Luca Cattani[2], Tiziano Manfredini[3], and Monia Montorsi[1,3]

[1] Department of Science and Methods for Engineering, University of Modena and Reggio Emilia, Via Amendola 2, 42122 Reggio Emilia, Italy
silvia.barbi@unimore.it
[2] Ask Industries S.p.A., Viale Ramazzini, 42124 Reggio Emilia, Italy
[3] Interdepartmental Center for Applied Research and Services in Advanced Mechanics and Motoring, University of Modena and Reggio Emilia, Via P. Vivarelli 10/1, 41125 Modena, Italy

Abstract. In this work a rational approach, such as Design of Experiments, has been used to design E-glass and S2-glass reinforced polyamide 6 composites. The models, derived by the multivariate analysis of the experimental tests, allowed deriving response surfaces in which the effect of reinforce's composition, content and shape on the thermo-mechanical have been related to composite's behavior during cycling loads and high temperatures. These composites find application in the developing of a sensor used in the automotive engine compartment where thermal and vibration effects must be taken in account to avoid premature failure. Thirty experiments were planned by Design of Experiments and analyzed through Analysis Of Variance to correlate reinforce's properties to coefficient of thermal expansion, Young Modulus and damping over temperature/frequency variation. Statically reliable models were calculated to obtain a numerical estimation of the overall quadratic and cubic interactions among reinforce's properties, explaining how matrix/reinforce interaction affects composite's properties. Nevertheless, the employment of S2-glass led to restrained coefficient of thermal expansion of the composites, reinforce's content of E-glass fibers over 30wt% is in a better agreement with the composite's overall requirements for this tailored application, due to restrained mechanical damping.

Keywords: Glass fiber · Polyamide 6 · Mechanical properties · Sensors · Design of Experiments

1 Introduction

Composites are widely employed for automotive applications where the obtainment of peculiar thermo-mechanical properties associated with a restrained weight is necessary [1]. In the car's engine area, where high performing components are needed due to wide temperature changes and high vibration stresses, component's damping is one of the main relevant issues because of its impact on system performance, safety and

C. Rizzi et al. (Eds.): ADM 2019, LNME, pp. 528–539, 2020.
https://doi.org/10.1007/978-3-030-31154-4_45

reliability [2]. Unexpected vibration may generate unwanted or unsafe operating conditions, high-noise levels and premature wear or stress fatigue failure. Therefore, a numerical estimation of the vibration response of the materials is essential to the effective composite's design and employment in this specific car's area [3]. Thus, to reduce the level of vibrations or noises in a temperature-frequency range required, polymers which offer significant restrained damping behavior in the temperature or frequency interval are chosen. For this reason polyamide based composites are largely used as automotive components, also due to theirs general low density and high thermal resistance [4]. One of the most efficient methods in order to further reduce damping is the design of structures where the viscoelastic nature of polymer is balanced through reinforce addition.

For this purpose, glass inclusions are often used as reinforce, either as fibers or particles, thanks to their resistance to high temperature, mechanical stresses, and their low cost, properties that are needed in application such as sensors or antennas encapsulation and precision assembly [5]. Several glasses have been considered as reinforce to meet these specific requirements, such as E-glass and S2-glass [6, 7]. In addition, polymer modifications are also due to the interaction between glass reinforce and polymeric matrix, promoting a local environment at the interfaces with enhanced strength, which favors the crystallization of the plastic material at the boundary, therefore changing locally the composite's structure [8]. Taking in consideration that fibers are the primary factors controlling the properties of fiber-reinforced composite materials, many specific elements could affect the final behavior of composites in their working conditions: the peculiar characteristics of reinforces and matrix, the content and the shape factor (length/diameter ratio) of reinforces, the dispersion method, reinforce's distribution and alignment into the matrix and the ability to transfer stresses across the interfaces between distinct phases [9]. For this reason most of the composites are far to be isotropic, leading to the need of specific model of material's design, in particular in an industrial context where fast and well-defined responses to precise requirements are needed [10, 11]. Recognizing the important role that the computational approach plays in composites design and analysis, many constitutive models were developed to be able to predict the material's performance based on different principles: continuum mechanics [12, 13], micromechanics [14] and computational (e.g. Monte Carlo) [15, 16]. With the same main goal Design of Experiments (DoE) can be used to generate models describing the complex correlation between input variables (subjected to manipulation) and output variables (which are to be examined downstream the experiment) from a predictive perspective using experimental data [17–19]. Analysis of variance (ANOVA) is generally employed to verify if the effects of the main factors and their interactions terms are statistically reliable [20].

In this work, glass reinforce's composition, shape factor and content were investigated as input variables of the DoE to obtain reliable models describing thermo-mechanical properties. For first, an experimental plan was established including all and only the experiments, in terms of composite's formulation, that are needed to calculate statistical reliable correlations. Thereafter, characterizations were performed on each composite of the experimental plan and the collected data were employed for the multivariate analysis, through which mathematical models were obtained correlating each material's property with relevant input variables. Coefficient of thermal expansions (CTE-α) was considered as a meaningful response that must be minimized to avoid stress due to the fast temperature changes typical of the car engine's area. The damping ability were determined by the intensity of the relaxation transition region on the tan δ graph at a specific frequency (110 Hz), to avoid any peak of resonance that could lead to a worse response of the sensor near the engine area. In addition, Young Modulus (E) was considered to evaluate a more complete behavior of the materials at room temperature. Finally, each model derived from the ANOVA analysis was combined into a desirability function (D) to generate a unique response surface area to meet the overall material's requirements [21].

2 Materials and Methods

2.1 Composite's Fabrication

PA6, provided by Ask Industries S.p.A. – Italy as granulate, has been used as matrix. E-glass and S2-glass reinforces were supplied by AGY (SC, USA) as chop with diameter (d) of 0.05 mm and length (L) of 0.50 mm, with a shape factor ratio (L/d) of 10. The two reinforces were grounded separately in agate mill for 1 h to reduce the L/d ratio from 10 to 1 (D0.99), obtaining two different glass particulates with the same composition of the starting reinforces. PA6 granules were treated in a vacuum oven for 24 h at 90 °C, then blended mechanically with glass inclusions into a two-roll mill (Haake, Rheomix 600) at 240 °C following the compositions shown in the experimental plan (Table 1). Standard dog-bone specimens (EN ISO 527-2, type 1A) were produced for tensile testing, and rectangular specimens (40 mm × 10 mm × 2 mm) were produced for dynamic testing by hot pressing at 230 °C and 5 Tons for 5 min (Carver, Mini-C Laboratory Press) [22]. With the same procedure, CTE testing specimens were fabricated with rectangular shape (15 mm × 5 mm × 5 mm). To overcome property's degeneration due to humidity uptake, all testing samples were dried at 90 °C for 5 days in a vacuum oven before characterization.

2.2 Characterization

Composite's morphology was observed through environmental scanning electron microscopy (ESEM Quanta FEI 200). An optical dilatometer (Expert System Solutions, Misura ODHT) was used to calculate the coefficient of linear thermal expansion (CTE-α) of PA6/glass composites [23]. Each specimen was heated up to 200 °C with a heating rate of 5 °C/min. CTE-α was calculated over a linear portion of the thermal curve (from 100 °C to 200 °C). TA Instruments Q800 Dynamic Mechanical Analyzer (DMA) was used to measure the dynamic response of the composites versus temperature and frequency variation, with particular focus on the damping property (tan δ). The rectangular specimens were tested in a single cantilever setup using a constant displacement amplitude in a multi-frequency procedure from 50 to 120 Hz. The tests were performed at five different temperatures of –20 °C, –10 °C, 0 °C, 10 °C and 20 °C ramped at 2 °C/min. For the tensile test, dog-bone shaped specimens were tested at an extension rate of 50 mm/min on a tensile tester (Instron, 5900 Universal Testing

Systems) with a load of 30 kN. Young's modulus (E) was calculated as the slope of the initial region of the stress strain curve from 0.05% to 0.25% of strain according to EN ISO 527-1 method [24].

3 Calculation

The Design Expert 8.0 (Stat-Ease) code was used both to set-up the experimental plan and to analyze the results. A full factorial design was selected to obtain predictive reliability on the effect of reinforce's commercial type (E-glass, S2-glass), volume (15–45%) and shape factor ratio L/d (1–10) on the thermo-mechanical properties of the composites. Therefore, the number of base experiments required for the selected experimental plan was equal to 2^3 (two level for three variables). Three replicates for each experiment were done for error estimation purposes. Central points, considered as the arithmetic average of the factor's levels, were added in triplicate to investigate the presence of curvature in the data analysis [17]. Table 1 shows the experimental plan consisting of 30 experiments.

ANOVA was employed to estimate the influence of each factor over the responses observing the p-values (α-level of 0.005) and F-tests [17]. The main assumptions of this statistical analysis are: (1) each input factor is independent and normally distributed; (2) the response's variation can be decomposed into different components to evaluate the effect of each factor, their interactions, and experimental error (or unexplained residual) [17]. The quality of the fit in terms of regression analysis and the predictive power of the model were evaluated by using the R^2 and Pred-R^2 respectively [18]. R^2 is the proportion of the dependent variable's variance predictable from the independent variables. In a similar way Pred-R^2 shows how well the model can predict the responses for new observations [18].

The overall desirability function (D) provides the most desirable response's values, taking into consideration all the analyzed responses at once (Eq. 1). For each response Y_i, a desirability function $d_i(Y_i)$ establishes the desirability value of the relative function with a value between 0 and 1, i.e. $d_i(Y_i) = 0$ represents a completely undesirable value of Y_i and $di(Y_i) = 1$ represents a completely desirable, or ideal, response value. The individual desirability functions $d_i(Y_i)$ are then combined using a mean, based on weight (w_i) assigned to each desirability function depending on how much each response has to match the tailored purpose. This mean gives the overall desirability function D [21]:

$$D = \frac{\prod_{i=1}^{k} d_i(Y_i) w_i}{k} \qquad (1)$$

Table 1. Experimental plan

Run	Code	Fibers	Volume (%)	Ratio L/d	CTE-α (10^{-5}/°C)	E (MPa)	tan δ
5	15EP	E	15	1	6.89	2605	0.022
19	15EP	E	15	1	6.92	2620	0.021
29	15EP	E	15	1	6.85	2589	0.019
7	15EF	E	15	10	7.80	3612	0.025
9	15EF	E	15	10	7.76	3582	0.025
13	15EF	E	15	10	7.84	3593	0.025
6	30EP	E	30	1	5.73	3523	0.044
8	30EF	E	30	10	5.95	3417	0.027
24	30EF	E	30	10	5.98	3556	0.027
10	45EP	E	45	1	4.51	3623	0.032
15	45EP	E	45	1	4.54	3635	0.029
16	45EP	E	45	1	4.50	3642	0.031
14	45EF	E	45	10	2.34	4000	0.062
25	45EF	E	45	10	2.36	3940	0.059
28	45EF	E	45	10	2.32	4034	0.063
12	15SP	S2	15	1	4.78	5175	0.081
18	15SP	S2	15	1	4.80	4190	0.075
20	15SP	S2	15	1	4.78	4100	0.082
1	15SF	S2	15	10	2.94	4730	0.064
3	15SF	S2	15	10	2.92	4840	0.067
27	15SF	S2	15	10	2.96	5950	0.068
11	30SP	S2	30	1	3.38	5550	0.052
22	30SP	S2	30	1	3.36	4560	0.046
17	30SF	S2	30	10	3.12	5556	0.078
2	45SP	S2	45	1	6.68	3670	0.013
23	45SP	S2	45	1	6.69	3550	0.054
26	45SP	S2	45	1	6.66	3730	0.014
4	45SF	S2	45	10	3.06	5990	0.025
21	45SF	S2	45	10	3.04	5930	0.027
30	45SF	S2	45	10	3.07	6128	0.027

4 Results and Discussion

Micrographs of composite's microstructures, obtained by ESEM analysis, are detailed in Fig. 1. The inclusions, glass fibers in Fig. 1a and glass particulates in Fig. 1b, are dispersed uniformly in the PA6 and no cluster or aggregation has been observed through ESEM analysis.

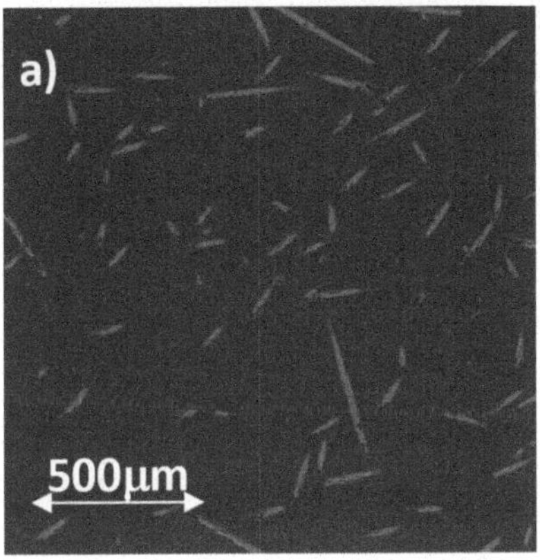

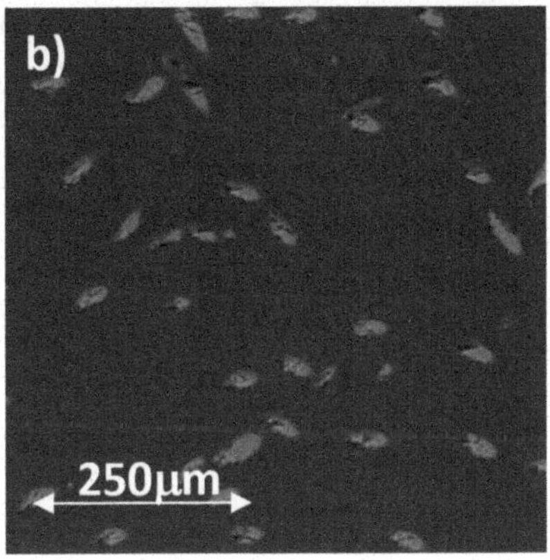

Fig. 1. ESEM micrographs representative of the composite's microstructure based on (a) glass-fibers (b) glass-particles.

It is worth noting that preliminary evaluation of the data collected during experimental sessions (Table 1) describe a quite complex system in which surely each single input variable plays a distinct effect on the final properties. In this condition a statistical approach, such as ANOVA, must be used to simplify the data analysis, correlating starting conditions and final composite behaviors. In Table 2 the ANOVA results have been reported confirming the statistical validity of the derived models and highlighting which factors and/or interactions among them play the main role on the response variables. All models are significant from the statistical point of view and the two F tests are satisfied (data not reported): F-value is higher than one with a p-value largely less than 0.001, indicating that, for each response analyzed, there is only a 0.01% chance that the calculated models occur due to noise [17, 18]. Furthermore, the lack of fit is not significant: no terms have been omitted in the models allowing a good predictive power for each analyzed property [25]. R^2 point out a reliable correlation for all the responses analyzed with values quite close to 1.00 and Pred R^2 is lower and in reasonable agreement with R^2 [18]. The normal distribution of the residuals has been analyzed (Fig. 2) confirming that each model can be used to quantify the relationships between factors and the final measured responses. ANOVA point out many significant factors (Table 2) confirming the complex nature of the correlations between thermo-mechanical properties and reinforce's parameters These relationships are better explained by the interaction factors than by a single factor, confirming the need of a multivariate statistical approach for the analysis and to derive the best formulation of these materials.

In Fig. 3 the graphical results about CTE-α have been reported showing very different trends varying from E-glass (Fig. 3a) to S2-glass (Fig. 3b). The range of CTE-α is much more grater employing E-glass, on the other hand S2-glass helps significantly to keep the CTE-α low in a wide range of conditions, as expected considering the higher resistance to heat attributed to S2-glass with respect to E-glass [26, 27]. In fact, employing E-glass fibers (Ratio = 10) in a restrained volume, a remarkable increasing of CTE-α is reached, on the opposite, no significant variations are detected

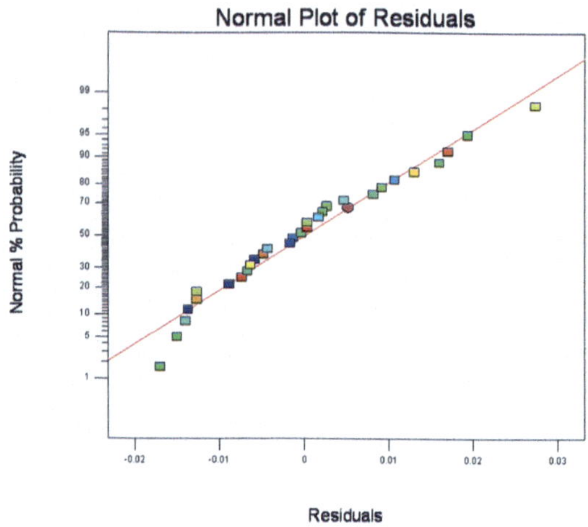

Fig. 2. Normal Probability plot for CTE-α

Table 2. Summary of ANOVA results and selected desirability parameters.

Responses	Significant factors	R^2	R^2 Pred	Target (importance)
CTE-α	Volume; Fiber; Ratio;	0.89	0.85	To minimize (5/5)
	Volume * Fiber			
	=(E-glass) + 8.18 * 10^{-5} – 8.12 * 10^{-7} * Volume + 2.23 * 10^{-6} * Ratio * 8.98 * 10^{-8} * Volume * Ratio =(S2-glass) + 2.88 * 10^{-5} + 8.31 * 10^{-7} * Volume + 3.33 * 10^{-7} * Ratio – 8.99 * 10^{-8} * Volume * Ratio			
tan δ	Volume; Fiber;	0.78	0.66	To minimize (5/5)
	Volume * Fiber			
	Volume * Ratio			
	=(E-glass) + 0.011 + 7.556 * 10^{-4} * Volume =(S2-glass) + 0.097 – 1.539 * 10^{-3} * Volume			
E	Volume; Fiber;	0.93	0.88	In range (3/5)
	Volume * Fiber;			
	Fiber * Ratio;			
	Volume * Fiber * Ratio			
	=(E-glass) + 2017 + 36 * Volume + 130 * Ratio – 2 * Volume * Ratio =(S2-glass) + 5199 – 34 * Volume – 45 * Ratio + 6 * Volume * Ratio			

using S2-glass fiber (Ratio = 10) in different volumes. Moreover, considering the singular effect, an increase of the shape ratio leads to a decrease of CTE-α for both types of reinforce glasses: higher shape ratio increases the molecular chain's constraints applied to the matrix due to an increased fiber/matrix interaction, resulting in a decreasing CTE-α [28].

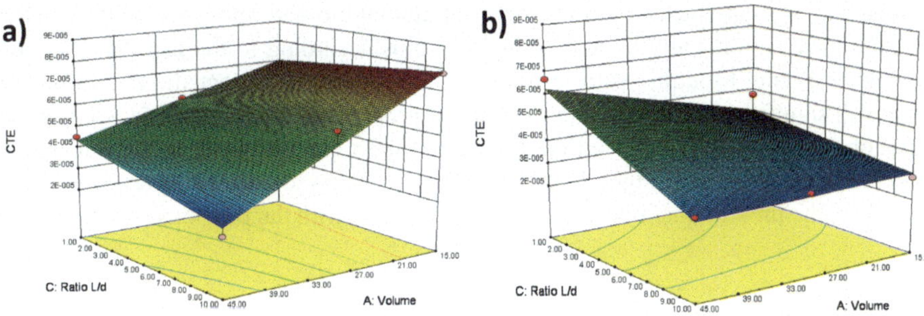

Fig. 3. Graphical models for CTE-α response: (a) E-glass fiber (b) S2-glass fiber.

As shown in Fig. 4, the magnitude of tan δ has been found to change significantly only if S2-glass is employed in substitution of E-glass, due to higher S2-glass fibers stiffness. Addition of high modulus fibers to the base polymer matrix reduces the damping characteristics of the composite as confirmed by previous literature [29]. Energy dissipation due to frequency variation took place mainly at the interface employing S2-glass, therefore, more energy is needed to promote polymeric chains rearrangement leading to an increase of the tan δ. As shown in Fig. 4b by decreasing reinforce volume fraction tan δ peak increases at the expenses of stiffness and strength as confirmed by literature [30, 31]. On the other hand, the shape ratio is the factor with the lower influence on tan δ: only a not-significant variation can be seen across the different shape ratio if compared with the differences due to different glass type or content of reinforce. The lowest tan δ is attributed to the E-glass containing composites in restrained volume (15wt%), with respect to the whole range studied in this work.

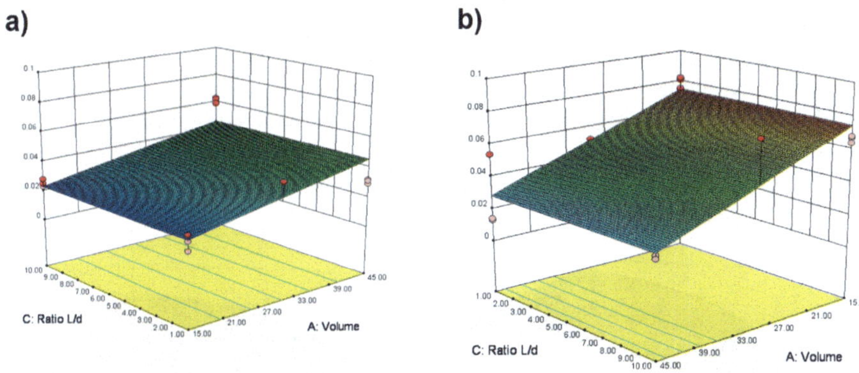

Fig. 4. Graphical models for tan δ response: (a) E-glass fiber (b) S2-glass fiber.

In Fig. 5, the Young Modulus (E) has been found to report very similar trend for E-glass (Fig. 5a) and S2-glass (Fig. 5b) in function of reinforce's volume and shape ratio. These two graphs shown the same trend even if the ranges of Young Modulus are different; For both type of inclusions, the Young Modulus is increased when higher volume and/or shape factor ratio is employed although composites produced by using S2-glass have significantly higher Young Modulus according to the fact that S2-glass has higher Young Modulus with respect to E-glass [32]. As expected, the reinforcing fibers will significantly improve the mechanical properties of composite if there is good adhesion between the matrix and the fibers, and if an enough percentage of fibers are longer than the critical length in order to avoid matrix deformation around fibers. Therefore, the present quantitative model in agreement with the previous literature [27].

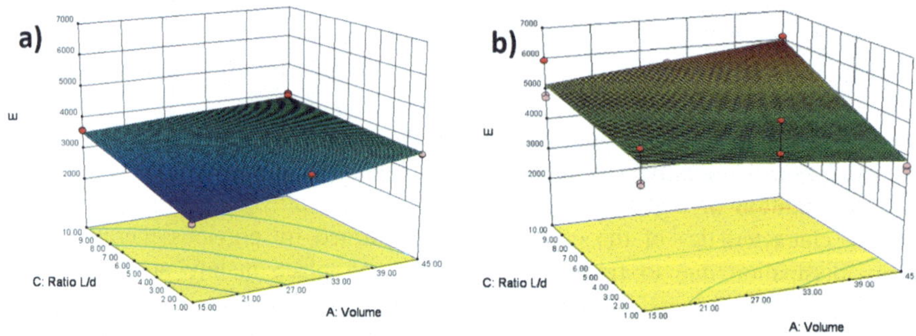

Fig. 5. Graphical models for E response: (a) E-glass fiber (b) S2-glass fiber.

Following target and importance of each single property, as shown in Table 2, D function (Fig. 6) has been derived to obtain the best reinforce parameters needed for the tailored application of these composites: the greatest importance has been attributed to the thermo-mechanical behavior during periodic load measured as CTE-α and tan δ. Figure 6 shows that D strongly changes moving from E-glass reinforce to S2-glass. The highest D, of about 0.8, is obtained employing E-glass with respect to the S2-glass, resulting equal to 0.6. Both functions are quite flat nearly the optimal solution, indicating that small variations around the best condition will not change significantly the overall desirability. Finally, among the E-glass fiber containing samples, the best performances are achieved using a content >30% in volume and a shape factor ratio L/d equal to 10, indicating that fiber should be strongly preferred to particle inclusion.

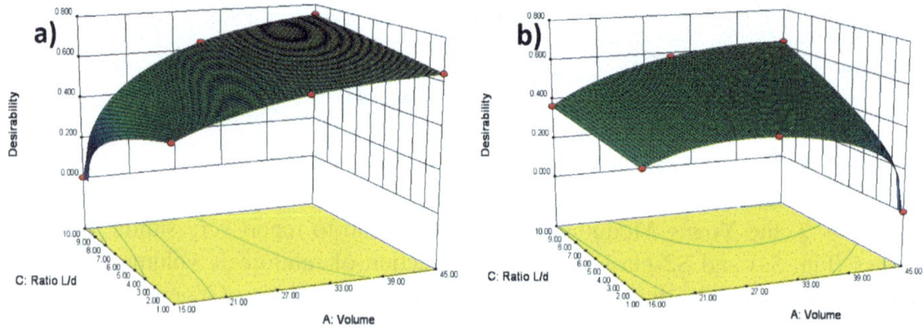

Fig. 6. Graphical models for overall Desirability function: (a) E-glass fiber (b) S2-glass fiber.

5 Conclusion

In this study Design of Experiments has been employed to develop isotropic PA6/glass composites for a specific purpose in automotive application having restrained CTE-α and damping over frequencies. Statistically reliable models ($R^2 > 0.78$), describing quantitatively the behavior of composites depending on the type and the content of reinforce have been obtained. It has been verified that, despite a restrained CTE-α and enhanced static mechanical properties by using S2-glass, the damping is lower with the employment of E-glass, due to the higher stiffness of S2-glass. No advantage has been observed by employing particulate instead of fiber glass for the specific purpose. Mathematical models show a good predictive power therefore they can be successfully used in materials design perspective. The best conditions codified by the D function clearly suggests an optimal volume of E-glass fiber in the range 30–45% of volume to promote restrained CTE-α and damping at 110 Hz. Compositions falling in this range were replicated and then tested in real condition close to engine in cars confirming the noise reduction and improved mechanical performances at higher temperature.

Acknowledgements. The authors would like to thank ASK industries S.p.A. for financial support and technical assistance. This work was supported by the Italian Ministry of Economic Development (MISE)'s FUND FOR THE SUSTAINABLE GROWTH (F.C.S) under grant agreement (CUP) B48I15000130008, project VASM ("Vehicle Active Sound Management").

References

1. Benaarbia, A., Chrysochoos, A., Robert, G.: Thermomechanical analysis of the onset of strain concentration zones in wet polyamide 6.6 subjected to cyclic loading. Mech. Mater. **99**, 9–25 (2016). https://doi.org/10.1016/j.mechmat.2016.04.011
2. Ullah Khan, S., Yin Li, C., Siddiqui, N.A., Kim, J.-K.: Vibration damping characteristics of carbon fiber-reinforced composites containing multi-walled carbon nanotubes (2011). https://doi.org/10.1016/j.compscitech.2011.03.022
3. Etaati, A., Mehdizadeh, S.A., Wang, H., Pather, S.: Vibration damping characteristics of short hemp fibre thermoplastic composites. J. Reinf. Plast. Compos. **33**, 330–341 (2013). https://doi.org/10.1177/0731684413512228
4. Santhanakrishnan Balakrishnan, V., Seidlitz, H.: Potential repair techniques for automotive composites: a review. Compos. Part B Eng. **145**, 28–38 (2018). https://doi.org/10.1016/j.compositesb.2018.03.016
5. Callister, W.D.: Composites. In: Material Science and Engineering An Introduction, 7th edn., pp. 577–620 (2006)
6. Yang, G., Park, M., Park, S.J.: Recent progresses of fabrication and characterization of fibers-reinforced composites: a review. Compos. Commun. **14**, 34–42 (2019). https://doi.org/10.1016/j.coco.2019.05.004
7. He, W., Gao, J., Liao, S., Wang, X., Qin, S., Song, P.: A facile method to improve thermal stability and flame retardancy of polyamide 6. Compos. Commun. **13**, 143–150 (2019). https://doi.org/10.1016/j.coco.2019.04.010

8. Ibáñez-Gutiérrez, F.T., Cicero, S., Carrascal, I.A., Procopio, I.: Effect of fibre content and notch radius in the fracture behaviour of short glass fibre reinforced polyamide 6: an approach from the theory of critical distances. Compos. Part B Eng. **94**, 299–311 (2016). https://doi.org/10.1016/j.compositesb.2016.03.064
9. Popescu, A., Hancu, L., Bere, P.: Research concerning the optimum extrusion temperature for reinforced polyamide. Appl. Mech. Mater. **371**, 394–398 (2013). https://doi.org/10.4028/www.scientific.net/amm.371.394
10. Güllü, A., Özdemir, A., Özdemir, E.: Experimental investigation of the effect of glass fibres on the mechanical properties of polypropylene (PP) and polyamide 6 (PA6) plastics. Mater. Des. **27**, 316–323 (2006). https://doi.org/10.1016/j.matdes.2004.10.013
11. Tognana, S., Salgueiro, W., Somoza, A., Pomarico, J.A., Ranea-Sandoval, H.F.: Influence of the filler content on the thermal expansion behavior of an epoxy matrix particulate composite. Mater. Sci. Eng. B Solid-State Mater. Adv. Technol. **157**, 26–31 (2009). https://doi.org/10.1016/j.mseb.2008.12.003
12. Nguyen, T.D., Jones, R.E., Boyce, B.L.: Modeling the anisotropic finite-deformation viscoelastic behavior of soft fiber-reinforced composites. Int. J. Solids Struct. **44**, 8366–8389 (2007). https://doi.org/10.1016/j.ijsolstr.2007.06.020
13. Nedjar, B.: An anisotropic viscoelastic fibre–matrix model at finite strains: continuum formulation and computational aspects (2007). https://doi.org/10.1016/j.cma.2006.09.009
14. Fukuda, H., Miyazawa, T.: Micromechanical approach to the tensile strength of unidirectional composites. Adv. Compos. Mater. **4**, 101–110 (1994). https://doi.org/10.1163/156855194X00231
15. Shojiro, O., Osamura, K.: Tensile strength of fibre-reinforced metal matrix composites with non-uniform fibre spacing (1989). https://doi.org/10.1007/BF02385736
16. Beyerlein, I.J., Phoenix, S.L.: Stress concentrations around multiple fiber breaks in an elastic matrix with local yielding or debonding using quadratic influence superposition. J. Mech. Phys. Solids **44**, 1997–2039 (1996). https://doi.org/10.1016/S0022-5096(96)00068-3
17. Montgomery, D.C.: Design and Analysis of Experiments, 8th edn. (2012). https://doi.org/10.1198/tech.2006.s372
18. Box, G.E.P., Hunter, J.S., Hunter, W.G.: Statistics for Experilllenters (2005)
19. Blondet, G., Duigou, J.Le, Boudaoud, N., Eynard, B.: An ontology for numerical design of experiments processes. Comput. Ind. **94**, 26–40 (2018). https://doi.org/10.1016/j.compind.2017.09.005
20. Colbourn, C.C., Dinitz, J.H.: Handbook of Combinatorial Designs, 2nd edn., 20065545, p. 1011 (2006). https://doi.org/10.1201/9781420010541
21. Derringer, G., Suich, R.: Simultaneous optimization of several response variables. J. Qual. Technol. **12**, 214–219 (1980). https://doi.org/10.1080/00224065.1980.11980968
22. BS EN ISO 527-5:2009: ISO 527-5:2009 Plastics — Determination of tensile properties. Part. 1, 527–1 (2009)
23. Sighinolfi, D.: Thermal expansion investigation on glasses used in ceramic industry. CFI Ceram. Forum Int. **87**, 33–35 (2010)
24. ISO, E.: ISO 527-1 - Plastics—Determination of tensile properties–Part 1: general principles (1996)
25. Leardi, R.: Experimental design in chemistry: a tutorial. Anal. Chim. Acta **652**, 161–172 (2009). https://doi.org/10.1016/j.aca.2009.06.015
26. Saravana Kumar, N., VinothKumar, G., VinothKumar, C., Prabhu, M.: Experimental investigation on mechanical behavior of E-Glass and S-Glass fiber reinforced with polyester resin **5**, 19–26 (2018)

27. Barker, R.E.: An approximate relation between elastic moduli and thermal expansivities. J. Appl. Phys. **34**, 107–116 (1963). https://doi.org/10.1063/1.1729049
28. Matykiewicz, D., Barczewski, M., Knapski, D., Skórczewska, K.: Hybrid effects of basalt fibers and basalt powder on thermomechanical properties of epoxy composites. Compos. Part B Eng. **125**, 157–164 (2017). https://doi.org/10.1016/j.compositesb.2017.05.060
29. Senthilvelan, S., Gnanamoorthy, R.: Damping characteristics of unreinforced, glass and carbon fiber reinforced nylon 6/6 spur gears. Polym. Test. **25**, 56–62 (2006). https://doi.org/10.1016/j.polymertesting.2005.09.005
30. Hadi, A.S., Ashton, J.N.: Measurement and theoretical modelling of the damping properties of a uni-directional glass/epoxy composite. Compos. Struct. **34**, 381–385 (1996). https://doi.org/10.1016/0263-8223(96)00005-0
31. Haddad, Y.M., Feng, J.: On the trade-off between damping and stiffness in the design of discontinuous fibre-reinforced composites. Compos. Part B Eng. **34**, 11–20 (2003). https://doi.org/10.1016/S1359-8368(02)00076-8
32. Wallenberger, F.T., Watson, J.C., Hong, L.: Glass fibers. In: ASM Handbook, ASM. International, Material Park, Ohio, vol. 21, pp. 27–34 (2001)

A Fiber Optic Strain Gage Sensor for Measuring Preload in Thick Composite Bolted Joints

Marannano Giuseppe[1(✉)] and Restivo Gaetano[2]

[1] Engineering Department,
University of Palermo, 90128 Viale delle Scienze, Palermo, Italy
giuseppe.marannano@unipa.it
[2] Mechanical Engineering Department, Santa Clara University,
500 El Camino Real, Santa Clara, CA 95053, USA
grestivo@scu.edu

Abstract. Mechanical fastening is a popular choice in joining composites because of the ability to transfer high loads and the ease of assembly and disassembly. In this study, the failure behavior of composite–aluminum single-lap bolted joint is investigated. In particular, the effects of varying the preload on the bolt are examined. In order to accurately measure the preload, a specialized sensor that uses a fiber Bragg grating sensor embedded in the bolt is proposed and created. This sensor is realized for the current tests but can be expanded to other applications. An experimental study of bolted single-lap joints varying the tightening torque value has been carried out and, in order to validate the experimental tests, several finite element analyses conducted in ANSYS environment with explicit solver are performed. Comparison between numerical results and experimental tests allows us to determine the effect of bolting parameters on the strength of joint and to verify the effectiveness of the use of the specialized preload sensor.

Keywords: Mechanical fastening · Composite materials · Preload sensor · Fiber Bragg sensor · Finite element analysis

1 Introduction

Laminated composite materials are used in greater frequency in all fields and they can provide lightness and strength advantages. The high strength and stiffness if compared to low weight contribute to design structures tailored to loads. After more than 50 years of basic research and development, composite joints are helping to considerably improve, even revolutionize, many technology and industrial sectors, such as energy and transportation. In fact, such type of joints is lighter and stronger than those of realized by using metallic adherents.

Among mechanical joints, bolted joints provide many advantages including the ability to be easily assembled and disassembled, and transfer high loads reliably. The advantages of composites can unfortunately be reduced due to the stress concentrations

C. Rizzi et al. (Eds.): ADM 2019, LNME, pp. 540–551, 2020.
https://doi.org/10.1007/978-3-030-31154-4_46

produced by drilling operation. Moreover, due to the anisotropy of composites, modeling the joints is much more complex than doing so with an isotropic material.

Many researchers [1–3] have studied both static and fatigue joint strengths by varying the bolted joint geometries. The aim of researches is to determine the effects of bolting parameters on the joint strength. Published works have mainly looked into the effects of preload or the amount of the load initially carried by the bolt. In fact, the bolted joint should be designed such that the clamp load is never overcome by the external tension forces acting to separate the joint. Once the two sides of the joint are separated, the bolt would be subjected to bending forces and this could lead to bolt failure or to the damage of the flanges.

Several studies have shown that the value of the preload could be related at the failure modes of bolted joints. Net tension failure, shear out or a combination of these failure modes may occur [1, 4, 5]. Increased tightening torque provides an improved connection between components, an increment of the stiffness of the overall joint [3–6] and a more gradual distribution of strains throughout the assembly. Moreover, experimental results have shown that under the same tightening torque, the stiffness increases with decreasing washer size [7]. Several authors studied the bearing failure of composite bolted joints with non-uniform bolt-to-washer clearance [8, 9]. The experimental results have shown that the relative location can affect the initial failure loads but not the ultimate failure loads. The measured failure loads were also used to validate the existing analytical models. Many authors used finite element analysis for studying composite bolted joints [10, 11]. These studies have shown that clearance and bolt preload have great influences on joint stress distributions around the hole and an optimized analysis scheme is desirable in order to obtain a comprehensive understanding on the effects of fasteners. Experimental results carried out on composite single-lap joint have moreover shown that the magnitudes of bearing strengths in bolted joints were strictly influenced by increasing value of applied preload as well as by ply sequence and fibers orientation. As a consequence, in order to accurately predict the behavior of the composite bolted joint, it is very important to evaluate the level of axial preload induced on the bolt.

Many manuals and textbooks propose several techniques for measuring preload. Among them, several require specialized tools [12], while some use more advanced techniques [13–16]. In order to determine the amount of preload on a joint, the simplest way is to use a measuring device to determine the elongation of the bolt. In this case, the preload can be easily calculated until the tensile elongation on the bolt is linearly related to the amount of load. The variance and inaccuracies associated with manual measurement errors have been found to be within a range of $\pm 5\%$. One of methods that allows to measure the preload is to realize a load transducer by means of a resistance strain gage (RSG) instrumented bolt. This involves creating a slot in the center of the bolt and then applying strain gages on the walls of the slot itself. These gauges are normally installed in a full bridge setup. This configuration removes variations due to bending moments and, also, an average of the strain can be found producing more accurate readings. The averaged strain can then be found to linearly vary with the application of preload. The main disadvantage is that this method removes a large amount of material reducing the resistant section of the bolt.

An additional method makes use of an instrumented load cell in order to measure the deformation of the transducer. These load transducers produce a linear relationship between the amount of deformation and the load but, unfortunately, these load cells are often larger adding weight and altering the mechanics of the joint. Once calibrated, both the load cells and RSG instrumented bolts can accurately measure the preload within 1 to 2%.

Recently, the use of ultrasonic velocity measurement is applied to determine the preload in a joint [13, 14]. This method uses a bolt with machined ends in order to allow accurate contacts. The ultrasonic wave is sent down the length of the bolt and its time of flight (TOF) is measured. After, the TOF for the loaded bolt is measured. The change in time of flight is then used to calculate the change in length. Because normal TOF in ultrasonic preload measurements is very susceptible to noise, the phase detection technique is used to upgrade normal TOF measurement. Like normal ultrasonic measurement, this technique uses TOF in order to determine the elongation of the bolt. Moreover, in order to produce a more accurate measure, this technique allows to measure the change in resonance. The effectiveness of this technique is confirmed by the good linear correlation of the experimental results. The main advantage of ultrasonic velocity measurement is that it does not require precisely machined bolts decreasing the cost.

In order to determine the bolt preload, a more recent technique uses 3D Electronic Speckle Pattern Interferometry (ESPI) [15]. ESPI is a full field technique that measures accurate displacements in all three directions. Low end values produce a certain amount of scatter resulting in less accuracy. Above 15 kN however, accuracy increases to ±10%. Advantage to this method is that it requires no specialized fasteners making it an effective tool that could be used in real time. The disadvantage is the calibration of all assembly materials.

Another full field method for use with measuring preload is Automatic Digital Image Correlation (ADIC) [16–18]. ADIC uses normal DIC with an automatic algorithm that measures deformation on washers. This algorithm first calculates the average strain in the washer axis direction and automatically converts the strain to a preload. Because the relation is again linear the program can continuously measure the preload as the strains are found. The performed experiments have shown that the maximum discrepancy was about ±4%. Much like ESPI, using ADIC does not require contact with the joint, but measures displacements in the washer instead of displacements on the surface. This means that once a certain type of washer is prepped for ADIC it can be used with different materials.

In this study, a fiber Bragg grating sensor (FBG) [19–21] is used. The design of the instrumented bolt consists of embedding a FBG gage in the bolt. This technique has considerable advantages compared to the techniques described previously.

2 Instrumentation of the Bolt

In the current study, a M12 grade 8.8 bolt is used. A longitudinal cut (width equal to 0.5 mm) up to axis of the bolt is realized by means of an electrical discharge machine (EDM). The cutting operation provides an extremely thin slot as seen in Fig. 1.

Fig. 1. Longitudinal EDM cut realized on the bolt

Instrumented bolts that use resistance strain gages require more material to be removed. In fact, this method creates weaker bolts that are not able to withstand the same loads. The studied solution removes much less material and, therefore, the FBG instrumented bolt will have better strength properties compared to a RSG instrumented bolt.

The EDM cut slot provides ample room allowing to fit a 0.1 mm diameter FBG. The FBG used in this study is specifically customized for this application. The gauge length is equal to 10 mm and the fiber also has a protective coating. The FBG, positioned so that it coincides with the axis of the bolt, is positioned by means of a specialized installation jig (Fig. 2a) made of extruded aluminum alloy.

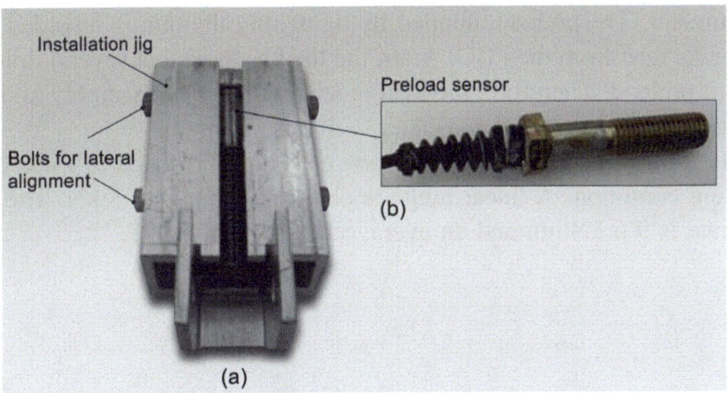

Fig. 2. (a) Installation jig; (b) FBG sensor

Four bolts allow the lateral alignment of the instrumented bolt. A slot on the top of the installation jig is created in order to allow at the FBG to be positioned into the bolt. Once aligned, the FBG is bonded with a cyanoacrylate adhesive. To ensure a strong bond, the slot of the bolt is previously cleaned with acetone. Finally, the clean, degreased, and etched surface is neutralized (treated with a basic solution) to give it the proper chemical affinity for the adhesive. It is also determined that the fragile intersection point between the FBG and bolt head needed to be protected. Heat shrink tubing and a strain relief connector commonly used in electrical connections are then applied in proximity to the bolt head. The tubing is shrunk onto the end to provide protection and to limit the bending of the cable during bolt preload. The completed sensor is shown in Fig. 2(b).

In order to calibrate the FBG sensor, a load cell transducer is used. The calibration and the experimental tests are conducted under controlled laboratory conditions. The first step in calibrating the sensor is assembling the joint with a load cell as shown in Fig. 3(a).

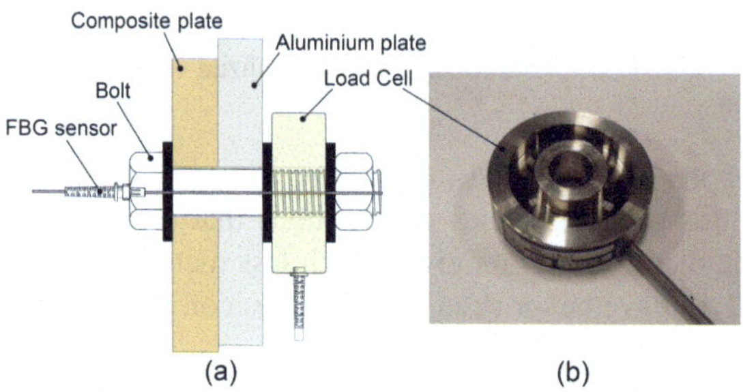

Fig. 3. (a) Setup for the preload sensor calibration; (b) load cell transducer

The accurate determination of the clamping load is performed by using a THC-1K-T5 kN through-the-hole type load cell transducer (Fig. 3b). The transducer is used in combination with a Hewlett-Packard 6200B DC power supply and a Hewlett-Packard 3468A multimeter. The preload, applied by tightening the nut, is recorded along with the wavelengths read from the FBG. After, the load is applied at several different levels up to 4 kN of preload. Therefore, the load is released and the assembly allowed to rest before additional calibrations.

The results from the calibration are shown in Fig. 4, related to three consecutive sets of loading condition. A linear range is observed from 0 to 3 kN. The average of calibration rate is 9.6 kN/nm and an average R^2 value of 0.997.

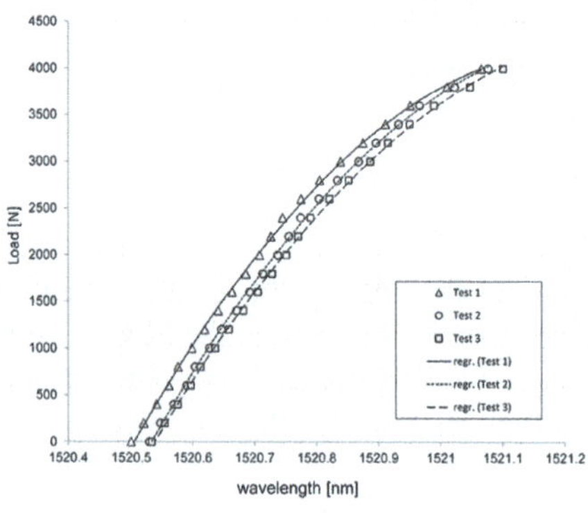

Fig. 4. Calibration of the preload sensor

3 Experimental Tests

In the present work, an experimental study of bolted single-lap joints varying the tightening torque value is carried out [22]. For destructive testing, nine specimens instrumented with Braggs fibers are tested in order to provide statistically significant results. Sample dimensions of the plates to be bolted are shown in Fig. 5(a). The thickness of plates to be joined is the same and equal to 12.5 mm.

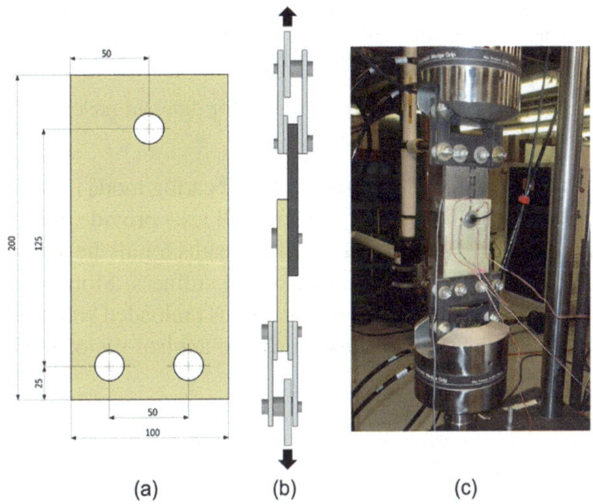

Fig. 5. (a) Plate dimensions and (b, c) test configuration for single-lap joint tests

The materials used for the two plates forming the bolted joint are 6082-T6 aluminum alloy [23, 24] and E-glass/epoxy composite panel, manufactured by hand lay-up and vacuum-bag techniques. The composite panel is cured for 1 h at 80 °C.

In order to obtain the desired thickness, sixty 0/90 woven fabric of E-glass/epoxy are used [25]. The final fiber volume fraction is calculated to be around 42%.

The values of preload are chosen based on the bolt yield characterization performed during the calibration phase. The preloads are set to 500 N, 2 kN, and 3.5 kN. The 500 N preload corresponds to a low amount of preload, 2 kN is in the middle of the values in the linear range of the curve in Fig. 4 while a preload of 3.5 kN is related to the start of nonlinear phase. In order to reduce the bonding effects due to the high thickness of the adherents, a load fixture for a single-lap joint configuration test is realized (see Fig. 5b, c). The specimens are centered by means of spacers and fixed by using several bolts.

Tensile tests are performed on a MTS 810 servo-hydraulic testing machine [26] with a 100 kN load cell (see Fig. 5c) and crosshead speed of 1 mm/min. During testing, the load on the fastener is monitored through the preload sensor (Fig. 3). Figure 6 shows the load-displacement curves for all experimental tests.

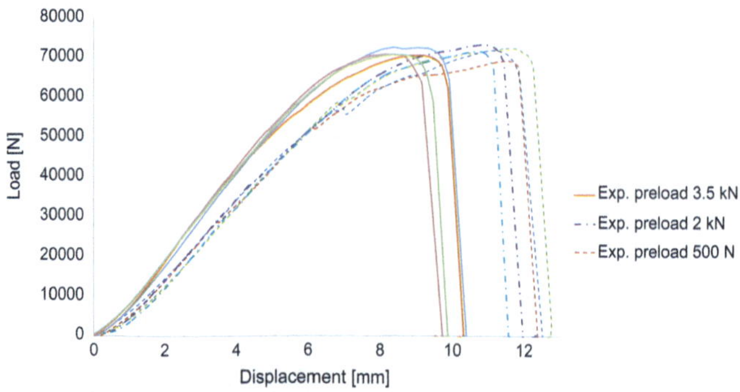

Fig. 6. Load *vs.* displacement results for different preload tests

All of the tests show that joint failure occurs in bearing mode [1, 4–7]. For each test, the load is monitored up to the yield of the bolt. All tests provide the same value of the ultimate load, approximately equal to 72 kN. The results for all the tests conducted using a preload equal to 3.5 kN show a closely grouped stiffness. Moreover, the stiffness is also noticeably higher compared to that of specimens preloaded with medium tightening torque value. Figure 7 shows test results for all the applied preload values.

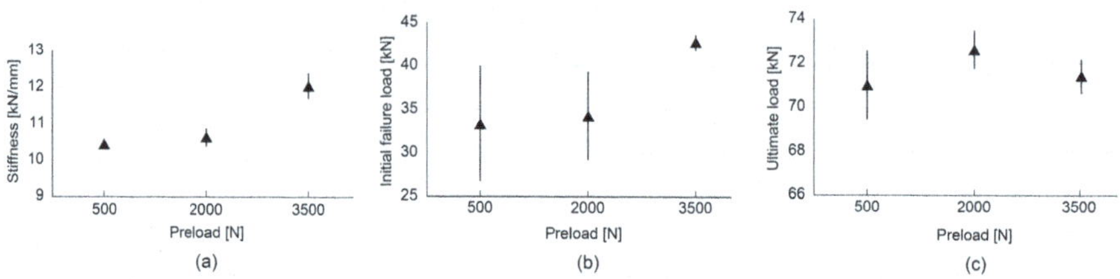

Fig. 7. Stiffness values (a), initial failure values (b) and ultimate failure values (c) plotted for different preload values

In particular Fig. 7 shows, respectively, the stiffness values (Fig. 7a), the initial failure load (Fig. 7b), the ultimate load of the joints (Fig. 7c). From Fig. 7(a) and (b) it is possible to observe that the stiffness and the initial failure load of the joints increase as the preload value increases. In the contrary, the ultimate load corresponding to the joint failure is less influenced by the amount of applied preload as it is influenced by the residual bearing strength of the composite.

4 Numerical Analysis

For an accurate determination of the effects of preload on the strength of the joint, several finite element analyses [27–29] are conducted in ANSYS APDL environment with explicit solver. The numerical study is performed in order to simulate the full damage process, up to the ultimate strength of the single-lap joint. In order to accurately predict the physical behavior of the joint, a detailed three-dimensional model is developed by parametric definition of all considered volumes and it is discretized by means of 8-node SOLID164 elements.

Figure 8(a) shows a mesh detail of the middle section of the joint, while Fig. 8(b) shows the mesh detail around the hole of the aluminum and composite plate.

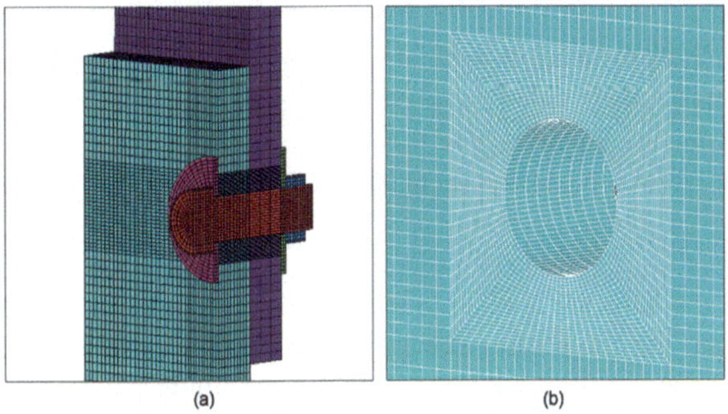

Fig. 8. (a) Mesh detail on the middle section of the joint; (b) mesh detail around the hole of the adherents

In particular, each part of the bolted joint elements is independently meshed. In order to adequately simulate the complex interactions between the volume parts constituting the single-lap joint, three different types of contact elements are used (Table 1).

Table 1. Contact elements used in the numerical analysis

Nodes to surface (NTS)	Surface to surface (STS)	Eroding surface to surface (ESTS)
Nut – bolt shank Washers – bolt shank	Composite plate - aluminum plate Nut - washer Bolt head - washer Washer - composite plate Washer - aluminum plate	Bolt shank - composite plate

Particular attention is taken in the simulation of the preloading phase. The aim is to model the load transfer between nut and bolt in a way that closely resembles the actual load exchange between the threads. To achieve this, the total preload is divided by the number of contact points and the obtained value is assigned to the internal nodes of the nut and to the external nodes of the bolt shank. Contact elements are introduced only to avoid interpenetrations and to transfer the load. Table 2 shows the orthotropic properties of the composite material.

Table 2. Orthotropic material properties and composite damage parameters

Density [kg/m^3]	1900
E_x, E_y [MPa]	38600
E_z [MPa]	7270
ν_{xy}, ν_{xz}	0.18
ν_{yz}	0.3
G_{xy}, G_{yz} [MPa]	4120
G_{xz} [MPa]	2040
KFAIL (Bulk modulus of material in compressive failure) [MPa]	830
S_{12} (In-plane shear failure stress) [MPa]	72
S_1 (longitudinal and transverse tensile strength) [MPa]	762
C_2 (transverse compressive strength) [MPa]	118

Loads and constraints are applied to the model in accordance with the experimental setup. In particular, the composite plate is fixed at the lower end and a moving speed of 1 mm/min is imposed to the upper nodes of the aluminum plate.

In accordance with the experimental tests, three numerical analyses are performed by varying the preload values. Figure 9 shows that the load-displacement curves obtained from FE analyses are in good agreement with those obtained experimentally.

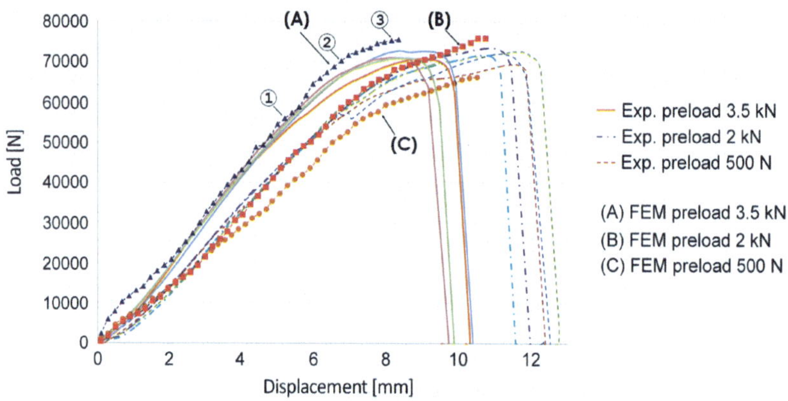

Fig. 9. Comparison between numerical results and experimental tests

It is possible to observe that increasing the preload value the maximum load at failure and the overall stiffness obtained from experimental tests are close to those obtained by numerical analyses.

The numerical results show that the damage increases in the composite plate up to failure. In particular, the damage starts on the edge of the hole in correspondence of the interface between the aluminum and composite plate, on the centerline of the joint, and it propagates in the axial direction subsequently affecting the entire composite thickness. Figure 10 shows a comparison between the damaged area of the composite laminate determined experimentally and that obtained by numerical analyses. The three images of Fig. 10(a) are also highlighted in the graph of Fig. 9 (points numbered from 1 to 3), relative to a preload value equal to 3.5 kN.

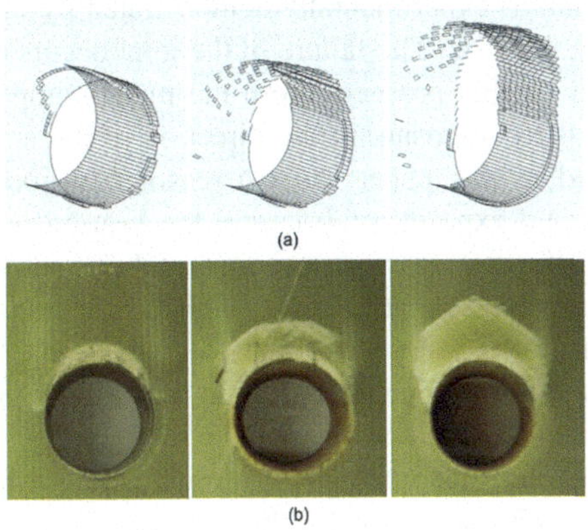

Fig. 10. Comparison of damaged area of the composite material. (a) FE analysis; (b) images of the specimen surface

The three images of Fig. 10(b) are recorded by stopping the experimental tests on three different crosshead position and photographing the extent of the damage area. The sizes of the damaged area experimentally measured are in good agreement with those obtained by numerical analyses.

5 Conclusions

In the paper, a specialized sensor for measuring preload is proposed and created. A fiber Bragg grating sensor (FBG) is bonded with a high strength epoxy adhesive inside the bolt, in a thin slot created by means of EDM technique. The proposed method, although it presents some experimental limitations due to the fragility, positioning and alignment of the fiber optic, presents the same advantages that could be obtained using electrical resistance strain gauges. It also has the advantage of keeping practically intact the bolt strength characteristics. The studied instrumented bolt is used in several experimental tests that are carried out on a single-lap bolted joint configuration. The specimens are obtained joining together a thick E-glass/epoxy composite laminate with a 6082 T6 aluminum plate, by varying the magnitude of the bolt preload. From these tests it is possible to observe that the preload improves stiffness of the joint and also increases the initial failure load of the bolted specimens. A high value of preload allows us to obtain high stiffness of the joint but, at the same time, it does not provide analogues enhancement of joint strength.

In order to verify the experimental results, several numerical analyses are conducted in ANSYS APDL environment with explicit solver. The load-displacement curves obtained by finite elements analyses show a good correlation with the experimental results and they highlight that the level of preload used for tightening the bolt provides an important contribution in the evaluation of the stiffness of the bolted joint.

The numerical model also allows to accurately predict the damaged area evolution in the composite material, as experimentally demonstrated by the comparison with the damaged surface during the tests. The failure of the joint occurs on the edge of the hole with a progressive damage that proceeds from the metal-composite interface towards the external surface of the composite, in the direction of the applied load.

The results obtained in this paper show a reasonably good correlation between finite-element analyses and experimental data in the bearing region of the composite laminate. The largest difference between experimental and numerical results is about 10%. In conclusion, in order to accurately predict by finite element method, the behavior of the bolted joint is very important to evaluate the level of axial preload induced on the bolt. Optical strain gauge based on fiber Bragg gratings is a solution of simple construction, easy calibration and sufficient accuracy. The use of the FBG instrumented bolt is therefore recommended in applications where the knowledge of the level of preload is a relevant factor for the accuracy of the results.

References

1. Kapti, S., Sayman, O., Ozen, M., Benli, S.: Experimental and numerical failure analysis of carbon/epoxy laminated composite joints under different conditions. Mater. Des. **31**, 4933–4942 (2010)
2. Cooper, C., Turvey, G.J.: Effects of joint geometry and bolt torque on the structural performance of single bolt tension joints in pultruded GRP sheet material. Compos. Struct. **2** (1–4), 217–226 (1995)
3. Choi, J., Chun, Y.: Failure load prediction of mechanically fastened composite joints. J. Compos. Mater. **37**(24), 2163–2177 (2003)
4. Liu, L., Zhang, J., Chen, K., Wang, H.: Combined and interactive effects of interference fit and preloads on composite joints. Chinese J. Aeronaut. **27**(3), 716–729 (2014)
5. Sen, F., Sayman, O., Ozcan, R., Siyahkoc, R.: Failure response of single bolted composite joints under various preload. Indian J. Eng. Mater. Sci. **17**, 39–48 (2010)
6. Minguez, J.M., Vogwell, J.: Effect of torque tightening on the fatigue strength of bolted joints. Eng. Fail. Anal. **13**, 1410–1421 (2006)
7. Khashaba, U.A., Sallam, H.E.M., Al-Shorbagy, A.E., Seif, M.A.: Effect of washer size and tightening torque on the performance of bolted joints in composite structures. Compos. Struct. **73**, 310–317 (2006)
8. Tong, L.: Bearing failure of composite bolted joints with non-uniform bolt-to-washer clearance. Compos. Part A Appl. Sci. **31**, 609–615 (2000)
9. Kelly, G., Hallström, S.: Bearing strength of carbon fibre/epoxy laminates: effects of bolt-hole clearance. Compos. Part B Eng. **35**(4), 331–343 (2004)
10. Camanho, P., Matthews, F.: Stress analysis and strength prediction of mechanically fastened joints in FRP: a review. Compos. Part A Appl. Sci. Manuf. **28**(6), 529–547 (1997)
11. Restivo, G., Marannano, G., Isaicu, G.A.: Three-dimensional strain analysis of single-lap bolted joints in thick composites using fibre-optic gauges and the finite-element method. J. Strain Anal. Eng. Des. **45**(7), 523–534 (2010)

12. Bickford, J.H.: An Introduction to the Design and Behavior of Bolted Joints. Marcel Dekker inc, New York (1995)
13. Joshi, S.G., Pathare, R.G.: Ultrasonic instrument for measuring bolt stress. Ultrasonics **22**(6), 261–269 (1984)
14. Jhang, K.Y., Quan, H.H., Ha, J., Kim, N.Y.: Estimation of clamping force in high-tension bolts through ultrasonic velocity measurement. Ultrasonics **44**, 1339–1342 (2006)
15. Nassar, S.A., Meng, A.: Optical monitoring of bolt tightening using 3D electronic speckle pattern interferometry. J. Press. Vessel Technol. **129**(1), 89–95 (2007)
16. Huang, Y.H., Liu, L., Yeung, T.W., Hung, Y.Y.: Real-time monitoring of clamping force of a bolted joint by use of automatic digital image correlation. Opt. Laser Technol. **41**(4), 408–414 (2009)
17. Ingrassia, T., Nalbone, L., Nigrelli, V., Ricotta, V., Pisciotta, D.: Biomechanical analysis of the humeral tray positioning in reverse shoulder arthroplasty design. Int. J. Interact. Des. Manuf. **12**(2), 651–661 (2018)
18. Ingrassia, T., Nigrelli, V., Ricotta, V., Nalbone, L., D'Arienzo, A., D'Arienzo, M., Porcellini, G.: A new method to evaluate the influence of the glenosphere positioning on stability and range of motion of a reverse shoulder prosthesis. Injury **50**, S12–S17 (2019)
19. Melle, S.M., Liu, K., Measures, R.M.: Practical fiber-optic Bragg grating strain gauge system. Appl. Opt. **32**, 3601–3609 (1993)
20. Yeager, M., Whitaker, A., Todd, M.: A method for monitoring bolt torque in a composite connection using an embedded fiber Bragg grating sensor. J. Intell. Mater. Syst. Struct. **29**(3), 335–344 (2018)
21. Woodruff, J., Marannano, G., Restivo, G.: Effects of clearance on thick, single-lap bolted joints using through-the-thickness measuring techniques. In: SEM Annual Conference and Exposition on Experimental and Applied Mechanics 2010, Indianapolis, Indiana, USA, pp. 128–136 (2010)
22. Fragapane, S., Giallanza, A., Cannizzaro, L., Pasta, A., Marannano, G.: Experimental and numerical analysis of aluminum-aluminum bolted joints subject to an indentation process. Int. J. Fatigue **80**, 332–340 (2015)
23. Marannano, G., Pasta, A., Parrinello, F., Giallanza, A.: Effect of the indentation process on fatigue life of drilled specimens. J. Mech. Sci. Technol. **29**(7), 2847–2856 (2015)
24. Marannano, G., Parrinello, F., Giallanza, A.: Effects of the indentation process on fatigue life of drilled specimens: optimization of the distance between adjacent holes. J. Mech. Sci. Technol. **30**(3), 1119–1127 (2016)
25. Marannano, G., Mariotti, G.V.: Structural optimization and experimental analysis of composite material panels for naval use. Meccanica **43**(2), 251–262 (2008)
26. Ingrassia, T., Lombardo, B., Nigrelli, V., Ricotta, V., Nalbone, L., D'Arienzo, A., D'Arienzo, M., Porcellini, G.: Influence of sutures configuration on the strength of tendon-patch joints for rotator cuff tears treatment. Injury **50**(2), S18–S23 (2019)
27. Cirello, A., Cucinotta, F., Ingrassia, T., Nigrelli, V., Sfravara, F.: Fluid–structure interaction of downwind sails: a new computational method. J. Mar. Sci. Technol. **24**(1), 86–97 (2019)
28. Mirulla, A.I., Bragonzoni, L., Zaffagnini, S., Bontempi, M., Nigrelli, V., Ingrassia, T.: Virtual simulation of an osseointegrated trans-humeral prosthesis: A falling scenario. Injury **49**(4), 784–791 (2018)
29. Mancuso, A., Pitarresi, G., Tumino, D.: Using FEM simulation to predict structural performances of a sailing dinghy. Int. J. Interact. Des. Manuf. **12**(3), 811–822 (2018)

How to Classify Compliant Mechanisms

Davide Russo(✉) and Antonio Caputi

DIGIP Department, University of Bergamo, 24044 Dalmine, Italy
davide.russo@unibg.it

Abstract. The design of compliant mechanisms is completely different from the design of all other mechanisms. It is assumed that traditional mechanisms are rigid or that they use rigid parts connected by mobile joints. As the name suggests, the compliant mechanisms are not rigid, but flexible; they exploit deformation to their advantage, thus also eliminating problems due to friction and play. An analysis of the state of the art of the scientific literature on compliant mechanisms revealed a chaotic proliferation of typologies, which differ in type, shape, material, size, application, etc. And the lack of a unique and complete classification. This problem has been addressed by proposing a series of classifications, designed according to the different inputs that can be had in the early stages of the design. These include a functional classification based on FBS ontology, used to define the state of the art and identify areas of investigation that are still unexplored. Finally the classification by degrees of freedom, rotations and translations allowed, which led to a system of organization of compliants families in the form of a matrix. The matrix represents a first step towards the future possibility of integrating this creation into a tool for modeling and optimizing the design by compliants. In this article an example of how to organize the emerging Lamina torsion mechanisms (LET) and the rotational joints in the matrix is proposed.

Keywords: Compliant mechanism · FBS · LET · Rotational joint

1 Introduction

A mechanism, as is known, is a mechanical device capable of transferring or transforming pairs, forces or energy.

The traditional mechanisms realize this aspect through the presence of rigid parts connected to each other through mobile joints, which allow their movement.

The compliant mechanisms perform the same function as the mechanisms, with the difference that they consist of a single body and owe their mobility to the bending of the members of which they are made, rather than to the joints [1].

The idea of compliant is intrinsic to the behavior of objects, but above all suggested by nature. Many living things use this behavior to perform the most common vital

C. Rizzi et al. (Eds.): ADM 2019, LNME, pp. 552–564, 2020.
https://doi.org/10.1007/978-3-030-31154-4_47

functions, such as the heartbeat. Heart valves can be considered as unidirectional compliant valve systems capable of sustaining millions of cycles by flexing their own fibers [2].

Among the many advantages of this type of design, the most important are: the reduction of costs as they consist of a single body and a single material, the increase in potential performance thanks to the absence of joints, the accumulation of energy that can be obtained by preloading the flexible parts and miniaturization [3]. The disadvantages, on the other hand, concern the difficulty of analysis and design, the movements of limited amplitude, or the resistance to bending of the conforming elements and the resistance to loads and temperatures [4, 5] and the low resistance to fatigue.

From the analysis of the state of the art of scientific literature on compliant mechanisms a growing interest has emerged in heterogeneous fields of application with a very wide range of use, from medical to war, involving components and mechanisms from nanoscale to macro scales [3]. However, this analysis also reveals the lack of a unique and comprehensive classification. This problem was addressed by proposing a classification based on the relationships between structure, function and behavior according to the FBS (function behavior system) methodology and other classifications based on granted movements and applied loads.

By breaking down and rearranging according to different classifications the most well-known mechanisms present in literature, maps have been produced that represent the starting point of a wider and more complete cataloging work. The results of this first screening has already led to the identification of investigations still unexplored on which there are no answers to any design requests.

The structure of the article is based on different criteria for classifying the mechanisms, which work at different levels of detail to meet the different needs of a designer. Section 2 collects functional classifications, such as an FBS-based organization. In Sect. 3 the FBS classification is used to create a conceptual map of the state of the art in terms of scientific publications in order to visualize how the different research areas of this topic and research opportunities are distributed. Section 4 proposes a more targeted classification, created starting from the number of degrees of freedom, the rotations and translations allowed. As previously the new organization follows an application tool (introduced in Sect. 5), in the form of a matrix and capable of hosting the different families of compliants, according to a more operational design logic. To demonstrate the potential of the instrument, the matrix was populated with LET mechanisms and rotational joints. Finally, the conclusions of the work.

2 Compliant Mechanisms Classification According to FBS

The first of the proposed classifications is based on a method of representing conceptual design, called FBS [6]. According to the FBS ontology, a system can be decomposed into function (F), behaviour (B), and structure (S). Practical definitions are proposed as it follows:

- Function (F): the function (F) of a technical system is the motivation/purpose of its existence, (i.e. what it is for). It should represent the designer's intention, given as the requirements that compliants must have. In this paper we classify mechanisms according to the following purpose: energy storage, bi-stable mechanisms, force multipliers and motions, transmission of constant force.
- Behaviour (B): it is a sequential change of states, what the system does to achieve the purpose expressed by the function (F). The behavioral level is based on the network of alternative behaviors (B) all deriving from the same functional concept. Our (B) level is built starting from the identification of the system function (F) and generating all possible movements/deformations by which it is possible to achieve the design purpose defined by the function (F). There are mechanisms designed in the plane that deform as a result of a stress in the same plane, or mechanisms that, although three-dimensional, act in the same plane of the deformation: under this category we collect all the so-called planar systems. Instead, there are mechanisms that from a solicitation in a plane react with a deformation in other planes. Depending on the way in which they deform and the configuration they assume at the end of the solicitation, they are called lems, orthoplanar, metamorphic, spherical. Finally, there are origami inspired mechanisms that in resting configuration are formed by planes that open up creating structures in space.
- Structure (S): describes the components of the object and their relationships. All transformations provided by behaviors (Bs) in order to achieve the design task (F) are realized thanks to the system structure (S). Therefore it is necessary to classify the mechanisms according to the geometry of the element that allows flexibility. In fact, some perform this behavior thanks to a reduction in thickness, creating a notch with the most diverse shapes and sizes, others use thin sheets that function as bent beams or thin segments that work by mechanical torsion. Always in the structural classification criteria are material changes, sizing of lengths and all those design parameters aimed at modifying local stiffness that are the basis of the optimization process. An organic framework of how to represent and manage these parameters is still under construction.

All types of compliant mechanisms can be subdivided according to function and behavior as described in Table 1.

Table 1. Subdivision of the main types of compliant mechanisms present in the literature based on function and behavior

Functional Classification	
Mechanisms with energy storage. Compliant mechanism which, if deformed, is able to store elastic potential energy, which can be released later to bring the mechanism back to the starting configuration with less implementation requirement [2,4].	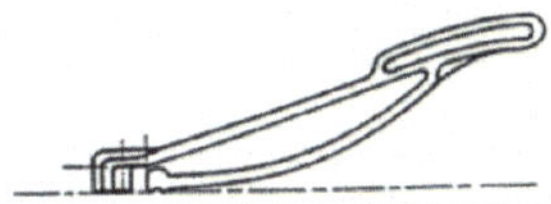
Bistable mechanisms. Used in applications where you want to reach two distinct positions of balance, keeping them without the use of energy inputs [3].	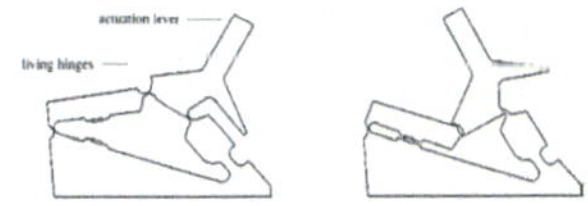
Force multipliers and motions. These effects can be achieved with zero clearance and high reliability, due to the distribution of mechanical stresses on the whole mechanism [3].	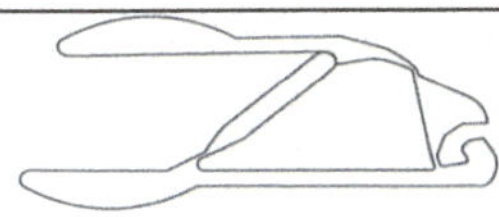
Mechanisms of transmission of constant force. Provide a constant output force for a wide range of input movements [4].	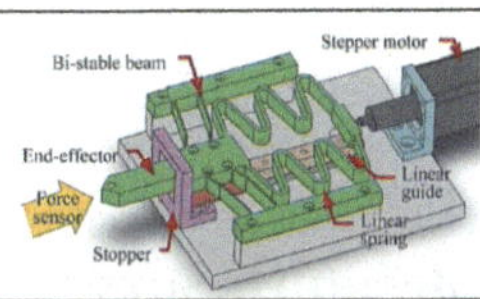
Behavioral Classification	
Contact-Aided Compliant Mechanisms. 2D or 3D structures able to confer movement to individual elements of which they are made, once a contact has occurred [7].	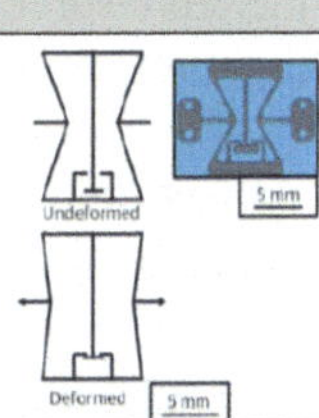
Lamina Emergent Mechanisms (LEM). 2D structures that have flexible segments arranged within a plane, whose peculiarity consists in the fact that the motion takes place outside the manufacturing plane [5,8].	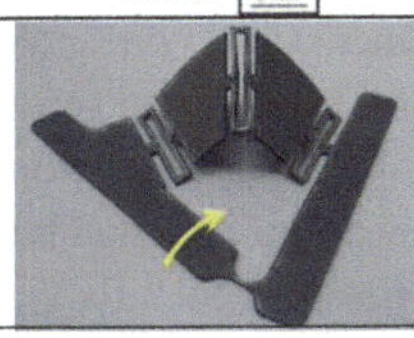
Ortho-planar mechanisms. 2D structures characterized by possessing all the connections simultaneously in a plane, while the motion occurs orthogonally to the plane itself [10].	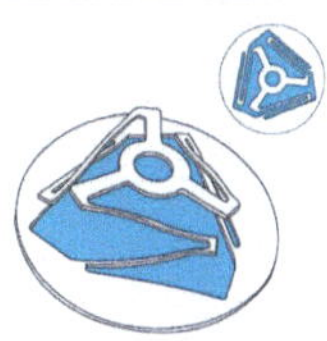
Planar mechanisms. 2D or 3D structures whose peculiarity is to keep the motion inside a plane [5,9].	

(*continued*)

Table 1. *(continued)*

Metamorphing mechanisms. 2D structures or Origami Inspired in which the actual number of connections and mobility change from one conformation to another [11].	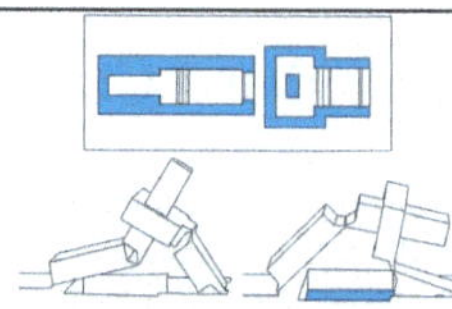
Spherical mechanism. Structures that combine the compactness and topological simplicity of LEMs with the three-dimensional motions of spherical mechanisms [10].	
Origami-Inspired compliant mechanism. Rigid-collapsible systems defined as mechanisms with variable stiffness and continuous transformable shape, and characterized by having regions separated by creases (or joints) that do not need to bend or twist during the bending motion [2,5].	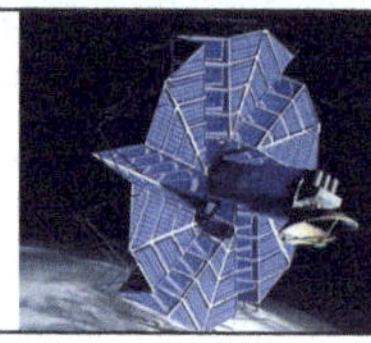

3 Literature Survey According to FBS Classification

A literature survey has been conducted on Scopus in order to estimate the coverage of each topic accord to this classification. The result of this survey has been proposed in form of an infographic conceptual map in Fig. 1. All relationships between the different compliants families are shown in a qualitative way, introducing radial graphs (Table 2).

These graphs, based on the recurrence in the bibliography, show which functions have been implemented for each behavior and function.

As we can see in Fig. 1, the most common conformal structure is the two-dimensional one, as it shows a greater behavioral versatility, while the three-dimensional and Origami Inspired ones are less studied. As for the functions, it is clear that the accumulation of energy is common to all the mechanisms, as a peculiarity of the compliant mechanisms, while to provide other functions there are only specific types able to satisfy this need.

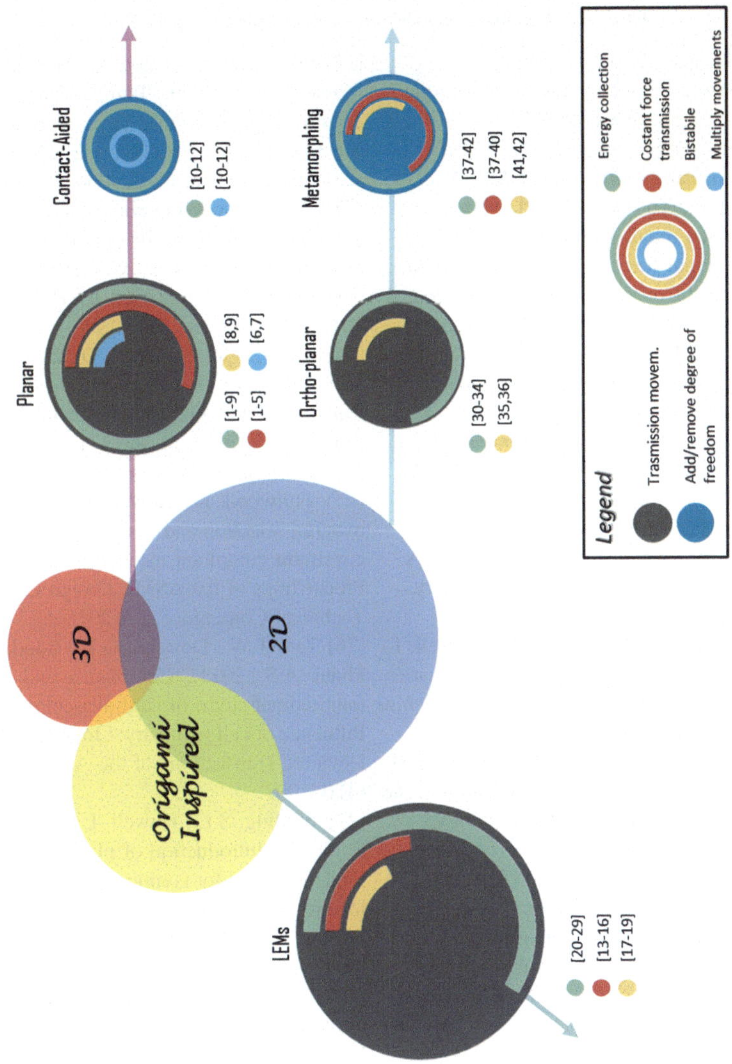

Fig. 1. Synthetic conceptual map representing the classification of compliant mechanisms [1–10].

Table 2. The list of papers inside conceptual map in Fig. 1

[1] Reddy, B.P., Zhou, H., 2017. "Synthesizing bidirectional constant torque compliant mechanisms." *ASME International Mechanical Engineering Congress and Exposition, Proceedings (IMECE)*, 4A-2017	[22] Jacobsen, J.O., Chen, G., Howell, L.L., Magleby, S.P., 2009. "Lamina Emergent Torsional (LET) Joint." Mechanism and Machine Theory, 44(11), pp. 2098-2109
[2] Wang, P., Xu, Q., 2017. "Design of a flexure-based micro-motion stage with constant output force." *Proceedings - IEEE International Conference on Robotics and Automation*, 7989343, pp. 2994-2999	[23] Liu, K., Shan, C.-C., Cao, Y., 2015. "Theoretical modeling and finite element analysis of planar tensile compliant joint." 2015 IEEE International Conference on Robotics and Biomimetics, IEEE-ROBIO 2015, 7419065, pp. 1991-1996
[3] Xu, Q., 2017. "Design of a constant-force microgripper mechanism for biological micromanipulation." *2017 IEEE 12th International Conference on Nano/Micro Engineered and Molecular Systems, NEMS 2017*, 8017055, pp. 418-421	[24] Xie, Z., Qiu, L., Yang, D., 2017. "Design and analysis of Outside-Deployed Lamina Emergent Joint (OD-LEJ)." Mechanism and Machine Theory, 114, pp. 111-124
[4] Wang, P., Xu, Q., 2018. "Design and modeling of constant-force mechanisms: A survey." Mechanism and Machine Theory, 119, pp. 1-21	[25] Sessions, J., Pehrson, N., Tolman, K., (…), Fullwood, D., Howell, L., 2016. "A material selection and design method for multi-constraint compliant mechanisms." Proceedings of the ASME Design Engineering Technical Conference, 5A-2016
[5] Tolman, K.A., Merriam, E.G., Howell, L. L., 2016. "Compliant constant-force linear-motion mechanism." Mechanism and Machine Theory, 106, pp. 68-79	[26] Tan, T.W., Douglas, G.R., Bond, T., Phani, A.S., 2011. "Compliance and longitudinal strain of cardiovascular stents: Influence of cell geometry." Journal of Medical Devices, Transactions of the ASME, 5 (4),041002
[6] Ouyang, P.R., Zhang, W.J., Gupta, M.M., 2008. "A new compliant mechanical amplifier based on a symmetric five-bar topology." *Journal of Mechanical Design, Transactions of the ASME*, 130(10), pp. 1045011-1045015	[27] Wilding, S.E., Howell, L.L., Magleby, S. P., 2012. "Introduction of planar compliant joints designed for combined bending and axial loading conditions in lamina emergent mechanisms." Mechanism and Machine Theory, 56, pp. 1-15
[7] Bharanidaran, R., Srikanth, S.A., 2016. "A new method for designing a compliant mechanism based displacement amplifier." *IOP Conference Series: Materials Science and Engineering*, 149(1),012129	[28] Chen, C., Lan, C., 2017. "An Accurate Force Regulation Mechanism for High-Speed Handling of Fragile Objects Using Pneumatic Grippers." IEEE Transactions on Automation Science and Engineering
[8] Stavenuiter, R.A.J., Birglen, L., Herder, J. L., 2017. "A planar underactuated grasper with adjustable compliance." Mechanism and Machine Theory 112, pp. 295-306	[29] Kuribayashi, K., Tsuchiya, K., You, Z., (…), Ito, T., Sasaki, M., 2006. "Self-deployable origami stent grafts as a biomedical application of Ni-rich TiNi shape memory alloy foil." Materials Science and Engineering A, 419(1-2), pp. 131-137
[9] Sarojini, D., Lassche, T.J., Herder, J.L., Ananthasuresh, G.K., 2016. "Statically balanced compliant two-port bistable mechanism." Mechanism and Machine Theory, 102, pp. 1-13	[30] Parise, J.J., Howell, L.L., Magleby, S.P., 2001. "Ortho-planar linear-motion springs." *Mechanism and Machine Theory*, 36(11-12), pp. 1281-1299
[10] Cirone, S.A., Hayes, G.R., Babcox, B.L., (…), Adair, J.H., Lesieutre, G.A., 2012. "Design of contact-aided compliant cellular mechanisms with curved walls." *Journal of Intelligent Material Systems and Structures*, 23 (16), pp. 1773-1785	

(*continued*)

Table 2. (*continued*)

[11] Aguirre, M.E., Frecker, M., 2011. "Design of a multi-contact-aided compliant mechanism." *Proceedings of the ASME Design Engineering Technical Conference* 6(PARTS A AND B), pp. 255-259	[31] Teichert, G.H., Jensen, B.D., 2013. "Design and fabrication of a fully-compliant mechanism for control of cellular injection arrays." *Production Engineering*, **7**(5), pp. 561-568
[12] Eastwood, K.W., Francis, P., Azimian, H., (…), Drake, J.M., Naguib, H.E., 2019. "Design of a contact-aided compliant notched-tube joint for surgical manipulation in confined workspaces." Journal of Mechanisms and Robotics, 10(1),015001	[32] Anderson, S., Jensen, B.D., 2009. "Viscoelastic damping of ortho-planar springs." *2008 Proceedings of the ASME International Design Engineering Technical Conferences and Computers and Information in Engineering Conference*, DETC 2008, 2 (PART A), pp. 323-330
[13] Jovanova, J., Frecker, M., 2017. "Two stage design of compliant mechanisms with superelastic compliant joints." ASME 2017 Conference on Smart Materials, Adaptive Structures and Intelligent Systems, SMASIS 2017, 2	[33] Qiu, C., Qi, P., Liu, H., Althoefer, K., Dai, J.S., 2016. "Six-Dimensional Compliance Analysis and Validation of Orthoplanar Springs." *Journal of Mechanical Design, Transactions of the ASME*, 138(4),042301
[14] Morgan, J., Magleby, S.P., Howell, L.L., 2016. "An approach to designing origami-adapted aerospace mechanisms." Journal of Mechanical Design, Transactions of the ASME, 138(5),052301	[34] Valentini, P.P., Pezzuti, E., 2016. "Computer-aided tolerance allocation of compliant ortho-planar spring mechanism." *International Journal of Computer Applications in Technology*, 53(4), pp. 369-374
[15] Balkcom, D.J., Mason, M.T., 2008. "Robotic origami folding." International Journal of Robotics Research, 27(5), pp. 613-627	[35] Carroll, D.W., Magleby, S.P., Howell, L. L., Todd, R.H., Lusk, C.P., 2005. "Simplified manufacturing through a metamorhic process for compliant ortho-planar mechanisms." *American Society of Mechanical Engineers, Design Engineering Division (Publication) DE*, 118 A(1), pp. 389-399
[16] Dai, J.S., Cannella, F., 2008. "Stiffness characteristics of carton folds for packaging." Journal of Mechanical Design, Transactions of the ASME, 130(2),022305	[36] Howell L. L., 2001. Compliant Mechanisms. Wiley, New York
[17] Alfattani, R., Lusk, C., 2017. "Design of a bistable origami reverse-fold using spherical kinematics." Proceedings of the ASME Design Engineering Technical Conference, 5B-2017	[37] Chen, Q., Lin, Q., 2013. "Analysis on metamorphic of the cable-driven parallel mechanism in wind tunnel test." *Applied Mechanics and Materials*, 336-338, pp. 1196-1203
[18] Dureisseix, D., 2012. "An overview of mechanisms and patterns with origami." International Journal of Space Structures, 27 (1), pp. 1-14	[38] Ding, X., Yang, Y., Lv, S., Dai, J., 2011. "Topology and kinematic performance analysis of Hoeken straight-line COPMM for micro-operation." *International Journal of Nanomanufacturing*, 7(5-6), pp. 544-558
[19] Huang, H., Deng, Z., Qi, X., Li, B., 2013. "Virtual chain approach for mobility analysis of multiloop deployable mechanisms." Journal of Mechanical Design, Transactions of the ASME, 135(11),111002	[39] Zirbel, S.A., Aten, Q.T., Easter, M., Jensen, B.D., Howell, L.L., 2012. "Compliant constant-force micro-mechanism for enabling dual-stage motion." *Proceedings of the ASME Design Engineering Technical Conference*, 4 (PARTS A AND B), pp. 191-198
[20] Xie, Z., Qiu, L., Yang, D., 2018. "Design and analysis of a variable stiffness Inside-Deployed Lamina Emergent Joint." Mechanism and Machine Theory 120, pp. 166-177	

(*continued*)

Table 2. (*continued*)

[21] Ku, J.S., 2017. "Folding thick materials using axially varying volume trimming." Proceedings of the ASME Design Engineering Technical Conference 5B-2017	[40] Li, D., Zhang, Z., Dai, J., Zhang, K., 2010. "Overview and prospects of metamorphic mechanism." *Jixie Gongcheng Xuebao/Journal of Mechanical Engineering*, 46(13), pp. 14-21 [41] Howell L. L., 2001. Compliant Mechanisms. Wiley, New York [42] Carroll, D.W., Magleby, S.P., Howell, L. L., Todd, R.H., Lusk, C.P., 2005. "Simplified manufacturing through a metamorhic process for compliant ortho-planar mechanisms." American Society of Mechanical Engineers, Design Engineering Division (Publication) DE, 118 A(1), pp. 389-399

4 Single Compliants Family Classification

The functional classification is not the only one possible. It is useful for determining the macro type of compliants but it is much less effective when choosing a device within the same family. For this purpose, those classifications that describe in more detail how compliants behave are much more interesting. In literature there many different alternatives. One example is based on the degrees of freedom and free movements [13], as shown in Table 3 where rotational joints family is divided in three macro classes dealing with the number of degrees of freedom. Although very effective in terms of classification, the major limitation of this classification is that the degree of freedom to choose a device is not enough.

To improve this limit in Table 4 Ivanov [14] proposes a dual-segment joint geometries classification according to the type of stress and applied load. Choosing among bending, torsion, tension, compression, we are more in line with the designer's way of thinking.

The two classifications have been chosen among the many present in the literature because both are useful even if they are not sufficient on their own to collect the entire range of compliants families. The idea behind this article is to integrate these two classifications and develop both criteria in order to create a new and more versatile one.

Table 3. Shows a classification of rotational joints

	**No translation allowed but only rotations**	
1degree of freedom	**Compliant Revolute Joints** These joints serve to create a purely rotary movement that is a bending around a single axis	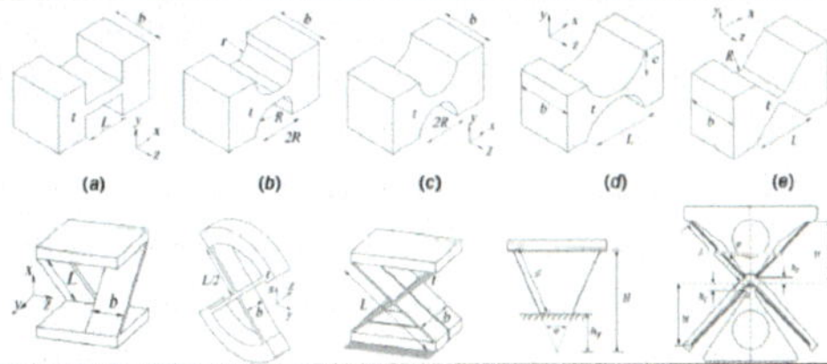
2 degrees of freedom	**Compliant Universal Joints** They are characterized by two rotational degrees of freedom and can therefore rotate around two axes	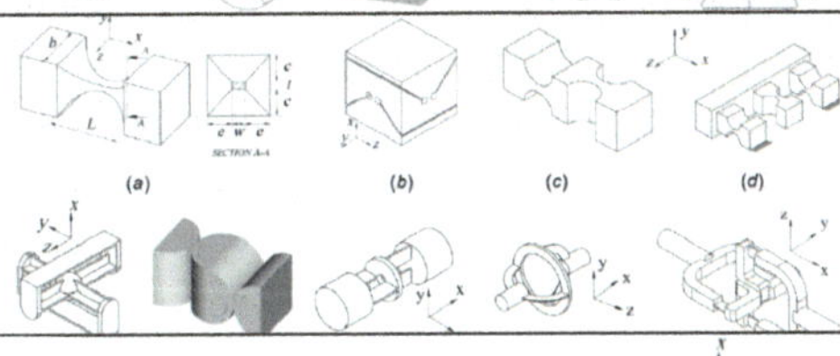
3 degrees of freedom	**Compliant Spherical Joints** They are characterized by three rotational degrees of freedom	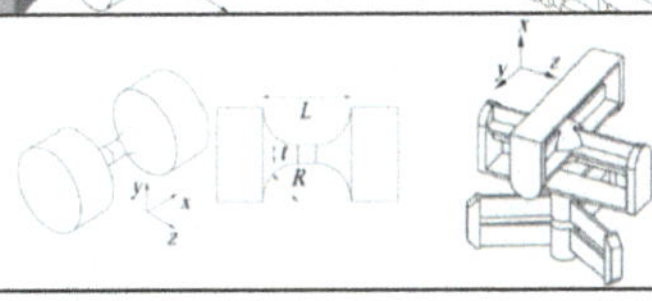

Table 4. Dual-segment joint geometries classified by Ivanov [14] according to the type of stress, rotation angle and applied load.

			Flexural Loading when dual-segment joint in		
Name	**φ**	**Geometry**	**Folding**	**Tension**	**Compression**
Bending-Orthogonal	0	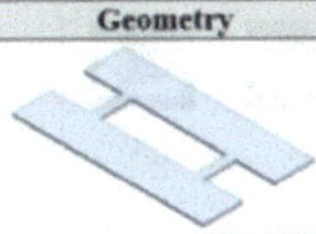	Bending	Tension	Compression
Mixed Tension Resistant	0-90	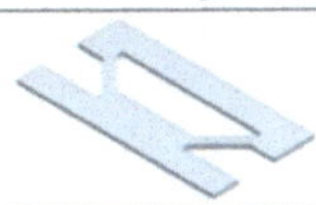	Mixed Torsion, Bending, and Tension	Mixed Tension and Bending	Mixed Compression and Bending
Torsion-Parallel	90	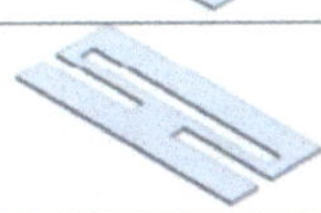	Torsion	Fixed-Clamped Bending	Fixed-Clamped Bending
Mixed Compression Resistant	90-180		Mixed Torsion, Bending, and Tension	Mixed compression and bending	Mixed tension and bending
Inverted Bending-Orthogonal	180	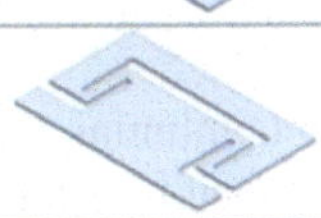	Bending	Compression	Tension

5 Proposal for an Universal Compliants Family Classification

The previous criteria for the organization of the structures can be combined together to create a single organizational framework capable of positioning entire families or individual components in an orderly and univocal manner. A matrix, shown in Fig. 2, contains as many lines as there are allowed primary movements (rotation and translation). They are produced according to the applied load (only flexion, lateral bending, torsion and all other combinations). Moreover each cell is characterized by a different color indicating the number of degrees allowed.

In this way the designer has an additional choice mode designed according to his constraints and the project specifications. To give a practical example of the usefulness of this tool we put representatives of Let Joints and Rotational Joints families present in the literature within the matrix.

As can easily be seen in Fig. 2, it appears to be incomplete in terms of content, as there is a vast presence of unoccupied cells. This means that many behavioral combinations have not been investigated, and to obtain a complete degree of filling, these unexplored spaces must be filled.

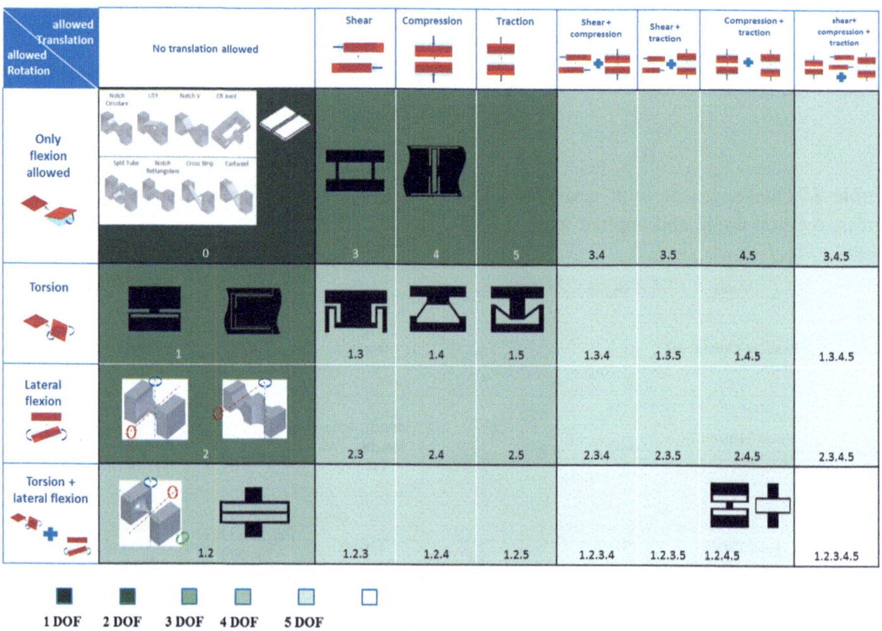

Fig. 2. Proposal for an universal compliants family classification

A future work could trivially consists in combining the existing LET joints in order to find the missing geometries. In fact if we want to create a new mechanism for filling an empty space cell we can hybridize already existent geometries. For example, taking into account cell 1.3.4, that is a torsional rotation, allowing a shear + compression displacement, we can combine one topology present in the shear-only column and in the torsion-only row. The result is visible in Fig. 3.

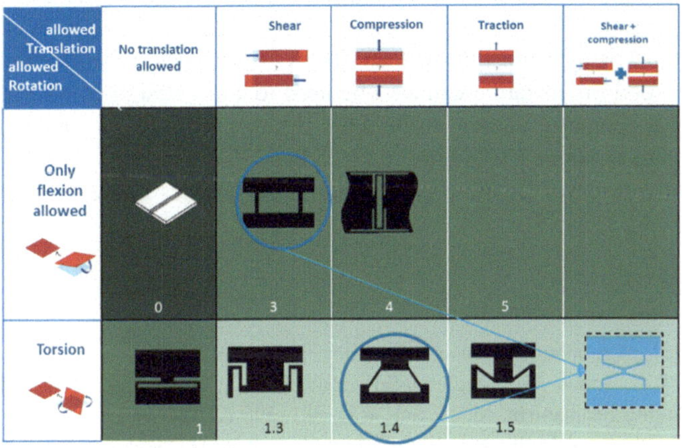

Fig. 3. An example of hybridization of cell 3 and 1.4 in order to create cell 1.3.4

According to this approach, Madotto and Casati [15], improved the coverage of Let joints increasing population from nine (already present in the literature) to sixteen mechanisms, namely precisely 50% of partial coverage. This allows the designer to have a greater number of topologies, extending the choice range.

The general idea is to populate the matrix by adding all the devices now present in literature. This allows the designer a broad overview of all the options he can draw on depending on what his project requirements are. A tool organized in this way does not require being an expert in the field.

The designer can now find in each cell at least one mechanism from which to take inspiration. The next step will be to move from a purely qualitative suggestion to one to quantitatively size the mechanism.

6 Conclusions and Future Developments

The first step to build a tool to push designers to use compliants instead of traditional joints is to find all the types of compliants that exist today and classify in order to have them ready according to the specifications of design required. The method by which they are classified is crucial. In literature most classification methods are not functional to the creation of a design tool but are useful as a means for teaching and archiving

In this work a selection was made of those methods that are most useful for a compliants design activity, i.e. functional methods such as FBS-based and those that work on allowed movements and forces.

A preliminary application of this approach has shown how research has proceeded in a chaotic way leaving entire areas uncovered. The uncovered areas can be filled by combining different topologies of compliants belonging to the same family or even to different families as long as they are organized according to the same functional scheme.

An example of hybridization of LET joints is proposed as an exemplary case to show how a good organization can also be used to show design combinations that do not yet have a geometry known in literature. More generally, the same classification tool was used to organize LET joints and rotational joint families, as a starting point to populate the representatives of all other compliants families in the same environment.

References

1. Howell, L.L.: Compliant Mechanisms. Wiley, New York (2001)
2. Merriam, E.G., Jones, J.E., Magleby, S.P., Howell, L.L.: Monolithic 2 DOF fully compliant space pointing mechanism. Mech. Sci. **4**(2), 381–390 (2013)
3. Oh, Y.: Synthesis of Multistable Equilibrium Compliant Mechanisms, pp. 4–8. Department of Mechanical Engineering, University of Michigan (2008)
4. Chen, Y.-H., Lan, C.-C.: An adjustable constant-force mechanism for adaptive end-effector operations. J. Mech. Des. Trans. ASME **134**(3), 031005 (2012)
5. Hegde, S., Ananthasuresh, G.K.: Design of single-input-single-output compliant mechanisms for practical applications using selection maps. J. Mech. Des. Trans. ASME **132**(8), 0810071–0810078 (2010)
6. Montecchi, T., Russo, D.: FBOS: function/behaviour–oriented search. Procedia Eng. **131**, 140–149 (2015)
7. Mankame, N.D., Ananthasuresh, G.K.: Contact aided compliant mechanisms: Concept and preliminaries. In: Proceedings of the ASME Design Engineering Technical Conference, vol. 5 A, pp. 109–121 (2002)
8. Jacobsen, J.O., Winder, B.G., Howell, L.L., Magleby, S.P.: Lamina emergent mechanisms and their basic elements. J. Mech. Robot. **2**(1), 1–9 (2010)
9. Dodgen, E., Stratton, E., Bowden, A., Howell, L.: Spinal implant development, modeling, and testing to achieve customizable and nonlinear stiffness. J. Med. Devices Trans. ASME **6** (2), 021010 (2012)
10. Parise, J.J., Howell, L.L., Magleby, S.P.: Ortho-planar mechanisms. In: Proceedings of the 2000 ASME Design Engineering Technical Conferences & Computers and Information in Engineering Conference, DETC2000/MECH-14193 (2000)
11. Dai, J.S., Jones, J.R.: Matrix representation of topological changes in metamorphic mechanisms. J. Mech. Des. Trans. ASME **127**(4), 837–840 (2005)
12. Wilding, S.E., Howell, L.L., Magleby, S.P.: Spherical lamina emergent mechanisms. Mech. Mach. Theor. **49**, 187–197 (2012)
13. Delimont, I.L.: Compliant Joints Suitable for Use as Surrogate Folds (2014). https://scholarsarchive.byu.edu/etd/4231/
14. Ivanov, I.: Methodical Development of a Parallel Kinematic Positioning System Based on Monolithic Structures With Flexure Hinges. Verlagshaus Mainz GmbH, Aachen (2016)
15. Casati, S., Madotto, L.: Metodo di progettazione dei Compliant mechanisms. Master's degree thesis at University of Bergamo (2018)

Condition Monitoring Techniques of Ball Bearings in Non-stationary Conditions

Matteo Strozzi(✉), Riccardo Rubini, and Marco Cocconcelli

Department of Sciences and Methods for Engineering, University of Modena and Reggio Emilia, Via Giovanni Amendola 2, 42122 Reggio Emilia, Italy
matteo.strozzi@unimore.it

Abstract. Frequently, the Industry suggests non-trivial problems and new fields of research for the Academy. This is the case of the ball bearing diagnostics in direct-drive motors. Direct-drive motors are brushless motors fully controlled by the drive system. Thanks to an encoder or a resolver mounted on the shaft, they can perform complex motion profiles, such as polynomials or splines, including reverse rotation of the shaft. The main advantage of direct-drive motors is the removal of cams or gearboxes afterwards motor with a consequent strong reduction of economic and maintaining costs. Indeed, their main drawback is the difficulty to make diagnostics on the bearings. Regarding bearing diagnostics, most of the techniques present in literature are based on the search of fault-characteristic frequencies in the vibration spectrum of the motor. These fault frequencies are linearly dependent on the rotational frequency of the shaft if it is supposed constant. However, in direct-drive motors the rotational speed changes continuously and consequently the fault frequencies are meaningless. The paper reports a brief overview of some techniques for the condition monitoring of ball bearings in non-stationary conditions used by the Authors in the case of a packaging machine working under variable speed. The techniques adopted include an improved version of the computed order tracking, the cross-correlation function and three supervised learning approaches: artificial neural networks, artificial immune systems and support vector machines.

Keywords: Condition monitoring · Non-stationary conditions · Ball bearings

1 Introduction

The issue of the condition monitoring of bearings for fault detection and diagnostics purposes results of great industrial interest and has been successfully dealt for decades with different methods based on the analysis of the vibration signals [1–13].

In particular, the vibrations induced by defective bearings have characteristics that make them recognizable by analysing the statistical distribution in the time domain and the periodicities in the frequency domain. The reliability of the current techniques of bearing monitoring has lifted up the objectives of the maintenance personnel who are engaged in the identification of sophisticated analytical procedures able to anticipate the location of defects in the service of predictive maintenance of machines.

C. Rizzi et al. (Eds.): ADM 2019, LNME, pp. 565–576, 2020.
https://doi.org/10.1007/978-3-030-31154-4_48

The widespread use, in the field of automatic packaging machines, of servomotors that replace some organs of powertrains to perform tasks requiring complex motion laws, invalidates the effectiveness of classical approaches, making non-Gaussian the distribution of the vibratory signal, even in healthy cases, and eliminating, within the spectra, the characteristic frequencies that distinguish the signals of damaged machines operating at constant speed.

Diagnostics of rotating machines in non-stationary conditions, e.g. bearings and gears, is a very attractive and promising research field. Recently, several works have been published in literature [14, 15] or presented in thematic Conferences [16–18], covering a wide range of applications, from manufacturing [19] to mining industry [20].

This paper reports the state of the art of the research carried out by the Authors in the areas of condition monitoring and diagnostics of ball bearings operating at variable speed. The Authors have proposed a series of solutions to the problem described, which are based on two different approaches of vibration signal analysis: (i) the modelling of the vibration characteristic signal of a rolling bearing, (ii) the application of expert systems. Considering the two previous approaches, five different diagnostics methodologies are proposed: computed order tracking [21], cross-correlation function [22–24], artificial neural networks [25], artificial immune systems [26] and support vector machines [27].

2 Methods Based on the Mechanical Model of the Defect

The knowledge of the physical phenomena that control the dynamics of the bearing allows to predict the expected vibration signal in both healthy and faulty condition of the bearing. The classical approach used to model the ball bearings is based on the similarity between ball bearings and epicycloid gears: the inner ring acts like the sun gear, the outer ring like the annulus gear, the spheres like the planet gears and the cage like the planet carrier. The kinematics relations that stand for the epicycloid gear allow to calculate the speed ratio between each bearing component as a function of the bearing geometry. Moreover, the modelling of the bearing allows to compute the well-known fault frequencies that are the basis of the diagnostic methods proposed in the literature in the last decades.

The ball-bearing fault frequencies are proportional to the rotational frequencies of the shaft (in the rest of the paper it's supposed that the inner ring of the bearing is integral with the rotating shaft, while the outer ring is fixed to the motor frame), then a constant value of the rotational speed of the shaft is an essential condition to perform the diagnostics of the bearing and it has been fulfilled for a long time by the working condition of asynchronous motors used in the Industry. As mentioned in the Introduction, the new concept of motors - the brushless servomotors - that have been started to be used in packaging machines no longer satisfies the constant-speed condition. It must be pointed out that the working condition of the motor is changed, but the physics of the bearing has remained the same, that is the mechanical model of the bearing and its defect can still be used with some tricks that are presented in this section.

In variable speed applications two ideal reference systems may be supposed in figurative sense: one is fixed in the variable speed domain and one in the constant speed domain. The vibration data are acquired in the variable speed domain while the algorithms developed in the literature belong to the constant speed domain. As a consequence, two different methodologies are possible: transform the vibration data in the constant speed domain or try to adapt the literature algorithms for the variable speed domain. In this section the Authors describe the Computed Order Tracking to tackle the first task and the Cross-Correlation Function to tackle the second task.

2.1 Computed Order Tracking

The choice of a technique based on the computed order tracking (COT) [28, 29] is justified because the scientific research in the field of monitoring of rolling bearings has always been focused, historically, to address the problems of machines operating at constant speed, by improving the signal analysis methodologies operating in the frequency domain.

The application of COT to a signal produced by the operation of a machine that must follow a complex law of motion - extrapolating the relationship between the magnitude of the parameter under observation (acceleration) and the angle of rotation of the shaft, by means of a re-sampling of the signal for constant angles of rotation - allows to get a signal in which, if you neglect the slippage between the surfaces in contact, the transient events induced by the presence of defects involving the elements of the bearings, appear periodically with a periodicity dependent on the geometry of the bearing. This processing procedure justifies the application, downstream from the COT, of the classical techniques of monitoring of the bearings.

The methodology proposed by the Authors [21] is covered by international patent [30] and can be summarized as follows:

- acquisition of the vibration signal of the support in the radial direction (radial ball bearings);
- identification of a resonance frequency of the system;
- extraction of the modulation signal of the resonant frequency selected in the previous step;
- application of the COT by referring to the signal of the crankshaft rotation speed acquired by an encoder;
- slip of the signal corresponding to rotations of the crankshaft in the opposite direction to that chosen as positive (inversion of the time axis);
- determination of the characteristic period of the defect based on the geometry of the bearing;
- selection, for each rotation period of the machine, of a signal section corresponding to an integer number of characteristic periods of the defect;
- application of the synchronous average of the signal sections selected, from the entire acquired time history, by following the criteria of the previous step;
- spectral analysis of the averaged signal.

In order to carry out a complete monitoring of the health conditions of the bearing, it is necessary to repeat this procedure for all possible types of defect. A flowchart of the procedure is reported in Fig. 1.

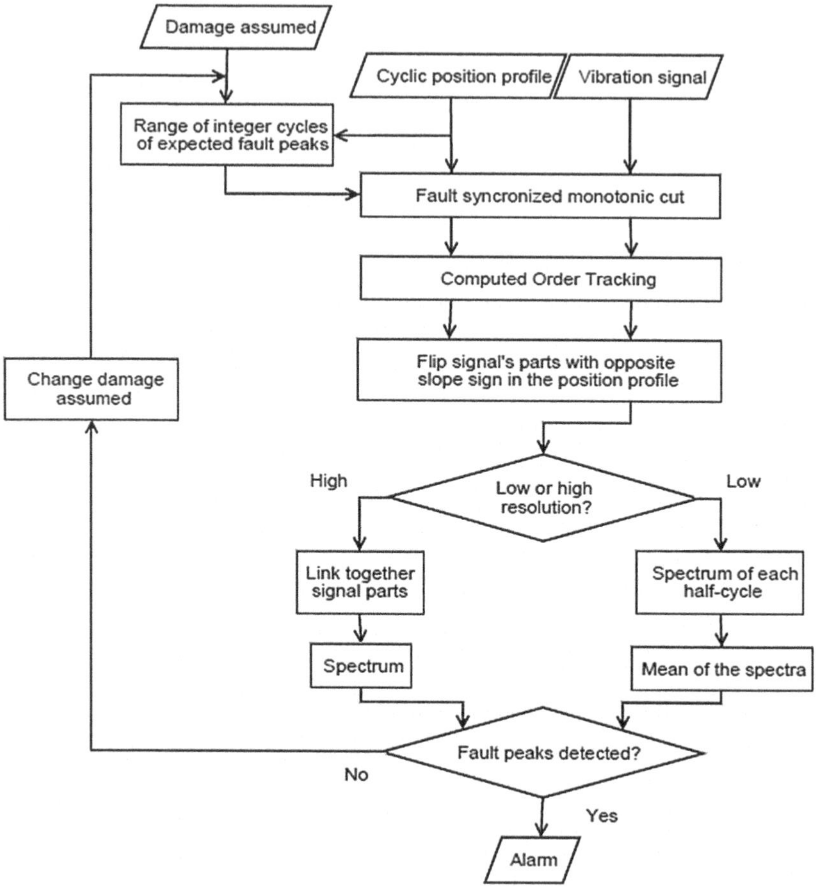

Fig. 1. Flowchart of the proposed procedure of computed order tracking.

2.2 Cross-Correlation Function

The use of servomotors, devices which transmit complex laws of motion to the rotating parts of the supports, invalidates the application of the more effective classical technique for the monitoring of rolling bearings, based on the detection of periodic events induced by the presence of surface defects on the components: the impulses, that arise in the vibration signal due to interaction with the damaged zones in operating conditions, occur with a complex time distribution dependent on the operating conditions, as well as the geometry of the supports.

The Authors have proposed a monitoring methodology [22–24] based on the construction of a simplified model of vibration signal, which looks like a train of Dirac impulses, which occupy on the time axis, unless the phase, the location of the events that would appear, if it was the defect to be monitored, under the expected operating conditions. This simplified model of signal must be compared with the one actually acquired from the supports, which is also affected by mechanical and electrical noise and the dynamic effects induced by complex laws of motion: accelerations and motion inversions.

In order to highlight, in the vibration signal, any fast transients that arise in the presence of localized defects, an analysis is performed in the time-frequency domain using the wavelet transform with impulse wavelet [31] as a mother function, particularly suitable for the detection of shocks.

With reference to Fig. 2, the proposed procedure can be summarized as follows:

- acquisition of the vibration signal of the support in the radial direction (radial ball bearings);
- application of Wavelet Transform with impulse wavelet;
- extraction of the filtered signal at a suitable frequency for the localization of possible pulses in the signal;
- construction of a model of the signal corresponding to a possible damage of the bearing located on a component, based on the law of motion imposed by the actuator: it consists of a Dirac pulse train;
- calculation of the cross-correlation function between the signals obtained in the previous steps (the actual signal and processed model);
- assessment of the degree of correlation by appropriate statistical parameters (kurtosis, maximum value, crest factor)

In order to carry out a complete monitoring of the health conditions of the bearing, it is necessary to repeat this procedure for all possible types of defect.

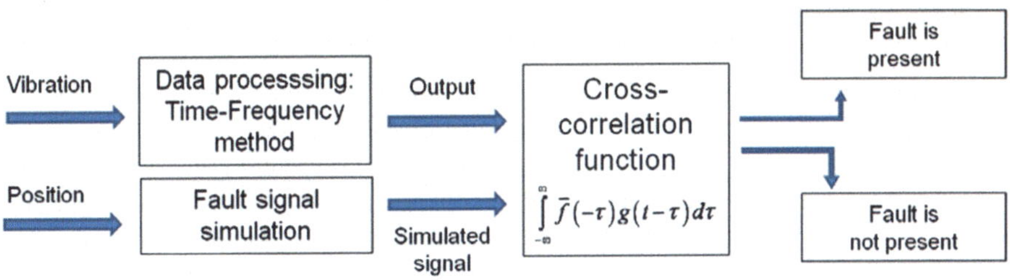

Fig. 2. Flowchart of the proposed procedure based on the cross-correlation function.

The suggested technique can be extended in order to use the electric current absorbed by the motor instead of the vibration signal. Although vibration data have demonstrated their effectiveness in diagnostics, there is a line of research in the literature that tries to use the current fluctuation absorbed by the motor as sensor of an incipient fault.

In fact, the presence of a defect induces fluctuations of the airgap between the stator and the rotor - that is a sort of vibration of the rotor - with a consequent presence of ripples in the current signal.

Although the view of an intrinsic sensor is quite attractive for the industries (external sensors are no more needed), the sensibility of the currents is very low compared to the one of the vibration signal. Starting from the idea that the current ripples are consequences of mechanical phenomena, like the impact of a sphere with a damage on the outer ring of the bearing, the Authors suggest to correlate the current data with the simplified model of vibration signal as shown in Fig. 3.

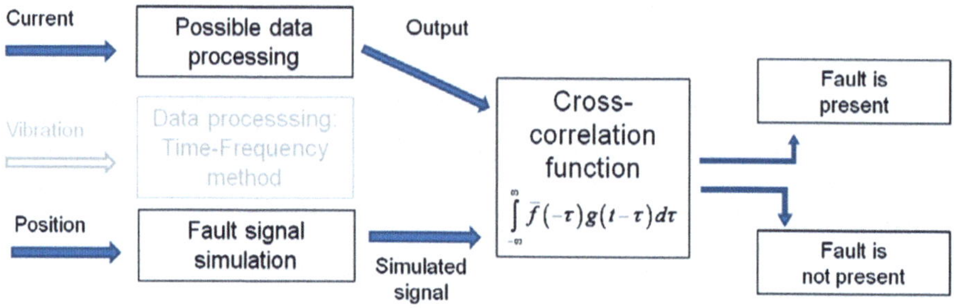

Fig. 3. Extension of the algorithm to the current measurement.

The simulated vibration signal acts like a map that highlights the similarity between the expected impulses in the vibration signal and the ripples in the current data through the cross-correlation function. The signal-to-noise ratio of the correlation can be improved with a pre-processing of the current data. For example, the presence of ripples in the current data can be highlighted calculating the gradient of the current data themselves.

3 Expert Systems

The adoption of expert systems for the recognition of signal patterns of damaged bearings [32–40] does not require the determination of a mechanical model. As is well known, in order to use the expert systems as efficient instruments of pattern recognition, it is necessary to educate them with information derived from actual operating conditions, by including all the possible manifestations of the phenomenon to be studied.

So, if one has the advantage of avoiding the simulation of the failure of the rolling bearings by proposing a signal model - which, to be efficient, should also consider the slippage of the rolling elements and the vibrations transmitted by other parts of the system - however, the application of expert systems requires a wide experimental phase and the identification of signal parameters that are sensitive to the different states of classification required.

For the identification of the presence of a particular defect in all the different operating conditions of a machine driven by servomotors, it is opportune to provide an indication of the hourly capacity of the machine (number of machine cycles per unit time, that in the paper will be addressed as *working speed* of the machine): this parameter can alter significantly the dynamics of the machine, acting on the shape of the vibration signal and the separation time between impulsive events induced by the surface damages.

3.1 Artificial Neural Networks

The Authors have obtained the maximum efficiency for bearing state recognition (100%) using an artificial neural network (ANN) feed-forward with single hidden layer consisting of 70 neurons, with hyperbolic tangent sigmoid transfer function and a training function based on the resilient back-propagation algorithm [25].

The network was provided, as inputs, by the root mean square value (RMS) and the kurtosis of the vibration of a large number of bearings, healthy or damaged on a particular component (natural or artificial damage), at different speeds. In order to obtain an adequate number of data for training and test phases, long acquisitions were made, then separated into parts corresponding to one machine cycle.

The combination of RMS-kurtosis, with the value of speed, allows to recognize the condition of the bearing, even in health cases with 0 h of operation: in this early stage of adjustment, the machine can vibrate with higher RMS value, but the signal remains stochastic, with a low value of kurtosis.

3.2 Artificial Immune Systems

The Authors developed a simple method inspired and derived from the mechanisms of the immune system to recognise bearing faults [26]. The proposed algorithm was a simplification of the original process, adapted to a particular case of a much bigger class of algorithms and methods named Artificial Immune Systems (AIS). These systems have proven to be useful and promising in many application fields [41, 42]. The AIS mimic the human body immune system that is able to recognize and react against never seen before virus. Figure 4 depicts the main idea of AIS: in (a), training dataset for healthy and faulted bearings is given in the space of the features (antigens). In (b), random antibodies are generated with a proper area of interest. Based on the major number of specific antigens, in (c) the antibodies are associated to a healthy or faulty status. Finally, once a new dataset is given in (d), its classification is given as the result of collisions with the antibodies. In particular, the proposed algorithm is based on the Euclidean distance minimization (EDM) method in the evaluation of binding between antigens. Applied to bearing diagnostics, the generic antigen used is created collecting together four features computed from the vibration signal: Kurtosis, Jerk-peak, RMS and the hourly capacity of the packaging machine the motor is mounted on.

A normalization with respect to the maximum computed value of the features is required in order to compare the antigens developed in training phase with the antigens under classification: experimental activity performed by the Authors considering this AIS based algorithm for bearing state recognition proved the maximum efficiency (100%) of the developed methodology.

3.3 Support Vector Machines

The Support Vector Machines (SVM) is a learning methodology based on the statistical learning theory [36–40]: it classifies a set of input into two categories.

During the training phase of the SVM the classification rule is generated by the specific kernel function selected, in order to obtain and optimize - on a plane or space

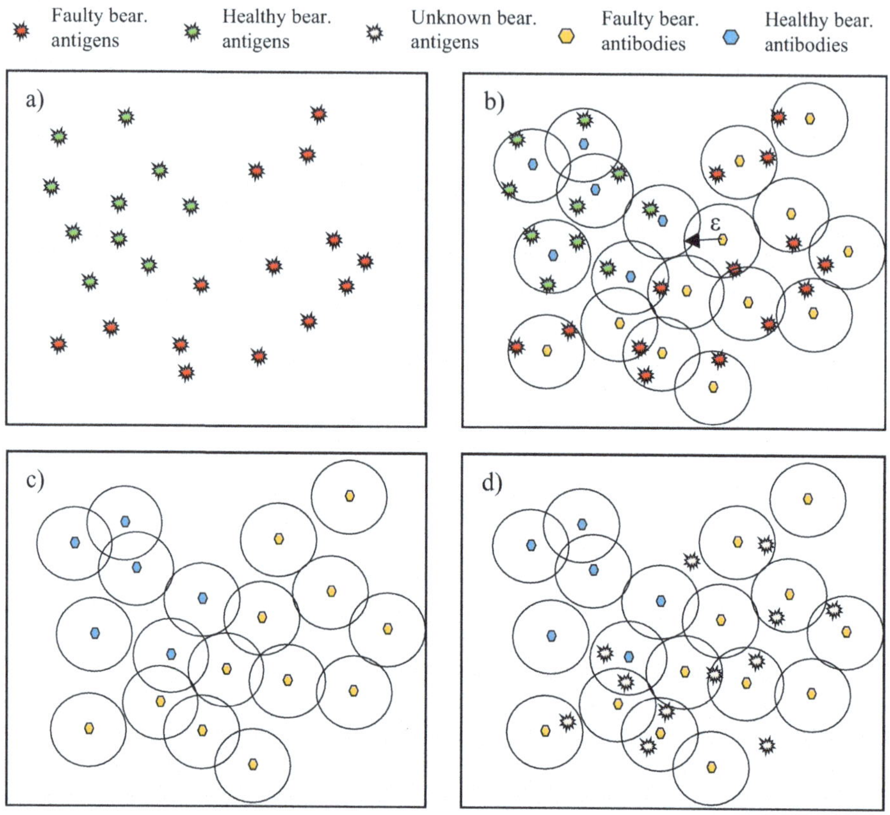

Fig. 4. Faulty and healthy bearing antigens (a). Generation (b) and separation (c) of the antibodies. The generated antibodies are used to classify the unknown antigens (d).

whose axes are represented by the input parameters - a curve or a surface boundary able to divide the two categories of outputs, maximizing its distance from the border cases.

The Authors used the algorithm of Professor Gunn [43] and the same input vector previously used for the ANN in order to make a comparison between the two expert systems. The results return the maximum efficiency for bearing state recognition (100% like in the ANN case) adopting an exponential radial-basis-function with sigma equal to five as kernel.

A second test has been done removing the speed information from the input vector, which is now made by the combination of RMS-kurtosis only. The results are 100% of success and it proves the capability of SVM as classifier. This second test has been applied also to ANN but with several errors in the recognition of the health of the bearing.

The differences between SVM and ANN are probably due to the capability of the SVM to pass the data to a multi-dimensional space through the use of the kernel function, where the single data can be divided in two categories by an hyperplane (see [43]). The ANN elaborate data in the input space and therefore they need an input vector with an increased dimension (RMS-kurtosis-speed instead of RMS-kurtosis only) from the beginning. Figure 5 shows a graphical representation of the features' space and the separation of the healthy (yellow) and faulty (blue) classification area.

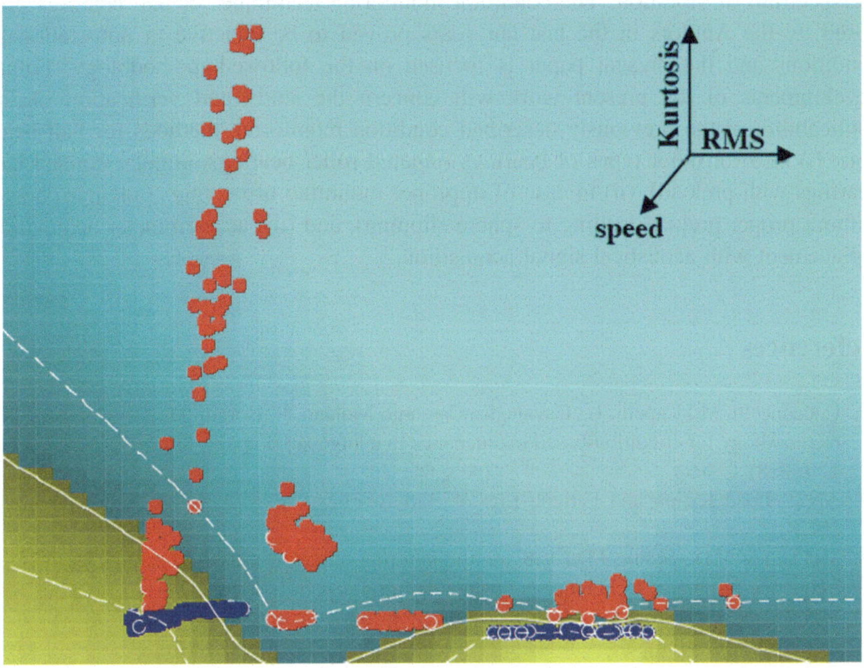

Fig. 5. Projection on the plane RMS-Kurtosis of the best Radial Basis SVM classification; the blue and red points represent the healthy and faulty patterns. White lines are the support vector machines (dotted) and the projection of separation hyperplane (solid).

The experimental activity made by the Authors proves that different expert systems give excellent results if the input parameters are suitably chosen. Often the expert systems seem to be an easy solution for pattern-recognition and classification tasks since they automatically build the weight matrix to classify new inputs, but it is just an appearance since the researcher still has the responsibility to define the right input vector that describes the physics of the process under consideration.

4 Conclusions

In this paper, the Authors run through their experience in the diagnostics of ball bearings in variable speed applications and outline the adopted solutions from a methodological point of view. All the methods are based on a specific test case, which is the condition monitoring of ball bearings for a brushless servo-motor in a packaging machine. Many mechanical devices, and in particular the automatic machines for packaging, continually increase the use of servo-motors which normally work at variable speed. Unfortunately, most of the literature on diagnostics assumes constant although arbitrary rotational frequencies of the motor. In fact, a continuous change of speed makes difficult to recognise the presence of fault characteristic frequencies within the spectrum of vibration. The condition monitoring techniques of ball bearings proposed by the Authors in the last ten years proved to be effective in non-stationary conditions and the present paper is focused on the followed methodology. Future developments of the present work will concern the study and verification of the applicability of the previously described condition monitoring methods for ball bearings: (i) also for other types of bearings (tapered roller bearings, angular contact ball bearings with preload), (ii) in case of improper mounting of bearings (angular contact without proper preload leading to sphere slipping), and (iii) accelerometer acquisition replacement with acoustical signal acquisition.

References

1. Cocconcelli, M., Capelli, L., Cavalaglio Camargo Molano, J., Borghi, D.: Development of a methodology for condition-based maintenance in a large-scale application field. Machines **6**, 17 (2018)
2. ANSI. Load Rating and Fatigue Life of Ball Bearings, ANSI/AFBMA Standard 9-1978 (1978)
3. McFadden, P.D., Smith, J.D.: Vibration monitoring of rolling element bearings by the high frequency resonance technique. A review. Tribol. Int. **17**(1), 3–10 (1984)
4. McFadden, P.D., Toozhy, M.M.: Application of synchronous averaging to vibration monitoring of rolling element bearings. Mech. Syst. Sig. Process. **14**(6), 891–906 (2000)
5. Harris, T.A.: Rolling Bearing Analysis. Wiley, Chichester (2001)
6. Rubini, R., Meneghetti, U.: Application of the envelope and wavelet transform analyses for the diagnosis of incipient faults in ball bearings. Mech. Syst. Sig. Process. **15**(2), 287–302 (2001)
7. Randall, R.B., Antoni, J., Chobsaaard, S.: The relationship between spectral correlation and envelope analysis in the diagnostics of bearing faults and other cyclostationary machine signals. Mech. Syst. Sig. Process. **15**(5), 945–962 (2001)
8. Taylor, J.I., Kirkland, D.W.: The Bearing Analysis Handbook: A Practical Guide for Solving Vibration Problems in Bearings. Vibration Consultants (2004)
9. Yu, D., Cheng, J., Yang, Y.: Application of EMD method and Hilbert spectrum to the fault diagnosis of roller bearings. Mech. Syst. Sig. Process. **19**(2), 259–270 (2005)
10. Li, H., Zheng, H.Q., Tang, L.W.: Faults monitoring and diagnosis of ball bearing based on Hilbert-Huang transformation. Key Eng. Mater. **291**, 649–654 (2005)
11. Li, H., Zhang, Y.: Bearing faults diagnosis based on EMD and Wigner-Ville distribution. In: Proceedings of the Sixth World Congress on Intelligent Control and Automation, Dalian, China, 21–23 June 2006 (2006)
12. Junsheng, C., Dejie, Y., Yu, Y.: A fault diagnosis approach for roller bearings based on EMD method and AR model. Mech. Syst. Sig. Process. **20**(2), 350–362 (2006)
13. Antoni, J., Randall, R.B.: The spectral kurtosis: application to the vibratory surveillance and diagnostics of rotating machines. Mech. Syst. Sig. Process. **20**(2), 308–331 (2006)
14. Randall, R.B., Antoni, J.: Rolling element bearing diagnostics - a tutorial. Mech. Syst. Sig. Process. **25**, 485–520 (2011)

15. Gelman, L., Harish Chandra, N., Kurosz, R., Pellicano, F., Barbieri, M., Zippo, A.: Novel spectral kurtosis technology for adaptive vibration condition monitoring of multi-stage gearboxes. Insight Nondestr. Test. Condition Monit. **58**(8), 409–416 (2016)
16. Barszcz, T.: Validation of vibration measurements for heavy duty machinery diagnostics. In: Proceedings of the First Conference on Condition Monitoring of Machines in Non-stationary Operations, Wroclaw, Poland, 7 March 2011 (2011)
17. Strozzi, M., Barbieri, M., Pellicano, F., Zippo, A.: Pitting and stress analysis of gears: a numerical and experimental study. In: Proceedings of the Thirteenth International Conference on Condition Monitoring and Machinery Failure Prevention Technologies, Paris, France, 10–12 October 2016 (2016)
18. Strozzi, M., Barbieri, M., Zippo, A., Pellicano, F.: Numerical simulation and experimental validation of normal strain distribution and pitting phenomenon in industrial gears. In: Proceedings of the Twenty-fourth International Congress on Sound and Vibration, London, UK, 23–27 July 2017 (2017)
19. Antoni, J.: Cyclic spectral analysis of rolling-element bearing signals: facts and fictions. J. Sound Vib. **304**(6), 497–529 (2007)
20. Bartelmus, W., Zimroz, R.: Vibration condition monitoring of planetary gearbox under varying external load. Mech. Syst. Sig. Process. **23**(1), 246–257 (2009)
21. Cocconcelli, M., Secchi, C., Rubini, R., Fantuzzi, C., Bassi, L.: Predictive maintenance of ball bearings for machines rotating with arbitrary velocity profiles. In: Proceedings of the International Conference on Noise and Vibration Engineering ISMA 2008, Leuven, Belgium, 15–17 September 2008 (2008)
22. Cocconcelli, M., Rubini, R.: Introduction to a simply and fast algorithm for variable speed bearing diagnostics. In: Proceedings of the Second Workshop in Memory of Ettore Funaioli, Bologna, Italy, 18 July 2008 (2008)
23. Cocconcelli, M., Secchi, C., Rubini, R., Fantuzzi, C., Bassi, L.: Comparison between time-frequency techniques to predict ball bearing fault in drives executing arbitrary motion profiles. In: Proceedings of the International Mechanical Engineering Congress and Exposition ASME 2008, Boston, MA, USA, 31 October–6 November 2008 (2008)
24. Cocconcelli, M., Rubini, R.: Correlation between the stator current signal and the kinematic model of the rolling bearing for the diagnostics. In: Proceedings of the XIX Italian Congress of Theoretical and Applied Mechanics, Ancona, Italy, 14–17 September 2009 (2009)
25. Cocconcelli, M., Rubini, R., Zimroz, R., Bartelmus, W.: Diagnostics of ball bearings in varying-speed motors by means of artificial neural networks. In: Proceedings of the International Conference on Condition Monitoring, Cardiff, UK, 20–22 June 2011 (2011)
26. Montechiesi, L., Cocconcelli, M., Rubini, R.: Artificial immune system via euclidean distance minimization for anomaly detection in bearings. Mech. Syst. Sig. Process. **76–77**, 380–393 (2016)
27. Cocconcelli, M., Rubini, R.: Support vector machines for condition monitoring of bearings in a varying-speed machinery. In: Proceedings of the International Conference on Condition Monitoring, Cardiff, UK, 20–22 June 2011 (2011)
28. Potter, R.: A new order tracking method for rotating machinery. Sound Vib. **24**, 30–34 (1990)
29. Fyfe, K.R., Munck, E.D.S.: Analysis of computed order tracking. Mech. Syst. Sig. Process. **11**(2), 187–205 (1997)
30. Cocconcelli, M., Bassi, L., Borghi, D., Rubini, R., Secchi, C.: Predictive rolling bearing maintenance. International World Patent, WO 2010/100253 A1 (2010)
31. Schukin, E.L., Zamarev, R.U., Schukin, L.I.: The optimization of wavelet transform for impulse analysis in vibration signals. Mech. Syst. Sig. Process. **18**, 1315–1333 (2004)

32. Larson, E.C., Wipf, D.P., Parker, B.E.: Gear and bearing diagnostics using neural network-based amplitude and phase demodulation. In: Proceedings of the 51st Meeting of the Society for Machinery Failure Prevention Technology, Virginia Beach, VA, USA, pp. 511–521 (1997)
33. Giuliani, G., Rubini, R., Maggiore, A.: Ball bearing diagnostics using neural networks. In: Proceedings of the 3rd International Conference on Acoustical and Vibratory Surveillance Methods and Diagnostics Techniques (1998)
34. Li, B., Chow, M.-Y., Tipsuwan, Y., Hung, J.C.: Neural-network-based motor rolling bearing fault diagnosis. IEEE Trans. Ind. Electron. **47**, 1060–1069 (2000)
35. Wang, C.-C., Kang, Y., Shen, P.-C., Chang, Y.-P., Chung, Y.-L.: Applications of fault diagnosis in rotating machinery by using time series analysis with neural network. Expert Syst. Appl. **37**(2), 1696–1702 (2010)
36. Boser, B., Guyon, I., Vapnik, V.: A training algorithm for optimal margin classifiers. In: Proceedings of the Fifth Annual Workshop on Computational Learning Theory, New York, NY, USA (1992)
37. Cortes, C., Vapnik, V.: Support-vector networks. Mach. Learn. **20**(3), 273–297 (1995)
38. Zhang, Z., Wenzhi, L., Shen, M.: Active learning of support vector machine for fault diagnosis of bearings. Lect. Notes Comput. Sci. **3973**, 390–395 (2006)
39. Rojas, A., Nandi, A.K.: Practical scheme for fast detection and classification of rolling element bearing faults using support vector machines. Mech. Syst. Sig. Process. **20**(7), 1523–1536 (2006)
40. Widodo, A., Yang, B.-S.: Support vector machine in machine condition monitoring and fault diagnosis. Mech. Syst. Sig. Process. **21**(6), 2560–2574 (2007)
41. Greensmith, J., Whitbrook, A., Aickelin, U.: Artificial Immune Systems. Handbook of Metaheuristics. Springer, Berlin (2010)
42. Dasgupta, D.: Artificial neural networks and artificial immune systems: similarities and differences. In: Proceedings of the IEEE International Conference of Orlando (1997)
43. Gunn, S.R.: Support vector machines for classification and regression. Technical report of the University of Southampton, Southampton (1998)

A CAE-Based Model of Aluminium Alloys Welded T-Joints for TEP Analysis

Federico Ruini, Fabio Pini(✉), and Francesco Leali

"Enzo Ferrari" Department of Engineering,
University of Modena e Reggio Emilia, Via Vivarelli 10/1, 41125 Modena, Italy
{federico.ruini, fabio.pini, francesco.leali}@unimore.it

Abstract. Continuous innovation in the field of high-end motor vehicle chassis demands optimization of the weight/stiffness ratio and to achieve high quality standards. The use of light materials, such as aluminum alloys, is therefore increasingly common in the design of the chassis, whose assembly process represents a technological challenge. Welding joining processes, and in particular robot-based welding, are widely used in automotive field despite causing distortions. To predict these deformations, finite element analyzes are performed, in particular thermo-elasto-plastic simulations, which are able to satisfactorily replicate the behavior of residual stresses and strains after cooling. However, such analyzes are computationally expensive making their application difficult to complex structures. This work would investigate an alternative solution to predict distortions that effectively returns the behavior of welded assemblies. A CAE-based model for TEP analysis of welded joints is proposed. As a case study, the T-welded junction between two aluminum alloy plates (T-Joint) was considered. The model is validated by a preliminary experimental campaign.

Keywords: Termo-elasto-plastic model · Welding process · CAE-based model · TEP analysis

1 Introduction

Aluminium alloys are increasingly used in the construction of automotive chassis as weight reduction is essential if an increase in performance or reduction in consumption is sought. The automotive chassis is commonly assembled by joining thin and fused structures through the welding process. The requirements of high quality standards over the welded joints have led to the increasingly frequent use of robots capable of guaranteeing excellent process repeatability. However, the large localized heat input causes distortion problems, particularly relevant for aluminium which has a higher thermal conductivity and expansion coefficient than steel alloys. In literature, the problem of distortion and its prediction has often been addressed with regard to large steel structures [1], mainly in naval application. The main techniques used to evaluate the deformations caused by welding are based on the theory of finite elements: they are the Thermo-Elasto-Plastic simulation methods (TEP methods) and methods based on

C. Rizzi et al. (Eds.): ADM 2019, LNME, pp. 577–588, 2020.
https://doi.org/10.1007/978-3-030-31154-4_49

residual deformations, Inherent Strain methods, commonly the Inherent Strain Stress (ISS) and Inherent Strain Equivalent Load (ISEL).

TEP analysis has proved to be very reliable in replicating the welding process, showing good matching between simulations and experimental results [2]. In TEP analyses, a thermal simulation and a mechanical simulation coupled together are performed sequentially. Thermal analysis returns the temperature field related to heat source of the welding operations. The temperature field complete the loading constraints for the subsequent mechanical analysis which provides the residual deformations of the post-welding cooling phase. The heat source, important for correctly describing the heat input, is mathematically modelled. The heat source model must be a mathematical description of the power distribution introduced into the melt bath, and must replicate the true shape of the isothermal surface of the weld pool and the associated heat flow. Furthermore, it must indirectly take into account the effects of the molten material in the bath, as they are not consistently modelled in finite element analyses. Goldak double ellipsoid is widely used in literature as heat source analytical model [3]. The Goldak model considers a Gaussian distribution of power density in the space applied to a double ellipsoid geometry: the shape and size of the model can simply be modified to adapt the penetration of both the arc and electron beam welding. The temperature distributions obtained using this mathematical model are sufficiently close to the experimental reality. From the constraint of the FEA model it is possible to simulate the effect of releasing the piece from the welding fixture, an important factor if you want to have an accurate simulation. Figure 1 shows the Goldak double ellipsoid while the two following Eqs. (1 and 2) respectively return the power distribution in the front and rear quadrant with respect to the double ellipsoid.

$$q(x,y,z,t) = \frac{6\sqrt{3}f_t Q}{abc\pi\sqrt{\pi}} e^{\frac{-3x^2}{a^2}} e^{\frac{-3y^2}{b^2}} e^{\frac{-3x^2}{a^2}} e^{\frac{-3(z+v(\tau+t))^2}{c^2}} \tag{1}$$

$$q(x,y,z,t) = \frac{6\sqrt{3}f_r Q}{abc\pi\sqrt{\pi}} e^{\frac{-3x^2}{a^2}} e^{\frac{-3y^2}{b^2}} e^{\frac{-3x^2}{a^2}} e^{\frac{-3(x+v(\tau-t))^2}{c^2}} \tag{2}$$

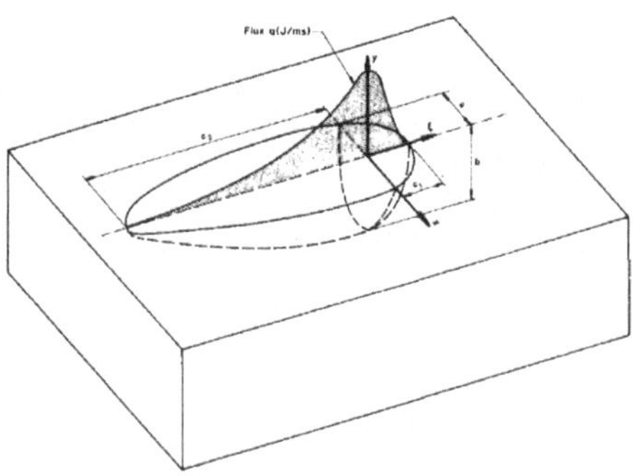

Fig. 1. Goldak double ellipsoid representation [3].

Where a, b and c are characteristic dimensions of the geometry of the model, f_f and f_r are fractions of the heat distributed between the front and the rear of the model, v is the welding speed in meters per second, τ is a time constant useful to define the position of the heat source at t = 0, and (x, y, z) are the coordinates of the heat source. Equation (3) provides the relation between the fractions of heat, while Eq. (4) returns the thermal power Q.

$$f_f + f_r = 2 \tag{3}$$

$$Q = IV\eta \tag{4}$$

Where I represents the input current value, V the voltage and η is the efficiency of the electric arc. The significant welding parameter Q_{input} is frequently used in connection with heat, and it is calculated by the relation (5).

$$Q_{input} = \eta IVv \tag{5}$$

TEP simulations give accurate results, even providing information on the quality of the weld itself. Conversely, they require a high computational effort leading to long simulation times even for relatively simple problems: a method often used is the local/global one, where a "local" TEP analysis of simple welded joints is performed whose results, in terms of stresses and residual deformations, are inserted into an overall elastic analysis of the assembly, the "global" model [4].

The thermal expansion of the material near the weld is blocked by the material in the lower temperature areas. The consequence of this phenomenon is the development of residual plastic deformations, to which residual stresses and material distortions are attributed. The transversal and longitudinal residual deformations are the main cause of the distortion. Therefore, if the entity and the position of the residual deformations were known, for each weld the deformation field could be applied as an initial condition of an elastic simulation of the finite elements.

Relating the temperature field with the deformation field without doing the thermal analysis allows to speed up the overall analysis, hence the inherent strain methods. The total volume Wx of the longitudinal residual deformation and the transverse residual deformation Wy are considered. Respectively, they are related to incoming heat Q by the relations (6 and 7).

$$Wx = KQ \tag{6}$$

$$Wy = \xi Q \tag{7}$$

The values K and ξ depend on the geometry of the model in question, for example for thick plates the Eqs. (8 and 9) are used. Respectively, α is the coefficient of thermal expansion, c is the thermal capacity, ρ is the density.

$$K = (0.255 - 0.335)\alpha/cp \tag{8}$$

$$\xi = (0.255 - 1.0)\alpha\rho \tag{9}$$

The residual longitudinal deformations εx and transversal εy are defined by the relations (10 and 11), assuming A as the area where the residual deformations are distributed.

$$\varepsilon x = Wx/A \tag{10}$$

$$\varepsilon y = Wy/A \tag{11}$$

The residual deformations calculated in this way are used to calculate the residual stresses: this is called ISS analysis [5]. In the ISEL model, on the other hand, the deformation caused by the welding process is assumed to be dependent on a compressive force, which generates a F_{Tendon} reaction force, commonly referred to in literature as "Tendon force", which acts on the welding path and influences the sign of residual deformations [6]. It depends on the heat supplied by the welding machine, Eq. (12). Analytical expressions report the prediction of distortion in typical configurations of common case studies, for example for the aluminium T-Joint considered the longitudinal and transverse forces are defined by relations (13 and 14).

$$F_{Tendon} = 200 \cdot Q_{input} \tag{12}$$

$$F_x = \int_A E_{YAlu} \cdot \varepsilon_x^* dA \tag{13}$$

$$F_y = \int_A E_{YAlu} \cdot \varepsilon_y^* dA \tag{14}$$

In relation (13), ε_x^* is the longitudinal residual deformation. While, the transverse deformation ε_y^* is related to the transverse distortion δ_y and length of the plate L (15).

$$\varepsilon_y^* = \frac{\delta_y}{L} \tag{15}$$

Thus, applying the forces as an initial load to a finite element model it is possible, with these steps, to calculate the distortion. The mathematics behind TEP methods demonstrate how this formulation is strictly accurate but the high time required to perform simulations using traditional FEM solvers makes TEP analysis difficult to apply on complex industrial cases. Conversely, IS methods being very useful since they require a decidedly lower computational cost and returns fastest results, but fail to reach the accuracy level of TEP simulations.

Consequently, the request is for more efficient approaches able to return the final distortion. In literature integrated computer aided tools have been developed profitably in different mechanical fields [7–9], but specific models must be implemented to effectively replicate the behaviour of the real processes.

Therefore, this work aims to set up a model to perform TEP analysis of welded joints with respect to the T-Joint configuration. For an effective response, the model relies on specific welding-driven CAE tools. An experimental campaign is carried out to set, verify the accuracy and test the efficiency of the T-Joint model. Accordingly, the next section describes the workflow used to identify the TEP CAE-based model. Section 3 describes the use case and the experimental setup. Section 4 focuses on the validation phase and provides the preliminary result gathered. Section 5 closes the paper with final remarks and an overview of the future activities required to return a robust validation.

2 CAE-Base Predictive Model for TEP Analysis

The model to predict distortions of welded joints is based on the characteristic parameters of the welding process, such as heat input, work angle, characteristic thicknesses and relative gap of the components to be welded. The workflow to derive the model is schematically illustrated in Fig. 2 and hereinafter described.

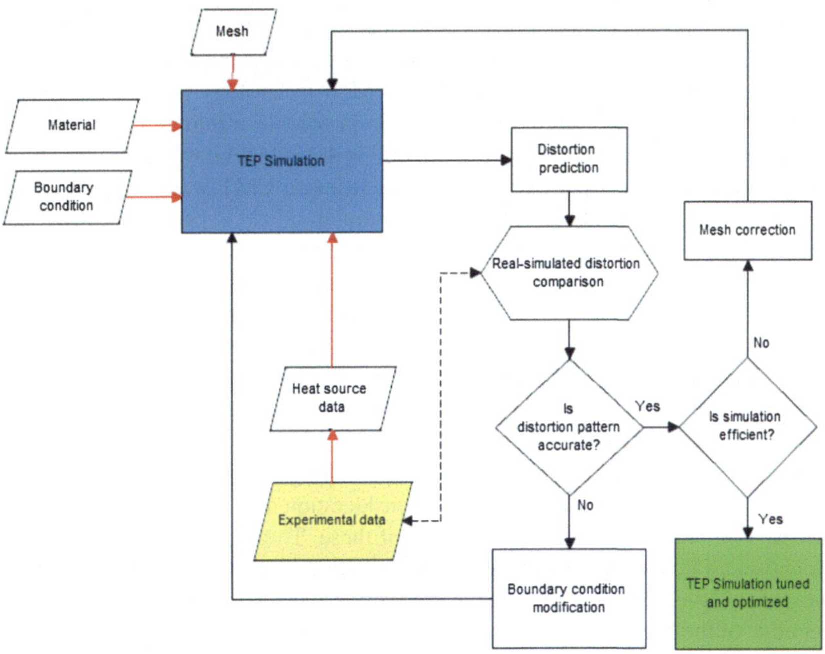

Fig. 2. Reference workflow to identifies the CAE-based model for TEP of welded joints.

The selected welding joint provides information about the material and defines the boundary conditions of the TEP analysis. To drive the TEP simulation, the other two inputs are the finite element model (mesh, constraint and material) and the heat source. The latter requires experimental welding test to identify the data set for the model.

The comparison between the simulation and experimental results allows us to understand the reliability of the model and to tuning its parameters. The first step assesses the accuracy of distortions while the second step evaluates the simulation efficiency in terms of completing time. In case of low accuracy, it is not possible to forward toward second step and the TEP simulation restarts with respect to modified boundary conditions. Conversely, the subsequent step optimizes the mesh quality through a sensitivity analysis until reaching a trade-off between simulation time and results. To stress the model, both the simulation and experiment are performed by varying the welding and model parameters.

3 Welding T-Joint Model and Experimental Setup

The model developed in this work refers to the T-Joint configuration, a typical welding joining adopted in the automotive field. It is a junction between two thin plates perpendicular to each other. In the literature, the vertical plate is called "*Web*" while the horizontal plate is called "*Flange*": the final welded assembly of both is called T-Joint.

The aluminium alloy chosen for the specimens is EN AW-6082-T6, a material used on high performance vehicles in the construction of space frame chassis and its subcomponents. The specimens are welded with MIG technology using S-Al-4043 as a filler material, a welding wire commonly used in robotic welding. The weld bead is 130 mm long. The ABB IRB 4400L/30 M2000 is the industrial robot that performs the welding operations, equipped with FRONIUS Robacta CMT welding torch. Furthermore, to prevent misalignments between the simulation and experiments, the fixturing for plates has been improved from manual adjusted modular solution toward T-Joint specific welding fixture. This solution provides unique positioning for the plates since follow the location principle 3-2-1, and consequently it returns better repeatability in the experiments, removes the uncertainty of locking point positions as well as recreates the expected configuration along subsequent tests. Figure 3 depicts the experimental setup which include the robot, the torch and the rigid fixture, and the model of the T-joint used, respectively on the left and right sides.

The experimental stage is performed by varying different welding and model parameters. The factors that are taken into consideration are the working angle, the gap between Web and Flange and the thickness of these. Three working angle levels have been chosen: 45°, that is typically indicated as optimal by welding manuals, and the two extremal levels 60° and 30°. The gap between flanges is the distance between the lower edge of the Web and upper surface of the Flange. Again, 3 levels are chosen also for the gap which are 0 mm (contact between plates), 0.5 mm, and 1 mm: this last level represents the weldability limit. Finally, the selected thickness for plates are 2 mm and 3 mm. Travel angle and power input are constant in these tests. Table 1 collects the variable input parameters and the selected values.

Distortion is measured by analysing 3D models acquired with a Zeiss Comet 6 8 M © scanner. In particular, acquisition and measurement are divided in two phases: the plates that make up the joint are scanned individually before welding, then the welded

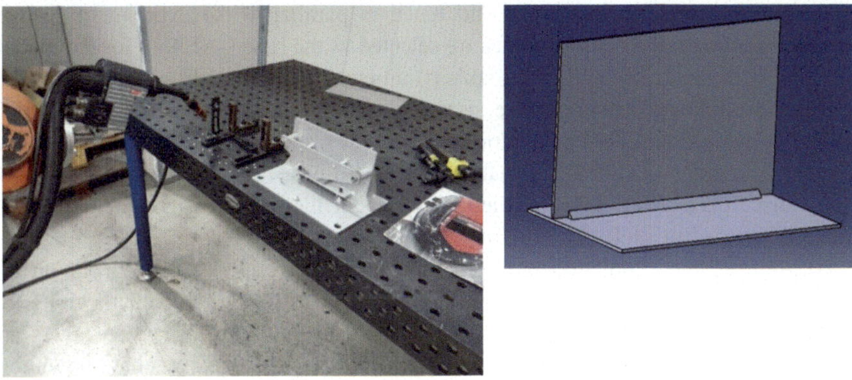

Fig. 3. Left side, the experimental setup with modular and rigid fixture systems; on the right, the T-Joint CAD model representation.

Table 1. Input parameters of the welding and their values

Input parameters	Level 1	Level 2	Level 3
Work angle	30°	45°	60°
Gap	0 mm	0.5 mm	1 mm
Plates thickness	2 mm	3 mm	/

T-Joint is scanned. In this way the true deformation caused by the heat source is detected. The distortion is quantified by analysing the 3D models of the T-Joint with a metrology software and by measuring the angular deflection of the Flange. As shown in Fig. 4, the lower corner of the joint is measured over 14 successive sections, allowing a clear visualization of the deformation pattern.

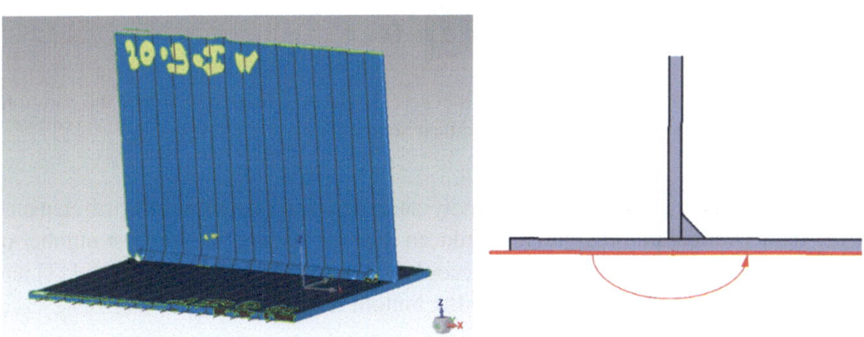

Fig. 4. On the left an acquisition of a welded T-Joint being measured. The section on which angular distortion is measured are shown. On the right the visualization of the angle measured.

The T-Joint model is evaluated and tuned in parallel to the experimental stage. Simufact Welding© has been the software selected as the base CAE tool for the model. To set the heat source correctly, Goldak's parameters are derived from the dimensions of the weld pool: the specimens are sectioned, lapped and then the welding pool is chemically etched with Keller solution to reveal the weld seam boundaries. The metallographic samples obtained are analysed using a Leica EZ4 stereomicroscope. The joint is discretized using cubic elements, chosen in literature for their better performances compared to shell elements in similar case studies [10].

To optimize the simulation model, a sensitivity analysis was carried out on the quality of the mesh. In a first phase the number of layers of finite elements along the thickness of the plates was considered. According to literature, at least 5 layers are required to correctly describe the deformation of the plate: the analysis has shown few differences over the 5 layers, highlighting differences only when the number of layers around the area of the weld bead is reduced too much. Consequently, 3 different areas have been defined which respectively identify the fusion zone, FZ, and heat effected zone, HAZ, and heat parent metal zones, PMZ. Figure 5 depicts the refinements areas while Table 2 collects the layer number of layers used along different simulations.

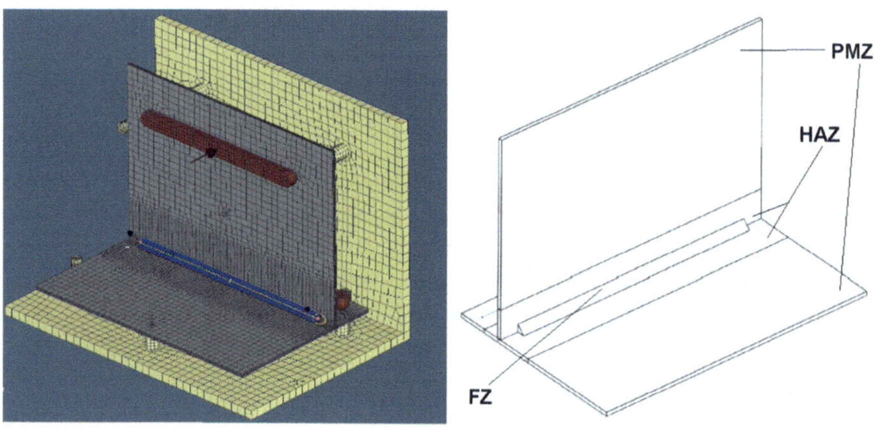

Fig. 5. Left side, simulation setup in CAE software. Right side, representation of the partitions of the model based on the distance from the heat source.

The subsequent phase of the sensitivity analysis concerns the number of elements of the model which are progressively reduced (always keeping the greatest number of elements around the welding bead, in accordance with the previous literature [11] and tests) evaluating the distortion angle of the bottom flange.

Table 2. Number of layers used to discretize plate thickness in the different areas.

Simulation	PMZ layers Web	HAZ layers Web	FZ layers	HAZ layers Flange	PMZ layers flange
1	5	5	5	5	5
2	5	5	10	5	5
3	5	10	10	10	5
4	10	10	10	10	10

4 Results

The CAE-based models of aluminium alloys welded T-Joints for TEP analysis demonstrates accuracy in predict the pattern of distortion caused by welding, but they succeed in quantifying the distortion angle only in some cases with errors ranging from 10% to 70%, overestimating the distortion effect.

The results obtained from two different T-Joint samples are provided in Figs. 6 and 7 with similar trends, but respectively there is a bad and a good matching between angular distortion.

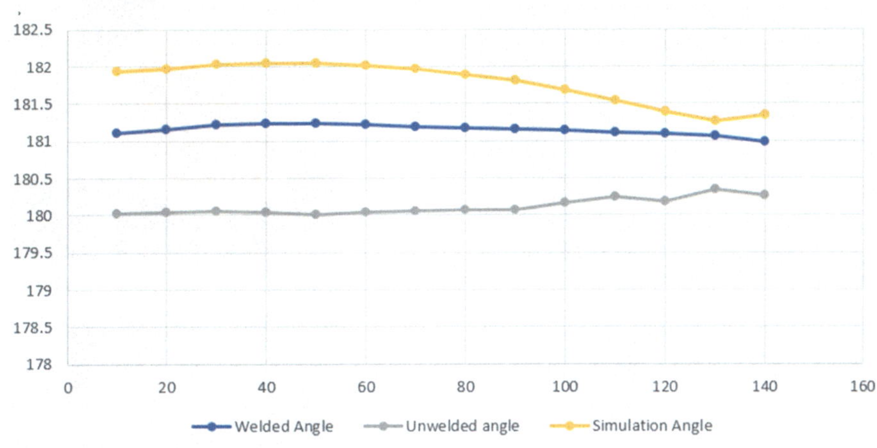

Fig. 6. Simulation and measurements on a T-Joint consisting in a 2 mm and a 3 mm plates, welded with a 45° work angle and 0.5 mm gap. The pattern predicted by the simulation is very similar to the measured one. However, the simulated angular distortion is bigger than the real one

Whit respect to deviation between the experimental T-Joints and related simulated models, the presence of the gap and the variation of the working angle proved to be critical, making it difficult to correctly represent the weld pool. This is the evidence that the model lacks of the parameters to replicate the softening effect related to aluminium alloy selected for the T-Joint [12]. Evidence of the phenomenon comes out also from analysis of the gap. As showed by the two pictures in the top part of Fig. 8, the model fails to capture the behaviour of the recovery of the gap. Conversely, in the absence of

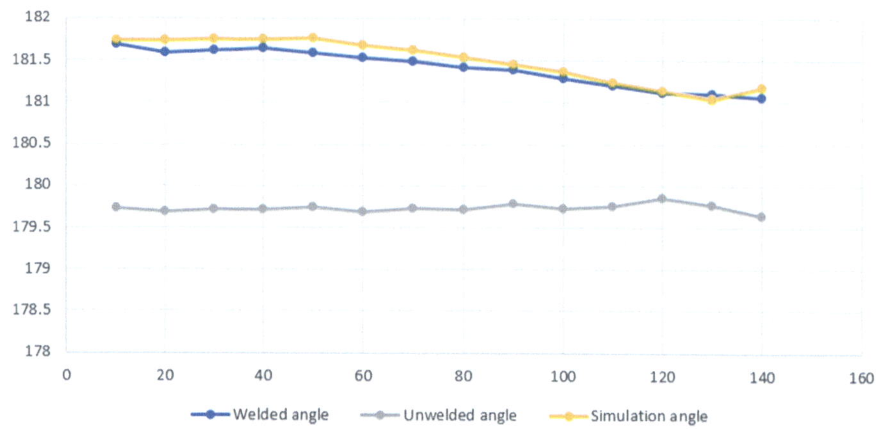

Fig. 7. Simulation and measurements on a T-Joint consisting in two 2 mm plates, welded with a 60° work angle and 0.5 mm gap. The pattern predicted by the simulation is very similar to the measured one and also the angular distortion is very accurate.

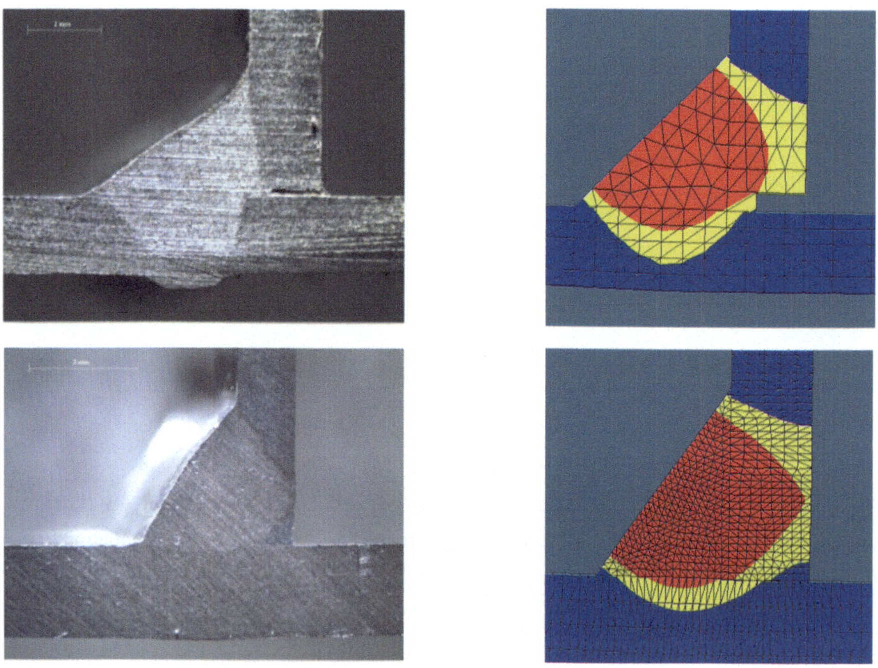

Fig. 8. Upper part of the figure: differences between real and simulated weld pool in a 0.5 mm gap joint. Lowest part of the figure: differences between real and simulated weld pool without gap.

gap and working angle conditions at 45°, the simulation replicates the shape of the weld pool with extreme precision.

The sensitivity analysis performed on the mesh is mainly based on the distance from the weld bead. For that areas the joint can be discretized differently, showing that progressively looser meshes cause deformations that are less different from each other than the critical areas near the heat source. Therefore, the use of looser meshes for the ends of the plates does not affect the goodness of the result allowing to improve the computational efficiency of the model, as often reported in literature [13]. Using a limited number of elements away from the bead while maintaining an adequate number close to the heath source results in an excessive distortion of the finite element: this introduces errors and must be avoided, although it is a limitation in the construction of the mesh. For the case study it has been quantified that between the 70,000 and 80,000 elements it is possible to considerably reduce the wall-time with minimal variations in the distortion angle. Figure 9 provides the results of sensitivity analysis with respect to the 4 simulations as described in Table 2.

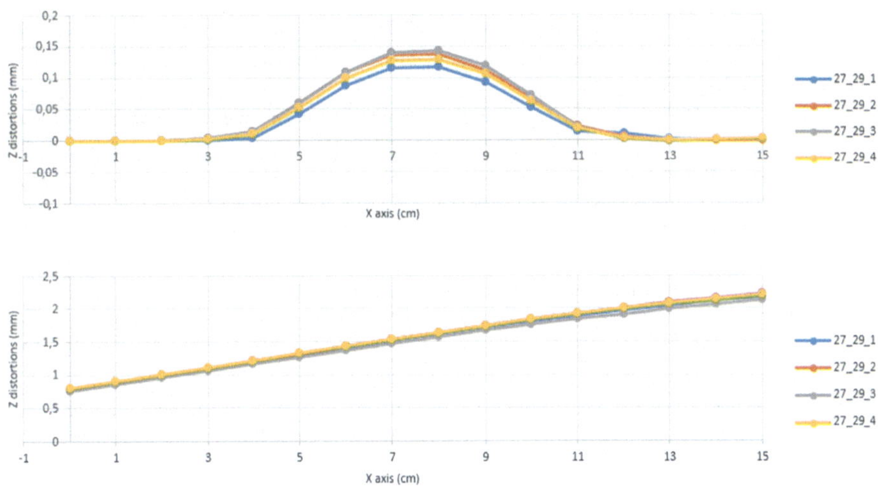

Fig. 9. Top: displacement pattern measured on the back edge of the flange. Bottom: displacement measured on the front edge of the flange. Difference are negligible except for simulation 1.

5 Conclusion

The current work proposes a CAE-based model to perform TEP analysis of aluminium alloy welded T-Joint. The aim is to identify an effective modelling strategy which returns a good tradeoff between accuracy and computational weight.

The proposed cases demonstrate the potential of CAE tools in obtaining data on the joint quality useful in the industrial field with acceptable wall-time. The sensitivity analysis performed on the mesh is effective reducing the simulation time of the optimized model up to a third, and being, for the same results, faster than the TEP models present in the literature. Conversely, further studies are needed to improve the accuracy of the results, in particular in relation with the correct modelling of the heat source in the software and material behaviour as well. In this regard, the extension of the experimental campaign would allow to investigate the influence of input parameters (gap, work angle, plate thickness) and their variation on distortion, thus defining a process tolerance.

References

1. Deng, D., Murakawa, H., Liang, W.: Numerical simulation of welding distortion in large structures. Comput. Methods Appl. Mech. Eng. **196**(45), 4613–4627 (2007)
2. Bonazzi, E., Colombini, E., Panari, D., Vergnano, A., Leali, F., Veronesi, P.: Numerical simulation and experimental validation of MIG welding of T-joints of thin aluminum plates for top class vehicles. Metall. Mater. Trans. Phys. Metall. Mater. Sci. **48**(1), 379–388 (2016)
3. Goldak, J., Chakravarti, A., Bibby, M.: A new finite element model for welding heat sources. Metall. Trans. B **15**(2), 299–305 (1984)
4. Duan, Y.G., Vincent, Y., Boiutout, F., Leblond, J.B., Bergheau, J.M.: Prediction of welding residual distortions of large structures using local/global approach. J. Mech. Technol. **21**(10), 1700–1706 (2007)
5. Murakawa, H., Deng, D., Ma, N., Wang, J.: Applications of inherent strain and interface element to simulation of welding deformation in thin plate structures. Comput. Mater. Sci. **51**(1), 43–52 (2012)
6. Lu, C., Wang, J.: Prediction of Welding Deformation with Inherent Strain Method Based On FEM. Shangai Huizhong Automotive Manufacturing Co., LTD report (2000)
7. Zardin, B., Borghi, M., Gherardini, F., Zanasi, N.: Modelling and simulation of a hydrostatic steering system for agricultural tractors. Energies **11**(2), 230 (2018)
8. Lee, D.C., Han, C.S.: CAE (computer aided engineering) driven durability model verification for the automotive structure development. Finite Elem. Anal. Des. **45**(5), 324–332 (2009)
9. Pini, F., Panari, D., Leali, F.: Computer-aided integrated method for effective development of robotic welding processes of car chassis assembly. DYNA **94**(2), 150–154 (2019)
10. Rezaei, A., Ranjbarnodeh, E., Iranmanesh, M.: Comparison of full 3D, shell/3D and inherent welding-induced distortion. J. Braz. Soc. Mech. Sci. Eng. **40**(6), 287 (2018)
11. Long, H., Gery, D., Carlier, A., Maropoulos, P.G.: Prediction of welding distortion in butt joint of thin plates. Mater. Des. **30**(10), 4126–4135 (2009)
12. Ferro, P., Berto, F.: Quantification of the influence of residual stresses on fatigue strength of Al-alloy welded joints by means of the local strain density approach. Strength Mater. **48**(3), 426–436 (2016)
13. Deng, D., Murakawa, H.: Prediction of welding distortion and residual stress in a thin plate butt-welded joint. Comput. Mater. Sci. **43**, 353–365 (2008)

Dynamic Modelling of Mechanical System for the Packaging Industry

Raffaele Di Canosa(✉) and Francesco Pellicano

Dipartimento di Ingegneria "Enzo Ferrari",
Università di Modena e Reggio Emilia, Via P. Vivarelli 10, 41125 Modena, Italy
{Raffaele.dicanosa, Francesco.pellicano}@unimore.it

Abstract. The development of robust and reliable systems in the packaging industry requires the creation and the validation of physical models that can estimate both the mission profile of the components and the sub-assemblies and to provide data related to the status of the system when operating out of the nominal working condition.

The work relies on the creation of a multibody model of a complex mechanism covering multiple phases of the production process in each operating cycle. The model has been developed with the commercial software MSC Adams® and validated on a test rig specifically manufactured for the purpose. Some components of the system were redesigned in order to behave as sensing elements and to provide data about the forces exchanged in the system, which were correlated with the virtual results.

The dynamic model will be used in further activities for integrating the response of the system when interacting with the other assemblies of the equipment, for simulating operating conditions that are out of the design ranges and collecting data from a system that is approaching failure.

Keywords: Multibody simulation · Experimental measurement · Complex dynamics

1 Introduction

The development of robust and reliable systems in the packaging industry requires the creation and the validation of physical models that are able to estimate both the mission profile of the components and the sub-assemblies and to provide data related to the status of the system, when operating out of the nominal working condition.

Through validated virtual models it might be possible to replicate issues detected during field operations, to investigate the behaviour of the equipment in case of anomalies and to study the impact of both the wear out of the components and the deterioration of system on the overall dynamic response of the equipment.

The design and modelling of complex machines requires a deep study of the dynamics involved in the process to optimize the production speed and improve the efficiency of the whole systems. Multibody modelling have the advance, once the model is experimentally validated, to optimize several process parameters, to investigate different process cycles and to study their effect on the quality of the formed

C. Rizzi et al. (Eds.): ADM 2019, LNME, pp. 589–597, 2020.
https://doi.org/10.1007/978-3-030-31154-4_50

products; in [1] Tanninen et al. optimized a parameter, force curve of the stroke, without reducing the production efficiency, moreover Golzar [2] studied the effect of temperature on the spring back of cellulose-based sheet in hot pressing using simulation results conformed to the experiments. Furthermore, thin shell structures, as the cylindrical paperboard that is processed in the present packaging machine, could present nonlinear dynamics phenomenon as noted in [3, 4] where the effect of temperature could arise instabilities and violent response in certain loading conditions. Have a deep knowledge of the system, is a keystone to forecast possible failures or damages that could occurs in particular operating conditions; Sapietová in [5] presented a research to study different type of misalignment of shaft simulating the rotating machine with MSC. ADAMS, considering parameters as deflection, velocity and acceleration and deriving the frequency spectrums to detect symptoms of failure states as with novel techniques as been studied by Gelman et al. [6] for a gearbox industrial application.

2 Model

The modelled system represents a mechanical assembly that is part of a packaging line, shown in Fig. 1, where a cylindrical paperboard slides through the packaging machine for the formation of the final product. The study rises from the need of understanding the complex dynamics shown, in former measurements, from the thin cylindrical structure during operation conditions.

Fig. 1. Picture of the packaging machine

The main function of the group is to drive the displacement of the cylindrical tube through several sub-phases, which are activated by both cams and electromechanical devices and that allows the creation of a sealed and folded package that is filled with liquid food.

The mechanism can be simplified in few elements, see Fig. 2: the chains (A), having the function of displacing the product, one driving wheel (B) in the right side of the mechanism, the tensioners (C) and the cams (D), that have the function of both constraining the displacement of the chain and activate some phases of the production process.

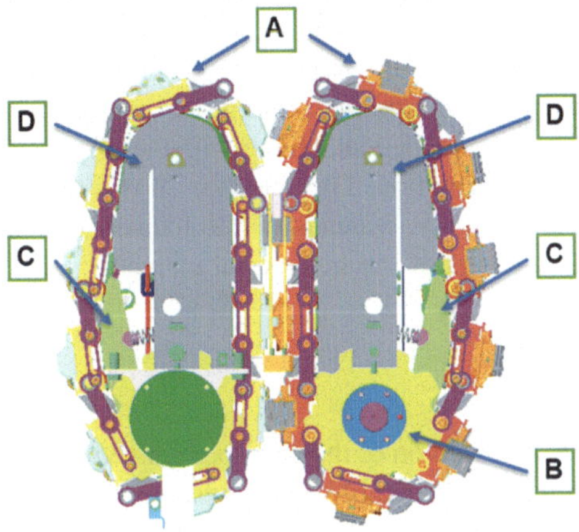

Fig. 2. Mechanism Model, front view

The chains are a twin paired system: they are rotating in opposite directions, moving the packages from the top to the bottom, one is connected to a driving wheel and it transfer the rotation to the other chain through the transversal elements, see Fig. 3. Those elements carry four ball bearings (E) that rotating on the cams which guide the displacement of the chain.

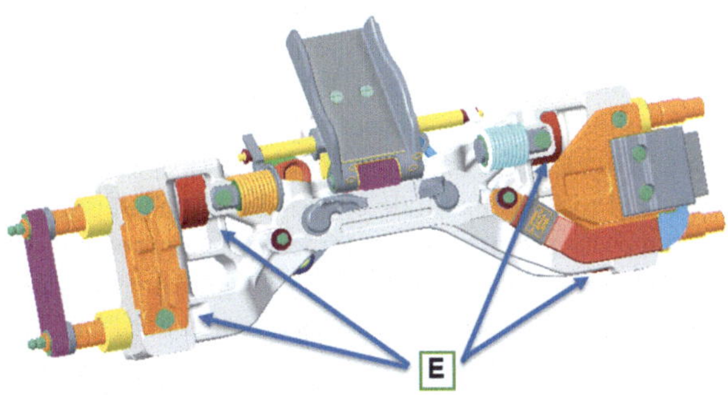

Fig. 3. Mechanism Model, transversal element detail

The main function of the tensioner is to recover the dynamic fluctuations of the chain, that are mostly created by the contact with the driving wheel and that are related to the polygonal effect during the rotation.

The cams ensure the contact between the transversal elements of the chain, that is fundamental for: transferring the motion, controlling the position of the product and driving the motion of the sub-assemblies installed in the transversal elements of the chain.

The multibody model was built with the commercial software MSC Adams®, considering in this part of the study only rigid bodies. The interactions among the elements of the system have been modelled as contact forces between solids, when occurring between the chains and the tensioner, wheel and cams, and as forces when generated by the contact with the product during the manufacturing phases.

The solid contacts were modelled according with the Impact Method, neglecting, in this phase, the frictional contribution. The selection of this type of contact was related to the intrinsic nature of the study, which aimed to focus on the stresses applied on the shaft carrying the ball bearings of the transversal element.

3 Test Rig

The test rig was a simplified version of the real operating equipment: all the elements of the chain were recreated as the standard version system in order to replicate the inertial properties. As in the virtual model, the product managed by the mechanism was replaced by springs, whose characteristics and installation position were calculated for achieving the identical stresses that are applied on the operative system. The target value of the forces and the torques were either data coming from former measuring campaigns or estimations provided by previous simulations.

The present work focuses on the resulting forces applied on the shaft carrying the ball bearings, thus an ad hoc sensor was developed for collecting the data related to the forces exchanged in the operative conditions and comparing them with the results obtained from the simulation.

The sensors were designed with the same geometry of the real shaft in order to preserve the coupling tolerances and interfaces, but a special geometry was obtained in the cross sections adjacent to the joints in the transversal element of the chain. The reduced resistance of those sections made the sensor sensitive to the shear stress applied by the ball bearing during the contact with the cams. The drilled sections were equipped with strain gauges, see Fig. 4.

In order to withstand to the magnitude of the loads applied during some phases of the working cycle and to preserve the geometrical requirements, the sensor was manufactured in the stainless steel AISI 630. The measuring capability of the sensor was limited only to one radial direction, that was oriented normally to the trajectory of the shaft. Additional reference surfaces obtained on the sensor allowed the possibility of installing the shaft in a position rotated of 90° for collecting the value of the force in two independent directions.

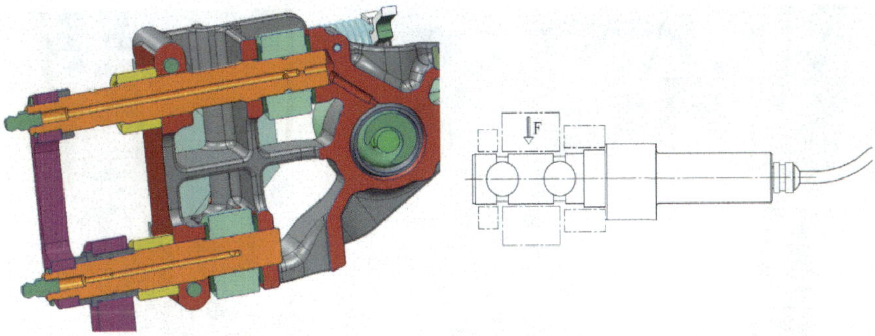

Fig. 4. Transversal element cross-section, shaft detail (a) and Sensor sketch (b)

The calibration curves of the sensors (Fig. 5) were obtained in the final assembled configuration, in order to compensate the manufacturing deviations of the interfaces or the errors caused during the installation.

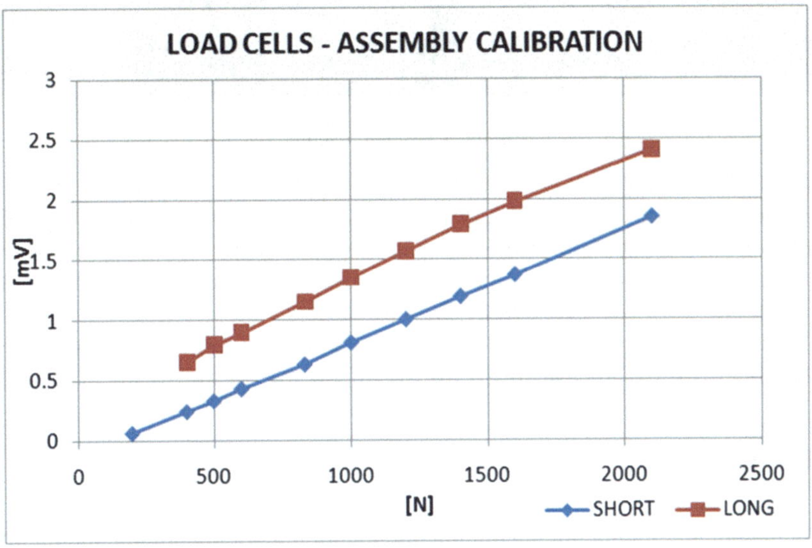

Fig. 5. Sensor calibration curves

In Fig. 6 a picture of the calibration setup has been shown; a wireless datalogger was required for collecting data from the operating system, since there was no access to the investigated shaft during the functioning of the equipment.

Due to the limited available space in the assembled system, restrictive dimensional constrains were prescribed for the data logger. The selected logger is the BeanDevice® DAQ connected with BeanGataway®, that has four channels available and the maximum sampling rate of 2 kHz.

Fig. 6. Calibration setup

In addition to the strain gauges, two triaxial accelerometers were installed both on the transversal element of the chain and on the frame of the testing equipment, with the purpose of detect unexpected hits between the elements of the chain and the cams or the frame, that were not measured by the load sensors.

An inductive sensor has been used to synchronizing the signals of the sensors with the working cycle of the chain.

4 Results

The analysis of the results is focused on the load, simulated and measured on the shaft carrying the ball bearings, which moves along the cam profile.

The Fig. 7 represents the simulated trend of the force during one complete cycle of the chain and the process phases can be clearly recognized: the shaft is not loaded in the beginning, than a peak is calculated when the bearing hit the tensioner and a load of low intensity is maintained during the contact with the tensioner profile, when leaving the tensioner contact the ball bearing is unloaded until it matches the cam and the further force values are growing since the transversal element is located in the upper position of the system and it is loaded by the weight of the entire chain, the next values refers to the load applied during the match with the paired element of the other chain, that is required both for completing a production phase and for transferring the motion between the chains.

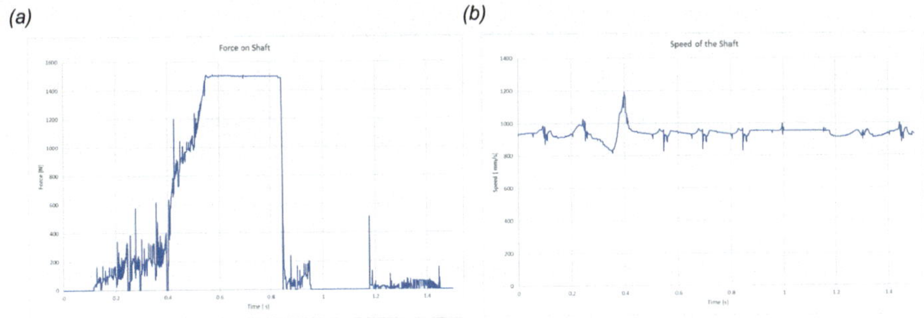

Fig. 7. Simulation of force trend (a) and speed trend (b) of the shaft in one operating cycle

The Fig. 7/b represents the tangential speed of the shaft along its trajectory, it is characterized by fluctuations that are correlated with the pitch of the chain, hence with the amount of transversal elements, that are 10, and with the trajectory of the chain, which describes a curve with a smaller radius in the upper position and it is subjected to a more intense acceleration.

The experimental data collected in the test rig showed a good matching with the values obtained by the simulation and provided some insight for tuning the model. The contact with the tensioner has a smaller intensity in comparison with the load applied during the production phase and it is hidden within the noises of the force signals. The growing trend of the force is detected in the upper positions of the trajectory of the transversal element and it is caused by the weight of the chain. Both these values as well as the plateau associated to the production phase are comparable with the simulated trends. In this section of the signals a fluctuation pattern related both to the tangential speed fluctuation and to the pitch of the chain is visible.

The peak, of the orange line, which is visible in the beginning of the plateau is generated by the interaction with the product and it is influenced by the manufacturing variability of the components, in the model it was represented as a localized force, hence lightly underestimated.

The comparison of the signals provided by the two shafts, highlighted that the load was not equally distributed among the four bearings, the causes of this gap was correlated both to the possible misalignment of the physical test rig and to the deformation of the transversal element, that was not considered in the multibody model. The hypothesis is confirmed by the presence of a peak after the highest plateau (Time 1.55 s – Fig. 8) only in the measurements collected on the long shaft; the peak corresponds to the one at about 1.2 s in the graphic in Fig. 7/a.

The measurements collected with the shafts rotated of 90° as well as the signal provided by the accelerometers did not highlight any relevant load applied on the shaft along the direction of the trajectory, confirming that the discontinuities between the cams were properly linked (Fig. 9).

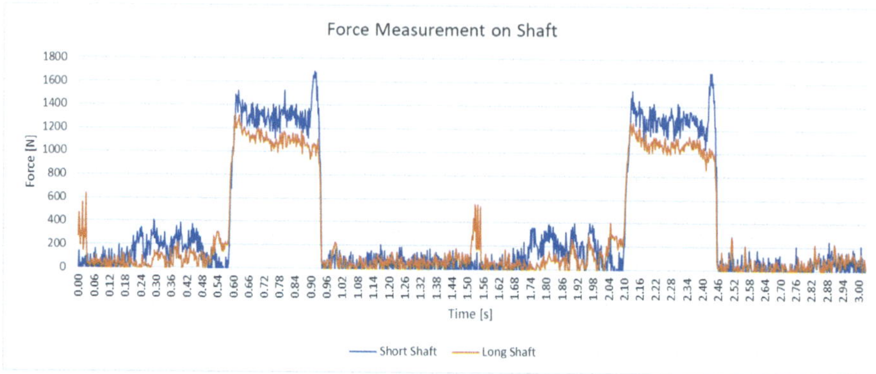

Fig. 8. Experimental measurement of Force on the Long and Short Shafts in 2 operating cycles

Fig. 9. Superposition of simulation and measurement results

5 Conclusion

The dynamic model created for simulating the mechanical system embedded into a production machine operating in the packaging industry provided a good representation of the real behaviour of the equipment.

The validation of the model was based on the correlation of the calculated and measured values of the forces applied on a shaft carrying a ball bearing. The process required the manufacturing of a test rig and special parts equipped with strain gauges.

The measurements on the test rig presented a good matching with the simulation, they provided insights for tuning the model and highlighted some deviations that might be generated on the system due to the manufacturing process of the mechanism.

The future development of the work will implement the optimization of the coefficients adopted for modelling the contacts and the implementation of flexible bodies in the chain.

References

1. Tanninen, P., Leminen, V., Matthews, S., Kainusalmi, M., Varis, J.: Process cycle optimization in press forming of paperboard. Packag. Technol. Sci. **31**(5), 369–376 (2018). https://doi.org/10.1002/pts.2331
2. Golzar, M., Ghaderi, A.: Effect of temperature on the spring back of cellulose-based sheet in hot pressing. Int. J. Adv. Manuf. Technol. **42**(7–8), 633–642 (2009). https://doi.org/10.1007/s00170-008-1626-y
3. Zippo, A., Barbieri, M., Iarriccio, G., Pellicano F.: Nonlinear vibrations of circular cylindrical shells with thermal effects: an experimental study. In: Nonlinear Dynamics (2019). https://doi.org/10.1007/s11071-018-04753-1
4. Zippo, A., Barbieri, M., Pellicano, F.: Temperature gradient effect on dynamic properties of a polymeric circular cylindrical shell. Compos. Struct. **216**, 301–314 (2019). https://doi.org/10.1016/j.compstruct.2019.02.098
5. Sapietová, A., Dekýš, V.: Dynamic analysis of rotating machines in MSC.ADAMS. Procedia Eng. **136**, 143–149 (2016). https://doi.org/10.1016/j.proeng.2016.01.188
6. Gelman, L., Harish Chandra, N., Kurosz, R., Pellicano, F., Barbieri, M., Zippo, A.: Novel spectral kurtosis technology for adaptive vibration condition monitoring of multi-stage gearboxes. Insight Non-Destructive Test. Condition Monit. **58**(8), 409–416 (2016). https://doi.org/10.1784/insi.2016.58.8.409

Experimental Methods in Product Development

How Do Design Changes and the Perception of Product Creativity Affect Value?

Yuri Borgianni[1(✉)], Lorenzo Maccioni[1], Guido Orzes[1], and Demis Basso[2,3]

[1] Faculty of Science and Technology, Free University of Bozen-Bolzano, Piazza Università, 3, 39100 Bolzano, Italy
yuri.borgianni@unibz.it
[2] Faculty of Education, Free University of Bozen-Bolzano, Viale Ratisbona, 16, 39042 Brixen, Italy
[3] Centro de Investigación en Neuropsicología y Neurociencias Cognitivas, Universidad Catolica del Maule, Talca, Chile

Abstract. A considerable part of the design literature focuses on creativity and puts forward means to enhance creativity. It is assumed that boosting creativity results in product improvements and benefits for many stakeholders, starting from the recipients of design deliverables. However, the actual outcomes of creative endeavors and, especially, creative products have never been assessed systematically. Within this overall goal, the present paper compares the results of two experiments in which the same participants were involved. Both experiments were meant to evaluate according to multiple dimensions couples of products, where an element of the pair was subjected to some variations. The experiments foresaw the use of eye-trackers to achieve additional behavioral information. As creativity or variations thereof (novelty, unusualness) were assessed, it is possible to infer dimensions of value that are affected by creativity in multiple settings. The outcomes show that creative products significantly give rise to increased exploration time and efforts to make sense out of objects, but this process leads nevertheless to difficulties in understanding the products and to the identification of some disadvantages. According to these preliminary results, the relationship between creativity and perceived value, as well as their measurement, is overall dubious and highly depending on circumstances. While the authors support the relevance of design creativity, the paper urges to consider value at the same time.

Keywords: Design creativity · Perceived value · Eye-tracking · Product evaluation · Fused Deposition Modelling

1 Introduction and Background

Especially in the early product development stages, creativity is attributed of outmost importance in the design field [1, 2]. While it is commonplace to consider design creativity as a source of advantage and improved performances, some contributions openly state how creativity is perceived as a driver of success [3], economic development [4] and innovation capabilities [5] also outside of academia. The change

C. Rizzi et al. (Eds.): ADM 2019, LNME, pp. 601–611, 2020.
https://doi.org/10.1007/978-3-030-31154-4_51

brought on by creative moves is supposed to explain the success of new value propositions [6] or new technological paradigms [7]. However, different transformations are responsible for higher or lower chances to thrive in the marketplace. In a retrospective analysis of award-winning creative mechanical products, [8] stresses how some potential innovations have never reached the marketplace, as well as the number of innovative characteristics is not conducive to the achievement of success. Within eco-design, although creative green products are overall appreciated, more established ones are likewise attributed of value, mostly in the utilitarian form, and this is supposed to hinder their replacement [9]. This is largely in line with [10], which shows that creative products perceived as meaningful are attributed of utilitarian value, while the simple display of novelty does not ensure consumers' appreciation, unless it is not accompanied by coolness. The distinction between utility-oriented and originality-oriented forms of creativity are focused on also in [11], even if, actually, both usefulness and novelty are considered as the pillars for a product to be considered as creative, e.g. [12].

Therefore, the link between creativity and capability of succeeding is seemingly more nuanced than a hypothetic direct relationship. Just a few studies have focused on said relationship as a core objective of their research. An exception can be represented by [13], in which a firm's self-assessment of its products results in a correlation between creativity and success, where the former is characterized in terms of both form and function creativity. However, too little evidence has been gathered so far to state the actual contribution of creativity to the determination of success. To the scope, an obstacle is clearly represented by the fact that success metrics are established on a case to case basis, few of those can be measured quantitatively, and products' market and financial outcomes are seldom available [14]. Another issue concerns the nature of creativity, which seems to be activated only if the task requires an action [15]. In addition, while creativity is exploited during product development and can be evaluated at the end of the design process, success can be judged just a posteriori, after social and market conditions could have undergone unpredictable changes. To this respect, it seems more appropriate to study the link between creativity and value, which can be seen as both the actual capability of creative transformations to be appreciated and a precondition for future success [16]. For the sake of clarity, value is here intended as the presence of perceived functional, emotional or hedonic benefits enabled by a designed product. The creation of value is inherently linked to design activities, and this kind of connection is stressed for the (re-)design of products [17], services [18] and processes [19]. Here, a fundamental pitfall is the possible interpretation of value and its variants, such as customer satisfaction or user experience. While mainstream research sees value as a broader interpretation of customer satisfaction [20], some studies consider utilitarian value only and indicate it as a component of creativity in lieu of usefulness [21]. In the present paper, the authors refer to a more articulated characterization of (the perception of) value, which includes emotional and affective dimensions besides functional and utilitarian ones [9]. In other terms, value is intended in the paper as the presence of perceived functional, emotional or hedonic benefits enabled by a designed product.

The objective of the paper is to add knowledge concerning the influence of (perceived) creativity or terms that can be associated to creativity on (perceived) value. The authors, as better explained in Sect. 2, benefit from different experiments with plainly different goals and experimental conditions. The comparison leads to check whether certain creativity-value links have the chance of being considered of general meaning or they are deliberately restricted to specific experimental conditions (Sect. 3). The main evidence and implications of the results are discussed in Sect. 4, while Sect. 5 draws conclusions and outlines future work.

2 Experiments and Sources of Data for the Present Investigation

In [9], the authors motivate why creativity can be considered as a dimension of value, if the latter is interpreted with its broadest meaning. Nevertheless, Sect. 1 has presented some findings that make this assumption (partially) invalid and motivate the need for more insightful studies. To this respect, it is useful to decouple creativity, along with its variants, and the other factors contributing to perceived value, so that their relationship can be analyzed. Undoubtedly, a large number of studies would be necessary to attain generalizable outcomes: the present paper intends to move a first step in this direction by benefitting from already available data. In particular, the sources of data are two recently conducted multiple-session experiments described in [9, 22], whose details are provided below and that have involved 43 participants with different backgrounds, although all involved in study or research at the University level.

In [9], the participants were asked to evaluate 40 products (20 pairs), each of them displayed at a time on a computer screen. The chosen products are freely accessible in the Appendix of [9]. Each pair was constituted by a reference product (baseline) and an alternative version (green), which exhibits the implementation of one or more eco-design principles. The evaluation consisted in assessments of the quality and expected performances of products, beyond an evaluation of their perceived novelty and creativity (a single variable). The experiment foresaw the use of some biometric instruments, increasingly diffused in design and product development research [23], and markedly eye-tracking, to explore inadvertent participants' reactions. Based on employed instruments, those unintended reactions might include emotions, attention, concentration, stress. More in details, eye trackers are increasingly diffused in engineering and design research used to determine and measure:

- Fixations, i.e. people's continuous observation of a specific point or region, which feature interest or attention meant to extract information from the area or element focused on [23];
- Saccades, i.e. eye movements between two different fixation points, whose trajectories do not include interesting elements for the scope of achieving relevant information [23];
- Pupil diameter, whose values and variations are proxies of attention, concentration [24], stress or workload [25].

In [22], 20 physical objects (10 pairs) were the subject of evaluation. Each pair was formed by a commercial product and a replica manufactured with a diffused

3D-printing technology, namely Fused Deposition Modelling (FDM). The products are displayed in Fig. 1.

Fig. 1. Products presented in [22], whose evaluations benefit the present study

Participants' assessments targeted appropriateness and the extent to which objects were considered fit to purpose, beyond the degree of unusualness. Also in this case, participants' visual behavior was monitored by means of eye-tracking technologies.

The following conditions can be mentioned that allow using the two experiments conjointly for the scopes of the present study.

- The same participants (43) were involved in the two experiments, whose sessions were alternated. This ensures that the outcomes of the comparison between the two experimental conditions is not affected by diverging personal inclinations. Therefore, said comparison can be overall referred as a within-subject test, by considering the different products/objects to be evaluated as two different treatments.
- In both cases, evaluations referable to creativity or, more in general, diversity, have been made. As already mentioned, the evaluation metric was "Novelty/Creativity" in [9], whereas, "Unusualness" was assessed in [22]. Undoubtedly, the two concepts do not overlap thoroughly, but their exploration can provide knowledge of the effects caused by the perception of something being deemed as diverging from standards. For instance, review works [26, 27] highlight the undisputed role played by novelty in evaluations of creative products and include unusualness as possible synonym, variation or nuance of novelty.
- Also with respect to value dimensions, the same terms and variables were not extracted in the two experiments. However, both the experiments included the elicitation of a variety of terms addressing the perception of quality and utility, presence of advantages/disadvantages, knowledge and understanding. In addition, variables are also available extracted by means of eye-tracking instruments during a free interaction with the products or their depictions. Beyond physiological, e.g.

pupil diameter, and behavioral measures, e.g. fixations and saccades, the use of eye-tracking technologies allowed the authors to extrapolate variables concerning the nature of participants' interactions, such as their duration and frequency.

- Some common measures were taken to enhance the validity of the results, for instance randomizing the order of presentation of products/objects, avoiding the presence of products/objects belonging to the same pair in the same experimental session, disclosing the objectives of the experiments just after all experimental phases had been completed.

It is worth clarifying that the aims of the present paper differ from those illustrating the two experiments in details and, here, the authors will make no reference to the effects of the main manipulated variables (eco-design principles and FDM 3D-printing) on value perception. As the focus is on creativity-related dimensions, these metrics will be considered irrespective of their attribution to baseline vs. sustainable products, as well as commercial objects vs. FDM-fabricated replicas. Thus, the considered samples of products (objects) analyzed by 43 people is 40 (20) with reference data extracted from [9] ([22]).

3 Statistical Analysis and Main Outcomes

3.1 Procedure to Obtain Statistical Evidence

The statistical analysis consisted in multivariate regressions linking value dimensions (dependent variables) and creativity assessments (regressor). This process was repeated for data extracted from both the two experiments by using the software R. Value-related variables include assessments, predominantly in the form of a 0 to 5 Likert scale, and measures obtained by means of eye-tracking. This implies that, according to the nature of those variables, ordinal or continuous regressions were leveraged. The considered dependent variables were restricted to those mutually ascribable across the two experiments. In other words, the authors did not take into account those variables that were present in one experiment only and could not be connected to anything similar in the other one.

The regressions gave rise to outcomes in terms of regression coefficients and p-values. As a common rule of thumb, the relationship between the dependent variable and creativity is considered significant when the corresponding p-value is lower than 0.05. Positive (negative) regression coefficients indicate the presence of a direct (inverse) relationship between the dependent variables and creativity – this applies to both continuous and ordinal regressions although the mathematical interpretation of the coefficient differs.

3.2 Most Meaningful Results

Table 1 presents a summary of the regression results that have involved equivalent or akin value dimensions significantly affected by the creativity-related variables in both dimensions. The first (and fourth) column indicates the definition of the dependent variable explored in [9] ([22]). The second (and fifth) column indicates the extracted regression coefficient for the creativity/novelty (unusualness) term, while the p-value of such a relationship is reported in the third (and sixth) column. As variables have different units of measurement and statistical distribution, beyond being either ordinal or continuous, the extent of the regression coefficients is not to be considered as the power of creativity measures to affect those dimensions. The seventh column indicates whether there is agreement (Yes or No) in the kind of effect (negative or positive) of the creativity-related term on the considered value dimensions.

Table 1. Value dimensions that have resulted significant in both the scrutinized experiments.

Value dimension as in [9]	Coeff. novelty creativity	P-value	Value dimension as in [22]	Coeff. unusualness	P-value	Agreement
Absence of disadvantages	–0.237	0.000	Absence of cons	–0.614	0.000	Y
Knowledge	–0.395	0.000	Comprehension	–2.603	0.000	Y
Preference	0.266	0.000	Liking	–0.145	0.001	N
Quality	0.098	0.016	Appropriateness	–0.604	0.000	N
Willingness to pay	0.494	0.000	Desire	–0.872	0.000	N
Overall involvement time	1514.300	0.000	Overall involvement time	5069	0.000	Y
Number of interactions	0.009	0.043	Number of interactions	0.101	0.007	Y
Number of saccades	4.853	0.000	Number of saccades	5.301	0.000	Y
Overall duration of saccades	302.940	0.000	Overall duration of saccades	261.9	0.000	Y
Number of fixations	2.175	0.000	Number of fixations	5.939	0.000	Y

In the Table, it is evident that significant variables overlap across the two experiments to different degrees. All those extracted from eye-tracking technologies are considered for the time intervals during which interactions with products took place. The other compared variables differ to a larger extent. The absence of disadvantages or cons were evaluated by participants in both cases with the same Likert scale. An ordinal scale and participants' assessment feature also Knowledge, Preference, Quality and Willingness to pay in [9], as well as Liking, Appropriateness and Desire in [22]. Conversely, Comprehension was assessed by the authors with an ordinal scale, based on participants' declared understanding of the name and/or the function of the object in [22]. However, the correspondence between Knowledge and Comprehension is deemed quite obvious, as they both refer to the understanding of the artefact, its functions, reasons and conditions for using it. The terms presented between the third and fifth row of Table 1 (all those reporting disagreement in the seventh column) do not overlap completely, although all of them feature perception of quality and appreciation. Overall, they are positively affected by creativity in [9] and negatively affected by unusualness in [22]. More in details, the reasons for comparing the mentioned variables, despite some inherent differences, are explained below.

- Preference features the chance of choosing the artefact over similar ones, while Liking is an indicator of perceived attractiveness of the object; the latter is plainly a precondition for the former, as emphasized in the Attention, Interest, Desire, Action model in [28].
- Quality indicates the adequateness of expected performances, while Appropriateness mainly refers to the object's fitness for its scopes; both terms are expectedly proxies of perceived utilitarian value and are compared for this reason.
- Willingness to pay and Desire both characterize a positive predisposition towards the object; the main difference stands in the former's additional element of availability to spend money for acquiring the product.

4 Discussions

4.1 Comments on the Results

The data presented in Table 1 show that participants who evaluated products high in creativity seldom perceived them as free of disadvantages. This might mean that the creative transformation process, while including novel elements, implies the introduction of some (real or perceived) limitations with respect to more traditional artefacts. The process of realizing what the product is for is viable to affect the perceived presence of weaknesses or inadequate aspects. Several indicators consistently show that the acquisition of relevant information is lengthier and more strenuous when forms of newness are in play. At first, the process leads less likely to the correct identification of a product and its functions. The longer times spent when interacting with products have, therefore, just partially helped participants to make sense of objects deemed as creative. For the sake of clarity, the mentioned lengthier interactions and examinations are inferable by the positive relationship between creativity and, markedly, involvement times and frequencies. These prolonged and repeated interactions have brought to the significant increase of both saccades (in terms of number and duration) and fixations (number). Then, the analysis of creative artefacts has included significantly more explorations of elements attracting attention for achieving information (or more repetitions thereof) and, more markedly, untargeted wide eye movements. While pupil diameter data could not be compared across the two experiments, the surge of saccades is a good proxy of increased efforts to perform the analysis of the products in light of the evaluation tasks included in [9] and [22]. By the way, further data analysis suggests that the higher number and longer duration of saccades (as opposed to fixations) is significantly affected by creativity also when percentage values are computed, i.e., when the duration-related measures are divided by the duration of the interaction events. The presentation of these values and their calculation process are omitted in the present manuscript for space reasons. The difficulties in interpreting and capturing the advantages of new and creative products emphasizes the importance of research in design affordances [29, 30].

As already outlined in Sect. 3.2, quality-related value dimensions are not congruent in terms of the effects creativity has on. In [9], more creative products have been considered significantly more attractive, adequate and worth of spending money for despite mentioned problems concerning their understanding and efforts to make sense out of them. The novelty of the objects displayed in [22] have played a reverse role. This can partially motivated by the manipulation operated in that experiment, where artefacts less similar to commercial products (likely more unusual) have been fabricated, somehow intentionally, with mean manufacturing technologies, i.e. FDM. This is also congruent to the view that creativity is related to action rather than simple perception [15]. It follows that participants, when required to evaluate creativity, produce a reliable estimation if they can rely on their (manual) experience with the object. Therefore, they could compare the present object and its features, with those in their memory. If a person does not have such an experience, is less likely to detect features like creativity issues in the present stimuli. Interest, therefore, is not sufficient per se, but the past knowledge about the stimuli need to be investigated as well. On the one hand, this aspect can be seen as a weakness of the present analysis. On the other hand, it shows that the possible interest and curiosity aroused by creative objects is not capable of compensating (real or perceived) performance gaps. The actual effect of the presumably worse performances caused by the recalled manipulation should be taken into account by means of more sophisticated statistical models, e.g. mixed regression functions.

4.2 Limitations

Beyond the stated circumstances applying to the experiment [22], which could have influenced the outcomes, and the aforementioned assumptions of similarity between terms featuring value dimensions (Sect. 3.2), the present study is affected by further limitations. In particular, the discussed outcomes could be affected by the following aspects and circumstances.

- Although both experiments required the evaluation of artefacts, the evaluation tasks differed in terms of (number of) investigated value dimensions, whose analysis might have required different exploration strategies.
- The experimental sessions exhibited unequal durations, especially because of the larger number of analyzed products in [9].
- The articulation of the tasks differed across the two experiments; for instance, a sort of quiz was launched during a first rapid exhibition of the objects in [22].
- The nature of the interaction was different in the two experiments, as pictures and physical objects were presented in [9] and [22], respectively. This could have biased understanding and exploration, while surely impacting on eye-tracking procedures (see below).
- A remote eye-tracker was used in [9], while eye-tracking glasses were employed in [22].

Many of these limitations are due to the experimental procedure, whose choices were made in order to make participants comfortable with respect to the devices and the task. Thus, a further series of experiments could evaluate whether results may depend on the task. Other ones are intrinsic limitations, which must be considered when running experiments of this kind. For example, the unequal duration depends on the kind of exploration: manipulative exploration requires more time but it is, overall, more informative than visual exploration.

5 Final Rcmarks

The literature on the role played by creativity in making products valuable and attractive is limited and largely inconclusive. As the design community is making considerable efforts to nurture creative processes, more knowledge of the actual benefits that this can engender is deemed useful for academia and industry. Undoubtedly, the present paper, while addressing this open issue, cannot provide a definitive answer to the highlighted research question. To this end, a larger number of products, tests, participants and testing conditions are required in future experiments. However, the exploitation of available data has allowed the exclusion of some circumstances (novel products are valuable per se) and the reinforcement of certain hypotheses (generalized difficulty to make sense of creative artefacts). On the one hand, the potential positive perception of creativity (here, substantially characterized by the novelty component) cannot compensate for performance lacks when one assesses the value of a product and its attractiveness. On the other hand, this exacerbates difficulties and efforts in understanding the product as a whole, its functions, and the performances that can be actually expected. This aspect has emerged as particularly critical in the authors' analysis, while, in parallel, research on the affordance of creative products is scarce and has arisen as an area worth investigating.

The reliability of the results of the study, here presented in a nutshell, might be affected by a series of methodological limitations that have been laid bare in Sect. 4.2. Those limitations are, nevertheless, not supposed to invalidate the findings, because of the rigor observed in the analyzed experiments and the conditions that make them adequate for the present investigation (Sect. 2). Moreover, in light of the fact that the research question has been just approached in the past, first indicative results can be useful for addressing future research directions. As for the authors, forthcoming work includes first more sophisticated statistical analysis to achieve more reliable outcomes and draw the first conclusions on the topic.

Acknowledgements. The study is conducted in the frame of the projects AMDAPA and EYE-TRACK funded by the Free University of Bozen-Bolzano with the calls CRC2016 and CRC2017, respectively.

References

1. Becattini, N., Borgianni, Y., Cascini, G., Rotini, F.: A TRIZ-based CAI framework to guide engineering students towards a broad-spectrum investigation of inventive technical problems. Int. J. Eng. Educ. **29**(2), 318–333 (2013)
2. Snider, C.M., Culley, S.J., Dekoninck, E.A.: Analysing creative behaviour in the later stage design process. Des. Stud. **34**(5), 543–574 (2013)
3. Yang, X., Helander, M., Dong, A.: Latent semantic analysis measures participant affiliation and team process quality. In: DS 58-9: Proceedings of ICED 2009, the 17th International Conference on Engineering Design, Human Behavior in Design, Palo Alto, CA, USA, 24–27 August, vol. 9, pp. 35–46 (2009)
4. Wood, B.M.: A leading role for design as an economic development enabler. In: DS 58-3: Proceedings of ICED 09, the 17th International Conference on Engineering Design, Design Organization and Management, Palo Alto, CA, USA, 24–27 August, vol. 3, pp. 37–48 (2009)
5. Livotov, P.: Competitive capability assessment of industrial companies within the framework of advanced innovation design approach. In: DS92: Proceedings of the DESIGN 2018 15th International Design Conference, Dubrovnik, Croatia, pp. 1903–1914 (2018)
6. Borgianni, Y., Cascini, G., Pucillo, F., Rotini, F.: Supporting product design by anticipating the success chances of new value profiles. Comput. Ind. **64**(4), 421–435 (2013)
7. Casagrande-Seretti, A., Montagna, F., Cascini, G.: A decision support model to assess technological paradigms. Int. J. Technol. Manage. **80**(1–2), 61–84 (2019)
8. Hölttä-Otto, K., Otto, K., Song, C., Luo, J., Li, T., Seepersad, C.C., Seering, W.: The characteristics of innovative, mechanical products—10 years later. J. Mech. Des. **140**(8), 084501 (2018)
9. Maccioni, L., Borgianni, Y., Basso, D.: Value perception of green products: an exploratory study combining conscious answers and unconscious behavioral aspects. Sustainability **11** (5), 1226 (2019)
10. Im, S., Bhat, S., Lee, Y.: Consumer perceptions of product creativity, coolness, value and attitude. J. Bus. Res. **68**(1), 166–172 (2015)
11. Chen, S., Benedicktus, R., Kim, Y., Shih, E.: Teaching design thinking in marketing: Linking product design and marketing strategy in a product development class. J. Mark. Educ. **40**(3), 176–187 (2018)
12. Sarkar, P., Chakrabarti, A.: Assessing design creativity. Des. Stud. **32**(4), 348–383 (2011)
13. Sehn, C.M., Jacques, J.J.D.: Correlations between successful consumer goods in the market and creativity in form and function attributes. In: DS 80-3 Proceedings of the 20th International Conference on Engineering Design (ICED 15), Organisation and Management, Milan, Italy, 27–30 July, vol. 3, pp. 309–318 (2015)
14. Maccioni, L., Borgianni, Y.: A product success scale for supporting research in engineering design. In: DS 92: Proceedings of the DESIGN 2018 15th International Design Conference, Dubrovnik, Croatia, pp. 69–80 (2018)
15. Withagen, R., van der Kamp, J.: An ecological approach to creativity in making. New Ideas Psychol. **49**, 1–6 (2018)
16. Yu, E.: Designing for value: insights from the emotional appraisal approach to understanding user value. Des. J. **21**(2), 185–207 (2018)
17. Boztepe, S.: Toward a framework of product development for global markets: a user-value-based approach. Des. Stud. **28**(5), 513–533 (2007)

18. Andreassen, T.W., Kristensson, P., Lervik-Olsen, L., Parasuraman, A., McColl-Kennedy, J.R., Edvardsson, B., Colurcio, M.: Linking service design to value creation and service research. J. Serv. Manage. **27**(1), 21–29 (2016)
19. Borgianni, Y., Cascini, G., Rotini, F.: Business process reengineering driven by customer value: a support for undertaking decisions under uncertainty conditions. Comput. Ind. **68**, 132–147 (2015)
20. Eggert, A., Ulaga, W.: Customer perceived value: a substitute for satisfaction in business markets? J. Bus. Ind. Mark. **17**(2/3), 107–118 (2002)
21. Grace, K., Maher, M.L., Fisher, D., Brady, K.: Data-intensive evaluation of design creativity using novelty, value, and surprise. Int. J. Des. Creativity Innov. **3**(3–4), 125–147 (2015)
22. Borgianni, Y., Maccioni, L., Basso, D.: Exploratory study on the perception of additively manufactured end-use products with specific questionnaires and eye-tracking. Int. J. Interact. Des. Manuf. (IJIDeM) **13**(2), 743–759 (2019)
23. Lohmeyer, Q., Meboldt, M.: The integration of quantitative biometric measures and experimental design research. In: Experimental Design Research, pp. 97–112. Springer, Cham (2016)
24. Sun, L., Xiang, W., Chai, C., Yang, Z., Zhang, K.: Designers' perception during sketching: an examination of Creative Segment theory using eye movements. Des. Stud. **35**(6), 593–613 (2014)
25. Maier, A., Baltsen, N., Christoffersen, H., Störrle, H.: Towards diagram understanding: a pilot study measuring cognitive workload through eye-tracking. In: International Conference on Human Behaviour in Design (2014)
26. Besemer, S.P., Treffinger, D.J.: Analysis of creative products: review and synthesis. J. Creative Behav. **15**(3), 158–178 (1981)
27. Ranjan, B.S.C., Siddharth, L., Chakrabarti, A.: A systematic approach to assessing novelty, requirement satisfaction, and creativity. AI EDAM **32**(4), 390–414 (2018)
28. Strong, E.K.: The Psychology of Selling and Advertising. McGraw-Hill book Company, Incorporated, New York (1925)
29. Pucillo, F., Cascini, G.: A framework for user experience, needs and affordances. Des. Stud. **35**(2), 160–179 (2014)
30. Flach, J.M., Stappers, P.J., Voorhorst, F.A.: Beyond affordances: closing the generalization gap between design and cognitive science. Des. Issues **33**(1), 76–89 (2017)

Improving the Efficiency of Design Protocol Analysis: An Approach to Speed Up the Coding Stage

Niccolò Becattini[1(✉)], Gaetano Cascini[1], Jamie O'Hare[2], and Jean-Francois Boujut[3]

[1] Politecnico di Milano, Milan, Italy
niccolo.becattini@polimi.it
[2] University of Bath, Bath, UK
[3] Grenoble INP, Grenoble, France

Abstract. This paper aims at checking the capabilities of tools to support coding activities for protocol design analysis to improve the efficiency of the process. The application of existing coding schemes to different protocols from real collaborative design sessions allows showing that the approach and the proposed tools can be flexibly adopted for various type of design tasks (product interface design and packaging design). The experiments to check the effectiveness of the proposed tools involved non-expert coders to create challenging experimental conditions and explore. This also allows exploring the viability of approaching the coding stage by non-expert coders in design protocol analysis, with a double goal: the positive results obtained show that is possible to release experts from time-demanding and HR consuming activities and, at the same time, obtain multiple parallel coded protocols for more reliable analysis. The outcomes are also discussed in terms of the implications on the development of similar supporting tools.

Keywords: Product interface design · Packaging design · Cognition · Creativity

1 Introduction

Design protocol analysis, in the last decades, has been one of the most used approach to study the behavior of designer(s), as it allows to unveil hidden or less evident aspects that are hard to capture during the design discourse, such as cognitive processes (e.g. [1]). It consists of a process of subsequent stages that enables the extraction of meaningful information from design observations, typically characterized by unstructured qualitative data. Its typical sequence consists in:

1. Recording the design session (protocol source);
2. Transcribing the protocol and annotating;
3. Segmenting the protocol into elementary design moves, according to the coding scheme;

C. Rizzi et al. (Eds.): ADM 2019, LNME, pp. 612–624, 2020.
https://doi.org/10.1007/978-3-030-31154-4_52

4. Coding the protocol segments according to the coding scheme;
5. Analyzing the coded protocol and drawing conclusions.

Such information, once coded by raters, consists of a set of structured coded data, which allows for both qualitative and quantitative analysis, depending on the applied metrics for the assessment at hand [2].

Most of these phases are extremely time consuming and their impact on research is significant as human involvement is currently necessary through all the stages. Literature provides figures for the expenses behind design protocol analysis. Someren et al. [2] estimates that transcribing may require about 10 times the duration of the protocol itself, which might also get worse, as several factors can hinder an efficient transcription: clarity of the protocol and of the speaking subject(s), voice overlapping… Blessing and Chakrabarti [3] estimate that the whole process takes up to 40 times the duration of the recorded protocol. Jiang and Yen [4], in turn, set this ratio to 100–1000 the protocol duration. Recently, some researchers addressed design protocol analysis stages with more efficient approaches. For instance, Becattini et al. [5] explored the support of computer, highlighting opportunities for a more efficient recording of the design sessions, their transcriptions and segmentation. The development of strategies to improve the efficiency of such analyses is then critical for their viability in research, potentially enabling a more convenient and frequent adoption.

The paper stems from existing coding schemes and aims at experimentally checking the effectiveness of support tools for the coding stage, which ease the interpretation of the verbalized contents discussed during a collaborative design session by non-expert coders. This approach can be considered replicable depending on the coding scheme to be implemented, but it is in principle applicable to all the design sessions with concurrent verbalization and production of visual design representation (sketches, models, 3D models, mock-ups, prototypes…). The involvement of non-expert coders allows for an easier involvement of HR in the protocol analysis process, so that multiple coders work in parallel and provide data with higher accuracy for the quantification of results.

The next section provides an overview of existing support tools and/or approaches that might improve the efficiency of design protocol analysis, highlighting their strength and weaknesses. Then, the coding scheme and the inherent support tools are respectively introduced and discussed. These are used analyze three different co-design protocols to allow for the experimental assessment of the approach and the related coding support tools. The results highlight the strength and the weakness of this approach and allow to discuss on implications for the development of similar tools or their integration in multi-modal evaluation computerized environments.

2 Tools and Instruments to Support Protocol Analysis

Table 1 collects representative tools or approaches among the alternatives to support design protocol analysis. A more detailed review of data visualization tools to support the analysis stage, here overlooked because less relevant to the research, is available in [6]. VizScribe [6] and Transana [7] are two examples of computer-based applications for qualitative data analysis (others are, e.g. NVivo, ATLAS.ti). These systems address one or more stages of protocol analysis; especially by enabling access to multimodal information from transcription to coding by facilitating the identification of potential patterns. Yet, they just allow the analyst to access a larger amount of information, but do not provide any specific support for information matching (i.e. decoding the protocol segment) and encoding. From this perspective, this might also trigger loss of attention/distraction or information overload, which might impair the coding. Among tools to support transcription, Dragon Naturally Speaking (www.nuance.com) is one of the most promising options for transcribing, as it requires short training (10/20 min) to automatically transcribe the recorded utterances.

For what concerns the stage of segmentation, Observer [8] allows for coding gestural interactions in real-time, making the coder an onlooker in the session. However, its adoption to reliably code verbal interactions in real time has not been proved yet. Nguyen et al. [9] approached the segmentation of the design discourse with EEG signal captured during design activities. This kind of pioneering approach, however, is still relatively new and correlates with verbal, gestural or documental interactions are not yet consolidated to make it sufficiently reliable. Among the tools supporting coding and analysis, Linkographer [10] allows analysts to define links between design moves, visualize inherent graphs and provide quantitative measurements for design relevant metrics based on link structure. Similarly, Becattini et al. [11] introduced an essential coding schemes to map navigation in and between problem and solution space, together with rules for the analysis of problem graphs (OTSM-TRIZ Network of Problems) according to creativity based metrics, showing significant time savings with reference to the whole analysis. As the coding phase is the stage that requires higher order reasoning (observation, interpretation, comparison with inherent decision making) and it strongly impacts on results and time/HR consumption, the authors decided to develop tools to assist coders in interpreting protocol segments.

Table 1. Existing tools and approaches to support Design Protocol Analysis

Tool or approach	Stage of design protocol analysis				Data for coding			Notes
	Transcription	Segmentation	Coding	Analysis	Verbal	Gestural	Documental	
VizScribe	x	x	x	x	x	x	x	Computer-based multimodal information access
Transana	x	x	x	x	x	x	x	
Dragon Naturally Speaking	x				x			Speech recognition software
Linkographer			x	x	x			Software to manage links among design moves
Network of problems			x	x	x			Semi-automatic analysis with Problem/Solution codes to map creativity
Observer		x	x			x		On the fly coding with concurrent observation
EEG-based segmentation		x						Use of bio-physiological signals

3 A Coding Scheme to Map the Superficial Characteristics of Products and Related Designers' Intentions

This research is part of the activities carried out within the SPARK project (H2020 ICT). It aims at providing co-designers with a real-like shared design representation, in the form of a mixed prototype: partially tangible (its shape) and partially virtual (its superficial characteristics rendered through projections), which proved to be particularly effective for product interface and packaging design. The system leverages Spatial Augmented Reality (projection, tracking, interaction technologies). This led to the development of a coding scheme that had already passed several refinements [12, 13] and that allows mapping different facets within the design discourse.

The coding schemes maps the projected contents as separate items (which typically correspond to file types). Items are contents that detail the superficial characteristics of a prototype (Table 2).

Table 2. The set of codes to map Items in verbalized interaction

Text	What is expressed by words
Image	Sketched or computer-generated picture (scalable)
Photograph	Photograph of a real object (non-scalable)
Logo	Graphic/textual representation of the brand identity
Icon	Representation of certification marks or similar
Background Motif	Texture characterizing the background of the design
System Parts	A part of the entire system (i.e. multiple items)
Whole	The whole design proposal as a single entity

The transformations items undergo during the design session are mapped in terms of parameters. Table 3 presents the description for the parameters considered relevant for the analysis of transformations of features of the superficial feature of prototypes.

Table 3. The set of codes to map Parameters in verbalized interaction

Position	Geographical location of the item (i.e. [x,y] coordinates)
Orientation	Degree of rotation of the item (i.e. θ coordinates)
Size	Change in item's dimensions (same aspect ratio)
Number	Number of Items onto the surface of the prototype
Presence	Introduction/removal of an item onto the prototype surface
Colour	Chromatic characteristics of an item
Reflectivity	Glossy or matt properties of an item
Material	The physical material that constitutes the item
Content	Subject represented onto the item (being it graphical or textual)
Shape	Dimensions, proportions of an item (i.e.: aspect ratio changes)
Sharpness	Resolution and definition of the item

The coding of designers' intentions is necessary to map cognition. It relies on existing codes, similarly to three "micro strategies" presented in [14]: Analysis of problems, Synthesis of solutions and Evaluation of solutions. However, the difference between Analysis of problems and Evaluation of solutions is weak for surface features of products, as most of the analysis of problems (e.g. accessibility, clarity in product interaction design) depends on what the prototype displays to mimic the solution. The code Synthesis keeps its original meaning. Moreover, as the SPARK platform should also ease the selection of alternatives to quickly design more effective solutions [15], the authors introduced an additional code: Choice. It maps the design moves where a decision is taken among two or more alternatives. Table 4 presents the description of these codes.

Table 4. The set of codes to map designers' intentions in verbalized interaction

Analysis	Designer interpreting the prototype, evaluating its matching with expectations and providing judgement about it (pros and cons)
Synthesis	Designer proposing a solution that can address one of the issues (explicitly emerged or not) concerning the prototype
Choice	Designer selecting between alternatives for the prototype
Other	Other spoken interactions that do not correspond to one of the above (i.e. non-design moves)

4 Tables and Decision Tree to Improve Coding Efficiency

Coding is essentially a decision-making process where the coder interprets what observed (decoding). Then, the coder matches its interpretation with relevant code(s) that appropriately describe the design move as interpreted (encoding the design move according to a scheme, e.g. Tables 2, 3 and 4). Its working memory processes a large set of information: from the interpreted observation to the variety of meanings the codes hold. A large number of alternatives for coding might bias its decision, also because of fatigue-related effects. Thus, the development of coding support tools should not be limited to enabling multimodal information access. This can provide a multifaceted perspective and a comprehensive viewpoint, but also potentially increases the load on working memory with the addition of potentially distracting information. To this purpose, support tools for coding protocols should aim at improving the reliability of the analysis and its effectiveness in terms of expired time and involved human resources.

Decisions can be both guided through appropriate structured procedures (e.g. decision trees) and with structured contents which helps information mapping and code-matching. Visually displayed information support decision-making processes in several contexts, from military [16] to environments for people with learning disabilities [17].

Coding schemes for items and parameters appear to be the more promising to provide visual support for coding. The coding scheme for intentions has just 3 alternatives, plus a code for residuals. Items and parameters, on the contrary, are characterized by a larger set of alternative codes, respectively 8 and 11, that make them more suitable to reduce working memory load, because of their intrinsic variety. The coding support tools, whose coding effectiveness is measured in this research, then are organized as

- Coding flowcharts with control questions to explore the decision tree and consequently assign codes to design moves;
- Coding tables with design session-related contents graphically displayed, structured and classified coherently with the coding scheme.

Coding tables, then, present contents characterizing the surface features of the prototype, as discussed and displayed during the design sessions. Within the design protocol analysis process, their development can take place right after the recording by expert coders, that can quickly structure design-related contents into a coherent version of codes interpretation together with visual contents (e.g. predefined coded items…).

Figures 1 and 2 display the coding tables developed as a support tools, respectively for Items and Parameters. On top of them, a decision tree allows for a quicker information processing (available at http://tiny.cc/5w8eaz).

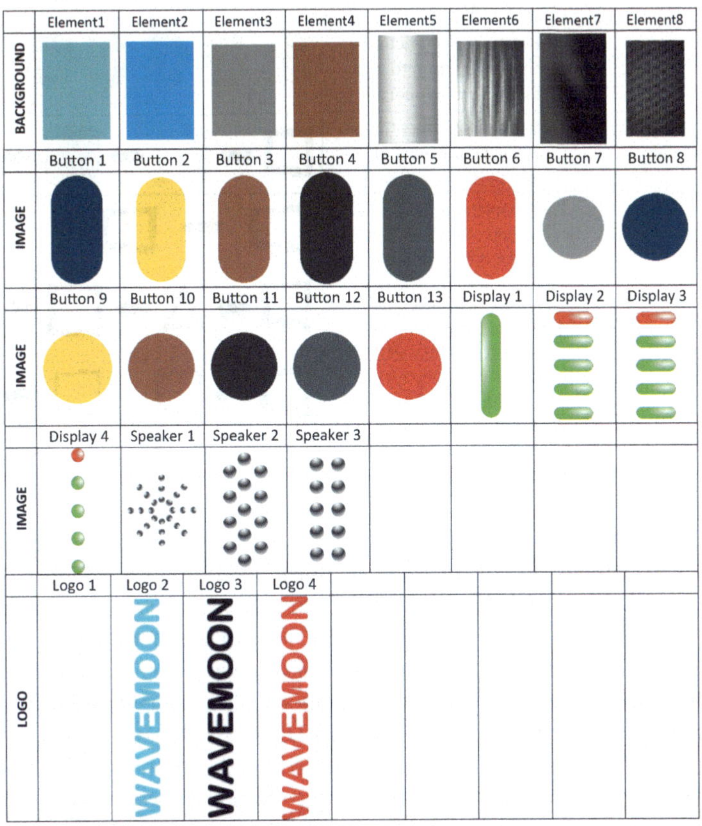

Fig. 1. Table to support the identification of items (for session E17 – see Sect. 5)

Reference Table:

Parameter Type	Definition	Example & Symbol
Position	The Parameter refers to the geographical location of the item on the design proposal or on one of its parts (i.e. [x,y] coordinates)	
Orientation	The Parameter refers to the degree of rotation of the item (i.e. θ coordinates)	
Size	The Parameter of the item describes refers to a change in its dimensions (without changing the aspect ratio)	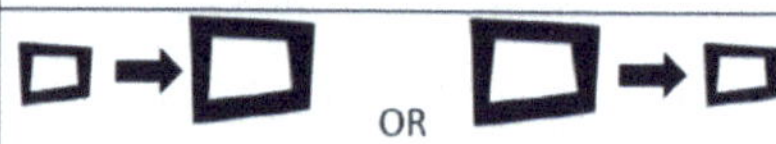
Number	The Parameter refers the number of Items of the same kind to be added on or removed from the design proposal	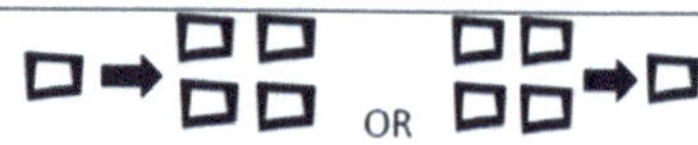
Presence	The Parameter refers to the introduction or the removal of an item within the design proposal	
Colour	The Parameter refers to the chromatic characteristics of an item	
Reflectivity	The Parameter refers to the capability of an item to show glossy or matt properties	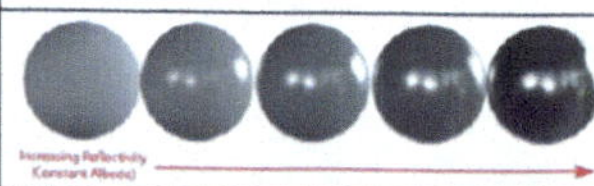
Material	The Parameter refers to the physical material that constitutes the item	
Content	The Parameter refers to the subject represented by the item (being it graphical or textual)	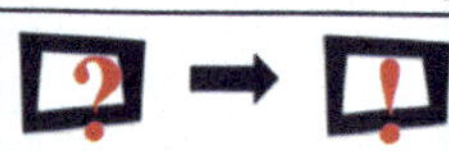
Shape	The Parameter refers to the representational characteristics (dimensions, proportions...) of an item (i.e: a change in the aspect ratio)	
Sharpness	The Parameter refers to the resolution and definition of the representation of the item.	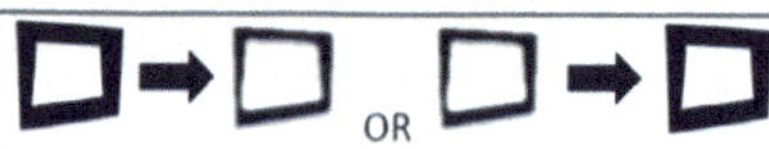

Fig. 2. Table to support the identification of parameters

The table in Fig. 1 directly presents the items as used in the design session, while the table for parameters (Fig. 2) presents graphical descriptions at a higher abstraction level.

There is no support table in addition to the decision tree for coding intentions, as this creates the opportunity to test the goodness of coding both with and without the tables with graphical contents. These two conditions allow comparing the robustness of analyzed protocols through inter-rater reliability (IRR) assessment, i.e. the convergence of multiple coders towards a coherent and similarly coded protocol, which is a crucial requirement to infer what has happened during the design session and studying the inherent creativity and cognition. Tables 1 and 2 have been used as coding support, consistently with the description of the experiment presented in the next section.

5 Experimental Application and Related Results

The first subsection describes design tasks, duration and usage of coding support tool. The second one motivates the involvement of non-expert raters for coding the protocols. The third subsection presents and discusses the results of the investigation with reference to the objectives of the research:

- to achieve a good inter-rater reliability among non-expert coders
- to reduce the overall time required for the analysis.

5.1 Description of Experimental Case Studies and Setting

All the co-design sessions (Table 5) were A/V recorded with multiple cameras and individual microphones to completely cover the scene. Each protocol was transcribed and translated into English to allow coders from different regions to understand contents and meaning of verbal interactions. For this experiment, the authors provided coders with pre-segmented protocols in the form of spreadsheets.

Table 5. Summary of case studies/design sessions characteristics

Case study ID	Design domain	Session time	Coding support tool	Session aim (design of…)
Estell 2017 (E17)	Product Design	71 min	- Decision tree - Tables with visual contents	Interface of a sport/gym device to measure workload parameters
Alce Nero 2017 (AN17)	Packaging Design	100 min	- Decision tree - Tables with visual contents	Baby food and baby rice packaging
Alce Nero 2016 (AN16)	Packaging Design	40 min	- Decision tree	Pouch shaped packaging for biscuits

5.2 Profile and Role of the Involved Coders

Three non-expert coders, who did also not observe the sessions nor participated in the development of coding schemes, received instructions to work independently not to bias the results. They received video mosaic of the recorded sessions (4 cams) and synchronized audio tracks for each co-designer, transcripts of the verbalizations with time codes. They were also provided with the above presented support tools for coding as for Table 4. An expert coder (not involved in coding) delivered a brief seminar (<60 min) on codes meaning, together with a tutorial to access and use the supporting material. The 3 coders (males) were selected among the university-allowed students to be recruited for paid activities in and for departments. All of them are in the MS in Mechanical Engineering at Politecnico di Milano. They had no experience nor previous knowledge on design protocol analysis and coding schemes in general. Non-expert coders, which proved to proficiently substitute expert evaluations in other domains (e.g. [18, 19]), represents a twofold opportunity. First, it creates harsher conditions to check the effectiveness of supporting tools. Second, it provides preliminary evidence on the effects of releasing experts from coding duties, so that they can mostly focus on the analysis of results, which tacit knowledge and experience are crucial.

5.3 Results of the Investigation

Consistently with the paper focus, the results of these experiments will not entail the creative processes emerged during idea generation, nor ideas themselves. This sub-section presents the results which measure the effectiveness of the coding support tools:

- Time required by stage of the design protocol analysis, to measure the impact coding support tools on the overall efficiency of the protocol analysis process (Table 6);

Table 6. Duration (in minutes) or relevant phases of design protocol analysis

Case study	Transcript	Tables	Segmentation	Coding	Tot.
AN16	600' (15:1)	–	300' (15:2)	675' (17:1)	1575' (40:1)
AN17	900' (9:1)	35' (1:3)	510' (5:1)	758' (15:2)	1963' (20:1)
E17	720' (10:1)	20' (1:3)	420' (7:1)	630' (9:1)	1790' (24:1)

- Inter-Rater Reliability, to measure the effectiveness of coding support tools to generate more uniform and statistically meaningful protocol data to analyze (Table 7).

Table 6 does not show time for the analysis as coded data appear in the same format for the set three coders and for every protocol, making analysis duration essentially the same. Table 6 shows that the time to transcribe is coherent with Someren et al. [2] (frequent speakers' voices overlaps increased transcription time for the AN16 session). The time to prepare coding tables have been considered just for the relevant case studies. Table 6 also shows that time require, phase by phase, are comparable at least in terms of order of magnitude. Figures in brackets, which display the ratio phase duration: protocol duration, show that the three sessions are, conversely, significantly different in essence. The duration of the whole process for AN16, which just benefited of decision trees, is consistent with Blessing and Chakrabarti [3]. A direct comparison between AN16 and AN17/E17 shows that the introduction of tools with visually displayed information has a significant impact on reducing time consumptions for coding. This kind of support substantially halved the time spent on this task

(−50%). Moreover, the lack of a graphical table for coding intentions suggest that its future introduction can further reduce the time spent for this activity, despite its expected impact might be proportional to the variety of the related coding scheme (3 + 1 codes for intentions vs 8 for items and 11 for parameters). The duration of segmentation here considers two contributions, one of which extends the typical sequence of design protocol analysis: beyond the time to segment the protocol, time consumptions here account also for the consolidation of the segmented protocol by expert coders before non-expert coders access it.

In terms of overall duration of design protocol analysis, the last column of Table 6 shows that the proposed approach impacts efficiency with significant time savings and a more efficient usage of resources, as the tools to support coding reduce time consumptions 40–50%. More cautiously, the author estimates that the impact on the coding stage is a more reliable indicator of efficiency. As coding takes, on average, 40% of the whole design protocol analysis, the expected time savings coding support tools should be around 20%. The definition and preparation of the tables to support coding determine a negligible increment of the workload (1–2%). The approach also proved to be applicable for different kind of design sessions, as the time savings are of the same order for coding of product interface protocols (E17) and packaging protocols (AN17).

Table 7 shows Fleiss' Kappa [22] values for inter-rater reliability (IRR), i.e. coherence between coded protocols as produced by the non-expert coders. This is a numerical evidence of the reliability of results with the proposed approach.

Table 7. Fleiss' Kappa values for the IRR among three independent coders

Session	Intentions	Items	Parameters
AN16	0,91	0,54	0,39
AN17	0,40	0,65	0,54
E17	0,20	0,60	0,53

As said, the coders received decision trees with rules and control questions to assign codes for each of the three coding schemes, but tables with graphical contents to support coding regard just items and parameters. Moreover, coders received supporting tables for coding that are tailored for the protocols of AN17 and E17 sessions, not for AN16.

An intra-protocol comparison of the results shows that IRR values are similar for both Items and Parameters in AN17 (0,65 items, 0,54 parameters) and E17 (respectively 0,60 and 0,53). These values are satisfactory, as these results just concerns the first coding round carried out by non-experts. Moreover, as these two protocols are different in nature, this result is a preliminary evidence is robust across design domains. The IRR values for intentions, on the contrary, show that decision trees are not sufficient for coherent coding among non-expert coders. This is confirmed by IRR value for AN16, where coders just benefited of decision trees (no graphical tables). The high value of IRR for intentions could depend on the nature of that design session (mostly design review, with small idea generation), which results in a "analysis-oriented" protocol.

The comparisons of IRR across sessions also highlight strong differences between purely text-based schemes (intentions) and those provided with graphical support (items and parameters). This suggests that the combination of information eases the interpretation (decoding/encoding) of verbal interactions. The protocols AN17 and E17 have higher values of IRR for items (0,65–0,60) and parameters (0,54–0,53) than AN16 (respectively 0,54 and 0,39). This also witnesses that graphics-based support tools for coding produce more reliable protocols to be analyzed, while the exclusive use of decision trees (intentions) produce less coherent results, as these IRR values vary significantly among the three protocols (from 0,90 to 0,20). Such experimental evidence might be explained with coders' stronger difficulties to interpret abstract concepts (as the cognitive processes behind intentions), compared to more concrete ones that can be displayed with graphical contents (items and parameters).

6 Conclusions

This paper proposes an original research aimed at the development of support tools to study design protocols in contexts of product interface design and packaging design (as well as other context where surface characteristics have a paramount role). In details, the proposed tools focus on support for the coding stage, as this is one of the most resource demanding stage of design protocol analysis. Tailored tools (graphical tables and decision trees) have been developed and tested, consistently with coding schemes that map verbal interactions in co-design sessions as (i) designers' intentions and (ii) the surface characteristics of a prototype. The paper stresses the usefulness of these tools when used by non-expert coders, who can significantly slash the time required for the analysis and release expert coders from time-consuming and low value activities.

The results of the first application of the proposed approach are promising, in that it provides evidences that the assistance of structured textual and visual contents for coding allows non-expert coders to interpret (decode/encode) design moves more uniformly, thus providing reliable protocols (evaluated through IRR scores) for analysis. Experimental results also show that graphically supported coding requires less time than traditional approaches (40–50% less time for coding, which results in a reduction of 20% for the whole protocol analysis process). The application of support tools for coding, especially with graphical pre-structured and classified contents, also provide repeatable results both for product interface design and packaging design.

As the codes having no graphical support (intentions) yielded worse results in terms of IRR, future studies should explore the development of tools that help interpretation and decoding/encoding more abstract characteristics of segments/design moves (e.g.: cognitive processes, designers' intentions, Function-Behaviour-Structure variables) through the integration of information. This would allow extending the approach to a wider set of design sessions that are to be analyzed with different kind of coding schemes.

This paper also explored the involvement of non-experts for coding design protocols, which is one the most time consuming stage. Beyond making the experimental conditions to validate the proposed tools more challenging, the relatively positive outcome detailed in Sect. 5 suggest that this could be a winning strategy to invest more effectively experts' time and HR on the analysis stage, as their deeper knowledge is crucial to infer conclusions on the protocol. Results show that graphical support tools for coding allow non-experts to provide reliably coded protocols, in a shorter time compared to what mentioned in [2, 3]. Nevertheless, the development of supporting tools introduces changes to the process of design protocol analysis that are expert-driven. Right after the execution of the protocol and before the coding activity starts, expert coders need to consolidate the segmentation of the protocol to be coded and collect and organize contents describing the surface characteristics of a prototype in visual tables that help decision-making. Despite this require additional time by experts (approximately one third of the protocol duration), the balance within the whole design protocol analysis process show a significantly improved efficiency.

With reference to existing tools to support design protocol analysis, these results show that providing access to information is just part of the support coders can receive, as cognitive support appears to increase efficiency. This implies that the future development of computer-based support tools for design protocol analysis, as shown in Sect. 2, should also include rules and strategies to make coding more repeatable and less fatiguing for working memory.

Acknowledgements. The authors would like to thank Tito Begnoni who contributed to data analysis. This project has received funding from the European Union's Horizon 2020 research and innovation programme under grant agreement No. 688417 (H2020_ICT_2015_SPARK_688417).

References

1. Hay, L., Duffy, A.H., McTeague, C., Pidgeon, L.M., Vuletic, T., Grealy, M.: Towards a shared ontology: a generic classification of cognitive processes in conceptual design. Des. Sci. **3**, E7 (2017)
2. Someren, M.V., Barnard, Y.F., Sandberg, J.A.: The Think Aloud Method: A Practical Approach to Modelling Cognitive Processes. Academic Press, London (1994)
3. Blessing, L.T., Chakrabarti, A.: DRM, a Design Research Methodology. Springer, London (2009)
4. Jiang, H., Yen, C.-C.: Protocol analysis in design research: a review. In: Rigor and Relevance in Design: IASDR 2009, Seoul, Korea, 18–22 October, pp. 147–156. International Association of Societies of Design Research (2009)
5. Becattini, N., Borgianni, Y., Cascini, G., Rotini, F.: Experiencing protocol analysis of computer-guided design tasks. In: DS 70: Proceedings of DESIGN 2012, the 12th International Design Conference, Dubrovnik, Croatia (2012)
6. Chandrasegaran, S., Badam, S.K., Kisselburgh, L., Peppler, K., Elmqvist, N., Ramani, K.: VizScribe: a visual analytics approach to understand designer behavior. Int. J. Hum. Comput. Stud. **100**, 66–80 (2017)
7. Halverson, E.R., Bass, M., Woods, D.: The process of creation: a novel methodology for analyzing multimodal data. Qual. Rep. **17**(11), 1 (2012)
8. Ben Guefrache, F., Masclet, C., Prudhomme, G., Cascini, G., O'Hare, J.: On the fly coding method for real-time capture of artefact-centric interactions in co-creative design sessions. In: Design 2018, Dubrovnik, 21–24 May 2018. https://doi.org/10.21278/idc.2018.0468

9. Nguyen, P., Nguyen, T.A., Zeng, Y.: Physiologically based segmentation of design protocol. In: DS 80-11 Proceedings of the 20th International Conference on Engineering Design (ICED 2015), Milan, Italy, vol. 11, 27–30 July 2015
10. Pourmohamadi, M., Gero, J.S.: LINKOgrapher: an analysis tool to study design protocols based on FBS coding scheme. In: DS 68-2: Proceedings of the 18th International Conference on Engineering Design (ICED 2011), Lyngby, Denmark, vol. 2, 15–19 August 2011
11. Becattini, N., Cascini, G., Rotini, F.: An OTSM-TRIZ based framework towards the computer-aided identification of cognitive processes in design protocols. In: Design Computing and Cognition 2014, pp. 99–117. Springer, Cham (2015)
12. Becattini, N., Masclet, C., Ben-Guefrache, F., Prudhomme, G., Cascini, G., Dekoninck, E.: Characterisation of a co-creative design session through the analysis of multi-modal interactions. In: International Conference on Engineering Design 2017, pp. 479–488 (2017)
13. Becattini, N., Cascini, G., O'Hare, J., Masclet, C.: Coding schemes for the analysis of ICT supported co-creative design sessions. In: Design 2018, Dubrovnik, May 2018
14. Mc Neill, T., Gero, J.S., Warren, J.: Understanding conceptual electronic design using protocol analysis. Res. Eng. Des. **10**(3), 129–140 (1998)
15. O'Hare, J., Dekoninck, E., Mombeshora, M., Martens, P., Becattini, N., Boujut, J.F.: Defining requirements for an augmented reality system to overcome the challenges of creating and using design representations in co-design sessions. In: CoDesign, pp. 1–24 (2018)
16. Wickens, C.D., Scott, B.D.: A comparison of verbal and graphical information presentation in a complex information integration decision task (No. EPL-83-1/ONR-83-1). Illinois univ at Urbana. Engineering-Psychology research lab. (1983)
17. Bailey, R., Willner, P., Dymond, S.: A visual aid to decision-making for people with intellectual disabilities. Res. Dev. Disabil. **32**(1), 37–46 (2011)
18. Kaufman, J.C., Baer, J., Cole, J.C., Sexton, J.D.: A comparison of expert and non-expert raters using the consensual assessment technique. Creat. Res. J. **20**(2), 171–178 (2008)
19. Fleiss, J.L.: Measuring nominal scale agreement among many raters. Psychol. Bull. **76**(5), 378 (1971)

Proof of Concept as a Multidisciplinary Design-Based Approach

Diego Paderno, Ileana Bodini(✉), and Valerio Villa

Department of Mechanical and Industrial Engineering,
University of Brescia, via Branze, 38, 25123 Brescia, Italy
ileana.bodini@unibs.it

Abstract. This work deals with the description of the activities developed in the PoC-BS laboratory. In particular, it is proposed first to provide a multidisciplinary definition of Proof of Concept - PoC - and then to illustrate the design-methods-based approach used in the laboratory for the development of different types of PoC.

In particular, the presentation of the integrated approach, for the realization of different types of PoC is provided through the description of some examples of activities carried out to support experimental research. In this work we describe: (i) the construction of a cosmic-ray telescope in collaboration with the group of Experimental Physics, (ii) the design and manufacture of devices for qualitative evaluation of the drying process of sludge generated by the inertization of waste incinerator fly-ash, carried out with the group of Chemistry for Technologies.

In conclusion, the paper focuses on the future prospects of the Laboratory, whose ambition is to consolidate its activity and to lay the foundations for a new way of service delivery, based on a fair acknowledgement of contributions from this type of laboratories to both educational and scientific research.

Keywords: Proof of Concept · Multidisciplinary approach · Design methods · Value added services for research

1 Introduction

Nowadays, the importance of carrying out so-called Proof of Concept - PoC - has become a fundamental task in different fields of application, including engineering, computer science, medical/pharmaceutical industry, economics, financial/business analysis and digital art.

Starting from the definition provided in the Oxford dictionary, PoC is an "evidence, typically deriving from an experiment or pilot project, which demonstrates that a design concept, business proposal, etc. is feasible." [1]. On the other hand, as suggested by the European Research Council in [2] and in different professional sites as [3], PoC indicates the possibility to bring ideas to a pre-demonstration stage where potential commercialization or societal opportunities may be identified. In particular, PoC is intended as "a demonstration, the purpose of which is to verify that certain concepts or theories have the potential for real-world application. PoC can be therefore considered an early stage prototype that is designed to determine feasibility, but does not represent deliverables".

C. Rizzi et al. (Eds.): ADM 2019, LNME, pp. 625–636, 2020.
https://doi.org/10.1007/978-3-030-31154-4_53

Providing a multidisciplinary definition of PoC, it is possible to highlight the following characteristics: the realization of a PoC device is strictly connected to the production of a model/prototype or to the development of a feasibility study (hardware or software) in order to understand and demonstrate if the starting concept related to a project/research, or even to a new business idea, is actually attainable.

Usually, PoC can be applied to a feasibility study: in this case, the nature of the proof is "partial", since a specific part of the project is investigated to the purpose of confirming the goodness of the research line. Here the clear intent is indeed not to build a preliminary version of the product, but to demonstrate its feasibility [4–16].

This work aims to provide a multidisciplinary definition of PoC, introducing a design-based methodology for the development of different types of PoC in a specific laboratory environment, the PoC-BS lab. The current lab configuration is based on a previous educational laboratory, equipped with traditional machinery for wood, plastics and non-ferrous metals, in order to make "3D-untrained" students be able to start working on tridimensional expression. CNC machinery and 3D-printers would be an interesting development to complete the possibilities of our lab.

1.1 The Model: Design, Representation, Experimentation and Communication

Being the evolution of an industrial design educational lab, the methodological approach to problem solving has been strongly influenced by the experience in modeling for design purposes. Even though the introduction of computer techniques, modeling has a fundamental role in design and remains an important mean of representation, experimentation and communication [17, 18].

Modeling types and applications are multiple, as well as techniques and materials. It is an expressive tool to shape an idea from early intuition. Study model is indeed an essential support for experimental validation, and its use highlights how a general problem could be separated and analyzed by specific studies of technical or morphological details. As an example, realizing volumetric models gives the opportunity to compare different formal hypotheses retaining 3D realism. With functional study models one can realize partial models and reproduce only the elements under investigation, or even outline the object adding details only in the parts of interest. Besides being a tool for validation and formal experimentation, the model gives an immediate feedback on the feasibility of the object, thus providing the designer with essential information for the success of the final product. Moreover, study models should be as fast and cheap as possible, to let the designer and the other professionals working on a project to quickly reach the following step, by finding different solutions, comparing them and excluding the unworkable ones.

In all the phases where professionals in different fields are involved, model represents a powerful mean of communication. Its 3D nature has a universal impact, and can be understood by all the professionals who contribute to the evolution and the realization of the project. In the presentation of an idea and in its development, it has a strong communication potential towards people belonging to different cultural worlds.

Along the design process, the model gets closer and closer to the final product.

1.2 Devices for Research Purposes

In research, standard scientific instruments used for qualitative and quantitative evaluation are frequently integrated with new cases devices, expressly built for specific needs or for preliminary observation in a more complex research path. Such artifacts may have a wide range of purposes, types, names and definitions.

The first distinction can be made according to the autonomy and purpose of the device. In some cases special accessory are built for making laboratory tests possible even on samples that would not be compatible with the laboratory standard equipment [19, 20]. Other times, stand-alone objects are realized for one specific test or to carry out qualitative analyses that can provide useful information for the future of the experimental process [21–23]. Another interesting aspect is the need to back up research when preliminary results drive the work towards a possible industrialization. This moment is often characterized by larger-scale tests with respect to early experimentation, but still not so large to suit the typical methods of industry [24–27].

Moreover, when a digital mock-up is developed by other scientific groups to carry out analyses and simulations [28–32], morphological and dimensional data are needed, the pragmatic approach to problem solving developed through lab activity can become an important support by defining the design scheme of the studied solutions.

2 Case Histories

2.1 The Cosmos Project

The Cosmos project, financed by the European Union in the framework of Life+ program, was born in 2010 to develop a technology for the inertization of fly-ash (residual light ash produced by combustion processes and made of small particles that can be carried by the smoke) generated by municipal solid waste incinerator and waste-to-energy facilities [33–38]. Once the inertization process had been optimized, converting the special waste into inert waste, a suitable application field has been had to be found for Cosmos, in order to perform a further transformation, from waste to secondary raw material.

An ambient-temperature process has been developed, based on the chemical reaction between fly-ash and an aqueous solution containing colloidal silica, later replaced by rice husk or Silica Fume after further experimental tests. At the end of the process, the material looks like a gray sand. Among the number of possible applications, it has been chosen to examine those envisaging the use of Cosmos as filler in plastic material industry.

The research carried out has shown the Cosmos compatibility with the extrusion process of thermoplastic granule highlighting the step of drying of the compound as the most critical of the process.

In order to perform a preliminary evaluation of the Cosmos behavior during drying, two different experimental set-ups were built. The problems encountered before the beginning of this stage had been the slowness of the process at ambient temperature and the compaction of the material into solid blocks. Two solutions were examined, with the idea of evaluating the devices qualitatively in both cases, to the aim of providing useful information for the construction of full-size facilities. The first solution consisted in a "dynamic" drying, obtained by keeping the material constantly stirred in order to avoid its compaction. It was clear that the duration and the effectiveness of this system had to be evaluated before structuring the process, i.e., before choosing whether to put together or to keep separate the stirring and the drying stages. The second option was based on a continuous process and on the maximization of the surface in contact with air. The downside of compaction could have been attenuated by the low thickness of the drying material layers. This would have made easier to perform a preliminary chipping in machines capable of providing a controlled granulometry.

In both cases, hot air at 80/90 °C was chosen as a heat source because of the existing hypothesis of building the inertization and drying facilities close to waste-to-energy incinerators, where large quantities of hot air are made available as a byproduct.

The first proof of concept consisted in a closed stirrer similar to a cement mixer, characterized by a constant flow of hot air. The device was aimed at the sole evaluation of the drying process, so the Cosmos was added immediately after mixing the components.

As shown in Fig. 1, the machine body is a cylinder made of PVC with an opening to load the material. An aluminum tube with 40-mm diameter inside the body is the rotation pivot of the cylinder and the hot air inlet/outlet duct at the same time. Mixing blades were placed inside the central body, integral with the rotation axis, to the aim of keeping the drying material constantly stirred.

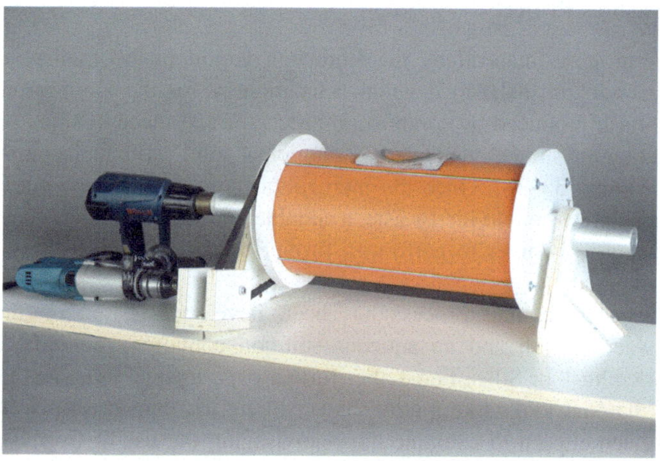

Fig. 1. Proof of concept for dynamic drying test.

The tests showed the low effect of the mixing blades. The progressive drying has indeed caused the formation of semi-solid blocks, which were compact enough to start rolling within the device body, and of a solid scab.

The second proof of concept, shown in Fig. 2, aimed at testing the effects of maximizing the contact surface with air, consisted of a reduced-height tunnel constantly flown through by hot air. The device has been built with a structure in semi-expanded PVC, where two glass panels have been fixed to obtain the top and the bottom of the tunnel. Two hot-air blowers have been fixed to the top panel structure by means of a converging manifold, so to distribute air flow suitably to the tunnel geometry. The bottom surface was mounted on an independent structure and could slide under the air flow, so to simulate the possible motion of the material with respect to the injection point of air. The tests showed interesting results. After about 90 min, the material looked solid but friable. Despite not being completely dried, it could be easily pulverized and sent to a further step of static drying. The availability of a larger air flow rate and of a de-humidifier in the pilot facility could have improved the process considerably. This type of drying has a downside in that it requires a larger facility, but it allowed to turn the material from sludge to a powder with no tendency to compaction. Moreover, it is suitable to "hybrid" plants with a static drying followed by a period of either storage in drying room or dynamic drying.

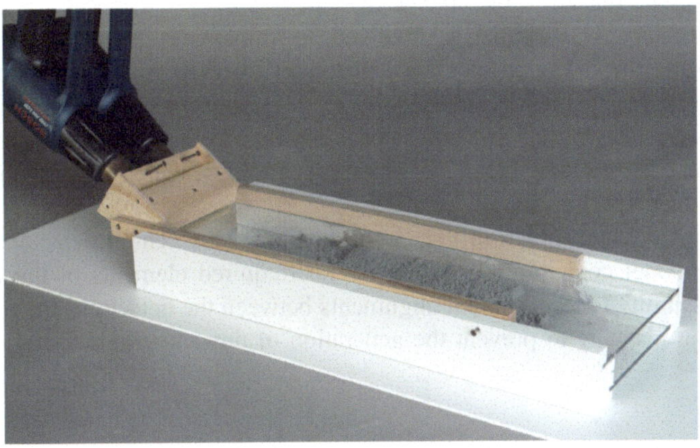

Fig. 2. Proof of concept for drying with increased air contact surface.

On the basis of test indications a pilot facility for 20-kg batches was developed. Stirring was performed by plowshares and a radial emulsifier was added to break compact conglomerates into pieces and integrate the action of the plowshares.

2.2 Cosmic Ray Detector

The detection of cosmic rays and the use of detector telescopes are one of the research themes developed by the Experimental Physics group of the University of Brescia [28–32]. Thanks to their high penetration, cosmic rays pass through the matter with negligible deviations, which makes this technology ideal to implement non-invasive solutions to monitor the stability of historical buildings or to identify the presence of dangerous materials within shipping containers (muon tomography).

The efficiency of detectors based on scintillating plastic fibers and silicon photomultipliers is currently under investigation. Scintillating fibers used, BCF-10 manufactured by Saint-Gobain, have a 3 × 3-mm^2 section, and consist of a polystyrene core containing a combination of fluorescent dopants that emit light if energetically excited, like in the case of a subatomic particle passing through the material. The core is coated with a PMMA cladding ensuring total internal reflection by refraction index difference. The passage of a particle through the fiber causes the excitation of scintillating molecules in the core and the emission of light, which is trapped within the fiber thanks to the total reflection between core and cladding, and is channeled to the endpoints of the fiber where it gets detected by silicon photomultipliers.

The ongoing research has led to the construction of a "detector telescope" based on three overlapping detectors, each of which containing 8 200-mm scintillating fibers. This device provided data for a comparison with the results of simulations. The first telescope consisted of three containing and masking elements, at the endpoints of which silicon photomultipliers with their electronic boards were mounted, in contact with fiber extremes. The encouraging results obtained during the experimental tests opened the possibility of building larger and more complex telescopes, thus highlighting any issue related to the change of scale.

The laboratory was primarily involved to implement an experimental set-up allowing testing the detection efficiency of longer fibers (1 m), in order to check the real entity of light attenuation along the fiber. The need for developing a customized device for such tests raised a practical problem due to the change of size. The first telescope had been made by FDM 3D printing technology, which is not suitable to print a five-times-bigger set-up.

In the new device, shown in Fig. 3 and in Fig. 4, the central detector is replaced with a new model containing 1-m-long fibers. Required elements of the new set-up were the possibility to test different alignments between the detectors, and the presence of an efficient masking to prevent the generation of unwanted light radiation.

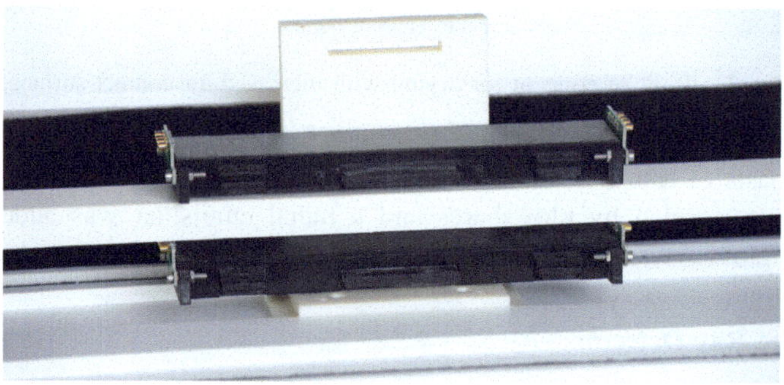

Fig. 3. Positioning of 1-m detector.

A new containing element was built for one of the three sets of scintillating fibers, using different techniques. In particular, the headers housing fiber endpoints and photomultipliers had been 3D printed, while the central body and the cover had been made in PVC by chip removal machining. The three elements were mounted on a U-shaped aluminum profile, to ensure a linear assembling and to confer stiffness to the container, while being supported only at the endpoints. The central detector was designed to be fixed, while the structures of the remaining two detectors were allowed to move, so to enable tests in different spatial configurations.

As concerns masking, the idea was to assemble all the components in a single device, with the additional function of allowing the 200 mm detectors to slide in parallel to the central detector and to the electronic board housing. To this aim, a hollow semi-expanded PVC base was created with openings for the flat cables to be connected to each of the detectors, and a rail to slide the old telescope base and the supports for the central detector. Hinges were mounted on the base to allow the flap opening of the cover, made of the same material and coated with a layer of black tufted polypropylene fabric, also used to upholster the zone where the cover leans against the base. This zone was recessed with respect to the edge to the aim of minimizing light entrance.

As concerns the assembling of the individual detector, black 3 × 1000-mm, 170 g/m^2 cardboard strips were prepared and put between fibers to limit bending and prevent cross-talking. Moreover, small steel plates, size, 3 × 3 × 0.6 mm, were cut and mirror-polished for a possible installation on the fiber endpoints opposite to photomultipliers, in order to reflect the light radiation transmitted in that direction.

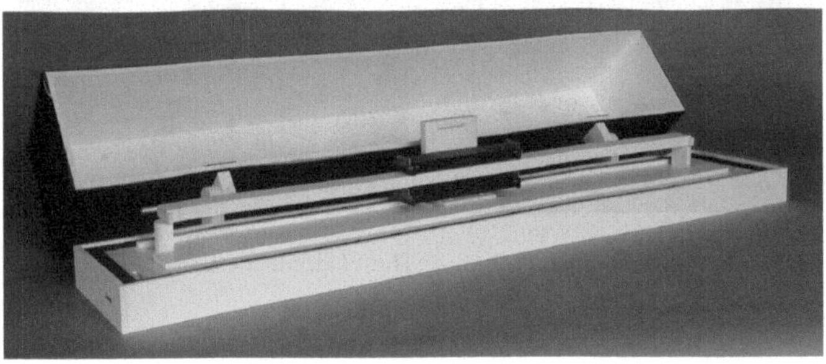

Fig. 4. Complete device.

Currently, the set-up is still under construction: the electronic equipment designed on the basis of the research group specifications should be delivered shortly.

In both telescope versions, a further problem was observed, which would be even amplified by a change to a larger scale experimentation: fiber cutting. The next stage of the research would envisage the construction of advanced detectors, made of two orthogonal fiber layers, for a total surface of 1 m^2. Considering three detectors for each telescope and given the fact that two telescopes are necessary to detect the angle of the incoming and outgoing rays relative to the investigated object, the number of required fibers to build a new set-up would be close to 4000 units, which is clearly incompatible with manual cutting and polishing. Fibers must be cut to size with a fair orthogonality (only slight inclination can be compensated with optical grease), at low temperature to keep the fiber constituents from melting and blending, and without damaging the cladding. Finally, the cut ends should be mirror-polished.

Cladding chipping turned out to be the most critical element. Cut tests have been performed with different technologies, revealing how this 0.1-mm layer is fragile, and how the slight concavity of the profile faces can make things even worse. Such concavity causes a small gap to be present between the base and the face during the cutting operations. In cutting or abrading tool exit zone this may chip the cladding, as shown in Fig. 5, thus creating points of dispersion for the light signal.

The solution found to cut and polish fibers with the equipment available in the Department has been the inspiration for a possible design development in view of a further change of scale. The problem has indeed been solved by holding the fibers being cut in a support made of two symmetrical plastic halves, as shown in Fig. 6 and in Fig. 7, each of which featuring milled half-fiber impressions, that can be closed together by means of screws.

Fig. 5. Cut and polished endpoint. The cladding looks slightly chipped.

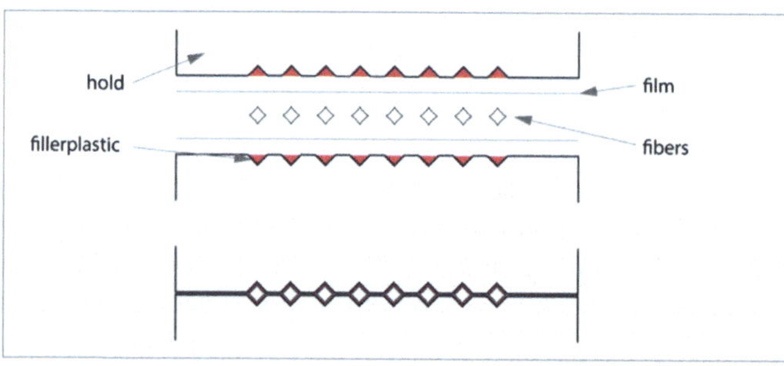

Fig. 6. Operation layout of the cutting/polishing device.

Impressions have been filled with plaster and a plastic film has been used to protect fibers from contact. In this way, the pieces being cut could be blocked in a support that was perfectly adherent to the profile faces.

The exceeding plaster and the protruding part of the fibers have been subsequently removed by abrasion, by means of a water-cooled lapping machine and several steps of gradually finer abrasives (the last of which had 1 μ grain size), until a mirror-polishing was obtained. The size of the support, far larger than fiber size, allowed to have a wide base surface ensuring a fair orthogonality.

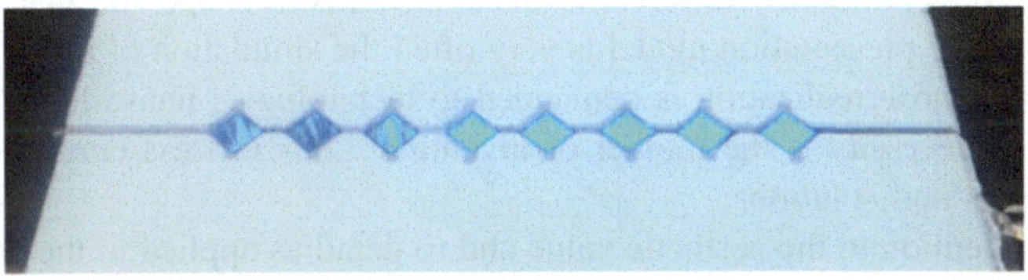

Fig. 7. Fibers within the support after polishing.

The use of a filler to solve the problem of fiber geometric irregularities inspired the idea to go beyond the current detector layout, based on a containing element serving as a masking device and acting as an interface between fibers and electronic components, and to think about a monobloc instead. This solution is still in the development phase.

3 Conclusions and Laboratory Development Outlook

The case histories described in this paper are just examples of the great variety of possible applications, and they point out how the interdisciplinary conceptual and practical contribution of such laboratories can impact on the activities of the different research groups. It is worth noting that only the "model" section has been dealt with, with the "photography" section barely mentioned. The generation of image and technical illustration represents indeed another impacting field in both teaching and research activities. Those who followed from the outside the evolution of the former "design" laboratories in the last years may think that this is the result of an extremely personal approach to the direction and orientation of the laboratory activities and, as such, difficult to "formalize". The different personalities of people "dwelling" in a structure certainly do influence how the work is organized, but the Authors believe that this vision should be completely reinterpreted, by considering the education, and therefore the mind-set, of those who contributed to the structure growth.

It is no coincidence that the laboratories described in this paper have been managed in the years by a designer, whose approach to problem solving is interdisciplinary by education, and by a teacher of technical design, the communal language of technical disciplines. The peculiarity of the laboratory is indeed how interdisciplinary it has proved to be in the years, and its great potential lies in the influence it can have on the evolution of the methodological research approach of experimental disciplines.

A further key factor for the development of the laboratory is that it has its roots in years of activities in teaching design. In particular, the experience in the construction of study models offers a valid approach even for complex instruments like those outlined in the case histories, since it represents a quick starting point to study and to solve a problem.

Similarly, the laboratory deals with a number of different materials, thus allowing to choose the most convenient one to develop a device according to the specific constraints. In particular, the availability of leftover material from various activities carried out for the Industrial Design program in the years has been of great help for the PoC-BS initial development. One of the most interesting aspects of the activity related to Industrial Design programs is, indeed, the wide range of solutions to the same theme, hence the need to build very different artifacts in terms of shape and material as well.

Moreover, the presentation model is very often the simulation of a real object, so it requires parts whose realization is demanded to technologies unavailable to the laboratory. Since "necessity is the mother of invention", this process enriches skills with new techniques and solutions.

Finally, attention to the aesthetic value and to detail as applied to the design model can bring formal accuracy to research devices, which is of great importance for communication purposes. The subsequent preparation of product picture and technical illustration is shared between design and engineering technical drawing, though with different shades, and it fulfills a primary role in supporting the research.

The laboratory activity can provide a significant contribution not only in the experimental phase, but also when the research has to be presented to the aim of funding request, participation to competitions or project reports. The dynamics is, to some extent, analogue to the design model: those who evaluate the project are usually technicians, but rarely experts of the specific discipline. Therefore, well-finished illustrations of the research, of the preliminary results, and of the specially-made tools, enable easier understanding of the research topic and of the tangible issues tackled, thus providing an essential competitive edge.

In conclusion, the activity of PoC-BS has wide room for growth in the framework of a complete evolution as an interdisciplinary structure to support research. However, this transformation cannot be started without a reflection on methods or the creation of rules to discipline collaborations and ensure the fair acknowledgement of the laboratory scientific contribution, if present.

Exploiting the laboratory as a mere cheaper, or even free-of-charge, alternative to outsourcing for the realization of research devices, would fail to give the right value to the skills developed in years of work and would lead to lose the creative potential expressed in the proposed case histories.

Lacking such potential may restrict the development of preliminary research tools to design disciplines alone, or at least to those entities that, despite not being focused on product design, include people with the individual ability of translating into practical solutions the problems and challenges of the specific research.

Only the formalization and application of a research method that regulates the development of tools to support the early stages of projects requiring experimental campaigns, can ensure the laboratory activity is actually interdisciplinary, and increase its application effectiveness.

Acknowledgments. Authors would like to thank Maria Cristina Ricossa for the help and the involvement in the PoC-BS lab.

References

1. Oxford dictionary: https://en.oxforddictionaries.com/definition/proof_of_concept
2. European Research Council: http://ec.europa.eu/research/participants/data/ref/h2020/other/guides_for_applicants/h2020-guide18-erc-poc_en.pdf
3. Techopedia: https://www.techopedia.com/definition/4066/proof-of-concept-po
4. Wu, W., et al.: Proof of concept test of a centrifugal particle receiver. Energy Procedia **49**, 560–568 (2014)
5. Redmond, J.A., et al.: The design and experimental validation of an ultrafast shape memory alloy resettable (SMART) latch. J. Mech. Des. 132/061007-1 (2010)
6. Toma, M., et al.: Design and proof-of-concept validation of a latched arch active seal. J. Mech. Des. 134/075001-1 (2012)
7. Santer, M., Pellegrino, S.: Concept and design of a multistable plate structure. J. Mech. Des. 133/081001-1 (2011)
8. Luo, Y., Hutapea, P.: Design of a bone transport device using smart material actuators. J. Mech. Des. 131/091005-1 (2009)
9. Murthy Arelekatti, V.N., Graves, C.W., Mkrtchyan, A.: Proof-of-concept evaluation of a low-cost and low-weight tractor for small scale farms. In: Proceedings of the ASME 2014 International Design Engineering Technical Conference & Computers and Information in Engineering Conference, EDETC/CIE 2014, August 2014, Buffalo NY, USA (2014)
10. Moon, Y., Seo, J.B., Choi, J.: Development of new end-effector for proof-of-concept of fully robotic multichannel biopsy. IEEE/ASME Trans. Mechatron. **20**(6), 2996–3008 (2015)
11. Trucco, H.A.: Developing ultra-light-weight aerial vehicle design and proof of concept - AC 2011-957: novel aeronautical engineering student project (2011)
12. Bliesner, W., et al.: Proof of concept and test validation of a 25 kilowatt dual shell stirling engine. Public Interest Energy Research (PIER) Program, Final Project Report CEC-500-2009-061 (2009)
13. Hutapea, P., Jacobs, K., Harper, M., Meyer, E., Roth, B.: Development of a proof-of-concept aircraft smart control system. Aeronaut. J. **113**, 1147 (2009)
14. Lammlein, D.H., et al.: Friction stir welding of small diameter pipe: an experimental and numerical proof of concept for automation and manufacturing. In: Proceedings of the Institution of Mechanical Engineers, Part B: Journal of Engineering Manufacture (2010)
15. Otten, J., et al.: Proof-of-concept of the shape memory alloy resettable dual chamber lift device for pedestrian protection with tailorable performance. J. Mech. Des. 135/061008-1 (2013)
16. Takai, S., Esterman, M., Midha, A.: An approach to study associations between design concepts and design outcomes. J. Mech. Des. 137/041101-1 (2015)
17. Critoph, R.E., Metcalf, S.J., Telto, Z.T.: Proof of concept car adsorption air conditioning system using a compact sorption reactor. Heat Transfer Eng. **31**(11), 950–956 (2010)
18. Paderno, D., Baronio, G.: Aspectos de integracion entre modelacion virtual y maqueta en la enseñanza del diseño. In: Atti del Congresso nazionale XX INGEGRAF, Valencia (Spagna) 4–6 Giugno 2008
19. Mendel, J.: A taxonomy of models used in the design process. Interactions **19**(1), 81–85 (2012)
20. Pola, A., et al.: Effect of globular microstructure on cavitation erosion resistance of aluminium alloys. Solid State Phenom. **256**, 51–57 (2016)
21. Uberti, S., Baronio, G., Cambiaghi, D.: Study & design of a special test bench for hydrostatic spindle housings. In: 11th International Design Conference, DESIGN 2010, pp. 1729–1740 (2010)

22. Borboni, A., et al.: Experimental kinematics of a special shape actuator. In: ASME 2014 12th Biennial Conference on Engineering Systems Design and Analysis, ESDA 2014, vol. 3 (2014)
23. Bodini, I., et al.: A novel optical apparatus for the study of rolling contact wear/fatigue based on a high-speed camera and multiple-source laser illumination. Rev. Sci. Instrum. **87**(8), 083701 (2016)
24. Bodini, I., Lancini, M., Pasinetti, S., Vetturi, D.: Techniques for on-board vibrational passenger comfort monitoring in public transport. In: 12th IMEKO TC10 Workshop on Technical Diagnostics: New Perspective in Measurements, Tools and Techniques for Industrial Applications, Florence, Italy (2013)
25. Uberti, S., Copeta, A., Motyl, B., Baronio, G.: An eco-innovation and technical contaminated approach for designing a low environmental impact off-road motorcycle. Int. J. Interact. Des. Manufact. **12**(1), 281–295 (2018)
26. Baronio, G., Copeta, A., Motyl, B., Uberti, S.: Gölem project: concept and design of a trekking/hiking wheelchair. Adv. Mech. Eng. **9**(11), 1–11 (2017)
27. Volonghi, P., Baronio, G., Signoroni, A.: 3D scanning and geometry processing techniques for customized hand orthotics: an experimental assessment. Virtual Phys. Prototyp. **13**(2), 105-116 (2018)
28. Luciano, V., Sardini, E., Serpelloni, M., Baronio, G.: An energy harvesting converter to power sensorized total human knee prosthesis. Measur. Sci. Technol. **25**(2), 025702 (2014)
29. Bodini, I., Bonomi, G., Cambiaghi, D., Magalini, A., Zenoni, A.: Cosmic ray detection based measurement systems. A preliminary study. Meas. Sci. Technol. **11**(18), 3537–3546 (2007)
30. Bonomi, G., et al.: Muon tomography as a tool to detect radioactive source shielding in scrap metal containers. Int. J. Mod. Phys. Conf. Ser. **27**, 1460157 (2014)
31. Pesente, S., et al.: First results on material identification and imaging with a large-volume muon tomography prototype. Nucl. Instrum. Methods Phys. Res., Section A: Accelerators, Spectrometers, Detectors Assoc. Equip. **604**(3), 738–746 (2009)
32. Bonomi, G., et al.: Measur. Sci. Technol. **30**(4), 045901 (2019)
33. Benettoni, M., et al.: Noise reduction in muon tomography for detecting high density objects. J. Inst. **8**(12), 12007 (2013)
34. Bontempi, E., et al.: A new method for municipal solid waste incinerator (MSWI) fly ash inertization, based on colloidal silica. J. Environ. Monit. **12**(11), 2093–2099 (2010)
35. Zacco, A., et al.: Use of colloidal silica to obtain a new inert from municipal solid waste incinerator (MSWI) fly ash: first results about reuse. Clean Technol. Environ. Policy **14**(2), 291–297 (2012)
36. Besco, S., et al.: Processing and properties of polypropylene-based composites containing inertized fly ash from municipal solid waste incineration. J. Appl. Polymer Sci. **130**(6), 4157–4164 (2013)
37. Rodella, N., et al.: Beyond waste: new sustainable fillers from fly ashes stabilization, obtained by low cost raw materials. Heliyon **2**(9), e00163 (2016)
38. Benassi, L., et al.: Chemical stabilization of municipal solid waste incineration fly ash without any commercial chemicals: first pilot-plant scaling up. ACS Sustain. Chem. Eng. **4**(10), 5561–5569 (2016)

Experimental Study on Nonlinear Random Excitation

Francesco Pellicano, Antonio Zippo(✉), Giovanni Iarriccio, and Marco Barbieri

Dipartimento di Ingegneria "Enzo Ferrari", Università di Modena e Reggio Emilia, Via P. Vivarelli 10, 41125 Modena, Italy
{francesco.pellicano,antonio.zippo, giovanni.iarriccio,mark}@unimore.it

Abstract. Experimental design methods are instruments for directing useful, time-effective and efficient experiments and an accurate strategy in experimental activities lead to successful results.

In the present paper is explained an experimental campaign focused on the random vibrations of circular cylindrical shells under thermal gradients across the shell thickness and broadband random loading to identify a particular phenomenon called synchronicity: the investigation is fully experimental.

Nuclear, aerospace and automotive are some of the engineering fields involved in this subject, and in these real environments non-deterministic excitations can be coupled with a thermal load; extreme thermal conditions can cause differences of the temperature inside and outside the shell, e.g. thermal exchangers.

Due to the importance of the subject, the literature on shell vibration is extremely wide, it is not analyzed here for the sake of brevity; however, it is to note that the number of papers containing experimental results is not large.

Under a random forcing, a system generally expects a random response, the statistical properties of the random response are correlated with the forcing through the transfer function in the case of linear systems, or more complicated relationships in the case of nonlinear systems. However, in some particular conditions (e.g. internal resonances, parametric resonances, …) the presence of nonlinearity in the systems can give rise to a surprising phenomenon, said synchronicity or entrainment.

In this work a shell subjected to a random base excitation is analyzed experimentally, the excitation is random (flat or limited frequency band), and takes advantage of previous setup and experimental techniques [3–5] developed by the present research team. The phenomenon of synchronicity is clearly observed for some particular thermal conditions: a strong transfer of energy from a broadband excitation signal to an almost harmonic response is experimentally observed, confirming the general findings of refs. [1, 2].

Keywords: Shells · Random vibration · Synchronization

C. Rizzi et al. (Eds.): ADM 2019, LNME, pp. 637–648, 2020.
https://doi.org/10.1007/978-3-030-31154-4_54

1 Introduction

Thin-walled structures have been a crucial role in several engineering fields as automotive (bodywork panels) or aerospace (fuselage and aircraft and satellite panels).

Those structures present often a strong nonlinear dynamic behaviour that leads designers and engineers to deal with it conditions to be able to forecast as accurate as possible the structural response due to external loadings, in [3] Zippo et al. studied experimentally the behaviour of the response of a thin circular cylindrical shell, it was shown that in tests performed with a sinusoidal axial forcing load clarify the role of the preload in enlarging the instability regions and the coexistence of more than one stable states when the shell is pre-loaded and excited with a moderately large sinusoidal excitation: the dynamic scenario is completely different if the frequency step is reduced from 1 Hz to 0.1 Hz.

The possible condition of critical temperature and thermal gradients through the thickness lead to a change in the mechanical properties of the structure with a consequent complexity in predicting the behaviour of the structures; from a deep analysis of the scientific literature it is possible to observe that there is a gap in the experimental studies and in the dynamic vibrational modelling of thin-walled structures subject to strong thermal gradients, which presents only a few publications in this field: the effect of temperature has been deeply studied in [6] where is clarified the role of thermal gradients, in particular, the damping and stiffness properties of the shell shown a complex behaviour that implies that using linear modelling could underestimate the dynamic response of the structure. In a previous study, Pellicano and Avramov [7] analyzed the dynamics of a circular cylindrical shell carrying a rigid disk on the top applying a numerical approach to obtain a complete analysis of the linear and nonlinear response. In a study published in Nonlinear Dynamics [4] is shown that the effect of temperature strongly affects the instability regions and the magnitude of the measured kinematic quantity and has been pointed out that high environmental temperature leads to a more complex shell dynamic.

Another critical condition in thin structures, of considerable interest for the automotive and aerospace fields, is the behaviour of these structures subjected to random loadings whose dynamics, characterized by strong nonlinearity, leads to the emergence of complex phenomena that cannot be predicted with current design tools, like finite element modelling. It is well known that when a chain of non-linear oscillators, used to model physical or biological structures, is subjected to intense periodic forcing, could present a "mode-locking" phenomenon that synchronizes the forcing load with the system response.

A similar phenomenon, but much less known and studied, can occur in the case of broadband random forcing [1] if certain conditions occur of internal resonance the random motions are found to be quasi-stationary, with steady oscillatory terms in the response moments. This suggests the possibility of entrainment of regular harmonic responses by the system.

This phenomenon, called synchronization of non-linear oscillators subjected to random forcing [2, 8], has been partially studied in the literature for its remarkable characteristic of conveying the spectral energy of a random forcing over specific frequencies, determining remarkable vibration amplitudes: Jansen et al. showed that a single, non-resonant periodic forcing is insufficient to activate the mode-locking phenomenon, but the supplement of a reasonable quantity of broadband noise allows transient mode-locking to the original periodic drive to occur. Similar studies involve the synchronization of chaos [9] referred to a process in which several chaotic systems adapt a certain property of their dynamics to a common behavior due to a coupling or a periodic or random forcing and analytical and numerical work [10] has been presented showing the synchronization induced by noise in chaotic systems.

Some experimental studies on very simple systems can be found in the literature, Ibrahim [12] presented a review article on experimental investigations of random excitation of dynamic systems; Cicek and Ertas [11] conducted a sets of experiments on a beam tip mass and pendulum system subjected to random excitation to determine the autoparametric interaction between the two modes of the system in the neighborhood of the autoparametric region, in particular the response behaves as ultra-narrowband process in the neighborhood of exact internal resonance that converge into a periodic response. A study carried out [13] by Roberts presents an analytical and experimental comparison with the broad band random excitation of a two degree of freedom vibratory system with non-linear coupling of autoparametric type which consists of a cantilever beam with at tip clamped another cantilever beam; attached to a primary vibratory system in such a way that response of the primary system imposes an axial motion on the cantilever, the system is a common structural arrangement, representing the fuselage-tail plane coupling in aircraft structures. All experiments present in the literature shows a test setup where the structures under investigations were excited within a frequency band around the main resonances and, as could be expected, the correspondent system response in the same bandwidth. In this work is presented the occurrence of the synchronicity coupled with a subharmonic response that arise 450 Hz below the excitation band. The main aim of this work is to demonstrate the occurrences of synchronicity in a complex system as a circular cylindrical shell with a top mass and to verify that this phenomenon is related to particular thermal conditions and forcing conditions, this showed an interesting scenario in the system response.

2 Setup

In the present work a random broadband loading has been applied at the base of a circular cylindrical shell with a top disk fixed at a electrodynamic shaker: the role of the disk is to impose a rigid body displacement at the top end of the shell and to exert an inertial axial load to the shell when a motion is imposed.

The shell under investigation has been 3d printed to avoid any geometrical imperfection and to allow the repeatability of the experimental campaign without geometrical errors: the specimen is made of Polyethylene terephthalate (P.E.T.) a thermoplastic polymer. The cylindrical sample present at the top end a rigid thick disk (the top disk) glued with special epoxy glue, resistant to high temperature, and its base is clamped to the fixture by means of a shaft collar that guarantees a uniform connection to the vibration table adapter.

In Table 1 are reported the geometrical and material parameters of the shell and the material properties of the fixture and top disk.

The system description is completed by Fig. 1 where a schematic representation of the system is shown: an accelerometer is located on the fixture, such accelerometer is used by the control system of the shaker when a closed loop strategy is selected, three triaxial accelerometers are located equally spaced on the top disk, they are used to measure all the six dofs of the rigid body, this allows to detect axisymmetric modes, asymmetric modes, beam-like modes and torsional modes.

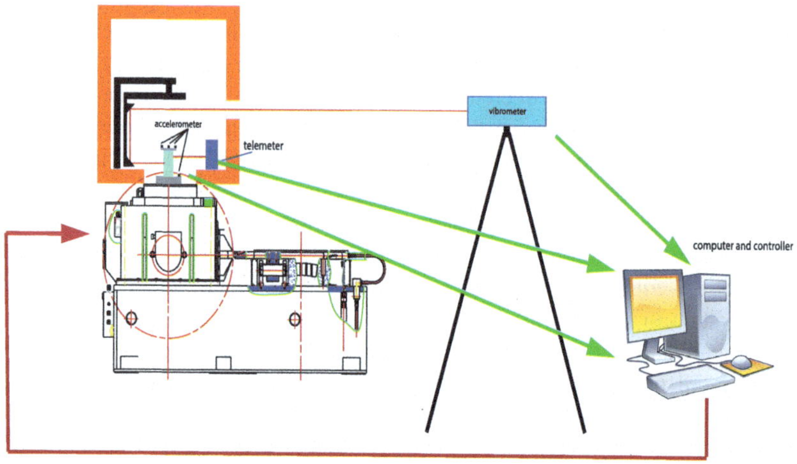

Fig. 1. Schematic view of experimental setup

Inside the shell a cartridge heater is present and is used to modify the temperature inside the shell and to create a thermal gradient across the shell wall; the specimen under investigation is mounted on a Es40 Dongling shaker coupled with a climate chamber that allow to set the outer temperature of the shell and hold it constant. The temperature ranges of the climate chamber are –70 °C up to +180 °C; the climate chamber is provided by an inspection hole from witch a laser vibrometer beam go through and is deflect by a mirror periscope to measure the velocity of a lateral point located on the opposite side respect to the hole.

Two thermocouples are used to measure the temperature, inside and outside the shell.

Moreover, a laser telemeter measures the displacement of a point of the shell out of phase of about 135°, see Fig. 2.

Fig. 2. Picture of the experimental setup

Table 1. Geometrical and material parameters of the shell, the fixture and top disk

Fixture and Top disk		
Material	Aluminium Alloy	
General	Mass Density	2700 kg/m^3
Stress	Young's Modulus	68.9 GPa
	Poisson's Ratio	0.33
Shell		
Material	P.E.T.	
General	Mass Density	1366 kg/m^3
Stress	Young's Modulus	3.05 GPa at 30 °C
	Poisson's Ratio	0.417
Dimension	Diameter	80 mm
	Thickness	0.35 mm
	Height	160 mm

3 Test Procedure and Methods

The experimental test campaign consists of a random forcing load applied through the electrodynamic shaker at the specimen, the Siemens LMS Scadas mobile has been used to generate a random signal that is controlled by means of the accelerometer placed at the base of the shell, for each test different increasing rms amplitude levels have been applied consecutively.

On the first, temperature is set for the climatic chamber and an adequate amount of time has been waited to stabilize the external temperature for about 1 h, after the inner temperature is raised slowly to reach the desired gradient, the system is left for another hour to reach a thermic stability.

When the temperature is stable a forcing random load is applied together with the thermal gradient across the shell thickness; such gradient will induce a continuous variation of elastic and dissipative properties through the thickness of the shell.

The test campaigns were carried out at different level of rms amplitude and at different bandwidth, in Fig. 3 is shown the experimental measured controlled base acceleration.

Afterwards, data was saved and imported into MATLAB where each level was analyzed computing spectrum, power spectral density, and spectrogram.

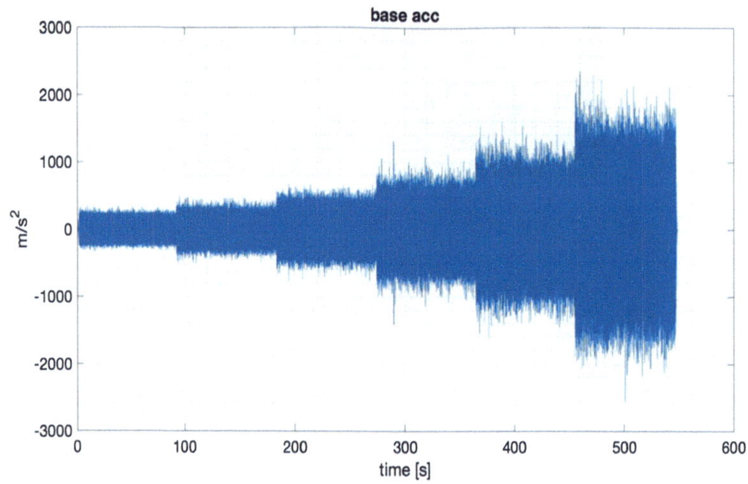

Fig. 3. Time history of base acceleration, test case 24 °C outer temp – 74 °C inner temp

4 Results

To sake of brevity only one test case will be shown in the following paragraph: the specimen under investigation is excited by a broadband random signal between 900 Hz and 4000 Hz together with a 50 °C thermal gradient from 74 °C in the inner surface and 24 °C in the outer surface; this combinations of conditions leads to a complex and unexpected scenario; in Figs. 4, 5 and 6 is shown the spectrogram of, respectively, the base acceleration, top acceleration and lateral velocity and is clearly visible that respect to the uniform frequency band of the base, the top acceleration and the lateral vibration of the shell shows an energy transfer in particular to the first axial-symmetric mode at

467 Hz. The spectrogram of base acceleration, see Fig. 4, clearly presents the energy condensed in the excitation band with a smaller spillover at higher frequencies due to the shaker dynamics for all the amplitude levels, in contrast to that the response of the top acceleration, see Fig. 5, presents a subharmonic response out of the excitation band together with a progressive reduction of energy content with a minimum at about 2900 Hz, this energy seems to be transferred at the lateral motion of the shell, see Figs. 6 and 12, with a complementary energy spikes where in top acceleration is present an energy fall, see Fig. 11. In Figs. 7, 8 and 9 the time histories of 0.025 s are shown, the piece taken corresponds at second 500 (at minute 8 and 20 s) of the whole recording, as can be noted the response of lateral velocity and the top acceleration show a contribution of an harmonic component, while the base acceleration is evidently random; more clear is power spectral density showed in Figs. 10, 11 and 12, calculated using 3 s of time history of the last amplitude level.

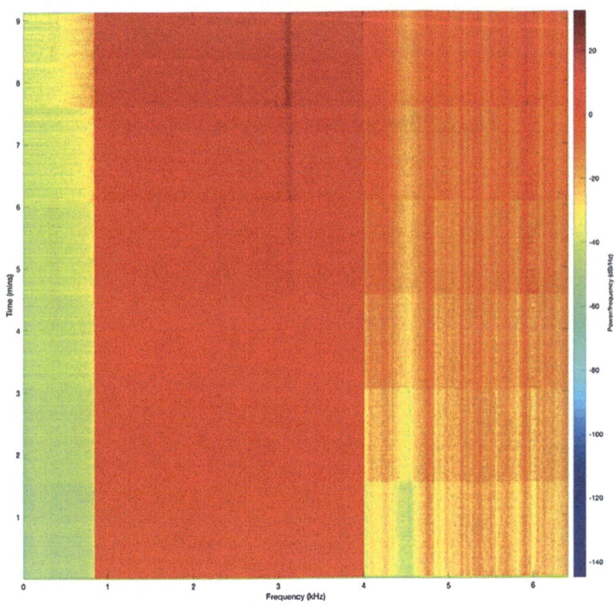

Fig. 4. Spectrogram of base acceleration

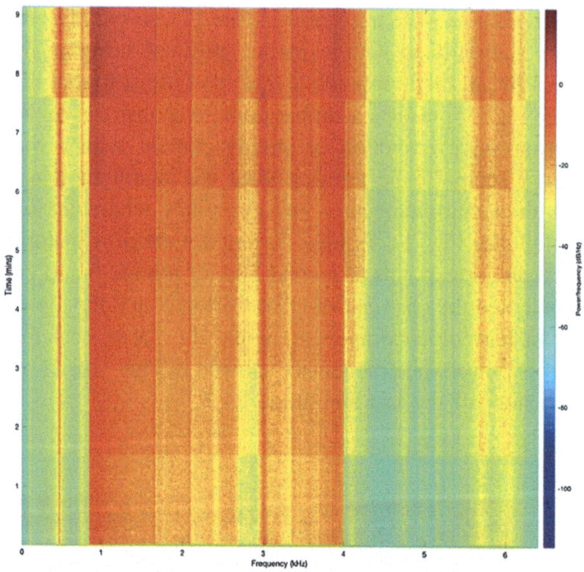

Fig. 5. Spectrogram of top acceleration

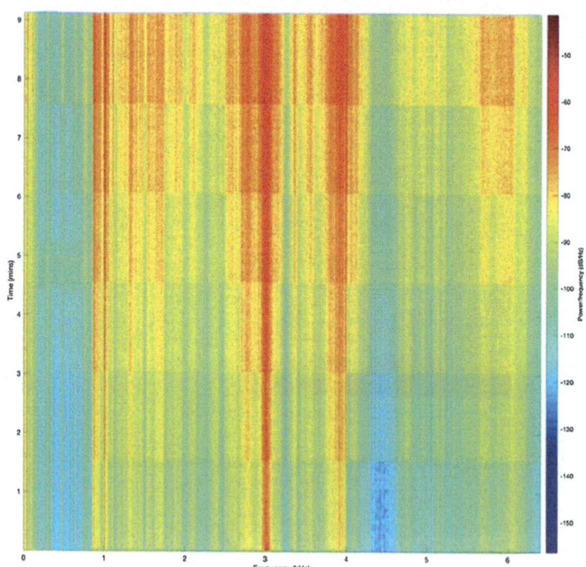

Fig. 6. Spectrogram of lateral velocity

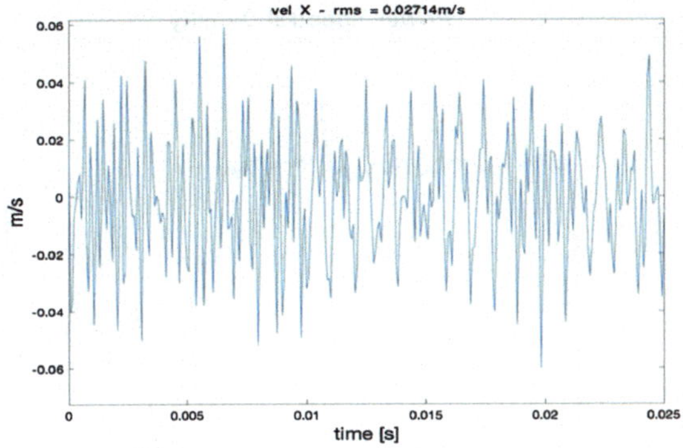

Fig. 7. Time history of lateral velocity

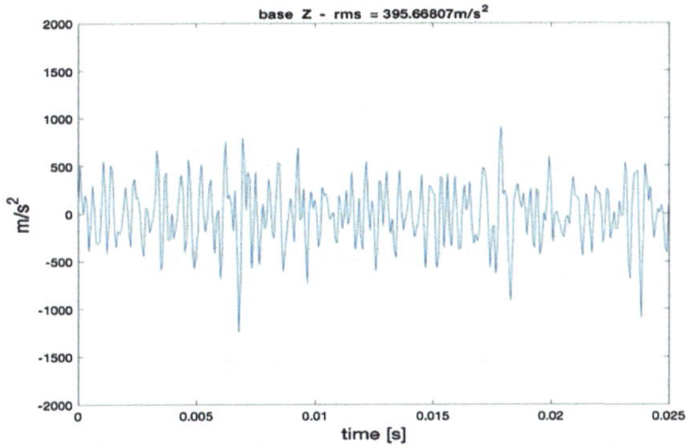

Fig. 8. Time history of base acceleration

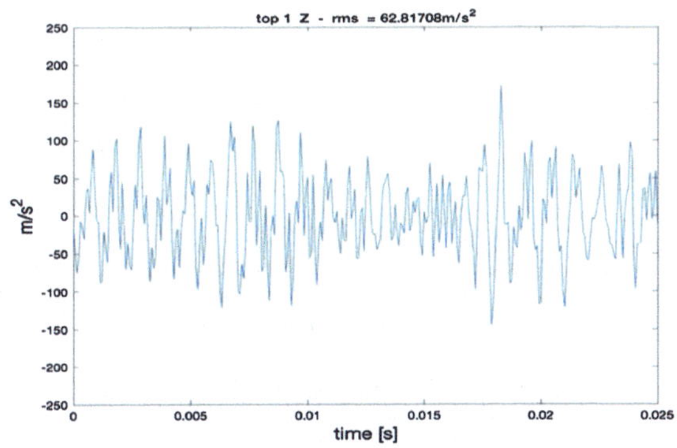

Fig. 9. Time history of top acceleration

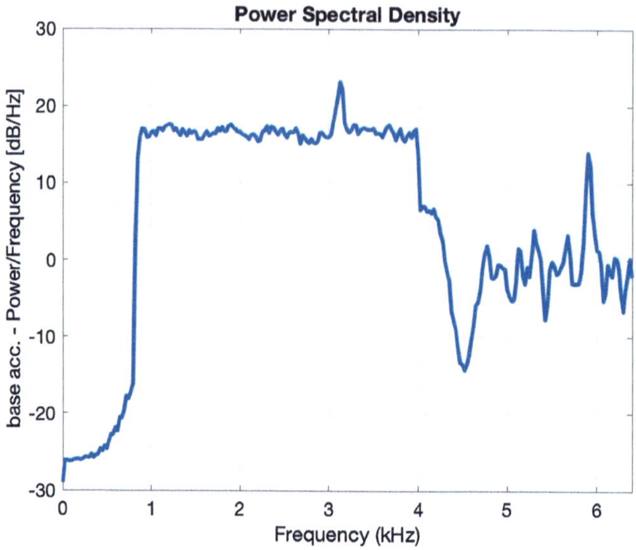

Fig. 10. PSD of driven base forcing load

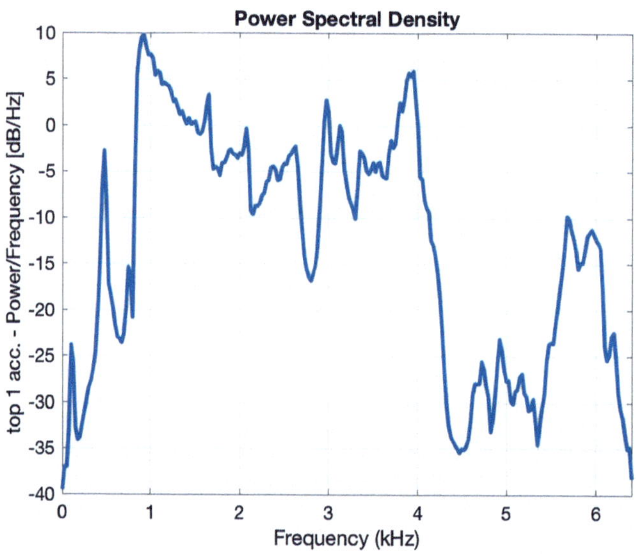

Fig. 11. PSD of response of the top disk

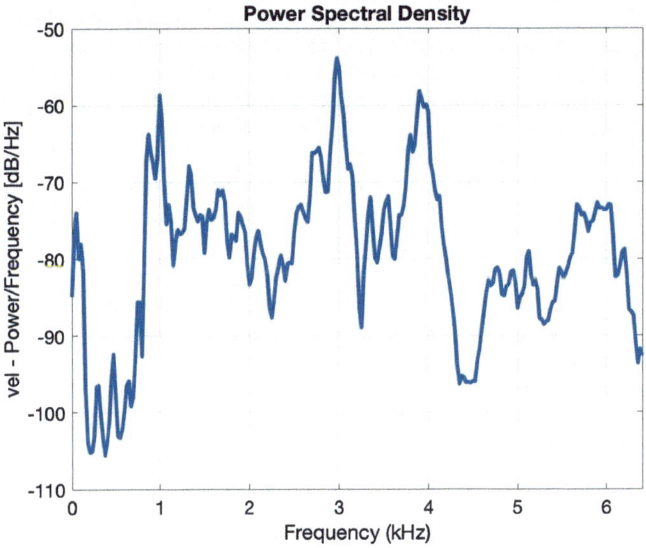

Fig. 12. PSD of lateral velocity

5 Conclusions

In the presented experimental study, a thin cylindrical shell has been driven with a random base excitation and the response has been analyzed: several thermal conditions has been tested (different thermal gradient has been applied) and numerous combinations of excitation bandwidth within amplitude levels were verified and data analyzed, in this paper is presented one of the scenario recorded where the excitation is random with a flat spectrum between 900 Hz and 4000 Hz. Two remarkable results can be pointed out: in the response is present an energy transfer to the first axial symmetric mode, that is 450 Hz far from the excitation band, and this phenomena is more visible in the top acceleration response, respect to the lateral vibration, that however show an articulated spectrum, more complex respect to the flat spectrum of the base acceleration, this subharmonic response has been noted, in previous work [3], where a saturation phenomenon occur under an harmonic excitation; this scenario is completed with the occurrence of synchronicity that hold a strong harmonic response under random excitation.

In future works a modelling of the system dynamics will be performed in order to better understand the mechanisms the lead to this scenario.

References

1. Ibrahim, R.A., Roberts, J.W.: Broad band random excitation of a two-degree-of-freedom autoparametric coupling. J. Sound Vib. **44**(3), 335–348 (1976). https://doi.org/10.1016/0022-460X(76)90506-X

2. Jensen, R.V.: Synchronization of randomly driven nonlinear oscillators. Phys. Rev. E: Stat. Phys., Plasmas, Fluids, Relat. Interdiscip. Top. **58**(6), R6907–R6910 (1998). https://doi.org/10.1103/physreve.58.r6907
3. Zippo, A., Barbieri, M., Pellicano, F.: Experimental analysis of pre-compressed circular cylindrical shell under axial harmonic load. Int. J. Non-Linear Mech. **94**, 417–440 (2017). https://doi.org/10.1016/j.ijnonlinmec.2016.11.004
4. Zippo, A., Barbieri, M., Iarriccio, G., Pellicano F., Nonlinear vibrations of circular cylindrical shells with thermal effects: an experimental study. In: Nonlinear Dynamics (2019). https://doi.org/10.1007/s11071-018-04753-1
5. Gelman, L., Harish Chandra, N., Kurosz, R., Pellicano, F., Barbieri, M., Zippo, A.: Novel spectral kurtosis technology for adaptive vibration condition monitoring of multi-stage gearboxes. Insight Non-Destructive Test. Condition Monit. **58**(8), 409–416 (2016). https://doi.org/10.1784/insi.2016.58.8.409
6. Zippo, A., Barbieri, M., Pellicano, F.: Temperature gradient effect on dynamic properties of a polymeric circular cylindrical shell. Compos. Struct. **216**, 301–314 (2019). https://doi.org/10.1016/j.compstruct.2019.02.098
7. Pellicano, F., Avramov, K.V.: Linear and nonlinear dynamics of a circular cylindrical shell connected to a rigid disk. Commun. Nonlinear Sci. Numer. Simul. **12**(4), 496–518 (2007). https://doi.org/10.1016/j.cnsns.2005.04.004
8. Jensen, R.V.: Synchronization of driven nonlinear oscillators. Am. J. Phys. **70**(6), 607–619 (2002). https://doi.org/10.1119/1.1467909
9. Boccaletti, S., Kurths, J., Osipov, G., Valladares, D.L., Zhou, C.S.: The synchronization of chaotic systems. Phys. Rep. **366**(1–2), 1–101 (2002). https://doi.org/10.1016/s0370-1573(02)00137-0
10. Toral, R., Mirasso, C.R., Hernández-García, E., Piro, O.: Analytical and numerical studies of noise-induced synchronization of chaotic systems. Chaos **11**(3), 665–673 (2001). https://doi.org/10.1063/1.1386397
11. Cicek, I., Ertas, A.: Experimental investigation of beam-tip mass and pendulum system under random excitation. Mech. Syst. Sig. Process. **16**(6), 1059–1072 (2002). https://doi.org/10.1006/mssp.2001.1475. ISSN 0888-3270
12. Ibrahim, R.A.: Nonlinear random vibration: experimental results. ASME. Appl. Mech. Rev. **44**(10), 423–446 (1991). https://doi.org/10.1115/1.3119487
13. Roberts, J.W.: Random excitation of a vibratory system with autoparametric interaction. J. Sound Vib. **69**(1), 101–116 (1980). https://doi.org/10.1016/0022-460X(80)90437-X. ISSN 0022-460X

Knowledge and Product Data Management

A Knowledge Repository to Support Ecodesign Implementation in Manufacturing Companies

Marta Rossi[1(✉)], Marco Marconi[2], Roberto Menghi[1], and Alessandra Papetti[1]

[1] Università Politecnica delle Marche, Via Brecce Bianche, 12, 60131 Ancona, Italy
marta.rossi@staff.univpm.it
[2] Università degli Studi della Tuscia, Largo dell'Università, 01100 Viterbo, Italy

Abstract. In last years, increasing attention on environmental matters is registered and companies are forced by legislations, normative and protocols to increase the environmental performances of their products. Observing the industrial context, it emerges there are several barriers for an effective implementation of eco-design strategies inside design departments. The paper presents a knowledge repository tool, which aims at both, providing a basic guide on environmental sustainability issues and favoring knowledge sharing among design departments. The tool is mainly based on a structured repository in which company materials, organized in different forms, are collected. The repository contains training section, well-organized guidelines, company specific knowledge and milestone. The implementation of the tool in two industrial companies is presented and results of usability and tool effectiveness discussed.

Keywords: Company knowledge · Knowledge sharing · Environmental sustainability

1 Introduction

In last years, an increasing attention on environmental matters is registered both at consumer level, national and European legislators. European legislations face this argument with several aims, from the reduction of energy consumption, to the identification of the most recyclable materials and the reduction of critical raw materials used in products. From the other side, also consumers gives a growing attention to environmental sustainability in products' purchasing. All these initiatives require that companies face the question of environmental sustainability with the aim to improve the related performances of their products. The household appliance sector, to which a lot of normative and regulations are referred, is particularly affected by these requirements, and therefore companies are forced to implement improving strategies on their products. In particular, within the broader scenery of the EU 20-20-20 Energy and Climate Package [1], the EU launched the eco-design directive [2] and the energy-labelling directive [3]. These actions formalize eco-design and minimum energy performance requirements with the objective to increase resource productivity and decouple economic growth from resource use and its environmental impact. All these

C. Rizzi et al. (Eds.): ADM 2019, LNME, pp. 651–661, 2020.
https://doi.org/10.1007/978-3-030-31154-4_55

initiatives require companies to applied ecodesign strategies, i.e. to include environmental aspects into the product development process, but often their attention is focused only on those aspects interested by legislations, which for house hold appliance sector, is often related to reduction or optimization of use phase consumption.

Several works in literature analyzed the problems related to the poor implementation of eco-design methods and tools in industrial context. The outcomes of these works show that several barriers exist, e.g. tools and methods often diverging from industry real needs [4, 6], the availability of a large number of tools make difficult the selection of the most suitable ones [7] and the implementation of methods and tools is time, effort and resource consuming [8, 9]. Hallstedt et al. [9] recognized the need of eco-knowledge sharing tools to stimulate production, efficient use, reuse and evaluation of company knowledge in the sustainability field, thus promoting the implementation of ecodesign strategies to improve environmental products' performances.

The present paper, after the investigation of ecodesign implementation inside industrial contexts (Sect. 2), proposes a methodology and a tool to support companies in the implementation of eco-design strategies through a structured repository of guidelines, best practices and training material on eco-design matters (Sect. 3). The implementation of the tool inside two manufacturing companies belonging to the household appliance sector is also presented in order to validate the methodology and to evaluate the tool usability and efficacy inside design departments (Sect. 4). Finally, discussion and conclusion are presented in Sect. 5.

2 State of the Art

The state of the art regards two main questions: the importance of knowledge sharing inside design departments and its correlation with environmental sustainability.

Design activities are strongly depending on knowledge, which can be defined as the sum of individual designers' education and company experience, reached on a specific product or family of products and used to design, develop and innovate them. The knowledge is recognized as a crucial element for the survival of industrial organizations [11, 12] and its sharing among different actors of the design process as an element for the success of the organization [13]. Iterative processes generally characterize design activities related to manufacturing products, where data exchange and collaboration represent two important needs to assure efficient and simultaneous engineering [14]. When the knowledge is transferred among individuals, groups, teams, departments, and organizations [15, 16] we are talking about knowledge sharing. The reuse of knowledge is particularly useful especially for companies, which operates in mature domains, due to the high quality of knowledge they are able to produce, the overlapping of new versions with the previous ones and the possibility to dedicate more time to innovations [16]. Focusing the attention on approaches which favor the knowledge sharing, several works have been realized both in academic and commercial fields. Among the academic ones, Briggs [18] and Chevalier [19] developed an approach to support the project development through consultation of data, but usually they are pilot projects, not able to support the generation and the sharing of the tacit knowledge [20]. If commercial tools are analyzed, the classical management systems usually allow to collect in a unique repository all the documents developed by a company, without allowing a smart classification of the information and knowledge that are contained inside those documents. They allow to increase the knowledge organization and to facilitate its storing, but not its useful sharing and effective re-using.

The eco-design literature has registered in the last years an important growth, focusing in particular on the development of tools and methodologies [21, 22], which, however, are not really internalized in industrial contexts [23]. What emerges is therefore, that most of companies usually do not dispose of knowledge to share, [24, 26] or have no competences and skills on environmental sustainability [26]. The literature analysis shows how several methods, tools or approaches have been developed by researchers to collect general environmental knowledge and provided support for designers. They are often proposed in the form of guidelines, i.e. procedures to orient a decision process towards given objectives [28]. Among them, some examples are represented by the works of Vezzoli and Sciama [29], Bischof and Blessing [30], Koh et al. [31], Rose et al. [32], including the well-known ten golden rules proposed by Luttropp and Lagerstedt [33] or the Ecodesign Pilot project [34] where a series of eco-design guidelines are collected. Usually, eco-design guidelines provide only general indications to designers and difficulties in translation into effective design choices are noticed [29, 31]. Most of these proposed solutions have a given knowledge, not stimulating its creation, customization and/or use and reuse inside design department.

Starting from these considerations the present paper proposed a methodology and tool to support companies in the implementation of eco-design strategies through a structured repository of guidelines, best practices and training material on eco-design matters. The tool, which aims at stimulating the knowledge consultation, creation and sharing, has been tested into two Italian manufacturing companies, which evaluated its functionalities and usefulness.

3 Methodology and Tool Functions

The methodology proposes in this paper guides designers through the design/redesign process and assist them in environmental design choices thanks to the consultation of eco-knowledge collected in a structured database (DB). The DB integrates contents related to eco-design and environmental sustainability.

Two complementary aspects characterize the methodology: creation and sharing of the environmental knowledge with the double aim to facilitate the problem solving and to support designers in acquiring competences on sustainability issues. The following five-steps methodology has been defined (Fig. 1):

- Step 1: Define the objective. Designers and/or managers define the reasons driving their work (e.g. increase the recyclability of a product);
- Step 2: Asses the level of knowledge. Designers and/or managers reflect on the competences, knowledge and company experiences (both present and past) on the project matter.

- Step 3: Acquire knowledge. This facultative step allows acquiring knowledge by the consultation of training materials or company's past experiences and best practices;
- Step 4. Define strategies to face the problem and to answer to the identified objective, thanks to the consultation of past experiences or eco-design guidelines;
- Step 5. Capitalization, i.e. derive materials for knowledge generation. Company in this way stores useful material related to the project, which will be useful in future. In case no successful results, company will register the reason related to the failing or possible alternatives.

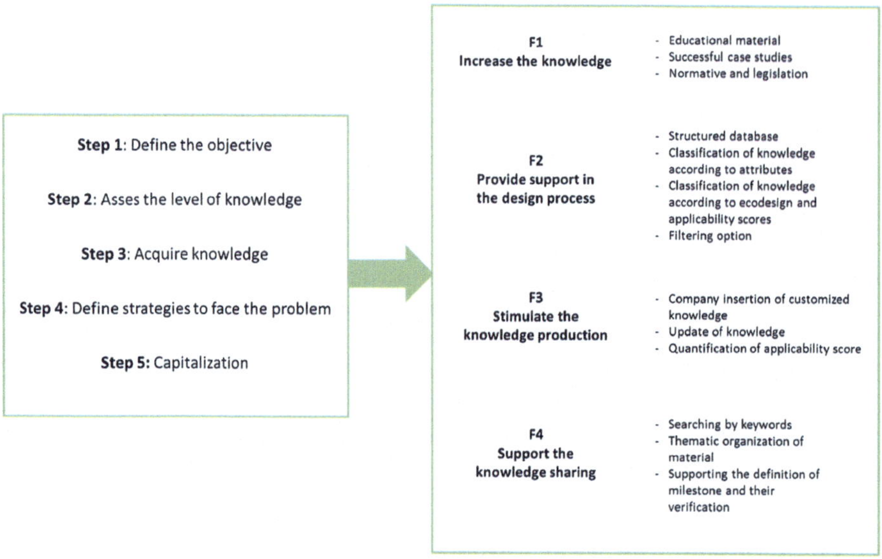

Fig. 1. Methodology

Based on the defined methodology the tool structure is derived. In particular, the tool is composed mainly by a structured DB, which allows the fulfilment of the following four functions:

- Training on environmental sustainability: designers and managers can consult specific material stored in the training section, with the objective to increase their competences by short lessons in the form of videos, presentations, documents and links to relevant materials;
- Support the design phase: designers consult the repository where guidelines are collected; they are supported in decision phase, by deriving design suggestion useful for the optimization of products under the environmental aspects;
- Stimulate the creation of customized knowledge: designers and mangers evaluate if and in what measure guidelines could be implemented in their company, giving them a further specificity for the specific context and insert specific knowledge through a guided procedure;

- Supports knowledge sharing activities, by the consultation of materials contained in the DB clustered according to thematic organization (e.g. typology of problem, key work, and product type), definition of project milestones and verification of their achievement.

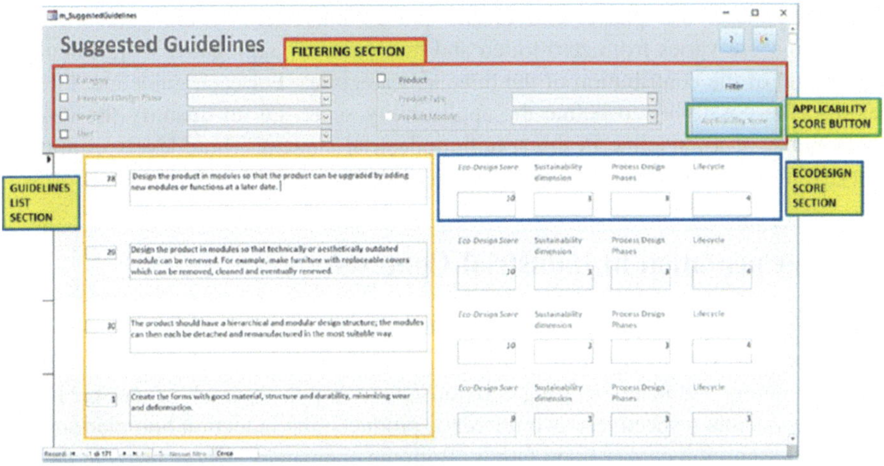

Fig. 2. Filtering main interface

In order to fulfil the abovementioned functions, the tool DB contains several types of material:

- General guidelines retrieved from the literature, simple lessons, presentations, and documents related to ecodesign, environmental sustainability and "green" strategies [35];
- Specific company guidelines derived from companies' specific experiences and analysis in the field of environmental sustainability (e.g. results from previous studies [36]);
- Specific company material uploaded inside tool DB after its practical use in design departments.

The tool was organized in few and simple interfaces allowing the user to choose the function he/she wants to select. The principal one, i.e. the filtering interface is shown in Fig. 2. In this section, the user can select parameters to filter guidelines contained in the DB content. Several filtering option are defined:

- Category of guidelines, this attribute allows to filter guidelines according to their category: general guidelines related to several products or family of products, or specific company guidelines related to specific products or components;
- Interested design phase, this attribute indicates what is the design phase in which the guidelines could be implemented;
- User, this attribute allows filtering guidelines according to the user typology (e.g. mechanical designer, electrical designers, etc.);
- Source, i.e. if the guideline has been retrieved from literature or from company specific knowledge and experiences.

The tool shows guidelines according to the filtering options and organizes them according to an ecodesign score. The ecodesign score evaluates guidelines according to their sustainability on three dimensions, i.e. sustainability dimension (environmental economic, and social sustainability), design phases (i.e. the design phase in which the guideline can be applied), and the life cycle phases (design phase, material and manufacturing phase, use phase, end of life phase) interested by the guideline. The ecodesign score varies from zero to ten and represents an aggregated score calculated on the sum of the contribution of the three subcategories. For each guideline, the user has also the possibility to define the applicability score, i.e. to quantify the level of applicability of a guideline in the specific industrial context. Three levels of applicability (from low, medium and high) have been defined for this second score.

4 Implementation in Industrial Contexts

After the tool development, the experimentation phase started. It was performed in two Italian companies, working in the field of household appliances; in particular, the first one is designing and manufacturing cooktops/ovens and the second one cooker hoods. Companies define respectively two reference products: the induction hob platform and a standard aspiration cooker hood family. Both the companies have realized in the past studies in the matter of environmental sustainability and in particular the first one has realized studies on environmental impact of their products and studies related to the virtual simulation of use phase in order to derive consumption profiles as in the energy consumption test. The e second one has realized environmental analyses in order to reduce the impact of material and manufacturing phases.

A one-day training has been performed before the tests started to train on the tool use both designers and managers involved in the test. The implementation and evaluation phases were organized in two steps, each one allowing obtaining a validation and evaluation of specific aspects. The first phase regards the tool customization: the tool DB was filled in with specific company knowledge, through a two-day activity collaboration with company designers and managers. In particular, designers and managers involved in the test phase were stimulated in the development of company specific guidelines, which summarize:

- the content of previous studies related to environmental sustainability if present;
- results of experimental tests conducted in laboratories (e.g. new material performances, energy efficiency tests);
- results of simulation procedures (realized to derive use phase profiles and related environmental impacts);

In each company, about 20 specific company guidelines were added into the DB through a dedicated interface. In this first step the companies have also evaluated the applicability of a selection of general guidelines yet contained into the tool DB (about 150 general guidelines), i.e. if and how those guidelines retrieved by the literature and contained in the tool DB could be implemented in their specific industrial context.

This evaluation has been realized by using an applicability score, which is then used to order guidelines during filtering phase according to their importance for the company. Three levels of applicability (from low, medium and high) have been defined. In this preliminary phase two designers and one manager with a long experience in the household appliances design have been involved.

The second stage of experimentation is based on the evaluation of the effectiveness and usability of the tool after it use for a period of two months, during redesign processes, inside the two design departments of the selected companies. The resource involved in this test phase were divided into four groups:

- Group 1: Expert designers with environmental skills: users belonging to this group (2 persons in each company) were a Mechanical Design Engineers and a Manager involved in innovation product projects and with strong expertise in eco-design and in environmental sciences.
- Group 2: Expert designers without environmental skills: users belonging to this group (3 persons in each company) were Mechanical Engineers and Designers without expertise in eco-design.
- Group 3: Young designers without experience: users belonging to this group are young mechanical engineers (3 persons in each company) recruited by company as lab engineers. They were involved in the training activities.
- Group 4: Young designer with environmental skills: users belonging to this group (2 persons in each company) were junior mechanical engineers, recruited by company as designers. They have basic expertise in eco-design and in environmental sciences.

Questionnaires in the Excel Format have been provided to the users. Forty-four questions were given, and two different typologies of questions have been defined:

- General, when the user provides an overall opinion on a metric (E.g., when you open the tool, is it clear what steps you can/should perform?)
- Specific, when the user provides an opinion on a specific aspect. (E.g., is it the tool aesthetically pleasing to the user?)

Usability and effectiveness questions have been divided into several groups: "Compatibility", "Consistency and standards", "Explicitness", "Flexibility and Control", "Functionality", "Informative Feedback", "Language and Content", "Navigation", "User guidance and Support", "Visual Clarity Description", "Impact on the traditional design process", "Personnel to involve". This approach has been chosen to obtain some general opinions about the tool. This allows the software developers to understand if globally the tool satisfies the user expectations or not. Each question has five possible options (sufficient, satisfactory, good, very good, excellent), ranking the tool usability and effectiveness. The following ranking criteria have been chosen for the quantification of user's opinions:

- Sufficient (score = 1);
- Satisfactory (score = 2);
- Good (score = 3);

- Very Good (score = 4);
- Excellent (score = 5);

The scores given by the users for each question have been summarized in a single medium value. In accordance with the software developer experience and with the company involved in the test, each aspect has the same weight so it possible to evaluate it using the arithmetic mean of the questions. Users have taken a medium period of one week to compile the questionnaire.

5 Result and Discussions

After the training session and first customization, users evaluated the applicability of general guidelines retrieved from literature and stored in the DB in the specific context of their companies. Guidelines present in the DB have been evaluated by users with experience (Groups 1 and 2) using an applicability score. Questionnaires with open questions have been prepared for this evaluation. The evaluation of guidelines applicability shows result contained in Table 1.

Table 1. Applicability evaluation results

First company	Second company
23 guidelines have been evaluated with a high applicability score (15%)	32 guidelines have been evaluated with a high applicability score (21%)
90 guidelines have been evaluated with a medium applicability score (60%)	78 guidelines have been evaluated with a medium applicability score (52%)
35 guidelines have been evaluated with a low applicability score (25%)	40 guidelines have been evaluated with a low applicability score (27%)

These first results shown the generality of guidelines retrieved from literature make most of them applicable in different industrial contexts. However if results of effectiveness evaluation are observed, designers underlined only company specific guidelines really help them during design choice, while general ones only provide suggestions without guiding the decision. The effectiveness of general guidelines and/or materials on ecodesign is on the contrary useful for users (Group 2) without knowledge on environmental sustainability matter, when training activities are performed. If these two results are combined together it is possible to understand expert users (Group 1 and 2) benefit from the consultation of company specific knowledge, both in terms of support in the decision phase, both in term of time and effort to retrieve information on company knowledge. Past projects and related result can be easily retrieved and capitalized.

The evaluation of the usability (Table 2) has provided interesting results, with opinions varying between "good" and "very good". The tool has been positively evaluated by the user in performing its functions. The most appreciated aspects were filtering options clarity, simplicity of the tool functions, and clarity of navigation.

On the contrary, the aspects which have obtained the lowest values were the about the syntax of the guidelines and the user guidance and support. The guidelines are gathered with the aim of completeness, but sometimes the syntax appears difficult to understand at first read and guidelines could be simpler in order to avoid redundancies in the information. The idea is to make the guideline more concise and add further section with deep information (e.g. additional materials or attachments).

Table 2. Usability and effectiveness evaluation results

Main topic used for the evaluation	Medium results first company	Medium results second company
Compatibility	3	3.5
Explicitness	4	4
Flexibility and control	3	3.5
Functionality	4	4
Informative feedbacks	4	3
Language and contents	3.5	3
Navigation	4	4
User guidance and support	3	3
Visual clarity	3.5	3
Impact on traditional design process	3.5	3
Personnel to involve	3.5	4

The results obtained in the evaluation of tool effectiveness and usability within the company appear therefore satisfactory. The tool provides results that can effectively support the designer work, and in the integration in the design process, it appears as useful tool to facilitate decisions and reduce decision time. The tool has been evaluated as usable without specific knowledge, admissible and compatible with the standard time of the traditional design process.

6 Conclusion

The paper describes the development of a tool that supports companies in the implementation of eco-design strategies through a structured repository of guidelines, best practices and training material on eco-design matters. Barriers that actually limit the implementation of eco-design strategies have been faced and a structured repository with multiple objectives have been proposed. It allows users to increase their competences on environmental sustainability through training materials, to be guided in the identification of possible solution strategies through the consultation of guidelines, while favoring the company knowledge storing and sharing by a guided procedure to catalogue best practices and company's analysis.

The first version of the tool has been implemented and customized in two manufacturing companies in order to evaluate its usability, effectiveness and integration level. First results have shown a good acceptance of the tool, its functionalities and interfaces. Future work will consist in updating of guidelines, with the objective to include also other aspect of sustainability, such as social aspect ones and to test the tool in other industrial contexts.

References

1. EU: European Commission Communication on the sustainable consumption and production and sustainable industrial policy action plan. COM, 2008, 397/3 (2009)
2. EU: DIRECTIVE 2009/125/EC of the European parliament and of the Council (2009)
3. EU: Regulation (EC) No 66/2010 of the European parliament and of the Council (2010)
4. ISO: Environmental Management Systems and Guidelines for Incorporating Ecodesign. ISO 14006. ISO, Geneva, Switzerland (2011)
5. Mathieux, F., Rebitzer, G., Ferrendier, S., Simon, M., Froelich, D.: Implementation of ecodesign in the European electr(on)ics industry - a an analysis of the current practices based on cases studies – an analysis of the current practices based on cases studies. J. Sust. Prod. Des. **1**(4), 233–245 (2002)
6. Blessing, L.: What is this thing called design research? In: International CIRP Design Seminar, Hong Kong (2002)
7. Araujo, C.S.: Acquisition of product development tools in industry: a theoretical contribution. Technical University of Denmark, Lyngby, Denmark (2001)
8. Van Hemel, C., Cramer, J.: Barriers and stimuli for ecodesign in SMEs. J. Clean. Prod. **10**, 439–453 (2002)
9. Hillary, R.: Environmental management systems and the smaller enterprise. J. Clean. Prod. **12**, 561–569 (2004)
10. Hallstedt, S.I., Thompson, A.W., Lindahl, P.: Key elements for implementing a strategic sustainability perspective in the product innovation process. J. Clean. Prod. **51**, 277–288 (2013)
11. Bonjour, E., Geneste, L., Bergmann, R.: Enhancing experience reuse and learning. Knowl.-Based Syst. **68**, 1–3 (2014)
12. Asrar-ul-Haq, M., Anwar, S.: A systematic review of knowledge management and knowledge sharing: trends, issues, and challenges. Cogent Bus. Manag. **3**, 1127744 (2016)
13. Witherspoon, C.L., Bergner, J., Cockrell, C., Stone, D.N.: Antecedents of organizational knowledge sharing: a meta-analysis and critique. J. Knowl. Manag. **17**(2), 250–277 (2013)
14. Mcharek, M., Toufik, A., Hammadi, M., Choley, J., Larouci, C.: Knowledge sharing for mechatronic systems design and optimization. In: Symposium on Information Control Problems in Manufacturing (2018)
15. Crossan, M.M., Lane, H.W., White, R.E.: An organizational learning framework: from intuition to institution. Acad. Manag. Rev. **24**, 522–537 (1999)
16. Ipe, M.: Knowledge sharing in organizations: a conceptual framework. Hum. Resour. Dev. Rev. **2**, 337–359 (2003)
17. Baxter, D., Gao, J., Case, K., Harding, J., Young, B., Cochrane, S., Dani, S.: A framework to integrate design knowledge reuse and requirements management in engineering design. Robot. Comput.-Integr. Manufact. **24**, 585–593 (2008)
18. Briggs, H.C.: Knowledge management in the engineering design environment. In: Structures, Structural Dynamics, and Materials Conference, Newport, USA (2006)

19. Chevalier, C., Pourry, F., Villenueuve, F., Du Pasquier, A.: The right knowledge management strategy for engineering analysis SME: a case study. In CIRP Design Conference, Bochum, Germany (2013)
20. Mehzer, T., Abdul-Malak, M.A., Ghosn, I., Ajam, M.: Knowledge management in mechanical and industrial engineering consulting: a case study. J. Manage. Eng. **21**, 138–147 (2005)
21. Baumann, H., Boons, F., Bragd, A.: Mapping the green product development field: engineering, policy and business perspectives. J. Clean. Prod. **10**, 409–425 (2002)
22. Bovca, M.D., P_erez-Belis, V.: A taxonomy of ecodesign tools for integrating environmental requirements into the product design process. J. Clean. Prod. **20**, 61–71 (2012)
23. Pigosso, D.C.A., Rozenfeld, H., McAloone, T.C.: Ecodesign maturity model: a management framework to support ecodesign implementation into manufacturing companies. J. Clean. Prod. **59**, 160–173 (2013)
24. Pigosso, D.C.A., McAloone, T.C., Rozenfeld, H.: Characterization of the state-of the- art and identification of main trends for Ecodesign tools and methods: classifying three decades of research and implementation. J. Indian Inst. Sci. **95**, 405–428 (2016)
25. Bonou, A., Skelton, K., Olsen, S.I.: Ecodesign framework for developing wind turbines. J. Clean. Prod. **126**, 643–653 (2016)
26. Dekoninck, E.A., Domingo, L., O'Hare, J.A., Pigosso, D.C.A., Reyes, T., Troussier, N.: Defining the challenges for ecodesign implementation in companies: development and consolidation of a framework. J. Clean. Prod. **135**, 410–425 (2016)
27. Ilgin, M.A., Gupta, S.M.: Environmentally conscious manufacturing and product recovery (ECMPRO): a review of the state of the art. J. Environ. Manage. **91**, 563–591 (2010)
28. Bonvoisin, J., Mathieux, F., Domingo, L., Brissaud, D.: Design for energy efficiency: proposition of a guidelines-based tool. In: International Design Conference, Dubrovnik, Croatia, pp. 629–638 (2010)
29. Vezzoli, C., Sciama, D.: Life cycle design: from general methods to product type specific guidelines and checklists: a method adopted to develop a set of guidelines/checklist handbook for the eco-efficient design of NECTA vending machines. J. Clean. Prod. **14**, 1319–1325 (2006)
30. Bischof, A., Blessing, L.: Guidelines for the development of flexible products. In: International Design Conference - Design 2008 (2008)
31. Koh, S.C.L., Demirbag, M., Bayraktar, E., Tatoglu, E., Zaim, S.: The impact of supply chain management practices on performance of SMEs. Ind. Manag. Data Syst. **107**(1), 103–124 (2007)
32. Rose, C.M., Beiter, K.A., Ishii, K.: Determining end-of-life strategies as a part of product definition. In IEEE International Symposium on Electronics and the Environment, Danvers, USA (1999)
33. Luttropp, C., Lagerstedt, J.: EcoDesign and the ten golden rules: generic advice for merging environmental aspects into product development. J. Clean. Prod. **14**, 1396–1408 (2006)
34. Ecodesign Pilot. http://www.ecodesign.at/pilot/ONLINE/ENGLISH/INDEX.HTM. Accessed Jan 2018
35. Rossi, M., Germani, M., Mandolini, M., Marconi, M., Mengoni, M., Morbidoni, A.: Ecodesign guidelines and eco-knowledge integration in product development process. In: International Conference on Engineering Design, vol. 5, pp. 161–170 (2013)
36. Favi, C., Germani, M., Landi, D., Mengarelli, M., Rossi, M.: Comparative life cycle assessment of cooking appliances in Italian kitchens. J. Clean. Prod. **186**, 430–449 (2018)

Engineering Methods in Human-Related Applications

Deep CNN for 3D Face Recognition

Elena Carlotta Olivetti[1(✉)], Jacopo Ferretti[2], Giansalvo Cirrincione[3], Francesca Nonis[1], Stefano Tornincasa[1], and Federica Marcolin[1]

[1] Department of Management and Production Engineering, Politecnico di Torino, 10129 Turin, Italy
elena.olivetti@polito.it
[2] Department of Electronics and Telecommunications Engineering, Politecnico di Torino, 10129 Turin, Italy
[3] Université de Picardie Jules Verne, Laboratory LTI, Amiens, France

Abstract. Three dimensional face analysis is being widely investigated since it appears as a robust solution to overcome the limits of two dimensional technologies. 3D methods allow to relate the recognition process on features not depending on lightning, head poses, make up and occlusions. This paper proposes a new approach to the problem consisting of a novel image representation, where specific facial descriptors replace the RGB traditional channels and a convolutional neural network performs the classification. We chose to use MobileNetV2, a relatively new network, as it has a low amount of parameters to train. The method has been evaluated on the Bosphorus database, and even though it is still a preliminary study, the results obtained with our method are extremely encouraging; the recognition rate achieved is 97.560% and it is comparable to the state of the art. This result, reached despite the fact that the Bosphorus database has a great number of subjects (105) but a low number of scans (4666), shows the effectiveness of this representation combined with convolutional neural networks.

Keywords: Deep Convolutional Neural Network · 3D face recognition · Shape index · Curvedness · MobileNetV2

1 Introduction

In the recent years, face recognition has passed from being a key-point in security and safety domains to revolutionize people's daily activities; smartphones and digital payment systems are clear examples of this aspect. Even if traditional 2D technologies have achieved excellent levels of precision, these are closely related to environmental conditions such as lightning, pose variations, face occlusions and make up. 3D face recognition has the purpose of overcoming these limits introducing the third dimension information, allowing the use of geometrical features to describe the facial surface. An additional improvement in recognition accuracy seems to be arguably favored by the integration between 3D methodologies and Deep Convolutional Neural Networks (CNN). Recent literature shows

C. Rizzi et al. (Eds.): ADM 2019, LNME, pp. 665–674, 2020.
https://doi.org/10.1007/978-3-030-31154-4_56

the research interest in this direction. In the work by Feng *et al.* [1], authors proposed the use of two Deep CNNs for feature extraction, respectively for 2D images converted to gray scale and for 3D grid models built from the point cloud data. The two feature layers are fused becoming the input of an Artificial Neural Network (ANN) recognition system. All the experiments have been conducted on CASIA-3D Face V1. The method has been experimented and compared with traditional approaches; both 2D and 3D DCNN based models led to higher recognition rates (93.750% and 85.938% respectively), achieving the better result with the feature layers fusion. No data augmentation (DA) has been performed. Even if based on a CNN, Kim *et al.* [2] propose a different approach involving a pre-trained network, together with a DA process. The 3D point clouds are converted into 2D depth maps used to fine tune VGG ("Visual Geometry Group" NN) Face. The size of the 2D depth maps is 224×224 with 3 channels, according to the input layer of the network. Furthermore, the size of the 3D dataset has been enlarged with expression generation, pose variation and random patches. FRGC and CASIA 3D database have been used as training data. Bosphorus, BU-3DFE and 3D-TEC have been used to evaluate performances. A Deep CNN model designed specifically for 3D face recognition is proposed by Gilani *et al.* [3]. In their FR3DNet, a 3D point cloud is used to generate a three channels image, in which the first channel is the depth image, the second and the third are obtained from the surface normals of the original point cloud; the three channels are rendered as an RGB image. The recognition accuracy achieved on Bosphorus database is 96,18%. Even in this study, a method to enlarge the dataset is proposed; labeled 3D face identities are generated for training. The adopted approaches of these previous works, along with ours, have been summed up in Table 1 in the Results section.

In this paper, we propose a methodology for 3D face recognition based on MobileNet CNN and on Shape index and Curvedness geometrical descriptors that is intended to be specific to work on 3D depth map. The goal of our research is to propose a new approach to face recognition, in which not only a CNN is incorporated in the recognition process, but a different representation of the face is experimented. In Sect. 2, the details of the database, of the geometrical descriptors (shape index and curvedness) and of the network are reported. In Sect. 3 the results of our research are presented.

2 Methodology

Our approach to the face recognition problem is schematized in Fig. 3. The proposal is to mix the traditional techniques, which represent images with three channels -one for each color-, with a novel representation in which the first channel is the face depth map, whilst the second and the third one are two geometrical feature maps that will be described in the following. Face representations obtained with this process -shown on the left side referring to Fig. 3- are then preprocessed and fed to a CNN -shown on the center- which perform the analysis, and finally passed to a classifier -shown on the right side- which provide the identity of the person.

2.1 Data Preparation

In the following sections the process of preparing data is presented. First, shape index and curvedness have been computed from Bosphorus database 3D depth maps. Then, a preprocessing of data has been performed to make them suitable for the CNN.

Database

The Bosphorus database contains 4666 images of 105 subjects, including different types of face occlusions and of head pose variations. For each of the 105 person, available scans range from 31 to 54. There are 45 women and 60 men; between male subjects, 33 have beard or short facial hair. A total of 34 expressions are collected, involving action units and the six basis emotions. Moreover, a gallery of different poses is provided (head inclinations and rotations). Hand, hair and eyeglasses occlusions are provided [4]. In our study, the whole database is involved.

Shape Index and Curvedness

In this work, we relied on the definitions of shape index and curvedness given by Koenderink and Van Doorn [5]. The shape index describes the shape of a surface; Koenderink and Van Doorn proposed a partition of the range [−1, 1] in nine categories corresponding to nine different surfaces, from spherical cup to spherical cap. The formal definition of shape index can be given as follows:

$$S = -\frac{2}{\pi} \arctan \frac{k_1 + k_2}{k_1 - k_2}, \qquad S \in [-1, 1], \quad k_1 \geq k_2, \tag{1}$$

where k_1 and k_2 are respectively the first principal curvature and the second principal curvature. We relied on their definitions given by Do Carmo in 1976 [11]. Since S does not give an indication of the scale of curvature present in the shape, the curvedness index of a surface is introduced as a measure of how highly or gently curved a point is. The formal definition is given as:

$$C = \sqrt{\frac{k_1^2 + k_2^2}{2}}. \tag{2}$$

Figure 1 reports the shape index and the curvedness mapped point-by-point on a facial surface.

These descriptors are reliable and complete, since they are able to accurately describe every part of the face, even focusing on the less easily identifiable zones [6]. Cirrincione *et al.* [7] assessed the discriminative capabilities of shape index and curvedness using statistical and neural techniques; what they found out is that, among the considered descriptors, the ones belonging to the shape index class are the most meaningful for classification task. Given to their completeness, new geometrical descriptors for face analysis have been defined on the form of S and C, achieving excellent results in describing 3D faces [8,9]. The relevance of 7 new geometrical descriptors has been validated with a new neural network,

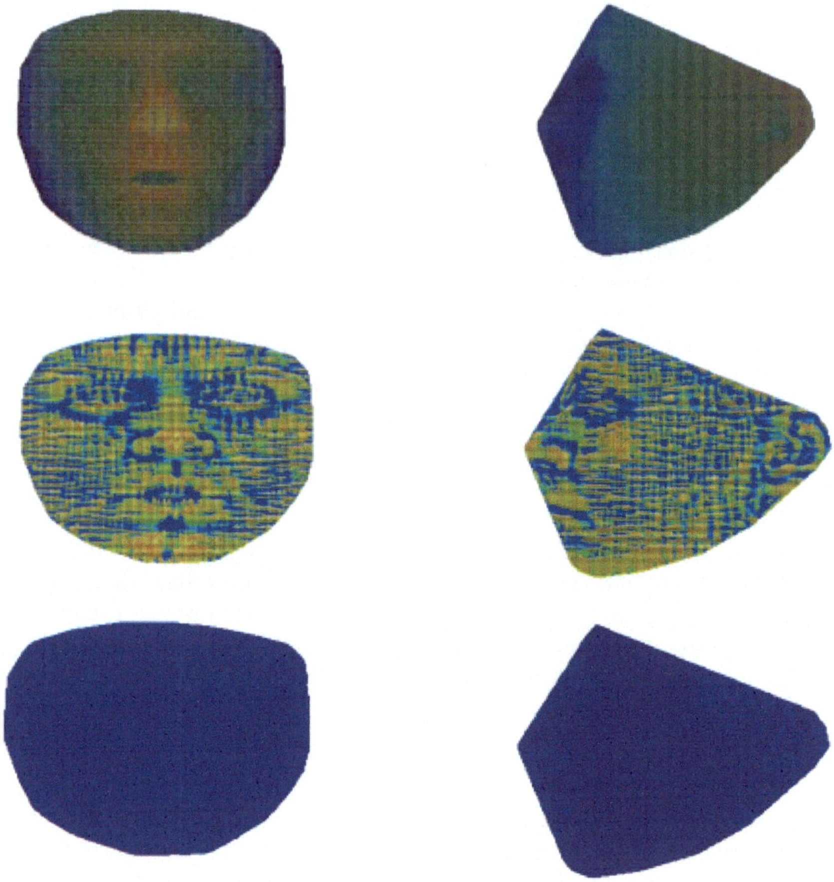

Fig. 1. Top to bottom, left to right. 3D depth map, shape index and curvedness of the same person; frontal view, lateral view

GH-EXIN, which has been used as a feature selection approach. The study stated the reliability of curvedness based geometrical descriptors in describing a facial surface [12].

Preprocessing
Preprocessing has been the last step required to adapt the data obtained from the above mentioned procedure to the neural network input. First, we resized the channels to a common dimension of 224 $\times$ 224, which is the input size of MobileNetV2. Then, we reduced inter-channel variability by normalizing each channel in the range $[0, 1]$. This way, each channel had the same impact in the network, and the variability between different maps has been minimized.

Data Augmentation
After few initial tests, we observed that it was not trivial to train a CNN with only 4666 images. Since we did not have any other source, we increased the

number of images with DA techniques. First, we created a copy of the original depth map with a clockwise rotation of 25° and a copy with a counterclockwise rotation of 40°, effectively tripling the database. Then, we horizontally mirrored each of the depth map, ending with six times the initial size of the database (27996 images).

2.2 CNN

A CNN is a class of NN where a filter, commonly called kernel, is passed (convoluted) along data in order to learn particular patterns.

In Fig. 2 an example is reported with two passages of a convolution where a kernel K, of height and width 3×3, is passed through the matrix IN to produce the matrix OUT. The elements $OUT_{i,j}$ of the matrix OUT are obtained superimposing the kernel on the corresponding element $IN_{i,j}$ on the input matrix, multiplying element-by-element, and finally summing the results. Because of how the convolution works, the output matrix has to be smaller of the input one and the final size is given by the input matrix size S_{IN} minus the kernel size S_K plus one: $S_{IN} - S_K + 1$ (in the example of Fig. 2: $7 - 3 + 1 = 5 \times 5$). For simplicity sake in the mentioned example the matrices represent one-channel images; in the case of three-channels images the kernel is applied on each channel and the results are summed. Therefore, the resulting image will be smaller in height and width, and will only have one channel.

In a CNN each Convolutional layer consists of many filters (kernels), resulting in a smaller output image with an amount of channels equal to the number of filters.

For our experiments we chose MobileNetV2 as deep CNN. The choice fell upon this network mostly because of its ease of training having a low amount

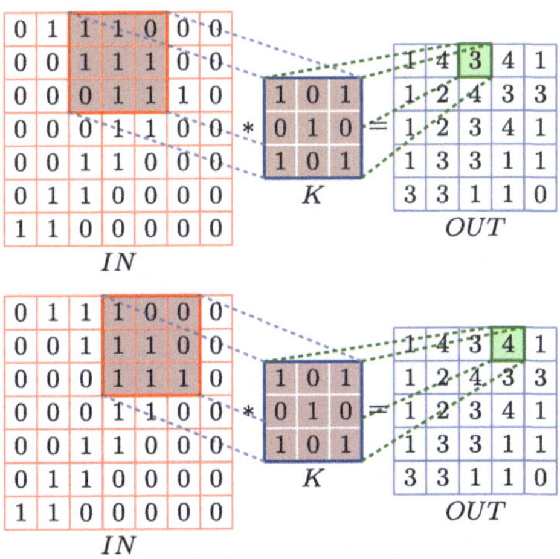

Fig. 2. Example of a convolutional layer.

of parameters to tune. MobileNet is a relatively new neural network specifically designed to work on mobile phones, hence the name. To do so, it was designed trying to obtain the best efficiency in terms of both memory utilization and number of required operations so that it could effectively work in limited hardware such as the one found on hand-held devices [10]. The strength of MobileNetV2 can be resumed in three main points.

Depthwise Separable Convolutions (DSC). DSCs are a close approximation of traditional 2D convolution with a considerably lower computational cost. They are performed applying a Depthwise Convolution, followed by a Pointwise Convolution. The final computational cost of DSC is: $h_i \cdot w_i \cdot d_i(k^2 + d_j)$, where h_i, w_i, and d_i are the dimension of the image, k is the kernel size, and s_j is the number of filters; whilst the computational cost of a classic 2D convolution is $h_i \cdot w_i \cdot d_i \cdot d_j \cdot k^2$.

Linear Bottlenecks. Experimental evidence suggests that using linear layers is crucial as it prevents nonlinearities from destroying too much information, assuming that the manifold of interest is low-dimensional.

Shortcut Connections Between the Bottlenecks. Shortcuts improve the ability of a gradient to propagate across multiplier layers. Moreover it is considerably more memory efficient.

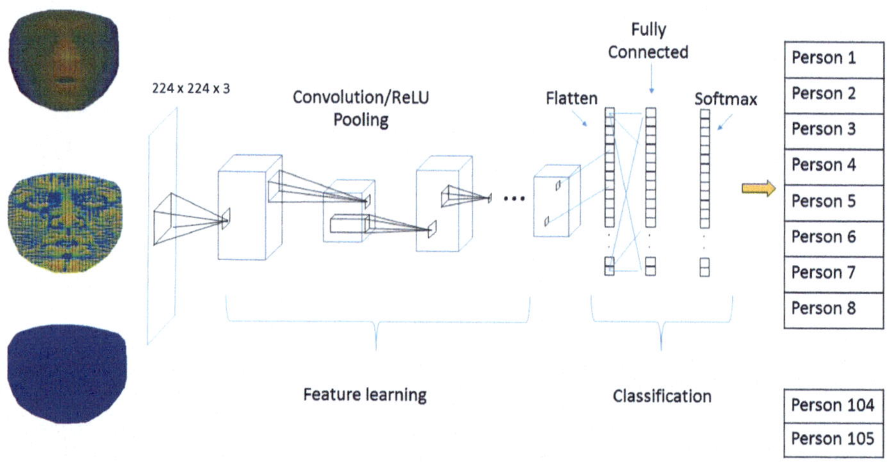

Fig. 3. Schematic overview of the adopted method

3 Results

In the following section, the results are presented. The first part of the section is focused on the outcomes of our experiments; in the second part, a comparison with the outcomes of the cited studies is reported.

All our experiments have been performed on Bosphorus public database, dividing it in a *training dataset*, used only to train the network, and a *validation dataset*, used only to assert the quality of the network. After a preliminary comparison of various network architectures, amongst which VGG16 and Inception V3, we decided to use MobileNetV2 as our CNN. The choice fell on this network mainly because of the significantly lower number of parameters -8.8 Million, against 134M and 27M of VGG16 and Inception V3- which required less time to train and execute.

At first, we have tried to use a pretrained network (transfer learning) based on imagenet. However, since the differences between our image type and the imagenet database were remarkable, this first experiment resulted in making the pretrained weights ineffective in analyzing our data. This led us to train the CNN from scratch. The preliminary experiments resulted in a recognition rate (RR) of about 70%, showing the necessity of enlarging the size of the database. After the data augmentation process, the database was six times its original size; accuracy increased to a percentage of more than 97%, a level comparable to the state of the art results. In Table 1 the accuracies of different methods are reported. For a deeper insight on our results we reported in Fig. 4 the confusion map computed on our validation dataset. In this Figure, the columns represent the label predicted by our model, the rows represent the actual label, whilst the color-map indicates the prediction ratio. Accordingly, if the cells on the principal diagonal are lit the network is classifying correctly.

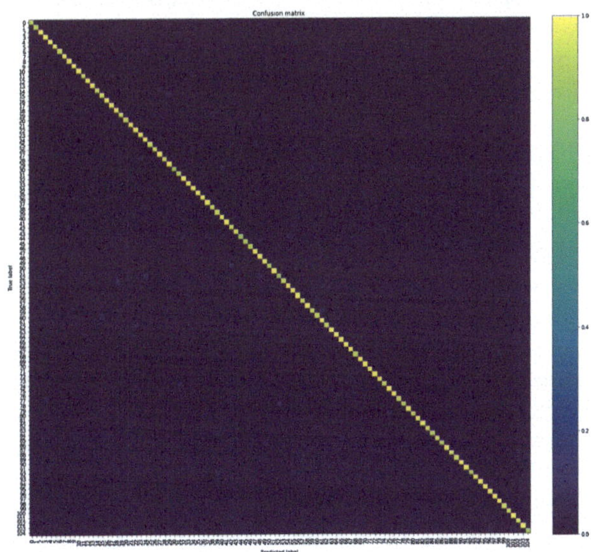

Fig. 4. Confusion map

The accuracy reached using MobileNetV2 (97.560%) with this new type of representation resulted to be higher than the one reported by Feng *et al.* for 2D feature extraction network (93.750%) and significantly higher than the RR obtained with the 3D feature extraction CNN (85.938%). The key point of their work was the fusion between the two methods, 2D and 3D, leading to a RR of 98.440%. Despite the excellent result, it is important to point out that the 2D contribution is related to an optimal environmental situation, while a purely 3D method would avoid decreasing of performances due to different conditions such as lightning and pose. Experiments have been conducted on CASIA-3D Face V1. The size of this database (4624 scans of 123 people) is comparable with that of the Bosphorus. In both studies, the number of classes (number of subjects to be identified) is significantly high in relation to the number of scans available. To effectively train a network many more data per subject are needed. In this perspective, the fact that our RR is 12% higher than theirs is to be considered as a proof of the validity of the methodology we propose.

The method proposing the fine tuning of VGG-16 and VGG Face CNNs led to an accuracy of 99.2% [2]; as pointed out by authors, the size of available 3D datasets is limited. A 3D DA technique is proposed, with the creation of facial expressions from a single image. Along with expression variation, data are augmented with the generation of random rotation matrices (resulting in different face poses) and with the application of random patches on the depth map (resulting in simulated face occlusions). Three different CNNs have been evaluated. Five different 3D face database have been involved: FRGC, CASIA 3D, Bosphorus, BU-3DFE and 3D-TEC. The first two were used as training sets, the remaining were used to evaluate the performance of the networks. For comparisons with our work, we focused on the results they obtained with fine tuned VGG Face on Bosphorus (that shows the highest rank-1 accuracy). The first difference from our approach is that augmented 3D models have been converted to 2D maps. It introduced one more step in the preprocessing of data.

It was not specified how data have been adapted to the input depth size of the network; since VGG Face is a CNN trained on RGB images, the input data for transfer learning must have the same depth size. This is a second difference, since our images had three channels replacing the RGB ones. Another difference is the training process; due to the lack of large scale 3D database, transfer learning seemed to be the better solution. In fact, in the case of VGG Face it gave an excellent result on Bosphorus (99.240%). With a training from scratch and with a relatively small database, our method reached an accuracy which is less than 2% lower. This is due, in our opinion, to the strength of the face representation by the 3D depth map, S and C.

In the work by Gilani *et al* a deep CNN specifically designed for 3D data was presented. The input was a three channels image; the first channel was represented by a depth map calculated by fitting a surface of the type z = (x,y) to the 3D point cloud. The second channel has been filled with surface of the form of $\theta = (x,y)$ and $\phi = (x,y)$, where θ and ϕ were the azimuth and the elevation angle of the surface normals of the 3D point cloud calculated in spherical coordinates. The image has been normalized in the range 0–255, rendering the representation as an RGB image. Authors proposed a method to generate labeled 3D face data to be used as training set. Starting from a property database of 1785 individuals, this data augmentation method enabled to obtain a final database of 3 169 275 faces belonging to 1 000 005 identities, with variety of poses and expressions. The 90% of the dataset was used for training and the remaining scans were used for validation. They also proposed a method for merging public databases in a single dataset called *LS3DFace*. RR obtained on this large dataset with FR3DNet is 95.510%. As in our study, a novel type of three channels image is proposed. However, there is a substantial difference, since in our representation the second and the third channels contain a univocal geometric description of the facial surface whose reliability has been validated by previous studies, as described in Sect. 2.

Considering that this method is still preliminary and requires some further implementation steps, the results are encouraging since our RR is slightly lower than the one obtained with a 3D data dedicated DCNN. In our view, this shows the strength of the face representation given by the shape index and the curvedness even with a reduced database.

Table 1. Recognition accuracy of different 3D methods. Training Database (DB training), test Database (DB testing), type of Neural Network (NN), Data Augmentation (DA) and Recognition Rate (RR) are reported. Comparisons have been made with our work on the results obtained on Bosphorus (typed in bold). For Gilani *et al.*, not all testing databases are reported in this table.

Ref	DB training	DB testing	NN	RR	DA
Feng *et al.*	CASIA 3D-Face V1	CASIA 3D-Face V1	DCNN with softmax classifier	85.938%	/
Kim *et al.*	FRGC	Bosphorus	VGG Face	**99.240%**	Facial expressions
	CASIA 3D	BU-3DFE		95.000%	
		3D-TEC		94.800%	
Gilani *et al.*	Property DB	LS3DFace	FR3D Net	95.510%	Labeled identities
		Bosphorus		**96.180%**	
Our method	Bosphorus	Bosphorus	MobileNetV2	**97.560%**	Rotations

4 Conclusions

3D face recognition has been deeply investigated in the last years since it seems to allow overcoming the limits of 2D in this field such as enlightenment, pose variations and face occlusions. The integration between 3D methods and deep learning algorithms is enhancing the accuracy of face recognition even in non ideal environmental situations. In this work, we propose a new approach based on the neural network MobileNetV2. Our idea is to use as network input data a novel type of three channels image different to traditional RGB images, in which the first channel is the facial 3D depth map, the second channel is the shape index and the third one is the curvedness. As far as we know, it is a novelty with respect to other works proposing 3D face recognition methods exploiting DCNN. In our opinion, the use of geometrical descriptors allows to enhance the accuracy of the network, even if it has been originally designed for 2D images. In fact, the RR we achieved (97%) shows the potentials of using the shape index and the curvedness instead of a simple RGB face image. Given their completeness, it has not been necessary to construct a 3D custom made DCNN to obtain high accuracy. Starting from the good results obtained with MobileNetV2, we are currently working on a second DCNN: Inception. In the light of the excellent results achieved with a non specific network, our future goal is to construct a deep CNN dedicated to the type of data we proposed in this work to overtake the accuracy reached with MobileNetV2. Moreover, we will consider to use only the geometrical descriptors as input data instead of pure facial depth maps, also considering other indices.

References

1. Feng, S., Guan, Y., Wu, M., Zhang, X., Ti, C.: 3D face recognition method based on deep convolutional neural network. In: Panigrahi et al. (ed.) Advances in Intelligent Systems and Computing, pp. 123–130 (2019)
2. Kim, D., Hernandez, M., Choi, J., Medioni G.: Deep 3D face identification. In: International Joint Conference on Biometrics (IEEE), pp 133–142 (2017). https://doi.org/10.1109/BTAS.2017.8272691
3. Gilani, S.Z., Mian, A.: Learning from millions of 3D scans for large scale 3D face recognition. In: Conference on Computer Vision and Pattern Recognition (IEEE), pp. 1896–1905 (2018). https://doi.org/10.1109/CVPR.2018.00203
4. Savran, A., Alyüz, N., Dibeklioğlu, H., Çeliktutan, O., Gökberk, B., Sankur, B., Akarun, L.: Bosphorus database for 3D face analysis. In: European Workshop on Biometrics and Identity Management, pp. 47–56. Springer (2008)
5. Koenderink, J.J., Van Doorn, A.J.: Surface shape and curvature scales. Image Vis. Comput. **10**(8), 557–564 (1992)
6. Vezzetti, E., Marcolin, F.: Geometrical descriptors for human face morphological analysis and recognition. Image Vis. Comput. **60**(6), 557–564 (2012)
7. Cirrincione, G., Marcolin, F., Spada, S., Vezzetti, E.: Intelligent quality assessment of geometrical features for 3D face recognition. In: Italian Workshop on Neural Nets, pp. 153–164. Springer (2017)
8. Marcolin, F., Vezzetti, E.: Novel descriptors for geometrical 3D face analysis. Multimed. Tools Appl. **76**(12), 13805–13834 (2017)
9. Marcolin, F., Violante, M.G., Moos, S., Vezzetti, E., Tornincasa, S., Dagnes, N., Speranza, D.: Three-dimensional face analysis via new geometrical descriptors. In: Advances on Mechanics, Design Engineering and Manufacturing, pp. 747–756. Springer (2017)
10. Sandler, M., Howard, A., Zhu, M., Zhmoginov, A., Chen, L.: MobileNetV2: inverted residuals and linear bottlenecks. In: Conference on Computer Vision and Pattern Recognition, pp. 4510–4520. Springer (2018). https://doi.org/10.1109/CVPR.2018.00474
11. Do Carmo, M.: Differential Geometry of Curves and Surfaces, pp. 4510–4520. Springer (1976)
12. Ciravegna, G., Cirrincione, G., Marcolin, F., Dagnes, N., Piccolo, E.: Assessing discriminating capability of geometrical descriptors for 3D face recognition by using the GH-EXIN neural network. In: Italian Workshop on Neural Nets (WIRN), p. 11 (2019)

Multiperspective Ergonomic Assessment Approach for Human Centered Workplace Design

Alessandra Papetti(✉), Martina Scafà, Agnese Brunzini, and Marco Mandolini

Università Politecnica delle Marche, Via Brecce Bianche 12, 60131 Ancona, Italy
{a.papetti,m.scafa,a.brunzini,m.mandolini}@univpm.it

Abstract. The transition toward smaller lot sizes in production requires the design of highly flexible processes where person and automation systems work together in a synergistic, safe and efficient manner. The new role of the operator requires researchers to study methods and tools able to evaluate the workers performance in order to maximize the comfort and quality of work. The virtual prototyping and simulation systems become fundamental to be able to design adequate production solutions, but before the tools it is necessary to study methods able to represent and consider the characteristics of the operators in the Industry 4.0-oriented factory environment appropriately. In this context, this paper proposes a method to assess the well-being of operators from different perspectives and, consequently, design operator-centered workstations. The first objective is preparatory in order to be able to fine-tune the second one, in a continuous improvement perspective. The approach has been successfully tested in collaboration with a food company.

Keywords: Human-Centered Manufacturing · Ergonomics · Operator 4.0

1 Introduction

The growing attention to the sustainable development issues in manufacturing is increasingly pushing companies to deepen all three pillars of sustainability: environmental, economic and social. Regarding this last aspect, since 2014, Zink [1] focuses on the needs of sustainable production system, defining that a correct design of a production system should focus on the human capital. Furthermore, the fourth industrial revolution defines not only a change in production but also a new way of interaction between human and the production system. For this reason, a special attention should be paid to physical and non-physical influencers to implement interventions that can induce positive change and minimize negative behavior for all operators [2].

Evidence from literature demonstrates that ergonomics has benefits in terms of shareholder value, productivity, and comfort, highlighting how participatory approaches contribute to success [3]. Although the importance of both operational performance and employee well-being for organizational success are intuitive, organizations

C. Rizzi et al. (Eds.): ADM 2019, LNME, pp. 675–685, 2020.
https://doi.org/10.1007/978-3-030-31154-4_57

often perceive a conflict between them [4]. From the literature review by Otto and Battaïa [5], it emerged the importance to adapt the ergonomics estimation methods to the needs of preventive planning. Nevertheless, few papers analyze the workers' experience in order to support human-centered product-process design.

Overall, studies in literature address the well-being of the operator [6] and his/her role in the Industry 4.0 [7] from different and complementary points of view but highlight the lack of holistic and proactive approaches. The scientific state of the art provides numerous methods and tools for biomechanical risk assessment [8] and illustrates how technology increasingly supports the design of human-centered manufacturing workstations [9]. It also underlines the importance and the boundaries of collaboration between human and automation and reiterates the need for a new factory model based on dynamic adaptation [10]. However, there are no comprehensive approaches that address multiple ergonomic domains to concretely assess the operator well-being in the working context. Furthermore, there is a need for new methods and tools to support the planning and evaluation of work, in a production context characterized by the presence of human workers and Cyber-Physical Systems, evermore necessary when the flexibility of production is required due to small lots.

In this context, the objective of the present paper is the development of a refined method that systematically points out current issues faced by workers in the current workplace from different perspectives. It would be a first step towards the definition of a human-centered plant model that pays particular attention to the quality of life and comfort of the operators within the workshop. The method is based on the best trade-off between production objectives (i.e. productivity, flexibility, etc.) and physical-cognitive well-being of individuals operating in the production context, measured through a sensorized environment able to monitor vital and environmental parameters within the plant. It also combines subjective and objective evaluations. The analysis of the data through knowledge-based rules allows identifying activities that have a greater impact on the operators and define the corrective actions to be implemented in order to modify the productive environments towards their needs, defining a dual functioning paradigm of human-machine integration. The goal is to define the optimal conditions for the operator 4.0 [11] that interacts with advanced technology systems (collaborative robots, augmented reality technologies, etc.) towards the standardization of advanced skills necessary for the management of the factory of the future.

2 Method

The proposed approach involves monitoring the operators' activities, as well as implementing corrective/improvement actions in order to make the factory environment more socially sustainable and in particular improve the design of workplaces. The idea is to integrate the analysis of the work environment and the needs of the operator within a factory that is evolving towards Industry 4.0, where new skills and abilities are required.

As shown in Fig. 1, the approach is divided into four main phases.

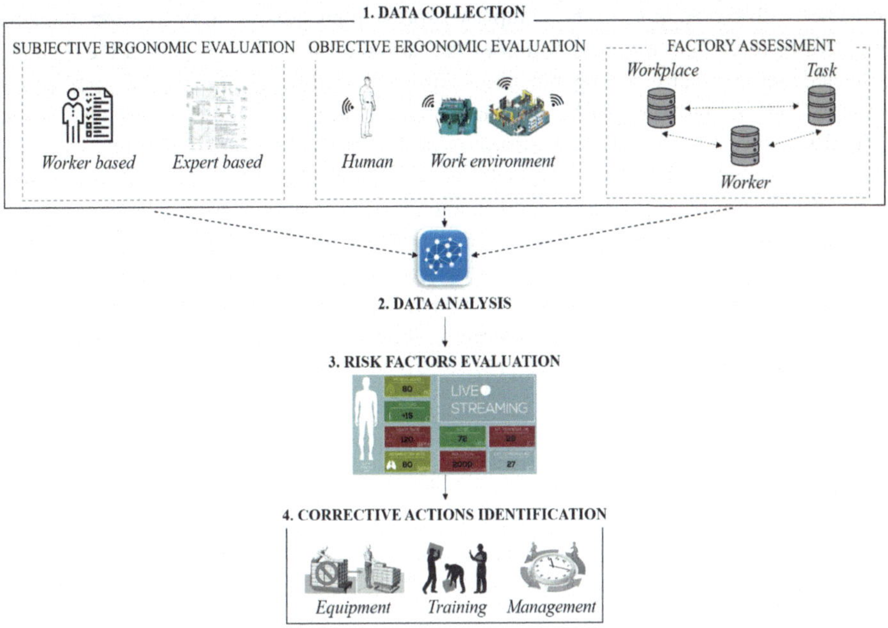

Fig. 1. Engineering approach for ergonomic assessment and improvement in the workplace.

The first phase is the data collection, which includes the creation of a data model of the factory environment and the monitoring of physical, cognitive and environmental parameters.

The data model consists of three different and communicating databases in which the most significant variables related to the three different aspects that influence the well-being of the operator are cataloged:

- Worker characteristics: demographic variables, anthropometry, functional skills, knowledge, personal needs, etc.
- Workplace characteristics: temperature, brightness, noise, workstation layout, etc.
- Task characteristics: type, duration, frequency, skills required, etc.

The correlation between these three domains, apparently independent of each other, is fundamental for an effective and reliable ergonomic analysis. For example, the activity of picking up a component, albeit light, may require significant physical effort for the operator according to the container placement and the worker's height. There is therefore a need to correlate the height of the worker with the layout of the workstation with the type of activity to be performed to verify the actual feasibility without generating frustration in the worker himself.

Different types of analysis are conducted to populate the three databases. First of all, an in-depth analysis of the plant is performed in terms of layouts, material flows, information flows and procedures. Interviews with the workers are conducted to collect subjective information to correlate with the results obtained from ergonomic analyses such as, for example, perceived exertion or level of satisfaction. Other data are collected with the support of the company process managers in order to know the best practices of reference in the execution of the various tasks. Finally, part of the data are collected automatically during the monitoring phase through the use of appropriate sensors. This allows data to be acquired more quickly and more accurately and to objectify a process of gathering information that emerges from literature to be still almost completely subjective, as it lacks standardized rules. The choice of the sensors to be used is the result of an in-depth technological scouting in compliance with the specifications of the analyses to be carried out and the company requirements [12]. Table 1 summarizes possible applications of smart devices for monitoring key parameters of an ergonomic evaluation, which are usable across a whole spectrum of industrial processes. However, the literature shows that there are only a few applications in real industrial environment [13].

Table 1. Smart devices for ergonomic evaluation

Smart devices	Parameters
Chest bands	Heart rate, respiratory rate, trunk posture, activity intensity level
Watches and bracelets	Heart rate, galvanic skin response, steps, activity intensity level
Bracelet with EMG sensors	Strength, posture of the limb
Clothing with EMG sensors	Heart rate, respiratory rate, strength, posture of multiple body segments
Glasses and eye trackers	Head posture, eye movements, blinks, attention
Trackers	Proximity, presence, localization, traceability of vehicles and assets (e.g. vibration)

The second phase is the management of the data collected in order to identify the critical issues to be addressed. In this phase, the raw data must be analyzed and transformed into a content that makes them easy to understand, interpret, reorganize, sort and manipulate to generate useful information. In other words, a descriptive analysis has to be conducted that is preliminary to the processing in order to clean the data from errors and/or anomalies, identify relevant information, evaluate their quality and significance, differentiate isolated events from recurring ones and being able to define hypotheses and limits. This type of analysis can add information relating to already known phenomena or could highlight new ones.

The next step is the processing and analysis of the information obtained in order to characterize the main points detected and identify possible relationships between the variables. This analysis aims to deepen the data to understand the causes of events and behaviors. In this phase, the data described in the data model are analyzed simultaneously with those detected by one or more sensors in order to:

- Exclude false alarms (e.g., an elevated heart rate could correspond to a high physical effort, but in the case of a person with tachycardia this consideration could lose significance);

- Verify the presence of a risk by using multiple sources (e.g., if the chest band detects a recurrent back flexion >60°, the video analysis could be used to confirm the adoption of an incorrect posture by the operator during the task execution);
- Identify possible risks by correlating several variables (e.g., a state of suffering of the eye could be correlated with inadequate lighting or excessive tiredness).

A series of algorithms, based on the knowledge-based rules derived from experience and the legislation currently in force in terms of health and safety in the workplace [14], allow the analysis of the data collected by the devices available. The algorithms have the function of identifying at the level of the raw data the thresholds for which such data have to be classified as alarm situations, or not, of the monitored activity. The algorithm identifies, therefore, the overcoming of the defined thresholds and, at the same time, classifies their severity. The logic used to implement the results obtained by the algorithms are oriented to the optimization of the work to make the workplace more streamlined and, at the same time, adequate for the worker.

At this point, the risk assessment is performed. Depending on the data collected and the knowledge management, the critical issues detected are associated to one or more of the following risk categories: posture; workstation; task; work organization; work environment and tools and devices. At the end of this phase, therefore, for each operator/station/task analyzed, the list of identified risks is available, sorted according to the intervention priority.

The last phase is the decision-making in which for each risk identified in the previous step a series of corrective actions capable of eliminating or mitigating it are identified [15], supporting the (re)design of a human-centered workstation. They may concern:

- Visual or auditory messages delivered to the operator through the use of monitors to suggest, for example, to correct the posture, to perform an operation in a different manner or to take a break. In the same way messages could be sent to improve the work environment, for example, to encourage a change in air quality or to suggest the use of a Personal Protection Equipment;
- Advanced components, tools and ergonomic solutions for the assembly, screwing, handling of loads in total safety, etc.;
- The ergonomic design of products and components so that they can be easily handled, tasks so that they can be carried out without significant physical and/or cognitive efforts and workstations so that they are comfortable for the operator respecting his needs;
- Measures to ensure the safety and health of the working environment;
- Organizational actions to support the correct allocation of resources and activities;
- Training of the operator with the aim of improving his well-being and avoiding the manifestation of occupational diseases resulting from unsatisfactory job performance or inappropriate use of an instrument.

Suggested corrective actions may require investments, long implementation times, organizational changes, etc. Therefore, the selection of the corrective action to be undertaken must be the result of a careful cost-benefit analysis. In fact, the same risk could be solved by different corrective actions. In this step, further parameters could be considered to suggest the best solution taking into account the trade-off between ergonomic, economic and productivity aspects. Once the corrective actions have been implemented and the design of the new workstation is realized, new monitoring can be carried out to verify their effectiveness.

3 Case Study

The method has been experimented in an Italian dairy farm with a particular attention to artisan quality and dairy tradition. Quality of its products is always connected to the social corporate values and responsibility, which entails a particular attention to operators' health and well-being.

The present case study focuses on physical ergonomics assessment in order to mitigate the relative risks. The ricotta storage workstation (Fig. 2), which involves two operators, has been reported in this paper in order to exemplify the approach, but the main processes have been analyzed.

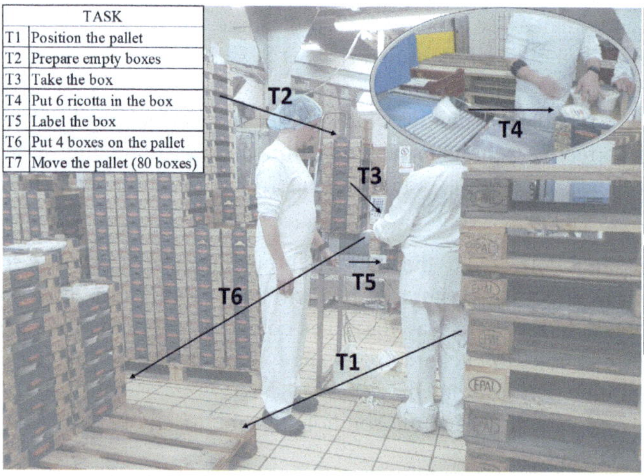

TASK	
T1	Position the pallet
T2	Prepare empty boxes
T3	Take the box
T4	Put 6 ricotta in the box
T5	Label the box
T6	Put 4 boxes on the pallet
T7	Move the pallet (80 boxes)

WORKSTATION	
Pallet height	15 cm
Work surface height	65 cm
Box dimensions	39x22x10 cm
Box weight (full)	1,8 kg

OPERATOR 1	
Gender	M
Age	42
Height	170 cm
BMI	25,95 kg/m^2

OPERATOR 2	
Gender	F
Age	53
Height	157 cm
BMI	30.83 kg/m^2

Fig. 2. Workstation layout and tasks.

Firstly, the main information related to the workstation, tasks and operators have been collected. The factory assessment included the workers' interviews to gain an understanding of the job tasks and demands and the observation of workers' movements and postures during several work cycles. The operators' perspective has been considered by evaluating the perceived exertion trough the Borg CR10 [16], a Category-Ratio (CR) scale anchored at number 10, that allows workers to subjectively rate their level of exertion during tasks execution. The selection of the postures to be evaluated and the relative methods was based on the most difficult postures and work tasks, the posture sustained for the longest period of time, or the posture where the highest perceived exertion occur. According to this preliminary evaluation, RULA [17] was applied to T1, T4 and T6 and NIOSH [18] to T6. Some parameters (back flexion, arm position, wrist twist, and muscle use) have been gathered from sensors to reduce the evaluator bias. The objective evaluation was carried out through two wearable devices: BioHarness 3.0 and Myo armband. The former is a physiological monitoring telemetry device that consists of a chest strap and an electronics module that attaches to the strap. The device stores and transmits vital sign data including heart rate (HR), breathing rate (BR), posture and activity that have been analyzed by OmniSense™ Analysis Software. The latter has 8 separate electromyography (EMG) sensors that allow measuring electrical activity from muscles and a 9-axis IMU (Inertial Measurement Unit) that senses the motion, orientation and rotation of forearm. Data are transmitted over a Bluetooth Smart connection. Raw EMG and IMU data have been utilized to evaluate the workers exposure to ergonomic risk factors associated with upper limbs musculoskeletal disorders (MSDs).

Two video cameras were positioned to have a complete view of the testing area. They were useful to capture workers' behaviors performing the working tasks, to detect any difference from the usual working pattern and interpret the variables trend appropriately.

The IoT framework also included a network connectivity; a Wi-Fi router to ensure data collection in real-time and remotely; a PC for data storage and a tablet for real time monitoring.

4 Results and Discussion

The analysis results are summarized in Table 2.

Table 2. Ergonomic evaluation results.

Task	Before			After		
	RULA	NIOSH	Borg CR10	RULA	NIOSH	Borg CR10
T1	6		3	–		3
T2			0			0
T3			1			1
T4	4		1	4		1
T5			0			0
T6	6	0,76	5	3	0,43	2
T7			2			2

According to the Borg CR10 score, the most critical tasks are T6 and T1. In the first case, it is associate to the leg fatigue caused by boxes lowering and in the second case to the arms exertion to lower the pallet. Similar results are shown by RULA scores, which highlight a medium risk requiring corrective actions to be planned and implemented in the medium term. It mainly depends on trunk and neck position. In particular, for T1, a +4 score was used both for the neck position (extension) and the back position (more than 60°). For T6, the assigned scores were +2 for the neck (0–20°) and +3 for the trunk (20–40°). They were collected by the chest belt (Fig. 3).

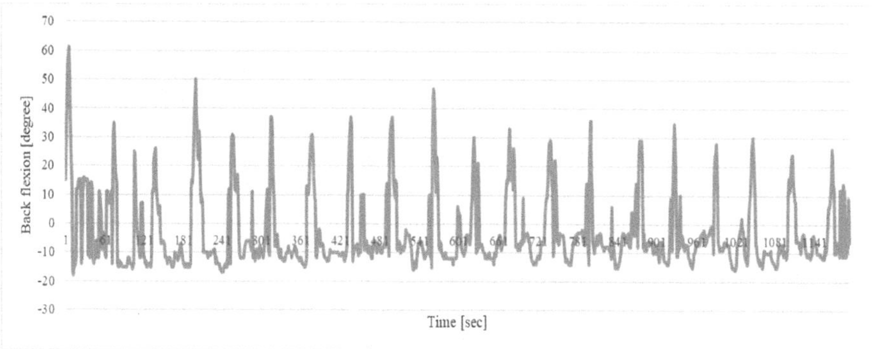

Fig. 3. Back posture during a work cycle.

No risk was associated to lowering activity according to NIOSH, as shown by the Lifting Index value less than 1.0. However, the individual multipliers have been analyzed to identify specific weaknesses in the workstation layout. Potential improvements are related to the horizontal location of the object relative to the body, the distance the object is moved vertically and the frequency and duration of lifting activity.

The analysis of physiological parameters allows evaluating the worker stress and fatigue. In particular, the BioHarness measures the heart rate variability (HRV) using the SDNN – the Standard Deviation of the Normal to Normal intervals, measured in milliseconds. It calculates an SDNN value over 300 heart beat detections and updates the SDNN value for each new detected heart beat. The SDNN is then reported once per second, showing the most up to date value. False and missed detections are filtered from the data and accounted for before the SDNN calculation is made. It resulted in the low-mid normal range, however, a decrease from 38 to 29 was observed after 2 work cycles. This result it was also confirmed by the physiological intensity that measures a person's cardiovascular output. After two work cycles it increased from 3 to 4 (where a score of 0 is a resting level whereas a score of 10 is equivalent to the individual working at their maximal effort).

Mechanical intensity, which is the measure of a person's kinematic output, resulted less than 2 during the whole shift (where 0 means that the subject is not putting any impact onto their musculoskeletal system whereas a score of 10 is equivalent to a sprinting level of impact and movement).

4.1 Workstation Re-design

According to the results, the main risk factors are related to the workstation layout and, consequently, to the awkward posture of operator's neck and back. At the same time, other potential improvements could concern the manual lifting and the task repetitiveness. Based on the classification proposed by authors in a previous work [15], different corrective actions have been selected and implemented to reduce the identified risks.

In terms of equipment, a lifter using automatic height adjustment for safe and easy movements has been introduced to improve work postures and reduce the force required to position the pallet and handling boxes (Fig. 4). It allows reducing the distance the object is moved vertically from 50 cm to 3 cm. Thanks to the lifter, the RULA total score is reduced from 6 to 3 (T6) and from 6 to 0 (T1), as shown in Table 2.

Fig. 4. Boxes handling at "as-is" (left) and "to-be" (right) workstation.

A training session was conducted to provide best-practices to better execute tasks and educate workers about the early warning signs of stress and fatigue. In the first case, a different boxes positioning at origin (i.e., 90° rotation) has been suggested, which allows reducing the horizontal multiplier factor of the revised NIOSH lifting equation. Indeed, the horizontal distance of the object is halved.

In the second case, the alternation between the two operators every two cycles has been recommended in order to improve the HRV and the physiological intensity. However, it has been given workers some control over the way they do their work to better satisfy their needs.

4.2 Human Experience Evaluation

Beyond the results obtained, the method application has been evaluated considering also the workers' opinions. For this aim, an ad hoc questionnaire has been developed and administered to evaluate the human experience related to the ergonomic assessment. It aimed to investigate the following issues:

- Devices' intrusiveness in terms of physical and emotional discomfort;
- Influence of monitoring devices on the usual way of carrying out the work;
- Awareness about ergonomic risks and the perceived usefulness of the analysis.

The questions were asked on a 7-point Likert scale from 1 (strongly disagree) to 7 (strongly agree). Figure 5 reports the answers of the two operators of the ricotta storage workstation. Overall, both operators expressed a positive judgment on the study. They perceived the benefits deriving from the ergonomic analysis and proved to be collaborative and proactive. They said they became even more aware of the ergonomic risks arising from their work and behavior. The use of wearable devices did not seem to cause operators' discomfort. However, they do not completely forget to wear them. Moreover, they said they felt under pressure just a little at first because they were under observation. One operator also stated he felt encouraged to assume a more correct posture.

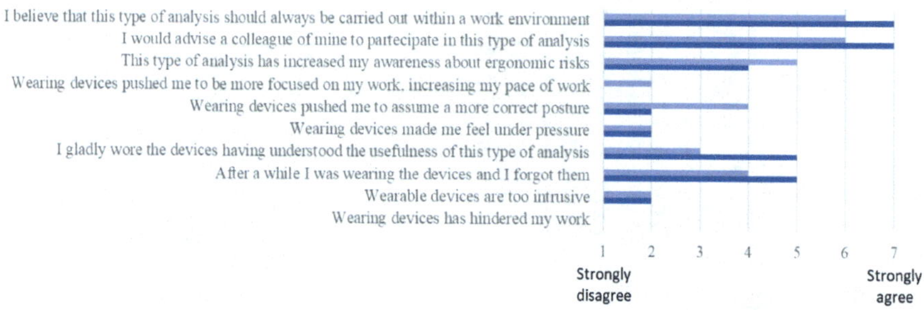

Fig. 5. Worker experience evaluation.

5 Conclusions

This paper deals with a topic that is currently very much felt in the context of business strategies related to Industry 4.0 and Lean Manufacturing: how to put into practice the Human-Centered Manufacturing approach aimed at being able to create high-quality and comfortable workplaces. The design of workstations could be based on design and simulation methods and tools that allow the evaluation of technical solutions based on the specific manufacturing operations and personal characteristics of the operators. An important phase, preliminary to the study and implementation of these methods and tools, is the identification of approaches to detect the physical and cognitive effort of the operator based on the activities performed. This evaluation can be the starting point to define the requirements for the design of the workstations themselves.

The 4-phases method of analysis, evaluation and correction of the operators' behavior illustrated in the paper has given good results from an experimental point of view. In fact, the case study dealt with in a company that produces food has made it possible to verify that through the application of the approach it is possible to identify the critical issues that put the operators in difficulty and the corrective actions to be applied. This achievement will be used, as future research work, to define specific Virtual Prototyping tools able to simulate the real scenarios and to allow the identification of main critical points.

References

1. Zink, K.J.: Designing sustainable work systems: the need for a system approach. Appl. Ergon. **45**(1), 126–132 (2014)
2. Genaidy, A.M., Huston, R.L., Dionysiou, D.D., Karwowski, W.: System-of-systems framework for improved human, ecologic and economic well-being. Sustainability **9**(4), 1–16 (2017)
3. Vink, P., Koningsveld, E.A.P., Molenbroek, J.F.: Positive outcomes of participatory ergonomics in terms of greater comfort and higher productivity. Appl. Ergon. **37**, 537–546 (2006)
4. Hoffmeister, K., Gibbons, A., Schwatka, N., Rosecrance, J.: Ergonomics climate assessment: a measure of operational performance and employee well-being. Appl. Ergon. **50**, 160–169 (2015)
5. Otto, A., Battaïa, O.: Reducing physical ergonomic risks at assembly lines by line balancing and job rotation: a survey. Comput. Ind. Eng. **111**, 467–480 (2017)
6. Mattsson, S., Fast-Berglund, A., Åkerman, M.: Assessing operator wellbeing through physiological measurements in real-time—towards industrial application. Technologies **5**(4), 61 (2017)
7. Peruzzini, M., Grandi, F., Pellicciari, M.: Exploring the potential of Operator 4.0 interface and monitoring. Computers & Industrial Engineering, in press
8. Alberto, R., Draicchio, F., Varrecchia, T., Silvetti, A., Iavicoli, S.: Wearable monitoring devices for biomechanical risk assessment at work: current status and future challenges—a systematic review. Int. J. Environ. Res. Public Health **15**(9), 2001 (2018)
9. Peruzzini, M., Pellicciari, M., Gadaleta, M.: A comparative study on computer-integrated set-ups to design human centred manufacturing systems. Robot. Comput.-Integr. Manuf. **55** (B), 265–278 (2019)
10. El Zaatari, S., Marei, M., Li, W., Usman, Z.: Cobot programming for collaborative industrial tasks: an overview. Robot. Auton. Syst. **116**, 162–180 (2019)
11. Romero, D., Wuest. T., Stahre. J., Gorecky. D.: Social factory architecture: social networking services and production scenarios through the social internet of things, services and people for the social Operator 4.0. In: Lödding, H., Riedel, R., Thoben, K.D., von Cieminski, G., Kiritsis, D. (eds.) Advances in Production Management Systems. The Path to Intelligent, Collaborative and Sustainable Manufacturing. APMS 2017. IFIP Advances in Information and Communication Technology, vol. 513. Springer (2017)
12. Papetti, A., Gregori, F., Pandolfi, M., Peruzzini, M., Germani, M.: IoT to enable social sustainability in manufacturing systems. Adv. Transdiscipl. Eng. **7**, 53–62 (2018)
13. Borgianni, Y., Rauch, E., Maccioni, L., Mark B.G.: User experience analysis in Industry 4.0 - the use of biometric devices in engineering design and manufacturing. In: IEEE International Conference on Industrial Engineering and Engineering Management, pp. 192–196 (2019)
14. http://www.hse.gov.uk/legislation/index.htm
15. Scafà, M., Papetti, A., Brunzini, A., Germani, M.: How to improve worker's well-being and productivity: a method to identify corrective actions. Procedia CIRP **81**, 162–167 (2019)
16. Borg, G.A.: Psychophysical bases of perceived exertion. Med. Sci. Sports Exerc. **14**, 377–381 (1982)
17. McAtamney, L., Corlett, E.N.: RULA: a survey method for the investigation of work-related upper limb disorders. Applied. Ergon. **24**(2), 91–99 (1993)
18. Waters, T.R., Putz-Anderson, V., Garg, A., Fine, L.J.: Revised NIOSH equation for the design and evaluation of manual lifting tasks. Ergonomics **36**(3), 749–776 (1993)

Towards a Non-invasive Pectus Excavatum Severity Assessment Tool Using a Linear Discriminant Analysis on 3D Optical Data

Michaela Servi[1(✉)], Rocco Furferi[1], Yary Volpe[1], Marco Ghionzoli[2], and Antonio Messineo[2]

[1] Department of Industrial Engineering, University of Florence, Florence, Italy
{michaela.servi,rocco.furferi,yary.volpe}@unifi.it
[2] Department of Pediatric Surgery, Meyer Children's Hospital, Florence, Italy
{marco.ghionzoli,antonio.messineo}@meyer.it

Abstract. Pectus Excavatum (PE) is a congenital anomaly of the thoracic cage, at the level of the sternal-costal plane, which consists of an inward angle of the sternum, in the direction of the spine. Its incidence, equal to 1 in 300–400 people, makes it the most frequent among thoracic malformations. For a proper care of patients suffering from PE, it is essential to monitor the progression of the disease and estimate its severity, in order to follow the course of the pathology over time and define the correct treatment. To this end, over the years, severity indexes, or chest indexes, have been introduced. Among them, in recent years, the so-called optical indices, calculated on optical scans of the patient's chest, are gaining ground. In fact, the optical scan is faster to apply and prevents the patient from being exposed to radiation. In this work the peculiar morphological features of PE have been examined by extracting various geometric parameters, in order to develop an outpatient support tool for the evaluation of the severity of the pathology and the monitoring of its progression over time. The extracted values have been compared against a ground truth obtained through five independent surveys collected from paediatric specialists. A Linear Discriminant Analysis was performed to determine the accuracy of the classification using the proposed geometric parameters, obtaining positive results.

Keywords: Pectus Excavatum · Depth camera · Severity index · Clinical support

1 Introduction

Pectus Excavatum, (PE) is a congenital anomaly of the thoracic cage, at the level of the sternocostal plane, which consists of a sternal inward angle, in direction of the spine [1]. In other words, the PE is an anterior cavity corresponding to the internal bending of the rib cartilages and can be either symmetrical or asymmetrical. Often there is an abdominal prominence of varying degrees depending on the size of the defect. This congenital form occurs three times more frequently in males and its incidence, equal to 1 in 300–400 people, makes it the most frequent of the thoracic malformations [2, 3].

C. Rizzi et al. (Eds.): ADM 2019, LNME, pp. 686–695, 2020.
https://doi.org/10.1007/978-3-030-31154-4_58

Although the disease is essentially asymptomatic in some cases the psychological and behavioral condition, which varies greatly from subject to subject, should not be neglected. Moreover, in severe cases, PE can exert pressure on the lungs and heart, and, if it is considered necessary, there is the possibility of intervening with a surgical treatment, with a traditional technique (Ravitch [4]) or with a more recent minimally invasive technique (Nuss [5]). Also, recent non-surgical correcting technique exists, which provides for the application of an external vacuum device, able to uplift the chest cartilages [6].

Since, as said, the pathology is likely to be asymptomatic with the possibility of worsening during growth, it is essential for the proper treatment of patients with PE to monitor the progression of the disease over time and estimate its severity. To this end, over the years, so-called gravity indices or chest indices have been introduced.

A chest index is a formula used to qualify or quantify the chest deformity [7, 8]. Severity indices are used to assess the gravity of the defect, define the treatment and eventually determine its conclusion; they can also be used to quantify post-operative changes. Severity indices can be divided into four main categories: Clinical, Chest X-ray, Chest CT/MRI scan and Optical.

Clinical indexes [7] are manually evaluated by the physician during the examination. Although easily applicable and low-cost, the indices in this class may be subject to errors and are not fully repeatable, as it is not possible to consider the patient's body mass and changes in age and height [9]. For this purpose, indexes referring to the patient's bone and internal cartilage can be calculated on X-rays, CT or MRI. Despite these indices are more repeatable and reliable, there is a great deal of variability in their interpretation, and they may also depend on the variation in age, sex or thoracic shape of the patient, as well as on the respiratory phase with which the tomographic image is acquired.

In recent years, so-called optical indices, calculated on external optical scans of the patient's chest [10–12], have been gaining ground. The reason for their popularity is that external optical scanning is faster to apply and prevents the patient from being exposed to radiation. In addition, they allow the accuracy of the measurement, which can be automated (and therefore not manually carried out) as in the case of clinical indices. Optical indices, however, may also suffer from variations in weight and shape of the individual patient.

Starting from a previous study in which an optical system for the monitoring and assessment of PE was developed [10], in this work we have tried to define and evaluate possible parameters that can be calculated on the optical scan, to better describe the chest deformation, such to implement a diagnostic aid system.

2 Materials and Methods

Over the past few years, the use of optical scanning techniques in the field of thoracic malformations has increased considerably. This is mainly due to the fact that common CT or MRI analyses, despite they may report precise values, suffer from some disadvantages. First of all, the exposure of the patient to radiation, in the case of CT, or the need to keep the patient still for a significant amount of time, in the case of MRI.

Furthermore, the result is highly dependent on human intervention since the identification of reference points for the calculation of the indexes is performed manually by scrolling through the tomographic images.

With the aim of creating a simple, repeatable and low-cost system [13] that can be used in common clinical practice, this work uses a chest optical acquisition system, capable of reproducing the external profile of the patient. Specifically, the acquisition system presented in [10] is employed, which allows a repeatable acquisition of the patient's chest (Fig. 1). 3D laser scanners, even if widely used in the past as body scanners, are not suitable for this application since it implies longer acquisition time and a post processing step to add colors on the 3D data [14, 15]; on the contrary, the new generation consumer grade depth sensors are particularly suited to deal simultaneously with both 3D model and colors (needed to better represent the deformity). The proposed system relies on a low-cost depth camera (i.e. the Microsoft Kinect v2 [16]) which allows the instantaneous and sufficiently accurate capture of the three-dimensional geometry of the patient's chest [17].

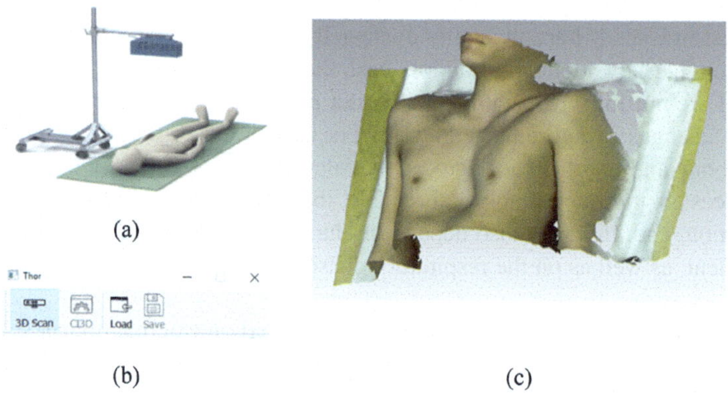

Fig. 1. (a) Three-dimensional chest scanner for the optical acquisition; (b) screenshot of the thorax scanner software interface and (c) three-dimensional image of a patient for clinical evaluation.

The scan is performed in the supine position and in the phase of maximum exhalation [18, 19]. The scanned chest is then processed automatically and transparently to the user to extract the reference plane XY (the floor), and to extract the parameters associated with chest depression, described below.

The first extracted index is the Three-Dimensional Correction Index (CI3D), already introduced in [10], which represents the adaptation of the Correction Index (CI) to the outer perimeter of the chest, as shown in Fig. 2.

The CI assesses the percentage of chest depth, independently of the chest width, unlike the more common Haller Index [20].

CI3D is computed as:

$$CI3D = \frac{C - D}{D} \qquad (1)$$

where C and D are the vertical distances of, respectively, the maximum and minimum sternal depression from the reference plane (Fig. 2).

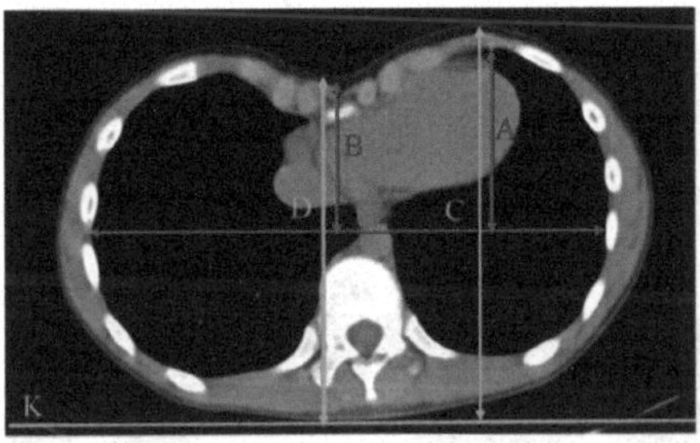

Fig. 2. Example of correction index (CI) calculation (red) and corresponding correction index three-dimensional (CI3D) calculation (green).

Two deformity indices are calculated in addition to CI3D: Asymmetry (AI3D) and Eccentricity (EI3D). These two measures, introduced in the literature [2, 7, 21] and usually calculated on CT and MRI, have also been adapted to optical scanning by considering external rather than intercostal perimeters.

The selection of these deformity indices is based on the fact that some characteristics of the thorax can produce different impacts also with regard to the success of the correction procedure [22]. In fact, many of these characteristics cannot be evaluated with the most widely used methods, i.e. the Correction Index and the Haller Index, which only take into account the position on the sagittal plane of the point of maximum depression. The Asymmetry Index determines the ratio between the right and left prominence of the chest, measured by considering the maximum points with respect to the sagittal plane, calculated on the section containing the point of maximum depression (Fig. 3).

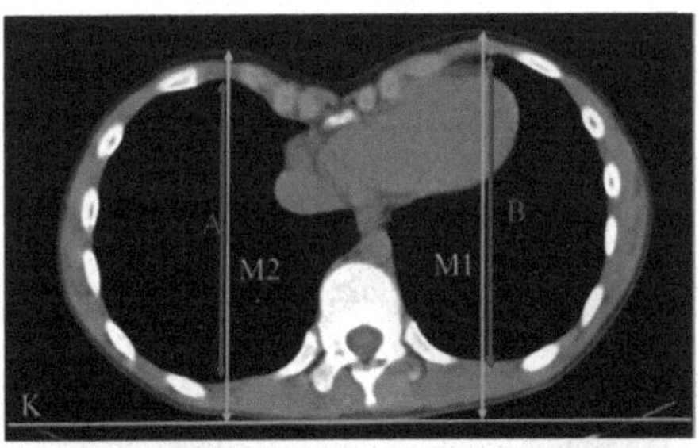

Fig. 3. Example of asymmetry index (AI) calculation (red) and corresponding asymmetry index three-dimensional (AI3D) calculation (green).

The Eccentricity Index represents the displacement, on the transversal plane, of the cavity and is calculated as the ratio between the two distances of the minimum point from the thoracic wall on the right and left sides (Fig. 4).

$$AI3D = \frac{M1}{M2} \tag{2}$$

$$EI3D = \frac{R}{L} \tag{3}$$

Where $M1$ and $M2$ are left and right highest points on the section containing the maximum depression; R and L are the left and right distances of the point of maximum depression from the external thoracic wall.

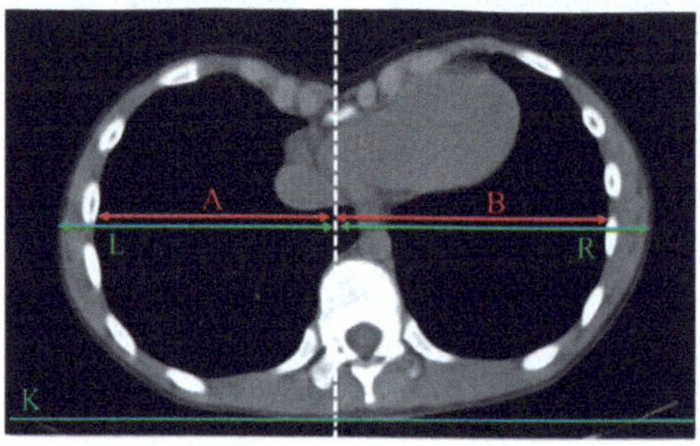

Fig. 4. Example of eccentricity index (AI) calculation (red) and corresponding eccentricity index three-dimensional (EI3D) calculation (green).

Two further measures have been introduced to describe the thoracic malformation: Area and Width. The idea is to describe the extent and shape of the cavity, not only its depth. The Width is calculated as the Euclidean distance between left and right highest points, in correspondence of the section containing the point of maximum depression (see Fig. 5). The Area is calculated by connecting the two highest points and approximating the area between the line thus obtained and the profile of the chest, always in correspondence of the section containing the minimum (yellow in Fig. 5).

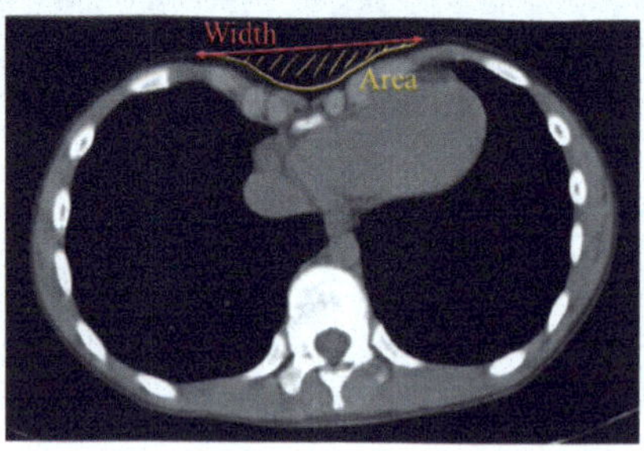

Fig. 5. Example of width and area computation on the external profile of the chest.

The described values are calculated on the 3D model of the patient's chest and are, for explanatory purposes, depicted on the CT image, to have a visual comparison between the traditional indices and their adaptation to the 3D model, calculated considering external rather than intercostal perimeters.

In particular, for each of these values, the section containing the point of maximum depression is calculated automatically. In detail, once the XY reference plane (i.e. the floor) is automatically identified in the 3D model, a number of transverse planes parallel to XZ with a reciprocal distance of 1 mm are automatically drawn (Fig. 6). The intersection of these planes with the 3D model of the chest allows the determination of a different Δ value (corresponding to the local sagittal depression) for each transverse plane. Of all the Δ values measured in the transverse planes, the section which gives the maximum value is considered.

Fig. 6. Sheaf of planes perpendicular to the ground plane cutting the chest for the automatic computation of the point of maximum depression.

3 Data Analysis

With the aim of creating a diagnostic tool that can support the physician's judgment during the outpatient examination, it was decided to evaluate the parameters extracted by assigning a class of severity to each analyzed case. Specifically, a severity clinical score (SCS) was obtained by consulting five doctors, experts in the field of thoracic deformities, to independently assess the defect of each patient according to the classes listed in Table 1. To each clinical case is assigned the mean value of the five SCS (MSCS). The severity values were mapped to six severity levels to allow a better discrimination of the different cases with respect to the basic severity levels (mild, moderate and severe).

50 new patients with mild (1) to extremely severe (5) diagnosis of PE were studied, aged between 2 and 17 years. For each patient the values of CI3D, Width, Area, AI3D and EI3D were extracted as described above and the MSCS was assigned.

Table 1. Values of severity for clinical assessment.

0	Absent
1	Mild
2	Moderate
3	Moderate to severe
4	Severe
5	Extremely severe

The data were evaluated with Linear Discriminant Analysis (LDA), a very common technique used for supervised classification problems [23]. The objective of the LDA is to project characteristics of a larger space on a smaller space. This can be achieved in three steps: the first step is to calculate the separability between different classes (i.e. the distance between the mean of the different classes), also called as variance between classes. The second step is to calculate the distance between the mean and the samples of each class, which is called variance within the class. The third step is to construct the lower dimensional space that maximizes the variance between classes and minimizes the variance in the class.

In first instance, the Pearson correlation coefficient between each index and the MSCS was calculated (Table 2).

Table 2. Pearson correlation coefficient between each index and the MSCS

CI3D	Width	Area	AI3D	EI3D
Corr: 0.87	Corr: 0.54	Corr: 0.80	Corr: −0.372	Corr: −0.154
p-value: 0.0005	p-value: 0.0005	p-value: 0.0005	p-value: 0.016	p-value: 0.336

As can be deduced from Table 2, among the selected values, the one that most describes the disease individually is the CI3D. Figure 7 shows the least-squares regression line predicting the clinical diagnosis from the evaluated CI3D values. The correlation was tested with the Pearson coefficient at a significance level of $\alpha = 0.05$. The null hypothesis was formulated as a lack of correlation between CI3D and MSCS values. As a result, a correlation coefficient of 0.87 was obtained and, due to a value of p lower than 0.005, the non-correlation hypothesis was rejected.

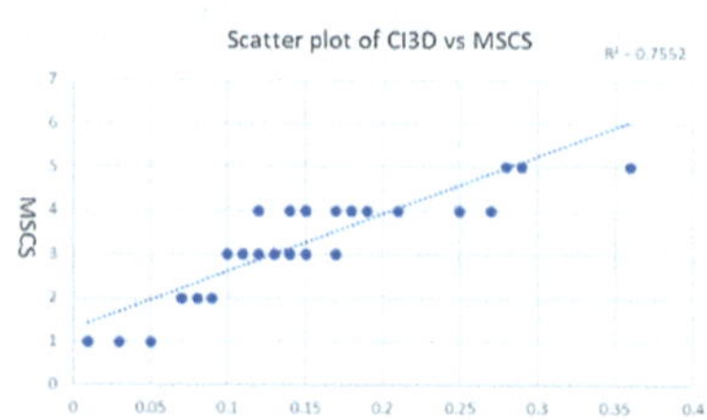

Fig. 7. Scatter plot and regression line of CI3D values and MSCS values.

Furthermore, LDA analysis performed on CI3D values reached 70.7% of correspondence with clinical judgments.

We then tested which of the extracted parameters could improve the correspondences rate, by performing LDA analysis with combination of inputs as described in Table 3.

According to the results presented in Table 3, the two most significant parameters are the Width, which, if combined with the CI3D allows to obtain a score of 87.8% (Fig. 8), followed by the Area.

Table 3. LDA analysis results with combined inputs.

CI3D	Width	Area	AI3D	EI3D
70.7%	56.1%	48.8%	46.3%	39.0%
	CI3D-Width	**CI3D-Area**	**CI3D-AI3D**	**CI3D-EI3D**
	87.8%	75.6%	73.2%	70.7%
		CI3D-Width-Area	**CI3D-Width-AI3D**	**CI3D-Width-EI3D**
		87.8%	85.2%	82.9%

The maximum percentage of precision achieved is of ~87%, which can be considered more than acceptable for the intended purpose. In addition, errors made by the classifier deviate from the assigned MSCS of only one class.

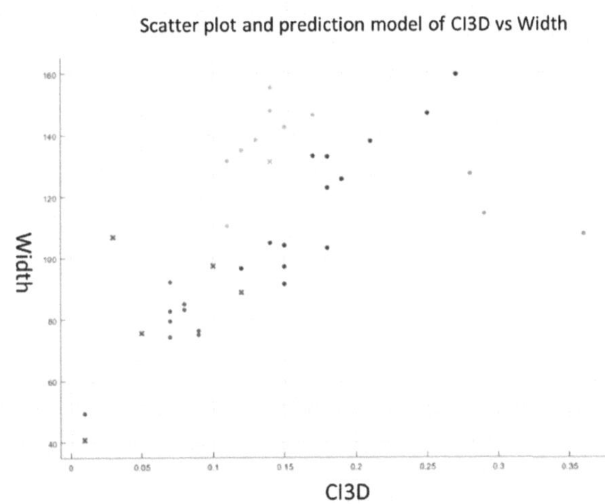

Fig. 8. Scatter plot and LDA prediction results of CI3D vs Width: colors indicates the assigned class; dot values are correctly classified, "x" values are classification errors.

4 Conclusions

In this paper, the geometrical structure of Pectus Excavatum (PE) was examined in order to develop an outpatient support tool for assessing the severity of the disease and monitoring its progression over time.

The study is based on the possibility of automatically retrieving chest indices from the optical scan of the patient's chest. The idea is to define which measure can assist in the characterization of the PE in order to meet the doctor's opinion with the utmost precision.

In addition to the already introduced Correction Index 3D (CI3D), two deformity indices, the Asymmetry Index 3D (AI3D) and the Eccentricity Index 3D (EI3D), have been used and adapted to the optical scan. It was also decided to introduce the Width, which represents the width of the cavity and the Area, calculated as the area enclosed by the cavity.

The calculation of all indices has been automated within the software presented in [10] that enable the clinical user to use it without the need for technical skills.

In the first instance, an analysis carried out on the CI3D value alone led to a correlation with the mean severity clinical score of 0.87, with p value lower than 0.005. Moreover, by applying the Linear Discriminant Analysis an accuracy of 70.7% is achieved.

The results obtained by the classification tool reached a maximum of 87.8% of accuracy. The score is significantly improved if compared to using the CI3D as the only measurement tool to assess the severity.

In this context, it is important to underline that the proposed measures are all computed on the section that contains the point of maximum depression and consequently they cannot consider any other defects of the chest (e.g. costal flaring), or even the vertical shape and extension of the cavity.

The tool is in fact designed to help, and not replace, the doctor's judgment in the diagnosis by providing objective measures regarding the condition of each individual patient.

References

1. Pectus Excavatum: historical background, clinical picture, preoperative evaluation and criteria for operation. In: Seminars in Pediatric Surgery, vol. 17, pp. 181–193 (2008)
2. Cartoski, M.J., Nuss, D., Goretsky, M.J., Proud, V.K., Croitoru, D.P., Gustin, T., Mitchell, K., Vasser, E., Kelly, R.E.: Classification of the dysmorphology of Pectus Excavatum. J. Pediatr. Surg. **41**, 1573–1581 (2006)
3. Goretsky, M.J., Kelly, R.E., Croitoru, D., Nuss, D.: Chest wall anomalies: Pectus Excavatum and pectus carinatum. Adolesc. Med. Clin. **15**, 455–471 (2004)
4. Fonkalsrud, E.W., Beanes, S., Hebra, A., Adamson, W., Tagge, E.: Comparison of minimally invasive and modified Ravitch Pectus Excavatum repair. J. Pediatr. Surg. **37**, 413–417 (2002)
5. Nuss, D., Kelly, R.E., Croitoru, D.P., Katz, M.E.: A 10-year review of a minimally invasive technique for the correction of Pectus Excavatum. J. Pediatr. Surg. **33**, 545–552 (1998)

6. Haecker, F.-M.: The vacuum bell for conservative treatment of Pectus Excavatum: the Basle experience. Pediatr. Surg. Int. **27**(6), 623–627 (2011)
7. Martinez-Ferro, M.: Indexes for pectus deformities. In: Kolvekar, S., Pilegaard, H. (eds.) Chest Wall Deformities and Corrective Procedures, pp. 35–60. Springer, Cham (2016)
8. Sujka, J.A., Peter, S.D.: Quantification of Pectus Excavatum: anatomic indices. Semin. Pediatr. Surg. **27**, 122–126 (2018)
9. Ewert, F., Syed, J., Wagner, S., Besendoerfer, M., Carbon, R.T., Schulz-Drost, S.: Does an external chest wall measurement correlate with a CT-based measurement in patients with chest wall deformities? J. Pediatr. Surg. **52**, 1583–1590 (2017)
10. Uccheddu, F., Ghionzoli, M., Volpe, Y., Servi, M., Furferi, R., Governi, L., Facchini, F., Lo Piccolo, R., McGreevy, K.S., Martin, A., Carfagni, M., Messineo, A.: A novel objective approach to the external measurement of Pectus Excavatum severity by means of an optical device. Ann. Thorac. Surg. **106**, 221–227 (2018)
11. Bliss, D.P., Vaughan, N.A., Walk, R.M., Naiditch, J.A., Kane, A.A., Hallac, R.R.: Non-radiographic severity measurement of Pectus Excavatum. J. Surg. Res. **233**, 376–380 (2019)
12. Sarwar, Z.U., DeFlorio, R., O'Connor, S.C.: Pectus Excavatum: current imaging techniques and opportunities for dose reduction. In: Seminars Ultrasound, CT MRI, vol. 35, pp. 374–381 (2014)
13. Carfagni, M., Furferi, R., Governi, L., Servi, M., Uccheddu, F., Volpe, Y., Mcgreevy, K.: Fast and low cost acquisition and reconstruction system for human hand-wrist-arm anatomy. Procedia Manuf. **11**, 1600–1608 (2017)
14. Uccheddu, F., Del Mastio, A., Remondino, F., Pelagotti, A., Cappellini, V.: Texture mapping of flat-like 3D models. In: 2009 16th International Conference on Digital Signal Processing, pp. 1–6. IEEE (2009)
15. Furferi, R., Governi, L., Uccheddu, F., Volpe, Y.: A RGB-D based instant body-scanning solution for compact box installation, pp. 819–828. Springer, Cham (2017)
16. Kinect - Windows app development. https://developer.microsoft.com/en-us/windows/kinect. Accessed 28 June 2019
17. Carfagni, M., Furferi, R., Governi, L., Santarelli, C., Servi, M., Uccheddu, F., Volpe, Y.: Metrological and critical characterization of the Intel D415 stereo depth camera. Sensors. **19**, 489 (2019)
18. Raichura, N., Entwisle, J., Leverment, J., Beardsmore, C.S.: Breath-hold MRI in evaluating patients with Pectus Excavatum. Br. J. Radiol. **74**, 701–708 (2001)
19. Birkemeier, K.L., Podberesky, D.J., Salisbury, S., Serai, S.: Breathe in… Breathe out… stop breathing: does phase of respiration affect the Haller index in patients with Pectus Excavatum? Am. J. Roentgenol. **197**, W934–W939 (2011)
20. Jaroszewski, D., Notrica, D., McMahon, L., Steidley, D.E., Deschamps, C.: Current management of Pectus Excavatum: a review and update of therapy and treatment recommendations. J. Am. Board Fam. Med. **23**, 230–239 (2010)
21. Kim, H.C., Choi, H., Jin, S.O., Lee, J.J., Nam, K.-W., Kim, I.-Y., Nam, K.C., Park, H.J., Lee, K.H., Kim, M.G.: New computerized indices for quantitative evaluation of depression and asymmetry in patients with chest wall deformities. Artif. Organs **37**, 712–718 (2013)
22. Sesia, S.B., Heitzelmann, M., Schaedelin, S., Magerkurth, O., Kocher, G.J., Schmid, R.A., Haecker, F.-M.: Standardized Haller and asymmetry index combined for a more accurate assessment of Pectus Excavatum. Ann. Thorac. Surg. **107**, 271–276 (2019)
23. Izenman, A.J.: Linear discriminant analysis. In: Modern Multivariate Statistical Techniques, pp. 237–280. Springer, New York (2013)

A Preliminary 3D Depth Camera-Based System to Assist Home Physiotherapy Rehabilitation

Francesca Uccheddu(✉), Lapo Governi, and Monica Carfagni

Department of Industrial Engineering, University of Florence, Florence, Italy
{francesca.uccheddu, lapo.governi, monica.carfagni}@unifi.it

Abstract. Total hip arthroplasty (THA) and total knee arthroplasty (TKA) have been recently heralded as the operations of the Century. Large improvements in mobility and patient-reported outcomes are typically observed compared with the small-to-moderate effects experienced with non-surgical interventions. Following surgery, physiotherapy-led exercise-based rehabilitation is often prescribed to yield better gait-related outcomes. Nevertheless, outpatient rehabilitation is expensive and heavily burden the national health service. When specific machines are not needed during the physiotherapy, patients, if assisted, can perform a home program. The purpose of this paper is to qualitatively investigate the applicability of a self-managed, home-based system for the automated evaluation of a home physiotherapy rehabilitation after TKA and THA. The system leverages the cost effectiveness and the versatility of a RGB-Depth camera system together with a commercial skeleton tracking system to analyse specific exercises. A novel computation of lower limb movements and related angles is proposed to evaluate the quality of the daily exercises. The laboratory experimental campaign, envisaged the analysis of the rotation angles of hips and knees; a lower limb schematic model is considered to estimate both knee and hip angles during ab/adductor and flex/extension movements. A novel real time calculation of the hip bone plane is proposed to assess the joint angles during specific exercises performance. A qualitative data analysis of each exercise has been performed. Results on the system usability in a domestic environment are reported as well as a visual comparison of the analysed output.

Keywords: RGB-D camera · 3D · Body tracking

1 Introduction

Total hip and knee arthroplasty have revolutionized the care of patients leading to functional recovery, and substantial improvements in quality of life. The demand for joint replacement surgeries is growing simultaneously with the ageing of the population physiologically prone to joint pain and disability who expect a longer and better quality of life; projection studies expect a sharp rise in hip and knee replacements over the next 30 years [1]. Evidence has shown that rehabilitation after surgery is important to

C. Rizzi et al. (Eds.): ADM 2019, LNME, pp. 696–706, 2020.
https://doi.org/10.1007/978-3-030-31154-4_59

prevent decline postoperatively and to restore a high functional level [2]; in the immediate post-operative phase, the adherence to the program is the most critical issue.

Recently, many orthopaedics medical centres, have been adopted the fast-track surgery protocol [3, 4] that involves an early mobilization of the patient after surgery, an early oral hydration and nutrition, thereby abolishing any kind of invasive procedure such as urinary catheterization or needle-tube beyond 48 h. This new approach is characterized by an extremely precise management of the whole perioperative period of candidate to total knee arthroplasty; patient education through informative material and meeting with medical staff is therefore crucial. Patient plays an active role starting from pre-operative phase thanks to his knowledge of his own recovery steps. During the entire perioperative period, short acting anaesthesia and excellent pain management assuring an early mobilization represent the real challenges. Accordingly, the role of a tool for measuring the functional recovery of the operated joint is crucial to both the patient to be motivated in performing the right number of exercises, and the doctor/physiotherapist that can follow up the patient remotely during the post-operative days.

The approaches to track limb motion of diverse precision, cost and complexity are many [5]; among others we find *Optical systems* that use visual data captured by one or more cameras to triangulate the 3D position of a set of points detected; high precision can be reached, e.g. of a few millimetres [6], but at a high cost [7]. A cheaper affordable solution of lower precision, e.g. of a few centimetres, is achieved by stereo vision systems [8] or depth cameras which use an integrated colour camera with an infrared depth sensor [9]. *Exo-skeletons* are rigid structures of jointed bars coupled with potentiometers that articulate at the joints of the body. Such systems offer real-time, high precision acquisition and are not being influenced by visual occlusion, quality and number of cameras. On the other hand, the main disadvantage is the movement limitation imposed by the mechanical constraints. Furthermore, commercial medical devices of this type have a prohibitive cost for most of the patients, not yet allowing their widespread use for home rehabilitation. Still these devices can be a useful acquisition in clinics and rehabilitation institutes [10]. *Inertial Motion Capture* technology is based on miniatured inertial sensors (IMUs), and sensor fusion algorithms. Most inertial systems use gyroscopes and accelerometers to measure rotational speed. Inertial motion systems capture the full six degrees of freedom body motion of a human in real-time. Such systems can be cheap, small, portable, and can capture large areas at a price of a lower positional accuracy and positional drift [11].

Within the context of home based low cost approaches, the optical systems implemented through depth cameras remain the more affordable solutions to monitor the patient exercise and to indirectly obtain more adherence to the program. Recently many low-cost, optical, body and hand tracking systems [12] have been employed successfully in the health care context to assess the impact of post-operative exercises on the patient range of motion. Among these, the Microsoft Kinect v1 device has been used to monitor people movements in rehabilitation [13, 14]. The Microsoft Kinect v2 is more robust and accurate as compared to Microsoft Kinect v1, and it has been recognized a viable tool for tracking human movement in clinical applications [15].

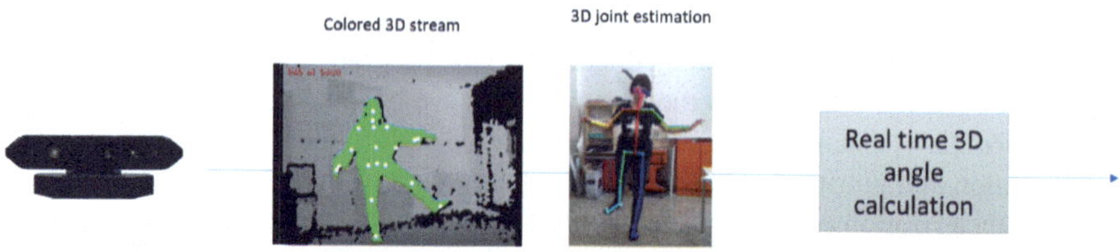

Fig. 1. Overview of the proposed system; an interactive Android box system served as a 3D skeleton tracker for demonstrating human 3D joints dynamic angles estimation.

Since Kinect sensors are no longer being manufactured, similar technology also recently appeared in other consumer products such as the Asus Xtion and the Occipital Structure Sensor and the ORBBEC Astra Pro 3D. Intel has also released the "Real-sense" family, a competitive triangulation-based range sensing technology [16, 17], implemented in standalone units by Creative, and in some Lenovo laptops. A 3D camera typically consists of a pair of colour and depth cameras, placed next to each other inside the device. The colour camera is a standard webcam whereas the depth camera generally uses infrared light for depth measurement.

In several 3D camera-based optical systems, since it is only needed to monitor the exercise repetitions number, the skeleton joints 3D coordinates are enough to execute the task. Authors in [13] and in [14] starting from a real time skeleton estimation, proposed to indirectly measure the joint rotation angles for elbows and knees. The knee joint, due to the anatomical movements constraints, can be reasonably considered as a cylindrical joint; thus, the rotation angle can be estimated by only knowing the 3D coordinates of the connected joints (i.e. hip-knee-ankle joints for the knee joint, shoulder-elbow-wrist joints for the elbow joint). On the contrary, the hip joint has more degree of freedom and can be approximated by a spherical joint; for this reason, the unsupervised evaluation of the hip rotation angle it is not trivial but instead requires a real time estimation of the plane on which the movement is performed.

The present article aims at presenting a novel approach to the real time measurement of lower limb joints rotation angles during post TKA and THA surgery rehabilitation; moreover, it aims at qualitatively investigating the feasibility of the proposed real-time 3D skeleton tracking system to perform an efficient monitoring of patient movements during the physiotherapy in a domestic setting.

2 Methodology

Figure 1 shows an overview of the proposed system; an interactive Android box system (i.e. TVICO by 3Divi, a rebranded Orbbec Persee combined with the Nuitrack SDK environment [18]) served as a 3D skeleton tracker for demonstrating human 3D joints dynamic angles estimation.

The system is connected via Wi-Fi to a remote laptop that starts and stop the joint tracking application. Such application allows to analyse, for each frame, the joints 3D coordinates and to calculate the hip and knee angles for each selected exercise. The camera system, since it transmits data via radio signal, is stand alone and can be freely positioned in the room without any need for specific structures or complicated cabling. The system relies on a simple representation of the human body as a stick figure, which consists of line segments linked by joints; Fig. 2 shows the configuration of the joints used in our algorithm.

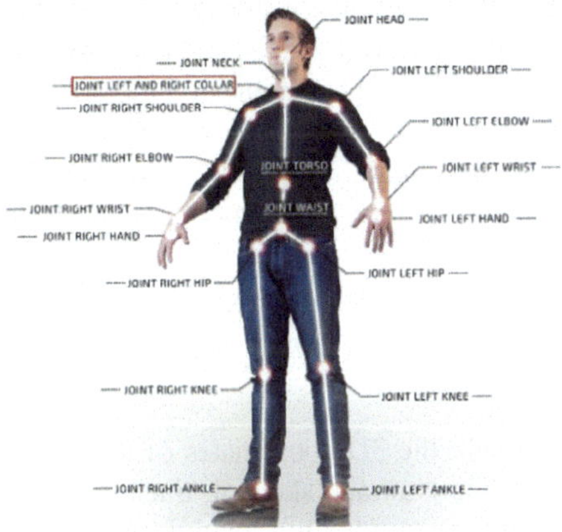

Fig. 2. The skeleton scheme as adopted in the Nuitrack skeleton tracker system

Figure 3 shows a sketch of the room settings. The height for sensor placement is about 1.0 m. It is put facing straight forward and parallel to the floor close to the supporting edge. The distance between the patient and the sensor is about 1.5 m.

This work has been carried out as part of the regional project ARCONTE, an informative system integrating a novel home-based patient movement monitoring system. The selected physical exercises analysed in this paper are among those prescribed to patients that underwent TKA and THA surgery in the Istituto Fiorentino di Cura e Assistenza (IFCA, Florence, that acts in the project as the clinical driver) adhering to the fast track protocol. The exercises for both hip and knee physiotherapy are the flexion and extension

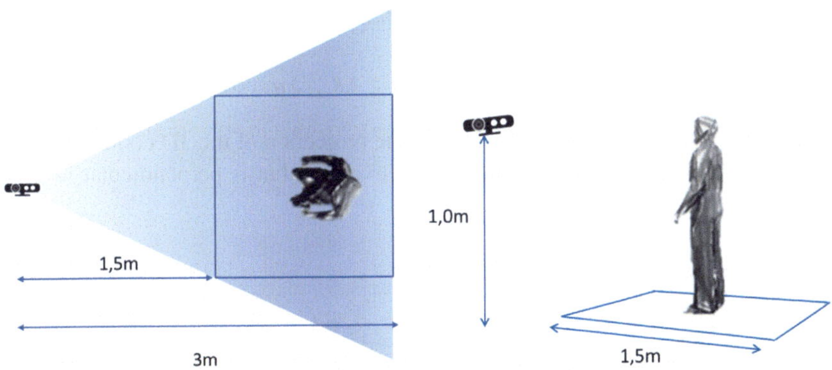

Fig. 3. Sensor placement in the experimental setup. On the left the view from top, on the right the sensor height from the floor and the "gaming area" in which the patient is freely to move during the exercises.

exercise and abduction and adduction exercise for left and right knee and hip joints with some variants as depicted in Fig. 4. The video frames reported in Fig. 4, show a superimposed coloured skeleton that has been manually depicted in order to help the reader to understand the joints positions in the 3D space.

The final output is a text file, with a row for each frame, whose columns representing the x-y-z coordinates of the tracked body joints, and the left and right hip and knees joints angles evaluated according to specific exercises (i.e. flex-extension or ab/adductor). Since the knee rotation [21] is physiologically only possible on the plane identified by the two long bones (i.e. the segments hip joint – knee joint and knee joint – ankle joint) the knee rotation angle $\boldsymbol{\theta}_{knee}$ to be evaluated during the flex/extension exercises is easily estimated as in Fig. 5a according to the following equation:

$$\cos(\boldsymbol{\theta}_{knee}) = \frac{< \overrightarrow{(\boldsymbol{H}-\boldsymbol{K})}, \overrightarrow{(\boldsymbol{K}-\boldsymbol{A})} >}{\left|\overrightarrow{(\boldsymbol{H}-\boldsymbol{K})}\right| * \left|\overrightarrow{(\boldsymbol{K}-\boldsymbol{A})}\right|} \tag{1}$$

where $\boldsymbol{H}, \boldsymbol{K}, \boldsymbol{A}$ are respectively the 3D coordinates of the (left or right) hip, knee and ankle joint, as provided by the Nuitrack skeleton tracker.

On the contrary the estimation of the hip rotation angle $\boldsymbol{\theta}_{hip}$ is more complex since the hip joint has more degrees of freedom. For this reason, it is needed to identify a plane for each exercise, on which to project the hip and the knee joints and to calculate the hip rotation angle. The flex/extension hip angle $\boldsymbol{\theta}_{hip_{flexextension}}$ is evaluated as in Fig. 5b, by projecting the joints on the sagittal plane, according to the following equation:

$$\tan\left(\boldsymbol{\theta}_{hip_{flexextension}}\right) = -(\boldsymbol{x}_H - \boldsymbol{x}_K)/(\boldsymbol{y}_H - \boldsymbol{y}_K) \tag{2}$$

where $(\boldsymbol{x}_H, \boldsymbol{y}_H)$ and $(\boldsymbol{x}_K, \boldsymbol{y}_K)$ are respectively the projection of the 3D coordinates of the right or left hip and knee on the hip plane (i.e. identified by the three points the waist joint, the right hip joint and the left hip joint as indicated in Fig. 2).

The ab/adductor hip angle $\boldsymbol{\theta}_{hip_{abadductor}}$ is evaluated as in Fig. 5c, by projecting the joints on the longitudinal plane according to the following equation:

$$\tan\left(\boldsymbol{\theta}_{hip_{abadductor}}\right) = -(\boldsymbol{x}_H - \boldsymbol{x}_K)/(\boldsymbol{y}_H - \boldsymbol{y}_K) \tag{3}$$

where $(\boldsymbol{x}_H, \boldsymbol{y}_H)$ and $(\boldsymbol{x}_K, \boldsymbol{y}_K)$ are respectively the projection of the 3D coordinates of the right or left hip and knee on the plane of the exercise that is perpendicular to the hip plane.

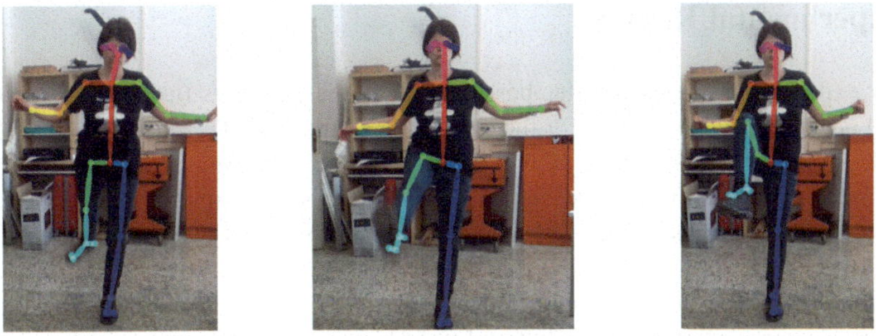

Frontal flex/extension type I | Frontal flex/extension type II | Frontal flex/extension type III

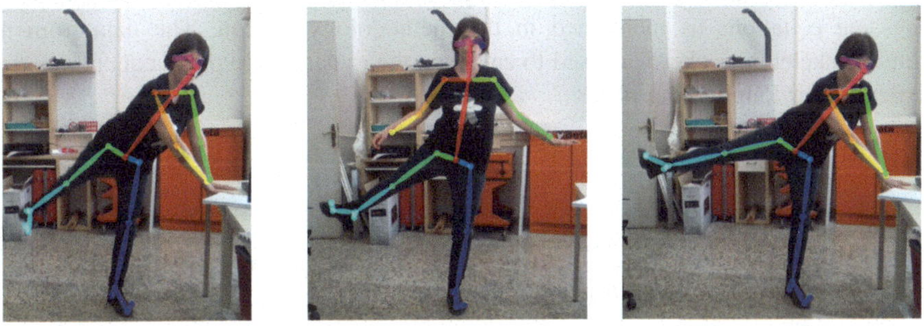

Flex/extension with hands on a table | Frontal ab/adduction | Ab/adduction with hands on a table

Fig. 4. The selected exercises to be performed after TKA and THA envisages a variety of flex/extension of the hip and the knee, and ab/adductor exercises of the same joints.

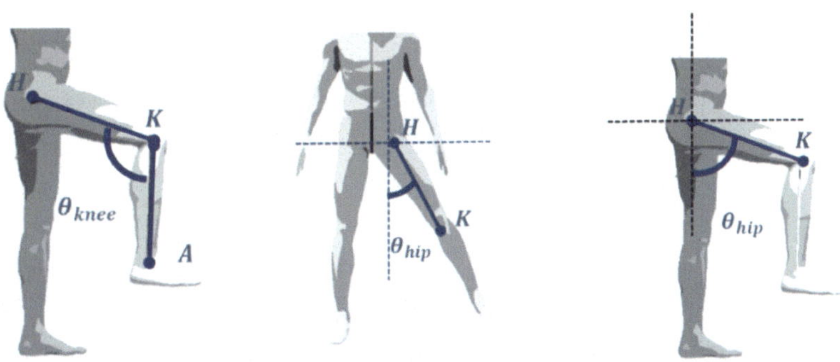

(a) (b) (c)

Fig. 5. A simplified human joint/bones model of the lower limbs during (a) flex/extension of the knee, (b) flex/extension of the hip, (c) ab/adduction of the hip.

3 Experimental Results

For our experiments we used a database of 6 dynamic exercises (as the most recommended post-operative exercises for the lower limbs after both TKA and THA [19, 20], also adopted by the specialists at the IFCA, Florence for patient in the fast track protocol); the aim of the analysis is the qualitative evaluation of a real time measurement of joint angles for the selected movements in terms of recovered range of motion.

The data from the RGB-D camera were recorded for about 7 s for each exercise, in terms of x, y, z coordinates of the left and right hip joint, the left and right knee joint, the left and right ankle joint and the waist joint. Joint angle calculation was done using the same equations as mentioned in the methodology section. The data from the Nuitrack skeleton tracker was used to calculate the dynamic hip plane equation and its perpendicular longitudinal plane.

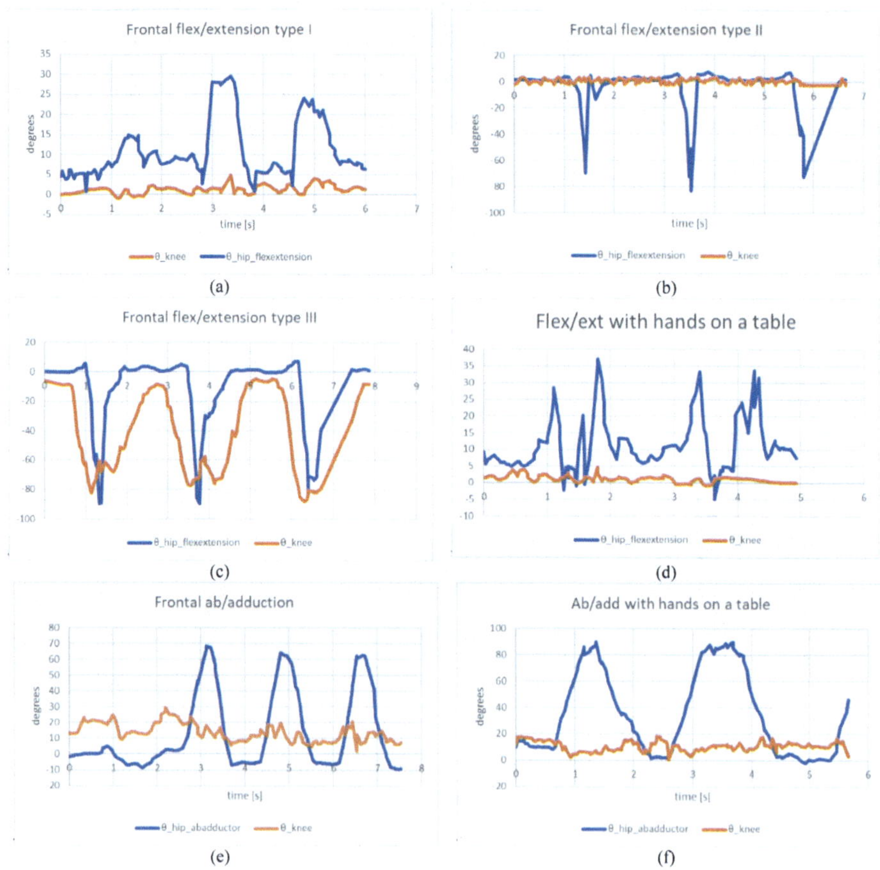

Fig. 6. Right hip and knee joint angles collected during the selected exercises. The Blue colour indicates the hip data and Orange indicates the knee data.

Figure 6 shows the right hip and knee joint angles collected during the selected exercises. The Blue colour indicates the hip data (i.e. flex/extension angle or ab/adductor angle) and Orange indicates the knee data (i.e. flex/extension angle). The graph shows change in angle in degrees over time in seconds. Graph 6a shows that the knee is almost still, while the hip is performing extensions repetitions; hip angles data are not free of noise mainly due to the position frontal to the camera, causing the knee/ankle joints to disappear during some frame as depicted in Fig. 7 top row. Graph 6b show hip flexion exercise in which the three movement repetitions of the hip are well represented as well as the knee angle (close to 0°). Graph 6c represents data from flexion of both hip and knee independently. Graph 6d represents the same exercise as in 6a framed from a different point of view and with the patient hands on a table. As depicted in Fig. 7 bottom row, a significative number of frames loses the knee joint thus provoking very noise and inaccurate data. Graph 6e shows a perfect hip adduction exercise; the knee angle oscillates on one hand due to the intrinsic noise of the system depth sensor/skeleton tracking algorithm, on the other hand because the knee inherits a residual drift motion from the hip movement. Graph 6f show the same exercise as in 6e with a different energy and velocity as it can be appreciated in the graph.

Fig. 7. Errors occurring during real time skeleton tracking; in the top row and in the bottom row, are depicted two movement sequence in which in the third frame is missing the lower limb joints (right knee and ankle).

4 Conclusions

This study focused on the qualitatively investigation of the applicability of a self-managed, home-based system for the automated evaluation of a home physiotherapy rehabilitation after TKA and THA. The system leverages the cost effectiveness and the versatility of a RGB-Depth camera system together with a commercial skeleton tracking system to analyse specific exercises. A database of 6 dynamic exercises has been evaluated in terms of knee and hip rotation angles. A lower limb schematic model is considered to estimate both knee and hip angles during ab/adductor and flex/extension movements and a qualitative data analysis of each exercise has been performed. Preliminary results show the ability of the proposed approach, differently to other Kinect based approaches, to fast estimate both the knee and the hip rotation angles during the flexo/extension ad ab/adduction exercises; results have also shown that not all the points of view are suited for tracking the patient skeleton during rehabilitation exercises, especially those exercises that move away or move closer the patient foot. Inaccurate rotation angles tracking data noise is always present in the analysed results; since the 3D tracking system relies on a kinematic-based tracking approach, a systematic drifting noise can be efficiently diminished by applying Kalman or particle filtering [22]. Furthermore, the system usability in a domestic environment is highly influenced by the specific patient pose; in fact, 3D body joints detectors performances remain very poor for non-frontal poses. On the other hand, 2D pose estimation is recently leveraging the power of Convolutional Neural Networks solutions by providing high quality pose estimation paid in terms of computational time; nevertheless, this computational limitation is mitigated by recent Graphical Processing Units architectures, that allows efficient real-time estimations. In future works a convenient double step algorithm can be considered to overcome the main limitation of using a pure 3D skeleton tracking algorithm. A 2D skeleton tracking system can be used to estimate the 2D body pose of the patient in the scene (i.e. using OpenPose [23] library), then the system should compute the 3D positions of each joint returned by the 2D tracker by exploiting the depth information coming from the RGB-D sensor used.

Acknowledgements. The author would like to thank Dr. Andrea Baldini and Dr. Marco Ponti for their helpful advice on describing the recommended exercises prescribed after TKA and THA surgery and various clinical issues examined in this paper, for their advice and comments during the experimental campaign. The author, however, bears full responsibility for the paper.

Funding. This work has been carried out within the project ARCONTE, thanks to the decisive regional contribution from the Regional Implementation Programme co-financed by the FAS (now FSC) and the contribution from the FAR funds made available by the MIUR.

Regione Toscana | FAS Fondo Aree Sottoutilizzate 2007-2013 | REPUBBLICA ITALIANA

References

1. Birrell, F., Johnell, O., Silman, A.: Projecting the need for hip replacement over the next three decades: influence of changing demography and threshold for surgery. Ann. Rheum. Dis. **58**(9), 569–572 (1999)

2. Di Monaco, M., Vallero, F., Tappero, R., Cavanna, A.: Rehabilitation after total hip arthroplasty: a systematic review of controlled trials on physical exercise programs. Eur. J. Phys. Rehabil. Med. **45**(3), 303–317 (2009)
3. Kehlet, H., Wilmore, D.W.: Evidence-based surgical care and the evolution of fast-track surgery. Ann. Surg. **248**(2), 189–198 (2008)
4. Fransen, B.L., Hoozemans, M.J., Argelo, K.D., Keijser, L.C., Burger, B.J.: Fast-track total knee arthroplasty improved clinical and functional outcome in the first 7 days after surgery: a randomized controlled pilot study with 5-year follow-up. Arch. Orthop. Trauma Surg. **138** (9), 1305–1316 (2018)
5. Gherardini, F., Mascia, M.T., Bettelli, V., Leali, F.: A co-design method for the additive manufacturing of customised assistive devices for hand pathologies. J. Integr. Des. Process Sci. 1–18 (2018)
6. Naddeo, A., Cappetti, N., Califano, R., Vallone, M.: Manual assembly work-station redesign based on a new quantitative method for postural comfort evaluation. Appl. Mech. Mater. **459**, 368–379 (2013)
7. Pfister, A., West, A.M., Bronner, S., Noah, J.A.: Comparative abilities of Microsoft Kinect and Vicon 3D motion capture for gait analysis. J. Med. Eng. Technol. **38**(5), 274–280 (2014)
8. Ferrara, P., Piva, A., Argenti, F., Kusuno, J., Niccolini, M., Ragaglia, M., Uccheddu, F.: Wide-angle and long-range real time pose estimation: a comparison between monocular and stereo vision systems. J. Vis. Commun. Image Represent. **48**, 159–168 (2017)
9. Mehta, D., Sridhar, S., Sotnychenko, O., Rhodin, H., Shafiei, M., Seidel, H.P., Xu, W., Casas, D., Theobalt, C.: VNect: real-time 3D human pose estimation with a single RGB camera. ACM Trans. Graph. (TOG) **36**(4), 44 (2017)
10. Talaty, M., Esquenazi, A., Briceno, J.E.: Differentiating ability in users of the ReWalk TM powered exoskeleton: an analysis of walking kinematics. In: 2013 IEEE 13th International Conference on Rehabilitation Robotics (ICORR), pp. 1–5. IEEE, June 2013
11. Seel, T., Raisch, J., Schauer, T.: IMU-based joint angle measurement for gait analysis. Sensors **14**(4), 6891–6909 (2014)
12. Gabel, M., Gilad-Bachrach, R., Renshaw, E., Schuster, A.: Full body gait analysis with Kinect. In: 2012 Annual International Conference of the IEEE Engineering in Medicine and Biology Society, pp. 1964–1967. IEEE, August 2012
13. Haas, D., Phommahavong, S., Yu, J., Krüger-Ziolek, S., Möller, K., Kretschmer, J.: Kinect based physiotherapy system for home use. Curr. Dir. Biomed. Eng. **1**(1), 180–183 (2015)
14. Phommahavong, S., Haas, D., Yu, J., Krüger-Ziolek, S., Möller, K., Kretschmer, J.: Evaluating the Microsoft Kinect skeleton joint tracking as a tool for home-based physiotherapy. Curr. Dir. Biomed. Eng. **1**(1), 184–187 (2015)
15. Capecci, M., Ceravolo, M.G., Ferracuti, F., Iarlori, S., Kyrki, V., Longhi, S., Romeo, L., Verdini, F.: Physical rehabilitation exercises assessment based on hidden semi-markov model by kinect v2. In: 2016 IEEE-EMBS International Conference on Biomedical and Health Informatics (BHI), pp. 256–259. IEEE, February 2016
16. Carfagni, M., Furferi, R., Governi, L., Servi, M., Uccheddu, F., Volpe, Y.: On the performance of the Intel SR300 depth camera: metrological and critical characterization. IEEE Sens. J. **17**(14), 4508–4519 (2017)
17. Carfagni, M., Furferi, R., Governi, L., Santarelli, C., Servi, M., Uccheddu, F., Volpe, Y.: Metrological and critical characterization of the Intel D415 stereo depth camera. Sensors **19** (3), 489 (2019)
18. https://nuitrack.com/. Accessed June 2019

19. Bruun-Olsen, V., Heiberg, K.E., Wahl, A.K., Mengshoel, A.M.: The immediate and long-term effects of a walking-skill program compared to usual physiotherapy care in patients who have undergone total knee arthroplasty (TKA): a randomized controlled trial. Disabil. Rehabil. **35**(23), 2008–2015 (2013)
20. Snell, L.D., Hipango, J., Sinnott, K.A., Dunn, J.A., Rothwell, A., Hsieh, C.J., DeJong, G., Hooper, G.: Rehabilitation after total joint replacement: a scoping study. Disabil. Rehabil. **40** (14), 1718–1731 (2018)
21. Naddeo, A., Cappetti, N., Vallone, M.: Assessment of ranges of rest postures of human lower limbs. Work **63**(2), 219–229 (2019). https://doi.org/10.3233/WOR-192916
22. Ferrari, A., Ginis, P., Hardegger, M., Casamassima, F., Rocchi, L., Chiari, L.: A mobile Kalman-filter based solution for the real-time estimation of spatio-temporal gait parameters. IEEE Trans. Neural Syst. Rehabil. Eng. **24**(7), 764–773 (2015)
23. Cao, Z., Simon, T., Wei, S.E., Sheikh, Y.: Realtime multi-person 2D pose estimation using part affinity fields. In: Proceedings of the IEEE Conference on Computer Vision and Pattern Recognition, pp. 7291–7299 (2017)

Design of a Customized Neck Orthosis for FDM Manufacturing with a New Sustainable Bio-composite

Rita Ambu[1(✉)], Alessandro Motta[2], and Michele Calì[3]

[1] Department of Mechanical, Chemical and Materials Engineering, University of Cagliari, via Marengo 2, 09123 Cagliari, Italy
ambu@unica.it
[2] Studio di Radiologia Motta S.r.l., Via Francesco Crispi 4, 95032 Belpasso, CT, Italy
alemotta.med@gmail.com
[3] Electric, Electronics and Computer Engineering Department, University of Catania, Viale A. Doria 6, 95125 Catania, Italy
mcali@diim.unict.it

Abstract. The interest in developing customized external orthopaedic devices, thanks to the advent of Additive Manufacturing (AM), has grown in recent years. Greater attention was focused on upper limb casts, while applications to other body's parts, such as the neck, were less investigated. In this paper the computer aided design (CAD) modelling, assessment and 3D printing with fused deposition modelling (FDM) of a customized neck orthosis are reported. The modelling, based on anatomic data of a volunteer subject, was aimed to obtain a lightweight, ventilated, hygienic and comfortable orthosis compared to the produced medical devices generally used for neck injuries. CAD models with different geometrical patterns, introduced for lightening and improving breathability, were considered, specifically, a honeycomb pattern and an elliptical holes pattern. These models were structurally assessed by means of finite elements analysis (FEA). Furthermore, an innovative composite material was considered for 3D printing. The material, Hemp Bio-Plastic® (HBP), composed by polylactic acid (PLA) and hemp shives, offers different advantages including lightweight, improved superficial finish and antibacterial properties. The results obtained in terms of design methodology and manufacturing by 3D printing of a prototype have shown the feasibility to develop customized cervical orthoses, with potentially improved performance with respect to cervical collars available on the market also thanks to the use of the innovative composite material.

Keywords: CAD · Neck orthosis · Additive Manufacturing · Bio-composite

1 Introduction

Additive Manufacturing (AM) techniques are increasingly considered as a viable manufacturing method for various applications. In fact, AM enables the production of complex geometries that were previously difficult or impossible to manufacture [1].

C. Rizzi et al. (Eds.): ADM 2019, LNME, pp. 707–718, 2020.
https://doi.org/10.1007/978-3-030-31154-4_60

In particular, the area relative to biomedical applications has some relevant potential benefits of using AM. These include tissue engineering bone scaffolds [2–4], preoperative planning 3D printed models [5, 6], patient specific instruments [7], and customized orthopaedic devices [8]. The latter involves the customized design and AM printing of external medical devices, or orthoses, used in the treatment of specific pathologies. An orthosis is generally designed to stabilize, immobilize, prevent deformity or protect against injury. The fabrication of a customized device can be, potentially, more advantageous than that of a prefabricated orthosis, both clinically and for the patient's comfort.

Recent research on personalized orthoses was primarily focused on upper limb casts [9–18] highlighting the feasibility of AM in this field. On the other hand, few investigations [19] have considered this manufacturing method as a viable option for the cervical region. This can be attributable to the complex functionality of a cervical orthosis which has to support, prevent immobilize or correct the spine. Moreover, the clinical aspects have to be combined with other factors such as the sensation of comfort and the aesthetic appearance.

In this paper, the design and manufacturing with fused deposition modelling (FDM) of a customized orthopaedic neck immobilization device are reported. The aim of the study is to evaluate the customized design and 3D printing of a neck orthosis highlighting the potential benefit of this methodology. Semi-rigid cervical collars are available on the market to protect the cervical spine after injuries and, postoperatively, to immobilize the spine following surgical reconstruction. A primary measure of their effectiveness is the capability to immobilize the spine. As known, some problems are usually associated with the use of these devices including muscle atrophy, skin rush or breakdown, pain, lack of hygiene, odor, heaviness and discomfort.

The design of the customized cervical orthosis was, therefore, conceived for an optimal performance by adopting a proper geometry for functionality, and, at the same time, reducing weight and improving breathability for a better comfort. Furthermore, the concept of sustainability, in terms of material, was introduced by considering for 3D printing an innovative composite material. The new composite is a 3D printing filament composed by polylactic acid (PLA) and hemp shives, patented by MICA s.r.l. (Ragusa-Italy) as Hemp Bio-Plastic® (HBP). The sustainable nature of the filament, its lightweight, improved mechanical performance [20] and superficial finish, and last but not least its antibacterial properties [21] were the main reasons for this choice.

The study was conducted starting from computer tomography (CT) scans of the anatomic area of the neck made available by a volunteer subject, that were used to reconstruct the part of interest. The next step consisted in the computer aided design (CAD) modelling of the customized cervical orthosis based on the reconstructed neck. Different geometrical patterns were introduced in the CAD model for lightening the structure and improving breathability. The models obtained were numerically assessed by means of Finite Elements Analysis (FEA) in order to globally evaluate the proposed geometrical configurations. Finally, a prototype of the designed cervical orthosis, made with the advanced composite, was manufactured with fused deposition modelling.

2 Design Procedure

The first step in the design procedure was relative to the acquisition and digital reconstruction of the neck (human neck). A digital representation of an outer part of the human body can be obtained by means of 3D photogrammetric or laser scanners. However, a 3D reconstruction can also be achieved from Computer Tomograpy (CT) scan acquisitions, if available. For this study, CT scan acquisitions (Fig. 1) were provided by a volunteer subject which underwent a CT Scan for personal issues.

These data were used to create the neck 3D model. Through the use of the Materialise's Interactive Medical Image Control System (MIMICS) software, the CT images (about 600 each analyzed case) were filtered to eliminate the parts that were not used (such as bones, internal organs, or clothes).

In particular, the "thresholding" filter was used for a first definition of the segmentation. The object was defined based on lower and upper thresholds. In the first case, the object of the segmentation had all the pixels in the images with a value which was greater than or equal to the threshold value; in the second case, the pixel value had to be comprised between both the threshold values to be part of the segmentation object.

The CT images are a pixel map of the linear attenuation coefficient of the X-rays in tissue. The pixel values were reduced so that the linear attenuation coefficient of the X-rays in air was equal to −1024 and that of water was equal to 0. This scale is called "Hounsfield scale" from the name of one of the tomography pioneers, Godfrey Hounsfield. The calculation of the polylines was made as the contours of the pixels in the axial view, so they were built automatically for the cavities inside the mask, as well.

In order to delete such cavities, considering the final interest in extrapolating the three-dimensional model only of the outer geometry of the neck, the "Cavity Fill From Polylines" filter was used, thus obtaining a new mask.

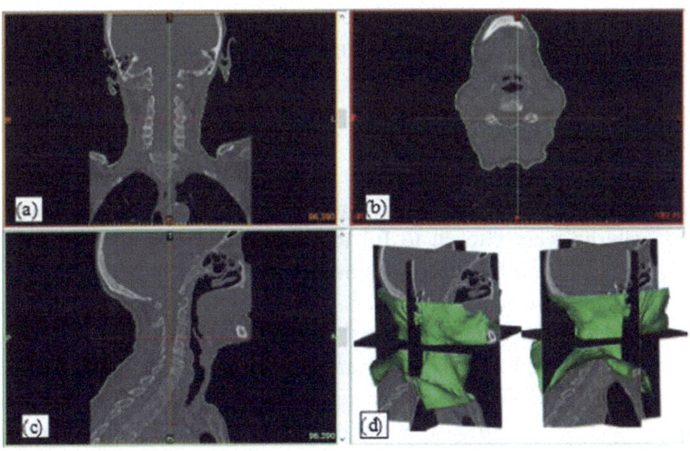

Fig. 1. CT Scans views: (a) frontal; (b) transversal; (c) lateral; (d) isometric.

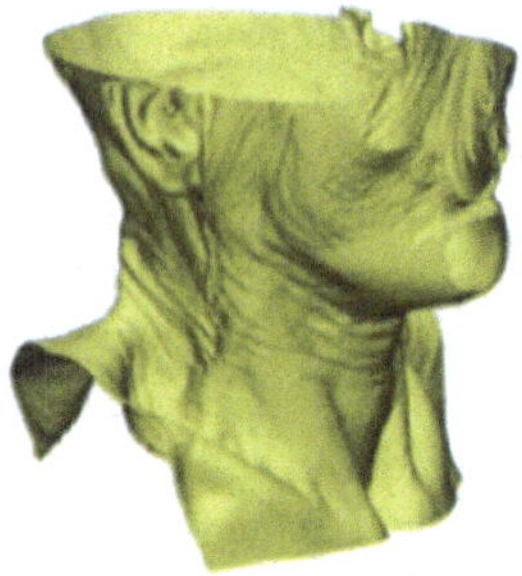

Fig. 2. 3D tessellated final surface.

Once the geometry was "cleaned", the part associated with the mask ("Calculate Part" operation) was calculated (Fig. 2). This part is a mesh of the neck in STL format.

This model was imported in a CAD environment and the automatically reconstructed surface was used to design the geometry of the neck orthosis.

The CAD model developed was intended to design a device with functionality similar to that of cervical semi-rigid collars available on the market. Generally, this kind of cervical collar is conceived as an orthosis that fits the neck, anywhere from the jaw to the chest. It is used to restrict movement and support the neck to allow healing when treating a cervical injury, fracture, or surgical procedure. On the other hand, soft cervical orthoses, which have a lighter geometrical design, help in the treatment of minor neck strains, support the neck and, control pain after an injury (e.g., whiplash).

For obtaining the model, subsequent sections of the neck were reconstructed, also introducing an offset with respect to the neck surface in order to allow the placement of a soft pad in the inner part of the orthosis to improve the comfort. Then, the surface obtained was thickened (4 mm) to obtain the ultimate architecture. The value of the thickness chosen was the average thickness usually adopted in prefabricated orthoses. A geometrical pattern was finally introduced in the obtained model.

In fact, as stated in the introduction, among the problems usually associated with the use of a cervical orthosis, there is the lack of breathability which can result in discomfort. Furthermore, since the orthosis needs be worn in a very visible part of the body, the aesthetic aspect is also relevant. Thus, the introduction of a pattern on the customized orthosis has several purposes, in addition to lightening the structure which is one of the main advantages allowed by the use of additive manufacturing.

Different patterns were chosen to fill the modelled structure. The first geometry considered was a honeycomb pattern with uniform cells. The cell size regularity allows to attain a more isotropic mechanical behaviour [22] with respect to a non-uniform pattern, such as a Voronoi pattern, frequently used for AM orthoses. A Voronoi pattern was also taken into account for this application; however, after a preliminary analytical analysis it was decided to discard this pattern since the relative model showed a low stiffness with respect to the stress that the neck orthosis has to withstand.

A zone devoid of the pattern was also provided to insert hinges for wearing purposes. The finally obtained honeycomb pattern orthosis, shown in Fig. 3, has a volume fraction of 0.79. This last parameter was chosen to globally characterize the geometrical pattern in terms of porosity, which can affect the structural performance of the orthosis.

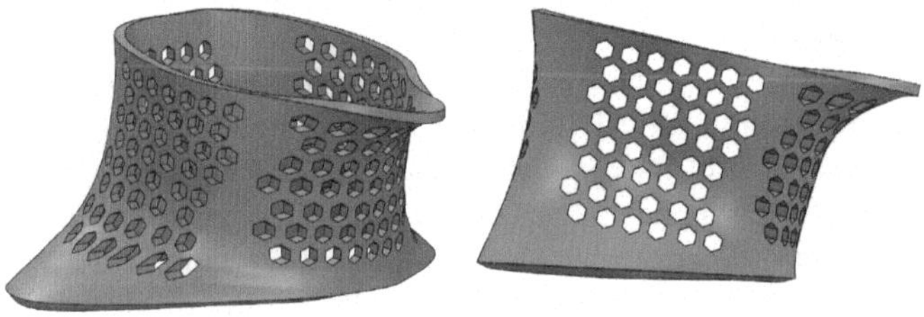

Fig. 3. CAD model of the honeycomb pattern cervical orthosis.

The other geometrical pattern considered, shown in Fig. 4, consisted of elliptical holes oriented at ±45°.

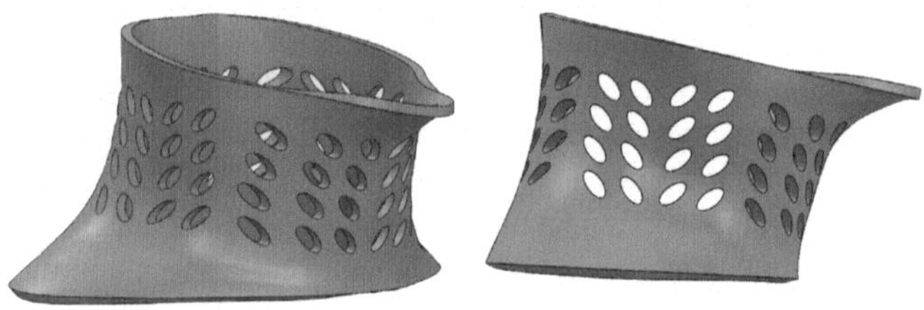

Fig. 4. CAD model of the elliptical holes pattern cervical orthosis.

This geometrical pattern, made of features with different orientations, provides orthotropic properties, keeping at the same time the porosity at a low value, being 0.87 the calculated volume fraction. Therefore, the investigation of this model, intrinsically different from the previous considered, is of interest in relation with the complex combination of forces acting on the neck orthosis.

3 Numerical Assessment of the Customized Cervical Orthoses

The assessment of the proposed models was done by means of finite elements analysis. The numerical simulations were used to obtain a global evaluation of the proposed architectures with the different patterns also comparing their performance in relation with some main parameters.

The composite material Hemp Bio-Plastic® (HBP), previously introduced, was chosen for the simulations. The mechanical properties were assumed as follows: elastic modulus 4420 MPa, flexural modulus 3833 MPa and Poisson's ratio 0.32. Each solid model was imported in a FE code and meshed by using eight nodes linear brick elements. Convergence testing was performed in order to minimize the influence of mesh density on the results. Finally, a model made of 1096142 elements was obtained for the honeycomb pattern while 1041519 were the elements of the model with the elliptical holes pattern.

A cervical orthosis is generally subjected to a combination of loads given the complex structure of the cervical spine and muscles of the neck. Investigations on cervical collars available on the market to quantify their capability of restricting the cervical spine motion during flexion, extension, axial rotation, and lateral bending were made using different experimental methods and also comparative analyses were reported in literature [23, 24]. FE numerical models of the cervical spine were also documented in literature mainly developed with the aim to study their injuries due to impact or similar [25, 26]. In this study a simplified analysis was done in order to assess the proposed architectures.

The models of the cervical orthoses were structurally analyzed under different loading conditions namely, extension, flexion and, lateral bending to simulate the action of the neck that the orthosis has to withstand.

The applied forces were derived from an experimental study [27] which reported the three-dimensional maximum moments measured during maximum voluntary contractions of neck muscles. In that study, measurements for both genders were reported; thus it was possible to extract the maximum moments applicable for this analysis. In particular, the lower limit values for women were considered, in relation with the characteristics of the volunteer subject under study, and, from these, the force to apply for each loading arrangement considered, was calculated.

The applied forces for each simulation and the boundary conditions are depicted in Fig. 5 relatively to the model with the honeycomb pattern. Analogous loads and boundary conditions were applied to the model with the elliptical holes pattern.

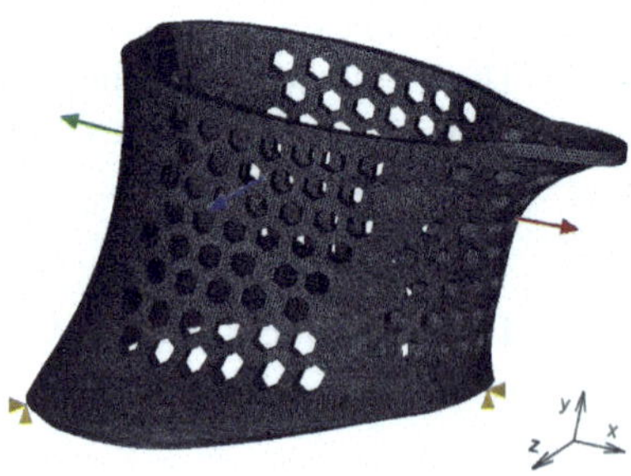

Fig. 5. FE model of the honeycomb cervical orthosis: loads and boundary conditions.

Figure 5 shows the approximated positions of the applied forces on the outer surface of the model for clarity of the presentation, but, indeed, they were applied on the inner surface. In particular, the green arrow in the figure is relative to the load considered for extension, the red arrow is relative to flexion, while the blue arrow is relative to lateral bending. The extension force (173.07 N) originates a bending moment in the plane x-y, as well as the flexion force (193.02 N), while the lateral bending force (153.84 N) acts on the plane y-z.

The first analyzed parameter was the total displacement resulting from each loading configuration considered for both models.

Figure 6 shows the isocolour representation of the total displacement, expressed in mm, obtained for each loading configuration, relatively to the model with the honeycomb pattern.

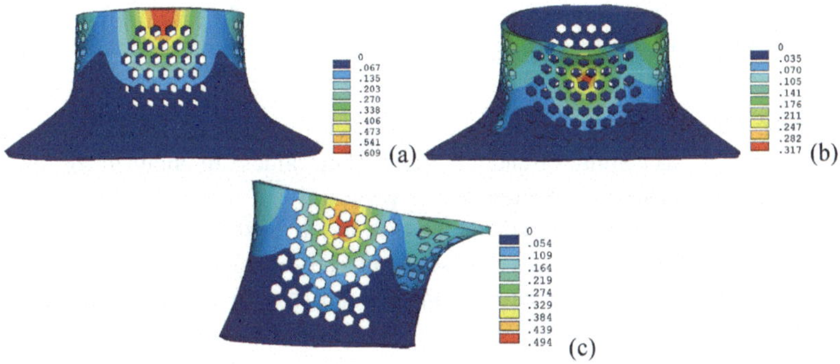

Fig. 6. Isocolour representation of the total displacement for the honeycomb pattern model and different loading configurations; (a) extension, (b) flexion, (c) lateral bending.

The figure shows that the displacements assume values that are quite limited, being under 1 mm, for the different loading conditions. A similar distribution was obtained for the other pattern considered, as can be observed from the isocolour representation shown in Fig. 7.

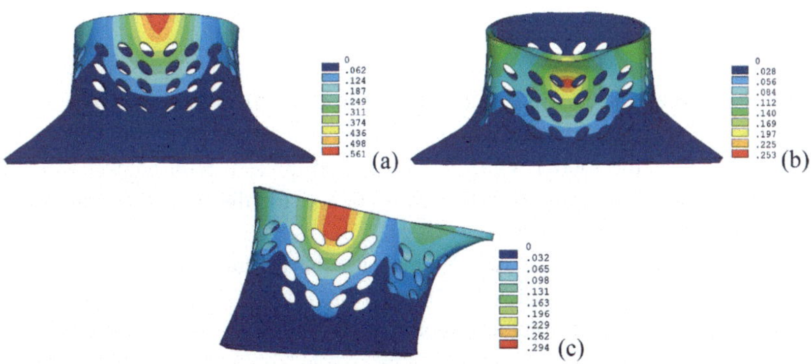

Fig. 7. Isocolour representation of the total displacement for the elliptical pattern model and different loading configurations; (a) extension, (b) flexion, (c) lateral bending.

The results obtained for the models with the different patterns can be compared in terms of maximum total displacement, as shown in the diagram of Fig. 8.

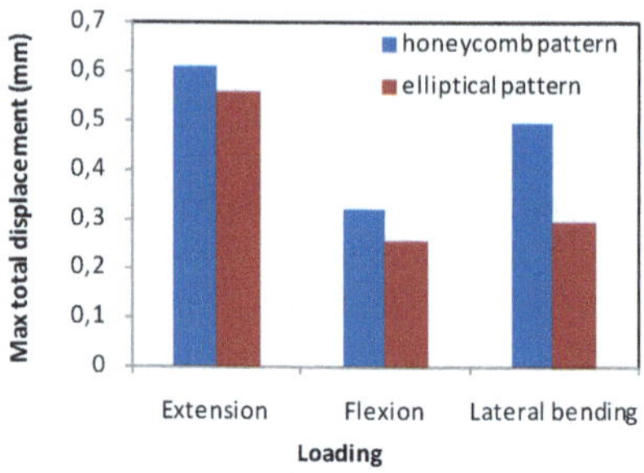

Fig. 8. Comparison between maximum total displacements of the two models obtained for different loading configurations.

From the figure it can be inferred that, for each model considered, the maximum total displacement has a similar trend and that higher values are obtained for the model of the cervical orthosis with the honeycomb pattern.

The other parameter considered for the evaluation of the two analyzed models is Von Mises stress. Figure 9 shows the isocolour representation of Von Mises stress, expressed in MPa, obtained for lateral bending and relative to the two different geometrical patterns considered.

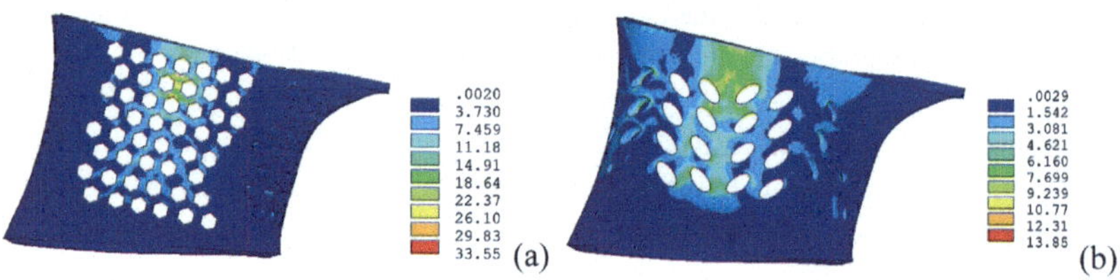

Fig. 9. Isocolour representation of Von Mises stress for lateral bending; (a) honeycomb pattern, (b) elliptical holes pattern.

From the figure it is possible to observe that Von Mises stress distribution has a similar trend for the two patterns considered. However, localized stress concentrations are more evident in the model with the honeycomb pattern with respect to the other pattern considered. Analogous results were obtained for the other loading configurations considered.

In this analysis, a parameter of relevance for the evaluation of the two patterned models is the maximum value of Von Mises stress. Figure 10 shows the values of this parameter obtained for each loading layout analyzed.

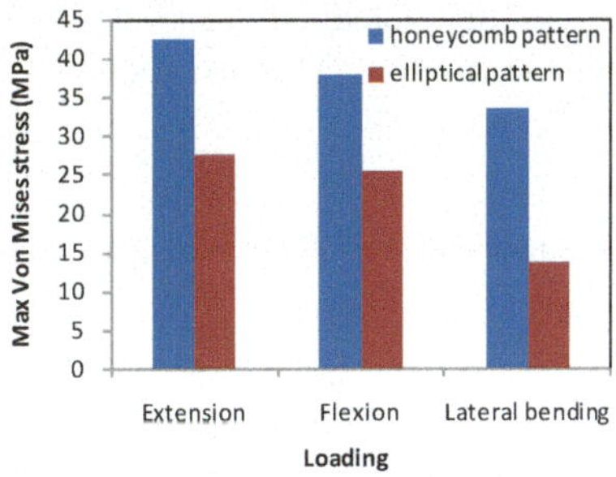

Fig. 10. Comparison between maximum Von Mises stress of the two models obtained for different loading configurations.

The diagram shows that, for each loading layout considered, higher values of the maximum Von Mises stress are obtained for the cervical orthosis model with the honeycomb pattern with respect to the elliptical pattern architecture.

However, the maximum stresses obtained for each loading configuration of the two models fall under the flexural maximum strength of HPB (69.7 MPa); accordingly, both patterns can be considered acceptable from a structural point of view.

4 Additive Manufacturing of a Prototype Cervical Orthosis

As a final step of this study, a prototype cervical orthosis was 3D printed. Between the two analyzed models, it was preferred to manufacture the model with the honeycomb pattern, even if it has shown a slightly global lower performance with respect to the other model assessed, since it has a lower volume fraction and this can be advantageous both for weight saving and for improved breathability.

The filament used for this case study is the HBP®, acronym standing for Hemp-bio plastic, previously introduced. The authors believe that, for its characteristic, this material could be used to create many external medical devices, besides the application proposed. In fact, it has shown many advantages, first of all in terms of mechanical properties, with a 30% of improved resistance compared to standard PLA alone, as stated by the producer [28]. This makes HBP® a good competitor with a largely used, oil-based thermoplastic, that is ABS (Acrylonitrile butadiene styrene). Furthermore, hemp shives allow to obtain lightweight composites; in particular, HBP®, thanks to its composition with hemp shives, shows an average of 20% less weight rather than standard ABS filaments [20]. The unique composition with hemp shives provide the HBP® antibacterial properties [21], a feature always preferred in medicine, hence the smooth and natural superficial finish of the printed part (allowing manual minor finish for imperfections) generate an hygienic surface. The material tested has not been proven with dermal contact standards; anyhow it is being tested for many ISO standards (including skin contact) and has already obtained food contact approval.

The prototype model was manufactured using the Printer D300 Technology® which uses the FDM process for printing. The characteristics of the printer are reported in Table 1.

Table 1. Characteristics of 3D printer D300 technology®.

Characteristic	Value
Frame	Delta
Print area:	Ø 250 × 300 mm
Max. bed temperature	100 °C
Nozzle size	0.6 mm
Max. nozzle temperature	260 °C
Max. Z-axis resolution	0.1 mm
X, Y, Z-axes accuracy	0.05 mm
Printer size	380 mm × 680 mm
Power input	AC 110–220 V AC, 50/60 Hz

Figure 11 shows the 3D printer and the prototype during an intermediate stage of manufacturing, whereas Fig. 12 reports the final prototype of the neck orthosis.

Fig. 11. An intermediate stage of the printing process.

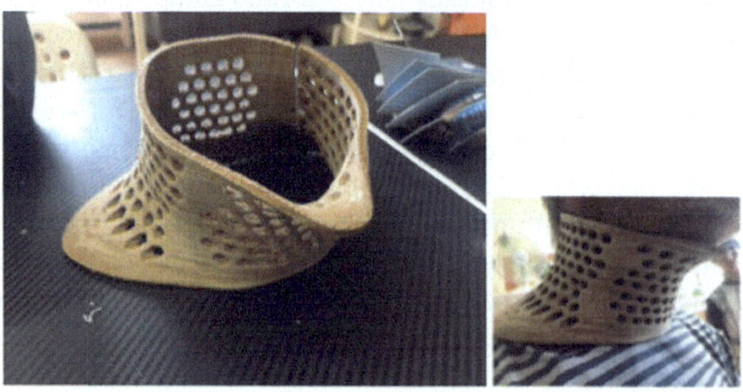

Fig. 12. Prototype of the neck orthosis.

5 Conclusions

In this paper the design, assessment and manufacturing with fused deposition modelling of a customized neck orthosis are reported. The procedure of reconstruction of the neck and subsequent modelling allowed to obtain an orthosis which potentially should accomplish similar functionality to that of semi-rigid collars available on the market but with an improved comfort. In fact, the device is customized, lighter and more breathable thanks to the geometrical pattern introduced in the structure. FE numerical analysis was applied to assess two intrinsically different geometrical patterns of the modelled orthosis and the differences in terms of main structural parameters of interest for this application were highlighted. Reduced stress concentrations were also observed in the model with the elliptical holes pattern with respect the other pattern considered. Finally, the manufacturability of a prototype neck orthosis with an innovative composite material was successfully verified. The use of a bio-based material can have additional advantages including satisfying the aesthetic demands thanks to its wooden appearance. Future research will be concerned with the improvement of the procedure by automating the steps for the development of the CAD model, in order to implement the AM neck orthoses on clinical trials in orthopaedic treatments.

Acknowledgements. This research was supported in part by European Union, European Social Fund, and the University of Catania research programme 2019/2021.

References

1. Gao, W., Zhang, Y., Ramanujan, D., Ramani, K., Chen, Y., Williams, C.B., Wang, C.C.L., Shin, Y.C., Zhang, S., Zavattieri, P.D.: The status, challenges, and future of additive manufacturing in engineering. Comput.-Aided Des. **69**, 65–89 (2015)
2. Ahmadi, S.M., Yavari, S.A., Wauthle, R., Pouran, B., Schrooten, J., Weinans, H., Zadpoor, A.A.: Additively manufactured open-cells porous biomaterials made from six different space-filling unit cells: the mechanical and morphological properties. Materials **8**, 1871–1896 (2015)
3. Ambu, R., Morabito, A.E.: Porous scaffold design based on minimal surfaces: development and assessment of variable architectures. Symmetry **10**(9), 361 (2018)
4. Afshar, M., Pourkamali Anaraki, A., Montazerian, H., Kadkhodapour, J.: Additive manufacturing and mechanical characterization of graded porosity scaffolds designed based on triply periodic minimal surface architectures. J. Mech. Behav. Biomed. Mater. **62**, 481–494 (2016)
5. Ganguli, A., Pagan-Diaz, G.J., Grant, L., Cvetkovic, C., Bramlet, M., Vozenilek, J., Kesavadas, T., Bashir, R.: 3D printing for preoperative planning and surgical training: a review. Biomed. Microdevice **20**, 65 (2018)
6. Speranza, D., Citro, D., Padula, F., Motyl, B., Marcolin, F., Calì, M., Martorelli, M.: Additive manufacturing techniques for the reconstruction of 3D fetal faces. Appl. Bionics Biomech. **2017**, 10 pages (2017). Article ID 9701762
7. Wong, K.C., Kumta, S.M., Sze, K.Y., Wong, C.M.: Use of a patient specific CAD/CAM surgical jig in extremity bone tumor resection and custom prosthetic reconstruction. Comput. Aided Surg. **17**(6), 284–293 (2012)

8. Pallari, J.H.P., Dalgarno, K.W., Woodburn, J.: Mass customization of foot orthoses for rheumatoid arthritis using selective laser sintering. IEEE Trans. Biomed. Eng. **57**(7), 1750–1756 (2010)
9. Lin, H., Shi, L., Wang, D.: A rapid and intelligent designing technique for patient-specific and 3D-printed orthopedic cast. 3D Print. Med. **2**(4), 1–10 (2016)
10. Buonamici, F., Furferi, R., Governi, L., Lazzeri, S., McGreevy, K.S., Servi, M., Talanti, E., Uccheddu, F., Volpe, Y.: A practical methodology for computer-aided design of custom 3D printable casts for wrist fractures. Vis. Comput., 1–16 (2019). https://link.springer.com/article/10.1007/s00371-018-01624-z
11. Kim, H., Jeong, S.: Case study: hybrid model for the customized wrist orthosis using 3D printing. J. Mech. Sci. Technol. **29**(12), 5151–5156 (2015)
12. Fitzpatrick A.P., Mohanned, M.I., Collins P.K., Gibson, I.: Design of a patient specific, 3D printed arm cast. In: The International Conference on Design and Technology, KEG, pp. 135–142. DesTech Conference Proceedings (2017)
13. Kelly, S., Paterson, A., Bibb, R.J.: A review of wrist splint designs for additive manufacture. In: Proceedings of 2015 14th Rapid design, Prototyping and Manufacture Conference, Loughborough, Great Britain, pp. 1–12 (2015)
14. Cortex: Exoskeletal Cast.http://www.evilldesign.com/cortex. Accessed 15 June 2019
15. A' Design Award & Competition, https://competition.adesignaward.com/design.phpID=34151. Accessed 18 June 2019
16. Zhang, X., Fang, G., Dai, C., Verlinden, J., Wu, J., Whiting, E., Wang, C.C.L.: Thermal-comfort design of personalized casts. In: UIST 2012 Proceedings, Quebec City, Canada, pp. 243–254 (2017)
17. Buonamici, F., Furferi, R., Governi, L., Lazzeri, S., McGreevy, K.S., Servi, M., Talanti, E., Uccheddu, F., Volpe, Y.: A CAD-based procedure for designing 3D printable arm-wrist-hand cast. Comput.-Aided Des. Appl. **16**(1), 25–34 (2019)
18. Blaya, F., San Pedro, P., López Silva, J., D'Amato, R., Soriano Heras, E., Juanes, J.A.: Design of an orthopedic product by using additive manufacturing technology: the arm splint. J. Med. Syst. **42**(54), 1–14 (2018)
19. Prates, A.: Self-adjusting orthoses design. Scripta-Ingenia **3**, 3–8 (2014)
20. Stevulova, N., Kidalova, L., Junak, J., Cigasova, J., Terpakova, E.: Effect of hemp shive sizes on mechanical properties of lightweight fibrous composites. Procedia Eng. **42**, 496–500 (2012)
21. Khan, A.K., Warner, P., Wang, H.: Antibacterial properties of hemp and other natural fibre plants: a review. BioResources **9**(2), 1–18 (2014)
22. Chen, C., Lu, T.J., Fleck, N.A.: Effect of imperfections on the yielding of two-dimensional foams. J. Mech. Phys. Solids **47**(11), 2235–2272 (1999)
23. Gavin, T.M., Carandang, G., Havey, R., Flanagan, P., Ghanayem, A., Patwardhan, A.G.: J. Rehabil. Res. Dev. **40**(6), 527–538 (2003)
24. Worsley, P.R., Stanger, N.D., Horrell, A.K., Bader, D.L.: Investigating the effects of cervical collar design and fit on the biomechanical and biomarker reaction at the skin. Med. Devices: Evid. Res. **11**, 87–94 (2018)
25. Toosizadeh, N., Haghpanahi, M.: Generating a finite element model of the cervical spine: estimating muscle forces and internal loads. Scientia Iranica B **18**(6), 1237–1245 (2011)
26. Osth, J., Mendoza-Vazquez, M., Sato, F., Swensson, M.Y.: A female head-neck model for rear impact simulations. J. Biomech. **51**, 49–56 (2017)
27. Vasavada, A.N., Li, S., Delp, S.L.: Three-dimensional isometric strength of neck muscles in human. Spine **26**(17), 1904–1909 (2001)
28. Kanesis. https://www.kanesis.it/products/#filaments. Accessed 26 June 2019

A Multi-layer Approach for the Identification and Evaluation of Collaborative Robotic Workplaces Within Industrial Production Plants

Ferdinando Vitolo[1], Agnese Pasquariello[1], Stanislao Patalano[1(✉)], and Salvatore Gerbino[2]

[1] Department of Industrial Engineering, University of Naples Federico II, Naples, Italy
stanislao.patalano@unina.it
[2] Department of Engineering, University of Campania Luigi Vanvitelli, Aversa, Italy

Abstract. Collaborative robotic solutions, where humans and robots share a common workspace performing tasks concurrently without physical safety barriers dividing them, are entering the 4.0 manufacturing market. Some proven and tested use cases of Human-Robot Collaboration have been implemented, but their identification process is often just based on the intuition of planning engineers. The purpose of this work is to propose a systematic approach for the identification of potential collaborative workstations within an industrial production plant. In order to do this a multi-layer modelling approach was used and enriched. The multi-layer approach defines the overall goal of the industrial process, the sub-processes that made it possible, the activity models that enables a flow of activities and, finally, a set of methods to carry out the activities. A morphological box of methods that can be used to achieve the specific goal of identifying suitable collaborative workplaces in an industrial plant, through a process of HRC potential analysis is, therefore, ready to be deeply investigated and used.

Keywords: Collaborative workplaces · Multi-layer approach · Evaluation criteria

1 Introduction

Manufacturing plants characterized by multi-stage production-line and based on medium batch are placed in a transition band between low and high batch production [1]. Low batch productions are usually manufactured by humans (low automation) which are characterized by high flexibility and long production time. In opposite, automated plants are characterized by low/poor flexibility and short production time; moreover, they require massive batches for cost affordable. Medium batch production needs high flexibility and short production time in order to be market leader; therefore, production line should implement the best human and robot skills. The collaboration

C. Rizzi et al. (Eds.): ADM 2019, LNME, pp. 719–730, 2020.
https://doi.org/10.1007/978-3-030-31154-4_61

between humans and robots seems to be the key approach for manufacturing improvement towards high flexibility and short production time [2].

The Human-Robot Collaboration (HRC) research field arises in the context of a new industrial automation trend, the so-called Industry 4.0 [3]. The main purpose of the collaboration between humans and robots is combining the advantages of robots which enjoy high levels of accuracy, speed and repeatability with the flexibility and cognitive skills of human workers. Nevertheless, safety must be assured when humans and robots share the same workspace at the same time. Usually, safety is guaranteed by advanced sensors [4]. Some of the many recommended advantages of Human Robot Collaboration include greater process efficiency, especially regarding quality and flexibility and improved ergonomics for human operators [5]. The benefits of HRC are not well-known; thus, many companies are not sure that HRC could actually improve their performance. There is also a fundamental fear in adopting solutions that do not include physical safety barriers and that require compliance with stringent standards and the use of advanced sensors. Furthermore, expertise and specialists related to collaborative systems still lack. Some proven and tested use cases of Human-Robot Collaboration have been implemented, but their identification process is often just based on an experience analysis and the intuition of planning engineers. Whereas the already provided ISO standards [6–8] are indispensable for the proper design of a new or existing collaborative workplace, the identification process of suitable workstations for the HRC has not yet been settled by standards or common guidelines. The purpose of this paper is to propose a systematic approach for the selection and the implementation of potential collaborative workstations within an industrial production plant. Although some practical methods will be presented, the objective of this work was not to directly provide a solution for identifying suitable workstations for HRC; on the contrary, the objective was to develop an overall approach for the modelling of a business process. The process was seen as a system made of a set of elementary activities carried out to achieve a certain objective, and then its instantiation consists into the specific case of analyzing the collaborative potential of existing workstations in a manufacturing company. Following the brief introduction to the problem of the current Sect. 1, an overview over the current state of the art related to collaborative workplaces is given with a focus on the references related to the problem of identifying the collaborative potential of a manufacturing system. Then, Sect. 3 presents the multi-layer approach for process modelling and its implementation. Finally, the Sect. 4 summarizes the proposed approach and highlights the overcoming of the current research state and the future developments of the work.

2 State of the Art

Although there is plenty of researches in the design field of collaborative robotic workplaces, in-depth analysis on the problem of identifying suitable workplaces for HRC in a manufacturing system it's not nearly as direct to find. Delang et al. [9] presents an approach to identify the individual best suiting workplace for an HRI scenario considering the individual requirements of the manufacturing companies, through an in-depth interview, and highlighting connections between them and the typical collaborative workstations, providing various benefits with a matrix diagram. The same authors identified, later, the previous approach as one of several ways to achieve the goal of the implementation of collaborative workstation. The proposed multi-layer approach [10] for service modelling ensures a manufacturing company identifying potential workstations with the most suited benefits for the individual requirements, choosing the most suitable methodology to analyze its production line. Other studies cope with the implementation of a quantifiable methodology for the task planning between human and robot. Schröter et al. [11] adopt capability indicators to identify which resource is suited most for a certain assembly operation task and then, using the deviation of the capability indicators of neighboring assembly tasks, it's possible to identify workstations yielding the highest potential benefits when automated as a hybrid workstation. Teiwes et al. [12] propose instead the use of MTM (Methods Time Measurements) analysis to assign an automation potential to different type of standardized movements in a generic workplace. An overall score is calculated for each real movement in the examined assembly line depending on the previous established automation potential and other parameters. A significant innovation is that proposed by Zanella et al. in [13] where the identification of HRC workplaces is based on the preliminary assignment of values to multiple Key Performance Indicators (KPIs). KPIs allows to filter the analysed workplaces according to a proper performance analysis in order to select only those cells which can take higher advantage by the application of HRC. A well-structured method to distinguish potential workplaces for HRC is proposed by Blankemeyer [14]. The proposed method firstly filters the set of existing workstations using enablers or inhibitors parameters for HRC and then focuses on the task allocation as in [11]. The overcoming is given by the possibility to establish also the degree of human-robot interaction (coexistence/cooperation/collaboration). In Table 1 advantages and limitations of the selected research works are given in order to collocate the proposed multi-layer approach in a well-defined scenario.

3 The Proposed Multi-layer Approach for Process Modelling

Taking a cue from the idea of proposing firstly a general modelling approach as in [10] but having a series of definitions belonging to several modelling domains as a basis, a multi-layer approach to process modelling is hereby considered (Fig. 1). The same four levels of modelling, as in [10], are still considered but with differences in their characterisations. In the present approach the four levels are: M3 (objectives), M2 (process), M1 (model), M0 (methods). A process could be defined as a set of elementary activities carried out to achieve a given objective [15]. At the M3 level of the present approach, the objective to be achieved and the boundary conditions are established, as a result of the process development. At the M2 level, a process meta-model establishes the architecture of the process by defining its possible sub-processes and respective sub-objectives. Defining inputs and outputs for each sub-process, their temporal and logical connections are highlighted. At the M1 level, the activity models, to be performed to achieve the sub-objectives of each sub-process, have to be identified. Each sub-process can therefore be modelled with an activity model [16]. Through the connections between sub-processes highlighted in the previous level, sub-models can

Table 1. Comparison among research works that addresses the topic of identifying suitable workstations for an implementation of HRC at the industrial plant.

	Advantages	Limitations
Delang et al. 2017	The identification of the possible collaborative workplace is based on the individual requirements and motivations of each company compared with the main benefits of HRI	The approach is mainly qualitative and does not consider any process parameters for the workplace identification
Delang et al. 2018	The HRI implementation is considered as a service and a general approach for service modeling is used for the scope. The multi-layer approach ensures a manufacturing company choosing the most suitable methodology to analyse its production line since the approach's output is a box of possible method	The idea of modelling a service, as it normally happens for systems, is to be appreciated but should be followed by means of a solid formalization, which instead is lacking or maybe is not well explained
Schröter et al. 2016	A well-defined quantitative analysis is made for task allocation between human and robots	This approach cannot be applied as a first step to identify potential collaborative workplaces while a first filtering of existing workplaces will be necessary. A distinction between all the form of interaction does not exist, therefore the true collaboration is not treated
Teiwes et al. 2016	The use of "categories" of MTM allows the estimation of an overall HRC potential of a production line based on different type of movements that are usually carried out in a manufacturing plant	It seems that the collaboration is perceived simply as the possibility to collocate a robot in a completely manual assembly line; technical parameters are not considered for the evaluation
Blankemeyer et al. 2018	The potential check considers both a first filtering of workplaces basing on enablers and inhibitors for HRC and a quantitative task allocation between human and robots and reference to various level of interaction	The research work does not show in detail how data can be collected in order to apply the two-stage methodology
Zanella et al. 2017	A strong connection between the identification of suitable collaborative workplaces and performance parameters is important for a not-time consuming analysis before the task allocation	The method applied is focused rather on the Criteria and KPs used for the evaluation than on the absolute value assigned to a specific KP or to the weights applied to specific set of KPs

be connected so that all activities will lead to the overall objective achievement. Basically, more sub-models could define an overall model of the process.

The way in which things could be carried out in order to achieve a sub-process goal are different, thus different models can be proposed for each sub-process. Finally, at the M0 level, for each sub-model, the actual methods used to carry out the activities formalised at the previous level are instantiated, i.e. by means of the application of recognised methods. Once again, in order to achieve a given sequence of activities, different methods could be used according to the level of details that one wishes to adopt yet. A pictogram of the approach is shown in Fig. 1.

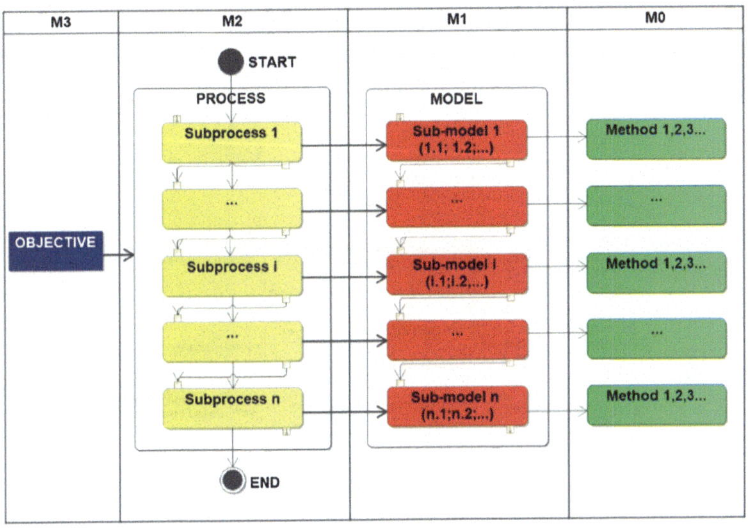

Fig. 1. Multi-layer approach framework.

3.1 The Characterisation of the Multi-layer Approach

M3-OBJECTIVE

To apply the described approach to the process of HRC potential analysis, first of all, it is necessary to define the objective that is to "introduce a collaborative workstation within a manufacturing system". Boundary conditions can be the cost limits that the company does not want to overcome and the benefits it would like to obtain (see Table 2).

Table 2. Definition of the objective and the boundary conditions in layer M3.

Objective	Introduce a collaborative workstation within a manufacturing system
HRC potential	
Costs	XXX €
Benefits	Improve ergonomics Improve accuracy of processes ...

M2-METAMODEL

To achieve the goal set in M3 without excessive planning efforts, five sub-processes, with an increasing level of details, on the manufacturing system have been considered (see Fig. 2). The *overall manufacturing system analysis* has the aim to determine whether there are the pre-requisites for a hybrid system.

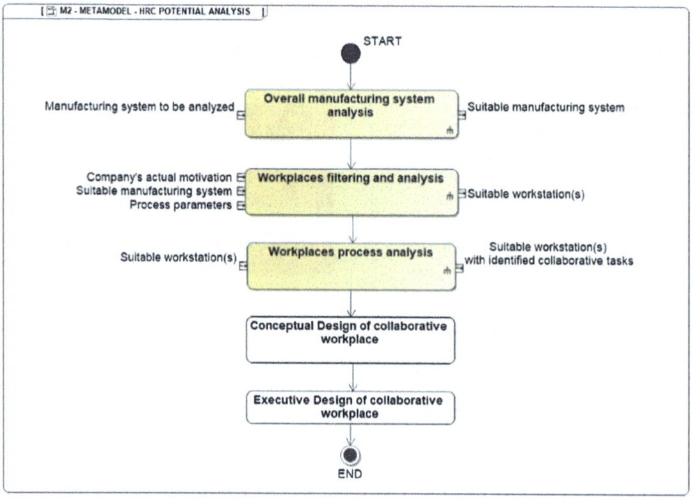

Fig. 2. Definition of the metamodel in layer M2.

The second sub-process investigates the HRC potential by *workplaces filtering and analysis* to initially recognize the workplace with the highest potential for the HRC application. The *workplace process analysis* focuses on individual activities performed at the selected workstations, their possibilities for automation and the characterization of tasks that can be carry out in collaboration with humans thus improving given requirements. The other two sub-processes, related to conceptual and executive design of the selected workplace, are not deeply investigated in the present work.

M1-MODEL

Once the sub-processes have been defined, it is necessary to identify the activities that could be carried out to reach the sub-objectives. A way to accomplish this goal is to adopt the activity models.

Sub-model 1. Overall Manufacturing System Analysis
According to the developed sub-model (see Fig. 3), the manufacturing system should be evaluated by using certain criteria. The activity of defining criteria will be always present in the activity diagram as a "process- independent step" because in order to evaluate something, first of all, it is necessary to define benchmarks. Once the evaluation criteria have been defined, the manufacturing system will be evaluated according to these criteria and if its characteristics correspond to those of a hybrid manufacturing system, it will be used to carry out a more in-depth analysis; otherwise it will be necessary to renounce the opportunity of an HRC in that specific manufacturing system.

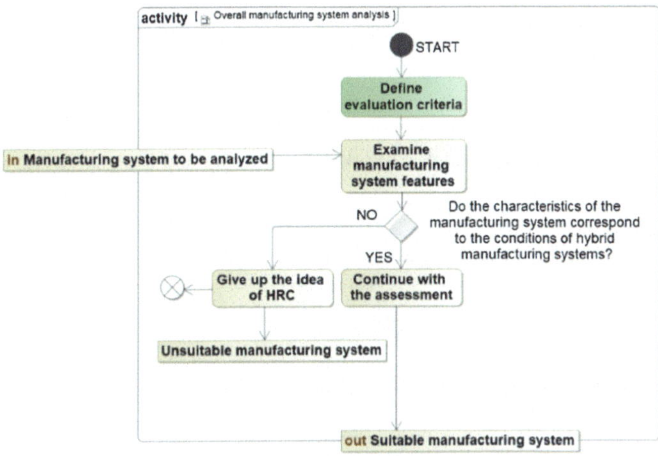

Fig. 3. Activity-model 1, overall manufacturing system analysis (M1).

Sub-model 2. Workplaces Filtering and Analysis
In order to select the workplace with the highest potential for HRC, workstations will be analysed with reference to the advantages that the HRC could provide for them. In fact, there could be process parameters, properties or boundary conditions of the workstations that could restrict the application of HRC while other characteristics could be improved using a collaborative robot. Once the criteria have been defined, the next step is the selection of the station to be analysed, the collection of data based on the previous criteria and the iteration of these two steps for all existing stations. If there are no more workstations to be analysed, the last step to be performed is the choice of the most suitable workstation for human-robot collaboration. The second activity model is shown in Fig. 4.

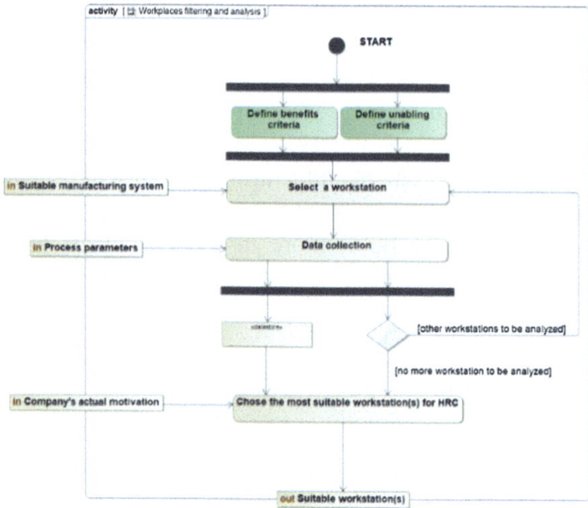

Fig. 4. Activity-model 2, Workplaces filtering and analysis (M1).

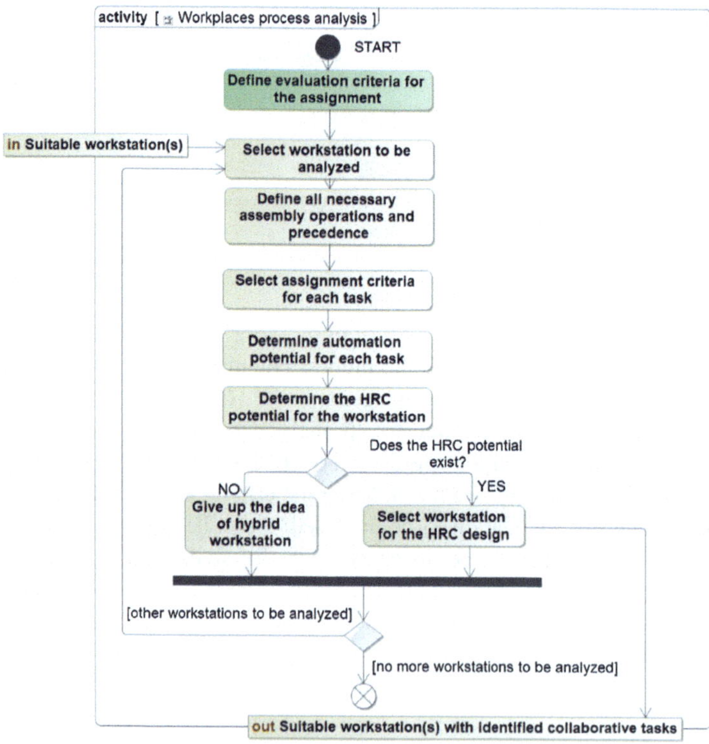

Fig. 5. Activity-model 3, Workplaces process analysis (M1).

Sub-model 3. Workplace Process Analysis
The workplace process analysis (see Fig. 5) starts with the definition of evaluation criteria for the task assignment. Then, it is necessary to define the sequence of the assembly tasks for the selected potential workstation and decide for each task if it is better to be carried out by humans or by robot, according to the selected criteria. Following this stage, the interaction points between humans and robots among the assigned tasks and the level of collaboration must be evaluated. These steps are repeated for each selected workstation.

M0-METHODS
Table 3 summarizes the identified methods which can be adopted for the identification of potential collaborative workstations. As prescribed by the models, for each step it becomes necessary to establish evaluation criteria. Those chosen are summarized in Table 4. Methods diverge in terms of level details (qualitative/quantitative methods), a necessary effort to carry out the results and the type of information requested.

Table 3. Methods summary in layer M0.

SUB-MODEL	Methods			
SUB-MODEL 1 Overall manufacturing system analysis	*Interviews*	*Studying documentation*	*Direct observation*	*Focus group*
SOURCE	Data gathering techniques [18]			
SUB-MODEL 2 Workplaces filtering and analysis	*Interviews Questionnaire Studying documentation Direct Observation Focus Group*	*Checklist*	*Matrix Diagram*	*Simple Multi-Attribute Rating Technique (SMART)*
SOURCE	Literature, [14] Data gathering techniques [18]	Literature, [19] Performance appraisal technique	Literature, [14]	Literature, [13] Multi Attribute Decision Ana lysis [20]
SUB-MODEL 3 Workplaces process analysis	*Calculation of capability indicator*	*MTM- US analysis with point scoring*	*Analytic Hierarchical Process (AHP)*	*Weighted Pro/Con*
SOURCE	Literature, [11, 14]	Literature, [12]	Multi Attribute Decision Analysis [20] Literature [21, 22]	Pro/Con Techniques for decision making [23]

Table 4. Selected evaluation criteria for each process step.

Overall manufacturing system analysis	Workplaces filtering and analysis	Workplaces process analysis
• Actual level of automation • Product volume • Desire to increment product volume • Desire to reduce unit cost • Product variety • Desire to introduce a product variant (flexibility) • Batches size • Product structure • Degree of customization • Desire to increment product customization	OPERATORS' POINT OF VIEW • Handheld weight (kg) • Lifting task's frequency/duration • Illness Rate • Overall operator satisfaction • Used ergonomic support • Need for human presence • Acceptance of HRC system • Repetitive motions • Uncomfortable positions • Vibrations • Ergonomic PRODUCTION POINT OF VIEW • Defect Rate • Process Time • Throughput • Downtime • Rework TECHNICAL POINT OF VIEW • Workstations size • Level of automation • Number of variants • Part supplying process • Preferences of the company	COBOTS' BENEFITS • Repeatability • Precise and accurate • High speed • Tireless • Can apply high force • Handling of heavy parts • Constant quality • Acceptable for unreasonable and monotonous tasks • Endurance COBOTS' LIMITS • Little flexibility • Handling complex and soft components • High investment and engineering costs • Reprogramming requires resources and expert knowledge • Requires huge resources for safety measures • Usually complex to use reasonable tasks • Limited ability to perceive its surroundings HUMANS' BENEFITS • problem solving, cognitive and soft skills • Ability of sense and locate things • Ability of adapting to unforeseen events • Flexibility and versatility • Immediate vision • Can handle complex parts HUMANS' LIMITS • Limited quality and process control • High running costs • Long-time process • Fatigue • Physical and mental performance limits

4 Conclusions and Future Work

The present paper illustrates a structured multi-layer approach for process modelling and its application to the process of HRC potential analysis in a manufacturing system. The multi-layer approach allows to define, respectively, the overall goal of a process at the M3 (objectives) level, the sub-processes that made it possible to achieve the defined goal at the M2 (process) level, the activity models at the M1 (model) levels that enabled a flow of activities to reach the sub-objective of the sub-processes and, finally, operative methods to carry out the activities at the M0 (methods) level. A preliminary morphological box of methods that can be used to identify a collaborative robotic workplace in a manufacturing system is, therefore, gained through the present work. Therefore, the contribute of the present work to the actual research state can be summarized in the identification of methods placed in the flow of the methodology and the possibility to constantly update those methods but also models and a meta-model in order to consolidate and enhance the approach. Future developments will address, in fact, the testing of selected methods to transform the approach in a robust methodology for HRC potential analysis. In particular, the proposed approach will be applied to two existing and different assembling lines in order to: (i) define the combination of methods to be used within SUB-MODEL1; (ii) set SUB-MODEL2 and SUB-MODEL3 methods to assembling lines and accomplish a comparison between them.

Acknowledgements. The present work was developed with the economic support of MIUR (Italian Ministry of University and Research) performing the activities of the project ARS01_00861 "Integrated collaborative systems for smart factory - ICOSAF".

References

1. Franciosa, P., Palit, A., Vitolo, F., Ceglarek, D.: Rapid response diagnosis of multi-stage assembly process with compliant non-ideal parts using self-evolving measurement system. Procedia CIRP **60**, 38–43 (2017)
2. Di Marino, C., Rega, A., Vitolo, F., Patalano, S., Lanzotti, A.: The anthropometric basis for the designing of collaborative workplaces. In: 2019 IEEE International Workshop on Metrology for Industry 4.0 and IoT, Naples (2019)
3. Gerbert, P. Lorenz, M., Rüßmann, M., Waldner, M., Justus, J., Engel P., Harnisch, M.: www.bcg.com. Accessed 09 Aprile 2015
4. Villani, V., Pini, F., Leali, F., Secchi, C.: Survey on human–robot collaboration in industrial settings: safety, intuitive interfaces and applications. Mechatronics **55**, 248–266 (2018)
5. Wilhelm, B., Manfred, B., Braun, M., Peter, R., Oliver, S.: Lightweight robots in manual assembly – best to start simply! Examining companies' initial experiences with lightweight robots. Fraunhofer-Gesellschaft (2016)
6. EN ISO 10218-1:2011, Robots and robotic devices – Safety requirements for industrial robots, Part1: Robots (2011)
7. EN ISO 10218-2:2011, Robots and robotic devices – Safety requirements for industrial robots, Part2: Robot systems and integration (2011)
8. ISO/TS 15066:2016, Robots and robotic devices – Collaborative robots (2016)

9. Delang, K., Bdiwi, M., Harsch, A.K., Putz, M.: Evaluation and selection of workstations for an application of human-robot-interaction in manufacturing. In: International Conference on Intelligent Robots and Systems; Workshop: Human-Robot Interaction in Collaborative Manufacturing Environments, Vancouver (2017)
10. Delang, K., Todtermuschke, M., Bdiwi, M., Putz, M.: An approach of service modeling for the demand-driven implementation of human-robot-interaction in manufacturing. In: 14th International Conference on Automation Science and Engineering (CASE), Munich, Germany (2018)
11. Schröter, D., Jaschewski, P., Kuhrke, B., Verl, A.: Methodology to identify applications for collaborative robots in powertrain assembly. Procedia CIRP **55**, 12–17 (2016)
12. Teiwes, J., Bänziger, T., Kunz, A., Wegener, K.: Identifying the potential of human-robot collaboration in automotive assembly lines using a standardised work description. In: 22nd International Conference on Automation and Computing (ICAC), Colchester (2016)
13. Zanella, A., Cisi, A., Costantino, M., Di Pardo, G., Pasquettaz, G., Giulio, V.: Criteria definition for the identification of HRC use cases in automotive manufacturing, In: 27th International Conference on Flexible Automation and Intelligent Manufacturing, FAIM2017, Modena, Italy (2017)
14. Blankemeyer, S., Recker, T., Stuke, T., Brokmann, J., Geese, M., Reiniger, M., Pischke, D., Oubari, A., Raatz, A.: A method to distinguish potential workplaces for human-robot collaboration. Procedia CIRP **76**, 171–176 (2018)
15. Rolland, C.: Modeling the requirements engineering process (1994)
16. Activity Diagram - Activity Diagram Symbols, Examples, and More. https://www.smartdraw.com/activity-diagram/
17. Robert, C., Tai, V.: Collaborative robots, the rise of a new generation. https://www.myforesight.my/2015/07/28/robert-cv-tai/
18. Preece, J., Yvonne, R., Helen, S.: Data gathering techniques in interaction design - beyond human-computer interaction, p. 214
19. Asplund, J., Brile, J.: Application of Human-Industrial Robot Collaboration. Master of Science Thesis in Production Engineering, Sweden (2017)
20. Rameshkumar Patel, M., Vijaykumar Bhatt, B., Pranav Vashi, M.: SMART-multi-criteria decision-making technique for use in planning activities. In: New Horizons in Civil Engineering, Surat, Gujarat, India (2017)
21. Heydaryan, S., Suaza Bedolla, J., Belingardi, G.: Safety design and development of a human-robot collaboration assembly process in the automotive industry. Appl. Sci. **8**, 344 (2018)
22. Tan, J.T.C., Duan, F., Kato, R., Arai, T.: Collaboration Planning by Task Analysis in Human-Robot Collaborative Manufacturing System (2010)
23. Decision Making Techniques - Decision Making Solutions. https://www.decision-making-solutions.com/decision_making_techniques.html

Accurate Liver 3D Reconstruction from MRE Images Using Shift-Compensated Volumetric Interpolation

Chiara Santarelli[1(✉)], Francesca Uccheddu[2], and Elisa Mussi[2]

[1] Department of Information Engineering, University of Florence, Florence, Italy
chiara.santarelli@unifi.it

[2] Department of Industrial Engineering, University of Florence, Florence, Italy
{francesca.uccheddu, elisa.mussi}@unifi.it

Abstract. In the last decade, recognition of the diagnostic potential arising from the analysis of the mechanical properties of biological tissues has led to research into methods for determining the mechanical properties of in vivo tissues. Magnetic resonance elastography is a newly developed imaging technique, based on the study of the phase contrast of a propagated acoustic transmission wave in tissues subjected to harmonic mechanical excitation. The acquired data allow to calculate the local quantitative values of the shear modulus and the generation of images that, in the form of colored maps, represent the elasticity of the tissue. The main limit of this method is the limited number of images that can be acquired during a scan. This is reflected in the accuracy of the 3D reconstruction of the anatomy and leads to coarser segmentation and modeling. In this article, Motion Compensated Frame Interpolation techniques are applied for effective axial interpolation of magnetic resonance elastography image sequences, with the main goal of achieving a refined 3D reconstruction.

Keywords: CT slices · Interpolation · 3D modeling · Magnetic Resonance Elastography · 3D liver reconstruction

1 Introduction

The mechanical characterization of human organs is of great impact on medical practice, particularly for diagnosis, surgical planning and training, tissue engineering and trauma research. Classic imaging methods, such as CT and MRI, allow to visualize an anatomical structure, verify its functionality and the presence of any pathologies, without giving any information about the mechanical properties of tissues. Measuring the elastic properties of tissues is of great interest because, not only do they differ between different physiological tissues, but it is also assumed that the tumor mass and the healthy tissue have totally different mechanical properties; the greater rigidity of the tissues, for example, may be related to a pathological state. Magnetic Resonance Elastography (MRE) is the ideal tool for determining certain mechanical properties of in vivo tissues. This method has been developed as an alternative, reliable and

C. Rizzi et al. (Eds.): ADM 2019, LNME, pp. 731–740, 2020.
https://doi.org/10.1007/978-3-030-31154-4_62

non-invasive method for the detection and staging of liver fibrosis as it has been suggested that changes in the mechanical properties of liver tissue may correlate with this disease; in particular, the rigidity of the liver is directly related to the various stages of the disease and increases with the progression of this disease [1, 2]. Coupling the data obtained from this elastographic analysis with a procedure for 3D reconstruction from images [3, 4], it is possible to obtain an anatomical 3D model with geometric and mechanical characteristics typical of the antomic organ under investigation. The technology of reconstruction and 3D printing in the clinical field allows to study and implement solutions customized to the needs of patients and is widely applied in medical diagnosis and surgical planning, also represents a valuable aid for surgeons in the preoperative and post-operative phase. Thanks to the three-dimensional reconstruction of anatomical districts, biomodels are created that faithfully replicate the pathology of the patient and that are used as a basis for preventive study of the operation. Difficulties, anatomical conditions, different depths and lengths of surgical instruments and instrumentation can be understood in advance, with the obvious advantage of having greater clarity, greater safety and speed of execution. Moreover, in the post phase, it allows to verify the correct execution of the operation and, in particular cases, the defects and possible corrections. The limitation of the low resolution of the MRE leads to an inaccurate 3D with low axial resolution. The proposed work is to apply the interpolation method to increase axial resolution and have an accurate 3D anatomical model.

2 Magnetic Resonance Elastography (MRE)

The most basic and immediate clinical tool for the diagnosis of diseases is palpation, since the pathological processes cause significant changes in the mechanical properties of the resulting tissues and therefore be of different consistency. In the medical field, "elastic" tissues are considered healthy, while "rigid" tissues are associated with pathological processes such as, for example, chronic inflammation or cancer. However, palpation as a diagnostic tool has its limitations because it is subjective and unable to identify the different tissue components, thus providing only qualitative analysis. A new imaging technique called Magnetic Resonance Elastography (MRE) has been developed to overcome these problems. This method is proposed as a non-invasive technique capable of quantifying the viscoelastic properties of in vivo tissues, i.e. the ability of the tissues to undergo deformation when subjected to pressure stress and to return to their original shape afterwards [5, 6]. The basic concept behind elastography includes three phases:

- generation of stress tension in a target tissue,
- tissue deformation analysis
- dynamic characterization of stress-deformation tissue.

MRE uses the propagation of mechanical shear waves (Gamma, 20–200 Hz) in conjunction with an MRI technique to probe the mechanical properties of tissues. Mechanical waves are propagated into the tissue by means of a driver (Fig. 1), and an additional cyclic motion coding gradient cyclic (MEG) is incorporated into a sequence

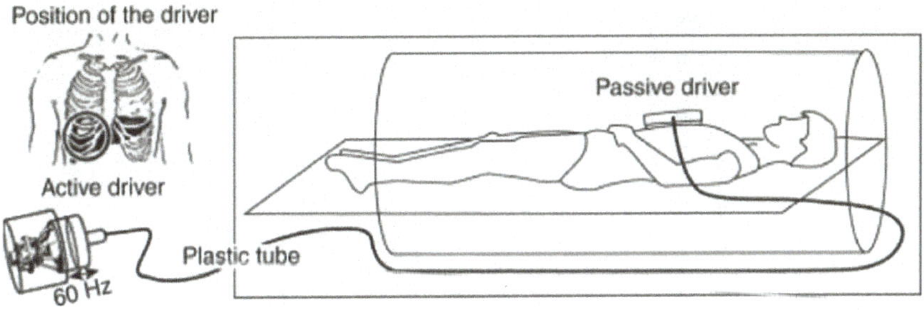

Fig. 1. MRE pneumatic system.

of magnetic resonance in order to directly represent the propagation pattern of the wave. To generate quantitative images representing the rigidity and other mechanical properties of the tissue, mathematical algorithms are used to process the images of the waves (Fig. 2).

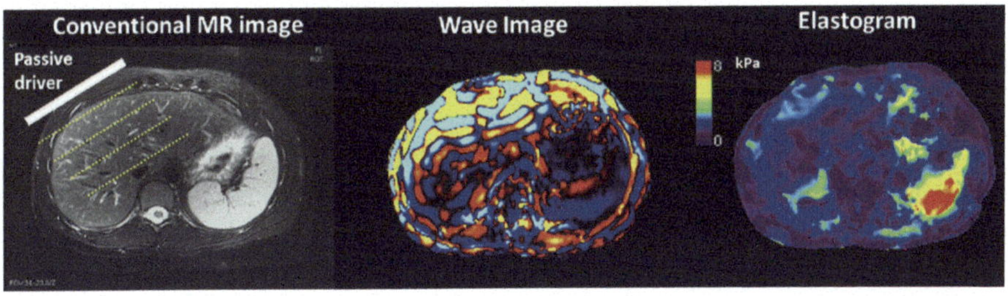

Fig. 2. MRE images of the abdomen. From right to left: RM image, wave image and Elastographic image.

The images of the mechanical properties calculated by MRE, are maps of rigidity where the rigidity of the tissues is expressed in kiloPascal (kPa) and displayed in color scale. As mentioned in the introduction, elastographic magnetic resonance imaging has been developed as an alternative method to liver biopsy, an invasive practice that makes it difficult to monitor the disease over time. Normally, the healthy liver tissue is very elastic compared to the diseased one, thus making elasticity a fundamental property in the diagnosis of liver fibrosis. Liver MRE is performed at a frequency of 60 Hz using active pneumatic pressure drivers, acquiring wave data with four phase offsets and using a modified direct inversion rigidity estimation algorithm with multiscale functionality. The passive driver, in this case, is applied to the abdomen and held in place by an elastic bandage. Figure 3 shows a graph describing liver stiffness in some volunteers with healthy liver and in patients with liver in different stages of fibrosis [7]. The graph is obtained by reporting on the abscissae the case of normal liver and the various phases of fibrosis F0–F4, while on the ordinate the corresponding mean stiffness values. As you can see, liver stiffness is directly related to the various stages of the disease and increases as it progresses.

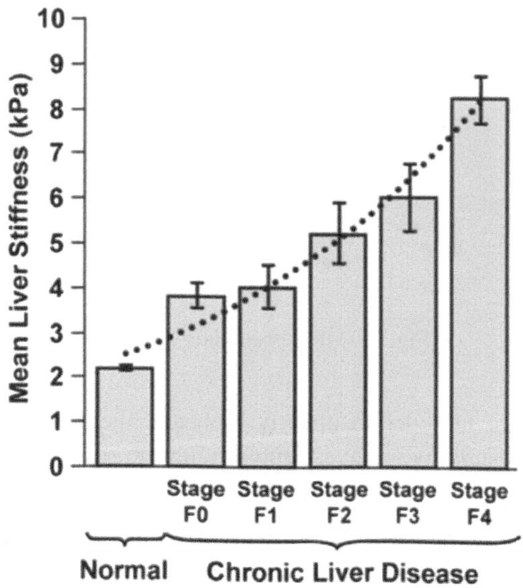

Fig. 3. Liver stiffness measures in the case of healthy and diseased patients divided into 5 different stages of fibrosis, which are referred to as F0, F1….F4.

One of the limits of this method is the low axial resolution, i.e. only a few slices can be obtained; this is due to the fact that the acquisition of images takes place during the apnea phase of the patient, which is very reduced because of the disease. In the figure the blue lines indicate the number of slices that will be acquired through MRE. As you can see, the portion of liver to be investigated is reduced and the slices are 11 mm spaced. To overcome this limit, a method of image interpolation has been proposed that aims to increase the axial resolution of the MRE sequence with the objective of obtaining a more accurate 3D reconstruction. This interpolation method will be explained below.

3 Optical Flow Based Interpolation

The objective of this method is improving the axial resolution in 3D medical sequences. This problem has analogies with other areas of signal processing. For example, in the field of video processing, the Motion Compensated Frame Interpolation (MCFI) technique is used to increase the number of frames of a video sequence, that is generating intermediate frames between adjacent ones. In this case, motion estimation is needed to track objects within the scene and predict their position in the interpolated frames. On the other hand, in the case of medical images sequences, even though the imaged anatomical parts are not subject to movement in adjacent slices, a spatial transform (a sort of morphing) may be foreseen to relate them in the different sections. This type of interpolation can be classified as scene-based or object-based [6, 7].

In scene-based methods, the interpolated pixel values are directly determined from the original images pixel intensities [8]. In object-based interpolation approaches, structural information is extracted and used to guide the interpolation process. In this work, techniques taken from MCFI [9–13] are applied to axial interpolation of tomographic sequences, having in mind, as a major objective, achieving more accurate 3D reconstructions. A scheme of the proposed method steps is presented in Fig. 4.

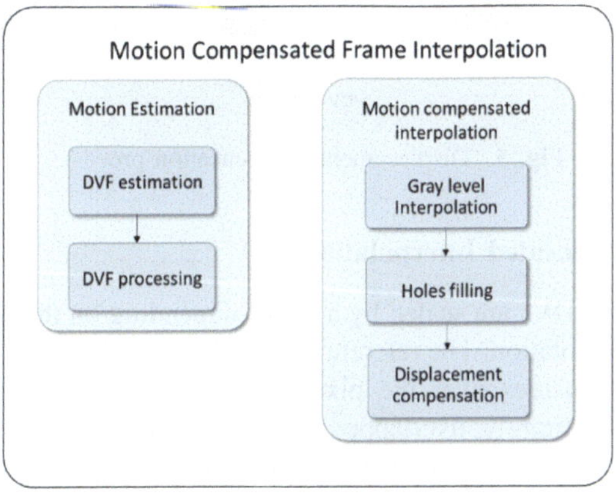

Fig. 4. Block diagram of the proposed method.

3.1 Motion Estimation

In the first step, we estimate the displacement vector field (DVF) between the pixels belonging to two adjacent sections. Due to the analogy with MCFI, the DVF is assimilated to a motion vector field (MVF) and the problem is solved by using algorithms taken from motion estimation approaches; more specifically, an Optical Flow (OF) technique has been employed. OF methods aim at estimating a dense field of motion vectors (one for each pixel of the images) that, applied to the pixels of one image (reference image), yields a prediction of a second image (target image). The Horn & Schunck algorithm, for example, solves this task by minimizing the energy of the prediction error with some regularity constraints. The DVF information obtained from the OF algorithm is then used on a patch basis to estimate the interpolated frames. In practice, the reference image is divided into small patches and a vector median filter (VMF) is applied to the displacement vectors related to a patch, resulting in a single displacement vector (DV) associated to the patch [14, 15]. The resulting vector is then scaled by a factor that depends on the position of the slice to be created by interpolation. The estimate of the DVF is performed forward and backward (i.e. swapping the reference images and the target images), so that two different scale factors are considered (see Fig. 5). In order to improve the results, a weighted VMF (WVMF) was also tested.

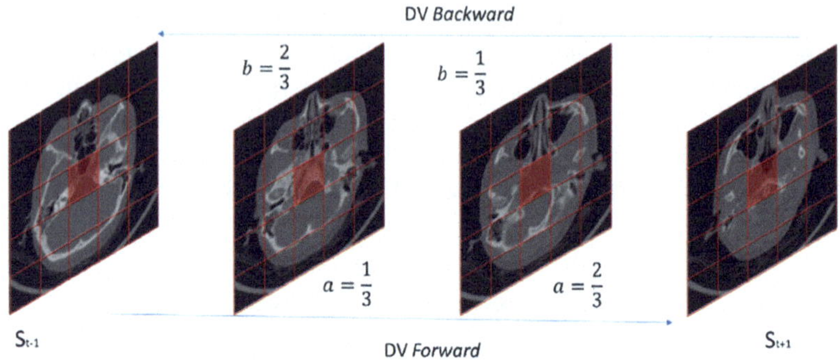

Fig. 5. Displacement compensation process.

3.2 Motion Compensated Interpolation

During this phase the DVs are scaled by a factor depending on the position of the slice to be created by interpolation. The reference patches are displaced by the corresponding scaled DVs and intensities over the pixel grid are achieved by means of bilinear interpolation. Since strongly overlapped patches in the reference image are used, overlapping displaced patches are produced and averaged to obtain the final interpolated section. Eventually, a filling "holes" procedure is adopted to provide intensity values to those pixels that were covered by no patches. To assess the performance of the proposed method, a high-resolution CT sequence was downsampled in the axial direction to simulate a low resolution one and, by using the DCSI algorithm, the missing slices were recovered to obtain an estimate of the original sequence. The experimental results show that the 3D models obtained from the downsampled/interpolated sequence are very close to those obtained from the original one.

4 Experimental Results

To assess the performance of the proposed method, a high-resolution CT sequence was downsampled in the axial direction to simulate a low resolution one and, by using the DCSI algorithm, the missing slices were recovered to obtain an estimate of the original sequence. The proposed algorithm was developed in the MATLAB environment and tests were performed by varying both the type of median filtering, the method for spatial interpolation and downsampling factor. In addition to the main functions that implement steps of the diagram in Fig. 4, a function has been built to translate the dynamic range of the images, so as not to have negative values and included in the range 0–255.

To evaluate the interpolation quality of this method, reconstructed images were compared with the original ones by means the root-mean-square error (RMSE), defined as

$$NRMSE = \sqrt{\frac{\sum_{m,n}(x_{true}[m,n] - x_{int}[m,n])^2}{\sum_{m,n}(x_{true}[m,n])^2}} \times 100 \qquad (1)$$

where x_{true} is the original image and x_{int} is the interpolated one. Table 1 shows the average RMSE values obtained for different choices of the system configuration parameters (spatial interpolation, DVF filtering type and downsampling factor).

Table 1. The average RMSE of each reconstructed sequence.

Spatial interpolation	DVF filtering	RMSE (m = 2)	RMSE (m = 3)	RMSE (m = 4)	RMSE (m = 5)
Bilinear	VMF	0,0318	0,0548	0,0808	0,1046
Bilinear	WVMF	0,0318	0,0548	0,0808	0,1046
Bicubic	VMF	0,0307	0,0541	0,0802	0,1043
Bicubic	WVMF	0,0302	0,0532	0,0791	0,1034

In addition, the mean value and standard deviation of the distance between the reference mesh and the reconstructed one were used to evaluate our method. The 3D model was by using MIMICS, a CE-certified medical software. The metrics allow the 3D models obtained from the original high-resolution CT sequences and the reconstructed one to be compared. Table 2 show the values resulting from this comparison and the comparison of the 3D models obtained from the downsampled CT vs. the original one is also reported.

Table 2. Average distance and standard deviation values by comparing the 3D model obtained from the original CT sequence with those reconstructed using the proposed method for m = 2 and m = 3.

Spatial interpolation	DVF filtering	Average distance m = 2	Standard deviation m = 2	Average distance m = 3	Standard deviation m = 3
Bilinear	VMF	0,0974	0,2151	0,2282	0,3922
Bilinear	WVMF	0,0955	0,2184	0,2234	0,3956
Bicubic	VMF	0,1000	0,2311	0,2322	0,4038
Bicubic	WVMF	0,0966	0,2267	0,2261	0,4060
3D model from down-sampled sequence		0,3376	0,5029	0,9038	0,9293

The computational cost of the software depends on the type of parameters set, downsampling factor, type of interpolation and filtering. The minimum value (8.53 s) is obtained by setting m = 2, bilinear interpolation and VMF filter, while the maximum value (115.86 s) is obtained by setting m = 5, bicubic interpolation and WVMF filter.

4.1 Real Low-Resolution MRE Sequence

As explained above, MRE is a method with low axial spatial resolution, therefore it is possible to apply the proposed method to obtain a more accurate 3D reconstruction. In this case an MRE sequence of the liver with such characteristics: number of slices = 4, slices distance = 11 mm spatial resolution 256 × 256, was interpolated. The reconstruction of the 3D models was carried out by setting the subsampling factor m = 4 in order to have the same number of slices contained in the sequence MR. The reconstruction method that gave the best results (bicubic-WVMF) during the test phase was chosen. Figures 6 and 7 show the 3D models corresponding to the original (low resolution) liver and to the interpolated sequence. As can be seen, the 3D model produced with the proposed method is much more regular then that obtained from the original data.

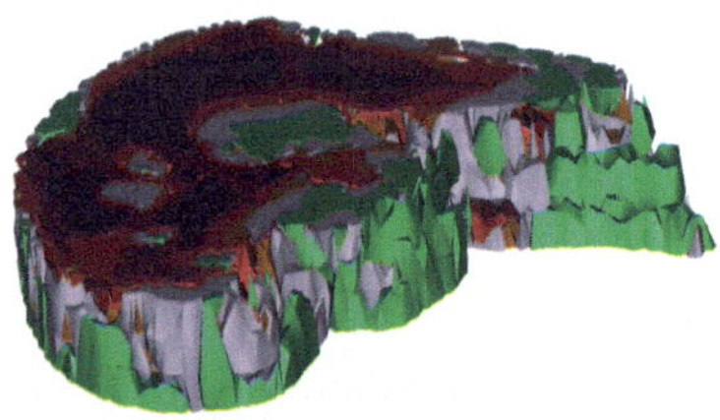

Fig. 6. 3D MRE original liver model.

Fig. 7. 3D MRE liver model reconstructed with the proposed method

5 Conclusion

Over the centuries, a wide range of tools have been developed to assess the mechanical properties of biological tissues. The first and immediate tool used by physicians in routine clinical examinations is palpation, as pathology is associated with changes in the mechanical properties of tissue that can be detected by touch. However, the sensitivity of this method is limited to the soft tissues located in the vicinity of the external surface of the body; in fact, there are many organs such as the brain or liver that are not accessible through palpation and therefore cannot be investigated through this simple diagnostic tool. Although advances in conventional imaging techniques, such as CT and MRI, make it possible to identify the pathologies of the less superficial organs, they are not able to directly evaluate their mechanical properties. Magnetic Resonance

Elastography (MRE) is an innovative imaging mode that visualizes and measures tissue elasticity allowing, in a non-invasive way, the determination of the tissue shear modulus in vivo. However, this method is affected by the limit of having a low axial resolution which leads to an inaccurate 3D reconstruction of the anatomical model. In this article a method for interpolating low resolution sequences has been proposed. The method takes into account the work in the field of motion compensated frame interpolation (MCFI) proposed for video processing and is based, fundamentally, on the estimation of the optical flow between adjacent slices. Different choices of the overall system configuration (post-processing of vector fields, interpolation functions) were compared in experimental tests using low-resolution CT scans obtained synthetically. These results guided the experiments on real MRE sequences, demonstrating that it is possible to obtain accurate 3D reconstructions.

References

1. Mariappan, Y.K., Glaser, K.J., Ehman, R.L.: Magnetic resonance elastography: a review. Clin. Anat. **23**(5), 497–511 (2010)
2. Rouviere, O., Yin, M., Dresner, M.A., Rossman, P.J., Burgart, L.J., Fidler, J.L., Ehman, R. L.: MR elastography of the liver: preliminary results. Radiology **240**(2), 440–448 (2006)
3. Uccheddu, F., Carfagni, M., Governi, L., Furferi, R., Volpe, Y., Nocerino, E.: 3D printing of cardiac structures from medical images: an overview of methods and interactive tools. Int. J. Interact. Des. Manuf. (IJIDeM) **12**(2), 597–609 (2018)
4. Nocerino, E., Remondino, F., Uccheddu, F., Gallo, M., Gerosa, G.: 3D modelling and rapid prototyping for cardiovascular surgical planning-two case studies. Int. Arch. Photogrammetry Remote Sensing Spat. Inf. Sci. 41 (2016)
5. Muthupillai, R., Lomas, D.J., Rossman, P.J., Greenleaf, J.F., Manduca, A., Ehman, R.L.: Magnetic resonance elastography by direct visualization of propagating acoustic strain waves. Science **269**(5232), 1854–1857(1995)
6. Manduca, A., Oliphant, T.E., Dresner, M.A., Mahowald, J.L., Kruse, S.A., Amromin, E., Ehman, R.L.: Magnetic resonance elastography: non-invasive mapping of tissue elasticity. Med. Image Anal. **5**(4), 237–254 (2001)
7. Yin, M., Talwalkar, J.A., Glaser, K.J., Manduca, A., Grimm, R.C., Rossman, P.J., Fidler, J. L., Ehman, R.L.: Assessment of hepatic fibrosis with magnetic resonance elastography. Clin. Gastroenterol. Hepatol. **5**(10), 1207–1213 (2007)
8. Furferi, R., Governi, L., Palai, M., Volpe, Y.: Multiple incident splines (MISs) algorithm for topological reconstruction of 2D unordered point clouds. Int. J. Math. Comput. Simul. **5**, 171–179 (2011)
9. Grevera, G.J., Udupa, J.K.: Shape-based interpolation of multidimensional grey-level images. IEEE Trans. Med. Imaging **15**(6), 881–892 (1996)
10. Joliot, M., Mazoyer, B.M.: Three-dimensional segmentation and interpolation of magnetic resonance brain images. IEEE Trans. Med. Imaging **12**(2), 269–277 (1993)
11. Krishnamurthy, R., Woods, J.W., Moulin, P.: Frame interpolation and bidirectional prediction of video using compactly encoded optical-flow fields and label fields. IEEE Trans. Circuits Syst. Video Technol. **9**(5), 713–726 (1999)
12. Choi, B.D., Han, J.W., Kim, C.S., Ko, S.J.: Motion-compensated frame interpolation using bilateral motion estimation and adaptive overlapped block motion compensation. IEEE Trans. Circuits Syst. Video Technol. **17**(4), 407–416 (2007)

13. Dikbas, S., Altunbasak, Y.: Novel true-motion estimation algorithm and its application to motion-compensated temporal frame interpolation. IEEE Trans. Image Process. **22**(8), 2931–2945 (2012)
14. Alparone, L., Barni, M., Bartolini, F., Cappellini, V.: Adaptively weighted vector-median filters for motion-fields smoothing. In: IEEE International Conference on Acoustics, Speech, and Signal Processing Conference Proceedings, vol. 4, pp. 2267–2270. IEEE (1996)
15. Alparone, L., Barni, M., Bartolini, F., Caldelli, R.: Regularization of optic flow estimates by means of weighted vector median filtering. IEEE Trans. Image Process. **8**(10), 1462–1467 (1999)

A Multi-disciplinary Assessments Tool for Human-Machine Interaction

Margherita Peruzzini[1(✉)], Fabio Grandi[1], Marcello Pellicciari[2], Giovanni Berselli[3], and Angelo Oreste Andrisano[1]

[1] University of Modena and Reggio Emilia, via Vivarelli 10, 41125 Modena, Italy
margherita.peruzzini@unimore.it
[2] University of Modena and Reggio Emilia, via Amendola 1, 42122 Reggio Emilia, Italy
[3] University of Genoa, Via all'Opera Pia 15/A, 16145 Genoa, Italy

Abstract. Human-centered design is based on the satisfaction of the user needs mainly related to performances, interaction, usability, accessibility, and visibility issues. However, the quality of the interaction process is hidden and usually difficult to detect. The paper proposes a multi-disciplinary assessment tool for the evaluation of the human-machine interaction, based on the collection of physiological data and anthropometrical performance data. Such a method can be used both within on-field tests and virtual simulations, supporting the spread of digital approaches in industry. The methodology allows objectifying how users interact with machine or interface items, thanks to the collection of the users' performance during task execution, the digitalization of collected data, and the evaluation of users' physical and mental workload. Such a system has been applied to an industrial case study focusing on agricultural machinery driving and control to support the system re-design in terms of interface features, commands' location and grouping, and positioning of additional devices.

Keywords: Human-centered design · Human-machine interaction · Digital factory · Ergonomics · Human factors

1 Introduction

In the industrial context, the study of human-machine interaction is a crucial activity and will be increasingly crucial for the design of efficient systems. Indeed, human work is drastically changing both in typology and contents (e.g., less physical load in favour of greater control activity and higher analysis of context data, interpretation of information, and decision-making tasks). In this scenario, the adoption of a human-centered approach is mandatory for the creation of successful interaction processes. The proper design of workspaces and user interfaces is definitely a strategic aspect, and it must include human factors in the system design in order to respond to physical, psychological, social and cultural needs of human beings [1]. Human factors specifically refer to research "regarding human psychological, social, physical, and biological characteristics, and

C. Rizzi et al. (Eds.): ADM 2019, LNME, pp. 741–752, 2020.
https://doi.org/10.1007/978-3-030-31154-4_63

working to apply that information with respect to the design, operation, or use of products or systems for optimizing human performance, health, safety, and/or habitability" [2]. As a consequence, Human-Centered Design (HCD) consists of the application of human-related information to the design of tools, machines, systems, tasks, jobs, and environments for safe, comfortable, and effective human use. As far as industrial system design, the optimization of posture, physical overload, perceived effort, discomfort, and physical fatigue is fundamental to satisfy the users' needs, prevent musculoskeletal disorders and stressing conditions [3]. In this context, the analysis of human factors has a central role in the understanding of human behaviours and performance interacting with systems, and the application of that understanding to design of interactions [4].

The paper presents a multi-disciplinary methodology to assess the users' performance during interaction processes and provides an assessment matrix as a tool to support design activities. The result is a global comfort index able to synthetize the user comfort detected by three dimensions (i.e., postural analysis, behavioural analysis, and physiological analysis) into a unique index, to provide an overall assessment. For data collection, a monitoring set-up integrating different methods for data acquisition of the human interaction, including physical, physiological, and visibility parameters, is used. The used set-up includes:

- Motion capture for real-time analyses of physical ergonomics, to measure body movements and positions of the different body parts (e.g., arms, hands, head) to objectify the human-machine interaction in terms of distance, instantaneous speed or acceleration) by applying well-known methods for comfort analysis (e.g., Dreyfuss);
- Use of sensors for real-time analyses of physiological parameters, which can provide a clear feedback on the user workload. The adopted sensors refer to electrocardiography for heart rate (HR) and heart rate variability (HRV) monitoring, breathing monitoring for breath rate (BR) analysis, and eye-tracking for eyes' fixation analysis and visual attention mapping;
- Human simulation software for the digitalization of human postures and movements and quick data post-processing.

The industrial case study focuses on the human-machine interaction on agricultural machinery, which has rapidly evolved in recent years in terms of technologies and on-board instrumentation in recent years to replace traditional means. It is a good example of work changing; indeed, the use of new, more technological machines are drastically reducing the users' workload, but at the same time requires a great mental and physical burden, demanding different skills and experience. This is due to the higher complexity of the tasks on one hand, and the greater technological development of system control and interfaces on the other, which made these machines more sophisticated and complex to use.

2 Assessment of the Human-Machine Interaction

The analysis of the quality of human-machine interaction is strongly related to the measure of the physical and psychological fatigue, which both affect the final user performance [1]. If ergonomics and usability aspects are not considered, the performance of the system will be poor and the effective working time will be reduced. The main factors affecting the final performance are numerous: from uncomfortable postures to higher task complexity and longer task duration, overload of information, time pressure, until awkward ambient conditions [5]. Besides more traditional physical measures, physiological analyses can be added. The basic idea is that physical activities stimulate certain physiological responses in human beings; these responses generate the human fatigue [6]. These investigations provide fundamental information on the user condition, even if high competence is required for the understanding and interpretation of physiological data. However, an automatic and non-invasive detection system is very useful, but it does not exist yet. Four main criteria can be distinguished in detecting the human stress, according to ISO 10075-3 [7]: psychological, physiological, behavioural, and biochemical. Psychological responses include the increase of strong emotions, usually negative, such as anger, anxiety, irritation and depression. From a physiological point of view, activity changes hormone levels in the body and provokes reactions such as sweat production, increased heart rate and muscle activation, faster breathing and increases blood pressure. Also the diameter of the pupils, eye movements and eye change rates can vary. A wide review of the main methods for human stress detection is presented by [8]. The electrocardiogram (ECG), based on the recording on the body surface of the electrical activity generated by heart, is one of the most used signals in stress detection research because it reflects directly the activity of the heart. The most typical and useful features computed with an ECG are probably the ones related to Heart Rate (HR) as the number of beats per minute, Heart Rate Variability (HRV) as the temporal variation between sequences of consecutive heart beats, and LF/HF ratio between low and high frequency signals. In addition, eye-related measures can be used to monitor the user performance. Indeed, changes in blink frequency (mean value, standard deviation, gaze spatial distribution) and pupil dilatation can be registered under stressful conditions by an infrared eye tracking systems or with image processing techniques applied to visual spectrum images of the eyes.

As far as the measurement of the physical stress, the most used method for assessing a user comfort level is the U.S. questionnaire National Institute of Occupational Safety and Health (NIOSH) [9], which allows the ergonomist to measure parameters concerning the level of musculoskeletal comfort in numerous parts of the body such as, for example, the intensity, frequency, and duration of the particular task. However, it is time consuming. Different approaches have been proposed in the past, including processing of video-recorded images [10] or implementation of gyroscopes or accelerometers [11], or motion capture cameras [12]. Video-based analysis is relative cheap, but video post-processing is time-consuming and requires several experts for an objective evaluation. The adoption of accelerometers and gyroscopes is highly accurate, but often expansive multiple sensors are required for three-dimensional analysis of a whole body, with the drawback of preventing or interfering with the body movement. Diversely, motion capture allows real-time acquisition of several body parts, representing a good compromise between accuracy and time spent for data analysis, even if the adopted systems (cameras) are suitable mainly for lab tests. Such methods can be easily integrated with postural risk assessment. The most known and used methods for risk assessment are the Rapid Upper Limb Assessment (RULA) used for the study of sedentary activities such as computer work, the Rapid Entire Body Assessment (REBA), ideal for quick job evaluations in feet and the Ovako Working Analysis System (OWAS), which allows a direct observation and sampling of individual processes by applying a system of coding the whole body posture and estimating, for each area of the body, the risks of damage.

From the analysis of the state of the art, two research hypotheses can be formulated:

- An increase of the mental workload means a worsening of driving comfort, and can be detected by a decrease in the HRV medium frequency band [0.07–0.15 Hz] in the power spectrum analysis;
- An increase of the mental workload, therefore a worsening of driving comfort, can be detected also through a monotonous increasing trend of the pupil diameter (PD).

The research aims at checking if these hypotheses are valid also for the specific area of investigation (i.e., agricultural machinery design) and defining which parameters can be used for user experience analysis in this field.

3 The Multi-dimensional Assessment Matrix

The purpose of this research is to present a methodology for a multi-dimensional analysis of the human-machine interaction, able to provide a quick assessment of the user comfort level. For the specific study, it was therefore decided to investigate and combine physical, behavioural and physiological aspects through the use of well-structured methodology and a lab-suitable monitoring system. It is a further development of previous researches focussed on the tractor's human-centered design [13] and driver's workload monitoring [14]. A proper matrix called "Assessment Matrix" is introduced to combine the different aspects and to provide a final overall user Comfort Index. The basic idea is to weave data from different dimensions in order to return a single output able to synthesize the user performance in terms of quality of interaction. The inputs are represented by:

1. Postural parameters, mainly based on the identification of the postures assumed by the operator during task execution, posture digitization, and evaluation by Dreyfuss 3D method [15]. This analysis also considers the impact of task conditions (e.g., repetition frequency, loads) on physical stress is considered;
2. Behavioural parameters, considering both visibility issues and eye-related measures (i.e., gaze plot and heat maps). This analysis considers the operator viewpoint and the distance of the focal point with respect to the overall dimensions of the workspace;

3. Physiological parameters, based on ECG data (i.e., HR and HRV analysis in time and frequency domain) and pupil diameter (PD) analysis.

The set of parameters is revised using a VBA-Excel, properly programmed to calculate, at each registered posture, the deviations of the point of view with respect to the workspace. Moreover, postural analysis is supported by virtual simulation using Siemens Jack software. A reference posture, called Zero position (characterized by the values of postural parameters as close as possible to the perfect comfort zone according to the Dreyfuss 3D database) has been implemented in the Jack software. Finally, behavioural and physiological data are synchronized with the specific postures assumed by the user characterizing a specific task. As a result, the Comfort Index is a function of time, which represents the level of comfort perceived by the operator throughout the job (Fig. 1).

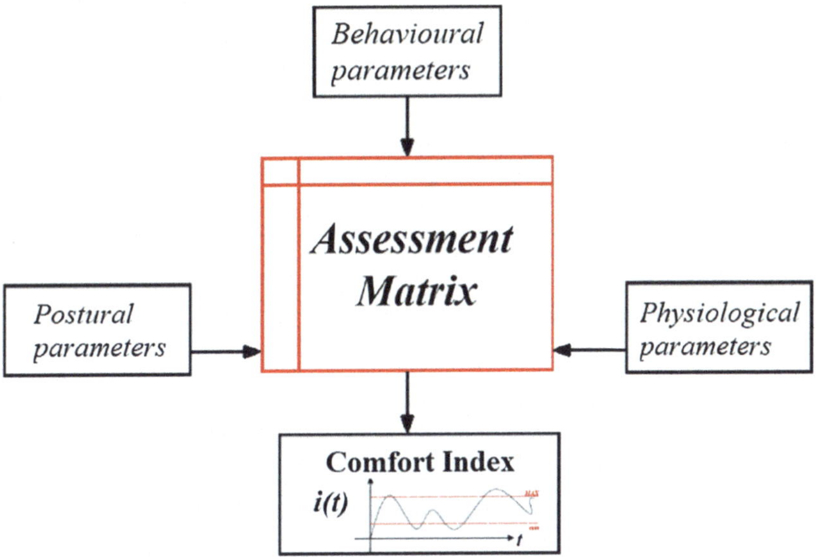

Fig. 1. Assessment Matrix

During testing, the proposed Assessment Matrix is used as described in Fig. 2. For each test, input data recording is activated by a trigger event (command activation monitored by CAN trace, telemetry applications, or any other data event). Therefore, the command activation is analysed and the different input data recognized: for each instant of time an active operation is identified, and data collected about postural, behavioural and physiological parameters can be synchronized. For each user interactions with the system, postural, physiological and behavioural parameters that identify the three dimensions of analysis are evaluated individually at the same time. For each time step, a routine assigns a weight, which graphically represents a numerical score, assigned according to the value of the dependent variables of that dimension with respect to pre-established limits. Such weights are defined as follows:

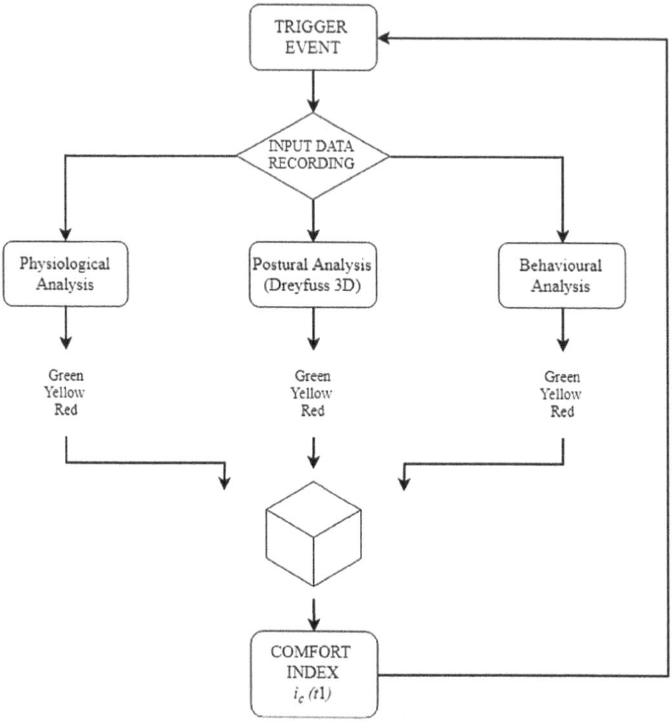

Fig. 2. Workflow to use the Assessment Matrix

- Green: good comfort (any significant difference from target value);
- Yellow: poor comfort (between optimal value and a minimum comfort value);
- Red: comfort limit value (below the minimum comfort threshold).

Starting from the input values, the matrix sums the contribution of each variable to the input and performs the arithmetic average. To overcome a linear trend that links the output to the various inputs, the matrix assigns corrective factors, as follows:

- For each Yellow weight, remove 0.5 from the final score;
- For each Red weight, remove 1 from the final score.

In this way a greater weight is attributed to the stressful phenomena that influence the final comfort value of the user. The matrix outputs a numeric value that represents the comfort index of a discrete moment of time.

Once obtained the value of the comfort index time function at the time instant *t1*, proceed with the detection of the subsequent activations in the cabin at time *t2, t3,...,* *tn*. In order to have a continuous time trend of the Comfort Index function, a linear interpolation between the points detected by the previous analysis is expected. This is because the activations, therefore the interactions between man and machine in the cabin, are in a finite number. The proposed tool provides a continuous time index representing the level of user comfort during human-machine interaction. Furthermore, since the comfort perceived by the user could deviate from the results obtained, self-report tests have been introduced to take into account also the user subjective evaluation.

4 The Industrial Case Study

The case study was based on the analysis of the human-machine interaction during the control of tractors. Indeed, driving and control agricultural machinery is a stressing activity. From a physical point of view, the driver undergoes a lot of movements which mainly involve the upper part of the body: steering, looking forward and backward, controlling the monitor or the vehicle's dashboard, using clutch, brake, control levers or joystick [16]. In the long term such activity could generate physical health problems in different upper parts of the body (i.e., arms, neck, shoulders, back, head). Furthermore, time pressure and uncomfortable external conditions can cause high mental workload. In this context, the cabin design including the seat, the control board and the interaction devices, plays an important role in the driver's comfort. A good human-machine design, such as well-accommodated operator enclosures, comfortable seat, achievable and easy to use commands, natural mapping and easy-to-use interface, can significantly enhance the user performance, with positive impact on productivity and safety [17].

The case study was developed in collaboration with CNH Industrial. Task analysis started by observing real users on the job within a tractor cabin (Fig. 3); after that, the virtualization of the cabin was useful to move the analysis within a virtual scenario and to verify different design solutions. The proposed methodology was applied to investigate the user comfort in order to support re-design actions. In particular, the user physiological, behavioural and physical data were collected by a proper monitoring set-up, and then synchronized and analysed in order to define the Comfort Index. A set of tasks were monitored during field tests and reproduced into the virtual scenario.

Fig. 3. Tractor cabin design used for experimental testing

For each task, a set of sub-tasks was identified and data collected. Correlating physiological parameters and their change over the time, with task execution, visual interaction by eye tracking, and command use allowed investigating the human-machine interaction and highlight the main criticalities.

The monitoring system used for experimental testing was composed by:

- Tobii Pro Glasses 2 eye-tracking to capture eye-related measures and study visual interaction;
- Zephyr BioHarness 3.0 sensor belt to record human physiological data, in particular to collect Heart Rate (HR), Heart Rate Variability (HRV), Breathing Rate (BR), Body activity (expressed in VMU, vectorial sum of activity counts in x-y-z directions), and the angular torso flexion;
- CAN data from the tractor interface and CANanalyzer software toolkit for data elaboration;
- a GoPro camera to record the scene.

Table 1 shows the monitored parameters during experimental tests, related to the three analysis dimensions (physical, behavioural, and physiological). 20 users were involved during testing (all males, from 20 to 45 years old). Figure 4 shows an example of physical data analysis on Joint angles and deviations for different body parts. Figures 5 and 6 shows example of HRV and pupil diameter (PD) analysis.

Table 1. Monitored parameters during experimental tests

Analysis dimensions	Tools and methods	Monitored parameters
Physical	Virtualization with JACK Stress analysis by Dreyfuss 3D	Joint angles and deviations for different body parts
Behavioural	Data collection with eye tracking Visibility analysis with Gaze plot and Heat Map	HV, HRV, BR average data in time and frequency domain (SDNN and R-R intervals)
Physiological	Data collection with sensor belt and eye tracking	PD (average data in time domain)

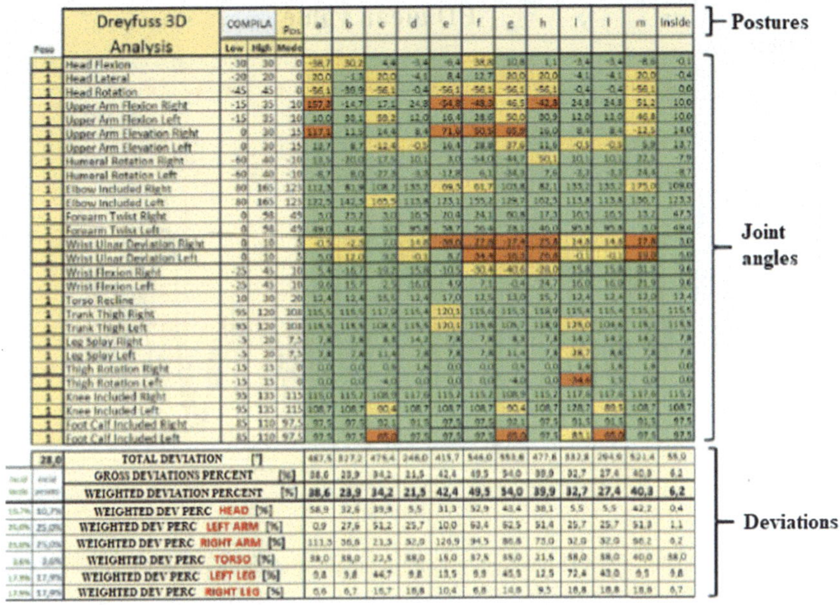

Fig. 4. Example of physical comfort assessment by Dreyfuss 3D

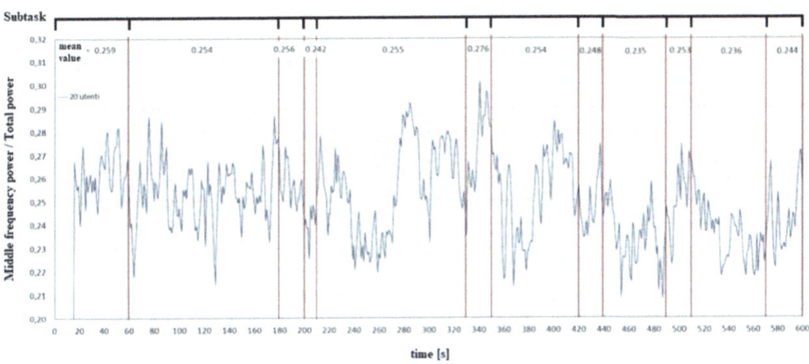

Fig. 5. Example of HRV data analysis on 20 users during task execution

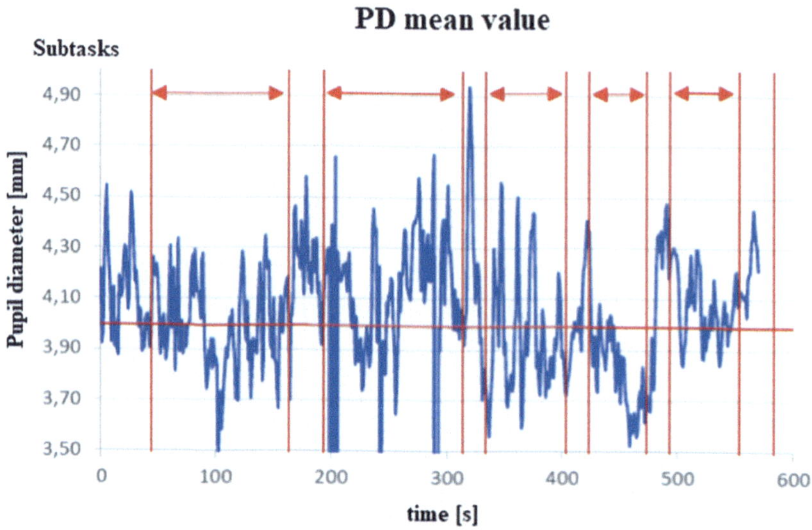

Fig. 6. Example of PD data analysis for a single user during task execution

5 Discussion of the Results

Considering the research hypotheses (presented in Sect. 2), the main results can be synthetized as follows:

- Heart rate (HR) and its variability (HRV) in time domain are well-correlated with the mental load, so their variation can be used to identify mental stress conditions;
- Standard deviation of the successive intervals (SDNN) in the sequence of beat-beat intervals (RR curve) turned out to be a fairly reliable metric in the evaluation of the user mental workload: low values correspond to poor ability to cope with stressful situations. This characteristic had indeed distinguished the commands used in situations of high psychophysical commitment from those less critical.
- The HRV analysis in frequency domain, considering the spectrum of the middle frequency band (MF) of the RR curve in the assessment of cognitive stress, showed good results. All these parameters have therefore been considered as inputs of the Assessment Matrix: it filters the input signals, associates to each signal certain values based on pre-set thresholds and produces an output continuous function of comfort;
- For the specific application, the PD values measurement demonstrated to have poor correlation with the mental workload. It is mainly due to the specific working conditions, in particular the lack of a controlled lighting environment in the cabin during the test, which greatly affects pupil dilation and invalidates the measurement;

- The use of questionnaires was useful to profile users from a psychological and non-psychological viewpoint, and provided support in data analysis. Indeed, subjective judgements were useful to assess the degree of comfort and to calibrate the matrix results;
- The overall comfort index has been proved to be a good measure of the degree of comfort during man-machine interaction over time.

6 Conclusions

The paper provided an innovative methodology to evaluate the user performance during human-machine interaction of complex machines (i.e., tractors). In particular, the proposed Assessment Matrix enabled the correlation of different variables and identified the most critical interactions. In particular, correlations of HR, HRV and PD with command activation during task execution allowed understanding the quality if the interaction thanks to the elaboration of a single ergonomic evaluation index through the multi-disciplinary Assessment Matrix. Finally, in order to support the application of the Assessment Matrix, classic evaluation tools such as heat maps and gaze plots were added as well as evaluation questionnaires. HR and HRV are demonstrated to be good indicators of comfort and workload to be used for agricultural machinery design, based on field tests, while PD is not well suitable due to the high variability in the environmental lighting conditions during tests.

Future works will be oriented to enlarge the type of input data, also including other physiological measures (e.g., skin conductance, EEG) and physical data by a direct interaction with motion capture technologies (e.g., magnetic full body motion capture systems).

Acknowledgements. This research has been carried out using the research facilities of the UNIMORE INTERMECH MO.RE. centre, funded with a grant from the Emilia Romagna High Technology Network within the project entitled “CREAM: human-CentREd Agricultural Machinery product and process design” (CUP E18C1500031000).

References

1. Ergonomics of human system interaction - Part 210: Human-centered design for interactive systems, International Organization for Standardization (ISO), ISO 9241-210 (2009)
2. Stramler, J.H.: The Dictionary for Human Factors/Ergonomics. CRC Press, Boca Raton (1993)
3. Pheasant, S.: Body Space: Anthropometry, Ergonomics and the Design of Work, pp. 121–123. Taylor & Francis, Philadelphia (1999)
4. Wilson, J.R.: Fundamentals of ergonomics in theory and practice. Appl. Ergon. **31**, 557–567 (2000)
5. Zhang, J.Y., Qiu, W.W., Fu, H.J., Zhang, M.T., Ma, Q.G.: Review of techniques for driver fatigue detection. Appl. Mech. Mater. **433–435**, 928–931 (2013)
6. Yerkes, R.M., Dodson, J.D.: The relationship of strength of stimulus to rapidity of habit formation. J. Comput. Neurol. Psychol. **18**, 459–482 (1908)

7. Ergonomic principles related to mental workload — Part 3: Principles and requirements concerning methods for measuring and assessing mental workload, International Organization for Standardization (ISO), ISO 10075-3 (2005)
8. Alberdi, A., Aztiria, A., Basarab, A.: Towards an automatic early stress recognition system for office environments based on multimodal measurements: a review. J. Biomed. Inform. **59**, 49–75 (2016)
9. Dempsey, P.G.: Usability of the revised NIOSH lifting equation. Ergonomics **45**(12), 817–828 (2002)
10. Tran, C., Trivedi, M.M.: Towards a vision-based system exploring 3D driver posture dynamics for driver assistance: issues and possibilities. In: Proceedings of the 2010 IEEE Intelligent Vehicles Symposium, pp. 179–184 (2010)
11. Yang, C.C., Hsu, Y.L.: A review of accelerometry-based wearable motion detectors for physical activity monitoring. Sensors **10**, 7772–7788 (2010)
12. Bashiri, B., Mann, D.D.: Automation and the situation awareness of drivers in agricultural semi-autonomous vehicles. Biosys. Eng. **124**, 8–15 (2014)
13. Grandi, F., Zanni, L., Peruzzini, M., Pellicciari, M., Campanella, C.E.: A transdisciplinary digital approach for tractor's human-centred design. Int. J. Comput. Integr. Manuf., 1–17 (2019, in press)
14. Peruzzini, M., Tonietti, M., Iani, C.: Transdisciplinary Design Approach Based On Driver's Workload Monitoring. J. Ind. Inf. Integr. (2019, in press)
15. Tilley, A.R., Dreyfuss, H.: The Measure of Man and Woman: Human Factors in Design. Wiley, Chichester (2001)
16. Peruzzini, M., Grandi, F., Pellicciari, M.: Benchmarking of tools for user experience analysis in Industry 4.0. Procedia Manuf. **11**, 806–813 (2017)
17. Hsiao, H., Whitestone, J., Bradtmiller, B., Whisler, R., Zwiener, J., Lafferty, C., Kau, T.Y., Gross, M.: Anthropometric criteria for the design of tractor cabs and protection frames. Ergonomics **48**(4), 323–353 (2005)

Understanding the Human Motor Control for User-Centered Design of Custom Wearable Systems: Case Studies in Sports, Industry, Rehabilitation

Teodorico Caporaso[1(✉)], Stanislao Grazioso[1], Dario Panariello[1,2], Giuseppe Di Gironimo[1], and Antonio Lanzotti[1]

[1] Fraunhofer Joint Lab IDEAS, Dipartimento di Ingegneria Industriale, Università degli Studi di Napoli Federico II, P.le Tecchio 80, 80125 Napoli, Italy
teodorico.caporaso@unina.it

[2] Dipartimento di Ingegneria Gestionale, dell'Informazione e della Produzione, Università degli Studi di Bergamo, Viale Marconi 5, 24044 Dalmine, BG, Italy

Abstract. This paper shows how studies on the biomechanics and neuroscience of human movements might be used for the design of wearable systems customized for humans. Such design is driven by key biomechanical and neuromuscular parameters extracted from accurate measurements made on the human body motion, as well as by subjective data collected from the end-users of the products through questionnaires. We present three case studies developed at ERGOS Lab: a wearable system for sports performance analysis; a synergy-based approach for industrial wearable robots; a soft wearable robotic glove for hand rehabilitation.

Keywords: Design methods · Biomechanics · Neuromuscular activity · Wearable technology

1 Introduction

Wearable systems are becoming widespread in many different fields, as sports and fitness, healthcare and wellness, safety and prevention in workplaces. They can be used for recording of human-related performance parameters for monitoring applications, or they can be used for aiding people in performing movements. In the first case, we refer to wearable devices [1], while in the second case, we refer to wearable robots [2].

The effective development of wearable systems require an extensive knowledge on the human behaviors. This knowledge might be acquired by understanding the human motor control, i.e. the process that involves brain, muscles, limbs and external objects to regulate human movements [3]. As the human motor control involves a cooperative interaction between the central nervous system and

C. Rizzi et al. (Eds.): ADM 2019, LNME, pp. 753–764, 2020.
https://doi.org/10.1007/978-3-030-31154-4_64

the musculoskeletal system, understanding this process requires a deep comprehension on the neuromuscular and biomechanical activities which generate the human motion.

In sport, understanding the human motor control could allow the development of wearable devices that can monitor and therefore maximize athletes' performance. In industrial workplaces, understanding the human motor control can bring to the development of custom wearable systems which guide the workers towards ergonomically correct postures [4], or can reduce the operator's workload during the activities [5]. Finally, in rehabilitation medicine, where are present deficiency in the execution of normal movements, understanding the human motor control is fundamental to restore the functions of the motor system.

As these systems have to be worn by humans, it is also important to extrapolate information from the end-users of the products, in particular related to their internal sensations and feeling with the technology. In this respect, a subjective way of analysis consists of submitting questionnaires to capture the users' emotions; this is of great importance for the development of human-centered devices [6,7]. Many design methods exist to obtain such kind of information; however, one of the most well-known approaches is represented by the Kansei engineering methodology. This approach allows capturing the internal sensations from the end-users and the identification of product quality elements satisfying both functional and emotional needs [8].

In this work we show how neuromuscular and biomechanical studies might generate human-related key design parameters for wearable systems. To valorize this strategy, we present three case studies conducted at ERGOS Lab, CeSMA (Centre for Advanced Metrology Services, University of Naples Federico II, Italy) in different domains of applications: sports, industry, rehabilitation. Then, we present an extensive discussion which can be used to guide further works in this context and we report some conclusions.

2 Method

In this work we show how key input parameters obtained from the human motor control system can be used to design of customized wearable systems for humans. In order to understand the basics of human motor control, biomechanics and neuroscience of human movements play a relevant role. The biomechanics of human motion studies the musculoskeletal system for the assessment of movements. Nevertheless, for a full understanding of the human motor control, it is important to take into consideration how the neural system acts in generating the movements: this is done through the study of neuroscience applied to human motion. The integrated approach for the assessment of human motor control is shown in Fig. 1. From one hand, this process outputs biomechanical parameters which can be implemented on wearable devices for monitoring of human performances. From the other hand, it can output some information as range of motion

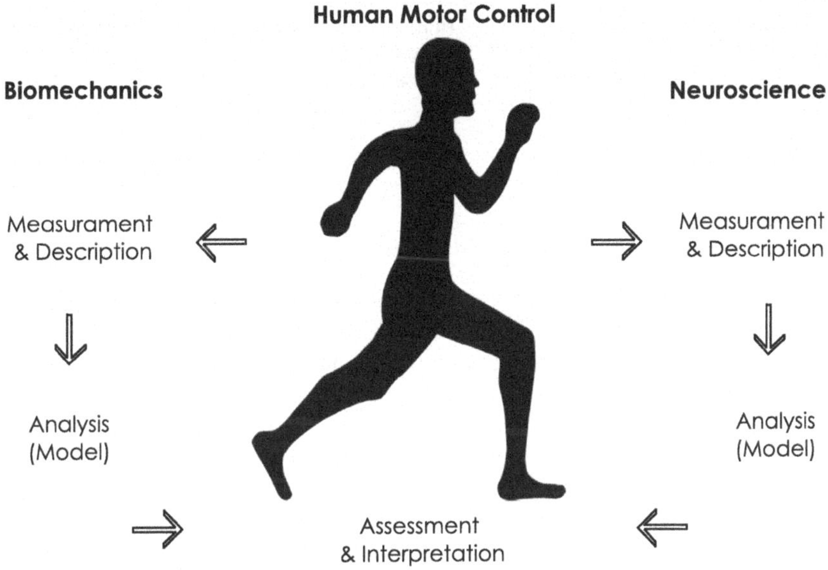

Fig. 1. Schematic diagram of the assessment of human motor control. On the left the main steps of biomechanics approach and on the right neuroscience ones.

of limbs and neuromuscular synergies of movement, which can be the basis for the development of wearable robots (i.e. exoskeletons) able to work closely to humans.

2.1 Biomechanics of Human Motion

The human movements are assessed through a biomechanics process involving several steps. First, two preliminary steps are required: (1) *measurement* of human movements (i.e. kinematic parameters acquired by an inertial sensor); (2) *description* of human movements (i.e. different graphical way to visualize the measurements). Then, the combination of measurements and description data into specific biomechanical models allows doing *analysis* of the movements [9]. The construction of these models is based on the combination of physical and biological principles. Typical models of the human skeletons used in the biomechanics community involve rigid-body segments connected by frictionless joints [10].

The use of these models with biomechanical experimental measurements (i.e. trajectories of specific land-markers, ground reactions forces and antropometrical characteristics), allows to assess the whole-body motion (i.e. during a walk). At the end of this analysis phase, a large number of information related to the movement can be available. The last stage for the assessment of the biomechanics of human motion is the *interpretation*. This stage aims at individuating specific key parameters able to summarize the most important information related to the human motion. These key parameters are interpreted in a custom way based on the main characteristics of the end-users.

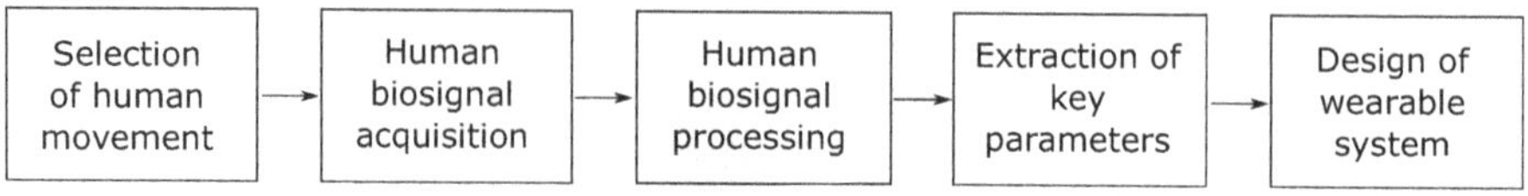

Fig. 2. A generic flow chart for bio-inspired design of customized wearable systems.

2.2 Neuroscience of Human Motion

The neuroscience of human motion allows to understand the role of human neural system for generating the motor actions. As done with biomechanics, we need to find proper measurement data related to the neural control system: with this respect, electromyography (EMG) allows recording the electrical activity produced by muscles as they are neurologically activated. Therefore, measurements are done using EMG sensors placed on the human body in correspondence of the muscles involved in the movements. After the measurement process, we need to describe the data. Specific models allows to analyze the processes underpinning this neuro-muscular activity. This activity is highly complex; however, reduced-order models exists which condense the information related to the neuro-muscolar activation process, as the concept of neuromuscular synergies. In the context of motor control, synergies are defined as groups of muscles jointly activated by a single central control signal [11]. Through the sequencing and superposition of a limited number of synergies, a large variation of complex movements can be achieved [12]. This concept can be implemented on robotic prostheses and exoskeletons, for simplifying the control of a multi degrees-of-freedom system through emulating only few synergies.

3 Examples of Applications

In this section we show with three practical examples how studies on the human motor control can be used for design of custom wearable devices. The three case studies refer to three different fields of applications: sports, industry, rehabilitation. Their development follows the design methodology illustrated in Fig. 2.

3.1 Sports

Biomechanics of human motion is used in the case study #1 for the development of a wearable system for sports performance analyses customized for elité athletes.

Case Study #1: Inertial Assistant Referee and Trainer for Race-Walking

Problem Statement: Currently, the most common infringement in race walking is the loss of ground contact (LOGC), which is evaluated during the race by the human eyes of judges (it lasts few hundredths of seconds). Therefore, a wearable system (composed by a measurement unit placed on the athletes' body and a management unit for judges) able to automatically assess the infringements in race-walking is highly desirable. On the other side, maximizing the performance (which, in this discipline, is strictly related to infringements) is also required. Starting from these motivations, we develop a system able to simultaneously detect the infringements and assess the performances.

Methodological Approach: The measurement unit is composed by an inertial system placed at the end of the athlete vertebral column (L5–S1). This placement was chosen according to two input data: (i) estimation of the 3D body center of mass during human locomotion [13,14]; (ii) users sensation extracted from specific questionnaires according to a Kansei engineering approach [15]. The measurement data are translated in biomechanical parameters customized for élite athletes for the assessment of infringement (i.e. estimation and classification of LOGC) and performance (i.e. step cadence, step length and smoothness) [16]. These parameters are plotted simultaneously on a radar chart, and shown by a mobile app for smartphone [17]. The design process of this wearable device is shown in Fig. 3.

Results: The system estimates the LOGC with an accuracy of more than 80% [18]. The management unit composed of a mobile app is considered intuitive from coaches (it allows to easily understand the strong and critical point of the athletes technique [16]) and judges (it helps in the assessment of sequence of LOGC [18]).

3.2 Industry

In industry, one of the most recent trend is related to the *humanization of factories*[1], which aims at developing bionic workplaces[2] where humans and machines can work together in a symbiotic way. In this respect, biomechanics and neuromuscular studies on the humans can play a relevant role for the design of assistive devices, as industrial exoskeletons [19,20], which can potentially reduce the workers' physical overloading.

Case Study #2: Synergy-Based Method for the Development of Soft Wearable Industrial Robots

Problem Statement: The current industrial exoskeletons present the following problems: (1) they are bulky; (2) they reduce the operative workspace available for workers and they can damage the surrounding objects; (3) they do not inherently guarantee a safe human-robot interaction; (4) their motion can be perceived as non-natural for the humans. The development of more effective

[1] https://www.robotics.org/.
[2] https://www.festo.com/.

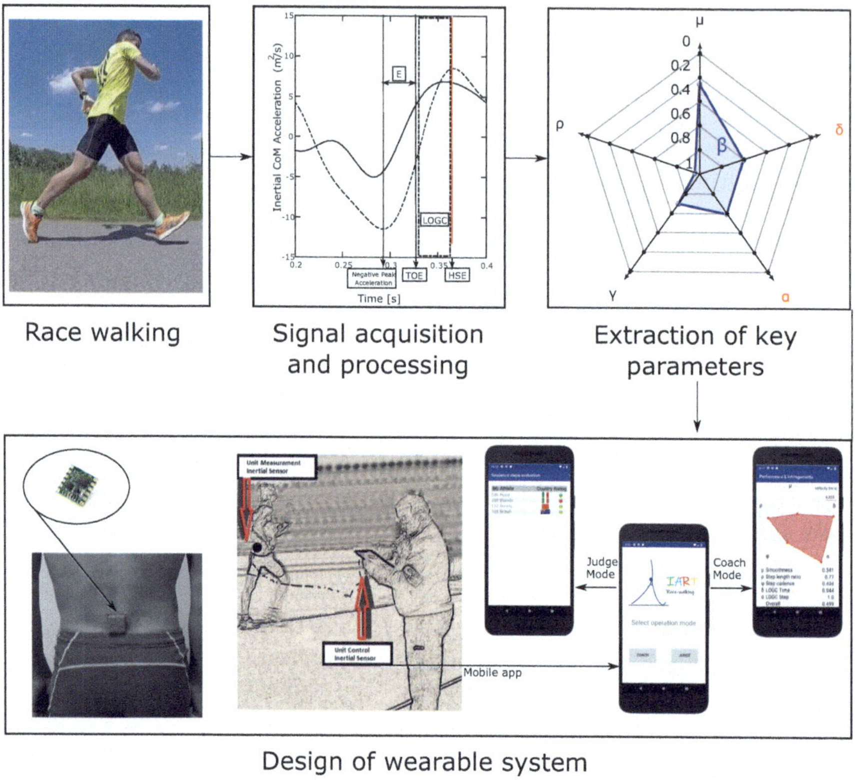

Fig. 3. Design of a wearable system for race walking. In the second rectangle, the correlation between the CoM accelerations and gait temporal events (TOE: toe-off event; HSE: heel strike event) for assessing the LOGC (using the threshold value E). In the third rectangle performance (μ, ρ, γ), infringement (α, δ) and overall (β) parameters.

assistive devices require a new radical design philosophy [21]. Therefore, we propose a strategy for design and control of industrial wearable robots which is based, from one side, on the extraction of neuromuscular synergies during the execution of industrial tasks, from the other side, on the use of soft robotics technology [22–24].

Methodological Approach: First, four industrial tasks of interest are selected. These tasks are those for which the exoskeleton has to provide assistance. Then, EMG sensors are placed on the worker's body according to a specific protocol, in order to capture the muscular activity. Three repetitions are performed for the same task, for a total of twelve experiments. From the EMG data, the muscle activation curves are reconstructed. Then, from these, the muscle synergies are extracted according to the non-negative matrix factorization (NMF) algorithm [25].

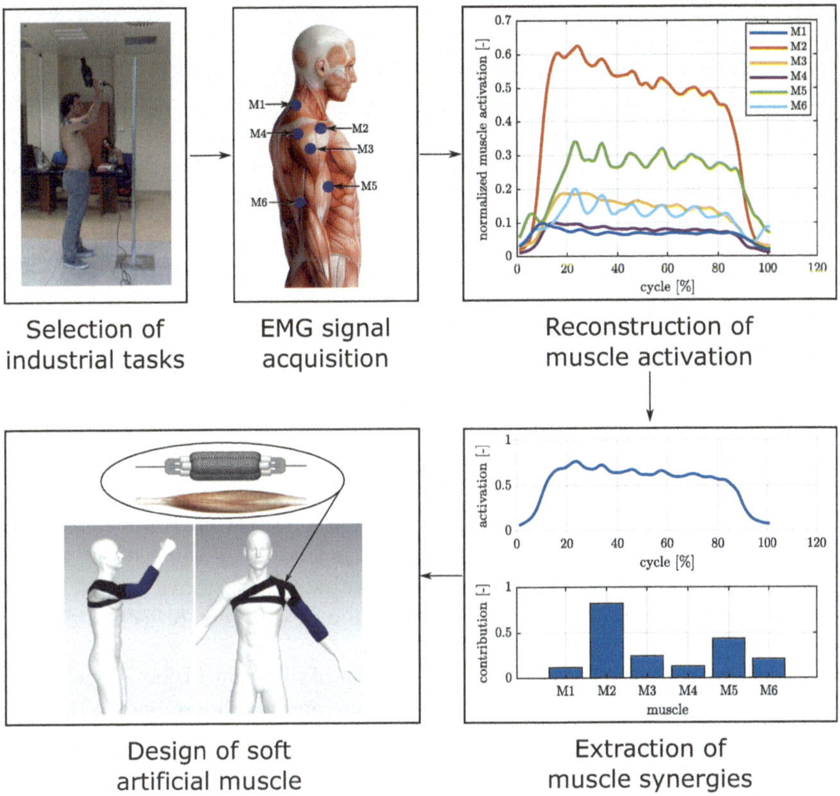

Fig. 4. The synergy – based approach for soft wearable robot. In the block "EMG signal acquisition" we have the following muscles: M1, upper trapezium; M2, anterior deltoid; M3, medial deltoid; M4, rear deltoid; M5, biceps brachii; M6, long head of the triceps.

Results: For the considered industrial tasks, the neuromuscular study outputs two major results: (1) the first synergy account for >98% of the total muscle activation [25], and therefore it is sufficient for reconstructing the overall movements; (2) the most activated muscle is the anterior deltoid. These results suggest that one artificial muscle (i.e. a soft actuator [26]) which replicates the first synergy and is placed in correspondence of the anterior deltoid muscle might reduce the physical overloading on the workers. Indeed, soft actuators can be designed to work as a biological muscle; furthermore, as they are pneumatically actuated and composed of soft materials, they are intrinsically safe for workers. The overall design process of soft wearable industrial robots is illustrated in Fig. 4.

3.3 Rehabilitation

In the recent year, there is a growing interest in using robotics technology to support physiatry. Rehabilitation robots must have the following features: safety, effectiveness, simplicity of use (for both patients and therapists). In the following, pneumatic actuation and soft materials are used together for the development of an intrinsically safe robotic glove [27] used for rehabilitation exercises.

Case Study #3: Design and Fabrication of a Soft Robotic Glove for Hand Rehabilitation

Problem Statement: Existing robotics solutions for hand rehabilitation are limited to medical centers, where the patients are under the supervision of physiotherapists. This happens because the existing solutions are usually heavy and they could be invasive for the patients. The widespread of soft robotic technology offers an alternative way for the development of systems which can interact with patients in a safer way [28]. Thus, a soft rehabilitation robot can be used also for home therapy, as it can be worn and actuated by the patient in a safe way. With this in mind, we develop a soft wearable robotic glove[3] which encloses the hand and allows a whole-body manipulation of different objects.

Methodological Approach: First, power grasping of different objects is selected as rehabilitation exercise. In order to develop a system which can enclose the hand and allows the grasping of different objects, we use soft bending actuators for augment the natural movements of each finger. These actuators are composed by two materials with different stiffness and they are designed such that they bend when the air flows in. Therefore, they can guide a finger towards the grasping pose. The typical range of motion of an healthy hand is used to compute the desired bending angle; indeed, anthropometric data of the natural fingers are used to size the length of each actuator, since they are custom fabricated. A mapping between bending angle and the input pressure is derived for actuation purposes. The robotic glove is safe for humans as it is composed of a soft material (elastomer) and it is actuated by a pneumatic source. The design process of the soft robotic glove is illustrated in Fig. 5. In order to test the capabilities of the system, a volunteer performed ten grasping experiments for each of the following objects: tennis ball, hard-disk case, plastic bottle, tape (for a total of 40 experiments).

Results: The grasping experiments involved three phases: (i) getting close to the object; (ii) grasping the object; (iii) keeping the object between fingers for 5 s. The experiments are considered to be successful if all the three conditions are successful. For the 40 experiments, we achieve a successful grasping in the 77.5% of the cases. We actuate the robotic glove in the same way for all the experiments: this underline how the system is robust with respect to different objects.

[3] https://softroboticstoolkit.com/.

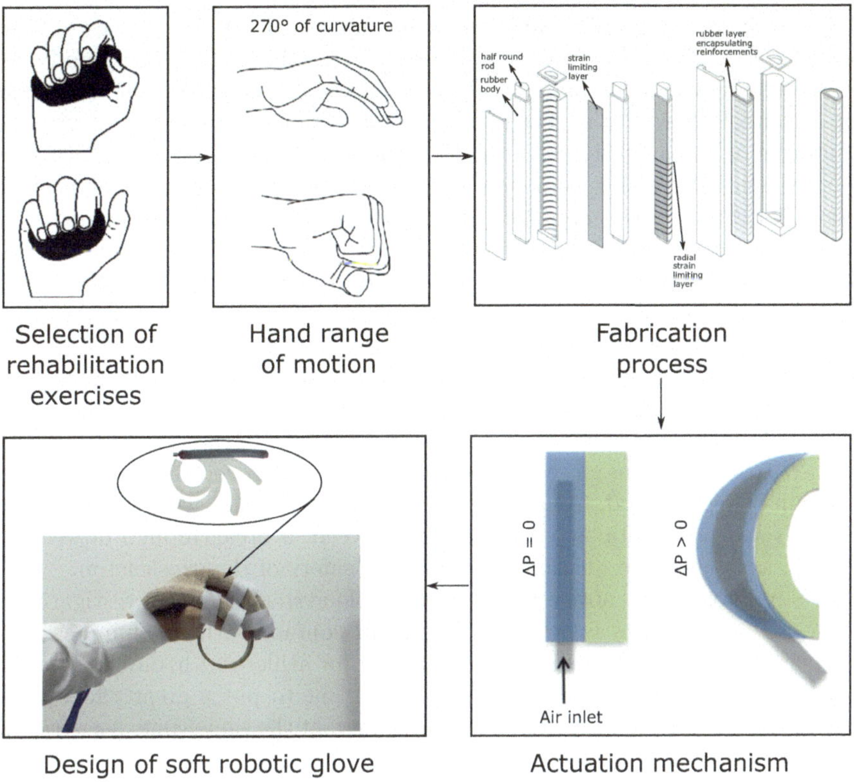

Fig. 5. Design of a soft robotic glove for hand rehabilitation

4 Discussion

The case studies presented in Sect. 3 are used to give practical examples of the more general design methodology shown in Fig. 2. This methodology have included five main steps: (i) selection of human movements, as industrial or rehabilitation tasks or sport gestures; (ii) acquisition of biosignals (biomechanical or neuromuscular measurements); (iii) processing of biosignals (through the definition of specific models); (iv) extraction of key biomechanical or neuromuscular parameters; (v) definition of design factors related to the previous key parameters and, finally, design of the system. A further improvement of the strategy could be including, in the design process, a 3D body scanning step useful for design of tailored assistive devices (in this case, only exoskeletons) by taking into consideration the real three-dimensional shapes of the subject [29–31]. In this work, the step in (iv) has been implemented and used in three different ways: (a) performance indicators used for monitoring of activities (Sect. 3.1); (b) neuromuscular indicators (i.e. synergies) used for defining the architecture and control strategy of a wearable robots (Sect. 3.2); (c) human-centered parameters used to design a wearable system (Sect. 3.3).

The proposed approach can be used to improve the design phases of many products, as it allows to obtain more customized, minimally invasive and effective systems devoted for humans. As a rule, a deep knowledge on the human motor control allows, from one hand, the development of effective devices for monitoring of human activities, from the other hand, the fabrication of wearable robots which behave and think similarly to humans. In the context of wearable devices, for example, this approach underlines the importance of deriving biomechanical indices customized for a particular category of end-user; moreover, the such developed indices result easy to be interpreted. As an illustrative example, authors in [32] have also proposed a system able to detect infringements in race-walking (as the one proposed in Sect. 3.1), but using two inertial sensors. However, their detection algorithms are not customized on the specific requirements of race-walking end-users (athletes, coaches and judges). In the context of wearable robots, neuromuscular studies could bring to the implementation of robot control algorithms closed to those generated by the central nervous system of human beings. Indeed, from the technology point of view, we believe that soft materials and pneumatic actuation could represent a breakthrough in wearable robotics, as it allows to guarantee an intrinsic safety of the exoskeleton.

It is importance to notice that the wearable systems under investigation in this work are products which are directly in contact with the humans for a long period during the day (i.e. about 3 h/day for athletes, 8 h/day for workers, 1–2 h/day for patients). Therefore, it is important to put a great effort in the design phase for the development of effective and, at the same time, user-friendly products.

In addition, as these systems have to be worn by humans, it is important to analyse the users' emotional needs. For example, in the case study #1, this aspect has been analysed through questionnaires developed according to the Kansei engineering method.

A further advance in this direction could be the development of objective techniques for quantifying the internal feelings and sensations. From the physiological point of view, the emotional life is largely housed in the limbic system, which also directs intentional movements. Therefore, applied neuroscience techniques (aims at correlating the internal sensations with measurements of electroencephalography, electromyography, heart rate and galvanic skin sensitivity) could offer rules and methods to quantify the Kansei outputs [33].

5 Conclusions

In this work we have presented some recent activities carried out at ERGOS Lab in the context of wearable systems. These activities show how studies on the human motor control might be useful for the design of user-centered systems, in different domains. These studies could be used in the future for improving many of existing and recognized design methods, as they can provide quantitative data related to humans behavior to be included in the overall design process of wearable systems.

References

1. Yang, G.Z., Yang, G.: Body Sensor Networks, vol. 1. Springer, London (2006)
2. Pons, J.L.: Wearable Robots: Biomechatronic Exoskeletons. Wiley, Hoboken (2008)
3. Rosenbaum, D.A.: Human Motor Control. Academic Press, Cambridge (2009)
4. Kim, W., Lorenzini, M., Kapıcıoğlu, K., Ajoudani, A.: ErgoTac: a tactile feedback interface for improving human ergonomics in workplaces. IEEE Robot. Autom. Lett. **3**(4), 4179–4186 (2018)
5. De Looze, M.P., Bosch, T., Krause, F., Stadler, K.S., O'Sullivan, L.W.: Exoskeletons for industrial application and their potential effects on physical work load. Ergonomics **59**(5), 671–681 (2016)
6. Caporaso, T., Grazioso, S., Vaccaro, D., Di Gironimo, G., Lanzotti, A.: User-centered design of an innovative foot stretcher for ergometers to enhance the indoor rowing training. Int. J. Interact. Des. Manuf. (IJIDeM) **12**(4), 1211–1221 (2018)
7. Franciosa, P., Gerbino, S., Lanzotti, A., Silvestri, L.: Improving comfort of shoe sole through experiments based on CAD-FEM modeling. Med. Eng. Phys. **35**(1), 36–46 (2013)
8. Lanzotti, A., Tarantino, P.: Kansei engineering approach for total quality design and continuous innovation. TQM J. **20**(4), 324–337 (2008)
9. Winter, D.A.: Biomechanics and Motor Control of Human Movement. Wiley, Hoboken (2009)
10. Caporaso, T., Di Gironimo, G., Tarallo, A., De Martino, G., Di Ludovico, M., Lanzotti, A.: Digital human models for gait analysis: experimental validation of static force analysis tools under dynamic conditions. In: Eynard, B., Nigrelli, V., Oliveri, S., Peris-Fajarnes, G., Rizzuti, S. (eds.) Advances on Mechanics, Design Engineering and Manufacturing, pp. 479–488. Springer, Cham (2017)
11. Latash, M.L., Scholz, J.P., Schöner, G.: Toward a new theory of motor synergies. Mot. Control **11**(3), 276–308 (2007)
12. d'Avella, A., Portone, A., Lacquaniti, F.: Superposition and modulation of muscle synergies for reaching in response to a change in target location. J. Neurophysiol. **106**(6), 2796–2812 (2011)
13. Di Gironimo, G., Caporaso, T., Del Giudice, D.M., Tarallo, A., Lanzotti, A.: Development of a new experimental protocol for analysing the race-walking technique based on kinematic and dynamic parameters. Procedia Eng. **147**, 741–746 (2016)
14. Hanley, B., Bissas, A., Drake, A.: Kinematic characteristics of elite men's and women's 20 km race walking and their variation during the race. Sports Biomech. **10**(02), 110–124 (2011)
15. Di Gironimo, G., Caporaso, T., Del Giudice, D.M., Lanzotti, A.: Towards a new monitoring system to detect illegal steps in race-walking. Int. J. Interact. Des. Manuf. (IJIDeM) **11**(2), 317–329 (2017)
16. Caporaso, T., Grazioso, S., Di Gironimo, G., Lanzotti, A.: Biomechanical indices represented on radar chart for assessment of performance and infringements in élite race walkers (2019)
17. Caporaso, T., Grazioso, S., Panariello, D., Di Gironimo, G., Lanzotti, A.: A wearable inertial device based on biomechanical parameters for sports performance analysis in race-walking: preliminary results. In: 2019 IEEE International Workshop on Metrology for Industry 4.0 and IoT, pp. 259–262. IEEE (2019)
18. Di Gironimo, G., Caporaso, T., Amodeo, G., Del Giudice, D.M., Lanzotti, A., Odenwald, S.: Outdoor tests for the validation of an inertial system able to detect illegal steps in race-walking. Procedia Eng. **147**, 544–549 (2016)

19. Xiloyannis, M., Dhinh, K., Cappello, L., Antuvan, C., Masia, L.: A soft wearable elbow exosuit: design considerations. In: Wearable Technology in Medicine and Health Care, p. 193 (2018)
20. O'Neill, C.T., Phipps, N.S., Cappello, L., Paganoni, S., Walsh, C.J.: A soft wearable robot for the shoulder: design, characterization, and preliminary testing. In: 2017 International Conference on Rehabilitation Robotics (ICORR), pp. 1672–1678. IEEE (2017)
21. Siciliano, B., Khatib, O.: Springer Handbook of Robotics. Springer, Cham (2016)
22. Kim, S., Laschi, C., Trimmer, B.: Soft robotics: a bioinspired evolution in robotics. Trends Biotechnol. **31**(5), 287–294 (2013)
23. Grazioso, S., Di Gironimo, G., Siciliano, B.: A geometrically exact model for soft continuum robots: the finite element deformation space formulation. Soft Robot. (2018)
24. Grazioso, S., Di Gironimo, G., Siciliano, B.: From differential geometry of curves to helical kinematics of continuum robots using exponential mapping. In: International Symposium on Advances in Robot Kinematics, pp. 319–326. Springer (2018)
25. Grazioso, S., Caporaso, T., Palomba, A., Nardella, S., Ostuni, B., Panariello, D., Di Gironimo, G., Lanzotti, A.: Assessment of upper limb muscle synergies for industrial overhead tasks: a preliminary study. In: 2019 IEEE International Workshop on Metrology for Industry 4.0 and IoT, pp. 89–92. IEEE (2019)
26. Abe, T., Koizumi, S., Nabae, H., Endo, G., Suzumori, K.: Muscle textile to implement soft suit to shift balancing posture of the body. In: 2018 IEEE International Conference on Soft Robotics (RoboSoft), pp. 572–578. IEEE (2018)
27. Polygerinos, P., Wang, Z., Galloway, K.C., Wood, R.J., Walsh, C.J.: Soft robotic glove for combined assistance and at-home rehabilitation. Robot. Auton. Syst. **73**, 135–143 (2015)
28. Polygerinos, P., Lyne, S., Wang, Z., Nicolini, L.F., Mosadegh, B., Whitesides, G.M., Walsh, C.J.: Towards a soft pneumatic glove for hand rehabilitation. In: 2013 IEEE/RSJ International Conference on Intelligent Robots and Systems, pp. 1512–1517. IEEE (2013)
29. Grazioso, S., Selvaggio, M., Di Gironimo, G.: Design and development of a novel body scanning system for healthcare applications. Int. J. Interact. Des. Manuf. (IJIDeM) **12**(2), 611–620 (2018)
30. Grazioso, S., Selvaggio, M., Caporaso, T., Di Gironimo, G.: A digital photogrammetric method to enhance the fabrication of custom-made spinal orthoses. JPO J. Prosthet. Orthot. **31**(2), 133–139 (2019)
31. Grazioso, S., Caporaso, T., Selvaggio, M., Panariello, D., Ruggiero, R., Di Gironimo, G.: Using photogrammetric 3D body reconstruction for the design of patient-tailored assistive devices. In: 2019 IEEE International Workshop on Metrology for Industry 4.0 and IoT, pp. 240–242. IEEE (2019)
32. Taborri, J., Palermo, E., Rossi, S.: Automatic detection of faults in race walking: a comparative analysis of machine-learning algorithms fed with inertial sensor data. Sensors **19**(6), 1461 (2019)
33. Schütte, S.T., Eklund, J., Axelsson, J.R., Nagamachi, M.: Concepts, methods and tools in Kansei engineering. Theor. Issues Ergon. Sci. **5**(3), 214–231 (2004)

3D Digital Surgical Planning: An Investigation of Low-Cost Software Tools for Concurrent Design

Francesco Buonamici, Lorenzo Guariento(✉), and Yary Volpe

Department of Industrial Engineering, University of Florence,
via di Santa Marta 3, 50139 Florence, Italy
{francesco.buonamici,lorenzo.guariento,
yary.volpe}@unifi.it

Abstract. Bone tumor resections have to be carefully planned in order to avoid intralesional cuts and thus lower the recurrence rate. Until the present-day bone resections have been performed freehand, using anatomical landmarks as reference points to retrieve planned resection planes over the very patient. Such method is highly prone to failure to the detriment of the patient's health, in fact survival rate for osteosarcomas is very low, ranging between 20% and 47%. The clinical outcome is highly dependent on the resections' accuracy; several emerging techniques proved to dramatically increase cutting accuracy, as well as survival rate, in particular Patient Specific Instruments (PSIs) and custom 3D printed metal prosthesis. The main limitation to a massive spread of this method is the large manufacturing time due to a lack of communication between surgeons and engineers about the surgical approach and the design constraints. This paper aims to compare several general-purpose low-cost software and to provide surgeons with an effective and easy to use platform to visually share information in a natural manner with engineers thus providing as many design constraints as possible, speed up the design process and avoid unfeasible results. Two surgeons from Azienda Ospedaliera Universitaria Careggi tested and evaluated a series of software. From this preliminary investigation Forger, a digital sculpting and texture painting application for iOS, resulted as the most user friendly and intuitive application among the test group.

Keywords: Human computer interaction · Computer aided surgery · Digital surgical planning · Concurrent design

1 Introduction

Surgical planning is a crucial step that strongly affects the clinical outcome of any surgical operation and plays a pivotal role dealing with oncological patients.

Primary bone tumors are rare conditions; they account for 3% to 5% of childhood cancers and for less than 1% of all cancers in adults [1]. Despite the low incidence, bone tumors are the third cause of tumor-related deaths in developed countries. In particular, pelvis osteosarcomas represent 7% to 9% of all osteosarcomas and the related survival rate ranges between 20% and 47% [2], mainly due to late diagnose in

C. Rizzi et al. (Eds.): ADM 2019, LNME, pp. 765–775, 2020.
https://doi.org/10.1007/978-3-030-31154-4_65

advanced stage with large tumor size. Pelvis surgery is particularly challenging and technically demanding because of its complex anatomical structure, surrounded by several vital structures as large blood vessels and organs.

In this context, safe resection margins are crucial to lower the local tumor recurrence and complications rate [3–5]. Resection planes are planned according to tumor staging [6], vital structures involved, access strategy, bone quality for implant fixation and desired resection margin [7, 8].

Recent advances in medical imaging and surgical equipment allow clinicians to accurately locate the lesion and plan the resection planes in order to spare, whenever possible, the limb's functionality and appearance by replacing the resected bone with a custom metal implant [9]. Metal 3D printing is becoming more and more popular to cure large bone resections due to the possibility to manufacture complex patient-specific implants.

During surgery, planned cutting planes are typically located exploiting anatomical landmarks, and resections are performed freehand. Cartiaux et al. proved that the freehand achievable accuracy of pelvic resections is not satisfactory [10], as experienced surgeons performed only 52% of pelvic resection with 10 mm margins from tumor within 5 mm tolerance [10]. Besides the high risk of intralesional cuts, such poor accuracy is not sufficient to guarantee a good fit of a patient-specific implant.

Recently, new techniques provided higher cutting accuracy, namely:

- Surgical Navigation System
- Patient Specific Instrument (PSI)

Surgical Navigation System allows a real-time tracking of the surgical equipment thanks to a 3D stereo camera system, and virtually displays the surgical instrument along with Computer Tomography (CT) and Magnetic Resonance (MR) images by performing an alignment of the two coordinate systems. Such instrumentation allows a real-time tracking of the cutting instrument and on-screen visualization of the resection planes, thus supplying surgeons a visual feedback of cutting accuracy.

Patient Specific Instruments, on the other hand, are 3D printed custom surgical jigs designed to perfectly fit the bony surface and guide surgical instruments to cut the bone as planned.

These technologies provide comparable cutting accuracy, but the latter proved to be more intuitive, faster and cheap, as it removes the need for a sophisticated instrumentation and a complex calibration process [11]. Despite its efficacy, several limits are still perceived during the PSI design process. Accordingly, this article will focus on the study of tools and methods capable of (i) optimizing PSIs design and (ii) promoting a proficient concurrent design that involves surgeons and engineers, in order to limit time losses.

Patient specific surgical guides are designed according to the specifications defined during surgical planning. In order to transfer the planned resections onto the patient, PSI has to perfectly fit the bone, taking account of the accessible area and exposed regions. In this framework, preoperative planning is a fundamental step to delineate the main constraints before the detailed project, which complies with the practical issues related to manufacturing, use and implantation; for example, the cutting guides have to be dimensioned accordingly to the specific blade employed and guarantee a good fit over the bone in the area of interest.

Currently the market offers several software specifically designed for preoperative orthopaedic planning that allow to simulate many surgical steps. The most popular planning software packages are reported in Table 1.

Table 1. Most popular preoperative planning software

Name	Developer
Mimics	Materialise [12]
Sectra	Sectra AB [13]
mediCAD	mediCAD [14]
PeekMed	Peek Health [15]
TraumaCAD	Brainlab [16]

Such programs are powerful tools but, in order to fully exploit their potentialities, users have to be well trained; for this reason preoperative planning process is mainly demanded to engineers and eventually validated by surgeons after many iterations, as engineers are not aware of all the clinical aspects to take account of.

As described in Fig. 1, the manufacturing process of PSIs includes several steps that ranges across several areas of competence; the overall process requires a whole set of tools and methods drawn from multiple specific technical fields. The process starts with diagnostic imaging segmentation, that is crucial to allow a precise 3D reconstruction of tumor, bone and surrounding tissues. Only through an accurate identification of the lesion and a valid planning of bone resections with adequate safe margins, it's possible to spare vital structures and, whenever possible, joints. Follows the design phase, that is driven by the specifications provided by surgeons, and eventually manufacturing and sterilization.

It's important to highlight that PSI and implant's design process involves clinicians and engineers; the first provide patient-specific details about the surgical approach and PSI and implant's prerequisites. Engineers, on the other hand, deal with the technical aspects of design and manufacturing. Both contributions are integrated through an iterative design approach, with engineers that typically propose different solutions, recursively examined and corrected by surgeons, until a final validation is reached. This method is evidently time-consuming. Moreover, the project is typically handled by third-party companies specialized in 3D printing of medical devices that are not in close contact with clinicians, which makes communication inefficient and slow.

Several iterations are often necessary to develop a satisfactory design due to communication issues between engineers and doctors about the surgical approach. The specific nature of each clinical case means that engineers are usually not aware of how surgery will be performed in detail; as a result, unfeasible PSI and implant solutions are occasionally produced when the surgical approach is not clearly assessed, in terms of access strategy, exposed bone stock and expendable soft tissues. Time is crucial in oncological surgery because tumors can rapidly evolve between the time when

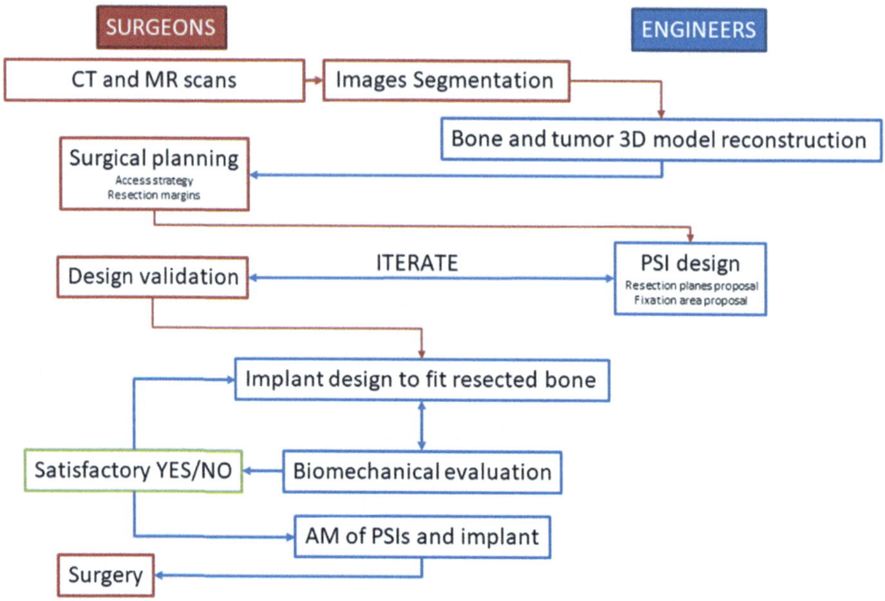

Fig. 1. Typical workflow for PSIs and custom implants' design and manufacturing

diagnostic imaging is performed and surgery [17]. PSIs and prosthesis should be designed and manufactured as quickly as possible, as bone resections are planned upon the 3D reconstruction of CT and MR scans, as pointed out in Fig. 1.

As previously mentioned, PSI and prosthesis' project needs to take account of several technical difficulties of both medical and engineering nature, so it is essential to involve both engineers and surgeons to share their specific know-how. Accordingly, a proficient communication and cooperation could dramatically shorten the design process and lead to better surgical outcomes.

Aim of this study is to identify the most efficient method for preoperative planning, in order to provide engineers and clinicians with a communication platform to share information and to avoid many iterations. For this purpose, we evaluated several approaches for preoperative planning such as Virtual Reality, 3D printing of the patient's bone and digital 3D planning were evaluated; among these possibilities we decided to explore the digital approach because is cheaper and more intuitive for surgeons. The ideal tool is a virtual 3D environment with few specific tools to visualize and manipulate the 3D anatomical models in order to point out any relevant information for engineers to accurately design the cutting guides, taking account of the patient-specific issues.

Follows a review of commercially available low-cost general-purpose software to promote the concurrent design between engineers and surgeons.

2 Preoperative Planning Software

Preoperative planning software can simulate most of the surgical procedures and bone resections, but typically require an engineering approach and a deep knowledge of digital environment.

To promote a cooperation between engineers and surgeons we tested general-purpose low-cost applications with the goal to find an easy and user-friendly software with the following features, defined in agreement with surgeons:

- Intuitive interface
- Visualization of multiple models
- Adjustable model's transparency
- Measurements
- Easy planes insertion
- Paintable mesh
- Touch screen or graphic tablet compatibility
- Exportable mesh with texture
- File compatibility with CAD software
- Ease of exchange
- Price under 1000€

This review aims to the identification of a virtual environment in which surgeons can autonomously manipulate 3D models and provide engineers all the design constraints for PSI and implant. These can usually be reduced to: (i) the exposed bony area, (ii) visible structures, (iii) resection planes and (iv) any other important information. All the information provided by the surgeon should be immediately retrieved in the CAD environment for detailed PSIs' design.

The adoption of an information sharing platform between two different fields is the first and most important step towards a proficient concurrent design. Such approach allows to freely exchange data and constitutes an important way of communications between the two areas. In order to guarantee the most intuitive interaction with the 3D visualization and modelling environment, several commercial software systems compatible with touch screen or graphic tablet have been tested.

The tested applications are presented in Table 2.

Table 2. Compared software

Name	OS	Price
Blender	Windows/MacOS	Free
Zbrush	Windows/MacOS	895 €
Sculptris	Windows/MacOS	Free
Forger	iOS	5 €
Armorpaint	Windows/MacOS	19 €
3Dcoat	Windows/MacOS	99 €
Polybrush	Windows	150 €

3 Materials and Methods

The above cited software packages were tested by two expert oncological orthopaedic surgeons from AOU Careggi. They were asked to plan a real case of tumor removal surgery with PSI that consisted of:

- 3D navigation (pan, rotate, zoom)
- perform measurements
- set resection planes, according to the desired safety margin
- mark the exposed bone for PSI fixation as well as critical areas with different colours
- export 3D mesh and texture

The same tasks were performed with the different software after a training session to get acquainted with the basic commands of every software.

Between the two surgeons who participated the test, one already had a basic knowledge of Blender and navigation of 3D environment, while the other one had no experience with any of the examined applications or other 3D modelling software.

Surgeons were eventually asked to express a qualitative evaluation about the ease of execution of every task. They assigned grades according to Table 3, and the overall execution time to qualitatively assess the ease of use of the tested programs was measured.

Table 3. Grading scale

1	**Poor**
2	
3	
4	**Average**
5	
6	
7	
8	**Good**
9	
10	**Excellent**

4 Results

The average grades assigned by surgeons to different tasks are shown In Table 4 and resumed in Fig. 2.

Table 4. Average Grades (P = Poor, A = Average, G = Good, E = Excellent; T1 = 3D navigation, T2 = Measurements, T3 = Plane's setting, T4 = Object's transparency, T5 = Mesh painting, T6 = Import/export)

		Software					
Tasks	Blender	Zbrush	Sculptris	Forger	Armorpaint	3Dcoat	Polybrush
T1	G	G	G	E	A	A	P
T2	P	A	No	G	P	E	P
T3	P	P	P	E	G	P	A
T4	G	G	No	G	E	G	E
T5	P	G	E	G	E	E	G
T6	E	E	E	E	G	E	A

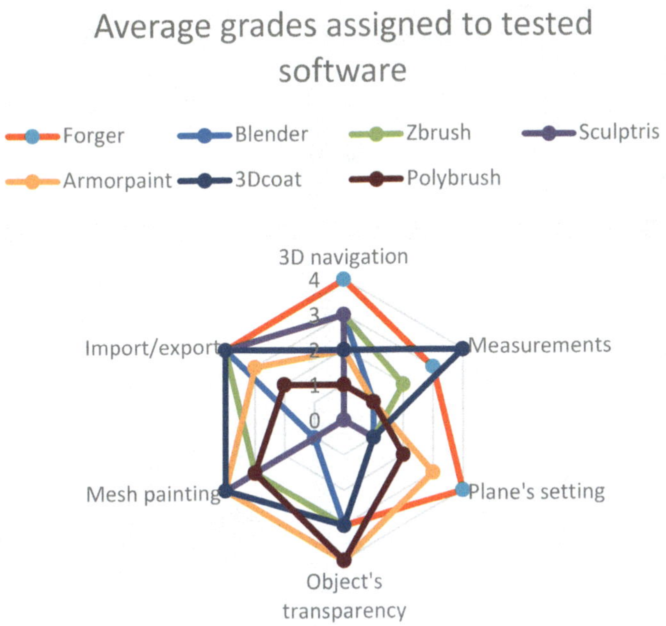

Fig. 2. Radar chart representing the average grades assigned by the surgeons to the submitted tasks

Table 5. Global score

	Forger	3Dcoat	Zbrush	Armorpaint	Polybrush	Blender	Sculptris
Score	3,5	3	2,7	2,8	2,2	2,2	2

From this preliminary investigation about these applications' usability, Forger resulted as the easiest and most user-friendly with the highest global score, as underlined in Table 5. The same weight has been assigned to every feature.

The user can easily pan, zoom and rotate the 3D models with an intuitive touch screen interface; planes can be imported into the scene and roto-translation transformations are interactively performed by dragging or rotating the objects, as shown in Fig. 3.

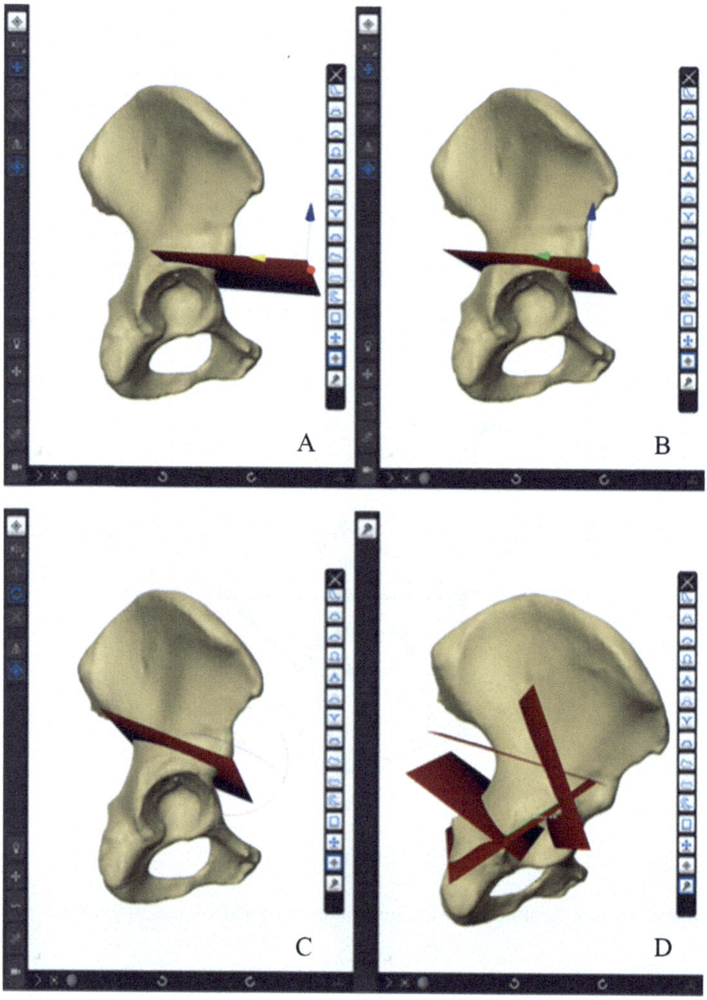

Fig. 3. Interactive planes' transformations in Forger. (A) and (B) illustrate translations, while (C) rotations. (D) represents the final result with four cutting planes.

Mesh painting is very natural, thanks to the combination of touch screen and Apple Pencil®, so different areas can be precisely highlighted or segmented using different colours. This software, developed for iPad, is lightweight and the exported .obj file with texture is compatible with most CAD software systems.

Table 6 shows the time in which the two surgeons completed the tasks.

Table 6. Time in which surgeons completed the preoperative planning for any of the tested software

Software	Surgeon 1 [minutes]	Surgeon 2 [minutes]
Blender	65	No
Zbrush	49	58
Sculptris	37	48
Forger	29	34
Armorpaint	47	52
3Dcoat	34	40
Polybrush	No	No

The whole test procedure was carried out in about 30 min using Forger. In this timespan, surgeons were able to prepare the 3D model of the bone with resection planes and colours to show accessible areas, exposed bone stock and point out critical areas.

Forger is easy to use, cheap, has all the features required, is very intuitive thanks to the touch screen provided by the hosting hardware and the exported files are compatible with the main CAD software packages. The main drawbacks are: models' transparency can only be enabled selectively once at a time and painting is not enabled when transparency is turned on. For more, this application does not allow to cut the model or show a section of the model.

Moreover, according to this preliminary investigation, 3D Coat proved to be a valuable alternative to Forger for preoperative planning, as it provides the same basic features and has been appreciated by the surgeons; 3D Coat, though, is less user friendly and more computationally demanding than Forger.

5 Conclusions

The advent of new technologies as Additive Manufacturing unlocked new methods for limb-sparing surgery, namely Patient Specific Instruments for accurate bone resection and metal custom implants. The project of such devices is typically demanded to third-party companies with a large loss of time, mainly due to communications issues between surgeons and engineers about the case-specific design constraints. Several iterations are often necessary to produce the definitive design of PSIs and custom prosthesis, so it is crucial to speed up the design process with dedicated digital tools. For this purpose, two experienced oncological orthopedics were asked to test a series of low-cost general-purpose software in order to find an efficient communication platform to allow an easy and effective exchange of knowledge and information between engineering and medical field, promoting an effective concurrent design.

Despite some limitations, Forger proved to be a very easy and intuitive software to visualize and paint 3D meshes, as well as to define resection planes; it resulted the best option between the tested software to share knowledge and information between surgeons and engineers for preoperative planning.

6 Future Developments

Future studies will be addressed to draft a usability questionnaire for a quantitative evaluation of the applications' performances and to extend the test to a wider group of surgeons.

A long-term study will assess how the adoption of the proposed method affects the clinical practice, in terms of time saving and surgical output's quality.

With the aim to establish a proficient concurrent design method the implementation of a web-based platform will be taken into account. This should allow surgeons to share diagnostic images, get the 3D reconstruction of the structures of interest, plan surgery and, eventually, validate the proposed design of PSI and implant, with great flexibility as no specific hardware is required. A web-based platform could enable a real-time communication though a live chat and a proficient interaction between users.

References

1. Kumar, N., Gupta, B.: Global incidence of primary malignant bone tumors. Curr. Orthop. Pract. **27**, 530–534 (2016)
2. Ozaki, T., et al.: Osteosarcoma of the pelvis: experience of the cooperative osteosarcoma study group. J. Clin. Oncol. **21**, 334–341 (2003)
3. Gouin, F., Paul, L., Odri, G.A., Cartiaux, O.: Computer-assisted planning and patient-specific instruments for bone tumor resection within the pelvis: a series of 11 patients. Sarcoma **2014** (2014)
4. Leng, S., et al.: Customized cutting template to assist sternotomy in pectus arcuatum. Ann. Thorac. Surg. **107**, 1253–1258 (2019)
5. Carfagni, M., et al.: Towards a CAD-based automatic procedure for patient specific cutting guides to assist sternal osteotomies in pectus arcuatum surgical correction. J. Comput. Des. Eng. **6**, 118–127 (2019)
6. Enneking, W.F., Spanier, S.S., Goodman, M.a.: A system for the surgical staging of musculoskeletal sarcoma. Clin. Orthop. Relat. Res. 4–18 (1980)
7. Kawaguchi, N., Ahmed, A.R., Matsumoto, S., Manabe, J., Matsushita, Y.: The concept of curative margin in surgery for bone and soft tissue sarcoma. Clin. Orthop. Relat. Res. **5**, 165–172 (2004)
8. Volpe, Y., et al.: Surgery of complex craniofacial defects: a single-step AM-based methodology. Comput. Methods Programs Biomed. **165**, 225–233 (2018)
9. Fan, H., et al.: Implantation of customized 3-D printed titanium prosthesis in limb salvage surgery: a case series and review of the literature. World J. Surg. Oncol. **13**, 1–10 (2015)
10. Cartiaux, O., Paul, L., Francq, B.G., Banse, X., Docquier, P.L.: Improved accuracy with 3D planning and patient-specific instruments during simulated pelvic bone tumor surgery. Ann. Biomed. Eng. **42**, 205–213 (2014)
11. Wong, K.C., Sze, K.Y., Wong, I.O.L., Wong, C.M., Kumta, S.M.: Patient-specific instrument can achieve same accuracy with less resection time than navigation assistance in periacetabular pelvic tumor surgery: a cadaveric study. Int. J. Comput. Assist. Radiol. Surg. **11**, 307–316 (2016)

12. Mimics Homepage. https://www.materialise.com/en/medical/software/mimics. Accessed 19 June 2019
13. Sectra Homepage. www.medical.sectra.com. Accessed 19 June 2019
14. mediCAD Homepage. www.hectec.de. Accessed 19 June 2019
15. Peekmed Homepage. http://www.peekmed.com. Accessed 19 June 2019
16. TraumaCAD Homepage. http://www.traumacad.com. Accessed 19 June 2019
17. Chen, Z., King, W., Pearcey, R., Kerba, M., Mackillop, W.J.: The relationship between waiting time for radiotherapy and clinical outcomes: a systematic review of the literature. Radiother. Oncol. **87**, 3–16 (2008)

CAD Modeling for Evaluating LVOT Obstruction in Transcatheter Mitral Valve Replacement

Salvatore Pasta[1(✉)], Stefano Cannata[2], Giovanni Gentile[2], Tommaso Ingrassia[1], Vincenzo Nigrelli[1], and Caterina Gandolfo[2]

[1] Università degli Studi di Palermo, Palermo, Italy
salvatore.pasta@unipa.it
[2] ISMETT IRCCS, Palermo, Italy

Abstract. The current generation of transcatheter heart valves (THV), as the Edwards SAPIEN 3 Ultra (S3), is not specifically designed for mitral position implantation and has intrinsic design geometry that may make mitral implantation suboptimal. This study aimed to develop a computed-tomography (CT) based CAD workflow for the preoperative planning of transcatheter mitral valve replacement (TMVR) by evaluating the resulting obstruction in the left ventricular outflow tract (LVOT). Specifically, the computational framework to reconstruct heart anatomy and virtually deploy the THV into mitral valve annulus was developed and successively applied to the cases of two patients who experienced annuloplasty ring failure. Planimetric assessment of the cross-sectional area of the neo-LVOT was quantified at different anatomic levels of implanted THV. Findings revealed the importance of the proposed CAD modeling workflow to enable more informative pre-operative assessments of the risk related to the development of the neo-LVOT obstruction and even to guide the Heart Team regarding device selection, sizing and intended positioning for TMVR.

Keywords: Transcatheter mitral valve replacement · CAD · Virtual simulation · Reverse engineering · Medical imaging

1 Introduction

1.1 Clinical Need

Evolution of catheter-based structural interventions has given patients less invasive alternatives to surgery; however, the current generation of transcatheter heart valves (THV) are not specifically designed for mitral position implantation and have intrinsic geometry that may make mitral implantation suboptimal. Indeed, while the term transcatheter mitral valve replacement (TMVR) is often used for implantation of a dedicated transcatheter mitral valve in a native mitral anatomy, aortic transcatheter heart valves (THVs) are frequently used for transcatheter valve replacement in degenerated bioprosthetic mitral valves (i.e., valve-in-valve), failed mitral annuloplasty rings (i.e., valve-in-ring), or patients with calcific mitral valve disease [1]. Interventional cardiologists are

C. Rizzi et al. (Eds.): ADM 2019, LNME, pp. 776–787, 2020.
https://doi.org/10.1007/978-3-030-31154-4_66

faced with unique challenges including valve deliverability, embolism, and notably left ventricular outflow tract (LVOT) obstruction. This condition is a prevalent procedural complication with reported incidence rates of 8.2% and 11.2% in valve-in-ring and valve-in-annular calcification procedures, respectively [2].

Transcatheter mitral valve replacement (TMVR) can lead to THV elongation in the outflow tract of the left ventricle, whereas the pre-existing "native" LVOT confined by the most basal septum and the intervalvular fibrosa remains unchanged (see Fig. 1) [3]. The newly created elongation due to the protrusion of the THV into left ventricle is confined posteriorly by the deflected anterior mitral valve (AML) leaflet in TMVR.

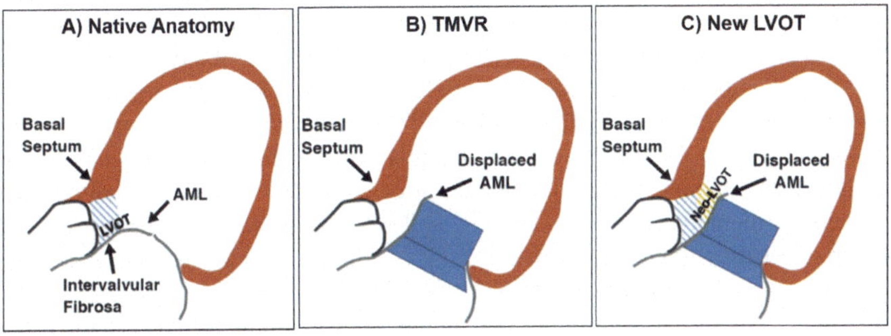

Fig. 1. (A) In the native anatomy, the LVOT is confined by the basal septum, the intervalvular fibrosa and the basal portion of the anterior mitral valve leaflet. (B) The TMVR device deflects the AML "septally," thereby elongating the outflow tract toward the left ventricle. This elongation is referred to as the "neo-LVOT", confined by the basal septum and the septally deflected anterior mitral valve leaflet.

Such a THV protrusion into LVOT likely alters hemodynamic into the left ventricle as well as the ascending aorta [4]. A small outflow tract geometry will create a region of turbulent flow distal to the narrowing, causing large drop in pressure beneath the aortic valve and ultimately determining sub-aortic stenosis when residual LVOT cross-sectional area is about 185 mm^2 [5]. Indeed, hemodynamically relevant obstruction can result in acute hemodynamic compromise and death.

1.2 Role of CAD Modeling and Virtual Simulation

Pre-procedural cardiac computed tomography (CT) has become the main tool for assessing anatomical suitability of the bioprosthesis for TMVR. This imaging technique can provide high-spatial resolution images that allows for a granular interrogation of the mitral valve apparatus and surrounding structures such as the left ventricle and outflow tract throughout the cardiac cycle. A complete pre-procedural CT evaluation for TMVR will usually include:

- mitral annular sizing for device selection,
- determination of LVOT obstruction risk with a specific device profile,
- surgical planning of transcatheter intervention.

In spite of high-resolution CT images, interventional cardiologists cannot account for CAD tools to pre-operatively plan the optimal deployment of THV in the mitral valve and therefore evaluate the obstruction into LVOT. In this setting, stereolithographic (STL) device representation outlines the dimensions and configuration of an unfolded THV in three-dimensional (3D) space and can be virtually superimposed onto the CT dataset using common CAD tools. These include reverse engineering [6, 7] for semi-automatic segmentation of heart anatomy, CAD operation for the evaluation of geometric features [7, 8] such as the left ventricle centerline, and finally the measurement of the cross-sectional area of the neo-LVOT.

In this study, we developed a computational CAD framework for the pre-operative virtual evaluation of the THV obstruction into LVOT to guide interventional cardiologists in the surgical planning of TMVR. Specifically, the case of two patients with mitral valve complications related to previously treated minimally-invasive annuloplasty ring bands are described. Successful TMVR was then accomplished by deployment of the Edwards SAPIEN 3 Ultra THV (Edwards Lifesciences, Irvine, CA, USA) with sizing of 26 mm. To accomplish this goal, we first generated a STL model of the SAPIEN 3 Ultra device and then reconstructed patient-specific anatomy by semi-automatic segmentation of left ventricle and left atrial heart chambers using the commercial medical imaging software Mimics (Mimics Innovation Suite v.21, Materialise, Belgium). Finally, the SAPIEN 3 Ultra THV was virtually superimposed onto the CT images by CAD operations and then the cross-sectional area determined by the neo-LVOT was quantified.

2 Methods

2.1 Transcatheter Heart Valve Model

The Edwards Lifesciences SAPIEN 3 Ultra is manufactured by a cobalt-chromium alloy frame with four cells for each row and twelve columns where three-sutured bovine pericardial leaflets are anchored. The S3 device is designed with an outer sealing skirt to minimize paravalvular leakage at inflow portion of device. The geometrical model of 26 mm S3 frame was acquired with a high-resolution micro-CT scanner (Skyscan 1172 with a resolution of 0.17 micron) and then a general reverse engineering approach was adopted to obtain the CAD model of the prosthetic heart valve (Fig. 2). Using Rhinoceros (Rhino v.5, Robert McNeel & Associates, USA), the point cloud obtained from the micro-CT scanner was used to generate a set of 3rd-order NURBS curves representing the cell unit of THV frame. Pattern and mirroring commands were adopted to reconstruct each row and column on a flat plane (Fig. 2A). A cylindrical surface with diameter and height equals to the nominal dimensions of SAPIEN 3 Ultra THV was generated and then wrapping of the flat THV frame onto the cylindrical surface object was performed to obtain the final THV frame geometry (Fig. 2B). The sealing skirt was modeled using a cylindrical surface from the bottom to the second row for the THV device. Prosthetic valve leaflets were modeled using a general 3D parametric geometry of the native aortic valve leaflet developed previously by our group [9].

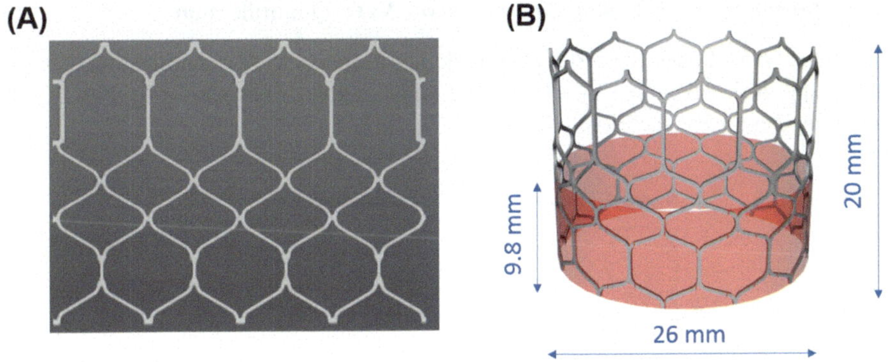

Fig. 2. (A) CAD representation of each row and column before wrapping onto a cylindrical surface and (B) final model of SAPIEN 3 Ultra THV stent frame.

2.2 Anatomic Heart Segmentation

Electrocardiographically-gated computed tomography (ECG-gated CT) scan was preoperatively performed for patient-specific anatomic reconstructions a using a 64 detector-row CT scanner. Specifically, ECG-gated has the advantage to effectively "freeze" cardiac pulsation and reduce motion artifacts due to heart beating when compared to conventional CT imaging. The entire scan was obtained during one breath-hold that was discretized using n.10 equally-distributed time phase as done previously [14]. Each time phase represents an image volume of information at specific point of the cardiac cycle. Images corresponding at diastole (i.e., minimum ventricular volume) were therefore processed by Mimics to reconstruct the left ventricle, left atrium and aortic root anatomy in terms of spatial position and dimension by applying different greyvalues and multiple masks. Volumetric CT images in standard DICOM format were imported in Mimics to show the sagittal, coronal and axial views of patient-specific heart. Using different Hounsfield unit thresholds, it was possible to discern the inner lumen of left ventricle, left atrial and the ascending aorta from surrounding tissues. Segmentation was performed by semi-automatic thresholding operation followed by manual mask editing and morphological operation including cropping, erosion and smoothing. The annuloplasty ring band was distinctly reconstructed using a different mask. Once the segmented regions were extracted, the heart model and annuloplasty ring band were independently exported as STL files for the modeling of THV in Rhino. Reconstructions were performed at end-systolic phase for which the left ventricular chamber is the smallest and thus the neo-LVOT is the worst.

2.3 Workflow of THV Modeling for Neo-LVOT Quantification

The framework of basic steps of CAD modeling for neo-LVOT quantification is based on the approach developed by Blanke and collaborators [3] as shown by Fig. 3. The following steps were therefore adopted:

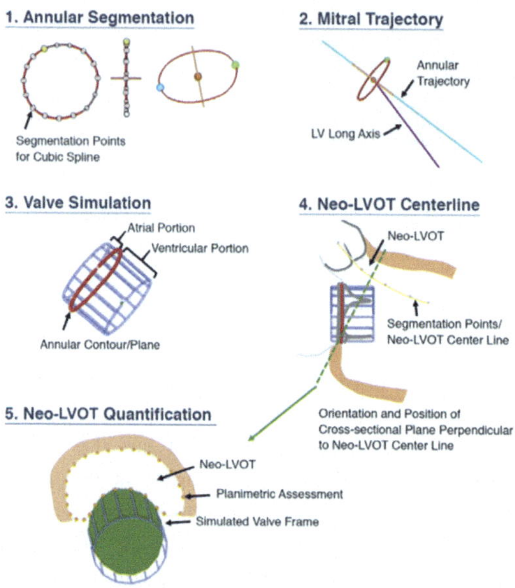

Fig. 3. Overview of basic steps for CAD modeling of THV as proposed by Blanke et al. [3]

1. **Annular segmentation;** for dedicated TMVR in a native mitral anatomy, the biomechanical landing zone of THV was the mitral annulus. Thus, a cubic spline curve interpolating the vertices of the STL file of each patient heart at mitral annular region was generated in Rhino.
2. **Mitral trajectory;** once the mitral annulus curve was developed, the curve center on annulus plane was determined, and the annular trajectory was generated by drawing a line perpendicularly to the annular plane and passing through the annulus center. This annular trajectory is commonly different from that described by the left ventricular long axis.
3. **THV CAD positioning;** the SAPIEN 3 Ultra longitudinal axis of STL file was then oriented parallel to the annular trajectory and coaxially to the cubic spline curve describing the mitral annulus. To ensure enough radial force between the THV and the myocardial wall, the SAPIEN 3 Ultra was positioned with 1/3 height into atrial portion while the remaining part was inside the ventricular portion. The level of implantation can be eventually modified according to clinical indications dictated by interventional cardiologists.
4. **Neo-LVOT centerline;** once virtual device position and orientation were confirmed, a geometric centerline was created along the curvilinear axis of the

neo-LVOT. This centerline should begin at the center of the aortic valve annulus and trace the center of the outflow tract to the smallest region of the neo-LVOT as determined between the THV and the left ventricular wall.

5. **Neo-LVOT quantification;** a set of area measurements were made at cross-sections orthogonal to the neo-LVOT centerline. Specifically, measurements of the area between the left ventricular wall and the SAPIEN 3 Ultra THV wall were done at proximal, mid and distal levels of the device.

3 Results

3.1 Case Study #1

A 70 years-old gentlemen was hospitalized at ISMETT IRCCS institution for mitral valve complications related to a previously treated minimally-invasive annuloplasty band ring. At baseline, echocardiography revealed dysfunction of the mitral valve and a slightly dilated left ventricular chamber. Pre-operative CT imaging was performed to assess the feasibility of TMVR and measure mitral valve annulus (diameter of 19.7 mm). The 26 mm SAPIEN 3 Ultra THV was taken into consideration for TMVR with decision taken by the Heart Team on the basis of clinical considerations of the risk profile of the patient. Figure 4 shows the anatomic reconstruction of both the heart anatomy and annuloplasty band ring in STL format while Fig. 5 displays the final position of THV and the neo-LVOT centerline after CAD modeling. It can be clearly observed that the presence of the SAPIEN 3 Ultra THV into the left ventricle reduces remarkably the configuration of native LVOT. Specifically, a 3D view from the ascending aortic root shows obstruction induced by the protrusion of THV into left ventricle portion, suggesting hemodynamic disturbance portending to high flow velocity and a pronounced pressure gradient (Fig. 6). Nevertheless, planimentric assessment of neo-LVOT by cross-sections with respect to neo-LVOT centerline revealed a proximal area of 415.5 mm^2, mid area of 386.9 mm^2 and distal area of

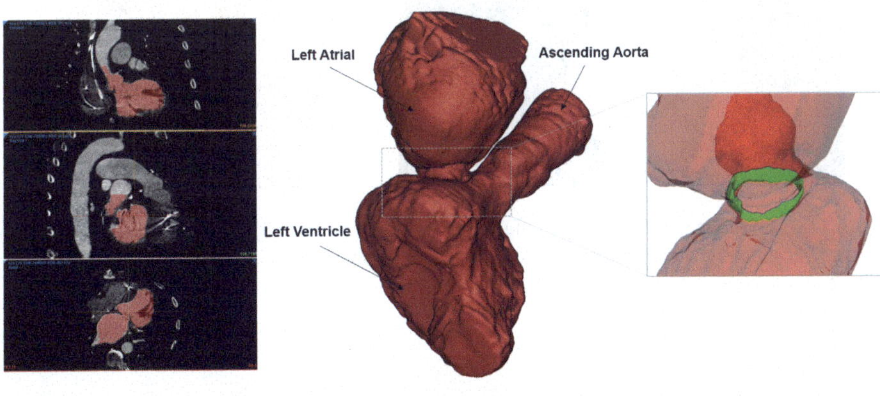

Fig. 4. Segmentation of patient Case #1 (left), STL 3D model of the heart (middle) in red color and inset of annuloplasty ring band (right) in green color

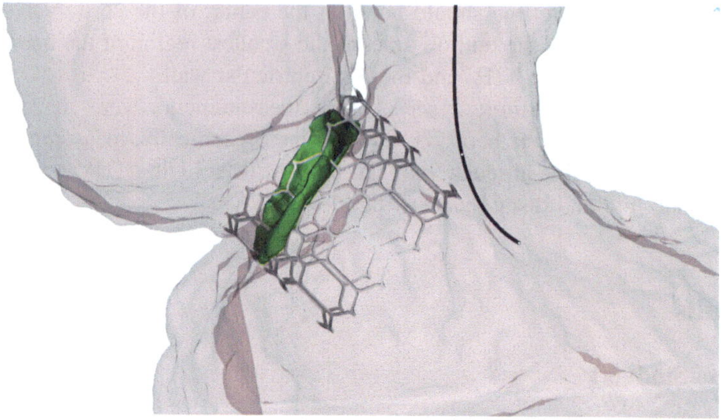

Fig. 5. CAD model showing virtual deployment of SAPIEN 3 Ultra valve (white color) and neo-LVOT centerline (black solid line) for Case #1

367.2 mm^2 (Fig. 7). These measurements show that the neo-LVOT area is higher than the clinical cut-off of 100 mm^2, which is considered predictor of adverse outcome related to neo-LVOT. Patient underwent TMVR procedure and post-operative echocardiographic assessment reveled successful deployment of SAPIEN 3 Ultra THV, although this should be confirmed by CT imaging.

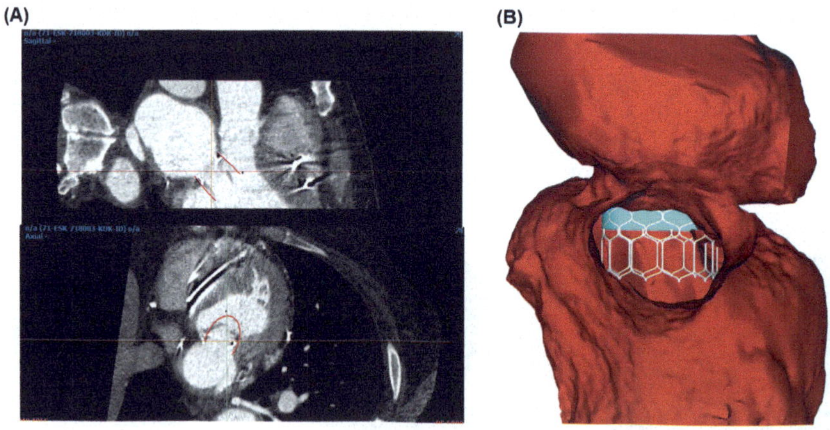

Fig. 6. (A) CT images reformatted along annular trajectory and (B) view from ascending aorta of Case #1 showing the obstruction into LVOT due to the presence SAPIEN 3 Ultra THV for TMVR (light blue = skirt and white = THV frame)

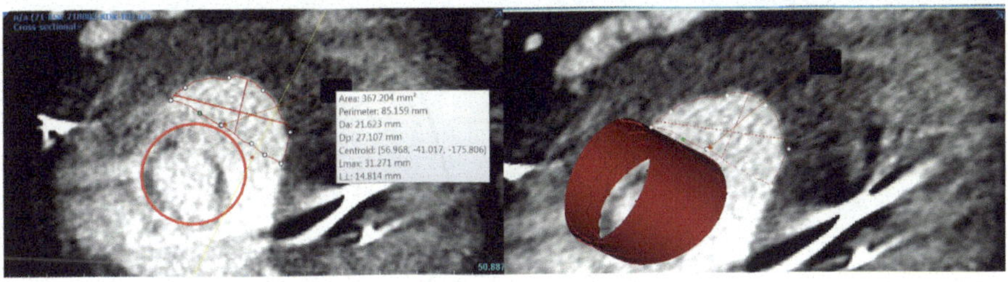

Fig. 7. Measurement of the cross-sectional area of neo-LVOT at distal end of SAPIEN 3 Ultra for Case #1

3.2 Case Study #2

A seventy-one-year old woman presented failed mitral annuloplasty rings for the treatment of a severe mitral regurgitation. This was initially found by echocardiography and then confirmed by high-contrast CT imaging after congestive heart failure symptoms. Clinical indications suggested to implant the 26 mm SAPIEN 3 Ultra THV, although the reduced mitral annulus dimension (diameter of 17.8 mm) and the absence of ventricular dilatation.

For Case #2, the anatomic reconstruction as well as neo-LVOT centerline and quantification of the cross-sectional area are shown in Figs. 8, 9, 10 and 11. We found that the neo-LVOT area at proximal, mid and distal levels of THV geometry were 366.2 mm^2, 294.9 mm^2, and 205.3 mm^2 respectively. These data suggest that TMVR for this patient could be problematic due to a remarkable reduction of the native LVOT configuration.

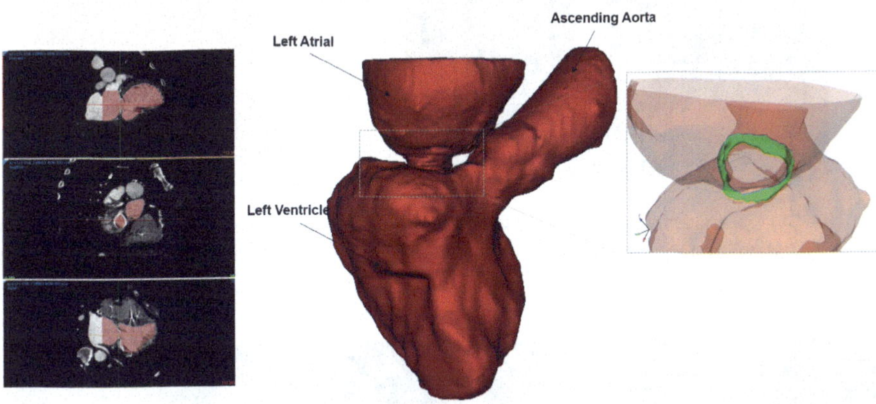

Fig. 8. Segmentation of patient Case #2 (left), STL 3D model of the heart (middle) in red color and inset of annuloplasty ring band (right) in green color

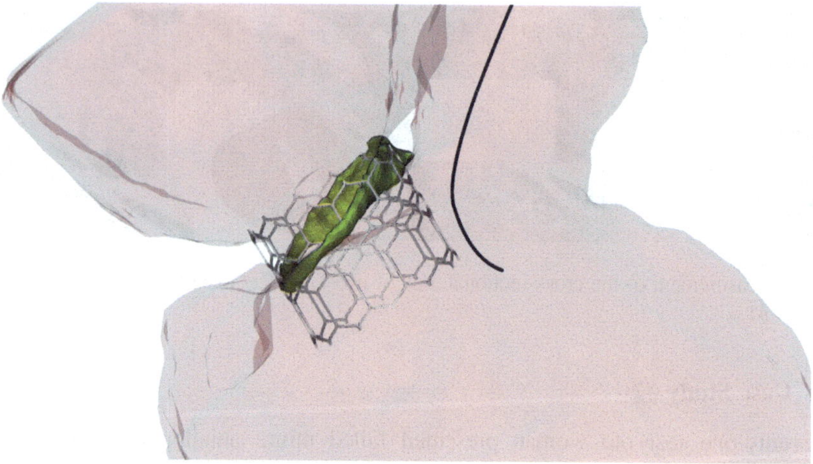

Fig. 9. CAD model showing virtual deployment of SAPIEN 3 Ultra valve (white color) and neo-LVOT centerline (black solid line) for Case #2

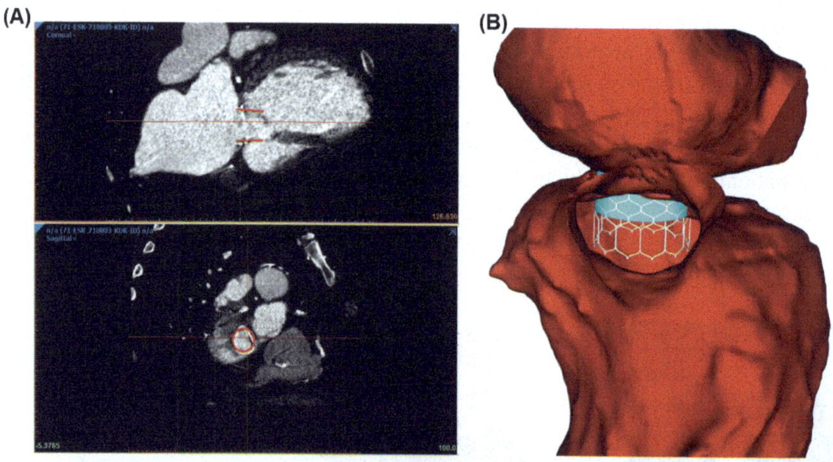

Fig. 10. (A) CT images reformatted along annular trajectory and (B) view from ascending aorta of Case #2 showing the obstruction into LVOT due to the presence SAPIEN 3 Ultra THV for TMVR (light blue = skirt and white = THV frame)

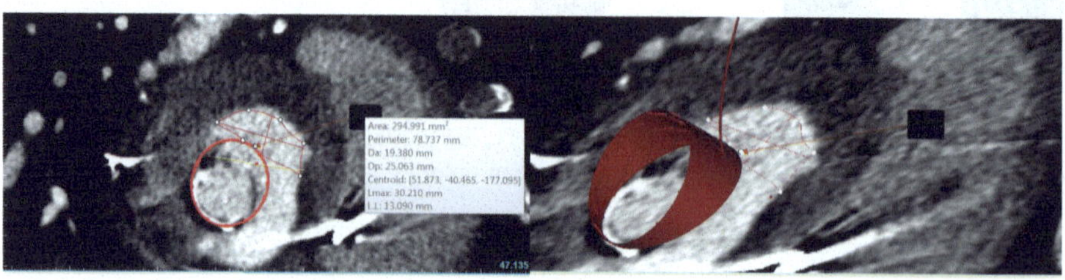

Fig. 11. Measurement of the cross-sectional area of neo-LVOT at mid-level of SAPIEN 3 Ultra for Case #2

4 Discussion

In this study, a computational CAD modeling framework was developed to pre-operatively quantify the obstruction induced by the THV protrusion into native LVOT for TMVR and then described the cases of two patients who experienced annuloplasty ring failure. The proposed CAD modeling framework allowed to inform clinicians on the residual cross-sectional area generated by the virtual deployment of THV into mitral annulus. Areas of less than 185 mm^2 at end-systole are considered to signify an increased risk of outflow obstruction and adverse events [1, 5]. We found that an area of 367.2 mm2 for Case #1 to likely suggest successful TMVR while Case #2 had neo-LVOT area of 205.3 mm2 to portend somewhat suboptimal deployment of SAPIEN 3 Ultra THV. This approach combining diagnostic imaging with CAD modeling represent a valuable tool to virtually simulate TMVR and different scenario to guide prophylactic intervention. We will integrate the proposed CAD modeling framework with computational fluid dynamic to assess whether the neo-LVOT alters aortic valve mechanics.

Successful TMVR depends on accurate sizing of the mitral annulus and avoidance of LVOT obstruction. Incorporation of computer-aided design into high-contrast CT imaging allows for *in-silico* device bench testing in patient-specific anatomy [10]. Modeling of proposed THV at different deployment heights and angles/depths into the left ventricle allows estimation of LVOT obstruction. A CT-based CAD prediction modeling tool is also helpful in optimizing patient selection for novel transcatheter techniques such the laceration of the anterior mitral leaflet to reduce the hemodynamic impact of LVOT obstruction into left ventricle [11].

From a hemodynamic perspective, our experience on computational fluid dynamics of cardiovascular diseases advises that the pressure drop across a narrowing is dependent not only on the smallest cross-sectional area of the narrowing, but also on the length of the narrowing [12–16]. In the context of TMVR, a neo-LVOT which has a small cross-sectional area at a discrete point will have different flow characteristics, that could be evaluated through computational fluid-dynamics simulations [17–19], than a neo-LVOT with small cross-sectional areas at a series of points along the outflow tract. Moreover, cross-sectional area alone may not fully account for the pressure gradient across the neo-LVOT. In the case of TMVR devices which have open-cell architectures such as the Edwards SAPIEN platform, there is potential for systolic flow to exert a pressure on the anterior mitral valve leaflet surface through the stent frame. This pressure would result in drag force that could push the anterior mitral valve leaflet towards the septum, thereby leading to THV migration and ultimately increasing the risk of adverse events related to LVOT obstruction.

Although virtual implantation of THVs in the setting of the aortic valve has been developed in modern angiography imaging systems to guide prophylactic intervention, the application to TMVR is limited by the fact that THVs are not designed for mitral valve repair. Angiographic scanner can be equipped with automated algorithms to allow the reconstruction of the aortic root centerline and to virtually place the THV in the aortic annulus. In the case of TMVR, the CAD modeling approach here described is more complex by not only requiring the assessment of mitral trajectory but even the evaluation of the neo-LVOT area by cross-sectional plane analysis. Moreover, the modeling accuracy and development of procedural standards need to be addressed for the application of the proposed CAD modeling framework in clinical practice.

5 Conclusion

We conclude that CT-based CAD modeling is a fundamental tool for the pre-operative planning of the risk related to LVOT obstruction of THV for TMVR and to guide the Heart Team discussions regarding device selection, sizing and intended positioning. Future studies will be undertaken to assess the computational fluid mechanics induced by the protrusion of THV into LVOT and to print 3D models [20] for providing useful insights into the size and shape of the predicted neo-LVOT for TMVR.

References

1. Wang, D.D., Eng, M., Greenbaum, A., Myers, E., Forbes, M., Pantelic, M., et al.: Predicting LVOT obstruction after TMVR. JACC Cardiovasc. Imaging **9**(11), 1349–1352 (2016)
2. Paradis, J.M., Del Trigo, M., Puri, R., Rodes-Cabau, J.: Transcatheter valve-in-valve and valve-in-ring for treating aortic and mitral surgical prosthetic dysfunction. J. Am. Coll. Cardiol. **66**(18), 2019–2037 (2015)
3. Blanke, P., Naoum, C., Dvir, D., Bapat, V., Ong, K., Muller, D., et al.: Predicting LVOT obstruction in transcatheter mitral valve implantation: concept of the Neo-LVOT. JACC Cardiovasc. Imaging **10**(4), 482–485 (2017)
4. Kohli, K., Wei, Z.A., Yoganathan, A.P., Oshinski, J.N., Leipsic, J., Blanke, P.: Transcatheter mitral valve planning and the Neo-LVOT: utilization of virtual simulation models and 3D printing. Curr. Treat. Options Cardiovasc. Med. **20**(12), 99 (2018)
5. Yoon, S.H., Bleiziffer, S., Latib, A., Eschenbach, L., Ancona, M., Vincent, F., et al.: Predictors of left ventricular outflow tract obstruction after transcatheter mitral valve replacement. JACC Cardiovasc. Interv. **12**(2), 182–193 (2019)
6. Ingrassia, T., Nalbone, L., Nigrelli, V., Ricotta, V., Pisciotta, D.: Biomechanical analysis of the humeral tray positioning in reverse shoulder arthroplasty design. Int. J. Interact. Des. Manuf. **12**(2), 651–661 (2018)
7. Ingrassia, T., Nigrelli, V., Ricotta, V., Nalbone, L., D'Arienzo, A., D'Arienzo, M., Porcellini, G.: A new method to evaluate the influence of the glenosphere positioning on stability and range of motion of a reverse shoulder prosthesis. Injury **50**, S12–S17 (2019)
8. Mirulla, A.I., Bragonzoni, L., Zaffagnini, S., Bontempi, M., Nigrelli, V., Ingrassia, T.: Virtual simulation of an osseointegrated trans-humeral prosthesis: a falling scenario. Injury **49**(4), 784–791 (2018)
9. Pasta, S., Gentile, G., Raffa, G., Scardulla, F., Bellavia, D., Luca, A., et al.: Three-dimensional parametric modeling of bicuspid aortopathy and comparison with computational flow predictions. Artif. Organs **41**(9), E92–E102 (2017)
10. Wang, D.D., Eng, M.H., Greenbaum, A.B., Myers, E., Forbes, M., Karabon, P., et al.: Validating a prediction modeling tool for left ventricular outflow tract (LVOT) obstruction after transcatheter mitral valve replacement (TMVR). Catheter Cardiovasc. Interv. **92**(2), 379–387 (2018)

11. Khan, J.M., Babaliaros, V.C., Greenbaum, A.B., Foerst, J.R., Yazdani, S., McCabe, J.M., et al.: Anterior leaflet laceration to prevent ventricular outflow tract obstruction during transcatheter mitral valve replacement. J. Am. Coll. Cardiol. **73**(20), 2521–2534 (2019)
12. Lee, J.J., D'Ancona, G., Amaducci, A., Follis, F., Pilato, M., Pasta, S.: Role of computational modeling in thoracic aortic pathologies - a review. J. Card. Surg. **29**(5), 653–662 (2014)
13. Pasta, S., Philippi, J.A., Tsamis, A., D'Amore, A., Raffa, G., Pilato, G., et al.: Constitutive modeling of ascending thoracic aortic aneurysms using microstructural parameters. Med. Eng. Phys. **38**(2), 121–130 (2016)
14. D'Ancona, G., Amaducci, A., Rinaudo, A., Pasta, S., Follis, F., Pilato, M., et al.: Haemodynamic predictors of a penetrating atherosclerotic ulcer rupture using fluid–structure interaction analysis. Interact. Cardiovasc. Thorac. Surg. **17**(3), 576–578 (2013)
15. Rinaudo, A., D'Ancona, G., Lee, J.J., Pilato, G., Amaducci, A., Baglini, R., et al.: Predicting outcome of acute aortic dissection with patent false Lumne by computational flow analysis cardiovascular. Eng. Technol. **5**(2), 176–188 (2014)
16. Pasta, S., Gentile, G., Raffa, G.M., Bellavia, D., Chiarello, G., Liotta, R., et al.: In silico shear and intramural stresses are linked to aortic valve morphology in dilated ascending aorta. Eur. J. Vasc. Endovasc. Surg. **54**(2), 254–263 (2017)
17. Cirello, A., Cucinotta, F., Ingrassia, T., Nigrelli, V., Sfravara, F.: Fluid–structure interaction of downwind sails: a new computational method. J. Mar. Sci. Technol. (Japan) **24**(1), 86–97 (2019)
18. Scardulla, F., Rinaudo, A., Pasta, S., Scardulla, C.: Evaluation of ventricular wall stress and cardiac function in patients with dilated cardiomyopathy. Proc. Inst. Mech. Eng. Part H J. Eng. Med. **230**(1), 71–74 (2016)
19. Scardulla, F., Rinaudo, A., Pasta, S., Scardulla, C.: Mechanics of pericardial effusion: a simulation study. Proc. Inst. Mech. Eng. Part H J. Eng. Med. **229**(3), 205–214 (2015)
20. Uccheddu, F., Carfagni, M., Governi, L., Furferi, R., Volpe, Y., Nocerino, E.: 3D printing of cardiac structures from medical images: an overview of methods and interactive tools. Int. J. Interact. Des. Manuf. **12**, 597–609 (2018)

A Reliable Procedure for the Construction of a Statistical Shape Model of the Cranial Vault

Antonio Marzola(✉), Michaela Servi, and Yary Volpe

Department of Industrial Engineering, University of Florence, 50139 Florence, Italy
{antonio.marzola,michaela.servi,yary.volpe}@unifi.it

Abstract. The application of CAx tools in surgery is representing a breakthrough for clinical practice, both in terms of effectiveness and costs. Working directly on the patient's own diagnostic images, this approach provides powerful tools for pre-operative simulation, complex-surgery planning, quantitative evaluation of asymmetry or dysmorphism and for the design of the patient-specific instrumentation. To exploit its full potential, methodologies are being developed to automatize and simplify the existing tools and strategies, in order to make them available also to less experienced CAx users, or directly to the surgeons.

With this aim, it is proposed a methodological procedure to automatically create a Statistical Shape Model of the cranial vault starting from a Training Set of pathologically unaffected adult crania. The Statistical Shape Model is useful as a template for a data-driven restoration of the physiological shape of the considered anatomy. The proposed procedure provides a reliable strategy for robust automatic detection of shape correspondence. Not requiring any user intervention, the number of samples in the Training Set can be increased at will to consequently increase the variability, and therefore the accuracy, of the resulting parametric model.

Keywords: Statistical Shape Model · Reverse engineering · CAx

1 Introduction

In the last few years, the astonishing evolution of Computer-Aided Technologies (CAx) software has made possible to process and analyze complex geometry. As a consequence, advanced CAx tools, primarily developed for engineering applications, has been introducing in non-traditional fields, as biology and anatomy. Especially in the field of medicine, this increasing ability to face with non-mathematical surfaces is proving its maximum potentialities: the possibility to virtually process 3D anatomical shape is a powerful tool to allow preoperative planification and simulation as well as to allow preoperative design of personalized prosthesis and surgical instruments.

Reverse Engineering (RE) techniques are applied to reconstruct the three-dimensional anatomical region of interest (ROI) starting from the patient's own diagnostic images made by Computerized Tomography (CT) or Magnetic Resonance

C. Rizzi et al. (Eds.): ADM 2019, LNME, pp. 788–800, 2020.
https://doi.org/10.1007/978-3-030-31154-4_67

Imaging (MRI). CAx software provide advanced tools to properly handle the so obtained 3D model allowing intervention planning and simulation, but also the design of custom medical devices (Fig. 1). This new high customized approach has proving to be really breakthrough, especially considering the possibility of manufacturing any complex geometry directly from the designed virtual model by means of Additive Manufacturing technologies [1]. For these reasons, in the last decade such an approach has been widely applied in the clinical practice, proving to be reliable both in terms of costs and effectiveness [2–5].

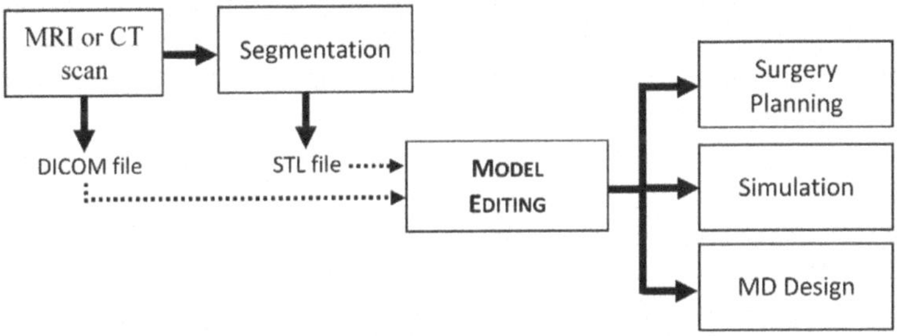

Fig. 1. Typical CAx framework in medical application. MD is for Medical Device.

In this scenario, there is now a great interest in automating the whole framework presented in Fig. 1, with the aim:

- of speeding it up, because time between the diagnosis and the surgical intervention is critical in some applications (as in oncology),
- of making the instruments available also to less experienced CAx users or directly to surgeons,
- of eliminating all time-consuming, repetitive and burden operations required of the user.

One of the most critical tasks in the process is the reconstruction of the anatomical regions to be treated, which often have malformations or missing parts. A correct reconstruction is essential to achieve an adequate surgery outcome, both for functional and, in some cases, aesthetical aims. At the same time, this task is undoubtedly the most complex, cumbersome and time-consuming for the user. For this reason, the related literature delivered a wide roster of alternatives to achieve a reconstruction that is reliable and as fast and simple as possible. Each reconstruction strategy is often properly designed for a specific ROI: namely considering the skull, a large number of very different approaches for reconstructing cranial vaults can be found [6]. This approaches can be categorized in three different strategies: *(i)* template-based approach uses a template as a prior knowledge which contains information about the expected shape for a data-guided reconstruction of the cranial bones to be restored, *(ii)* the surface-based approach uses a mathematical surface that interpolates the edge of the defect and *(iii)* slice-based approach which deform a mathematical curve to wrap the contour of the healthy bone slice-by-slice.

Such a wide interest on the cranial vault restoration is justified by the need to obtain a very accurate reconstruction: face is the most important mean of communication and social interaction [7] and a correct restoration will lead to relevant social and psychological benefits for the patient [8, 9].

Among the existing approaches, the template-based seems to be the most adequate to lead to an acceptable outcome, allowing a data-driven reconstruction able to fill defective region in a sensible way since it incorporates knowledge encoded in the template model. Despite that, there are still some critical issues that limit its application, i.e. the high user interaction required for a proper registration between the template and the defective model and the strict dependence of the reconstruction on the template.

To date, the best way to deliver an adequate template able to overcome the just discussed shortcomings is by means of statistical analysis methods, to gather and interpret information from a collection of samples belonging to the same class of shape (e.g. cranial vaults). A statistical representation of normal shape variations for the considered class makes possible to take into account the high complexity as well as the wide interpersonal variability of the anatomical structures. Statistical Shape Model (SSM) [10–12] represents the most established statistical tool in medical image analysis [13–15]. SSM describes a set of samples (forming the Training Set) by a Mean Shape and a series of Modes of Variation (MoV). The MoVs resume all possible shape deformation, i.e. the way in which each point of the Mean Shape can be distributed in space; such a distribution is learned from the initial Training Set. A more comprehensive description of the SSM is provided in the Sect. 2.

SSM "has been widely proposed in orthopedics, especially for anatomical bone modelling, joint cinematic analysis, staging of morphological abnormality, and pre- and intra-operative reconstruction" [16]. Furthermore, SSM has already proving its reliability as a template for cranial vault reconstruction [15].

In the other hand, the numerous and very different methods found in the literature for the definition of an anatomical SSM prove that obtaining a reliable SSM is not straightforward [14]. Actually, due to the wide anatomical variability and the extreme complexity of the hard tissues in the cranial region, the methods often fail to detect the shape correspondences. As an example, Fig. 2 shows a typical mismatching between the points of the outer and inner crust on a cranial vault model. The model is the result of a non-rigid pair-wise registration between two unaffected adult crania. The mismatching is mainly due to a physiological difference in size of the starting models, which cause incorrect correspondences. Establishing the point correspondences between all shapes is a fundamental request to be able to properly gather the information starting from the Training Set. Points correspondence is a crucial assumption to apply standard multivariate statistics to model a probability distribution over shapes because it provides properly organized data. As a consequence, achieving well-established correspondences is the major factors influencing the template quality, but, at the same time, it generally represents the most challenging part in the SSM construction [14].

In this work a methodological procedure to compute an SSM using a pair-wise registration based on an automatic detection of shape correspondences is presented. The procedure is properly designed to obtain a parametric model of the human cranial vault. Being automatic, it makes possible to easily enlarge the number of samples in the

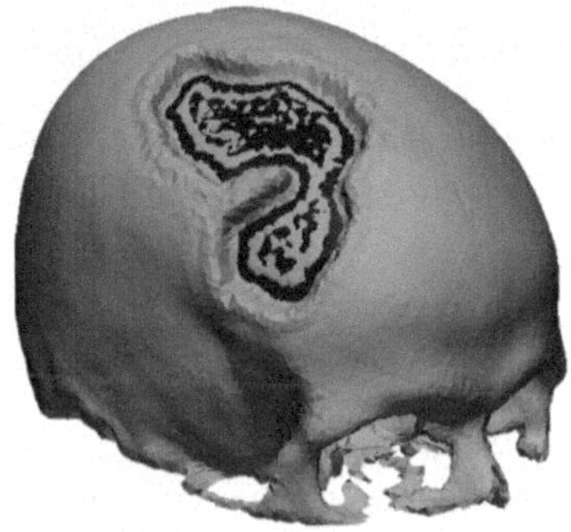

Fig. 2. Example of the erroneous result consequents to a non-correct correspondence: the mismatching between external and internal surface of the skull.

Training Set, increasing consequently the variability of the parametric model and therefore its accuracy.

2 Statistical Shape Model

The straightforward solution to represent complex shapes is by means of a discrete description based on a set of points distributed across the surface, usually referred to as landmarks [14, 17]. The resulting discrete model is called Point Distribution Model (PDM) [18]. In early approaches, landmarks refer to salient feature points and can be manually selected. To date, working with complex 3D models requires a dense set of points for an adequate representation of the shape addressed. Therefore, they are not located at as per the common definition but are extracted from the whole volumetric model by proper algorithms (e.g. Marching Cube). Such a high number of data makes difficult to gather and analyze the information. To reduce the problem's dimensionality without losing generality, the typical approach to SSM construction is based on Principal Component Analysis (PCA) [19].

SSM assumes that the space of all possible shape deformations can be learned from a set of M samples forming an appropriate Training Set $\{\Gamma^1\ \Gamma^2 \ldots \Gamma^M\}$. Each shape Γ^i is represented as a PDM, with a discrete set of landmarks $\mathbf{x}^i$.

$$\Gamma^i = \left\{\mathbf{x}_k^i | \mathbf{x}_k \in R^3, k = 1, \ldots, N\right\} \tag{1}$$

Where N is the number of points used to describe the shape and $\mathbf{x}_k^i$ contains the Cartesian coordinates of the point k. Γ^i is a 3 × N matrix.

For the application of the PCA, each shape Γ^i is rearranged by stacking the coordinates of each point k in a large (3N elements) column vector. Considering the global reference system in x-, y- and z-axis:

$$\vec{\mathbf{x}}^i = \left[x_1^i\, x_2^i \ldots x_N^i\, y_1^i\, y_2^i \ldots y_N^i\, z_1^i\, z_2^i \ldots z_N^i\right] \tag{2}$$

This way, the column vector represents the 3N variables of single observation Γ^i.

The whole Training Set is then organized in a matrix whose rows are the observations (i.e. the M shapes):

$$\mathrm{TS} = \left[\vec{\mathbf{x}}^1\, \vec{\mathbf{x}}^2 \ldots \vec{\mathbf{x}}^M\right] \tag{3}$$

TS represents the raw dataset on which apply the PCA. It contains all the variations described by the training samples: rows of TS are the observations and columns are the variables. But such a data manipulation is not enough for the straight application of the PCA, since must be strictly assumed that the points are in correspondence, i.e. the points in all training samples are located at corresponding position. In other words, it means that the k-th landmark x_k^i and x_k^j of two shapes Γ^i and Γ^j represents the same point of the shape. Since the manual landmarking approach is not suitable working with complex 3D shape, due to the large amount of points required to achieve an adequate superimposition, several methods for the automatic detection of shape correspondences have been proposed [14]. All the proposed automatic algorithms perform a registration between PDMs: the straightforward solution is to choose an arbitrary shape as reference and warping it to all others by using a non-rigid Iterative Closest Point-based approach. The main advantage in using ICP-based algorithms is that they treat correspondence as a variable to be estimated and not as an input as in the Procrustes-based algorithms. To overcome the bias introduced by choosing an arbitrary reference shape, the procedure can be repeated a second time with the Mean Shape obtained by the SSM [20] or by applying a symmetric version of the ICP (as presented in [21]).

Once the dataset TS is properly defined, the Mean Shape can be obtained by simply averaging over all M samples:

$$\bar{\mathbf{x}} = \frac{1}{\mathrm{M}} \sum_{i=1}^{\mathrm{M}} \vec{\mathbf{x}}_i \tag{4}$$

For the PCA, the corresponding covariance matric S is given by:

$$\mathrm{S} = \frac{1}{\mathrm{M}-1} \sum_{\mathrm{i}=1}^{\mathrm{M}} (\vec{\mathbf{x}}_i - \bar{\mathbf{x}}_i)(\vec{\mathbf{x}}_i - \bar{\mathbf{x}}_i)^T \tag{5}$$

An eigendecomposition on S delivers the principal Modes of Variation φ_{m} (eigenvectors of the matrix S) and their respective variances λ_{m} (eigenvalues of the matrix S) estimated on each points $\mathbf{x}_k$ of the initial dataset. It is now possible to approximate new valid shapes, i.e. belonging to the same family of the Training Set, by a linear combination of the first c Modes of Variation:

$$\mathbf{x} = \bar{\mathbf{x}} + \sum_{m=1}^{c} \alpha\sqrt{\lambda_m}\varphi_m \tag{6}$$

Where:

- $\bar{\mathbf{x}}$ is the Mean shape as define in Eq. (4).
- λ_m and φ_m are, respectively, the eigenvalues and eigenvectors resulting from the eigendecomposition (Eq. (5)), ordered by their variances so that $\lambda_1 \leq \ldots \leq \lambda_{M-1}$.
- $\alpha \in R^M$ is a coefficient vector able to constrain the allowed variation to plausible shape; usually, α is delimited in the range [−3; +3], which corresponds to a deformation that is 3 standard deviations away from the mean ($\alpha = 0$ means that the model is equal to the Mean shape).
- c represents the number of significative eigenvalues. In the presented application, c is defined as:

$$\frac{\sum_{i=1}^{c} \lambda_i}{\sum_{j=1}^{M-1} \lambda_j} = 0.98 \tag{7}$$

As Eq. (6) shows, SSM provides a parametric representation of the considered family of shapes; the variance is learned by a finite Training Set of significative models. In the medical field, having such a parametric representation of an anatomical healthy ROI means to have a deformable model useful as a template able to estimate a coherent undefective shape starting from the known parts of the structure to be restored. As discussed in the Sect. 1, the main limitation of this approach still remains the correspondences definition, despite there are many alternative methods proposed in the related literature: in the specific application of skull, the typical approach based on the ICP registration has proved to be inadequate if brutally applied on the 3D model as it is obtained from the CT data (Fig. 2). Trying to overcome these shortcomings, the procedure presented in the subsequent Section has been developed with the aim to make the correspondences definition via ICP as robust as possible.

3 The Proposed Procedure

The proposed procedure starts from an adequate number of PDMs of the cranial vault. Instead of PDMs, STL files can be provided as input. STL files can be easily obtained directly from the segmentation process. To facilitate the subsequent operations, each model has to be roughly finished (removal of outliers and the redundant regions). No further operations are required. In the following, the proposed procedure is described step by step. The flow chart of the proposed procedure is presented in Fig. 4.

3.1 Alignment and Scaling

To ensure a well-defined correspondence, the proposed procedure uses a pair-wise approach to deform a reference shape (henceforward called Reference) onto each sample (henceforward called Target) of the Training Set. Such a deformation is achieved by means of a non—rigid registration algorithm based on an Iterative Closest Point (ICP) algorithm, as described in detail in [22]. Using an ICP based approach allows to find the optimal non-rigid transformation from Reference to each Target, notwithstanding Reference and Targets are PDMs with different number of points. This way, a new Training Set is defined, that still describes the same variance of the initial population but containing samples that are the Reference properly deformed onto each Target. This ensures to have the same number of point N for all the Training shapes, and it usually ensures also an adequate correspondence. Among the existing approaches, the algorithm proposed in [22] has proved to be reliable in our application if the points correspondence has been correctly fulfilled: the rough application of this algorithm has leaded to questionable results when complex shape, as cranial vaults, are involved (Fig. 2). To overcome this shortcoming and to make the procedure more reliable, a different pre-alignment between the Reference and each Target is performed before the application of the ICP.

First, a roughly alignment is done by representing each sample of the Training Set in its Principal Component Space: it means that each shape is translated to superimpose its centroid onto the origin of the global reference system and rotated to align its principal axes of inertia with the x-, y- and z- axis of the global reference system.

To refine the alignment, the sagittal plane of each skull (calculated by the algorithm presented in [23]) is aligned with the y-plane (i.e. the plane orthonormal to the y vector). As a result, all the samples are aligned as shown in Fig. 3. In addition, a new Reference model is created as follows: for each Target i the original Reference is scaled in the three dimensions a, b and c (defined as shown in Fig. 3) by values equal to s^i_a, s^i_b and s^i_c. The s values are calculated as in Eq. (8):

$$s^i_\xi = \frac{\xi^i_{BB}}{\xi^R_{BB}}, \quad i = 1, \ldots, M - 1 \tag{8}$$

where ξ_{BB} indicates the dimension a, b or c of the Bounding Box (as defined in Fig. 3) belonging to the shape indicates by the superscript index: i is for the considered Target and R for the original Reference. As a result, the dimensions of the Bounding Box of each deformed References i is equal to the Bounding Box of the corresponding Target i.

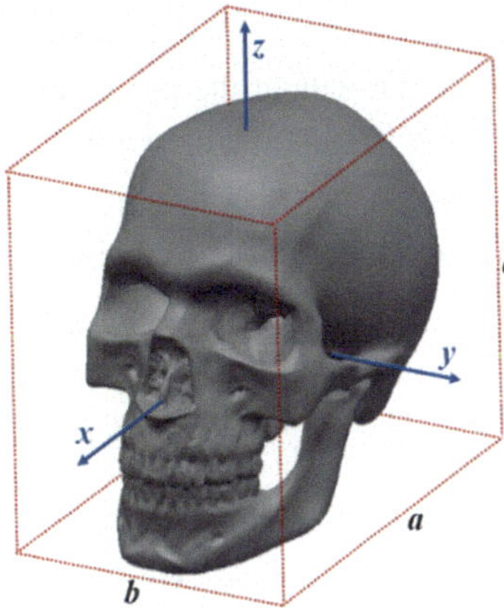

Fig. 3. The coordinate frame of the pre-aligned Training Set. The centroid of each model is superimposed to the origin of the axes. The red dotted line represents the Bounding Box of the sample. The dimension a, b and c are defined as reported.

3.2 External Crust Selection

The previous step ensures a consistent orientation of the shapes which are now aligned in a common coordinate frame. Despite that, this it is not still sufficient to ensure a reliable point-to-point matching during the subsequent ICP because of the high complexity and high variability of the population data. To reduce the possibility of mismatching, the internal and external crusts of each cranial vault are automatically isolated using a ray-tracing algorithm [24] as explained in [25]. The aim is to apply the ICP separately to the two crusts by treating them as different surfaces. This way, the data sets to be matched are much simpler than the initial model. The two parts will be re-merged after establishing the correspondence, and the SSM will be calculated on the complete cranium.

3.3 Non-rigid ICP

Non-rigid ICP algorithm described in [22] has been implemented to warp the external and internal crust of the scaled Reference *i* onto, respectively, the external and internal crust of each Target *i*. This algorithm performs an initial rigid alignment carried out by a bidirectional ICP algorithm [26], that also delivers the correspondence between the scaled Reference and the considered Target, followed by a non-rigid transformation modelled as a sum of Gaussian Radial Basis Function [27].

Applying this algorithm on the Training Set modified as described in the previous Sections has allowed to obtain an adequate warping (Fig. 5). As a result, a new Training Set is generated in which Targets are represented by the Reference properly deformed. As said in Sect. 3.1, this new Training Set still contains the same variance of the initial population, but the data are now organized appropriately for the creation of the SSM.

3.4 Correspondence

Having regard to its importance throughout the procedure, a further check is made to ensure that correspondences have been correctly defined. Given the vectors:

- $\mathbf{x}^i$ containing the points' coordinate of the new Target i, and
- $\mathbf{x}^R$ containing the points' coordinate of the Reference,

a k-Nearest Neighbor (k-NN) algorithms is iteratively applied to finds the nearest neighbor (i.e. the closest point) in $\mathbf{x}^i$ for each query point in $\mathbf{x}^R$. k is imposed equal to 1. Then, each Target vector $\mathbf{x}^i$ is sorted so that the j-th point is the nearest local neighbor of the j-th point of the Reference.

To avoid double assignments (i.e. more than one $\mathbf{x}^i_j$ defined as local neighbor of the same $\mathbf{x}^R_j$), the Target's points already assigned are removed from the list of possible subsequent matches. For each Target, the algorithm is iteratively applied as long as d_c, i.e. the median of the Euclidean distances calculated between the points in correspondence of the two considered PDMs, converges (Eq. (9)).

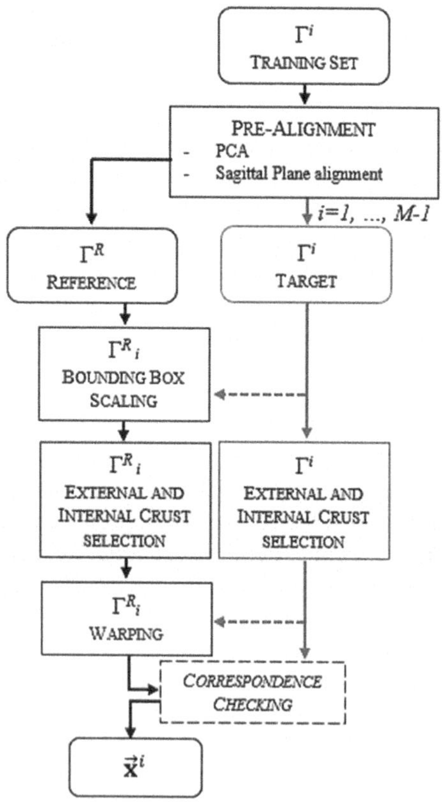

Fig. 4. The flow chart of the proposed procedure.

$$d_c = median\left(\sum_{j=1}^{N} dist\left(x_j^R, x_j^i\right)\right), \quad i = 1, \ldots, M-1 \tag{9}$$

After the data have been sorted properly to fulfil the correspondence, the internal and external crust of both Reference and Targets are re-merged as a single model to create the TS dataset (as described in Eq. (3)). The SSM of the cranial vault can be then created starting from the just created TS matrix.

Figure 4 shows the flow chart of the proposed procedure.

4 Results

The presented approach has been applied to a training set contains 100 STL models of pathologically unaffected adult crania (43 female, 57 male, average age = 32.6 years). The STL models have been obtained from anonymized diagnostic images (Computed Tomography or Magnetic Resonance Imaging) using properly designed tools in Materialise Mimics® software (Materialise NV, Leuven, Belgium) for image segmentation and for exporting the STL model. To facilitate the following operations, each of the so obtained STL models are roughly finished (removal of outliers and the redundant regions). In addition, in order to maintain the calculation cost limited, the number of points of each model is decimated up to about 150000–200000 points.

The presented procedure has been implemented in Matlab ® (MathWorks, Inc.) and applied on the 100 STL. As a result, the SSM of the cranial vault has been automatically provided. To assess the effectiveness of the proposed procedure, each model of the TS was compared with the corresponding cranial vault contained in the initial training set. The comparison is carried out by a deviation analysis between corresponding models (i.e. the deformed Reference and the correspondent Target). Figure 5 shows an example of the deviation measured between two corresponding samples.

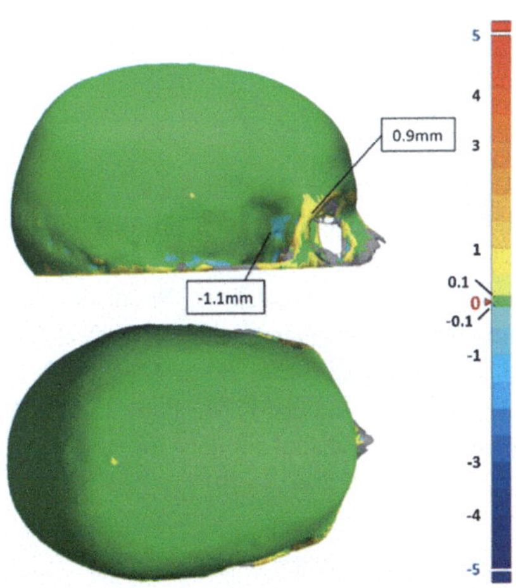

Fig. 5. Deviation (in mm) between the deformed Reference *i* and the correspondent Target *i*.

The results have shown a deviation between −0.5 and 0.5 mm, except in areas with high curvature where the correspondence is still not reliable, and the deviation increases to values between −2 and 2 mm. This error is however limited and relegated to small areas. Furthermore, in contrast with the outcomes obtained without the proper data alignment presented in this work, no model in the TS has showed inaccurate geometry. The result is, then, fully acceptable.

5 Discussion and Conclusions

The increasingly powerful and specialized CAx tools for handling complex geometries, such as anatomical ones, are representing breakthrough for clinical practice.

Especially in surgery, the capability to make reliable simulations as well as to design the actual implant at preoperative stage using the patient's own diagnostic images has proved to be very worthwhile both in terms of effectiveness and costs. To make the most of the potential shown by this approach, tools and methodologies are being developed that are increasingly suited to dealing with the difficulties inherent in anatomical structures. With this aim, a procedure able to ensure a reliable automatic detection of shape correspondence among cranial vaults' 3D discrete models has been proposed. The procedure allows to automatically create a parametric model useful as a template for a data-driven reconstruction of defective or deformed cranial vault avoiding the typical mismatching shown in Fig. 2. Not requiring any manual operation, the number of samples in the Training Set can be increased at will, without time-consuming or heavy and tedious operations required of the user. Therefore, this allows to have a Training Set with a larger population and, consequently, a more precise template being more able to reproduce the wide interpersonal variability of the anatomical structures.

The proposed procedure is properly developed for the cranial vault. Because of the high complexity of the anatomy structures it is difficult to provide a methodology reliable for all applications, despite this should be the objective of the future work.

References

1. Martelli, N., Serrano, C., van den Brink, H., Pineau, J., Prognon, P., Borget, I., El Batti, S.: Advantages and disadvantages of 3-dimensional printing in surgery: a systematic review. Surgery **159**(6), 1485–1500 (2016)
2. Tarsitano, A., Battaglia, S., Crimi, S., Ciocca, L., Scotti, R., Marchetti, C.: Is a computer-assisted design and computer-assisted manufacturing method for mandibular reconstruction economically viable? J. Cranio-Maxillo-Fac. Surg. **44**(7), 795–799 (2016)
3. Rogers-Vizena, C.R., Sporn, S.F., Daniels, K.M., Padwa, B.L., Weinstock, P.: Cost-benefit analysis of three-dimensional craniofacial models for midfacial distraction: a pilot study. Cleft Palate-Craniofacial J. **54**(5), 612–617 (2017)

4. Volpe, Y., Furferi, R., Governi, L., Uccheddu, F., Carfagni, M., Mussa, F., Scagnet, M., Genitori, L.: Surgery of complex craniofacial defects: a single-step AM-based methodology. Comput. Methods Programs Biomed. **165**, 225–233 (2018)
5. Bartalucci, C., Furferi, R., Governi, L., Volpe, Y.: a survey of methods for symmetry detection on 3d high point density models in biomedicine. Symmetry **10**(7), 263 (2018)
6. Buonamici, F., Furferi, R., Genitori, L., Governi, L., Marzola, A., Mussa, F., Volpe, Y.: Reverse engineering techniques for virtual reconstruction of defective skulls: an overview of existing approaches. Comput. Aided Des. Appl. **16**(1), 103–112 (2018)
7. Martini, M., Klausing, A., Messing-Jünger, M., Lüchters, G.: The self-defining axis of symmetry: a new method to determine optimal symmetry and its application and limitation in craniofacial surgery. J. Cranio-Maxillofac. Surg. **45**(9), 1558–1565 (2017)
8. Grammar, K., Thornhill, R.: Human facial attractiveness and sexual selection: the role of averageness and symmetry. J. Comp. Psychol. **108**(3), 233–242 (1994)
9. Gangestad, S.W., Thornhill, R., Yeo, R.A.: Facial attractiveness, developmental stability, and fluctuating asymmetry. Ethol. Sociobiol. **15**(2), 73–85 (1994)
10. Cootes, T.F., Taylor, C.J., Cooper, D.H., Graham, J.: Active shape models-their training and application. Comput. Vis. Image Underst. **61**(1), 38–59 (1995)
11. Cootes, T.F., Edwards, G.J., Taylor, C.J.: Active appearance models. IEEE Trans. Pattern Anal. Mach. Intell. **23**(6), 681–685 (2001)
12. Blanz, V., Vetter, T.: A morphable model for the synthesis of 3D faces. In: SIGGRAPH 1999, pp. 187–194. ACM Press (1999)
13. Heimann, T., et al.: Comparison and evaluation of methods for liver segmentation from CT datasets. IEEE Trans. Med. Imaging **28**(8), 1251–1265 (2009)
14. Heimann, T., Meinzer, H.P.: Statistical shape models for 3D medical image segmentation: a review. Med. Image Anal. **13**(4), 543–563 (2009)
15. Fuessinger, M.A., Schwarz, S., Cornelius, C.P., Metzger, M.C., Ellis, E., Probst, F., Semper-Hogg, W., Gass, M., Schlager, S.: Planning of skull reconstruction based on a statistical shape model combined with geometric morphometrics. Int. J. Comput. Assist. Radiol. Surg. **13**(4), 519–529 (2018)
16. Cerveri, P., Belfatto, A., Manzotti, A.: Pair-wise vs group-wise registration in statistical shape model construction: representation of physiological and pathological variability of bony surface morphology. Comput. Methods Biomech. Biomed. Eng. **22**(7), 772–787 (2019)
17. Bookstein, F.L.: Landmark methods for forms without landmarks: morphometrics of group differences in outline shape. Med. Image Anal. **1**(3), 225–243 (1997)
18. Cootes, T.F., Taylor, C.J., Cooper, D.H., Graham, J.: Training models of shape from sets of examples. In: Hogg, D., Boyle, R. (eds.) BMVC 1992. Springer, London (1992)
19. Jolliffe, I.T.: Principal Component Analysis, 2nd edn. Springer, New York (2002)
20. Vos, F.M., de Bruin, P.W., Aubel, J.C.M., Streekstra G.J., Maas, M., van Villet, L., Vossepoel, A.: A statistical shape model without using landmarks. In: ICPR 2004, vol. 3. IEEE (2004)
21. Brett, A.D., Taylor, C.J.: A method of automated landmark generation for automated 3D PDM construction. Image Vis. Comput. **18**(9), 739–748 (1999)
22. Audenaert, E.A., Van Houcke, J., Almeida, D.F., Paelinck, L., Peiffer, M., Steenackers, G., Vandermeulen, D.: Cascaded statistical shape model based segmentation of the full lower limb in CT. Comput. Methods Biomech. Biomed. Eng. **22**(6), 644–657 (2019)

23. Di Angelo, L., Di Stefano, P., Governi, L., Marzola, A., Volpe, Y.: A robust and automatic method for the best Symmetry Plane detection of the craniofacial skeletons. Symmetry **11** (2), 245 (2019)
24. Möller, T., Trumbore, B.: Fast, minimum storage ray-triangle intersection. J. Graph. Tools **2** (1), 21–28 (1997)
25. Marzola, A., Governi, L., Genitori, L., Mussa, F., Volpe, Y., Furferi, R.: A semi-automatic hybrid approach for defective skulls reconstruction. Comput. Aided Des. Appl. **17**(1), 190–204 (2019)
26. Besl, P.J., McKay, N.D.: A method for registration of 3-D shapes. IEEE Trans. Pattern Anal. Mach. Intell. **14**(2), 239–256 (1992)
27. Carr, J.C., Beatson, R.K., Cherrie, J.B., Mitchell, T.J., Fright, W.R., McCallum, B.C., Evans, T.R.: Reconstruction and representation of 3D objects with radial basis functions. In: SIGGRAPH 2001, pp. 67–76. ACM Press (2001)

A New Approach to Evaluate the Biomechanical Characteristics of Osseointegrated Dental Implants

Vito Ricotta(✉), Tommaso Ingrassia, Vincenzo Nigrelli, and Marco Zicari

Dipartimento di Ingegneria, Università degli Studi di Palermo, Viale delle Scienze, 90128 Palermo, Italy
vito.ricotta@unipa.it

Abstract. Tooth loss is a common pathology that affects many people. Dental osseointegrated implants are the ideal solution to restore normal functionality in partially or completely edentulous patients. The not perfect osseointegration and the fixture fracture are the main causes of failure for these kinds of implant. To avoid these drawbacks, several studies have been conducted to analyse the behaviour of dental implants.

Aim of this work is to analyse the biomechanical behaviour of three different endosseous dental implants. For this purpose, a new numerical model has been developed to simulate different levels of osseointegration and to evaluate the stress values on the bone at different times. In this way, it can be investigated the possibility of anticipating the use of dental implants that usually is delayed three months after surgery. Obtained results confirm the validity of the proposed approach and can provide useful guidelines for dentists.

Keywords: Dental implant · Osseointegration · FEM · Virtual simulation · CAD

1 Introduction

Periodontitis, caries, trauma, developmental defects and genetic disorders represent the most common causes of tooth loss [1].

Before dental implants were introduced, dentures or bridges were used to treat these pathologies in edentulous patients. Unfortunately, these solutions did not guarantee high performances, stability and duration.

Nowadays, dental implants are the ideal solution for lack of dentition, being considered the best alternative to natural teeth [2, 3].

It is possible to use dental implants to substitute single or multiple teeth. As a single-tooth replacement, an implant is freestanding, not attached to adjacent teeth and, therefore, much easier to maintain. For patients who have lost multiple teeth, dental implants can provide support for a stable, healthy and aesthetic dentition that is not removable. Perhaps the most important aspect of implants is that they can restore a high quality of life and greatly boost a patient's self-esteem.

C. Rizzi et al. (Eds.): ADM 2019, LNME, pp. 801–811, 2020.
https://doi.org/10.1007/978-3-030-31154-4_68

Based on their structural and functional characteristics, dental implants can be classified as:

- *endosseous implants*: having a pin (fixture) deeply inserted into the bone tissue (also called "root-form");
- *subperiosteal implants*: used when the width and depth of the bone are not such as to allow the insertion of endosseous pins. They are designed to be inserted on the upper part of the bone, but always under the gum;
- *transosseous implants*: with this type of implant, it is possible to treat patients in which bone atrophy is particularly severe without resorting to invasive surgeries (with staples, single pivot, and needle).

Today, the root-form is the most frequently used dental implant. These kinds of implants consist of three main components: fixture, abutment and prosthesis (Fig. 1).

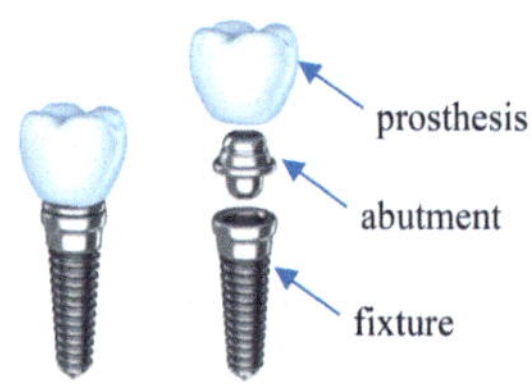

Fig. 1. A root-form implant and its main components.

The fixture is a cylinder (or cone)-shaped metal pin that is surgically embedded into the bone (jaw or mandible). The abutment is placed above the mucosal surface and is attached to the fixture using a screw. The prosthesis is fixed to the abutment either through cement or by means of a screw [1].

A good osseointegration, defined as the direct structural and functional connection between the bone and the implant [4, 5], guarantees the implant stability. Two kinds of stability could be considered in dental implants. Primary stability characterizes the mechanical engagement of the implant just after its insertion, while secondary stability is the result of (long-term) bone regeneration and remodelling (biological process) around the inserted implant. Primary and secondary stabilities are closely related [6, 7]. Usually, failure of dental implants occurs due to not perfect implant osseointegration or fixture fracture [5, 8–10].

To reduce the number of implant failure, several studies have been conducted to analyse the behaviour of dental implants [11–14]. In this field, the numerical methods are widely used [14, 15] for the investigation of the biomechanical behaviour of bone-implant components and simulating their mechanical interaction [16–23].

Aim of this work is to study the biomechanical behaviour of three different endosseous dental implants and the development of a new parametric numerical model [16, 24] to simulate different levels of osseointegration in order to evaluate the stress distribution on the bone-implant interface at different times, also to evaluate an early use of the dental implants [25, 26].

2 CAD Modelling

To setup FEM analyses of the dental implants, the 3D models of a mandible and the three types of analysed fixtures have been created by means of reverse engineering tools and 3D modelling parametric software.

2.1 CAD Modelling of the Fixtures

Three different types of fixture have been studied: a conical-shaped with triangular thread, a cylindrical-shaped with square thread and a cylindrical-shaped with triangular thread (Fig. 2). The main dimensions of the fixtures have been obtained from scientific literature and catalogues. All three types of analysed fixtures have the same maximum diameter and the same length.

Fig. 2. CAD models of three different fixtures: conical with triangular thread (left), cylindrical with square thread (centre) and cylindrical with triangular thread (right).

2.2 3D Acquisition and CAD Modelling of the Mandible

The CAD model of a mandible has been created using DICOM images of computerized tomography (CT) scan data. The 3D model has been reconstructed by semi-automatic segmentation procedures using the commercial medical imaging software Mimics. The final 3D solid model of the mandible is shown in Fig. 3.

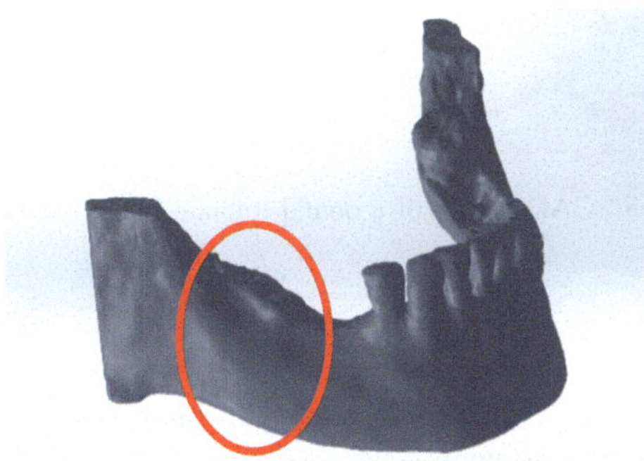

Fig. 3. CAD model of the mandible.

To reduce the computational costs of the FEM analyses, it was decided to study only a limited part of the mandible. For this reason, the 3D model of the bone has been sectioned in the area where the dental implant is installed (Fig. 3).

The mandibular bone has been subdivided in two main regions: a 2 mm thick (external) part of cortical bone and an internal one made of cancellous bone (Fig. 4). The separation surface between cortical and cancellous regions has been obtained by imposing an offset to the external surface.

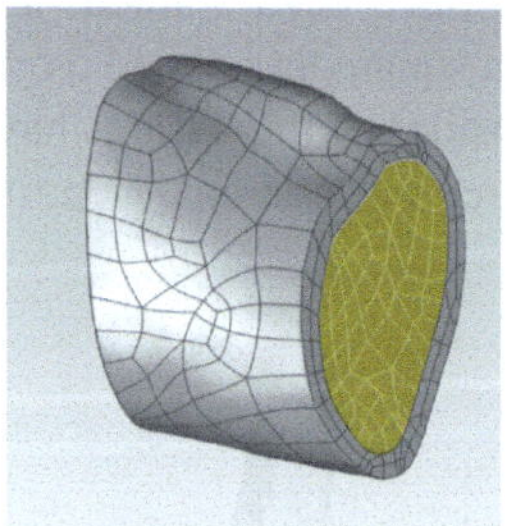

Fig. 4. CAD model of the analysed portion of the mandible: cortical bone (in grey) and cancellous one (in yellow).

Finally, the assemblies of the CAD models of the bone and the fixtures have been made following the surgical guidelines typical of dental surgery (Fig. 5). The interfaces among the different objects have been modelled by boolean operations to make them perfectly complementary.

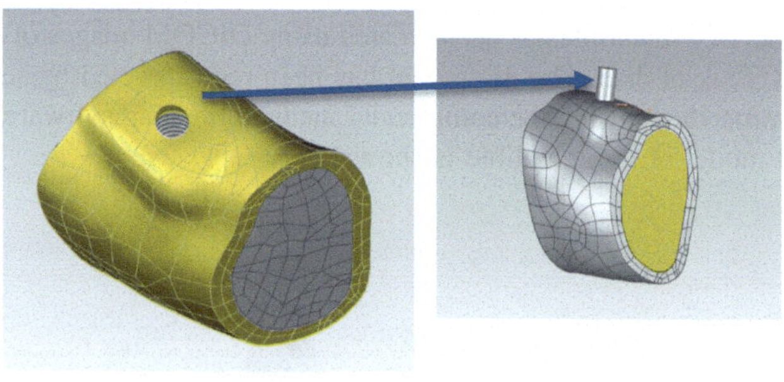

Fig. 5. CAD model of a dental implant-bone assembly.

3 FEM Analyses

FEM analyses have been performed to study the biomechanical behaviour of three different endosseous dental implants.

Two different forces [16, 25] have been applied to the implant: a compression load of 400 N and a shear load of 150 N distributed on the surface of the upper surface of the fixture (Fig. 6).

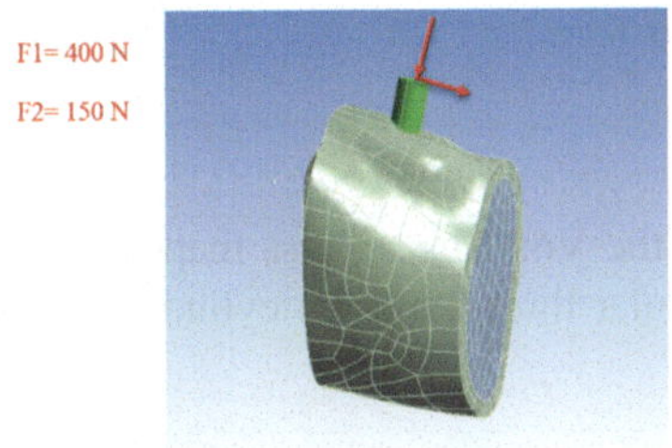

Fig. 6. Applied loads.

After a convergence analysis, it was chosen to use about 1.000.000 tetrahedral elements to mesh the implant-bone assembly (Fig. 7). To improve the quality of the mesh, the crests of the cylindrical fixtures with triangular thread have been truncated.

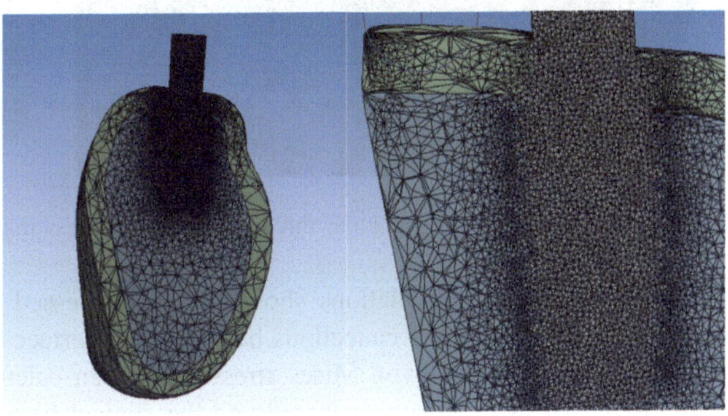

Fig. 7. Mesh of model.

To obtain very accurate results also in critical areas, the mesh of the assembly has been refined at the screw-bone interfaces and where small curvatures of the geometry occur.

Two fixed constraints [27] have been applied to the extreme surfaces of the analysed portion of the mandible. The contact between the bone and the implant has been considered as bonded.

The material used for the fixture is the titanium alloy Ti6Al4V; the bone structure was considered isotropic (both for cortical and spongy bone) as reported in [28–30]. Table 1 shows the main mechanical properties of the used materials.

Table 1. Mechanical properties of the materials.

Materials	Young modulus [MPa]	Poisson ratio
Ti6Al4V	110000	0.3
Cortical bone	13400	0.3
Cancellous bone	1370	0.31

Results

Figures 8, 9 and 10 show the Von Mises stress maps on the bone respectively for the conical fixture with triangular thread and the cylindrical fixtures with square and triangular thread.

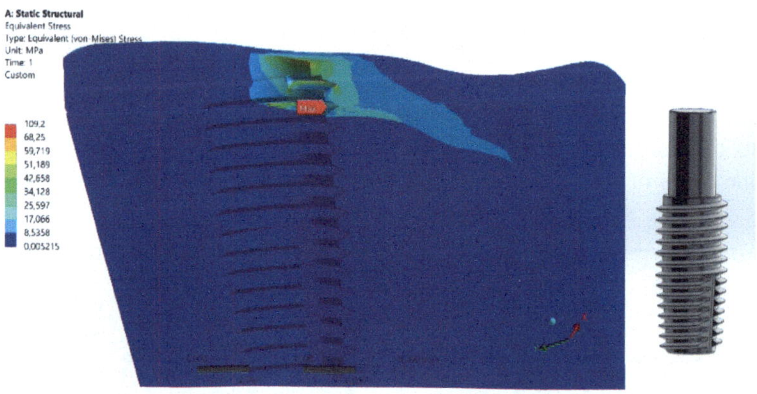

Fig. 8. Conical fixture with triangular thread: Von Mises stress map.

The results of the three FEM simulations show the most stressed parts are the cortical areas around the screw and the cancellous bone-screw interfaces.

The lowest and highest values of von Mises stress have been calculated respectively for the cylindrical fixture with square thread and the conical fixture with triangular thread. In all the cases analysed, the maximum von Mises stresses are lower than the ultimate tensile stress of the cortical bone (150 MPa).

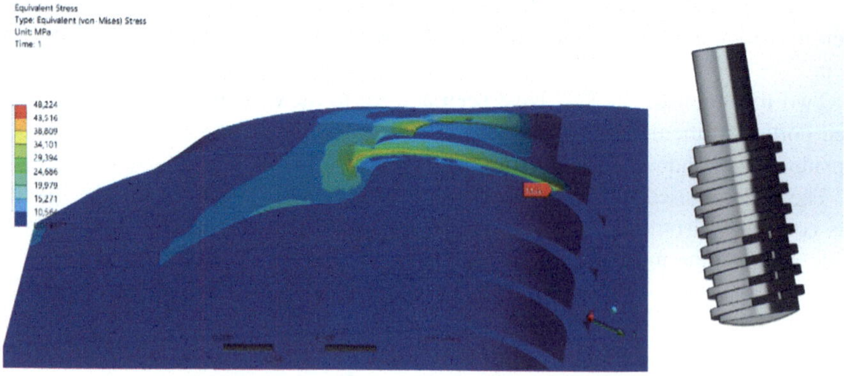

Fig. 9. Cylindrical fixture with square thread: Von Mises stress map.

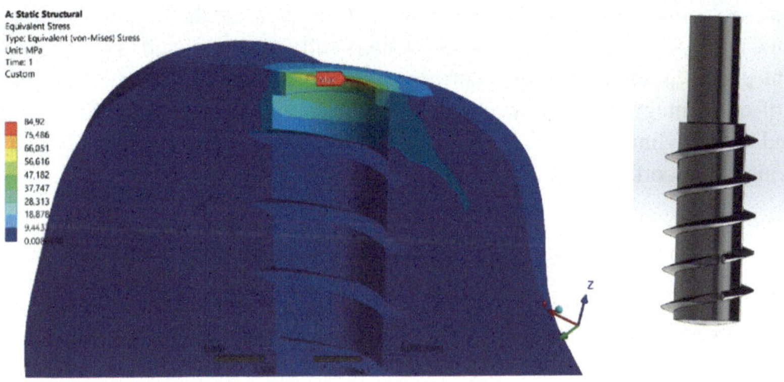

Fig. 10. Cylindrical fixture with triangular thread: Von Mises stress map.

4 New Numerical Model to Simulate Different Levels of Osseointegration

To evaluate the von Mises stress distributions at different times after the surgery, a new parametric numerical model [16, 24] has been developed. In particular, the solid model of the bone-implant interface has been suitably subdivided into two different cylindrical regions ("inner" and "outer") to which different mechanical characteristics can be assigned (Fig. 11).

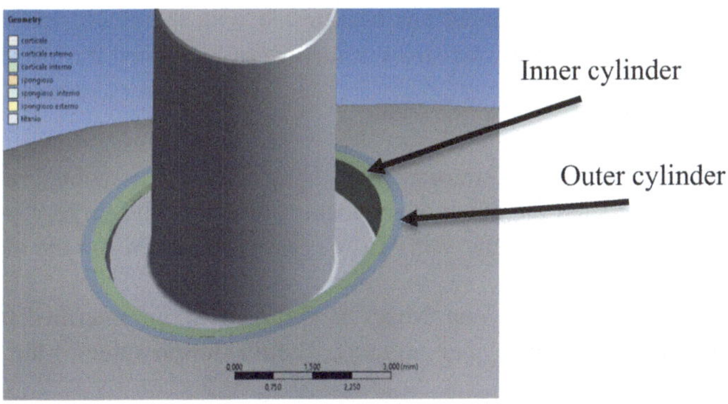

Fig. 11. FEM model with two cylindrical parts having different mechanical characteristics.

By assigning different and increasing mechanical characteristics (in terms of Young modulus) to these cylindrical parts, different levels of osseointegration can be simulated. In particular, lower values of the Young modulus can be used to simulate the initial level of osseointegration, just after the surgery, when the mechanical characteristics at the bone-implant interface are poor. Higher and higher values of the Young modulus can be used to simulate successive levels of osseointegration, until the complete osseointegration is obtained (about 90 days after the surgery), when good mechanical properties occur because the bone is fully reconstructed. To simulate three different phases after the surgery, corresponding to the initial, intermediate and final levels of osseointegration, the mechanical characteristics shown in Table 2 have been assigned to the cylindrical parts and three FEM simulations have been performed using these values for cortical and cancellous bones.

Table 2. Mechanical properties of the cylindrical parts at different levels of osseointegration.

Level of osseointegration	Initial (just after surgery)	Intermediate (after 45 days from surgery)	Final (after 90 days from surgery)
Inner cylinder - Cortical bone: value of the Young modulus [MPa]	670 (5% of the final value)	6700 (50% of the final value)	13400
Inner cylinder - Cancellous bone: value of the Young modulus [MPa]	68 (5% of the final value)	685 (50% of the final value)	1370
Outer cylinder - Cortical bone: value of the Young modulus [MPa]	6700 (50% of the final value)	10720 (80% of the final value)	13400
Outer cylinder - Cancellous bone: value of the Young modulus [MPa]	685 (50% of the final value)	1096 (80% of the final value)	1370

The developed numerical model has been tested with the cylindrical fixture with square thread, for which the lowest stress values were calculated during the previous numerical simulations.

Results

Figure 12 shows the Von Mises stress map for the first level of osseointegration. In this case, it has been simulated the effect of the loading just after the surgery. As it can be noticed, the maximum value of the Von Mises stress is 136 MPa and the most stressed area is the cortical area.

Figure 13, instead, shows Von Mises stress map at an intermediate level of osseointegration (about 45 days after surgery). The maximum value of the Von Mises stress is 92 MPa.

The obtained results demonstrate that it could be evaluated anticipating the use of the dental implant. Considering the ultimate tensile strength of the cortical bone is equal to 150 MPa [5] and that after 45 days from the surgery the maximum von Mises stress calculated on the cortical bone is about 92 MPa, the dental implant could be used quite safely.

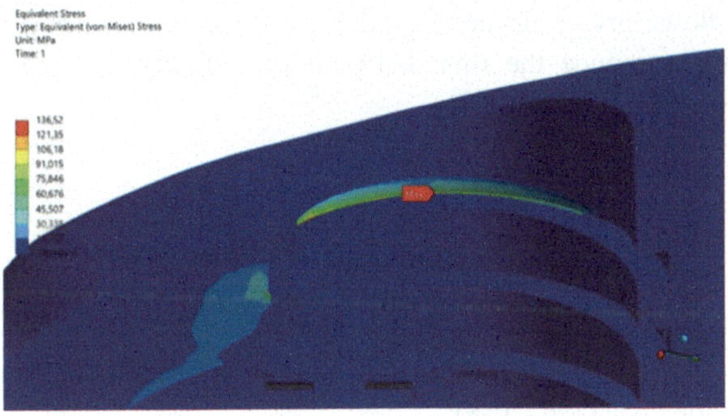

Fig. 12. Cylindrical fixture with square thread: Von Mises stress map just after the surgery.

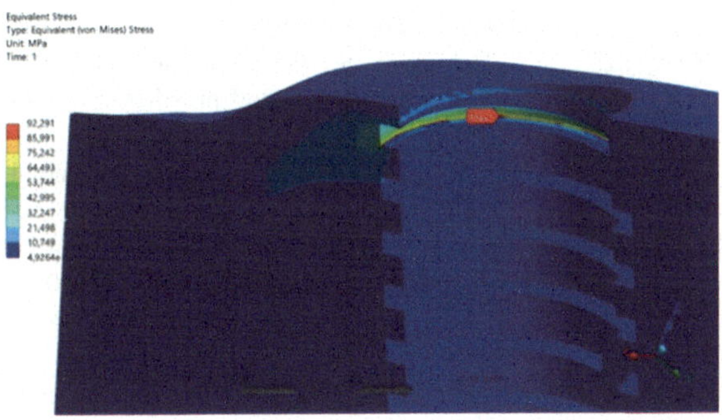

Fig. 13. Cylindrical fixture with square thread: Von Mises stress map 45 days after the surgery.

5 Conclusions

In this work the study of the biomechanical behaviour of three different endosseous dental implants and a new numerical model to simulate different levels of osseointegration have been presented.

The first step of the work was related to the modelling of the mandible and the fixtures of the analysed dental implants. After, the CAD models of the mandible and the dental implants have been virtually assembled following, step-by-step, the surgical guidelines.

FEM analyses performed on the three analysed endosseous dental implants have demonstrated that the best results, in terms of lowest von Mises stress calculated on the bone, are obtained for the cylindrical fixture with square thread.

The new numerical model to simulate different levels of osseointegration has been tested on the cylindrical fixture with square thread. The obtained results have demonstrated that it could be possible to evaluate anticipating the use of this kind of dental implant of 45 days.

The developed FEM models could be used also customizing the values of the mechanical characteristics of the bone depending on the specific patients. The imposed mechanical characteristics, in fact, could be defined depending of the porosity of the bone. All that allows giving accurate guidelines for dentists during rehabilitation of patient, allowing to reduce the time between the surgery and the full use of the implants.

References

1. Zohrabian, V.M., Sonick, M., Hwang, D., Abrahams, J.J.: Dental implants. Semin. Ultrasound CT MRI **36**, 415–426 (2015)
2. Block, M.S.: Dental implants: the last 100 years. Am. Assoc. Oral Maxillofac. Surg. J. Oral Maxillofac. Surg. **76**, 11–26 (2018)
3. Bicudo, P., Reis, J., Deus, A.M., Reis, L., Vaz, M.F.: Mechanical behaviour of dental implants. Procedia Struct. Integr. **1**, 026–033 (2016)
4. Shah, F.A., Thomsen, P., Palmquist, A.: Osseointegration and current interpretations of the bone-implant interface. Acta Biomater. **84**, 1–15 (2019)
5. Mirulla, A.I., Bragonzoni, L., Zaffagnini, S., Bontempi, M., Nigrelli, V., Ingrassia, T.: Virtual simulation of an osseointegrated trans-humeral prosthesis: a falling scenario. Injury **49**(4), 784–791 (2018)
6. Dorogoy, A., Rittel, D., Shemtov-Yona, K., Korabi, R.: Modeling dental implant insertion. J. Mech. Behav. Biomed. Mater. **68**, 42–50 (2017)
7. Sundell, G., Dahlin, C., Andersson, M., Thuvander, M.: The bone-implant interface of dental implants in humans on the atomic scale. Acta Biomater. **48**, 445–450 (2017)
8. Croitoru, S.M., Popovici, I.A.: R&D on dental implants breakage. Appl. Surf. Sci. **417**, 262–268 (2017)
9. Liaw, K., Delfini, R.H., Abrahams, J.J., Liaw, K.: Dental implant complications. Semin. Ultrasound CT MRI **36**, 427–433 (2015)
10. Tabrizi, R., Behnia, H., Taherian, S., Hesami, N.: What are the incidence and factors associated with implant fracture? Am. Assoc. Oral Maxillofac. Surg. J. Oral Maxillofac. Surg. **75**, 1866–1872 (2017)
11. Bicudo, P., Reis, J., Deus, A.M., Reis, L., Vaz, M.F.: Performance evaluation of dental implants: an experimental and numerical simulation study. Theor. Appl. Fract. Mech. **85**, 74–83 (2016)
12. Roy, S., Dey, S., Khutia, N., Chowdhury, A.R., Datta, S.: Design of patient specific dental implant using FE analysis and computational intelligence techniques. Appl. Soft Comput. **65**, 272–279 (2018)
13. Wang, K., Geng, J., Jones, D., Wei, X.: Comparison of the fracture resistance of dental implants with different abutment taper angles. Mater. Sci. Eng. C **63**, 164–171 (2016)
14. da Costa Valente, M.L., de Castro, D.T., Macedo, A.P., Shimano, A.C., dos Reis, A.C.: Comparative analysis of stress in a new proposal of dental implants. Mater. Sci. Eng. C **77**, 360–365 (2017)
15. Ingrassia, T., Nalbone, L., Nigrelli, V., Pisciotta, D., Ricotta, V.: Influence of the metaphysis positioning in a new reverse shoulder prosthesis. In: Advances on Mechanics, Design Engineering and Manufacturing. Lecture Notes in Mechanical Engineering. Springer, Cham (2017). https://doi.org/10.1007/978-3-319-45781-9_47
16. Dhatrak, P., Shirsat, U., Sumanth, S., Deshmukh, V.: Finite element analysis and experimental investigations on stress distribution of dental implants around implant-bone interface. Mater. Today Proc. **5**, 5641–5648 (2018)

17. Ingrassia, T., Nigrelli, V., Ricotta, V., Nalbone, L., D'Arienzo, A., D'Arienzo, M., Porcellini, G.: A new method to evaluate the influence of the glenosphere positioning on stability and range of motion of a reverse shoulder prosthesis. Injury **50**, S12–S17 (2019)
18. Rinaudo, A., D'Ancona, G., Lee, J.J., Pilato, G., Amaducci, A., Baglini, R., et al.: Predicting outcome of acute aortic dissection with patent false Lumne by computational flow analysis. Cardiovasc. Eng. Technol. **5**(2), 176–188 (2014)
19. Mancuso, A., Pitarresi, G., Tumino, D.: Using FEM simulation to predict structural performances of a sailing dinghy. Int. J. Interact. Des. Manuf. (IJIDeM) **12**(3), 811–822 (2018)
20. NarendraKumar, U., Mathew, A.T., Iyer, N., Rahman, F., Manjubala, I.: A 3D finite element analysis of dental implants with varying thread angles. Mater. Today Proc. **5**, 11900–11905 (2018)
21. D'Ancona, G., Amaducci, A., Rinaudo, A., Pasta, S., Follis, F., Pilato, M., et al.: Haemodynamic predictors of a penetrating atherosclerotic ulcer rupture using fluid–structure interaction analysis. Interact. Cardiovasc. Thorac. Surg. **17**(3), 576–578 (2013)
22. Fragapane, S., Giallanza, A., Cannizzaro, L., Pasta, A., Marannano, G.: Experimental and numerical analysis of aluminum-aluminum bolted joints subject to an indentation process. Int. J. Fatigue **80**, 332–340 (2015)
23. Santiago Jr., J.F., Verri, F.R., de Faria Almeida, D.A., de Souza Batista, V.E., Lemos, C.A. A., Pellizzer, E.P.: Finite element analysis on influence of implant surface treatments, connection and bone types. Mater. Sci. Eng. C **63**, 292–300 (2016)
24. Cirello, A., Cucinotta, F., Ingrassia, T., Nigrelli, V., Sfravara, F.: Fluid–structure interaction of downwind sails: a new computational method. J. Mar. Sci. Technol. (Japan) **24**(1), 86–97 (2019)
25. Aumnakmanee, S., Yodpiji, N., Jantong, N., Jongprasithporn, M.: Finite element analysis of dental implant prosthetics. Mater. Today Proc. **5**, 9525–9534 (2018)
26. Ingrassia, T., Lombardo, B., Nigrelli, V., Ricotta, V., Nalbone, L., D'Arienzo, A., D'Arienzo, M., Porcellini, G.: Influence of sutures configuration on the strength of tendon-patch joints for rotator cuff tears treatment. Injury **50**, S18–S23 (2019)
27. Ingrassia, T., Nalbone, L., Nigrelli, V., Ricotta, V., Pisciotta, D.: Biomechanical analysis of the humeral tray positioning in reverse shoulder arthroplasty design. Int. J. Interact. Des. Manuf. **12**(2), 651–661 (2018)
28. Camarda, L., Pitarresi, G., Moscadini, S., Marannano, G., Sanfilippo, A., D'Arienzo, M.: Effect of suturing the femoral portion of a four-strand graft during an ACL reconstruction. Knee Surg. Sport. Traumatol. Arthrosc. **22**(5), 1040–1046 (2014)
29. Marcian, P., Wolff, J., Horackova, L., Kaiser, J., Zikmund, T., Borak, L.: Micro finite element analysis of dental implants under different loading conditions. Comput. Biol. Med. **96**, 157–165 (2018)
30. Linetskiy, I., Demenko, V., Linetska, L., Yefremov, O.: Impact of annual bone loss and different bone quality on dental implant success – a finite element study. Comput. Biol. Med. **91**, 318–325 (2017)

Biomechanical Analysis of a New Elbow Prosthesis

Vito Ricotta[1(✉)], Laura Bragonzoni[2(✉)], Giuseppe Marannano[1], Lorenzo Nalbone[3], and Andrea Valenti[1]

[1] Dipartimento di Ingegneria, Università degli Studi di Palermo, Viale delle Scienze, 90128 Palermo, Italy
vito.ricotta@unipa.it

[2] Alma Mater Studiorum, Università di Bologna, Bologna, Italy

[3] Ambulatorio di Ortopedia e Traumatologia, Azienda Ospedaliera Universitaria Policlinico Paolo Giaccone di Palermo, 90100 Palermo, Italy

Abstract. Total elbow arthroplasty (TEA) is an effective and frequently used treatment for patients with debilitating elbow pathology. Total elbow prostheses have lagged behind those of the knee, hip and shoulder for different reasons, such as the high failure rate of the early designs. Concern remains regarding the longevity of TEA implants, especially in younger patients. The main cause of revision of the implant is usually related to the phenomenon of aseptic loosening mainly due to the cement-bone interface failure.

Aim of this work is the biomechanical analysis of a new elbow prosthesis to investigate the mechanical behaviour at the cement-bone interface. For this reason, a musculoskeletal model has been developed by modelling the forces of the muscles and after FEM analyses have been performed.

Obtained results confirm the validity of the implemented model and can provide guidelines for surgeons regarding the implant configurations with the aim to reduce the aseptic loosening.

Keywords: Total elbow arthroplasty · FEM · Reverse engineering · CAD

1 Introduction

The elbow joint allows the movements of flexion, extension and rotation of the forearm. Injuries to this joint can lead to the reduction of these movements and a subsequent reduced functionality of the arm. The elbow is composed of the ulno-humeral, the radio-humeral and the proximal radio-ulnar joints (Fig. 1) [1].

The clinical cases in which an elbow arthroplasty may be necessary are various, but all are characterized by serious damages of the articular surfaces. Total elbow arthroplasty (TEA) is an effective and frequently used treatment for patients with severe elbow pathology [2]. TEA was initially designed to treat rheumatoid arthritis of the elbow, but the indications for this procedure have been expanded and refined in recent years. TEA is now used for inflammatory arthritis, post-traumatic arthritis, primary osteo-arthritis, haemophilic arthropathy, chronic instability, revision arthroplasty, distal humeral non-unions and tumor reconstruction [3–5].

C. Rizzi et al. (Eds.): ADM 2019, LNME, pp. 812–823, 2020.
https://doi.org/10.1007/978-3-030-31154-4_69

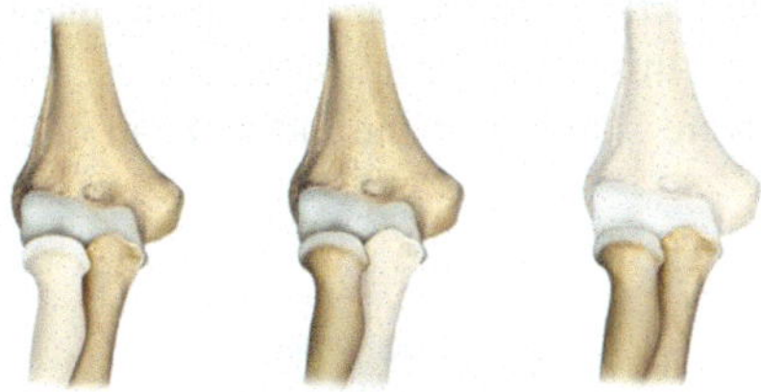

Fig. 1. From left to right: ulno-humeral, radio-humeral and proximal radio-ulnar joints.

Nowadays, the main elbow arthroplasty are of two types: unlinked and linked. Unlinked prostheses mainly rely on the surrounding soft tissue to provide stability to the elbow, even if the articular geometry of the unlinked prostheses also can contribute to the stability [6]. Linked prostheses, instead, have an intrinsic stability and are, therefore, indicated when significant osseous or ligamentous deficiency exists. The decision to use a linked or an unlinked implant is generally made preoperatively. This is usually based on different factors, i.e. the patient's level of function, age, diagnosis and status of the articulation. The advantage of an unlinked prosthesis is that the elbow soft tissues share some of the load, which may decrease the forces on the cement-bone and cement-prosthesis interfaces, and that could decrease the rate of aseptic loosening and minimize the surfaces wear. Linked implants, instead, offer the advantage of not relying on soft tissues for stability, are inherently stable but they usually experience greater forces that could lead to earlier aseptic loosening and bearing wear. Developments in elbow joint replacements have lagged behind that of the knee, hip and shoulder [7–9] arthroplasties for different reasons, such as the high failure rate of the early designs [10]. TEA remains a challenging and technically demanding surgical procedure. Reported complications and failure rates are up to 62% and 90%, respectively. In comparison with arthroplasty of the hip and knee, the complication and failure rates of TEA are relatively high [2, 11]. As mentioned, the main cause of revision of the implant is related to the phenomenon of aseptic loosening that is the detachment of the prosthetic component from the bone in which it is implanted [2, 12–15].

Aim of this work is the biomechanical analysis of a new elbow prosthesis to investigate the mechanical behaviour at the cement-bone interface, in order to provide guidelines for surgeons regarding aseptic loosening. For this purpose, a new parametric numerical model [16, 17] has been developed and tested with an innovative elbow prosthesis. The present work was divided into three main steps. The first step consists of the 3D acquisition, reconstruction and CAD modelling of bones and an elbow prosthesis. In the second step, the musculoskeletal model was created. Finally, FEM analyses have been performed.

2 Case Study: Linked Prosthesis LINK Endo-Model®

In this work, the innovative total elbow prosthesis LINK® Endo-Modell® by Link (Fig. 2) has been studied. This prosthesis led to a significant improvement of the surgical technique. This kind of prosthesis integrates a fully constrained hinge for the replacement of the ulno-humeral joint and an additional unconstrained replacement of the radio-humeral joint. In this way the muscles forces are widely distributed and movements with more severe slanting of the radius head or eccentric rotational shifts are possible.

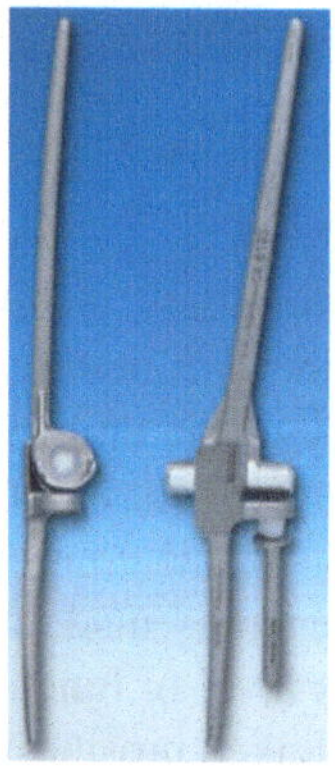

Fig. 2. LINK® Endo-Modell®

2.1 3D Acquisition and CAD Modelling

The shapes of the prosthetic components have been acquired with a triangulation-based 3D laser scanner by Hexagon metrology. This system has adjustable line lengths and it is able to acquire point cloud at high speed (150000 points per second) with a good accuracy level (0.013 mm).

After 3D acquisition, the point clouds have been post-processed and, finally, converted into CAD solid models (Fig. 3).

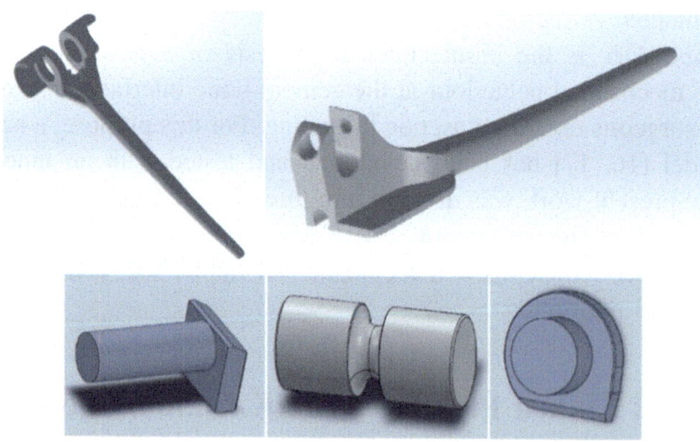

Fig. 3. CAD model of all components of the elbow prosthesis.

2.2 Virtual Assembling

All prosthetic components have been assembled using a 3D parametric CAD software (Fig. 4).

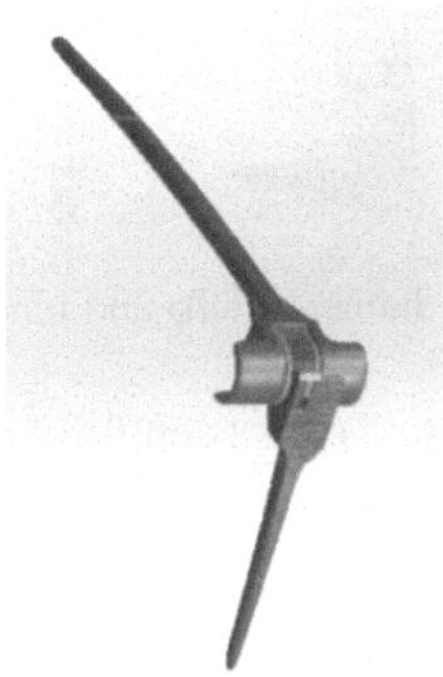

Fig. 4. CAD model of the elbow prosthesis.

To perform the virtual biomechanical study of the elbow joint, the digitalized prosthetic components have been assembled with the CAD models of the bones, obtained through DICOM images.

The assembly of the CAD models of the bones and the prosthesis has been made following, step-by-step, the surgical guidelines of a total elbow arthroplasty [5] (Fig. 5).

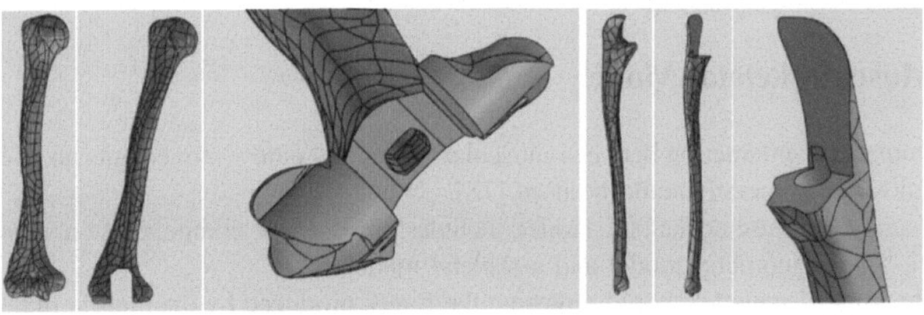

Fig. 5. CAD models of humerus and ulna

The cement has been modelled by generating solids whose cross-section, circular, followed the average line of the stem of the prosthetic elements (Fig. 6).

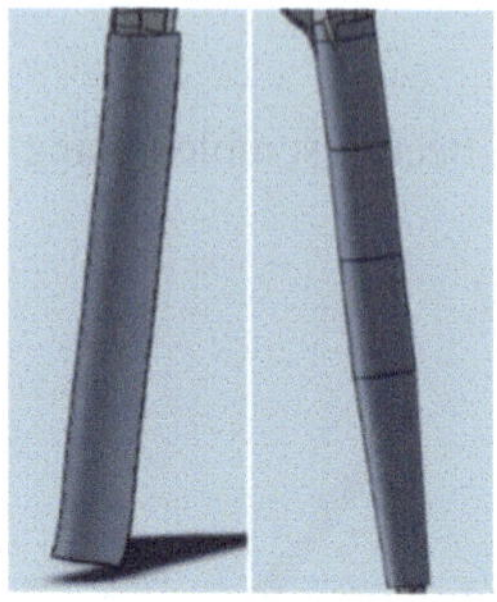

Fig. 6. Cement in the humeral (left) and ulnar (right) components.

The final assembly model (Fig. 7) has been fully parameterized in order to be able to modify it in a very simple and fast way during the virtual simulations.

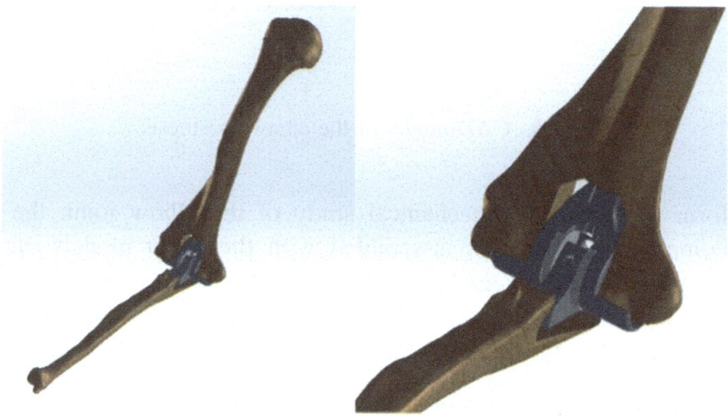

Fig. 7. CAD models of the elbow joint

3 Musculoskeletal Model

To simulate the interaction between muscular forces and elbow movements an effective musculoskeletal model should be used [18].

Usually, a musculoskeletal model includes three main components: a muscular model, a joint geometry model and a skeletal model.

The muscular model should represent the forces produced by the muscle fibres, the joint model must reflect the articulation kinematics while the skeletal model must correctly represent the inertial characteristics and the constitutive relations afferent to the bones.

A generic muscle can be modelled as a set of fibres connected to two tendons, able to generate force through the muscular contraction. This force is transmitted to the bones at the insertion points of the tendon [19].

Concerning muscle force modelling, two methods can be used: the straight-line approach, which connects the insertion points of a muscle, and the centroid line approach, represented by a line that passes through the locus of the cross sectional centroids of the muscle. In this work, it was decided to use the first approach.

As regards the value of the muscle force, the Physiological Cross-Sectional Area (PCSA) is used, that is the transversal area of the muscle perpendicular to its fibres [20, 21]. This value, if pennate muscles are considered, is different from the Anatomical Cross Section Area (ACSA), which is the transversal area perpendicular to the muscle axis (Fig. 8).

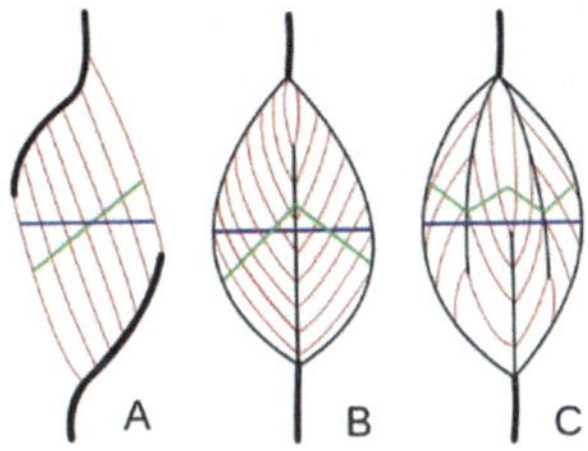

Fig. 8. Pennate muscle fibre arrangements. The green lines represent PCSA; the blue lines represent ACSA.

The PCSA can be evaluated as [20, 21]:

$$PCSA = \frac{M_m}{L_m \cdot \rho} \tag{1}$$

where M_m is the muscle mass, L_m is the muscle length and ρ is the density of the muscle.

Multiplying the PCSA by the maximum normal stress that a muscle can exercise, the maximum force during an isometric contraction can be calculated. The maximum stress value present in the literature varies from 0.2–0.35 MPa [19].

To create the musculoskeletal model to analyse the elbow joint, muscles forces related to anconaeus, biceps brachialis and triceps brachialis were considered in static load conditions, due to an isometric contraction.

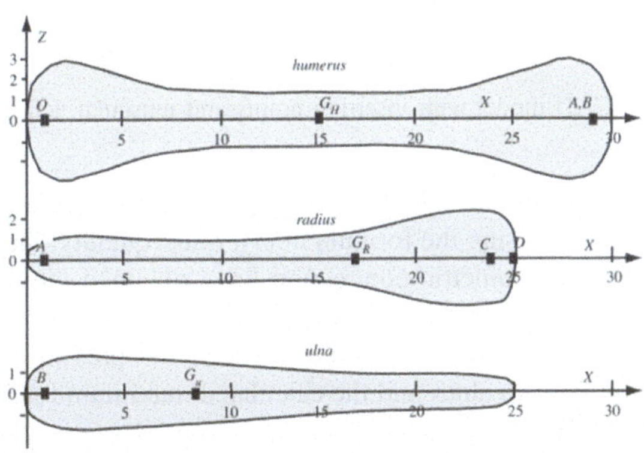

Fig. 9. Main dimensions of the upper limb bones in a 1.73 m tall man [21]

To calculate the muscle forces it was necessary to derive the muscle length.

This has been done using suitable tables [21] that, considering the main dimensions of the bone of a medium sized 1.73 m tall man (Fig. 9), report the coordinates of the insertion points of the muscles to be considered in the numerical analyses.

In Table 1 are reported the calculated muscle lengths.

Table 1. Muscle lengths and, in brackets, the corresponding number of the insertion points shown in Fig. 10.

Muscle and insertion points	Anconaeus (9)	Biceps brachialis (8)	Biceps brachialis (8 bis)	Triceps brachialis (10)
Length [mm]	64	285	250	222

In Fig. 10 the complete CAD model with insertion points and muscular action lines is shown.

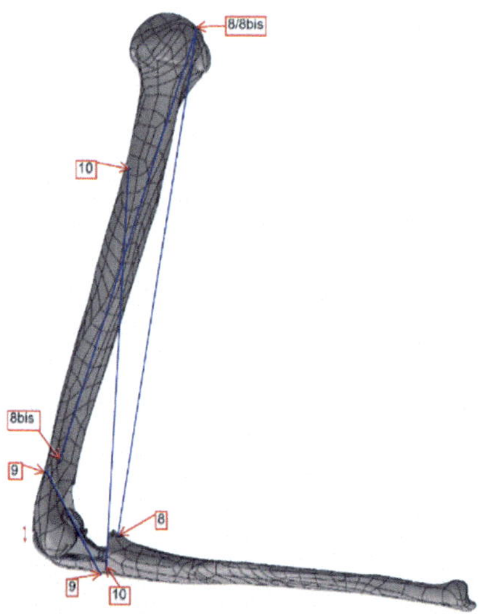

Fig. 10. CAD model with insertion points and muscular action lines.

After calculating the muscle lengths, it was possible to calculate the PCSA value for every analysed muscle using the formula in (1). Subsequently, the maximum forces of the muscles during an isometric contraction have obtained multiplying the PCSA values by the maximum normal stress explicable by the muscles. An average value, equal to 0.275 MPa, of the maximum normal stress values present in the literature has been considered. The PCSA values and the calculated maximum forces of the analysed muscles are shown in Table 2.

Table 2. PCSA and forces used for the FEM analysis of the elbow joint.

Muscle	PCSA [mm^2]	Maximum force [N]
Anconeus (9)	170	47
Biceps brachialis (8)	220	60
Biceps brachialis (8 bis)	250	69
Triceps brachialis (10)	490	−135

4 FEM Analysis

In order to investigate the mechanical behaviour at the cement-bone interface, FEM analyses [22, 23] have been performed using the commercial software Ansys Workbench. Regarding the mechanical properties of materials, the bone is considered an anisotropic material transversely isotropic, while cement polymethyl methacrylate (PMMA) and prosthesis (Cr-Co-Mo alloy) are considered isotropic. In Tables 3 and 4 are reported, respectively, the mechanical properties of the bone and of the other materials.

Table 3. Mechanical properties of bone material.

E_1 [GPa]	E_2 [GPa]	E_3 [GPa]	n_{12}	n_{13}	n_{23}
18	18	22,8	0,28	0,3	0,3

Table 4. Mechanical properties of PMMA and materials of the components of the elbow prosthesis.

Material	E [GPa]	n
PMMA	2,28	0,3
Co-Cr-Mo alloy	220	0,29
Polyethylene	0,5	0,3

Figure 11 shows the FEM model of the elbow joint.

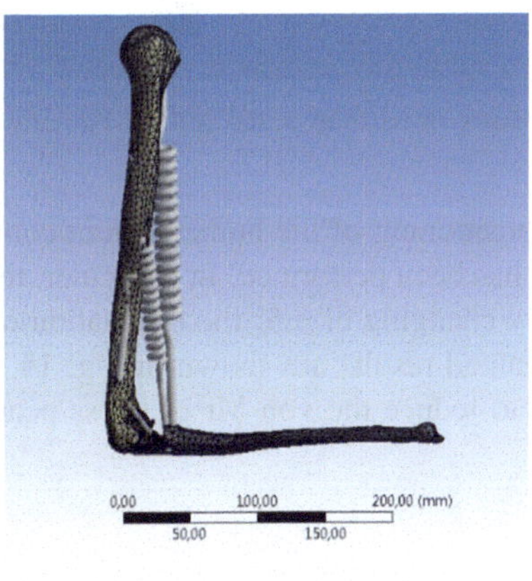

Fig. 11. FEM model of the elbow joint.

4.1 Results

Figure 12 shows Von Mises stress maps of the cement on the ulna side. The maximum calculated value is about 4 MPa and is lower than the fatigue strength [24–26] of cement, whose value in literature range from 6 to 10 MPa (>10 million cycles) [27].

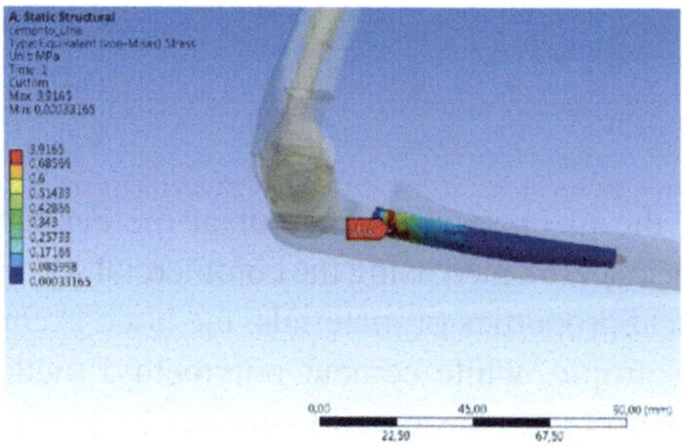

Fig. 12. Von Mises stress map on the cement of the ulna.

Figure 13 shows the Von Mises stress map at the cement-humeral stem interface. It can be noticed a very high value of the stress that could cause early mobilization of the prosthesis due to cement failure.

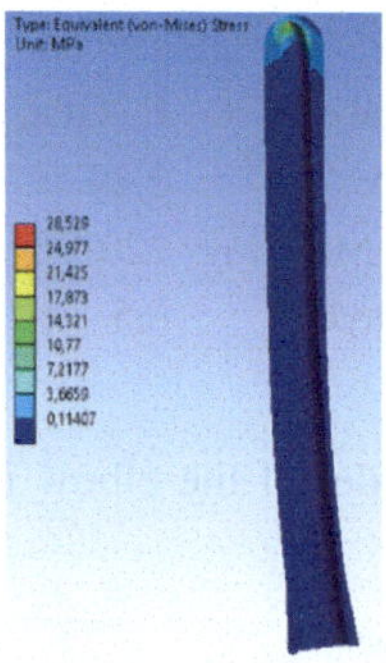

Fig. 13. Von Mises stress map at the cement-humeral stem interface.

To understand if the positioning of the humeral stem could affect the stress on the cement, a new simulation has been performed. In particular, the position of the humeral stem has been modified by changing of $\pm 5°$ the orientation of the humeral stem along the sagittal stem. The obtained results are shown in Fig. 14. It can be noticed that an inclination of +5° allows to reduce the von Mises stress peak.

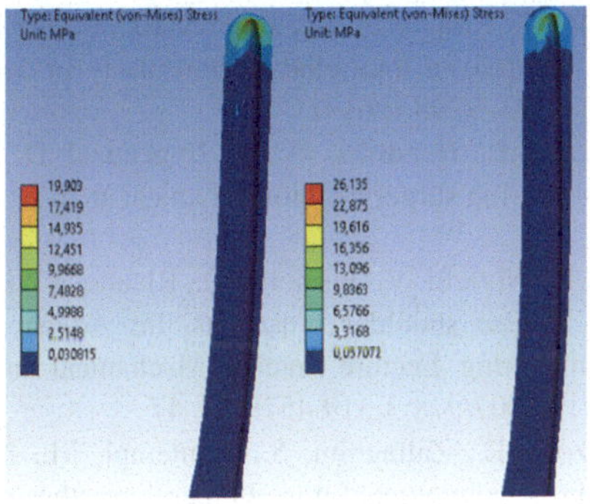

Fig. 14. Von Mises stress map at the cement-humeral stem interface by varying the inclination of the stem along the sagittal plane of +5° (on the left) and −5° (on the right).

5 Conclusions

In this work, a new numerical model of an elbow prosthesis to investigate the mechanical behaviour at the cement-bone interface is presented. To this aim, a musculoskeletal model has been implemented to simulate the interaction between muscular forces and skeleton movements. The first step of the work was related to the modelling, by means of reverse engineering techniques, of the prosthetic components and the bones and their virtual assembling. After, a musculoskeletal model has been implemented, the muscle forces and their insertion points have been calculated and, finally, FEM analyses of the elbow joint have been performed. Some modifications of the positioning of the prosthesis components have been also investigated to understand how they can affect the stress at the cement-prosthesis interface. Obtained results demonstrate that modifying the inclination of the humeral stem the peak stresses on the cement can decrease and that could reduce the number of early failures of the prosthesis. The developed numerical model could also effectively used to investigate how changes on geometry of the components of the prosthesis affect the stress on the cement and the bones, so to provide useful information to orthopaedic surgeons.

References

1. Felstead, A.J., Ricketts, D.: Biomechanics of the shoulder and elbow. Orthop. Trauma **31**(5), 300–305 (2017)
2. Geurts, E.J., Viveen, J., van Riet, R.P., Kodde, I.F., Eygendaal, D.: Outcomes after revision total elbow arthroplasty: a systematic review. J. Shoulder Elb. Surg. **28**, 381–386 (2019)
3. Marsh, J.P., King, G.J.W.: Total elbow arthroplasty. Oper. Tech. Orthop. **23**, 253–264 (2013)
4. Schoch, B.S., Werthel, J.-D., Sánchez-Sotelo, J., Morrey, B.F., Morrey, M.: Total elbow arthroplasty for primary osteoarthritis. J. Shoulder Elb. Surg. **26**, 1355–1359 (2017)

5. Mansat, P., Bonnevialle, N., Rongières, M., Bonnevialle, P., the Bone, Joint Trauma Study Group (GETRAUM): The role of total elbow arthroplasty in traumatology. Orthop. Traumatol. Surg. Res. **100**, S293–S298 (2014)
6. De Vos, M.J., Wagener, M.L., Hendriks, J.C.M., Eygendaal, D., Verdonschot, N.: Linking of total elbow prosthesis during surgery; a biomechanical analysis. J. Shoulder Elb. Surg. **22**, 1236–1241 (2013)
7. Ingrassia, T., Nalbone, L., Nigrelli, V., Pisciotta, D., Ricotta, V.: Influence of the metaphysis positioning in a new reverse shoulder prosthesis. In: Advances on Mechanics, Design Engineering and Manufacturing. Lecture Notes in Mechanical Engineering. Springer, Cham (2017). https://doi.org/10.1007/978-3-319-45781-9_47
8. Mirulla, A.I., Bragonzoni, L., Zaffagnini, S., Bontempi, M., Nigrelli, V., Ingrassia, T.: Virtual simulation of an osseointegrated trans-humeral prosthesis: a falling scenario. Injury **49**(4), 784–791 (2018)
9. Ingrassia, T., Lombardo, B., Nigrelli, V., Ricotta, V., Nalbone, L., D'Arienzo, A., D'Arienzo, M., Porcellini, G.: Influence of sutures configuration on the strength of tendon-patch joints for rotator cuff tears treatment. Injury **50**, S18–S23 (2019)
10. Lenoir, H., Micallef, J.P., Djerbi, I., Waitzenegger, T., Lazerges, C., Chammas, M., Coulet, B.: Total elbow arthroplasty: influence of implant positioning on functional outcomes. Orthop. Traumatol. Surg. Res. **101**, 721–727 (2015)
11. Stone, M.A., Singh, P., Rosario, S.L., Omid, R.: Outpatient total elbow arthroplasty: 90-day outcomes. J. Shoulder Elb. Surg. **27**, 1311–1316 (2018)
12. Frostick, S.P., Elsheikh, A.A., Mohammed, A.A., Wood, A.: Results of cementless total elbow arthroplasty using the discovery elbow system at a mean follow-up of 61.8 months. J. Shoulder Elb. Surg. **26**, 1348–1354 (2017)
13. Kodama, A., Mizuseki, T., Adachi, N.: Macroscopic investigation of failed Kudo type 5 total elbow arthroplasty. J. Shoulder Elb. Surg. **27**, 1380–1385 (2018)
14. Ingrassia, T., Nigrelli, V., Ricotta, V., Nalbone, L., D'Arienzo, A., D'Arienzo, M., Porcellini, G.: A new method to evaluate the influence of the glenosphere positioning on stability and range of motion of a reverse shoulder prosthesis. Injury **50**, S12–S17 (2019)
15. Ingrassia, T., Nalbone, L., Nigrelli, V., Ricotta, V., Pisciotta, D.: Biomechanical analysis of the humeral tray positioning in reverse shoulder arthroplasty design. Int. J. Interact. Des. Manuf. **12**(2), 651–661 (2018)
16. Rinaudo, A., D'Ancona, G., Lee, J.J., Pilato, G., Amaducci, A., Baglini, R., et al.: Predicting outcome of acute aortic dissection with patent false Lumne by computational flow analysis. Cardiovasc. Eng. Technol. **5**(2), 176–188 (2014)
17. Cirello, A., Cucinotta, F., Ingrassia, T., Nigrelli, V., Sfravara, F.: Fluid–structure interaction of downwind sails: a new computational method. J. Mar. Sci. Technol. (Japan) **24**(1), 86–97 (2019)
18. Wang, T., Chellai, R.: A novel adaptive control of a human musculoskeletal arm model. Comput. Electr. Eng. **69**, 732–739 (2018)
19. Millard, M., Emonds, A.L., Harant, M., Mombaur, K.: A reduced muscle model and planar musculoskeletal model fit for the simulation of whole-body movements. J. Biomech. **89**, 11–20 (2019)
20. Bolsterlee, B., Vardy, A.N., van der Helm, F.C.T., (DirkJan) Veeger, H.E.J.: The effect of scaling physiological cross-sectional area on musculoskeletal model predictions. J. Biomech. **48,** 1760–1768 (2015)
21. Pennestrì, E., Stefanelli, R., Valentini, P.P., Vita, L.: Virtual musculo-skeletal model for the biomechanical analysis of the upper limb. J. Biomech. **40**(6), 1350–1361 (2007)

22. Baron Saiz, C., Ingrassia, T., Nigrelli, V., Ricotta, V.: Thermal stress analysis of different full and ventilated disc brakes. Frattura Integrità Strutt. **34**, 608–621 (2015). https://doi.org/10.3221/IGF-ESIS.34.67
23. Khoo, T.A., Normahira, M., Revati, R., Abdul Majid, M.S., Mohd Nasir, N.F.: A finite element analysis of elbow joint in daily activities. ARPN J. Eng. Appl. Sci. **11**(8), 5139–5145 (2016)
24. Marannano, G., Pasta, A., Parrinello, F., Giallanza, A.: Effect of the indentation process on fatigue life of drilled specimens. J. Mech. Sci. Technol. **29**(7), 2847–2856 (2015)
25. Restivo, G., Marannano, G., Isaicu, G.A.: Three-dimensional strain analysis of single-lap bolted joints in thick composites using fibre-optic gauges and the finite-element method. J. Strain Anal. Eng. Des. **45**(7), 523–534 (2010)
26. Marannano, G., Parrinello, F., Giallanza, A.: Effects of the indentation process on fatigue life of drilled specimens: Optimization of the distance between adjacent holes. J. Mech. Sci. Technol. **30**(3), 1119–1127 (2016)
27. Yang, D.T., Zhang, D., Arola, D.: Fatigue of the bone/cement interface and loosening of total joint replacements. Int. J. Fatigue **32**, 1639–1649 (2010)

Additive Manufacturing

Adoption of Additive Technologies by Florence Industries: Designing a Survey Session

Francesco Saverio Frillici(✉), Lorenzo Fiorineschi, Rocco Furferi, and Federico Rotini

University of Florence, via di Santa Marta 3, 50139 Florence, Italy
francescosaverio.frillici@unifi.it

Abstract. Although the wide diffusion and technological development of Additive Technologies, it is still unclear to what extent the related potentialities are actually exploited. The work described in this paper aims at developing an on-line survey to be administered to industrial practitioners from different types of firm to elicit the information required to better understand the role of additive technologies and/or prototypes. In particular, we developed a preliminary version of the survey, and tested it with a limited sample of academic participants. The followed procedure, which includes the administration of a NASA Task Load Index questionnaire to participants, allowed to rapidly receive important feedbacks to support the development of a robust survey. Once administered to the expected final participants, the survey is expected to provide information about which are the main technologies used by different industrial sectors, how firms currently select 3D printers, and how they are currently used for design purposes.

Keywords: Additive Manufacturing · 3D printing · Prototypes · Survey

1 Introduction

Additive Technologies (AT) are currently spreading, thanks to the versatility of their applications and to the development of many technological variants, which allows the diffusion in different contexts. Although the original use of AT was related to the production of prototypes, nowadays is claimed to be a real alternative and/or an integrated manufacturing process for the final production of some products. End-product AT manufacturing processes are currently acknowledged under the name of Additive Manufacturing (AM). Recently, research topics related to AM reached a terrific increase of interest within scholars, while the number of contributions mentioning "Rapid Prototyping" (RP) (actually not limited to AT), encountered a sensible diminution [1]. These trends somehow communicate that the attention toward design-related topics concerning AT is now mainly focused on manufacturing aspects, while those related to prototyping encounter a descending phase. Nevertheless, the spreading of low cost or "entry level" technologies, AT-based RP can play an important and concrete role in modern product design and development, providing a relatively cheap and rapid support to the well-known fundamental activity of building physical prototypes for design purposes.

C. Rizzi et al. (Eds.): ADM 2019, LNME, pp. 827–836, 2020.
https://doi.org/10.1007/978-3-030-31154-4_70

To better understand how firms and industrial practitioners actually exploit the benefits of AT, we designed an investigation activity focused on the Florence Metropolitan Area, aimed at extracting fundamental information for the purpose. More precisely, although limited to a restricted area, the investigation is expected to provide detailed information about the following issues:

- Which kind of technologies are used by different firm types
- How AT are exploited within the product design process
- Which are the main parameters affecting the selection of specific AT

To obtain the related information, it is crucial to get in contact with a number of firms. However, it is acknowledged that industry-academia communication is often difficult, due to different needs and perspectives [2]. Therefore, it is extremely important to find an agile way for elicit the required information, without discouraging the industrial practitioners with too extended and articulated interviews. To this aim, online surveys are often used by scholars, also for AT related investigations [3–7]. Nonetheless, in order to obtain a comprehensive set of data, it is important to correctly design the survey, by considering some critical tips. Accordingly, the objective of the work described in this paper was to design the survey and to receive first important feedbacks for further optimizations. In particular, Sect. 2 reports a short background about acknowledged AT and prototype related surveys. Then, Sect. 3 shows the specific guidelines and tips followed to develop the survey. Section 4 illustrates the first version of the survey, and presents a set of structured feedbacks obtained from a pre-test session. Thanks to these first results, a discussion is performed in Sect. 5, highlighting limits and formulating hypothesis about the potential impact and the future developments. Eventually, Conclusions are reported in the last section.

2 Surveys as Investigation Tools for Understanding Prototypes and AT

The interest toward the use of AT in design processes led scholars to perform many investigations, some of which in form of surveys and/or interviews. Lauff et al. [3] recently performed a pilot study, including surveys and interviews, whose results highlight that, while both students and professionals agree that prototypes can be useful for testing functionality of concepts, students rarely mention the possible use of prototypes for supporting activities like communication, decision making and learning. Lauff et al. [8] also exploited the information extracted from specific surveys and interviews to extract a new definition of prototype: i.e. “a physical or digital embodiment of critical elements of the intended design, and an iterative tool to enhance communication, enable learning, and inform decision-making at any point in the design process”.

Deininger et al. [4] used semi-structured interviews to collect data about how respondents use prototypes. The results confirm that respondents perceive prototypes as tools for testing and evaluating functionality of concepts, but differently from what observed by Lauff et al. [3], all participants reported that prototypes have been used in their projects also for communication purposes.

Böhmer et al. [7] performed an investigation about how students use physical prototypes, the related effect in design thinking processes, and about what affects prototyping results. The observed results led the same scholars to conclude that it is fundamental to train students in developing a correct prototyping strategy to make them capable of exploiting the potentialities of prototypes.

Koohgilani et al. [9] used a basic online questionnaire to gather information about the students' and practitioners' views of prototyping techniques. Among the obtained results, it emerged only a slight preference toward rapid prototyping technologies.

Even if not used as main investigation tools, surveys and/or interviews have been often employed by researchers to find preliminary or additional data. For example, questionnaires were administered to identify the major difficulties perceived in project-related mechanical design courses [10]. It emerged that prototyping activities are perceived as difficult tasks, especially when working within time and budget constraints. In particular, one of the hardest tasks appeared to be the development of feasible prototyping strategy.

Also in the work of Greenhalgh [11], a survey has been used to understand participant's perception about the different techniques for manufacturing prototypes. In the mentioned work, it has been highlighted that the construction method to build the physical prototype (hand-made or 3D printed) influenced the students' design outcomes.

Interviews have been used also in the work of Yu et al. [12], to investigate the differences between industrial designers and engineers. One of the most interesting results from this study, is that while engineers tend to use prototypes for testing purposes, industrial designers are more willing to use them for exploratory purposes. Moreover, while engineers often adopt high fidelity prototypes, industrial designers largely exploit rough low fidelity prototyping.

Pradel et al. [5] exploited an online survey to better understand how industrial and design practitioners are actually experienced about how to design products for AT series production. They observed that the majority of the respondents remained sceptical about the use of AT for series production, due to cost and technical performances. Moreover, the few respondents reporting experience in designing components for AT series production, referred to plastic as the main adopted material. Moreover, it also emerged that designers continue to view the main benefits of AT if exploited for prototyping during product development.

The above-presented background highlights that the adoption of surveys can be a valid tool to elicit information about the use of AT and/or prototypes. Moreover, the results obtained by Pradel [5], indicate that when investigating around AT, it is still important to comprehensively consider the aspects of prototypes and prototyping processes.

3 Designing the Survey

According to the main questions claimed in Sect. 1, we have developed a survey by following suggestions and guidelines available in literature. First of all, we identified the "descriptive" surveys as the preferred type. Indeed, descriptive surveys are used to collect information about the phenomenon to be investigated [13], by reporting factual data or opinions that can lead to more comprehensive studies [14].

It is also important to discern between cross-sectional studies and longitudinal studies [15]. In the first, the data are collected within a single and relatively short time period, while the longitudinal studies collect data at more than one point in time (even across multiple years). Accordingly, and due to time schedule imposed by project, we necessarily opted for a cross-sectional study.

Then, to formulate the questions, the following important requirements have been considered [13]:

- Reliability: Ambiguous or vague wording must be avoided
- Response rate: Non-response to questions needs to be minimized. Intrusive, insensitive, irrelevant or repetitive questions as well as those that are difficult to understand and answer or have insufficient response categories can produce frustration and non-response.
- Same meaning for all respondents: If different respondents interpret a question in different ways, they effectively answer different questions, thus making the survey analysis largely meaningless.
- Relevance: Each question must earn its place in your survey. For each question, ask yourself whether it really is necessary.
- Exhaustiveness (inclusiveness): There must be sufficient response alternatives so that all respondents can answer the question
- Exclusiveness: The alternate responses should be mutually exclusive so that only one response for each variable is applicable to any respondent

Moreover, the following steps and tips have been followed to build the survey [14, 16]:

- State the problem
- Plan all stages of the project
- State the Research Question
- Review the literature
- Develop the Survey Instrument
- Use the minimum number of items needed to obtain the data desired, and resist the temptation to gather data not pertinent to the research question
- Long instruments tire respondents and can lower response rates.
- Do not ask anything you do not really need to know.
- Give definitions where appropriate and avoid jargon or slang unless needed.
- Not leading. "Don't you think that. . ." is almost always a bad way to start a question. Let them tell you what they think.
- Do not ask questions they cannot answer or do not know

Besides these guidelines, we have also felt the need to perform a pre-test on the survey, before administering it to industrial practitioners. Indeed, in order to better exploit the limited time slot that industrial practitioners are supposed to concede for our investigation, it is important to limit any loss of time and to optimize the investigation tool. To that purpose, a first version of the survey (see Fig. 1), has been administered to 12 participants from academia, asking them to "simulate" the industrial participants.

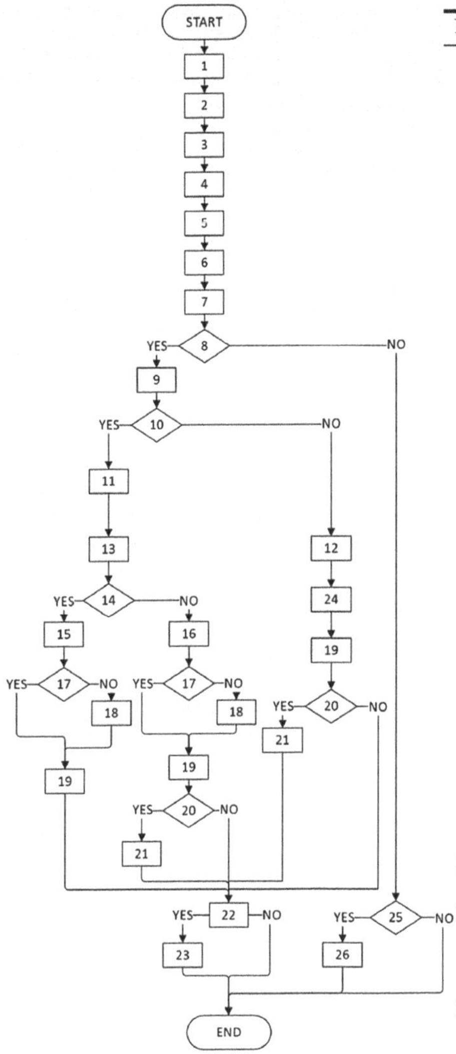

Nr.	Question content
1	What are the dimensions of the company in which you work?
2	What is your role within the company?
3	How many years have you worked in the same industrial sector?
4	How many years was the company founded?
5	In which sector is your company located?
6	What are your main customers?
7	What is your definition of a prototype?
8	Are prototypes used in your company?
9	Where do the prototypes are made?
10	Does your company use additive manufacturing technologies (3D printing) for prototyping?
11	Which additive manufacturing technologies (3D printing) do you use?
12	Why is additive manufacturing not used for prototyping?
13	Where the prototypes realized for additive manufacturing are made?
14	Does your company use exclusively AT for prototyping?
15	Why do you only use AT for prototyping?
16	What other prototyping technologies does your company use?
17	Is AM used exclusively for prototyping?
18	For what additional purposes, in addition to prototyping, is AM used?
19	In which phase of the product development cycle do you use prototypes?
20	Do you use different prototyping technologies at different design stages?
21	Which prototyping technologies do you use in the initial or ideation phase, in the intermediate or feasibility analysis phase, in the final or detailed phase?
22	Does the level of similarity of a prototype with respect to the final product depend on the progress of the project and/or on the design phase?
23	How much does the "shape" (or "data", "functionality", "interactivity", "prototyping technique", "material", "performance", "environment") parameter impact on the similarity between a prototype and the finished product?
24	What other prototyping technologies does your company use?
25	Does your company use AT for an internal business that is not prototyping?
26	For which purposes are the AT used?

Fig. 1. Schematic representation of the survey and its branching logic.

In order to extract information about the perceived difficulties, we also asked them to provide answers to an additional questionnaire developed by following the logic of the NASA Task Load Index (TLX) [17, 18], i.e. a multidimensional assessment tool that rates the perceived workload for different activities, among which, data entry tasks [17]. The TLX can be viewed at https://forms.gle/2c9oDU8enBTn1Kzm7.

4 The Designed Survey: First Impressions

Thanks to guidelines and tips introduced in Sect. 3, and other similar ones available in literature [15, 19, 20], the survey represented in Fig. 1 has been developed. The online version of the survey is available at https://forms.gle/qhCxb1NWD4KQkCNi6. The reader is invited to take vision of the survey and provide suggestions, comments and/or highlight potential criticalities (e.g. through the TLX).

The most critical results obtained with the performed pre-test and the related TLX questionnaire (compiled by 9 respondents), are shown in Figs. 2, 3 and 4 (the other TLX parameters led to good results, so we omitted them to accomplish the length limit imposed for the paper).

As shown in Fig. 2, most of the participants declared a medium mental effort spent in answering the questions, while Fig. 3 shows that perceived level of satisfaction of the survey's is quite good, but could be improved. However, Fig. 4 reports that there is a non-negligible heterogeneity in the perceived level of frustration, and this indicates that there is something that it is not clear or not adequately motivated.

Accordingly, three participants highlighted that the matrix-type format and the specific formulation of Question 23 in Fig. 1, make the last part of the survey too onerous and difficult to understand. More precisely, they suggested to reformulate the question in clearer way, and to reformulate the Likert scale levels as well. Additionally, it was posted a comment asking to insert the option "other" to Question 2 in Fig. 1.

Fig. 2. Mental effort perceived by participants

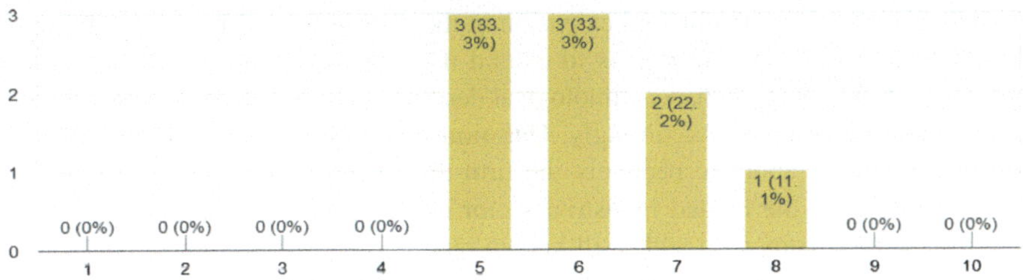

Fig. 3. Perceived level of satisfaction, according to introduced survey's objective.

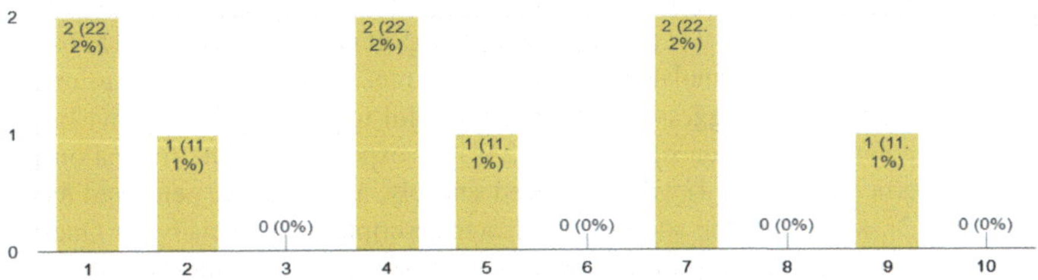

Fig. 4. Frustration perceived by participants during the survey.

5 Discussions

According to the objective introduced in Sect. 1, the work described in this paper led to a preliminary structured survey, focused on some of the most important parameters affecting the exploitation of AT and prototypes in general. In particular, the survey was focused on the identification of the technological variants of AT acknowledged by industrial staffs, and their use in industry. When referring to prototyping activities (i.e. still the most diffused use of AT [5]), a series of questions concerning the perceived meaning of the term "prototype", its role within the design process and the main "fidelity" dimensions [21] have been used to elicit detailed information. Indeed, it is expected that by comparing the AT variants selected by respondents with the answers to Question 23 (Fig. 1), it would be possible to evaluate possible links between the needed fidelity level of the prototype, and the specific AT variant. Therefore, one of the future outcomes expected for the survey, is that of providing important information that could pave the way for the development of an heuristic guideline, to support (at least in part) the complex task of the selecting the most suited AT [22–24] for prototyping purposes. Indeed, different AT have different performance in terms of surface finishing, tolerances, material strength, etc. which affect the fidelity parameters used in the survey. Accordingly, the respondents will be asked to participate at more comprehensive interviews, aimed at understanding why the specific AT has been selected, and which are the advantages that drove the respondent's company to that particular technology.

Moreover, thanks to Question 7 in Fig. 1, it will be possible to extract the personal definitions of the term "prototype", and then to perform comparisons with those acknowledged in literature [25–29] (some of which extracted with similar procedures [8]).

Indeed, besides the first limited extension of the distribution of the survey (Florence Metropolitan Area), a future step is to extend it to the whole Italy, with the aim of capturing a more comprehensive ethnological description of the actual use of AT in the various industrial sectors. Accordingly, Question 1 and Question 5 in Fig. 1 allow to extract information about respectively the firm dimension (micro-enterprise, SME or big enterprise) and the related industrial sector (e.g. automotive, mechanics, fashion, etc.). This data, if compared with well known reports (e.g. Wohlers [30]), could allow to obtain detailed information about how AT are used in Italian firms for prototyping and design purposes.

However, before starting with the real investigation activities, the issues highlighted by the TLX respondents must be faced and resolved first. Especially concerning Question 23 in Fig. 1, since it is perhaps the most important one, but revealed to be also the most complex to be formulated. The performed pre-test session allowed to receive important feedback and suggestions, extremely useful to improve the survey. In particular, after the modification of Question 23, it is expected that the frustration perceived by participants (Fig. 4) will be reduced sensibly, as well as the perceived mental effort (Fig. 2). Moreover, the adoption of clearer descriptions (especially for Question 23) is expected to further improve the perceived satisfaction of the survey's objectives (Fig. 3). Once performed the required corrections on the survey, a new pre-test session will be performed to check whether the performed amendments and optimization are sufficient, according to the TLX framework. Then, the next step is to try to extract first "illustrative" results from the respondents of the pre-test to check also the post-processing phase. Once reached a satisfactory quality level for both the survey and the post-processing procedure, the on-line survey will be eventually administered to industrial practitioners, managers and staff in general.

The "take-home" of this paper is a structured procedure and hints from literature, which can be conveniently re-used by peers involved in qualitative investigations where surveys constitute the main investigation tool.

6 Conclusions

This paper aimed at providing important indications and suggestions about how to develop a survey for specific AT-related investigations. Notwithstanding the limits that characterize this work and which make impossible to provide generally valid indications, the instructions and the procedure provided in this work can be conveniently re-used by peers. Especially concerning the pre-testing procedure based on the NASA Task Load Index, it revealed to be very useful to optimize our survey. Therefore, even though not originally conceived for such a kind of applications, it can be a valid instrument to rapidly elicit important information about the actual performance of the questionnaire to be developed. Accordingly, it allowed us to rapidly identify an important criticality in one of the most important questions of the survey, together with a quantification of the related negative effect.

Thanks to the outcome of this work, it will be possible (after the required optimizations) to proceed toward the next step of the project, i.e. the administration of the questionnaire to the industrial practitioners in the Florence Metropolitan Area.

Nevertheless, a further step is expected, i.e. to extend the survey to the whole Italy, in order to obtain a more comprehensive and extended set of data to be compared with acknowledged definitions and methods. In this way, it is expected to provide an important contribution for better understanding the actual potentialities of AT, especially in terms of their role in the design process, with particular attention to prototyping activities.

Acknowledgments. The work described in this paper has been partially financed by Fondazione CR Firenze, within the project TAM-TAM: "Tecnologie Additive per l'innovazione di prodotti, processi e Metodi nelle aziende del Territorio dell'Area Metropolitana" (Additive technologies for product, process and method innovation in the Florence's metropolitan area).

References

1. Carfagni, M., Fiorineschi, L., Furferi, R., Governi, L., Rotini, F.: The role of additive technologies in the prototyping issues of design. Rapid Prototyp. J. **24**, 1101–1116 (2018). https://doi.org/10.1108/RPJ-02-2017-0021
2. Bruneel, J., D'Este, P., Salter, A.: Investigating the factors that diminish the barriers to university-industry collaboration. Res. Policy **39**, 858–868 (2010). https://doi.org/10.1016/j.respol.2010.03.006
3. Lauff, C., Kotys-Schwartz, D., Rentschler, M.E.: Perceptions of prototypes: pilot study comparing students and professionals. In: Volume 3: 19th International Conference on Advanced Vehicle Technologies; 14th International Conference on Design Education; 10th Frontiers in Biomedical Devices, V003T04A011 (2017). https://doi.org/10.1115/DETC2017-68117
4. Deininger, M., Daly, S.R., Sienko, K.H., Lee, J.C., Street, H.: Novice designers' use of prototypes in engineering design. Des. Stud. **51**, 25–65 (2017). https://doi.org/10.1016/j.destud.2017.04.002
5. Pradel, P., Zhu, Z., Bibb, R., Moultrie, J.: Investigation of design for additive manufacturing in professional design practice. J. Eng. Des. **4828**, 1–36 (2018). https://doi.org/10.1080/09544828.2018.1454589
6. Böhmer, A., Richter, C., Hostettler, R., Schneider, P., Plum, I., Böhler, D., Lindemann, U., Conradt, J., Knoll, A.: Think.Make.Start. - an agile framework. In: International Design Conference - Design 2016, pp. 917–926 (2016)
7. Böhmer, A.I., Kayser, L., Sheppard, S., Lindemann, U.: Prototyping as a thinking approach in design. Insights of problem-solving activities while designing a product. In: International Conference on Engineering, Technology and Innovation (ICE/ITMC 2017), pp. 955–963. IEEE (2017). https://doi.org/10.1109/ICE.2017.8279985
8. Lauff, C.A., Kotys-schwartz, D., Building, F., Rentschler, M.E.: What is a prototype? What are the roles of prototypes in companies? J. Mech. Des. 1–51 (2018). https://doi.org/10.1115/1.4039340
9. Koohgilani, M., Powell, J., Underwood, G.: RP or not RP, that is the co-creation question. In: International Conference on Engineering and Product Design Education, 3 & 4 September 2015, pp. 246–251 (2015)
10. Holland, D., Conor, J.W., Bennett, G.J.: Tools for assessing student learning in mechanical design courses. In: DS 76: Proceedings of E&PDE 2013, the 15th International Conference on Engineering and Product Design Education, Dublin, Ireland, pp. 064–069 (2013)

11. Greenhalgh, S.D.: The effects of 3D printing in design thinking and design education. J. Eng. Des. Technol. (2016). https://doi.org/10.1108/JEDT-02-2014-0005
12. Yu, F., Pasinelli, M., Brem, A.: Prototyping in theory and in practice: a study of the similarities and differences between engineers and designers. Creat. Innov. Manage. **27**, 121–132 (2018). https://doi.org/10.1111/caim.12242
13. DeVaus, D.: Survey research. In: Greenfield, T., Greener, S. (eds.) Research Methods for Postgraduates. Wiley (2016). https://doi.org/10.4135/9781849209984
14. Passmore, C., Dobbie, A.E., Parchman, M., Tysinger, J.: Guidelines for constructing a survey. Fam. Med. **34**, 281–286 (2002). https://doi.org/10.1109/MOBISECSERV.2015.7072880
15. Christensen, L.B., Johnson, R.B., Turner, L.A.: Research methods, design, and analysis (2014)
16. Janes, J.: On research survey construction. Libr. Hi Tech. **17**, 321–325 (1999). https://doi.org/10.1108/07378839910289376
17. Hart, S.G.: Nasa-Task Load Index (NASA-TLX); 20 years later. Proc. Hum. Factors Ergon. Soc. Ann. Meet. **50**, 904–908 (2006). https://doi.org/10.1177/154193120605000909
18. Hart, S.G., Staveland, L.E.: Development of NASA-TLX (Task Load Index): results of empirical and theoretical research. Adv. Psychol. **52**, 139–183 (1988). https://doi.org/10.1016/S0166-4115(08)62386-9
19. Krosnick, J.A., Visser, P.S., Lavrakas, P.J.: Survey research. Ann. Rev. Psychol. **50**, 537–567 (1999). https://doi.org/10.1146/annurev.psych.50.1.537
20. Fink, A.: How to Ask Survey Questions. SAGE Publications Inc., Thousand Oaks (2003)
21. McCurdy, M., Connors, C., Pyrzak, G., Kanefsky, B., Vera, A., Field, M.: Breaking the fidelity barrier: an examination of our current characterization of prototypes and an example of a mixed-fidelity success. In: Proceedings of the SIGCHI Conference on Human Factors in Computing Systems, pp. 1233–1242 (2006)
22. Vimal, K.E.K., Vinodh, S., Brajesh, P., Muralidharan, R.: Rapid prototyping process selection using multi criteria decision making considering environmental criteria and its decision support system. Rapid Prototyp. J. **22**, 225–250 (2016). https://doi.org/10.1108/RPJ-03-2014-0040
23. Borille, A., Gomes, J., Meyer, R., Grote, K.: Applying decision methods to select rapid prototyping technologies. Rapid Prototyp. J. **16**, 50–62 (2010). https://doi.org/10.1108/13552541011011712
24. Byun, H.S., Lee, K.H.: A decision support system for the selection of a rapid prototyping process using the modified TOPSIS method. Int. J. Adv. Manuf. Technol. **26**, 1338–1347 (2005). https://doi.org/10.1007/s00170-004-2099-2
25. Schrage, M.: Cultures of prototyping. In: Bringing Design to Software, pp. 191–205. ACM, New York (1996). https://doi.org/10.1145/229868.230045
26. Houde, S., Hill, C.: What do prototypes prototype? In: Helander, M., Landauer, T., Prabhu, P. (eds.) Handbook of Human Computer Interaction, pp. 1–16. Elsevier Science B.V., Amsterdam (1997). https://doi.org/10.1016/B978-044481862-1.50082-0
27. Hallgrimsson, B.: Prototyping and Model Making for Product Design. Laurence King Publishing Ltd., London (2012). https://doi.org/10.1017/CBO9781107415324.004
28. Ullman, D.G.: The Mechanical Design Process, 4th edn. McGraw Hill, New York (2010)
29. Jensen, L.S., Özkil, A.G., Mortensen, N.H.: Prototypes in engineering design: definitions and strategies. In: International Design Conference - Design 2016, pp. 821–830 (2016)
30. Wohlers, T.: Wohlers report. https://wohlersassociates.com/2019report.htm

Properties Enhancement of Carbon PA 3D-Printed Parts by Post-processing Coating-Based Treatments

Sandro Barone, Paolo Neri, Sara Orsi, Alessandro Paoli, Armando V. Razionale, and Francesco Tamburrino(✉)

Department of Civil and Industrial Engineering, University of Pisa, Largo L. Lazzarino 1, 56122 Pisa, Italy
francesco.tamburrino@ing.unipi.it

Abstract. In recent years, Fused Deposition Modelling (FDM) has become one of the most attractive Additive Manufacturing (AM) techniques, due to the advantages in the production of complex shapes with a wide range of materials and low investment costs. The thermoplastic polymers used for FDM technology are characterized by low mechanical properties if compared to those of composites and metals. This issue is usually overcome by reinforcing the thermoplastic polymer with chopped fibres or particles. Moreover, a second issue arises, which is represented by the water absorption with a relevant impact on mechanical properties and dimensional stability of printed models. In this paper, an experimental study is presented with the aim at evaluating the water absorption influence on mechanical properties of Carbon PA (Polyamide matrix reinforced with Carbon Fiber at 20%) specimens fabricated with the FDM technique. Two post-processing treatments, based on the use of acrylic spray and photosensitive resin, have been also proposed to improve the behaviour of Carbon PA printed parts. Results of water absorption tests and tensile tests demonstrated a significant improvement in terms of weight stability and mechanical properties by adopting the proposed post-processing treatments.

Keywords: Fused Deposition Modeling · Carbon fiber · Coating treatment · Water absorption · Mechanical characterization

1 Introduction

Fused Deposition Modeling (FDM) is one of the most attractive Additive Manufacturing techniques. FDM process is based on the extrusion of a thermoplastic material by a temperature-controlled head with a nozzle. The extrusion is carried out in accordance to programmed patterns and the process is repeated layer by layer. When the material is extruded onto a previous layer, it is in a semi-molten state and its surface re-melts the previous layer creating a polymer bond [1, 2].

3D printers relying on FDM technology represent the most widespread and low-cost AM systems. A huge amount of thermoplastic semi-crystalline or amorphous polymers, which can be printed with the FDM process, have been developed in the last few years. However, their properties cannot cover a wide range of applications,

C. Rizzi et al. (Eds.): ADM 2019, LNME, pp. 837–847, 2020.
https://doi.org/10.1007/978-3-030-31154-4_71

especially if high mechanical properties, humidity resistance or temperature resistance are required. The most common solutions adopted to overcome these issues rely on the development of composite materials made by a thermoplastic polymer and a reinforcement made of chopped fibres or particles. In this regard, different material typologies have been designed and tested in the last few years [3–5]. Currently, the most common approach is the use of a single printing head, which extrudes a polymeric filament containing a volume fraction of fibres (carbon, glass or wood) or particles (metallic or ceramic) which flow with the melted plastic matrix [3, 6].

In this paper, Carbon PA has been studied. Carbon PA is a composite material, based on a polyamide matrix reinforced with carbon fiber at 20%, supplied by Roboze® company. This material shows attractive physical and mechanical properties, which could make it an appropriate solution for a wide range of different applications in the automotive, aerospace, and electrical fields. However, the polyamide matrix has a significant disadvantage, which has been widely analysed and discussed in the literature. It is a highly hygroscopic semi-crystalline polymer (moisture sensitive), which is a crucial issue to be considered for parts design, mechanical performance prediction and optimization. The moisture absorption is a variable that has a significant impact on the processability of the material and on end-use performances (mechanical properties, dimensional stability, etc.). In particular, the water absorption acts as a plasticizer, affecting material properties such as strength, stiffness, and ductility [7, 8]. The characteristic behaviour of the polyamide matrix results in a decay of the potential end-use performances of Carbon PA.

The aim of this work is to experiment two different coating treatments in the reduction of the water absorption of Carbon PA specimens: one based on the use of an acrylic spray and the other on the use of a photosensitive resin. Some existent literature studies addressed their attention to water absorption of polymeric models printed by FDM technology, showing quite significant results of post-processing treatments in terms of water absorption reduction, but no improvements of the mechanical properties [9, 10]. In this paper, water absorption tests and tensile tests have been carried out in order to assess the effectiveness of the proposed post-processing treatments in the enhancement of the mechanical properties of Carbon PA parts. Outcomes are beneficial to enlarge the range of application of FDM printed parts to new and more advanced fields, where high performances and reliability of the components are essential.

2 Materials and Methods

2.1 Materials and Specimen's Fabrication

Carbon PA specimens have been investigated in this study. A filament of Carbon PA was processed with a Roboze® One FDM machine, with a nozzle diameter of 0.6 mm. The geometry of the specimen was chosen in accordance to the ISO 527-2:2012 (specimen type 1-A) [11]. Shape and dimensions of the specimen are shown in Fig. 1.

The 3D model of the specimen was created in Solidworks®, exported as STL file and then imported into Simplify 3D®, a slicer software, used to create the G-code needed for the 3D printing process. Table 1 reports the most significant parameters selected to

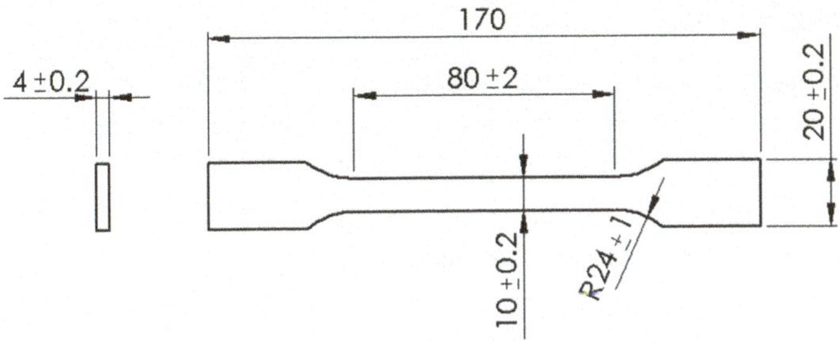

Fig. 1. Schematic representation of the ISO 527 Type 1A tensile specimen (dimensions in mm).

minimize the porosity of the fabricated Carbon PA samples. The "xyz" orthogonal orientation, defined by the ISO/ASTM 52921:2013 [12], and a raster orientation of "+45/−45", have been used (Fig. 2) in order to reduce printing time and increase mechanical properties [13]. A total number of 45 specimens have been fabricated.

Table 1. Process parameters used to print the specimens with the Roboze® One FDM machine.

Parameter	Description	Value
Air gap	Space between rasters	0
Slice height (mm)	Layer height	0.2
Extrusion width (mm)	Width of the deposited raster	0.55
Infill density (%)	Density of the printed part	100
Extrusion temperature (°C)	Nozzle temperature	300

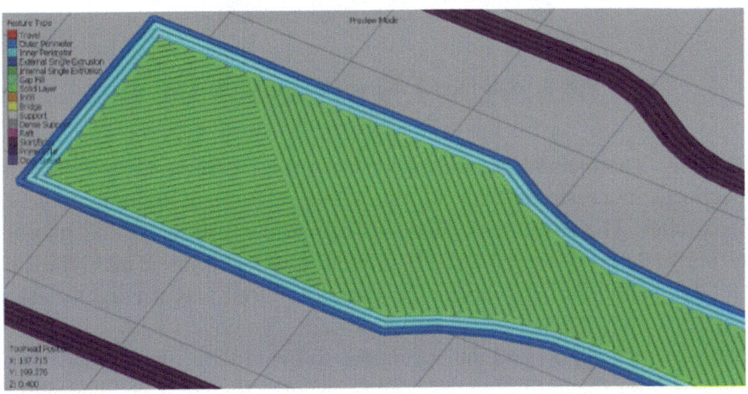

Fig. 2. Orthogonal "xyz" orientation (ISO/ASTM 52921:2013) and raster orientation "+45/−45" adopted for the specimens printing.

2.2 Coating Treatment Process

The printed specimens were coated with the aim at sealing the parts, reducing open porosity and water absorption and improving, at the same time, their mechanical properties. Two different coating treatments were used:

- "ACRIL COLOR" clear acrylic spray, supplied by Arexons®;
- "UV clear firm resin", supplied by Photocentric Group®.

The coating treatment, for both products, was always preceded by uniform heating at 100 °C (1 h) to remove water content from the specimens. 30 specimens were coated: 15 with acrylic spray and 15 with UV resin.

Acrylic Coating. The acrylic coating treatment consisted in manually spraying the specimens once clamped. The coating was applied twice in order to guarantee a fairly uniform coating on the specimens. The final specimen thickness resulted just slightly varied with respect to the uncoated samples, although the spraying procedure was hand-made: a thickness increase of 0.14 ± 0.006 mm was found by a micrometre. A representation of the acrylic coating process is schematized in Fig. 3.

Fig. 3. Scheme for the acrylic coating process.

Resin Coating. The method developed for UV resin coating relies on the immersion of the specimens in a resin bath for 30 min with a subsequent draining process carried out using a roller. The specimens, then, undergo a UV curing for 150 min at 80 °C. A supporting tool was designed and 3D printed in order to insert and hold steady the specimens during the curing process. This auxiliary tool was needed to stabilize the specimens, thus guaranteeing a repeatable process. Moreover, a layer of silicone was placed between the support and the specimens to avoid their glueing, thus facilitating their removing after the curing process. The specimen thickness resulted only slightly affected by the treatment process: a thickness increase of 0.085 ± 0.012 mm with respect to the uncoated samples was found by using a micrometre. Figure 4 schematizes the UV resin coating treatment process.

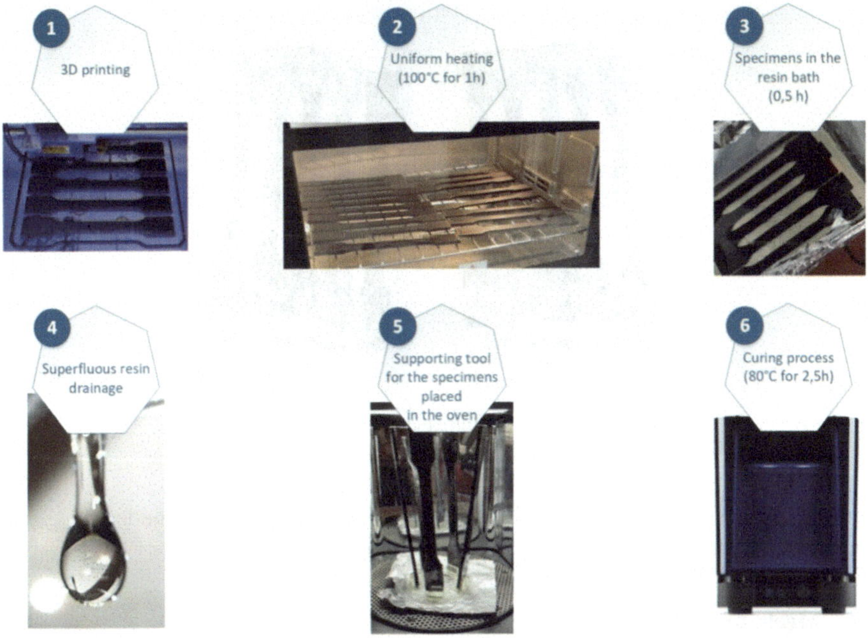

Fig. 4. Scheme for the resin coating process

2.3 Water Absorption

The standard test method for water absorption of plastics described in the ASTM D570 – 98 (2018) [14] was used as a guideline. A variation on the specimen geometry was considered since it was selected in accordance with the ISO 527-2:2012 in order to perform tensile tests on specimens subjected to water absorption. The ASTM D570 generally uses disks, sheets, or rods. The experimental equipment consisted in an analytical balance with a reading capacity of 0.001 g (JJ-B Series by G&G), an oven capable of maintaining a uniform temperature of 50 ± 3 °C, containers and distilled water. The specimens were placed in a container with distilled water maintained at a temperature of 23 ± 1 °C, kept completely immersed (Fig. 5) and weighted after 2 h, 24 h and 168 h (long-term immersion). The procedure involved a final reconditioning process, with the specimens dried in the oven for 24 h at 50 ± 3 °C, then cooled in a desiccator and finally weighed. Water absorption tests were carried out for 10 specimens for each typology ("as printed", coated with acrylic spray and coated with UV resin). Moreover, 5 specimens out of 10 were subjected to the reconditioning process. The maximum water content (CI) and the open porosity (OP) have been defined as [9]:

$$CI(\%) = \frac{w_{sat} - w_{dry}}{w_{dry}} \cdot 100 \quad (1)$$

$$OP(\%) = \frac{w_{sat} - w_{dry}}{\rho_{water} \cdot V} \cdot 100 \quad (2)$$

Fig. 5. Specimens immersed for absorption tests.

where w_{dry} is the dry weight, w_{sat} the weight achieved at the end of the test, ρ_{water} the water density (at 23 °C) and V the specimen volume.

2.4 Mechanical Tests

Tensile tests were carried out for 15 specimens for each typology ("as printed", coated with acrylic spray and coated with UV resin). In particular, 5 specimens were tested without water immersion, 5 specimens after a long-term immersion, and 5 specimens after a long-term immersion and subsequent reconditioning process.

The choice to perform the tensile tests not after 2 h and 24 h of water absorption, but only after 168 h, was determined by the intention to take into account the most critical condition in term of CI. Stress-strain curves have been plotted for each specimen or set of specimens, to derive some observations on failure mechanisms. The parameters determined by the tensile tests are Young's modulus E, yield stress σ_Y and strain ε_Y and the ultimate stress σ_u and strain ε_u.

3 Experimental Results

3.1 Absorption Behaviour

Water absorption tests were conducted to assess the effectiveness of the coating treatments (acrylic and UV resin) in the reduction of water content, CI, and open porosity, OP. The results are compared to those obtained with the specimens as printed.

Table 2 and Fig. 6 report the mean percentage of the weight variation and evidence that both CI and OP significantly decrease in the case of the coated specimens. In particular, the most critical condition in terms of water immersion is characterized by a reduction of the maximum water content from 4.04 ± 0.3% for the specimens as printed to 2.06 ± 0.6% for the acrylic coated specimens to 1.66 ± 0.35% for the UV resin coated specimens. In general, it is worth noting that the reduction in water absorption is considerable for coated specimens and the sealing performance of UV resin is slightly higher than the one of the acrylic treatment. OP values follow the trend of CI values. The lowest open porosity value is the one obtained for UV resin coated specimens, while the highest open porosity value is the one relative to as printed specimens. Results demonstrate the beneficial effect connected to the proposed post-processing treatments. Furthermore, qualitative observations on OP, have been performed by means of a digital microscope on the cross-section of some specimens subjected to tensile tests (Fig. 7). The difference in term of porosity is evident for coated specimens, particularly after water absorption (second row of Fig. 7).

Table 2. Water absorption properties of specimens untreated and treated with acrylic and resin coatings.

Treatment	2 h		24 h		168 h		Reconditioned	
	CI (%)	OP (%)	CI (%)	OP (%)	CI (%)	OP (%)	CI (%)	OP (%)
Untreated	0.75 ± 0.092	0.89 ± 0.106	1.74 ± 0.129	2.06 ± 0.138	4.04 ± 0.304	4.78 ± 0.336	1.19 ± 0.112	1.42 ± 0.144
Acrylic	0.17 ± 0.049	0.21 ± 0.059	0.71 ± 0.224	0.87 ± 0.265	2.06 ± 0.602	2.51 ± 0.711	0.53 ± 0.098	0.65 ± 0.118
Resin	0.15 ± 0.048	0.17 ± 0.056	0.53 ± 0.148	0.63 ± 0.174	1.66 ± 0.346	1.97 ± 0.406	0.72 ± 0.256	0.85 ± 0.300

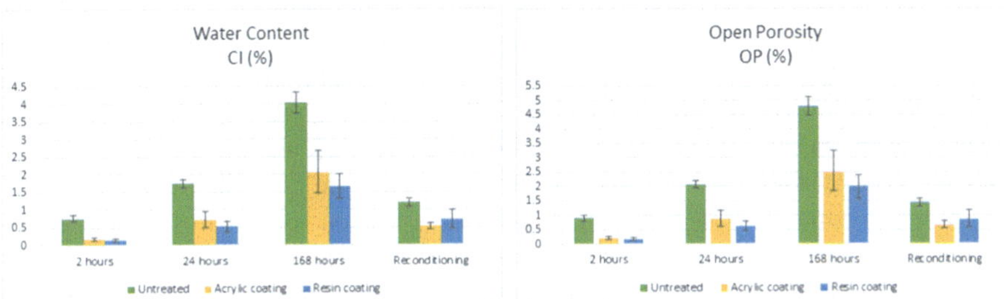

Fig. 6. Mean and standard deviation values for water content (CI) and open porosity (OP).

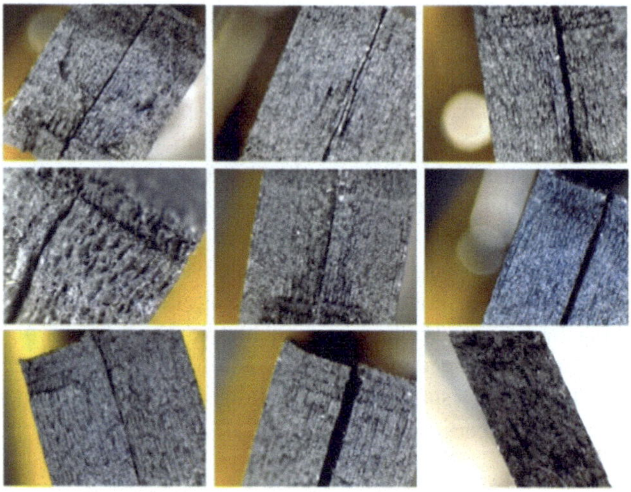

Fig. 7. Detailed views of the fracture surfaces evidencing OP. The first two rows evidence a higher content of voids for the untreated samples with respect to the coated ones.

3.2 Tensile Behaviour

Tensile tests were carried out to assess the mechanical behavior of coated specimens, comparing their properties to the untreated ones. In particular, the Young's modulus E, the yield stress σ_Y, the yield strain ε_Y, the ultimate stress σ_u and ultimate strain ε_u, were considered (Table 3). It is worth noting that the Young's modulus E, for the long-term immersion condition, is roughly two times higher for the acrylic coated specimens (5.57 $\pm$ 0.75 GPa) and even higher for UV resin coated specimens (6.95 $\pm$ 0.76 GPa) with respect to untreated samples (2.65 $\pm$ 0.76 GPa). An enhancement for the coated specimen properties can be found also for the yield stress σ_Y, and the ultimate stress σ_u, while the yield strain ε_Y and the ultimate strain ε_u values are only slightly affected. The general improvement of the mechanical behavior of coated specimens can be ascribed to the reduction of absorbed water and to the contribution given by the coating materials. The first aspect is proved by the comparison between coated and uncoated specimens after water absorption while the second is evidenced by analyzing the specimens characterized by a low water content (i.e., "as printed", "as printed and coated", and reconditioned). In this latter case, enhanced mechanical properties can be ascribed to the contribution of the coating material and to a reduced porosity.

Uncoated and coated specimens also show different stress-strain curves (Fig. 8). The failure for uncoated ones occurs at lower stress values, but higher plasticity and a higher strain (%) is present. The higher plasticity for uncoated specimens can be motivated by the higher content of water and the higher porosity [15]. The behaviour of the coated specimens is different. In general, their failure occurs at significantly higher stress values, but the plasticity is decreased, and brittle fracture occurs. For long-term immersion specimens, coated samples present higher ultimate stress values if compared to uncoated samples, but higher plasticity values than coated samples not subjected to water absorption. Water absorption results, discussed in Sect. 3.1, evidence a correlation between the plasticized failure behaviour and the samples' water content. Finally, some examples of fracture for "as printed", acrylic coated and UV resin coated specimens are reported in Fig. 9. The difference in terms of fracture is evident for coated specimens, particularly after water absorption (second row of Fig. 9). The coated samples exhibit a brittle fracture with respect to untreated samples, which are characterized by a more ductile fracture.

Table 3. Results obtained for tensile tests.

Treatment		E (GPa)	σ_Y (MPa)	ε_Y (%)	σ_u (MPa)	ε_u (%)
Uncoated	*Absorption*	2.65 ± 0.76	20.30 ± 7.63	0.01 ± 0.003	32.11 ± 6.50	0.045 ± 0.025
	Reconditioned	7.62 ± 0.27	67.84 ± 7.83	0.011 ± 0.001	79.22 ± 6.00	0.015 ± 0.001
	As printed	8.54 ± 0.31	74.62 ± 1.98	0.011 ± 0.001	91.04 ± 2.59	0.011 ± 0.001
Acrylic	*Absorption*	5.57 ± 0.75	52.97 ± 7.37	0.011 ± 0.003	72.73 ± 5.92	0.059 ± 0.078
	Reconditioned	9.39 ± 0.61	96.39 ± 6.11	0.012 ± 0.001	115.55 ± 5.21	0.018 ± 0.002
	Only coated	9.66 ± 0.17	111.77 ± 1.32	0.014 ± 0.001	117.97 ± 3.68	0.015 ± 0.001
Resin	*Absorption*	6.95 ± 0.76	68.65 ± 9.01	0.012 ± 0.001	76.71 ± 8.15	0.016 ± 0.002
	Reconditioned	8.80 ± 1.61	75.64 ± 17.16	0.011 ± 0.003	84.66 ± 12.54	0.014 ± 0.002
	Only coated	9.06 ± 0.34	89.17 ± 3.33	0.012 ± 0.001	90.75 ± 5.84	0.012 ± 0.001

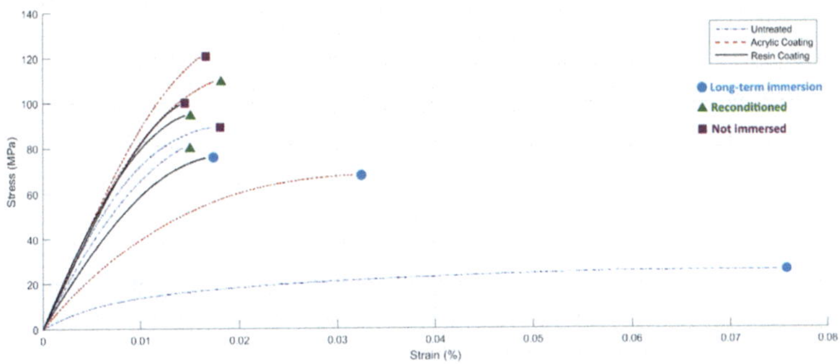

Fig. 8. Stress-strain curves for "untreated", acrylic coated and UV resin coated specimens. A single representative curve is plotted for each treatment typology.

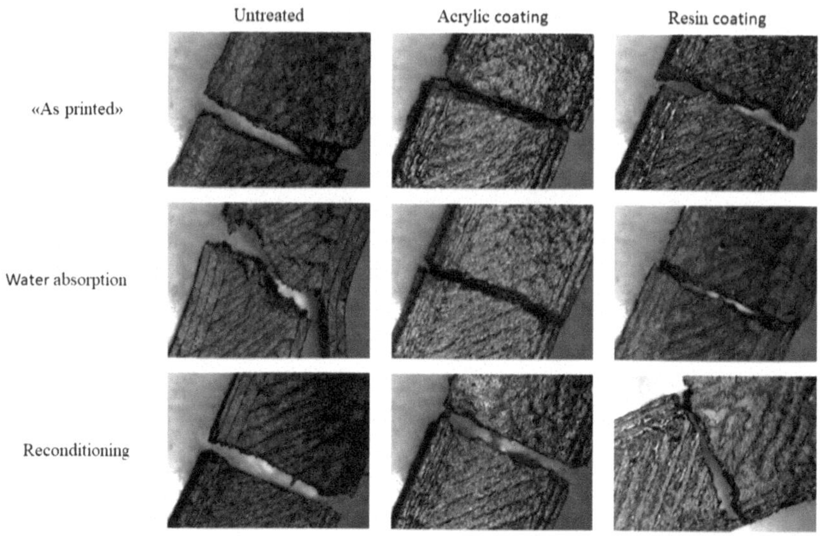

Fig. 9. Detailed views of the fracture surfaces for "as printed", acrylic coated and UV resin coated specimens.

4 Conclusions

The use of composite materials to fabricate mechanical parts by FDM technologies is increasingly widespread. This trend is motivated by the need of using flexible and low-cost technology to produce components characterized by high performances for advanced applications. This growing demand is not always accompanied by a complete control of material properties over time, particularly when subjected to environmental influences. In this paper, a post-processing method, based on coating treatments, has been developed to reduce water absorption of Carbon PA specimens. The aim is at increasing performance stability over the time and mechanical properties of this material, which suffers for the high hygroscopicity of its polyamide matrix. Obtained results are significant and suggest that coating-based post-processing treatments can be an effective solution when reliable and advanced mechanical properties are requested.

Future developments will address issues related to dimensional stability (e.g., swelling, warping, etc.). Moreover, further research activities could investigate the mechanical characterization of the analyzed material by considering different CI values, the selection of additional coating materials and the application of the described experimental approach to test other hygroscopic 3D printing materials.

References

1. Patterson, A.E., Bahumanyam, P., Katragadda, R., Messimer, S.L.: Automated assembly of discrete parts using fused deposition modeling. Rapid Prototyp. J. **24**, 249–260 (2018)
2. Tamburrino, F., Graziosi, S., Bordegoni, M.: The influence of slicing parameters on the multi-material adhesion mechanisms of FDM printed parts: an exploratory study. Virtual Phys. Prototyp. **14**(4), 316–332 (2019). https://doi.org/10.1080/17452759.2019.1607758
3. Ligon, S.C., Liska, R., Stampfl, J., Gurr, M., Mulhaupt, R.: Polymers for 3D printing and customized additive manufacturing. Chem. Rev. **117**, 10212–10290 (2017)
4. Masood, S.H., Song, W.Q.: Development of new metal/polymer materials for rapid tooling using fused deposition modelling. Mater. Design **25**, 587–594 (2004)
5. Singh, R., Singh, S., Fraternali, F.: Development of in-house composite wire based feed stock filaments of fused deposition modelling for wear-resistant materials and structures. Compos. Part B-Eng. **98**, 244–249 (2016)
6. Ning, F.D., Cong, W.L., Hu, Y.B., Wang, H.: Additive manufacturing of carbon fiber-reinforced plastic composites using fused deposition modeling: effects of process parameters on tensile properties. J. Compos. Mater. **51**, 451–462 (2017)
7. Jia, N.Y., Fraenkel, H.A., Kagan, V.A.: Effects of moisture conditioning methods on mechanical properties of injection molded nylon 6. J. Reinf. Plast. Comp. **23**, 729–737 (2004)
8. Parodi, E., Peters, G.W.M., Govaert, L.E.: Prediction of plasticity-controlled failure in polyamide 6: influence of temperature and relative humidity. J. Appl. Polym. Sci. **135**, 45942 (2018)
9. Miguel, M., Leite, M., Ribeiro, R., Deus, A.M., Reis, L., Vaz, M.F.: Failure of polymer coated nylon parts produced by additive manufacturing. Eng. Fail. Anal. **101**, 485–492 (2019)
10. Leite, M., Varanda, A., Ribeiro, A.R., Silva, A., Vaz, M.F.: Mechanical properties and water absorption of surface modified ABS 3D printed by fused deposition modelling. Rapid Prototyp. J. **24**, 195–203 (2018)
11. ISO: Plastics - Determination of tensile properties - Part 2: Test conditions for moulding and extrusion plastics, vol. ISO 527-2:2012, p. 11 (2012)
12. ISO/ASTM: Standard terminology for additive manufacturing – Coordinate systems and test methodologies, vol. ISO/ASTM 52921:2013 (2013)
13. Fatimatuzahraa, A.W., Farahaina, B., Yusoff, W.A.Y.: The effect of employing different raster orientations on the mechanical properties and microstructure of Fused Deposition Modeling parts. In: 2011 IEEE Symposium on Business, Engineering and Industrial Applications (ISBEIA), pp. 22–27 (2011)
14. ASTM: Standard Test Method for Water Absorption of Plastics. vol. ASTM D570 - 98 (2018) (2018)
15. Miri, V., Persyn, O., Lefebvre, J.M., Seguela, R.: Effect of water absorption on the plastic deformation behavior of nylon 6. Eur. Polym. J. **45**, 757–762 (2009)

Sensor Embedding in a 3D Printed Flexure Hinge

Francesco Rosa(✉), Diego Scaccabarozzi, Simone Cinquemani, and Francesco Bizzozero

Dipartimento di Meccanica, Politecnico di Milano, via La Masa, 34, 20156 Milano, MI, Italy
francesco.rosa@polimi.it

Abstract. Additive Manufacturing has opened privileged roads towards rapid-realization of highly customized goods. The most spread AM technology is the *Fused Deposition Modeling* (FDM), that allows production of plastic prototypes and finished parts.

Among these opportunities, the FDM technology offers the possibility of printing objects made up of different materials (*Multi-Material-Deposition*) to create objects characterized by local variations of compliance. This capability can be exploited to generate flexible joints that connect more rigid portions of the same object, enabling finite relative movement between portions of the object. In other words, it is possible to realize monolithic and joint-less structures that rely on the compliance of specifically designed zones to govern motion and forces transfer, i.e. the so called *Compliant Mechanisms.*

This paper presents the design process of such a flexible junction, and how it is possible to embed a distributed sensor (i.e. an optical fiber) within it in order to measure the junction deformation and, then, read the relative movement of the two elements connected to it. The control of this type of systems, in fact, is always problematic because of the intrinsic 'uncertainty in the elasticity and mass density of the links', that can make open loop control systems prone to instability.

Furthermore, the presented and discussed embedding technique can be also adopted to embed sensors in other types of applications.

Keywords: Additive Manufacturing · Sensor embedding · Smart Flexure Hinge · Compliant Mechanism

1 Introduction

The better possibilities to fully exploit the capabilities of Additive Manufacturing (*AM*) technologies occur when it is possible to design and develop a new product from scratch, abandoning the traditional strategies and the typical solutions, in favor of new solutions.

The so called *Compliant Mechanisms* (CM) are an example of such a design approach.

C. Rizzi et al. (Eds.): ADM 2019, LNME, pp. 848–859, 2020.
https://doi.org/10.1007/978-3-030-31154-4_72

CMs are single-part flexible mechanisms that transform input forces and/or displacements into output forces and displacements only by means of the elastic deformations of specifically designed small portions featured to be compliant [4,5]. In other words, CMs are monolithic and joint-less structures that rely on the compliance of specifically designed zones to govern motion and forces transfer.

With respect to traditional systems, CMs can integrate different functions into fewer parts, a feature that enables them to accomplish complex task with less parts, reducing (or avoiding) assembly operations and costs, and increasing reliability.

The realization of a robust control of this systems is still an open issue, because of the *uncertainty in the elasticity and mass density of the links* [2]. This issue is also confirmed by Linß et al. [3], that, after having compared numerical and experimental force - displacement relationships of several types of aluminum (AW 7075) flexure hinges, concluded that *the error of the calculated results is in the range of less than 2% to less than 16% for the suggested parameter.*

A possible approach to overcome this limitation is to directly measure their deformations during functioning, thus allowing correction and improving the control system reliability.

This paper is focused on the opportunity to monitor a *Flexural Hinge* (FH) [6,7], i.e. a portion of a CM which connect or transmit load between two "components" while allowing limited relative movement between them by deflecting elastically [1], introducing a novel solution, i.e. a *smart FH* (SFH): a FH with an embedded distributed sensor capable to monitor hinge deflection in order to allow its real-time and continuous monitoring to implement a closed loop control system. In particular, we will focus on the design of this SFH, trying to highlight the crucial aspects, that may be transferable into the design process of other similar parts.

In order to simplify the development of the SFH, we started studying how to manufacture it using plastic materials. Coming to the manufacturing technology, in this development stage, we will realize the SFH using the *Fused Modeling Deposition* (FDM) 3D-printing technique. It is, in fact, a relatively cheap and easy-to-use technology, and it allows to realize multi-material objects. Furthermore, thanks to the wide variety of printable materials, more combinations of connections can be studied. This technology has, in fact, been already used to realize compliant components.

In this first study, with the aim of avoiding, if possible, to introduce additional constraints for the designers, we searched for a solution that does not require specific printing parameters (e.g. speed, temperature, ...) nor the addition of particular artifacts (like brims). We therefore used the printing parameters suggested by filament supplier (Main parameters values are: printing speed: 50 mm/s, bed temperature: 40 °C, extruder temperature: 230 °C, first layer thickness: 0.3 mm, other layer thickness: 0.2 mm).

2 A Brief Overview on 3D-Printed Compliant Mechanisms

One of the great advantages of AM technologies consists in widening designer freedom, allowing the realization of complex geometries without the need of special tool machines and fixtures, provided that the parts are designed taking into account the manufacturing constraints of these technologies.

CMs are a bright example of these wider design possibilities. Figure 1 shows a 3D-printed complaint crimping mechanism made entirely with *PolyFlex*™, a highly flexible material, easily printable via FDM. In this specific example the flexibility is given by FHs geometry and position in the structure, that has been realized with a single material.

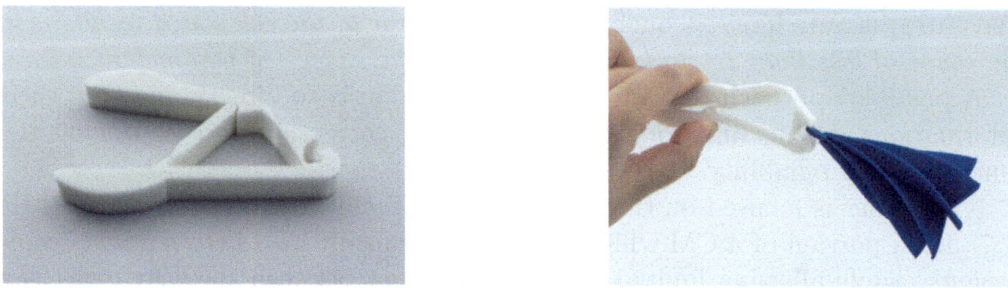

Fig. 1. 3D-printed complaint crimping mechanism [8]

An alternative to the design shown above (Fig. 1) is to create a compliant mechanism combining materials with different stiffness characteristics. An example of such a solution has been developed by Doran [9] to create a sort of glove entirely made of 3D-printed components. Doran conceived his system designing the rigid parts to be made in ABS and the deformable ones (finger articulations) in *NinjaFlex*®, a flexible material, whose mechanical characteristics are similar to the previously cited *PolyFlex*™.

A CM with similar mechanical functionality, but different design, has been presented by Rosa [10]. This particular CM was used to replicate the behavior of a human hand. In particular, a FH was placed in correspondence of each finger knuckle. The flexible component was made in *ThermoPlastic Polyurethane* (TPU), while the rigid ones were made in *PolyLactic Acid* (PLA).

Since the available multi-material solutions can be designed to obtain larger displacements between the rigid portions with smaller deformations of the flexible part of the FH, we selected this design approach, in order to ease the deformation measuring task.

We therefore used a dual extruders *Sharebot NG* FDM printer, that is capable to print two materials in the same layer.

3 Embeddable Sensors for Deformation Measuring

There is a wide variety of applications for which embedding sensors into parts is desired [11]. First, enclosing the sensible element inside the structure protects the sensor itself from hostile environments and extend its lifetime [12]. Second, an embedded sensor is in direct contact with the zone of the system that has to be monitored, even if this part is not directly and immediately and/or easily accessible [13].

Embedding *Optical Fibers* (OF) is the typical solution to measure part deformations.

At a glance, two main solutions are available: *Fiber Bragg Grating* (FBG) and *Optical Backscatter Reflectometry* (OBR) sensors [14]. Both these techniques (FBG and OBR) rely on the fact that the refraction index changes because of fiber mechanical deformations. This change results in a shifting (both in time and in frequency domain) of the Rayleigh back-scattered spectrum diffused by the “inhomogeneities”. This signal can be read and interpreted to determine part local strains. The main difference is that FBG reads deformations only in correspondence of gratings preliminarily realized via photo-incision, while OBR uses the whole fiber as a sensor, making possible to read the deformations along all the OF.

One of the first demonstrations of fiber embedded in an ABS part manufactured via FDM has been done using *Fiber Bragg Grating* (FBG) sensors: Zubela et al. [14] have developed a procedure to embed *Silica* and *Polymer* optical FBGs into 3D-printed patches. The procedure was similar for both this types of FBGs: the printing operation was paused at the end of the realization of the embedding zone; the fiber was then fixed into the zone with a super-glue and, after some time (to let the glue cure), the printing process was resumed.

Other studies have been performed by embedding the FBG fiber into ABS 3D-printed parts. For example, Kantaros et al. [15] embedded FBG fibers to study the effect of layer thickness and deposition orientation on residual strains.

Among the measuring techniques based on optical fibers, we selected the OBR one, instead of the more commonly adopted FBG systems. OBR systems, in fact, allow to use all the OF as a distributed sensor, while the FBG reads the information only in correspondence of gratings realized in the fiber. Using OBR systems, it is thus possible to obtain a real distributed measurement. Furthermore, there is no need to engrave any sort of gratings through photo-incision, like for the FBG. In OBR, the whole fiber is the sensor, since it relies on variation of the refraction index along the wire.

While the fibers employed in OBR systems are cheap and commonly available, sensors for FBG interrogators are generally designed according to customer needs and requires costly technological operations. Furthermore, the measure spatial-resolution of the employed system is at least one order of magnitude higher than the FBG one. The continuous nature of OBR sensor allows also to better appreciate local deformations of FH in order to define its position.

The high chemical and mechanical compatibility of optical-fibers with many materials make them embeddable in many materials. Particular care will be paid to avoid its over-heating and the possibility to mechanically damage it with the nozzle.

In particular, we used a LUNA® [16–19] system based on the OBR technology that reads the data coming from an optical-fiber to determine the local strains of the fiber itself. The sensor is hence a common optical fiber, that is transformed into a strain-sensor directly by the interrogator, according to the LUNA® OBR-technology.

4 Optical Fiber Embedding

The first step consisted in defining how to embed the LUNA® OF inside a 3D-printed part. The definition of the embedding strategy required the parallel development of an adequate geometry of the FH as well as of its manufacturing process.

The main goal of this approach is the realization of an area, enclosed inside the specimen, to receive the fiber optic sensor, i.e. a full-infill zone where the deposed fiber can fully adhere to the part being printed.

As expected, from initial tests and measurements, we have observed that limiting interactions between hot-printing nozzle and the thin optical wire is crucial. In particular, what has been found to be critical is the filament direction of deposition in the layer that covers the optical fiber. Using a direction perpendicular to the optical fiber to depose the material represents the major risk of damaging and/or moving the fiber from its embedding zone. To overcome this problem, the layer that covers the fiber has to be deposited parallel to the fiber.

4.1 First Tests

The first (and simpler) idea was to depose the fiber directly between two normal layers, without preparing a special embedding zone. The first consequence of this decision is that it is necessary to impose a 100% infill in all the part, in order to avoid cavities in the embedding area so that the fiber can be completely attached to the part. Secondly, all the filaments of the layer that covers the fiber has to have the same orientation of the fiber, in order to guarantee that the fiber will not be moved away from the embedding zone and that it will not undergo to dangerous shear stresses.

The advantage of this solution is its simplicity. The part geometry can be freely designed, and it is possible to obtain the desired direction of deposition by simply modifying the Infill Angle directly in Slic3r®. In fact, if this parameter is set equal to 0° or 90°, the software automatically generates one layer with transverse orientation and the subsequent one with a longitudinal direction of deposition. It will be therefore enough to depose the fiber on a traverse layer; a longitudinal one will be automatically used to cover it. The drawbacks of this solution (100% infill density) are long printing time (and hence a greater risk of thermally damaging the fibre) and too high stiffness of the part, or difficulty in controlling its stiffness.

Some tests have demonstrated that this drawbacks are too severe. This solution was therefore abandoned.

4.2 Micro-holes

A second solution with a lower infill (15%) has therefore been conceived. Nevertheless, the fiber embedding zone has to be realized with a 100% infill to guarantee a proper adhesion. Practically, it is necessary to realize two different infills in the same part. This has been obtained with the *Micro-holes* technique, presented in [20]: the trick we can use to *increase the amount of infill or perimeters* only in a peculiar zone of a part *is to simply pierce the area with a group of cylinders. This will cause the SW to print a bunch of perimeter lines around the cylinders which in effect will fill in the area more densely.* The upper left picture of Fig. 2 shows the complete sample, while the top right picture of Fig. 2 shows the longitudinal filament pattern, located at the center of the part, in which the fiber has been embedded. The two lower pictures of Fig. 2 show the strains measured after manufacturing. It is evident that these strains are very high and that the fiber is no longer able to read the deformations near to the specimen ends. Possible reasons of these results can be traced back to:

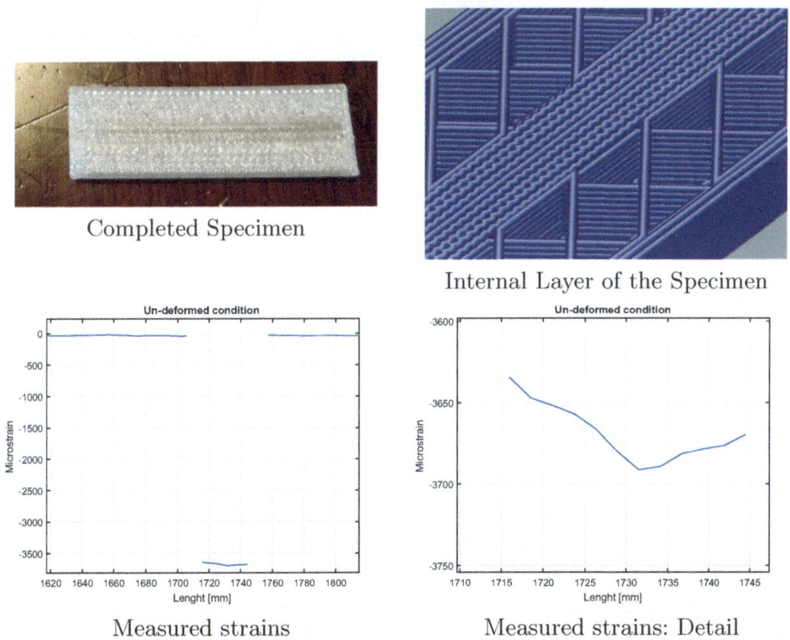

Fig. 2. 15% infill TPU specimen, Micro Holes technique

- thermal shrinkage of the specimen after its detachment from printing bed;
- too long interaction with the hot printing nozzle.

The actual printing time was, in fact, approximately equal to 55 min, much higher than expected. This is mainly due to the presence of the micro-holes: their realization extremely slows down the printing process. For these reasons, the micro-holes technique has been discarded.

4.3 Rectangular Cuts

With the aim of overcoming the problems encountered with the Micro-Holes technique, we developed the *Rectangular Cuts* solution, consisting in designing, directly in the CAD model, two small cuts confining the area of the embedding zone. Figure 3 shows the final geometry of these cuts. To find the minimum width of the cuts, an iterative procedure has been followed. This operation is needed because a too small cut width will not be detected by the slicer software, and hence their presence will not result in an adequate embedding zone. On the other hand, too wide cuts will actually create cavities in the part. The minimum rectangular cut width has been identified to be 0.1 mm.

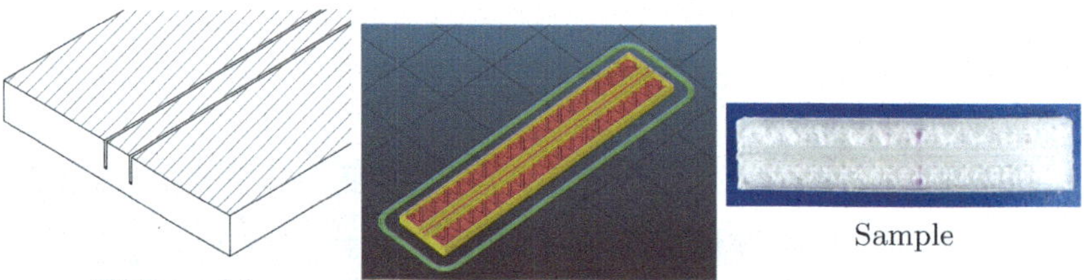

CAD model, half section | Layers preview (Slic3r®) | Sample

Fig. 3. CAD, Slic3r® and sample of sample realized with Rectangular Cuts

5 Flexure Hinge

5.1 The Design

In the light of the results obtained in preliminary tests, a satisfactory design of a 3D-printed SFH has been obtained starting from the FH architecture presented

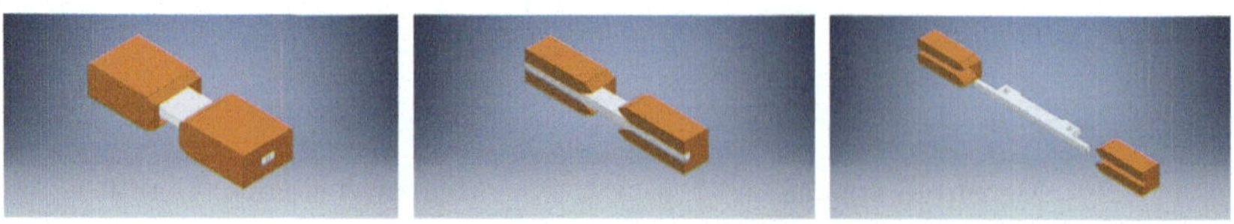

Assembly | Longitudinal section | Exploded view

Fig. 4. CAD model of the Smart FH

in [10]. The design considerations that lead us to design compliant mechanism made of two materials with a satisfactory OBR optical fiber embedding are summarized hereafter.

1. The TPU flexible part has been designed so that it is possible to embed the fiber in a single material. This solution allows to avoid discontinuities and shear actions of printing head during fused material deposition. When the SFH will be incorporated in a mechanism, a simple duct for the fiber have to be designed in the adjacent parts of the CM.
2. The embedding zone has been created along the whole FH (see Fig. 4). It has been designed with the *Rectangular Cuts* technique, that guarantees the required longitudinal filament direction of deposition, together with the shortest interaction time between optical fiber and hot printing nozzle.
3. Thickness of the central TPU part has been increased. This solution is used to create a better fiber embedding and to avoid possible shifting or undesired motion of the OF during the covering procedure. Nevertheless, a direct effect of this action, is an increase on the bending stiffness of the compliant joint. To preserve the symmetry of the stiffness behavior, material has been added both on the upper side and the lower side. To limit the bending stiffness of the compliant joint, the thickness has been increased just in the correspondence of the embedding zone and not on the whole surface (see Fig. 4). By properly defining infill and thickness, it is possible to obtained the desired bending stiffness of the FH.
4. The rigid parts has been designed with a broad fillet to avoid stress concentrations in that zones during cooling, as well as when the FH undergoes to wide deflections. The radius of this fillet corresponds to the radius of the bent flexible part when the angle between the two rigid parts reaches 90°.

5.2 Fiber Embedding and Strains Monitoring

The embedding procedure has been carried out imposing a pause to the *Sharebot-NG* 3D printer at a height equal to 4, 70 mm, i.e. 1.1 mm above the mid-height of the specimen. This pause has been manually programmed by manually editing the *G-Code* generated by Slic3r®. Placing the fiber at such an height allows to obtain a strain measurement compatible with the capabilities of LUNA® system. Figure 5 shows the FH during printing and immediately after the fiber embedding procedure. The sensor wire, fixed on both extremities of the specimen, has been positioned on a ThermoPlastic Polyurethane layer with a longitudinal infill. Then, the printing is resumed and the FH is completed.

During the manufacturing of the SFH the strains were monitored, in order to evaluate if the discontinuities that have been encountered in previous samples were avoided. Figure 6 shows the strains measured in two peculiar moments. The left diagram shows the strains after a slow cooling of the FH inside the printer and still attached to printer bed. This behavior is related to the different thermal expansion/contraction of the two materials and to the constrain of the bed. In fact, just after the end of the FH printing, the measured strains are quite smaller

Overall Central Zone Detail

Fig. 5. Smart Flexure Hinge during printing, after fiber embedding

along all the SFH. As the specimen cools down, the compression strains of the two PLA portions increase, while the central TPU portion shows tensile strains. The diagram on the right shows the strains immediately after that the sample is detached from the printer bed. It shows that the peculiar geometry of the FH and the presence of materials with different thermal expansion/contraction behaviors does not allow for a complete relaxation of the strains in the zone where the fiber is embedded, even if measured strains are about 30% lower with respect to the previous printing strategy.

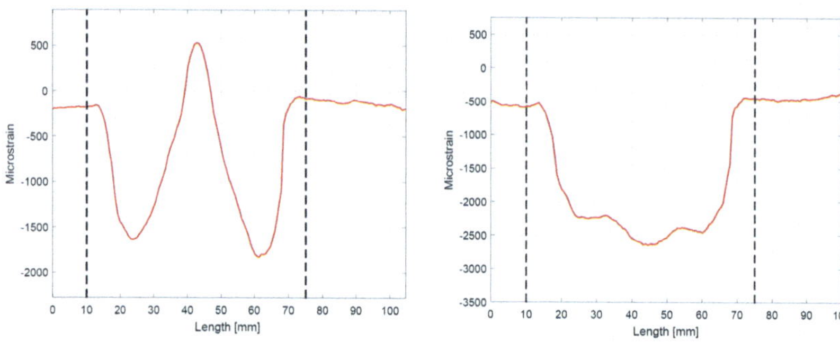

Cooled and attached Detached

Fig. 6. Measured strains in the FH. Vertical dotted lines shows the actually embedded trait of fiber.

6 Smart FH Testing and Calibration

The SFH has been tested on *angular-jigs* that allows to measure the strains imposing a predefined angle between the two rigid parts. Practically, each of the two PLA portions of the SFH has been firmly constrained to one of the leg of a V shaped frame, so that they are forced to form a predefined angle. In particular, the used jigs allow to impose the following angles: 10°, 15°, 20°, 25°, 30°, 35°, 40°, 45°, 50°, 60° and 90°. Tests have been performed acquiring five measurements at each of these angles, each day for 5 days; 275 measurements have therefore been collected and analyzed. The maximum bending strain has been related to the SFH bending angle to finally retrieve the joint orientation. Experimental results have been fitted by a double exponential function (Eq. 1), where: $A = -7174$, $B = -0.03701$, $C = 6310$ and $D = 0.005829$. It is worth underlying that these values are valid only for the tested SFH.

$$Y_{int} = A \cdot e^{B \cdot x} + C \cdot e^{D \cdot x} \tag{1}$$

It has been possible to measure values of strain up to approximately 11000 $\mu\varepsilon$, corresponding to the 90°, rotation.

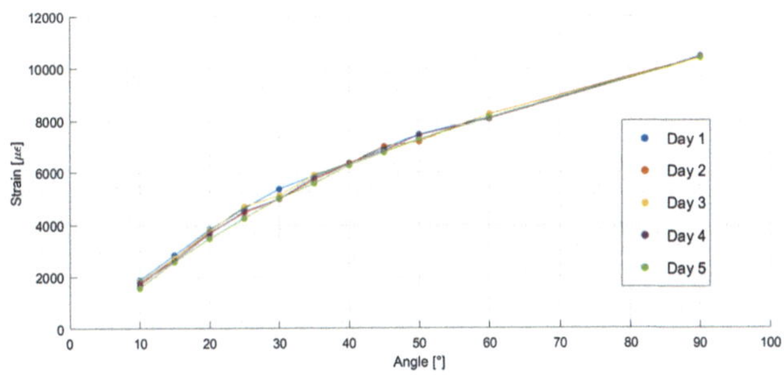

Fig. 7. Results of last days of analysis on the SFH

As evidenced in Table 1, the resulting standard deviation[1] is less than 6% in the worst case (10° input). This confirms good measurement repeatability for the performed test campaign. Figure 7 shows the experimental points superimposed to the interpolating curve. The performed tests demonstrate a quite good measurement capability of 3D-printed SFH to retrieve the joint orientation and therefore giving the possibility to use it for closed loop control systems and strategies.

Table 1. Standard deviation

Rotation [°]	10°	15°	20°	25°	30°	
Std. Deviation [$\mu\varepsilon$]	119	97	131	174	162	
Rotation [°]	35°	40°	45°	50°	60°	90°
Std. Deviation [$\mu\varepsilon$]	130	39	91	124	86	26

[1] The tabulated standard deviation has been computed starting from sample variance at each input.

7 Conclusions

The paper investigated the possibility to embed sensors based on optical fiber technologies into 3D printed compliant mechanisms. A process based on geometrical constraints and manufacturing parameters has been outlined to efficaciously embed an OBR optical fiber into a FH to read its deflection.

The main aspects that have to be considered to achieve good measurement accuracy are:

- a proper slow cooling of the SFH to prevent excessive residual strains;
- avoiding that the hot nozzle crosses the OF when deposing material and/or it operates too long in a zone close to the OF;
- an accurate deposition of the OF so that it can well adhere to the part being printed;
- paying attention to the strain relaxation/accumulation that could alter part residual strains.

Under a general methodological perspective, this experience demonstrated once more that, when designing for Additive Manufacturing, it is fundamental to account also for the manufacturing capabilities and properties of the selected technology (FDM in this case) even in the early stages.

Acknowledgments. This research activity has been financially supported by the *Dipartimento di Meccanica* of the *Politecnico di Milano* within the project: *3D Printed Mechanisms with Distributed Sensing and Actuation.*

References

1. Neale, M.J.: Tribology Handbook - Chapter A25. Elsevier, Oxford (1996). ISBN 978-0-7506-1198-5. https://doi.org/10.1016/B978-0-7506-1198-5.X5000-0
2. Shojaei Barjuei, E., Boscariol, P., Vidoni, R., Gasparetto, A.: Robust control of three-dimensional compliant mechanisms. J. Dyn. Sys. Meas. Control **138**(10), 101009 (2016). https://doi.org/10.1115/1.4034019
3. Linß, S., Schorr, P., Zentner, L.: General design equations for the rotational stiffness, maximal angular deflection and rotational precision of various notch flexure hinges. Mech. Sci. **8**(1), 29–49 (2017). https://doi.org/10.5194/ms-8-29-2017
4. CMR Awarded Research Grant from National Science Foundation. https://www.compliantmechanisms.byu.edu/. Accessed 04 Apr 2018
5. Howell, L.L., Magleby, S.P., Olsen, B.M.: Handbook of Compliant Mechanisms. Wiley, Chichester (2013)
6. Dassa, L., Cambiaghi, D., Legnani, G.: Mechanisms based on flexure articulations. Ph.D. thesis, Universitá degli studi di Brescia, Ciclo XXI
7. Lobontiu, N.: Compliant Mechanisms. Design of Flexure Hinges. CRC Press LLC, Boca Raton (2003)
8. PolyMaker.com: Project Update no. 9: compliant mechanisms printed with PolyFlex!. http://3dfilaprint.com/project-update-9-compliant-mechanisms-printed-with-polyflex/. Accessed 04 Apr 2018

9. Doran, B.: 3D printing rigid and flexible material on the sigma dual extruder 3D printer and simplify 3D. www.youtube.com/watch?v=9hix87v8LDo&t=100s. Accessed 04 Apr 2018
10. Rosa, F., Bordegoni, M., Dentelli, A., Sanzone, A., Sotgiu, A.: Print-in-place of interconnected deformable and rigid parts of articulated systems. In: 27th International Conference on Flexible Automation and Intelligent Manufacturing, FAIM 2017, Modena, Italy, pp. 555–562, June 2017. https://doi.org/10.1016/j.promfg.2017.07.149
11. Sbriglia, L.R., Baker, A.M., Thompson, J.M., Morgan, R.V., Wachtor, A.J., Bernardin, J.D.: Embedding sensors in FDM plastic parts during additive manufacturing. In: Topics in Modal Analysis & Testing, vol. 10 (2016). https://doi.org/10.1007/978-3-319-30249-2_17
12. Shemelya, C., Banuelos-Chacon, L., Melendez, A., Kief, C., Espalin, D., Wicker, R., Krijnen, G., MacDonald, E.: Multi-functional 3D printed and embedded sensors for satellite qualification structures. In: IEEE SENSORS, Busan, South Korea, pp. 1–4, November 2015. https://doi.org/10.1109/ICSENS.2015.7370541
13. Gunes, O.: Failure modes in structural applications of fiber-reinforced polymer (FRP) composites and their prevention. In: Developments in Fiber-Reinforced Polymer (FRP) Composites for Civil Engineering, pp. 115–147. Woodhead Publishing Series in Civil and Structural Engineering (2013). ISBN 978-0-85709-234-2. https://doi.org/10.1533/9780857098955.1.115
14. Zubela, M.G., Sugdena, K., Webba, D.J., Saez-Rodriguez, D., Nielsenc, K., Bangc, O.: Embedding silica and polymer fiber Bragg gratings (FBG) in plastic 3D-printed sensing patches. In: Micro-Structured and Specialty Optical Fibers IV, Proceedings of SPIE, vol. 9886, 98860N, SPIECCC code: 0277-786X/16/$18 (2016). https://doi.org/10.1117/12.2228753
15. Kantaros, A., Karalekas, D.: Fiber Bragg grating based investigation of residual strains in ABS parts fabricated by fused deposition modeling process. In: Materials & Design, vol. 50, pp. 44–50, September 2013. https://doi.org/10.1016/j.matdes.2013.02.067
16. LUNA® Onepager: Introducing the ODiSI-B 5.1 for Strain & Temperature. www.lunainc.com. Accessed 04 Apr 2018
17. Samiec, D.: Distributed fiber-optic temperature and strain measurement with extremely high spatial resolution. In: Photonik International (2012). https://lunainc.com/wp-content/uploads/2013/04/photonik_intl_2012_010.pdf
18. LUNA® Datasheet: High-Definition Fiber Optic Strain Sensors. www.lunainc.com. Accessed 04 Apr 2018
19. LUNA® Final brochure: Fiber Optic Sensing. www.lunainc.com. Accessed 04 Apr 2018
20. 3D Verkstan: Designing for 3D printing. https://support.3dverkstan.se/article/38-designing-for-3d-printing. Accessed 04 Apr 2018

A Virtual Design Process to Produce Scoliosis Braces by Additive Manufacturing

Davide Felice Redaelli[1(✉)], Fabio Alexander Storm[2], Emilia Biffi[2], Gianluigi Reni[2], and Giorgio Colombo[1]

[1] Politecnico di Milano, Via La Masa, 1, 20156 Milan, Italy
davidefelice.redaelli@polimi.it
[2] Scientific Institute, IRCCS E. Medea, Via Don Luigi Monza, 20, 23842 Bosisio Parini, LC, Italy

Abstract. The present work aims to present the current virtual design process of scoliosis braces and propose an improvement by considering the internal geometry. Starting from the external scanning it is possible to apply only corrections that do not consider the interaction with the human body. For this reason we propose to embed the 3D model reconstruction of the skeleton of the patient from the medical images. This would bring to a better virtual design process that could help avoiding the current need for a physical positive mold. The final goal of such a change is to pass from the current thermoforming production to the use of additive manufacturing. In the paper, we briefly analyze the choice of the most convenient 3D printing technology and the selection of a proper material that can be comparable to the ones used for thermoforming. Finally, a case study is presented to test the assumptions regarding both the design and manufacturing processes.

Keywords: Scoliosis brace · Virtual design · 3D printing

1 Introduction

1.1 Scoliosis and Brace Treatment

Scoliosis is defined as the abnormal spine curvature, in respect to the physiological spine curve. The prevalence of scoliosis in early childhood is about 1–2% [1], among adolescents it reaches about 5% [2] and it has been shown to spread out in elder subjects [3]. The main index for establishing if a person has scoliosis is the deformation on the frontal plane, whose severity is defined by the Cobb angle, *i.e.* the maximum angle between the vertebrae, as depicted in Fig. 1. Moreover, the line that connects the center of the vertebrae defines the type of curve (C shape = single curvature, S shape = double curvature, etc.).

We can thus distinguish between mild scoliosis, when the Cobb angle is below 25°, moderate scoliosis if it is between 25° and 45°–50°, and severe scoliosis if above. In case of severe scoliosis, the orthopedic physician usually opts for surgery in order to stabilize the spine with metal rods, avoiding the further progression of the deformation,

C. Rizzi et al. (Eds.): ADM 2019, LNME, pp. 860–870, 2020.
https://doi.org/10.1007/978-3-030-31154-4_73

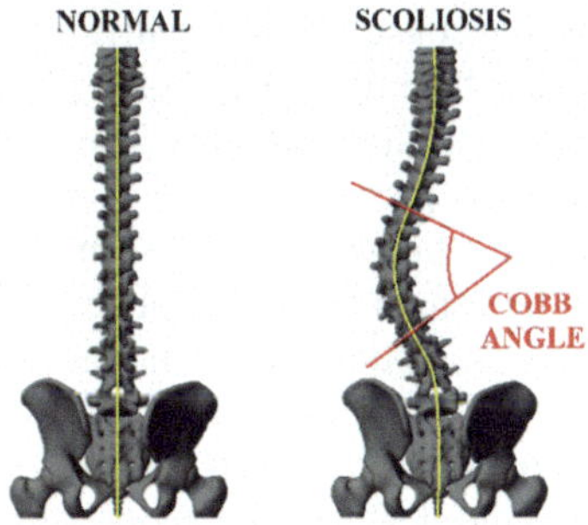

Fig. 1. Scheme of a normal spine (left) and a spine affected by scoliosis with the indication of how the Cobb angle is measured (right) in the frontal plane.

but limiting the natural flexibility of the trunk. For this reason, it is very important to start treating the disease as early as possible, also when the severity level is low [4].

The typical treatment for mild and moderate scoliosis is composed of physical exercise, both for stretching and for strengthening the muscles, and of the use of an external device, called scoliosis brace. There exist different versions and models of this orthotic device, from the very rigid and old Milwaukee brace, to the newest and extremely soft SpineCor. If the first type is now less common because of the limited comfort and low quality of life of the patients, the last seems to have a limited effectiveness and is appropriate only to light scoliosis, under 20° of the Cobb angle [4]. The most common back braces are thus the ones in between, that are jacket-style plastic braces. These are typically made of a rigid and lightweight plastic plate that covers the torso from hips to armpits.

The aim of the device is to support the chest and, in most of the cases, also to realign the spine by applying a correction based on the tree point bending system. The type of load depends on the extent of deformation, on the location along the spine and on the characteristics of the patient. The orthopedist can decide for the proper brace and asks the orthotist to apply the patient specific correction to the models.

The full process is usually long, from a full day to a week depending on the technology, and still expensive. 3D printing is considered to be a good solution for this kind of unique lot production, in opposition to the current thermoforming where it is needed to create every time a mold that will not be used after. The main problem of passing to this manufacturing process is related to the modelling phase that is still based on the manual experience of the technicians. For this reason, we present an improvement for the virtual sculpting of the back brace in comparison to the current process.

1.2 As-Is Manufacturing Process

As presented in [5], the manufacturing process that is currently adopted for the custom made orthosis is based on the thermoforming of a plastic plate around the patient's positive mold. This model can be obtained in different ways according to the available technology and to the experience of the orthotists.

The traditional method is to reconstruct the negative shell mold of the patient using plaster of Paris bandages. This is then filled with liquid plaster, left to dry and solidify, and finally sculpted manually by an expert orthotist. The entire production takes several days due to the drying time of the plaster.

The "modern" method regards the use of a 3D scanner to acquire the shape of the patient's body and pass it to the computer. A specialized software is adopted for a first phase that consists on virtually sculpting and preparing the CAM file to control the computer numerical control (CNC) machine. This machine, typically a lathe or robot arm, produces the positive model milling a cylinder of polyurethane foam. This procedure avoids the creation of a physical negative model, thus reducing the overall time to a full working day in best cases, by removing the part required for the plaster to solidify. Nevertheless, the present virtual sculpting step is not complete and requires a further manual sculpting phase of the milled positive mold and, at the same time, it adopts a material that is not recyclable nor reusable.

1.3 To-Be Manufacturing with Additive Manufacturing

The idea of using Additive Manufacturing (AM), also known as 3D printing, has been proposed by other authors before [6]. 3D printing has also been used for various other applications of medical fields, mainly for the surgery planning or for educational purposes [7]. Moreover, it has been tested in the field of orthotics and prosthetics (O&P) for cases of wrist orthosis, pelvis splint and lower limb prosthesis socket, where the shape modelling was limited or simplified [8, 9].

In order to pass to such a manufacturing process, the orthotist should adopt a fully virtual sculpting phase, abandoning the creation of the physical positive model required by the thermoforming. This could lead to difficulties in imagining the 3D result of the sculpting phase for the orthotist when working only on a computer.

1.4 Proposed Approach for Virtual Modelling

Several specialized software are currently used for modelling the orthosis in the orthopedic centers, *e.g.* Rodin4D Neo or Vorum Canfit [10, 11]. They enable importing also X-ray images to keep the reference of internal organs, mainly skeleton, and some of them allow also to re-project some reference points from the X-ray images to the 3D model acquired with the scanner.

The main problem today, when modelling the orthosis with these software, is that orthotists do not take into account the 3D geometry of the internal organs, that create the overall interaction of the patient's body with the orthosis itself. This could enhance the reliability of the final model of the brace obtained completely through virtual sculpting.

The aim of this work is to present the virtual design process of a back brace for a patient with scoliosis, required to produce it with additive manufacturing techniques. The proposed improvement regards the addition of the internal geometry of the skeleton reconstructed by medical images. This model must be aligned to the external geometry acquired with 3D scanners in order to better shape the brace, considering the interaction with the underlying structures. Moreover, we briefly analyze the choice of the 3D printing technology and the materials that allowed to 3D print the back brace for a case study.

2 Methods and Tools

2.1 Virtual Design Process

As written before, for the initial steps of the process there exist already 3D scanners adopted for the orthopedic use. In a previous work [12], we evaluated the accuracy of some of these instruments in comparison to other general purpose devices and we verified that the Structure Sensor by Occipital has an accuracy that is adequate for the application.

In the following paragraphs we present the current modelling process, based on the personal experience of the orthotist, and the additional step that we can embed to consider the internal geometry of the patient's body.

Experience Based Modelling

The model that comes out of the 3D scanner is tessellated (surface mesh STL/OBJ); hence, the best type of modelling software is the one that uses the dynamic mesh, such as Autodesk Meshmixer. Moreover, other software commercialized for orthopedic applications have been developed in this direction, *e.g.* the abovementioned Rodin4D Neo and Vorum Canfit. All these software packages enable the direct deformation of the tessellated surface with commands similar to the manual sculpting operations that the orthotist would perform on the physical positive model. For example, we can find commands to reduce the volume, add more material and smooth the surface. In addition to these, the orthopedic oriented software allow a finer control and a better visualization with a direct comparison between the initial model and the final one, giving also numerical values about the applied changes, part that is missing in the general purpose software Meshmixer.

Finally, all the programs enable the technician to select and crop the edge of the orthosis. The problem encountered with Rodin4D Neo is the absence of the command to apply the thickness, necessary to pass from the ideal inner surface of the brace to the solid model. This step is necessary for passing the model to the software that realize the slicing required for creating the commands of the 3D printer, because it cannot be an open surface but a closed one surrounding the volume we want to print.

This step does not always perform well in all the software. For example in Fig. 2, we can see that a spike appears once the thickness is applied in certain cases when using Meshmixer. This probably happens because of the difficulty of the software in evaluating the normals of the triangles along the border, but at the same time it is not possible to smooth the surface in this region.

For this reason, after the sculpting phase was completed in the first software, we preferred to pass the brace model as an open surface version in Rhinoceros in order to cut the edge and apply the thickness in a more stable way.

It has to be noticed that, while Meshmixer is able to work directly on the surface model deriving from the 3D scanner, Rodin4D Neo performs a quick reconstruction of the shape by approximating a certain number of horizontal sections. This operation is

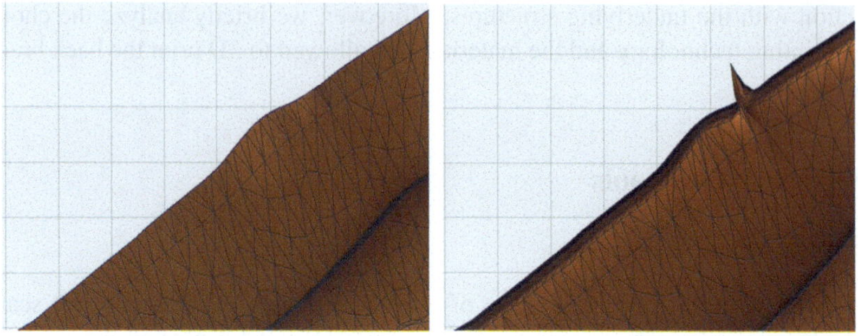

Fig. 2. Example of spike appearing on the border of the open surface (left) when the thickness is applied in Autodesk Meshmixer (right).

required by the software to apply other special operations that are not available in Meshmixer. These operations are related to correcting the whole model according to a flexible axis, representing the spine, and to stretching the model along this axis to preserve the volume inside the brace. Both correspond respectively to the spine deformity correction and to the spine traction.

Skeleton Model

In addition to the model of the skin, acquired with such scanner, we propose the use of the patient specific 3D model of the skeleton, not only of the spine, but also of the pelvis and the ribcage. The former gives an indication of the deformation, while the latter enables modelling the brace considering the interaction with the patient's body.

At present, in order to obtain such model, a different medical examination is required. The model could be created by segmenting a CT scan or, alternatively, an MRI.

On the one hand, the problem of aligning organ models of the medical images with the 3D scanned model could be simpler than considering only the 2 planar X-ray images, because we could also segment the skin from the same CT scan or MRI and keep it as a partial reference. On the other hand, the CT scan and the MRI are usually performed over a long period (20–40 min) and the patient has to lay on the mobile rigid table of the machine. As a result, the reconstructed skin model is always flattened on the back.

Instead, the 3D scan of the skin is performed with the patient standing straight in order to evaluate the proper anatomical shape. This difference in the patient's posture can also lead to a reconstruction of the skeleton that does not match to the skin model acquired with the 3D scanner. Moreover, it is usually hard to have this kind of medical examination for each patient, mainly if the patient is a child, because of radiation limits. Indeed, the scoliotic patients usually undergo only a biplanar X-ray imaging examination, except for particular cases, *i.e.* accidents, extreme pathologies or before a surgery.

Thus, we can state that the ideal situation would be to acquire simultaneously both the internal organs and the skin model with the patient standing straight. This posture reflects better the condition of use (not for the night-time and over correcting

Charleston brace), and using a low dosage technology, as presented for the biplanar X-ray images with the EOS system [13].

2.2 Polymer 3D Printing

Because of the big volume of the back brace and because of the print speed, the most convenient printer is the one based on the technology Fused Filament Fabrication (FFF). The other technologies available for 3D printing, Selective Laser Sintering (SLS) and Stereo Lithography Apparatus (SLA), typically produce smoother parts, but they have smaller volumes, they are much slower and the machines are also more expensive. Therefore, except for the reduced waste (that is comparable between these technologies), they would not be advantageous when compared to the thermoforming process.

Among the other advantages of the FFF 3D printers, we can mention the wide choice of materials available on the market and the low initial investment required for both machine and materials. The powders for SLS and the liquids for SLA are indeed more expensive than the filaments. For example, the standard filament is PLA an average cost of 30–40 $/kg, while the common polyamide powder for SLS is about 45–75 $/kg and the standard SLA resin is around 50 $/L [14].

Regarding the choice of the materials, the back braces are usually worn over a t-shirt, so there is no direct contact with the patient's body. However, in an orthopedic center, there is always the need to produce other types of orthoses, such as the wrist splints, that are applied in direct contact with the human body. In these cases, the use particular materials with a certification for "skin contact approval" are required. These are already available for the FFF technology and are considered to be an interesting option for the future of orthopedic centers.

Materials Evaluation

A first choice among the available materials can be performed comparing the mechanical properties of the materials used for the traditional braces, reported in Table 1. As mentioned before, there exist different types of braces that present different characteristics in terms of stiffness. Rigid models, as the Milwaukee brace, adopt polycarbonate (PC) while more flexible models, considered semirigid, use polyethylene (PE) and polypropylene (PP).

At the same time, depending on the chosen material, we need to take care about the melting temperature that determines the type of extruder that the printer needs. For these reasons, the most common applications found in literature [7] present the use of polylactic-acid (PLA) and acrylonitrile butadiene styrene (ABS), or also polyamide (PA) in case of SLS. However, these materials are rigid and fragile, comparable to PC.

For more flexible braces, we should thus consider materials more similar to PE and PP, as the available alternative PET-G. These particular filaments are typically more expensive and present more problems of initial adhesion to the build plate. Consequently, they could reduce the final quality of the product. Further research on this topic is required in order to select a proper solution for the O&P applications.

Table 1. Comparison between the properties of the materials used for current orthoses and the ones that could be substituting them in 3D printing with FFF technology. Source CES EduPack 2018, except * melting temperatures of PC and ABS [15].

	Material	Young modulus [GPa]	Elongation to rupture [%]	Melting temperature [° C]
Thermoforming	Polycarbonate (PC)	2–2.44	70–150	250–343*
	Polyethylene (PE)	0.621–0.896	200–800	125–132
	Polypropylene (PP)	0.896–1.55	100–600	150–175
3D printing	Polylactic-acid (PLA)	3.3–3.6	3–6	145–177
	Acrylonitrile butadiene styrene (ABS)	1.1–2.9	1.5–100	180–274*
	Polyethylene Terephthalate (PET)	2.76–4.14	30–300	212–265

Printing Setup

The setting of the printer and the orientation of the orthosis in the printer are critical for the final result. Among all the parameters available in the slicing software, for example Ultimaker Cura and Slic3r, the most important ones regard the printing speed, the choice of the layer height and the temperatures for the nozzle and for the printing bed.

On a side, it is important to understand the limits about the fusion and the suggested bed temperature associated to the chosen material in order to exceed the melting point and provide good adhesion of the first layer. On the other side, the combination with the nozzle size, the layer height and the printing speed will determine whether the filament is able both to flow through the nozzle and to stick to the previous layer.

General indications are usually provided by the filament producer and can be found in literature or in specialized websites with tips from experts and other users. However, a certain number of tests is suggested to verify the characteristics of a sample that can be representative of the final project for the shape and the orientation, which determines the nozzle path.

3 Results

3.1 Model from Virtual Design Process

Considering all the information described above, we started with the reverse engineering step, by reconstructing the geometry with the Structure Sensor. The sculpting phase was left to a trained orthopedic technician that based his design on the long personal experience. Once the final shape was defined, the model was imported in Rhinoceros in order to apply the uniform thickness. Finally, the STL model was passed to Ultimaker Cura, where the analysis of the overhangs was automatically performed and the proper settings, also of the required supports, were applied. The output file is then the G-code containing all the commands for the printer.

3.2 Skeleton Model from DICOM

In order to obtain a 3D model from the medical images, known as DICOM files, we performed the segmentation of a CT scan. This process consists in setting a threshold for the greyscale values of the images to define the portion of the images that are part of the desired volume. The software, 3D Slicer [16], can then apply this filter to all the 2D images and connect the different segmented areas to create the final volume.

Figure 3 illustrates the first result, applying a threshold value equal to 100. Depending on the type of CT scan (machine brand and model, presence of contrast, etc.) we could obtain different results changing the threshold level [17]. In this case, the CT scan was segmented with the intensity threshold approach in its basic form of visual selection of the value by the operator. Moreover, we can notice that, with such approach, the hard tissues are segmented all together, without distinction for the component of the sliding bed.

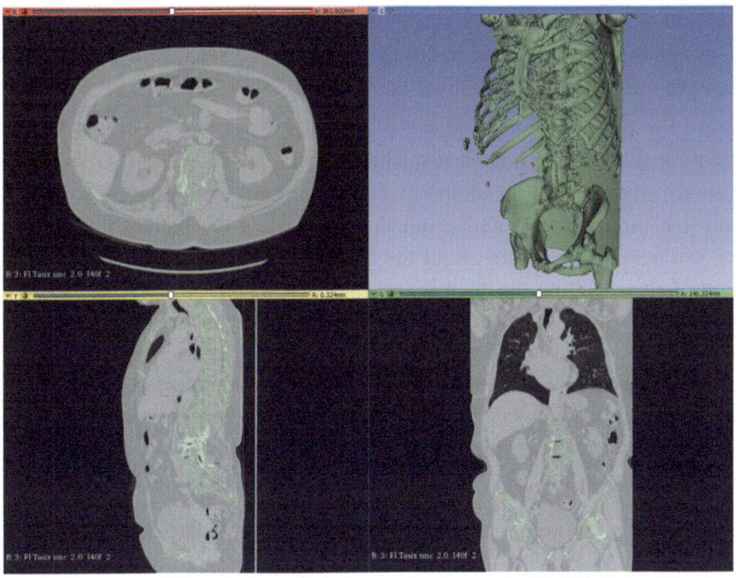

Fig. 3. Reconstruction of the internal organs 3D model (top right view) from a CT scan (other views) applying a threshold filter (green selection in the views).

After a cleaning phase of the model, it is possible to isolate the skeleton that can be then imported in the other modelling software. Figure 4 depicts the example of the previous model once it has been cleaned and it is ready for being used in a software like Meshmixer.

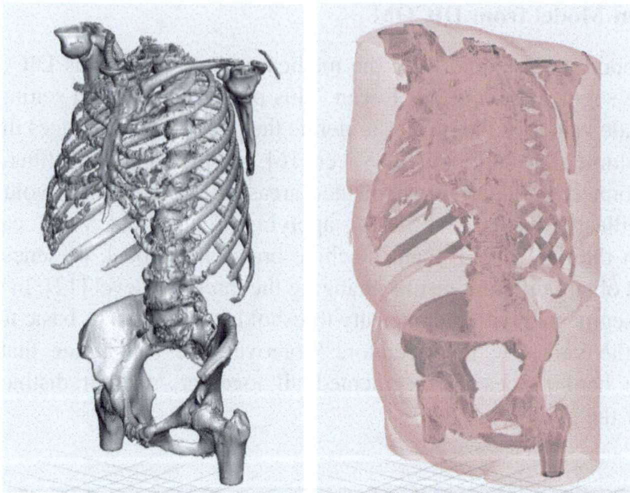

Fig. 4. Segmented skeleton model (left) and aligned skin model segmented from the same medical examination over the skeleton (right).

In a similar way, we extracted also the skin model from the medical images that will be already aligned with the skeleton. From the grey scale images in Fig. 3, we can understand that the back part will result flat because of the patient laying on the bed. Moreover, we have to take care about the dimensions of the patient and the setting of the volume of the machine not to miss any portion of the acquisition.

Once the skin model is reconstructed from the CT scan, we have to align the skin model acquired by 3D scanning. This operation can be accomplished by partially aligning the models considering only the front part. As mentioned, the CT scan model of the skin is flat on the back because of the bed and it cannot be considered for shaping the brace. Another alternative approach could contemplate the use of markers placed in anatomical reference points, for example iliac spines and xiphoid process, as it is done in other biomedical fields, *i.e.* for biomechanical analyses.

3.3 Printed Back Brace

The orthosis was printed in the vertical and upside-down position in order to minimize the required supports. Indeed, the top border of the orthosis (under the breast of the patient) was straighter than the lower border (around the hips) (Fig. 5 – left). The chosen material was the PET-G, because of the better trade-off between interlayer adhesion, elongation to rupture and surface finishing, in respect to PLA and ABS. The nozzle temperature was set to 235 °C, which is in line with the range 230–250 °C recommended by the producer of the filament. The build plate temperature was set to 90 °C, higher than the suggested range 0–70 °C, because of previous tests on the material. We used a nozzle of 1.2 mm that could create the final thickness of 2.2 mm in just two complete lines per layer.

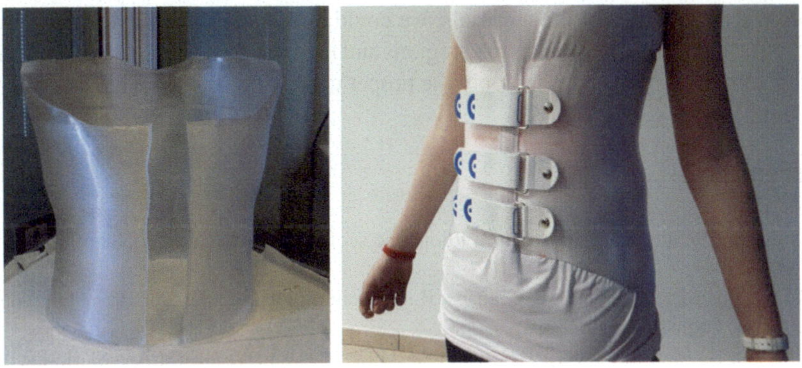

Fig. 5. 3D printing of the brace (on the left) and patient wearing the brace after the postprocessing operations performed by the orthotist (on the right)

After the model was printed, a series of standard operations was performed by the orthotist to verify the post-processability of the product. These operations regarded the support removal, the smoothing of the border and the drilling of holes for applying the straps used for closing the orthosis on the patient's body (Fig. 5 – right side). The time for printing the brace was 4 h 33 min, that can be compared to the modern process

4 Conclusions and Discussion

The work showed the feasibility of the approach in distinct steps. The virtual design process has already all the tools for sculpting completely a back brace. Orthotists still prefer to pass to a physical positive model and sculpt it manually because of their expertise, but the future process should consider remaining in the virtual environment till the final version of the orthosis, in order to manufacture it by 3D printing.

The current design process uses tools that can be improved by adding the information that we can obtain by the medical images. In particular, it is important to embed a part of reconstruction of the internal organs of the patients to be inserted in the skin model acquired by 3D scanning. Having such information would improve the resulting model of the scoliosis brace, considering the interaction of the device not only with the skin but also with the underlying skeleton. This could bring to the final version of the brace, avoiding the positive physical mold, necessary to 3D print it directly.

As we presented in the results, printing a complete back brace is feasible and the overall time for the production is comparable to the current one, but further research is required in terms of material choice and printing setup. For example, mechanical tests should prove the real characteristics of the printed product with the chosen parameters. We could better understand how changing temperature, speed, nozzle size and layer height could influence in the properties of the real back brace and compare them to the other braces that are currently manufactured by thermoforming.

These results could be then used in relation to the interaction with the patient's body. Stiffer materials will need more care when used in regions close to the bones or, from a different point of view, the model will have to be more precise, considering the bones that are under the skin. Nevertheless, a softer result could be more adaptable to these regions.

Knowing the properties of the materials, we could also balance the final stiffness of the brace reducing the thickness in defined regions and increase it in others, for example in the loading zones where we have to apply the proper correction for the spine.

References

1. Trobisch, P., Suess, O., Schwab, F.: Idiopathic scoliosis. Deutsches Ärzteblatt Int. **49**, 875–884 (2010)
2. Konieczny, M.R., Senyurt, H., Krauspe, R.: Epidemiology of adolescent idiopathic scoliosis. J. Child. Orthop. **7**, 3–9 (2013)
3. Kebaish, K.M., Neubauer, P.R., Voros, G.D., Khoshnevisan, M.A., Skolasky, R.L.: Scoliosis in adults aged forty years and older: prevalence and relationship to age, race, and gender. Spine **36**, 731–736 (2011)
4. KidsHealth from Nemours. https://kidshealth.org/en/parents/scoliosis-brace.html. Accessed 28 June 2019
5. Redaelli, D.F., Biffi, E., Colombo, G., Fraschini, P., Reni, G.: Current and future manufacturing of chest orthoses, considering the case of osteogenesis imperfecta. In: ASME 2018 International Design Engineering Technical Conferences and Computers and Information in Engineering Conference, ASME, Quebec City, vol. 1B (2018)
6. Weiss, H.R., Tournavitis, N., Nan, X., Borysov, M., Paul, L.: Workflow of CAD/CAM scoliosis brace adjustment in preparation using 3D printing. Open Med. Inform. J. **11**, 44–51 (2017)
7. Ventola, C.L.: Medical applications for 3D printing: current and projected uses. Pharm. Ther. **10**, 704–711 (2014)
8. Palousek, D., Rosicky, J., Koutny, D., Stoklásek, P., Navrat, T.: Pilot study of the wrist orthosis design process. Rapid Prototyp. J. **20**, 27–32 (2014)
9. Colombo, G., Comotti, C., Redaelli, D.F., Regazzoni, D., Rizzi, C., Vitali, A.: A method to improve prosthesis leg design based on pressure analysis at the socket-residual limb interface. In: ASME 2016 International Design Engineering Technical Conferences and Computers and Information in Engineering Conference, ASME, Charlotte, vol. 1A (2016)
10. Rodin4D. http://rodin4d.com/en/Products/rectification/r4d-cadcam-software. Accessed 28 June 2019
11. Vorum. https://vorum.com/cad-cam-prosthetic-orthotic/canfit-design-software/. Accessed 28 June 2019
12. Redaelli, D.F., Gonizzi Barsanti, S., Fraschini, P., Biffi, E., Colombo, G.: Low-cost 3D devices and laser scanners comparison for the application in orthopedic centres. In: ISPRS TC II Mid-term Symposium "Towards Photogrammetry 2020", ISPRS, Riva del Garda, vol. XLII-2 (2018)
13. EOS imaging. https://www.eos-imaging.us/us/professionals/eos/eos. Accessed 28 June 2019
14. All3DP. https://all3dp.com/2/how-much-do-3d-printer-materials-cost/. Accessed 28 June 2019
15. MatWeb Material Property Data. http://www.matweb.com/index.aspx. Accessed 28 June 2019
16. Dslicer. https://www.slicer.org/. Accessed 28 June 2019
17. Rathnayaka, K., Sahama, T., Schuetz, M.A., Schmutz, B.: Effects of CT image segmentation methods on the accuracy of long bone 3D reconstructions. Med. Eng. Phys. **33**, 226–233 (2011)

High Density AlSi10Mg Aluminium Alloy Specimens Obtained by Selective Laser Melting

Federico Uriati[1], Filippo Da Rin Betta[1], Paolo Ferro[1], Stefano Rosso[2], Gianpaolo Savio[2], Gianmaria Concheri[2], and Roberto Meneghello[1](✉)

[1] Department of Engineering and Management, University of Padova, Stradella San Nicola 3, 36100 Vicenza, Italy
roberto.meneghello@unipd.it

[2] Department of Civil, Environmental and Architectural Engineering, University of Padova, via Venezia 1, 35131 Padua, Italy

Abstract. Additive manufacturing technology offers new design possibilities compared to traditional casting processes applied to metallic materials. Not only there are no limits in shape, but a higher microstructure control is allowed compared to traditional processes. Irrespective of the sample dimensions, the solidification defects induced by SLM process depend only on process parameters and do not vary from zone to zone of the component like in a casting component: the higher the casting dimensions and thickness variations, the lower the microstructure homogeneity resulting from different cooling conditions inside the casting itself.

The effect of process parameters on porosity, in selective laser melted AlSi10Mg aluminium alloy, is carefully analysed with the aim to find optimal conditions that guarantee the maximum material density and the best mechanical properties.

Keywords: Additive manufacturing · Aluminium alloy · Selective laser melting · Porosity · Mechanical properties

1 Introduction

Additive manufacturing (AM) first emerged in 1987 with stereolithography (SL) from 3D Systems, a process that solidifies thin layers of ultraviolet (UV) light-sensitive liquid polymer using a laser. Starting from that period, different AM processes developed so that, in 2010, the American Society for Testing and Materials (ASTM) group "ASTM F42 – Additive Manufacturing", formulated a set of standards that classify the range of Additive Manufacturing processes into seven categories (Standard Terminology for Additive Manufacturing Technologies, 2012) [1]: VAT Photopolymerisation, Material Jetting, Binder Jetting, Material Extrusion, Powder Bed Fusion, Sheet Lamination, Directed Energy Deposition. Additive manufacturing offers several advantages throughout the design workflow including little lead time, few constraints, little-skill manufacturing. Furthermore, it is a low environmental impact process because it is characterised by less waste and energy saving. As a matter of fact,

C. Rizzi et al. (Eds.): ADM 2019, LNME, pp. 871–878, 2020.
https://doi.org/10.1007/978-3-030-31154-4_74

when compared with traditional manufacturing processes, additive manufacturing can significantly reduce energy usage by using less material and eliminating steps in the production process [2].

A significant problem with additive processes stands up in failure assessment. The material properties of AM components change depending on the fabrication process making it dependent on the geometry of the components. This means that every change in the geometry will change the way that the AM machine performs its fabrication routine affecting the properties of the resulting solid [3–5]. Currently, the main challenges for laser applications in metal powder-based fusion AM processes include the formation of defects (e.g., porosity), low surface finish quality, and spatially non-uniform properties of material.

Among the aluminium alloys, AlSi10Mg is certainly the most used for fabricating components through selective laser melting (SLM) process. Such alloy is often used in components of motor racing, automotive industry and general engineering. It has historically been a right choice for lightweight and thin walled casting parts or any components with a complex geometry subjected to high loads.

AlSi10Mg parts issued from powder-bed additive layer manufacturing (ALM) processes exhibit higher or at least, comparable, mechanical properties than casted parts [6]. In particular, ultimate tensile strength of the as-built Aluminium ALM parts is always higher than those obtained by High pressure die casting (HPDC) in both as-casted and in the aged condition.

Static and fatigue properties of additive manufactured AlSi10Mg samples obtained by SLM were investigated by Brandl et al. [7]. Mechanical properties were highly influenced by the building direction; however, the combination of 300 °C platform heating and peak-hardening were found to increase the fatigue and static resistance, neutralizing at the same time the building direction effect. Unfortunately, in that work, process parameters optimization was not considered.

With the objective to increase the process efficiency, in 2011 Buchbinder et al. [8], tried to increase the build rate of AM AlSi10Mg parts by using a 1 kW laser, while reaching a 99.5% material density. They demonstrated the possibility to extend the built rate from approximately 5 mm^3/s to 21 mm^3/s. In a recent paper, Han et al. [9] investigated the selective laser melting AlSi10Mg cellular lattice strut in terms of molten pool morphology, surface roughness and dimensional accuracy. The results showed that the average width and depth of the molten pool, the lower surface roughness and dimensional deviation decrease with the increase of scanning speed and hatch spacing. In that work, the laser power was kept constant (200 W). The influence of laser power, scan speed, scan spacing and island size on porosity development in AlSi10Mg alloy builds has been investigated by Read et al. [10]. They found that the laser power, scan speed, and the interaction between the scan speed and scan spacing have the major influence on the porosity development in the builds. They also demonstrated the higher strength and elongation properties of SLMed samples compared to those of die cast samples of similar composition.

Aboulkhair et al. [11] investigated the effect of scanning speed (in the range of 250–1000 mm/s) and hatch spacing (from 50 to 250 μm) on porosity in AlSi10Mg parts processed by SLM. The laser power was kept constant and equal to 100 W. Surveying the windows of parameters, the best combination was found to be a speed of

500 mm/s and hatch spacing 50 μm when using a layer thickness of 40 μm and employing the pre-sinter scan strategy yielding a relative density of 99.8%

In their work, Hasmuni et al. [12] focused their study on porosity, influenced by the hatch distance, in fabricated SLM AlSi10Mg parts. The hatch distance was varied from 0.1 to 0.15 mm while the laser power, the scanning speed and the layer thickness were kept constant and equal to 350 W, 1650 mm/s and 30 μm, respectively. The lowest porosity amount was obtained by using a hutch distance of 0.13 mm.

The several papers present in literature about SLM applied to AlSi10Mg alloy confirm, from one hand the high interest of academic and industrial world on this topic, and on the other the need to still deepen the strong and complex interaction between process parameters, microstructure and mechanical properties that characterises the SLM process. This study is the initial of a wider work aimed to contribute in covering this gap by analysing the influence of laser power and exposure time on the porosity amount in SLMed parts.

2 Materials and Methods

The analysed material is the AlSi10Mg hypoeutectic alloy. Its chemical composition is collected in Table 1.

Table 1. Chemical composition of the alloy AlSi10Mg (wt%)

Al	Si	Mg	Fe	N	O	Ti	Zn	Mn	Ni	Cu	Pb	Sn
balance	9–11	0.25–0.45	<0.25	<0.2	<0.2	<0.15	<0.10	<0.10	<0.05	<0.05	<0.02	<0.02

The particle size of the AlSi10Mg powder is in the range of 15–45 μm.

The samples were obtained by selective laser melting with a Renishaw 400AM. The building square-based platform has a side length equal to 250 mm. The laser is characterised by 400 W maximum beam power with a diameter of 70 μm. Argon was used as protective gas against powder oxidation. Both cylindrical (diameter and length equal to 10 mm) and dog bone specimens were produced following the Standard ASTM E606 (Fig. 1).

In order to reduce possible process parameters combinations, the following inputs were kept constant during the tests:

- Layer thickness (d), 30 μm;
- Spot diameter (Φ), 70 μm;
- Platform temperature, 170 °C;
- Point distance (s) equal to hatch distance (h), 80 μm;
- Building direction: sample axis;

In Fig. 2 the laser scanning strategy and the definition of point distance and hatch distance are shown.

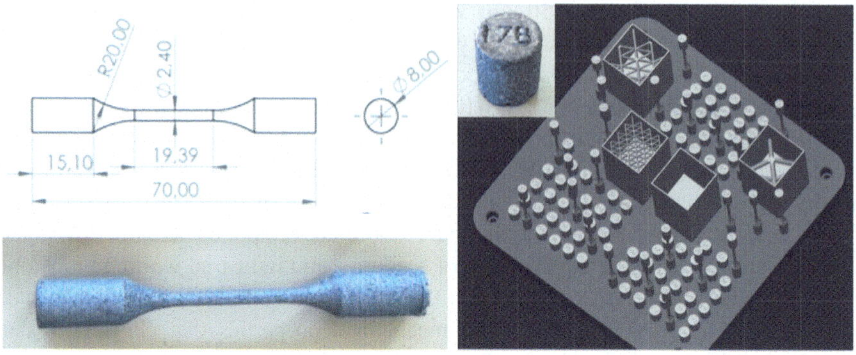

Fig. 1. Geometry of the specimens (dimensions in mm) and their position on the building platform

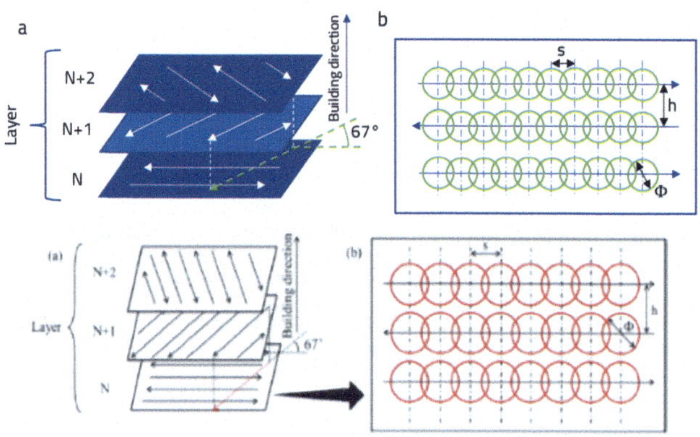

Fig. 2. Scanning strategy (a); spot-to-spot fabrication process, where 's' is the point distance, 'ø' is the laser beam spot and h is the hatch distance (b).

With the aim to minimize the porosity inside the material, the laser beam power and exposure time values were changed starting from those suggested by Renishaw (275 W and 40 μs). Three cylindrical samples for each couple of process parameters shown in Fig. 3 were carried out.

In order to optimize the material at disposal, samples for tensile tests were produced with the couples of parameters marked in red in Fig. 3. It is noted that even if it was not possible to change the scan speed (v_s), this last parameter can be approximated by Eq. (1), being s the point distance:

$$v_s \approx \frac{s}{exposure\ time} \tag{1}$$

The microstructure was analysed using a scanning electron microscope (Quanta 2580 FEG, FEI, Boston, MA, USA). Light microscopy pictures were also taken with Leica DMC 2900 microscope.

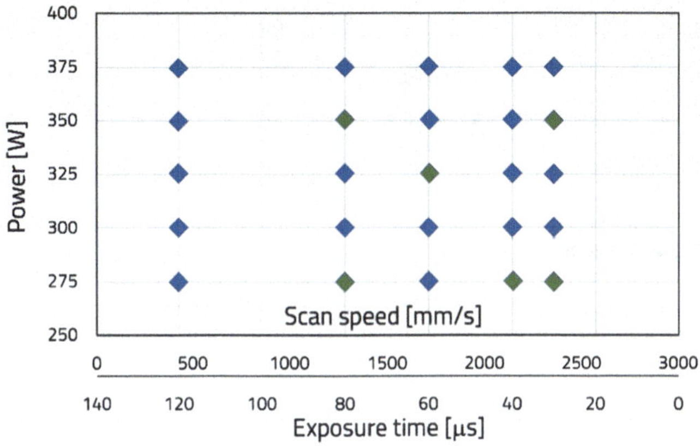

Fig. 3. Process parameters tested in the experiment

With the help of its dedicated software for image analysis, the microscope is driven to obtain the overall specimen image by means of the combinations of several micrographs carried out by scanning the entire surface of the specimen. The final micrograph is then processed for the automatic counting of the area interested by porosity.

The tensile tests were carried out by using the MTS Acumen 3 and following the Standard ISO 6892-1:2016. Figure 1 shows the tensile specimen geometrical parameters. Samples obtained with the power and exposure time equal to 350 W and 80 μs, respectively, were not tested because of their high deformation resulted after the building process.

3 Results and Discussion

3.1 Microstructure

Pores are easily formed during the SLM process. Figure 4 shows the pores morphology; gas porosity and balling were the most common defects observed. The microstructure (Fig. 4c) is characterised by α-Al dendrites inside a eutectic matrix constituted by α-Al and Si-particles.

The mean value of the area percentage occupied by pores as a function of the range of tested process parameters is shown in Fig. 5, where the bubble areas are proportional to the porosity percentage detected in the cylindrical samples.

It is observed that minimum values of porosity are reached in the lower right corner of the graph (Fig. 5). Similarly, by using the concept of energy density (Ed), given by [13]:

$$Ed = \frac{P}{v_s \cdot h \cdot d} \quad (2)$$

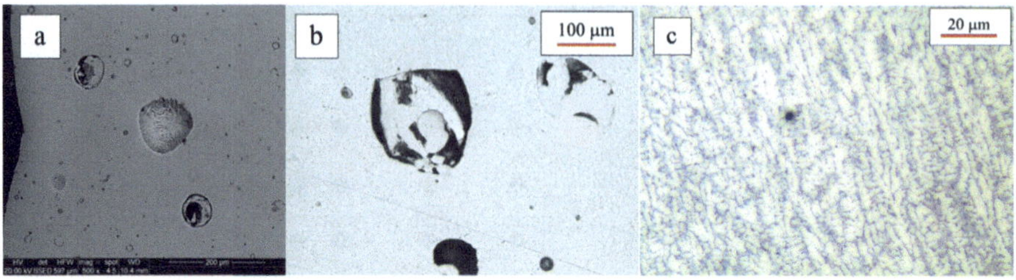

Fig. 4. Pores morphology (a), 'balling' (b) and microstructure morphology (c) induced by SLM process (Power = 350 W; exposure time = 120 μs).

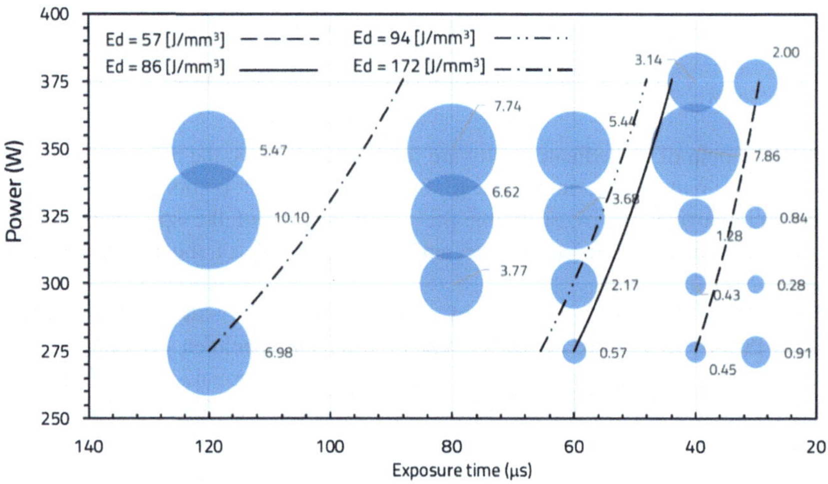

Fig. 5. Porosity percentage (%) as a function of process parameters (laser power, exposure time) and energy density (Ed)

It can be easily noted that samples characterised by the highest density, stay in the zone crossed by iso-energy density curves having the lowest values, in the range of tested parameters (Fig. 5). Nevertheless, as observed by Prashanth et al. [13], the applicability of Eq. (2) is still at stake, even though it has been widely used in literature for optimizing the SLM parameters [12]. As a matter of fact, the Eq. (2) may not properly represent the effective energy transferred to melt the powder bed, and thus it needed to be improved involving the hatch parameters and the material properties. With the aim to reinforce that idea, the graph energy density versus porosity is plotted in Fig. 6. It shows the positive trend of porosity (continuous line) versus energy density but with a high scattering (R2 = 0.6). Despite this, results obtained agree with those published by Read et al. [9] who found the minimum pores fraction induced by a critical energy density equal to approximately 60 J/mm^3.

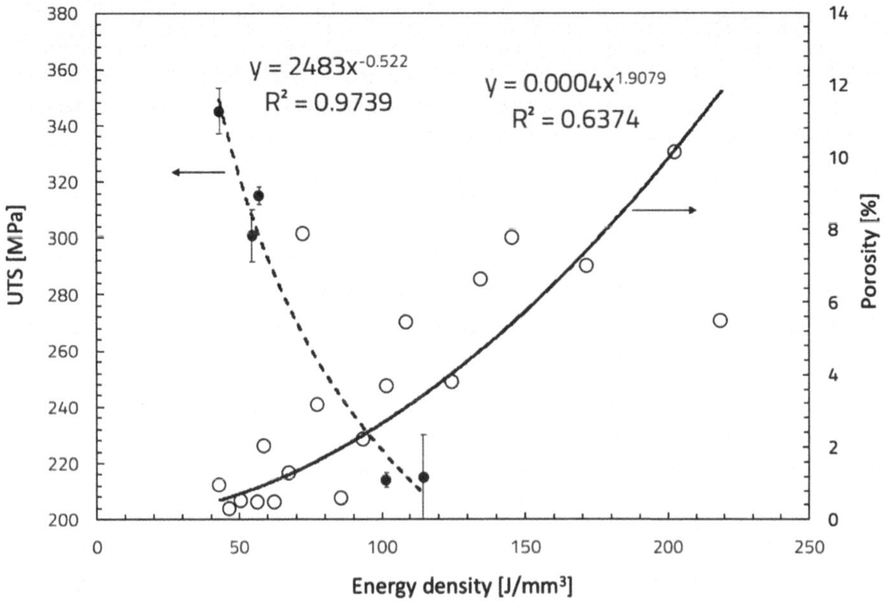

Fig. 6. Porosity and UTS versus energy density

3.2 Tensile Tests

As expected, the material tensile strength is linked to its microstructure morphology, with particular reference to defects. The higher the porosity percentage, the lower the ultimate tensile strength (dotted line in Fig. 6). A negative trend characterises the UTS versus the energy density. This is attributed both to stress concentration effect and nominal cross section area reduction induced by the porosity itself.

4 Conclusions

The influence of process parameters on porosity, in SLMed AlSi10Mg samples, was studied. The laser power and the exposure time has been varied from 275 W to 375 W and 30 μs to 120 μs, respectively. Tensile tests were finally carried out on samples produced with different process parameters. Results were described as a function of energy density and are summarised, within the limitations of the study, as follows:

- gas porosity is the main defect detected in the microstructure,
- the mean pore diameter value and its standard deviation decreases as the energy density decreases,
- minimum values of porosity percentage are obtained with the minimum values of energy density, between approximately 60 and 50 J/mm^3,
- the higher the porosity percentage, the lower the UTS.

Acknowledgments. This work was partially funded by the research grant "FSE 2105-37-11-2018: Hybrid manufacturing strategies through selective laser sintering for the customization of series components" from Regione Veneto, by the research project "FESR D23D18000160007" of Tesolin S.p.A., funded by Regione Friuli Venezia Giulia and by a Ph.D grant of Fondazione Cassa Di Risparmio di Padova e Rovigo (CARIPARO).

References

1. ASTM F2792-12a: Standard Terminology for Additive Manufacturing Technologies (2012)
2. Gibson, I., Rosen, D., Stuker, B.: Additive Manufacturing Technologies: 3D Printing, Rapid Prototyping, and Direct Digital Manufacturing (2015)
3. Razavi, S.M.J., Ferro, P., Berto, F., Torgersen, J.: Fatigue strength of blunt V-notched specimens produced by selective laser melting of Ti-6Al-4V. Theor. Appl. Fract. Mech. **97**, 376–384 (2018). https://doi.org/10.1016/j.tafmec.2017.06.021
4. Razavi, S.M.J., Ferro, P., Berto, F.: Fatigue assessment of Ti-6Al-4 V circular notched specimens produced by selective laser melting. Strength Mater. **7**, 291 (2017). https://doi.org/10.1007/s11223-017-9884-2
5. Razavi, S.M.J., Bordonaro, G.G., Ferro, P., Torgersen, J., Berto, F.: Fatigue behavior of porous Ti-6Al-4V made by laser-engineered net shaping. Materials (Basel), **11** (2018). https://doi.org/10.3390/ma11020284
6. Kempen, K., Thijs, L., Van Humbeeck, J., Kruth, J.P.: Mechanical properties of AlSi10Mg produced by selective laser melting. Phys. Procedia **39**, 439–446 (2012). https://doi.org/10.1016/j.phpro.2012.10.059
7. Brandl, E., Heckenberger, U., Holzinger, V., Buchbinder, D.: Additive manufactured AlSi10Mg samples using selective laser melting (SLM): microstructure, high cycle fatigue, and fracture behavior. Mater. Des. **34**, 159–169 (2012). https://doi.org/10.1016/j.matdes.2011.07.067
8. Buchbinder, D., Schleifenbaum, H., Heidrich, S., Meiners, W., Bültmann, J.: High power selective laser melting (HP SLM) of aluminum parts. Phys. Procedia **12**, 271–278 (2011). https://doi.org/10.1016/j.phpro.2011.03.035
9. Han, X., Zhu, H., Nie, X., Wang, G., Zeng, X.: Investigation on selective laser melting AlSi10Mg cellular lattice strut: molten pool morphology, surface roughness and dimensional accuracy. Materials (Basel), **11** (2018). https://doi.org/10.3390/ma11030392
10. Read, N., Wang, W., Essa, K., Attallah, M.M.: Selective laser melting of AlSi10Mg alloy: process optimisation and mechanical properties development. Mater. Des. **65**, 417–424 (2015). https://doi.org/10.1016/j.matdes.2014.09.044
11. Aboulkhair, N.T., Everitt, N.M., Ashcroft, I., Tuck, C.: Reducing porosity in AlSi10Mg parts processed by selective laser melting. Addit. Manuf. **1**, 77–86 (2014). https://doi.org/10.1016/j.addma.2014.08.001
12. Pakkanen, J., Lombardi, M., Lorusso, M., Trevisan, F., Fino, P., Manfredi, D., Aversa, A., Ambrosio, E., Calignano, F.: On the selective laser melting (SLM) of the AlSi10Mg alloy: process, microstructure, and mechanical properties. Materials (Basel). **10**, 76 (2017). https://doi.org/10.3390/ma10010076
13. Prashanth, K.G., Scudino, S., Maity, T., Das, J., Eckert, J.: Is the energy density a reliable parameter for materials synthesis by selective laser melting? Mater. Res. Lett. **5**, 386–390 (2017). https://doi.org/10.1080/21663831.2017.1299808

Scale and Shape Effects on the Fatigue Behaviour of Additively Manufactured SS316L Structures: A Preliminary Study

Stefano Rosso[1(✉)], Roberto Meneghello[1], Gianmaria Concheri[2], and Gianpaolo Savio[2]

[1] Department of Management and Engineering, Laboratory of Design Tools and Methods in Industrial Engineering, University of Padova, Stradella S. Nicola, 3, 36100 Vicenza, Italy
stefano.rosso.3@phd.unipd.it

[2] Department of Civil, Environmental and Architectural Engineering, Laboratory of Design Tools and Methods in Industrial Engineering, University of Padova, Via Venezia, 1, 35131 Padua, Italy

Abstract. The freedom in geometry given by additive manufacturing allows to produce cellular materials, also called lattice structures, with unit cells and mesoscale features that are impossible to obtain with traditional manufacturing techniques. The geometric modeling of lattice structures still presents issues such as robustness and automation, but, with a novel modeling approach based on subdivision surface algorithm, these troubles were limited. Furthermore, the subdivision method smooths surfaces, avoiding sharp edges at nodal points and increasing performances in fatigue properties. The aim of this work is twofold; a. The subdivision surface method is validated through fatigue tests on specimen additively manufactured by selective laser melting technology in SS316L stainless steel; dynamic tests were carried out on two types of lattice structures based on cubic cell: one obtained with a traditional modeling method, one obtained with a subdivision surface approach. b. Additional tests on bulk cylindrical samples, allowed to propose a preliminary model that describes the fatigue behaviour of additively manufactured lattices as a function of the bulk material properties, considering the shape and scale effects coming from stress concentration factor, increased area, surface roughness and porosity of the part. Results show that the subdivision surface approach improves the fatigue life of lattice structures, as expected. More, the lattices have a worse fatigue life compared to the bulk samples due to the scale and shape effects, that results in a higher sensibility to surface and internal defects related to the manufacturing process.

Keywords: Additive manufacturing · Lattice structure · Fatigue behaviour · Scale effect · Shape effect

C. Rizzi et al. (Eds.): ADM 2019, LNME, pp. 879–890, 2020.
https://doi.org/10.1007/978-3-030-31154-4_75

1 Introduction

Additive manufacturing (AM) and its layer-by-layer approach allow producing parts with extreme freedom in geometry. More, AM presents other benefits if compared to conventional techniques: design can be optimized for function and customized for different users, design of the final part can quickly change, small production batches become more feasible and economical, and material wasting is reduced [1]. Looking at the market, it is also possible to see an exponential rise in metal AM systems sales, supported by the significant growth of the revenues from metals for AM [2]. Furthermore, the layer-wise approach of AM technology enables the production of cellular materials [3], also called lattice structures. In addition to the fact of being lightweight, lattices present other interesting properties: they are stiff in relation to their low mass, they are good energy absorbers thanks to the possibility to undergo large deformations, good acoustic insulators thanks to the internal porosity and good heat exchangers due to their large surface area [4]. While the static behaviour of metal lattice structures produced by SLM has been extensively studied, the fatigue behaviour still needs to be analyzed deeper [5]. In this scenario, the need for characterization of materials tailored for AM has been highlighted in [6] and [7] and is an essential task that has to be addressed in order to guarantee that AM parts have the same mechanical performance of parts produced with more traditional subtractive technologies. Another important point for the characterization of additively manufactured parts is the understanding of the influence of surface texture and porosity on mechanical properties, considering also the sensitivity of these two parameters to the size and shape of the geometry [8].

Greitemeier et al. [9] studied the effect of surface roughness on fatigue life of AM Ti6Al4V samples, showing that the fatigue properties highly depend on this parameter, and formulating a model based on an equivalent initial flaw size (EIFS). Masuo et al. [10] identified the major factors that influence the fatigue strength of AM materials as microstructure, build direction [11, 12], defects (i.e. surface roughness, pores, lack of fusion, shrinkage and inclusions) and residual stress, and proposed a method for estimating the effect of defects with the $\sqrt{area}$ parameter. Wycisk et al. [13] analyzed and simulated the fatigue life and crack propagation considering AM defects, such as porosity and surface roughness.

Nevertheless, further studies that investigate the effect on fatigue behaviour of typical AM defects on parts with respect to the wrought material are needed, especially when dealing with complex geometries such as lattice structures.

The aim of this study is twofold. a. A previously presented method for modeling lattice structures based on a subdivision surface algorithm [14] was tested to verify its actual effectiveness in enhancing the fatigue life of the parts with respect to a more standard method; b. A preliminary model was proposed in order to describe the sensitivity of AM parts to typical AM defects, coming from the technology, i.e. porosity and surface roughness, and from the modeling phase, i.e. sharp edges, especially when dealing with scale and shape effects. To do so, three types of specimens were additively manufactured with selective laser melting (SLM) technology in 316L stainless steel (SS316L): bulk parts, lattice structures modeled with a traditional approach, and lattice structures modeled with the subdivision surface method.

2 Materials and Methods

2.1 Specimens

Three types of specimens were designed. Bulk cylindrical (BC) specimens were adapted from ASTM E606/606M – 12 standard [15], adopting the dimensions in Fig. 1. The diameter of the central uniform section was reduced to 2.8 mm to obtain an area close to 6 mm^2, similar to the resistant area of 6 beams with unitary area as in the lattice samples.

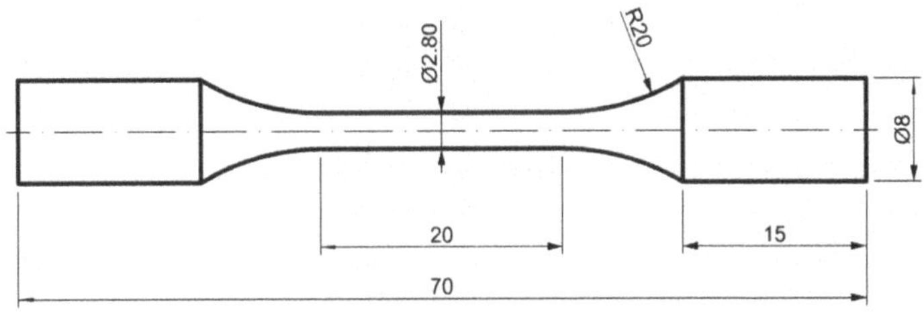

Fig. 1. Bulk cylindrical specimens dimensions.

Then, two types of lattice structures were modeled, cylindrical (CY) and Catmull-Clark (CC); both of them are made up from a regular repetition of a simple cubic cell; the number of instances is 3, 2 and 7 along x-, y-, z- axes, respectively. The cylindrical beams around the wireframe have been modeled with two different approaches; in the CY unit cell, the beams approaching nodal points are simply intersected, generating sharp edges; the CC unit cell, instead, is modelled adopting the Catmull-Clark subdivision surface algorithm [14, 16]: starting from a quadrilateral mesh, the initial vertices are iteratively averaged following the subdivision rule, splitting into four each quad face and resulting in a smooth surface, especially at nodal points, without the need for additional filleting operations. Figure 2 shows the CY and CC unit cell, and a lattice specimen. The dimension of the beams of the lattice specimens are obtained in order to ensure a cross-section area of 1 mm^2 for each beam, and cell size of 3 mm; the beams of the CY specimens have a diameter of 1.128 mm, while the CC has a beam dimension ranging between 1.116 mm and 1.150 (the section of a CC beam is not a circle). Moreover, all the specimen types have a similar cross-section area, 6.16 mm^2 for BC, 6 mm^2 for CY and CC lattices; this allows to concentrate on the mechanical fatigue behaviour of specimens originating from different modeling approaches. More, the adopted dimensions respect technological constraints, such as minimum hole size for de-powdering and minimum struts size.

A total of 36 specimens, 12 for each specimen type, were manufactured in 316L stainless steel by SLM technology, using the AM400 machine by Renishaw. Renishaw SS316L-0409 powder was used, and its main properties are reported in Table 1. The SS316L is an austenitic stainless steel and presents good weldability, machinability and high corrosion resistance; it is widely used in cyclically loaded components in a variety of engineering sectors such as automotive, construction, refinery, oil and gas, and chemical and petrochemical industries [17].

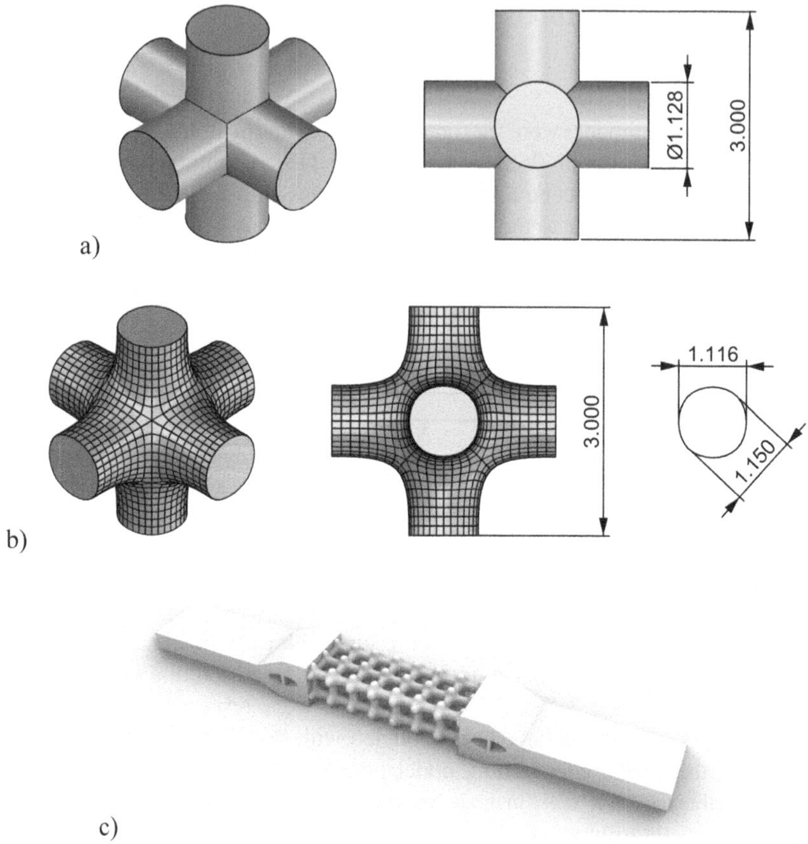

Fig. 2. (a) Cylindrical unit cell, (b) Catmull-Clark unit cell, (c) lattice specimen.

Table 1. Renishaw SS316L-0409 stainless steel powder properties.

Powder composition	
Element	Mass (%)
Iron	Balance
Chromium	17.6
Nickel	12.7
Molybdenum	2.32
Manganese	1.31

Particle size*	
Percentile	Diameter [μm]
$d_{10\%}$	18.8
$d_{50\%}$	29.4
$d_{90\%}$	45.5

*Laser diffraction analysis

All specimens were manufactured adopting the process parameters summarized in Table 2, "Interior" column; only for BC parts, different process parameters were used at the borders. The build direction was parallel to the longitudinal axis of the specimens and to the loading direction. The samples were tested as-built, i.e. no post-processing operations were performed, except for powder evacuation and supports removal.

Table 2. Process parameters.

	Interior	Borders
Laser power [W]	200	110
Exposure time [μs]	80	100
Point distance [μm]	60	20
Layer thickness [μm]	50	50
Layer rotation angle [°]	67	67
Scan pattern	Stripe	Contour

2.2 Experimental Tests

Specimens were weighed, and measured by the optical coordinate measuring machine (CMM) OGP SmartScope Flash 200: diameters of the central section were measured for BC samples and diameters of the beams were measured for CY and CC ones; a routine was created in Zone3 software, in which the beams are optically detected and measured; measurements were taken both in the front (3 × 7 cells) and side (2 × 7 cells) face of the lattice. Surface roughness along building direction was measured by a Talysurf i-Series. Measurement was taken in the central section of 2.80 mm diameter for the BC samples, while for the lattice structures specimens, due to the difficulty to measure the beams of the lattice, measurements were taken on the flat surfaces at the extremity of the samples, where they are gripped to the testing machine. The following parameters were used for roughness measurement and analysis: evaluation length = 12.5 mm, sampling length = 2.5 mm, according to ISO 4287 [18] and ISO 4288 [19] standards.

For CY and CC cells, numerical analyses were done to calculate the curvature and the stress concentration factor k_t, defined as $k_t = \sigma_{peak}/\sigma_{nom}$, where σ_{nom} is the applied force divided by the strut section area and σ_{peak} is the maximum Von Mises stress on the cell. Minimum and maximum surface curvature were analysed in Rhinoceros 6 using AdvMesh tool from Rhino Open Projects. k_t was obtained in ANSYS® Release 16 by the finite element analysis (FEA) of a simplified model allowing to study a single cell, as if it was surrounded by others [20].

Then, fatigue tests were executed on an MTS Acumen 3 Electrodynamic Test System equipped with a 3 kN load cell and an MTS 634.31 F extensometer. Tests were performed at 10 Hz frequency with 0 stress ratio (R = 0, i.e. $\sigma_{max}/\sigma_{min} = 0$, where σ is the stress). Each type of specimen was tested at 4 different stress levels, 473.2 MPa, 405.6 MPa, 338 MPa and 270.4 MPa, equal to 70%, 60%, 50% and 40% of the ultimate tensile strength (UTS), 676 MPa, according to Renishaw's SS316L datasheet [21]. Data were acquired with a timed trigger at 64 Hz to store entire cycles signals and with a peak-and-valley trigger to store minimum and maximum signals at every cycle. No artificial cooling was applied. The estimated Stress – Number of cycles (S-N) curves, also known as Wöhler curves, were statistically computed according to standard ISO/DIS 12107 [22].

3 Results and Discussions

Figure 3 shows one printed sample for each type of specimen. Due to the lack of post-processing procedures such as sandblasting or chemical etching [23], in the magnified view of the lattice structure of CY and CC specimens, it is possible to observe some partially melted powder, especially at bottom surfaces.

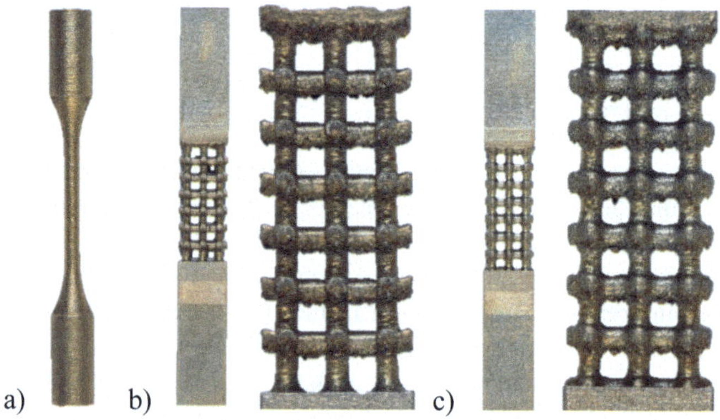

Fig. 3. SLM SS316L specimens: (a) BC sample, (b) CY sample, (c) CC sample.

Table 3 presents the average values and the standard deviations for mass and diameter dimension of the specimens, compared with nominal values. Nominal mass is obtained adopting a density of 7.99 g/cm^3, according to the datasheet, and calculating the volume of the geometric model. For CY and CC models only the diameters of the beams aligned with the building direction were reported. Table 3 also shows the roughness parameters Ra, Rz, and Rt of specimens. The mass and dimensions of the specimens are always higher than the nominal value; this is probably due to the absence of offset compensation preset in the build preparation software. The lack of a contour strategy for the lattice specimens affects the quality of their surfaces in terms of size and roughness [24]. Moreover, the more complex shape of the CY and CC samples could result in higher dimensional errors.

Table 3. Mass, diameter and roughness measurements.

	Mass [g]			Diameter [mm]			Roughness [μm]					
							Ra		Rz		Rt	
	Mean	SD	Nominal	Mean	SD	Nominal	Mean	SD	Mean	SD	Mean	SD
BC	16.09	0.15	15.93	2.829	0.012	2.800	5.5	0.5	30.8	2.7	49.2	6.9
CY	13.72	0.11	12.85	1.259	0.006	1.128	7.7	0.8	49.6	6.2	60.0	8.0
CC	14.06	0.08	13.20	1.265	0.007	1.116	8.4	1.0	55.0	6.7	69.2	11.8

Figure 4a presents the curvature analysis for CC (subdivided) and CY (circular section) specimens. CY geometry does not have fillet radius but during the manufacturing of the parts, sharp edges are slightly filleted; an approximated fillet radius of 0.1 mm was measured by the CMM. The minimum and maximum curvature analysis highlights that while the CY geometry has a C^1 surface (tangent continuous surfaces), the Catmull-Clark subdivision algorithm produces a surface with curvature continuity (C^2 surface), except at extraordinary vertices where they are C^1 [16, 25] (for a quadrilateral mesh an extraordinary vertex is a vertex in which the number of incident edges differs from 4). Thanks to its curvature continuity, CC cells surfaces are smoother than CY ones. This also reflects on the stress concentration factor k_t; in Fig. 4b, k_t is plotted as a function of the Fillet radius/Cell dimension ratio. CC cell has a constant k_t = 1.15 deriving from the subdivision algorithm, while CY cell has a decreasing behaviour as the ratio increases. This is due to the beneficial effects that large fillet radii have on the peak stresses. Nevertheless, as far as the ratio increases, k_t of CY will not reach the k_t of CC.

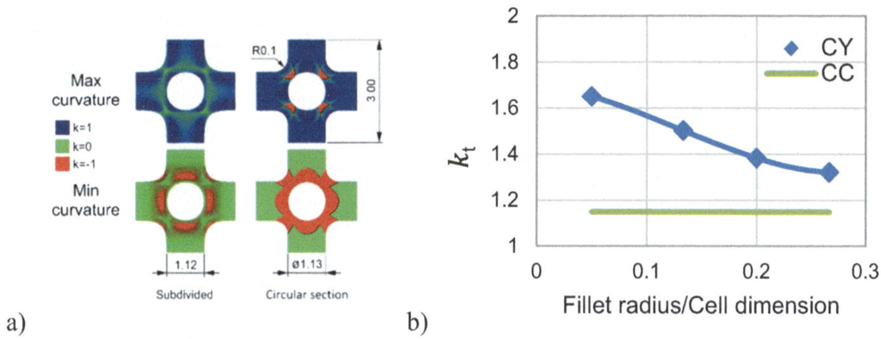

Fig. 4. (a) Curvature analysis for CC (subdivided) and CY (circular section) specimens. (b) stress concentration factor k_t as a function of fillet radius/Cell dimension ratio.

Figure 5 shows the results of fatigue tests as Stress – Number of cycles (S-N) curves (Wöhler curves) obtained according to ISO/DIS 12107 standard. The curves were obtained by the model equation:

$$\log_{10} N = b_0 + b_1 \cdot \log_{10} S \quad (1)$$

that can be rewritten as:

$$N = k \cdot S^{b_1} \quad (2)$$

where $k = 10^{b_0}$.

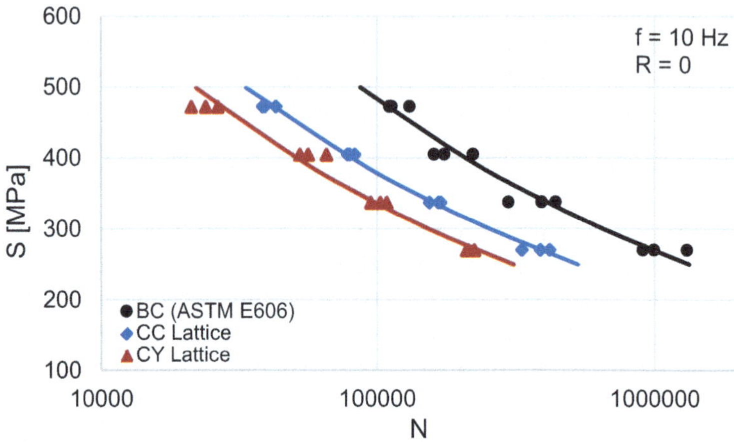

Fig. 5. S-N curves, according to ISO/DIS 12107 standard.

In Table 4 the parameters b_0 and b_1 are reported, together with the standard deviation SD and the correlation coefficient R^2, that is close to 1 for all of the specimen types, that means the model well describes the behaviour of the samples.

Table 4. Statistical parameters for S-N curve, according to ISO/DIS 12107.

Specimen type	b_0	b_1	SD	R^2
BC	15.58	−3.94	0.076	0.96
CC	15.26	−3.98	0.032	0.99
CY	14.63	−3.81	0.064	0.97

CC lattice specimens fail about one and a half times later with respect to CY ones. This is due to the different modeling approach adopted. As previously analyzed, the lattice structure modeled with the Catmull-Clark subdivision method presents smoother surfaces than the CY, especially at nodal points where the CY model has sharp edges, except for the small fillet radii resulting from the manufacturing technology; the lower stress concentration factor improves the fatigue behaviour. So the surface subdivision method results in lattices with an enhanced fatigue life, that are modeled in an efficient way, and that can be exploited in different applications and fields, for example, scaffolds in biomedical field, lightweight and energy absorber structures in the automotive industry, and bearing structures in industrial field.

The BC samples fail at a higher number of cycles if compared to lattice structures; indeed, they resist more than twice the CC lattices. This is explained at least by 3 aspects:

- the stress concentration factor, k_t, is negligible for the BC sample;
- the effective resistant area is more sensitive to the porosity in a small section;
- the cross-section profile of the sample is greater for the lattice sample and increases the opportunity of cracks initiation.

k_t, mainly related to the shape of the sample, is higher in CY lattices than in CC ones and, especially, in BC models can be neglected thanks to the axisymmetrical shape that does not present sharp edges, but filleted edges with large radii instead.

The cross-section and the porosity are related to the scale effect, that appears in this work due to the "segmentation" of one bigger element in several smaller ones. Let d be the diameter of a pore and let suppose that a pore is present in a beam, which could be the weak link in the chain of a lattice. The effective area will be:

$$A_{eff} = A_n - \frac{\pi d^2}{4} \tag{3}$$

where A_n is the nominal area. The coefficient k_A that keeps in consideration the sensitivity to the porosity depending on the number N of beams can be defined as a function of the effective areas:

$$k_A = \frac{\sigma_N}{\sigma_1} = \frac{A_{eff,1}}{A_{eff,N}} = \frac{A_n - \frac{\pi d^2}{4}}{A_n - \frac{\pi N d^2}{4}} = \frac{D^2 - d^2}{D^2 - N d^2} \tag{4}$$

where D is the diameter of a single beam of area A_n, σ is the stress applied, and subscripts 1 and N indicate how many sections are present in the model.

The cross-section profile of the lattice structure is greater due to the absence of a contour strategy that results in a higher roughness, and to the shape of the sample: to obtain the same area with more beams, the length of the profile increases. Since the relation between area A and perimeter p of a circumference can be written as $p = 2\sqrt{A\pi}$, if the same area is divided in N equal smaller areas, the ratio between the perimeter of a single bigger beam p_1 and the perimeters of N smaller beams p_N will be:

$$\frac{p_1}{p_N} = \frac{2\sqrt{A\pi}}{2\sqrt{A\pi N}} = \frac{1}{\sqrt{N}} \tag{5}$$

The coefficient k_P that describes the scale effect of the cross-section profile will consider the ratio between the roughness parameters Rt_1/Rt_N as well, and can be described as follows:

$$k_P = f\left(\frac{1}{\sqrt{N}}, \frac{Rt_1}{Rt_N}\right) \tag{6}$$

To better understand Eq. (6), further studies are needed to confirm these hypotheses, which are not yet confirmed in the literature. Indeed, studies exist that state that larger parts experience lower fatigue limits with respect to smaller parts [26].

Combining all the coefficients, Eq. (2) can be re-written as:

$$N = k \cdot k_p \cdot (k_t \cdot k_A \cdot S)^{b_1} \tag{7}$$

Equation (7) allows to describe the fatigue behaviour of geometries in which the circular cross-section of a part has been divided in smaller cross-sections, keeping in consideration the scale and shape effects; this perfectly suits the situation where lattice structures made of struts are compared to the bulk material. The development of an accurate and effective model in fatigue life of lattice structures is fundamental in the design phase, allowing reliable simulations of the mechanical behaviour. For instance, this model can be introduced in homogenization methods, reducing the complexity in the simulation of trabecular components [27].

4 Conclusions

In this work, three types of specimen were modeled and produced by AM SLM technology in SS316L: bulk specimens (BC), regular lattice structures modeled with a traditional modeling method (CY), and regular lattice structures modeled adopting a previously proposed modeling approach based on a subdivision surface algorithm (CC). All the parts were fatigue tested to better understand the fatigue behaviour of the geometric modeling approach.

In particular, the comparison between CY and CC modeling approaches highlights that CC samples fail at a higher number of cycles, thanks to a modeling method that results in smooth surfaces, avoiding sharp edges and reducing the negative consequences of shape effects. The comparison between bulk and lattice specimens allows to understand the scale effects on porosity and surface roughness: when the diameter of a beam decreases, the percentage of area occupied by a porosity is bigger if compared to the same porosity in a beam with a bigger diameter, and this weakens the structure; moreover, when the area of a beam is subdivided into several smaller beams, the total perimeter of a cross-section increases, enhancing the probability of points for crack initiation. Thanks to these observations, confirmed by experimental tests, it was possible to propose a preliminary model that takes into consideration the scale and shape effects and describes the fatigue behaviour of AM parts characterized by the reduction of the elements dimension and by the increase of the shape complexity, allowing to predict the fatigue behaviour of lattice structures.

The proposed model is still in an early stage and further studies are needed to complete and validate it through an extensive fatigue test campaign, with samples with different shapes and element sizes. Nevertheless, it is already possible to assert that the surface subdivision modeling method allows overcoming issues in lattice structure modeling highlighted in literature, such as robustness and scalability. Finally, the development of a model in fatigue life of lattice structures is fundamental in the design phase, where can be implemented in Computer Aided Engineering software, reducing the complexity of the analyses.

Acknowledgments. This work was partially funded by Fondazione Cassa Di Risparmio di Padova e Rovigo (CARIPARO) and by Regione Friuli Venezia Giulia "FESR D23D1800-0160007" to Tesolin S.p.A.

References

1. Holmström, J., Partanen, J., Tuomi, J., Walter, M.: Rapid manufacturing in the spare parts supply chain. J. Manuf. Technol. Manag. **21**, 687–697 (2010). https://doi.org/10.1108/17410381011063996
2. Wohlers, T., Campbell, I., Diegel, O., Kowen, J., Caffrey, T.: Wohlers report 2019 (2019)
3. Gibson, L.J., Ashby, M.F.: Cellular Solids. Cambridge University Press, Cambridge (1997). https://doi.org/10.1017/CBO9781139878326
4. Savio, G., Rosso, S., Meneghello, R., Concheri, G.: Geometric modeling of cellular materials for additive manufacturing in biomedical field: a review. Appl. Bionics Biomech. **2018**, 1–14 (2018). https://doi.org/10.1155/2018/1654782
5. Van Hooreweder, B., Kruth, J.-P.: Advanced fatigue analysis of metal lattice structures produced by selective laser melting. CIRP Ann. **66**, 221–224 (2017). https://doi.org/10.1016/J.CIRP.2017.04.130
6. SASAM: Additive Manufacturing: SASAM Standardisation Roadmap (2015)
7. AMSC: Standardization Roadmap for Additive Manufacturing (version 1.0) (2017)
8. Yadollahi, A., Shamsaei, N.: Additive manufacturing of fatigue resistant materials: challenges and opportunities. Int. J. Fatigue **98**, 14–31 (2017). https://doi.org/10.1016/J.IJFATIGUE.2017.01.001
9. Greitemeier, D., Dalle Donne, C., Syassen, F., Eufinger, J., Melz, T.: Effect of surface roughness on fatigue performance of additive manufactured Ti–6Al–4V. Mater. Sci. Technol. **32**, 629–634 (2016). https://doi.org/10.1179/1743284715Y.0000000053
10. Masuo, H., Tanaka, Y., Morokoshi, S., Yagura, H., Uchida, T., Yamamoto, Y., Murakami, Y.: Influence of defects, surface roughness and HIP on the fatigue strength of Ti-6Al-4V manufactured by additive manufacturing. Int. J. Fatigue **117**, 163–179 (2018). https://doi.org/10.1016/J.IJFATIGUE.2018.07.020
11. Croccolo, D., De Agostinis, M., Fini, S., Olmi, G., Vranic, A., Ciric-Kostic, S.: Influence of the build orientation on the fatigue strength of EOS maraging steel produced by additive metal machine. Fatigue Fract. Eng. Mater. Struct. **39**, 637–647 (2016). https://doi.org/10.1111/ffe.12395
12. Bača, A., Konečná, R., Kunz, L.: Influence of build direction on the fatigue behaviour of Ti6Al4V alloy produced by direct metal laser sintering. Mater. Today Proc. **3**, 921–924 (2016). https://doi.org/10.1016/J.MATPR.2016.03.021
13. Wycisk, E., Solbach, A., Siddique, S., Herzog, D., Walther, F., Emmelmann, C.: Effects of defects in laser additive manufactured Ti-6Al-4V on fatigue properties. Phys. Procedia **56**, 371–378 (2014). https://doi.org/10.1016/J.PHPRO.2014.08.120
14. Savio, G., Meneghello, R., Concheri, G.: Geometric modeling of lattice structures for additive manufacturing. Rapid Prototyp. J. **24**, 351–360 (2018). https://doi.org/10.1108/RPJ-07-2016-0122
15. ASTM: ASTM E606/E606M - 2012 – Standard Test Method for Strain-Controlled Fatigue Testing (2012)
16. Catmull, E., Clark, J.: Recursively generated B-spline surfaces on arbitrary topological meshes. Comput. Des. **10**, 350–355 (1978). https://doi.org/10.1016/0010-4485(78)90110-0

17. Elangeswaran, C., Cutolo, A., Muralidharan, G.K., de Formanoir, C., Berto, F., Vanmeensel, K., Van Hooreweder, B.: Effect of post-treatments on the fatigue behaviour of 316L stainless steel manufactured by laser powder bed fusion. Int. J. Fatigue **123**, 31–39 (2019). https://doi.org/10.1016/J.IJFATIGUE.2019.01.013
18. ISO International Organization for Standardization: ISO 4287:1997 - Geometrical Product Specifications (GPS) – Surface texture: Profile method – Terms, definitions and surface texture parameters (1997)
19. ISO International Organization for Standardization: ISO 4288:1996 - Geometrical Product Specifications (GPS) – Surface texture: Profile method – Rules and procedures for the assessment of surface texture (1996)
20. Savio, G., Rosso, S., Curtarello, A., Meneghello, R., Concheri, G.: Implications of modeling approaches on the fatigue behavior of cellular solids. Addit. Manuf. **25**, 50–58 (2019). https://doi.org/10.1016/j.addma.2018.10.047
21. Renishaw: Data sheets - Additive manufacturing - SS 316L-0407. https://www.renishaw.com/en/data-sheets-additive-manufacturing–17862
22. ISO International Organization for Standardization: ISO/DIS 12107 – Metallic materials – Fatigue testing – Statistical planning and analysis of data (2017)
23. Van Hooreweder, B., Apers, Y., Lietaert, K., Kruth, J.P.: Improving the fatigue performance of porous metallic biomaterials produced by selective laser melting. Acta Biomater. **47**, 193–202 (2017). https://doi.org/10.1016/j.actbio.2016.10.005
24. Yadroitsev, I., Yadroitsava, I., Bertrand, P., Smurov, I.: Factor analysis of selective laser melting process parameters and geometrical characteristics of synthesized single tracks. Rapid Prototyp. J. **18**, 201–208 (2012). https://doi.org/10.1108/13552541211218117
25. Zorin, D.: Subdivision zoo. Subdiv. Model. Animat. 65–104 (2000)
26. Murakami, Y.: Metal Fatigue: Effects of Small Defects and Nonmetallic Inclusions. Elsevier, Boston (2002)
27. Savio, G., Curtarello, A., Rosso, S., Meneghello, R., Concheri, G.: Homogenization driven design of lightweight structures for additive manufacturing. Int. J. Interact. Des. Manuf. (2019). https://doi.org/10.1007/s12008-019-00543-0

Additive Manufacturing Challenges and Future Developments in the Next Ten Years

Antonio Bacciaglia, Alessandro Ceruti(✉), and Alfredo Liverani

Department of Industrial Engineering (DIN), University of Bologna, Bologna, Italy
{antonio.bacciaglia2,alessandro.ceruti, alfredo.liverani}@unibo.it

Abstract. A significant growth of Additive Manufacturing technology has been noticed in the last few decades due to its well-established advantages. This production process based on the idea of adding material layer by layer, instead of removing it, has several advantages such as: time reduction in the design-to-manufacturing cycle, capability to produce complex shapes in a single piece thus reducing the connections, and weight saving just to mention the main significant. On the same wave of enthusiasm, the research community is significantly focused on this technology and many contributions that are focused on the whole range of aspects connected to this manufacturing way are already available. However, Additive Manufacturing is in a development phase, and the design tools, the material portfolio and the production methodologies are still far from the optimum typical of mature technologies. This paper critically reviews the Additive Manufacturing advantages and weaknesses which nowadays limits its wider application; a discussion about how these problems could be solved is included, together with an outlook of the challenges the practitioners must face off from nowadays up to a time window of ten years.

Keywords: Additive Manufacturing · Design tools · Topological optimization · Technology review · Future challenges

1 Introduction

Additive manufacturing (AM) is becoming an available strategy to produce parts for a wide range of applications. AM describes a new way to produce components based on the idea of adding material layer by layer, in opposite to the traditional manufacturing processes based on casting or material removal with milling, turning and drilling processes where several manufacturing restraints must be respected. The design of a component must follow a set of rules when the traditional manufacturing is adopted: undercuts, tool paths, geometry, thickness, geometry in general. On the other hand, when AM is adopted, the designer can shape the geometry of the body with a reduced set of restraints. The literature expresses this concept with the expression: "What You See Is What You Build" [1]. So, when traditional machining is adopted the

C. Rizzi et al. (Eds.): ADM 2019, LNME, pp. 891–902, 2020.
https://doi.org/10.1007/978-3-030-31154-4_76

manufacturing drives the shapes, but when AM is adopted the shape drives the manufacturing. AM is also known to the large audience with the popular name of "3D printing" or "Rapid Prototyping (RP)". These latter two terms are a bit restrictive and improper, because they do not describe all the AM potentiality: nowadays AM is used not only to produce both aesthetical or functional prototypes (like suggested by the term RP), but also structural parts to be installed in products available to end-users.

The additive manufacturing market is increasing year after year and it represents one of the three main manufacturing processes, along with subtractive processes and formative ones [2], which can be nowadays used to manufacture components.

The scope of this paper is to critically review the AM advantages and factors which nowadays limits its wider application, focusing on the possible challenges and future developments of this technology in the next years. This work is organized as follows: Sect. 1 will give an historical background along with a brief mention of the different AM technologies. Section 2 describes the AM advantages and criticalities while Sect. 3 contains a discussion about the future developments and challenges that the scientific community has to face off. Finally, Sect. 4 summarizes the results and suggests direction where to aim future researches in this field.

1.1 Historical Background

Looking to the past, the Stereolithography (SLA) invention by Charles Hulls in 1983 can be considered the first application where an additive manufacturing processes has been introduced. The Hull's concept is based on the idea to create three-dimensional objects layer by layer, in which a computer-controlled moving energy source, i.e. a laser beam, is used to build up the required structure, from a liquid photo-sensible polymer that hardens on contact with laser light.

However, even if this AM process was invented, something was still missing: creating a conceptual model, a 3D component or functional prototype ready to be manufactured, was very expensive in terms of money and time. Moreover, designers and engineers were using quite new design tools for that time such as Computer Aided Design or Manufacturing software (CAD/CAM), but there was still a gap on the communication between the design software and the first Rapid Prototyping System, SLA 1. Therefore, Charles Hulls and 3D Systems developed the STL file format - that is still in use - to transfer the component's shape information from the CAD software to the 3D printers in order to produce it [3].

From that moment, a huge effort from the scientific community was oriented to the research and development of different AM processes involving different materials in order to get a wide range of material properties for a heterogeneous set of users. Juts to mention one, in the 1990s the development of the electron beam melting (EBM) [4] and laser beam melting (LBM) [5] paved the way to the manufacturing in AM of titanium, aluminium alloys, Inconel and steel parts from powders. In the 2000s, the research community focused the attention towards the software tools necessary to support AM: thanks to the more affordable prices of computers and the increase of computational power it was possible to handle complex shapes and to apply optimization techniques. The late 2000s and the more recent years saw the commercialization of AM machines based on the Fused Deposition Modelling (FDM) technique in many research groups because of more affordable prices.

1.2 AM Process Terminology and Classification

Several technologies are now available for users interested in AM. Thanks to ASTM standards [6], all the processes, materials and technologies are classified in a rigorous way, as it can be seen in Fig. 1. Each AM process has its peculiarities, advantages and limitations as discussed in papers like [7], and a detailed description of each technology is beyond the scope of this work. However, some interesting contributions for each method are suggested to the reader. The more diffused AM technologies are:

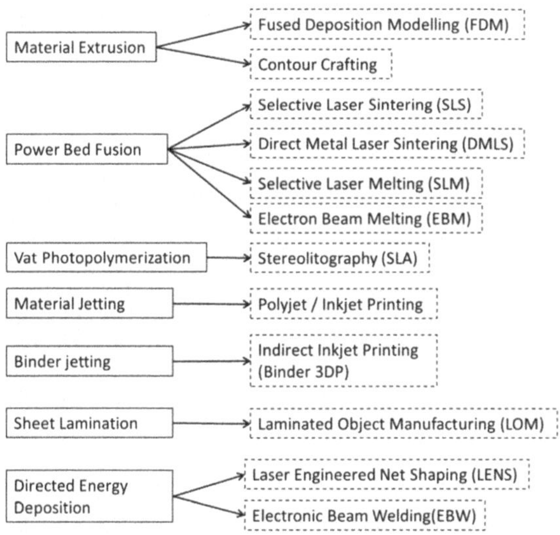

Fig. 1. ASTM terminology for AM processes (adapted from [6])

- Fused Deposition Modelling (FDM) [8]
- Polyjet - Inkjet printing [9]
- Selective laser sintering (SLS) [10], Selective laser melting (SLM) [11] and Electron beam melting (EBM) [4]
- Laser engineering net shaping (LENS) [12]
- Stereolithography (SLA) [13]

Each one of these technologies is based upon different operating philosophies, production strategies, materials properties and so on. This reflects on the production time, costs, final component quality and mechanical properties. For these reasons the designer has to properly choose the best AM process, depending on the specific application in a trade-off between the pros and cons of each single technology.

FDM can be used for low strength parts where the external roughness isn't important; Stereolithography (SLA) is a more expensive technology for parts with high details which can be obtained with polished surfaces and better structural properties respect to FDM (after curing with heat and UV the solids obtained polymerizing resins with a laser beam). Selective laser sintering (SLS), LBM and EBM are expensive technologies but effective in producing high strength metal parts.

However, there is a main design flowchart [14] to follow that is in common to all the available technologies, even if some small differences could be noticed (see Fig. 2).

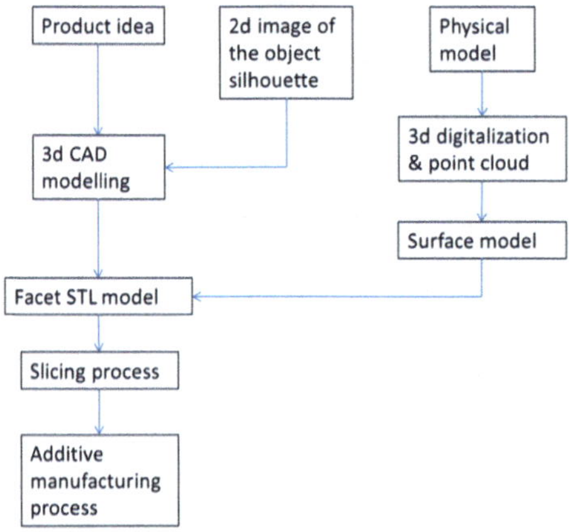

Fig. 2. Additive Manufacturing design flowchart from component idea to 3D production.

2 Additive Manufacturing Advantages and Weaknesses

2.1 Advantages

As previously said, several advantages reflect the increasing amount of AM usage as main production process.

Design Flexibility. Rapid prototyping offers an extreme flexibility on the component design, thanks to the reduction of manufacturing constraints in terms of shapes and geometries: with AM the design drives the shape, while the manufacturing drives the shape when traditional processes are chosen. This flexibility is given by the synergy of CAD software with the rapid prototyping machines, referred as Digital Manufacturing. Using 3D printing and CAD software packages, the designer can produce in reduced times components with high structural efficiency, with a reduced effort in respecting manufacturing restraints.

Minimization of Internal Process Logistic. 3D printing can reduce the time needed for the design-to-manufacturing cycle, together with the minimization of internal process logistics due to significant less operations with respect traditional manufacturing processes, but with similar final-product performance [15]: in case of small lots, the need for mould in casting can waste time and money. Similarly, the tooling pre-setting and CNC machines programming can increase the cost of components when numbers are low. AM can be used also to manufacture prototype parts for quality/functional/aesthetic tests well before the production, thus saving money, efforts and time. This is particularly important for companies where the cost of changes in the products' design cycle increases with the advance of the development phase.

Product Customization. Due to its extreme design flexibility, custom shapes can be produced; this is an important aspect for the automotive industry, especially for the production of spare parts for luxury, racing (Fig. 3) and classic cars, and in the biomedical industry, where each patient needs specific product characteristics such as shapes, materials, and in general when high customization and low volume production rate are involved.

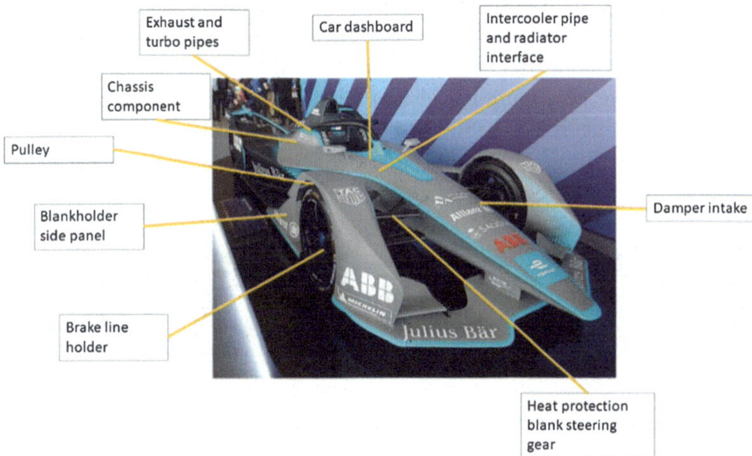

Fig. 3. AM application in automotive industry

Material Customization. Another important advantage is the possibility to customize the material used to produce a component, or in other words the possibility to create custom mixes of powders; to embed fibres in a plastic filament; to increase the material strength with carbon fibres or wood fibres; and to use bio-compatibles materials. On the other hand, if a material that suits the need isn't available, it is possible to produce a component with the required material property by changing the internal porosity with the adoption of cellular structures, or functionally graded materials.

In literature, there are different types of cellular structures like foams (stochastic structures), honeycombs and lattices (periodic structures) [16]. Lattices are bio-inspired, lightweight structures composed of simple small elements, which form a representative volume called "unit cell" that is repeated thousands of times in space to fill the internal volume of a component (Fig. 4).

From Fig. 4 it can be seen the complexity of the shapes of a lattice structure that may rise several problems if a traditional manufacturing process is chosen. However, thanks to the AM flexibility, it is possible to design and manufacture optimized structures made of elements with integrated electrical wiring (through hollowed structures) or topologically optimized structures.

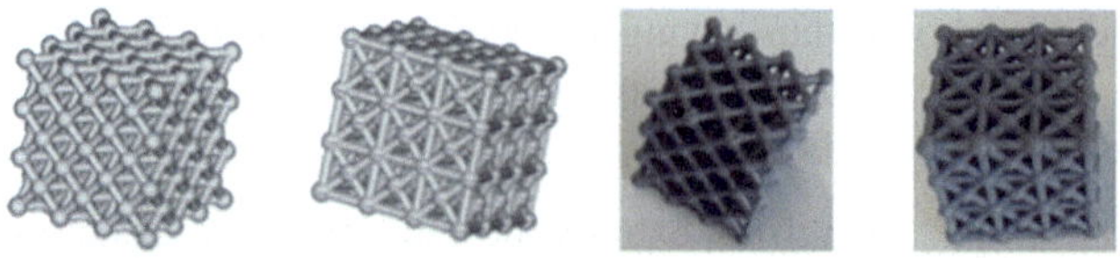

Fig. 4. Uniform lattice CAD models in the left and 3D printed in the right with SLA technique

Topology Optimization. Topology optimization (TO) is a numerical procedure where the shape of a body is automatically changed depending on a fitness function that in most of the cases is identified in the compliance, stress or maximum displacement. TO requires the boundary conditions and body constraints, the loads position orientation and magnitude, the locations of holes or forbidden areas which must be left empty, the maximum material volume fraction which is needed to avoid a fully dense solution [17]. After an iterative procedure based on Finite Element Analysis (FEA), the output of the algorithm is a material distribution which minimizes the fitness function (Fig. 5): this leads to low weight structures with homogeneous stress in the material.

Fig. 5. Cantilever beam optimized through TO algorithm

The resulting geometry is far from the common simple shapes which can be sketched by a human designer, and it is impossible to manufacture it using traditional processes. On the other hand, AM and TO can gather enormous economic advantages: reduced quantity of waste material, time savings, and weight reductions can be achieved easily. This is of straightforward importance in the design of vehicles and transports in general where reduced weight means less power required and less pollution. From literature, rapid prototyping has been already adopted for non-critical components. In aerospace, AM is often coupled with TO or lattice structure implementation to produce complex shapes which can be manufactured only with AM [18].

Less Economical Effort. Another advantage of AM is that with a single machine companies can manufacture parts with a wide range of dimensions, geometries and materials. This is a key point for small companies that would like to enter in the AM production market. Thanks to AM flexibility, small companies could invest in AM machines a smaller amount of capital with respect to a more expensive entire supply chain based on traditional machining. In addition, AM enables the possibility for companies to manufacture in house some tools needed for product manufacturing or for validation tests.

Assembly Connection Minimization. AM shows the capability to generate complex shapes in one single piece instead of an assembly of multiple parts, that is advantageous to reduce the time needed for proper bolting connections or part welding. Furthermore, one-piece parts are more reliable with respect to an assembly of bolted parts: connections are critical in vibrating environments and money has to be spent in maintenance operations for inspections in order to detect failures.

Multi-material Capability. At last, it is worth noting that with the development of the Additive Manufacturing File format (AMF) and 3MF, new capabilities could be available to machines, such as the use of several materials in a singles component. This multi-material, multi-structure freedom can give a highly competitive advantage to the additive processes respect to the traditional ones.

2.2 Weaknesses

On one hand AM has great opportunities and advantages, but on the other hand this technology has also some drawbacks and weaknesses that will be discussed in this section.

Anisotropic Material. First, the resulting component after an AM process is usually highly anisotropic because of the process itself: to provide an example, in FDM the material is deposed in each single layer (supposed to be in the XY plane) and the piece grows in the Z direction with voids between layers. This could arise some problems in structural applications, and this is true in several kinds of AM processes.

Surface roughness can be another problem with some AM techniques. If an FDM machine is not particularly accurate, it is possible to see each slice with the naked eye (stairs effect).

CAD Software Still Show Large Limitations. Even if the advantages of lattice structures are clear, common design tools available to the engineer, show limitations such as because of the implementation of a B-rep representation technology: it is not well suited for lattices and more in general in all the applications when the external surface is quite complex [19]. The software tools used in design, topological optimization, STL export, slicing, simulation and AM manufacturing are separate each other and an end-to-end workflow capability is still missing. Moreover, designers are not ready to handle in a correct way the Design for AM and *skills gap problems* can arise.

Limited Material Portfolio and Limited Building Sizes. Limited material portfolio available for the designers compared to the traditional manufacturing processes is another important constraint in the AM applications. Moreover, the limited size of AM produced components should be noticed as well: as a matter of fact, 3D printers with larger building volume are equipped with 1 m^3 printing chamber, that could be a limiting factor for all the applications in which components with large dimensions are involved.

Inspection and Maintenance. From a Design for Assembly point of view, making a complex assembly in one single piece produced in AM could be advantageous, except when the component has to be repaired: as-printed assemblies usually can't be disassembled, thus creating some problems in case of maintenance and repair operations. When the whole component has to be replaced with a higher economic impact, a pros & cons analysis should be carried out depending on the single application.

High Costs. The AM technology is associated to high costs when metallic parts are produced: this is the main showstopper limiting the spreading of this manufacturing process in case of structural parts. The so-called upfront capital expenditures for AM machines are quite high respect to traditional milling and turning machines (800 k€ for a SLM AM machine compared to 200 k€ for a traditional CNC). In AM the costs per unit is constant when the volume production rates increase, making the traditional manufacturing more attractive. The raw material cost is another important aspect affecting economic issues related to AM: following [20] these costs are always higher than the raw material necessary for casting or moulding.

High Production Times. Problems of time needed to produce components in AM is noticed when serial production in short times is requested. So, for each single application it is important to estimate the even breakpoint value for which AM is more convenient with respect to traditional processes. Fortunately, several tools for cost estimation are available in the internet.

Slow Certification Process. High variation in material properties, changes in raw materials (e.g. powders) with supplier, properties repeatability, availability from supplier, storage, ageing contributes to slow/unavailable certification process. Moreover, AM is a relatively new technology and a limited set of historical data is available in order to understand the material behaviour during the operative life when fatigue analysis plays an important role. So, in applications where reliability, repeatability and high product quality are mandatory (i.e. aerospace sector) structural parts in AM can't be certified. Even if international organizations like FAA, EASA, SAE are planning standards and regulations, the application of AM is still focused on non-safety-critical components for automotive and aerospace. A similar scenario can be noticed in the biomedical field [21], where AM is used in experimental way and in pioneering studies.

Security Issues. Finally, security issues are involved in the AM production because 3D models are shared between design offices and manufacturers: ways to avoid tampered data, stolen/modified 3d models must be implemented through the development of secure software with data exchange validation procedures.

3 Future Challenges and Developments

From the overview provided in the previous chapters, the reader could already capture some cues for future development and maturation of the AM technology. As a matter of fact, there are different fields where the research community should address efforts in the next ten years in order to increase the AM technology readiness and offering AM as an appealing alternative respect to traditional manufacturing processes.

Going in deep in this analysis, the actual CAD tools are inadequate to model lattice structures and still show strong limitations: current CAD systems are usually based on boundary representation technology (B-rep) that is not well suited for lattice structures, and in general where the external shapes are quite complex. For this reason, it is expected in the next 10 years that the research community could develop a new representation technology: the higher computational power available nowadays with respect to some decades ago can push this technology change. An extended use of voxel-based representation methods could be advantageous both to obtain fast geometry manipulation, both to carry out topological optimization of large structures.

However, if a voxelization method is used an iso-surface transformation algorithm should be used in order to get smooth external surfaces (which affect aesthetical issues and fatigue behaviour). The current algorithms supporting smoothing still have some limitations that in the future have to be solved: just to provide an example, if the output of the topological optimization results in a body with a layer of voxel in some points, nowadays algorithms returns a plane in that region, and a solid where layers with more than one voxel lie: the resulting body can't be manufactured. Future decade CAD systems will probably offer powerful iso-surface tools so that optimization could be carried out with a voxel representation, allowing a further voxel to smooth surfaces (and vice versa) conversion.

The research on the self-supporting structures is another recurring issue labelled as critical in several research contributions. When dealing with AM technologies where the material support can't be melted or easily removed such as the SLA (Fig. 6), there is a strong need of efficient algorithms embedded in the pre-processing software. Before starting the printing, the component should be automatically modified and oriented in order to reduce the use of support structures, thus aiming to self-supporting structures. Another strategy which could be adapted is based upon the development of 3D printers allowing the use of easily removable materials for the supports. This technology is already available in FDM machines, but this must be extended to other technologies as well.

Fig. 6. Lattice structure (left) and propeller (right) manufactured in SLA with support material that can't be removed easily.

Next decade design tools for AM structures will handle a multi-objective optimization approach where the final shape is obtained considering requirements on stress, maximum deformation, resonance frequency, reduction of support material, thus proposing not only a single shape, but a set of shapes which the designer could select depending on the final application.

A higher efficiency in the AM process could be achieved changing the standards of parts representation for AM: the STL file format where a body is represented through the approximation of its external surface with triangles is now obsolete. In next years, it will be probably substituted by the AMF or 3MF [22] file formats assuring multi-materials and multi-colours capabilities in the production process.

To fully exploit AM capabilities, the research community will develop new materials which could be used in machines: the use of carbon filaments in FDM or mixed nylon/carbon fiber wires are an example of enabling technologies under investigation now that will be available in next years. But the range of possible materials which can be used in AM is broad: plastics, nylon, carbon/glass fibers, metallic powders, resins are some examples. And this is a non-inclusive list: the Italian Company 3DWASP is developing a large-scale delta 3D printer capable of manufacturing an entire small building by extruding mud or concrete by the nozzle. Similar studies aimed to use regolith and AM to build space stations on the moon have been carried out by the European Space Agency (ESA). The use of AM in civil engineering could be a probable scenario in the next years.

The development of guidelines for the Redesign for Additive Manufacturing (RfAM) of parts produced using traditional processes parts is another research topic worth of being investigated: it is mandatory to rethink the design rules in order to exploit all the AM potentiality. In Universities, courses of Design for AM and AM technologies will be offered in Engineering Schools to complement traditional technology and manufacturing classes.

In the future the building volumes available in the AM machines should increase as much as possible to allow the manufacturing of large parts in one-piece (e.g. body bones [23], cars, large parts of aircraft). Currently, parts' size affects the production time which can increase significantly and could limit the possible AM industrial applications, especially where a large and fast-production rate is needed. To provide the reader with an example, using a current FDM machine with a cubic building volume with 600 mm long edges a topologically optimized beam with a footprint of 580×60 mm could take more than one week using a 30% infill and a 0.1 mm layer height (Fig. 7). Such a production time doesn't satisfy industry needs neither for small production lots, nor for aesthetic prototypes. For this reason, efforts should be spent towards multi speed algorithms that could set the printing speed based on the body local complexity.

However, regulations and standards will play a decisive role in the development speed of AM in the next 10 years. Aerospace, automotive and medical sectors need standards to drive the use of AM parts towards critical structural applications. On the other hand, regulatory agencies need historical data and experience to understand how to deal with technologies, so that it is expected that standards will gradually widen their applicability from non-critical applications up to the most complex design scenarios.

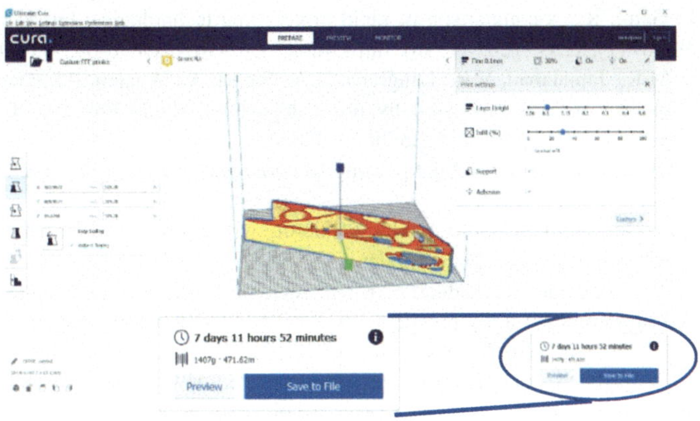

Fig. 7. Example of production time needed for large parts using FDM technology, simulated with CURA software

4 Conclusions

Additive Manufacturing is a recent technology which is now in the development phase and several studies are necessary to push it to a maturity phase. AM shows dramatic advantages respect to traditional manufacturing processing such as casting, milling, and lathing: production time reduction, customization, lightweight structures, assembly parts without connections are some examples. On the other hand, AM is an expensive technology when metallic powders, resins, or carbon reinforced wires are used. File format to connect CAD models with AM machines, new materials, new optimization and assisted sketching tools, improved slicing and speed setting algorithms are some examples of issues which will be addressed in the next 10 years to boost AM capabilities. Apart from the efforts from the academic community and AM machines and software manufacturers, national authorities play a fundamental role: the issuing of regulations in automotive and aerospace industry where design, manufacturing and testing of AM components is specified is mandatory to push AM towards highly critical structural applications.

References

1. Gibson, I., Rosen, D., Stucker, B.: Additive Manufacturing Technologies. Springer, New York (2015)
2. Gebhardt, A.: Understanding additive manufacturing (2011)
3. Wohlers, T., Gornet, T.: History of additive manufacturing (2014)
4. Arcam History: Arcam AB (2015). http://www.arcam.com/company/about-arcam/history/
5. Shellabear, M., Nyrhila, O.: DMLS-development history and state of the art. In: Laser Assisted Netshape Engineering 4, Proceedings of the 4th LANE, pp. 21–24 (2004)
6. ASTM F2792-12A: Standard Terminology for Additive Manufacturing Technologies. ASTM International, West Conshohocken, PA (2012)

7. Pham, D., Gault, R.: A comparison of rapid prototyping technologies. Int. J. Mach. Tools Manuf. **38**(10–11), 1257–1287 (1998). https://doi.org/10.1016/s0890-6955(97)00137-5
8. Agarwala, M.K., Jamalabad, V.R., Langrana, N.A., Safari, A., Whalen, P.J., Danforth, S.C.: Structural quality of parts processed by fused deposition. Rapid Prototyp. J. **2**(4), 4–19 (1996). https://doi.org/10.1108/13552549610732034
9. Le, H.: Progress and trends in ink-jet printing technology. J. Imaging Sci. Technol. **42**, 49–62 (1998)
10. Beaman, J.J., Deckard, C.R.: Selective laser sintering with assisted powder handling. https://patents.google.com/patent/US4938816A/en
11. Olakanmi, E., Cochrane, R., Dalgarno, K.: A review on selective laser sintering/melting (SLS/SLM) of aluminium alloy powders: processing, microstructure, and properties. Prog. Mater Sci. **74**, 401–477 (2015). https://doi.org/10.1016/j.pmatsci.2015.03.002
12. Griffith, M.L., Harwell, L.D., Romero, T., Schlienger, E., Atwood, C.L., Smugeresky, J.E.: Multi-material processing by lens. In: University of Texas at, pp. 387–393 (1997)
13. Cooper, K.: Rapid Prototyping Technology. CRC Press, Boca Raton (2001)
14. VDI-Guideline 3404: Additive Fabrication-Rapid Technologies (Rapid Prototyping) – Fundamentals, Terms and Definitions, Quality Parameter, Supply Agreements (2009)
15. Leal, R., Barreiros, F.M., Alves, L., Romeiro, F., Vasco, J.C., Santos, M., Marto, C.: Additive manufacturing tooling for the automotive industry. Int. J. Adv. Manuf. Technol. **92** (5–8), 1671–1676 (2017)
16. Azman, A.H., Vignat, F., Villeneuve, F.: Design configurations and creation of lattice structures for metallic additive manufacturing. In: 14eme Colloque National AIP PRIMECA, La Plagne (2015)
17. Bendsoe, M.P., Sigmund, O.: Topology Optimization - Theory, Methods and Applications. Springer, Heidelberg (2004)
18. Ceruti, A., Marzocca, P., Liverani, L., Bil, C.: Maintenance in aeronautics in an industry 4.0 context: the role of augmented reality and additive manufacturing. J. Comput. Des. Eng. (2019). https://doi.org/10.1016/j.jcde.2019.02.001
19. McMillan, M.L., Furg, M., Leary, M., Brandt, M.: Programmatic generation of computationally efficient lattice structures for additive manufacture. Rapid Prototyp. J. **23**(3), 486–494 (2017)
20. Wohlers, T.: Wohlers Report 2015: Global Reports. Wohlers Associates, Belgium (2015)
21. Campbell, I., Bourell, D., Gibson, I.: Additive manufacturing: rapid prototyping comes of age. Rapid Prototyp. J. **18**(4), 255–258 (2012)
22. Cicely, E.: New directions for additive manufacturing: workshops, agreements aid standards work and more. Stand. News **44**(5), 30–33 (2016)
23. Frizziero, L., Liverani, A., Donnici, G., Osti, F., Neri, M., Maredi, E., Trisolino, G., Stilli, S.: New methodology for diagnosis of orthopedic diseases through additive manufacturing models. Symmetry **11**, 542 (2019). https://doi.org/10.3390/sym11040542

Investigating the Relationships Between Additive Manufacturing and TRIZ: Trends and Perspectives

Barbara Motyl[1](✉) and Stefano Filippi[2]

[1] DMIF, University of Udine, 33100 Udine, Italy
barbara.motyl@uniud.it
[2] DPIA, University of Udine, 33100 Udine, Italy

Abstract. As Industry 4.0 is the driving force behind the manufacturing industry in this period, increasing importance is being attached to its enabling technologies, formerly known as the nine pillars. In particular, within this context, interest in Additive Manufacturing – AM is constantly growing. For this reason, the authors have considered as appropriate to carry out a literature search to assess whether there is any relationship between AM processes and product design and development practices after 2011. Particularly, the authors were interested in considering the use of additive technologies in relation to the TRIZ – theory of inventive problem solving - and its tools. To this end, it has been decided to carry out a search among the publications contained in the Scopus Database. 115 potential papers have been identified for analysis. After a careful selection, 14 papers, containing information on AM and TRIZ useful for our survey, were selected. Reading these works, it was possible to identify current research trends related to the use of creative problem solving strategies such as TRIZ and its combined use with AM, ranging from the definition of particular integrated methodologies of product development to the definition of guidelines for Design for AM or technological predictions, the state of the art of current research, highlights possible levers of intervention and potential developments and future trends.

Keywords: Additive Manufacturing · TRIZ · Industry 4.0 · Research trends

1 Introduction

Industry 4.0, formerly known as the fourth industrial revolution, consists in the massive use of Cyber- Physical Systems-CFS and in the diffusion of digital or smart factories [1]. In this framework, several changes have to be expected for the industrial world such as the introduction of novel business opportunities and novel service-based, real-time enabled CPS platforms with the arise of new social infrastructure for the workplaces. The shared vision about the Industry 4.0 framework considers the massive use of smart networked systems and IoT focusing on the creation of smart products, procedures, and processes. Thus, smart factories constitute the key feature of this framework. In the smart factory, human beings, machines, and resources communicate with each other as naturally as in a social network. Smart products will allow both manufacturers and users

C. Rizzi et al. (Eds.): ADM 2019, LNME, pp. 903–911, 2020.
https://doi.org/10.1007/978-3-030-31154-4_77

to know many details of how they were made and how they are intended to be used as they actively support the manufacturing and production processes. Consequently, Industry 4.0 should be implemented in an interdisciplinary manner and in close cooperation with the other key areas and using different technologies drivers. These drivers are formerly known as the nine pillars of the technological advancement, and they comprise: Big Data; Autonomous Robots; Simulation; Universal System Integration; Industrial IoT; Cybersecurity; Cloud Computing; Additive Manufacturing and Augmented Reality [2, 3]. Given this context and the growing importance attributed now to Additive Manufacturing -AM technologies after 2011, it seemed appropriate to carry out a literature search to understand the current situation about AM and the possible research gaps concerning the use, applications, and diffusion of this kind of technologies. In particular, this work wants to investigate if there are relationships or particular implications between the AM technologies and the product design and development processes. Particularly, the authors were interested in considering the use of TRIZ, a well-known and structured creativity method, with a wide possibility of related tools, in combination or in synergy to the AM processes in general or referring to Design for AM (DfAM) strategies.

The paper is organized as follows. After the introductory section, in Sect. 2 there is an overview of AM and TRIZ while in Sect. 3 it is described the research method adopted. Then, in Sect. 4, the selected papers are analyzed, and the related observations are reported. Finally, Sect. 5 presents the conclusions and possible future developments.

2 Overview of Additive Manufacturing and TRIZ

It seems appropriate to introduce some basic concepts in relation to AM and TRIZ theory.

2.1 Additive Manufacturing

AM extends the spectrum of traditional manufacturing processes, as parts or complete products are produced by adding material layer by layer in contrast to traditional subtraction processes. This opens up innovative design opportunities and advances in product performance and customization, for example through geometric freedom and the use of highly integrated structures impossible to produce with machine tools. While until recently AM was used exclusively for rapid prototyping, it is now possible to achieve direct production of finished and very precise objects. This is thanks to the improvement of existing technologies and to the different and increasingly performing characteristics of the new materials available. On the other hand, one of the contrasting aspects to the spread of these techniques is given by the designers' lack of experience and knowledge about the capabilities and limitations of these manufacturing technologies in the designing of parts to be incorporated into traditional design processes. In fact, one of the most evident trends that can be seen in this field is the effort to try to define specifications and guidelines for DfAM [4–8].

2.2 TRIZ - The Theory of Inventive Problem Solving

TRIZ - the theory of inventing problem solving - was developed by G. Altshuller in 1946 to support engineers and scientists in solving problems using the knowledge of former inventors, derived from the analysis of thousands of patents and proposing innovative solutions in a systematic way [9–11]. TRIZ guides problem-solvers by suggesting only directions of research to find solutions, leaving room for designers' creativity [12]. With TRIZ, inventors can generate more ideas than before, faster than before, and select the best ones. TRIZ offers a wide range of tools to analyze and solve problems in different perspectives and its systematic approach guides people through problem solving activities avoiding a random exploration of the space of solutions. Some of the most known and used tools are the 40 principles of innovation, the contradiction matrix and the evolution trends [10, 11, 13].

3 Research Method

To accomplish this search, the authors started from asking the following three questions in relation to TRIZ and AM or 3D Printing (3DP).

- Q1. Are there examples of use of TRIZ and its tools for solving problems related to the use of AM or 3DP technologies?
- Q2. How are these technologies applied? to what type of problems (design or process optimization, DfAM purposes)?
- Q3. Is there any relation between the use of AM and TRIZ techniques? Any example of combined use? Any example of innovative use?

In order to answer these questions, the authors have carried out a review of the scientific literature that will be illustrated in the following paragraphs. This literature review exploited the Scopus Database [www.Scopus.it]. The authors used the following search string *"TRIZ* AND *Additive Manufacturing" or "TRIZ* AND *3D Printing"*. 3D Printing term was used to consider all the possible implication connected to additive processes in general, and to find more publications as possible. Also, Industry 4.0 was used as a milestone for filtering publications published later than 2011 to consider only those papers that may have been affected by its effects. In addition, the authors considered only papers or conference contributions (proceedings or book chapters) written in English and fully available (for fully available authors intend both papers available from university catalogs and open access papers). This way, excluding any duplication, 115 papers have been identified. First, all the authors established the following exclusion criteria: to consider only articles dealing with AM and TRIZ issues in the text, and not only in the references section. This way, the number of interesting papers was decreased from 115 to 31. Subsequently, the 31 articles were examined, by all the authors, considering abstracts, keywords and carefully reading the whole text. Thus, this led to the selection of 14 articles, eliminating those submitted by the same authors who had very similar content or the description of the same study cases in relation to the topics (AM and TRIZ) of interest to our search. All these papers appear in the references section while Table 1 summarizes them, highlighting the topics they focus on and their implications with TRIZ and AM.

Table 1. The 14 papers selected by the literature search.

Ref.	Focus of the work	Implications with TRIZ	Implications with AM
[14]	To propose a DfAM method using TRIZ for the void conceptual design of microwave waveguide components	TRIZ is used for concept generation	Definition of DfAM guidelines
[15]	To define a new concept of 3D printed vibratory scaffold for cell culture	TRIZ is used to identify and analyze problems and contradictions, to generate 3DPVS conceptual solutions	Evaluation of the generated solutions in function of 3DP fabricating constrains
[16]	To present AM advantages related to the concept design of high heat flux components aimed at use in a fusion reactor. Report two examples of novel cooling geometry concept design	TRIZ and systems engineering techniques are used in the concept generation phase	Use of AM with new materials (tantalum) to realize the new designed product concepts
[17]	To develop a customized ear cap for medical treatment using TRIZ + AM techniques	TRIZ is used, in a collaborative way with AM, as idea generation method. Authors use the 2010 expanded TRIZ matrix	The FDM technique is used to realize a prototype of the customized designed product
[18]	To develop an instrument (a pen) capable of drawing dotted lines	TRIZ is used to solve the design problem of a new product concept, using functional decomposition and Inventive principles analysis	The FDM technique was used to realize a prototype of the new designed pen
[19]	To use Quality Function Deployment - QFD and TRIZ methods to validate a design method for direct open molds by means of a new strategy: hybrid manufacturing method that employ subtractive and additive techniques efficiently combined	TRIZ is used to find innovative solutions. This work uses the Ideal Final result schematization and the functional diagram to highlight useful or non-useful functions to trim. TRIZ is implemented using TEchoptimizer software	AM is integrated in a new and hybrid manufacturing process to produce direct open molds
[20]	To create a new TRIZ matrix specifically developed for DfAM purposes. Two case studies are presented: a credit card holder and Shotgun Barrel Support Fixture	The TRIZ matrix can help effective design decision making for AM in an early design process as well as a redesign process	A DfAM worksheet is provided to enable users to find specific design solutions for certain AM techniques
[21]	To evaluate and classify the applicability of the 40 inventive principles of TRIZ in function of DfAM and of Laser-beam based Additive Manufacturing – LAM or Additive Manufacturing process characteristic – AMPC	Examples for three TRIZ principles (i.e. asymmetry, nested doll, blessing in disguise) are evaluated in the context of DFAM, outlining their advantages in functionality and performance	Two examples of DfAM applied to TRIZ are presented: the design of a high-pressure hydraulic valve system optimized for stiffness and the design of a Counter-flow heat exchanger profiting from increased surface area and high surface roughness
[22]	To develop a design methodology that integrates axiomatic design and TRIZ with consideration of AM environment. Report a case study on a hydraulic pump to validate the proposed framework	Use of an integrated approach for design problems between axiomatic design and TRIZ (used to generate innovative solutions)	Definition of a DfAM framework that uses TRIZ and Axiomatic Design in the conceptual design phase
[23]	To develop a design approach for part redesign using a biomimicry inspired design methodology in combination with AM design rules to overcoming common thinking patterns. Present a case study of redesign of a star tracker's housing	TRIZ is used in the biomimicry-based design approach for problem solving, using the problem analysis and product functional decomposition in useful or harmful functions	AM technique metal-based L-PFB is selected as manufacturing process for the star tracker housing

(*continued*)

Table 1. (*continued*)

Ref.	Focus of the work	Implications with TRIZ	Implications with AM
[24]	To define a TRIZ-based systematic design approach that incorporates database-enhanced biomimetic part design specifically for AM. Presents a case study of design optimization of a gear wheel to validate the method	Use of TRIZ-based biomimetic part design and functional analysis for the optimization of a design problem	Use of an integrated CAD/CAE-tool to provide data for the AM-machines
[25]	To apply technology forecasting using the technical evolutionary theory, based on the TRIZ trends of evolution. Trends are graphically presented as an S-curve, showing how the main characteristic(s) of the system change over time. The application of fuzzy logic to positioning a technology on the S-curve is illustrated in the case study concerning technologies for making small diameter holes in superalloy components and compared to next generation AM technologies	Use TRIZ trends of evolution as a technology forecasting method	The next generation additive manufacturing technologies are compared to the traditional technologies for drilling holes and positioned on an S-curve to consider their grade of development and possible use
[26]	To develop an integrated methodology based on QFD and TRIZ for the design of a tracheal stent to palliate an airway obstruction. The result is then 3D modelled, and a parameterization tool is used to modify the stent's general dimensions to fit a specific patient. The final conceptual design was printed using AM Fab@Home technology	Use of the QFD and TRIZ contradiction matrix to solve conflicts and to find all possible solutions generating a large number of concepts. The combination of QFD and TRIZ presented herein was used to theoretically guide the development of a new stent design suitable	The Fab@Home technology is not capable of processing many types of materials; however, by following the TRIZ principles, work is being done to process biocompatible materials. It is also important to consider the packaging, sterilization and labelling at an early stage, as these are an integral part of the whole device
[27]	To define a method for technology forecasting and decision-making that guide technology development based on co-classification analysis and co-word analysis to extract implicit or unknown patterns and topics and employ main path analysis to discover significant clues about technology hotspots and development prospects. To illustrate this hybrid approach 3D printing, is evaluated referring to various technologies and processes used to synthesize a three-dimensional object	Semantic-TRIZ, tech mining can help obtain better understanding of developmental trends over a relatively short time series	The process to identify patent evolution pathways and some insights in evolution pathways for 3D printing are presented in "Case study: 3D printing"

4 Data Analysis and Discussion

The analysis of the content of Table 1 leads to the following considerations.

At the moment, the literature proposes a limited number of examples of the use of TRIZ and AM, even combined.

- In the 14 articles considered, there is a clear prevalence of the use of TRIZ as a tool for problem solving and generation of concepts and solutions, including creative ones, alongside the use of AM techniques to create the prototype of such a solution (six out of fourteen articles [15–19, 26]).
- In four documents [14, 20–22] the TRIZ is used to define guidelines or indications for the DfAM. In particular, two of these papers [20, 21] deal with redefining or customizing the TRIZ contradiction matrix according to the problems related to DfAM.
- In two cases [25, 27], there is the innovative use of the TRIZ as a technological forecasting tool to predict the trend of technologies and deduce the possible trend also in the case of AM techniques.

In at least two cases [23, 24], there is the use of TRIZ plus biomimetic design for optimized component design.

In order to have a better picture of this situation in Fig. 1, the 14 papers have been collected by dividing them according to the type of research application considered for the use of AM and TRIZ.

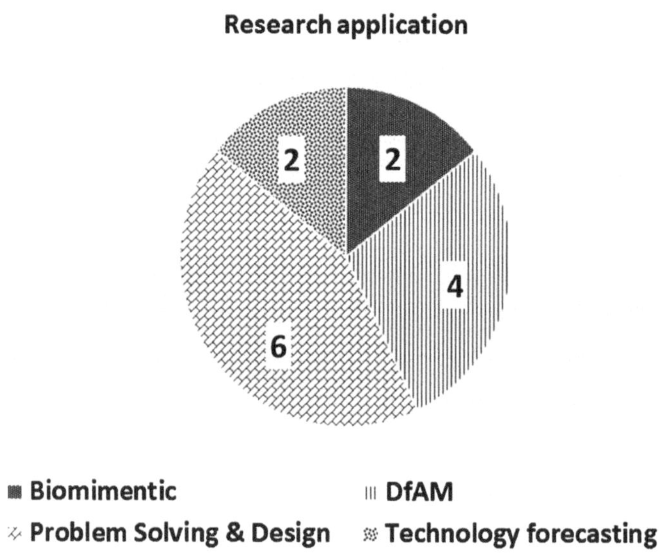

Fig. 1. Kind of research application of the 14 papers analyzed.

These considerations help in answering to the questions expressed at the beginning of Sect. 3. The answers are as follows.

- A1: Our research has shown the presence of 14 works out of 115 (equal to 12% of the starting head) that report and describe the use of the TRIZ and/or any of its tools in relation to issues related to the AM.
- A2: Mainly the applications described in these 14 works describe the use of the TRIZ for the resolution of problems related to the design of products to be made with AM technologies. These are mainly works that define the methods of design/product development that use the TRIZ also in combination with other methods such as Quality Function Deployment (QFD) and Axiomatic Design and especially for the generation of innovative concepts related to the optimization of the design of products of different types, from industrial products to particular applications of medical type. Interesting are also the cases of works that use the TRIZ for the definition of the DfAM guidelines, up to the customization of the matrix of contradictions for the solution of particular problems related to the AM.
- A3: Particularly interesting and innovative are the works concerning the use of TRIZ (in particular, the Trends of Evolution) as a method of technological forecasting to deduce and predict the trend and evolution of new technologies.

From these observations, it can be deduced that there is a link between the field of AM and the use of TRIZ methodologies especially related to the use of the latter within product development methodologies and in particular for the phases of analysis of design problems and for the development of innovative concepts. The combined use of the two methodologies is particularly noticeable in the work concerning the definition of guidelines for the DfAM.

5 Conclusions and Future Developments

From these observations, it can be deduced that there is a link between AM applications and the use of TRIZ methodologies moreover for the analysis and resolution of design issues. The work done so far has made it possible to respond quickly enough to the questions that we had set, to investigate whether there are links between the world of AM and systematic methods of innovation such as the TRIZ. In particular, the search and the analysis of the literature have been carried out, taking into account the scientific publications contained in the Scopus Database. This analysis is certainly not exhaustive and can be further refined and deepened using more structured investigation techniques, consulting further Databases. At the moment, however, the first results obtained have shown that there are several rather recent works that have addressed these issues and new aspects related to the use of these methodologies, such as the increasing attention to the methods of DfAM in support of designers or the use of TRIZ as a method of technological forecasting. It is planned to extend this search in the future in order to obtain a more complete picture of the situation. In particular, authors plan to extend the research also investigating the synergic use of other creative methods and to characterize better the type of use in relation to the phase of design and/or product development process.

References

1. Kagermann, H., Helbig, J., Hellinger, A., Wahlster, W.: Recommendations for implementing the strategic initiative INDUSTRIE 4.0: securing the future of German manufacturing industry. https://www.acatech.de/publikation/securing-the-future-of-german-manufacturing-industry-recommendations-for-implementing-the-strategic-initiative-industrie-4-0/. Accessed 20 June 2019
2. Rüßmann, M., Lorenz, M., Gerbert, P., Waldner, M., Justus, J., Engel, P., Harnisch, M.: Industry 4.0: the future of productivity and growth in manufacturing industries. https://www.bcg.com/it-it/publications/2015/engineered_products_project_business_industry_4_future_productivity_growth_manufacturing_industries.aspx. Accessed 20 June 2019
3. Motyl, B., Baronio, G., Uberti, S., Speranza, D., Filippi, S.: How will change the future engineers' skills in the Industry 4.0 framework? A questionnaire survey. Procedia Manuf. **11**, 1501–1509 (2017). https://doi.org/10.1016/j.promfg.2017.07.282
4. ISO/ASTM DIS 52900 - Additive manufacturing – General principles – Fundamentals and vocabulary. https://www.iso.org/standard/74514.html. Accessed 19 July 2019
5. ISO/ASTM 52910:2018 - Additive manufacturing – Design – Requirements, guidelines and recommendations. https://www.iso.org/standard/67289.html. Accessed 19 July 2019
6. Campbell, I., Bourell, D.L., Gibson, I.: Additive manufacturing: rapid prototyping comes of age. Rapid Prototyp. J. **18**(4), 255–258 (2012). https://doi.org/10.1108/13552541211231563
7. Kumke, M., Watschke, H., Vietor, T.: A new methodological framework for design for additive manufacturing. Virtual Phys. Prototyp. **11**(1), 3–19 (2016). https://doi.org/10.1080/17452759.2016.1139377
8. Rosen, D.W.: Research supporting principles for design for additive manufacturing. Virtual Phys. Prototyp. **9**(4), 225–232 (2014). https://doi.org/10.1080/17452759.2014.951530
9. Altshuller, G., Al'tov, G., Altov, H.: And Suddenly the Inventor Appeared: TRIZ, the Theory of Inventive Problem Solving. Technical Innovation Center, Inc., Worceste (1996)
10. Orloff, M.A.: Inventive Thinking Through TRIZ: A Practical Introduction. Springer, Heidelberg (2006)
11. Rantanen, K., Domb, E.: Simplified TRIZ: New Problem Solving Applications for Engineers and Manufacturing Professionals. St Lucie Press, Boca Raton (2002)
12. Motyl, B., Filippi, S.: Integration of creativity enhancement tools in medical device design process. Procedia Eng. **69**, 1316–1325 (2014). https://doi.org/10.1016/j.proeng.2014.03.124
13. Filippi, S., Motyl, B., Ciappina, F.M.: Classifying TRIZ methods to speed up their adoption and the ROI for SMEs. Procedia Eng. **9**, 172–182 (2011). https://doi.org/10.1016/j.proeng.2011.03.110
14. François, M., Segonds, F., Rivette, M., Turpault, S., Peyre, P.: Design for additive manufacturing (DfAM) methodologies: a proposal to foster the design of microwave waveguide components. Virtual Phys. Prototyp. **14**(2), 175–187 (2019). https://doi.org/10.1080/17452759.2018.1549901
15. Yuan, H., Xing, K., Hsu, H.-Y.: Generate basic conceptual solutions for 3DPVS via utilizing TRIZ. Bio-Des. Manuf. **2**(2), 76–95 (2019). https://doi.org/10.1007/s42242-019-00039-6
16. Hancock, D., Homfray, D., Porton, M., Todd, I., Wynne, B.: Exploring complex high heat flux geometries for fusion applications enabled by additive manufacturing. Fusion Eng. Des. **136**, 454–460 (2018). https://doi.org/10.1016/j.fusengdes.2018.02.097
17. Mawale, M.B., Kuthe, A., Mawale, A.M., Dahake, S.W.: Development of an ear cap in chronic suppurative otitis media using additive manufacturing and TRIZ. Proc. Inst. Mech. Eng. [H] **232**(7), 733–738 (2018). https://doi.org/10.1177/0954411918780126

18. Chang, T.-Y., Lu, H.-P., Luor, T.-Y., Wu, C.-Y.: Research and innovative design of a drawing instrument - a case study of dashed line pen. In: Proceedings of 4th IEEE International Conference on Applied System Innovation 2018, ICASI 2018, EAI, Bratislava, SK, pp. 712–715 (2018). https://doi.org/10.1109/icasi.2018.8394357
19. Frizziero, L., Francia, D., Donnici, G., Liverani, A., Caligiana, G.: Sustainable design of open molds with QFD and TRIZ combination. J. Ind. Prod. Eng. **35**(1), 21–31 (2018). https://doi.org/10.1080/21681015.2017.1385543
20. Gross, J., Park, K., Okudan Kremer, G.E.: Design for additive manufacturing inspired by TRIZ. In: ASME 2018 International Design Engineering Technical Conferences and Computers and Information in Engineering Conference. American Society of Mechanical Engineers, p. V004T05A004 (2018). https://doi.org/10.1115/detc2018-85761
21. Kretzschmar, N., Chekurov, S.: The applicability of the 40 TRIZ principles in design for additive manufacturing. In: Annals of DAAAM and Proceedings of the International DAAAM Symposium, vol. 29, no. 1, pp. 888–893 (2018). https://doi.org/10.2507/29th.daaam.proceedings.128
22. Renjith, S.C., Okudan Kremer, G.E., Park, K.: A design framework for additive manufacturing through the synergistic use of axiomatic design theory and TRIZ. In: IISE Annual Conference and Expo 2018, pp. 551–556 (2018)
23. Vogel, D., Geismayr, L., Langer, M., Leslabay, P., Schlick, G.: Combining additive manufacturing and biomimetics for the optimization of satellite structures. In: Proceedings of the 69th International Astronautical Congress, IAC 2018 (2018)
24. Kamps, T., Gralow, M., Schlick, G., Reinhart, G.: Systematic biomimetic part design for additive manufacturing. Procedia CIRP **65**, 259–266 (2017). https://doi.org/10.1016/j.procir.2017.04.054
25. Malinin, L.: Technology selection based on main parameters of value and fuzzy logic. Int. J. Bus. Innov. Res. **11**(3), 431–443 (2016). https://doi.org/10.1504/IJBIR.2016.078876
26. Melgoza, E.L., Serenó, L., Rosell, A., Ciurana, J.: An integrated parameterized tool for designing a customized tracheal stent. CAD Comput. Aided Des. **44**(12), 1173–1181 (2012). https://doi.org/10.1016/j.cad.2012.07.004
27. Huang, Y., Zhu, D., Qian, Y., Zhang, Y., Porter, A.L., Liu, Y., Guo, Y.: A hybrid method to trace technology evolution pathways: a case study of 3D printing. Scientometrics **111**(1), 185–204 (2017). https://doi.org/10.1007/s11192-017-2271-8

Optimizing the Nozzle Path in the 3D Printing Process

Manuel Iori and Stefano Novellani(✉)

DISMI, University of Modena and Reggio Emilia, Via Amendola 2, 42123 Reggio Emilia, Italy
{manule.iori,stefano.novellani}@unmimore.it

Abstract. In this paper, we define the 3D printing routing problem, the problem of finding the optimal path of the nozzle in a fused deposition modeling 3D printing system, so as to minimize the time required to create on object. We formally model the problem with an integer linear programming formulation and then solve it via heuristic algorithms. We test the algorithms on a set of large-size real-life instances, comparing them with one of the most widely used open source software for the problem. We show that large time reductions can be obtained. We finally propose a set of interesting directions for future research.

Keywords: Additive Manufacturing · 3D Printing · Rural Postman Problem · Heuristics · Optimization

1 Introduction

Three-dimensional Printing (3DP), also known as *Additive Manufacturing*, is a production technology that allows to construct three-dimensional objects with complex shapes by adding one or more materials layer by layer. Since its rise in the '80s, different techniques for 3DP have been developed, including *Stereolithography*, that makes use of liquid polymers hardened by UV beams layer by layer, *Selective Laser Sintering*, where a laser melts the powder, layer by layer, in a selective way, and *Fused Deposition Modeling* (FDM), also known as *Fused Filament Fabrication*. FDM involves extruding thermoplastic material (such as ABS plastic or polycarbonate) through heated nozzles: the filament is pulled by a feed roller, it is heated to a temperature above the material's melting point, and it is then extruded by the nozzle. Once a layer is completed, the platform where the material has been deposited is lowered (or the nozzle is raised) to the next layer level, and a new layer is added above the previous one. From now on, when mentioning 3DP we will refer to the FDM technique.

In this work, we focus on the optimization of the 3DP process, in particular we concentrate on the optimization of the path that the nozzle has to follow to produce the three-dimensional object. The objective is to minimize the printing time while respecting given printing constraints. The large amount of time

C. Rizzi et al. (Eds.): ADM 2019, LNME, pp. 912–924, 2020.
https://doi.org/10.1007/978-3-030-31154-4_78

needed to produce objects with 3DP is one of the main disadvantages of this technology, representing a limit to its use for mass production, so minimizing its printing time is a very relevant objective. Before introducing models and heuristic algorithms, we describe the 3DP process. We focus on the aspects that are related with the optimization pursued in this paper, and refer the reader interested in structural properties of the materials and technical details of the production process to, e.g., Campbell et al. [2] and Akella [1].

The manufacturing process of a product through 3DP follows the typical scheme depicted in Fig. 1. The process starts with the definition of the 3D model, which can be created by a computer-aided design (CAD) software or by directly scanning an object that we want to reproduce. The second step is to reproduce the three-dimensional object in a polygonal mesh, normally made of triangles. The typical file used to report a polygonal mesh is in a .stl format. When the model is represented as a mesh, the corresponding .stl file is passed to a slicing software that divides the mesh in layers. The following step is the most interesting to our work and it is to decide the path to be followed by the tool for reproducing the three-dimensional object. This is usually obtained by dedicated software/freeware by the use of quick heuristic algorithms (see, e.g., Cura [27]). The solution obtained is expressed as a sequence of commands in a g-code format, which is then passed to the printer in order to produce the final 3D object (see, e.g., Campbell et al. [2]).

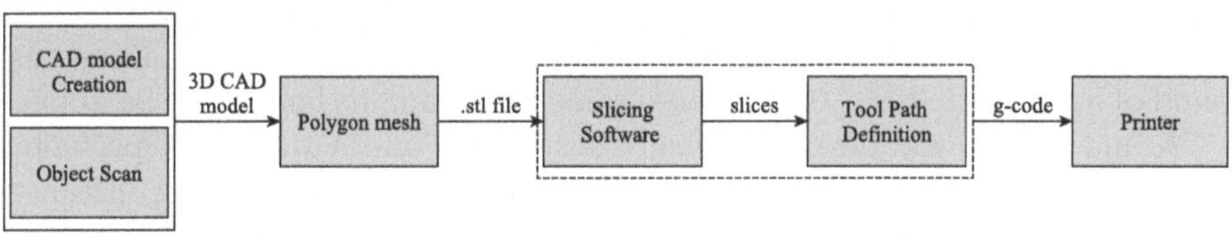

Fig. 1. Scheme of the 3DP process.

In the 3DP process, there are two relevant aspects that can be optimized. The first one is about the quality of the printed object and in particular its surface. The second one refers to the printing time that can be minimized, and it is strictly connected to the tool path. In this paper, we account for this second aspect by focusing in minimizing the distance traveled by the nozzle during the 3DP process, thus indirectly minimizing the time of the print.

In the following, we present a formal description of our 3DP optimization problem, which can be seen as a generalization of the well-known rural postman problem (see, e.g., Eiselt et al. [8]). We then propose a brief literature review and a mathematical model (Sects. 2 and 3). To solve the problem, we introduce some heuristic algorithms and produce a collection of instances derived from three-dimensional objects on which we perform computational tests (Sects. 4 and 5). Last but not least, we list a set of features and future research direction to be followed in the field that represents the intersection between optimization and 3DP.

2 Literature Review

The problem of defining the tool path in a 3DP process is a generalization of the *Rural Postman Problem* (RPP), a particular version of the more general class of *Arc Routing Problems* (see, e.g., Corberán and Laporte [4]). Given a multigraph $G = (V, A)$, where V is the set of vertices and A the set of arcs (that can be directed, indirected, or both), each associated with a non-negative cost, and given a subset $R \subseteq A$ of required arcs, the RPP is the problem of finding a minimum-cost closed path that traverses all arcs of R at least once. The RPP is NP-hard if the set of required arcs is not strongly connected (Laporte [19]).

The RRP arises in many real-world situations such as snow plowing, garbage collection, etc. The reader can find several contributions related to these applications in Eiselt et al. [8].

Many exact algorithms, mostly cutting planes and branch-and-cut algorithms, have been proposed in the last years for the RRP and its generalizations (see e.g. Laporte and Ghiani [14], Fernández et al. [9]). According to the survey by Corberán and Prins [6], the largest RPP instances that have been solved to optimality are characterized by up to 1000 vertices, 3080 edges and 204 required components and are solved in about one hour of computing time on a standard PC by using the branch-and-cut described in Corberán et al. [5] for the *windy general routing problem* (which contains the RPP).

Probably, the most famous heuristic algorithm for the RPP is the one by Frederikson [11] (FA), which is built on the well-known heuristic for the *Traveling Salesman Problem* (TSP) by Christofides [3], and, similarly, presents a worst case bound of 3/2. Hertz et al. [16] proposed some improving techniques to be applied to FA and to a newly developed heuristic. Groves and Vuuren [15] presented an effective local search framework for the undirected RPP solving instances with up to about 5000 vertices and 30000 edges. Ghiani et al. [13] proposed a constructive heuristic for the *undirected RPP* with a local post-optimization whose results can compete with FA. Holmberg [17] proposed some heuristics for the RPP built on FA by changing the order of the algorithmic components, and included two post-processing heuristics to improve the solution.

We report now a set of variants of the RPP that are interesting for 3DP. Dror and Langevin [7] introduced the *directed clustered RPP*, a generalization of the RPP in which each subset of arcs to be served has to be completed before serving another subset. These subsets do not have to be visited in a given order. The authors proposed an enumerative solution approach by transforming the directed clustered RPP into a *Generalized TSP*, that is a modification of the TSP in which nodes are partitioned into clusters and exactly one node from each cluster is visited in a cycle. The largest instance they solved is made of 581 vertices, 770 arcs, and 44 required arcs. Ghiani and Improta [12] considered the problem of drawing and decorating metal surfaces by means of a laser plotter that works as a 2D printer, minimizing the total length of the non-drawing moves. The so-called *Laser-plotter beam Routing Problem* (LRP) is modeled as an RPP with additional constraints that aim at minimizing the spot created on the surface every time a new line is drawn and at reducing the number of the shutter openings to reduce maintenance costs. They solved the LRP by transforming it into an RPP. By using the branch-and-cut presented in Ghiani and Laporte [14], instances with up to 225 vertices were solved. Orazi et al. [22] optimized the path of the scanning head in laser texturing of free form surfaces. They modeled the problem as a TSP and solved it by using the heuristic methods provided by the Concorde software.

The following recent works have considered optimization in the 3DP process. Kulkarmi et al. [18] gave an overview on the most important 3DP tasks: the orientation problem, the determination of supports, the slicing problem, and the path definition problem. They proposed possible future research directions. In our work we propose new future research directions specific for the path optimization problem. Panesar et al. [23] designed an optimization strategy for the definition of the internal configuration for structural purpose. Zhang et al. [26] and Wang et al. [24] studied the best printing direction to diminish the possible flaws and improve the surface quality given by the external support needed when printing an object. Leochowicz et al. [20] optimized the path of a 3D printer: they considered one layer and model it as a set of pixels, each of those that intersect the printing area must be visited. They propose a closest neighbor type of greedy, a second algorithm where 2-opt exchange is applied to the closest neighbor solution, and a simulated annealing algorithm to accept non-improving solutions. To fasten the algorithms they decomposed the considered layer in a set of sublayers. Wojcik et al. [25] also considered one layer divided in a set of pixels, they proposed four greedy algorithms to visit the pixels that intersect the printing area: in a zig-zag way, by following the contour of the figure, in a spiral way, or with a fractal strategy. Both of these papers present computing times for one layer that are comparable to ours for the entire set of layers. Fok et al. [10] also optimized the 3D Printing path: they show the difference between the closest neighbor algorithm used by the famous software Cura [27] and an algorithm based on the Christofides one [3]. Our contribution with respect to these works is that we formally define for the first time the arc routing problem linked to the 3D printing path optimization, we consider all the layers, and we use the edges to be printed and not the pixels (that can represent a simplification of the object to print). Moreover, among the proposed algorithms, we also compare the closest neighbor algorithm to a clustered one and we show that we obtained good results with very fast computing times even if considering all the layers at once. We propose a set of new heuristic algorithms that could improve largely the solutions provided by Cura.

3 The 3D Printing Routing Problem

We call *3D printing Routing Problem* (3DRP) the generalization of the RPP that is related to the tool path definition in a 3DP process. The 3DRP represents the *basic problem* of optimizing the tool path of a 3DP process, but some additional characteristics might arise across the 3DP process. We provide a brief description of those features in Sect. 6.

In the 3DP tool path, the nozzle must start from its starting position (vertex 0) and return to the same position when the print is terminated. The graph is divided into clusters to be visited in a sequence, each representing a layer to be printed. On each layer, the set of *compulsory edges* represents the set of edges on which the nozzle is required to depose material, while the other edges are called *optional edges*. No layer can be started before terminating the previous one, and the nozzle cannot move back to previous layers. Then, being that only optional edges are present between two layers, one and only one optional edge can be used to move from one layer to the next one. Moreover, no edge between two non-subsequent layers exists.

Formally, the problem is defined on an undirected graph $G = (V, E)$, where V is the set of vertices and E is the set of edges. Moreover, it is defined a subset $R \subseteq E$ of edges that must be visited at least once. In our case, R contains the compulsory edges, where the material is extruded (printed) the first time they are used, whereas $E \setminus R$ contains the set of optional edges. Let us define *layer* as a set of vertices characterized by the same altitude. We call $\mathcal{L}$ the set of all layers, where a layer $L_h \in \mathcal{L}$ is one layer of print, so $\mathcal{L} = \cup_{h=1}^{|\mathcal{L}|} L_h$, and $L_h \cap L_k = \emptyset, h \neq k, L_h, L_k \in \mathcal{L}$. The graph G is clustered in several subgraphs, each induced by a layer plus the starting and ending points of print, vertex 0, and all the edges connecting these components. Thus, $V = \mathcal{L} \cup \{0\}$.

Let variable x_{ij} give the number of times edge $(i, j) \in E \setminus R$ is traveled, and the number of times edge $(i, j) \in R$ is traveled in addition to the first time. Let c_{ij} be the shortest distance between i and j, $(i, j) \in E$. The 3DRP aims at minimizing the path of the nozzle starting from vertex 0, traveling all the compulsory edges at least once, respecting the order of the layers, and returning to the vertex 0. Let us define $\Omega = \{S \subseteq L_h, \bar{S} = L_{h+1} \cup L_h \setminus S, \text{for } h = 1, \ldots, |\mathcal{L}| - 1\} \cup \{S \subseteq L_{|\mathcal{L}|}, \bar{S} = \{0\} \cup L_{|\mathcal{L}|} \setminus S\}$. Then, for each $S \in \Omega$, let us also define $R^1(S) = \{(i, j) \in R : i \in S, j \in \bar{S}, j > i\}$, $R^2(S) = \{(i, j) \in R : i \in S, j \in \bar{S}, j < i\}$, $E^1(S) = \{(i, j) \in E \setminus R : i \in S, j \in \bar{S}, j > i\}$, $E^2(S) = \{(i, j) \in E \setminus R : i \in S, j \in \bar{S}, j < i\}$. The basic 3DRP can be modeled as

$$\min \sum_{(i,j)\in R} c_{ij}(1 + x_{ij}) + \sum_{(i,j)\in E\setminus R} c_{ij}x_{ij} \tag{1}$$

$$\sum_{j\in R:j>i} (1 + x_{ij}) + \sum_{j\in R:j<i} (1 + x_{ji}) + \sum_{j\in V:j>i} x_{ij} + \sum_{j\in V:j<i} x_{ji} = 2z_i, \quad i \in V \tag{2}$$

$$\sum_{(i,j)\in R^1(S)} (1 + x_{ij}) + \sum_{(i,j)\in R^2(S)} (1 + x_{ij}) + \sum_{(i,j)\in E^1(S)} x_{ij} + \sum_{(i,j)\in E^2(S)} x_{ij} \geq 1 \quad S \subseteq \Omega \tag{3}$$

$$\sum_{j\in L_1} x_{0j} = \sum_{j\in L_{|\mathcal{L}|}} x_{j0} = 1 \tag{4}$$

$$\sum_{(i,j):i\in L_h, j\in L_{h+1}} x_{ij} = 1 \quad h = 1, ..., |\mathcal{L}| - 1 \tag{5}$$

$$x_{ij} \in \mathbb{N}_0 \quad (i, j) \in E \tag{6}$$

$$z_i \in \mathbb{N}_0 \quad i \in V \tag{7}$$

The objective function (1) aims at minimizing the total costs, i.e., distances. Constraints (2) state that the number of edges incident to a vertex are even or zero. In (3), we impose the connectivity of the solution. Constraints (4) state that one and only one edge must leave and return to the vertex 0. Constraint (5) imposes that one and only one edge can be used between one layer and the next one. The variables are defined as integer and non-negative in (6) and (7).

4 Heuristic Algorithms

The 3DP process typically involves from thousands to millions of edges to be printed and traveled by the nozzle. Thus, real 3DRP instances are impossible to solve exactly, because of the huge number of variables and constraints needed. For this reason, we developed four heuristic algorithms.

Closest 3DP. The first algorithm we present to solve the 3DRP starts from vertex 0 and moves from one edge to the closest unused compulsory edge (we enter in the edge by the closest vertex with respect to the current one and then we travel the edge to arrive in its second vertex, which will be the new current vertex). If the second vertex of the current edge is also the first vertex of the next edge, we link them directly, otherwise we insert a non-printing edge. When the compulsory edges of the current layer have been traversed at least once, we move to the next layer by using the cheapest edge. We repeat this until the last layer is completed, and then connect the last visited vertex to vertex 0 to end the algorithm.

Clustered 3DP. Let us first clarify that each layer is divided into different subpolygons and each subpolygon can be formed by at most three clusters of compulsory edges: the outer part (the border of the subpolygon), the inner part (that follows the outer part internally, to strengthen the structure), and the filling part (that represents the edges used to fill the subpolygon), see Fig. 2. The second algorithm we present is called *Clustered 3DP* algorithm, and follows a similar idea with respect to the Closest 3DP algorithm, while making use of the different clusters just presented. It selects a cluster on a layer (choosing among outer, inner, and filling) and travels all the compulsory edges of the cluster, before going to another cluster. It starts from vertex 0 and selects the closest vertex of the compulsory edges on layer L_1. This action not only defines the next edge, but also the first cluster to be completed before moving to the next one.

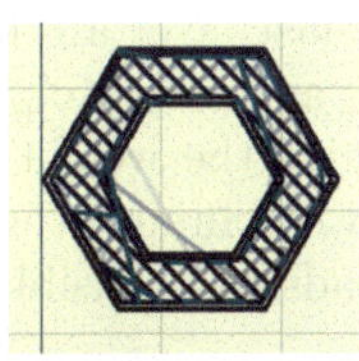

Fig. 2. An example of polygon with outer, inner, and filling parts.

When the current cluster is completed, the algorithm moves to the closest vertex, which defines the next edge and the next cluster. The algorithm continues until all clusters of the current layer are completed and then the closest vertex of the next layer is chosen. When the last layer is completed we travel back to vertex 0 and the algorithm terminates.

Look Ahead 3DP. The name of the third algorithm that we propose is inspired by the fact that it makes use of heuristic information on the future possible decisions. It starts from vertex 0 and then selects the three closest vertices among those of the compulsory edges of layer L_1. A particular cost is then computed by running the Closest 3DP algorithm from each of the three vertices till the end of the layer. The cost is computed as follows: we sum up the traveling cost of the non-printing edges needed to terminate the layer by the Closest 3DP, the vertex of the three selected ones that furnishes with the cheapest cost is used. This process is iterated until all compulsory edges of the current layer are used at least once. We then use the cheapest edge to move to the next layer and repeat the procedure until the last layer. To conclude the algorithm, we move back to the vertex 0.

Shortest Path Based 3DP. The forth proposed algorithm is called *Shortest Path Based 3DP* and is divided into two phases, the first one builds a supporting graph to be used in the second phase. The second phase computes the Dijkstra algorithm on the newly built supporting graph to obtain a solution for the 3DRP. In the first phase we define the supporting graph $G' = (V', E')$, where $V' \subseteq V$ is the set of vertices and E' the set of edges that we describe in the following. For each layer we select a random set of vertices among those of the compulsory edges (R), called *starting vertices*, from which we can start the layer. The union between the starting vertices, vertex 0, and its duplication, vertex $0'$, defines V'. The set E' is defined by: the edges going from each starting vertex of one layer to each starting vertex of the next layer, the edges going from vertex 0 to the starting point of the first layer L_1, and the edges going from the starting points of the last layer $L_{|\mathcal{L}|}$ to vertex $0'$. Starting from each starting vertex we terminate the layer as if using the Closest 3DP algorithm ending up in one and only one vertex that we call *ending vertex*. We call c'_{ij} the cost of each edge of E'. The costs c'_{ij} is computed as the summation of the distance between a starting vertex and the corresponding ending vertex, whose cost is computed by the Closest 3DP algorithm, and the distance between the corresponding ending vertices and the starting vertices of the next layer. The starting vertices of the first layer are linked to the vertex 0 and a cost $c'_{0j} = c_{0j}, j \in L_1$ is associated to them. The ending vertices of the last layer are linked to the vertex 0', that has the same coordinates of vertex 0, and a cost $c'_{j0'} = c_{j0}, j \in L_{|\mathcal{L}|}$ is associated to them. The costs $c'_{ij}, (i,j) \in V'$ give the weight of the edges of the supporting graph G'. In the second phase, we compute the shortest path form 0 to $0'$ on the supporting graph G'. This leads to a feasible solution for the 3DRP.

5 Computational Results

We coded our algorithms in C++ and tested them on an Intel Core i3-2100 with 3.10 GHZ. We collected 30 instances on the Internet from the 3DP design community thingiverse.com. The instances have been processed by our algorithms and with one of the most famous software used for slicing polygonal meshes and for path definition, Cura [27]. We used the instances as input for Cura that takes the polygonal meshes of the instances, slice them, and defines a path of the nozzle in a code called g-code. The g-code represents a heuristic solution for the related 3DRP in a language that the 3D printer can understand. By processing the obtained g-code we can obtain the set of layers, the set of compulsory edges to be printed, the corresponding vertices and their coordinates. These pieces of information define the data that we used as instances for our algorithms. In Table 1, we report the name of the instance, the number $|\mathcal{L}|$ of layers, the number $|R|$ of compulsory edges, and the total distance given by Cura.

Let Γ be the cost of the non-printing edges used in a 3DRP solution. Note that the compulsory edges are the same for all the solutions, and thus we report only the non-printing cost. Let Γ_{Cura} be the value of Γ proposed by Cura, in mm, and Γ_A the value Γ defined by the solution obtained thanks to one of our algorithms. Let $\%impr$ be the percentage improvement between them, computed as $\%impr = 100 * (\Gamma_A - \Gamma_{Cura})/\Gamma_{Cura}$. In Table 1, we report the results of the proposed Closest 3DP, Clustered 3DP, and Look Ahead 3DP algorithms showing their percentage improvement and their computing time in seconds. Extended results can be found in the PhD thesis by Novellani, summarized in [21].

From Table 1, one can observe that the Closest 3DP can improve the solution value for 28 out of 30 instances by 22.91% on average. The average solving time is very promising being lower than one second. With Clustered 3DP algorithm we can improve the solution value with respect to Cura for 27 out of 30 instances, with an average improvement of 12.15% in a time that is about half a second on average. The Look Ahead (LA) 3DP algorithm results show that the solution value is improved for every instance with respect to Cura, by 36.73% on average. To reach these results, the LA needs to run for about 39 min on average, which is a relevant solving time if compared with the previously proposed algorithms. These very good solutions pay their quality in terms of higher solving times.

In Table 2, the average results for the Shortest Path Based (SPB) 3DP algorithms are shown, averaged on all the instances. In the left part, we report the results for the SPB, where the number of starting vertices (SV) for each layer is fixed to 10, 20, 30, 50, 60, and 100; in the right part, we present the results for the SPB where the SV are fixed with respect to a percentage of the vertices of each layer (1%, 2%, 5%, 10%, 15%, 20%, and 25%). One can see that for both SPB types the average improvement increases with the number of starting vertices explored. The improvement is proportional also to the computing times. The SPB with fixed numbers of starting vertices can reach a 27.38% gap on average in the best case, that is represented by the starting vertices set to 100. This value is obtained in about 80 s. One can see that this version works similarly to the version with percentage starting vertices when evaluating the average improvement, but needs consistently more seconds to terminate. The algorithm that shows the best results in terms of improvement of the solution value is LA. However, LA needs higher computing times with respect to the other algorithms. The SPB furnishes, in its two versions, good results in acceptable solving times.

Table 1. Computational results for Closest, Clustered, and Look Ahead

Instance			Cura	Closest		Clustered		Look Ahead	
Name	$\vert\mathcal{L}\vert$	$\vert R\vert$	Γ	$\%impr$	$t(sec)$	$\%impr$	$t(sec)$	$\%impr$	$t(sec)$
30grams	219	8693	11687.56	5.66	0.00	2.84	0.00	17.38	0.31
angel	153	353671	147351.1	10.12	6.67	8.51	3.56	29.02	22664.21
block_4×4	37	28694	15263.46	40.68	0.14	19.86	0.09	48.68	122.82
brickb	332	107355	132462.2	−1.00	0.59	−1.03	0.47	12.67	765.48
bunny	41	21333	7569.886	16.88	0.08	9.68	0.05	31.25	45.67
chamfer	20	12700	11813.21	3.17	0.09	−3.12	0.06	15.39	77.05
Cityscape	342	488409	293773.5	9.95	7.44	5.02	3.64	30.59	17105.26
gear_b	85	19275	6167.369	3.86	0.03	7.19	0.03	14.26	10.42
geoboard	82	52244	51043.13	3.93	0.28	2.89	0.24	15.14	303.95
hilbert2	99	26703	14938.52	11.71	0.03	11.71	0.08	21.50	13.39
Hilton_Chicago	73	7179	6081.438	18.09	0.00	14.28	0.00	30.21	1.06
ingranaggio	39	52569	16516.57	22.22	0.42	4.34	0.33	36.70	758.60
miwin1	132	130267	95751.26	13.26	1.11	0.90	0.58	32.22	1645.93
orso_bis	52	35772	41535.23	60.24	0.16	38.17	0.11	73.46	131.84
pesce_bis	52	65262	77239.71	64.16	0.53	31.46	0.36	70.89	1012.54
polysoup	3	3018	2042.693	71.14	0.02	43.72	0.02	78.06	21.67
portab	32	47293	34371.95	13.53	0.44	12.01	0.45	32.08	1208.38
portaingr	51	69664	17318.14	10.54	0.92	7.57	0.49	29.48	3032.91
RFin	188	35898	9401.675	10.32	0.05	0.42	0.03	30.41	17.16
RostockBottom	44	18120	27514.87	31.54	0.13	13.92	0.14	46.74	128.41
RostockTop	50	21132	22071.05	23.36	0.17	10.80	0.13	36.72	369.54
Rpenta	369	433954	203950.3	28.52	3.13	18.71	2.05	46.30	5144.17
RShowerHead	274	146602	140072.8	24.85	0.52	17.41	0.45	40.69	472.31
tardisflat2	170	24695	23625.85	39.63	0.03	9.14	0.03	47.73	6.00
tetra	160	20839	13115.34	−4.81	0.03	−0.89	0.03	16.62	6.94
trail	125	268372	244951.1	48.29	4.19	45.74	2.77	59.47	14259.08
wbuilding	100	3364	2643.942	2.86	0.00	1.21	0.02	16.70	0.08
wine_fixed	39	20276	29576.97	38.03	0.09	12.84	0.06	48.10	62.18
wine1	39	19145	29808.14	40.19	0.08	1.59	0.06	51.12	46.33
WitchCastle	112	56334	12736.63	26.33	0.22	17.49	0.17	42.37	187.03
Average				22.91	0.92	12.15	0.55	36.73	2320.69

Table 2. Results for the Shortest Path Based 3DP algorithm.

#SV	%impr	t(sec)	%SV	%impr	t(sec)
10	23.50	8.59	1%	18.54	24.96
20	25.20	16.32	2%	21.10	48.74
30	26.04	24.16	5%	24.02	118.13
40	26.43	31.85	10%	25.70	181.53
50	26.63	39.83	15%	26.29	200.25
60	26.89	47.51	20%	26.59	208.47
100	27.38	80.94	25%	27.00	212.81

6 Conclusions and Future Research Directions

We studied a generalization of the RPP related to the definition of the tool path in a 3DP process, called 3DRP. We provided a collection of 3DRP instances and solved them with four heuristic algorithms. We compared them with one of the most famous software for 3DP path definition, Cura [27]. We obtained very good results, consistently improving the Cura solution values within limited computing times. The Closest and the Clustered algorithms can improve the solutions in less than a second for most of the instances; the Look Ahead algorithm can improve the solutions by a 36% but requires long times; on the other hand we believe that the Shortest Path Based algorithms provide the best trade off between improvements and computing times. Following the established line of research by Lechowicz et al. [20], Wojick et al. [25], and Fok et al. [10], the obtained data derive from a simulation of the real print. Only a small subset of instances has been printed by using the Cura algorithm and our best performing ones and no visual evidence of non-conformity with the original can be reported; however, in the future we intend to test the algorithms by performing real prints of all the objects.

Other versions of the 3DRP with other secondary features can be considered. In the following we report an overview of these features.

Quality Constraints. The 3DP process can produce some blemish on the external surface of the printed product. Indeed, when traveling the optional edges after ejecting material, the nozzle can eject some leftovers and ruin the quality of the external surface. Additional constraints could be considered to avoid it. *Subpolygons.* Several separated polygons, called *subpolygons*, can be present in the same layer. To diminish blemish on the external surface, we can impose a new constraint: to complete a polygon before stepping to the next one, by taking one and only one optional arc between a subpolygon to another on the same layer. *Corner.* To preserve the quality of the surface from the blemishes, we can consider not only to leave the polygon just once, but also to impose the nozzle to pass only through a vertex of the outer part of the same subpolygon in several layers, called *corner*. This will make it easier to remove the blemishes after the print. In some other cases the accumulation of material has to be avoided: this can be done by impose that the first vertex of a layer (or of a subpolygon) must have different coordinates if compared to the first vertex in the previous layer. *Temperature.* The 3DP process includes the melting of the material before extruding it from the nozzle. After being melt, the material conserves a high temperature for a certain amount of time. Thus material cannot be extruded on the top of the previous layer until its temperature is not under a certain value, otherwise flaws or bending can happen. We can impose this by setting that an edge cannot be completed before an underlying edge had the time to cool down.

Costs. We modeled the cost of an edge as its euclidean distance and the time to travel an edge as linear to the distance. We could, alternatively, model them with an acceleration, a constant speed, and a deceleration. Moreover, the deceleration (acceleration) could depend on the angle between the current edge and the following (previous) one visited. This can lead to different modeling options, considering non-linear distances between the vertices or considering the speed of the nozzle as a variable.

Other Features. *Layers with different thickness.* Sometimes, to deal with structural problems we can print some parts of the same layer with a different thickness. *Supporting material.* Some 3D objects can have undercuts, thus a supporting material is needed when printing. In order to minimize the supporting material, the positioning of the object can be a variable to account for. *Filling.* In this work we consider the filling part of a certain polygon as an input, but to decide the filling, in terms of distances between the edges of the filling, their direction, and even a particular pattern can change the optimization of the print or be part of it. *Colors.* Some printing objects can have different set of materials and/or colors to be used. Indeed we can see the supporting material part of this case. When printing an object made of different colors or materials we can decide to print first the polygons of one material and then the others by deciding a sequence. Otherwise, we can decide to mix the use of colors to make the printing time shorter. Then the number of nozzles becomes relevant: if we have more materials than nozzles we must consider the time spent in changing material and going back to the vertex 0 to clean the nozzle, otherwise we must consider the distance between the nozzle of one color and the others.

All these features have a strong impact on the speed of the print and on the quality of the surface, both very important characteristics in 3DP. We believe that they will be the focus of many future research works in optimization.

References

1. Akella, S.: Hot off the press: the technology behind 3D printing. Professor Scarlatos HON 301 (2012)
2. Campbell, T., Williams, C., Ivanova, O., Garrett, B.: Could 3D printing change the world? Technologies, Potential, and Implications of Additive Manufacturing. Atlantic Council, Washington, DC (2011)

3. Christofides, N.: Worst-case analysis of a new heuristic for the travelling salesman problem. Technical report, DTIC Document (1976)
4. Corberán, Á., Laporte, G.: Arc Routing: Problems, Methods, and Applications. SIAM, Philadelphia (2014)
5. Corberán, Á., Plana, I., Sanchis, J.M.: A branch & cut algorithm for the windy general routing problem and special cases. Networks **49**(4), 245–257 (2007)
6. Corberán, Á., Prins, C.: Recent results on arc routing problems: an annotated bibliography. Networks **56**(1), 50–69 (2010)
7. Dror, M., Langevin, A.: A generalized traveling salesman problem approach to the directed clustered rural postman problem. Transp. Sci. **31**(2), 187–192 (1997)
8. Eiselt, H.A., Gendreau, M., Laporte, G.: Arc routing problems, Part II: the rural postman problem. Oper. Res. **43**(3), 399–414 (1995)
9. Fernández, E., Meza, O., Garfinkel, R., Ortega, M.: On the undirected rural postman problem: tight bounds based on a new formulation. Oper. Res. **51**(2), 281–291 (2003)
10. Fok, K.Y., Ganganath, N., Cheng, C.T., Tse, C.K.: A 3D printing path optimizer based on Christofides algorithm. In: IEEE International Conference on Consumer Electronics-Taiwan (ICCE-TW), pp. 1–2 (2016)
11. Frederickson, G.N.: Approximation algorithms for some postman problems. J. ACM (JACM) **26**(3), 538–554 (1979)
12. Ghiani, G., Improta, G.: The laser-plotter beam routing problem. J. Oper. Res. Soc. **52**(8), 945–951 (2001)
13. Ghiani, G., Laganà, D., Musmanno, R.: A constructive heuristic for the undirected rural postman problem. Comput. Oper. Res. **33**(12), 3450–3457 (2006)
14. Ghiani, G., Laporte, G.: A branch-and-cut algorithm for the undirected rural postman problem. Math. Program. **87**(3), 467–481 (2000)
15. Groves, G.W., Van Vuuren, J.H.: Efficient heuristics for the rural postman problem. ORiON J. ORSSA **21**(1), 33–51 (2005)
16. Hertz, A., Laporte, G., Hugo, P.N.: Improvement procedures for the undirected rural postman problem. INFORMS J. Comput. **11**(1), 53–62 (1999)
17. Holmberg, K.: Heuristics for the rural postman problem. Comput. Oper. Res. **37**(5), 981–990 (2010)
18. Kulkarni, P., Marsan, A., Debasish, D.: A review of process planning techniques in layered manufacturing. Rapid Prototyping J. **6**(1), 18–35 (2000)
19. Laporte, G.: Modeling and solving several classes of arc routing problems as traveling salesman problems. Comput. Oper. Res. **24**(11), 1057–1061 (1997)
20. Lechowicz, P., Koszalka, L., Pozniak-Koszalka, I., Kasprzak, A.: Path optimization in 3D printer: algorithms and experimentation system. In: 4th International Symposium on Computational and Business Intelligence (ISCBI), pp. 137–142 (2016)
21. Novellani, S.: Models and algorithm for the optimization of real-world routing and logistics problems. 4OR **14**(3), 331–332 (2016)
22. Orazi, L., Montanari, F., Campana, G., Tomesani, L., Cuccolini, G.: CNC paths optimization in laser texturing of free form surfaces. In: Procedia 9th CIRP Conference on Intelligent Computation in Manufacturing Engineering, vol. 33, pp. 440–445 (2015)
23. Panesar, A., Brackett, D., Ashcroft, I., Wildman, R., Hague, R.: Design optimization strategy for multifunctional 3D printing. In: 25th International Solid Freeform Fabrication Symposium, pp. 592–605 (2014)
24. Wang, W.M., Zanni, C., Kobbelt, L.: Improved surface quality in 3D printing by optimizing the printing direction. Comput. Graph. Forum **35**(2), 59–70 (2016)

25. Wojcik, M., Pozniak-Koszalka, I., Koszalka, L., Kasprzak, A.: Experimentation system for path planning applied to 3D printing. In: Advances in Applied Digital Human Modeling and Simulation, pp. 291–30 (2017)
26. Zhang, X., Le, X., Panotopoulou, A., Whiting, E., Wang, C.C.: Perceptual models of preference in 3D printing direction. ACM Trans. Graph. (TOG) **34**(6), 215 (2015)
27. Ultimaker: Cura software (2015)

A Build-Time Estimator for Additive Manufactured Objects

Luca Di Angelo(✉), Paolo Di Stefano(✉), and Emanuele Guardiani(✉)

Department of Industrial and Information Engineering and Economics, University of L'Aquila, L'Aquila, Italy
{luca.diangelo,paolo.distefano}@univaq.it, emanuele.guardiani@graduate.univaq.it

Abstract. Additive Manufacturing is a very time consuming technology. An estimation of the build time is fundamental to:

- Evaluate the production cost in budgeting process.
- Make use of optimization methods, which use as parameter the build time, for determining optimal build direction.

In both these cases a fast and valid build time estimator, which can work with a few input data deducible from geometric model, is required.

In the proposed paper a reliable parametric-based method to determine the build time for additive manufactured objects is provided. The implemented method is based on a back-propagation artificial neural network, which gives the possibility to implement the complex functions that elapse some driving build-time factors and the build time. The neural network training is based on data provided by a properly developed analyzer of the list of commands given to AM machines, which performs an analytical estimation of the build time. The implementation of the proposed methodology is illustrated and some comparisons between the real and estimated build-time are provided, then the results are critically analyzed.

Keywords: Build time estimation · Adaptive model · Process planning · Additive Manufacturing

1 Introduction

Additive Manufacturing (AM) is a technology for fabricating functional prototypes and small batches of part. It allows to realize components characterized from a complex shape and provides important advantages related to the cost saving, compared with other manufacturing processes, and the reduction of the Time to Market [1, 2]. But AM is also a very time consuming technology. Due to the limitations of the technologies currently in use the speeds of the growing process need to be limited for maintaining a good quality of the manufactured part. Consequently AM is generally applied only to small batches of parts. In this sense is very important to provide an estimation of the build time before the part is realized, in order to evaluate the production costs.

C. Rizzi et al. (Eds.): ADM 2019, LNME, pp. 925–935, 2020.
https://doi.org/10.1007/978-3-030-31154-4_79

This information is essential for the sellers during the budgeting process for providing the part cost to the customer and also for planning the production. Indeed many optimization methods for AM have been suggested during the last years [3–12], whose intents are to increase the profits and also the quality of the manufactured objects. In these methods, whose final output is the optimal build direction of the interested part, the build time is generally required as input. Consequently, a reliable way for evaluating this parameter is necessary.

Two possibilities are offered by the literature to estimate the build time: by parametric-based methods [13–21] or by detail analysis-based methods, such as [24]. In parametric-based methods the build time is expressed as a function of the driving build-time factors. These last can be evaluated starting from the geometrical model and before defining the manufacturing process, which can be very useful for the budgeting process. The parametric-based methods are fast and easy-to-use but, on the other side, their accuracy is poor. In detail-analysis based method instead the estimation of the build time is very accurate but they commonly require a full geometrical information of the interested part, which could not be available in some situations.

In this paper the use of a parametric-based method is illustrated. Most of the driving factors of the build time are identified and analyzed. Then these factors are used as input of an artificial neural network, which allows to identify the correlation between the driving build-time factors and the build time. A similar work was proposed by the Authors in a previous activity research, but it evidenced some limitations related to the calculation of some driving factors due to new typologies of hatching that have been introduced during this recent years (Fig. 1). Moreover network learning was very slow since, not being available an accurate tool for estimating the build time, the real value of this last was required. Both limitations have been passed thanks to the use of the new methodologies which will be afterwards described.

2 Proposed Method

In parametric-based methods the build time t_f is expressed as

$$t_f = t_f(T, DTF_1, DTF_2, \ldots, DTF_n) \qquad (1)$$

The parameter T is representative of the technology in use, while DTF_i identifies the *i-th* driving build-time factor. A very difficult step is represented by the individuation of these factors. In order to avoid cross-correlation the driving build-time factors should be independent each other. Moreover, to make the method general-purpose, they should be independent from the specific technology used for manufacturing the part.

In the proposed method the use of the following five driving build-time factors is suggested:

$\frac{V_{mat}}{L}$ Volume of part/Layer thickness
p_{mat} Total length of part's perimeter
n_{r-mat} Number of repositioning required for part's hatching

$\frac{V_{sup}}{L}$ Volume of supports/Layer thickness
n_{r-sup} Number of repositioning required for supports hatching

In Table 1 each factor is compared with the most widespread AM technologies to establish which parameters should be used for a specified process.

The chosen factors can be evaluated relatively easily starting from the geometrical model of the part. Moreover they are quite independent between them, as it was demonstrated in [20].

The driving build-time factors are evaluated by a method implemented into a Matlab script, which process a.stl representation of the object.

The volume of the part V_{mat} is computed using the methodology proposed by [22].

Table 1. Correlation between the driving build-time factors and typically commercial rapid prototyping technologies.

Driving build-time factors	Technologies							
	FDM	*SLA*	*SLS*	*LOM*	*3DP*	*MJM*	*EBM*	*SLM*
MATERIAL								
$\frac{V_{mat}}{L}$	✓	✓	✓	✓	✓	✓	✓	✓
p_{mat}	✓	✓	✓	✓			✓	✓
n_{r-mat}	✓	✓	✓		✓	✓	✓	✓
SUPPORT								
$\frac{V_{sup}}{L}$	✓	✓		✓	✓	✓		
n_{r-sup}	✓	✓		✓	✓	✓		

Some technologies, as evidenced in Table 1, require a deposition of the layer contour. A very accurate estimation of the total layers' contour is provided by the formula (2) suggested by [23], in which the perimeter is calculated as the "projection" of the triangular facets of the geometrical model divided by the layer distance.

$$p_{mat} = \frac{\sum_{j=1}^{n_T} S_j \sqrt{1 - (z \cdot \boldsymbol{n}_j)^2}}{L} \tag{2}$$

Where:

n_T Number of triangular facets of geometrical model
S_j Area of *j-th* triangular facet
z Normal vector of building direction
$\boldsymbol{n}_j$ Normal vector of *j-th* triangular facet

Another parameter which significantly affects the build time is related to the number of repositioning of the deposition tool. During each repositioning an acceleration and a jerk phase are required, causing an increasing of the build time [24]. The number of repositioning is strictly related with the typology of hatching used. In Fig. 1 some of the most common hatching structures are reported.

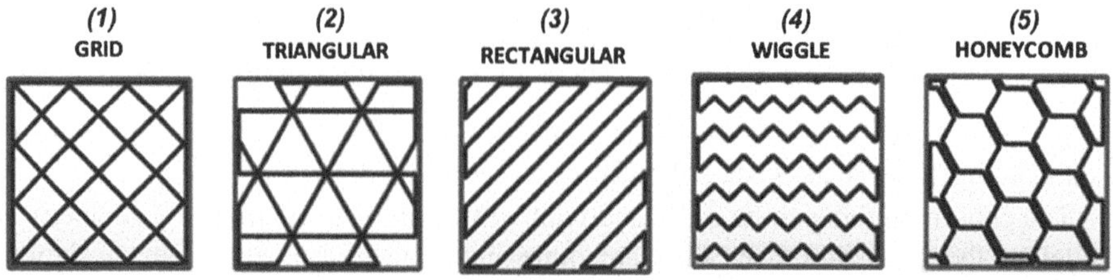

Fig. 1. Most in use typologies of hatching for AM.

For the structures 1, 2 and 3, in order to evaluate the number of repositioning n_{r-mat}, an extension of the methodology already in use for the rectangular hatching is proposed [21].

$$n_{r-mat} = \sum_{k=1}^{2} \left(\frac{\sum_{j=1}^{n_T} S_j \sqrt{1 - (z \cdot \boldsymbol{n}_j)^2}}{H_{mat} L} \left| \tau_{k,c} \cdot \boldsymbol{n}_{j,xy} \right| \right)$$

Where:

H_{mat}	Hatching distance
$\tau_{k,c}$	Normal vector referred to hatching direction
$\boldsymbol{n}_{j,xy}$	Projection onto the stratification plane of the normal vector of *j-th* triangular facet

The phase $\boldsymbol{\theta}$ of the vector $\boldsymbol{\tau}_{k,c}$ assumes different values depending on hatching typology, in accordance with Table 2.

Table 2. Relationship between k, phase θ and hatching typology.

Hatching typology	*k = 1*	*k = 2*
Grid (1)	$\boldsymbol{\theta}(\tau_{1,c}) = \tau$	$\boldsymbol{\theta}(\tau_{2,c}) = \tau + \frac{\pi}{2}$
Triangular (2)	$\boldsymbol{\theta}(\tau_{1,c}) = \tau$	$\boldsymbol{\theta}(\tau_{2,c}) = \tau + \frac{\pi}{3}$
Rectangular (3)	$\boldsymbol{\theta}(\tau_{1,c}) = \tau$	$\tau_{2,c} = \boldsymbol{0}$

For the structures 4 and 5 instead, the number of repositioning is not proportional to the external perimeter p_{mat} but rather to the hatching area A_i of the *i-th* layer. In that case the expressions reported in Fig. 2 are used for evaluating n_{r-mat}:

Wiggle	*Honeycomb*
$n_{m-rip} = \sum_{i=1}^{\frac{b_{z-mat}}{L}} 2\frac{A_i}{Dh}$	$n_{m-rip} = \sum_{i=1}^{\frac{b_{z-mat}}{L}} 4\frac{A_i}{\frac{3}{2}\sqrt{3}\,D}$

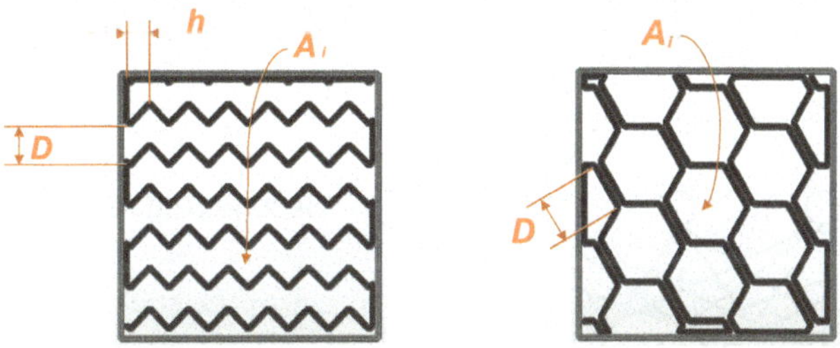

Fig. 2. Formulation of for wiggle and honeycomb hatching.

A formulation for determining the same parameters with reference to the supports should now be provided. But no information about the support structures is contained into the geometrical model and, moreover, these last are strictly dependent on user choices and technology in use. So a strategy, based on voxelization, has been implemented into a Matlab script for determining the geometrical information referred to the supports [24]. The total volume of supports V_{sup} is provided by the sum of the volume of the voxel elements:

$$V_{sup} = \sum_{i=1}^{N_v} V_{v,i}$$

Where:

N_v Number of voxel elements
$V_{v,i}$ Volume of the *i-th* voxel element

While the total number of repositioning n_{r-sup} required for the supports is obtained considering the total perimeter p_{sup} of the supporting structure.

$$p_{sup} = N_v\left(4\frac{d^2}{L}\right)$$

Where:

d Grid dimension of voxel element

Then the same methodology proposed for the material can be used, depending on the kind of hatching chosen.

Now a relationship between the proposed driving build-time factors and the build time needs to be founded. Generally speaking, the function (1) is very complex and unknown. So an Artificial Neural Network (ANN) is used to meet the request. Those mathematical instruments represent a powerful solution for generating a complete representation of a domain not entirely known.

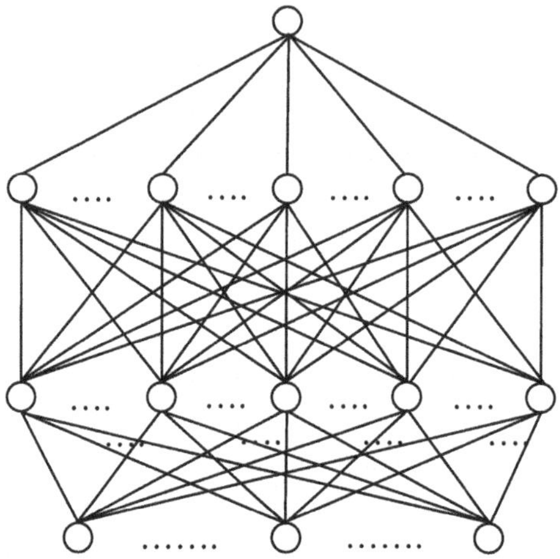

Fig. 3. Configuration of the artificial neural network.

A neural network, being able to reproduce non-linear functions, provides the possibility of considering all the complex dependencies which elapse between the driving build-time factors, offering, for example, the opportunity to predict the effects of the acceleration and jerk phases. Moreover a neural network is characterized by an adaptive capability; if the estimation provided by the network for a prototype results to be unsatisfactory, this last can be added to the set of samples used for training the network. In this way the accuracy of results increases over time.

For our purpose, a typical back-propagation artificial network with two hidden layer (Fig. 3) has been used. The parameter T, which identifies the AM technology in use, is provided in input to the network in addition to the driving build-time factors. That parameter should work as a logical value and it is chosen between the positive integer numbers depending on the technology in use ($T_{FDM} = 1, T_{SLA} = 2, \ldots, T_{SLM} = 8$). In order to train the network, the real build time of the test samples is required. In [21] the learning process was very slow because, not being disposable a reliable way for evaluating this quantity, each test case required to be physically realized. Therefore, in order to maintain the method economically and temporally sustainable, few samples could be provided for network learning. For example no other build direction except the optimal one was used and this represented a significant limitation of the network's ability to generalize the results.

In the proposed method, by the use of a detail-analysis based method [24], now an extremely accurate evaluation of the build time, based on the use of GCode, is applied for evaluating very precisely the real build time. The use of this methodology requires that some specific information concerning the manufacturing process should be provided, for example the magnitude of accelerations and jerks, or the infill percentage of part and support. The necessity to know this information represents a limit of the method itself in comparison with the proposed one, but the use of it allows us to obtain very quickly an evaluation of the build time. In this way many test cases, using different orientations or shapes, can be used as reference for the learning process, increasing the method accuracy. In Fig. 4, the workflow followed for obtaining the real build time of the samples used for network learning is illustrated.

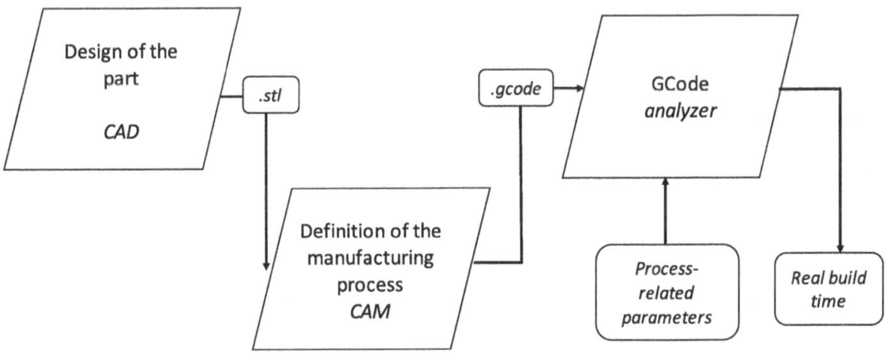

Fig. 4. Workflow for obtaining the build time using the proposed detail-analysis based method.

3 The Results

In order to train the ANN, seventy-seven test cases, which are characterized from different shape, dimension, geometrical and topological complexity, have been used taking as reference the FDM technology. Some of these test cases have been evaluated more than once by changing building direction and/or scaling its geometry. Therefore the ANN was trained by using ninety training instances.

A first experiment (A) was conducted for identifying the best number of nodes to use for the two hidden layers of ANN. As evidenced by [25], an optimal number of nodes can be founded only trough an empirical investigation and it is affected by the complexity of the problem. If the number of nodes is too low, the accuracy of the neural network will low too. Otherwise if the number of nodes is increased beyond a threshold level, then the method will lost its capability to generalize the results.

Six different ANN configurations were trained, using, respectively, 25, 50, 65, 75, 85 and 100 nodes. Then the samples reported in Fig. 5 were submitted to the ANNs to verify their capability to evaluate the build time. These samples belong to the seventy-seven models used for training ANNs, but they were rotated and/or scaled and then submitted to neural network. The estimation error, shown in Fig. 6, evidence that the optimal case is provided when 75 nodes are used for each hidden layer. In that configuration the area of the polygonal figure that corresponds to 75 nodes is minimum.

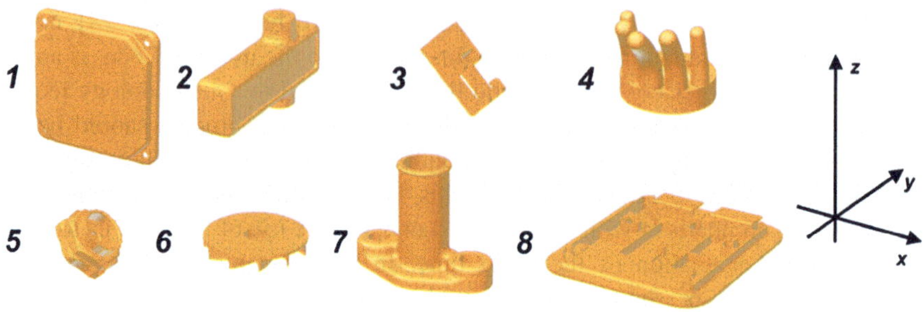

Fig. 5. Validation samples used for experiment (A).

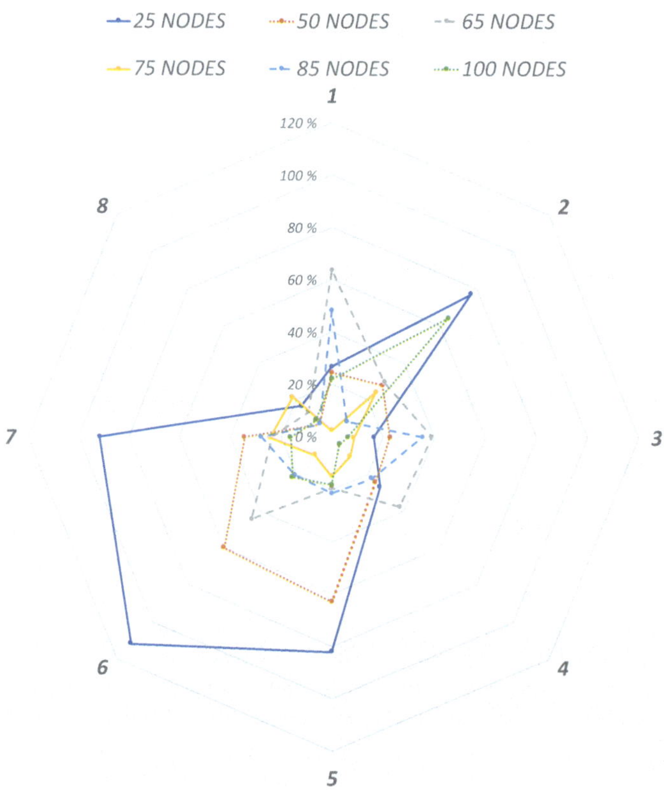

Fig. 6. Relationship between number of nodes and percentage error in build time estimation of experiment (A) for the eight proposed samples.

In order to analyze the behavior of the ANN in regard to new geometrical models, not used in the training phase, another experiment (B) was conducted. Six samples, reported in Fig. 7, were analyzed and their build time estimated. The results, reported in Table 3, evidenced a very satisficing estimation of the build time when 100 nodes are used.

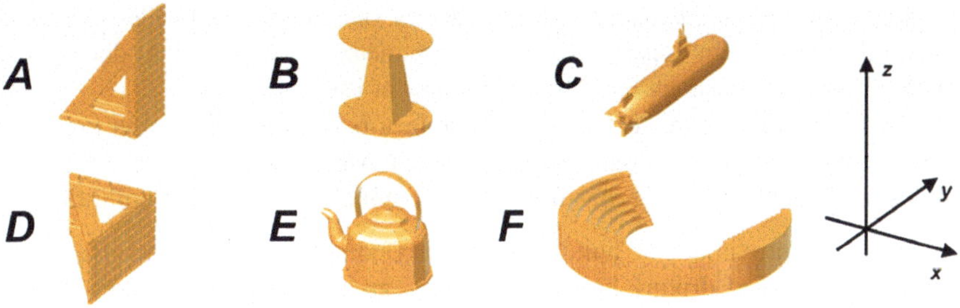

Fig. 7. Validation samples used for experiment (B).

Table 3. Results of experiment (B).

Model	75 NODES		100 NODES		REAL BUILD TIME
	Estimation [min]	Error	Estimation [min]	Error	[min]
A	123	18%	106	3%	104
B	241	39%	159	8%	173
C	239	146%	89	8%	97
D	131	3%	131	3%	135
E	553	15%	423	12%	479
F	490	36%	748	2%	762

The difference between the behavior of the neural network towards already-trained samples, in which only the building direction or the size have been modified, and new samples has been attributed to a mathematical instability of ANN. When too many neural connections are present into the hidden layers, small variations of the driving build-time factors cause a significant change of the build time. In the case of experiment (A) many of the driving build-time factors does not vary a lot when the build direction changes. For example one the most significant driving factors, such as the volume, is not affected by build direction. This does not occur in the experiment (B) when new geometrical models are analyzed. In that experiment an increasing number of neurons seems to improve the accuracy of the network. This is demonstrated by the fact that, when 75 nodes are used, the average error for the build time estimation is equal to 43% of the real build time, while for the case of 100 nodes the average error is only 6%.

4 Conclusions

In this paper a parametric-based approach for evaluating the build time for AM technologies has been presented. A parametric estimation of this quantity could be very useful for a fast and privacy-aimed evaluation of the costs required for manufacturing an object. Respect to the published methods by Authors, in this paper the following new elements are introduced:

- Among eight original driving build-time factors, only the five with a low correlation each other have been considered. A new parameter T has been introduced in formula (1) which considers the AM technology in use.
- A new formulation of the driving build-time factors is proposed in order to take into account the different typologies of hatching for AM.
- A fast analytical procedure was used for determination of the real build time.

In order to approximate the build time, the very complex and non-linear function of the previously defined driving build-time factors, an ANN with two hidden layers is used.

With a first experimentation the optimal number of neurons for the two hidden layers was defined. The net training was performed with ninety test cases opportunely chosen.

The obtained results show a good performance with a mean error of 6% in the build time estimation for objects for which the ANN has never been trained.

In a further work may be interesting to use an ANN with more neural connections which may lead to an even better estimation of the build time. Furthermore the availability of a reliable detailed-analysis based method for evaluate the build time could be exploit for improving even more the quality of the test cases. For example, as suggested by the results of experiment (A), a better specialization of the ANN for different build directions of a same object is required.

References

1. Rajenthirakumar, D., Jagadeesh, K.A.: Analysis of interaction between geometry and efficiency of impeller pump using rapid prototyping. Int. J. Adv. Manuf. Technol. **44**, 890–899 (2009)
2. Vinodh, S., Devadasan, S.R., Maheshkumar, S., Aravindakshan, M., Arumugam, M., Balakrishnan, K.: Agile product development through prototyping technologies: an examination pump-manufacturing company. Int. J. Adv. Manuf. Technol. **46**(5–8), 663–679 (2010)
3. Alexander, P., Allen, S., Dutta, D.: Part orientation and build cost determination in layered manufacturing. Comput. Aided Des. **30**(5), 343–356 (1998)
4. Byun, H.S., Lee, K.H.: Determination of optimal build direction in rapid prototyping with variable slicing. Int. J. Adv. Manuf. Technol. **28**(3–4), 307–313 (2006)
5. Byun, H.S., Lee, K.H.: Determination of the optimal build direction for different rapid prototyping processes using multi-criterion decision making. Robot. Comput. Integr. Manuf. **22**(1), 69–80 (2006)
6. Armillotta, A., Cavallaro, M., Minnella, S.: A tool for computer-aided orientation selection in additive manufacturing processes. In: Proceedings of the 6th International Conference on Advanced Research in Virtual and Rapid Prototyping (2013)
7. Thrimurthulu, K.P.P.M., Pandey, P.M., Reddy, N.V.: Optimum part deposition orientation in fused deposition modeling. Int. J. Mach. Tools Manuf. **44**(6), 585–594 (2004)
8. Canellidis, V., Giannatsis, J., Dedoussis, V.: Genetic-algorithm-based multi-objective optimization of the build orientation in stereolithography. Int. J. Adv. Manuf. Technol. **45**(7–8), 714–730 (2009)

9. Singhal, S.K., Prashant, K.J., Pandey, P.M., Nagpal, A.K.: Optimum part deposition orientation for multiple objectives in SL and SLS prototyping. Int. J. Prod. Res. **47**(22), 6375–6396 (2009)
10. Phatak, A.M., Pande, S.S.: Optimum part orientation in rapid prototyping using genetic algorithm. J. Manuf. Syst. **31**(4), 395–402 (2012)
11. Brika, S.E., Zhao, Y.F., Brochu, M., Mezzetta, J.: Multi-objective build orientation optimization for powder bed fusion by laser. J. Manuf. Sci. Eng. **139**(11), 111011 (2017)
12. Jaiswal, P., Patel, J., Rai, R.: Build orientation optimization for additive manufacturing of functionally graded material objects. Int. J. Adv. Manuf. Technol. 1–13 (2018)
13. Kamash, T., Flynn, D.: Build time estimator for stereo-lithography machines—a preliminary report. Prototype Express **2**, 2 (1995)
14. Kechagias, J., Maropoulos, S., Karagiannis, S.: Process build-time estimator algorithm for laminated object manufacturing. Rapid Prototyping J. **10**(5), 297–304 (2004)
15. Nezhad, A.S., Vatani, M., Barazandeh, F., Rahimi, A.: Build time estimator for determining optimal part orientation. Proc. Inst. Mech. Eng. Part B J. Eng. Manuf. **224**(12), 1905–1913 (2010)
16. Chen, C., Sullivan, P.: Predicting total build-time and the resultant cure depth of the 3D stereo-lithography process. Rapid Prototyping J. **2**(4), 27–40 (1996)
17. Ruffo, M., Tuck, C., Hague, R.: Empirical laser sintering time estimator for Duraform PA. Int. J. Prod. Res. **44**(23), 5131–5146 (2006)
18. Campbell, I., Combrinck, J., Barnard, D.L.: Stereo-lithography build time estimation based on volumetric calculations. Rapid Prototyping J. **14**(5), 271–279 (2008)
19. Munguía, J., Ciurana, J., Riba, C.: Neural-network-based model for build-time estimation in selective laser sintering. Proc. Inst. Mech. Eng. Part B J. Eng. Manuf. **223**(8), 995–1003 (2009)
20. Zhang, Y., Bernard, A.: Generic build time estimation model for parts produced by SLS, in high value manufacturing. In: Advanced Research in Virtual and Rapid Prototyping: Proceedings of the 6th International Conference on Advanced Research in Virtual and Rapid Prototyping, Leiria. CRC Press (2010)
21. Di Angelo, L., Di Stefano, P.: A neural network-based build time estimator for layer manufactured objects. Int. J. Adv. Manuf. Technol. **57**(1), 215–224 (2011)
22. Zhang, C., Chen, T.: Efficient feature extraction for 2D/3D objects in mesh representation. In: Proceedings 2001 International Conference on Image Processing (Cat. No. 01CH37205), vol. 3, pp. 935–938 (2001)
23. Di Angelo, L., Di Stefano, P.: Parametric cost analysis for web-based e-commerce of layer manufactured objects. Int. J. Prod. Res. **48**(7), 2127–2140 (2010)
24. Di Angelo, L., Di Stefano, P., Guardiani, E.: A build time estimator for additive manufacturing. In: Proceedings 2019 International Workshop on Metrology for Industry 4.0 and Internet of Things, pp. 327–332, IEEE catalog number CFP19N49-USB (2019)
25. Sukthomya, W., Tannock, J.: The training of neural networks to model manufacturing processes. J. Intell. Manuf. **16**, 39–51 (2005)

3D Printed Materials for High Temperature Applications

Antonio Marzola, Elisa Mussi(✉), and Francesca Uccheddu

Department of Industrial Engineering, University of Florence, Florence, Italy
{antonio.marzola, elisa.mussi, francesca.uccheddu}@unifi.it

Abstract. Additive Manufacturing technologies are proving to be reliable alternatives to traditional ones. Because of their capability to allow short-run production with very low changeover costs and minimal setup, additive manufacturing technologies represent the best choice in case of low-volume highly customized productions. Furthermore, they represent the only alternative to manufacture lattice components, characterized by a complex meso-structure. To date, the main limitation of additive manufacturing is represented by a limited selection of printable materials. In addition, despite lots of information can be found about the mechanical characterization of the available printable materials, there is not as much information regarding their mechanical and geometrical behavior in high-temperature applications. Especially for thermoplastics, the lack of information limits the possibilities for the selection of adequate material, affecting the effectiveness of the design. In this work, a close inspection of the specific literature has been done, with the aim to gather information on the available printable materials suitable for high-temperature applications.

Keywords: High temperature · Additive manufacturing · 3D printed materials

1 Introduction

In the last decades, additive manufacturing (AM) technologies have had a disruptive effect on the traditional manufacturing process thanks to their flexibility in manufacturing operations [1]. Although they were developed specifically for rapid prototyping, additive technologies are now commonly employed for the actual components production, finding great uptake in a wide range of industrial sectors. The main breakthroughs introduced by AM can be summarized as follows: no manufacturing constraints to build extremely complex and highly optimized components (e.g. lattice structures), production of fast prototypes as well as actual components in short-run production with very low changeover costs and minimal setup, possibility to reuse waste material (i.e. powder, resin), lack of need for any tools and molds [1, 2] being able to manufacture directly from 3D CAD models.

As a consequence, AM allows to reduce both the production and materials costs, creating a more sustainable production and consumption system [3].

C. Rizzi et al. (Eds.): ADM 2019, LNME, pp. 936–947, 2020.
https://doi.org/10.1007/978-3-030-31154-4_80

Because of their potentialities, AM technologies are used in a wide industrial and medical applications spectrum: electronics, automotive, aerospace, tool making, personalized medicine and oil and gas sector [4, 5].

In Electronics, geometrically appropriate prototypes can be manufactured reducing the development time from weeks to hours [6]; Selective Laser Melting (SLM) technology can be employed for substrate manufacturing, system integration and packaging concepts in electronic systems, particularly in molded interconnect device [7].

In Automotive and Aerospace industry, topological optimized components with extremely complex cross-sections can be manufactured for the purpose to lighten the vehicles while ensuring safety [8]; Selective Laser Sintering (SLS) or Electron Beam Melting (EBM) are commonly used to produce the actual components [9].

In Medicine, additive manufacturing is able to deliver highly customized Medical Devices (MD) designed on the patient's own diagnostic images (such as Computerized Tomography or Magnetic Resonance Imaging), as well as allowing the construction of realistic physical models which enable physicians to detect the best surgical strategy [3, 6, 7, 10]. To date, AM machines are properly designed for MD fabrication based on EBM or SLM technologies to manufacture metal implants. Selective Laser Sintering (SLS) or Material Extrusion (ME) technologies are used to build thermoplastic parts useful as training models as well as patient-specific surgery instrumentations (e.g. cutting guides).

Oil and Gas industry is characterized by a unique and low-volume production. The use of AM technologies can provide benefits such as time reduction, possibility to produce complex shape, improvement of supply chain efficiency [11].

Despite the mentioned strengths, there are some limitations that still restrict the actual use of AM technologies to few selected industrial and medical applications. Related literature highlights the limited selection of materials available to be used in AM, especially for low-cost technologies [12]. In addition, 3D printed components generally suffer from anisotropic mechanical properties and limited surface quality and aesthetics (stair stepping effect) [3, 13–15]. The mechanical properties and anisotropic behavior of 3D printed components are strictly dependent on the printing pattern [16]; surface quality, on the other hand, can be improved in a series of mechanical post-processes. It should be noted that, in addition to the technical and technological problems reported, there are also open legal questions as product validation, intellectual properties and liability [3].

Despite a considerable number of literature studies [9, 12], regarding AM manufacture technology, mechanical properties and surface quality characterization of 3D printed parts, there is very few information about the use of AM for High Temperature (HT) applications. As per common definition, the term "HT application" refers to a maximum working temperature between 30% and 70% of the melting temperature of the considered material.

Temperature is a fundamental variable to be considered, because it can cause important changes to the mechanical and geometric characteristics of the components. The relationship between the temperature changes and component characteristics highly depends on the component's material. Therefore, in HT applications the choice of materials is a key component of an effective design, but it is not a trivial task. It is more relevant considering 3D printed parts because of the limited selection of materials available.

The present work aims to gather the available information on the printable materials (both plastics and metals) suitable for high temperature applications. In the following sections, all the available 3D printed materials are reported, focusing on the most commonly used in industry because of their cost and availability on the market.

2 3D Printed Materials for HT Applications

Current additive technologies are able to process refractory metals alloys, such as cobalt, nickel and tungsten (used for the production of components for HT applications), tool steels and stainless steels, titanium, aluminum and copper-based metals or light alloys (for aerospace and biomedical applications).

Historically, metals have been the most successful materials in structural applications due to their excellent mechanical properties, such as resistance, rigidity and fracture toughness, but also due to their high machinability and low cost. Steel and other traditional metal alloys are still the most widely used materials today, but they show their limits when specific mechanical characteristics are required; such limits are often perceived in performance-driven design, as in the automotive, aviation and aerospace industries. In these sectors, the main limit of conventional metals is their weight, which is incompatible with the design requirements of vehicles and, above all, aircraft or components for aerospace systems. It is therefore necessary to switch to lighter metallic materials, mainly aluminum and magnesium alloys, even accepting lower mechanical performance, which often result inadequate. Metal matrix composite materials, especially those with aluminum, magnesium or titanium-based matrixes, are of great engineer interest because they combine lightness with high mechanical properties which, in some cases, can be maintained even when operating at high temperatures (600–700 °C). However, the use of this class of advanced materials is limited by the cost and difficulties of their production: in fact, all the processes used so far to produce composite material products with a metal matrix have a number of problems mainly related to the reactivity of the matrices/interfacial reinforcements. For this reason, additive manufacturing is necessary to produce parts even of complex shape in composite material, with short delivery times, low costs, good performance of the final material.

Despite metals are historically the best choice for an HT application, also some thermoplastic material can be employed in some cases. In HT applications which do not require high mechanical performances, thermoplastic polymer could be a cheaper but still effective solution. As an example, patient specific instrumentation (e.g. cutting guides) are usually made with polymeric materials, since strong mechanical characteristics are not required. On the other hand, it is mandatory for to be biocompatible but also sterilizable. Sterilization imposes temperatures close to thermoplastic melting temperature (about 130 °C) even if for a short amount of time (about 30 min). In Sect. 2.2 the sterilization processes, which play a major role for the application of plastic materials for MD, are treated more in details.

In recent years, innovation in the world of 3D polymer printing concerns new applications to be tested, new materials adapted to the technology, new hardware, but also other technologies, such as 4D printing, where the fourth dimension is time. “4D printing” means the processing, by means of additive technologies, of intelligent and programmable materials, i.e. materials capable of changing shape and properties (shape memory matrices). 4D materials are able to perform specific actions - such as self-assembly over time - following an external stimulus such as heat, vibration, gravity, magnetism or electricity. In the case of shape memory polymers, for example, rigid materials at room temperature take on a certain flexibility and deformability at higher temperatures: by exploiting this behavior it is possible to give different shapes by working the component at high temperature and fixing the new shape simply by cooling down, knowing that the component will summary the original shape as soon as it is heated following the shape memory.

Nowadays, the limited number of materials available on the market for AM poses some limits on the use of such a technology for some industrial sectors. The present work focuses on materials reliable for high temperature applications, i.e. materials with geometrical and mechanical characteristic stability at the considered working temperature. Metals, metal-alloy and thermoplastic materials will be analyzed highlighting their mechanical and thermal properties. Composite materials are not considered because the addition of the reinforcements (such as silicon carbide, aluminum, glass particulates) improves the mechanical and structural properties of materials but cannot modify the thermal properties, that still remain the same of the least performing material. Usually, thermal properties are imposed by the matrix [17].

In the following sections we present an overview of metals, alloys and thermoplastic materials employed in HT applications. A separate section focuses on the sterilization problem for materials for MD, which is a hot topic due to the wide use of additive manufacturing in medicine.

2.1 Metals and Alloys

Inconel 625 is a nickel-based superalloy available for powder bed fusion processes and directed energy deposition process. It is a non-magnetic material characterized by an excellent fatigue, oxidation and corrosion resistance. It resists pitting and cracking when exposed to chloride and has a high resistance to alkalis and acids. In addition, Inconel 625 has exceptional strength and toughness for a wide range of temperature from cryogenic to 982 °C. For these reasons Inconel 625 is commonly used in aerospace, chemical, petrochemical and marine sectors. The melting temperature of Inconel 625 ranges from 1290 to 1350 °C [18]. The price of the powder is about 135 €/kg.

Inconel 718 is a nickel-chromium-based alloy and is and age-hardened version of 625. It has good tensile strength, fatigue strength, rupture strength and good weldability besides an excellent oxidation resistance, high resistance to the corrosive effect of hydrochloric and sulfuric acid. The alloy can be used permanently at temperatures to 700 °C. The mechanical properties of Inconel 718 make its employment ideal for parts that must perform well in high-temperature, high-stress environments. Indeed, Inconel 718 is used to produce metal parts of gas turbines, jet engines, pumps, measurement probes in energy technologies and parts for aerospace and petrochemical applications.

The melting temperature of Inconel 718 ranges from 1210 to 1344 °C [19]. The price of the powder is about 135 €/kg.

Cobalt-chromium alloys have high temperature, wear and corrosion resistance, very high hardness, high strength and high ductility. In addition, cobalt-chromium alloys are biocompatible materials. Due to high hardness and biocompatibility these materials are mainly used in the medical industry to produce implants and prostheses. Nevertheless, Co-Cr alloys are also used to produce components for application in high temperature environments, such as in jet-engines, gas and wind turbines [19]. The alloy can be used permanently at temperatures to 650 °C. The melting temperature of Co-Cr alloys ranges from 1260 to 1482 °C. The price of the powder is about 220 €/kg.

The AISI 4140 is another type of material suitable to operate at high temperature that is made with technology L-PBF (Laser Powder Bed Fusion) [20]. AISI 4140 is an alloy of steel, commonly used in a wide range of applications for heavy industry (aerospace and automotive applications). The mechanical properties, such as yield strength, tensile strength, wear resistance, elongation and toughness of the material are excellent compared to those of traditionally produced steel. The alloy is able to resist creep in temperature up to 538 °C and to maintain its properties being exposed at comparatively high working temperatures for long periods.

Titanium alloy, Ti-6Al-4V, is a lightweight material that is widely used in space vehicle structures; it provides an excellent combination of high strength, low density, high modulus, low thermal expansion coefficient, and higher operating temperature than aluminium alloys [21]. Ti-6Al-4V alloy therefore has excellent mechanical properties, high corrosion resistance and low density. It covers more than 50% of the industrial production of titanium alloy. Table 1 shows a summary of 3D printed metal and alloys materials reporting the related technology and price for each material.

Table 1. Technologies, melting temperature and price of 3D printed metal and alloys material for HT applications. the acronyms are to be understood. PBF (Powder-Bed Fusion), DED (Directed Energy Deposition), ME (Material Extrusion).

	AM Technologies	Melting temperature (°C)	Price (€/kg)	Biocompatibility
Metals and alloys				
Inconel 625 [18]	PBF/DED	1290–1350	135	✗
Inconel 718 [19]	PBF/DED	1210–1344	135	✗
Cobalt-chromium alloys [19]	PBF/DED	1260–1482	220	✓
AISI 4140 [20]	PBF	1416	Not available	✗
Ti6Al4V [21]	PBF/DED	1604–1660 °C	400	✓

2.2 Thermoplastic Materials

PEEK (Polyetheretherketone) is an efficient semi-crystalline thermoplastic alternative to metallic implantable materials due to its excellent biocompatibility combined with its excellent mechanical properties (in terms of strength and stiffness). PEEK has the ability to maintain its stiffness even at high temperatures and is suitable for continuous operation at temperatures of 260 °C. It is widely used in aerospace applications, oil and gas, and in semiconductors. It shows excellent resistance to a wide range of chemicals agents, such as those used in the oil and gas, automotive, aerospace, industrial and chemical processing industries [22].

PEKK (Polyetherketoneketone), is another thermoplastic material used in high temperature applications. It has peak melting temperatures in the range from 359 °C to 362 °C [23], compared oto amorphous thermoplastics, has a wider temperature range in continuous use, and remains performing also at temperatures above the glass transition temperature; it also has better creep strain resistance [24]. It is used in space vehicles operating under extreme environmental conditions with the aim of protecting electronics from the risk of extreme temperatures and electrostatic discharges.

ULTEM 9085 resin is a high performance thermoplastic suitable for the ME process with a high strength-to-weight ratio. It has a good strength, that allows products autoclave sterilization, optimal durability and resistance to heat and chemicals that make it suitable for prototypes or functional production parts. ULTEM 9085 is found in many aircraft and aerospace products because it meets stringent safety requirements.

ULTEM 9085 is FST classified, which means that it meets flame, smoke and toxicity standards and once processed, the FST characteristics of the material remain unchanged. ULTEM 9085 also has high tensile (72 MPa, ASTM D638) and flexural strength (115 MPa, ASTM D790) [25].

PA2200 is a nylon powder, biocompatible according to EN ISO 10993-1 and USP/level VI/121°C. Laser-sintered parts in PA 2200 are characterised by a high strength and stiffness, good chemical resistance and excellent long-term constant behaviour [26]. In comparison to standard polyamide PA2200 is characterized by higher crystallinity and higher melting point (184 °C) as result of specific production process. The powder price is approximately 80 €/kg.

PC-ISO is another ME biocompatible thermoplastic material. It is commonly used in applications such as food and drug packaging and medical device manufacturing thanks to high tensile and flexural strength and biocompatibility. The material price is approximately 121 €/kg [27].

ABS-M30i is an ME biocompatible material characterized by high strength. It is particularly expensive (320 €/kg) compared to other thermoplastic material [28].

ULTEM 1010 resin is a biocompatible and high-performance thermoplastic material. It ensures highest heat resistance (glass transition temperature of 215 °C), chemical resistance and tensile strength of any ME thermoplastic material. It is used for applications such as food production and custom medical applications and for many industrial tooling applications that require the combination of strength and thermal stability thanks to a very low coefficient of thermal expansion [29]. The material price is approximately 195 €/kg. Table 2 shows a summary of 3D printed thermoplastic materials reporting the related technology and price for each material.

Table 2. Technologies, melting temperature and price of 3D printed thermoplastic materials for HT applications. the acronyms are to be understood. PBF (Powder-Bed Fusion), DED (Directed Energy Deposition), ME (Material Extrusion). *Due to amorphous nature, materials do not display a melting point, the glass transition temperature is reported.

	AM Technologies	Heat Deflection (ASTM D648)	Melting temperature (°C)	Price (€/kg)	Biocompatibility
Thermoplastics					
PEEK [22]	ME	160	359–362	500–700	✓
PEKK [24]	ME	150	359–363	600	✗
ULTEM 9085 [25]	ME	153	*186	260	✗
PA 2200 [26]	PBF	150	184	80	✓
PC-ISO [27]	ME	133	*161	121	✓
ABS-M30i [28]	ME	96	*108	320	✓
ULTEM 1010 [29]	ME	216	*215	195	✓

2.3 Materials for Surgical Sterilization

In the medical field, the AM technologies are widely used to produce patient-specific solutions in a short time with high precision. Medical industry is the third largest market of the AM, with approximately 16% of the overall revenue, following automotive and consumer electronics [30]. Nowadays, AM is used in a large variety of medical applications including anatomical models, tissue engineering scaffolds, dentistry, drug delivery, implants and surgical guides (Fig. 1).

3D printers are used to create customized anatomical model for preoperative planning, training, and/or education [31] which allow a better understanding of the complex patients-anatomy to predict potentials weaknesses of the surgery before entering surgery room.

In tissue engineering area, bioprinting technologies produce precise, well-defined micro- and nanoscale 3D structure consisting of living cells, biomaterials and biomolecules able to repair and replace damaged tissues [32].

The strong growth of the personalized medicine is making the impact of AM also strong in pharmaceutical technology. 3D printing gives the opportunity to prepare different kind of dosage forms, multidrug devices and drug formulations for personalized therapy [33].

Finally, AM allows to design and produce customized implants or surgical guides [34] for maxillofacial, reconstructive and neurosurgery applications. The use of AM to manufacture a custom implant has shown decreased complication rates, improved accuracy [35] and reduction of surgery steps [36] compared to the traditional approaches.

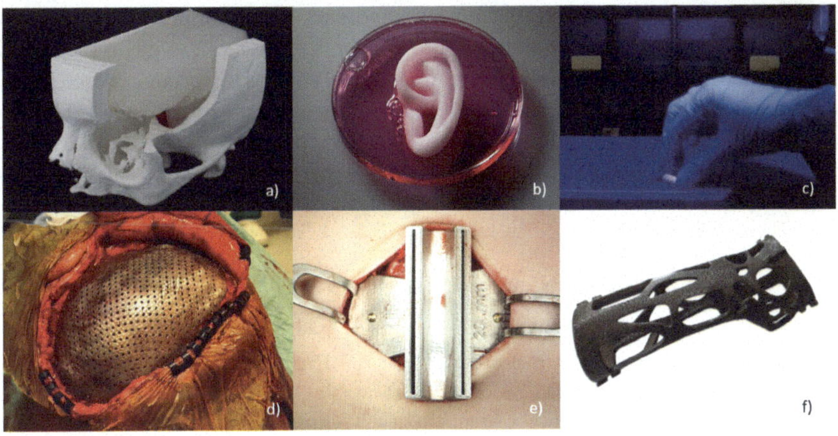

Fig. 1. Medical applications of AM technologies: (a) anatomical model for preoperative planning, (b) ear printed with bioprinter, (c) personalized drug, (d) titanium plate for cranioplasty, (e) cutting guide to assist sternal osteotomies, (f) orthoses model.

According to the application, the materials must meet specific requirements including biocompatibility and sterilizability. Biocompatibility is intended as the ability of a material to perform with an appropriate host response in a specific application [37].

The sterilization process consists in the use of physical or chemical procedures to destroy all microbial life, including large numbers of highly resistant bacterial endospores. There are different methods of sterilization that can be classified in three categories: (i) high temperature/pressure sterilization, (ii) chemical sterilization and (iii) radiation sterilization [38]. However, three sterilization techniques are commonly used to sterilize medical equipment: autoclave, ethylene oxide gas and hydrogen peroxide gas plasma.

Steam autoclaves was the first method utilized for sterilization of medical products. The basic principle of steam sterilization is to expose each item to direct steam contact at a specific temperature, pressure and for a precise time which ensure the microbicidal activity. The two common settings are 30 min at 121 °C in a gravity displacement sterilizer or 4 min and 132 °C in a pre-vacuum sterilizer [39]. The advantage of steam sterilization is that it is nontoxic to the patient, medical staff, or environment; on the other hand, steam sterilization can damage heat-sensitive materials.

Ethylene oxide gas (ETO) is a colorless gas that is also microbicidal. It is used as a low-temperature sterilization method (below 60 °C) which consists in five phases: preconditioning and humidification, gas introduction, exposure, evacuations and air washes. The main disadvantages associated with ETO are the lengthy cycle time, the cost, and its potential hazards to patients and staff; the main advantage is that it allows to sterilize heat- or moisture-sensitive medical equipment without deleterious effects.

Another low-temperature sterilization method uses hydrogen peroxide gas plasma. The devices are placed in a chamber under a deep vacuum where gas molecules are excited by electromagnetic waves generating gas plasma. This produces free radicals that interact with cell components disrupting the metabolism of microbes. The cycle duration is 55 min at a maximum temperature of 55 °C.

In Table 3 are reported the main advantages and disadvantages of sterilization methods.

Table 3. Sterilization methods: advantages and disadvantages [39].

	Advantages	Disadvantages
Steam autoclaves	– Short processing time – Process simplicity – Efficacy – Non-toxicity	– High operating temperature (121 °C or 132 °C) – Possible corrosion
Ethylene oxide gas	– Low operating temperature – High microbicidal activity	– Long processing time (10–24 h) – Toxic, flammable and carcinogenic gas
Hydrogen peroxide gas plasma	– Low operating temperature – Medium processing time	– Very expensive – Damages Nylon based materials

As shown in Table 2, the steam autoclave process represents the best alternative in terms of cost-effectiveness and timesaving. However, such a sterilization process requires high temperature which cannot be withstand by all the material listed above. Table 4 reports the suitable sterilization process for each of the treated materials. Since the metal components can withstand with the required temperatures of the considered sterilization processes, they are not reported. All the thermoplastics in Table 3 are biocompatible.

Table 4. 3D printed biocompatible materials with their suitable sterilization processes

	Steam autoclaves	Ethylene oxide gas	Hydrogen peroxide gas plasma
PA 2200 [26]	✓	✓	Not available
PC-ISO [40]	✓	✓	✓
ABS-M30i [40]	✗	✓	✓
PEEK [41]	✓	✓	✓
ULTEM 1010 [25]	✓	Not available	Not available

3 Conclusions

Additive manufacturing is rapidly changing the design concept and manufacturing industry. The customization and fast production of prototype are just a few advantages that make AM the best choice for several applications.

On the other hand, the limited selection of materials for AM is perhaps one of the main constraints for its use in some applications. The present paper aimed to report the most common thermoplastic materials adequate for HT application, with a particular focus on medical sector.

Specifically, the presented work provides an overview of the materials available to produce AM parts for applications in HT. A close inspection of commercially available materials has confirmed the limited availability of 3D printable materials reliable for HT applications. It is well known that metallic materials have mechanical performances at high temperature and higher pressure than thermoplastic materials. However, the costs of materials and technologies are higher. It is possible to predict that in the next years research will address to identify innovative AM materials able to overcome the present limits such as costs and usability in extreme applications.

References

1. Spalt, P., Bauernhansl, T.: A framework for integration of additive manufacturing technologies in production networks. Procedia CIRP **57**, 716–721 (2016)
2. Nordin, N.A.B., Johar, M.A., Ibrahim, M.H.I., Marwah, O.M.F.: Advances in high temperature materials for additive manufacturing. IOP Conf. Ser. Mater. Sci. Eng. **226**(1), 012176 (2017)
3. Ford, S., Despiesse, M.: Additive manufacturing and sustainability: an exploratory study of the advantages and challenges. J. Cleaner Prod. **137**, 1573–1587 (2016)
4. Guo, N., Leu, M.C.: Additive manufacturing: technology, applications and research needs. Front. Mech. Eng. **8**, 215–243 (2013)
5. Macdonald, E., Salas, R., Espalin, D., Perez, M., Aguilera, E., Muse, D., Wicker, R.B.: 3D printing for the rapid prototyping of structural electronics. IEEE Access **2**, 234–242 (2014)
6. Beltrametti, L., Gasparre, A.: Industrial 3D printing in Italy. Inrt. J. Manuf. Technol. Manag. **32**, 43–64 (2018)
7. Syed-Khaja, A., Walawski, D., Stoll, T., Franke, J.: Is selective laser melting (SLM) an alternative for high-temperature mechatronic integrated devices? Methodology, hurdles and prospects. In: 12th International Congress Molded Interconnect Devices (MID). IEEE (2016)
8. Beyer, C.: Strategic implications of current trends in additive manufacturing. J. Manuf. Sci. Eng. (2014)
9. Wong, K., Hernandez, A.: A review of additive manufacturing. ISRN Mechanical Engineering (2012)
10. Furferi, R., Guariento, L., McGrevy, K.S., Mussi, E., Parri, N., Uccheddu, F., Volpe, Y.: 3D printing-based pediatric trainer for ultrasound-guided peripheral venous access. In: 15th Mediterranean Conference on Medical and Biomedical Engineering and Computing. Springer (2019, in press)
11. Burns, M., Wangenheim, C.: Metal 3D printing applications in the oil & gas industry. In: SPE Middle East Oil and Gas Show and Conference. Society of Petroleum Engineers (2019)

12. Gao, W., Zhang, Y., Ramanujan, D., Ramani, K., Chen, Y., Williams, C.B., Zavattieri, P.D.: The status, challenges, and future of additive manufacturing in engineering. Comput. Aided Des. **69**, 65–89 (2015)
13. Singh, S., Singh, M., Prakash, C., Gupta, M., Mia, M., Singh, R.: Optimization and reliability analysis to improve surface quality and mechanical characteristics of heat-treated fused filament fabricated parts. Int. J. Adv. Manuf. Technol. **102**, 1–16 (2019)
14. Di Angelo, L., Di Stefano, P., Marzola, A.: Surface quality prediction in FDM additive manufacturing. Int. J. Adv. Manuf. Technol. **93**, 9–12 (2017)
15. Borgianni, Y., Maccioni, L., Basso, D.: Exploratory study on the perception of additively manufactured end-use products with specific questionnaires and eye-tracking. Int. J. Interact. Des. Manuf. **12**, 743–759 (2019)
16. Cantrell, J., Rohde, S., Damiani, D., Gurnani, R., DiSandro, L., Anton, J., Young, A., Jerez, A.: Experimental characterization of the mechanical properties of 3D-printed ABS and polycarbonate parts. Rapid Prototyping J. **23**(4), 811–824 (2017)
17. Ghita, O., James, E., Davies, R., Berretta, S., Singh, B., Flint, S., Evans, K.E.: High temperature laser sintering (HT-LS): an investigation into mechanical properties and shrinkage characteristics of Poly (Ether Ketone) (PEK) structures. Mater. Des. **61**, 124–132 (2017)
18. Jinoop, A., Paul, C., Bindra, K.: Laser-assisted directed energy deposition of nickel super alloys: a review. Proc. Inst. Mech. Eng. Part L J. Mater. Des. Appl. (2019)
19. Demir, A., Previtali, B.: Additive manufacturing of cardiovascular CoCr stents by selective laser melting. Mater. Des. (2017)
20. Jamshidinia, M., Sadek, A., Wang, W., Kelly, S.: Additive manufacturing of steel alloys using laser powder-bed fusion. Adv. Mater. Process. **173**(1), 20–24 (2015)
21. Rawal, S., Brantley, J., Karabudak, N.: Additive manufacturing of Ti-6Al-4V alloy components for spacecraft applications. In: 6th International Conference on Recent Advances in Space Technologies (RAST), pp. 5–11. IEEE (2013)
22. Vaezi, M., Yang, S.: Extrusion-based additive manufacturing of PEEK for biomedical applications. Virtual Phys. Prototyping **10**(3), 123–135 (2015)
23. Kishore, V., Chen, X., Ajinjeru, C., Hassen, A., Lindahl, J.: Additive manufacturing of high performance semicrystalline thermoplastics and their composites. In: Proceedings of the 26th Annual International Solid Freeform Fabrication (2016)
24. https://www.stratasys.com/materials/search/antero-800na. Accessed 24 June 2019
25. https://www.stratasys.com/it/materials/search/ultem9085. Accessed 24 June 2019
26. http://www.e-manufacturing.it/downloads/PA2200.pdf. Accessed 24 June 2019
27. https://www.stratasys.com/it/materials/search/pc-iso. Accessed 24 June 2019
28. https://www.stratasys.com/-/media/files/material-sp-sheets/mss_fdm_absm30i_en_0318a.pdf. Accessed 24 June 2019
29. https://www.stratasys.com/it/materials/search/ultem1010. Accessed 24 June 2019
30. http://wohlersassociates.com/press71.html. Accessed 24 June 2019
31. Mussi, E., Furferi, R., Volpe, Y., Facchini, F., McGreevy, K.S., Uccheddu, F.: Ear reconstruction simulation: from handcrafting to 3D printing. Bioengineering **6**(1), 14 (2019)
32. Gu, B., Choi, D., Park, S., Kim, Y., Kim, Y.J., Kim, C.H.: 3D bioprinting technologies for tissue engineering applications. In: Cutting-Edge Enabling Technologies for Regenerative Medicine, pp. 15–28. Springer (2018)
33. Jamróz, W., Szafraniec, J., Kurek, M., Jachowicz, R.: 3D printing in pharmaceutical and medical applications – recent achievements and challenges. Pharm. Res. **35**, 176 (2018)
34. Leng, S., Bici, K., Facchini, F., Volpe, Y., Uccheddu, F., Furferi, R., Messineo, A.: Customized cutting template to assist sternotomy in pectus arcuatum. Ann. Thorac. Surg. **107**(4), 1253–1258 (2019)

35. Oh, T.S., Jeong, W.S., Chang, T.J., Koh, K.S., Choi, J.W.: Customized orbital wall reconstruction using three-dimensionally printed rapid prototype model in patients with orbital wall fracture. J. Craniofac. Surg. **27**(8), 2020–2024 (2016)
36. Volpe, Y., Furferi, R., Governi, L., Uccheddu, F., Carfagni, M., Mussa, F., Genitori, L.: Surgery of complex craniofacial defects: a single-step AM-based methodology. Comput. Methods Programs Biomed. **165**, 225–233 (2018)
37. Ksander, G.A., Alto, P.: Definitions in biomaterials, progress in biomedical engineering. Ann. Plast. Surg. **21**, 291 (1988)
38. Govindaraj, S., Muthuraman, M.S.: Systematic review on sterilization methods of implants and medical devices. Int. J. ChemTech Res. **8**, 897–911 (2015)
39. https://www.cdc.gov/infectioncontrol/guidelines/disinfection/sterilization/steam.html. Accessed 24 June 2019
40. Perez, M., Block, M., Espalin, D., Winker, R., Hoppe, T., Medina, F., Wicker, R.: Sterilization of FDM-manufactured parts. In: Proceedings of the 2012 Annual International Solid Freeform Fabrication Symposium, Austin, pp. 6–8 (2012)
41. Sastri, V.R.: Plastics in Medical Devices: Properties, Requirements, and Applications, 2nd edn. Elsevier (2013)

Optimization Design Strategy for Additive Manufacturing Process to Develop 3D Magnetic Nanocomposite Scaffolds

Antonio Gloria[1], Marco Domingos[2], Saverio Maietta[3], Massimo Martorelli[3](✉), and Antonio Lanzotti[3]

[1] Institute of Polymers, Composites and Biomaterials – National Research Council of Italy, V.le J.F. Kennedy 54 – Mostra d'Oltremare Pad. 20, 80125 Naples, Italy

[2] School of Mechanical, Aerospace and Civil Engineering, University of Manchester, Manchester, UK

[3] Department of Industrial Engineering, Fraunhofer JL IDEAS – University of Naples Federico II, P.le Tecchio 80, 80125 Naples, Italy

massimo.martorelli@unina.it

Abstract. In the current research, an optimization design strategy for additive manufacturing processes based on extrusion/injection methods was extended to the fabrication of poly(ε-caprolactone) (PCL)/iron oxide (Fe_3O_4) scaffolds for tissue engineering. The attention was focused on four parameters: process temperature (PT), deposition velocity (DV), screw rotation velocity (SRV), slice thickness (ST). Specifically, PCL/Fe_3O_4 scaffolds were manufactured varying iteratively one parameter, while maintaining constant the other three parameters. A further insight into the influence of process parameters on the morphological features and mechanical properties of PCL/Fe_3O_4 scaffolds was provided.

Keywords: Design for additive manufacturing · Magnetic nanocomposite scaffolds · Mechanical and morphological properties · Tissue engineering

1 Introduction

Today, the possibility to develop living constructs integrated with the surrounding tissues represents a great challenge in the field of tissue engineering [1–4].

However, static cell culture systems show many limitations especially in terms of nonhomogeneous cell-matrix distribution and inadequate diffusion, and research efforts have been devoted to the design of different kinds of scaffolds and bioreactors.

As frequently reported [4–6], scaffolds should possess a set of morphological features as well as chemical and biochemical cues for promoting and controlling specific biological events at different levels.

On the other hand, a bioreactor should be ideally able to supply appropriate levels of nutrients, oxygen and growth factors, and to provide suitable mechanical stimuli for cell migration and scaffold colonization [4–7].

C. Rizzi et al. (Eds.): ADM 2019, LNME, pp. 948–958, 2020.
https://doi.org/10.1007/978-3-030-31154-4_81

However, over the past years the important role of reverse engineering [8], additive manufacturing [4, 6, 7], computer-aided design (CAD), modeling/simulation processes and experimental tests [9–11] has been widely stressed in the literature. The design of several kinds of polymeric and composite devices for tissue engineering is strongly related to the above reported methodologies and innovative materials [4–7, 12].

In this context, 3D additive manufactured scaffolds should represent promising candidates for tissue engineering applications. They would possess a structure with interconnected pore network, controlled architectural features and pore size distribution, tailored mechanical and mass transport properties. In addition, a magnetic structure may be also employed as its own bioreactor [13].

The potential to magnetically switch-on/switch-off the scaffold could be considered to deliver biomolecules and stem cells, as well as to stimulate cell adhesion, proliferation differentiation [13].

Many studied investigated the effect of a magnetic field on the biological systems [14–17]. In this context, iron oxide-based phases such as magnetite or maghemite have been extensively investigated.

3D additive manufactured PCL/iron oxide (PCL/Fe_3O_4) magnetic scaffolds were already designed and fabricated through 3D fiber deposition technique, and the effect of the amount of Fe_3O_4 nanoparticles on the biological, mechanical and magnetic performances was also studied [18]. Process and instrument parameters clearly play a crucial role in determining the mechanical and morphological properties of additive manufactured scaffolds. For this reason, the influence of process parameters on such features was previously analyzed for 3D additive manufactured poly(ε-caprolactone) (PCL) scaffolds obtained through a bioextruder [19]. The attention was especially focused on process temperature (PT), deposition velocity (DV), screw rotation velocity (SRV) and slice thickness (ST) to find the best set of parameters to fabricate PCL scaffolds with improved properties and reproducibility [19]. However, in processing composite materials difficulties are generally greater than those found for the neat polymers.

Accordingly, benefiting from previously obtained results on the neat PCL structures [19], in the current research an optimization design strategy for additive manufacturing processes based on extrusion/injection methods was adopted to develop magnetic nanocomposite PCL/Fe_3O_4 scaffolds for tissue engineering.

2 Materials and Methods

2.1 Design and Fabrication of 3D Additive Manufactured PCL/Iron Oxide Scaffolds

Poly(ε-caprolactone) (PCL, Mw = 65000, Sigma-Aldrich, St. Louis, MO Aldrich) pellets were dissolved in tetrahydrofuran (THF, Sigma-Aldrich, St. Louis, MO) with stirring at room temperature. Polyvinylpyrrolidone (PVP)-coated iron oxide (Fe_3O_4, 99.5%, 25 nm, 0.2% w PVP-NanoAmor) nanoparticles and ethanol were then added to the solution.

A PCL/Fe_3O_4 weight ratio (w/w) of 90/10 was employed and the dispersion of nanoparticles in the polymer solution was optimized employing an ultrasonic bath (Branson 1510 MT, Danbury, CT).

The nanocomposite PCL/Fe_3O_4 pellets were processed using a 3D fiber deposition technique to manufacture 3D block-shaped scaffolds (7.0 mm in length - *L*, 7.0 mm in width – *W*, 8.0 mm in height – H_0) with a 0/90° lay-down pattern.

The PCL/Fe_3O_4 pellets were placed in a metal chamber and heated using a unit on the arm of a bioextruder system similar to that employed for the fabrication of PCL scaffolds [19].

The material was extruded through a nozzle with an inner diameter of 400 μm, and the continuous filaments/fibers were deposited. Specifically, additive manufactured PCL/Fe_3O_4 scaffolds were built by extruding and depositing the filaments along specific directions according to the adopted lay-down pattern. A constant filament distance (FD) of 1,000 μm was used (see Fig. 1).

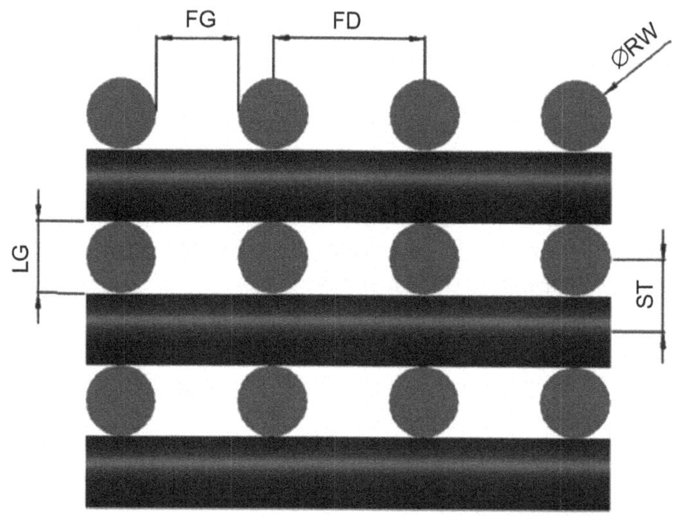

Fig. 1. A schematic view of the cross-section in the XZ plane: road width (RW), slice thickness (ST), filament gap (FG), layer gap (LG), filament distance (FD). The image was adapted from Domingos et al. [19].

The nanocomposite scaffolds were fabricated by using three different values of each parameter: process temperature (120, 130 and 140 °C), screw rotation velocity - SRV (20, 25 and 30 rpm), deposition velocity - DV (8, 10 and 12 mm/s) and slice thickness - ST (350, 380 and 400 μm).

The structures were manufactured varying iteratively one parameter, while maintaining constant the other three parameters.

2.2 Mechanical Compression Tests

Mechanical compression tests were performed on the designed 3D PCL/Fe_3O_4 scaffolds at a rate of 1 mm/min up to a strain of 0.4 mm/mm, using an INSTRON 5566 testing system equipped with a 1 kN load cell.

The "apparent" stress (σ) and strain (ε) were evaluated as reported below using a procedure reported in a previous work [19]:

$$\sigma = \frac{F}{L \cdot W} \quad (1)$$

$$\varepsilon = \frac{\Delta H}{H_0} \quad (2)$$

where F and ΔH are the force measured by the load cell and the height variation of the scaffold, respectively.

The compressive modulus was determined from the slope of the initial linear portion of the stress-strain curve.

For each set of manufacturing parameters, five specimens were tested.

3 Results and Discussion

The internal architecture, filament gap (FG), road width (RW) and pore size together with structural features and pore spatial distribution determine the performance of the scaffolds. Some works in the area of the design and fabrication of additive manufactured scaffolds have already highlighted the potential to obtain a wide range of "road" width (RW) by varying both process and instrument parameters at a fixed nozzle size [19].

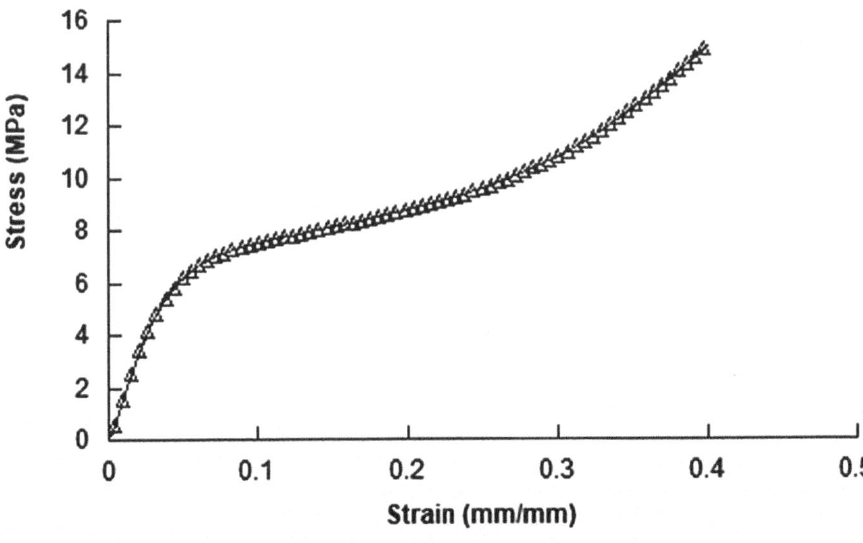

Fig. 2. An example of stress-strain curve obtained from compression tests on PCL/Fe_3O_4 scaffolds tested at a rate of 1 mm/min up to a strain of 0.4 mm/mm.

In the current study, an approach already employed to fabricate additive manufactured PCL scaffolds [19] was adopted and all the parameters were suitably settled to achieve a value of RW equivalent to the inner nozzle diameter (400 μm), trying to maintain the highest reproducibility with a minimal fabrication time without negatively affecting the structural stability of the scaffolds.

Results from compression tests displayed stress-strain curves (see Fig. 2) which were consistent with those reported for 3D additive manufactured scaffolds [18, 19].

3.1 Process Temperature (PT)

As a first step, the values of DV, SRV and ST were fixed (DV = 10 mm/s, SRV = 30 rpm and ST = 400 μm) while varying the temperature (120, 130 and 140 °C).

An increment of temperature from 120 to 140 °C provided an increase of RW (Table 1).

On the contrary, porosity decreased (Table 1) as pore width (FG) and pore height (LG) also decreased.

The results confirmed that PT directly affected the morphological features and, hence, RW and porosity values, the mechanical properties (i.e., compressive modulus and maximum stress) were likely to be influenced.

An increment of PT from 120 to 140 °C caused the thickening of the scaffold filament and a decrease of porosity. Values of compressive modulus and maximum stress are reported in Table 1.

Table 1. RW, porosity, compressive modulus and maximum stress of 3D PCL/Fe_3O_4 scaffolds obtained for different PT values (DV = 10 mm/s, SRV = 30 rpm and ST = 400 μm). Data are reported as mean value ± standard deviation.

PT (°C)	RW (μm)	Porosity (%)	Compressive modulus (MPa)	Maximum stress (MPa)
120	454 ± 6	60.2 ± 1.5	125.6 ± 12.2	14.2 ± 1.8
130	480 ± 6	54.3 ± 1.3	130.3 ± 13.7	15.3 ± 2.0
140	500 ± 5	51.2 ± 1.3	134.1 ± 15.2	14.0 ± 2.1

The obtained findings suggested that an increase of temperature above 120 °C did not greatly affect the compressive modulus and maximum stress, even if differences in terms of porosity and RW were found between the values obtained at 120 °C and those at 130 and 140 °C.

3.2 Deposition Velocity (DV)

The influence of DV on the morphological and mechanical properties was analyzed using different DV values (8, 10, 12 mm/s) and fixing the values of PT (120 °C), SRV (30 rpm) and ST (400 μm).

An increase of DV from 8 to 12 mm/s caused the filament thinning, providing scaffolds with wider pores and higher porosity (Table 2).

DV seemed to influence the mechanical performance of the nanocomposite structures. In particular, a high deposition speed should negatively affect the mechanical properties of the additive manufactured PCL/Fe_3O_4 scaffolds. As consequence of a fast deposition process in the XY plane the filaments of the scaffold resulted too stretched (lower values of RW and higher porosity). The lowest values of compressive modulus and maximum stress were found for DV of 12 mm/s (Table 2).

Table 2. RW, porosity, compressive modulus and maximum stress of 3D PCL/Fe_3O_4 scaffolds obtained for different DV values (PT = 120 °C, SRV = 30 rpm and ST = 400 μm). Data are reported as mean value ± standard deviation.

DV (mm/s)	RW (μm)	Porosity (%)	Compressive modulus (MPa)	Maximum stress (MPa)
8	489 ± 6	54.1 ± 1.2	132.0 ± 13.1	13.4 ± 1.7
10	454 ± 6	60.2 ± 1.5	125.6 ± 12.2	14.2 ± 1.8
12	434 ± 6	63.8 ± 1.4	92.3 ± 9.8	9.7 ± 0.8

3.3 Screw Rotation Velocity (SRV)

The amount of the extruded/deposited material is clearly related to the screw rotation velocity.

In this case the above reported principle was still employed as three different SRV values (20, 25 and 30 rpm) were used while maintaining constant the values of PT (120 °C), DV (10 mm/s) and ST (400 μm).

When SRV was varied from 20 to 30 rpm, an increment in RW was evident due to the higher amount of extruded material.

On the contrary, higher SRV values provided a reduction of pore height (LG), pore width (FG) and porosity (Table 3).

An increment in the SRV value clearly resulted in the thickening of the filament and structures characterized by lower porosity.

The highest values of compressive modulus and maximum stress were achieved for SRV of 30 rpm (Table 3).

Table 3. RW, porosity, compressive modulus and maximum stress of 3D PCL/Fe_3O_4 scaffolds obtained for different SRV values (PT = 120 °C, DV = 10 mm/s and ST = 400 μm). Data are reported as mean value ± standard deviation.

SRV (rpm)	RW (μm)	Porosity (%)	Compressive modulus (MPa)	Maximum stress (MPa)
20	380 ± 7	71.0 ± 1.8	68.1 ± 7.0	9.4 ± 0.9
25	435 ± 7	63.5 ± 1.5	84.1 ± 9.7	9.8 ± 0.9
30	454 ± 6	60.2 ± 1.5	125.6 ± 12.2	14.2 ± 1.8

3.4 Slice Thickness (ST)

As reported in the literature [19], the possibility to tune the ST allows for the direct control of the pore height (LG). However, the ST also influences the fiber diameter, pore width and porosity.

From a theoretical point of view, a reduction of the ST causes a compression of the filaments in adjacent layers, providing changes in the geometry of the filaments, which become more elliptical. This effect leads to an increase of the filament diameter as well as to a reduction of filament gap (FG) and porosity.

As expected, by reducing ST values from 400 to 350 μm a decrease of the pore height was obtained, thus leading to an increase of the filament diameter and, hence, a reduction of pore width and porosity (Table 4).

Changes in terms of ST would affect the compressive modulus and maximum stress.

In general, an increment of the material amount in the structure reduces the porosity, as consequence of a decrease in pore height and pore width related to higher RW values and represents the rationale of an improved mechanical behavior. However, values of compressive modulus and maximum stress are summarized in Table 4.

Table 4. RW, porosity, compressive modulus and maximum stress of 3D PCL/Fe_3O_4 scaffolds obtained for different ST values (PT = 120 °C, DV = 10 mm/s and SRV = 30 rpm). Data are reported as mean value ± standard deviation.

ST (μm)	RW (μm)	Porosity (%)	Compressive modulus (MPa)	Maximum stress (MPa)
350	493 ± 5	47.1 ± 1.4	142.0 ± 13.8	16.3 ± 1.5
380	478 ± 5	52.8 ± 1.3	137.7 ± 14.0	15.2 ± 1.4
400	454 ± 6	60.2 ± 1.5	125.6 ± 12.2	14.2 ± 1.8

3.5 Optimization of Process Parameters

The aim of the current research work was to find the optimal set of process/instrument parameters for the development of 3D PCL/Fe_3O_4 scaffolds with minimal fabrication time, however avoiding to compromise the structural integrity and reproducibility features.

For this reason, all the parameters were settled to obtain a value of RW equivalent to the inner nozzle diameter (400 μm).

It has been frequently reported that additive manufacturing techniques based on extrusion/injection methods allows for the direct fabrication of 3D porous structures with controlled architectural features, pore size and distribution, as well as tailored mechanical and mass transport properties [19].

From the obtained results, the effect of parameters such as LT, DV, SRV and ST on the RW (hence, on pore size and porosity), compressive modulus and maximum stress was analyzed.

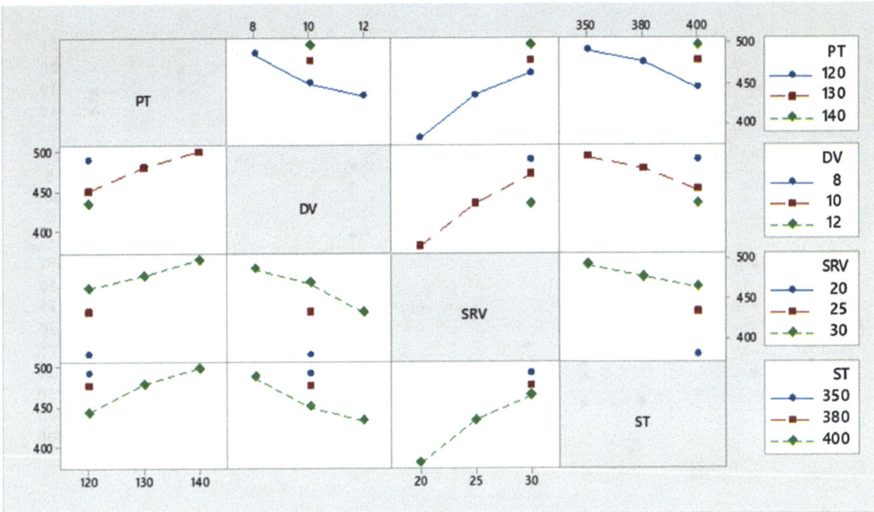

Fig. 3. Interaction plot for RW (μm).

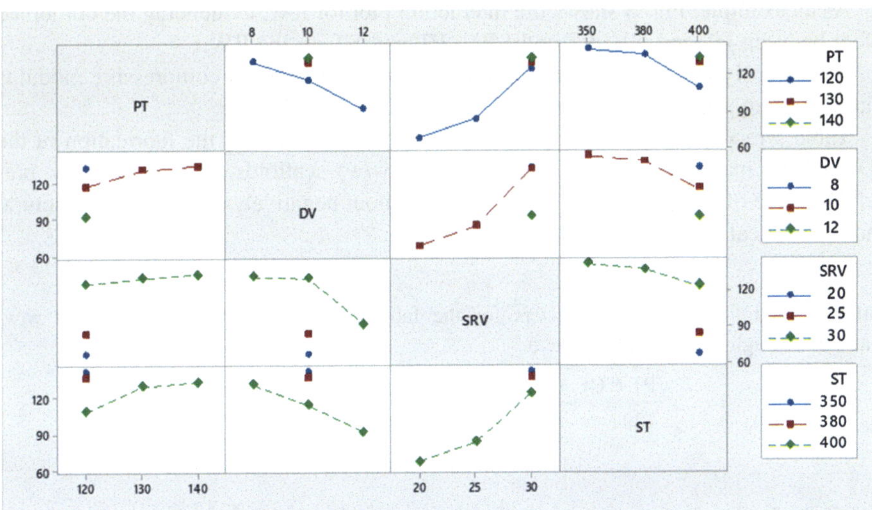

Fig. 4. Interaction plot for compressive modulus (MPa).

Furthermore, most of the parameters were interdependent, however differently showing a direct impact on the morphological features and mechanical performances of the designed 3D PCL/Fe_3O_4 scaffolds.

Figures 3, 4 and 5 report the interaction of different pairs of parameters (i.e., pairs of two parameters) on the analyzed morphological features (i.e., RW) and mechanical properties (i.e., compressive modulus, maximum stress) of the additive manufactured nanocomposite scaffolds.

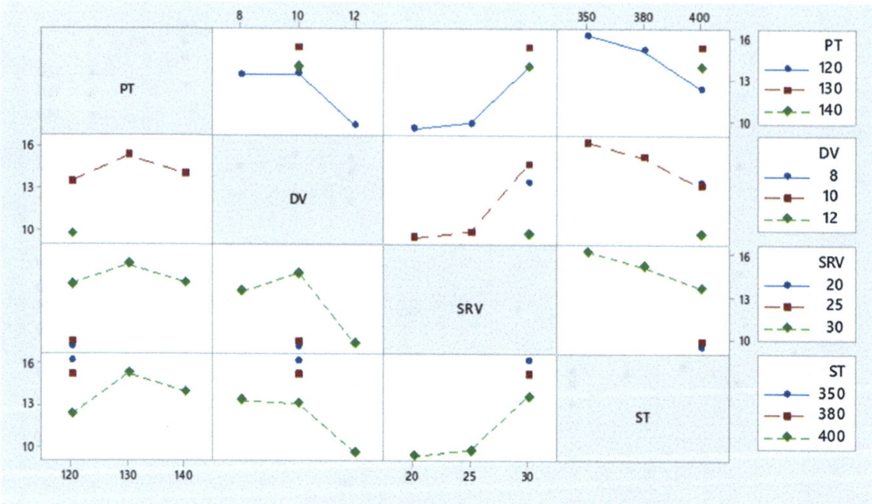

Fig. 5. Interaction plot for maximum stress (MPa).

As an example, Fig. 3 shows the interaction plot for RW, evidencing the combined effect between PT and DV, PT and SRV, PT and ST on the RW.

Analogously, Figs. 4 and 5 illustrate the interaction plots for compressive modulus and maximum stress, respectively.

Table 5 reports the optimized set of parameters allowing for the fabrication of the 3D additive manufactured PCL/Fe_3O_4 (90/10 w/w) scaffolds characterized by pre-defined value of RW, pore size and porosity, without negatively affecting the structural and mechanical features.

Table 5. Optimized set of parameters for the fabrication of 3D PCL/Fe_3O_4 (90/10 w/w) scaffolds through a bioextruder system.

PT (°C)	DV (mm/s)	SRV (rpm)	ST (μm)
120	10	30	400

4 Conclusions

The current research extended to 3D PCL/Fe_3O_4 scaffolds an optimization design strategy for additive manufacturing processes based on extrusion/injection methods, which was already reported to fabricate 3D PCL scaffolds.

The obtained results showed that DV and SRV have the highest impact on morphological features (i.e., RW) and mechanical properties (i.e., compressive modulus, maximum stress).

Taking into account the difficulties in processing nanocomposite materials (i.e., PCL/Fe_3O_4), which are generally greater than those related to the neat polymers (i.e., PCL), the employed optimization design strategy plays a fundamental role and may be extended to further kinds of materials.

References

1. Rijal, N.P., Adhikari, U., Khanal, S., Pai, D., Sankar, J., Bhattarai, N.: Magnesium oxide poly(ε-caprolactone)-chitosan-based composite nanofiber for tissue engineering applications. Mater. Sci. Eng., B **228**, 18–27 (2018)
2. Inzana, J.A., Olvera, D., Fuller, S.M., Kelly, J.P., Graeve, O.A., Schwarz, E.M., Kates, S.L., Awad, H.A.: 3D printing of composite calcium phosphate and collagen scaffolds for bone regeneration. Biomaterials **35**, 4026–4034 (2014)
3. Costa-Pinto, A.R., Reis, R.L., Neves, N.M.: Scaffolds based bone tissue engineering: the role of chitosan. Tissue Eng. Part B Rev. **17**, 331–347 (2011)
4. Hollister, S.J.: Porous scaffold design for tissue engineering. Nat. Mater. **4**, 518–524 (2005)
5. Ng, K.W., Hutmacher, D.W., Schantz, J.T., Ng, C.S., Too, H.P., Lim, T.C., Phan, T.T., Teoh, S.H.: Evaluation of ultra-thin poly(ε-caprolactone) films for tissue-engineered skin. Tissue Eng. **7**, 441–455 (2011)
6. Moroni, L., de Wijn, J.R., van Blitterswijk, C.A.: Three-dimensional fiber-deposited PEOT/PBT copolymer scaffolds for tissue engineering: influence of porosity, molecular network mesh size and swelling in aqueous media on dynamic mechanical properties. J. Biomed. Mater. Res. A **75**, 957–965 (2005)
7. De Santis, R., D'Amora, U., Russo, T., Ronca, A., Gloria, A., Ambrosio, L.: 3D fibre deposition and stereolithography techniques for the design of multifunctional nanocomposite magnetic scaffolds. J. Mater. Sci. Mater. Med. **26**, 250 (2015)
8. Martorelli, M., Ausiello, P., Morrone, R.: A new method to assess the accuracy of a cone beam computed tomography scanner by using a non-contact reverse engineering technique. J. Dent. **42**(4), 460–465 (2014)
9. Giordano, M., Ausiello, P., Martorelli, M., Sorrentino, R.: Reliability of computer designed surgical guides in six implant rehabilitations with two years follow-up. Dental. Mater. **28**(9), e168–e177 (2012)
10. Ausiello, P., Ciaramella, S., Fabianelli, A., Gloria, A., Martorelli, M., Lanzotti, A., Watts, D. C.: Mechanical behavior of bulk direct composite versus block composite and lithium disilicate indirect Class II restorations by CAD-FEM modeling. Dent. Mater. **33**(6), 690–701 (2017)
11. De Santis, R., Gloria, A., Maietta, S., Martorelli, M., De Luca, A., Spagnuolo, G., Riccitiello, F., Rengo, S.: Mechanical and thermal properties of dental composites cured with CAD/CAM assisted solid-state laser. Materials **11**, 504 (2018)
12. Sun, H., Zhu, F., Hu, Q., Krebsbach, P.H.: Controlling stem cell-mediated bone regeneration through tailored mechanical properties of collagen scaffolds. Biomaterials **35**, 1176–1184 (2014)
13. De Santis, R., Russo, A., Gloria, A., D'Amora, U., Russo, T., Panseri, S., Sandri, M., Tampieri, A., Marcacci, M., Dediu, V.A., Wilde, C.J., Ambrosio, L.: Towards the design of 3D fiber-deposited poly(ε-caprolactone)/iron-doped hydroxyapatite nanocomposite magnetic scaffolds for bone regeneration. J. Biomed. Nanotechnol. **11**(7), 1236–1246 (2015)

14. Xiaolan, B., Hadjiargyrou, M., Di Masi, E., Meng, Y., Simon, M., Tan, Z., Rafailovich, M.H.: The role of moderate static magnetic fields on biomineralization of osteoblasts on sulfonated polystyrene films. Biomaterials **32**, 7831–7838 (2011)
15. Ross, S.M.: Combined DC and ELF magnetic fields can alter cell proliferation. Bioelectromagnetics **11**, 27–36 (1990)
16. Hashimoto, Y., Kawasumi, M., Saito, M.: Effect of static magnetic field on cell migration. Electr. Eng. Jpn. **160**, 46–52 (2007)
17. Kotani, H., Kawaguchi, H., Shimaoka, T., Iwasaka, M., Ueno, S., Ozawa, H., Nakamura, K., Hoshi, K.: Strong static magnetic field stimulates bone formation to a definite orientation *in vitro* and *in vivo*. J. Bone Min. Res. **17**, 1814–1821 (2002)
18. De Santis, R., Gloria, A., Russo, T., D'Amora, U., Zeppetelli, S., Tampieri, A., Herrmannsdörfer, T., Ambrosio, L.: A route toward the development of 3D magnetic scaffolds with tailored mechanical and morphological properties for hard tissue regeneration: Preliminary study. Virtual Phys. Prototyp. **6**(4), 189–195 (2011)
19. Domingos, M., Chiellini, F., Gloria, A., Ambrosio, L., Bartolo, P., Chiellini, E.: Effect of process parameters on the morphological and mechanical properties of 3D Bioextruded poly (ε-caprolactone) scaffolds. Rapid Prototyp. J. **18**, 56–67 (2012)

Determination of Adhesive to Be Applied in the Fabrication of Prototypes Using FDM Techniques of 3D Printing in Component Parts Using ULTEM™ 1010

Miguel Suffo(✉)

Department of Mechanical Engineering and Industrial Design,
High Engineering School, University of Cádiz,
Campus Río San Pedro s/n 11510, Puerto Real, Cádiz, Spain
miguel.suffo@uca.es

Abstract. It is common in the 3D printer or additive manufacturing techniques of small or medium format, to manufacture prototypes or products in parts and then join them to complete a functional set to a specific industrial purpose. This technology is discarded as a valid alternative to many large-format parts or mechanical assemblies. However, adhesive technology offers the additive manufacturing an opportunity to avoid dependence on the dimensions of the prototype. When selecting a commercial adhesive it is important to follow the manufacturer's instructions that inform about its basic composition and its use on some materials. However, these specifications are not applicable to materials commonly used in additive manufacturing. Polyetherimide (ULTEM™ 1010), is one of these materials due their relative novelty generate uncertainty when using them in large-format parts or assemblies because the manufacturer's adhesives do not have tests enough provide guarantees. The problem is even more serious if the mechanical assembly is composed for the other complex parts that make up a kinematic device. The unions joint are stressed in the normal movement mechanism and some study about bond stress or adhesion strength is required. In this paper, a tensile test are carried out with 6 different adhesive technologies from the same manufacturer and the best response is determined experimentally. Through a second tensile test on 12 specimens in Fused Deposition Modeling technology with different orientations it is demonstrated who this parameter is determinant according adhesion strength parameter. Finally, with the best adhesive obtained, a first prototype mixer agitator has been carried out for the biomedical industry.

Keywords: Adhesives · Additive manufacturing · Adhesion strength · ULTEM™ 1010

1 Introduction

Additive Manufacturing (AM) offers the industry the possibility of generating prototypes or functional models of low cost, high availability of inputs and, of a good quality and resistance [1, 2]. One of the few disadvantages of the 3D printing technology is the

C. Rizzi et al. (Eds.): ADM 2019, LNME, pp. 959–969, 2020.
https://doi.org/10.1007/978-3-030-31154-4_82

reduced dimension of the printer drawer, except some recent large-format printers whose costs are not affordable in many cases [3, 4]. This disadvantage can be solved using joining techniques such as threaded elements, or, by means of fixed adhesive unions, but these techniques are more unknown in type of polymer commonly used in Fused Deposition Modeling (FDM) technology [3]. The literature is not profuse in research works related to the use of adhesives for these polymers such as PEI (ULTEM), PEEK, PC, PLA, etc. [5]. Reference [6] performs some tensile tests on different plastics materials boned with commercial adhesives, determining the adhesion strength and the friction resistance. This reference highlights the difficulty of achieving a conclusive result with certain plastics due to the lack of knowledge of its composition. FDM, also known as Fused Filament Fabrication (FFF), Modeling Extrusion (ME), Fused Layer Manufacturing (FLM) or 3DP are AM process or technology that generate parts whose mechanical properties depend on the process parameters [7] and the orientation of the piece with respect to the printer drawer [5, 8].

More experimental contributions are needed in this field so that the industry is able to apply this technology of adhesives in the designs of parts and mechanical assemblies reducing uncertainty. For the selection of the best adhesive for a plastic material manufactured by FDM techniques, we must involve more variables than those based solely on the adhesion strength that support. Some of these variables can be: the medium in which these parts will work, their temperature, relative humidity, orientation in the manufacture and, the cost of the support material, among others.

A simple application of this type of adhesives in the design and manufacture of a mixer agitator for the mixing of biphasic fluids used in the biomedical or food industry is proposed. Due to the corrosive environment a non-metallic technological solution is preferred; because of the length of the axis, it goes beyond the limits of the printer drawer. In this paper, a series of commercial adhesives will be tested on different specimens manufactured by biocompatible ULTEM1010 which is used in the aeronautics, automotive and biomechanical industries [9]. Additionally, this material is flame retardant and has a high thermal stability [10, 11]. From the result obtained in the tensile tests, the optimum adhesive will be determined to join the different parts of the prototype.

2 Materials and Methods

2.1 Materials

Specimens used in the tests was polyetherimide (PEI) material, particularly ULTEM^{TM} 1010 filament biocompatible from the Savic supplier and adapted for Fortus 450MC machine. In its manufacture, it was used a T14 type nozzle (0.254 mm); for the support material it was used the same material, but by means of a T16 type nozzle (0.254 mm). Specimens were manufactured from the 3D model of a dimensional specimens: 90 × 25 × 5 mm according to the UNE-EN 1465: 2009 standard (ISO 4587), as shown in Fig. 1.

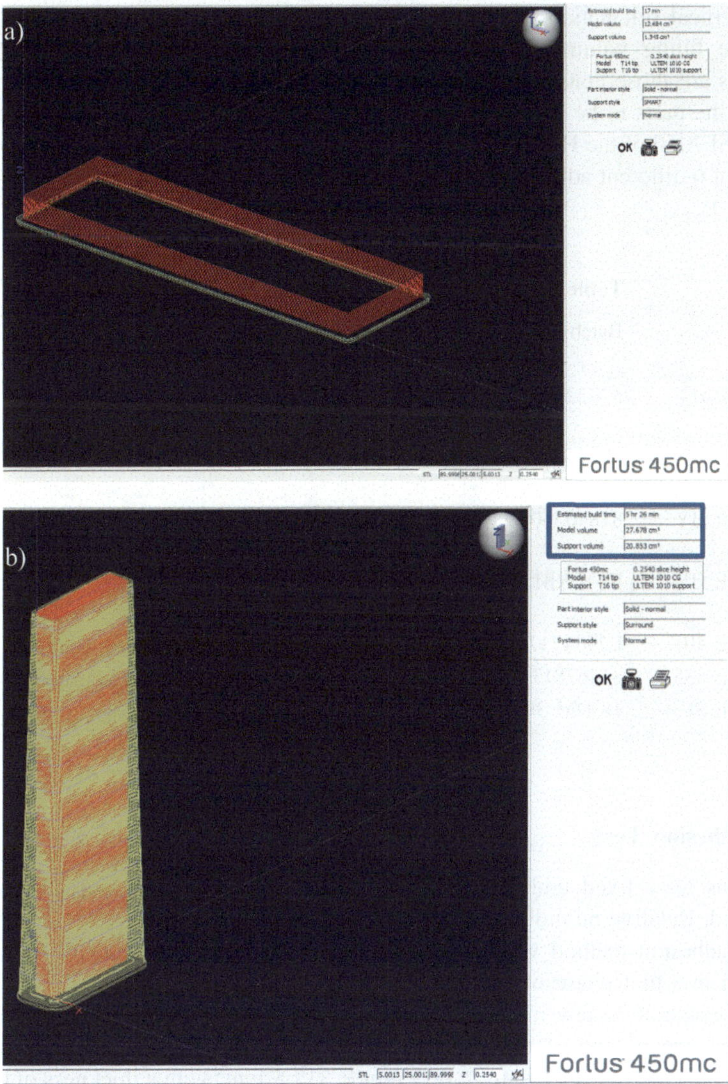

Fig. 1. Simulation of parameters before manufacture of prismatic parts used such as specimens using Insight 3D v10.6 Printing Software. (a) specimens in horizontal orientation; (b) specimens in vertical orientation.

In a first series of tests, 12 specimens manufactured in horizontal orientation were used. In a second series of tests, 12 specimens were manufactured in horizontal direction and another 12 specimens in vertical orientation to the manufacturing plane. Figures 1a and b show the orientations of the specimens horizontally and vertically, respectively and, the simulation was carried out using Insight 3D v10.6 Printing Software. In these figures the cost of support material in yellow color is also observed, in the vertical orientation this cost is twenty times higher than horizontal, although twice the filling volume of the specimen is required in vertical. The supporting structures are those considered essential by the device software to guarantee the stability of the parts to be manufactured. Table 1 lists the 6 adhesives plus the cleaning product SF7070 of the Henkel Ibérica manufacturer. On the specimens it will be tested a series of 6 different adhesive technologies with various chemical compositions listed in Table 1.

Table 1. Adhesives and equipments used in the tests

Product	Batch	Expiry date	Description
LOCTITE HY 4080	83022936	10/19	Structural hybrid adhesive (Cianoacrylate/Acrylic)
LOCTITE AA 3295	L280521264	02/20	Acrylicstructuraladhesive
LOCTITE HY 4070	81840560	10/19	Hybrid quick-fixation adhesive (Cianoacrylate/Acrylic)
LOCTITE HY 4060	ET10X81	01/20	Structural hybrid adhesive (Cianoacrylate to/epoxi)
LOCTITE 401	L290471769	01/21	Instantadhesive (Cianoacrylate)
LOCTITE 435	84467817	08/20	Instanttoughadhesive (Cianoacrylate)
LOCTITE SF 7070	18 DM 397	04/20	Cleaner

2.2 Adhesion Test

Specimens were fixed under the following environmental conditions: Temperature: 21 °C and, Relative humidity: 28.8%.

The adhesion method was adapted to UNE-EN 1465:2009 standard (ISO 4587), consisted in a first phase of specimens cleaning through the 7070 liquid cleaner and leave to evaporate a few minutes. In a second phase, the adhesive was added in the specimens one on top of the other in simple overlapping (12,5 mm overlap). The adhesion area measures 25 mm × 12,5 mm = 312,5 mm^2 with a thickness of 0,1 mm; 31.25 mm^3 is the quantity of adhesive applied. The adhesive density is 1,1 g/cm^3 then, the adhesive mass applied is aprox. 35 mg. The specimens allowed to cure for 24 h at room temperature and relative humidity and were subjected to tensile shear tests. Once the best adhesive was found, it was used in a second phase to carry out a series of tests with 12 pairs of specimens in the following configurations: 4 Horizontal-Horizontal pairs (H-H); 4 Vertical-Vertical pairs (V-V) and; 4 Horizontal-Vertical pairs (H-V).

2.3 Tensile Test

Once the adhesions of the specimens manufactured in horizontal orientation were made, they were subjected to tensile tests to determine the adhesion strength or resistance to friction. The tests were carried out on a Zwick Roell Z20 machine, at a speed of 2 mm/min. The method was also adapted to the aforementioned UNE-EN 1465: 2009 (ISO 4587) standard.

2.4 Mixer Agitator Design

The best adhesive product will be applied on a prototype of the mixer agitator whose 3D model is shown in Fig. 2. The prototype (1150 mm height × 25 mm diameter) is composed of an impeller of 6 equal prismatic blades evenly distributed around the base. As shown in Fig. 2a, the prototype has been divided into 3 parts to conduct the test plus two parts will be attached by cylindrical reinforcements with a height of 50 mm and a diameter of 10 mm. The upper part of the agitator is prepared to fit in a motor-gearbox kit through an 85 mm length key. The device will be joined by the adhesive selected according to the previous tests.

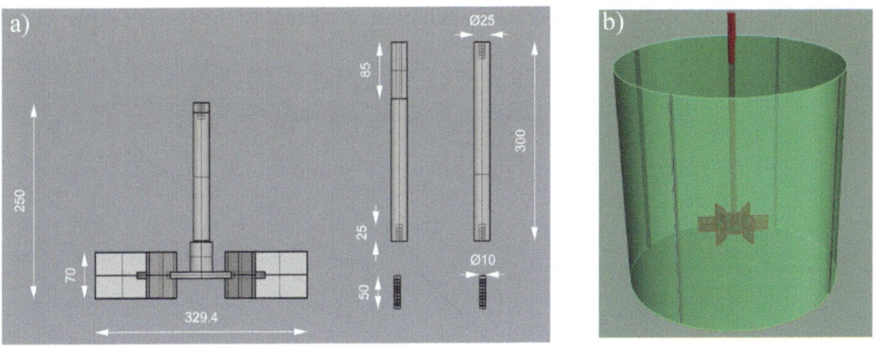

Fig. 2. 3D model of the mixer agitator to be incorporated into the reactor. (a) bill of parts with simplified dimensions; (b) reactor and mixer agitator 3D assembly with a capacity of 1 m^3.

3 Results and Discussion

The results obtained from the first tensile test are presented classified based on the materials tested in the study, in this way, it is possible to establish comparatives that relate the types of failures with the applied adhesives. Table 2 shows the quantitative results and the types of failures obtained in the first series of tests corresponding to the specimens manufactured only in horizontal orientation. As it is showed in the results of Table 2, all the test specimens showed adhesive failure, except in the case of 401 product in which the substrate based on ULTEM failed. In addition, the shear strength of this material was superior to that of the other adhesives, approximately double. Therefore, the adhesive selected for this union was 401.

Table 2. Results obtained from the first series of tensile tests with specimens manufactured in horizontal orientation

Product	Force [N]	Shear Strength σ [MPa]	Type of failure	Specimens
4080	540,60	1,48	Adhesive	
3295	191,89	0,61	Adhesive	
4070	488,41	1,54	Adhesive	
4060	382,6	1,19	Adhesive	
401	1166,34	0,71	Substrate	
435	806,32	2,36	Adhesive	

In Fig. 3 plots the force-displacement curves results of the first tensile test, in which curves presented by each of the adhesion behavior subjected to the test are represented.

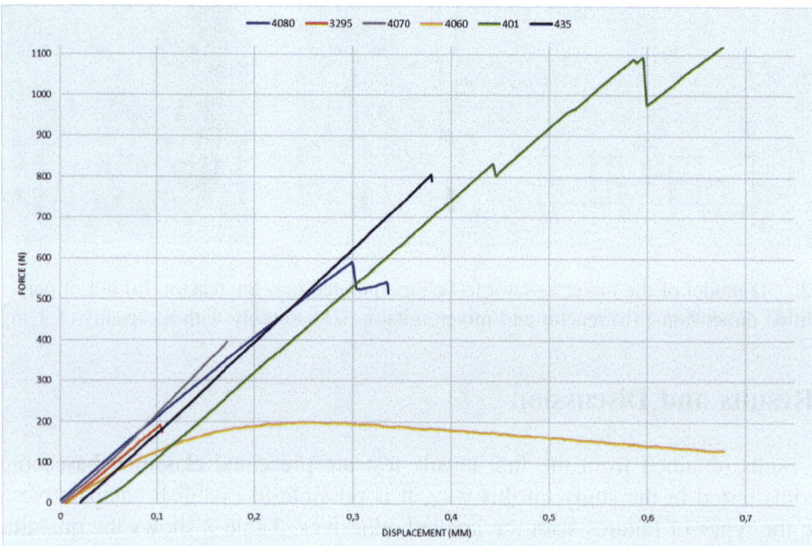

Fig. 3. Results obtained from the tensile test for specimens manufactured in horizontal orientation.

All the adhesives have identical linear behavior in all the deformation except the model 4060 which has an elasto-plastic behavior maintained in time, but with a first failure faster than all the others. It is interesting to observe the behavior of this hybrid adhesive compared with 4070 and 4080, both being cyanoacrylate mixes. It is possible that the cyanoacrylate/epoxy combination does not achieve a stable chemical bond with the ULTEM 1010 but to determine this hypothesis a microscopic study is necessary.

Table 3. Results obtained from the second series of comparative tensile tests applying only the 401 adhesive.

Product	Force [N]	Shear Strength σ [MPa]	Type of failure	Specimens
1 V-V	1370	4,35	Substrate	
2 V-V	1740	5,32	Substrate	
3 V-V	1450	4,51	Adhesive/Substrate	
4 V-V	1690	5,27	Substrate	
Average		***4,86 ± 0,5***		
5 H-H	892	2,8	Adhesive	
6 H-H	1020	3,06	Adhesive	
7 H-H	1720	5,5	Adhesive/Substrate	
8 H-H	1140	3,79	Adhesive	
Average		***3,79 ± 1,22***		
9 V-H	1640	5,31	Substrate	
10 V-H	1700	5,54	Substrate	
11 V-H	1730	5,19	Substrate	
12 V-H	1760	5,62	Substrate	
Average		***5,42 ± 0,2***		

Based on the results obtained, a second tensile test was carried out using a combined series of specimens manufactured in horizontal and vertical orientation with the same geometry, under the same environmental conditions and using the identical adhesive surfaces as before (12.5 mm), but only applying the 401 adhesive. Table 3 shows the results obtained from the test that served to establish a comparative series of adhesion between test specimens at different orientations: H-H; V-V and; H-V.

The average shear strength obtained for the 401 over ULTEM in the three combinations applied continues to be much higher than the other adhesives tested previously. The difference in the mode of failure that is observed in some specimens is due to the fact that the resistance value of the ULTEM substrate is very similar to that shown by the 401 adhesive. It is observed in Table 3, most the failures cases in H-H unions are "adhesives" types and only one specimen results in "substrate" failure type. Likewise, the values of Shear Strength (Adhesion Strength) are slightly inferior to those obtained in the joints involving vertical manufactured specimens (V), as indicated in references [1, 2]. In these joints involving vertical manufactured specimens (V-V and H-V), substrate failures (ULTEM) are obtained in all cases. It should be noted that in horizontal and vertical manufactured specimens joints, "substrate" failure always occurs in the vertical manufactured specimen. The theoretical tensile strength in this manufacturing orientation of the specimen provided by the ULTEM 1010 manufacturer is about 75 MPa compared to that obtained in the adhesion test (5.42 MPa). Despite the differences in the type of theoretical test applied to obtain this result, the maximum average value of the shear resistance is much lower than the theoretical one. There are significant discrepancies between the theoretical value of this parameter and the actual value obtained by the habitual users of the same polymeric material in AM. This result coincides with others reported previously with similar uncertainty considering this type of materials [9]. When FDM is used with this polymeric material, unknown parameters such as porosity, anisotropy, surface topology or adhesion are incorporated, which influence the final prototype obtained [12, 13].

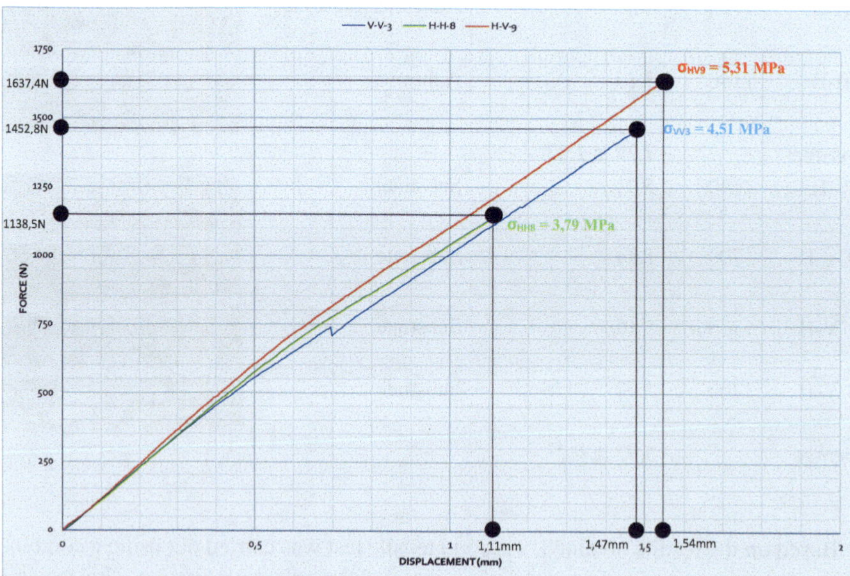

Fig. 4. Force-Displacement curves obtained from the tensile test for specimens manufactured in the combined series.

Figure 4 shows the most representative Force-Displacement curves of the specimens used in the second combined series of tensile tests.

The curves represented in the graph are those that approximate the most to the average value obtained in each combined series: H-H; V-V; H-V. Thus, in blue color, the curve that most closely approximates to the averaged value of the 4 replicas made in the H-H oriented specimens is shown. In green color, the curve that most closely approximates to the averaged value of the 4 replicas made in the V-V oriented specimens is shown. In red color, the curve that most closely approximates to the averaged value of the 4 replicas made in the H-V oriented specimens is shown.

In spite of the good results obtained by the vertically oriented specimens, which may be attributed to the roughness of the surfaces in this orientation that provides better adherence [5, 14], the economic cost of manufacturing should be assessed. The volume of supporting material needed triples the manufacturing cost compared to those oriented in Horizontal while the verticals provide an improvement in tensile strength of only 15%, like [4]. The most advisable solution will be to select the combination of mixed H-V orientations, which offer the best compromise alternative in the "cost - strength" ratio.

For the union between the components shown in Fig. 5a, 2 internal reinforcements were manufactured in horizontal orientation and the 3 parts were made in vertical orientation.

Fig. 5. First prototype manufactured; (a) Separates parts; (b) Prototype assembled with 401 adhesive

The result showed in Fig. 5b represents the first prototype that is currently working at a rate of 8 h a day and 5 days a week, during 640 h in total without any failure.

4 Conclusions

Even though our study has some limitations, it has been determined that it is possible to manufacture large prototypes with small-medium scale machines from smaller parts that can then be bonded in adhesive techniques. The dependence on the use of different adhesive technologies applied to a material commonly used in several strategic industrial sectors, such as ULTEM 1010, has also been proven. According to the results of resistance in tensile tests, the 401 model composed of Cianoacrylate applied to this material, has been behaving in a similar way. Experimental results from this study offer an opportunity for designers who need to use a type of adhesive in the prototypes of this type of material and dimensions. As an example of studio application, A device agitator mixer is being currently adapted to measure the torque of different configurations of the agitator mixer and considering different materials used in AM. Additionally, a research field on adhesive technologies and polymeric materials commonly used in AM is created.

Acknowledgements. The author wishes to thank the Henkel group for products supply.

References

1. Ian, B., Deakin, G., Rosen, D.: Additive Manufacturing Technologies: 3D Printing, Rapid Prototyping, and Direct Digital Manufacturing, 2nd edn. no. 3, pp. 193–198 (2015)
2. (2015) ISO/ASTM 52900. Additive manufacturing, General principles. Terminology (2017)
3. Popescu, D., Zapciu, A., Amza, C., Baciu, F., Marinescu, R.: FDM process parameters influence over the mechanical properties of polymer specimens: a review. Polym. Test. **69** (April), 157–166 (2018)
4. Roschli, A., et al.: Designing for big area additive manufacturing. Addit. Manuf. **25** (November), 275–285 (2019)
5. Byberg, K.I., Gebisa, A.W., Lemu, H.G.: Mechanical properties of ULTEM 9085 material processed by fused deposition modeling. Polym. Test. **72**(October), 335–347 (2018)
6. Troughton, M.J.: Handbook of Plastics Joining : A Practical Guide (2008)
7. Zaldivar, R.J., Mclouth, T.D., Ferrelli, G.L., Patel, D.N., Hopkins, A.R., Witkin, D.: Effect of initial filament moisture content on the microstructure and mechanical performance of ULTEM ® 9085 3D printed parts. Addit. Manuf. **24**(October), 457–466 (2018)
8. Alharbi, N., Osman, R., Wismeijer, D.: Effects of build direction on the mechanical properties of 3D-printed complete coverage interim dental restorations. J. Prosthet. Dent. **115** (6), 760–767 (2016)
9. Chuang, K.C., Grady, J.E., Draper, R.D., Patterson, C., Santelle, T.D.: Additive Manufacturing and Characterization of ULTEM Polymers and Composites (2019)
10. Wu, H., Sulkis, M., Driver, J., Saadecastillo, A., Thompson, A., Koo, J.H.: Multifuncional ULTEM TM 1010 composite filaments for additive manufacturing using Fused Filament Fabrication (FFF). Addit. Manuf. **24**(August), 298–306 (2018)
11. Jiang, W., Sundarram, S.S., Wong, D., Koo, J.H., Li, W.: Polyetherimide nano-composite foams as an ablative for thermal protection applications. Compos. Part B Eng. **58**, 559–565 (2014)

12. Fayazbakhsh, K., Movahedi, M., Kalman, J.: The impact of defects on tensile properties of 3D printed parts manufactured by fused filament fabrication. Mater. Today Commun. **18** (December 2018), 140–148 (2019)
13. Sun, S., Li, M., Liu, A.: A review on mechanical properties of pressure sensitive adhesives. Int. J. Adhes. Adhes. **41**, 98–106 (2013)
14. Motaparti, M., Taylor, K.P., Leu, G., Chandrashekhara, M.C., Castle, K., Matlack, J.: Effects of build parameters on compression properties for ULTEM 9085 parts by fused deposition modeling. In: Proceedings of 27th Annual International Solid Freeform Fabrication Symposium, no. 2016, pp. 964–977 (2016)

Assessment of Design for Additive Manufacturing Based on CAD Platforms

Enrico Dalpadulo(✉), Fabio Pini, and Francesco Leali

"Enzo Ferrari" Department of Engineering,
University of Modena e Reggio Emilia, Via Vivarelli 10/1, 41125 Modena, Italy
{enrico.dalpadulo, fabio.pini,
francesco.leali}@unimore.it

Abstract. The aim of this paper is to analyze some critical issues in the Design for Additive Manufacturing workflow and evaluate the introduction of CAD platforms as backbone tools to shorten product development time and raise its efficiency. It is focused on the design of components to be printed by Powder Bed Fusion metal Additive Manufacturing. Even though the use of additive technologies firmly joins a CAD mathematical model and the actually printed component, the workflow from the concept to the definitive job may result in many sequential steps which have complex and slow relationships. Currently, at the state of art for the production of components specifically designed to be produced by additive manufacturing, there are issues both with the adoption of STL as interchange files and the not reversible sequence of tasks. For example, if a problem occurs in the part re-design during component industrialization, usually one must restart the work from the beginning. Thus, an improvement of the design workflow that could shorten time to product and improve both product performances and process quality and reliability, is necessary. In particular, the use of CAD platforms that integrates CAD and CAE tools has been investigated. An automotive case study, originally made by traditional subtractive technology (CNC milling), has been re-designed with topology optimization in order to be printed by Selective Laser Melting process with benefit of weight reduction. Design and industrialization tasks have been tested with respect to the selected integrated CAD platform, and potential improvements have been evaluated.

Keywords: Design for Additive Manufacturing · Topology optimization · CAD platform · Automotive

1 Introduction

Additive Manufacturing, AM, is a process of joining materials to make objects from 3D model data, usually layer upon layer, as opposed to subtractive manufacturing methodologies [1]. The idea below this process is that a bi-dimensional shape is easy to create independently from the geometrical complexity of a component. Thus, it is possible to imagine every type of object and build it in a direct and fast way [2–4]. AM was born in the '80 and it has been massively used for visualizing product concepts or building prototypes (Rapid Prototyping) in the last decades. Subsequently the evolution

C. Rizzi et al. (Eds.): ADM 2019, LNME, pp. 970–981, 2020.
https://doi.org/10.1007/978-3-030-31154-4_83

of machines and materials enabled the construction of parts that can both fit in assemblies due to the improvements in accuracy and perform functionality due to the better mechanical properties. This process is the achievement of the so called F3 rule: form, fit and function. Nowadays the innovation is the extension of the technology at a real manufacturing system for end use parts, especially for small batch sizes of products. Consequently, the Design for Additive Manufacturing, DfAM, is receiving an increasingly attention in the industrial settings focused on development of high performances customized products.

The basis of DfAM workflow is an iterative refinement of the starting model of the part with respect to *Design* and *Industrialization* constraints to respectively improve the part shape and the associated printing process. With respect to schematic representation of Fig. 1, four main phases compose the approach, as listed below.

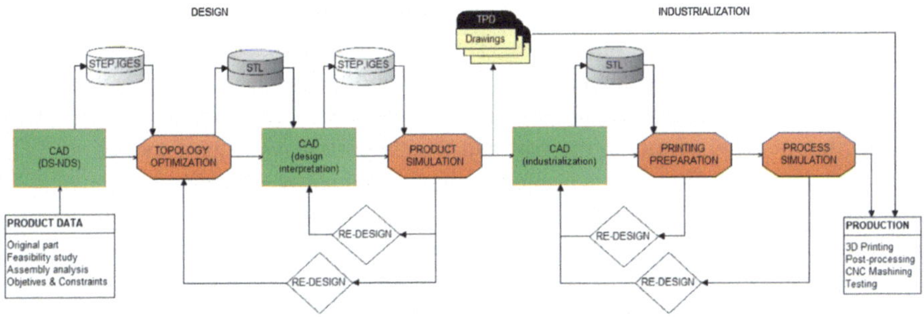

Fig. 1. Design for Additive Manufacturing workflow.

Topology Optimization and Product Simulation. The optimization phase is to search for the optimal shape of the part. Inputs are the original model of the part, the feasibility study and assembly analysis, as well as the objectives and constraints of the project. Any critical issue has to be recognized and it has to be taken into account for the design inputs. Then the design space and the non-design space of the project have to be created. This often starts from the original geometry of the part and the analysis of the assembly where it has to fit and work. These elements have to be modeled in the CAD environment letting all the couplings and the assembly fitting to be granted. To get the best topology optimized part, the design space must be as large as possible, according to the project constraints, so that the material distribution can be calculated with the maximum freedom. The non-design space is necessary to define the functional surfaces and the volumes where the design must not be modified. These CAD models are exported in a neutral interchange file and used to run the topology optimization, setting up the simulation model, with specific targets and constraints.

Typically, this is performed with specific CAE finite-elements based software [5]. The result of this step is generally a geometry described by a cloud of points. Some operations can be necessary to manipulate it and perform triangulation, or eventually it can be exported as STL file via software.

The resulting geometry drives the reconstruction in the CAD environment, using it as a support for the manual free-shape modeling of the topology optimized component, which is one of the most time-consuming activity in the workflow.

Product Simulation evaluates the structural performance, examines possible critical issues, and then defines the elements to be re-designed. So the CAD model must be exported in a neutral interchange file and has to be verified by finite elements analysis in a CAE environment. Usually some iterations between the design and product simulations and in worse cases also between design and topology optimization are needed. Final models and technical product documentation can now be produced.

Printing Preparation and Process Simulation. The third and fourth phases aim to set-up and improve the AM production process. The printing of re-designed part requires further oversizes, the definition of correct printing pose, (location and orientation) and support structures, as well as slicing analysis. Usually, part geometries of CAD models are exported in STL file and imported in CAM printing preparation software to perform these operations. The introduction of solid supports requires importing the STL file containing parts oriented and the build plate, generated in the printing preparation software, in the CAD environment for support design. After introduction of solid supports, the part and the supports have to be exported as STL. Then the industrialization task is needed again. Subsequently, all geometries have to be re-exported as STL to undergo the process simulation phase to analyze part thermal behavior and predict its warping and distortions.

Often some elements of parts have to be re-designed to achieve a correct industrialization and if that happen, the CAD model which should be parametric can be modified, but then it has to be exported in a STL format and the industrialization must be done from the beginning. After some iterations, CAD models and drawings are reviewed, final industrialization is ready to be sent to the machine and the part construction can start. Finally, all heat treatments and post-processing operations, such as part cutting, supports removal and surface finish treatments have to be done. All the functional surfaces then have to be machined to improve surface quality and ensure fitting and functionality, and at last parts can be tested.

1.1 Critical Issues

Troubles in the product development could be many and interest the singular phase itself and also the whole workflow. The former type of issues could be referenced as local to each phase and those that have the main drawbacks are listed as follows.

- *Topology Optimization issues*. Manual geometry interpretation of topology optimization results can be highly time consuming [6] and tools for automatic parametric model construction can be improved.
- *Product Simulation issues*. It could be interesting to implement DFM [7] with manufacturability constraints and parameters for build direction, to take account process anisotropy, minimization of supports or part deformations or post-processing operations. There are efforts in including in topology optimizations algorithms of build orientation, overhangs, and the minimization of support structures [8].

- *Printing Preparation issues.* Industrialization tasks such as part orientation and support generation can be difficult, especially with the introduction of oversize or CAD-designed support structures, which could bring saving of material, energy and build time.
- *Process Simulation issues.* Process simulations, even with difficult calibration, could predict part warping and defects, but it is still difficult to modify part geometries or industrialization tasks or machine parameters to avoid any kind of issue.

With respect to the whole workflow, the main issues are related to the file exchange and data management due to the use of many specific software to perform the described tasks, from concept definition to final job creation. Sundry CAD-based tools help users in part design, CAE ones in topology optimization, product and process simulation and validation, CAM ones in part industrialization. Typically, STL file is used for data management as interchange file between different software. In particular, it has 3 main applications:

- It is used in design operations with CAD software, as a support for modelling parts starting from results of topology optimizations. The cloud of points defined by a density cutting of finite elements based on selected iso-values can be exported as STL and the free-shape design can be performed overlapping the shape on the result.
- It is used for industrialization tasks in 3D printing preparation software, for interchange of geometry of parts to be produced. In that environment it can undergo manipulations such as multiplying, scaling, sectioning, orienting, adding features such as support structures.
- It is used for additive manufacturing process simulation, as input file to import an industrialized job in the simulation software. It is generally the input for generation of voxel meshes for modeling parts, supports and the building plate of a job.

STL file is actually a surface mesh that approximates a mathematical geometry representation. Its name is the abbreviation of STereoLithography, the first additive manufacturing process developed by 3D System in the early '80s. It is a triangular mesh, where each element is defined by the position in space of the three vertex and a normal vector. The data can be stored in binary representation, but also in ASCII format to be more easily readable or eventually editable. The STL file has always deviations from the ideal mathematic, moreover it can present many errors, such as unit changing, problems in vertex and edges connection, intersecting or overlapping triangles, degenerated facets or bad aspect ratio triangles, bad edges, holes. The adoption of this format can lead to issues in file generation, fixing and editing in CAD environment. Manual repairing is a highly time-consuming operation; luckily automatic fixing tools are now very common but sometimes they are not completely effective or they can lead to problems in geometry. STL models cannot benefit of CAD manipulation and parametric edition but are often necessary as design support for part development and re-design.

Evidence of above mentioned issues emerges from case studies in design and construction of topology optimized components that make use of the described workflow. Gebisa and Lemu applied the method for the re-design of the GE jet engine bracket [9]. Orme and Madera applied it for the re-design of many aerospace case studies for SSTL's Technology's and SpaceIL's missions: a star tracker camera bracket, an edge insert, a thruster mount for a lunar lander vehicle [10]. In particular, Süß et al. implemented the method to re-design the main gearbox bracket of a helicopter, showing issues in STL file manipulation and data management between the design tasks [11].

The additive manufacturing ISO standard 17209-4 [11] explains that "STL file format (…) has established itself as a quasi-industry standard format for transferring data to additive manufacturing technology" ([11], subsection 4.2.2), but it also states that "the STL data format is unsuitable for exchanging data between CAD/CAM systems because the geometry is irreversibly faceted" ([11], subsection 4.3.1). In addition, it warnings users that the representation must be faultless and that there could be troubles in data management based on STL files. There are considerations about guidelines to follow for triangulations and it even asserts that "data set repairs can be very time consuming and costly and therefore require individual approval" ([11], subsection 4.3.1). In fact, according to the standard, STL file has to be used only to send the geometry information for the slicing operation and additive fabrication process. Figure 2 depicts the scope of STL as reported in the ISO standard.

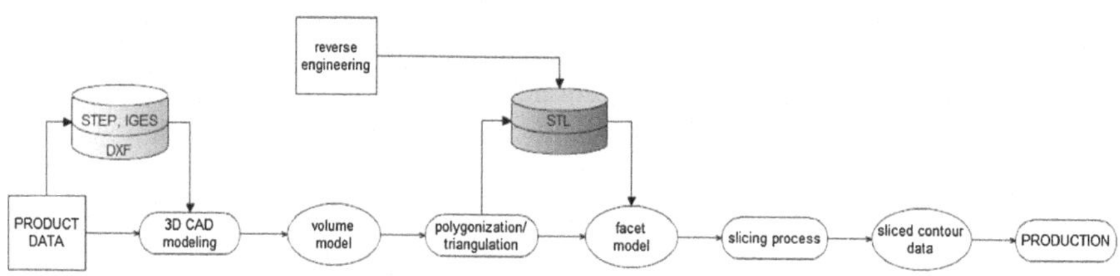

Fig. 2. From ISO 17209-4 [11].

2 Data Consistency Through a Computer Aided Platform

The present work proposes the introduction of CAD platforms as backbone tools to remove the issues related to the file exchange and data management along the DfAM workflow. The founding idea is to improve the state-of-the-art workflow removing the needs to use STL file and to exploit product-process platforms which integrate CAE and CAD applications to enable product design and drafting, as well as product and process simulation and analysis. The aim is to shorten product development, but also improve quality and reliability and thus to implement the Design for X idea of gaining competitiveness by modifying product, process and systems [12]. The phases for the development of additive manufacturing components are fundamentally the same. Nevertheless, data exchange and export to interchange files between the tasks is not required. Nowadays new design platforms include a plethora of applications fully integrated in one environment which allow an unambiguous management of data throughout all the product lifecycle. Consequently, it could be possible to define a single model to work on, from part concept to production.

The different applications integrated in the CAD-based environment are switched to perform the different tasks, keeping parametric and associative properties, which are intrinsic in the CAD model. As a consequence, design intent in feature definition to keep intelligence in the model for all the operations needed in product and process development becomes fundamental.

The phases of CAD platform-based DfAM can be summarized as follow. The first step is to import product data or to create the basic elements to set up the design space. Subsequently, the model is used to create partitions, non-design space, functional features and elements to set up the optimizations. Geometry interpretation could be required to transform the data in a parametric model, or automatic tools for shape generation could be then exploited. The model already set for the optimization process could be directly used for product simulation or eventually with specific analysis-oriented set up. If re-design is required, it can be done directly on the model. The same model can be also introduced in assemblies for particular analysis or in the printing preparation environment, with the machine and the build platform, in order to be handled for industrialization tasks. Eventually, the final data can be exported in STL, but also in other formats suitable for additive manufacturing: AMF (standardized Additive Manufacturing File), 3MF (XML-based data), MTT (metadata) files for direct production, or also CLI (Common Layer Interface) and SLC (SLiCe) which are basically representations of the slicing.

Consequently, the DfAM workflow is strictly based on the effectiveness of CAD platform used which relies on software structure, feature functionality, partitioning and logical sequence, and methods required to perform specific operations. One of these platform is the Dassault Systèmes 3DExperience [13]. With respect to additive manufacturing, it lets the integration of CAD environment with generative design and product simulation, 3D printing preparation and process simulation and can lead to design and manufacturing time optimization.

3 Case Study - Topology Optimized Wheel Knuckle

A Formula SAE wheel knuckle identified as a case study to assess the introduction of the 3DExperience CAD platform as a support tool for DfAM. The part, originally made by traditional subtractive technology (CNC milling – Fig. 3(a)), has been re-designed in order to be printed by selective laser melting (SLM) process with benefit of weight reduction. A first topology optimized wheel knuckle had been designed, following the state of the art design for additive manufacturing workflow, see Fig. 3(b). The first prototype development required the use of many specific software for the different tasks. In particular, Dassault Systèmes Solidworks had been used for CAD design and re-design operations, Altair Optistruct for topology optimization, MSC Marc for product simulation, Materialise Magics for printing preparations, PTC Creo for CAD operations connected with industrialization (raw part, solid designed supports).

The case study has been re-designed for additive manufacturing to validate all *Design* and *Industrialization* tasks in the Dassault Systèmes 3DExperience CAD

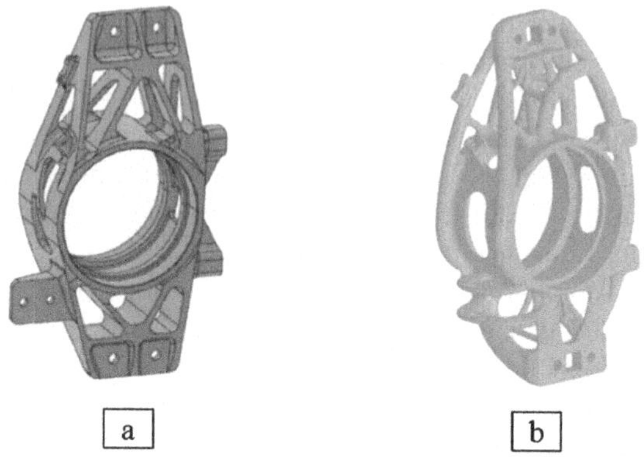

Fig. 3. Original component model (a) and first topology optimized model (b).

platform. Non suspended masses have to be minimized in order to guarantee comfort and in particular cornering performance, due to the better tarmac-tyre contact. The re-design for additive manufacturing of the component benefits of the limitations of technological constraints at functional surfaces and the possibility of part integration. The original part was made by machining of 7075 Ergal with 2.81 kg/cm^3, elasticity modulus of 72 GPa, yield stress of about 500 MPa, weight 436 g, but had stress concentration of about 300 MPa. The design objective was a mass reduction of 10%, with constraints related to part stiffness, reduction of stress concentrations, part couplings and kinematic fitting in the wheel-suspension assembly. In order to define the design space, some technical improvement have been introduced in the model, such as radial supports for brake caliper, double bearing housings instead of one, exciter slot, integrated steering rod connection.

An assembly analysis has been performed, to guarantee couplings with upper and lower wishbone brackets and elements of steering and braking systems, with the feasibility of assembling operations (space for fasteners, tools and movements) and room for the envelope of the steering rod and the wishbone to be taken into account. The design space model has been partitioned and functional surfaces and non-design space volumes have been detected. Figure 4(a)–(b) respectively depict the pose of original wheel knuckle with respect to mounting assembly, and the related design space.

In order to perform topology optimization, finite elements model is set up by meshing the design space and defining loads and restraints for the simulation. Three load cases have been introduced, to simulate car bump or vertical jump, cornering at maximum speed, braking at maximum deceleration. Vertical, longitudinal and lateral forces are applied at the wheel center point, whereas braking forces are applied at the brake caliper pads contact surfaces. The topology optimization has been set at the same time with the three load cases, with the target mass reduction and the constraint of maximizing part stiffness. Product analysis and simulations have been the next step.

Structural validation has been performed, to check part stiffness and part stresses compared to the three load cases. In addition, digital mock-up (DMU) of the component inserted in the wheel-suspension group assembly has been performed with assembling operations and check for clash. Final model has been obtained with the

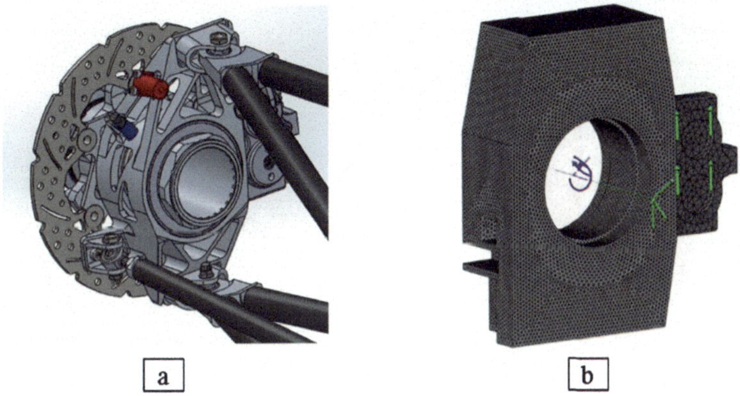

Fig. 4. Assembly analysis for design space definition (a) and finite elements model with mesh and set-up of load cases (b).

introduction of functional features and surfaces so that the part could work properly in the assembly. To proceed toward the industrialization of the component, the right and the left wheel knuckle raw model had to be created, oversizes on all the surfaces to be machined and studs on spots to be drilled. A PBF (powder bed fusion) machine environment for the SLM 280HL has been created, construction parameters have been set and the build plate has been inserted. Part orientation has been calculated with an optimization algorithm to both minimize support areas on the parts and support volume. To bear considerable areas, solid designed supports have been created. Afterwards vectorial line, volume and cone supporting structures are added to complete the task, checking the operation with the slicing analysis and completing the job preparation for the printing with specimens.

4 Results

The re-designed part with the 3DExperience CAD platform is considered. The result of topology optimization inclusive of the three load cases is shown in Fig. 5(a). An automatic shape generation tool has been used to obtain a concept shape setting 0.52 iso-value for density. At the end of this step a manipulable solid model has been created as output of the topology optimization (Fig. 5(b)–(c)). The final part obtained with some design improvements and the introduction of missing functional features is shows in Fig. 5(c).

The concept has been analyzed by finite element simulations of the three load cases. Thanks to the branched natural shape, with the absence of sharp edges, the main stresses are all above 200 MPa, the stiffness is comparable with the original component, the weight of the concept part is about 382 g. Figures 6(a)–(b) show the stresses on the component with respect to the cornering load case and also the digital mock-up of the concept part into the wheel and front suspension assembly.

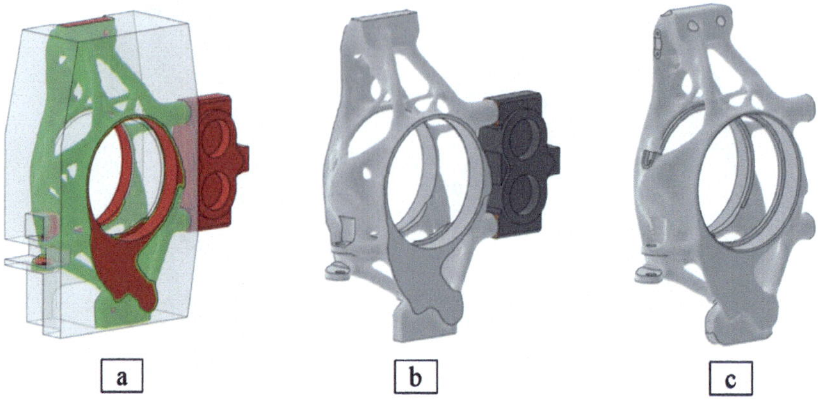

Fig. 5. Topology optimization result (a), generated concept shape (b), final design model (c).

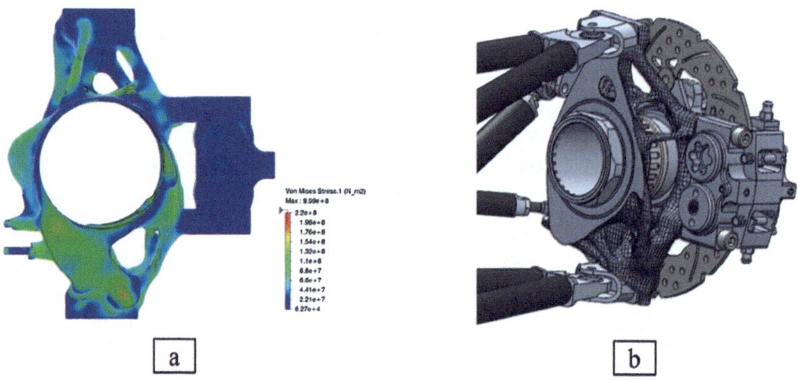

Fig. 6. Concept model finite elements analysis (a) and digital mock up analysis (b).

The weight for the component rises at 389 g. Figure 7 shows part industrialized, with CAD designed solid supports, that lead to savings of material, energy and build time for supports up to 40%. Results of the topology optimized part, designed in the 3DExperience CAD platform can be analyzed and a comparison with the original optimized component, designed following the standard method can be done. In particular, considering almost the same values about stiffness and stress concentration level far below the one of the CNC-milled part, the weight of the parts can be compared, see Table 1.

Some considerations can be extracted from CAD platform-based DfAM. Topology optimization has been calculated at the same time with respect to three load cases. An easily manipulable solid model has been obtained as output of this step, instead of a cloud point to be exported as STL and to be used as a support for geometry reconstruction. Design, optimization and product validation tasks have been made in the same environment due to the integration with the CAE applications, on the same parametric model, without need of data interchange, also for re-design. This model

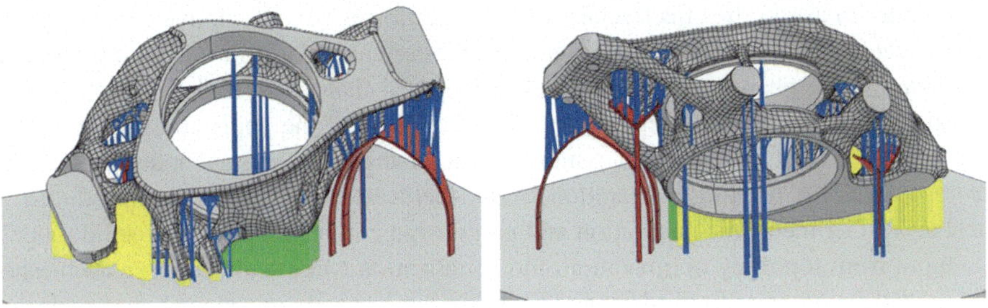

Fig. 7. Industrialized job for the left raw part.

Table 1. Weight comparison between the components.

CNC-milled	AM (state of the art)	AM (3D Experience platform)
436 g	401 g	389 g

could be also used for industrialization, due to the integration with the CAM environment, exploiting CAD parametricity and associativity for part re-design. In this way even CAD designed solid supports could be introduced directly in the virtual printing environment. Also, industrialization has been feasible without the use of STL files meshes, without issues in file generation, manipulation or reparation.

Consequently, it is possible to design topology optimized components entirely in the CAD platform, by switching between specific applications and working on a single model. STL file is not necessary as interchange element, as modeling support, for re-design, for printing preparation. Therefore, the use of integrated product-process CAD platforms can evidently improve the DfAM workflow simplifying operations and relations between tasks and can shorten product development workflow reducing time to product.

5 Conclusions

Some features and potentials of the implementation of AM as technology to build high performance components have been introduced. At the state of the art, different phases have to be performed by the use of different CAD and CAE specific software, adopting many interchange files in the sequence of operations and in required iterations. Two kinds of issues have been recognized, the first at local level, concerning the operations about each phase, the second at global level, regarding the entire workflow.

CAD platforms might represent effective choice to overcome troubles in the DfAM, with the aim of shortening product development time and optimizing product and process performance. In particular, these platforms are based on the use of integrated product-process applications and consequently they simplify and reduce data management and exchange.

In order to assess the effectiveness of CAD platforms, an automotive case study has been developed within Dassault Systèmes 3DExperience platform, for the *Design* and *Industrialization* phases. At global level, it emerged that every re-design step is simplified by the parametricity and associativity between all the applications. The use of STL files could not be required anymore as interchange element, as modeling support, for re-design, for printing preparation. Also, there are some potentials introduced at local level. Geometry reconstruction and part design can be simplified by solid models as output from topology optimization. Industrialization can avoid dealing with meshes and related fixing operations, designed supports can be easily introduced, part re-design can be very quick. It is possible to perform only one file exportation of the final industrialized job, at the end of the workflow, for parts production. Possible troubles in data management can be minimized, the entire product development workflow can evidently be shorted, the lead time connected with the design and industrialization operations can be reduced.

Further investigations will be carried on tuning the printing process simulation by introducing material thermo-mechanical data, machine building and scanning parameters, to predict part construction behavior. As a result, build issues and warping could be detected and *Industrialization* can be improved in order to get better results in terms of part tolerance and material quality.

References

1. ASTM F2792-12a, Standard Terminology for Additive Manufacturing Technologies, (Withdrawn 2015), ASTM International, West Conshohocken, PA (2012)
2. Gherardini, F., Mascia, M.T., Bettelli, V., Leali, F.: A co-design method for the additive manufacturing of customised assistive devices for hand pathologies. J. Integr. Des. Process Sci. **22**(1), 21–37 (2019)
3. Buonamici, F., Carfagni, M., Furferi, R., Governi, L., Saccardi, M., Volpe, Y.: Optimizing fabrication outcome in low-cost FDM machines. Part 1-metrics. Manuf. Technol. **18**(3), 372–378 (2018)
4. Bici, M., Brischetto, S., Campana, F., Ferro, C.G., Seclì, C., Varetti, S., Maggiore, P., Mazza, A.: Development of a multifunctional panel for aerospace use through SLM additive manufacturing. In: 11th CIRP Conference on Intelligent Computation in Manufacturing Engineering, pp. 215–220. Elsevier B.V., Gulf of Naples (2017)
5. Bici, M., Broggiato, G.B., Campana, F.: Topological optimization in concept design: Starting approach and a validation case study. In: Lecture Notes in Mechanical Engineering, pp. 289–299 (2017)
6. Süß, M., Schöne, C., Stelzer, R., Kloeden, B., Kirchner, A., Weissgaerber, T., Kieback, B.: Aerospace case study on topology optimization for additive manufacturing. In: Fraunhofer Direct Digital Manufacturing Conference DDMC, Berlin, pp. 37– 41 (2016)
7. Boothroyd, G., Dewhurst, P., Knight, W.: Product Design For Manufacture And Assembly. CRC Press, Boca Raton (2011)
8. Orme, M., Gschweitl, M., Ferrari, M., Madera, I., Mouriaux, F.: Designing for additive manufacturing: lightweighting through topology optimization enables lunar spacecraft. J. Mech. Des. **139**(10), 100905 (2017)

9. Gebisa, A., Lemu, H.: A case study on topology optimized design for additive manufacturing. In: IOP Conference Series: Materials Science and Engineering, vol. 276, no. 1, p. 012026 (2017)
10. Orme, M., Madera, I., Gschweitl, M., Ferrari, M.: Topology optimization for additive manufacturing as an enabler for light weight flight hardware. Designs **2**(4), 51 (2018)
11. UNI EN ISO 17296-4:2017 - Additive manufacturing - General principles - Part 4: Overview of data processing (2017)
12. Huang, G.Q.: Design for X: Concurrent Engineering Imperatives. Chapman & Hall, London (1996)
13. Dassault Systèmes. https://www.3ds.com/about-3ds/3dexperience-platform/. Accessed 07 Jan 2019

Author Index

C. Rizzi et al. (Eds.): ADM 2019, LNME, pp. 983–986, 2020.
https://doi.org/10.1007/978-3-030-31154-4

Z

The manufacturer's authorised representative in the EU is Springer Nature Customer Service Centre GmbH, Europaplatz 3, 69115 Heidelberg, Germany. If you have any concerns regarding our products, please contact ProductSafety@springernature.com

Printed and bound by CPI Group (UK) Ltd, Croydon, CR0 4YY
25/09/2026
02222888-0001